GENERAL AND COMPARATIVE PHYSIOLOGY

PRENTICE-HALL BIOLOGICAL SCIENCE SERIES
William D. McElroy and Carl P. Swanson, *editors*

GENERAL AND COMPARATIVE PHYSIOLOGY

Second Edition

WILLIAM S. HOAR

Department of Zoology
The University of British Columbia
Vancouver, Canada

PRENTICE-HALL, INC., *Englewood Cliffs, New Jersey*

Library of Congress Cataloging in Publication Data

HOAR, WILLIAM S.
General and comparative physiology.

(Prentice-Hall biological science series)
Bibliography: p. 774
1. Physiology, Comparative. I. Title.
[DNLM: 1. Physiology, Comparative. QP31 H679g]
QP31.2.H6 1975 591.1 74-17118
ISBN 0-13-350272-4

Cover illustration: "Verbum," a lithograph
by Maurits C. Escher, a Netherlands artist
who died in 1972. (Escher Foundation, Haags
Gemeentemuseum, The Hague.)

Printed in the United States of America

10 9 8 7 6 5 4 3 2

Prentice-Hall International, Inc., *London*
Prentice-Hall of Australia, Pty. Ltd., *Sydney*
Prentice-Hall of Canada, Ltd., *Toronto*
Prentice-Hall of India Private Limited, *New Delhi*
Prentice-Hall of Japan, Inc., *Tokyo*

TO MY STUDENTS—PAST, PRESENT, AND FUTURE

CONTENTS

TWO ENVIRONMENTAL RELATIONS 317

THREE NERVOUS INTEGRATION AND ANIMAL ACTIVITY 457

FOUR REPRODUCTION AND DEVELOPMENT 693

PREFACE

The objectives and philosophy of the current revision can best be stated by quoting from the preface of the first edition: "Physiological literature is now rich in details pertaining to every major group in the animal kingdom. Zoologists, who must acquire a working knowledge of this information, are faced with the difficult task of remembering diverse facts and incorporating them into a meaningful scheme. The details often seem isolated or pertinent to only one animal and, although they may have real meaning for the specialist, the beginner frequently finds little interest in them. Yet the numerous facts of comparative physiology are not really unrelated. They are all a part of the story of evolution. The processes have been discovered and recorded as isolated facts, but they came into existence and acquired meaning as steps and stages in the progressive adaptation of animal life to varied habitats and changing environments. They are historical details in the organization of protoplasm for the varied activities of animal life.

"This book is written with the conviction that a story of phylogeny in animal functions can now be sketched and that this will provide a framework into which the many details of physiology can be interestingly fitted. It is written for students who must acquire a working knowledge of functional biology whether their special field is to be animal physiology or any one of the many other branches of zoology.

"No attempt has been made to write a detailed treatise. A wealth of well-established information—related particularly to human, medical, and cellular physiology—is not included. These topics are comprehensively covered in many excellent texts. Likewise, numerous monographs on the different animal groups provide a ready source of detailed physiological information. What seems to be less generally available is a synthesis of the major trends in physiological adaptation. This has been attempted here. It is hoped that the student will find it a structure into which the accumulated and ever-increasing body of facts concerning animal physiology can be easily and interestingly fitted."

In planning the revision, I rejected several possible rearrangements of the material. To me, the organization of the first edition is logical in a story of phylogeny; moreover, I feel that the many users of the previous edition will appreciate knowing approximately where to look for particular topics. Old books are like old friends; one comes to know what to expect from them and when to anticipate the responses.

Many parts of the text have been rewritten and most sections have been expanded; additional topics have been included; there are almost 200 new illustrations; the list of references has more than doubled. As in the first edition, the reference list includes the names of some pioneers who made fundamental contributions in comparative physiology but concentrates on a selected list of critical reviews; through these reviews "students can most speedily locate the many pertinent original references." Again, the index shows the taxonomic position for the scientific names of animals; the hypothetical phylogenetic tree at the end of the preface should assist readers unfamiliar with animal taxonomy.

This revision has required far more time than I anticipated. The back of the monster was broken in the Radcliffe Science Library at Oxford University while I enjoyed a Senior Research Fellowship from the National Research Council of Canada; I am indebted to Professor Niko Tinbergen, Professor J. W. S. Pringle, and Dr. A. E. Needham for the many courtesies which they extended during my stay in Oxford. The task was completed in the Woodward Library at the University of British Columbia, where the staff gave every assistance. My colleagues in our Department of Zoology were always responsive to my pleas for help; I am grateful to them all. I am particularly indebted to C. P. Hickman, Jr. of Washington and Lee University for his careful reading of the entire manuscript; I wish I could have taken advantage of all his constructive suggestions. Finally and most sincerely I acknowledge the cheerful help of my wife Myra who spent many many long hours questioning my statements and checking all parts of the manuscript at every stage. This made my work less dreary and resulted in a much better text for my students—past, present, and future—who inspired the book and to whom it is fondly dedicated.

Vancouver, Canada William S. Hoar

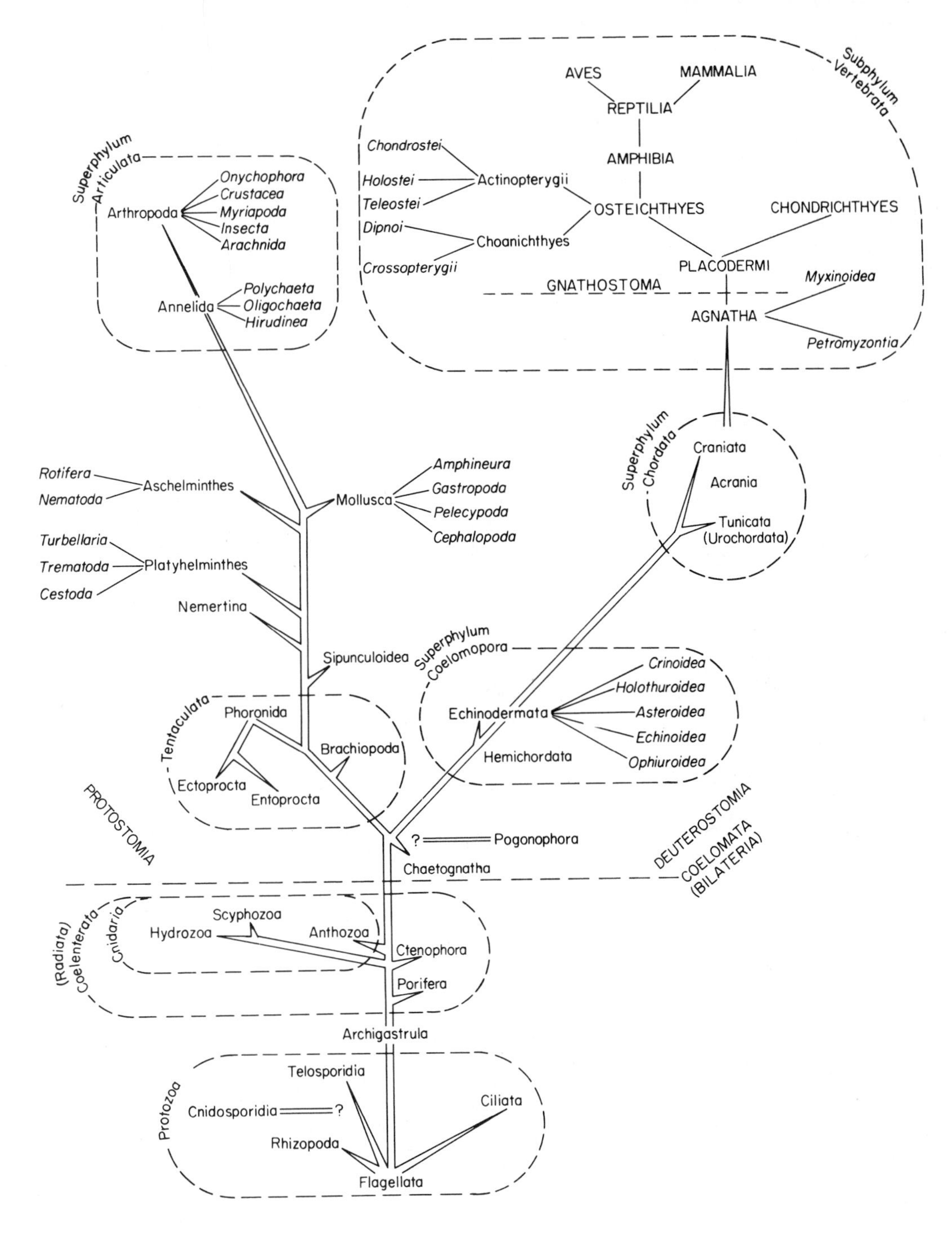

A hypothetical phylogenetic tree. [Invertebrate relationships based on Marcus after Kerkut, 1960: Implications of Evolution. Pergamon Press, Oxford. Vertebrate relationships based on Romer, 1970: The Vertebrate Body. 4th ed. Saunders, Philadelphia.]

ONE

SOURCES OF ENERGY AND ITS DISTRIBUTION

1

THE ORIGIN OF ANIMALS AND THEIR ENVIRONMENT

The Russian biochemist Oparin, about 1922, first developed those ideas which have become basic to the modern concepts of the origin of life. Oparin (1936) emphasized two alternatives. One might postulate that life arose in a world much like ours, or one might assume that it originated under very different conditions. Until about a generation ago only the first alternative seemed likely. Oparin argued that all the evidence pointed to the second alternative and that the earth, in its youth, was probably very different. He maintained that, although spontaneous generation is now impossible, it might have been inevitable in the earth which knew no life. The living environment has evolved with living organisms. They are inseparable, and our present world, filled with so much life, precludes a spontaneous generation.

This argument now seems sound, and it is likely that the world and the animals which live in it have evolved together. Geochemists and biologists agree that there has been a series of irreversible steps while the earth cooled, generated an atmosphere rich in hydrogen and produced the first simple organic compounds which gradually increased in complexity and formed the self-duplicating living systems (Fox, 1965; Calvin, 1969; Miller and Orgel, 1974). Each step has changed conditions and reduced the likelihood of a repetition of earlier events. Darwinian evolution has been documented by convincing facts for over a hundred years, but it is only in the past two decades that scientists have found evidence for the evolution of the organic from the inorganic, the biochemical from the organic, and the living from the biochemical.

The duration of the prebiologic period is incomprehensible. Gaffron (1960) suggests that it required three billion years for the development of the first cells and that Darwinian evolution, by comparison, has lasted only about a third as long. Some estimates are even greater. Molecular paleontologists have recently extracted hydrocarbons from shales which were probably formed more than three billion years ago. The presence of normal hydrocarbons, alkanes, and isoprenoids characteristic of present-day lower plants, strongly suggests a high degree of biological organization in the Archaean (Calvin, 1969). Although there is no real agreement concerning the precise time when life first appeared, scientists are unanimous that the early evolutionary period was longer than that with which Darwin was concerned. Three major stages are thought to have preceded conditions suitable for the life of our familiar plants and animals (Gaffron, 1960; Calvin, 1969). These were probably anaerobic or nearly so. The materials listed in the second, third, and fourth columns of Table 1.1 may be characteristic of the groups of substances which dominated these three important eras.

Table 1.1 Schematic Representation in Chemical Terms of the Formations Which Must Be Accomplished from the Atoms to Produce the Structure of a Cell. Compounds in Columns 2, 3, and 4 May Be Characteristic of the Substances Which Dominated Three Stages of Biological Evolution Prior to Life

Atom	Molecule	Polymer	Cell
Hydrogen	Acid ⟶	Lipid	Molecular aggregation
Carbon	Sugar ⟶	Cellulose, starch, etc.	+
Oxygen	Base ⟶	Nucleic acid ⟶	Autocatalysis
Nitrogen	Amino acid ⟶	Protein	+
			Membrane assembly

Source: Calvin, 1969: Chemical Evolution. Clarendon Press, Oxford.

ANAEROBIC STAGES IN TERRESTRIAL EVOLUTION

Era of Excess Hydrogen and the Formation of Micromolecules

Life had its origin in highly reducing conditions. The prebiologic atmosphere probably contained water, hydrogen, ammonia, and methane (Fig. 1.1). Small amounts of carbon dioxide may have been present (Keosian, 1966), but methane, not carbon dioxide, furnished the carbon for the first organic compounds. Hydrogen was abundant; free oxygen was absent. The primary source of energy for organic syntheses in the beginning was, as now, the sun. The major components of this energy budget were photochemical and ionizing radiations with lesser contributions from volcanism, meteorite impact, and lightning (Miller and Urey, 1959). These

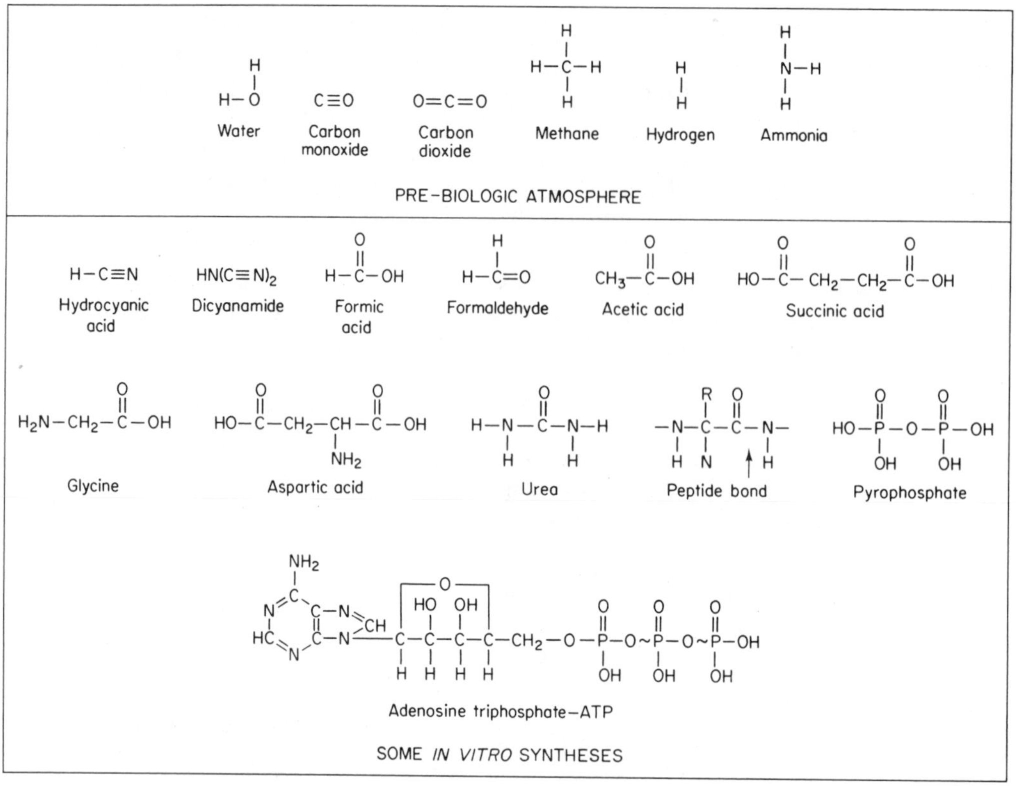

Water ($H-O-H$) | Carbon monoxide ($C\equiv O$) | Carbon dioxide ($O=C=O$) | Methane (CH_4) | Hydrogen ($H-H$) | Ammonia (NH_3)

PRE-BIOLOGIC ATMOSPHERE

Hydrocyanic acid ($H-C\equiv N$) | Dicyanamide ($HN(C\equiv N)_2$) | Formic acid ($H-C(=O)-OH$) | Formaldehyde ($H-C(H)=O$) | Acetic acid ($CH_3-C(=O)-OH$) | Succinic acid ($HO-C(=O)-CH_2-CH_2-C(=O)-OH$)

Glycine ($H_2N-CH_2-C(=O)-OH$) | Aspartic acid ($HO-C(=O)-CH_2-CH(NH_2)-C(=O)-OH$) | Urea ($H-N(H)-C(=O)-N(H)-H$) | Peptide bond | Pyrophosphate ($HO-P(=O)(OH)-O-P(=O)(OH)-OH$)

Adenosine triphosphate–ATP

SOME *IN VITRO* SYNTHESES

Fig. 1.1 The prebiological atmosphere and some likely prebiological compounds. Except for ATP, compounds shown in the lower group have been identified in *in vitro* systems.

were the conditions which are thought to have favored the first fortuitous combinations of carbon, hydrogen, oxygen and nitrogen.

Miller (1953, 1955) was the first to test this hypothesis. In his classic experiments, mixtures of methane, ammonia, hydrogen and water were subjected to electrical sparks in a closed system for about one week and the chemical changes in the apparatus were monitored by radiochemical and chromatographic techniques. Small quantities of glycine, alanine, aspartic acid, and α-amino-n-butyric acid were readily identified. These exciting experiments stimulated investigators in many laboratories (Calvin, 1969; Miller and Orgel, 1974). Syntheses of all the building blocks required for the formation of the complex organic substances have now been effected under abiological conditions, using various mixtures of gases and energy sources thought to have been present during the prebiotic period. Identification of many different amino and fatty acids, urea, aldehydes, sugars, and several of the purine and pyrimidine bases was followed by the demonstration of more complex molecules, such as adenosinetriphosphate (Ponnamperuma, 1965) and the porphyrins (Miller and Orgel, 1974). It is now apparent that nonbiological processes could have produced the essentials of such vital compounds as chlorophyll, hemoglobin, the cytochromes and the nucleosides. Oparin postulated an immensely long period for these syntheses before living material appeared and before the photosynthetic activities of plants filled the air with oxygen and the evolving animals began to oxidize plant materials as a source of energy. Urey (1952) calculates that the primitive oceans may have been a 10 per cent solution of organic compounds.

Changing Prebiologic Atmosphere and the Formation of Macromolecules

Formation of precursor subunits during the first era was followed during the second by the appearance of macromolecules in a gradually changing environment. The atmospheric conditions during this period are assumed to have remained essentially anoxic. Progressively more hydrogen escaped from the earth, and traces of oxygen began to accumulate from the direct decomposition of water and, later, from the action of living organisms. The ozone, formed from the oxygen by the intense ultraviolet radiation, gradually escaped into the upper atmosphere to form a thickening curtain between the short-wave ultraviolet and the earth. Under this ozone blanket organic compounds became more diversified and complex. The direct formation of the amino and aliphatic acids from methane and ammonia would decline, but the more elaborate and complex organic materials, for example, the nucleosides, porphyrins and polypeptides, often decomposed by intense ultraviolet radiation, could persist. The details are unknown; but certain classes of organic compound must have appeared since this stage is thought to have terminated with the appearance of life in the true sense of the word.

Many important reactions involved in the formation of macromolecules are dehydrations. Simple sugars are combined through glycosidic linkages to form polysaccharides (Fig. 3.3); amino acids are united by peptide bonds to produce proteins (Fig. 3.6); fatty acids combine with alcohols through the ester linkage to form fats (Fig. 3.5); and anhydride linkages join molecules of phosphoric acid to form pyrophosphate bonds (Fig. 7.2); similar dehydration condensation reactions produce polynucleotides such as ribonucleic acid (Figure. 1.2) which is concerned with the translation of the genetic code in the synthesis of specific proteins. Calvin (1967, 1969) discusses various schemes that might have effected these combinations before the evolution of high-energy phosphate compounds which usually play this role in living organisms. He favors the energy store in a carbon-nitrogen triple bond such as hydrogen cyanide. Hydrocyanic acid and dicyanamide appear in many of the reaction mixtures which have been used to synthesize relatively small organic molecules (Fig. 1.1). Cyanamide and its derivatives are capable of absorbing a water molecule to form a C═O and a N—H group together with a variety of products. Several plausible theories have been detailed but all are rather speculative and will not be described here.

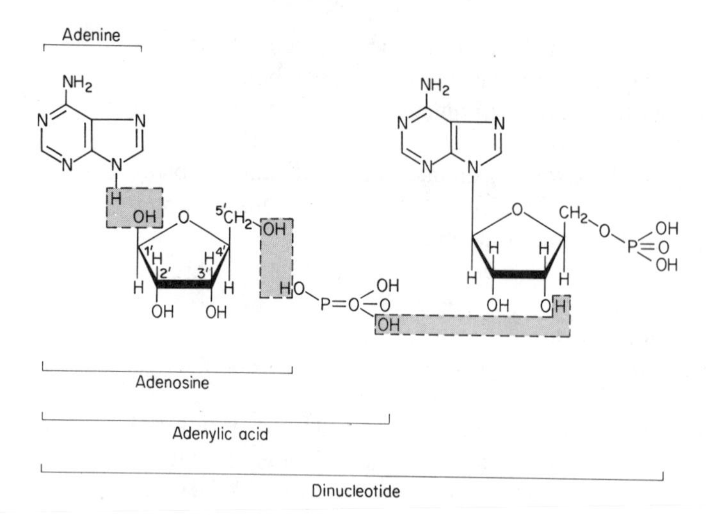

Fig. 1.2 Dehydration condensation reactions in the formation of ribonucleic acid. DNA lacks the OH on the 2′ carbon of ribose. The shaded blocks show positions of dehydration-condensation reactions. [From Calvin, 1969: Chemical Evolution. Clarendon Press, Oxford.]

Life is much more than a suspension of organic molecules and it is obvious that the stage was not yet set for it with the appearance of complex organic polymers. Several important biological processes must first have appeared: in particular, a system of biological catalysts to trigger and control specific chemical reactions, a ready source of biochemical energy and reliable mechanisms for self-duplication and for the preservation of the most valuable components of the system.

Biochemical processes depend on the action of ENZYMES. These are organic catalysts which lower the activation energy of specific reactions and permit life processes to occur in a regulated manner at moderate temperatures. Some of their properties are considered in the next chapter. Here it is noted that enzymes are either proteins or proteins combined with complex non-protein groups (prosthetic groups). Enzyme-catalyzed reactions depend on transitory complexing or interlocking of the substrate with specific areas of the enzyme molecule (the active sites). Although about 1,000 different enzymes have been identified, this is not really a very large number in view of the great diversity of organisms and their many different metabolic processes. This relatively small number argues for a great conservatism in biochemical evolution and supports the hypothesis of a phylogeny of complex and diversified forms from very simple systems of life, with a continuous interaction and interdependence of microorganisms, plants and animals. The evolution of enzymes, regulating metabolic pathways along specific lines, may have been the first step in establishing the biochemical uniformity essential to life (Eakin, 1963; Dixon, 1966).

In the present world, living organisms utilize chemical energy through a group of "high-energy" phosphate compounds, the most important of which is ADENOSINE TRIPHOSPHATE, usually referred to as ATP. This compound has also been synthesized in laboratory experiments under conditions which simulate those that might have preceded life. The structure of ATP (Fig. 7.2) and many details pertaining to the transformation of energy in living systems will be summarized in Chapter 7. At this point, it is noted that the hydrolysis of the terminal phosphate bond of ATP is associated with the exchange of about 8500 calories of free energy per mole; ADENOSINE DIPHOSPHATE (ADP) is formed by the reaction. Much of the chemistry of life has to do with the formation of energy reactive anhydrides of phosphoric acid from the free energy of the environment—either directly through radiant energy from the sun or indirectly from the carbon compounds. Biochemical work is then made possible through the hydrolysis of these pyrophosphate bonds. ATP is like a fully-charged storage battery whose energy can be transformed into mechanical work, light or heat; when a muscle contracts or a gland secretes, or when chemical syntheses are performed, the source of energy is usually the pyrophosphate bond of ATP. The universality of the ATP/ADP mechanism suggests that these compounds appeared early in the prebiologic world. It seems reasonably clear that the process of storing, transforming, and mobilizing energy by the ADP carrier system is older than life itself and has persisted, relatively unchanged, for a billion years or more.

A third prerequisite for life was the evolution of macromolecules concerned with the transfer of information. The many characteristics of living organisms are coded in the nucleic acids which, through self-duplication, pass this code on from one generation to the next. The nucleic acids also provide templates for assembling specific amino acids in the syntheses of proteins.

Deoxyribonucleic acid (DNA) and its closely related co-worker, ribonucleic acid (RNA), are chain compounds composed of four nitrogenous "bases," pentose and phosphate groups (Fig. 1.2 and 22.1). These are strung together in definite sequences to form a code for the duplication of complex organic compounds. The advent of the DNA/RNA machinery replaced the haphazard development of organic systems. When superior units appeared with the power of self-duplication, their advantage would be tremendous. Many of the characteristics of living systems probably preceded cellular organization and were present in self-duplicating macromolecules. The nucleic acids have now been intensively studied by comparative biochemists interested in genetics and evolution; many important clues to the interrelationships of organisms and their probable lines of phylogeny have emerged (Sueoka, 1965; Ohno *et al.,* 1968).

Emergence of Living Systems

Just where the line is drawn between the "living" and the "non-living" depends on the definition of life. Students of biopoesis or life-making have not always agreed on a definition (Keosian, 1966); however, it is universally recognized that the organisms of concern to animal physiologists require a definite structural basis for the orderly operation of their biochemical processes. In short, the macromolecules are enclosed in cell membranes while special internal membranes provide active surfaces for biochemical reactions; it can be assumed that the first living units must have been isolated from their environment by membranes—probably composed of lipoproteins—and that there was a gradual evolution of an internal cellular architecture.

Numerous attempts have been made to form discrete systems under abiologic conditions. Oparin's (1936) COACERVATES and Fox's (1960, 1965) PROTEINOIDS have been the most successful. However, even though enzymes and substrates have been incorporated into some of the coacervate droplets, these structures are static and soon come to an equilibrium with the environment. It has not yet been possible to achieve the stability found in living systems, where an unstable structure is maintained through the utilization of energy and materials from the environment.

Many problems associated with the origin of life remain to be investigated (Miller and Orgel, 1974). For present purposes it seems safe to assume that the first living organisms were small and simple in structure; they may have been of bacterial size but less rigidly organized. If the assumptions concerning the atmosphere are correct, all of them were anaerobic. Alcoholic fermentations as described for bacteria and yeast (Fig. 7.9), are anaerobic and produce sufficient ATP to meet the demands of these simple organisms. However, the energy yield per molecule of food is a relatively small part (2 per cent to 5 per cent) of that available to the aerobic organisms. Butyric acid fermentation is considered a very primitive reaction. It often occurs in the decomposition of organic material in bogs and swamps. Such processes are assumed to have occurred also in the primitive anaerobic world.

The energy for life in this phase may not have been entirely chemical. Solar radiation may have been vital, even before the chlorophyll-containing organisms made use of it for synthetic purposes. As Gaffron and others have emphasized, the rate of evolution would have slowed down considerably during this stage if it had depended entirely on the energy stored in organic compounds. Although the action of the short, more energetic components of solar radiation was greatly reduced, photochemical energy could have been abundantly available through various pigmented compounds. Iron and copper salts and many well-known dyes produce photochemical effects by absorption of energy in the visible spectrum. Transfer of hydrogen and varied oxidation reactions can be greatly accelerated by light-excited dyes. The porphyrin compounds, all of which are colored, may have played a vital part in photochemical reactions before one of them (the magnesium porphyrin, chlorophyll) initiated the era of photosynthesis.

PHOTOSYNTHESIS AND THE ORIGIN OF AN AEROBIC WORLD

The advent of photosynthesis marked the beginning of a fourth stage in the evolution of life—the stage with which we are primarily concerned in comparative animal physiology. In this, plants utilize radiant energy to produce ATP and then to synthesize carbon compounds and store chemical energy for themselves and for the animals which feed on them. "Each year the plants of the earth combine about 150 billion tons of carbon with 25 billion tons of hydrogen, and set free 400 billion tons of oxygen" (Rabinowitch, 1948). This liberation of oxygen gradually converted an anaerobic world into an aerobic world and set the stage for Darwinian evolution. If the constant flow of carbon dioxide into the plants and oxygen into the environment were to cease, the earth would rapidly revert to its primeval anaerobic condition; the free carbon dioxide would soon be tied up in carbonates and the oxygen would form inert oxides.

Although many of the biochemical details of photosynthesis are still unresolved, the major events are now known. Investigators made slight progress for centuries until modern tracer techniques of radiochemistry and chromatography disclosed the intricate flow pattern of carbon, hydrogen, oxygen, and phosphorus and until electronmicroscopy revealed the molecular architecture of the photosynthetic machines (Gabriel and Fogel, 1955; Hill, 1965). The major steps are outlined at this point but many of the details will be more understandable after the processes of energy transfer have been considered in Chapter 7. The intention here is to emphasize the significance of photosynthesis in the evolution of animals and their environment.

Photosynthesis, like other biological processes, evolved gradually and it is unlikely that there was ever a sharp break between the non-photosynthetic and the photosynthetic world, between the anaerobic and the aerobic. In the microbial world (Thimann, 1963), the green (Chlorobacteriaceae) and the purple to red (Thiorhodaceae) sulfur bacteria utilize radiant energy to reduce CO_2 and to oxidize H_2S, forming sugars, represented by (CH_2O) in the following scheme:

$$CO_2 + 2H_2S \xrightarrow{\text{light}} (CH_2O) + H_2O + 2S$$

At high concentrations of H_2S, sulfur granules tend to accumulate in the cells of the purple species and are deposited extracellularly in the green ones. As the supply of H_2S becomes exhausted, the accumulated sulfur is also oxidized; sulfate is the end product as shown by the following summary reaction:

$$2CO_2 + H_2S + 2H_2O \xrightarrow{\text{light}} 2(CH_2O) + H_2SO_4$$

Some of the purple sulfur bacteria can utilize hydrogen directly or obtain it from a variety of hydrogen donors such as simple organic acids represented by H_2X in the second equation below:

$$CO_2 + 2H_2 \xrightarrow{\text{light}} (CH_2O) + H_2O$$

$$2H_2X + CO_2 \xrightarrow{\text{light}} 2X + (CH_2O) + H_2O$$

These green and red sulfur bacteria are anaerobic and strictly dependent on sunlight as a primary source of energy. Similar forms probably existed at an early stage in the origination of life. Their photochemical processes are referred to as PHOTOREDUCTIONS to distinguish them from PHOTOSYNTHESIS which occurs in green plants and results in the elaboration of sugars through the reduction of CO_2 with electrons supplied from water:

$$CO_2 + H_2O \xrightarrow{\text{light}} (CH_2O) + O_2$$

Van Niel (1935) emphasized the generalized nature of the light reaction in all of these photochemical reactions. The process may be represented by an oxidation-reduction reaction as follows, where X represents oxygen in the green plants, while it represents other substances such as sulfur or organic groups in the photosynthetic bacteria.

$$CO_2 + 2H_2X \longrightarrow (CH_2O) + H_2O + 2X$$

This concept has been amply confirmed through tracer carbon techniques with a chromatographic identification of the intermediate products (Arnon, 1960). There is still speculation about biochemical details, but the major events have been outlined. For the animal physiologist, the most significant of the processes is that in which solar energy liberates oxygen from water into the atmosphere and passes the hydrogen ions and electrons to a series of carriers, thus forming reduced pyridine nucleotides and ATP. This process not only permitted the evolution of animals but also that of the environment in which they live.

Synthesis of Phosphate Bonds and Reduced Pyridine Nucleotide

Whether in sulfur bacteria or green plants, the photosynthetic processes must have been preceded by the evolution of many macromolecules concerned with the transfer of electrons—substances such as ferredoxin (Eck and Dayhoff, 1966), the cytochromes and the pyridine nucleotides, all of which will be identified in Chapter 7. It also required the organization of specialized cellular structures (chloroplasts or

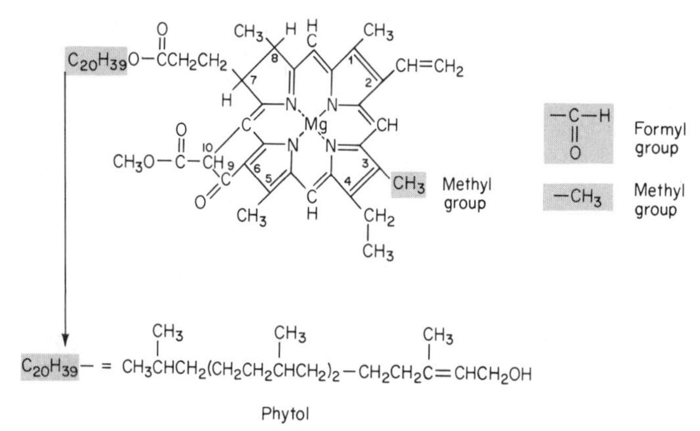

Fig. 1.3 Structure of chlorophyll *a*. The methyl group shown in the formula for chlorophyll *a* is replaced by a formyl group in chlorophyll *b*. [From Lehninger, 1971: Bioenergetics, 2nd ed. Benjamin, Menlo Park, California.]

their counterparts) containing chlorophyll and other pigments arranged in an orderly manner (Lehninger, 1971). The chlorophyll molecule which is the target for the action of light is a magnesium porphyrin (Chapter 6 and Fig. 1.3) capable of emitting electrons when excited by radiant energy. There are several different forms of chlorophyll with different absorption spectra and electron emitting capacities. The indications are that the chlorophylls found in green plants were preceded phylogenetically by the more primitive chlorophylls in bacteria and algae.

Two distinct processes can be studied in all photosynthetic organisms. The first is photochemical in nature and leads to the phosphorylation of ADP (PHOTOPHOSPHORYLATION) while the second occurs independently of light (DARK REACTION), requires a battery of special enzymes and utilizes ATP (and reduced NADP) to incorporate CO_2 into sugars (FIXATION OF CARBON DIOXIDE).

Three patterns of photophosphorylation are of interest in the present context. In all of them, absorption of light by chlorophyll leads to the emission of high-energy electrons which are accepted by substances like ferredoxin or plastoquinone (Arnon, 1960; Arnon *et al.,* 1965; Fogg, 1972); these substances are reduced as a consequence of acquiring additional electrons. Reduced ferredoxin and plastoquinone have high electron pressures (Chapter 7) and are thus able to pass electrons along to other electron transport substances (cytochromes) arranged in a series with progressively lower electron pressures. The substances forming the links in this chain are, in succession, first oxidized and then reduced; as the energy drops to lower levels it is channeled into the phosphorylation of ADP or the reduction of pyridine nucleotides (NADP), both of which are essential for the synthesis of sugars from CO_2 and H_2O.

Electrons emitted from chlorophyll must be quickly replaced. In CYCLIC PHOTOPHOSPHORYLATION, the same electrons are returned to the chlorophyll by the carrier system; in BACTERIAL NON-CYCLIC PHOTOPHOSPHORYLATION, electrons are replaced through the oxidation of such substances as succinate or thiosulphate; in NON-CYCLIC PHOTOPHOSPHORYLATION OF PLANTS, water becomes the electron donor and oxygen is released into the atmosphere. This latter process, which occurs only in the green plants, depends on more complex chlorophyll systems than those utilized by the other two, neither of which leads to the formation of oxygen. The cyclic and bacterial types might have contributed substantially to the energy budget of life while the earth was still anaerobic and before the evolution of plants.

The important steps in the cyclic process, as visualized by Arnon (1966), are shown in Fig. 1.4a. Each pair of electron equivalents which leaves the chlorophyll loses about 50,000 calories of energy before it returns to chlorophyll again (Lehninger, 1971). At two points this energy flow is coupled to the ADP/ATP phosphorylation and the plant increases its supply of chemical bond energy accordingly.

In the bacterial non-cyclic system (Fig. 1.4b), phosphorylation occurs at only one of the links but the process also results in the reduction of nicotinamide adenine dinucleotide phosphate (NADP), one of the basic compounds required for the fixation of CO_2. If succinate or thiosulfate is involved, the oxidation yields electrons and

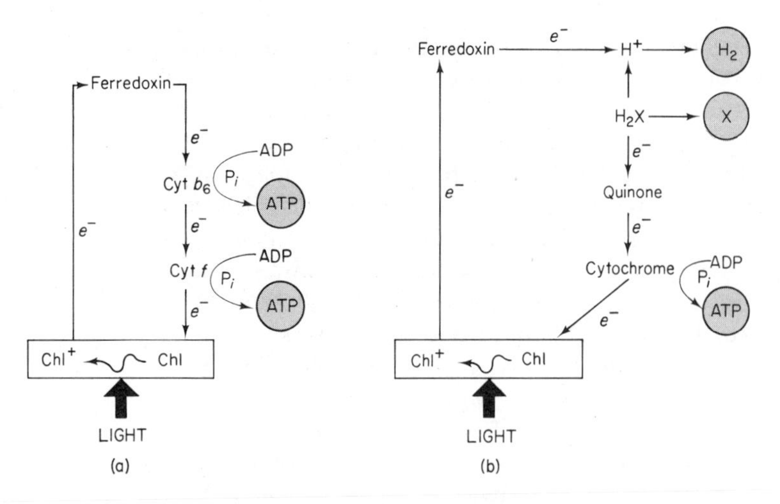

Fig. 1.4 Two schemes for photophosphorylation in primitive organisms: (a) cyclic; (b) non-cyclic bacterial, without evolution of oxygen. In each case, chlorophyll (Chl) becomes excited by absorption of light and donates a high-energy electron (e^-) to the photosynthetic system. In (a) the same electron is returned to the chlorophyll through a series of carriers; in (b) electrons are donated by a substance such as succinate or thiosulphate (H_2X) which becomes oxidized; the resulting hydrogen is released as gas in some bacteria or used to reduce NADP in others. [Based on Arnon, Tsujimoto, and McSwain, 1965: Nature, *207*: 1367.]

hydrogen ions; in succeeding reactions, electrons are restored to chlorophyll and used also to reduce NADP according to the following schemes:

$$\begin{array}{c} \text{COOH} \\ | \\ \text{CH}_2 \\ | \\ \text{CH}_2 \\ | \\ \text{COOH} \end{array} \xrightarrow{\text{succinic dehydrogenase}} \begin{array}{c} \text{COOH} \\ | \\ \text{CH} \\ \| \\ \text{CH} \\ | \\ \text{COOH} \end{array} + 2\text{H}$$

Succinic acid → Fumaric acid

and

$$\text{NADP}^+ + 2\text{H}^+ + 2e^- \longrightarrow \text{NADPH} + \text{H}^+$$

Oxidized NADP → Reduced NADP

The anaerobic photosynthetic bacteria that utilize these reactions may also reduce nitrogen to form ammonia—a reaction of considerable importance in the synthesis of nitrogen compounds.

The non-cyclic photophosphorylation of green plants depends on chlorophyll *b* while both the cyclic and the bacterial types can operate with chlorophyll *a*. Chlorophylls *a* and *b* differ both in light absorbing and electron emitting properties (Goodwin, 1966; Gregory, 1971). Under the influence of light, chlorophyll *b* transfers electrons from OH^- (water) *via* plastoquinone and chlorophyll *a* to ferredoxin; phosphorylation of ADP is thus coupled with the oxidation of the OH^-. This step, known as the PHOTOLYSIS OF WATER, is still not well understood and has proven to be the most difficult of the many complicated processes to unravel. The scheme shown in Fig. 1.5 is still provisional.

Cyclic photophosphorylation may be coupled with the non-cyclic process in the higher plants which contain both chlorophyll *a* and chlorophyll *b* (Fig. 1.5); highly efficient energy transfers are assumed between these chlorophylls in white light. The end result is the production of ATP, oxygen and reduced ferredoxin which may be oxidized to form reduced NADP in a manner comparable to that shown above for bacterial non-cyclic phosphorylation; one of the H^+ arises from the hydrolysis and the other from the photolysis of water while the reduced ferredoxin supplies the electrons. The process yields ATP and NADPH in the ratio of 1:1. As noted in the

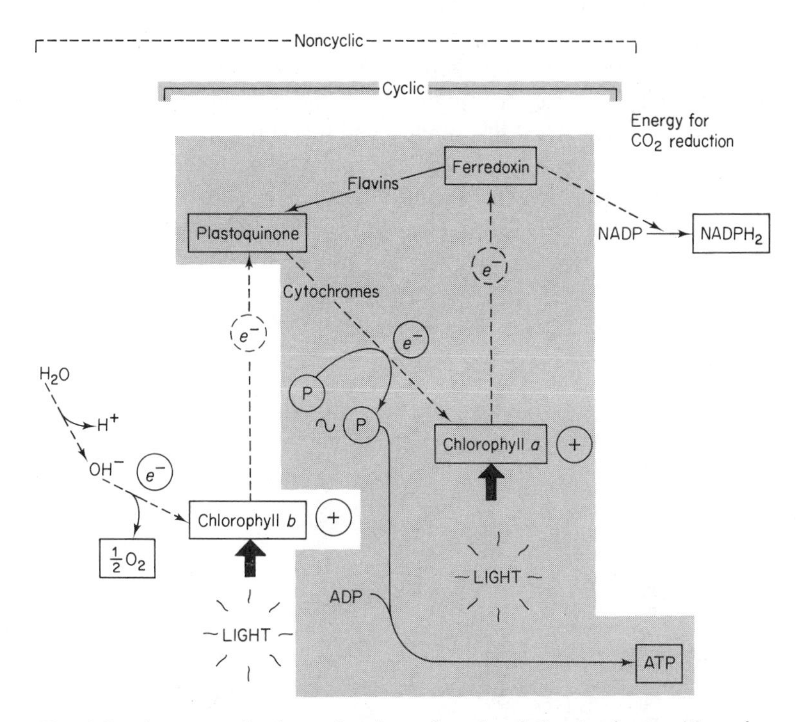

Fig. 1.5 A proposed scheme for photophosphorylation in plants with evolution of oxygen and formation of ATP and $NADPH_2$. The connection between the cyclic and non-cyclic systems is still hypothetical (Gregory, 1971). NADP is reduced by electrons and H^+ from the photolysis of water. [Based on Bennett and Frieden, 1966: Modern Topics in Biochemistry. Macmillan, New York.]

next section, two moles of NADPH and three of ATP are required for one mole of CO_2 assimilated to make sugar. Arnon (1966) assumes that the non-cyclic process will operate until the ATP supply is exhausted and sugar synthesis ceases; at this point the cyclic system comes into play and the electrons from reduced ferredoxin begin to cycle within the chloroplasts to increase the supply of ATP; the incorporation of CO_2 into sugars is then resumed. Thus, these two complementary and parallel pathways jointly supply the power for the fixation of carbon dioxide. This scheme is still hypothetical (Gregory, 1971).

Fixation of Carbon Dioxide

In this process, ATP and NADPH which were produced in the light are utilized to reduce atmospheric carbon dioxide and then, to synthesize the carbohydrates, fats, and proteins characteristic of plant protoplasm. This second system is an enzyme-operated process which does not require sunlight but can proceed just as well in the dark. The two systems—one dependent on light and the other independent of it—must have evolved separately and then interlocked fortuitously (Calvin, 1962a).

The important steps in carbon fixation are summarized in Fig. 1.6, but the details will be more meaningful when the section on tissue respiration has been studied in Chapter 7. At this point it should be noted that carbon dioxide, water, phosphate in the form of ATP, and hydrogen in reduced pyridine nucleotide are fed into an enzyme pool which synthesizes a three-carbon compound (glyceraldehyde-3-phosphate or PGAL). The cycle of chemical changes (the RIBULOSE DIPHOSPHATE

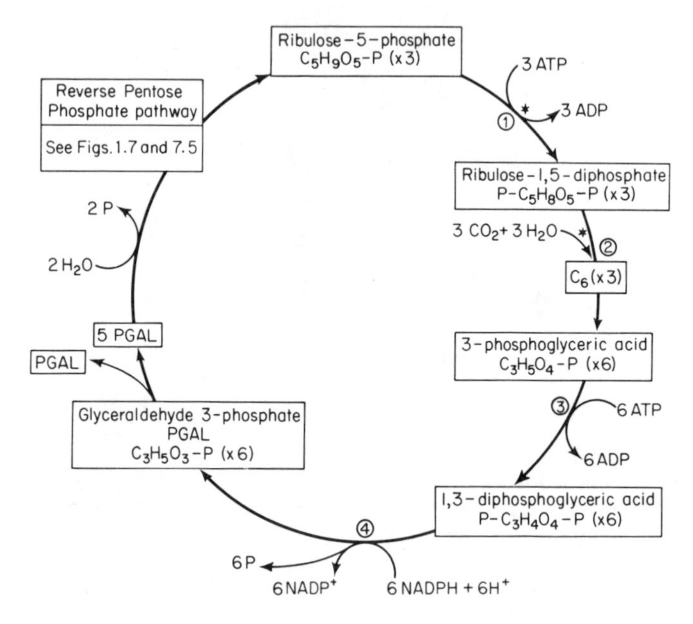

Fig. 1.6 Carbon dioxide fixation. Asterisk, reactions peculiar to photosynthesis while others are also found in tissue respiration. *Enzymes*: (1) phosphoribulokinase and Mg^{++}; (2) ribulose diphosphate carboxylase or carboxydismutase; (3) 3-phosphoglycerate-1-kinase and Mg^{++}; (4) glyceraldehyde-3-phosphate dehydrogenase.

CYCLE OF CALVIN CYCLE) may be considered to start and end with ribulose-5-phosphate, and with each turn of the "wheel" one unit of this three-carbon compound, PGAL, is synthesized from inorganic sources. In Chapter 7 it will become evident that the PGAL thus synthesized enters into other enzyme pools which lead directly to complex carbohydrates, fats and proteins and that the compounds and the enzymes (with two exceptions) which form the carbon dioxide fixation cycle (Fig. 1.6) are also present in tissue respiration. In general, the carbon dioxide fixation cycle elaborates the carbon compounds, and tissue respiration breaks them down in reversed reactions.

Attention is directed to the link between PGAL and ribulose-5-phosphate (Fig. 1.6). At this point, a complex series of reactions rearranges five units of PGAL (a 3-carbon compound) into three units of pentose (a 5-carbon sugar); the sixth unit of PGAL is available for syntheses of more complex substances. The many intermediate steps in this important link have been traced with radioactive carbon and it has been shown that additional carbon compounds, besides the C_3 and C_5 fragments, are formed as intermediates and become available for metabolism (Bassham, 1962; Calvin, 1962b). In short, the processes involved in carbon dioxide fixation provide a series of small carbon compounds for biosynthetic purposes. A diagram of these events with respect to carbon is shown in Fig. 1.7 where the squares show only the number of carbon atoms in the carbohydrate molecules concerned. Reference will again be made to these important pathways in Chapter 7.

Autotrophic and Heterotrophic Ways of Life

In our world, two main streams of life are quite distinct. The plants (AUTOTROPHIC or self-nourishing organisms) utilize inorganic materials to elaborate a

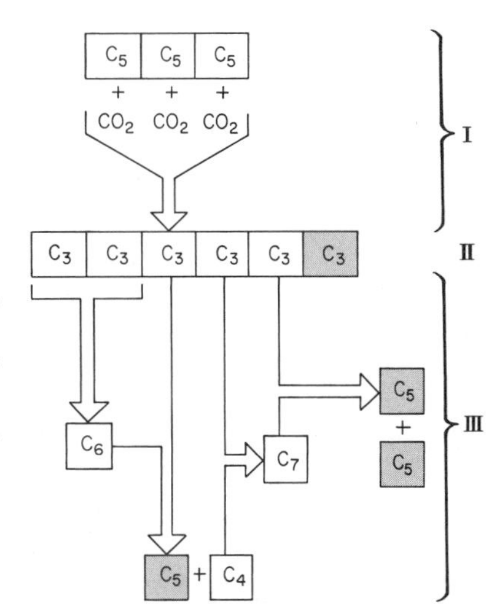

Fig. 1.7 Balance of carbons in the carbon-reduction cycle of photosynthesis. I, carbon dioxide fixation where three molecules of ribulose and three molecules of CO_2 combine to form six molecules of PGAL (II); III, reverse pentose pathway to recycle the three molecules of ribulose and yield C_4, C_6, and C_7 carbon fragments. [From Awapara, 1968: Introduction to Biological Chemistry. Prentice-Hall, Englewood Cliffs, N.J.]

great variety of organic compounds; the animals (HETEROTROPHIC organisms) exploit the organic compounds elaborated by the plants. The distinction is not so sharp among the more primitive forms of life and must have been even less evident during early periods of evolution. A great range of metabolic machinery has been found in present-day bacteria; investigation of their metabolism has provided many useful clues to biochemical evolution (van Niel, 1954; Starkey 1962; Pfenning, 1967).

Chemosynthetic bacteria These are forms which derive their primary energy from the oxidation of inorganic compounds rather than from sunlight. The oxidizable materials include sulfur, hydrogen sulfide, thiosulfate, ferrous and manganous salts, methane, carbon monoxide or formaldehyde, hydrogen gas, ammonia, and nitrite (Thimann, 1963; Rittenberg 1969). The energy yields are small (of the order of 25 to 120 kilocalories per mole) but sufficient to elaborate the familiar carbon compounds of protoplasm from carbon dioxide, ammonia or other nitrogen source and water.

This remarkable group of organisms ranges in metabolic capacities from those which obtain all of their energy from inorganic compounds to the heterotrophs; there are also many intermediate forms which combine the capacities of both groups (Rittenberg, 1969). The nitrifying bacteria *Nitrosomonas* and several species of colorless sulfur bacteria (*Thiobacillus*) belong to the first category. Special enzyme systems such as ammonia dehydrogenase in *Nitrosomonas* catalyze the oxidation of the substrate (ammonia in this case) with the energy channeled into the ADP/ATP system. Fixation of CO_2 then occurs by means of the ribulose diphosphate cycle in a manner similar to that described for photosynthesis (Fig. 1.6). In contrast, the strictly heterotrophic species like *Desulfovibrio desulfuricans* lack the ribulose diphosphate cycle and must acquire a supply of preformed simple organic compounds.

Photosynthetic bacteria The photosynthetic bacteria are particularly interesting since they contain pigments (chlorophylls, carotenoids) and make direct use of the sunlight as well as of a variety of oxidizable materials, including such inorganic substances as the sulfur compounds (Pfenning, 1967). In addition to the sulfur anaerobes referred to previously, the purple (or red or brown) bacteria (Athiorhodaceae) are non-sulfur organisms requiring organic compounds as hydrogen donors and certain growth factors (vitamins) for their metabolism. In the dark they live aerobically and oxidize a variety of organic materials; in the light they utilize radiant energy and synthesize organic compounds photochemically. Hydrogen is obtained from such donors as acetate, propionate or butyrate (Thimann, 1963). These photoreductions operate according to the anaerobic non-cyclical scheme shown in Fig. 1.4b).

Photoreduction to photosynthesis The step from photoreduction to photosynthesis may be bridged by the type of metabolic machinery found in the single-celled green alga *Scenedesmus* (Gaffron, 1944; Hill, 1965). When adapted to anaerobic conditions, *Scenedesmus* develops an active hydrogenase and displays the capacities of a chemoautotroph. When in the dark at low oxygen concentrations, these anaerobically adapted cells, like the bacteria *Bacillus pycnoticus* or *Hydrogenomonas,* reduce carbon dioxide in the presence of hydrogen; the hydrogen is oxidized to water in this process as follows:

$$6H_2 + CO_2 + 2O_2 \longrightarrow (CH_2O) + 5H_2O$$

If illuminated under anaerobic conditions, photoreduction occurs; CO_2 is utilized and hydrogen taken up as illustrated for the red sulfur bacteria on page 9. If the light intensity is increased beyond a certain point, the hydrogenase system is suppressed, photolysis occurs, oxygen is liberated, and normal photosynthesis takes place.

These photochemical processes in bacteria and algae suggest that the phylogeny of chlorophyll resulted in a sequence of organisms. The most primitive ones may have used chlorophyll only to increase their supply of synthetic energy in the form of ATP; more specialized forms went a step further to convert CO_2 into organic matter; finally the green plants dispensed with oxygen acceptors and produced free oxygen (Thimann, 1963). With the advent of an oxygen-rich environment, the stage was set for aerobic living and the evolution of animals.

THE ANIMAL AND ITS ENVIRONMENT

The animal and its environment have evolved together. Before any animals appeared an atmosphere of H_2, NH_3, and CH_4 was converted into one of N_2, CO_2 and O_2, and a planet of bare rock was gradually covered with thick layers of pulverized earth, rich in soluble nitrates, sulfates, phosphates and organic materials. Then the developing cover of green plants provided oxygen for respiration and shelters for the animals. The waters were filled with oxidized materials, both animate and inanimate, and the air became a transport medium for bacteria, seeds, and spores, and a highway for insects, birds and some mammals. The animal, in its evolution as in its daily living, is inseparable from the environment.

The physiologist may look at animals in many ways. In one sense, they are biochemical machines, acquiring elements from the environment and arranging these in a series of electron transfers to provide the necessary energy for the operation of their machinery. They must acquire fuel, distribute it, burn it and remove the waste. These problems are considered in Part One of this book. An animal usually operates in an oscillating environment. Cold periods are followed by warmer periods, day is followed by night, and the organism is geared to adjust and to regulate, to tolerate and to resist these changes. The mechanisms related to these environmental compensations are discussed in Part Two. To the layman, perhaps, the most characteristic thing about animals is the way in which they move and behave, the manner in which they see and hear and react to other animals or to objects in their environment. These activities depend on a battery of sensitive receptors, a group of specialized effector organs and an elaborate integrating nervous system. The physiology related more specifically to animal activity is dealt with in Part Three. Finally, life is constant renewal; the old must be replaced by the new. Only through replacement by a somewhat variable progeny has an evolutionary change been made possible and the formation of increasingly more complex and specialized animals become a reality. The physiology of reproduction and development is considered in the final section.

Any of these topics might be treated in various ways. One might emphasize the many functional differences or stress the similarities in patterns of metabolism and machinery. Superficially, the differences are more evident, and the pioneers in comparative physiology were particularly concerned with their tabulation and description. However, with the accumulation of information it has become evident that there are many fundamental similarities in metabolism and that the basic life processes are essentially the same wherever they are found.

2

REGULATORY MECHANISMS

Comparative physiology owes one of its most basic concepts to the renowned French physiologist of the nineteenth century, Claude Bernard. As early as 1859 he pointed out that complex organisms live in two environments—an external environment which is the same for both animate and inanimate objects and an internal fluid environment which surrounds the cells and tissues of the body, is characteristic of the animal species and remains relatively constant. Claude Bernard argued that the animal did not really live in the *milieu extérieur* but in the liquid *milieu intérieur*—which bathes the tissues and serves as a medium for the exchange of foods and wastes and for the distribution of chemical messengers of many kinds.

Bernard made many contributions to physiology, but the outstanding one was his appreciation of the vital role of this internal environment. He wrote, "It is the fixity of the '*milieu intérieur*' which is the condition of free and independent life," and, "All the vital mechanisms, however varied they may be, have only one object, that of preserving constant the conditions of life in the internal environment" [Bernard (translation), 1957; Cannon, 1929, 1939]. The individual cells and tissues of the complex organisms exist in a remarkably constant environment and, during the hundred years since Bernard, it has become evident that many of the familiar physiological phenomena are essentially mechanisms directed to the maintenance of this steady state. W. B. Cannon (1939) applied the term "homeostatic" to these coordinated physiological processes which maintain steady states. By HOMEOSTASIS, Cannon did not mean a static or stagnant condition, but one which varied only within narrow and precise limits. These limits of variability and the means of regulation form a large part of the classical physiology with which Bernard and Cannon and a host of other investigators have been concerned.

Concepts of the internal environment and homeostasis were first developed by mammalian physiologists with particular reference to the autonomic nervous system and the hormones which regulate metabolism (Adolph, 1961). Comparable controls have now been studied in the lower vertebrates and in many of the invertebrates and, although the variability is regularly greater in the lower forms, it is characteristic of all physiological processes to operate within definite limits. Rates of metabolism, blood sugars, moisture, electrolyte content and many other factors show relatively minor fluctuations. This concept of the regulated physiological rate can be extended to the cellular and enzyme level, for it is just as characteristic of the enzyme reactions as it is of the complex processes which ultimately depend on them. The genetic system preserves a constancy in the lines of protoplasm, and this is expressed in the physiology and biochemistry as well as in the morphology. It extends from the enzymatic to the cellular level and from the cell to the organ-system and on to the social level of organization.

Prosser (1955) has made a useful distinction between physiological adjustment or conformity and physiological regulation. This distinction, developed in connection with an organism's responses to altered environmental conditions, is also valuable in a discussion of homeostatic mechanisms. The less specialized animals operate over a much wider range of internal environmental conditions than the more complex ones. Blood sugars are more variable; the buffering action of the blood is less; the osmotic value of body fluids varies, and temperatures change with those of the environment. Here ADJUSTMENTS of the internal environment occur, and the tissues function over wide ranges. The internal environment, in contrast, is relatively stable in the higher vertebrates with an almost constant temperature and with the blood sugar, electrolytes and other constituents more precisely REGULATED. The evolution of large, active and more specialized animals has been associated with a phylogeny in regulatory mechanisms, from simple adjustments to precise regulations. There are, nevertheless, limitations in these adaptive processes for the animal which adjusts as there are for the animal that regulates. At the molecular level, the most frequent limitation is probably in the range of enzyme activity. In unicellular forms, the metabolic controls are entirely adjustments within a complex system of interconnected enzyme reactions; in multicellular animals, the many organs and organ-systems are integrated through nervous and endocrine mechanisms, although these too are limited by enzyme-controlled processes.

FACTORS REGULATING ENZYME ACTIVITY

Even the simplest manifestations of life involve a great complexity of enzymatic reactions. The fermentation of sugar (glucose), for example, requires a dozen well-known enzymes with associated coenzymes and activators (Chapter 7). These catalysts are highly specific, and the reactions always occur in orderly sequence. Characteristically, the control which they exercise is not on the nature of the reaction but on its rate. The kinetics of enzyme activity are discussed in textbooks of biochemistry and enzymology, and only a brief summary of the major rate-controlling factors is attempted here.

Enzyme-Substrate Complex

To begin with, enzymes are pure proteins or proteins associated with a complex organic group (prosthetic group). They associate themselves reversibly with the materials on which they act (substrates) in a relationship (THE ENZYME-SUBSTRATE COMPLEX) which is intimate and specific. It may involve both the protein portion and its prosthetic group. This union is so intimate and specific that it is often likened to a complex lock and its individual key. While in this association, the substrate is in some manner activated, subjected to molecular strain, or otherwise altered, so that the reaction path is smoothed and the end products form rapidly. Thus, the enzyme reduces the amount of energy required to form the activated complex.

The hypothesis of an enzyme-substrate combination has long been supported by an impressive amount of indirect evidence, and more recently it has been possible to identify certain intermediate compounds and demonstrate the reality of the complex (Baldwin, 1967; Yost, 1972). Biochemists agree that the enzyme-substrate complex is an essential first step in enzyme action; this means that reactions of this sort are strictly limited by the amount of enzyme present or, to put it another way, by the specific amount of reacting enzyme surface. When these enzyme surfaces, or their reactive centers, are tied up with substrate, reaction velocity will reach a plateau.

It is important to understand that these active enzyme centers may also be tied up by compounds similar in structure to a substrate but incapable of forming the end products. In this manner, essential reactions may be blocked. The phenomenon, known as COMPETITIVE ENZYME INHIBITION, is extremely important, both theoretically and practically. Theoretically, it has been used to investigate the elusive enzyme-substrate complex since some of the enzyme-inhibitor complexes are more readily isolated and identified than the normal substrate products (Baldwin, 1967). Practically, a multitude of important metabolic poisons which work on this principle are now used in medicine, in insect toxicology and in research (Adams, 1959). Two examples will suffice at this point. The sulfa drugs bear a marked structural similarity to an important biochemical, *p*-aminobenzoic acid (P.B.A.). So similar are they that in the presence of sulfanilamide and related compounds, bacteria (which require P.B.A. for their metabolism) may fail to obtain sufficient P.B.A. for growth. Their multiplication is thereby checked to the benefit of the sufferer with certain infectious diseases. Other interesting examples are found in some of the nerve poisons. Transmission of nerve impulses at many synapses and myoneural junctions is dependent on the release of acetylcholine. Only a brief period is required for stimulation, and the acetylcholine is then destroyed by the enzyme cholinesterase. If acetylcholine accumulates through failure of the cholinesterase mechanism, a continual stimulation with serious consequences may be expected. This is precisely what takes place when physostigmine, a poison once used in West African ordeal trials, is present in the system. Physostigmine, and the synthetic substitute neostigmine, combine with cholinesterase about 10,000 times more readily than acetylcholine and thus form competitive inhibitors of the enzyme cholinesterase.

Kinetics of Enzyme Action

Since the enzyme-substrate combination is essential for enzyme activity, relative amounts of the two substances forming the complex are of major significance in the reaction kinetics. Figure 2.1 is based on data from the hydrolysis of soluble starch (substrate) in reaction mixtures containing pancreatic amylase. Enzyme activity is expressed as mg of reducing sugar produced from the starch. When the amount of enzyme is constant (Fig. 2.1a), activity increases regularly with additional amounts of substrate up to a maximum where all the enzyme is (at that moment) associated with substrate. This is the sort of relation expected if an enzyme-substrate combination is requisite to the hydrolysis of starch by amylase. When the amounts of enzyme are varied, in the presence of excess substrate (Fig. 2.1b), activity rises in direct relation to the available enzyme, again in accordance with the enzyme-substrate concept.

These relationships were first analyzed quantitatively by Michaelis and Menten (1913) who assumed that an enzyme-substrate complex (ES) was an essential intermediate step in all enzyme reactions and that end products can only be formed by way of this complex. The overall reaction may be written as follows:

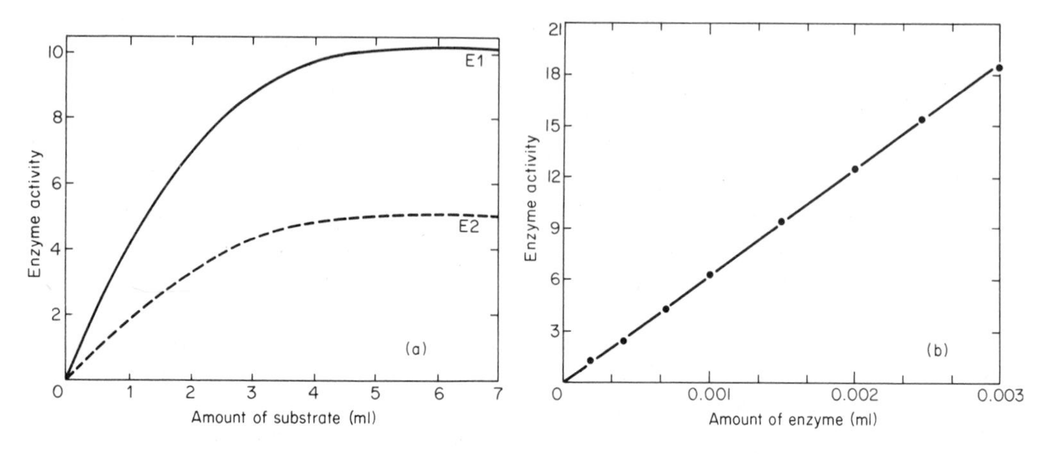

Fig. 2.1 Some factors controlling the rates of starch digested by pancreatic amylase. Enzyme activity measured as mg reducing sugar formed. (a) Amount of substrate varied with constant amounts of enzyme—E1 (upper) and E2 (lower). (b) Optimum amount of substrate and variable amounts of enzyme. The substrate was 1 percent soluble starch; the enzyme was a sample of duodenal contents at optimum pH and chloride. [From Myers and Free, 1943: Am. J. Clin. Path., *13*: 45.]

$$\mathrm{E} + \mathrm{S} \underset{k_2}{\overset{k_1}{\rightleftharpoons}} \mathrm{ES} \underset{k_4}{\overset{k_3}{\rightleftharpoons}} \mathrm{P} + \mathrm{E}$$

where E is the enzyme; S is the substrate, and P the reaction product; the k values represent the velocities of the reactions indicated by arrows. Enzyme reactions will attain a maximum velocity (V_{max}) when all the enzyme in the system is present in the complex ES. In theory, V_{max} is a useful parameter for the comparison of enzyme velocities under different conditions. In reality, however, it is frequently impossible to measure either V_{max} or the maximum amount of ES formed; anomalies often appear at high substrate concentrations and, moreover, it becomes difficult to decide when maximal velocities occur with the rate increasing by very small increments as V_{max} is reached. Further, in living cells the substrate concentrations are usually low and V_{max} may have little biological meaning. A more easily assessed value and one with greater biological significance is the velocity at half saturation ($\frac{1}{2}V_{max}$). At this point, half of the enzyme is free while half is in the ES complex. The Michaelis-Menten constant, referred to as K_m, is the *substrate concentration* at which the velocity is half maximal (expressed in moles per liter). The derivation of the Michaelis-Menten equation is given in texts of biochemistry and cell physiology (for example, Lehninger, 1970, and Yost, 1972); the relationships may be readily visualized from Fig. 2.2. Because of the practical problems of measuring maximum velocities, the data are usually transformed algebraically to a linear form which permits ready plotting. One of the most useful transformations is shown in Fig. 2.2 where the reciprocals of the velocities are plotted against the reciprocals of the substrate concentrations; the intercept on the horizontal axis is $1/K_m$ while the intercept on the vertical axis is $1/V_{max}$. K_m values are important parameters in comparative enzyme studies; they vary in different enzymes and also for the same enzyme under different reaction conditions. An enzyme with a large K_m value is one that is saturated at high substrate

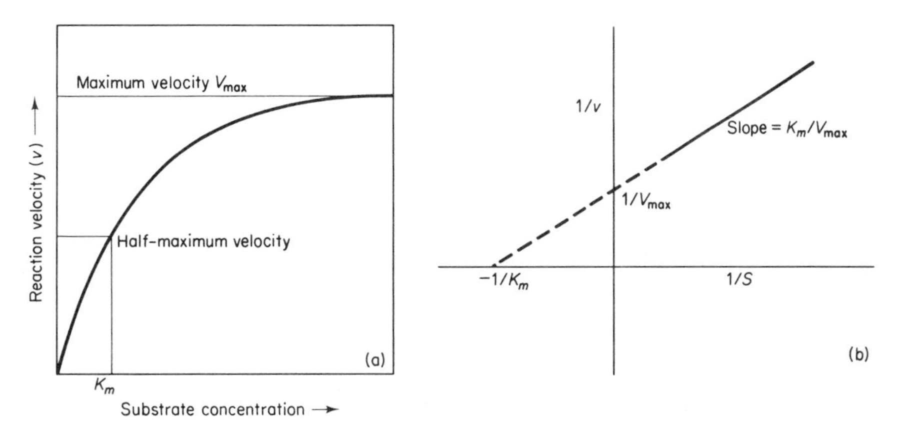

Fig. 2.2 Relation between substrate concentration (S) and reaction velocity (v) when the concentration of enzyme is constant. (a) The Michaelis-Menten plot. (b) The Linweaver-Burk plot of the reciprocal of substrate concentration ($1/S$) against the reaction velocity ($1/v$) to give a straight line. Further description in the text.

concentrations while a small K_m value indicates saturation at low substrate concentrations (Figs. 2.2 and 2.5).

Factors influencing the velocity of enzyme reactions By international agreement, the activities of enzymes are expressed in terms of enzyme units. An ENZYME UNIT is the amount of enzyme required to transform 1 micromole of substrate in one minute at 25° under optimal conditions of measurement. The SPECIFIC ACTIVITY, defined as the number of enzyme units per mg protein, is a measure of enzyme purity.

The velocities of enzyme reactions are controlled by several factors in addition to the relative concentrations of enzyme and substrate. Temperature, hydrogen ions, and the presence of activators or inhibitors are particularly significant in both *in vitro* and *in vivo* systems. Enzyme reactions, like other chemical activities, depend on molecular motion and this is, in part, temperature controlled. There is usually a well marked optimum (Fig. 2.3). Further, enzymes are proteins, and proteins, because of their amino and carboxyl groups, are amphoteric electrolytes (ampholytes) which dissociate either as acids or bases, depending on the pH of the solution. Consequently, the chemical properties of the enzyme and its ability to form reactive enzyme-substrate complexes may be expected to vary over a range of pH with the development of a well marked optimum (Fig. 2.4). These simple rules are not without exception. Some substrates are neutral and electrical charge plays no part in the catalysis; papain and invertase, for example, are independent of pH over a wide range. The complexity of temperature effects will be discussed in Chapter 10.

Under natural conditions, these optima may only have biological meaning when related to an appropriate time scale. Berrill's (1929) studies of digestion in an ascidian, *Tethyum,* illustrate this beautifully (Fig. 2.3). When *in vitro* tests of hydrolytic activity were carried out for only a short three-hour period, the optimum

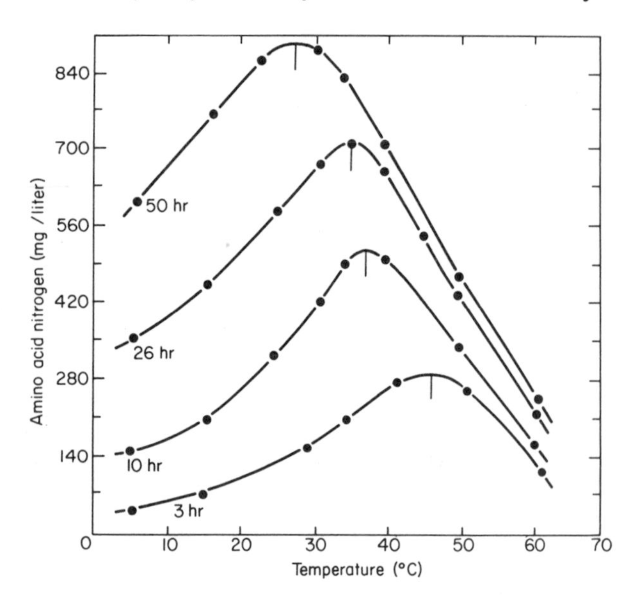

Fig. 2.3 Influence of temperature on the digestion of gelatin by proteinases of the ascidian *Tethyum.* Further description in the text. [From Berrill, 1929: J. Exp. Biol., *6*: 286.]

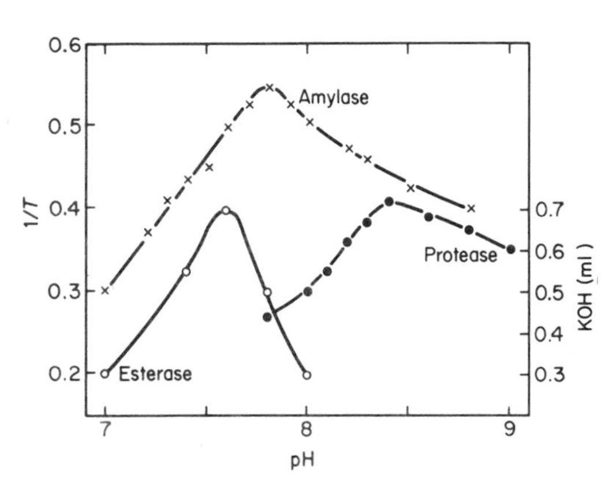

Fig. 2.4 Effects of pH on the action of the digestive enzymes in the midgut of an echiuroid *Ochetostoma erythrogrammon*. $1/T$, the reciprocal of time taken by the amylase-starch mixture to become colorless with iodine; ml KOH, titer of 0.01 N KOH per 0.1 ml digest of protease-gelatine and per ml digest of esterase-benzyl *n*-butyrate. [From Chuang, 1963: Biol. Bull., *125*: 465. Permission of the Managing Editor.]

temperature for the formation of reaction products was as high as 45°C. If, however, the *in vitro* tests were run for 50 hours, then the maximum development of reaction products occurred at 25°C. Since it requires about 50 hours for the food to pass through this animal's gut, the biological optimum for digestion is, in fact, 25°C and not 45°C.

Although many enzymes operate as simple proteins, others are only active when associated with specific COFACTORS. These cofactors are non-protein in nature and may be loosely grouped as inorganic ion ACTIVATORS, COENZYMES, and PROSTHETIC GROUPS. Many examples of ion activation will be found in succeeding chapters; some of the common ions involved are K^+, Mn^{++}, Mg^{++}, Ca^{++}, and Zn^{++}. Coenzymes are complex organic molecules which are usually heat stable and loosely bonded to the enzyme; they may be readily separated from it by relatively simple methods such as dialysis. The prosthetic groups, on the contrary, are firmly bonded to the protein portion of the enzyme and form an integral part of the complex molecule; in other ways they are similar to coenzymes. Many coenzymes operate in tissue respiration (Chapter 7) and serve to transfer electrons; several well-known vitamins are components of these molecules (thiamin, riboflavin, nicotinamide). The iron porphyrin group of cytochrome is a good example of a prosthetic group (Fig. 7.7).

Inhibition also plays a part in the regulation of enzyme activity. Enzyme reactions are reversible and the presence of reaction products may affect velocity in several different ways (McGilvery, 1970). Further, the binding of the substrate to a complex enzyme may affect the rate at which additional substrate is tied up, and it may alter the hyperbolic Michaelis-Menten curve (Fig. 2.2a) to one that is sigmoid in shape (compare the oxyhemoglobin curves in Fig. 6.3). In a comparable manner, some enzymes appear to have secondary binding sites as well as those concerned with substrate binding. These additional sites (ALLOSTERIC SITES) may be occupied by modulators which either increase the rate of the reaction (POSITIVE MODULATORS) or depress it (NEGATIVE MODULATORS). Enzymes that behave in this way are called REGULATORY ENZYMES (Fig. 2.5). They are complex proteins consisting of several subunits and usually display sigmoid substrate-saturation curves. Textbooks of

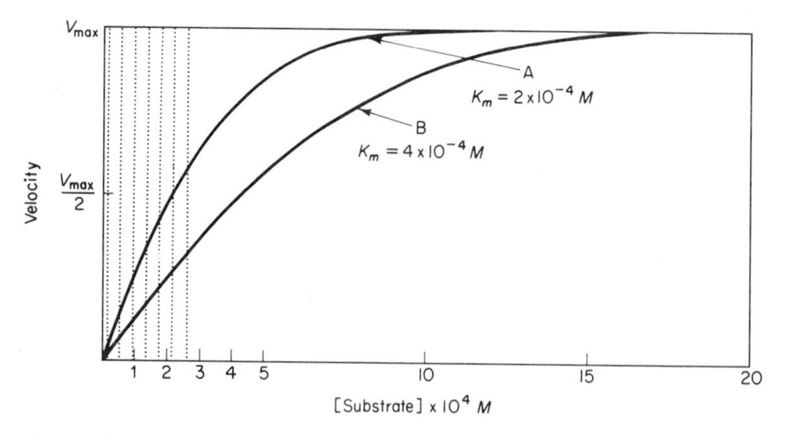

Fig. 2.5 The effect of varying enzyme-substrate affinity, defined as the reciprocal of the Michaelis constant (K_m) on enzyme activity. Enzymes A and B catalyze the same reaction and exhibit the same activity at saturating (V_{max}) concentration of substrate. The enzymes have a twofold difference in K_m. At physiological substrate concentrations (indicated by stippling), enzyme A is much more active than B. In terms of enzyme regulation theory, B may represent a deactivated state of the enzyme and A may represent an activated state. B may be converted to A by the binding of a positive modulator while A may be converted to B by a negative modulator. [From Hochachka and Somero, 1971: Biochemical Adaptation to the Environment. *In* Fish Physiology, Vol. 6, Hoar and Randall, eds. Academic Press, New York.]

biochemistry and enzymology should be consulted for details. The important point for the animal physiologist is that enzyme reactions may be adjusted in different ways by cofactors, end products and modulators to meet variable needs.

Some enzymes occur in multiple molecular forms within the same organism or even within the same cell. Slight differences in amino acid sequences or molecular configuration make a certain form of the enzyme more suitable for a particular situation. Thus, lactate dehydrogenase (LDH) may occur in five or more different forms (ISOZYMES or ISOENZYMES). They all catalyze the same reaction but have different K_m values, amino acid sequences, and immunological properties; one basic type predominates in heart muscle while another is more prevalent in skeletal muscle; it appears that the conditions in these two tissues are sufficiently different to make one particular type preferable. Adjustments in enzyme activity through modification in isozymes are known to occur both phylogenetically and during physiological compensation to environmental change (Hochachka and Somero, 1971, 1973).

These, then, are the familiar controls of enzyme reactions. Regulation of life processes at the cellular level depends on variable amounts of enzymes, substrates, activators, inhibitors and byproducts, reacting under the modifying influence of temperature and pH. Reserves of inactive enzyme molecules (proenzymes) may be present, awaiting some form of activation. For example, the protein splitting enzyme produced by the vertebrate gastric mucosa is secreted as the proenzyme pepsinogen; this is activated, in the first instance, by the hydrochloric acid in gastric juice. This process is referred to as UNMASKING and involves the removal of a polypeptide, leaving the active protease, pepsin. Several similar examples will be given in the discussion of digestive enzymes. Proenzymes may be present intracellularly as well as in the digestive juices. This is one of the important ways in which stores of enzymes are maintained. Similarly, reserves of enzymes may be brought into action through the removal of inhibitors or by the existence of coenzyme mechanisms, modulators, or specific activating ions.

The variety, complexity and precision of enzyme control mechanisms are currently subjects of active biochemical research. Regulation is achieved not only through modulators and feedback controls, but also through more complex mechanisms involving INDUCTION or complete REPRESSION of certain enzyme systems. In short, organisms possess a multitude of CONSTITUTIVE enzyme systems which are always present and demonstrable but, at the same time, may show the capacity to develop appropriate enzyme systems (INDUCIBLE ENZYMES) after the addition of a substrate (Yost, 1972).

Enzyme Regulation in Cellular Metabolism

The regulation of enzyme activity occurs at all levels in phylogeny and is basic to life itself. There are many familiar examples. Animals regularly process variable quantities of different sorts of foods. From a diversity of amino acids, sugars and fats, the individual cells select according to their particular requirements; the rates of processing (hence enzyme activities) are adjusted according to the supplies available and cellular demands for fuel, storage or synthesis. Enzyme adjustments also occur during environmental adaptation (Hochachka and Somero, 1973). In many poikilotherms, the enzyme-substrate affinities are modified in accordance with both sharp diurnal fluctuations and long-term seasonal changes in ambient temperature; isozymes may also show adaptive changes during temperature compensation (Chapter 10). Alterations in enzyme kinetics have likewise been associated with some other environmental variables and examples will be found in chapters devoted to these topics (Part Two); one further instance will be noted here. Several different amphibians alter liver and kidney arginase activities in accordance with the available water, so that urea production is favored in dry habitats, while ammonia is excreted when moisture is abundant (Chapter 8 and Table 2.1).

The classical examples of selective enzyme induction were first found in the bacterial world. *Escherichia coli,* an abundant bacterium in the gut, can meet all its dietary requirements if supplied with glucose as a source of carbon and ammonia as a source of nitrogen. From these starting materials, this bacterium synthesizes the variety of amino acids, nucleotides, vitamins, and so forth which are essential to its life. Biosynthetic pathways with the operating enzymes are known, and biologists

Table 2.1 Effect of Dehydration for 100 hr on the Kinetic Properties and Activity of *Rana pipiens* Kidney Arginase

	Hydrated	Desiccated
K_m	5.9 mM	9.3 mM
V_{max}	5.1×10^{-4} mM urea/1 min.	8.3×10^{-4} mM urea/1 min
Sp. act	2.9×10^{-4} mM urea/ 1 min per mg protein	2.8×10^{-4} mM urea/ 1 min per 1 mg protein
	9.1×10^{-6} mM urea/ 1 min per mg tissue	1.3×10^{-5} mM urea/ 1 min per 1 mg tissue

SOURCE: Boernke, 1973: Comp. Biochem. Physiol., 44B: 647.

no longer consider it remarkable that these substances are produced in the correct proportions for the life of the bacterium and its future generations. The interesting point to be noted here is that if a preformed amino acid such as tryptophan is added to the culture medium of *E. coli,* the endogenous synthesis of this particular amino acid ceases. This is not merely an equilibrium attained in a reversible reaction, since a great many chains of enzymatically controlled processes are involved in such a synthesis. Formation of other amino acids (methionine, proline, arginine) has been repressed in a similar way. Regulatory mechanisms of this type keep the different enzyme systems of a cell in step with one another. In an organism such as *E. coli,* which can synthesize a variety of amino acids, the accumulation and wasting of important compounds is avoided, and a balanced enzyme production maintains a steady or constant biochemical make-up. Similar phenomena have been demonstrated for the catabolic as well as the anabolic enzymes in bacteria (Prosser, 1958; Davis, 1961a).

Enzyme induction has also been studied in the higher animals. One of the best known examples is the induction of tryptophan pyrrolase (tryptophan peroxidase) in the liver of the rat (Knox, 1951). The activity in this enzyme may increase by as much as ten-fold in 5 to 6 hours following a dose of tryptophan equivalent to about twice that consumed in the normal daily diet of the rat. The substrate tryptophan is, in this case, the inducer. The enzyme may also be induced by an injection of the adrenocortical hormone, hydrocortisone. This hormone as an inducer can produce a three-fold increase within the same time period. This is a significant demonstration of regulatory processes operating in a higher animal by way of enzyme induction.

Enzyme regulation of cellular metabolism is fundamental to life. The examples just given show that it is an ongoing process during daily living and that it occurs during adaptation to environmental changes as well as in the evolution of the species. Many specific examples will be found in subsequent chapters; the principles are emphasized here. There are now many excellent reviews (Atkinson, 1966; Boyer, 1970; Weber, 1972; Hochachka and Somero, 1973; two symposia edited by P. W.

Hochachka and published in the *American Zoologist,* vol. 3, numbers 1 and 3, 1971).

REGULATION AT THE ORGAN-SYSTEM LEVEL

The complex multicellular organism is more than a collection of enzyme-regulating cells. The grouping of specialized cells into tissues, each adapted for a particular job, the association of tissues into organs designed to perform the major functions, and the arrangement of organs into a whole and unified animal require a different type of integration. For this, intercellular communication is essential. In final analysis, this too depends on enzyme reactions within individual cells but at a different level: it involves feedback mechanisms from metabolizing tissues, specialized coordinating chemical materials, communication by circulating fluids and the transmission of nerve impulses. The substances concerned may be broadly referred to as CHEMICAL MESSENGERS. They are extremely varied both biochemically and physiologically; there is no entirely satisfactory classification of them (Turner and Bagnara, 1971). For present purposes, they are grouped under TRANSMITTERS, HORMONES, and the hormone-like substances including PARAHORMONES and PHEROMONES.

TRANSMITTER SUBSTANCES are produced in nerve cells relatively unspecialized for secretory purposes. They are released at the ends of the fibers, travel only very short distances before being enzymatically destroyed and act on other neurons, muscles or glands in intimate contact with the nerve endings (Fig. 2.6). They are, then, short-range and short-lived materials. Acetylcholine, adrenaline and noradrenaline are well recognized members of this group. In addition, there is circumstantial evidence for a similar action by a few other chemicals, such as 5-hydroxytryptamine and gamma-aminobutyric acid, which are widely distributed in nervous tissues.

NEUROSECRETORY SUBSTANCES are produced by neurons which are specialized for secretion rather than conduction. Usually the endings are associated with special storage centers called NEUROHEMAL ORGANS. The chemicals are released from the storage centers; they are relatively stable and act at greater distances and over longer periods (Fig. 2.7). They control such varied phenomena as molting and chromatophore activity in arthropods or water balance and urine production in the verte-

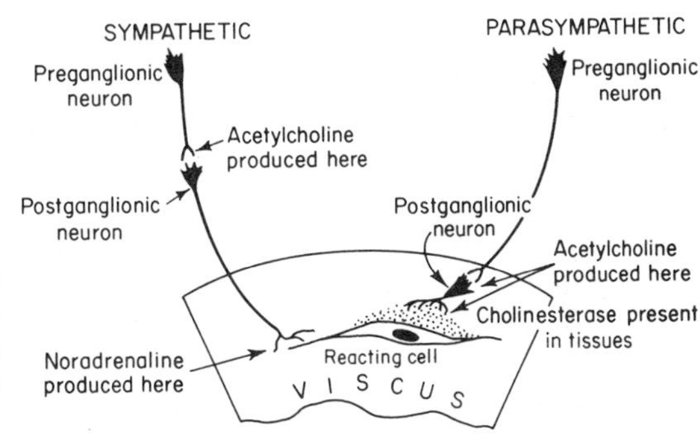

Fig. 2.6 Diagram showing the arrangement of the vertebrate autonomic fibers. [From Myerson, 1938: Am. Med. Assoc. J., *110*: 101.]

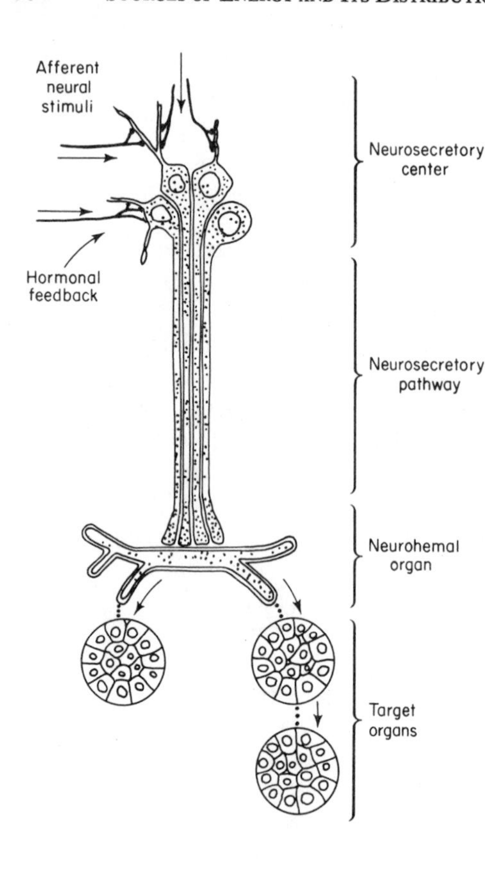

Fig. 2.7 Generalized diagram of the neurosecretory system applicable both to invertebrates and vertebrates. Afferent nerves release transmitters on specialized neurons which may conduct impulses but are primarily concerned with the production of hormones (neurosecretory substances); these enter the circulation to act on distant target organs. [From Scharrer and Scharrer, 1963: Neuroendocrinology. Columbia U. P., New York.]

brates. Neurosecretory substances may serve directly in processes of integration or act by way of a second endocrine structure which has a non-nervous origin (for example, the Y-organ of crustaceans). Transmitters and neurosecretory substances are sometimes grouped together as NEUROHORMONES.

The endocrine systems of the more advanced invertebrates and all of the vertebrates include several hormone-producing glands of non-nervous origin. Familiar examples are the anterior pituitary, the thyroid and the adrenal cortex; these organs may or may not be controlled by transmitters or neurosecretory materials. A distinction between the autonomic nervous system and the endocrine system is convenient for descriptive purposes, but it should be appreciated that the integrative homeostatic mechanisms are neuroendocrine and the nervous and endocrine systems are intimately related and often interdependent.

Throughout the animal kingdom a measure of regulation is achieved by feedback stimuli from tissue metabolites. Carbon dioxide, one of the universal byproducts of metabolism, is the most important of these metabolites. It acts directly on the cardiac and vasomotor centers in the vertebrate medulla to participate in the control of breathing and blood pressure. Comparable regulation occurs in many invertebrate animals. The respiratory activities of the annelid *Tubifex,* breathing movements of the pulmonate gastropods, the ventilation of the *Octopus* and the pumping movements of crustacea may be cited as examples of complex activities which are adjusted in part by the level of carbon dioxide. Several other physiologically important agents are known to be formed in active tissues. The polypeptide *bradykinin,* for example, is released by active sweat and salivary glands; it is a powerful vasodilator that enormously increases the blood flow locally and thus promotes secretion of sweat and saliva. In a comparable manner, low blood oxygen stimulates the production of more red blood cells by triggering the formation of a glycoprotein *erythropoetin* in the kidney (Fisher, 1972). Other "near hormones" are histamine, the kinins (Schachter, 1969), angiotensin (Chapter 8), the prostaglandins (Pike, 1971; Kirim, 1972), and the thymus hormone (Chapter 6). These substances, many of which act very locally, are sometimes grouped together as PARAHORMONES or TISSUE HORMONES.

Finally, chemical messengers are also important in regulating behavior. The PHEROMONES are species-specific chemical agents which are released into the environment and evoke behavioral, developmental or reproductive responses. They are well-known sex attractants, and they may also be involved in the synchronization of sexual cycles, trail marking, recognition of other individuals and certain other intraspecific reactions. Many examples will be found in the sections devoted to chemoreception (Chapter 17).

Neurotransmitters

Chemical transmission at synapses, so controversial a quarter of a century ago, is now firmly established. A nerve impulse triggers the release of a specific chemical at the axon terminus; this chemical, in turn, excites or inhibits a second neuron or effector organ. Exceptions occur only at the relatively infrequent tight junctions where transmission is electrical (Chapter 21). Although comparatively few different species of animals have been carefully studied, it is apparent that the actual number of neurotransmitters is not great in relation to the diverse forms of life; further, the particular functions of a transmitter may be quite different in the various animal groups. To mention but one example, acetylcholine—one of the best known transmitters—is firmly established as the active agent at the vertebrate neuromuscular junction but, throughout the arthropods, is entirely ineffective at the ending of motor nerves and serves as a transmitter for sensory nerves instead (Florey, 1967).

The recognized or suspected neurotransmitters are acetylcholine, several different amines, a few amino acids, and ATP. Their localization in tissues has been greatly facilitated since about 1960 by the introduction of techniques of subcellular fractionation (Smith, 1972) and fluorescence histochemistry. Pharmacological agents which mimic or block transmitters are also valuable but not reliable as the sole evidence of a transmitter. There are now many excellent reviews of the literature on physiology and pharmacology of neurotransmitters (for example, Florey, 1967; Burnstock, 1969, 1972; Phillis, 1970).

Acetylcholine Acetylation of the nitrogenous alcohol choline gives rise to acetylcholine, now recognized as a neurotransmitter in all major groups of animals (Florey, 1967). The biosynthesis is catalyzed by choline acetylase and readily reversed by acetylcholinesterase as indicated in the following reaction:

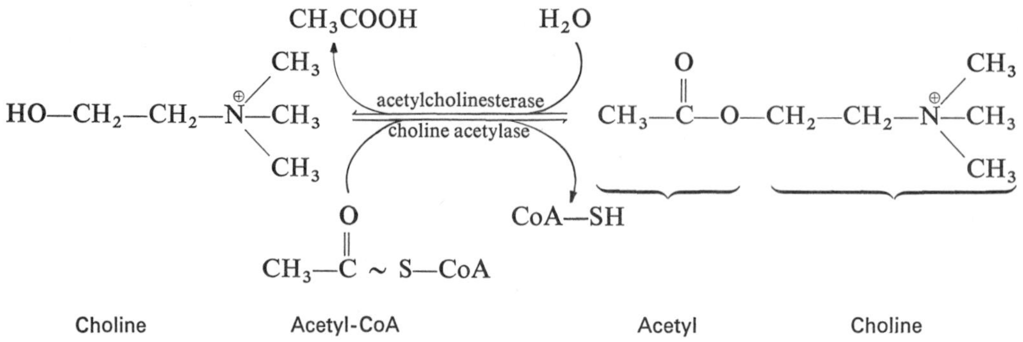

$$\underset{\text{Choline}}{HO{-}CH_2{-}CH_2{-}\overset{\oplus}{N}(CH_3)_3} + \underset{\text{Acetyl-CoA}}{CH_3{-}\overset{O}{\overset{\|}{C}} \sim S{-}CoA} \underset{\text{acetylcholinesterase}\ (+H_2O \rightarrow CH_3COOH)}{\overset{\text{choline acetylase}\ (\rightarrow CoA{-}SH)}{\rightleftharpoons}} \underbrace{CH_3{-}\overset{O}{\overset{\|}{C}}}_{\text{Acetyl}}{-}O{-}\underbrace{CH_2{-}CH_2{-}\overset{\oplus}{N}(CH_3)_3}_{\text{Choline}}$$

Acetylcholine satisfies all criteria for a transmitter substance. Its most widely recognized action sites are the endings of vertebrate motor nerves, the endings of the autonomic preganglionics and the parasympathetic postganglionics. It appears also to be the transmitter of motor nerves in all advanced invertebrate animals except the arthropods where it seems to operate at the endings of sensory nerves (Florey, 1967). Acetylcholine is concerned with various visceral functions in both invertebrates and vertebrates; details will be found in later chapters.

Claude Bernard launched the history of transmitter pharmacology in 1857 when he discovered that nerve stimulation failed to excite muscles that were poisoned with a plant alkaloid, curare, even though the muscle remained responsive to direct stimulation. Bernard correctly concluded that this substance blocked excitatory events at the neuromuscular junction. It is now known that curare acts by competing with ACh for receptor sites on the motor end plate. With highly purified preparations of curare (*d*-tubocurarine) muscular excitability is gradually reduced in accordance with the dose; thus, this substance has found a valuable place in clinical medicine as well as physiology. Its paralytic effects were recognized and exploited by native peoples of South America long before Bernard localized its action in the motor end plate. It occurs in a number of different plants and was being widely used as an arrow poison when the first waves of Europeans reached South America. The history of this interesting substance is recorded in many places (references in Goodman and Gilman, 1970).

Physostigmine, also called eserine, is another plant alkaloid used in studies of ACh physiology. It is an anticholinesterase and increases muscular excitability by competitively inhibiting the breakdown of ACh by the esterase. This substance, extracted from the Calabar beans growing along the banks of streams in tropical

West Africa, is sometimes called the "ordeal poison" because of its use in witchcraft trials. It has been used therapeutically for almost a century. Neostigmine is a close synthetic relative.

The physiologist has at his disposal a wide range of drugs for the investigation of acetylcholine. Hemicholinium-3 prevents ACh synthesis in motor nerves; botulinum toxin, one of the most potent poisons in the world, prevents its release. Several different substances, termed AGONISTS, act like ACh on the postsynaptic membrane; well-known examples are carbachol, succinylcholine, decamethonium and nicotine. In addition to eserine and its relatives, there are several synthetic organophosphorus compounds that act as anticholinesterases and have found uses as "nerve gases" and insecticides (parathion and malathion). Curare does not block ACh transmitter activity in the autonomic nervous system (postganglionic parasympathetics) nor does nicotine act as an agonist. The plant alkaloids muscarine (from mushrooms) and pilocarpine (from certain tropical shrubs) have the nicotine-like action in the autonomics while the action of atropine (from deadly nightshade) is comparable to curare. Thus, physiologists speak of the "nicotinic action" or the "muscarine action" of ACh, depending on the nature of the effective blocking agent. Brief accounts of this area of pharmacology are found in Florey (1966, 1967) and Aidley (1971); there is a comprehensive treatment in the monograph edited by Goodman and Gilman (1970).

Catecholamines Very early in the century it was suggested that sympathetic neurons function by releasing adrenaline (epinephrine). This hypothesis was based on obvious similarities between the physiological effects of sympathetic stimulation and the action of adrenaline, known to be the product of the adrenal medulla (Elliott, 1905). W. B. Cannon and his associates established the validity of this concept although they termed the transmitter SYMPATHIN rather than adrenaline in order not to prejudge the issue (Cannon and Uridil, 1921). These early investigators were greatly puzzled by the fact that tissues such as smooth muscle were sometimes excited by adrenaline while at other times its action was inhibitory; consequently, a concept of two sympathins with opposing activities was widely held for many years. It remained for von Euler (1946) to show that noradrenaline (norepinephrine) and not adrenaline was the major transmitter of mammalian sympathetic neurons. Ahlquist (1948) solved the other part of the puzzle by showing that in many situations there are two different types of adrenergic receptors (termed alpha and beta) rather than two different transmitters. In mammals, activation of α-receptors usually leads to excitation while stimulation of β-receptors has an inhibitory effect.

It is now known that three biochemically related amines are derived from tyrosine in many different animals and serve as neurotransmitters in both autonomic and central nervous systems. These three amines (dopamine, adrenaline and noradrenaline) are jointly referred to as catecholamines because of the presence of catechol, a dihydric phenol (Fig. 2.8). Dopamine is found in high concentrations in the brains of mammals and snails; it is presumed to serve as a transmitter (Woodruff,

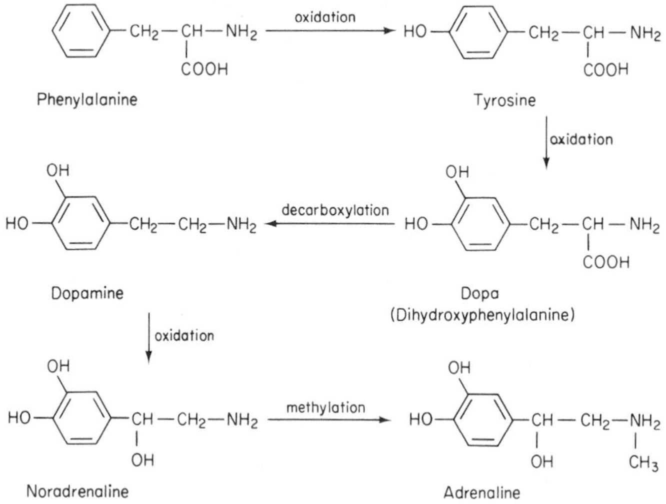

Phenylalanine —oxidation→ Tyrosine

Tyrosine —oxidation→ Dopa (Dihydroxyphenylalanine)

Dopa —decarboxylation→ Dopamine

Dopamine —oxidation→ Noradrenaline

Noradrenaline —methylation→ Adrenaline

Fig. 2.8 Catecholamines. The three oxidations are catalyzed by hydroxylases; tyramine (not shown) is formed through the decarboxylation of tyrosine; DOPA is a precursor of melanin (see Fig. 20.17).

1971). Adrenaline is the hormone of the adrenal medulla while noradrenaline predominates in sympathetic neurons of all groups of vertebrates except the teleost fishes and amphibians, where adrenaline is the transmitter. The patterns of transmitter distribution vary greatly in different groups of animals; there are no obvious phylogenetic trends although the literature contains some interesting speculation (Pscheidt, 1963; Burnstock, 1969).

A number of sympathomimetic drugs are used pharmacologically; the best known are isoprenaline, tyramine, ephedrine and amphetamine. Several physiologically different antagonists have also been extremely valuable in separating α-and β-receptor sites. Thus, α-receptors are blocked competitively by the alkaloids yohimbine and ergotoxin, and irreversibly by dibenamine and phenoxybenzamine; β-receptors are selectively blocked with dichloroisoproterenol (DCI) and its derivatives propranolol and sotalol. Goodman and Gilman (1970) should be consulted for details of adrenergic and other drugs.

Other biogenic amines Many different amines have been isolated from the tissues of both plants and animals. In addition to those already mentioned, serotonin [5-hydroxytryptamine (5-HT)] and histamine have long been suspected of transmitter activities. Both have been found in many different tissues at all phyletic levels (Blum, 1970); they are physiologically important in several metabolic and vascular reactions not related to transmitter action. However, they also occur in places where they might serve as neurotransmitters. Moreover, it is considered significant that

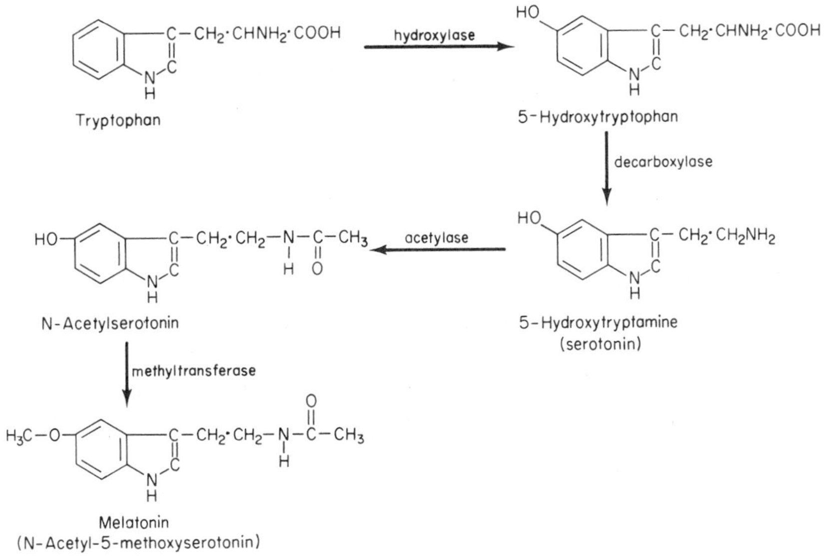

Fig. 2.9 Formation of serotonin and melatonin from tryptophan.

several well-known psychotogenic and sedative agents show biochemical affinities to one or another of them; for example, some of the hallucinogenic effects of LSD (lysergic acid diethylamide) may relate to the indole moiety that occurs in several psychotogenic drugs as well as 5-HT (Fig. 2.9), while the sedative qualities of the antihistamines are well known (Goodman and Gilman, 1970).

The history of serotonin stems from studies of a potent and bothersome vasoconstrictor substance which appears during mammalian blood clotting. About 1950, the culprit was identified as 5-hydroxytryptamine. Several lines of research then converged to show that this amine was identical with a histochemical material abundant in the enterochromaffin tissues of the gut, and known to biologists as *enteramine.* The ensuing spate of research established the presence of 5-HT in many vertebrate tissues, especially in the enterochromaffin cells (about 90 per cent of the total), the blood platelets and the brain. It occurs also in many invertebrates at all phyletic levels and is particularly abundant in the nervous system of molluscs (Welsh, 1970; Endean, 1972); the venoms of wasps and scorpions are rich in 5-HT as are also stinging nettles and cowhage (itching powder). The pineal gland is the site of intense 5-HT metabolism leading to the formation of melatonin (Fig. 2.9), a hormonal substance concerned with melanophore blanching (Chapter 20) and reproductive physiology in mammals (Reiter, 1970, 1973). Although the evidence for transmitter activity is less complete for 5-HT than it is for the catecholamines, most writers find the evidence adequate and agree that serotonin has neural functions as well as its recognized metabolic and vascular ones.

Histamine is formed as follows by the decarboxylation of the amino acid histidine:

$$\text{(imidazole ring: HN, N)}-CH_2-\underset{\displaystyle NH_2}{\underset{|}{CH}}-COOH \xrightarrow[\searrow CO_2]{\text{decarboxylase}} \text{(imidazole ring: HN, N)}-CH_2-CH_2-NH_2$$

Histidine Histamine

It is destroyed enzymatically through a specific histaminase. Because histamine actively dilates small blood vessels and increases capillary permeability, it has been intensively studied in connection with several different tissue reactions—particularly anaphylaxis, allergy and tissue damage. Among the vertebrates, Mast cells are the best source of histamine, although smaller amounts occur in other tissues including the brain. Its subcellular distribution in nervous tissues is consistent with the possibility of a transmitter action. There is now a voluminous literature on histamine—including its possible role in the central nervous tissues and the action of antihistaminic drugs (Kahlson and Rosengren, 1971).

Amino acids The first persuasive evidence that some amino acids serve as neurotransmitters came from comparative studies of crustacean muscles. Gamma-aminobutyric acid (GABA) applied to the muscles of certain crustaceans (Astacura) inhibit their contractions and duplicate the mechanical and electrical activity seen during normal inhibition; the effects of GABA are blocked by picrotoxin. These experimental results were obtained about 1960 but even before this, it was shown that GABA is identical with an inhibitory substance which had been extracted from mammalian brains. There is now convincing evidence that GABA is an inhibitory transmitter not only in crustaceans and insects but in other invertebrates and higher vertebrates (literature reviewed in Florey, 1966, 1967; Elliott, 1970; Goodman and Gilman, 1970; Pitman, 1971; Endean, 1972).

GABA is formed by the decarboxylation of glutamate and this substance may also be a neurotransmitter. The chemical affinities of these two amino acids are shown as follows:

$$HOOC-CH_2-\underset{\displaystyle NH_2}{\underset{|}{CH_2}}-COOH \xrightarrow{\nearrow CO_2} \underset{\displaystyle NH_2}{\underset{|}{CH_2}}-CH_2-CH_2-COOH$$

Glutamic acid

γ-aminobutyric acid
GABA

Glutamate has been found in the nervous systems of many animals at different levels in phylogeny. It is thought to be a transmitter in arthropod excitatory motor neurons and may operate in the central nervous systems of higher forms including man. Other amino acids for which there is some suggestive, although rather slight, evidence of transmitter function are aspartic acid and glycine. The literature may be traced through the reviews already cited.

Purines The most recent arrivals in the family of chemical messengers are some of the purine compounds. The action of one-methyl adenine in triggering the spawning of starfishes will be considered in Chapter 23. This purine, formed in the follicle of the ovary, is not a neurotransmitter. However, adenosine triphosphate has recently come to the fore in studies of the vertebrate autonomic and now appears to be well established as the transmitter of some postganglionic nerves.

For many years there has been a suspicion that, contrary to classical concepts, there are autonomic nerve fibers which release neither acetylcholine nor catecholamines. Since about 1960, the rapidly expanding evidence has focused on ATP as a neurotransmitter in the gut wall of vertebrates—particularly the inhibitory fibers of the stomach in all vertebrates and the entire intestine of the mammal. The cell bodies of these PURINERGIC nerves are located in Auerbach's plexus and controlled by preganglionic vagal fibers. Although the evidence is less complete, some purinergic nerves are probably also present in the urinary bladder, in the lung and in parts of the cardiovascular system. Burnstock (1972), who has been responsible for much of the experimental work, has summarized the literature. The evidence satisfies the requirements usually sought in studies of neurotransmitter activity.

How transmitters work Some transmitters may modify the ion-pumping machinery in cells, but most of them seem to open additional ion valves or gates in the membranes where they act (Lundberg, 1958; Miledi, 1969). The consequent change in distribution of ions excites the effector or postsynaptic cell. There are still many uncertainties about the molecular details; at best, the story can only be told in general terms. The most completely understood preparation is the ACh-mediated spread of excitation from vertebrate nerve to muscle (Chapter 19). Physiologists have been able to monitor electrical events at both junctional membranes (nerve and muscle), measure the passage of ions and transmitters, modify transmitter action by altering the extracellular environment, and utilize specific drugs that selectively block the passage of ions such as Na^+ and K^+. Some of the evidence will be presented later (Chapters 15 and 19) and can be readily traced through recent monographs and reviews (for example, Phillis, 1970; Rubin, 1970; Katz, 1971; Pappas and Purpura, 1972).

A second area of understanding has recently emerged and may eventually form part of a unified picture of transmitter physiology The compound ADENOSINE 3′, 5′-MONOPHOSPHATE, usually referred to as CYCLIC AMP, is a crucial link in mechanisms by which several hormones operate; it is also a part of the physiology of the catecholamine transmitters (Robison *et al.*, 1971). Cyclic AMP is formed from ATP by splitting off pyrophosphate. The enzyme adenyl cyclase catalyzes the reaction; phosphodiesterase destroys cyclic AMP (Fig. 2.10). Adenyl cyclase occurs widely in animal tissues but is particularly abundant in plasma membranes of certain ef-

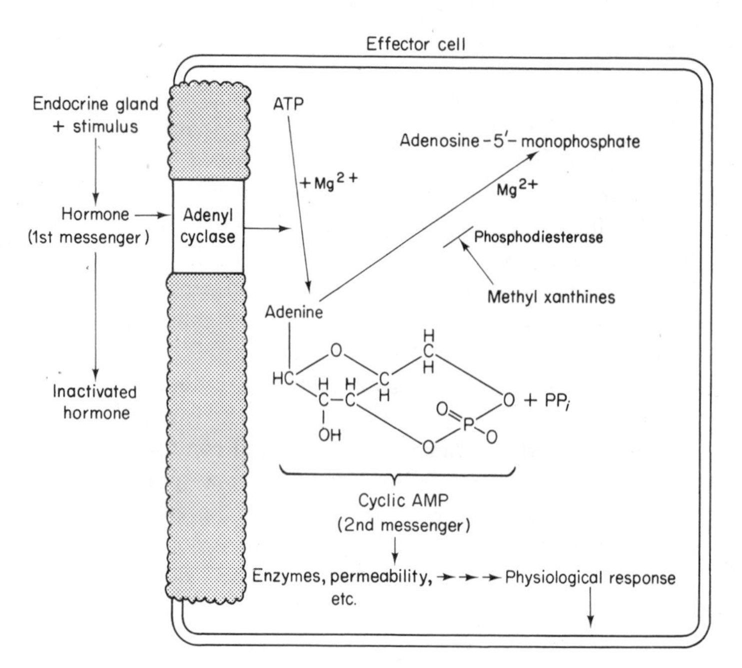

Fig. 2.10 The "second messenger hypothesis." Methyl xanthines block the phosphodiesterase reaction; further description in the text. [From Frieden and Lipner, 1971: Biochemical Endocrinology of the Vertebrates. Prentice-Hall, Englewood Cliffs, N.J. After Sutherland, Robison, and Butcher, 1968: Circulation, *37*: 280.]

fector cells. It is known to be released by several different hormones and by catecholamines; the intracellular cyclic AMP thus generated triggers the physiological response characteristic of the particular receptor cell. This concept, known as the "SECOND HORMONE HYPOTHESIS," has provided a unifying picture in several areas of hormone activity; it is important in the present concept since the release of cyclic AMP is related to the stimulation or inhibition that occurs at β-adrenergic receptors. This is currently an active area of research and it seems probable that cyclic AMP plays an important role in synaptic transmission, as well as peripherally in hormone activity (McAfee *et al.*, 1971; McAfee and Greengard, 1972). The comparative physiologist may be expected to play an important part in studies of this area of neurochemistry.

AUTONOMIC NERVOUS SYSTEM

Langley (1921) introduced the concept of an autonomic nervous system to describe the nerves controlling the visceral effectors for digestion, circulation, excretion and other more or less involuntary functions. More precisely, it was the motor innervation of smooth muscle, cardiac muscle, and glands in mammals.

The presence of ganglia was a basic feature in Langley's description of the autonomic nervous system. Whereas the motor fibers associated with the voluntary muscles have their cells located in the central nervous system, those associated with the autonomic have their cell bodies in ganglia located at some distance from the central nervous system (Fig. 2.11). These ganglia are in turn connected by nerve fibres to the central nervous system. Thus, a PREGANGLIONIC FIBER arising in the central nervous system forms a synapse in a ganglion with the POSTGANGLIONIC FIBER, which is in contact with the visceral effector organ (Figs. 2.6 and 2.11).

Langley distinguished two divisions of the vertebrate autonomic—the SYMPATHETIC and the PARASYMPATHETIC nervous systems. In the mammal these two divisions are distinct anatomically, physiologically and pharmacologically. The sympathetic arises from the thoracico-lumbar regions of the central nervous system, the parasympathetic from the cranial and sacral regions (Fig. 2.12). Typically, each visceral organ receives both sympathetic and parasympathetic fibers; one is excitatory while the other is inhibitory. Pharmacologically, excitation is mediated through two different transmitters; although all preganglionics release acetylcholine, there are two different postganglionic transmitters. Sympathetic effects are usually mediated by noradrenaline or adrenaline (ADRENERGIC NERVES), while the parasympathetic effects are due to acetylcholine (CHOLINERGIC NERVES). Since Langley's

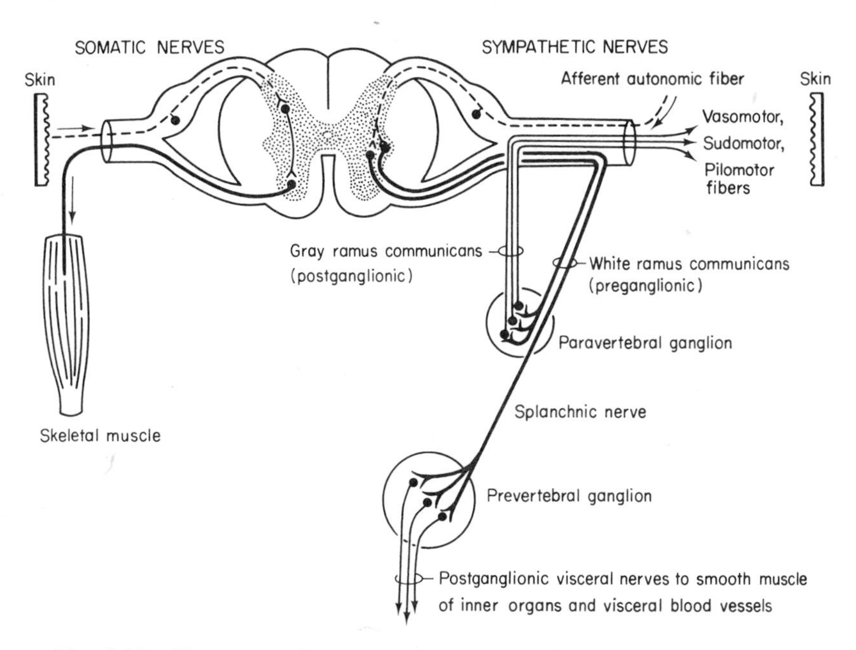

Fig. 2.11 The autonomic sympathetic outflow showing the preganglionic fibers by heavy lines and the postganglionics in lighter lines. [From Pick, 1970: The Autonomic Nervous System. Lippincott, Philadelphia.]

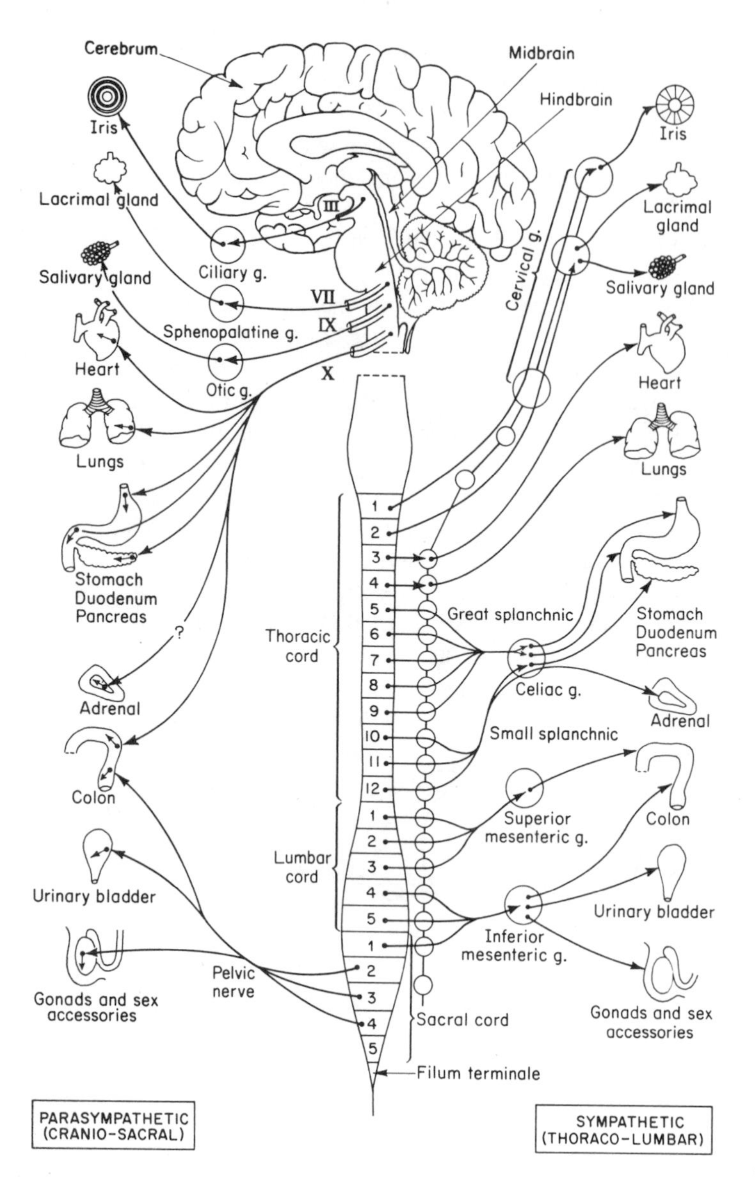

Fig. 2.12 Diagram of the mammalian autonomic nervous system. [From Turner and Bagnara, 1971; General Endocrinology, 5th ed. Saunders, Philadelphia.]

day it has become apparent that the mammalian sympathetic to some organs contains cholinergic postganglionics; these may be few in relation to the more conspicuous adrenergic fibers (lung, vasodilators to muscle) or numerous as in sweat glands. The situation is much more varied in lower vertebrates, with cholinergic sympathetics in several organs and some purinergic nerves as well (Burnstock, 1969,

1972). Details are noted in later sections; at this point the classical pattern as found in the mammal is emphasized.

The autonomic, as originally described, was a purely motor system. However, it seems reasonable and avoids misconceptions if the associated viscero-sensory components are also included and many descriptions of the autonomic nervous system now embody both sensory and motor elements. Pick (1970) reviews the history of this system and gives many details of comparative and clinical physiology.

Visceral Nervous System of Invertebrates

Throughout the invertebrates, nervous regulation of visceral functions depends on nets of neurons dominated by ganglia. In a broad sense this is also true of the vertebrates and may represent a primitive arrangement from which the autonomic nervous system, as described by Langley, evolved. However, the classical vertebrate pattern of presynaptic neurons, arising in a highly centralized nervous system and connecting with postsynaptics in peripheral ganglia, does not exist in the invertebrate world and, consequently, it is better to reserve the term "autonomic" for the system originally described. Several different terms have been applied to the visceral nervous systems of invertebrates. Bullock and Horridge (1965) refer to the STOMODEAL NERVOUS SYSTEM as that part of the peripheral nervous system which is distributed to the gut of higher invertebrates, with connections to the brain and anterior ventral ganglia. Other familiar terms are STOMATOGASTRIC and ENTERIC nervous systems.

During phylogeny, the regulation of visceral muscular activity seems to have appeared before other controls such as secretion (Nicol 1952, 1967). From flatworms to protochordates, nerve nets are found in special areas of muscle such as that of the pharynx, proboscis, rectum, gut, heart and copulatory organs. The nature of the neurotransmitters is almost entirely speculative. The present evidence is based largely on the pharmacological reactions to drugs commonly used in studies of visceral function. Responses to transmitters such as ACh, 5-HT and the catecholamines suggest that the same transmitters operate in both invertebrates and vertebrates even though the controls to comparable organs may be entirely different (Fänge, 1962; Florey, 1967). A pattern of dual controls with separate excitatory and inhibitory transmitters also appears to be a part of the visceral regulation in invertebrates as well as vertebrates. However, the present findings suggest that there are many differences in the controls of closely related groups and that generalizations are not yet justified. Monographs by Bullock and Horridge (1965) and the treatise edited by Florkin and Scheer (1967–72) summarize the recorded information.

Vertebrate Autonomic Nervous System

The detailed arrangement of the autonomic system is shown in Fig. 2.12. Three groups of ganglion cells may be conveniently distinguished. The well-marked chains of PARAVERTEBRAL GANGLIA on either side of the spinal cord are sympathetic; these are the loci of synapses between pre- and postganglionic fibers supplying the iris, salivary glands, sweat glands, blood vessels, heart and respiratory organs. The gut, from the stomach to the rectum, as well as the genital organs, are innervated by postganglionic sympathetic fibers which arise in the PREVERTEBRAL or COLLATERAL GANGLIA. These collateral ganglia may be looked upon as vertebral ganglia moved into closer relationship with the viscera which they innervate. Finally, TERMINAL GANGLIA, associated with the parasympathetic system, are located within the organs involved. The nerve plexuses of the digestive tract are familiar examples. It will be noted that, in the mammal, the preganglionics are frequently very long in the parasympathetic division (the vagus nerve, for example), and the postganglionics very short, while the reverse is true of the corresponding sympathetic components. This, however, is not true of the lower vertebrates. Sympathetic fibers to portions of the digestive tract and bladder of fish and some amphibians synapse with plexuses of terminal fibers in the viscera concerned, after the manner of the mammalian parasympathetic fibers.

Evolution of the autonomic nerves The phylogeny of the vertebrate autonomic is uncertain. The visceral systems of invertebrates are entirely different and offer few evolutionary clues. Likewise, studies of the visceral controls in amphioxus have proved fruitless. Although this small animal has a simple chordate nervous system which strongly resembles the spinal cord in larvae of primitive craniates, its visceral nervous system is fundamentally different from that of the craniates (Bone, 1960). In amphioxus, viscero-motor cells are located segmentally in the ventral region of the cord (*below* the somatic motor system); their long axons emerge from the cord with the dorsal root nerves and pass directly to the atrial and rectal musculature without intervening ganglia. A plexus of conspicuous nerve cells in the wall of the atrium and gut is *sensory*; long axons from these cells pass into the cord also by way of the dorsal roots (Bone, 1958); this plexus bears no relation to that found in the vertebrate gut which is motor and postganglionic. Thus, in amphioxus the visceral innervation is conspicuously sensory while the motor components dominate in the craniates; this may provide a clue to the evolutionary history. The visceral nervous system in amphioxus is associated with a highly specialized ciliary feeding apparatus; the vertebrate autonomic may have evolved with jaws and the habit of periodic feeding on chunks of food (Bone, 1958, 1961). This latter habit demands a coordinated sequence of activities in the gut—some enzymatic and some muscular; other radical changes occurred when gill bars acquired a respiratory function. Although the origins of the autonomic may always remain speculative, it seems quite possible that its evolution has been associated with these innovations typical of jawed vertebrates.

Comparative physiology of the vertebrate autonomic The gross anatomy of the autonomic system is basically the same throughout the vertebrates. The major

Table 2.2 Summary of Anatomical Organization of Vertebrate Autonomic Nervous System. (Roman numerals, cranial nerves; +, component or innervation present; −, absent)

Animal or group	Parasympathetic		Sympathetic			
	Cranial	Sacral	Origin Dorsal or ventral root	Ganglia: Segmental chain	Ganglia: Prevertebral	Gray rami
Hagfish	X	−	Ventral	−	−	−
Lamprey	III, VII, X	−	Dorsal & ventral	−[1]	−	−
Elasmobranch	III, VII, IX, X	−	Mostly ventral	Loose connections	−	−
Teleost	III, X	−	Mostly ventral	+	−	+
Anura	III, VII, IX, X	Rudiment	Mostly ventral	+	+	+
Amniota	III, VII, IX, X[2]	+	Ventral	+	+	+

[1] Collections of nerve cells along the cardinal veins may represent sympathetic ganglia (Campbell, 1970).

[2] Fibers from the accessory nerve (number XI) also contribute to the cranial outflow in the mammal (Burnstock, 1969).

variations are shown in Table 2.2 In detail, the parasympathetic system widened its control during phylogeny through additional cranial components and the acquisition of a sacral territory in the land vertebrates. The cranial expansion is related to the evolution of the digestive and respiratory systems (salivary glands, lungs, separation of stomach, large and small intestine with distinctive glands); the sacral component seems to have appeared with the cloacal bladder in the amphibians and expanded with the differentiation of the cloacal region into the separate urinogenital and rectal passages typical of land vertebrates. Sympathetic fibers probably emerged dorsally from the spinal cord of primitive craniates. However, as the neurons of the sympathetic ganglia moved away from the cord and became organized into chains, the roots acquired more ventral connections (Burnstock, 1969). Not shown in Table 2.2 is the intimate association of the sympathetic and vagal fibers in the vagosympathetic nerve of all lower vertebrates. The two components become distinct in the reptiles although a few sympathetic fibers still travel with the vagus nerve of the mammal.

The comparative physiology is much more variable than the anatomy (Burnstock, 1969). The gut and derivatives, such as the respiratory and urinogenital systems, show the most evidence of phylogenetic experiments which probably preceded the mammalian autonomic. In this regard, a comparison of cardiac and gastrointestinal controls is interesting. The primitive heart is supplied with inhibitory cholinergic fibers from the vagus; the excitatory cardiac control depends on catecholamine-containing cells in the lower craniates and on sympathetic adrenergic neurons in the more advanced forms. This pattern of cholinergic vagal inhibition and adrenergic excitation has remained largely unchanged throughout the vertebrate series. By contrast, the gut shows many differences at the various levels. In primitive craniates, adrenergic visceral controls depended on diffusely scattered chromaffin cells, while at a somewhat higher level (frog bladder, toad lung, lizard gut) there are numerous intramural adrenergic nerves which have migrated away from the central nervous system and lack the recognized ganglionic connections. This evolutionary experiment seems to have been abandoned in the higher vertebrates with well-organized ganglia and a clear-cut nervous control from the spinal cord.

During phylogeny the postganglionic sympathetic transmitter seems to have changed from ACh to noradrenaline. Thus, below the reptiles, visceral muscles of organs such as the stomach and lungs are excited by cholinergic sympathetics; some place between the amphibians and reptiles, this cholinergic excitatory activity is taken over by the vagus, which in lower forms has a non-adrenergic inhibitory action. With this change, the sympathetics become adrenergic and inhibitory, as in the mammalian gut. Reptiles and birds show an intermediate condition with nerves to the gut which are morphologically sympathetic but contain a mixture of cholinergic and adrenergic fibers. This gradual shift from a cholinergic to an adrenergic sypathetic control of the gut explains certain peculiarities long recognized in autonomic controls of the mammalian viscera; departures from the classical picture are relics of phylogeny and not deviations from the normal.

The recognition of postganglionic purinergic nerves in several vertebrate tissues has added a new dimension to the concepts of the autonomic system. Both excitatory and inhibitory fibers have now been described in certain visceral and cardiovascular tissues (Burnstock, 1972). The known distribution of these nerves suggests that they may have become more important as the more elaborate and precise controls of higher vertebrates evolved. Thus, purinergic nerves have been found only in the stomach of lower vertebrates but extend throughout the gut of mammals; in the stomach and distal rectum they are controlled by cholinergic preganglionics but in the mammalian large intestine they are regulated by intramural cholinergic nerves (Fig. 2.13).

These notes underline the physiological diversity in autonomic controls. Many evolutionary experiments have been tried—particularly with the transmitters; the specialization of the gut and its derivatives appears to have been a strong force in the evolution of this system. At this stage, there are obvious hazards in making generalizations and Burnstock's (1969) caution against oversimplification should be emphasized.

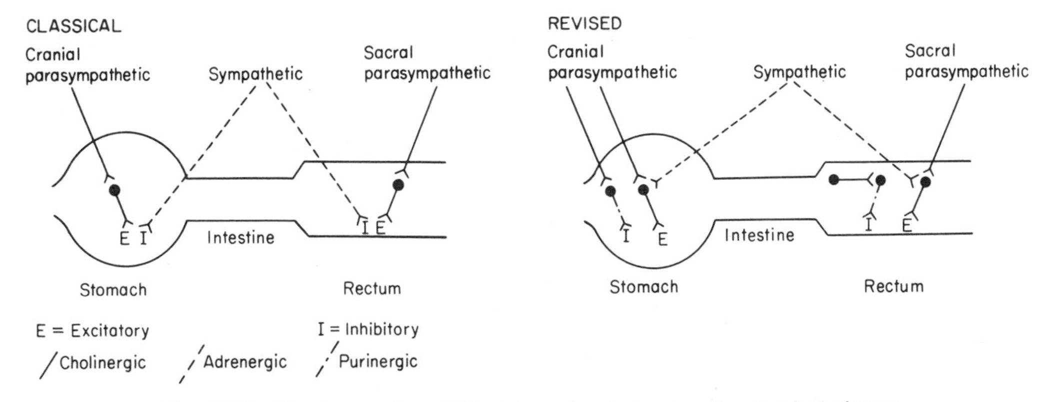

Fig. 2.13 The innervation of the tetrapod gut showing the classical picture (left) and the present concepts (right). Note the purinergic and intramural neurons in the revised concept. See original for further details and qualifications. [From Burnstock, 1972: Pharm. Rev., *24*: 560.]

ENDOCRINE SYSTEM

Hormones are special chemicals which are elaborated in restricted areas of the organism. They diffuse or are transported over variable distances to adjust metabolism, control remote effectors or regulate morphogenesis. They are effective in minute quantities. Their regulatory action is sometimes one of excitation and sometimes one of inhibition; consequently the word *hormone*, which comes from the Greek root meaning "to excite" is really a misnomer. The term endocrine (*endon*, within and *krinen*, to separate) was likewise coined before the true significance of this group of chemical coordinators was known. Vertebrate physiologists first recognized the endocrine gland as a ductless organ which discharged its secretions directly into the blood. Particular emphasis was placed on the vascularity and specialized anatomy. Now, however, endocrines are known to occur in animals with scanty circulating fluids, while very vascular organs such as the brain and the intestinal wall not only produce hormones but at the same time perform quite different physiological activities. As a matter of fact, the first internal secretion conclusively demonstrated was the hormone SECRETIN produced by the wall of the gut, and discovered in 1902 by the celebrated British physiologists Bayliss and Starling (Gabriel and Fogel, 1955). Since 1902 a host of chemicals has been added to the list of hormones.

The classical methods of investigations have been: removal of the endocrine tissue, either by surgical or chemical techniques, the study of the resulting disturbances in metabolism, and the restoration of the animal to normal physiological state by replacement of the endocrine tissue or the administration of tissue extracts (Fig. 23.9). Histophysiological studies have also contributed much, and the biochemist has now added substantially to the physiology by purifying or synthesizing many of the better known hormones; the organic chemist and pharmacologist have played their parts by providing many different chemicals which serve as blocking agents to the synthesis, release or action of specific hormones.

Almost a century of investigation has now shown that, at all levels in phylogeny, animal hormones are involved in a host of regulatory processes—both metabolic and metamorphic. More recent studies at the biochemical and cellular levels, however, indicate that these many processes are being governed at only two or three molecular sites (Clegg, 1969; Turner and Bagnara, 1971). At least one of the actions of insulin is to regulate the transfer of glucose into cells such as muscle and fat; a capacity to regulate transfer of essential materials into cells is a basic function of some hormones. The second group of functions takes place at the level of intracellular enzyme systems. Several hormones in addition to the catecholamines operate by way of the adenyl cyclase system already discussed (Fig. 2.10); vertebrate hormones which appear to operate this way in at least some places are TSH, ACTH, LH, MSH, vasopressin, glucagon, insulin, calcitonin, and the parathyroid hormone. Another intracellular enzymatic process, regulated at least in part by hormones, is the synthesis of messenger RNA (Chapter 22); through this channel hormones modify protein synthesis, differentiation and growth. Several of the steroid hormones and the growth hormone seem to exert their control at this level. In final analysis, all hormones may be operating through the modulation of enzyme systems. However, the intracellular targets (cell membranes, adenyl cyclase and RNA) are sufficiently distinct to warrant separate consideration.

Endocrinology is now an established segment of the biological sciences. It has contributed greatly to theoretical understanding in physiology and has found a secure place in the applied areas. The viewpoint of the comparative endocrinologist has been ably put by Bern (1972); Turner and Bagnara (1971) give a selected list of the important books in this field.

Phylogeny of Endocrine Tissues

The nerve cell was the grand evolutionary solution to problems of rapid intercellular communication. Its phylogeny was fundamental to the development of an active life; this matter is discussed in Chapter 21. Long-term adjustments that depend on hormones are of equal importance. The most valuable insurance of biological success is a capacity to take maximum advantage of favorable seasons for reproduction, development and growth and to be able to adjust adaptively to seasonal changes in moisture, temperature, and sometimes salinity. The first step in the cyclical and seasonal regulation of physiological processes depended on nerve secretions (neurosecretions) which persisted longer than the transmitters and acted over a wider territory; neurotransmitters act directly on another neuron or effector cell but neurosecretions usually diffuse or are transported some considerable distance (Fig. 2.14). It is only among the higher invertebrates (molluscs and arthropods) and the vertebrates that endocrine glands, other than those of a neurosecretory

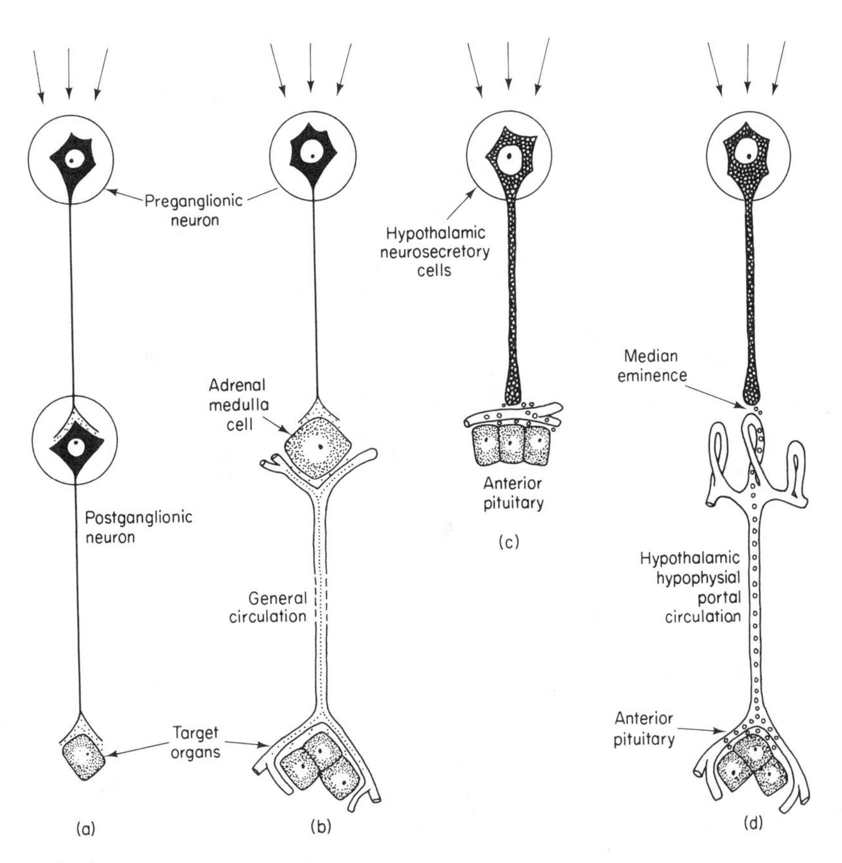

Fig. 2.14 Several discharge patterns for neurotransmitters and neurosecretions. (a) Sympathetic system where ACh is discharged at the preganglionic ending and noradrenaline at the postganglionic. (b) In the adrenal medulla, postganglionic neurons have transformed into endocrine gland cells which are excited directly by ACh. (c) Discharge of neurosecretory material into vascular channels which are very close to or directly on the target cells in teleosts and directly on them in the corpus allatum of insects. (d) Neurosecretions discharged into an elaborate portal system for transport to the target cells. [From Scharrer, 1965: Arch. Anat. Micr., *54*: 361, 364.]

nature, occur; the evidence suggests that the first hormones were neurosecretions and that these regulated processes which are strongly affected by seasonal changes: migration, reproduction, growth, metamorphosis, and tissue repair.

There are only a very few places where nerves directly regulate endocrine tissues (Scharrer, 1965, 1970). The adrenal medulla provides an example among the vertebrates but is scarcely typical since its tissues arise from the nervous system (Fig. 2.14b); an adrenergic hypothalamic control of some of the activities of the pars intermedia provides a better example (Chapter 20), as does also the nervous control of the optic glands in cephalopods (Chapter 23). However, direct nervous control is exceptional; neurotransmitters do not persist long enough to provide a reliable mechanism for regulating endocrine secretions which, by their nature, operate over prolonged periods. The neurosecretory cell provides the link between external and internal environmental cues and the secretory activity. There are several architectural patterns (Scharrer, 1965). The neurosecretory fibers to the corpus allatum of the insect discharge directly onto the target tissues; this, however, appears to be an unusual arrangement; neurosecretions usually reach the target tissues through circulatory channels which may be rather short as in the anterior pituitary of the teleost fish (Fig. 2.14c) but are usually much longer, as in the portal system of the mammalian pituitary (Figs. 2.14d and 2.18). Very frequently, numerous neurosecretory terminals expand into a neurohemal organ where secretions accumulate within the endings until required (Figs. 2.7 and 2.18); these neurohemal organs are recognized in several of the higher invertebrates (Highnam and Hill, 1969) as well as the vertebrates.

A regulated supply of hormones Regulation, which depends on a chain of different secretions, demands an appropriate system of internal controls. The best known of the endocrine control systems are those of the vertebrate pituitary hormones. Three different schemes are recognized and may be broadly representative of neuroendocrine systems of communication in the animal world.

The most direct control is by peripheral stimuli which act on the endocrine-producing cells. The hormones of the neurohypophysis belong to this group. Oxytocin, vasotocin and vasopressin are synthesized in the hypothalamic neurosecretory cells and released from their terminals in the neurohypophysis (neurohemal organ) in response to changing levels of electrolytes, temperature, tactile stimuli (suckling, coitus) and other peripheral stimuli.

Less direct controls occur when there are endocrine way stations between the neurosecretory cells and the terminal targets. The anterior pituitary is such a way station and the secretory activity of its cells is regulated in two somewhat different ways. One group of anterior pituitary hormones acts directly on target processes such as growth, melanosome activity and the prolactin effects. A second group works through additional hormone links between the pituitary and the target processes; thus, the thyroid, adrenal cortex and gonads form the final link in a third-order neuroendocrine chain with preceding links in the neurosecretory centers of the hypothalamus and the anterior pituitary. These two patterns appear to be associated with different control systems. When there are three links in the chain, the final one provides a feedback of information to the hypothalamus; thus, rising levels of thyroid hormone suppress the output of a hypothalamic releasing (regulating) factor which in turn controls the secretion of thyroid stimulating hormone TSH. A comparable regulation of adrenal cortical and gonadal hormone production occurs by way of correspondiing hypothalamic releasing (regulating) factors and anterior pituitary trophic hormones. This type of direct feedback is not found when there is a single link between neurosecretion and target process; the output of growth hormone, prolactin and melanophore stimulating hormone seems to depend on two hypothalamic secretions: a releasing (regulating) factor and a release-inhibiting factor (Schally *et al.*, 1973). Dual control systems with opposing excitatory and inhibitory substances are probably widespread in the animal world; they occur in the autonomic nervous system as well as the hypothalamic-pituitary system and are also found in regulatory hormones of some invertebrates, for example, the gonadal and molting hormones of crustaceans (Chapter 23; Adiyodi and Adiyodi, 1970). The whole field of neuroendocrine communication and control is currently one of the most active areas of endocrinology (E. Scharrer, 1965; B. Scharrer, 1970; Schally *et al.*, 1973).

The Hormones of Vertebrates

The functions of the hormones are considered in later chapters with the regulation of the different organ systems. As a background, the vertebrate endocrine tissues and their secretions will now be identified. Consideration of the corresponding invertebrate structures is included in chapters devoted to reproduction, development and growth (Chapters 23 and 24), since the better known areas of invertebrate endocrinology are concerned with these processes. In passing, it is also noted that the vertebrate endocrine system appears to be much more complex than that of invertebrates. The invertebrate systems are predominantly neurosecretory with endocrine tissues of non-nervous origin present only in the more advanced groups (particularly the arthropods). By contrast, vertebrate endocrine systems include six glands which develop from the anterior portion of the digestive tube and three or four which arise from the nephrogenic tissues in the posterior part of the coelom; these glands are in addition to the neurosecretory centers that dominate many of their activities.

Evolution of the vertebrate endocrine system There has been considerable speculation concerning the phylogeny of this system of controls (Barrington, 1964, 1968a; Hoar 1965a). The sites of evolution are suggested in Fig. 2.15—a fanciful diagram of the anterior end of an archetypal vertebrate. The diagram is based on structures and processes recognized in such primitive chordates as amphioxus and the larval cyclostomes; it assumes that the endocrine glands were preceded by groups of cells with very different functions. Attention is directed to three major areas of evolutionary change.

The first of these is the nervous system. Like nervous systems throughout the entire animal world, the archetypal system probably contained many neurosecretory cells. Indeed, studies of present-day lampreys strongly support this hypothesis since neurosecretion (and gliasecretion) is widespread in their brains and spinal cords (Sterba, 1972). In higher vertebrates, neurosecretory cells are restricted to two or three sharply circumscribed regions. Tetrapods, in line with increasing cephaliza-

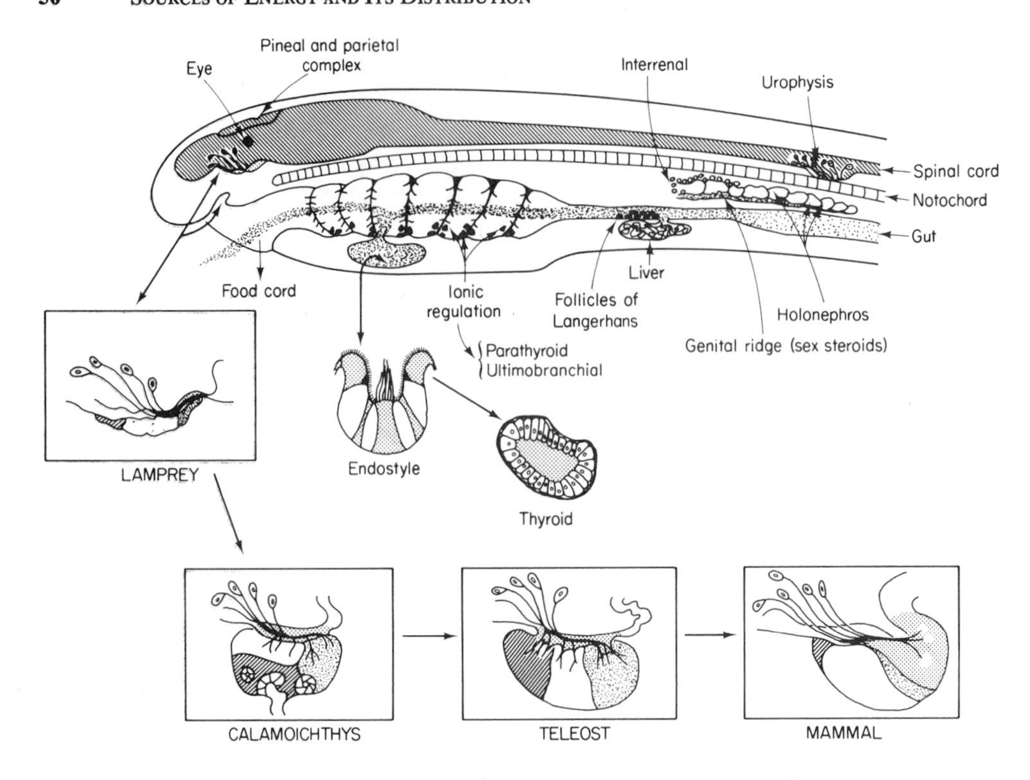

Fig. 2.15 Hypothetical organization of the evolving endocrine system in the archetypal vertebrate with the course of pituitary and thyroid phylogeny suggested below. [From Hoar, 1965a: Trans. Roy. Soc. Canada, *3* (ser. 4): 194.]

tion, have only two areas of neurosecretion, both confined to the brain. The PINEAL COMPLEX develops from the epithalamic region of the diencephalon and appears to have endocrine functions associated with color change (Chapter 20) and reproductive functions (Chapter 23); groups of neurosecretory cells in the hypothalamic region of the diencephalon (the supraoptic and paraventricular nuclei of mammals, the preoptic nuclei of lower forms) elaborate a number of related but distinct peptide hormones. One group of these hypothalamic substances (the RELEASING OR REGULATING FACTORS) controls the output of the anterior pituitary secretions; the other group is stored in a neurohemal organ (the NEUROHYPOPHYSIS) or posterior lobe of the pituitary, to be released as required for such varied activities as ion regulation, water balance, and the contraction of smooth muscle. The UROPHYSIS, in the caudal region of the spinal cord of jawed fishes, forms a third region of neurosecretion. This structure appears to have evolved through a gradual localization of widespread neurosecretory cells into a restricted area with a neurohemal organ at the ends of their fibers (Fridberg and Bern, 1968). In some of the teleosts it becomes a small discrete body which may be readily removed surgically from the caudal region of the spinal cord. No trace of it has been found in the tailed amphibians or higher vertebrates. Its functions are not yet well established but pharmacological evidence suggests the presence of more than one hormone with functions akin to those of the neurohypophysis—osmo-ion regulation and contraction of smooth muscles, such as those of the urinary bladder and blood vessels (Bern, 1969; Lederis, 1972).

The anterior end of the food tube formed a second site of endocrine phylogeny. The archetypal vertebrate was a filter feeder; copious sticky secretions were produced in the mouth and gill regions and constantly wafted posteriorly by cilia which clothed the gill bars and walls of the passages. This sticky mucus trapped food particles from the water circulating through the pharynx. Presumably, these secretions produced some molecules which acquired important digestive and metabolic functions. With the evolution of jaws and the assumption of predatory habits and periodic feeding, the need for sticky food traps disappeared but the requirements for regulatory molecules intensified; this may have served as a basis for the evolution of several of the glands which develop in this region. The ontogeny of the THYROID GLAND provides the best support for this hypothesis. The thyroid of the jawed vertebrates develops from a groove that appears at an early stage in the floor of the pharynx; in lampreys it forms at metamorphosis from a larval organ, the ENDOSTYLE. The endostyle is a ciliated groove or pouch (Fig. 2.15), whose major function is to produce mucus for food trapping in protochordates and larval lampreys; it also has the capacity to bind iodine to the amino acid tyrosine (Barrington, 1968a), and this becomes the major function of the vertebrate thyroid. A similar pattern of evolution is suggestive in several other endocrine organs which develop in this area: thus, (1) diffusely scattered cells in the gastrointestinal epithelium secrete GASTROINTESTINAL HORMONES, such as secretin and gastrin, which regulate the flow of digestive juices; (2) pockets of glandular epithelium in the region of the developing pancreas form the FOLLICLES OF LANGERHANS which control blood glucose concentrations by two hormones, insulin and glucagon; (3 and 4) gill pouch epithelium, which in the gills of adult fishes not only exchanges gases but also secretes and absorbs salts, is the source of the PARATHYROID and ULTIMOBRANCHIAL GLANDS concerned with the regulation of calcium and phosphorus. These four glands manufacture peptides which regulate digestion and metabolism; together with the thyroid which secretes specialized amino acids, they may have evolved from cells which, in the first instance, produced mucus for filter feeders and, coincidentally, contained molecules with regulatory potentialities.

The sixth gland arising from the primitive food-gathering tube has the most complex history of all. A pit or pouch of epithelium in the roof of the embryonic mouth pushes dorsally to establish an intimate association with the hypothalamus which is just above it. This proliferation of epithelium forms the ADENOHYPOPHYSIS; together with the NEUROHYPOPHYSIS which grows down from the diencephalon it becomes the HYPOPHYSIS OR PITUITARY GLAND. Something of the curious phylogenetic history of this gland is evident in the morphology of adult fishes (Fig. 2.15); at least one modern fish (*Calamoichthys*) has a persistent hypophysial pouch that opens into the mouth and shows evidence of exocrine as well as endocrine secretion; again, in teleost fishes, the adenohypophysis is little more than a bulbous mass attached to a rather flat neurohypophysis through which neurosecretory fibers penetrate directly, without the portal system which is characteristic of other vertebrates (Perks, 1969). Adaptive forces in the phylogeny of the pituitary gland are entirely speculative but it will be argued later that two of the most significant may have been the regulation of electrolytes by way of prolactin and the control of color changes by intermedin.

The nephrogenic tissues form a third area of endocrine evolution (Fig. 2.15). In the embryo, these tissues arise in association with the coelom—a cavity which in very primitive forms serves excretory and reproductive functions. The endocrine cells that arise from these tissues manufacture a variety of biochemically related steroid substances (Fig. 2.21). Those synthesized in the gonads coordinate reproductive functions; those formed in the adrenal tissues seem to have been first concerned with electrolyte regulation in the nearby nephric tissues but have assumed additional functions in the higher vertebrates. Both groups of tissues are regulated from the neurosecretory centers by way of the trophic hormones of the adenohypophysis; in turn, they regulate the activities of the neural centers through the feedback of information.

In summary, the endocrine glands of vertebrates have evolved from scattered neurosecretory cells, from several mucus-secreting and trapping tissues of a primitive filter feeder, and from that part of the mesoderm which forms the gonads, kidneys, and coelomic cavity. The most ancient functions of this system may have been ion regulation and protective coloration; two major forces appear to have shaped its subsequent history. One of these was the requirement for the seasonal regulation of reproduction, development and growth; the other was the evolution of jaws and the abandonment of non-selective filter feeding for the life of a selective, and often predacious, feeder with major problems of regulating digestive and metabolic functions. On this long evolutionary journey there has been a strong morphological tendency toward the formation of compact vascular glandular organs from scattered endocrine cells; this sequence is very evident in such organs as the adrenal medulla and cortex and the thyroid gland, all of which are scattered and diffuse in most fishes but are compact in the tetrapods. In the higher vertebrates, two or more organs may join forces to share a common vascular supply; the adrenal glands of higher vertebrates consist of two distinct anatomical and physiological tissues (medulla and cortex) while the mammalian thyroid contains the ultimobranchial as well as the thyroid tissue and in addition is anatomically very closely associated with the parathyroids. The advantages in endocrine physiology are obvious.

Neurohypophysis Investigations of the neurohypophysial hormones commenced in this century. Several different workers, using relatively simple techniques, showed that crude extracts produced four different effects when injected into mammals. These effects, revealed in the following order, were: elevation of blood pressure (the vasopressor effect shown in 1895), contraction of uterine smooth muscle (the uterotonic activity demonstrated in 1905), milk ejection found in 1910, and an antidiuretic activity shown in 1912–13. It is now recognized that these activities result from two basic effects: (1) an action on smooth muscle (uterine, vascular and myoepithelial in lactating glands) and (2) an effect on semipermeable membranes such as those of the kidney tubules. The first three activities depend predominantly on a basic peptide known as OXYTOCIN while the membrane effects are largely due to a neutral peptide, the antidiuretic hormone ADH or VASOPRESSIN. More than half a century of biochemical and pharmacological research has revealed five other peptides and it is not unlikely that additional ones remain to be discovered (Acher *et al.*, 1972). These seven peptides are very similar biochemically (Fig. 2.16) with many physiological properties, all of which relate to the two kinds of effects demonstrated almost seven decades ago. They are formed by the neurosecretory cells of the hypothalamus and combined with proteins called NEUROPHYSINS for storage in the terminals of the cells (Ginsburg, 1968). The history of the endocrinology of the neurohypophysis has often been summarized (Perks, 1969; Turner and Bagnara, 1971; Wolstenholme and Birch, 1971).

These hormones have molecular weights of about 1000 and are all octapeptides

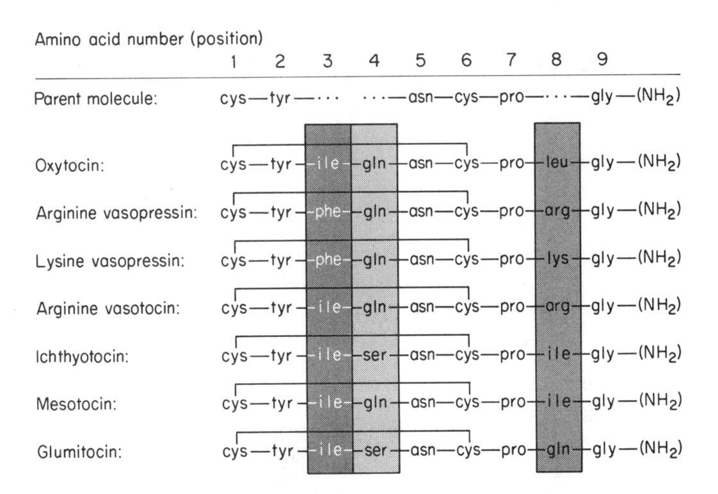

Fig. 2.16 The molecular structure of the seven known natural neurohypophysial hormones and the probable parent molecule from which they originated. One gene duplication and a series of subsequent single substitutions in position 3, 4, or 8 could produce two molecular "lines." Thus, substitution of isoleucine for glutamine in position 8 transforms glumitocin into ichthyotocin, and substitution of glutamine for serine in position 4 transforms ichthyotocin into mesotocin. The mammalian hormone oxytocin appears with the substitution of leucine for arginine in position 8 of vasotocin. Arginine vasopressin appears by the substitution of phenylalanine for isoleucine in position 3 of vasotocin. [From Frieden and Lipner, 1971: Biochemical Endocrinology of the Vertebrates. Prentice-Hall, Englewood Cliffs, N.J.]

with two molecules of cysteine being counted as one molecule of cystine. The disulfide linkage in cystine forms a pentapeptide ring with an attached side chain as indicated in the figure. The known hormones differ only in the amino acids at positions 3, 4, and 8 (Fig. 2.16). The vasopressins and vasotocin are basic peptides with their strongest effects on membrane permeability—not only membranes of the kidneys but also frog skin, cloaca and bladder (Chapter 11). The remaining four peptides (Fig. 2.16) are neutral with predominant effects on smooth muscle, particularly that of the reproductive organs (Heller, 1972). The overlap in pharmacological properties of these two groups of peptides is emphasized. Thus, both lysine vasopressin and oxytocin affect milk ejection in the rabbit and contraction of the rat uterus but oxytocin is about eight times more effective in the former assay and four times more effective in the latter. Among other things, this overlapping of properties argues for a biochemical evolution from one parent substance.

Additional evidence of a single line evolution is suggested by the phylogenetic distribution (Table 2.3). Most workers feel that arginine vasotocin is the parent substance since it is evidently the only octapeptide in the neural lobe of the cyclostomes and occurs throughout the vertebrate series, although it seems to be present only during fetal life in the higher mammals (Vizsolyi and Perks, 1969; Perks and Vizsolyi, 1973). Since the strongest effects of arginine vasotocin are on hydromineral balance, the speculation is that these hormones were first concerned with the regulation of water and electrolytes. A somewhat different line of argument is based on the developmental history of the pituitary gland; this ontogeny suggests that the first function of the neurohypophysis may have been to regulate the adenohypophysis. The hormones of the neural lobe show biochemical affinities with the releasing (regulating) factors of the adenohypophysis (Schally *et al.*, 1973). Among fishes, the neurohypophysis is broadly connected with the adenohypophysis, frequently interdigitating with it (Fig. 2.15 and 2.17). This anatomical arrangement is very appropriate for a direct delivery of information from the brain to the anterior lobe and this may have been the first function of the neural component of the gland. Among the tetrapods the neurohypophysis forms a distinct neural lobe,

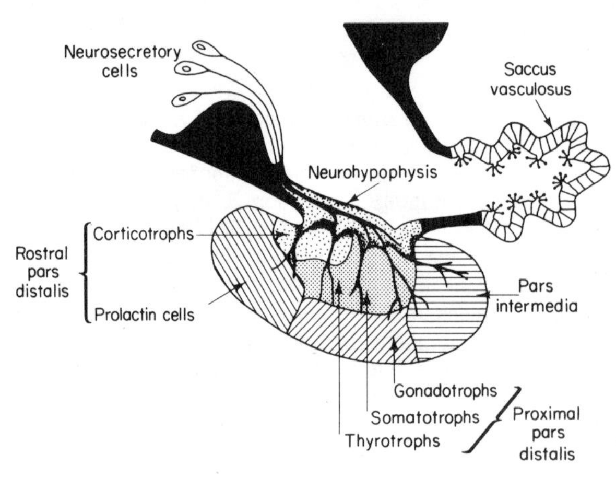

Fig. 2.17 Median sagittal section of the pituitary of a teleost fish showing different areas of the adenohypophysis where the various cell types are located. [Diagram based on several different sources as detailed by Ball and Baker, 1969: *In* Fish Physiology, Vol. 2, Hoar and Randall, eds. Academic Press, New York.]

Table 2.3 Distribution of Neurohypophysial Hormones (+, presence demonstrated; −, not found)

	Vasotocin (8-arginine oxytocin)	Vasopressin[1]	Oxytocin	Mesotocin (8-isoleucine oxytocin)	Isotocin or ichthyotocin (4-serine, 8-isoleucine oxytocin)
Cyclostomes	+	−	−	−	−
Cartilaginous fish:					
holocephalians	+	−	+	−	−
elasmobranchs[2]	+	−	−	−	−
Bony fish:					
lungfish	+	−	+	+	−
rayfinned fish	+	−	−	−	+
bichir	+	−	−	−	+
Amphibians	+	−	+	+	−
Reptiles	+	−	+	+	−
Birds	+	+(?)	+	−	−
Mammals	+[3]	+	+	−	−

1 Arginine vasopressin (3-phenylalanine, 8-arginine oxytocin) is the basic peptide in most mammals; the Suiformes provide a notable exception; lysine vasopressin (8-lysine) is the dominant factor in domestic pigs while peccaries and wart hogs may have either AVP or LVP or a mixture of the two, indicating that single allelic genes determine the structure (Ferguson, 1969).

2 Glumitocin (4-serine, 8-glutamine oxytocin) has been identified as the neutral peptide in skates while dogfishes have two peptides not yet firmly identified (A. M. Perks, personal communication); Acher *et al.* (1972) suggest valitocin (8-valine oxytocin) and aspartocin (4-asparagine oxytocin).

3 As yet identified only in some fetal mammals.

the pars nervosa, containing the terminations of many hypothalamic neurosecretory fibers and serving as a storage area (neurohemal organ) (Fig. 2.18). The evolution of a distinct neural lobe is thought to be related to the water balance demands of terrestrial living even though a number of very different activities (lactation, for example) have evidently come under its control. It should be noted that the vascular relationships of the tetrapod pituitary are such that the adenohypophysis receives chemical information from the hypothalamic nuclei (via the portal vessels, Fig. 2.18) even after the establishment of a neural lobe. Thus, in the higher vertebrates,

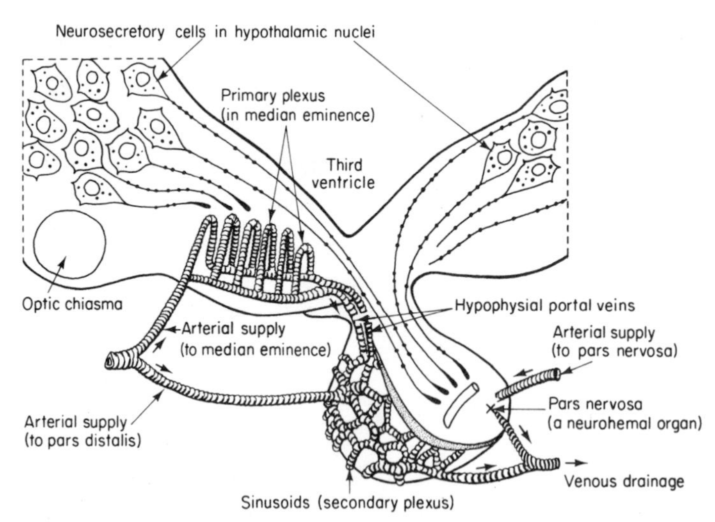

Fig. 2.18 Connections between the hypothalamus and pituitary gland of a mammal. Some of the neurosecretory fibers, with their stores of oxytocin and vasopressin, terminate near the blood vessels in the pars nervosa; others discharge their secretions (releasing factors or hormones) in close proximity to the capillary loops in the median eminence; releasing factors pass by the hypophysial portal vessels to their target cells in the adenohypophysis. [From Turner and Bagnara, 1971: General Endocrinology, 5th ed. Saunders, Philadelphia.]

secretions from the hypothalamic nuclei control the adenohypophysis locally and exercise a more remote control over distant organs by neurosecretions which are first stored in the neural lobe.

These two lines of argument concerning the evolution of the neurohypophysial hormones are not mutually exclusive. In primitive forms, neurosecretory cells in the forebrain may have regulated water and electrolyte balance directly or in association with an anterior pituitary substance such as prolactin which is actively concerned with electrolyte regulation in lower forms (Chapter 23). In modern fishes, both water and electrolyte balance are strongly affected by hormones of the neurohypophysis (Maetz, 1968). It may also be pertinent that the caudal neurosecretory system of fishes (urophysis) also seems to be involved in electrolyte balance. There are many stimulating discussions of the evolution of the neurohypophysial hormones (Sawyer, 1967, 1968; Vliegenthart and Versteeg, 1967; Perks, 1969).

Adenohypophysis This part of the pituitary secretes six or more protein hormones, each apparently elaborated by a different type of cell (Fig. 2.17). Since these hormones are proteins it is not surprising that they often show considerable species specificity. Although interspecific testing with pituitary fractions usually elicits some response, the efficacy is much greater in closely related species. Size and complexity of the protein molecule are probably responsible for the observed differences. Thus, beef thyrotropin with a relatively small molecule (about 10,000 molecular weight) elicits responses in thyroidal cells of all groups of vertebrates from the cyclostomes to the mammals, but the beef gonadotropins with molecular weights four to ten times greater are often relatively ineffective in lower vertebrates.

The growth promoting hormone SOMATOTROPIN (STH) has been studied in different vertebrates from fish to man and has displayed both metabolic and morphogenetic activities. The former are discussed in Chapter 7 and the latter in Chapter 24. The thyroid stimulating hormone or THYROTROPIN (TSH) regulates iodine metabolism by way of its action on the thyroid gland. The ADRENOCORTICOTROPHIC HORMONE or CORTICOTROPIN (ACTH) plays a comparable role with respect to the adrenal cortex. The melanophore stimulating hormone, INTERMEDIN, (MSH) is considered with chromatophores in Chapter 20. GONADOTROPINS—the follicle-stimulating hormone (FSH) and the luteinizing hormone (LH)—are discussed with reproduction in Chapter 23; PROLACTIN, which has been termed the "jack of all trades," is dealt with in Chapters 23 and 24.

Biochemists have shown that these proteins fall into three groups. The gonadotropins and TSH are glycoproteins made up of two subunits, each of mol wt about 15,000 (Chapter 23); prolactin and STH (mol wt about 22,000) are very similar and in some species have not been biochemically separated (Chapter 24); ACTH and MSH are much smaller molecules (mol wt about 5000) which share a common core of seven amino acids. In fact, the family of hormones to which ACTH belongs contains two melanophore stimulating factors known as α-MSH and β-MSH, as well as two lipotropic polypeptides known as β-LPH and γ-LPH; the latter substances are fat mobilizing factors. The ACTH/MSH chains range in length from 13 amino acids (α-MSH of pig, cow, and horse) to 39 amino acids in ACTH; the LPH molecules are about twice the size, ranging up to 90 residues. The interesting points are that all of these substances share a central core of 7 amino acids while the α-MSH molecule is identical with the first 13 amino acids in ACTH (Frieden and Lipner, 1971).

These biochemical affinities and the multitude of functions ascribed to the anterior pituitary hormones invite speculation concerning phylogeny. At this point, the subject will be dropped with the following observations: (1) The functions of prolactin, particularly those concerned with ion balance, were clearly established before STH had assumed distinct hormonal effects (Chapter 24). (2) Some of the best known functions of the thyroid gland in lower vertebrates relate to reproduction, and TSH and the gonadotropins may have evolved together as seasonal regulators of reproduction and development (Sage, 1973); it may also be significant that the TSH/LH/FSH hypothalamic releasing factors are biochemically very similar (Guillemin and Burgus, 1972). (3) ACTH, as well as MSH, has a strong stimulating effect on melanosomes; melanophore control is established in the most primitive living vertebrates while their pituitaries are not yet in full command of either the gonads or the adrenal cortex (Turner and Bagnara, 1971; Larsen and Rothwell, 1972). It is perhaps safe to speculate that the first functions of the evolving adenohypophysis concerned ion regulation and melanophore control and

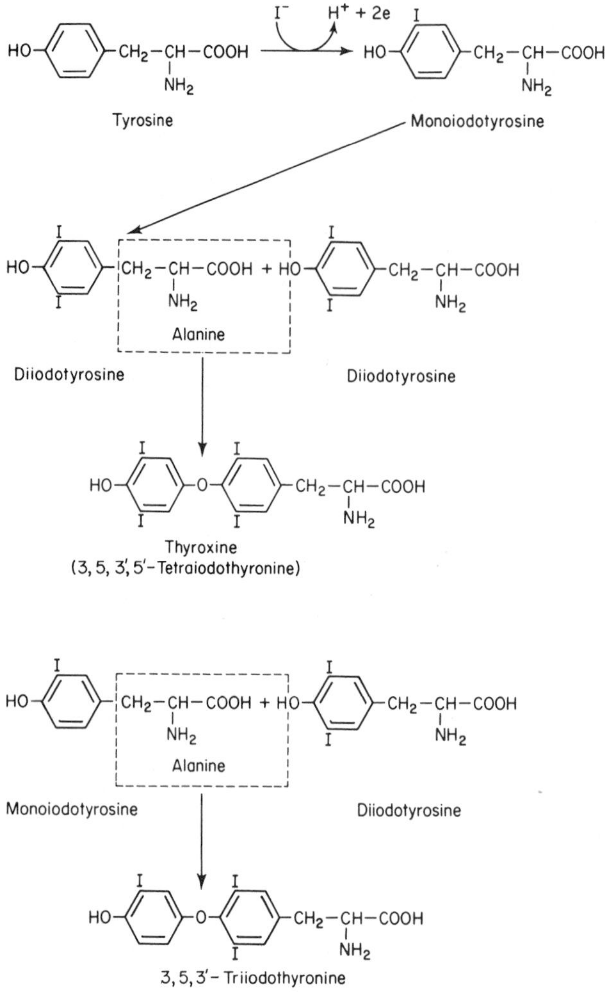

Tyrosine
Monoiodotyrosine
Alanine
Diiodotyrosine
Diiodotyrosine
Thyroxine
(3, 5, 3', 5'-Tetraiodothyronine)
Alanine
Monoiodotyrosine
Diiodotyrosine
3, 5, 3'- Triiodothyronine

Fig. 2.19 Biosynthetic pathways for the thyroid hormones. [From Turner and Bagnara, 1971; General Endocrinology, 5th ed. Saunders, Philadelphia.]

that these controls evolved by way of neurosecretory fibers arising in the floor of the forebrain.

Thyroid hormones Iodine combines readily with the amino acid tyrosine to form a series of iodinated compounds as shown in Fig. 2.19. The first of these to be described in natural materials was 3,5-diiodotyrosine, named iodogorgoic acid by Drechsel, its discoverer, when he found it in the gorgonian corals in 1896. Iodoproteins are widely distributed, and at least one of the iodinated amino acids has now been identified in every major group of animals—except the Protozoa and the Echinodermata—and in a number of marine plants (Gorbman and Bern, 1962). Bromine, iodine's nearest relative in the halogen series, also forms a tyrosine compound in some of the corals.

Iodination of protein occurs readily under certain *in vitro* conditions. Casein, for example, when incubated with iodine yields monoiodotyrosine (MIT), diiodotyrosine (DIT) and thyroxine (Tx or T_4). Some proteins, because of their tyrosine content—perhaps also because of the structure of their molecules—have greater affinities than others for iodine. In nature, some non-thyroidal tissues are particularly rich in iodoproteins—for example, the notochord, the tunic of tunicates, the radula of molluscs, the jaws and mucus of polychaetes (Fletcher, 1970). Since iodoproteins evidently form easily in some kinds of protein, their presence in such locations as tunic, jaws, radula and notochord might be purely fortuitous and of no physiological significance.

Gorbman, following this line of argument, postulates that at some point in evolution these iodinated amino acids—which occurred at first by chance—gradually assumed an indispensable role as one of the vertebrate hormones (Gorbman and Bern, 1962). Iodine trapping cells became specialized and eventually formed an organ for the synthesis, storage and regulated delivery of the hormone in accordance with the demands of the organism. The ubiquity of the iodoproteins, the failure to demonstrate any function for them in the invertebrates, and the increasing importance of thyroid hormones among the higher vertebrates all support this hypothesis. The development of the lamprey thyroid from the endostyle of the ammocoete has long been considered evidence of thyroid phylogeny from this mucus trapping structure. The endostyle is an organ of active iodine metabolism and forms both triiodothyronine (T_3) and thyroxine (T_4). It should be noted that although iodotyrosine compounds are widespread in the animal kingdom, only the chordates seem to synthesize T_3 and T_4 (Barrington, 1968a). Gorbman and Creaser (1942) did the pioneer work on iodine metabolism in the ammocoete endostyle, while Barrington (1959) first investigated the protochordates. It is now recognized that certain cells in the endostyle of larval lampreys bind iodides and form thyroid hormones (Barrington and Sage, 1972). It is entirely possible that the endostyle synthesizes T_3 and T_4 and that these substances pass with the mucus into the digestive tract and play some part in the metabolism, before the formation of thyroid follicles when the ammocoete changes into an adult lamprey.

The thyroid physiologist has now traced many of the steps by which iodine is incorporated into the proteins of the follicular colloid and then converted into hormone and released into the blood. His task has been greatly aided by three techniques. Standard histological methods, the oldest of these, are particularly useful; the height of the follicular epithelium and the staining characteristics of the colloid reflect rather faithfully the functional condition of the gland. Radiochemical techniques have also been valuable since radioiodine is relatively inexpensive, safe and easy to use. Several components of iodine metabolism can be precisely traced in this way. The technique is rapid and shows a good correlation with the histology. Finally, there are several chemicals (goitrogens or antithyroid substances) which block specific steps in the iodine cycle and have been especially valuable in studying animals, such as the teleost fishes, where the thyroid is diffuse and cannot be removed surgically.

The follicular epithelium pumps iodides and other essential compounds, such as amino acids, into the follicle. Within the follicular colloid, the iodide is then oxidized to iodine and this is followed by an iodination of tyrosine which is one of the amino acids in thyroglobulin. These biosyntheses lead to the formation of a protein thyroglobulin with a mol wt of about 700,000. T_4 and T_3, the active hormones, are released into the circulation as required, after digestion of the thyroglobulin by proteases produced by the follicular cells; they circulate in combination with the plasma proteins. This curious sequence of protein binding and intrafollicular digestion may reflect the phylogeny of the follicles from the endostyle; at least superficially the steps of protein binding and subsequent digestion are similar in both organs.

The pituitary hormone TSH stimulates all phases of iodine metabolism: the trapping of iodide, the iodination of tyrosine, and the synthesis of thyroxine. It also accelerates the formation of colloid, the proteolytic degradation of thyroglobulin, and the release of thyroid hormone. Several antithyroid agents (GOITROGENS) have found wide use in studies of thyroid physiology and in the diagnosis and treatment of certain thyroid disorders; trapping of iodide is blocked by such substances as thiocyanate, perchlorates, and iodates, while the thiocarbamides (thiourea, thiouracil, and others) prevent the iodination of tyrosine.

The circulating hormone is predominantly T_4. However, its partner T_3 is metabolically more active and the presence of T_3 in lesser amounts may indicate a rapid turnover; there is at least a suspicion that T_4 is converted to T_3 prior to usage (Sterling *et al.*, 1973). The role of thyroid hormones at the molecular and cellular level has been intensively studied for many years. It now seems certain that one of its major functions is a stimulation of the RNA machinery—hence, protein (enzyme) synthesis (Eaton and Frieden, 1969; Tata, 1969; Frieden and Lipner, 1971).

The thyroid is linked to the central nervous system by TSH which increases both production and secretion of hormones. Stimuli from the external or internal environment, including the "feedback" control of the thyroid hormones, regulate TSH production by way of the hypothalamic releasing hormone (TRH). TRH occurs in only minute amounts. About two million pig brains and five million sheep brains have yielded sufficient material to show that TRH is a tripeptide composed of glutamic acid, histidine and proline (pGlu-His-Pro-NH_2). Its isolation and characterization along with the other hypothalamic hormones form one of the most exciting chapters in modern biochemistry (Guillemin and Burgus, 1972; Schally *et al.*, 1973). This hypothalamic-pituitary control is probably an ancient one phylogenetically since the iodine-accumulating cells of the ammocoete endostyle seem to be responsive to TSH (Barrington and Sage, 1972).

Parathyroid and ultimobranchial glands One of life's most critical requirements is the regulation of inorganic ions. The integrity of cell membranes, the activities of enzymes, and protoplasmic excitability depend on a delicate balance of specific ions. Ion regulatory problems were solved in primitive vertebrates with semipermeable membranes and good ion pumps. This is possible if an ambient environment, containing adequate ions, is continually circulating over and through the animal; many of the invertebrates, like the jawless fishes, regulate in an open system through controlled transfers across the skin, gills, gut and kidney tubules (Chapter 11; Urist, 1963; Copp, 1969a). The increasing size and complexity of the jawed vertebrates, as well as the penetration of new habitats, (very high or low salinity waters and particularly the land) created extra demands which were met with the appearance of ion-regulating hormones. Those of the pituitary (prolactin and vasotocin) have already been noted; in addition, steroid hormones of the interrenal gland took on the job of regulating important monovalent ions, while special glands developed in the gill region to control calcium and phosphate.

Calcium and phosphorus are strictly regulated in animal tissues—and for very good reasons. Calcium is concerned with membrane permeability and plays a critical role in neuromuscular and synaptic transmission; it also activates several enzyme systems and is indispensable for the clotting of blood. Phosphorus itself is highly toxic and occurs only as phosphate; the importance of the organic phosphates in energy transfers and the position of phosphate in the DNA/RNA system has already been noted (Chapter 1). Calcium and phosphate are found in very definite ratios in many tissues; they are often stored together (bone, scales) and regulated hormonally in a balanced system, controlling rates of absorption, storage, and excretion.

Two hormones are strictly concerned with the regulation of calcium and phosphate. The first of these to appear in phylogeny was CALCITONIN (CT) a peptide consisting of 32 amino acids which is secreted by the ULTIMOBRANCHIAL GLANDS. These glands are found only in the jawed vertebrates. In mammals, calcitonin lowers blood calcium; it increases calcium uptake by bone cells (osteoblasts) and thus antagonizes the action of the second hormone PARATHORMONE (PTH) which stimulates bone-resorption cells (osteoclasts). Functions of calcitonin in the jawed fishes are uncertain (Copp, 1969a; Chan, 1972) but cannot be related to storage and resorption in bone since calcitonin is present in fishes with purely cartilaginous skeletons as well as the bony ones; moreover, bones and scales evolved in the jawless fishes (Romer, 1963) prior to the ultimobranchial glands. Calcitonin may first have been concerned with cation regulation at cell membranes—perhaps the removal of excess calcium in very saline waters.

Parathormone is a straight chain polypeptide with about 80 amino acids, secreted by the PARATHYROID GLANDS. These glands do not occur in fishes and seem to have evolved with terrestrial living and the problem of storing and mobilizing reserves of calcium salts. In mammals, the action of parathormone is opposite to that of calcitonin (Fig. 2.20). In the presence of parathormone (either *in vivo* or *in vitro* tests), osteoclasts are stimulated to dissolve calcium salts from bone; in addition, the kidney is markedly stimulated to increase phosphate excretion and to lower the calcium excretion slightly. The magnitude of these effects varies in different vertebrates (see symposia edited by Cortelyou, 1967 and Hoar and Bern, 1972).

Historically, this area of endocrinology commenced during the latter part of the nineteenth century when the lethal effects of "thyroidectomy" in cats and dogs were traced to the parathyroids and not the thyroids. Collip (1925) prepared the

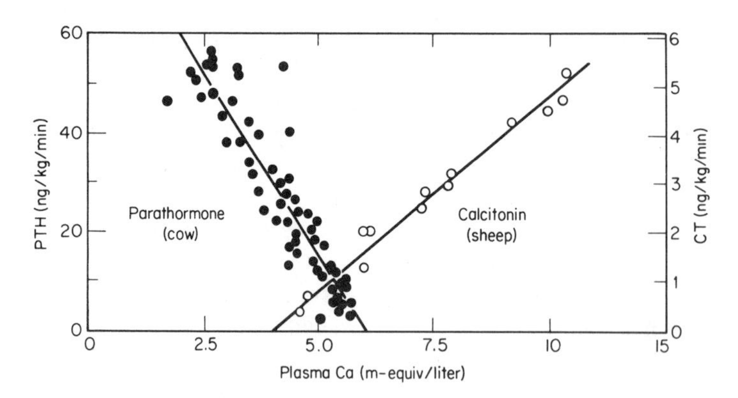

Fig. 2.20 Effect of plasma calcium on the secretion rate of parathormone (PTH) and calcitonin (CT). [From Copp, 1969: J. Endocrinol., *43*: 154.]

first extracts which counteracted the symptoms of parathyroidectomy. Collip's hormone—parathormone or parathyroid hormone—remained the only calcium phosphate regulating hormone for almost 40 years until a second fraction (calcitonin) with opposite effects was demonstrated by Copp in 1962 (reviews by Copp, 1969 a-c; Rasmussen *et al.*, 1970). The task of locating the source of calcitonin was complicated by the fact that its presence was first discovered in mammals where the secretory cells are diffusely scattered throughout the thyroid. During mammalian development, cells derived from the ultimobranchial body move into the developing thyroid, where they can be subsequently distinguished as the PARAFOLLICULAR or C-CELLS; in other tetrapods, the ultimobranchial glands remain as distinct structures. This story is now one of the classics in endocrinology and shows some of the dividends of comparative studies. These glands also provide an interesting chapter in comparative anatomy and embryology since they both are derived from pharyngeal or gill pouch epithelium; their phylogenetic history is tied to that of the thymus, tonsils, aortic and carotid bodies—all of which are derivatives of pharyngeal endoderm.

Several factors, in addition to CT and PTH, are known to be concerned with calcium and phosphate metabolism. Absorption from the gut depends on vitamin D; metabolism is strongly influenced by female sex steroids, probably because of the demands for egg production; in some fishes, the CORPUSCLES OF STANNIUS regulate ions and appear to produce a hormone that lowers blood calcium. These small corpuscles develop from the wall of the nephric ducts and are known only in holostean and teleostean fishes. Chan (1972) cites the relevant literature and discusses the possible interaction of calcitonin and the hormone of the Stannius corpuscles in regulating blood, muscle and bone calcium. The action of CT and PTH as well as these other factors vary markedly in different species of vertebrates (Greep, 1963; Turner and Bagnara, 1971; symposia previously cited). This is not surprising since the availability of calcium depends so strongly on habitat and diet while the demands for calcium change during development (bone, scales, teeth), season and reproductive condition (egg formation, lactation).

Chromaffin tissues and the adrenal medulla These tissues are a part of the autonomic sympathetic as well as of the endocrine system. Morphologically, the chromaffin cells, like the sympathetic ganglia, arise from the neural crests and are, in fact, postganglionic neurons which have lost (or almost lost) their fibers. They discharge their secretions directly into the blood instead of passing them along axons to the effector organs as in the neurohumors of the autonomic nervous system proper. In the most primitive vertebrates (cyclostomes) numerous small masses of these chromaffin cells (paraganglia) are scattered, in close proximity to the cardinal veins, in every segment of the body from the second branchial to near the end of the postanal region (Chester Jones, 1957). The number is reduced and the masses become progressively larger and more compact in higher vertebrates.In the elasmobranchs there is still a paired, distinct chain of paraganglia; but in the teleosts and higher forms they become associated with, and embedded in, the cortical tissues until, in the mammals, the chromaffin tissue forms a distinct and readily separable mass (the adrenal medulla) within the adrenal cortex. Nevertheless, it retains its preganglionic sympathetic nerve fiber connections. In this it is exceptional among the vertebrate endocrine glands. In general, the vertebrate endocrine glands are activated chemically, but the adrenal medulla receives its cues from the sympathetic nerves.

Histochemically, the granules in these cells become brown when treated with certain oxidizing agents; chromium salts are most frequently used, and for this reason, the tissues are called "chromaffin." This term is the preferred one for the lower vertebrates while the same tissues in the higher vertebrates—particularly the mammals—are usually called "medullary." The chrome reaction is another mark of the relation between these cells and the autonomic ganglia which show the same staining. In both cases the reaction is due to the presence of two catecholamines, noradrenaline and adrenaline (Fig. 2.8). The cells which produce them may be cytochemically separated by appropriate staining methods. The chromaffin or medullary tissues produce varying amounts of both amines in different animals. In most mammals, adrenaline predominates, but there are exceptions, and as much as 80 per cent noradrenaline is sometimes found (whales). It should be added that adrenaline and noradrenaline are of widespread occurrence in the animal kingdom. Chromaffin cells are found in the nervous system of many invertebrates (annelids, molluscs, arthropods), and both adrenaline and noradrenaline have been identified.

Interrenal gland and the adrenal cortex. The mesoderm of the roof of the coelom has an interesting history. Laterally, the nephrogenic tissue gives rise to the tubules and ducts of the excretory system; medially, as the genital ridge, it produces the gonads. Between these two areas strands of cells are budded off to form the interrenal or adrenal cortical tissues essential to the life of vertebrate animals. In the cyclostomes these buds of cells retain their scattered distribution in the tissues around the cardinal vein, along most of its length (Chester Jones, 1957; Gorbman and Bern, 1962). In the elasmobranchs they form compact structures of variable shape between the kidneys where they were quite appropriately named the "interrenals" in early descriptions. The progressivly more intimate association which occurs between these and the chromaffin tissues in vertebrate phylogeny was mentioned in the previous section.

Physiologically, the cortical tissue has affinities with the organs which develop from neighboring anlagen. On the one hand, hormones produced by both the gonads and interrenals belong to the same biochemical group (steroid hormones) and show many common biosynthetic pathways. In fact, several of the same steroids may be isolated from both organs. On the other hand, with respect to the nephric anlage, it has been suggested that the interrenal gland may have been developed by a response (evolutionary) to the demands for ionic regulation by the kidney (Chester Jones and Phillips, 1960). It has been argued that ionic regulation is one of the primary functions of the vertebrate kidney. It is now known that cortical steroids are concerned with electrolyte balance in vertebrates ranging from fish to men (Chapter 11). There is a suggestive connection, and the speculation may be extended to bring the corticosteroids into the group of materials essential for the evolution of the vertebrates.

The cortical tissues are centers for the manufacture of steroids; about fifty different variations of this molecule have been identified in the mammalian adrenal. Not all are hormones. Many represent biochemical steps in the synthesis of the active hormones (Figs. 2.21 and 23.8). It is possible to block the biosynthetic pathways in the adrenal cortex by the use of chemicals (amphenone B and metopirone, for example). The technique has the same experimental potentialities as chemical thyroidectomy in studies of the thyroid or the production of diabetes with alloxan (Gaunt *et al.*, 1965). It has been particularly useful in the lower forms where the cortical tissues are diffuse and cannot be removed surgically.

Corticosteroids have now been identified in the plasma of all major groups of vertebrates (Idler *et al.*, 1971). However, only microgram amounts occur in lampreys and hagfishes and it is questionable whether corticosteroids are physiologically significant in these animals (Weisbart and Idler, 1970). Significant quantities of corticosteroids occur in the blood of all jawed vertebrates. There is now good evidence that they are produced by the interrenal or adrenal cortex and that the pituitary regulates their production—in whole or in part, through the corticotropic hormone ACTH. The interesting point is that although most of the steroids shown in Fig. 2.21 can usually be identified in vertebrate plasma, many of them are only found in small amounts while one or a few molecules are prevalent (Idler, 1971; Idler *et al.*, 1972). The sharks and rays differ from all other vertebrates in the presence of a hydroxylase that converts corticosterone to 1 α-hydroxycorticosterone, the only corticosteroid found in significant amounts in these animals. It is curious that the closely related ratfishes (Holocephali) lack this enzyme and appear to form 11-deoxycortisol and cortisol as dominant steroids. Cortisol is also prominent in the more primitive actinopterygean fishes (sturgeon, bowfin) while cortisol and cortisone are conspicuous in the higher bony fishes. The lungfishes are interesting in showing significant amounts of aldosterone along with cortisol, thus showing

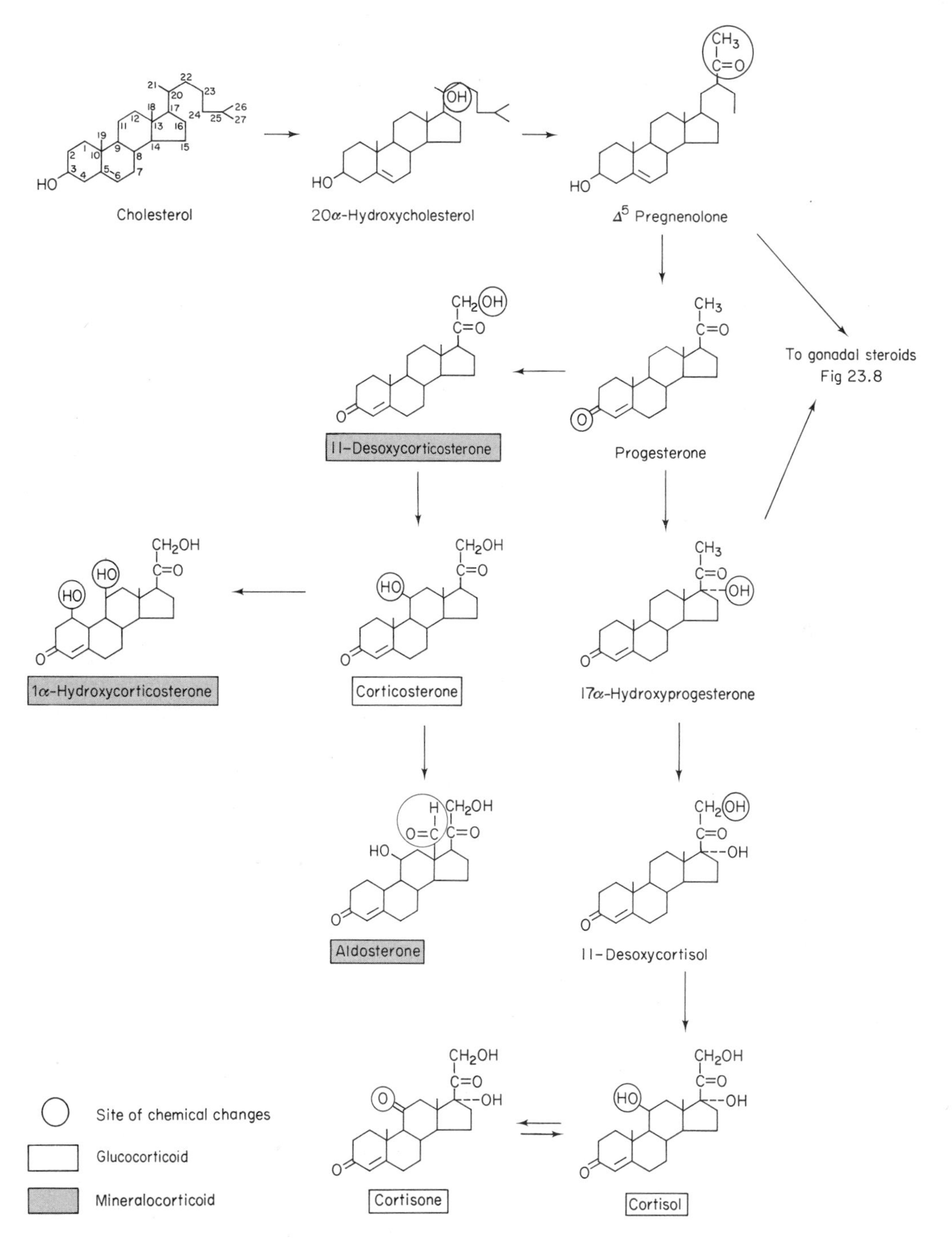

Fig. 2.21 Adrenocortical steroids of vertebrate animals with the physiological effects which they show in mammals. Note that the 1α-hydroxylation is characteristic only of sharks and rays. Further description in the text.

themselves to be in the company of the tetrapods, all of which have significant amounts of aldosterone. In amphibians, reptiles and birds, aldosterone and corticosterone, with their intermediate 18-hydroxycorticosterone, are the major corticosteroids while in most mammals (primate, sheep, cat, dog and others) cortisol is predominant with smaller but significant quantities of cortisone, corticosterone, aldosterone and 18-hydroxycorticosterone. Rodents and lagomorphs differ from most mammals in secreting mainly corticosterone, 18-hydroxycorticosterone and aldosterone (Idler, 1971). These systematic differences suggest a common pattern of steroid synthesis throughout the vertebrates with a phylogeny in which various groups have capitalized on one or a few different enzymes, perhaps in a purely fortuitous way or because some peculiar biochemical or ecological condition favors a particular enzymatic pathway.

Two major physiological groups of corticosteroids are recognized—one concerned with gluconeogenesis or the formation of glucose from the breakdown products of protein and fat (GLUCOCORTICOIDS) and the other with a primary action on electrolyte balance (MINERALOCORTICOIDS). This dichotomy into gluco- and mineralocorticoids is not a sharp one; compounds which are most active in gluconeogenesis may have some electrolyte effect, and vice versa. In mammals, cortisol (hydrocortisone) is by far the most active of the glucocorticoids although cortisone and corticosterone are also well-known representatives of this group. These compounds occur throughout the vertebrate series (Chester Jones and Phillips, 1960). Aldosterone is the most active of the mineralocorticoids. Deoxycorticosterone (DOC) is likewise active but much less potent than aldosterone. Both gluco- and mineralocorticoid effects appear to have been established at an early stage in the evolution of jawed vertebrates. Because of surgical difficulties, the evidence with respect to fishes is rather unsatisfactory and is mainly based on the pharmacological action of steroids. It seems likely, however, that both effects are physiological in fishes (Maetz, 1968; Chester Jones *et al.*, 1969) although dependence on these substances, especially in electrolyte regulation, may have become more marked in the terrestrial forms. As might be expected from comments on distribution and phylogeny, the various steroids may act quite differently in major phyletic groups. Further comment is reserved for the discussions of metabolism (Chapter 7) and electrolyte regulation (Chapters 8 and 11).

Different mechanisms control the secretion of gluco- and mineralocorticoids in higher vertebrates. ACTH is the prime regulator of glucocorticoid production, although there is also a mild effect on the secretion of mineralocorticoids. In the zonated mammalian gland, ACTH has its greatest action on the zona fasciculata and reticularis, but in other vertebrates, where zonation is largely absent, the effects are not localized. At the next higher level of integration, a polypeptide secreted by the hypothalamus, and known as the corticotropic releasing factor CRF or hormone CRH, controls pituitary activity through a negative feedback to maintain a delicate balance of ACTH production. Many factors affect the demands for glucocorticoids and hence for ACTH. Some of these variations are in accordance with normal physiological reactions (age, breeding biology) while others are pathological. Selye (1956, 1961 and earlier) first pointed out that a variety of situations collectively termed STRESS may create excessive metabolic demands that exhaust the pituitary-adrenal mechanisms. "Stress" situations range all the way from trauma and adverse environmental conditions to excessive social stimulation. Glucocorticoids are secreted in accordance with changing requirements but at a certain point the machinery may be overworked and a series of pathological states may develop. Since Selye's early definition of the GENERAL ADAPTATION SYNDROME in mammals, comparable states of adrenal and pituitary exhaustion have been noted in other groups of jawed vertebrates.

Mineral corticoid production is largely independent of the pituitary. In mammals, aldosterone is a powerful salt-retaining hormone. A decline in blood volume (hence pressure) or a fall in the plasma sodium stimulates production of aldosterone by the glomerulosa cells in mammals. The main receptor mechanisms reside in the juxtaglomerular (JG) apparatus of the renal corpuscle (Fig. 8.23). Stimulation of the JG system increases the output of an enzyme RENIN, which cleaves a decapeptide (angiotensin I) from a plasma globulin angiotensinogen (mol wt about 30,000) formed in the liver; a serum-converting enzyme then forms angiotensin II (eight amino acids), one of the most potent pressor substances known. Thus, both sodium absorption from the glomerular filtrate and blood pressure are restored (Chapter 8). These mechanisms appear to be confined to the tetrapods and the teleost fishes (Ogawa *et al.*, 1972).

Endocrine tissues of the gastro-intestinal endoderm The lining of the gut and its derivatives secrete a variety of peptide hormones concerned with the regulation of digestive activities and carbohydrate metabolism. Those connected with digestion proper—usually called the GASTROINTESTINAL HORMONES—are elaborated by mucosal cells in the wall of the stomach and intestine and, passing by way of the blood, act on the stomach, intestine, pancreas, gall bladder and liver. The hormone secreting cells have not been identified, but it is evident that they are scattered in the mucosa and are never organized into compact glandular masses. The information which comes mostly from studies of the higher vertebrates will be summarized in Chapter 3.

Embryologically, the pancreas and liver develop as evaginations of the gut endoderm. Both these glands elaborate secretions concerned with digestion, and these activities are regulated by the gastrointestinal hormones. In addition, the liver plays a major role in metabolism. It is strategically located to receive blood loaded with nutritive molecules from the digested food. These it metabolizes, stores and delivers on demand to the tissues of the body. Glucose is one of the key metabolites, and its movements and transformations are in large measure controlled by endocrine tissues—the FOLLICLES or ISLETS OF LANGERHANS—which also develop as specialized outgrowths of the gut endoderm.

The embryology seems to reflect the phylogeny (Barrington, 1962; Epple, 1969; Falkmer and Patent, 1972). In the larval lampreys (ammocoetes), distinct follicles of Langerhans are embedded in the submucosa of the anterior intestine. Ducts are not present, and the secretions pass directly into the blood. Barrington (1942, 1972) has shown that these follicles are concerned with carbohydrate metabolism and comparable to the islets of other vertebrates where the cells are arranged in nests—the islets of Langerhans. In some teleost fishes and in a few snakes the islets are grouped in several small but distinct globular masses (principal islets) in the region of the gall bladder, but in most vertebrates they are scattered throughout the exocrine tissue of the pancreas.

There are three types of cells in the islet tissues of the jawed vertebrates. The alpha or A-cells secrete GLUCAGON, a polypeptide which mobilizes blood sugar (hyperglycemic action), while the beta or B-cells produce INSULIN which has an opposite (hypoglycemic) action and lowers blood sugar. In addition, there is a third cell type—the delta or D-cells of uncertain physiology. Proportions of these three cells vary considerably in different groups of vertebrates (Falkmer and Patent, 1972), probably in accordance with diet and metabolic demands; for example, islets of the urodele amphibians contain very few A-cells. These three cell types stain distinctly; moreover, they are selectively affected by different poisons. Thus, B-cells are destroyed by alloxan and the A-cells by cobaltous chloride.

The first potent extracts of insulin were prepared in 1921 by Banting and Best (Banting and Best, 1922; Best, 1959). This discovery almost immediately alleviated the sufferings of many men and women with diabetes mellitus—a condition characterized by an inability to utilize or store glucose. The hyperglycemia is associated with widespread metabolic disturbances in the metabolism of fat and protein as well as carbohydrate. The clinical uses of insulin preceded by several decades an understanding of its chemistry and cellular physiology.

Insulin is a polypeptide of mol wt about 6000. Bovine insulin consists of an α-chain of 21 amino acids connected by disulfide bonds to a β-chain of 30 amino acids. In 1955, F. Sanger established the sequence of molecules in these chains; this was the first determination of macromolecular primary structure. The history of insulin may be traced through Best's 1959 paper and a symposium published in the *American Journal of Medicine* in 1966 (vol. 40, pp. 651–772). The insulin molecule appears to act at more than one point to affect the metabolism of cells. Its best known action is to increase the glucose permeability of cell membranes; insulin also increases RNA formation to affect the synthesis of protein. Further comment is reserved for sections on metabolism (Chapter 7).

The hyperglycemic factor, glucagon, was not recognized as a hormone until about three decades after the isolation of insulin. The glucagon molecule is smaller than that of insulin with 29 amino acid residues and a mol wt of about 3500. Its action is now well known. Glucagon stimulates the breakdown of glycogen (glycogenolysis) by activating liver glycogen phosphorylase which catalyzes the formation of glucose-1-phosphate from glycogen. This action, however, is not direct but by way of the adenyl cyclase system (Fig. 2.10). This is now one of the best known sequences for the action of a hormone on an enzyme reaction.

Phylogenetic studies indicate that insulin is a more ancient protein than glucagon. The arguments are based on demonstrations of insulin or proinsulin in a number of invertebrates and on the probable absence of glucagon (and A-cells) in the cyclostomes (Falkmer and Patent, 1972); it may also be significant that insulin appears to have more varied regulatory functions than glucagon. Insulin-like molecules appear to have been synthesized in the digestive mucosa of multicellular animals since a very early stage in evolution (Falkmer, 1972).

A COORDINATED SYSTEM OF REGULATION

For descriptive purposes it has been necessary to dissect the regulatory machinery of animals and separate out the various parts. Actually, enzymes, transmitters, neurosecretory substances and hormones operate as an integrated group of chemicals. In addition, there are the PARAHORMONES which act like hormones but differ in their mode of formation or distribution. These are the metabolites, like carbon dioxide which regulates the breathing centers of the brain, the secretagogues important in digestion, metabolic products such as histamine, and special secretions like erythropoetin, the kinins, renin, and the prostaglandins. The PHEROMONES also form a part of this chemical interrelation; unlike the hormones proper which operate within the animal that produces them, these work on other members of the same species. They have been called "social hormones" and include substances responsible for olfactory attraction between the sexes, alarm substances which warn other members of the species of dangers, and "markers" used to establish territories and trails or to mark rich sources of food.

Many links in the chain connecting these chemicals into a coordinated integrating system are now known. This is a neuroendocrine system, and its activities are frequently controlled by centers in the brain. The hypothalamus is only one of the centers involved in the vertebrates; the PINEAL ORGAN, which develops from the epithalamus, has been implicated in the control of chromatophores in some lower vertebrates (Chapter 20) and in the regulation of sexual cycles in birds and mammals (Wurtman *et al.*, 1968; Fenwick, 1970; Reiter, 1970, 1973). The urophysis in the tail of fishes is also a part of the neuroendocrine system.

This neuroendocrine relationship is basic to the physiology of most animals and important both in regulating metabolism through the various feedback mechanisms and relating the animal to its oscillating and changing environment. Seasonally altered photoperiods and temperatures act through the neuroendocrine system to trigger reproduction, growth and other physiological phenomena at the most appropriate times.

The other important level of recent understanding concerns the molecular or cellular site of neuroendocrine action. The examples already given emphasize important roles for these chemical coordinators at the cell membrane, in activating RNA synthesizing machinery and in releasing adenyl cyclase to form cyclic AMP, a substance known to trigger several physiological reactions (Fig. 2.10; Pastan, 1972). At the molecular level, these may all be effects on basic enzyme mechanisms.

3

NUTRITION

The term nutrition, as used here, includes most of the many processes involved in nourishing individual animal cells. It is sometimes used in a more restricted sense as synonymous with food or nutriment, but the older definition of the word, which implies also the processes of feeding, digestion and assimilation, is more appropriate for the comparative physiologist. In final analysis, it is the individual cell which requires for its nourishment a steady flow of energy-rich materials from the environment.

Living organisms may be loosely classified as AUTOTROPHIC or HETEROTROPHIC. The autotrophic organisms are able to synthesize all essential organic compounds from inorganic mineral sources. They include the chemosynthetic bacteria (CHEMOTROPHS) and the chlorophyll-bearing green plants (PHOTOTROPHS). The heterotrophic organisms, however, require organic substances as food and have limited synthesizing abilities. Almost all animals are heterotrophic and obtain energy for their life processes by swallowing, devouring, or engulfing other animals and plants and by breaking down the complex organic compounds which they contain through hydrolytic or other processes. A much smaller group of animals, including many parasitic forms, are SAPROZOIC and absorb relatively complex organic compounds through their body surfaces. This is considered a secondary method of nutrition, although, at one stage in evolution, the "soaking up" of organic compounds from the environment probably preceded the nutritional arrangements which are now found in animals and plants.

NUTRITIVE REQUIREMENTS

It seems likely that primitive ancestral animals were chlorophyll-bearing organisms with a full range of biosynthetic capacities. Although animal origins, like the origin of life itself, remain in the realm of speculation, the usual suggestion is that animal phylogeny has been from primitive plant to animal and from the obligate phototroph to the heterotroph. This view is consistent with the greater synthesizing abilities of the simpler forms of animal life. With increasing complexity, there is increasing dependence on special nutritive factors in the form of vitamins, essential amino acids and other complex organic groups.

Synthesizing abilities are controlled by genes, and a failure to manufacture any compound must depend on changes in the genetic material. This was first convincingly demonstrated by Beadle and Tatum (1941) in classical experiments on the genetic control of biosyntheses in the bread mold *Neurospora.* Because the cells of the mycelia are haploid and thus contain only one set of genes for any particular character, the genetics is reduced to very simple terms. For example, if a yellow and a white strain of bread mold are crossed and the resulting spores cultured separately, half of them will give white and half of them yellow molds. There are no dominant and recessive individuals to be sorted out in back crosses.

The red bread mold proved especially suitable for these experiments; it can be grown in pure cultures on a medium of nitrate, sulfate, phosphate and various other inorganic substances together with some sugar and biotin. From these simple starting materials, the mold synthesizes about 20 amino acids, 9 water-soluble vitamins of the B-complex and other equally elaborate groups. Every so often a mutant spore appears which is incapable of growing in the medium of salt, sugar and biotin but will grow on the addition of a mixture of complex organic compounds such as occur in an extract of yeast. The ability to synthesize one or more of these complex molecules has evidently been lost. Beadle (1948) increased the rate of mutation by using X rays or other mutagenic agents, and he isolated strains which were unable to synthesize certain particular molecules. These isolations were made by patiently testing strains, which failed to grow in the minimal salt-sugar-biotin medium, with media which had been supplemented by a single purified chemical such as one of the B-vitamins or an amino acid. A study of the genetics of these mutant strains then provided clear evidence for the genetic control of the biosyntheses. For example, when a strain that grows only if pyridoxine is added to the salt-sugar-biotin mixture is crossed with the wild type, it is found that half of the resulting spores can manufacture pyridoxine and half of them cannot. These experiments support the theory that the ability to synthesize materials such as pyridoxine is controlled by single genes.

The evolution of animal life involved not only the loss of chlorophyll and the photosynthetic ability, but also a loss of the ability to synthesize many enzymes whose reaction products are critically important for life. It is assumed that organisms growing in nutritively rich media occasionally produced mutants lacking the genes required for the production of certain enzymes. These mutants could survive and compete with other individuals if they lived in environments capable of supplying the products derived from the missing enzymic reactions. The protozoa provide many examples of closely related species with nutritive requirements which vary in accordance with the materials available to them in their environments.

Nutrition Among the Euglenoidina

Among the euglenoid Mastigophora, there is an almost continuous series from obligate photoautotrophs to obligate heterotrophs, with some strains capable of shifting back and forth from one mode of nutrition to the other (Hutner and Lwoff, 1955; Dogiel, 1965; Florkin and Scheer, 1967). In the light, *Euglena gracilis* is photoautotrophic and able to manufacture most of the compounds essential to its life. In the dark, *E. gracilis* may lose its chlorophyll and become saprozoic. Related genera likewise show variable modes of nutrition. *Polytoma* is a colorless and purely saprozoic flagellate; *Monas* lacks the colored pigments but possesses a pit or gullet for the entrance of food into the body and thus exemplifies the holozoic condition. Some of the colored forms also possess gullets and may combine the holozoic and the holophytic methods. *Ochromonas* and other chrysomonads combine all three modes of nutrition. All of the known methods of nutrition are evidently exploited by these primitive animals. With the loss of photosynthesis and a definite commitment to the holozoic way of life, there is a progressive loss of synthesizing abilities and a trend toward the obligate heterotroph.

There are many interesting differences in the capacity of the phytoflagellates to utilize inorganic nitrogen. Like most of the phytoflagellates, *Euglena stellata* can satisfy its nitrogen requirements with nitrate, and this is considered primitive. *E. anabaena,* on the other hand, cannot use nitrate but must obtain nitrogen from ammonia or an amino acid. Other species may require complex organic groups such as amino acids (*E. deses*) or peptones (*E. pisciformis*) and cannot rely on either NO_3 or NH_3 (Hollande, 1952; Dogiel, 1965). Amino acid requirements are also variable; some forms are able to grow on a single amino acid, while others may require a dozen or more different ones (Kidder, 1967). In their phylogeny, these organisms seem to have become progressively more dependent on the complex organic nitrogen compounds, and many forms require a rather specific group of ESSENTIAL AMINO ACIDS just like the higher vertebrates. They have lost the ability to manufacture these food factors because they have become specialized to exploit an environment which regularly contains them.

The chlorophyll-bearing protozoa are frequently described as photoautotrophs, implying synthesizing capacities comparable to the higher plants. The fact is that very few of them are completely autotrophic. As culture methods have been refined, it has become increasingly clear that most of the phytoflagellates cannot synthesize some essential enzymes and coenzymes (vitamins) from simple inorganic constituents. The importance of vitamin B_1 or thiamine in protozoan nutrition has been studied intensively and provides an instructive example. This substance is a coenzyme concerned with the release of energy from carbohydrates. Thus it enters into a chain of biochemical reactions which is a part of all life. It seems likely that animals, specializing in holozoic ways of nourishment, lost the ability to form this molecule at an early stage in evolution. Structurally, thiamine is a combination of two biochemical units, the pyrimidine ring and the thiazole ring (Fig. 7.14). The construction of each ring presents distinct biochemical problems and, although a few protozoa can manufacture the completed thiamine unit, most species must be provided with it or its two major blocks. Lilly (1967) in a summary of protozoan growth factors notes that only a very few of the phytoflagellates (*Chlamydomonas moewusi, Cryptomonas ovata* and *Polytoma uvella*) do not require thiamine or its two major building blocks. Some flagellates and one amoeba can manage with only one component of thiamine, either the thiazole or the pyrimidine portion. The loss of the power to synthesize thiazole and pyrimidine appears to be one of the earliest and most constant characteristics of animal evolution (Lwoff, 1947). Mutations which gave rise to forms incapable of manufacturing these essential compounds would be lethal, unless the mutant organisms lived in environments rich in foodstuffs which contained these compounds. Once the mutant is established as a new species, the odds of reversion to a state in which these compounds could again be synthesized are very small, albeit finite. Most likely the species will be forever chained to an environment in which its vitamin needs can only be satisfied from outside sources.

Components of an Adequate Diet

There is a constant exchange of materials between living organisms and their non-living environment. The energy requirements are met through oxidation of a series of carbon compounds—particularly the carbohydrates, fats and proteins. Growth, tissue repair and the synthesis of various secretions demand the amino acid units of protein. A number of chemical substances which are essential for metabolism cannot be manufactured by animals and must be obtained from outside sources; these dietary supplements are the VITAMINS. Finally, life demands a diversified group of minerals for the regulation of osmotic pressure and acid-base balance, the formation of skeletal materials, and the activation of numerous enzyme reactions.

The sources of energy Carbohydrate usually forms the major source of energy. In the well-balanced human diet, 55 per cent to 70 per cent of the calories are derived from carbohydrate. However, all animals may utilize other compounds as energy sources. Some green flagellates, which depend largely on photosynthesis, metabolize simple organic compounds such as acetate or butyrate. These animals are said to be FACULTATIVE because they can utilize a variety of organic acids when these are available. There is considerable variation in facultative abilities among different forms. Simple carbon groups, such as acetate, can probably be used by most animals. Fats and proteins are also sources of energy and may replace carbohydrate in the diet. Fats are stored when food is abundant, but when required they are readily converted into energy, and many animals (the migrating salmon, for example) can live for months on reserves of accumulated fat. Biochemically, the basic energy-yielding units of carbohydrate, fat and protein are readily interconverted (Chapter 7). The essential materials for animal tissues are the monosaccharides, simple triglycerides and a variety of amino acids. The metabolic machinery can readily shuffle these groups to obtain energy from any one of them or to store them in the form of fat (Chapter 7).

Essential amino acids A minimal amount of protein is imperative, since its structural units, the amino acids, are required for the construction of new tissues and the repair of old ones. Moreover, the amino acid requirements are usually very specific because of genetic limitations in the abilities to synthesize these complex molecules. The genetical work on *Neurospora* has shown that in this organism the synthesis of amino acids such as lysine, tryptophan, or proline depends on a single gene. Nutritional studies of several species of protozoa suggest that primitive animals lost the ability to synthesize these groups through random mutations in suitable environments.

All multicellular animals and most protozoa require some specific amino acids as dietary factors. These are referred to as ESSENTIAL AMINO ACIDS. Rose (1938) did much of the pioneer work on the rat and later (1949) showed that adult men could remain in nitrogen balance only if the following amino acids were included in their diets: lysine, trytophan, phenylalanine, threonine, valine, methionine, leucine and isoleucine. The same acids are required by mice, chickens, several species of fishes (Neuhaus and Halver, 1969; Cowey *et al.,* 1970), many different protozoa (Kidder, 1967), and insects (House, 1965); it is probable that all multicellular animals and most of the protozoans must obtain some preformed amino acids from their environment. From the level of the phytoflagellates, animals in general seem to carry just about the same set of enzymatic deficiencies related to amino acid syntheses (Kidder, 1967). Amino acids are the fundamental building blocks of animal tissues, and the enzyme systems required for their manufacture occur only in the plant world. Davis (1961a) points out that the number of enzymatic reactions involved in the biosynthesis of these essential amino acids is relatively large. Thus, non-essential acids like alanine or glycine have only one enzyme in the main anabolic pathway while essential acids such as threonine or arginine require six or seven, and the most complex ones (tryptophan, phenylalanine, isoleucine) have ten to fifteen such enzymatic reactions in their biosynthesis.

Many years ago it was recognized that there are not only essential and non-essential amino acids but also those which are semi-essential (Beaton and McHenry,

1964). Synthesis of arginine, for example, occurs in the rat but is not sufficiently rapid to permit optimum growth. Even the simplest amino acid, glycine, is probably not synthesized rapidly enough to provide for optimum growth of chicks. Clearly there are complete inabilities to synthesize some amino acids and there are limits to the rate of synthesis of others. Moreover, these limitations depend not only on the taxonomic position but also on the physiological condition of the animal.

In general, the naturally occurring amino acids are L-isomers. In fact, D-isomers of amino acids and L-isomers of sugars have long been regarded as artifacts when found in nature; the naturally occurring proteins are composed of L-amino acids while all the sugars are D-isomers. This biological curiosity has given rise to considerable speculation among biochemists, physiologists and students of biopoesis (Chapter 1). Further complexities have more recently been recognized. It has been clearly demonstrated that the D-isomers of some of the amino acids are toxic when included in the diets of certain insects (House, 1965). However, small amounts of the D-isomers of some amino acids are regularly present in the tissue of certain animals; these were first identified in insects, molluscs and annelids but, with improved techniques, have also been found in mammals. It has been suggested that these naturally occurring D-isomers may play very specific roles in metabolism and provide particular functional groups which do not enter the regular metabolic pool (Corrigan, 1969). In annelids, for instance, D-serine contains chemical constituents required for the synthesis of lombricine (page 224) but serine in this form cannot be incorporated into tissue proteins.

Essential lipids All animal tissues contain some fat or lipid material. In addition to that present as depot or reserve energy material, there are lipids in every cell, forming essential components of the plasma membrane and such important cytoplasmic constituents as the mitochondria. Animals can readily synthesize fats from carbohydrates or proteins but, as in the case of the amino acids, there are limitations with respect to certain chemical groups. This is particularly true of certain unsaturated fatty acids and the more complex lipid materials such as cholesterol.

The rat is unable to synthesize linoleic, linolenic and arachidonic acids. These are all polyunsaturated fatty acids in which the location and configuration of the double bonds are evidently of the utmost importance. A similar fatty acid requirement has been demonstrated in several other mammals, in some birds and, among the insects, in some Lepidoptera and a few Orthoptera (House, 1965; Levinson and Navon, 1969). No absolute requirements for polyunsaturated fatty acids have been demonstrated among the Protozoa but several different ciliates grow better in media supplemented with saturated acids such as stearic or oleic, and in a few cases growth has been stimulated by linoleic and arachidonic acids (Holz, 1964; Dewey, 1967). Although relatively few animals have as yet been investigated, it seems that the ability to synthesize the necessary lipids is much more widespread than the ability to form the necessary amino acids. This appears to be related to differences in biochemical complexity of the two classes of compounds.

The steroids are more complex biochemically than the unsaturated fatty acids.

The cholesterol molecule (Fig. 2.21) is ubiquitous as a constituent of cell membranes; its four-ring nucleus forms the skeleton of such important compounds as the vertebrate sex hormones and adrenocortical hormones.

Although the higher vertebrates can synthesize cholesterol from acetic acid or its active derivatives, this is not true of many invertebrates. A dietary requirement for cholesterol or a similar compound has been demonstrated in many of the protozoa, at least one coelenterate, *Rhizostoma*, the earthworm *Lumbricus*, a free-living nematode, several of the carnivorous molluscs, some crustaceans and many insects (Wootton and Wright, 1962; Holz, 1964; van den Oord, 1964; House, 1965; Dewey, 1967; Zandee, 1967; Rothstein, 1968). Only the insects and protozoans have been systematically studied. Although there is frequently an ability to rearrange certain groupings within the steroid molecule, the requirements are always rather specific.

The purines and pyrimidines These heterocyclic nitrogenous compounds are components of the nucleotides which form a part of such basic molecules as DNA and ATP. Dietary requirements for purines and pyrimidines are extremely rare in the animal kingdom. The first organism shown to require purine was the ciliate *Tetrahymena pyriformis* (Kidder and Dewey, 1945). It now appears that all ciliates and zooflagellates require both purine and pyrimidine. Specific requirements with respect to the ring structures vary but there are distinct limitations to the capacities of these organisms to shift about the different groups in the purine or pyrimidine ring (Kidder, 1967).

The vitamins The story of the vitamins is one of the most interesting chapters in the history of medicine and biochemistry. Much of it is written around two of the ancient scourges of mankind, scurvy and beri-beri. Scurvy was all too well known to the Crusaders of the thirteenth century and probably to generations of men from the beginning of human history. Wherever men must exist for prolonged periods without fresh fruits and vegetables, widespread small hemorrhages develop, the joints become swollen, the teeth loosen in their sockets, bones break easily and wounds fail to heal. Death is the inevitable result. In times of war and famine, on lengthy sea voyages, and in institutions with restricted diet, men and women were familiar with the symptoms long before an Austrian physician, Kramer, recommended citrus fruit for a prophylaxis in 1520. Another 200 years or more were to pass before James Lind in 1753 produced his classic report on this disease and presented such forceful evidence for the efficacy of citrus fruits that the British Navy introduced limes and lemons into their rations (1795). Thus, their sailors earned the name "limey," and their ships were rid of scurvy (Drummond and Wilbraham, 1939). The wharf area of London became "Limehouse," but the physiology of the disease was really no better understood than in the days of the Crusades.

In 1928, the active component of the citrus fruit juice (ascorbic acid) was isolated by Szent-Györgyi, to be followed by its crystallization in 1932. By this time it was clearly recognized that ascorbic acid is essential for the formation of the intercellular cement, and that without it the defective connective tissues fail to support bones, joints and teeth, while the ruptured capillary walls produce hemorrhages and the other characteristic pathologies of scurvy. Imperfect formation of the cementing substance is caused by a failure in collagen synthesis due to a deficiency in the enzyme proline hydroxylase which converts proline into hydroxyproline (Harper, 1973). However, most of the basic functions of ascorbic acid are still unresolved and the final chapter in the history of this vitamin is still to be written.

At present the history of beri-beri has a somewhat more satisfactory conclusion, since the identification of the missing dietary factor (thiamine) was followed by the demonstration of its role as a coenzyme involved in the oxidation of pyruvate. Without thiamine, the chain of reactions in carbohydrate oxidation stops at the pyruvic acid stage and this material accumulates. Carbohydrate metabolism is depressed; fat metabolism is accentuated. The nervous tissues suffer most; neuritis develops and leads to a paralytic condition and also to gastrointestinal and cardiovascular symptoms which, if not relieved, are fatal. Beri-beri was best known in the Orient where constant diets of polished rice prevailed. The polishing and preparation of the cereal remove a compound which, in nature, is formed only by plants. In 1882, Takaki reduced the incidence of the disease in the Japanese Navy by varying the diets, but it remained for a Dutch investigator, Eijkman in 1897, to show that the disease could be produced experimentally in chickens and cured with a water-soluble extract of rice polishings. Thiamine was crystallized in 1926, and its role as cocarboxylase was recognized in 1937 (Gabriel and Fogel, 1955).

The discovery and identification of many of the other food factors is just as interesting as that of ascorbic acid and thiamine. Intense research during the first forty years of the present century has explained the functions of many of these accessory food factors. Discussions of their chemistry and place in animal physiology can be found in texts of biochemistry and nutrition. Wagner and Folkers (1964) provide a brief historical account and a compilation of vitamin activities and nomenclature with discussions of the current literature. Our knowledge is almost entirely based on mammals, birds, and a few species of protozoans, insects and fishes. Although the nutritional requirements of the majority of animals are unknown, it is likely that the vitamin concept extends through all groups from the unicellular to the most complex multicellular forms. Thiamine and many other accessory food factors are just as truly "vitamins" for the protozoa as they are for man. There are, however, many variations and further studies are bound to reveal curious differences in species requirements. Ascorbic acid, for example, has often been recorded as a vitamin for only primates and the guinea pig. Careful investigations have, however, shown a vitamin C requirement for some protozoans (Lilly, 1967), a few insects (Levinson and Navon, 1969), and a considerable number of birds. The more primitive species among the birds (Galliformes, Anseriformes and others) synthesize ascorbic acid in the kidneys only. In the more highly evolved Passeriformes, some species form the vitamin in the liver as well; in others, the synthesis is confined to the liver while many species (16 of 28 species examined) fail to synthesize the vitamin at all (Chaudhuri and Chatterjee, 1969). There is an evident phylogenetic trend in this group of animals and this may also apply to other animal groups. Loss of the enzyme concerned with ascorbic acid has evidently occurred quite frequently. This raises the question of a definition.

The term VITAMIN does not refer to compounds with biochemical similarities but to a group of substances with similar general functions in metabolism. Vitamins are organic compounds, required in catalytic amounts by animal tissues. The organism cannot synthesize them and requires a constant supply; the ultimate source is the plant or bacterial world. Vitamins are obtained in food or supplied as byproducts of intestinal bacteria. Strictly speaking, any particular compound—ascorbic acid, for example—may or may not be a vitamin depending on the synthesizing abilities of the particular animal concerned. Certain amino and fatty acids have already been mentioned as dietary essentials. These are not vitamins, however, since they usually enter into the construction of tissues or form building blocks in secretions and other essential biochemical entities. Vitamins, on the other hand, act as catalysts. It should be noted, however, that this distinction based on catalytic and structural roles becomes rather blurred when the problem is looked at from the comparative and molecular levels. There are several dietary factors which are structural in the sense that they form essential constituents of enzyme (catalytic) molecules. Some of the trypanosomes will only grow in media containing blood and this has been shown to be due to a nutritional requirement for porphyrin (Kidder, 1967); porphyrins are basic constituents of the cytochromes present in all active cells and concerned with the release of energy. Again, the distinction is not so satisfactory for some of the lipid materials. The group of sterols known as vitamin D has long been recognized as essential for mammalian bone formation and the proper utilization of calcium and phosphorus (Loomis, 1970). Insects and some protozoa require cholesterol and cannot use other sterols such as vitamin D (calciferol). Cholesterol is sometimes classed as an insect vitamin; it certainly enters into the structure of many tissues whether or not it plays a catalytic role.

On the basis of their solubility, vitamins are usually considered in two groups, and in general these two groups discharge rather different functions. Most of the water-soluble factors are universally vitamins since they perform the same functions wherever they occur; they are catalytic factors and, in consequence, form vital links in the chains of biochemical reactions characteristic of all life. Thiamine, for example, is required wherever sugars are oxidized aerobically to release energy. The fat-soluble vitamins, on the other hand, play more specialized roles in certain groups of animals and in particular types of activities. They function in the formation of a blood-clotting factor in the vertebrates (vitamin K), the absorption of calcium and phosphorus from the vertebrate intestine (vitamin D), or in the formation of a visual pigment (vitamin A). Thus, most of the water-soluble vitamins are a part of all metabolizing tissue, but the fat-soluble vitamins are involved in particular tissues and special physiological activities.

Minerals In Table 3.1 the elements have been listed in the approximate order of their abundance in living tissues. The first four (oxygen, carbon, hydrogen, and

Table 3.1 Composition of Animal Tissues

Elements		Higher animals	Higher plants	Per cent: In man	Per cent: In tissues generally
Oxygen	Major elements (always present and essential)			65.0	> 1.0
Carbon				18.0	
Hydrogen				10.0	
Nitrogen				3.0	
. . .					
Phosphorus				1.0	0.05 to 1.0
Calcium				1.5	
Potassium				0.35	
Sulphur				0.25	
Sodium				0.15	
Chlorine				0.15	
. . .					
Magnesium				0.05	
Iron			←	0.004	< 0.005
. . .					
Manganese	Trace elements (always present)	← Essential for higher animals	← Essential for higher plants	0.0003	
Copper		←	←	0.0002	
Iodine		←		0.00004	
Cobalt		←			
Zinc		←	←		
Selenium		←			
Molybdenum		←	←		
Boron			←		
Silicon		← Perhaps required by higher animals	←		
Nickel					
Aluminum					
Fluorine		←			
Barium		←			
Strontium		←			
Chromium					
Tin					
Lead					
Titanium					
Rubidium					
Lithium					
Arsenic					
Bromine					
Vanadium	Occasionally present				
Silver					
Gold					
Cerium					
etc.					

SOURCE: Compiled from Oser (1965) and Underwood (1971).

nitrogen) form more than 95 per cent of protoplasm. Appropriately combined, they are the structural basis of living material. The element carbon unites readily with itself and with other members of this quartet to form the long chains and the complex rings which are the molecular framework of protoplasm. Carbon has been referred to as "the great juggler of matter" (Ducrocq, 1957), and no other element produces such an infinite variety of complex organic groupings. Hydrogen and oxygen, combined as water, form 70 per cent or more of active tissue; water is the most plentiful single compound of protoplasm. Nitrogen, in the NH_2 group attached to a carbon atom, characterizes the amino acids which are the structural units of protein and one of life's most abundant and specific biochemical entities.

The next seven elements in Table 3.1 are also included in those referred to as "major." They justify this title because of their occurrence in relatively high percentages (usually more than 0.1 per cent), and also because most of their activities are concerned with processes which are widespread or universal in living material. Phosphorus, through its high energy bonds, is involved in all major energy exchanges of life. It is also a constituent of the phospholipids which are a part of cell membranes and certain other equally indispensable structures. Further, in many animals phosphorus is a component of the skeleton and an important part of the phosphate buffer system. Calcium is associated with phosphorus in several of these activities but, in addition, discharges its own special functions in clotting reactions (blood clotting, milk coagulation and cell surface precipitation reactions) in muscle contraction and in numerous enzyme activities. The movements of sodium and potassium ions through the cell membranes are constant features of the stimulus response phenomena (Chapter 15); sodium chloride is the most important single compound in the control of osmotic phenomena. Sulfur forms a part of the essential amino acids cystine and methionine and also has a number of special functions in the—SH group. This is by no means a complete catalogue of the duties of the major elements. It does, however, emphasize their ubiquitous distribution and the type of role which they play in the universal construction and functioning of tissues.

Magnesium and iron are sometimes placed in a group with the major, and sometimes with the minor constituents. Quantitatively, they are much less abundant than those at the top of the list but far more abundant than those usually grouped as trace elements. Their activities are both structural and catalytic. Structurally, they are components of such universal and indispensable materials as chlorophyll (magnesium) and hemoglobin (iron); catalytically, like the trace elements, they are active in numerous enzyme systems.

Many of the trace elements are now known to be as essential to life and as universally distributed as the major elements. Like the vitamins they are not placed together because of chemical affinities but because they are all required in such minute amounts and because they are usually concerned with specific biochemical compounds and physiological activities. (Bowen, 1966). It should be noted that some elements which are found only in traces at the vertebrate level of phylogeny may be much more abundant and play indispensable roles in some of the invertebrates (vanadium in tunicates, copper in the blood of crustaceans, and silicon in the skeleta of radiolarians and sponges).

The list of essential trace elements will probably grow as research continues (Frieden, 1972). One of the more recent additions to the list is cobalt, now known to be a part of vitamin B_{12} or cyanocobalamine. This material is required by many animals ranging from certain of the protozoa (Lwoff, 1951) to the vertebrates (Underwood, 1971). It may be universally required. It is interesting that this cobalt-containing compound, unlike other members of the B-vitamin complex, is not manufactured by the higher plants nor by yeasts. It is synthesized only by bacteria, and in some way the higher animals have become dependent on bacteria for it. Without B_{12} blood formation is impossible in the higher vertebrates.

As components of enzyme systems, the trace elements are involved in two somewhat different ways. Many of them form an actual part of the structure of the enzyme—or more particularly, its prosthetic group. For example, xanthine oxidase is a molybdo-flavoprotein, and the prosthetic group of catalase is an iron porphyrin compound. On the other hand, the trace elements may activate enzyme systems. Aminopeptidases and dipeptidases, for example, are activated by traces of specific metals such as Mn, Zn or Mg, and enzyme activity is lost following dialysis (Baldwin, 1967). However, if the trace element enters into the structure of the enzyme there is no loss of activity with dialysis.

It should also be emphasized that these elements are often physiologically interrelated, and the functions of one cannot be easily divorced from one or more of the others. This is not surprising if it is remembered that life consists of interrelated chains of highly specific enzymatic reactions and that this specificity extends also to their prosthetic groups and activators. These interactions may occur in a variety of ways. Molybdenum may not be excreted sufficiently if dietary copper and sulfate are low. Cattle on forage rich in molybdenum then develop a severe diarrhea ("peat scours"), and grazing lands which appear rich and luxuriant may in this way be lethal. Yet traces of molybdenum are indispensable and seem to form a part of the enzyme xanthine oxidase. Soil rich in selenium may produce a different but equally serious condition called "alkali disease" or "blind staggers" in livestock. The condition has been recognized since the days of Marco Polo (Rosenfeld and Beath, 1964) but the biochemical nature of its toxicity (which is comparable to that of arsenic) is still not understood. However, selenium and sulfur are chemically related in several ways and it is believed that the sulfur of the two essential amino acids cystine and methionine is partially replaced by selenium, and certain enzyme systems (succinic dehydrogenase) are also inactivated, perhaps through the removal of sulfhydryl groups (Underwood, 1971). Selenium toxicity was recognized for more than a century before it was known that higher animals actually require this element in trace amounts. Evidence indicates that selenium is a component of the enzyme glutathione peroxidase which is important in degrading small amounts of H_2O_2

formed in metabolism; thus, it may protect certain biological structures (membrane lipids, for example) from oxidative disintegration (Rotruck *et al.,* 1973). In the vertebrates iron-deficient diets lead to a condition of anemia; but, even if the dietary iron is abundant, blood formation does not occur normally in the absence of minute amounts of copper. Enzymes containing copper are probably involved in the synthesis of the hemoglobin units.

The list could be extended. Bone formation depends on an appropriate calcium/phosphorus ratio; diets high in potassium create extra demands for sodium in mammals (sodium chloride hunger); excess magnesium may produce narcosis, and an appropriate balance of anions and cations (ion antagonism) is essential for the integrity and normal responsiveness of all tissues (Heilbrunn, 1952; Blair-West *et al.,* 1968). These examples will suffice to illustrate the intricacies of the interactions and to emphasize that physiological investigation of one element may require the simultaneous study of several others. A study of copper requirements, for example, is of little value without a knowledge of molybdenum metabolism in the ruminant or without an understanding of the zinc metabolism of the rat (Underwood, 1971). Since the biochemistry is most likely to be explained in terms of interacting systems of enzymes, similar interlocking relationships may be expected in all forms of life.

COLLECTION OF FOOD

Only plants and saprophytic animals such as intestinal and blood parasites can soak up or directly absorb the materials necessary for their life processes. The active procurement of food is basic to animal life, and much of animal evolution in terms of specialized anatomy and physiology is an adaptation for the capture of different kinds of food and its preparation for the cells of the body.

Yonge (1928), in a classical paper published many years ago, grouped feeding mechanisms into three major categories according to the type of food utilized: (1) MECHANISMS FOR DEALING WITH SMALL PARTICLES—MICROPHAGY (pseudopodia, cilia, tentacles, mucus, setae, muscles); (2) MECHANISMS FOR DEALING WITH LARGE PARTICLES OR MASSES—MACROPHAGY (swallowing inactive food, scraping, boring, seizing the prey and then swallowing or chewing it or digesting it externally); and (3) MECHANISMS FOR TAKING IN FLUIDS OR SOFT TISSUES (piercing and sucking, sucking only, absorption through the body surface). His article provides a detailed and well-illustrated description of the feeding mechanisms of the invertebrates. Jennings (1965) and Nicol (1967) have dealt in a similar way with both the vertebrates and the invertebrates.

Morton (1967) adopts a functional classification in accordance with the nature of the food rather than its size and considers five categories of animals: (1) HERBIVORES and OMNIVORES with digestive systems generalized for the treatment of bulky foods of uncertain nutritive value often containing a large non-digestible component, resulting in considerable wastage; (2) DEPOSIT FEEDERS that pass large amounts of the substratum through their guts; this source of food is often of low and variable nutritive content; (3) CARNIVORES which rely on concentrated sources of animal protein; the mobility and the uncertainty of the food has resulted in highly specialized organs for capturing, dismembering, and swallowing it; (4) FILTER FEEDERS which continuously strain small particles from large volumes of water, trapping the particles in mucus films or screens of setae and utilizing elaborate devices for sorting and transporting them; (5) FLUID FEEDERS with special organs for piercing and sucking juices from animals or plants and often possessing large fluid storage spaces in otherwise simple digestive tracts. Some animals do not fit neatly into any of these classifications. They exploit whatever foods are available in their environment or depend on several different habitats. The brittle star *Ophiocomina nigra,* for example, is at times a microphagous feeder utilizing a mucus net spread between the spines to capture food particles which are then formed into a bolus by the action of ciliary currents and tube feet. This animal is also a macrophagous feeder capturing food with its arms and tube feet or browsing on the bottom (Fontaine, 1965). In the brief comments that follow, the collection of food is considered under only two headings: NONSELECTIVE FEEDING and SELECTIVE FEEDING.

Nonselective Feeding

In general, the sessile animals and sedentary feeders, such as a bivalve mollusc, an *Amphioxus,* or ammocoete larva, depend on a varied group of filtering and trapping devices. Cilia and setae create complex water currents which bring particles of food onto the feeding surfaces where the particles are trapped in mucus. A moving belt or cord of mucus, propelled by cilia, carries the food particles into the digestive tract. These feeding currents may be elaborate and the structures may be highly specialized. Tentacles often form trapping organs. In some of the sea cucumbers, the sticky tentacles which trap small organisms are one by one thrust into the pharynx to wipe off the food (Hyman, 1955). Hydra, and some other coelenterates, also make use of a tentacle trap; in this case the trailing appendages are armed with nematocysts and these are discharged by tactile and chemical stimuli to pierce, poison, and hold the prey (Lenhoff, 1968, 1969; Lindstedt *et al.,* 1968).

Among the lower groups of animals, where filter feeding is widespread, the most prevalent technique is the ciliated mucus field (Morton, 1960; Jørgensen, 1966). The secretion of mucus, its circulation onto an external surface, its movement back into the body and finally its partial disintegration or digestion for reutilization require work. This is a price of living just as certainly as are the muscular efforts of those animals which actively search for their daily requirements. Usually the mucus in the esophagus forms a cord which is gradually rotated through the gut. While it moves along as a firm rotating rod or food string, its viscosity is altered by the changing pH in different areas of the gut. At the lower pH which occurs in the stomach, the mucus becomes less viscous and the food particles are released to be taken up by phagocytosis or to be digested extracellularly in the intestine. In the posterior part of the gut, where the pH is again higher, the mucus forms a firm fecal rope for discharge. These mechanisms reach their greatest complexity in the molluscs, even though the most primitive members of this phylum browse on algae by means of a radula.

Filter feeding, although characteristic of sessile and sedentary animals, is by no means confined to this way of life. Small active copepods are filter feeders as well as the sockeye salmon (*Oncorhynchus nerka*), the huge basking shark (*Ceteorhinus*), the whalebone whale (Mysticeti), and some marine birds like the broad-billed prions or whalebirds *Pachyptila* of the antarctic (Nicol, 1967). Feeding and digestive processes show many convergent adaptations (Morton, 1960); there is no particular phylogenetic trend in filter feeding mechanisms among the more specialized animal groups where it is exceptional and always a secondary modification.

Filter feeders are nonselective and must take what comes to them. They can control the situation only by operating or ceasing to operate the filter, and many of them are receptive to chemicals or other stimuli which warn them when filtering conditions are hazardous. Another type of nonselective feeder is the deposit feeder which passes the external medium through its body and takes from it whatever can be digested and absorbed. Many annelids, some echinoderms and the hemichordates feed in this way. The environment must be rich in nutritive material, and it seems likely that these forms absorb some part of the food directly in a saprophytic manner. These nonselective feeders are omnivorous and take whatever comes their way, provided it is sufficiently small to be caught in their filters or traps. Grinding organs may be present internally, but the piercing and cutting mouth parts characteristic of the selective feeders are absent.

Selective Feeding

Selective feeders have varied techniques for the capture and utilization of bulky foods or for the removal of juices from the bodies of animals and plants. The phylogeny which permitted the exploitation of varied sources of food has shaped the anatomy of many structures and altered the physiology as well. Mammalian feeding mechanisms provide an interesting example of the former (Davis, 1961b). Mammals have the unique ability to masticate food and reduce large objects to sizes readily managed by the digestive system. The jaws of the reptilian ancestors were primarily organs of prehension, and the evolution of efficient mastication involved not only changes in bones and muscles but also in the air passages to the lungs. Breathing must continue during mastication, and the air passages are separated and guarded by the soft palate, epiglottis and palatopharyngeal folds—characteristic mammalian structures. Such radical differences in cranial and buccal architecture are apparently directly related to these changes in feeding habits.

In all major groups of animals there are structures of highly specialized anatomy and physiology associated with the basic demand for food. In the Mollusca, one of the most successful and diversified animal phyla, the primitive feeding device is the RADULA. This unique organ is a straplike structure armed with sharp teeth and mounted like a rasp or file on a tongue of tissue; its to-and-fro movement abrades food particles from plants or rocks and rakes them into the mouth (Fig. 3.1a). In the large carnivorous gastropod *Conus geographus* and several related species, the radular teeth have been modified into numerous poison darts; these are housed in reserve as a sheaf of arrows within the radular caecum (Fig. 3.1b). *Conus* preys on fish that are located chemotactically and captured by the discharge of several of these harpoons which convey a neurotoxin along their partly closed shaft; the neurotoxin is secreted by a modified salivary gland. The paralyzed prey is engulfed by a highly distensible proboscis where partial digestion rapidly softens the fish for passage into the foregut (Hyman, 1967; Morton, 1967).

The emphasis on special types of foods may produce changes not only in the physiology of digestion and nutrition but also in such different functions as sexual maturation and social behavior. Many of the insects, for example, depend on only one specific type of food. The female mosquito must have a meal of blood to produce eggs and the number produced sometimes depends both on the amount and kind of blood (Roeder, 1953). *Culex pipiens* is reported to lay twice as many eggs per mg of canary blood as per mg human blood. The reproductive cycle of the rabbit flea *Spilopsyllus cuniculi* is directly controlled by its host's reproductive hormones which it obtains through feeding on pregnant rabbits (Rothschild, 1965; Rothschild *et al.,* 1970). Nutrition in the honey bee determines whether the female larvae will develop into sterile workers or sexually perfect queens. There are many such curious adaptations.

Neurosensory and neuromuscular specializations are also associated with selective feeding. The evolution of an active animal, with the phylogeny of a complex nervous

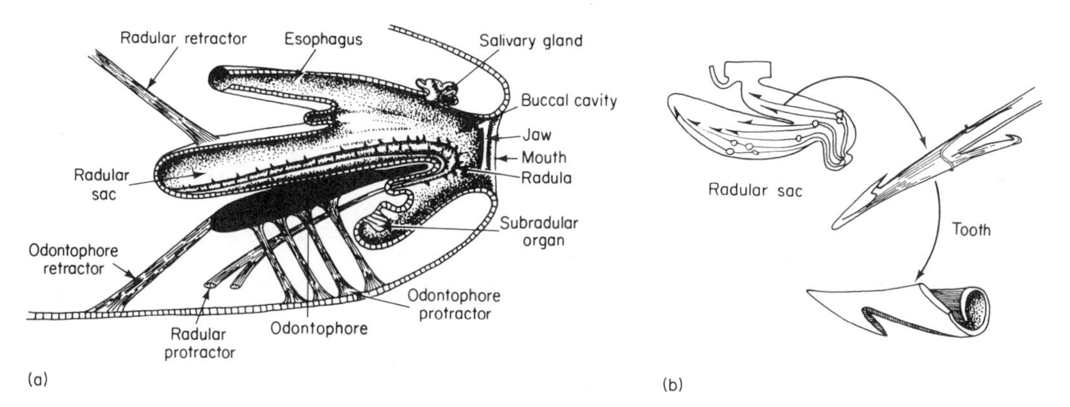

Fig. 3.1 Molluscan feeding apparatus. (a) Typical buccal cavity with radular apparatus; the radula is pulled to and fro like a toothed belt over the odontophore, a protrusible tongue of firm tissue. [From Meglitsch, 1972: Invertebrate Zoology, 2nd ed. Oxford U. P., New York.] (b) Radular apparatus of *Conus* specialized to form a sheaf of poison arrows. [From Morton, 1967: Guts. Arnold, London.]

system and a battery of sense organs, is both directly and indirectly related to the struggle for food. The location and capture of particular foods depend on sensory organs and the animal's ability to become conditioned to visual, chemical and other stimuli. Simple learning phenomena such as conditioning and the evolution of many other capacities of the central nervous system are, in a very real way, responses to the pressures for food and the specialization of the selective feeder.

DIGESTION

The word DIGESTION comes from two Latin words meaning "to carry" (*gerere*) and "apart" or "asunder" (*dis*). In this process, complex foods are broken down into the monosaccharides, amino acids, fatty acids, glycerol and several other constituents. Only these much simpler units can be utilized by cells or incorporated into living protoplasm. Digestion is an essential physiological activity in all animals, whether they feed on minute food particles (microphagous) or on large plants and animals (macrophagous). Some of the internal body parasites, such as the tapeworms, can dispense with a digestive system and absorb food predigested by their hosts. However, these parasitic conditions are secondary and not particularly relevant in the evolution of physiological processes.

In the macrophagous animal, digestive processes are both mechanical and chemical. Mechanical difficulties are substantial when animals utilize the higher plants with heavy cellulose walls or prey on animals with hard exoskeletons. The final processes of digestion are always chemical, but there is frequently an essential pretreatment of the food, either before or after it is taken into the body.

Mechanical Treatment

The biochemical disintegration of the cellulose molecule evidently presents difficulties. In most animals, cellulose passes directly through the digestive canal, and only the noncellulose portions of the plant tissue are digested and utilized. This means that cellulose walls must be crushed or otherwise broken in preparation for the enzymatic treatment of the cell contents.

Three groups of animals, with three different sorts of machines, are responsible for the primary utilization of most of the bulky plant material (Ramsay, 1968). The amphineuran and gastropod molluscs form one of the three groups. With their ribbon or strap-like radulae, set with numerous chitinized and readily replaceable teeth, they scrape away small particles of food from larger plants. Insects such as the locusts with their grinding and cutting mandibles form a second great group of plant feeders; the herbivorous mammals with their corrugated and grinding molar teeth form the third. Radulae, insect mandibles and molar teeth are all highly specialized for the mechanical destruction of cellulose plant walls.

There are several other structures concerned with cellulose disruption, but they turn over a relatively small amount of plant material. The valves of the mollusc have been modified to pulverize cellulose in the wood borers (*Teredo*). The insect mouth parts form stylets in plant bugs which puncture cell walls and suck out the contents. The Lantern of Aristotle in the Echinoidea is an effective crushing organ, and some sea urchins are mainly herbivorous, chewing algae and scraping seaweeds from the rocks with Aristotle's Lantern.

The mechanical treatment of the food continues within the bodies of many animals. Practically all digestive systems have gut musculature and often it is highly specialized. Heavily developed areas of muscle are sometimes assisted in their action by hardened surfaces. In certain forms, these hardened surfaces are the stones or grit taken in with the food (gizzards of earthworms, some birds and reptiles); in other cases, concretions of calcium salts develop in the stomach to serve a similar purpose (gastroliths of crayfish); or chitinized lining areas may develop in the anterior part of the gut (gastric mill of crayfish and comparable structures in some molluscs).

Movements of the Gut Contents

Digestion and absorption take place while the food is slowly moved through the gut. The propulsive force is almost always provided by cilia or a special gut musculature; sometimes both cilia and gut musculature cooperate in this activity (Campbell and Burnstock, 1968). The Nemathelminthes are exceptional and depend on the somatic musculature, for both locomotion and the maintenance of activities of the digestive tract; cilia are absent in this group and visceral musculature either completely or almost completely lacking (Hyman, 1951).

Muscular action usually plays a subsidiary part in the ciliary feeders (Morton, 1960). Transport of food particles takes place in a mucus rod or cord (ERGATULA). The cilia which rotate this food string are localized in one area of the gut, and in this way other areas are free to perform different functions such as the sorting of different sized food particles (molluscan stomach) or digesting them (molluscan intestine). The rotating movement itself provides an important stirring and circulating action in the gut. The rotating style as a propulsive force in the movement of the gut contents is most highly developed in the Mollusca but is also a characteristic feature of most other ciliary feeders, both invertebrate and vertebrate. Some of the sedentary polychaete annelids, however, provide exceptions. Within the gut, cilia are short or absent, and the propulsive force is provided by the contractions of both visceral and somatic musculature. Ciliary feeding in these animals is regarded as secondary in their phylogeny (Morton, 1960).

Cilia would be of little use when the food is bulky or hard; animals which live on such diets depend on well-developed layers of visceral muscle. This includes many annelids, molluscs and echinoderms, and all the arthropods and vertebrates except the larval lampreys and anuran tadpoles (Campbell and Burnstock, 1968). The trend from ciliary to muscular transport is well illustrated in the Mollusca. The microphagous species (amphineurans, most lamellibranchs, some gastropods) rely heavily on ciliary movement with muscular activities confined to the stomach where they play a part in the filling and emptying of the digestive diverticula (Fig. 3.2). The macrophagous feeders have a well-developed musculature throughout the gut with several specialized sphincters along the course; transport is wholly muscular among the cephalopods.

The arrangements of gut musculature are extremely variable. Throughout the vertebrates there are thick inner circular and outer longitudinal layers in the main wall of the tube with thinner layers, similarly arranged, just beneath the mucosa. The situation is less uniform among the invertebrates. In many of them the main inner layer is longitudinal and the outer layer circular (squid, sea cucumber, some insects), but in some (oligochaetes, certain insects) the disposition of the layers is like that of the vertebrate. Circular muscle may predominate with little or no longitudinal muscle (*Aplysia*); an oblique layer may also be present (squid stomach); or the muscle may be striated rather than smooth (some molluscs, arthropods, a few fishes such as the tench *Tinca*). These isolated examples testify to the variability in muscle arrangement but do not exhaust the patterns which the invertebrates have attained (Andrew, 1959).

Spontaneous activity is characteristic of the gut musculature of the annelids, molluscs, echinoderms, arthropods, and vertebrates. In other invertebrate Metazoa transport is dependent on cilia and the contractions of the body wall (Campbell and Burnstock, 1968). The type of muscular activity most characteristic of the hollow viscera is referred to as PERISTALSIS. Peristalsis has long been defined as a wave of contraction preceded by a wave of relaxation. In reality, the relaxation is often inconspicuous or absent, and peristalsis appears as a wave of contraction in the circular muscle which sweeps along the hollow viscus for a certain distance before dying out. Such a wave is easily initiated experimentally by mechanical stimulation. It may start at any point and proceeds for variable distances. Peristalsis is described in both invertebrates and vertebrates but there is no evidence of common controls at both levels of phylogeny. Antiperistalsis has also been described and appears to occur frequently in lower forms like the annelids, where it is thought to play a part in the general circulation; in these animals, vascular channels and digestive tube are closely associated anatomically.

In addition to peristalsis there are several other movements of the vertebrate visceral musculature. A TONUS RHYTHM in the various muscular layers produces slow alterations in the size of the gut. Superimposed on the tonus rhythm a strong RHYTHMIC SEGMENTATION is sometimes seen—most frequently in the duodenum and jejunum—and is essentially a mixing process. PENDULAR MOVEMENTS, involving the longitudinal as well as the circular musculature, produce rhythmic to-and-fro movements of a loop of digestive tube and tend to force the food from one end of the loop to the other, creating a rapid mixing. The activity of the vertebrate muscularis mucosa may also be important in mixing and transporting food, but in the small intestine it probably serves especially to move the villi about in the food.

Chemical Action

In the body of an animal, as in the laboratory, a successful chemical reaction depends both on the characteristics of the reaction vessel and the nature of the chemical reagents. The locus of enzyme activity will be considered prior to a discussion of the chemical reactions.

Intracellular digestion Digestion is intracellular in the protozoa, and this was presumably the situation in the most primitive animals. The protoplasm of the single-celled animal captures its food, digests it in a food vacuole, discharges wastes, and incorporates the simple sugars, amino acids and other molecules. The hydrolytic enzymes of a cell are thought to be contained in special packages called LYSOSOMES (de Duve and Wattiaux, 1966). These are surrounded by membranes which separate them from the cytoplasm in the living cell to prevent the autodigestion which occurs so quickly after death. According to current hypotheses, intracellular digestion is an aspect of lysosome function. It is still not clear just how the digestive enzymes enter the food vacuoles. Presumably, the lysosomes are in some manner discharged into them (Müller, 1967).

Digestion is wholly intracellular only in the Protozoa and the Porifera. In other phyla an extracellular digestion either supplements the intracellular mechanisms or completely replaces them. Many different variations in this combination of processes have been described. (Yonge, 1937; Barrington, 1962). Only a very few broad generalizations can be made, and the notes which follow are mostly from Barrington (1962).

The phylogenetic trend is toward the extracellular process; the most highly organized of the invertebrates (the cephalopods among the molluscs and the insects among the arthropods) rely entirely on this mode of digestion. Intracellular digestion with supplementary extracellular processes, is characteristic of the Coelenterata, Platyhelminthes, Nemertea, Annelida, Mollusca and some of the minor phyla. With few exceptions, such as protein digestion in the arachnids, digestion is entirely extracellular in the Arthropoda, the Nematoda and the Echinodermata; in the latter group it is probably strongly aided by phagocytic amoebocytes. Simple multicellular forms, like the acoelan worm *Convoluta,* depend entirely on intracellular digestion, clearly the most primitive process phylogenetically. It seems to be a short step from digestion within amoeboid cells to the release of enzymes from them into a space or onto a surface; such a simply organized animal as *Hydra* carries out a preliminary hydrolysis of its food before taking up the macerated remains in the phagocytic cells which line the gastrovascular cavity. Size of food may not always be an obstacle to intracellular digestion. The land planarian *Orthodemus terrestris* feeds on slugs by protruding its pharynx and disintegrating the body of its prey with a strong sucking action, aided by proteases secreted by the acidophilic gland cells of the pharynx. The tiny food particles eroded in this manner are sucked into the gut and taken up by phagocytosis

(Jennings, 1962, 1968). *Polycelsis cornuta,* a freshwater triclad, attacks large items of food in the same manner. Extraintestinal digestion is also found in some other invertebrate phyla and may be combined with intracellular or a mixture of intra- and extracellular digestive processes. Some arachnids, for example, exude proteases onto their prey and suck up the semidigested material for further digestion which is partially intracellular. Some echinoderms and insects like the water beetle *Dytiscus* also rely on extraintestinal digestion, but in these animals the later stages, which take place within the body, are also extracellular.

As a group, the molluscs display the most varied combinations of intracellular and extracellular digestion. Even the bivalves, which are purely microphagous feeders, exemplify an array of specializations with some species shifting over completely to the extracellular mode. The stomach of the bivalve receives the food particles, sorts them, and looks after their digestion. It is assisted in the latter process by two special structures which arise from it. One of these is a distal tube, the STYLE SAC, which in most species contains the CRYSTALLINE STYLE. This peculiar device is found only among the molluscs; it is most characteristic of the lamellibranchs but occurs also in some of the more advanced herbivorous gastropods. The evolution of the style sac in connection with ciliary feeding is discussed by Morton (1952, 1960). To the physiologist it is a thick gelatinous rod loaded with enzymes and rotated by strong cilia which force it gradually into the stomach where it rubs against the horny gastric shield to release its enzymes and to stir up the stomach contents. Amylases are the most abundant enzymes in the crystalline style, but lipases seem also to be present. When the style is poorly developed, the style sac probably produces enzymes which are involved in extracellular digestive processes.

The midgut glands or DIGESTIVE DIVERTICULA form the other special development from the stomach. These follicular glands communicate with the stomach by a system of ducts and are lined by epithelial cells which discharge several functions—elaboration of enzymes, phagocytosis, absorption, food storage and excretion. All of these functions are thought to have been discharged by the same type of cell in primitive forms but present-day species show at least a partial segregation of functions (Owen, 1966). Four patterns of diverticula are illustrated in Fig. 3.2. In the more primitive situations (Fig. 3.2a and b), food particles are pumped into flask-like diverticula where they are phagocytized by the lining cells, digested intracellularly and absorbed In the more advanced forms (Fig. 3.2c and d), numerous tubules and an extensive series of ducts greatly increase the epithelial surfaces; movement of particles and fluids is effected by cilia. There may be a two-way movement (Fig. 3.2c) with intracellular digestion in the tubules and extracellular digestion in the stomach. In the most specialized situation (Fig. 3.2d), the epithelia are completely glandular and digestive juices are wafted by cilia into the stomach where digestion is entirely extracellular. Thus, there seems to have been a transfer of both digestion and absorption from the diverticula to the stomach. This change is thought to be related to the nutritive habits of the Nuculidae (Fig. 3.2d) which feed on detritus; detritus contains

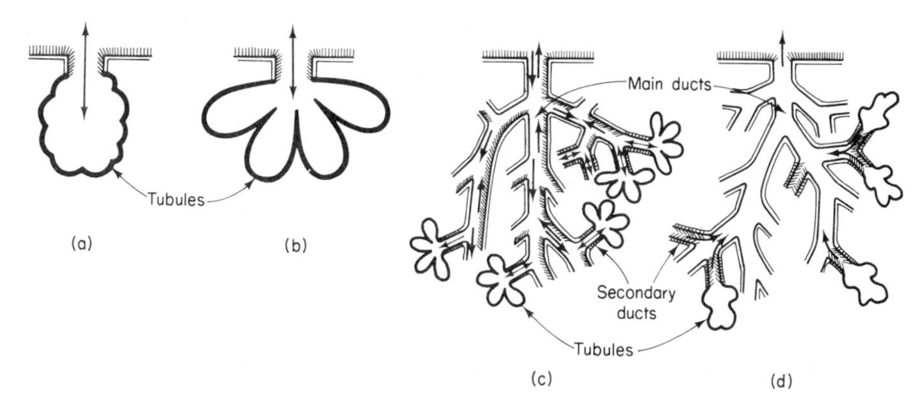

Fig. 3.2 Types of digestive diverticula present in the Lamellibranchia, shown diagrammatically: (a) hypothetical primitive condition found in many larval lamellibranchs; (b) in septibranchs, some eulamellibranchs and the wide diverticula of the Nuculanidae; (c) *Anisomyaria* and most Eulamellibranchia; (d) Nuculidae. Double-headed arrows represent movement resulting from muscular activity; single-headed arrows represent ciliary movements. [From Owen, 1956: Quart, J. Micr. Sci., *97*, 564.]

much indigestible material and phagocytosis of it would be uneconomical (Owen, 1956). This offers a clue to one of the factors probably responsible for the shift to extracellular digestion in macrophagous animals.

In a few animals intracellular digestion is partially dependent on an extensive phagocytosis. A host of wandering amoebocytes or phagocytic cells procure food particles, digest them and pass their products to the other cells of the body. The best examples are found among the lamellibranchs and the echinoderms where the three major classes of digesting enzymes (carbohydrate, fat and protein-splitting) have been identified within the wandering cells. Yonge's (1937) review is most frequently quoted in this connection. In the lamellibranchs, he described the migration of amoebocytes both into the digestive tract and into the mantle cavity (from palps and gills) where they capture food particles and transport them back into the body of the animals while digesting them. Although accessory digestive processes by amoebocytes within the gut are usually recognized, the contribution of phagocytosis from the mantle cavity has been questioned and the matter is still unresolved. Owen (1966) reviews the pertinent literature.

Pequignat (1966) has described a curious mode of "skin digestion" in several of the echinoderms. This depends on migration of spherule-coelomocytes onto the surface of the animal where they release hydrolytic enzymes which, in association with epidermal secretions, predigest a variety of materials on the surface of the animal. Digested products are absorbed directly through the epidermis or carried into the gut. Direct absorption of such dissolved organic matter as amino acids and sugars by the epidermal cells of echinoderms appears to be well-established and a certain amount of digestion of larger molecules by epidermal secretions seems likely (Ferguson, 1969). Similar claims for supplementary dietary contributions by direct absorption of amino acids and sugars have been made for several of the annelids (Stephens, 1963; Taylor, 1969) although the significance of the absorption to the nutritional requirements of the animals has been questioned (Johannes *et al*, 1969; Efford and Tsumura, 1973).

Extracellular digestion Many new avenues of evolution were open to the first organisms which secreted some of their digestive juices onto larger and more complex items of food. Such organisms were not only able to utilize bulky foods but could devote fewer cells to digestive processes and specialize in producing enzymes suitable for peculiar food substances. Digestion could be carried out more rapidly and less time had to be devoted to the actual business of feeding. It seems likely, however, that most of the advantages could not have been realized without the essential anatomical development of a tubular digestive tract, open at both ends. Animals like coelenterates and flatworms, which lack an anus, have not evolved separate areas for storage, digestion and absorption of food nor for the separation and discharge of those portions of the food which are not usable. Likewise, animals which depend almost entirely on intracellular processes have only achieved a modest specialization in cell types. Their digesting cells usually contain all the different types of enzymes and there are no particular masses of cells (glands) devoted specifically to the digestion of starch, protein or other types of food.

The digestive tract and its "enzyme chain" Yonge (1937) described five regions of the digestive system of the metazoan: (1) reception, (2) conduction and storage, (3) digestion and internal triturition, (4) absorption, (5) conduction and formation of feces.

This arrangement has permitted numerous anatomical specializations with concomitant physiological advantages. The region for reception is often associated with devices for food maceration (teeth), for paralyzing struggling prey (salivary toxic enzymes), for initiating digestion (salivary enzymes), or for lubricating the food (mucus-secreting buccal glands). Buccal glands may have more specialized functions in some groups. The bloodsucking insects and leeches produce an anticoagulant. Some carnivorous gastropods secrete a strong acid which dissolves calcareous shells. All those animals which carry on extraintestinal digestion produce proteolytic enzymes for the hydrolysis of connective tissue to produce the fluid material which they suck into their digestive tracts.

A separate region for storage has been even more important phylogenetically, since this permits the utilization of foods which are not continuously available. The leech may take several months to digest a single meal of blood. The herbivorous animal spends many hours masticating the food which it gathers hurriedly and stores temporarily in its stomach. The production of acid in this region of the vertebrate gut may have arisen as an adaptation for killing prey and for checking bacterial activity;

the production of an enzyme (pepsin) which is active in acid medium may have followed later. The special storage region also permits the gradual release of macerated and partially digested food into the main area of digestion.

In the third region, the enzymes rapidly reduce the food to an absorbable form. The terminal regions of the gut also present a number of specialized features; several of these which are related to absorption will be considered below. In such herbivorous animals as the rabbit, there are colonies of symbiotic bacteria which provide vitamins as byproducts of their metabolism and assist digestion through fermentation processes.

Vonk (1937) has discussed the "enzyme chain" in vertebrate digestion and stressed its physiological significance. In general, digestion occurs in only one area of the invertebrate gut, while the vertebrate gut—with the exception of *Myxine* (Adam, 1963)—has different enzymes localized in specific areas. In certain mammals (man, monkey, pig) carbohydrate digestion may start in the mouth by the salivary amlyases or ptyalin. In all vertebrates the pancreas splits starches into oligosaccharides in the anterior part of the duodenum. In subsequent steps these are hydrolyzed by oligosaccharases and maltases, mostly secreted by the gut lining. Here then is a chain of carbohydrases starting in the mouth and continuing into the intestine. Similarly, gastric or pancreatic enzymes initiate the protein hydrolysis, but the final separation of the dipeptide links occurs farther along the duodenum through the action of dipeptidases formed by the intestinal cells. Vonk suggests that this localization of enzymes in the vertebrates is associated with a more intense metabolism and a greater sensitivity to changes in blood sugar or amino acid levels. In this way food products enter the blood gradually and the animal is not inundated after a meal. This may or may not be a significant point. The greater efficiency provided by groups of cells devoted to the production of one or a few enzymes may be important, and this arrangement would be more effective if those promoting the hydrolysis of the major links were secreted before those concerned with the final cleavage.

In contrast to the vertebrates, the same glandular areas produce a mixture of the different kinds of enzymes in the invertebrates; moreover, the digestive epithelia are usually absorptive as well as secretory. The hepatopancreas (midgut gland) of the crustacean provides a good example. This diverticulum from the midgut secretes enzymes, absorbs digested food and stores fats, carbohydrates, probably proteins and minerals. The epithelia of the coelenterate enteron or the turbellarian gut, the pyloric caeca of the starfish, the digestive caecum of *Amphioxus* and some of the glands of the mollusc are equally versatile. In many cases the same cells seem to be capable of performing these multiple activities (Barrington, 1962).

Some of the more specialized invertebrates (insects and cephalopods) have achieved a partial separation of functional areas for the secretion of different enzymes and the absorption of digestive products. The octopus and the squid provide the best known examples (Barrington, 1962; Bidder, 1966). There are two morphologically distinct midgut glands, referred to as "liver" and "pancreas." These supply enzymes to the stomach and caecum where digestion occurs progressively as the food moves along through these areas into the intestine; digestion may also occur within the "liver" of the octopus (Bidder, 1966). In *Loligo,* but not in *Octopus,* absorption seems to have been completely separated from the digestive glands. In comparison with the other invertebrates, digestive processes are extremely rapid in the predacious cephalopods and approach those of the vertebrates in their efficiency.

In the higher vertebrates, each area of the gut is concerned with a special activity. Digestive enzymes are produced in discrete glands as well as in the wall of the gut; absorption occurs predominantly in certain areas of the intestine, and a special organ (the liver) takes over the job of storing materials. This is a decidedly more efficient arrangement, since larger animals require more food and a more precise homeostatic control over metabolic processes.

Digestive Enzymes

The actual enzymes concerned with digestion are conveniently considered according to the three major food items: carbohydrases or glycosidases acting on carbohydrates, lipases and esterases acting on fatty materials, and proteinases concerned with proteins.

Carbohydrases Dietary carbohydrates are potential sources of the simple hexose sugar glucose, which plays the key role in metabolism and is the direct source of much of the energy in all animals. Both plant and animal foods may contain free glucose, but most of the dietary carbohydrate is in a complex form and consists of numerous glucose or similar units joined by a condensation reaction, with the elimination of one water molecule for each linkage formed. Before these hexose units can be incorporated into the body the glycosidic bonds (Fig. 3.3) must be hydrolyzed, and this reaction is catalyzed by a family of highly specific and varied enzymes, the carbohydrases or glycosidases.

Many different carbohydrases have been reported from animal tissues (Barrington, 1962; Bernfeld, 1962). The key to this complex situation is the structure of the carbohydrate molecule itself, as is readily apparent from the architecture of glucose (Fig. 3.3). This molecule has a skeleton of 6 carbon atoms and, because 4 of these are asymmetrical, there is potentially a family of 16 isomers (8 belonging to the D-series and 8 to the L-series). Actually, the open-chain form occurs only in traces; the ring forms are the active units, and these may be linked into molecules ranging from disaccharide size (sucrose with a molecular weight of 342) to the starches with molecular weights up to 500,000 (rice starch) and cellulose with even larger particles. Thus, a potentially great variety of slightly different molecules is possible, and many different ones do occur in biological systems. The carbohydrases, as a group, are all concerned with the hydrolysis of the glycosidic bond between the different monosaccharide units, but the isomeric arrangements are often sufficiently different to require specific enzymes; this accounts for the variety of these catalysts found in nature. Bernfeld (1962) reviews their distribution in both plants and animals.

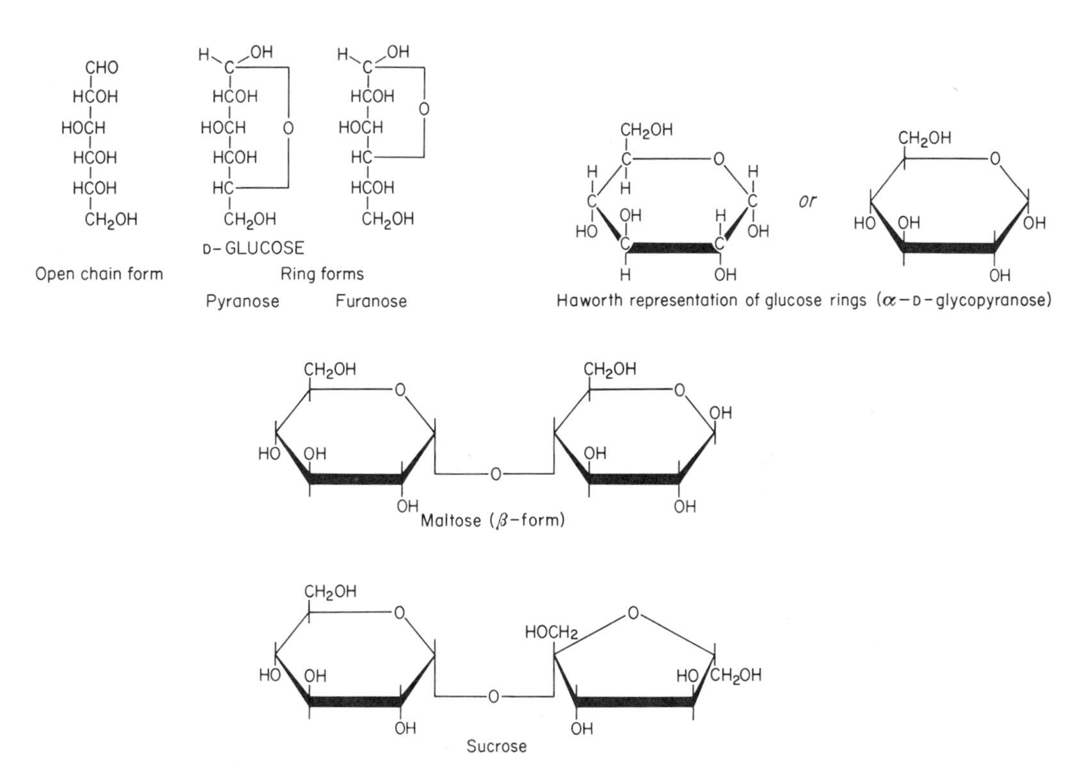

Fig. 3.3 The structure of carbohydrates. Above, several methods of representing the glucose molecule; below, the glycosidic linkage in a disaccharide, represented by the general formula R-CH-OR. [For further details, see Guthrie and Honeyman, 1968: An Introduction to the Chemistry of Carbohydrate, 3rd ed. Clarendon Press, Oxford.]

The carbohydrates of particular interest in animal nutrition, together with their associated enzymes, are listed in Table 3.2. More precise classifications of the enzymes, based on the structural arrangements of the hydrolyzed linkages, are found in texts of biochemistry and enzymology. Many of the enzymes listed in Table 3.2 are not, in reality, single protein molecules but classes or families of enzymes. This is particularly true of the amylases and polysaccharases.

Probably no animal requires all the carbohydrases shown in Table 3.2. The occurrence of digestive enzymes seems to be correlated with feeding habits rather than phyletic position. It follows that those carbohydrases which most frequently occur will be concerned with the digestion of the most abundant dietary carbohydrates. Thus, amylases and maltases are of almost universal occurrence, while inulases are probably absent in animals (Bernfeld, 1962) and cellulases occur rather infrequently. Likewise, among the disaccharases, sucrases are more frequently found than lactases and trehalases. It seems likely that the variety of naturally occurring carbohydrates has increased immensely since the dawn of animal life and that the animals, which

Table 3.2 Some Carbohydrates of Interest in Animal Physiology

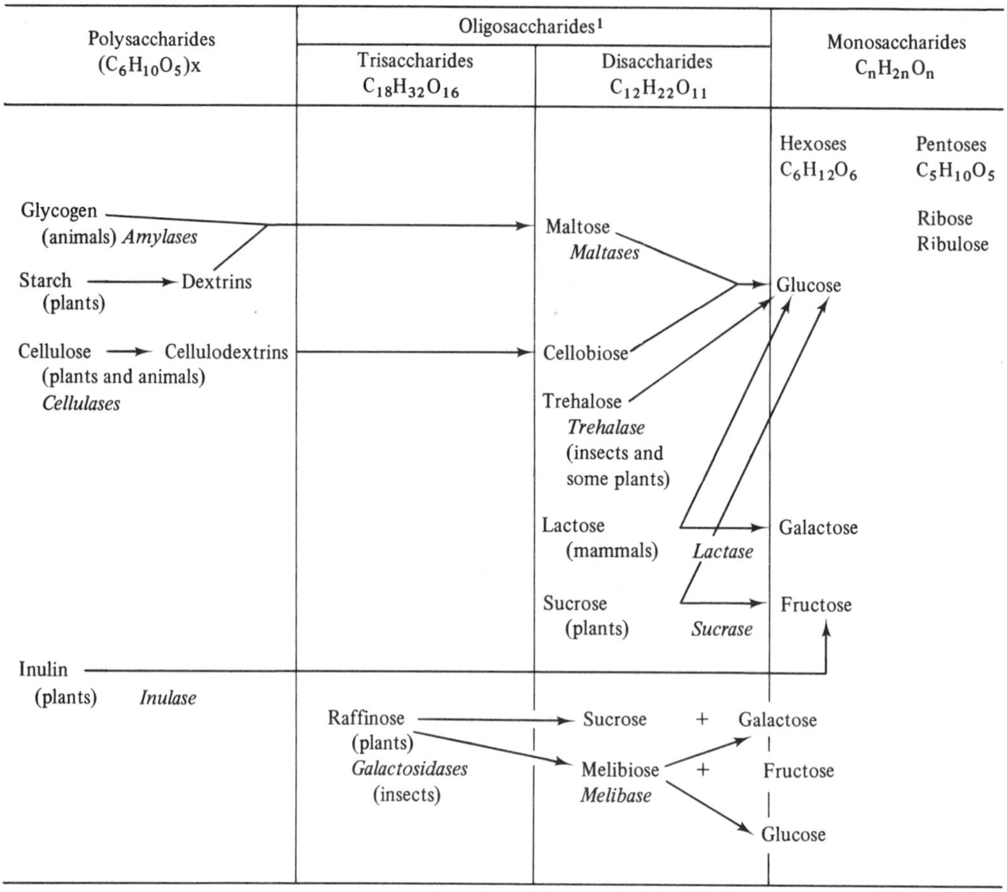

Polysaccharides $(C_6H_{10}O_5)x$	Oligosaccharides[1]: Trisaccharides $C_{18}H_{32}O_{16}$	Oligosaccharides[1]: Disaccharides $C_{12}H_{22}O_{11}$	Monosaccharides $C_nH_{2n}O_n$: Hexoses $C_6H_{12}O_6$	Monosaccharides $C_nH_{2n}O_n$: Pentoses $C_5H_{10}O_5$
Glycogen (animals) *Amylases*		Maltose *Maltases*		Ribose Ribulose
Starch → Dextrins (plants)			Glucose	
Cellulose → Cellulodextrins (plants and animals) *Cellulases*		Cellobiose		
		Trehalose *Trehalase* (insects and some plants)		
		Lactose (mammals) *Lactase*	Galactose	
		Sucrose (plants) *Sucrase*	Fructose	
Inulin (plants) *Inulase*				
	Raffinose (plants) *Galactosidases* (insects)	Sucrose +	Galactose	
		Melibiose *Melibase* +	Fructose	
			Glucose	

[1] The root *oligo-* means "few" and there is no sharp point of division between the oligosaccharides and polysaccharides

have evolved with the plants, have acquired enzymes for utilizing the more varied plant foods.

The vertebrates possess a more limited group of carbohydrases than the invertebrates. However, it should be emphasized that relatively few animals have been examined carefully for all types of digestive enzymes, and there may be many exceptions to these broad generalizations. For example, it is claimed that certain species of teleost fish can digest hemicelluloses as well as xylan and algin (Barrington, 1957). These substances are probably broken down enzymatically by several invertebrates but by very few vertebrates (Horiuchi and Lane, 1966; Harnden, 1968). However, some of the earlier reports of polysaccharases are now known to be based on the activities of associated cellulolytic bacteria. The occurrence of cellulases has, however, been established in some representatives of all phyla except the Chordata (Elyakova, 1972). The molluscs and arthropods in particular, have many cellulose-digesting species. In contrast, cellulases are extremely rare among echinoderms and seem to be entirely absent in the protochordates and the vertebrates.

The utilization of cellulose by animals merits an additional comment. It does form an important item in many diets, but its digestion is usually by symbiotic organisms rather than by a specialized system of digestive enzymes. Many insects feed on woody plants, and this group of animals has found a variety of ways to deal with the complex carbohydrate molecules. In some cases the plant cells are pulverized and the starches and sugars that they contain are digested, but the cellulose is of no nutritive value (powder post beetles, Lyctidae); in other groups of insects, hemicellulases (bark beetles, Scolytidae) and cellulases (wood-boring beetles, Cerambycidae and Anobiidae) are actually present and the woody plant material is enzymatically hydrolyzed; in still others (termites and roaches) symbiotic bacteria and flagellate protozoans find shelter and an abundance of macerated cellulose in the digestive tract of the insect and in return release products of their anaerobic metabolism or fermentation (mainly lower fatty acids) which serve the insect as an energy source.

The stomach of the ruminant is the most specialized organ for the digestion of plant material (Fig. 3.4 and Comline *et al.,* 1968). The anterior division forms a huge fermentation chamber within which bacteria, yeasts and protozoa reduce cellulose and related carbohydrate to a usable form (Annison and Lewis, 1959; Barnett and Reid, 1961; Hungate, 1966). The cellulolytic bacteria are the most important, and the role of the protozoa is probably insignificant or secondary. The bacteria are anaerobes or facultative anaerobes and release a variety of fatty acids (formic, acetic, propionic, butyric, succinic, lactic), depending on the cellulose and associated starches and sugars in the diet. These acids are absorbed directly from the rumen and

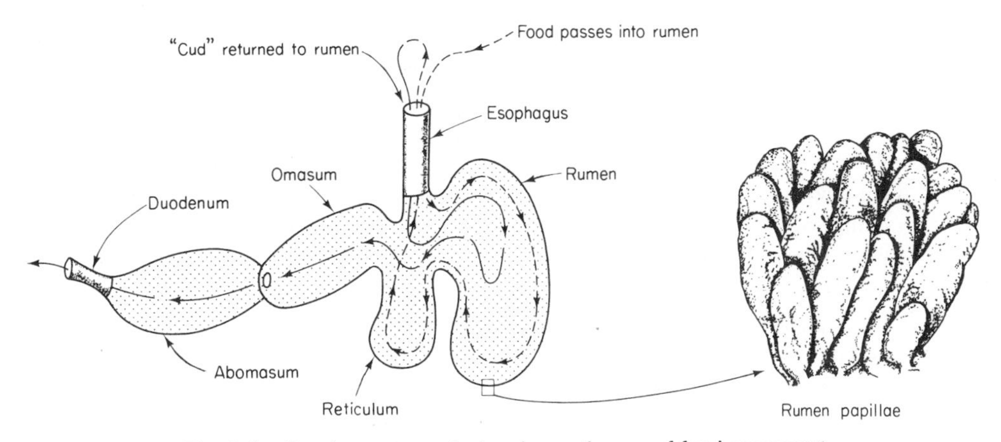

Fig. 3.4 Ruminant stomach showing pathways of food movement.

enter immediately into various metabolic pathways. Considerable amounts of carbon dioxide and methane are produced by the fermentation. Most of this gas is eructed but some is passed via the lungs. The operation of this complicated fermentation chamber involves several specializations of the gut and is dependent on the maintenance of a constant and suitable environment for the bacteria. A very copious supply of saliva, rich in bicarbonate, maintains the fluid content and pH. The rapid absorption of the fatty acids is also important and prevents their accumulation which would inhibit the activities of the bacteria. The volume is regulated by the passage of the semifluid material into the omasum at intervals. Gastric juice is secreted in the abomasum, and here the animal's own digestive enzymes come into play. The actual process of RUMINATION, or the regurgitation and remastication of the food, is important in slowing down the passage of food through the fermentation chambers as well as insuring rupture of cellulose walls and through mixing with saliva.

The ruminants do not hold a monopoly on fermentation as an adjunct to digestion. Ruminant-like fermentation is present in several groups of animals at different levels in phylogeny (Moir, 1968). Among the non-ruminant mammals, STOMACH FERMENTATION occurs in an expanded cardiac stomach of the hippopotamus and in a greatly elongated stomach of the kangaroo; CECAL FERMENTATION occurs in the extended intestinal cecum of the horse, pig or rabbit. COPROPHAGY, a fecal reingestation, is a physiologically related phenomenon common among rodents and rabbits where slowly digested foods are given additional treatment by repassage through the gut. The importance of these processes has been demonstrated experimentally. For example, rabbits prevented from eating "soft feces" which have a much higher protein content than hard feces (which are not eaten) show a 50 per cent reduction in nitrogen balance (Moir, 1968). The importance of these microbial populations in the utilization and metabolism of protein (especially in the recycling of ammonia and urea) will be discussed in a later section.

Lipases and esterases The lipids or fatty substances are a varied group of organic compounds, utilizable by living organisms and characterized by their insolubility in water, their solubility in certain organic solvents, and the ester linkage which is actual or potential in all of them (Fig. 3.5). The enzymes which hydrolyze lipids are actually esterases since it is the ester linkage which is broken to produce acids and alcohols. It is customary, however, to refer to the enzymes concerned with the triglycerides (esters of fatty acids and glycerol) as LIPASES while the term ESTERASES is reserved for enzymes which act on compounds like ethyl butyrate (simple esterases) and more complex lipids (phospholipids, cholesterol and waxes) where the cleavages require more specific enzymes (cholesterol esterase, for example).

Two essential processes are involved in the digestion of fat. Enzymatic hydrolysis by lipase is secondary to an emulsification by agents which reduce the surface tension at fat/water interfaces. In the vertebrates, bile salts from the liver serve as initial emulsifying agents and lipase can only act on the finely divided and emulsified droplets. Once fat hydrolysis is initiated, the digestion products—particularly monoglycerides and fatty acids (Fig. 3.5.)—also act as emulsifiers, resulting in extremely fine

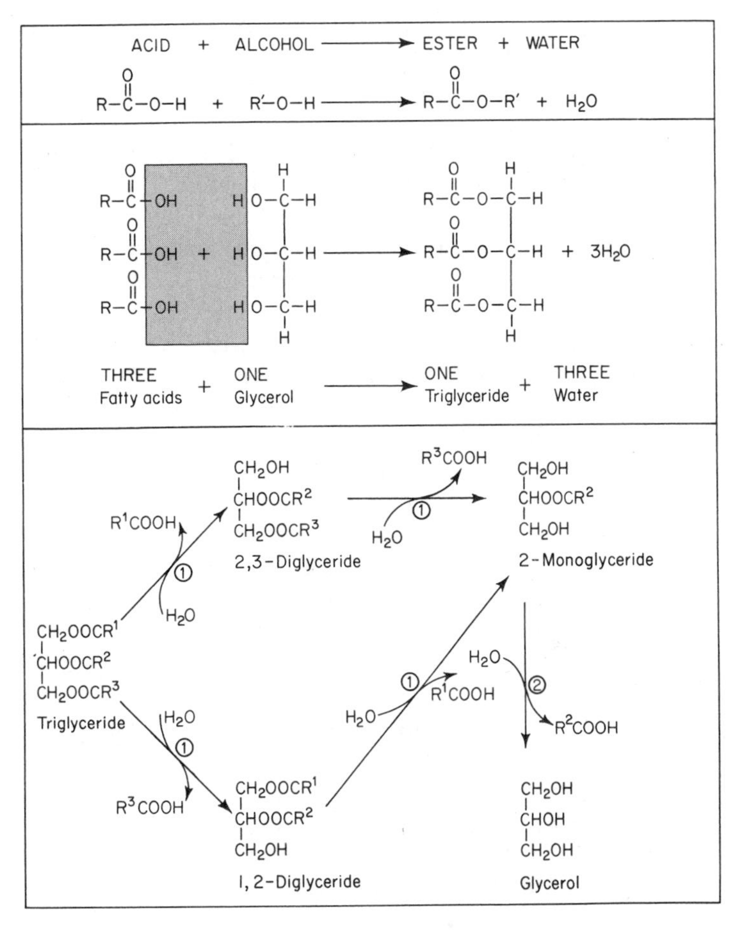

Fig. 3.5 Structure of a triglyceride (above) and the probable reaction sequence in its hydrolysis by pancreatic enzymes (below). Enzymes operating in positions 1 and 2 are different.

droplets that present much larger surfaces for enzymatic activity. Vonk (1962, 1969) has investigated the emulsifiers in several groups of invertebrates (sea cucumbers, crustaceans, molluscs, annelids) where a completely different family of biochemical substances is utilized. Vonk (1969) notes that the bile salts of the vertebrates and the crustacean emulsifiers both contain taurine but in the vertebrates this is conjugated with steroids (cholic and other bile acids) while in the crustaceans the conjugation is with fatty acids. The vertebrate type of bile acid (Haslewood, 1968) is not found among the invertebrates and Vonk speculates on its probable superiority in effecting very small particle size and increased solubility of the emulsified fats. It may be significant that many of the invertebrates are now known to be incapable of synthesizing cholesterol from which the bile acids are derived; the evolution of cholesterol biosynthesis among the vertebrates may be related to their demands for bile acids and the utilization of fat.

Some physiologists believe in the direct absorption of very finely emulsified fats (Baldwin, 1967; Glass, 1968). Careful studies in mammals indicate that if the droplets are fine enough (0.5 μ or less and negatively charged), they may pass directly into the intestinal cells and into the circulation. Bile salts alone do not produce this degree of emulsification, but in association with fatty acid and monoglyceride (from the enzymatic hydrolysis of the neutral fat) such a degree of dispersion is attained. Direct absorption of tiny fat droplets probably occurs also among lower vertebrates and invertebrates although there is very little information, even for such a carefully investigated group as the insects (Gilbert, 1967).

Steps in the hydrolysis of a triglyceride are shown in the lower panel of Fig. 3.5. In mammals, pancreatic lipase effects the hydrolysis of the triglycerides and diglycerides but the final step of monoglyceride hydrolysis depends on a different pancreatic enzyme (Harper, 1973). This last step occurs very slowly and, as a consequence, much of the monoglyceride is not hydrolyzed but absorbed directly. In short, fats are taken up from the digestive tract as triglycerides, diglycerides, monoglycerides, or fatty acids and glycerol.

Biologists have usually considered the lipases to be less varied and less widely distributed than the carbohydrases and the proteinases (Vonk, 1937). If this is true, it may be related both to capacities for direct absorption of emulsified fat (or partial products of its digestion) and to the comparatively few types of linkage requiring hydrolysis in fat digestion. However, few animals have as yet been carefully studied; esterases occur in multiple forms in both the digestive juices and the tissues and are highly species-specific (Holmes *et al.*, 1968; Reid, 1968); our concepts may be substantially modified by further investigation.

The digestion of the more complex esters, which occur in the compound lipids (phospholipids and cerebrosides), the derived lipids (sterols, etc.) and waxes (esters of fatty acids with alcohols other than glycerol) has been demonstrated in many groups of animals. The wax moth, *Galleria*, is able to utilize beeswax, probably through a combination of bacterial and enzymatic action in the digestive tract (Gilbert, 1967). This same animal has a cholesterase which converts cholesterol into its fatty acid esters. Lecithinases are common in many kinds of tissues, including the digestive mucosa of some animals. The presence of curious enzymes, such as cerase, in association with the digestion of unusual foods has often been considered evidence of the ready production of adaptive hydrolases by animals. However, careful investigation has frequently shown symbiotic microorganisms rather than digestive hydrolases, and physiologists are now more cautious in proclaiming the presence of cellulases, cerases and chitinases in multicellular animals (Barrington, 1962). An example of this is the honey guide (genus *Indicator*), a small African bird which has the curious behavior of leading men and other animals to the nests of bees. This they do by very noisy activities which may attract animals from considerable distances. When the bees' nest has been destroyed by the cooperative mammals, the honey guides then gorge themselves on the wax, but it is an intestinal microflora which hydrolyzes the wax, and not the enzymes produced by the bird itself (Friedmann and Kern, 1956).

Esterases from the digestive epithelium have sometimes been adapted to non-digestive functions. The salivary or buccal glands provide an interesting example. These organs, derived from gut epithelium, may produce amylases, proteases, or lipases. They play a minor part in digestion. Even among mammals, where amylases occur regularly, the digestion that takes place in the mouth is preliminary and often non-essential; the lubricating mucus of the saliva is more important than its enzymes. Lipases and proteinases, when found in buccal gland secretions, often serve specialized non-digestive functions. In certain reptiles they are potent constituents of venom. Some snake venoms contain a phospholipase capable of hydrolyzing lecithin to produce a powerful hemolytic agent, lysolecithin; others produce proteolytic enzymes which act as thromboplastic materials capable of inducing extensive intravascular blood clotting. Venoms often contain a mixture of toxic materials, including many enzymes which are not associated with digestive processes (cholinesterase, hyaluronidase, phosphatases, etc.) and other compounds (proteins, amines, etc.) which are not enzymatic in nature (Russell and Saunders, 1967; Bücherl *et al.,* 1968). The buccal glands of some amphibians and reptiles produce several of these toxic materials. A few of the same substances are found in spider venoms and insect poisons (McCrone, 1969). In these animals, buccal gland secretions are important in paralyzing the prey and sometimes also in initiating an extraintestinal digestion.

Proteinases Proteins are organic substances of high molecular weight, composed of numerous amino acids united by the peptide linkage. On hydrolysis the SIMPLE PROTEINS yield only amino acids while the conjugated proteins yield, in addition, non-amino groups. The non-protein portion of the conjugated protein is called the prosthetic group. A peptide linkage between two simple amino acids is shown in Fig. 3.6. During digestion the peptide links are hydrolyzed one by one to split off amino acids or groups of amino acids.

The most familiar proteolytic enzymes are shown in Table 3.3. They are highly specific and should be thought of as groups or families of enzymes rather than as specific chemical entities. There is no single pepsin molecule in the same sense as a

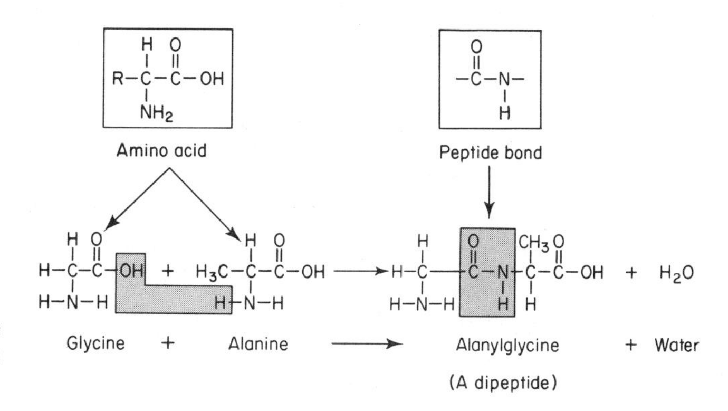

Fig. 3.6 Two amino acids unite through a peptide bond to form a dipeptide.

NaCl or HCl molecule. The specificity is of three sorts. Like all proteins, these enzymes show species specificity, and it has been demonstrated that pepsins from various species of vertebrates, for example, act differently on certain substrates and have different reaction optima. In addition, the proteases show a marked preference for particular peptide links. The presence of specific groups in the vicinity of the peptide bond makes it sensitive to the catalytic activity of only one of the proteases (Fig. 3.7). Finally, there is a stereospecificity, and the enzymes listed in Table 3.3. are specific for only the L-form of the amino acids.

Table 3.3 Protein Digesting Enzymes of Animals. Further Details of Specificity Are Given in Fig. 3.7. (I–IV, cathepsins of some texts; *A–C*, cathepsins now characterized)

		"Zymogen" $\xrightarrow[\text{Autocatalyst}]{\text{Activator}}$ Enzyme	Preferred peptide link
Endopeptidases (proteinases)	I – – *A*	Pepsinogen $\xrightarrow[\text{pepsin}]{\text{HCl}}$ Pepsin	Carboxyl group of dicarboxylic amino acid *to* amino group of aromatic amino acid
	II – – *B*	Trypsinogen $\xrightarrow[\text{trypsin}]{\text{enterokinase}}$ Trypsin	Carboxyl groups of arginine or lysine
	C	Chymotrypsinogen $\xrightarrow{\text{trypsin}}$ Chymotrypsin	Carboxyl group of aromatic amino acids (side opposite pepsin)
Exopeptidases (peptidases)	III	Aminopeptidase (Mn, Mg, Zn)	Terminal amino acid with free amino group
	IV	Carboxypeptidase (Zn)	Terminal amino acid with free carboxyl group
		Tripeptidase	Tripeptides
		Dipeptidase (Mn, Mg, Zn)	Dipeptides

SOURCE: Fruton and Simmonds (1958).

The digestive proteases may be broadly grouped into ENDOPEPTIDASES and EXOPEPTIDASES (Table 3.3). The former, usually referred to as PROTEINASES in older literature, are concerned with the hydrolysis of very specific and central peptide links of the protein molecule (Fig. 3.7); the latter, often called PEPTIDASES in older literature, catalyze the removal of terminal amino acids. Endopeptidases and exopeptidases occur as both intracellular and extracellular enzymes. The three intracellular endopeptidases—cathepsin *A, B,* and *C*—are the counterparts of pepsin, trypsin, and chymotrypsin in the extracellular group. In addition, cells contain intracellular

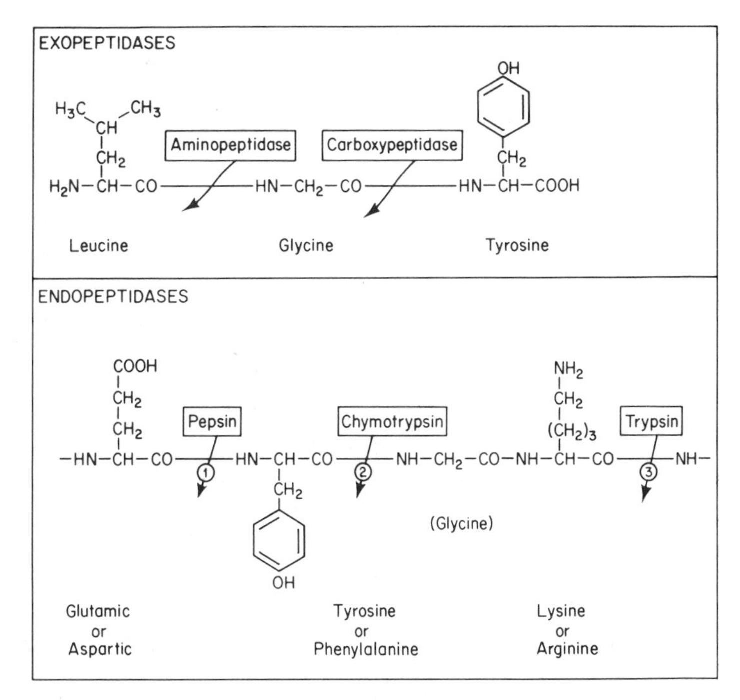

Fig. 3.7 The specific peptide bonds associated with activities of different proteases. (1) Pepsin acts at links between the carboxyl of a dicarboxylic acid and the amino group of an aromatic amino acid. (2) Chymotrypsin acts on the carboxyl group of an aromatic amino acid (tyrosine or phenylalanine). (3) Trypsin acts on the carboxyl of lysine or arginine.

exopeptidases corresponding to the four major groups of extracellular exopeptidases listed in Table 3.3. It should be emphasized that these intracellular enzymes have not yet been crystallized and are recognized only as broad groups.

An endopeptidase of the trypsin type, an aminopeptidase, and a carboxypeptidase have now been identified in all groups where careful investigations have been carried out. These three enzymes would be adequate for the digestion of proteins, although the process might not be as rapid or as complete as the vertebrate digestive process with its additional specific enzymes. Current information is fragmentary and little of it is based on modern biochemical techniques. However, there are indications that families of proteolytic enzymes evolved at a very early stage in animal phylogeny and that comparable groups of digestive proteases occur throughout the animal kingdom (Neurath *et al.,* 1967). Studies of the serine proteases supply the most pertinent data.

The SERINE PROTEASES, of which trypsin and chymotrypsin are representatives, are characterized by a tetrapeptide sequence of "glycine-asparagine-serine-glycine" with the serine as the functional component of the active site. Other members of the group include elastase concerned with the digestion of elastin, thrombin involved in blood clotting, and plasmin which effects a lysis of blood clots; these enzymes are considered homologous; they all appear to operate by the same mechanism and to have evolved from the same ancestral gene by a process of gene duplication (Neurath *et al.*, 1967). The striking similarities between trypsin and chymotrypsin were revealed by the determination of amino acid sequences which showed that 40 per cent to 50 per cent of the sequence is common to both enzymes (Neurath, 1964; Dixon, 1966). Now, it is particularly significant in the present context that three serine proteases have been extracted from the gastric filaments of the sea anemone *Metridium senile* and that one of them has properties very similar to mammalian chymotrypsin (Gibson and Dixon, 1969). These findings suggest that the evolution of the serine proteases commenced with the phylogeny of multicellular animals and that the active components of the enzyme have remained unchanged.

In many animals the hydrolysis of protein by the acid proteases, like pepsin and cathepsin I, is linked with the action of strong acid in killing or capturing live prey. In *Paramecium* and other ciliates, the food vacuole takes a fixed course from cytostome to anal pore while its contents first become progressively more acid until the pH reaches about 1.4 (equivalent to about 0.3% HC1); it then becomes increasingly alkaline to pH 7.5 or 8.0. During the acid phase, bacteria are killed and the food is presumably prepared for alkaline hydrolysis but it is not known whether or not proteases are operating at the low pH (Jennings, 1968). In several non-segmented worms, acid treatment of food is clearly associated with preliminary proteolysis as well as the killing of prey. Jennings (1968) and Jennings and Gibson (1969) describe the details in various species. The carnivorous freshwater triclad *Polycelsis* uses a highly muscular pharynx to suck out the body contents of small oligochaete worms and larval insects; this food then receives preliminary extracellular digestion by acid endopeptidases (optimum pH 5.0). There follows a phagocytosis of partially digested food and the hydrolysis is completed within food vacuoles that become progressively more alkaline as digestion takes place. *Linus ruber,* a carnivorous intertidal rhynchocoel, swallows its food alive and holds it for some time in the foregut while acid secretions (presumably HC1) which are devoid of enzymes kill the prey; subsequent steps in digestion are comparable to the process described for *Polycelsis.* In phylogeny, acid secretions used only to kill food—or capture it, as in boring of calcareous shells by some gastropods (Carriker *et al.*, 1967)—may have preceded acid hydrolysis, but the two processes appear to have been linked at an early stage in evolution by way of the acid proteases. Acid proteases of the pepsin-rennin variety, which are unmasked by HC1, are known only among the vertebrates. The acidity of mammalian gastric juice is extremely high, with a pH of about 1.0 equivalent to 0.2 per cent to 0.5 per cent HC1. The importance of the enzyme carbonic anhydrase in the production of HC1 by the gastric parietal cells is indicated in Fig. 3.8. Cells rich in carbonic anhydrase are associated with acid secretion in the flatworm *Linus,* suggesting a similar mechanism among very primitive animals (Jennings, 1968).

Extracellular digestive enzymes, in contrast to the intracellular ones, usually require an activator—either in the form of another enzyme, an inorganic compound, or an ion (Table 3.3). Enterokinase, for example, triggers a series of reactions when it promotes the change of trypsinogen to trypsin, since trypsin is not only a proteinase but also an activator for trypsinogen, chymotrypsinogen and carboxypeptidase. In-

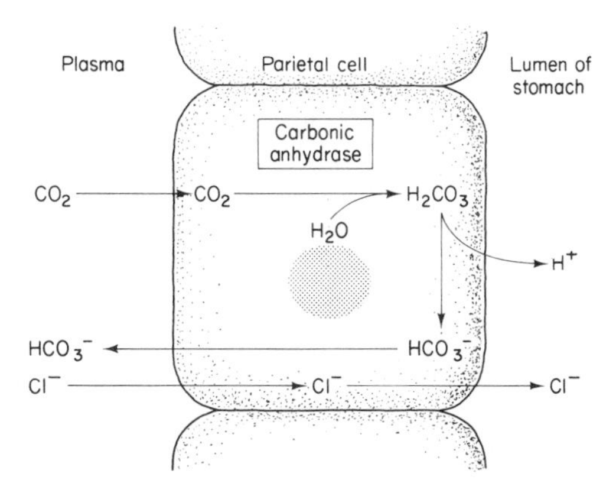

Fig. 3.8 Formation of hydrochloric acid by the parietal cells of the vertebrate stomach. [From Harper, 1973: Review of Physiological Chemistry, 14th ed. Lange Medical Publications, Los Altos, California.]

active zymogen granules similar to those in the mammalian pancreas have been identified by electronmicroscopy in the gastric filaments of the sea anemone *Metridium* and an activating enzyme mechanism demonstrated for them (Gibson and Dixon, 1969). The activation or unmasking of extracellular enzymes following their secretion is a part of the homeostatic or control mechanism of extracellular digestion.

The digestion of the very hard, resistant scleroproteins such as collagen (connective tissue), keratin (wool) and fibrous protein (silk) presents a special challenge which only the insects in the animal world seem to have met (Barrington, 1962). A keratinase, capable of digesting wool, is found in several insects and is well known in the clothes moth. Silk too can be utilized by the clothes moth and certain dermestids, but the mechanism has not been investigated (Roeder, 1953). A collagenase has been reported in the blow-fly larva *Lucilia* where its presence in the excreta promotes the extraintestinal digestion of fibrous tissue in meat; collagenases are also utilized by larvae of the bot fly *Hypoderma* which migrate through the tissues of their hosts by enzymatic digestion of the connective tissues (Wigglesworth, 1972).

Other digestive enzymes Various other molecules which are not pure carbohydrate, lipid, or protein are present in any varied diet and are digested by some animals. Chitin, for example, is a hexoseamine containing amino groups in addition to hexose sugar. Several molluscs, arthropods, and a few vertebrates are said to have chitinases (Jeuniaux, 1961; Alliot and Bocquet, 1967; Sedallian, 1968). Other hexoseamines are chondroitin in cartilage and the constituents of mucoproteins. Conjugated proteins like the nucleoproteins (Chapter 22) have several specific enzymes associated with their digestion in the vertebrates, and other animals may be expected to have comparable catalysts.

ABSORPTION

The monosaccharides, amino acids, and other products of digestion, whether arising in a food vacuole or in a complex digestive tube, must be passed on to the tissues before they can perform essential cell functions. The process by which they are transferred from the locus of digestion is referred to as absorption. In the higher animals this is a transfer from the digestive tract to the circulatory fluids. Similar mechanisms may be involved in the transfer from blood to the metabolizing cell, but this is not usually included in a discussion of absorption.

Two distinct questions have been investigated, viz., the locus of absorptive activity and the mechanics of transferring molecules through cells and cell membranes.

Locus of Transfer

In intracellular digestion the same cells are obviously concerned with digestion and absorption if, in fact, the latter is considered to operate. With the development of extracellular digestive enzymes and one-way traffic through a tubular gut, separate areas are often devoted to absorption and are specialized accordingly. This is particularly true in the vertebrates. Among the invertebrates, enzyme secreting and absorbing cells are often in close proximity, and the same areas of epithelium are frequently concerned with both processes. With finely divided or liquid foods this is not an unreasonable arrangement. In some invertebrates (crustaceans and insects, for example) the same cells have been shown to operate in cycles, at one time secreting enzymes and at another absorbing digested foods (Barrington, 1962; Jennings, 1968).

In many of the larger invertebrates there is an extensive series of diverticula or caeca from the stomach or midgut region—the digestive gland or so-called "liver" of the mollusc, the hepatopancreas of the crustacean and the pyloric caeca of the starfish. These glandular developments greatly increase the available surfaces and serve a combined function of secreting and absorbing. Digestion within these structures is often both intracellular and extracellular. Food, which is reduced to particulate size in the anterior gut or stomach region, is moved into the tubules of these glands where final digestion occurs and the absorption of digested materials takes place. The circulation of food within the diverticula is maintained by cilia or muscle fibers or both. In Crustacea the contraction of the circular muscle of the diverticula forces digestive juices from the diverticula into the stomach; their relaxation and the contraction of the longitudinal muscles draw digested or partially digested material into the diverticula—sometimes after a preliminary sorting or straining. In the lamellibranch or the starfish the exchange of materials between the gut and the digestive gland depends entirely on the activity of long cilia.

Among the invertebrates, several molluscs show a strong tendency to localize absorptive activities and, thus, to separate them from enzyme secretion. Although absorption is usually confined to the midgut glands, in at least one group of lamellibranchs (Nuculidae), it seems to have been entirely relegated to the stomach and intestine while the gut diverticula operate as an extracellular digestive organ (Owen, 1956). Similar trends are evident in the gastropods, but some cephalopods have made the greatest progress toward a separation of digestive and absorptive activities (Barrington, 1962; Bidder, 1966). Cephalopod digestion shows many of the advances found in the vertebrates; in consequence, they achieve a rapid and efficient flow of nutrients to the body tissues.

The diverticular glands of the invertebrate gut are often concerned with storage of reserve food as well as with digestion and absorption. The midgut gland or hepatopancreas of the crustacean (van Weel, 1974) and the pyloric caeca of the starfish (Ferguson, 1969) have been shown to store reserves of carbohydrate and fat which disappear during starvation. In most of the higher invertebrates, one and the same gland secretes enzymes, absorbs digested foods, and stores the body reserves. Conversely the vertebrate liver, which develops as a diverticulum of the gut, is a major organ of food storage and has no capacities for digestive enzyme production.

In all vertebrates, a portion of the wall of the small intestine is concerned with absorption. Again, this is not an exclusive property of a single area and some absorption may occur at almost all points in the gut. Many teleost fish have an elaborate development of pyloric caeca which are embryologically comparable to the vertebrate pancreas; physiologically they are equivalent to the digestive glands of some of the invertebrates, since they are concerned with both the secretion of enzymes and the absorption of digested foods (Barrington, 1957). Food storage, however, seems unlikely in pyloric caeca; this occurs in liver and body fat as in other vertebrates.

The wall of the vertebrate intestine is variously folded and ridged to provide an extensive surface for absorption. The "spiral valve" is an elaborate series of folds which seems to have arisen phylogenetically from a single longitudinal ridge or typhlosole like that of the earthworm or the ammocoete larva; it occurs in some cyclostomes, in elasmobranchs and certain other groups of fishes. Among teleosts, amphibia and reptiles an intricate network of ridges, often bounding deep tubular crypts, adds greatly to the absorptive area of the intestine. Comparable ridges in birds and mammals are covered with a velvet-like pile of minute absorptive villi (Fig. 3.9). These are highly specialized absorptive organs with a core containing a network of capillaries derived from blood vessels in the wall of the gut. Each also contains a central lymph capillary or LACTEAL which begins blindly under the epithelium at the tip of the villus

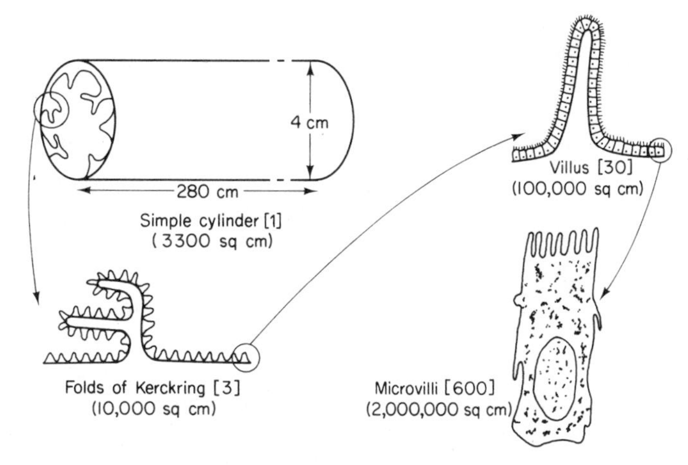

Fig. 3.9 Three mechanisms for increasing the surface area of the human small intestine. Numbers in brackets indicate increase in surface area relative to cylinder. Bases for calculations are given in the original. [From Wilson, 1962: Intestinal Absorption. Saunders, Philadelphia.]

and drains into the main lymphatic channels of the gut wall. The consensus is that lipids pass primarily into the lacteals while the sugars and amino acids are absorbed directly into the capillary blood. Both the villus and the intestinal fold contain smooth muscle, and the rhythmic movements thus produced are important in bringing the villi into contact with the intestinal contents and in maintaining circulation in the lacteals, lymphatics and small blood vessels.

Mechanics of Transfer

The processes involved in the absorption of digested foods may be conveniently grouped into three broad categories: (1) bulk transfer, (2) diffusion and (3) active transport. Monographs and textbooks of cell physiology should be consulted for more elaborate classifications and quantitative treatments (Conner, 1967; Curran and Schultz, 1968).

Bulk transfer In 1883, at a rather low point in his scientific career, the Russian zoologist Elie Metchnikoff introduced a few rose thorns beneath the delicate and transparent skin of a starfish larva. The excitement that followed his observation of the accumulation of amoebocytes about the foreign material sparked experiments that were to extend over a quarter of a century and take him from his native Odessa to the Pasteur Institute in Paris and a Nobel laureate in 1908 (Hirsch, 1965). Metchnikoff first used the term PHAGOCYTOSIS to describe the ingestion of particulate matter by cells. A phenomenon, comparable to phagocytosis, but involving the ingestion of fluid droplets by tissue culture cells was termed PINOCYTOSIS by Lewis in 1931. The historical details of these pioneer experiments have frequently been reviewed (Conner, 1967; Gosselin, 1967). These two processes of bulk transfer have now been studied intensively in many organisms ranging from amoeba to man. The similarities between them are more striking than the differences. In both, there is a wrinkling of the surface membrane at the point of contact with the foreign substance; the wrinkling is followed by "membrane flow" around the particle or fluid droplet to enclose the foreign substance in a vesicle or vacuole of the cell. Rabbit macrophages showing active pinocytosis may infold 2 per cent to 20 per cent of their cell surfaces every minute (Gosselin, 1967).

These processes are particularly important in food absorption by lower forms. Protozoa may depend entirely on "membrane flow" for their supply of organic nutrients. It permits many of the Lamellibranchiata and the Echinodermata (Yonge, 1937; Wagge, 1955) to digest food particles (diatoms, for example) which are too large to enter the tubules of the digestive diverticula. Yonge (1937) suggests that this may account for the fact that a highly specialized group such as the Lamellibranchiata has evolved so successfully with only a limited degree of extracellular digestion.

Diffusion In simple diffusion or PASSIVE TRANSPORT, materials move through cells, membranes and intercellular spaces because of differences in the concentration gradient. The substances are in aqueous solution, and their migration is the result of the random motion of solute molecules. Most living membranes have very special permeability properties which restrict the movements of many molecules and ions; consequently, a study of diffusion through living cells involves not only the concentration gradients or physical forces of diffusion but also the osmotic properties of the cell membranes or tissues. In some cases, however, the cell membranes and tissues *seem to* have little effect on the migration of the molecules; absorption is primarily due to concentration gradients so that there is a direct relationship between the concentration of the penetrating substance on the outside and inside of the cell membrane. The extent of "passive" transport varies with the type of cell, its physiological condition and the properties of the penetrating molecule. With these reservations, the absorption of quite a variety of materials may be said to depend on passive transport or diffusion (Holter, 1961).

Many drugs, poisons, alcohols, acid amides and other extraneous substances which pass from the digestive tract into the blood do so in accordance with their concentration and show no evidence of active transport. Some of the vitamins (ascorbic acid, for example) seem to enter the circulation by simple diffusion; others, like vitamin B_{12}, require elaborate transport systems (Matthews, 1967).

Several special cases of diffusion are of particular interest to the comparative physiologist since they emphasize once more the versatility of physiological organization. Two will be noted here. The first is the absorption of volatile fatty acids from the stomach of the ruminant. It has already been pointed out that the fermentation processes in the rumen produce large amounts of volatile fatty acids and ammonia and that the ready and rapid absorption of these materials is essential to the maintenance of constant conditions of pH, ions and other ecological factors of importance to the rumen flora. There is no evidence of active transport. Volatile fatty acids and ammonia diffuse readily into the blood capillaries and are quickly metabolized so that the diffusion gradient is continuously steep (Annison and Lewis, 1959).

The second example is related to the specialized nature of insect metabolism where the major carbohydrate of the hemolymph is the disaccharide trehalose and not the monosaccharide glucose which is the circulating blood sugar of vertebrates. In the locust *Schistocerca gregaria*, the uptake of glucose from the caeca and the anterior part of the midgut is directly related to its conversion into trehalose by the fat bodies strategically located to effect a rapid metabolism of the absorbed monosaccharide. The concentration gradient is maintained by the speedy conversion of monosaccharide into disaccharide; the process is unaffected by metabolic inhibitors and shows none of the features of active transport which moves glucose from the vertebrate intestine. Treherne (1967) refers to the process in the locust as a *facilitated diffusion mechanism* and relates it to the specialized nature of insect metabolism. Different mechanisms for glucose absorption probably exist in certain insects since some may have as much as 80 per cent glucose in the hemolymph carbohydrate (third instar larva of the dipterous parasite *Agria affinis*). Amino acid absorption also seems to occur by diffusion in the locust; in this case the gradient is established by the rapid uptake of water from the midgut (Treherne, 1967).

In association with the digestive and absorptive process, many insects (and some polychaetes) possess a curious molecular strainer, the PERITROPHIC MEMBRANE on the inside of the gut. This is a thin membranous sleeve, lining the midgut and anterior part of the hindgut and separating the gut epithelium from the mass of digesting food. It is an extremely delicate structure (0.5 μ or less in thickness) composed of chitin with a small amount of protein. Colloidal particles are unable to penetrate, but the digestive enzymes, the digested food, and the simpler molecules of partly digested food pass freely through it (Roeder, 1953; Wigglesworth, 1972). It protects the delicate epithelium of the gut from the abrasive action of hard foods. Many animals achieve a similar protection through copious secretions of thick mucus. In the hagfish *Myxine,* the mucoid secretions of the gut epithelium form a delicate but tough peritrophic membrane or sac within which the enzymes act (Adam, 1963).

Active transport During ACTIVE TRANSPORT, molecules may pass from the gut against a diffusion gradient. In other words, glucose, amino acids and other substances may move from very dilute solutions through the intestinal cells and membranes into more concentrated body fluids. Active metabolic work is required but neither the transport mechanism nor the coupling of this with the energy generator of the cell is fully understood. Current theories assume the activity of special carrier molecules which form a complex with the transported material on one side of the membrane and release it on the other (Yost, 1972; Curran and Schultz, 1968).

Glucose, the major product of carbohydrate digestion, is actively transported in the vertebrates. This transport is normally against a concentration gradient since the major circulating blood sugar of the vertebrate is glucose (about 100 mg per 100 ml). In spite of intensive investigations extending over many years, the transport mechanism remains uncertain (Crane, 1960, 1968). Many pertinent facts have, however, been demonstrated. It is now known that only hexoses with a specific molecular architecture (hydroxyl group at carbon-2) are moved by active transport; that the specific rates of transfer vary for the different hexoses (galactose moves faster and fructose more slowly than glucose) while pentoses, lacking the essential molecular configuration, penetrate by simple diffusion. A competition between chemically related penetrating molecules (glucose, galactose, 3-0 methyl glucose) argues strongly for the presence of a specific carrier or carriers while the dependence of transport on the presence of sodium has suggested an involvement of the sodium pump. For a long time, phosphorylation was considered an essential step but physiologists are no longer satisfied with this concept. Current theories, developed by Crane (1960, 1968), link the absorption of glucose with the transport of sodium. Two separate processes appear to be involved. At the brush border or mucosal side of the cell, a carrier molecule couples with glucose and sodium—perhaps also with potassium; the complex thus formed is readily permeable to the cell membrane and diffuses through it along a concentration gradient (FACILITATED OR CARRIER DIFFUSION). Once inside the cell, the carrier molecule separates from the rest of the complex; the cation pump then removes the sodium while the glucose concentrates inside the cell. This intracellular glucose will diffuse across the serosal membrane into the portal blood when it becomes more concentrated than the blood sugar. Thus, it is the operation of the sodium pump which actually requires energy in the form of ATP, while glucose moves by a special diffusion process.

Absorption of undigested protein probably occurs to a limited extent at all levels in phylogeny. In some cases, the direct absorption is of biological significance. Newly born mammals regularly acquire some of their immune substances as undigested protein from their mother's milk. Direct absorption of small amounts of protein in the adult may, unfortunately, create bothersome immunological reactions (allergies). Protein of nutritional significance, however, is absorbed only after its hydrolysis to amino acids. Again, the mechanisms are uncertain although active transport is known to occur; a coupling with the sodium pump has also been suggested for the transfer of amino acids. Sodium-dependent transport processes occur widely in the animal world (Schultz and Curran, 1970).

According to the classical LIPOLYTIC THEORY of fat absorption, a complete hydrolysis of the triglycerides into fatty acids and glycerol is followed by their absorption into the intestinal mucosa cells where they are resynthesized and pass as fine droplets or chylomicrons into the lymphatic system (lacteals). The alternate PARTICULATE THEORY claims a direct absorption of at least a portion of the finely emulsified fat droplets. (Frazer, 1946; Glass, 1968). Thus, in the presence of bile salts, long-chain fatty acids as well as monoglycerides, some diglycerides, and triglycerides form minute fat MICELLES a few millimicrons in diameter; these penetrate the cell membrane where they are resynthesized into triglycerides and packaged in delicate envelopes of phospholipid and protein to form tiny droplets (1 μ to 4 μ in diameter) known as CHYLOMICRONS which then pass into the lacteals.

Fat absorption has been investigated in relatively few animals and substantially more evidence will be required before any generalizations are possible. Current evidence, however, favors the lipolytic theory. In a careful electron microscopical study of fat absorption in the rat, Cardell *et al.* (1967) found that fatty acids and some monoglyceride diffused directly into the mucosal cells of the intestine where they encountered an extensive smooth endoplasmic reticulum (SER). Within the SER, triglyceride was resynthesized and chylomicrons formed; these were then delivered into the intercellular spaces and thence moved into the lacteals.

Many different factors are known to regulate absorption in the mammal. The action of bile salts in lipid absorption has been mentioned. Pyridoxyl phosphate participates in amino acid transport. The D-vitamins stimulate the uptake of calcium ions from the intestinal tract; the level of mucosal ferritin controls the uptake of iron. These mechanisms are detailed in textbooks of human and medical physiology.

COORDINATION OF DIGESTIVE ACTIVITIES

The regulated passage of food through the gut provides sufficient time for effective enzyme action and proper absorption before the wastes (feces) are discharged. The necessary coordinating mechanisms are much less elaborate in the continuous feeder than in the periodic feeder. In many invertebrates and particularly in the filter feeders, there is normally a continual flow of food particles into the body, a constant secretion of enzymes, with incessant digestion and absorption. The continuous feeder thus avoids the problems of storage between meals and the potential hazards of fluctuations in the blood levels of digested foods. A periodic feeder, however, may obtain large volumes of food during a relatively brief period, and these must be digested, absorbed, and stored to avoid excessive flooding of the tissues with nutrients. At the upper level of phylogeny, a complex interaction of the autonomic nervous system and endocrines initiates enzyme secretion, regulates the discharge of accumulated juices, and governs the motility of the gut and the passage of food through it. Even the filter feeder, however, may face periods of starvation and, although the controls are less elaborate, digestive activity varies somewhat in accordance with demands at all levels in phylogeny. In the bivalve mollusc, the crystalline style (source of amylases) may actually disappear during periods of starvation; in *Helix* the digestive gland cells show a secretion cycle which operates at varying speeds according to demand (Owen, 1966); at the coelenterate level, extracellular protease secretion has been shown to be stimulated by the presence of food and depressed by its absence.

Both humoral and neural mechanisms are active in the coordination of digestive functions, but their relative importance varies in different animals. Primitive forms are often continuous feeders, and the enzyme producing cells secrete constantly or are cyclically active in accordance with the supply of food or in response to secretagogues in the form of partially digested food. It would appear that the neuromuscular control necessary to mix and move the food through the gut and to discharge the wastes is a more acute problem in these lower forms than the problem of regulating enzyme secretion.

The Visceral Autonomic System

Characteristic of the digestive canal of the higher invertebrates and the vertebrates is the presence of one or more nerve nets. In the vertebrates a MYENTERIC PLEXUS OF AUERBACH between the two thick layers of smooth muscle is primarily responsible for the movements of the gut, while MEISSNER'S PLEXUS in the submucosa controls the activities of the enzyme producing glands. In the decapod crustacean there is one well-developed nerve plexus resembling that of Auerbach (Vonk, 1960), and similar nerve nets have been described in many other arthropods, in annelids and molluscs. Other phyla of invertebrates lack the special nerve nets of the visceral sympathetic or enteric system. There are, however, definite nerves to the specialized muscular areas of the pharynx, stomach, and rectum. A circumenteric nerve ring anteriorly, and sometimes also posteriorly, with branching nerves is a rather constant feature (Hyman, 1951–59). In the more primitive groups these may be thought of as nerve nets which have been concentrated into circumenteric nerve cords.

Among the invertebrates, the enteric or stomatogastric supply is predominantly motor to the muscles of the gut (Campbell and Burnstock, 1968). In the annelids, both inhibitory (adrenergic?) and excitatory (cholinergic?) fibers have been described; the annelids, however, may be exceptional in this regard since only excitation has been convincingly demonstrated following stimulation of the stomatogastric in other invertebrates. In *Arenicola,* a rhythmically spontaneous activity of the gut musculature (especially the proboscis and esophagus) is associated with a burrowing life in soft mud (Wells, 1950). This rhythm depends only on pacemakers in the enteric net and is quite independent of the central nervous control. Spontaneous activity is probably present in other invertebrates as well. The annelids may also provide an exception to the nervous control of enzyme secretion; secretion of proteolytic enzymes has been noted following stimulation of the enteric nerves in the earthworm but not in crustaceans (Vonk, 1960) and insects (Barrington, 1962).

The vertebrate autonomic nervous supply to the gut is typically double with parasympathetic and sympathetic fibers responsible for opposing activities. In the bird and mammal, stimulation of the vagus nerve (parasympathetic) increases enzyme secretion (especially in the salivary glands and stomach) and heightens gut motility while the sympathetic effect is inhibitory. Normally, a balance between the two systems maintains proper muscle tone and glandular activity (anteriorly) in relation to feeding. Recent evaluations of the experimental work on the lower vertebrates (Campbell and Burnstock, 1968; Campbell, 1970) indicate that the vagal outflow is primarily inhibitory in forms below the anurans and thus has a directly opposite effect to the parasympathetic of higher forms. Likewise, the spinal autonomic differs at the two levels of vertebrate organization; primitively it seems to have been responsible for both excitatory and inhibitory effects with the excitatory elements predominating in the lower forms. Our knowledge of the autonomic of lower forms is based on relatively few investigations and generalizations are extremely hazardous (Chapter 2 and Burnstock, 1969).

Nervous control of digestive secretion has not been demonstrated in the lower vertebrates and, as in the invertebrates, the autonomic control of the muscle is the more constant feature of the visceral system. Actually, even in the mammal there are variations in the different regions of the gut; a double balanced innervation is characteristic, but it is not safe to go beyond this in generalizing.

Gastrointestinal Hormones

Vertebrate physiologists have described a series of hormones produced by the gastrointestinal epithelium and concerned with regulating the secretions of the digestive glands. A comparable system has not been found in the invertebrates; it may well be absent although there is some evidence of a hormonal control of enzyme secretion in certain insects which feed periodically (House, 1965; Langley, 1967; Wigglesworth, 1972). Among most invertebrates, a chemical regulation of feeding and digestive activities is often evident, but this depends on the presence of food or partially digested food products (secretagogues) and not on chemical messengers produced by the animal itself. In *Hydra,* for example, the presence of certain amino acids and glutathione released from injured prey induces an elaborate series of feeding activities (Lenhoff, 1968; Lindstedt, 1971; Reimer, 1971). Even in the most complex invertebrates, processes of digestion, secretion and storage take place side by side, and a nervous control of gut motility, with some direct response of the enzyme secreting cells to the presence of food (secretagogues) is evidently adequate. The vertebrates, however, must mobilize a series of digestive juices and regulate the passage of food through several different areas at rates which permit sufficient time for the appropriate enzymes to act and for the digested foods to be absorbed. The demands on particular enzyme secreting tissues vary in accordance with the type of food eaten. To meet this complexity of demands, the vertebrate has a sequence of hormones (Fig. 3.10) as well as a highly organized autonomic control.

The first of the gastrointestinal hormones to be discovered was SECRETIN. Actually, secretin, isolated from the intestine by Bayliss and Starling (1902), was the first substance to which the term "hormone" was applied. These investigators showed that the presence of HCl stimulated the intestinal mucosa to liberate into the blood a compound (secretin) which initiated the release of pancreatic juices. They showed later (1903) that a similar mechanism existed among the lower vertebrates. In 1906, Edkins suggested that the stomach might also be responsible for the elaboration of a

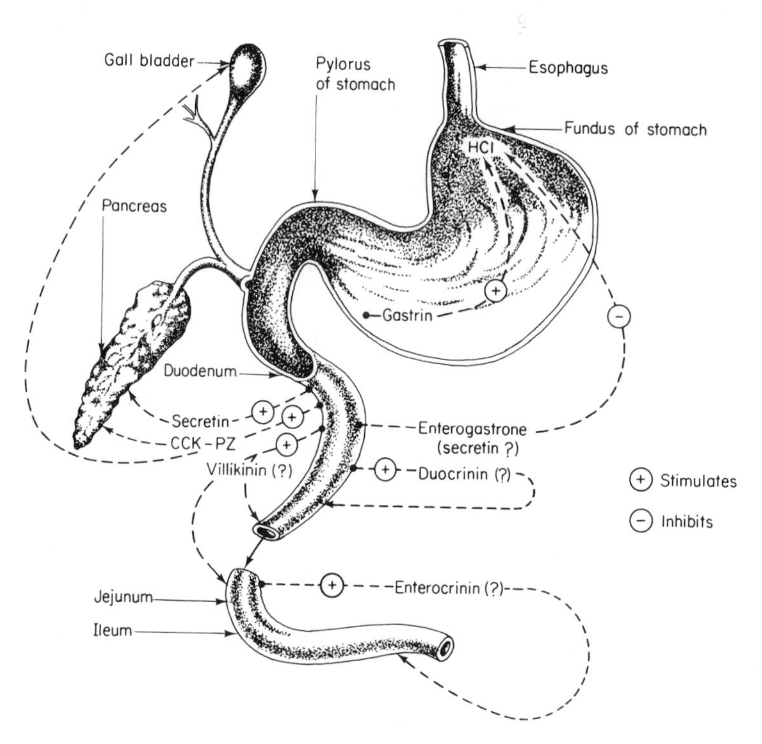

Fig. 3.10 Gastrointestinal hormones of the mammal. The polypeptides *Cholecystokinin* and *Pancreozymin*, long considered separate hormones, are now thought to be a single factor, represented above as CCK-PZ.

hormone and, since that time, no less than seven distinct hormonal factors or effects have been described in the mammal (Frieden and Lipner, 1971; Turner and Bagnara, 1971). The biochemist has now firmly established three distinct polypeptides in this group of hormones and physiologists have shown that several effects formerly attributed to distinct factors are caused by single hormones. The gastrointestinal hormonal factors are shown in Fig. 3.10. GASTRIN, SECRETIN, and CCK-PZ are now on firm ground both biochemically and physiologically. In addition, there is considerable physiological support for VILLIKININ; ENTEROCRININ and DUOCRININ remain rather speculative.

GASTRIN, secreted by the gastric mucosa, is responsible for the balanced production of HCl in the stomach. Local nervous mechanisms, initiated by the presence of food, cause gastrin production, and this in turn stimulates the parietal cells. The gastrin mechanism is responsible for the control of the volume of HCl; the presence of HCl itself serves as a feedback to inhibit gastrin secretion. Thus, the first link in the hormonal control of digestive activity provides a proper concentration of acid for peptic activity within the stomach.

A second substance concerned with gastric activity is released from the intestinal mucosa into the blood when fats and HCl enter the intestine. This factor, known as ENTEROGASTRONE, inhibits the secretion of gastric juice—particularly HCl. In this manner, the presence of partially digested foods in the intestine gradually slows down the activity of the stomach. Highly purified preparations of secretin have now been shown to exert all the effects of enterogastrone and it seems unlikely that enterogastrone is a distinct factor (Johnson and Grossman, 1968).

The endocrine control of the pancreas has now been found to be more complicated than visualized by Bayliss and Starling in 1902. Instead of one hormone, secretin, there are two—SECRETIN and PANCREOZYMIN. Secretin which is released under stimulus of low pH (HCl), digested fat or bile, initiates the production of a copious pancreatic juice; this juice is low in enzymes but rich in salts which may be important in neutralizing the acid chyme. Pancreozymin, secreted under the stimulation of partially digested protein (HCl to a lesser extent), induces the flow of an enzyme-rich secretion. Thus, the partially digested food from the stomach triggers the release of fluid, but the important proteolytic enzymes are only produced when they are required by the presence of peptone and proteoses.

Pancreozymin now appears to be biochemically identical with CHOLECYSTOKININ, a hormonal factor concerned with the contraction of the gall bladder. The peptide having the action of these two factors is referred to as CCK-PZ.

Rapid hydrolysis of fats depends on their prior emulsification by the bile salts and secretions of the liver. The arrival of fat from the stomach initiates the release of CCK-PZ by the intestinal mucosa, and this elicits contraction of the gall bladder. In this way, an accumulated store of bile is promptly added to the intestinal content in response to the fat which requires it for digestion. Secretion of bile by the hepatic cells depends on a different set of stimuli. The hormone secretin and the bile salts themselves, when reabsorbed into the blood, are responsible for this activity (Combes, 1964).

The presence of a gall bladder permits an animal to meet a sudden demand for large amounts of bile, and its anatomical development probably bears some relation to the diet. It is absent in the horse, deer, rat, and several other animals (Swenson, 1970).

It has been claimed that the stomach chyme in the upper intestine stimulates the release of additional hormones which activate mechanisms further down the intestine in preparation for the food which has entered its upper end. Evidence for these claims is still rather incomplete. Enterocrinin and duocrinin are said to release enzymes at different points in the small intestine; a hormone VILLIKININ seems to stimulate the motility of the absorptive villi.

In the intestine the control is entirely humoral; at the level of the salivary glands it is entirely nervous. The presence of food and the acts of feeding promptly activate nervous mechanisms for the immediate initiation of digestion, but as the food progresses along the gut, more slowly operating humoral controls regulate the different digestive glands in accordance with the foods eaten and their passage from one area to another. The interplay of nervous and humoral mechanisms—basic to most homeostatic regulation—is beautifully shown in the coordination of all the activities of the alimentary canal.

The gastrointestinal hormones are known to occur in some of the primitive fishes (Nilsson, 1970; Nilsson and Fänge, 1970) and have apparently had a very long evolutionary history. The amino acid sequence in secretin bears a marked resemblance to that of glucagon (Chapter 2). Both substances are concerned with the physiology of the pancreas, a gland which arises embryologically from the intestine. Gastrin and CCK-PZ also show marked biochemical as well as physiological similarities (Turner and Bagnara, 1971). These polypeptide hormones must have evolved as the primitive gut developed specialized physiological areas for digesting the foods, and as the liver and pancreas attained the status of separate glandular organs. However, much more comparative work will be necessary before it is possible to suggest evolutionary pressures which may have been responsible for the phylogeny of separate hormonal factors.

4

THE EXCHANGE OF GASES

Aerobic life, characteristic of animals, demands a steady flow of oxygen into the cells and a ready removal of the carbon dioxide which arises from their metabolism. At the cellular level, physical forces of diffusion alone effect these exchanges. No convincing evidence has ever been presented for active transport or secretion of either oxygen or carbon dioxide by cells. Diffusion is quite adequate at the cellular level, but in a large multicellular animal gas exchange is only possible because of a number of special adaptations which bring oxygen into close proximity with the cells; some of these adaptations are respiratory and some are circulatory.

Two main factors have shaped the structures and refined the processes of gaseous exchange. The first of these is associated with the increasing size of animals during phylogeny. Since rates of diffusion are relatively slow and since in many environments the oxygen supply is marginal, greatly extended surfaces for diffusion are usually required. The extent of the surface may be enormous in proportion to the mass of the organ. It has been calculated that human lungs have a surface area of from 50 to 90 square meters, somewhere between thirty and fifty times the external surface area of the body.

The second major factor is related to terrestrial life and aerial respiration. Soft protoplasmic extensions which characterize the respiratory organs of aquatic animals collapse and dry out in air. The evolution and success of terrestrial forms have, in part, depended on the development of respiratory surfaces inside the moist bodies of the animals. These surfaces are perpetually wet. Whether the animal is aquatic and extends its branchial membranes into the water or terrestrial with internal gas-filled cavities, oxygen and carbon dioxide are always dissolved in a water layer at the surface of the cell. Although the thickness of the water layer varies, respiration is, in a sense, aquatic in both cases.

Unless gases are piped to the individual cells, as they are in the tracheates, mere extension of surfaces could never solve the problem of gas exchange for a bulky animal. Diffusion of gases in protoplasm is slow, and a constant movement of the external environment over the outer surface of these membranes (ventilation) must be coupled with an equally efficient and continuous movement of gases inside the membranes. Thus, the evolution of organs for both ventilation and a blood circulatory system has accompanied the expansion and elaboration of the respiratory epithelium. The continuous circulation of gases—both on the inside and outside of the respiratory epithelium—is indispensable to a system which depends on concentration gradients and convection currents. The varied pigments involved in gas transport are discussed in Chapter 6. The machinery responsible for ventilation, as well as the respiratory membranes themselves, will be considered here. Krogh's (1941) monograph, published more than thirty years ago, remains an excellent summary of the pertinent information. More recently, Hughes (1963), Jones (1971a) and Steen (1971) have covered essentially the same ground in monographs of similar scope. Irving (1964) has provided a useful review in the *Handbook of Physiology*.

INTEGUMENTARY RESPIRATION

Gases move slowly through protoplasm. Calculations based on the metabolic demands of animals and on rates of diffusion of gases in protoplasm show that simple diffusion cannot satisfy the oxygen demands of organisms much larger than 1 mm in diameter (Krogh, 1941). The precise size limits depend not only on the rate of metabolism but also on the shape of the animal; any departure from the spherical will increase the surface area relative to the mass. Thus, a number of the smaller metazoa and the larvae of much larger ones exceed the 1 mm diameter range. Giant land planarians (Terricola) may be 50 cm long, but their flat elongated bodies result in very large surfaces in relation to the mass. Coelenterates and sponges often reach even larger sizes with modest metabolic demands and relatively short diffusion distances. A jellyfish is largely water with only about 1 per cent organic dry material; a sponge maintains a circulation of water by cilia over the surfaces of cells which line an intricate series of canals and spaces; tissues of some of the coelenterates have been shown to operate at very low oxygen tensions. Modifications such as these permit relatively large animals to effect sufficient exchange of gases without gas-transport pigments or specialized areas of respiratory epithelia.

Theoretically, the addition of an efficient circulatory system should allow adequate gaseous exchange through the integument. Metabolism is nearly proportional to body surface (Chapter 7), and if the surfaces are readily permeable and vascular, they alone should suffice. The earthworm, the leech, and some larval fishes are among the many animals which meet their oxygen demands in this way. Oligochaetes, for example, may reach much larger sizes than the 50-cm turbellarians referred to in the last paragraph; it seems likely that the circulation of hemoglobin through a highly vascular hypodermis is one reason for the difference (Johansen and Martin, 1966). Even larger animals, such as amphibians and fish, may rely on cutaneous respiration during emergencies or use it continuously as a supplement to gills or lungs. The eel can exchange 60 per cent of its respiratory gases through a highly vascular skin, and this is enough to permit prolonged activity in moist air at temperatures of 15°C or less (Randall, 1970b). Cutaneous respiration is adequate for an eel or a frog submerged in water if metabolic demands are not elevated by temperature or excessive activity. In air, frogs also normally exchange large amounts of oxygen and carbon dioxide through the skin. Cutaneous respiration is important in all amphibians. In *Ambystoma* approximately 80 per cent of the carbon dioxide is released through the skin at all temperatures above 5°C; below this temperature the lung and buccopharyngeal exchange of carbon dioxide is too small to be measured. The cutaneous oxygen consumption in this salamander increases with temperature in a linear manner to about 50 per cent of the total at 15°C and then declines to about 35 per cent of the peak value at 30°C (Whitford and Hutchinson, 1963). Foxon (1964) tabulates comparative values of relative capillary area in lungs, skin and buccal cavity as a measure of the contribution of each of these to respiratory exchange. The buccal cavity is apparently of minor importance since the capillary surface is always small, ranging from less than 1 per cent to 3 per cent. In a number of species the capillary areas are about equal in lungs and skin, but there is considerable variation, and the integumentary contribution ranges from as low as 20 per cent in dry-skinned forms, such as some toads, to about 76 per cent in the urodele *Triturus alpestris*.

Cutaneous respiration can only be successful, however, if the surfaces are moist, thin and readily permeable. This imposes a serious restriction in many habitats. Soft coverings are vulnerable to predators and prone to abrasion. Problems of electrolyte and water balance are magnified in proportion to the extent of permeable surface. The hazards of desiccation in the terrestrial environment are obvious. The addition of chitin to the cuticle, for example, drastically reduces the cutaneous exchange of gases, and none of the Crustacea which rely on cutaneous respiration has reached the size of the non-chitinized annelids. The compromise between moist, readily permeable surfaces and a protective impermeable outer covering has been met by the development of body surface extensions in the form of gills or lungs—usually housed in moist chambers specialized for this particular purpose.

BRANCHIAL RESPIRATION

Any appendage of the body primarily concerned with the exchange of gases may be called a gill. Gills are typically the respiratory organs of aquatic animals and range from cirri and simple epithelial extensions, which only supplement cutaneous respiration, to elaborate structures consisting of thousands of highly specialized lamellae, enclosed in a gill cavity which is ventilated by a continuous flow of water.

External Gills

External gills are phylogenetically more primitive. There is a multitude of structural variations, and only a few examples from this vast array will be included here. The echinoderms, like other major groups of invertebrates, have experimented with a variety of gill structures. The body of the asteroid is rather uniformly clothed with hollow, papillate, body-wall extensions from the coelom (papulae), while in most echinoids five pairs of small branched structures (the gills) surround the peristomial region. In both cases, exchange of gases through these structures is supplementary to the exchange which occurs through the tube feet or podia. Among the annelids the locomotory appendages often supplement the integument to meet the respiratory demands (*Nereis*); or there may be paired, segmentally arranged branchial tufts along the sides of the body (as in the burrowing polychaete, *Arenicola*); or the branchial tufts may be developed only anteriorly (as in the tubulous polychaetes) where they appear as plumes and feather-like structures or masses of tangled filaments.

Among some of the amphibia the length of the external gill filaments has been shown to change with the oxygen content of the water. A functional response to low oxygen tensions has been reported for both *Salamandra* and *Rana* (Krogh, 1941). In some fishes (Elasmobranchii, Dipnoi, Polypteridae) external gills are present only in the larvae and precede the development of the adult branchial apparatus. In the male lungfish, *Lepidosiren*, masses of filamentous respiratory appendages grow on the pelvic fins at the time of reproduction when the male attends the nest. He makes periodic excursions to the surface, fully aerates his blood by the lung, and returns to supply oxygen to the developing eggs and larvae in the nest by way of the highly vascular external pelvic gills. There are many other curious examples of external gills. They attest the primitive nature and varied form of this method of gaseous exchange.

Internal Gills

Exposed gills have several obvious disadvantages. In some habitats delicate appendages are subject to abrasion and may attract predators; they also increase resistance to locomotion. Further, the movement of water over their surfaces is intermittent, depending on the gentle movements of the appendages or on the water currents in their surroundings. The withdrawal of gills into a cavity permits the streamlining of the body, provides maximum protection for the delicate epithelia, and favors the specialization of a pumping system for their ventilation. Internal gills are found in some members of all the groups of large aquatic animals—whether sessile or active. Their many forms are described in textbooks of general zoology, and only the principles associated with their phylogeny and operation will be considered here.

In evolution there seems to have been a compromise between the attainment of a sufficiently large respiratory surface and the housing of this in limited spaces. To meet these demands a continuous flow of water is maintained through the branchial apparatus at rates which depend on the demand for oxygen. In addition, the arrangement of capillaries in the gills provides maximum contact between circulating blood and inflowing water; sometimes this is achieved with a system of multicapillary loops in contact with the respiratory epithelium but, more frequently, maximum gradients are developed through a counter-current exchange with blood and water flowing in opposite directions. Further refinements in efficiency involve variations in the number of gills and in the transport capacities of the blood pigments. Adaptations, both morphologically and physiologically, meet the demands of varied habitats.

Among sessile animals, such as bivalves, tunicates and some of the echinoderms, the circulation of water through the branchial chamber often depends on ciliary activity. A supply of food particles as well as oxygen is carried by this gentle stream. On the other hand, active, free living forms as well as many tube-dwelling and semisessile animals depend on muscular activity and frequently combine ventilation with locomotion—sometimes also with feeding. The coordinated action of many groups of muscles periodically aerates the branchial chambers or constantly pumps water through them. Integration depends on ganglionic and brain centers and may include a sensory system with specialized mechano- and chemoreceptors. Several examples will illustrate the trends in specialization.

A highly successful group, such as the Crustacea, displays a variety of gills ranging from an exposed series of relatively simple leaflets to masses of intricately divided filaments housed in regularly ventilated chambers; in some small forms and in larvae, integumentary respiration suffices and respiratory appendages are absent (Wolvekamp and Waterman, 1960). The more primitive gilled species, such as the fairy shrimp *Branchipus* or the brine shrimp *Artemia*, have vascularized lamellae on the eleven or more pairs of thoracic appendages. These appendages are also equipped with a brush of filtering hairs, and the currents which they maintain supply food as well as oxygen. In the Anostraca, these lamellate gills are fully exposed, while they are covered by a carapace in other orders of the Branchiopoda. In the Decapoda, among more specialized crustaceans, the gills are housed in a branchial chamber which, in forms like the crab, communicates with the external environment only through small slits (one above each leg) and a main inhalant and exhalant passage on each side. The exhalant passage is the largest and is situated in front of the mouth where the scaphognathite (of the second maxilla) moves to and fro as a bailer to draw water out of the exhalant passage and thus create currents over the gills. The mechanics of circulation in the chamber are such that the gills are

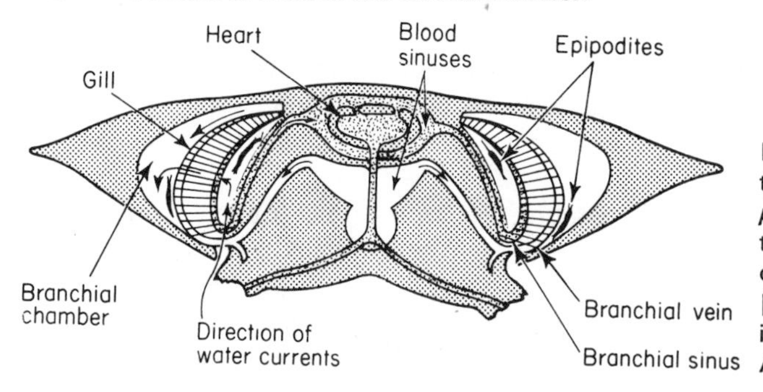

Fig. 4.1 Branchial chambers and the blood supply of the gills in *Cancer pagurus* (diagrammatic cross section). Arrows indicate directions of water currents and blood flow. [From Waterman, 1960: The Physiology of the Crustacea, Vol. 1. Academic Press, New York.]

thoroughly bathed, and water is brought close to the blood (Fig. 4.1). The extent of gill epithelium is related to habitat and respiratory demands (Gray, 1957 and Chapter 12).

The gills of the mollusc are organs of the mantle cavity and this space provides ready-made housing for them. In the sessile and relatively inactive forms, ventilation of the mantle depends on cilia and in this way gas exchange is linked with feeding; this primitive arrangement is also found in the tunicates and *Amphioxus* as well as many other invertebrates. In the active cephalopods the respiratory currents are created by muscles and ventilation is combined with locomotion. In species which lack a shell (*Octopus* and squid), two sets of opposing muscles (the longitudinal and circular) alternately expand and constrict the mantle. As the mantle space enlarges, water is drawn in around the edges of the mantle and circulates over the gills. Contraction of the circular muscles first constricts the edges of the mantle around the neck and then forces the water out in a jet from the funnel (Fig. 4.2). In the Shelled *Nautilus* the flow of water and somewhat gentler swimming move-

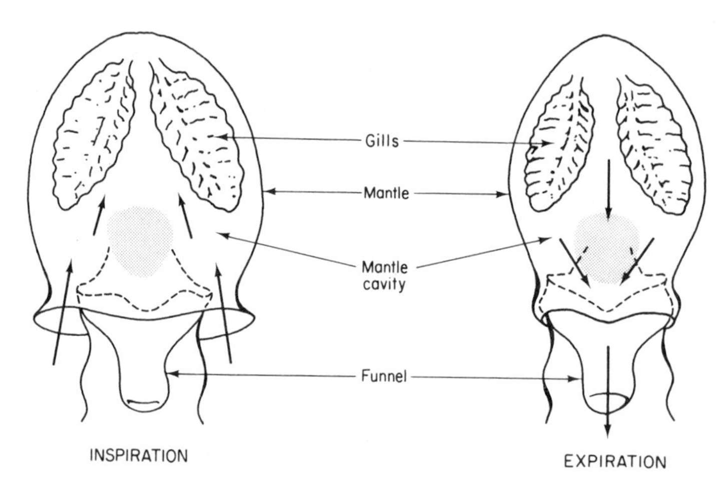

Fig. 4.2 Respiratory water flow in an octopus. Shaded area indicates presumed dead space. [From Johansen and Lenfant, 1966: Am. J. Physiol., *210*: 910.]

ments depend only on contractions of the funnel—a modified foot. Perhaps because of the less vigorous means of ventilation, the *Nautilus* has four pairs of gills while the more active forms operate with two pairs.

Branchial respiration attains its greatest efficiency in the aquatic vertebrates. Several reviews are available (Hughes and Shelton, 1962; Randall, 1970b; Shelton, 1970). Only the teleost fish is considered here. Its branchial apparatus exemplifies the specialized features of a highly efficient gill system: extensive and protected respiratory surfaces, a continuous circulation of water over them, and a countercurrent flow of oxygenated water and deoxygenated blood.

A series of gill filaments is spread like a curtain to separate two chambers, the oral or buccal cavity and the opercular cavity (Fig. 4.3). While a fish is breathing quietly, the tips of the hemibranchs on neighboring gill arches meet as shown in Fig. 4.3 to form a complete screen through which water flows from the oral to the opercular cavity. During activity, delicate muscles move the paired filaments on the branchial arches rhythmically and permit some of the water to bypass the interlamellar septa and flow freely from one chamber to the other; in this way resistance is greatly decreased (Saunders, 1961). The movement of the gill filaments may be

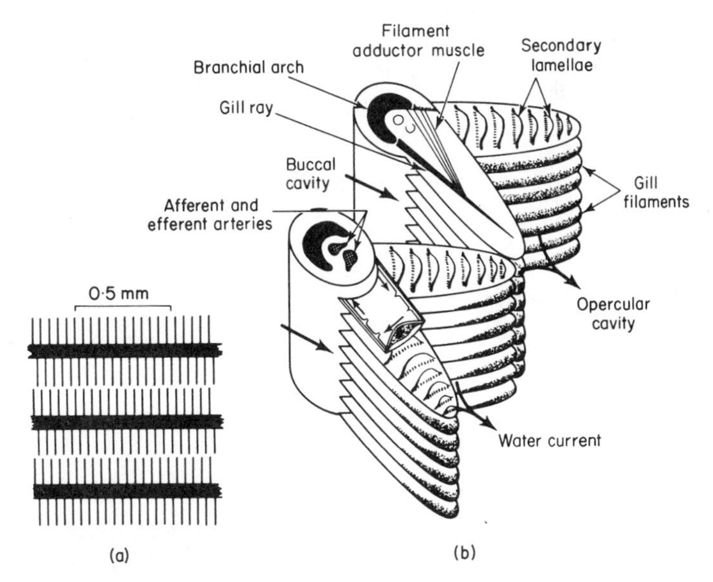

Fig. 4.3 (a) Diagram of part of the sieve provided by the filaments and secondary folds of the tench. The section passes through three filaments and shows the secondary lamellae projecting alternately above and below the surface of each filament. The water flows at right angles to the plane of the page. (b) Diagram of two gill arches and the double row of filaments attached to each of them in a teleost fish. The tips of the filaments of adjacent arches are shown in contact with one another. [From Hughes and Shelton, 1962: Adv. Comp. Physiol. Biochem., *1*: 294.]

especially important during swimming when the mouth is open and water flows directly through it. Some fish (mackerel, for example) do not actively ventilate the gill chamber but, by continuously swimming, maintain a steady current over the gills. In fact, this may be a most efficient means of gas exchange since a constant rate of flow can be maintained over the gills by the degree to which the mouth is opened and since no energy is required to operate respiratory muscles. Salmon, for example, while swimming vigorously or holding position in rapidly flowing water, temporarily suspend active breathing and in this way achieve a high ventilation volume at low respiratory cost (Randall, 1970b); specialists in "ram" ventilation such as the tunas have lost the capacity for active breathing with structural changes in the gill filaments and lamellae (Muir and Kendall, 1968).

Branchial irrigation normally depends on the skeletal muscle of the jaws, gill arches, and operculum. The mechanics may be explained in terms of a series of pumps: a suction pump drawing water into the mouth, a pressure pump in the buccal cavity pushing water through the gills and a suction pump posteriorly drawing water over them (Fig. 4.4). As the mouth closes in the breathing rhythm, folds of mucous membrane on the inner surface of the jaws (the oral valves) come into position to prevent an outflow of water anteriorly. Pressure in the oral cavity rises, and as the space is constricted the water is forced back over the gills into the opercular cavity. This is the action of the pressure pump. Opercular movements are also involved. First, the abduction of the opercula takes place while the branchio-

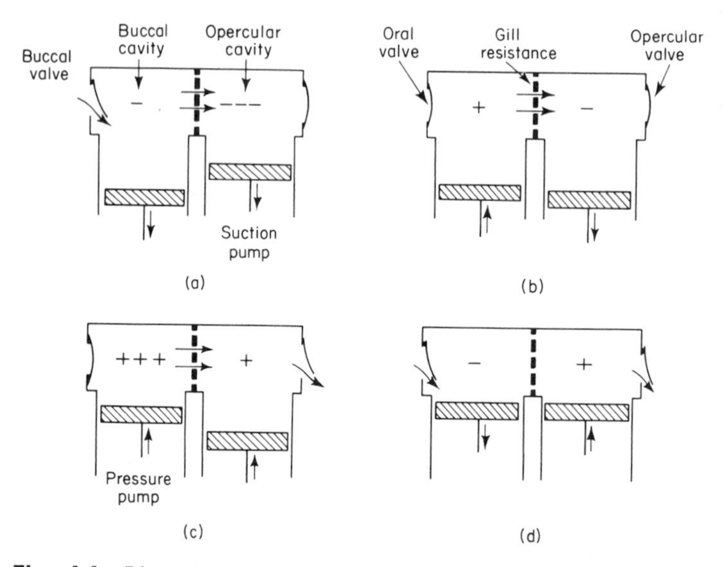

Fig. 4.4 Diagram illustrating the double pumping mechanism for ventilating the gills of fishes. The two major phases of the cycle are (a) in which the suction pumps are predominant and (c) when the buccal pressure pump forces water across the gills. The two transition phases (b) and (d) each take up only one-tenth of the whole cycle. [From Hughes and Shelton, 1962: Adv. Comp. Physiol. Biochem., *1*: 302.]

stegal membranes are closed and thus prevent the entrance of water posteriorly. In this way water is drawn from the oral into the opercular cavity (the suction pump). Then, with the adduction of the opercula, the branchiostegal valves open and the water flows out posteriorly. G. M. Hughes and his associates have carefully investigated the physiology of the respiratory muscles in several fishes, including elasmobranchs as well as teleosts (Hughes and Ballintijn, 1965, 1968; Shelton, 1970).

The breathing movements of a trout are recorded in Fig. 4.5. It is evident that the pressure is higher in front of the gills (oral cavity—pressure pump) than it is behind the gills (opercular cavity—suction pump) for all, or almost all, of the respiratory cycle. Thus, a gradient is maintained from mouth to operculum and, although breathing is rhythmical, the flow of water over the gills is continuous

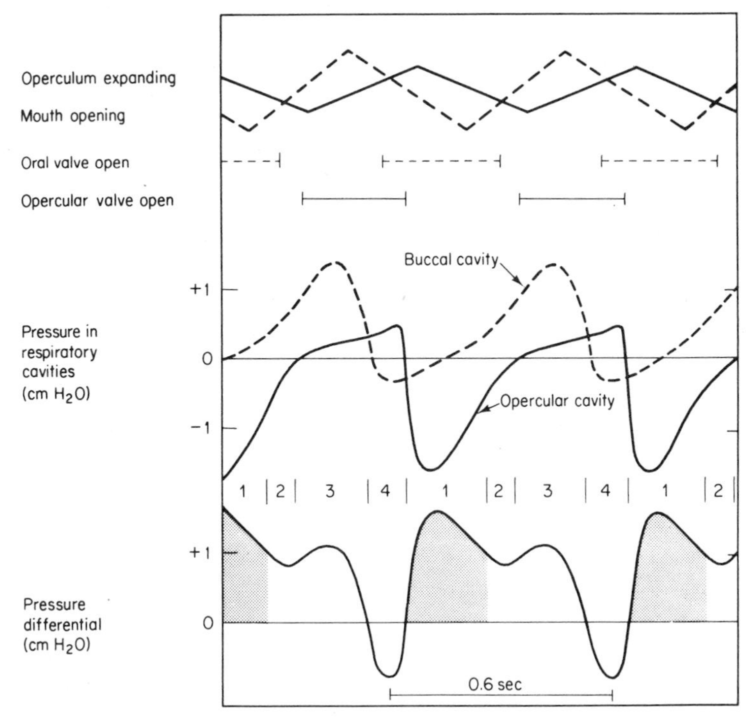

Fig. 4.5 Movement of the mouth and operculum with associated pressure changes in the buccal and opercular cavities of the trout. Broken lines, buccal side; continuous lines, opercular side of system; upper record, opercular expansion and mouth opening cycles which are out of phase; second pair of lines, relation of oral and opercular valve opening. The pressure changes are shown as separate records of actual observed pressures on either side of the gill resistance and below as the difference between the two sides; shaded areas show the periods when water flow is mainly due to the opercular suction pump. [From Jones, 1971: Comparative Physiology of Respiration. Arnold, London. After Hughes and Shelton, 1958: J. Exp. Biol., *35*: 812.]

(Hughes and Shelton, 1958; Saunders, 1961). Under different physiological and ecological conditions, the teleosts show many functional and structural modifications of the gill pumps. The amount of water pumped may be substantially altered by varying the frequency of the rhythm or the volume pumped per breathing cycle; the relative contributions of the two pumps vary in different species. Such bottom-dwelling forms as flat fishes and the dragonet rely much more heavily on the opercular suction pump than on the oral pressure pump (Shelton, 1970).

Gaseous exchange is still further improved in the teleost fishes by the counter-current flow of oxygenated water and deoxygenated blood. Counter-current exchangers find a use in several physiological operations and are also well known to engineers and laboratory workers. The principle is simple. Two channels in close proximity carry fluids in opposite directions. If the channel walls are freely permeable to any particular material and if the channels are long enough, equilibria will be established in the concentration of the permeable materials. The diagram in Figure 4.6 shows a hypothetical situation in which the numbers may represent concentrations of oxygen, sodium chloride, temperature or some other identity which moves freely through the wall. If deoxygenated blood were travelling in tube A and oxygenated water in tube B, then the maximum exchange of oxygen might be expected, and this is precisely the arrangement which occurs in teleost gills (Fig. 4.6). The system is extremely efficient, and in some teleosts 85 per cent

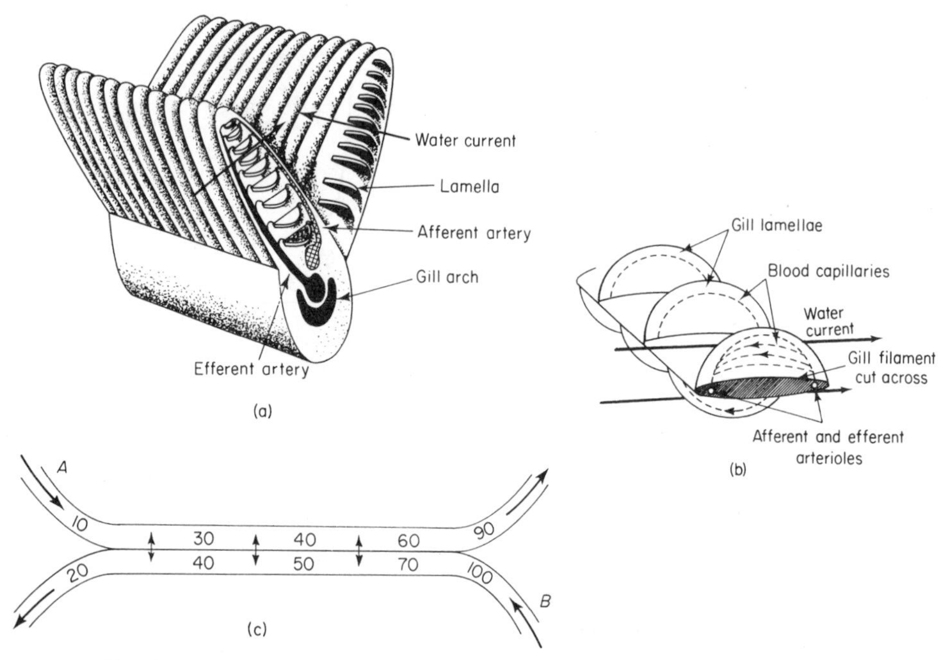

Fig. 4.6 Counter-current exchange system in the teleost gill. (a) Blood flow from the afferent to the efferent branchial arteries through the minute vessels in the secondary folds or lamellae as shown in (b). [From Hughes, 1961: New Scientist, *11*: 346.] (c) Diagram of a counter-current exchanger.

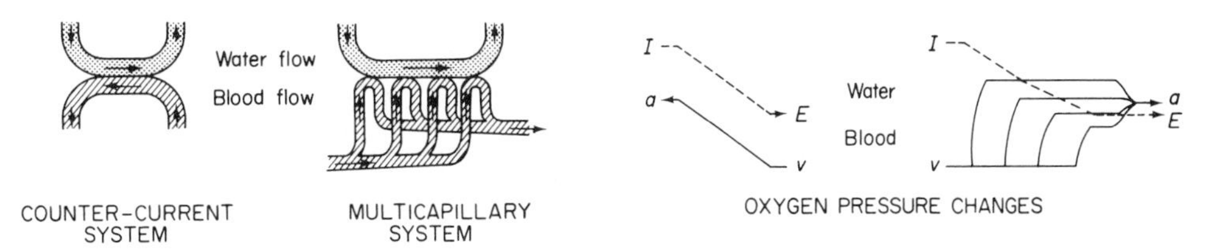

Fig. 4.7 Diagrams showing two systems facilitating oxygen exchange in fish gills and the changes in oxygen pressure of water flowing over the gills and blood flowing through them. *I*, inspired water Po_2; *E*, expired water Po_2; *a*, arterial Po_2; *v*, venous Po_2. [From Piiper and Schumann, 1967: Resp. Physiol., *2*: 146. Permission North Holland Publishing Co., Amsterdam.]

of the oxygen may be extracted from the water passing over the gills (Saunders, 1962). Exchangers of this type are found in the temperature control system of Arctic animals, in the swim bladder of fishes, and in the kidney. They will be referred to again when the physiology of these structures is considered.

Two functionally different circulatory paths have been described in the gills of several teleosts (Steen and Kruysse, 1964). The species represented are so widely separated taxonomically as to suggest that this is a universal feature of the teleost gill. In addition to the respiratory route whereby blood flows through the lamellae, there is also a system of non-respiratory vessels in the central portion of the filament. These companion vessels are probably comparable to the systemic vessels in other organs and serve a nutritive rather than a respiratory function (Gannon *et al.*, 1973).

Although counter currents of blood and water predominate in the teleost gill and also in the gill of the hagfish (Johansen, 1960), the elasmobranchs rely largely on parallel flows and achieve effective oxygenation of the blood by a serial multicapillary arrangement of vascular loops in intimate contact with the respiratory epithelium (Fig. 4.7). The multicapillary system is also the dominant mechanism in the mammalian placenta (Metcalfe *et al.*, 1967; Steen, 1971).

The oxygen uptake capacity of the gills differs in various species of fish and in relation to such factors as gill area, rate of ventilation, and the oxygen-carrying capacities of the blood and water. These factors and further comment are given in Chapter 12.

FROM AQUATIC TO AERIAL BREATHING

The aquatic environment was the first habitat of animal life. The more primitive members of all phyla are aquatic, and several phyla have no truly terrestrial representatives. However, all major phyla—including the purely aquatic ones such as the coelenterates and echinoderms—have representatives which use atmospheric air, even though they do so by way of their aquatic breathing organs. Air is the superior source of oxygen (210 ml per liter as compared to 5 to 10 ml per liter in fresh water), and if the additional hazards of terrestrial life (a much less stable environment, subject to desiccation) can be met, the potentialities of active life and aerobic living can be more completely realized. Carter (1931), in an outstanding review, finds that the evolutionary migration from water to land has occurred many times—more frequently than any other change which requires such profound modifications in animal organization. He describes the modes of aerial respiration in the different groups of animals and stresses the evolutionary trends.

There are two aquatic habitats where the ability to utilize atmospheric oxygen has distinct advantages, and these have probably been the geographic points of origin for the major terrestrial groups. One of these is the littoral area which is subject to periodic fluctuations in water level, through either tidal rhythms or seasonal droughts. The other is the area of stagnant water, especially the swampy, shallow tropical bogs and pools where water stagnates and where the dissolved oxygen reaches a low level. It is this latter environment which Carter believes was most forceful in the evolution of terrestrial groups.

Many animals make use of atmospheric oxygen and yet operate an aquatic mode of breathing. This may be done by living near the surface where there is relatively more dissolved oxygen; or the animal may actually come out into the moist air and still rely on cutaneous or gill respiration. Air may also be taken into a body cavity and mixed with the water circulating around the respiratory epithelia. For example, the sea cucumber *Holothuria tubulosa*, which is the only known echinoderm using atmospheric air, rises to the surface (when in stagnant water) and draws air into the cloaca where it is used to oxygenate the water in the respiratory tree. A few of the crustaceans and fishes rely on comparable mechanisms, and from this it is a short step to the development of a moist, vascular respiratory epithelium—usually in some area of the branchial cavity. The mantle of the gastropod, the epithelia of the branchial chambers of crustaceans and fishes, and parts of the gut may thus operate in the exchange of oxygen. These accessory developments are often elaborate and associated with distinctly altered mechanisms for ventilation.

Many of the physiological problems associated with life on the land are evident in the semiterrestrial vertebrates. The air-breathing fishes (Johansen, 1970) and the amphibians (Hughes, 1967a) display a spectrum of specialized adaptations in functional morphology and emphasize the acute problems of excretion (both carbon dioxide and nitrogen) which air breathing creates. When animals turned their backs on the aquatic environment it was necessary to find new avenues for both the acquisition of oxygen and the elimination of the inevitable wastes; air is a far superior source of oxygen but carbon dioxide and ammonia are extremely soluble and much more easily removed in water. The eel *Anguilla* provides a good example of some of the difficulties of the primitive air breather (Berg and Steen, 1965, 1966). On land the eel gulps air and holds it in the branchial chambers while the oxygen in this air gradually falls and the CO_2 rises; in this way about one-third of the oxygen required on land is obtained through the gills while the remainder diffuses through the vascular skin. Although eels make extended journeys onto land, acid metabolites gradually accumulate and about an hour may be required in water to pay off the debts of an excursion of 20 hours in air at a temperature of 15°C. Likewise, the teleost *Symbranchus marmoratus,* an inhabitant of stagnant swamps and rivers in South America, depends on its gills for air breathing. These creatures frequently crawl through the moist grass and regularly aestivate for several months in mud burrows during drought but their physiological problems are similar to those of the eel (Fig. 4.8). Unlike *Anguilla* and *Symbranchus,* most primitive air breathers rely on alternate areas of respiratory epithelium (gill chambers, stomach, air bladders) for oxygen acquisition while retaining the old routes for CO_2 elimination. Obligate air breathers like the lungfishes *Lepidosiren* and *Protopterus* normally absorb less than 10 per cent of their oxygen from the water but eliminate 70 per cent to 80 per cent of their carbon dioxide by the aquatic routes (Johansen, 1970).

Hughes (1967a) emphasizes the importance of integumentary respiration in the evolution of the land vertebrates. Periodic excursions to the surface for air can provide an abundance of oxygen but the elimination of CO_2 cannot be effective without regular ventilation. Thus, the first air breathers retained the old routes for the elimination of CO_2 and as the lungs developed and the gills were lost, the skin first took over the extremely important role of CO_2 excretion. Both the absence of scales and the production of mucus, which reduces water loss, are particularly evident in such fishes as eels and gobies which make excursions on to land. The phylogenetic pressures toward the development of a dry skin and regularly ventilated moist lungs are obvious and essential in the evolution of insulated warm-blooded vertebrates. Theoretical steps in the phylogeny of the machinery for gas exchange are shown diagrammatically in Fig. 4.9a. Carbon dioxide tensions are always higher

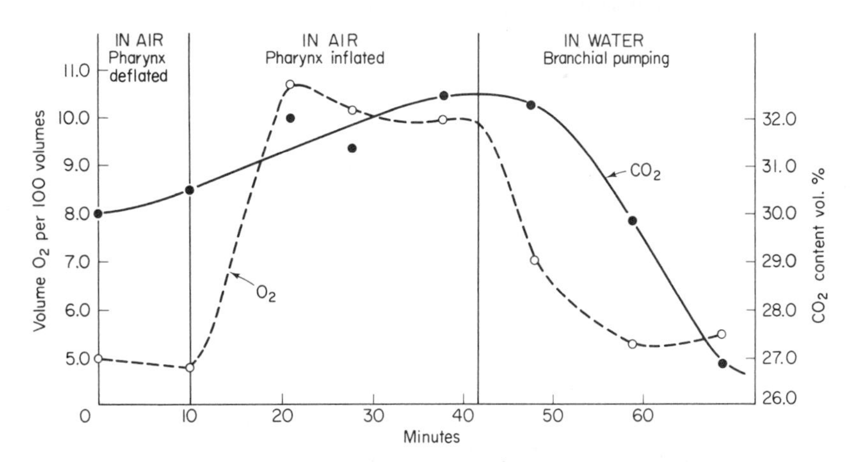

Fig. 4.8 Arterial oxygen and carbon dioxide content in *Symbranchus marmoratus* during air breathing and water breathing. The fish were initially out of water but later transferred back to the aquarium. [From Johansen, 1966: Comp. Biochem. Physiol., *18*: 388.]

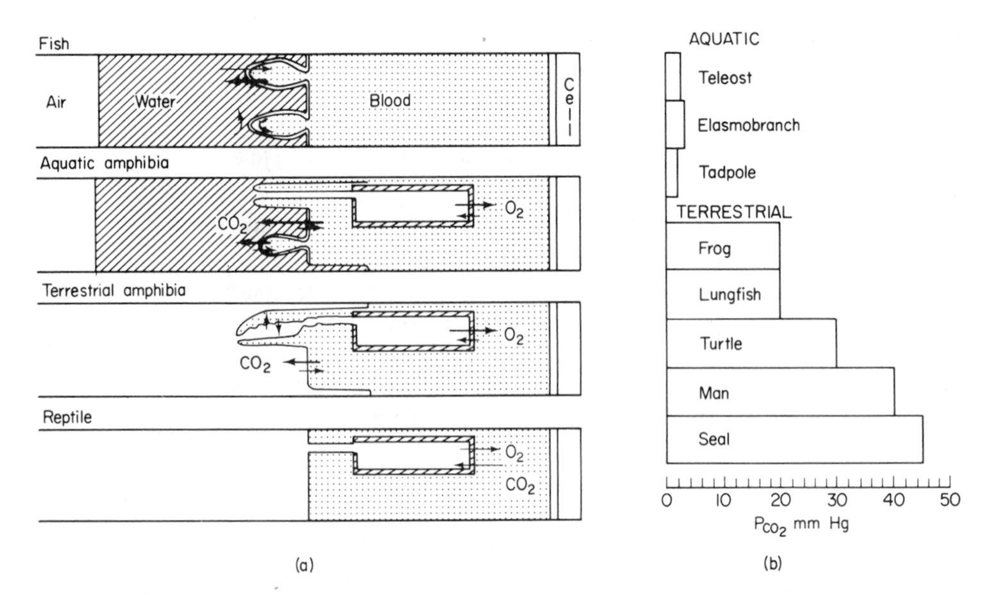

Fig. 4.9 Pathways for gas exchange in representative vertebrates at different stages in the phylogeny of air breathing. Main sites of oxygen intake and carbon dioxide release are indicated by arrows. [From Hughes, 1967a: *In* Development of the Lung. Ciba Foundation Symp. Churchill, London]. (b) Partial pressure (P_{CO_2}) in body fluids of aquatic versus terrestrial animals. [From Robin and Murdaugh, 1967: *In* Development of the Lung. Ciba Foundation Symp. Churchill, London.]

in the body fluids of terrestrial vertebrates than they are in the aquatic species (Fig. 4.9b); this is related to their high rates of metabolism as well as to modes of gas exchange. In Chapter 6 (page 202), this interesting topic will be considered again in relation to acid/base balance.

Cutaneous respiration, the oxygenation of water surrounding the gills, and the special adaptations of gills and gill chambers have never led to marked evolutionary advances in the air breathing machinery. With the possible exception of the gastropod mantle, it may be argued that the evolution of the advanced terrestrial groups has depended on the innovation of special air sacs (lungs) and air tubes (trachea). These are basic to the colonization of the land by the vertebrates and the insects.

LUNGS

A lung is a vascularized air sac. The exchange of gases between the environment and the body tissues depends on an intermediate circulating fluid. Morphologically, there is slight distinction between minute tubular lungs, such as the book lung of a spider (Fig. 4.10) or the tracheal lung of the chilopod *Scutigera*, and the tracheae of insects; but the gaseous exchange mechanisms are different, since oxygen

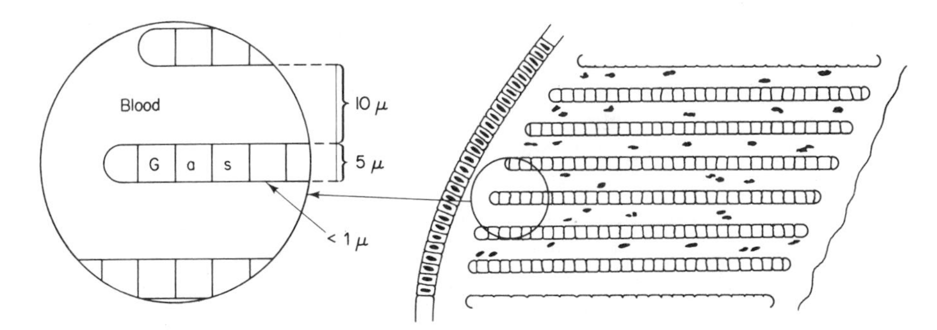

Fig. 4.10 Schematic drawing of the book lung of a spider. A series of blood-filled, parallel, thin plates are separated by thin layers of air. Measurements in left diagram are for the giant South American tarantula. [From Steen, 1971: Comparative Physiology of Respiratory Mechanisms. Academic Press, New York.]

is transported from lungs by a circulating fluid while it is piped directly to the tissues through tracheal tubes (Fig. 4.10 and 4.13).

Lungs, thus defined, are found in several of the terrestrial invertebrate groups: pulmonate snails, scorpions, and some spiders, chilopods and isopods. The efficiency of these lungs is restricted by the lack of a ventilating system; the to-and-fro movements of gases depend only on diffusion. In some forms the size of the lung aperture is altered in relation to activity, temperature and environmental carbon dioxide, but this provides the only control on the system. The movement of fresh air into the lung opening is so slight that none of the larger terrestrial animals, with their high rates of metabolism, depends on the diffusion type of lung.

Ventilation Lungs

Ventilation lungs are characteristic of the terrestrial vertebrates. Coordinated muscle movements create a rhythmic exchange of air, so that gradients remain relatively high in the air spaces even at considerable distances from the external environment. Embryologically, the vertebrate lungs arise from a median groove in the floor of the pharynx. Their morphological phylogeny is "none too certain" (Romer, 1970). Any sacculation of vascular epithelium, open to the air, would be useful to the emerging terrestrial vertebrates. It is probable that primitive fishes experimented with a variety of such structures. Among the present-day forms, highly vascular mucous membranes of the mouth and anterior gut, outpocketings and elaborate convolutions of the branchial cavities and sacs which may open dorsally, laterally or ventrally into the gut suggest the varied nature of this experimentation.

In a physiological sense, the phylogeny seems more evident. The amphibian lung, for example, merely supplements the cutaneous and oral exchange of gases, and the respiratory physiology of the successful terrestrial vertebrates has evidently evolved from this sort of beginning. Three interconnected trends are evident: progressive enlargement of the respiratory epithelium, specializations in the machinery for ventilation, and the provision of an efficient circulation. These trends reach a somewhat different climax in mammals and birds.

The mammalian lung Primitive air-breathing vertebrates possess sac-like lungs with internal walls which are occasionally smooth (some of the urodeles) but usually ridged to form air pockets that provide a modest increase in respiratory surface. In the more advanced reptiles (some lizards, turtles and crocodiles), the ridges or septa become intricately subdivided to convert the lung into a sponge-like structure with numerous respiratory air sacs or alveoli and a main central air tube running through it. This increase in the respiratory surface reaches a climax in the mammals. Minute, closely packed alveoli cluster around the alveolar ducts which lead into the respiratory bronchioles. In man there are an estimated 750 million of these alveoli with diameters averaging about 166 μ (Davson and Eggleton, 1968).

The limitations in this line of phylogeny are set by the increasing bulk of the lungs and the difficulties of effectively ventilating the tiny air sacs. For instance, even after a forced expiration, between 20 per cent and 35 per cent of the air still remains in the human lungs; this leads to an alveolar carbon dioxide tension of about 40 mm Hg which is greater than usually found in the most stagnant pond waters.

The bird lung The birds reduced the bulk of the lungs and eliminated the dead air spaces by the evolution of a special system of non-respiratory air sacs. Like several other avian structures, the lung of the bird is the most specialized organ of its type. Maximum differences in oxygen tension can never be achieved with the closed alveolar system of mammals. In the bird lung, the alveoli are replaced by air capillaries through which it is theoretically possible to circulate atmospheric air. The operation of this system depends on the presence of non-respiratory air sacs, connected with the lungs and serving as reservoirs. The morphological phylogeny of the air sacs can be traced from thin-walled, non-respiratory areas in the lungs of some reptiles. Their significance in the reptiles is unknown, but in the birds they have provided the basis for the evolution of the highly efficient respiratory system required in a flying homeotherm.

The physiological unit of the bird lung is a five- or six-sided cylinder about 1.0 mm in diameter, containing a central air passage—the parabronchus—with numerous radiating tubes—the air capillaries (Fig. 4.11a). These minute air capillaries (about 10 μ in diameter) coil and branch in intimate contact with the blood capillaries (Fig. 4.11b). In species with high respiratory demands such as pigeons, the air capillaries of one unit anastomose with those of other units; thus, the many different units may be interconnected. In less active species (chickens) connections are absent (Salt and Zeuthen, 1960), but even when absent, the maximum diffusion distances from the parabronchi are not more than 0.3 to 0.4 mm. At both ends the

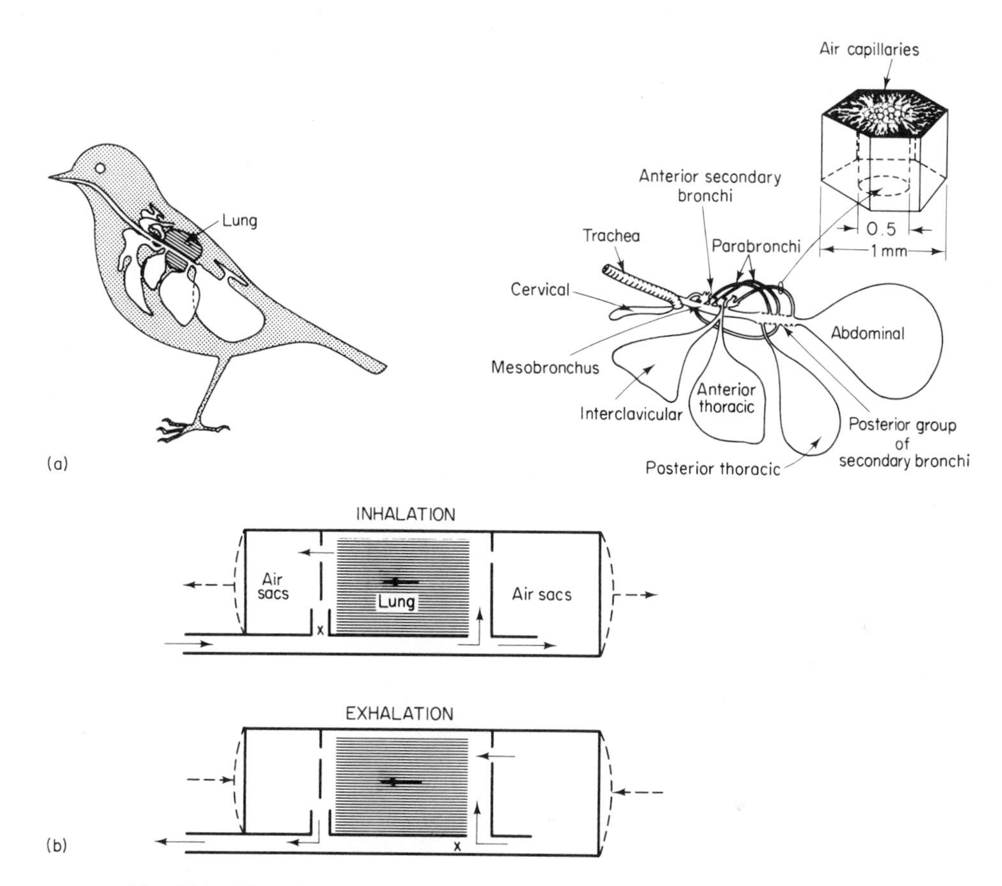

Fig. 4.11 The avian lung. (a) Main respiratory passages and air sacs from the left side with one of the units at upper right showing central parabronchus and the net of air capillaries which radiate from it. [Left diagram, from Schmidt-Nielsen, 1972: How Animals Work. Cambridge U. P., London, based on Salt, 1964: Biol. Rev., *39*: 125. Right diagram, based on Hughes, 1963: Comparative Physiology of Vertebrate Respiration. Heinemann, London.] (b) Air streams in bronchi, air sacs, and lung during breathing. Arrows show directions of volume changes; lung volume changes are small relative to those of the air sacs; lung volume diminishes during inhalation; the lungs expand during exhalation. [From Schmidt-Nielsen, 1972: How Animals Work. Cambridge U. P., London.]

parabronchi join the larger bronchial tubes, and these in turn connect with the air sacs or the outside world through the trachea (Fig. 4.11). The air sacs hold between 75 per cent and 90 per cent of all the air in the respiratory system.

There is a fundamental physiological difference between the tidal movements of air in the alveoli of the mammalian lung and the unidirectional stream passing through the parabronchi of bird lungs. This has long been recognized, but the precise mechanisms were speculative for centuries. Some of the most vexing problems have now been clarified through recent technical advances which permit the introductions of tracer "breaths" of pure oxygen and the precise monitoring of its flow in the major passages and the air sacs (Schmidt-Nielsen, 1971, 1972).

The system is shown diagrammatically in Fig 4.11b. As indicated in this diagram, the air sacs form distinct anterior and posterior groups with respect to the parabronchi. An important morphological correlate, not shown, is the presence of hinged ribs and sternum (Hughes, 1963; Schmidt-Nielsen, 1972) which effect large changes in thoracic volume during breathing. Another essential correlate is the long trachea, which creates an important dead air space; in birds such as swans and cranes, the trachea is an elongated, coiled, S-shaped tube in the sternum. How then does this system work to the advantage of the bird and why have these flying machines capitalized on it while bats, which are also very good flyers, do quite well with mammalian type lungs?

In outline, inspired fresh air—together with the "dead" air in the trachea and bronchial passages—moves directly into the posterior sacs with only small amounts passing into the lungs posteriorly *and none* going into the anterior sacs. Both groups of air sacs expand during inhalation while the lung volume diminishes; this means that some air moves from the lungs into the anterior sacs during inspiration. However, the main flow of air through the lungs occurs during exhalation when the thoracic volume decreases; at this time, air flows from the posterior sacs into the parabronchi and on through and out of the anterior sacs. Thus, in contrast to mammals, air passes through the lungs mainly during exhalation. It is still not clear how the flow is regulated at points marked x in the diagrams; the architecture of the air channels or the action of the musculature may be important. Recent evidence favors a counter-current flow of fresh air and blood to effect maximum gas exchanges. The tracheal dead space of expired air is important in preserving carbon dioxide levels in the blood. If only fresh air flowed over the respiratory surfaces, the blood CO_2 would soon be almost completely depleted with adverse pH and other changes; the importance of CO_2 in physiological regulation has already been mentioned. The second question raised in the previous paragraph concerns the evolutionary pressures which may have shaped the bird lung. This topic is discussed in Chapter 12 where it is related to the problems of high-altitude flying.

Mechanics of ventilation Studies of lungfishes and amphibians indicate that the air breathing machinery of land vertebrates evolved from the buccal force pump of fishes (McMahon, 1969). A frog fills its lungs by air compression rather than suction as in mammals. A sequence of muscular activities has been described (Gans *et al.*, 1969): (1) with nares open and glottis closed, the buccal cavity is flushed out with fresh air by an oscillating series of contractions of the buccal floor; (2) less frequently, the glottis opens *while the buccal floor is compressed* and the lung gases (which are at pressures above atmospheric) are blown out through the nares; (3) the nostrils then close while the glottis remains open and the buccal floor rises to force air into the lungs at pressures above atmospheric; (4) the glottis then closes, the nares open and the cycle is repeated. Relatively little mixing of gases occurs when air is expelled rapidly through the dorsal part of the buccal chamber.

Since the buccopharyngeal cavity is lined with a moist, vascular epithelium, the buccal oscillations may be expected to contribute to the exchange of gases. This may be especially significant at higher temperatures (Whitford and Hutchinson, 1963). In *Ambystoma*, the volume of air moved through the lungs increases about three-fold between 10° and 25°C, while that moved through the buccopharyngeal cavity increases twenty-five-fold.

In the higher vertebrates the breathing action is that of a suction pump (negative pressure mechanism) rather than a force pump. The situation is best illustrated by the mammal where the lungs are enclosed in the pleural cavities bounded by the thoracic cage and the dome-shaped diaphragm posteriorly. Inspiration is the active process. Impulses originating in the respiratory center of the brain activate the intercostal muscles and the diaphragm. The dome-shaped diaphragm flattens, the ribs are elevated, the space is enlarged and air flows into the lungs. As the muscles relax and the space is restricted, the elastic recoil of the lungs results in a passive expiration of the air.

Circulation combined with ventilation In birds and mammals a double heart provides for a separate pulmonary circulation followed by a systemic circulation of the blood. In this way fully oxygenated blood is supplied to the tissues. The branchial system of fishes also provides fully oxygenated blood to the tissues but, with only a single pump for both respiratory organs and body tissues, the blood pressures are greatly reduced in the oxygenated supply lines. The comparative anatomy again suggests that the evolving terrestrial forms tried several arrangements before the appearance of the double heart. In a few species (*Saccobranchus*, Fig. 4.12) the air-breathing organs are in parallel with the gills. In theory, this should be quite satisfactory for the amphibious way of life if there is a suitable mechanism for changing from one scheme to the other in association with the habitat. In the lungfishes, an almost completely double circulation is established with two atria, a partial interventricular septum and a twisted valve in the conus (Foxon, 1955; Johansen and Hanson, 1968). There is no interventricular septum in the amphibians but a limited separation of venous and arterial blood is effected by the anatomical arrangement of the principal vessels leaving the heart. Reptilian hearts have a ventricular septum, but even in the most advanced species, a small gap is found in it near the arterial trunks (Romer, 1970). A complete anatomical separation of the ventricles occurs only in the birds and mammals. The phylogeny of the ventricular septum is uncertain (Foxon, 1964).

TRACHEAE

The tracheal system of the terrestrial arthropods serves the dual functions of bringing air into the body and distributing it to the cells; consequently, the respiratory functions of the transport system are no longer required. The pattern of

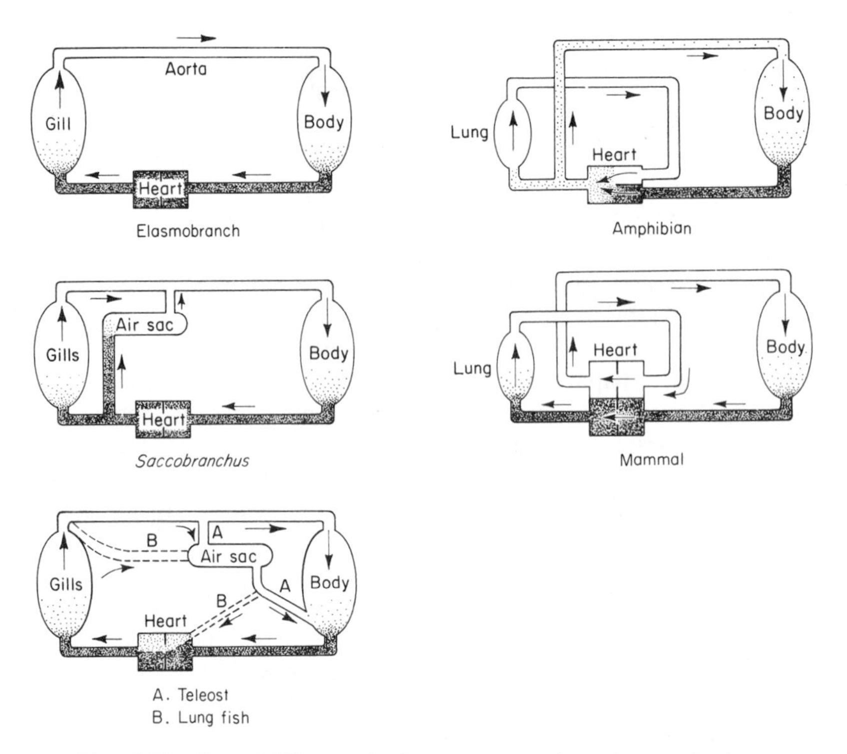

Fig. 4.12 Several different circulatory patterns of vertebrate animals.

tracheae and air capillaries (tracheoles) is similar to the pattern of blood vessels in other highly successful groups of animals; although, of course, there is not a complete circuit within the system of air tubes. Carter (1961) likens this system of air tubes to a highly divided lung, a lung in which the alveoli have extended throughout the intercellular space as a ramiform network of tubes; Weis-Fogh (1964) compares its functional design in flying insects with the avian lung.

Air enters the tracheal system through the spiracles. In primitive tracheates (*Peripatus*, some Apterygota), the apertures are permanently open, and there is no control of the movement of air through them. In most forms, however, the spiracles are provided with valves of one type or another, operated by muscles and sometimes provided with filters (Miller, 1964).

The tracheal tubes which lead from the spiracles develop as invaginations of the body surface, and their walls have the general structure of the integument. In the larger tracheae, thickenings (taenidia) in the cuticular layers form a chitinized spiral which permits a stretching of the tubes but prevents their collapse; taenidial support is also present in the smaller tubes although complete helices are absent. The walls of the tracheae become progressively thinner as the tubes become smaller. At a diameter of 2 to 5 μ, the tracheae pass into the air capillaries or tracheoles. These minute channels, less than 1 μ in diameter, are the physiologically important units in gas exchange, even though it seems unlikely that this process is confined entirely to them. Wigglesworth (1930) described a to-and-fro movement of fluid in the terminal tracheoles, associated with the activity of the organs which they supply (Fig. 4.13). His observation has often been confirmed, although the precise mechanism is still debated. Wigglesworth ascribed it to the formation of tissue metabolites during activity and the withdrawal of water osmotically from the tubes into the interstitial spaces. These minute air capillaries make numerous and intimate contacts with the cells. Contrary to older views, they do not enter the cells but only indent or sink into the plasma membranes of cells with high oxygen demands. In this way, tracheoles may become functionally, if not morphologically, intracellular and gaseous oxygen is brought very close to the mitochondria (Miller, 1964, 1966a).

The phylogeny of the tracheae is no more certain than that of the lungs. In this case it is probable that the emerging terrestrial forms experimented with several simple arrangements of air sacs and tubes. Living forms testify to the probable variety. In *Peripatus*, spiracles, which are permanently open, lead into pits which communicate with bundles of minute tracheae. In some of the terrestrial isopods (*Porcellio, Armadillidium*) tuft-like invaginations of the integument of the exopodites of the pleopods form branching air tubes known as pseudotracheae. Among the arachnids a series of forms suggests an evolution of lung books in the terrestrial scorpions and some spiders from the gill books of the aquatic arachnids (*Limulus*), and subsequently a replacement of the lung books by tracheae. Almost every combination of lung books and tracheae is found among the spiders, from species which depend on lung books to species entirely dependent on tracheae (Kerkut, 1963). However, even among the arachnids some groups (Acarina) are not in this series

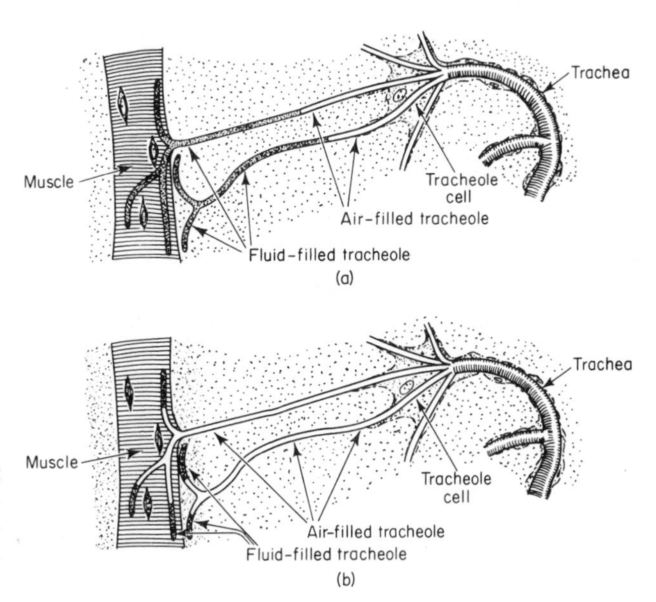

Fig. 4.13 Diagram of movement of liquid in the tracheoles. (a) Resting. (b) Active. [From Wigglesworth, 1930: Proc. Roy. Soc. London., *B106*: 231.]

and show a different kind of tracheal system, while among the insects, where the tracheal system attains its most complete development, there is no evidence of an evolution by way of the lung books. It appears that ectodermal invaginations have formed air tubes in many different anatomical locations, and the segmental ones of the evolving insects were most adaptable. The range of adaptation is spectacular. The tracheal system copes not only with the requirements of sluggish larvae but also with the excessive demands of small flying machines; at the same time, it has been modified for cutaneous respiration and utilized in the formation of beautifully elaborate gills in many aquatic species (Hinton, 1968; Wigglesworth, 1972).

Ventilation of the Tracheal System

There is no active ventilation in many of the tracheates. The Onychophora, Myriapoda, tracheate Arachnoidea, and the larvae and pupae of all insects depend on simple diffusion for the exchange of gases in the air tubes (Krogh, 1941). Calculations have shown that this is adequate in small and less active forms. However, in the adult insects which sometimes have very high rates of metabolism, there is ventilation of the system and control of the volume of air and the direction of its flow through the system. Three specializations will be considered: the spiracular control, the unidirectional flow of air and the action of air sacs.

Spiracular control is usually present even when there is no ventilation. The hazard of a fully open system of spiracles is evident in *Peripatus*. Although this animal has a dry and much less permeable skin than the earthworm, it loses moisture twice as fast through its open spiracles scattered over the surface of the body—several in each segment (Kerkut, 1963). Moreover, *Peripatus* loses water forty times as fast as a caterpillar which is also unable to ventilate its tracheae but is able to control its spiracles. A DIFFUSION CONTROL of respiration by the opening and closing of the spiracles is thus effective in the important compromises between water loss and the requirements for air. Wigglesworth (1972) described it in the flea. At rest only two pairs of spiracles operate. They open and close rhythmically at rates which depend on the temperature. If the animal is active, eight pairs of thoracic spiracles remain open until the activity has ceased and the metabolism returns to the resting condition. With a less forceful stimulation of the metabolism, such as occurs during digestion or while the eggs are ripening, the two pairs of spiracles may remain continuously open and the others may rhythmically open and close or, if metabolism so demands, remain open all the time. Thus, although this animal is unable to ventilate its system, there is an adequate control for the conservation of water and the regulation of the air supply in accordance with demands. Spiracular control is regulated peripherally by CO_2 and centrally by nervous mechanisms (Miller, 1964, 1966a).

Active ventilation is brought about through movements of the body walls which rhythmically compress the air spaces and force the air out of them. There are several different kinds of movement but the most common are the dorsoventral flattening of the abdomen (grasshoppers, beetles) and the telescoping movements of the abdominal segments (bees and flies). In the insects, contrary to mammals at rest, expiration is the active process; as the terga and sterna are pulled together or the segments are telescoped, the pressure of the rigid exoskeleton on the soft organs of the body forces air from the tracheal system. With the reverse movement, air again flows into the system.

In many forms the flow of air is unidirectional and thus the dead air space is reduced to a minimum. In the grasshopper, for example, it has been demonstrated that the thoracic spiracles are used primarily for inspiration while the air flows out through the abdominal spiracles. However, whether the flow is tidal or streams through the system, only the larger trunks are actively ventilated, and the smaller branches always depend on diffusion and the movements of fluids in the tracheoles.

The efficiency of ventilation is greatly improved in some species by the air sacs. These are balloon-like dilations of the trachea which range in size from tiny vesicles among the leg muscles of some insects to huge sacs in the main air tubes of the Hymenoptera. Their importance in ventilation is unquestioned. As in the birds, air sacs in the insects may serve functions other than respiration, but their primary activity is associated with ventilation (Miller, 1964).

REGULATING MECHANISMS

Sluggish and sedentary animals may temporarily suspend gaseous exchange and live anaerobically. Their rates of metabolism are low and their tissues capable of tolerating high concentrations of acid metabolites. Bivalves, tube-dwelling worms, and a host of invertebrates which live along the sea shores have margins of safety and discontinue the exchange of gases at low tide. Ventilation may be elevated for three or four hours after the flood while waste metabolites are eliminated and a supply of oxygen is accumulated for the next emergency (Chapters 5 and 12). In contrast, active animals such as the vertebrates, many of the arthropods and the cephalopods cannot withstand anoxia for even a brief period and have special machinery to insure a steady flow of water or air over the respiratory epithelium.

It has already been emphasized that the flow of gases *through* the epithelia is a matter of simple diffusion and depends on rapid distribution to the tissues from the internal surface and steady renewal of the air or water on the external surface. Regulation on the tissue side is considered in later chapters. The control of ventilation is discussed here.

Respiratory currents depend on either cilia or skeletal muscles. The former, when used for ventilation, operate mechanisms which provide food as well as oxygen. Unless conditions become intolerable and the animal suspends most of its activities, they work continuously and automatically. A nervous control of ciliated epithelia has been described in several organs (Chapter 19), and the activities of these cilia-driven respiratory currents may also be under nervous control. Their responsiveness to transmitter substances such as acetylcholine and 5-hydroxytryptamine suggests this (Gosselin, 1961). However, a large element of automaticity is certainly present, as evidenced by prolonged and continuous activity of small pieces of ciliated epithelium (lamellibranch gill) when removed from the animal and completely divorced from central nervous control.

The situation is quite different when ventilation depends on skeletal muscle. Whether the animal combines the activities of feeding and ventilation by creating water currents with its appendages (*Branchipus*, *Daphnia*), ventilates a tube or burrow (*Arenicola*), pumps water over gills, or moves air into lungs or tracheal tubes, the muscles involved lack automaticity and depend on a flow of impulses from neurons in central ganglia or nerve cords. The phylogeny of oscillatory activity in central neurons which control the respiratory muscles seems to be fundamental to all these controls. Such pacemaking systems have been described in many different animals.

The Pacemakers of Ventilation

The rhythmic ventilation of the *Arenicola* burrow exemplifies a relatively simple situation found in some marine burrowing forms. Wells (1949) traced the rhythmic activities of these animals in U-tubes of sea water by a float connected with a writing lever. The irrigation of *Arenicola's* burrow is caused by the wave movements in the musculature of the trunk. Wells (1950) also noted rhythmic contractions when he removed intact worms from their tubes, pinned them on cork plates submerged in sea water and recorded the worm's movements by a hook passing under the middle of the body and attached to a light lever. Bursts of activity corresponded to the irrigation cycles. Similar results were obtained with longitudinal strips of the body wall containing the ventral nerve cord; the brain was not necessary. Wells concluded that there is a pacemaker in the ventral nerve cord and that this controls the rhythm, although its action can be modified, like that of the heart, by a number of conditions.

In the insects also, physiologists have obtained evidence for pacemaking neurons concerned with the rhythm of ventilation (Miller, 1966a). Inhibition of specific breathing movements occurs when certain nerve ganglia are sectioned; stimulation of specific ganglia modifies breathing, while the rhythmic changes in action potentials of certain ganglia correspond to the respiratory activities. Experiments of this sort, with several species of insects, have shown the presence of a segmental control of the spiracles and respiratory rhythm with higher coordinating centers (pacemakers) in the thoracic or anterior abdominal ganglia. It has also been shown that the isolated nerve cord produces rhythmic bursts of action potentials which occur with the same frequency as the normal resting movements of ventilation (Fig. 4.14).

It has long been known that the hindbrain of the vertebrate contains groups of neurons which are essential for rhythmical breathing. Lesions in various parts of the brain show that the breathing centers are located in the medulla. In the skate, for example, only the medulla is involved in motor control. There are bilateral

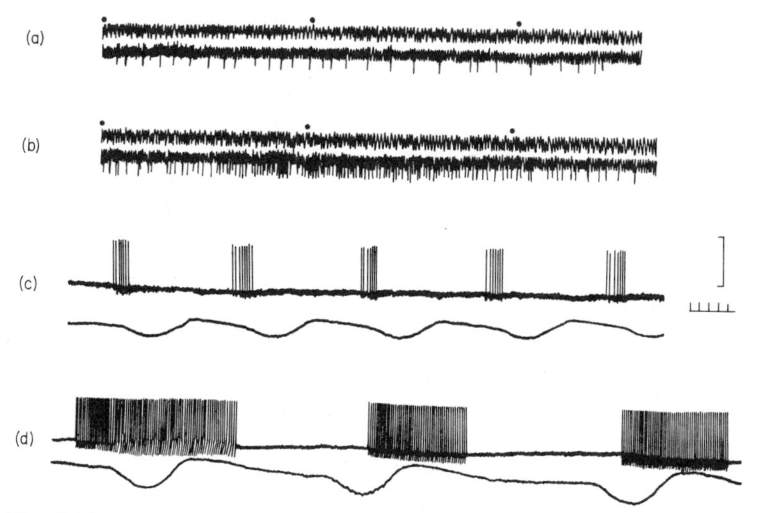

Fig. 4.14 Nervous control of ventilation. (a) Oscilloscope records from the isolated second abdominal ganglion of the desert locust; (b) same after blowing CO_2 at the ganglion; time, 50 cycles/sec (upper trace) and 1.0 sec (dots). [From Miller, 1960: J. Exp. Biol., *37*: 233.] (c) and (d) Rhythmic bursts of action potentials from respiratory neuron (c) and reticulomotor neuron (d) in *Squalus acanthias*. Upper traces, cell discharge; lower traces, respiration; upper movement of the beam = inward movement of first gill flap. Calibrations: time = 100 msec; amplitude = 10 mV. [From Satchell, 1968: Comp. Biochem. Physiol., *27*: 836.]

groups of neurons; the two sides operate independently with control for the spiracle and first gill arch clearly separated from that which regulates the last four gill arches (Healey, 1957). Similarly, in the mammal there are several centers or subcenters in the hindbrain (the pneumotaxic and apneustic center in the pons and the medullary centers) concerned with ventilation (Ruch and Patton, 1965). Lesions in these different areas show that their neurons are involved in the regulation of breathing.

Further evidence of centrally located pacemakers for respiration is found in the electrical activity associated with these neurons. More than forty years ago, Adrian recorded rhythmic action potentials from the isolated hindbrain of the goldfish and found that this corresponded to respiratory movements (Adrian and Buytendijk, 1931). In anaesthetized fish it has also been possible to insert microelectrodes and record oscillating potentials which correspond to breathing movements. These early experiments have been carefully repeated in recent years (Fig. 4.14), and the results are consistent with the earlier findings of an autonomous activity in the respiratory neurons of teleost fishes (Healey, 1957; Hughes and Shelton, 1962; Shelton, 1970). Numerous experiments of the same sort have been carried out on the higher vertebrates.

Thus, evidence from several lines and many different groups of animals indicates that ventilation depends on autonomous pacemakers in groups of central neurons. The mechanism which generates this rhythmic respiratory activity is uncertain although several theories have been considered (Ruch and Patton, 1965; Miller, 1966a; Shelton, 1970).

The activities of these pacemakers are modified by reflexes, set in action by sensory impulses arising primarily from chemical and pressure stimuli. In addition, both pacemakers and the effector organs of ventilation (cilia or muscles) sometimes show direct responses to thermal or chemical changes in their surroundings.

Chemical Regulation of Respiration

Structures concerned with ventilation have evolved in response to the demands for oxygen and the necessity of removing waste metabolites, particularly carbon dioxide. It follows then that levels of oxygen and carbon dioxide (through feedback mechanisms) provide reliable cues for the regulation of ventilation. This has been abundantly demonstrated in a wide variety of animals at many different levels in phylogeny. Only a few examples can be given here.

Tubifex, a minute, thread-like oligochaete worm, shows a series of interesting behavioral responses to changes in the gas tensions of its environment (Krogh, 1941). These small animals and some of their immediate relatives live in the mud at the bottom of stagnant pools. With a depletion of the oxygen they extend their bodies by more than ten times and, by wriggling, improve the circulation of the surrounding water. Some species avoid asphyxiation in extreme conditions by swimming toward the surface. Many other animals make equally effective responses. The bivalve pumps more water through its mantle cavity; the movement of the respiratory appendages of the crustacean is more rapid; the pumping action is augmented in the tracheal system of the insect; ventilation of the gills of fishes and the rate of breathing of birds and mammals have all been shown to change in relation to the oxygen and carbon dioxide tensions of the environment. Many curious examples are known. The male three-spined stickleback (*Gasterosteus aculeatus*) for example, incubates the eggs by creating a water current through his nest with a rapid movement of the pectoral fins (fanning) and the intensity of this fanning activity is dependent on the CO_2 produced by the developing embryos.

These varied responses are appropriate to the life of the animal but, in detail, may be quite different even in closely related species. Thus, although carbon dioxide frequently stimulates respiratory activities, it may be without any effect as in some of the crustaceans (Wolvekamp and Waterman, 1960), or it may depress respiratory activity. When the bivalve is forced to close its shell and when acid metabolites accumulate in the tissues, the cilia—on which respiration depends—become inactive. This is a characteristic response of cilia to lowered pH and provides an appropriate adjustment in an animal which must conserve energy when its filtering machinery is not operating.

Even in the same animal entirely different reflexes may sometimes appear with changes in tensions of the same gas. *Erythrinus unitaeniatus*, a fresh-water fish found in swamps of northern South America, provides an interesting example (Carter, 1957). This animal depends on its gills when the water is well-oxygenated but can use its air-bladder as a lung and breathe at the surface when the water is stagnant. Its complete oxygen demand may be met from the air or the water, but under certain conditions both sources of oxygen are used, and the animals make periodic excursions to the surface so that the air sac supplements the gills (intermediate respiration). Willmer (1934) was able to modify the breathing behavior experimentally. If the oxygen of the water fell below 1.5 ml per liter, the gill opening closed and the animal became a terrestrial breather; this same response was obtained when the CO_2 was elevated above 35 ml per liter. These reactions are obviously appropriate but Willmer also found that the animal closed its gill openings and breathed air when the CO_2 tension fell below 5 ml per liter. He concluded that a certain minimum level of CO_2 was necessary for activity of the respiratory center (as it is in the mammal) and that if the blood becomes too alkaline there is no excitation of the branchial breathing apparatus and the animal uses air. Thus, it is evident that the changing levels of CO_2 in the blood of this animal not only control the rate of respiration but also control a series of reflexes which signal gill breathing at certain levels and lung ventilation at others.

The carotid and aortic bodies Although the responses to changes in oxygen and CO_2 are well known, the details of the controls have not often been analyzed. These controls are best understood in the mammal, but even here there are still some uncertainties after years of careful study (Comroe, 1964; Porter, 1970).

If one rebreathes a small volume of air by holding a tight rubber bag around the mouth and nose, ventilation is soon noticeably accelerated. When the CO_2 rises to 3–5 per cent and the O_2 falls to about 17 per cent, this increase is marked. If the expired CO_2 is absorbed by some agent such as soda lime, then the effects are not noted until the O_2 falls to about 14 per cent; if the oxygen tension is kept at the normal level of 21 per cent or higher but CO_2 is permitted to accumulate, then hyperpnea (increased ventilation) again develops as in the rebreathing of normal air. Simple experiments such as these long ago demonstrated the excitatory nature of CO_2 on the rate of breathing, and it was concluded that this response was of more significance in removing excess CO_2 than it was in meeting an oxygen emergency (Haldane, 1927; Haldane and Priestley, 1935). The importance of oxygen in the regulation of respiration was not fully appreciated until the classical experiments of Heymans and his associates focussed attention on the carotid and aortic bodies as chemical receptor areas of great significance, not only in ventilation but also in cardiovascular responses (Heymans and Neil, 1958). In 1938, Heymans was awarded a Nobel prize for these discoveries.

The carotid bodies of the mammal are small nodules of vascular and neurosensory tissue (diameter, 2–5 mm in the human adult) lying near the fork of the common carotid artery, supplied by carotid blood and receiving fibers from the cervical sympathetic, the glossopharyngeal and vagus nerves. The aortic body situated near the arch of the aorta is similar in structure and function. These bodies have not been found in the fishes and amphibia, but homologous cells have been described in some reptiles; the carotid body is well known in birds although its location differs slightly from that of the mammal (Adams, 1958). Embryologically, these tissues arise from the aortic arches and their functional phylogeny may be based on receptors in the branchial vessels of fishes.

If the carotid and aortic bodies are perfused with a physiological saline solution, predictable effects are noted with alterations in the oxygen and carbon dioxide tensions or in the pH of the perfusion fluid. Ventilation increases markedly with lowered oxygen tension and, to a lesser degree, with elevated CO_2 and depressed pH. These responses disappear with denervation of the perfused bodies, thus indicating the reflex nature of the respiratory responses. This is also evidenced in the changing pattern of action potentials of the intact nerves in response to alterations in the perfusate. The receptors are particularly sensitive to oxygen, and although responses to elevated CO_2 and depressed pH can be demonstrated, this is only possible at relatively high levels of acidity. Whereas the respiratory centers of the mammal often respond to CO_2 changes of as little as 1 mm Hg partial pressure or even less, it requires something of the order of 10 mm to excite the centers by way of the peripheral chemoreceptor system. Thus, in the intact animal, the carotid and aortic bodies contain the important receptor cells for lowered oxygen, while the effects of elevated bicarbonate ion and other acid metabolites act directly on the centers in the medulla.

Peripheral chemoreceptors have been described in other groups of vertebrates, but the details are not well known. In several fishes increased respiration has been recorded when the CO_2 rises or the O_2 decreases in the water flowing over the gills, but neither the peripheral receptors nor the reflex pathways have been identified. The literature has been reviewed and the problem carefully discussed by Hughes and Shelton (1962) and Shelton (1970).

The emerging terrestrial vertebrates must have found CO_2 receptors increasingly more reliable as detectors for the regulation of breathing. In the aquatic environment, carbon dioxide dissolves rapidly into the water and the blood CO_2 levels are always low (Fig 4.9). Primitive air breathers, such as the eel, are insensitive to changing CO_2 while exposed to air but increase their ventilation sharply if the oxygen is lowered and cease breathing in pure oxygen. Likewise, the urodele *Amphiuma*, which lives an aquatic life but must surface at intervals (about once per hour at 15°C) to inflate its lungs, is relatively insensitive to manipulations of the CO_2 content of its lungs; it can, however, be readily induced to surface more frequently by lowering its lung oxygen and to remain submerged longer than normal either in an atmosphere of pure oxygen or by artifically introducing fresh air into its lungs (Toews, 1969). Terrestrial forms, however, lack ready means of removing carbon dioxide; blood (and lung) CO_2 levels are higher and relatively constant; small changes provide reliable cues for breathing.

Mechanoreceptors and Ventilation Reflexes

In 1868, Hering and Breuer, two Austrian scientists, reported an inhibition of respiration with distension of the lungs. This reflex, called the Hering-Breuer reflex, was carefully analyzed a few years later by Head, using a slip of rabbit diaphragm which can be freed sufficiently to record its contractions without disturbing the circulation or innervation (Porter, 1970). His experiment is interesting. It will be recalled that ventilation in the mammal consists of an inspiration involving contraction (hence posterior movement) of the dome-shaped diaphragm and an elevation of the ribs to enlarge the thoracic cavity; air flows into the lungs. Head found that the contractions of the strips of rabbit diaphragm were inhibited by distension of the lungs; alternatively, as long as the lungs were collapsed the diaphragm muscle remained contracted. In short, there is a reflex control of breathing which depends on stretch receptors in the lung; when stimulated, these discharge volleys of vagal impulses to the inspiratory centers of the medulla, and inspiration is inhibited. Action potentials in the vagus have been recorded in relation to this stretching of the lung. A rhythmic control is maintained with an excitation of the respiratory muscles which are periodically inhibited by the Hering-Breuer reflex. In addition to the pressoreceptors of the lungs, there is a system of stretch receptors in the adventitia of the aorta, common carotids and carotid sinus which plays a prominent part in the regulation of blood pressure but also modifies respiratory rhythm. The carotid sinus system is discussed with the control of mammalian circulation in the next chapter. Here it is noted that respiration is adjusted with blood pressure by means of this system of tension receptors.

Such detailed investigations have not yet been carried out on other animals, but there is evidence of comparable mechanisms in several different groups. Stretch receptors, associated with the Hering-Breuer type of reflex, have also been described in some fishes and higher invertebrates. Work on the dogfish indicates that the inflation of the pharynx reflexly inhibits inspiration. The experimental evidence comes from sectioning or stimulating the IX, X, and prespiracular branch of the VII cranial nerves, from inflation of air-filled balloons in the pharynx and from the recording of action potentials from the branchial nerves (Satchell, 1959; Satchell and Way, 1962). Finger-like (branchial) processes lining the internal openings of the gill pouches are richly supplied with proprioceptors responsible for the sensory discharge via the vagus to inhibit respiration. Satchell (1960) also described a coordination of the respiratory and cardiac rhythms (branchiocardiac reflex) which seems to adjust the flow of blood in the gills so as to effect maximum oxygenation. A similar relationship has been described in the teleost. Hughes and Shelton (1962) and Shelton (1970) discuss the physiological significance of this reflex.

The extent of lung inflation in the primitive air-breathing amphibian *Amphiuma* is also controlled by a volume detection mechanism in the lungs (Toews, 1969).

Ventilation of the mantle cavity of the cephalopod is again a highly coordinated process depending on rhythmic contractions of skeletal muscle. Neurons are located in the palliovisceral ganglia and a well-defined nervous control, including a giant fiber system, regulates the contractions of the mantle. The anatomy has been carefully described and a Hering-Breuer type of reflex has been shown to operate (Ghiretti, 1966b). Crustaceans, both aquatic and terrestrial, have been investigated and shown to alter their ventilation rates in response to hypoxia (Johansen *et al.*, 1970; Taylor *et al.*, 1973; Taylor and Butler, 1973) and hypercapnia (Cameron and Mecklenburg, 1973). Receptors have not been identified. The literature is cited in these recent papers.

In summary, more than a hundred years of careful study of mammalian ventilation has demonstrated a series of pacemaking neurons in the hindbrain which are indispensable for breathing. These generate oscillating volleys of impulses which, in turn, rhythmically stimulate the diaphragm and intercostal muscles. A series of interlocking and stabilizing mechanisms control the rate and precision of oscillation through information arising in effector organs (stretch receptors in the lungs) and in the blood (oxygen, carbon dioxide and the acid metabolites). The comparative physiologist finds equivalent pacemaking neurons and rhythmically controlled reflexes in the lower vertebrates and the more specialized invertebrates. Rhythmic movements of ventilation seem to be regulated by similar machinery at different levels in phylogeny from the polychaete worms to the mammals.

5

THE INTERNAL FLUID ENVIRONMENT AND ITS CIRCULATION

In many small animals the nutritive molecules, respiratory gases and waste metabolites diffuse readily through the intercellular spaces. No special arrangements are required for their transport. Likewise, some rather large animals, because of their primitive organization or low rates of metabolism, lack a circulatory system. The coelenterates and flatworms have achieved considerable size and complexity with little more than a highly branched and ramifying gut or gastrovascular cavity which combines some of the functions of a circulatory system with the digestive machinery; the echinoderms have such a low rate of metabolism that an active circulation does not appear to be essential. However, the majority of multicellular animals because of bulk, activity and the associated metabolic demands require a continuous and reliable circulation of body fluids, which are specialized to transport nutrients and gases to tissues remote from the source of supply. The coelenterate or platyhelminth plan would never suffice for such an elaborate organ as the vertebrate eye or kidney. At any rate, it is hard to imagine how such specialized cells and tissues could be associated with masses of closely applied gastrovascular tubes to form an efficient photoreceptor or compact excretory organ. The evolution of a transport system was essential to the phylogeny of the highly complex organ-systems of the higher animals.

VASCULAR CHANNELS

Embryologically, the hemal channels appear first as a series of spaces in the mesenchymal tissue. These coalesce, extend and gradually differentiate into the vascular system characteristic of any particular species. Mesenchymal cells included in these spaces and channels become the first blood cells. Phylogenetically, the primitive vascular system must have become organized in a similar manner. Many of the present-day flatworms do not possess vascular channels or even sinuses and lacunae in the mesoderm but transfer nutrients, respiratory gases, and wastes by simple diffusion. In others, however, a definite system is present, consisting of several longitudinal tubes on each side of the body which give off blind branches to the intestinal crura, the reproductive organs, and the suckers (Fig. 5.1). These tubes are lined with flattened mesenchyme and contain a lymph with some cellular elements arising from the neighboring mesenchyme. There is no special means of circulating the fluid, but it moves gently to and fro with the activities of the animal and thus expedites the diffusion of food, gases, and excretory products (Martin and Johansen, 1965).

Hemal spaces and channels are not the only mesodermal cavities of the multicellular animal. Within the massive layers of mesoderm a series of spaces are necessary, not only for the circulation of fluids, the distribution of fuel and gases, and the removal of wastes, but also to provide space within which large visceral organs can achieve an independence of movement, and through which excretory products and genital cells may find ready exits. These spaces or body cavities are related in one way or another to the vascular channels. Two patterns are found in the animal kingdom. In one, the major perivisceral space—referred to as the PRIMARY BODY CAVITY—is, in fact, a persistent blastocoel which has not been obliterated by the expanding mesoderm. This space, in an animal such as the arthropod, becomes an enlarged blood sinus, the hemocoel. The coelomic spaces proper (cavities in the mesoderm) are restricted to the gonads and excretory organs (Fig. 5.2). In the second pattern, an extensive SECONDARY BODY CAVITY (true COELOM) develops between two layers of mesoderm (mesothelium) and expands to obliterate the primary body cavity or blastocoel (Fig. 5.2). The coelomic spaces thus formed are only indirectly connected with the vascular channels and are intimately related to

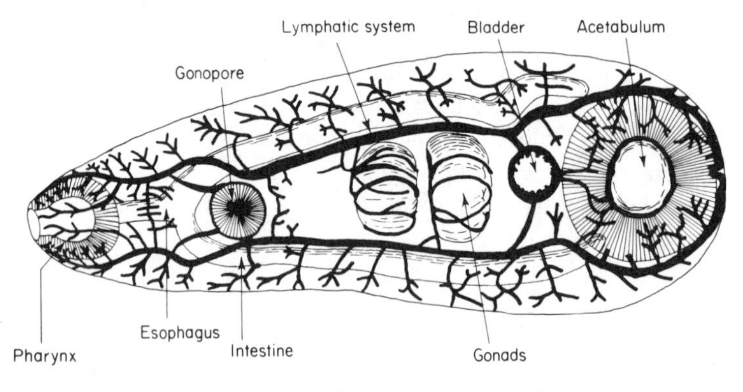

Fig. 5.1 The lymph system of *Cotylophoron*, a digenetic trematode. [Based on Willey, 1930: J. Morph., *50*: 33.]

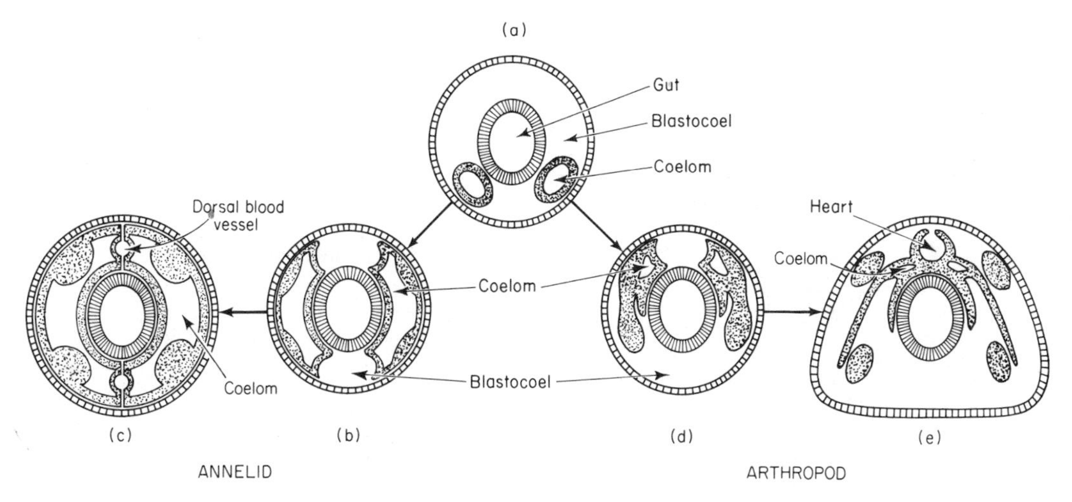

Fig. 5.2 Cross-sectional diagrams to show the relations of the body cavities in annelids and arthropods. (a) Early embryonic condition; (b) and (c) the blastocoel persists only in the dorsal and ventral blood vessels and the coelom becomes the secondary body cavity; (d) and (e) the coelom remains small and the blastocoel persists as the hemocoel or primary body cavity. [From Ramsay, 1968: A physiological approach to the Lower Animals, 2nd ed. University Press, Cambridge.]

the urogenital drainage systems (Chapter 8). In animals belonging to this second group, hemal channels are organized in the mesoderm as a system of continuously connected closed channels.

Thus, two distinct arrangements of circulatory channels characterize the highly organized groups of animals. Arthropods, most molluscs, and several groups of lesser animals have an OPEN SYSTEM in which there are no small blood vessels or capillaries connecting the arteries with the veins. Nets of minute sinuses or capillaries are found in some places such as the wings of insects or the gills of crayfish. However, these are interposed at some point in a blood sinus (insect wing) or located at the ends of arteries (cerebral ganglion and green gland of crayfish) or along certain veins (Fig. 4.1). They do not form a closed network between arteries and veins, and the arterial blood, sooner or later, passes into sinuses (large spaces) or lacunae (small spaces), so that the circulating fluids bathe the major organs and tissues. From these tissue spaces the fluids slowly work their way back into the open ends of the veins or the ostia opening into the heart. In these open systems the organs lie directly in the blood-filled hemocoel or in a primary body cavity. In the second type of system, the animals (vertebrates, annelids and a few other groups of invertebrates) have a completely CLOSED SYSTEM of blood channels. A continuous network of minute capillaries unites the smaller arteries with the veins. Fluids filter through the walls of the capillaries into the tissue spaces and thus transfer nutrients to the cells (Fig. 5.3 and 5.10).

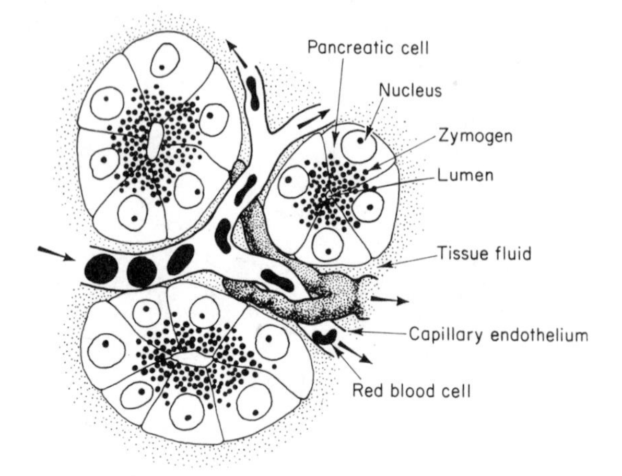

Fig. 5.3 The interrelations of the three vascular fluids in most vertebrates. Fine stipple, tissue fluid; clear, blood; close stipple, lymph as seen in a section of pancreas. [From Finerty and Cowdry, 1960: A Textbook of Histology, 5th ed. Lea and Febiger, Philadelphia.]

VASCULAR FLUIDS

In this discussion the body fluids will be referred to as TISSUE FLUIDS or LYMPH, BLOOD and HEMOLYMPH. Invertebrates which lack a circulatory system possess tissue fluids or lymph (Latin, *lympha*—clear water) surrounding their cells and forming minute lakes of watery solution with relatively low protein content, some salts, nutritive materials and wastes. Primitive blood cells (largely phagocytic in nature) float in this lymph or move through the tissue spaces. Animals with a closed circulatory system maintain a distinct separation between the blood (a tissue composed of cells and fluid plasma) and the tissue fluids. The latter are formed by filtration through the walls of the capillaries into the intercellular spaces under hydrostatic pressures. The filtrate is essentially non-colloidal (about 0.85 per cent protein compared with 7.0 per cent plasma protein in man) and is in part returned to the circulatory system by a special system of channels, the LYMPHATICS. In the higher vertebrates, these lymphatics commence blindly in the tissues and empty into the veins. Thus, a distinct separation between the interstitial fluids and the contents of the lymphatics (lymph) is maintained (Fig. 5.3). The invertebrates and fishes other than the teleosts lack proper lymphatics, although sinuses in the tissues may connect with the veins and foreshadow the development of this system. Some teleosts and amphibians exemplify intermediate conditions with many of the terminal lymphatics arising from tissue spaces. In these animals, tissue fluids and lymph proper are sometimes in direct continuity.

These varied arrangements make it difficult for the comparative physiologist to maintain a rigid distinction between tissue fluid and lymph. In all cases the lymph is derived from tissue fluids, whether it enters the lymphatics by diffusing through their walls or flows in through terminal openings. Further, in the open circulatory system, the distinction between blood and lymph or tissue fluid breaks down completely. The same fluids move through vascular channels and tissue spaces, and the term hemolymph seems justified. Hyman (1951), however, refers to this also as lymph and includes the coelomic fluids and contents of other fluid spaces under the same term. In this book the terms blood, hemolymph, tissue fluid and lymph will be retained as useful terms for somewhat different body fluids.

The morphology of blood cells is described in textbooks of comparative histology and anatomy; the composition of the fluids can be found in reference books of biochemistry. These topics will not be considered here; mechanisms concerned with the circulation of the fluids will be outlined in this chapter and several specialized characteristics of vascular fluids in the next.

HEMODYNAMICS

Hemodynamics is the study of blood flow and blood pressure. Many of the principles involved are the purely physical principles of the movement of viscous fluids (water, plasma, lymph) or "plastic" fluids (blood, a suspension of cells) through tubular channels. These physical principles are elaborated in textbooks of mammalian physiology. Here the concern is less with the fluid and how its physical properties affect the flow but more with the specializations of the tubular system and the various ways in which it has been modified to meet the demands of an efficient circulation.

Locomotion and the associated movements of the body provide sufficient agitation of the body fluids in a coelenterate, a flatworm and in some of the small animals in more advanced phyla. It is also the main propulsive force in some relatively large animals (certain leeches and echinoderms); the pressure in the body cavity of a holothurian increases from a low of about 1 mm Hg to a high of 25 or 30 mm Hg during activity (Table 5.1). Moreover, at all stages in phylogeny, locomotion is an important adjunct in hemodynamics; the massaging action of muscles on thin-walled veins and lymphatics is essential to the venous return in the appendages of higher vertebrates.

In most animals, locomotion is only subsidiary to a group of mechanisms which actively force the blood through special channels. A gentle stirring of fluids in a series of diffuse channels would be wholly inadequate to maintain the necessary exchanges in a bulky animal depending on different kinds of highly specialized tissues. From the nemertean worm to the vertebrate animal, the blood vessels are usually reinforced with muscle. In the primitive circulation, widespread propulsive force, due to the peristaltic activity of these muscles, moves the blood in a rather haphazard manner through its vascular channels. In a specialized circulation, the contractile force is localized in one or more rhythmically pulsating areas (hearts); the flow is directed throughout the system by a series of valves, and the pressures are steadily maintained (both during cardiac contraction and relaxation) since the pumps work in a closed series of highly elastic tubes.

Physically, the movement of viscous fluids through extremely small and lengthy tubes requires considerable force. Flow resistance increases with the length and the complexity of the tubular net. Pressure is also essential in the discharge of several associated physiological demands. For example, kidneys usually operate partly as filters; it is the high pressure in the vascular channels which forces the non-colloidal molecules through the lining of the blood vessels and associated membranes into the capsular spaces (Fig. 8.19). Likewise, the intercellular fluids, which ultimately carry the nutritive molecules to the cells, are driven through the capillary endothelia by the hydrostatic pressures in the blood channels (Fig. 5.10). These demands have become progressively more acute in the evolution of large active animals such as the cephalopod molluscs among the invertebrates and the birds and mammals among the vertebrates. The phylogenetic trend is toward mechanisms which provide a continuous blood flow at constant high pressures. Elaborate controls are necessary for the coordination of many interrelated physiological events. The details have been carefully investigated in mammals. In contrast, there are relatively few studies of the lower vertebrates (Johansen and Martin, 1965) and only fragmentary data for the invertebrates (Martin and Johansen, 1965). Several mechanisms responsible for the circulation of the blood will now be considered.

Peristalsis and "the Ebbing and Flowing" of Vascular Fluids

Galen (A.D. 131–201), the towering biological authority of antiquity, failed to understand a circulation of the blood. He visualized an ebbing and flowing in the great blood vessels of the body—between the heart and the viscera where NATURAL SPIRITS were formed; between the heart and the brain which was the locus of ANIMAL SPIRIT formation; and between the heart and the VITAL SPIRITS which came from the outside world through the windpipe and the lungs (Singer, 1959). Fourteen centuries passed before William Harvey (1578–1657), using simple demonstrations and logical arguments, corrected these ideas and proved the existence of a completely closed circulation. Harvey argued from his observations of the beating heart, from simple experiments on the direction of the flow of blood in the veins and from calculations of the volume of blood ejected from the left ventricle, that a closed and complete circulation was logical even though he could not see the connecting links. He drew on his knowledge of the lower vertebrates with a single ventricle (fish, frog) to support his assumption that blood is transferred from veins to arteries by the heartbeat. Thus, he combined comparative anatomical and physiological observation and used quantitative data to test the validity of his hypothesis (Graubard, 1964). His book, published in 1628, remains one of the outstanding scientific contributions of all time and a landmark in comparative as well as in medical physiology.

Since Harvey's time the circulation of the blood has not been seriously questioned in most animals. However, the ebbing and flowing of blood does occur in some primitive forms, such as the annelid *Nereis* with a somewhat Galenic circulation in its parapodia, which are important organs of gas exchange (Ramsay, 1968).

Vascular Pumps

As in any physical system, so also in the animal body, the efficient circulation of fluid through a series of pipes depends in large measure on the pumping arrangements. In both cases there are obvious advantages with larger pumps for more elaborate and longer systems and in the use of centralized pumping stations with auxiliary pumps at strategic points where additional flow or more reliable control is required.

Open and closed circulatory systems require distinctly different pumping devices. In an open system, the blood is discharged into the hemocoel which is comparable to a lake or well. The force is dissipated and not available for refilling of the pump. Actually, the pressure is absorbed in raising the internal body pressure which is maintained by a tightly fitting shell or exoskeleton. However, the important point is that during diastole (heart relaxation or expansion) the pressures within the vessels which return blood to the heart are equal to the internal body pressures and cannot aid in refilling the pump. The operation of this system requires a SUCTION PUMP, analogous to the simple lift pumps used to raise water from a well (Fig. 5.4). As the pressure rises with the contraction of the heart (or the downstroke of the pump), the intake valves close and fluid is discharged; filling is effected by the enlargement of the space within the pump (its upward stroke—outlet valves closed and inlet valves open) with aspiration of fluid from the hemocoel (lake or well). The flow is intermittent, depending on the effective (upward) stroke of the pump.

In the closed circulatory system, on the contrary, a FORCE PUMP or PRESSURE PUMP maintains a steady flow throughout a continuous network of pipes (Fig. 5.5). In this case, an elastic reservoir (or compressible air pocket in the comparable mechanical device) is added to maintain a positive pressure within the system while the pump is being filled. Although additional factors operate in the hydrodynamics of cardiac refilling, this low pressure venous reservoir (pressure about 5 mm Hg in man) provides the major force (Davson and Eggleton, 1968).

The tubular hearts of arthropods The arthropods show a high level of organization in which circulation is maintained in an open system. Some of the smaller representatives (*Cyclops*, for example) do not possess a heart or a proper circulation of the blood. The hemolymph is only gently stirred by the movements of the gut and body musculature. The majority of the arthropods, however, depend on the rhythmical contractions of the dorsal blood vessel or a heart which is a specialized area of the dorsal vessel. These hearts are single chambers with heavily reinforced walls containing striated muscle (Fig. 5.4). They may be tubular and extend for a considerable length of the body (*Branchipus*, *Artemia*, Insecta) or pulsating muscular sacs as in the crustaceans (Maynard, 1960). In any case, the major artery is directed anteriorly and carries the hemolymph forward; it returns through lateral ostia and sometimes posterior veins (Fig. 5.6). Valves are generally present in the main arteries and arterial junctions (Maynard, 1960).

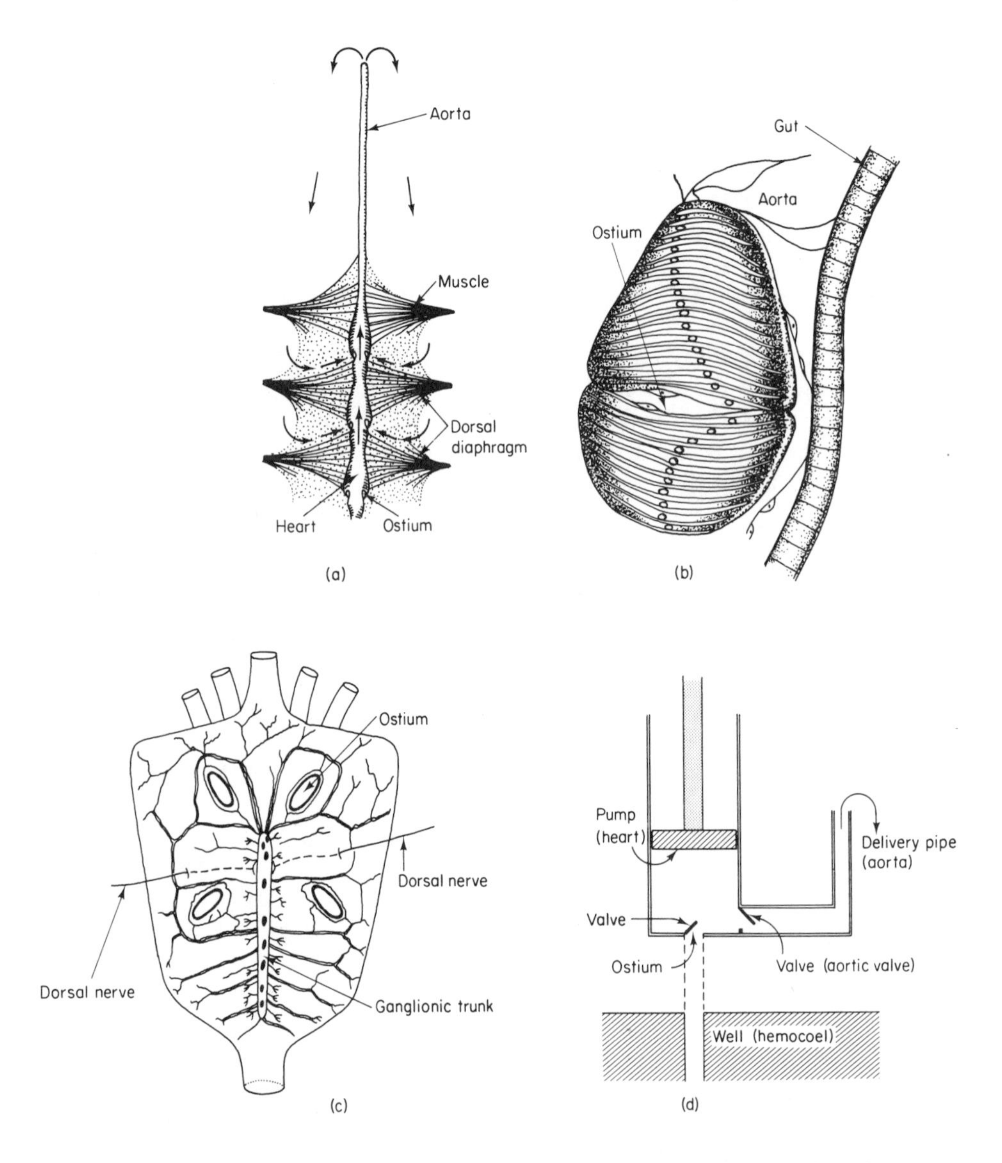

Fig. 5.4 Suction pumps. (a) Tubular heart of an insect represented by the grasshopper *Dissosteira*; (b) and (c) sac-like hearts of crustaceans represented by the cladoceran *Daphnia* and the spiny lobster *Palinurus*; (d) diagram of a common suction pump. [Insect heart based on Snodgrass, 1935: Principles of Insect Physiology. McGraw-Hill, New York; *Daphnia* from Maynard, 1960: *In* The Physiology of Crustacea, Vol. 1., Waterman, ed. Academic Press, New York; and *Palinurus* from Bullock and Horridge, 1965: Structure and Function of the Nervous System of Invertebrates, Vols. I and II. Freeman, San Francisco after Alexandrowicz, 1932: Quart. J. Micr. Sci., *75*: 192.]

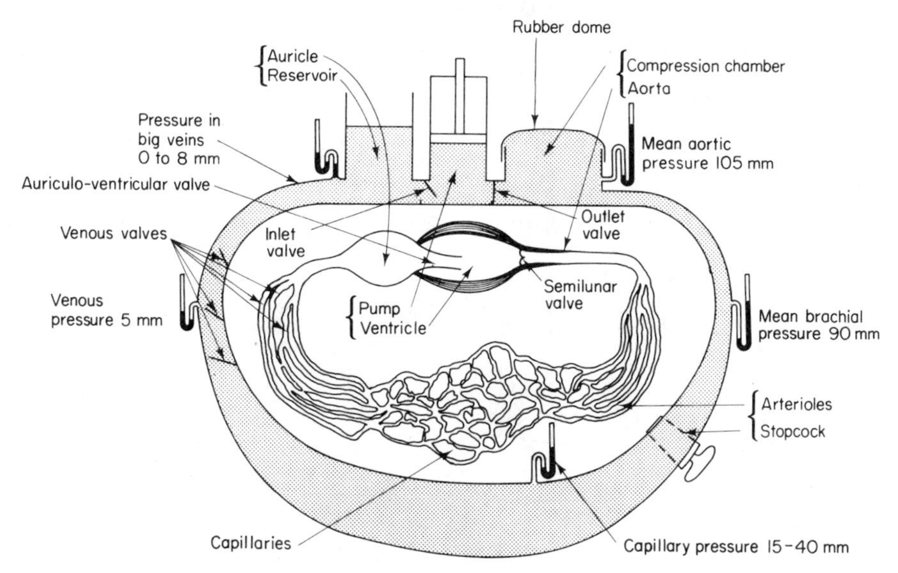

Fig. 5.5 Principle of the force pump or pressure pump in the circulation of the blood. [From A. W. Smith. 1948: The Elements of Physics, 5th ed. McGraw-Hill, New York.]

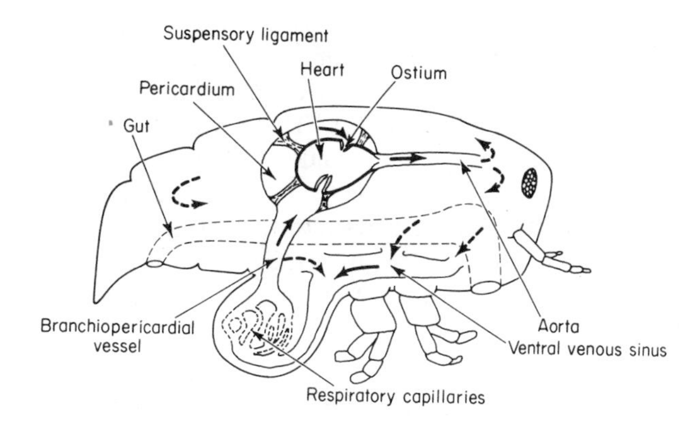

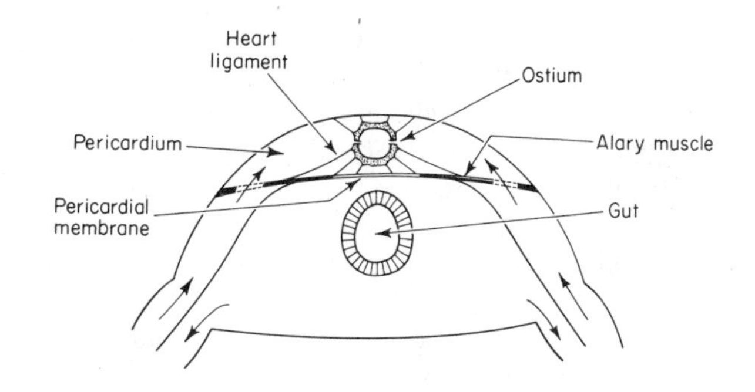

Fig. 5.6 Diagrams of the crustacean circulation. Solid arrows (upper figure), hemolymph flow in vessels; broken arrows, flow in unbounded sinuses. [Upper, longitudinal section from Maynard, 1960: *In* The Physiology of Crustacea, Vol. 1., Waterman, ed. Academic Press, New York. Lower, cross section from Ramsay, 1968: A Physiological Approach to the Lower Animals, 2nd ed. University Press. Cambridge.]

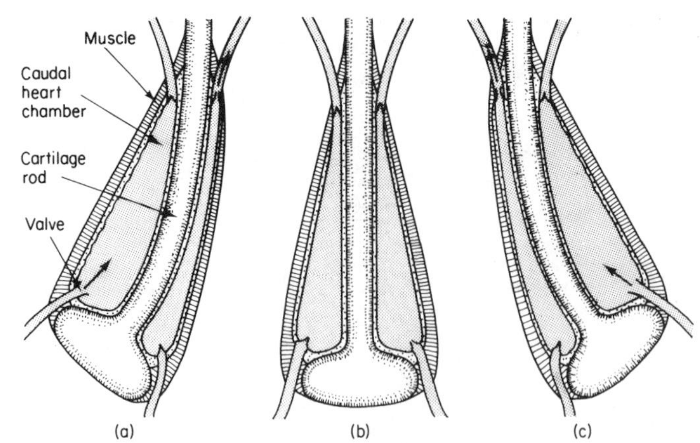

Fig. 5.7 Caudal heart of the hagfish. Heart is located about 1 cm from the end of the tail. A cartilaginous rod lies between two valved chambers; paired muscles, attached to the rod, pass along the outside of the chambers. Contraction of left muscle (a) expels blood collected in the right chamber and allows the left chamber to fill; the opposite contraction (c) empties the left chamber and fills the right one. The caudal heart beats irregularly. [for further details, consult Greene, 1900: Am. J. Physiol., *3*: 366–82.]

The heart is suspended in a fluid-filled pericardial sinus. Its hemodynamics require the filling of a perforated bag with fluid from the container which houses it. The mechanisms are entirely different from those which operate in chambered hearts of molluscs and vertebrates. The arthropod heart is suspended by a series of elastic ligaments or striated muscle fibers which are attached to the exoskeleton and to a heavy pericardial septum bounding the pericardial space ventral to the heart (Fig. 5.6). This septum is a tough connective tissue membrane; it may contain some muscle (alary muscle). The contraction of the heart places these elastic ligaments and muscles under considerable tension. As the heart relaxes in diastole, the tension of the ligaments pulls the ostia open and stretches out the walls of the heart so that the hemolymph flows into the organ (action of the suction pump). The rhythmic contraction of the myocardial muscle empties the heart; a passive refilling occurs during relaxation when the organ is stretched open by these elastic suspensory ligaments.

Auxiliary hearts are often present in the arthropods. These differ from the auxiliary hearts of molluscs and vertebrates (except *Myxine*, Fig. 5.7) in that their activity usually depends on specialized extrinsic muscles rather than on an intrinsic cardiac musculature. In some Crustacea, for example, there are marked local distensions of the blood vessels, called blood pumps; these are compressed by contractions of somatic muscles which have their origins and insertions outside the heart and run through the heart or its wall or lie in close proximity to it. These somatic muscles are secondarily adapted to the problems of circulation; in some cases they contract rhythmically (Maynard, 1960). Accessory pulsatile ampullar organs are also common in some groups of insects. They have been described in connection with the circulation in the wings (Diptera, Odonata), the antennae (Orthoptera), and the legs (Hemiptera). In some cases their musculature is intrinsic; in other cases it is extrinsic. Their rhythm is independent of the main heart and sometimes also of the accessory hearts in the same animal.

Chambered hearts The continuous output of blood at high pressure can best be achieved by increasing the thickness of the muscular cardiac walls. These thick walls, however, must be stretched somewhat when the organ is filled; this in itself creates a mechanical problem which is solved by coupling a thinner-walled receiving chamber (atrium) with a thicker-walled systemic pumping chamber (the ventricle). A sinus venosus may also be present, as in the frog or fish, and appears to operate as both a venous reservoir and an additional pump (Johansen, 1965). A system of valves prevents backflow; the ventricle is put under considerable tension in filling, and this improves the efficiency of the contracting muscle (see below). The coupling of a thin-walled primer pump with a thicker-walled pumping chamber is characteristic of both the molluscs and the vertebrates (Fig. 5.8).

Most of the molluscs operate an open circulatory system with a centrally located vascular pump consisting of an auricle and ventricle (Fig. 5.8). In the octopus and its allies, the system is closed and two auxiliary pumps (branchial hearts) are utilized to insure a circulation of blood through the gills (Fig. 5.9). A few vertebrates also require booster pumps in addition to a strategically located multichambered heart. The primitive cyclostome *Myxine* has a portal, a cardinal, and a caudal heart in addition to the systemic pump and also depends on contractions of the gill musculature to keep the blood moving in a partially open, very low pressure system (Johansen, 1960, Fig. 5.7). The frog requires several additional pumps (lymph hearts) to return the lymph to the venous system. Accessory hearts are absent in most of the higher vertebrates (page 156); birds and mammals operate two strong pumps in a central pumping station—one for the systemic circulation and the other for the pulmonary circuit (Fig. 4.12).

Venous pressure has already been identified as a major factor concerned with the refilling of the heart in a closed circulatory system. The hydrodynamics of refilling

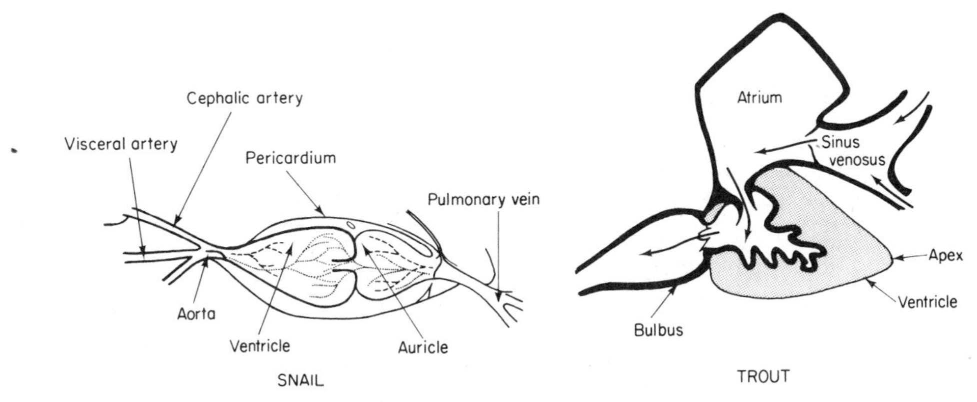

Fig. 5.8 Diagrams of chambered hearts. [The snail *Helix pomatia* after Ripplinger, 1957: Ann. Sci. Univ. Besançon, Zool. et Physiol., *8*: 21. Trout after Randall, 1968: Am. Zool., *8*: 180.]

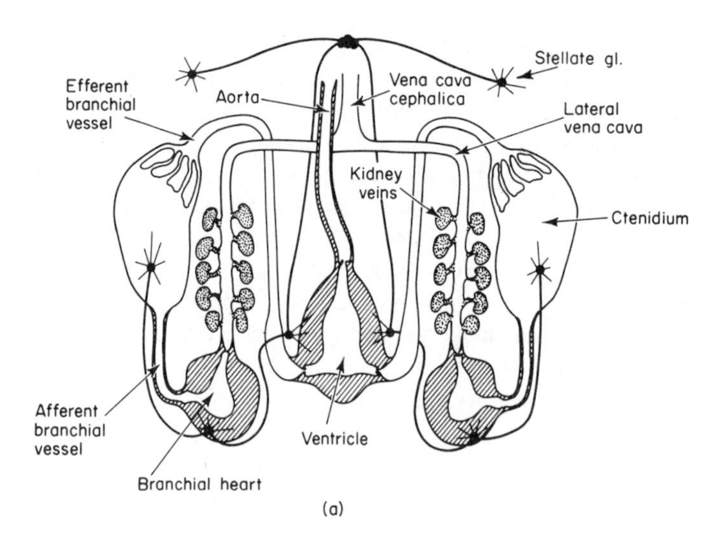

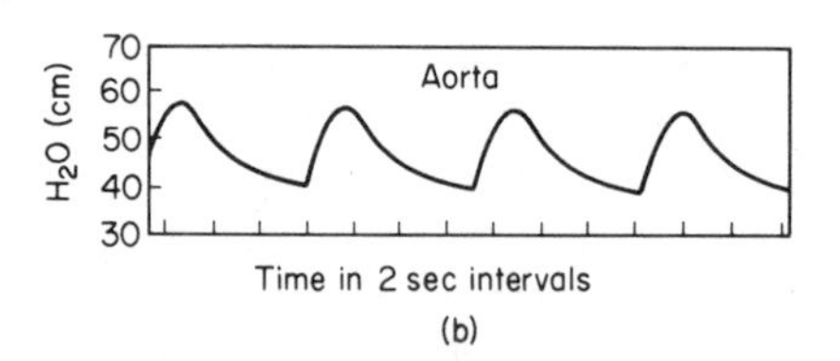

Fig. 5.9 Diagram of the central vascular system in the octopus (a) and a pressure record from the aorta cephalica (b). [From Johansen and Martin, 1962: Comp. Biochem. Physiol., *5*: 164.]

the molluscan heart, which (the cephalopods excepted) operates in an open circulatory system, are different and not completely understood (Hill and Welsh, 1966). The pericardial space, however, has long been considered important (Krijgsman and Divaris, 1955; Ramsay, 1968). Chambered hearts, whether molluscan or vertebrate, are housed in a special chamber of the coelom bounded by a tough connective tissue membrane, the pericardium. The expulsion of blood from the ventricle during contraction (systole) may be expected to create a negative pressure within the pericardium which will favor the filling of the thin-walled sinus and atrium during the subsequent period of relaxation (diastole). The action of the pericardium in filling the snail heart has recently been established by comparing the pressures in auricle, ventricle, and pericardial cavity throughout the cardiac cycle (H. D. Jones, 1970, 1971). An action of the pericardium is also recognized in some of the lower vertebrates (Randall, 1970a).

Fish in general and elasmobranchs in particular possess a tough pericardium which closely adheres to the external supporting structures to form a semirigid box capable of supporting negative pressures (Johansen, 1965). Intrapericardial pressures, measured in several elasmobranchs, are never above atmospheric and usually fluctuate around −5 mm H_2O. The pressure slowly rises as the ventricle fills and pulsates with its contractions; rupture of the pericardium depresses the arterial blood pressure. This evidence of an aspiratory effect caused by negative intrapericardial pressures is convincing in the elasmobranchs. A comparable action is probably present in other lower vertebrates although subatmospheric pressures have not been recorded; among amphibians a definite drop in pressure that occurs during ventricular systole could increase the filling rate (Johansen, 1965). In mammals and birds, the presence of a diaphragm and separate thoracic cavity provide a similar action associated with breathing. In man, the RESPIRATORY PUMP effects negative pressures of from −5 mm to −10 mm Hg during the respiratory cycle; this alternating suction on the walls of the great veins acts like an accessory pump to suck blood into the heart with each inspiration (Davson and Eggleton, 1968). Respiratory movements, which are usually synchronized with cardiac activity, are probably also important in filling the hearts of lower forms (Johansen and Martin, 1962).

The lymphatic circulation In the vertebrates, with their closed systems of vascular channels, the return of the lymph to the main circulation may require special assistance. The problem can be presented most easily by reference to Starling's classical hypothesis summarized in Figure 5.10.

The protein content of the blood is always greater than that of the tissue fluids and produces a colloidal osmotic pressure of about 36 cm H_2O. At the arterial end of the capillary the hydrostatic pressure, which depends on the cardiac pump, is 44 cm H_2O, and this results in a net filtration pressure of 8 cm H_2O. At the venous end of the capillary the hydrostatic pressure is much lower (17 cm H_2O), and the net pressure of 19 cm. is in the reverse direction and returns fluid to the capillary. In addition, there is a certain mechanical resistance (tissue pressure) to the flow of fluids from the capillaries into the tissue spaces. This diagram emphasizes the significance of blood pressure (cardiac pump), osmotic pressure of the fluids and tissue pressures which vary with the activity of the muscles. In some vertebrates these forces alone provide for the fluid exchanges between capillaries and tissue spaces, but in most vertebrates a lymphatic system is also required. Lymphatics are absent in cyclostomes and elasmobranchs but present in the bony fishes and tetrapods (Romer, 1970). A few fishes (eels and *Silurus*), all the amphibians and reptiles, all bird embryos and some adult birds have lymph hearts which improve the circulation of the lymph and aid its return to the venous channels. Most adult birds and

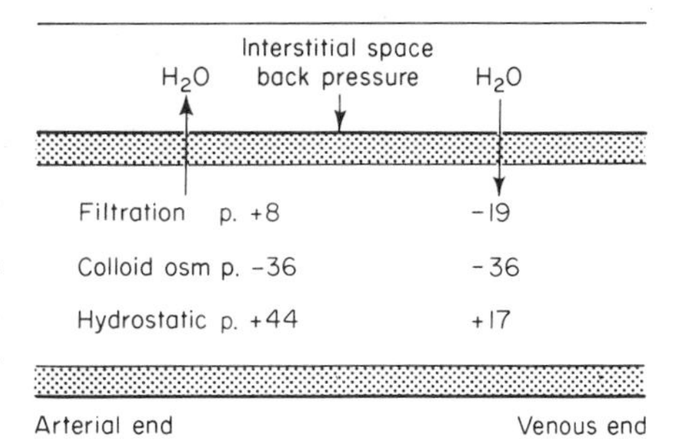

Fig. 5.10 Diagram of fluid exchange across the wall of a capillary. Note that pressures are in cm H_2O. [From Young, 1957: The Life of Mammals. Clarendon Press, Oxford, based on Davson, 1951: A Textbook of General Physiology. Churchill, London.]

all mammals depend on the massaging action of muscles and the pressure of neighboring pulsating arteries and other organs. This massaging of thin-walled vessels provided with numerous valves to direct the flow constitutes the so-called "lymphatic pump."

The hagfishes show a unique condition among vertebrates with arterial blood actually flowing into tissue spaces ("red lymphatics") from vascular papillae on the branchial and carotid arteries (Brodal and Fänge, 1963). The several accessory hearts mentioned previously are essential to the circulation in this semiclosed system.

Cardiac output The quantity of blood pumped by the heart depends on the volume ejected at each contraction (STROKE VOLUME) and the rate of pulsation. In a healthy young man this amounts to about 5.6 liters per min or 80 ml/kg/min (MINUTE VOLUME). In 1870, a German physiologist, A. Fick, described a simple method of estimating cardiac output from measurements of the oxygen consumption (or carbon dioxide production) and the difference between the oxygen (or carbon dioxide) contents of blood entering and leaving the heart (Fick, 1872). Thus:

$$\text{Cardiac output (ml/min)} = \frac{O_2 \text{ absorbed by lungs (ml/min)}}{\text{arteriovenous } O_2 \text{ difference (ml/liter blood)}}$$

If blood entering the right atrium contains 160 ml O_2/liter while that leaving the left ventricle holds 200 ml/liter, it follows that 40 ml O_2 have been picked up by each liter of blood which circulated through the heart and lungs. If oxygen is being removed from the inspired air at the rate of 200 ml/min then it is obvious that 5 liters of blood must have been pumped through the heart and lungs to do the job. Details will be found in textbooks of medical physiology. Several kinds of flow meters are also available but most of the measurements recorded in the literature are based on Fick's principle or some modification of it. It should be noted that the cardiac output of a higher vertebrate is the volume of blood pumped by *either* the right or the left ventricle—not both.

A cardiac output of about 5.5 liters/min in man corresponds to a mean blood flow of about 9 ml/min/100 gm of tissue. However, blood flow varies greatly in different organs, and values tabulated by Davson and Eggleton (1968) range from 1.3 and 2.7 in skin and muscle respectively to a high of 2000 ml/min/100 gm in the carotid body. The cardiac output of birds, with their relatively large and rapidly beating hearts, is high while that of the poikilotherms is relatively low. Recorded values for domestic birds range from about 200 to 400 ml/kg/min (Jones and Johansen, 1972) while those for fishes (only one ventricle) are of the order of 5 to 100 ml/kg/min (Holeton and Randall, 1967; Randall, 1970a). Values have been recorded for only a very few invertebrates; the cardiac output of a 750-gm rock lobster *Palinurus* was calculated as 60 ml/min (0.8 ml/Kg/min) and comparable figures for an 18-Kg *Octopus dofleini* are 320 ml/min and 17.7 ml/Kg/min (Maynard, 1960; Martin and Johansen, 1965).

It has already been noted that cardiac output can be altered by adjusting either the stroke volume or the rate. In phylogeny, various groups of animals seem to have emphasized these two possibilities somewhat differently. Among the lower vertebrates the major adjustment during vigorous exercise is an increase in stroke volume (Fig. 5.11); in mammals, however, there may be little change in stroke volume while major adjustments are made in the rate. Dogs, for example, when running on a treadmill at 3 mph on a 5 per cent grade showed a two- to three-fold increase in heart rate but the aortic flow per stroke was only slightly elevated (Rushmer and Smith, 1959). These differences are considered in several recent reviews (Johansen, 1965, 1971; Randall, 1968, 1970a).

Valves and Stopcocks

An intricate system for the circulation of any fluid will require valves at appropriate points to direct the flow and stopcocks to control its volume. The vascular channels of all animals with a true circulation are generously supplied with these; they become particularly vital to the maintenance of the sustained high pressures and continuous flows of vertebrate animals. Among the invertebrates, valves are common in the region of the heart. In an annelid, flap-like valves ensure the forward movement of the fluids in the long dorsal vessel and prevent its reversal where the

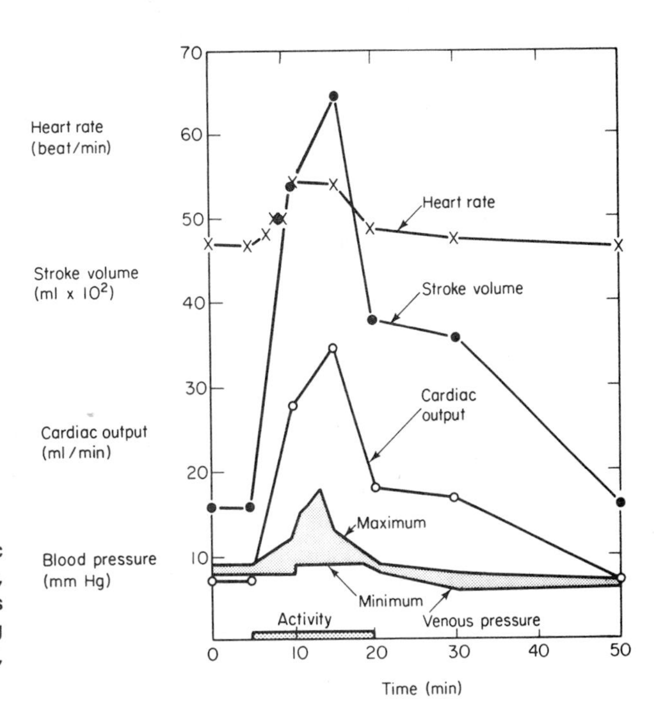

Fig. 5.11 Changes in cardiac output, stroke volume, heart rate, and maximum and minimum venous pressures in the rainbow trout during moderate swimming. [From Randall, 1968: Am. Zool., *8*: 184.]

non-contractile vessels enter the dorsal vessel. Several types of valves have been described in the crustaceans (Maynard, 1960) and are found where arteries arise from the heart as well as in the arteries and venous sinuses. Semilunar valves, consisting of two membranous flaps, are common. When stretched, these flaps form pockets. Reversal of flow fills the pockets and expands them so that the two components come together and close the vessel. Single flap-like valves are also present. In these, the reversal of blood flow stretches the flap to close the lumen of a small vessel.

Chambered hearts, whether vertebrate or invertebrate, possess valves as an essential feature of the primer pump arrangement. The semilunar type of valve attains a high degree of development in the vertebrate heart where two or three such pockets guard the atrioventricular openings, and the flaps of tissue are supported by fibers of muscle (papillary muscle) and connective tissue (chordae tendineae). Paired flap-like pocket valves in many of the veins and in the lymphatics of the vertebrates are essential to the movement of their fluids. In a few animals the arteries in organs remote from the heart, such as the tail of the shark (Birch *et al.*, 1969) are also provided with valves.

Several devices operate as variable resistance stopcocks in circulation. The ostia of an "open heart" like that of the crustacean are controlled by special sphincter-like valvular arrangements. Arterioles also control blood flow by sphincter-like muscles in their walls. Although some muscle is characteristic of many blood vessels from the level of the nemertean up the phylogenetic series, it is in the vertebrates that the development becomes marked in the smaller peripheral vessels and functions as an efficient stopcock arrangement to control the flow to different parts of the capillary bed. The muscle of the arterioles is actively controlled by both nervous and chemical factors to meet the variable circulatory demands of the tissues. These arteriolar stopcocks control the peripheral resistance and thus become one of the major factors in the maintenance of the high and constant blood pressures of the vertebrates.

Elasticity of Blood Vessels and a Sustained Pressure

Fluid flow depends on the construction of the pipes as well as on the nature of the pump and the valves in the system. If the walls are rigid, fluids move through the system in accordance with the action of the pump; if forced in as a series of jets, the fluids emerge in the same way. On the other hand, elastic tubes which are stretched under pressure recoil when that pressure is released and exert a continued force on the fluid during the interval between the strokes of the pump. In this way, a sustained and continuous flow is maintained even though the fluid enters the tubes as a series of jets. Such a mechanism can operate efficiently only in a closed system; the sustained high pressures of the vertebrates depend on the marked elasticity of the main arteries as well as the peripheral resistance, referred to above, and the action of the pump. In an open circulation there is practically no peripheral resistance. Flows tend to be intermittent; the pressures are variable and low (Table 5.1).

Table 5.1 Blood Volumes and Pressures of Representative Animals. (Pressures expressed as systolic/diastolic (e.g., 120/80); or limits of systolic (90–100); or mean pressures)

Species	Blood volume[1] % body weight (dye method)	Blood pressure	
		Vessel	Pressure mm Hg
Vertebrata			
Man	7.1–7.8	Radial artery	120/80
Dog	6.0		112/56
Rat	8.5		130/91
Frog	9.5		30/20
Eel, *Anguilla*	2.9	Ventral aorta	25–60
Salmon	3.5–9.8	Ventral aorta	30/22
Dogfish, *Squalus*	6.8	Ventral aorta	30/24
		Dorsal aorta	17/16
Annelida			
Earthworm, at rest		Body cavity	1.5
Earthworm, active			10.0
Arenicola, at rest		Body cavity	0.73
Arenicola, active			11.0
Mollusca			
Mytilus	51		1.9/0
Octopus	28	Aorta	44/22
Arthropoda			
Lobster, at rest	17	Ventricle	13/1
Lobster, active			27/13
Carcinus	32.6	Sternal artery	9.6
Crayfish	25.1	Cheliped sinus	7.4
Cockroach	15.7–17.5		
Locusta		Dorsal aorta	6.3/2.3

SOURCE: Data from Prosser (1973).

[1] Blood volumes recorded in the literature vary with the technique used for determination and show marked differences in closely related groups of animals (for example, in four species of marine elasmobranchs—Thorson, 1958). Values for invertebrates are volumes of extracellular space; blood volume of octopus (dye method) is 5.9.

Résumé

Throughout the animal kingdom the contractions of the somatic musculature play an indispensable part in the circulation of body fluids. In higher forms, locomotion is an adjunct to other hemodynamic mechanisms; in many primitive animals, the intermittent movements of the lymph and the variable blood pressures are entirely dependent on it.

A continuous circulation and sustained blood pressure require a constant pumping of the fluids through a system of tubes or spaces. In a few forms, this pumping depends on widespread peristalsis in the tubes but, almost universally, specialized contractile areas of the tubes (hearts) provide the major hemodynamic force. Increasing animal size, high levels of activity, and steady metabolic rates become possible only when the tissues are constantly perfused with nutrients and when the important capillary nets, responsible for gaseous exchange and the removal of wastes, are continuously serviced. The open plan of circulation does not permit such high and continuous pressure as the closed type and, partly for this reason, animals with open systems are limited in their size and/or activity. The *Octopus*, a large and active representative of the Mollusca (a phylum characterized by an open circulation), has an essentially closed circulation; the circulation in the intricate capillary net of the insect wing depends on special booster pumps.

The chambered heart is a feature of most circulations of the closed type and is primarily responsible for the high pressures attained. In the chambered heart, one or more thinner-walled receiving chambers serve as booster pumps to fill the thick-walled ventricle which provides the main pumping force. In some forms, the pericardium is an important adjunct to this pump.

The higher vertebrates exemplify the climax in circulatory efficiency. Separate pumps for the respiratory and systemic capillary nets insure high and continuous pressures in all parts of the system. The high systemic pressures, associated with the double circulation, insure steady perfusion of tissue juices into the intercellular spaces while a system of lymphatics (characteristic of animals with higher pressures) assists in the return of the lymph to the main circulation. Diastolic pressures are maintained by the elasticity of the vessels, and the relatively high pressures throughout the cardiac cycle permit a continuous operation of such vital functions as respiratory exchange and kidney filtration.

HOMEOSTASIS AND THE CIRCULATION OF BODY FLUIDS

To meet the requirements of active cells, located far away from sources of fuel and points of waste disposal, the transport system must not only operate continuously but must be capable of adjustments in accordance with variable demands. This is achieved through the constant, but rate-variable, activity of the pump (SUSTAINED CARDIAC OUTPUT) and the adjustments in the caliber of the blood vessels (PERIPHERAL RESISTANCE). Homeostatic mechanisms for the operation of the pump appeared early in animal phylogeny, but the precise control of peripheral resistance became essential and was established only in the vertebrates (particularly the homeothermic members of this group) and in the largest and most active invertebrates.

Cardiac Rhythm

Peristaltic activity, characteristic of smooth muscle in the larger vessels of embryonic and phylogenetically primitive circulations, is succeeded by more localized pacemaking areas in the heart. Groups of specialized muscle or nerve cells control the rhythmic activity of the muscle and are the centers through which the temperature, neurohumors and other factors operate to vary the cardiac output. Although such pacemakers are conspicuous in many hearts, it should be remembered that the rhythm is inherent in the cardiac muscle and that the pacemaking areas are established during differentiation. This fact was not always appreciated; for many years the origin of the heartbeat (whether myogenic or neurogenic) was a topic of intensive research and prolonged arguments.

Both types of control are now recognized. All vertebrates have myogenic hearts; but one cannot so easily generalize for invertebrates. The tunicate heart—particularly interesting because of its regularly reversing beat—is myogenic with a pacemaker near each end (Kriebel, 1968a). Molluscan hearts are also considered to be myogenic (Hill and Welsh, 1966). Among annelids and arthropods, however, some species possess neurogenic controls while in others the rhythm is myogenic. Martin and Johansen (1965) point out that neurogenic controls are most characteristic of the elongated tubular hearts found in the larger representatives of these phyla; larval stages and smaller representatives with more compact hearts frequently have a myogenic cardiac control. It is possible that the rhythm of a long tubular structure can be more effectively regulated from compact nervous centers with rapidly conducting nerves.

Martin and Johansen (1965) also speculate on the phylogenetic origin of the heart and its pulsations and suggest that traces of the evolutionary trail may be found among the annelids. In some simple freshwater oligochaetes (family Aeolosomatidae), the main dorsal blood vessel stems from a posterior sinus which lies within the dorsal musculature of the gut. Thus, the gut is a functional part of the circulatory system and the peristaltic activity of the intestine is responsible for the movement of the blood. In more advanced species (some of the Naididae), the dorsal blood vessel is distinct from, but still attached to the gut throughout its entire length; the contractions of gut and blood vessel occur in synchrony. In other species of Naididae and in the Tubificidae, the dorsal blood vessel is physiologically dissociated from the gut with a localized myogenic pacemaker. The suggestion is that the rhythmically contractile musculature of the dorsal blood vessel in the lower annelids was derived from the digestive tract and that the rhythm was myogenic. More specialized annelids, like *Arenicola* and *Lumbricus*, as well as larger and more advanced arthropods have neurogenic cardiac controls. These, however, appear to be associated with special problems of coordination; similarly specialized pacemaking areas of muscle have developed in the myogenic hearts of the higher vertebrates.

Pacemakers of myogenic hearts In the hearts of vertebrates, pacemaking activity resides in a system of specialized muscle cells. These impulse-conducting cells are often histologically different from the general cardiac muscle fibers, but the extent of differentiation varies remarkably. The system is most distinctive in some birds (hummingbird, chick) and in mammals such as the platypus, spiny anteater, sheep, cow and pig; it is either poorly formed or lacks differentation in the snake, turtle, bat, rat, guinea pig and rabbit. The hearts of primates, cats and dogs are intermediate in this series (Truex and Smythe, 1965). Claims have also been made for the presence of histologically specialized myogenic pacemakers in fishes, amphibians and reptiles (Prakash, 1957). However, relatively few forms have been examined and the phylogeny is uncertain (Davies and Francis, 1946; Cranefield, 1965). In any case, it should be emphasized that even in the chick with its highly distinctive impulse-conducting cells, cardiac rhythm is established during embryonic development prior to the appearance of this specialized system and that all hearts, whether myogenic or neurogenic in the adult, are basically myogenic (Hecht, 1965).

In the bird or mammal, pacemaking impulses arise in the sinuatrial node (S–A node), a small mass of specialized cells measuring about 2 cm by 2 mm in man and located in the right atrium near the entrance of the great veins (Fig. 5.12). This is the true pacemaker; localized changes in temperature, surgical manipulations, and electrical stimulations have shown that the wave of excitation associated with cardiac activity spreads from this point. It is picked up by a similar mass of tissue (the atrioventricular node or A–V node) situated in the right atrium of the mammal near the ventral part of the interatrial septum. Thence, excitation spreads along the atrioventricular bundle into the right and left bundle branches of the ventricles

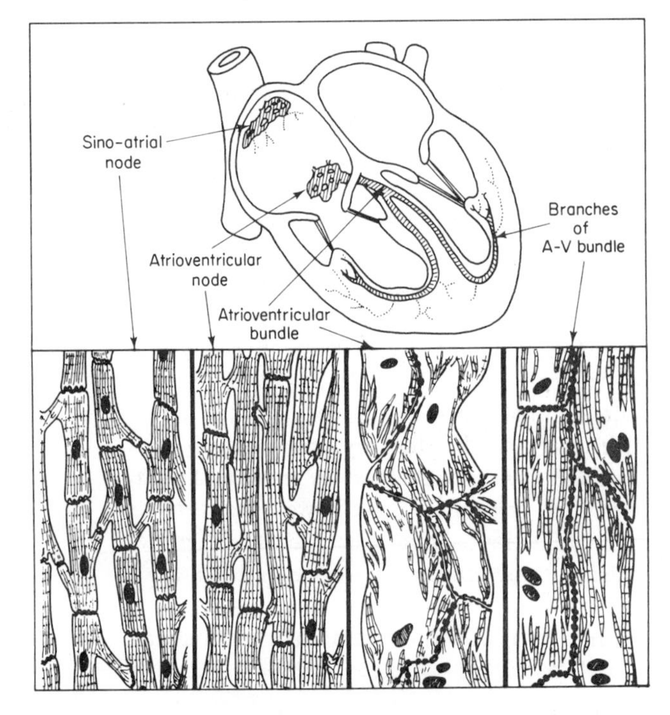

Fig. 5.12 Schematic diagram of the impulse-conducting system of the beef heart. Diagrams of histology shown below are based on camera lucida drawings at magnification of about 130×. [From Rhodin, Del Missier and Reid, 1961: Circulation *24*: 350]

and from these, by fine ramifications (the Purkinje fibers), into the mass of cardiac tissue. The system is similar in the bird, except for the extensive development of conducting fibers in the left as well as in the right atrium.

It is assumed that the cardiac pacemaking system of the homeotherm has, during phylogeny, been derived from the sinus venosus. This hypothesis is supported by embryology and by the presence of pacemaking tissue in the sinus venosus of many poikilotherms. Embryologically, the sinuatrial node is derived from the primordial sinus venosus; physiologically, it can be demonstrated that the sinus venosus of amphibians and reptiles (frog or turtle) is the area from which excitation spreads throughout the entire organ. Even though the cardiac muscle of the frog or turtle will show rhythmic contractions when divorced from the sinus venosus, under normal conditions excitation, as measured by electrocardiograms, spreads from the sinus; experimental manipulations, such as altered temperatures or stimulation of this area, will affect the entire organ.

However, the sinus venosus and the sinuatrial node are not the only pacemakers of vertebrate hearts. In teleost fishes the floor of the atrium and the atrioventricular junction usually contain the pacemaking cells, although in the eel the sinus venosus is also involved. In elasmobranchs, the sinus venosus, the atrioventricular junction, and truncus arteriosus all show pacemaking activity (Mott, 1957; Randall, 1970a).

Pacemaking cells are probably widely distributed in the myogenic hearts of many invertebrates, as they are in such spontaneously acting smooth muscle as that of the ureter and the intestine of the vertebrate (Burnstock *et al.*, 1963; Florey, 1966). However, localized pacemakers also occur in myogenic invertebrate hearts and have been studied in the tunicate (Kriebel, 1968a) and in some molluscs (Krijgsman and Divaris, 1955; Hill and Welsh, 1966). A trend in the specialization of pacemaking muscle cells localized in particular regions of the heart seems to have commenced early in animal evolution and reached a climax in some birds and mammals.

The rhythm of neurogenic hearts Classical studies of the pacemaking activity of a neurogenic heart were performed on *Limulus* by A. J. Carlson, about 70 years ago (Krijgsman, 1952). At that time, this heart seemed almost to have been designed to decide the controversy which was current between myogenic and neurogenic theories of cardiac rhythm. A series of ganglia on its dorsal surface can readily be removed without destroying the cardiac muscle. By appropriate experiments Carlson showed that the rhythm of the adult heart resided in these ganglia and that changes in temperature, stimulation of inhibitory nerves, and other manipulations would modify the muscular rhythm through their effects on these pacemaking ganglia. Subsequently, it has been shown that each of the myocardial cells is innervated by at least six motor neurons—three from its own segmental nerves and three from adjacent or more distantly located segmentals (Abbott *et al.*, 1969). Comparable pacemaking ganglia have now been studied in some other neurogenic hearts (Maynard, 1960; Watanabe *et al.*, 1967a, b), but demonstrations of their activities are more difficult because of the anatomical distribution of the ganglia.

The spread of excitation and the rhythmic activity of the heart creates an orderly sequence of changes in electric potential. Kolliker and Johannes Müller (1858) first demonstrated electrical activity in the beating heart (Fulton, 1955). They were able to detect movements in the crude coil-type galvanometer available at that time but depended on a biological experiment for proof of "animal electricity." The sciatic nerve of a frog gastrocnemius nerve-muscle preparation was looped around the ventricle of the frog, and two contractions of the gastrocnemius were noted with each heartbeat. It remained for Einthoven (1903), a Leiden professor, to develop the first suitable electrocardiograph for the measurement of these changes in electrical potential. Einthoven's instrument is now superseded by precise electronic recording devices which have become indispensable for research and for clinical diagnosis of cardiac pathology. The records (electrocardiograms, ECG—or EKG in German terminology) show consistent variations from the normal pattern in association with a variety of experimentally induced or naturally occurring pathological states.

Understanding of bioelectric phenomena has come a long way since Galvani (1737–98) first demonstrated the presence of "animal electricity" in his pioneer experiments with the gastrocnemius-sciatic nerve preparations of the frog (Verworn, 1899). It is now known that the ECG depends on one of the fundamental properties of living cells. Each and every cell, as long as it is alive, shows an electric potential difference across the surface membrane. This transmembrane potential depends on active metabolic processes which create an unequal distribution of ions between the interior and exterior of the cell. Its genesis will be considered in Chapter 15. At this point it is noted that the "resting" cardiac muscle cell of the vertebrate maintains a transmembrane potential of between 70 and 100 mv and that stimulation is associated with a drop in this potential to zero and then a momentary reversal in the sign to about 20 mv in the opposite direction. Thus, the rhythmically contracting muscle cells show an oscillation in depolarization and repolarization of their surface membranes.

Intracellular recordings Individual cardiac muscle cells have been explored with intracellular electrodes and characteristic differences were found in the action potentials of the pacemaking, conducting, and contractile cells (Fig. 5.13). The pacemaking cells, like pacemaker cells in other spontaneously active muscles, show action potentials preceded by a slowly developing depolarization, referred to as a PREPOTENTIAL OR PACEMAKER POTENTIAL (Fig. 5.13a). A recording from a spontaneously beating cell in the sinus venosus of a frog will show a gradual rise in potential during diastole from a negative value of about 55 mV to about 40 mV after which the rate of depolarization rapidly increases until the inside of the fiber becomes positive to about 10 mV; the membrane is then recharged in a slow manner which is characteristic of cardiac muscle. The membrane potential is never stationary and this is in marked contrast to non-pacemaking cells, whether conducting (Purkinje tissue) or contractile, where the pacemaker potential is absent (Fig. 5.13b). Electrical resistance between neighboring cardiac cells is low and excitation spreads from the rhythmically active pacemaker cells through the heart which behaves electrically as a syncytium.

Pacemaker potentials in ganglia of neurogenic hearts are similar in character to

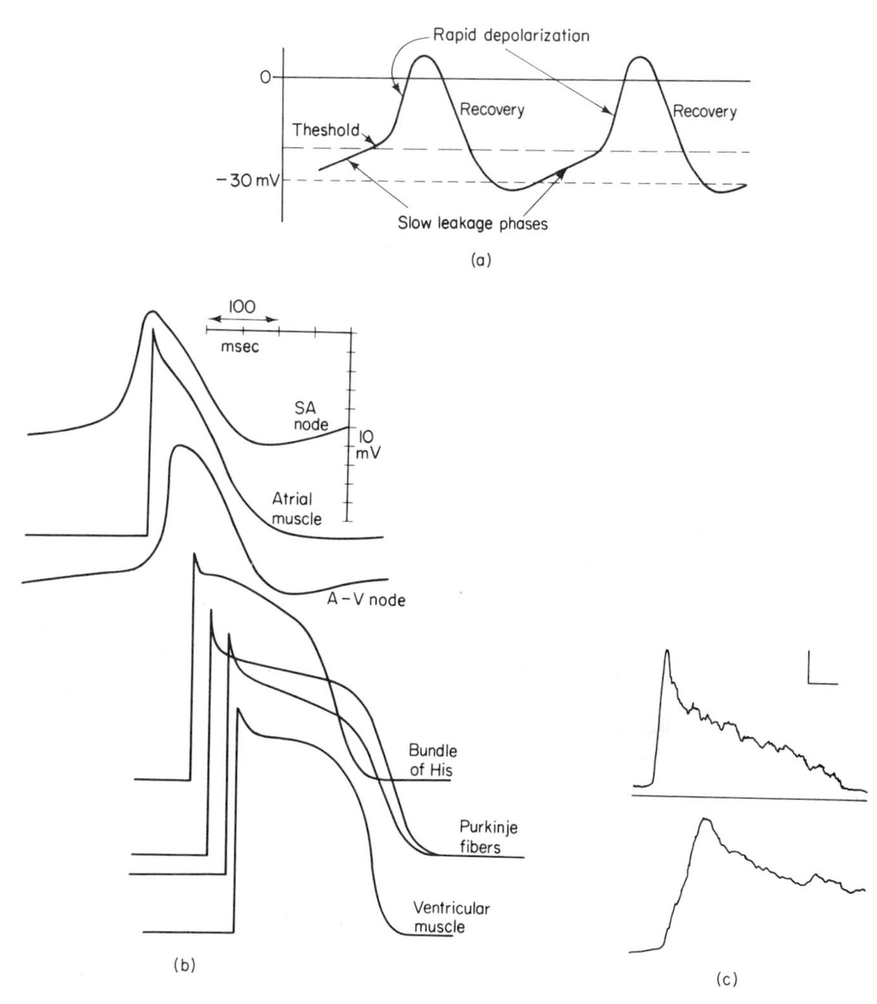

Fig. 5.13 Action potentials in cardiac muscle cells. (a) Microelectrode recording from a pacemaker cell. (b) Schematic diagram showing form of action potentials recorded from seven different sites in the vertebrate heart. Note the sequence of activation at the various sites as well as the differences in amplitude, configuration, and duration of the action potentials. [From Hoffman and Cranefield, 1960: Electrophysiology of the Heart. McGraw-Hill, New York.] (c) Intracellular electrical recording from *Limulus* myocardial cell at different speeds. Calibration: vertical, 10 mV; horizontal, in upper trace = 250 msec; in lower trace = 62.5 msec. [From Abbott, Lang and Parnas, 1969: Comp. Biochem. Physiol., *28*: 151.]

those recorded in vertebrate pacemakers. Intracellular recordings from contractile cells are, however, quite different in neurogenic and myogenic hearts. There may be a superficial similarity with a sharply rising upstroke and a plateau-like phase in myocardial recordings of the neurogenic (*Limulus*) heart (Fig. 5.13c), but when this response is examined on an extended time scale, the rising phase shows a succession of small junction potentials which summate to give the sustained depolarization. As previously noted, there is a multisegmental innervation to each myocardial cell and impulses arising in several ganglionic neurons produce a succession of these small junction potentials which summate, leading to a synchrony in contraction along the length of the heart (Abbott *et al.*, 1969). The neurogenic type of cardiac activation is in strong contrast to that seen in the myocardial cells of vertebrates with the overshooting spike and plateau (Fig. 5.13a).

Electrocardiograms The synchronous activity of many cells in a beating heart, whether myogenic or neurogenic, produces quite sizable potentials which can be recorded on the surface of the animal's body. The pattern of electrical potentials is characteristic for each type of heart (Fig. 5.14). In general, a marked wave of depolarization may be expected with the contraction of each chamber. Thus, a single wave is recorded for each contraction of the sac-shaped or vesicular heart of the crab (Fig. 5.14a) and the long tubular heart of the insect (Fig. 5.14b), while two extremely slow but well marked waves occur during the cardiac cycle of the double chambered heart of the clam (Fig. 5.14c). Vertebrate hearts generate more complex sequences of waves since there are always three or more chambers

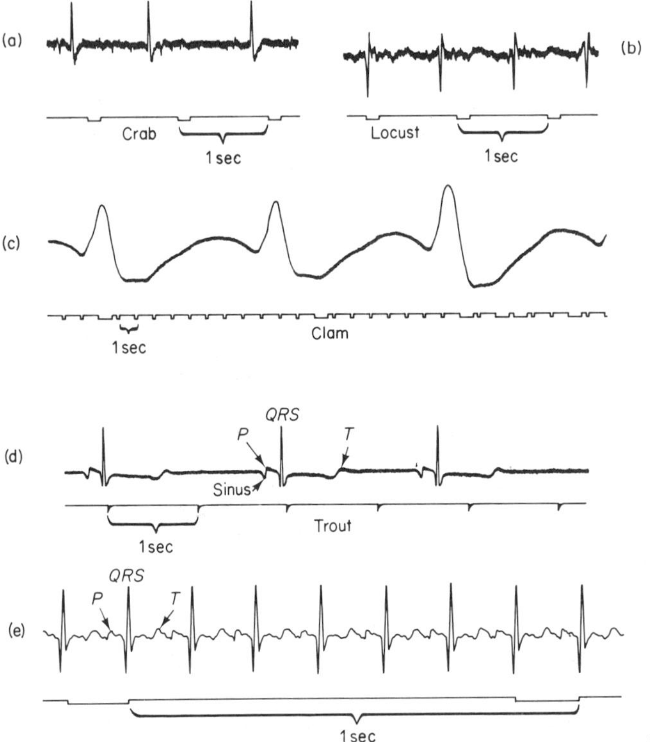

Fig. 5.14 Several forms of electrocardiograms. [Courtesy of D. R. Jones and Associates.]

involved, and since the repolarization of thick masses of muscle may also create characteristic waves. The ECG of man, so frequently recorded in the diagnoses of cardiac disease, is best known. It is typical of the double hearts of higher vertebrates where the atria contract together, as do also the ventricles, so that only two major components are associated with the cardiac contractions (Fig. 5.14e). These waves are, by convention, labeled *P*, *Q*, *R*, *S*, and *T*. The *P* wave and the *QRS* complex are caused by the depolarization of the muscle in the atria and the ventricles respectively. These waves of depolarization precede the contraction of the chambers. The *T* wave is a repolarization wave of the ventricle and marks the end of ventricular systole. The *P–R* interval repsesents the time required for excitation to spread from the pacemaker to the ventricle. The *QRS* wave associated with the spread of depolarization over the ventricle is a complex, dependent on numerous electrical changes impinging on the surface from the intricate arrangement of muscular units. The intervals and patterns of waves may be altered in a predictable manner by different experimental procedures and in various cardiac pathologies.

Cardiac activity in lower vertebrates, with separate sinus venosus and conus arteriosus, often shows additional waves associated with the contractions of these chambers. The spread of excitation in the sinus (*V* wave), which precedes the *P* wave, has been recorded in some fish and amphibians (Fig. 5.14d and Randall, 1970a).

Factors Modifying Cardiac Rhythm

Several factors are known to modify the rate and force of the heart either through an action on the pacemakers or on the cardiac muscle itself. The human heart may contract at anywhere from about 50 to 130 or more times per minute, depending on the age, health, activity, or emotions of the individual. Factors which modify cardiac rhythm may be conveniently grouped as chemical, mechanical, thermal and nervous.

Chemical control A consideration of the effects of the mineral environment should certainly begin with a description of Sydney Ringer's classical experiments on the influence of inorganic constituents of the blood on ventricular contraction. These were recorded in a series of papers published between 1880 and 1886 in the *Journal of Physiology* (particularly Volume 4) and are summarized in many textbooks of vertebrate physiology. Bayliss (1920) reproduced some of the original records.

Ringer perfused the frog heart with an isotonic solution of sodium chloride and found that it ceased beating after a short time and came to rest in a relaxed condition (diastole). If a small amount of calcium was then added, the beat was temporarily restored, but with excess calcium the heart was gradually arrested in a contracted condition (calcium rigor). Traces of potassium abolished the toxic action of calcium without destroying its ability to "neutralize" the sodium effect. By itself, the action of potassium chloride was like sodium chloride and favored relaxation. A balanced solution containing these three ions was thus shown necessary to maintain a rhythmic activity for any length of time. Ringer seems to have been the first to note that single cation solutions are extremely toxic.

We know that Ringer made a fundamental physiological discovery and that the activity of all tissues and cells is dependent, not only on the osmotic content of the medium but also on its specific ions. Single cation solutions are always toxic. A balance of ions is essential. This is the phenomenon of ION ANTAGONISM and may be demonstrated with many different tissues. Ringer's specific findings for the frog heart are in general true for other vertebrates, but many differences will be found in the effects of the particular ions if hearts of invertebrate animals are compared. For example, although high potassium favors a diastolic arrest and high calcium a systolic arrest in the frog heart, the reverse is true in many invertebrates, while the sodium effects are frequently, although not always, the same in both vertebrates and invertebrates (Prosser and Brown, 1961). The general rule is that a balance of monovalent and divalent ions is as essential to the heart as it is to other types of tissue; if certain ions are present in excess or not properly antagonized by other ions the rhythm will be affected. At the cellular and molecular levels, these ion effects are related to the cell membrane potentials (Chapter 15) and to the special role which calcium plays in contractile processes (Chapter 19).

The pH may also be important. In general, acid metabolites and a low pH favor relaxation while the reverse is true of alkalis (Mountcastle, 1968). Marked changes in pH destroy the rhythm. However, the heart, whether vertebrate or invertebrate, is often rather insensitive to experimental changes in the alkalinity or the acidity of the medium. This can probably be attributed to the low permeability of hydrogen or hydroxyl ions into cells (Heilbrunn, 1952).

In addition to the mineral constituents, there are many organic substances, both natural and synthetic, which modify cardiac activity. Several of the naturally occurring ones are considered with the transmitter substances in the discussion of nervous control.

Mechanical effects Muscle cells contract more forcefully when partially stretched. The improved performance of the partially distended heart is a reflection of this general property of muscle cells—whether they be smooth, skeletal or cardiac; adequate filling is an essential requisite to effective cardiac output. Starling (1918) first emphasized this in his studies of the vertebrate heart and expressed it as the oft-quoted "LAW OF THE HEART"—*the energy of contraction, however measured, is a function of the length of the fibers.* Starling's pioneer experiments and a great many of those that followed were carried out on heart-lung preparations from dogs. Experiments on lower vertebrates, particularly the frog, followed and it was widely concluded that, within physiological limits, the output per beat is directly proportional to the diastolic filling. The greater the volume of blood in the heart at the beginning of systole, the greater the amount of blood ejected per beat, although beyond a certain critical stretch the contractions will become weaker (Fig. 5.15a). It is now recognized that these findings were applied without justification to the hearts of intact mammals. Contrary to earlier views, the size of the mammalian heart and its stroke volume do not increase under conditions such as exercise which

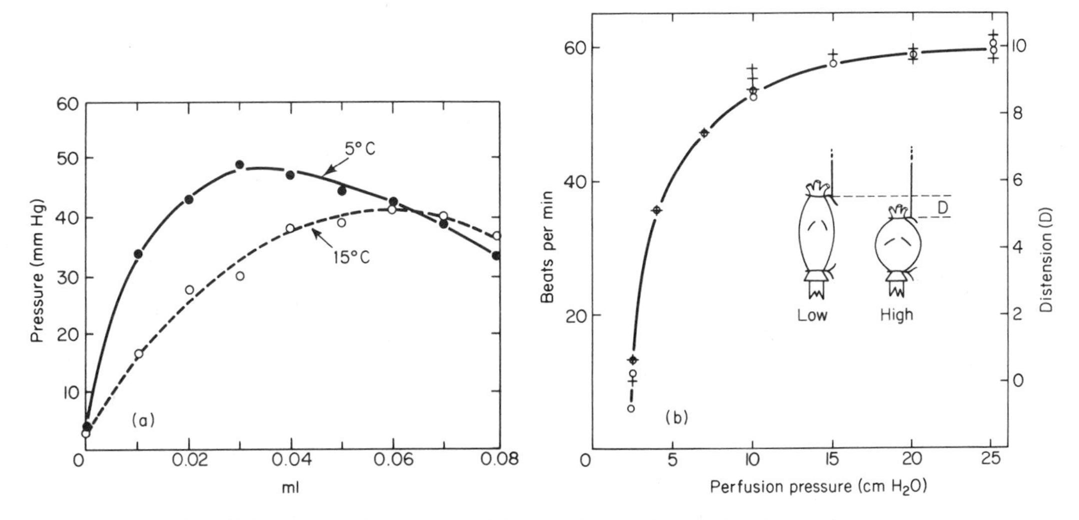

Fig. 5.15 Some factors modifying cardiac activity. (a) relation between tension or pressure developed by a frog heart contracting isometrically and its initial contained volume. [From Doi, 1920: J. Physiol., *54*: 224.] (b) Distension and heart rate in a malacostracan. Abscissa, perfusion pressure; left ordinate and open circles, heart rate; right ordinate and crosses, arbitrary measure (shown in inset) of distension (D) of heart resulting from perfusion pressure. [From Maynard, 1960: *In* The Physiology of Crustacea, Vol. 1., Waterman, ed. Academic Press, New York.]

augment venous return. An adjustment in rate looks after the increased flow. This difference between the mammalian heart and that of lower vertebrates has already been noted; fish hearts, for example, obey Starling's law whether isolated or in an intact animal during exercise (Randall, 1970a). These differences presumably depend on the more highly developed nervous and humoral controls of the mammalian heart. The adjustments in rate rather than stroke volume may safeguard the heart from overdistension. Starling's law still comes into operation in the intact mammal to adjust automatically the outputs of the two ventricles (Davson and Eggleton, 1968; Gordon *et al.*, 1972).

In some kinds of hearts the degree of inflation affects the frequency as well as the amplitude of contraction (Fig. 5.15b). In crustaceans the stretching of the cardiac tissues probably stimulates the ganglionic neurons in the organ, and these operate reflexly to control the frequency. In this case the inflation of the organ not only stretches the muscle and thus directly modifies its contractility but also activates ganglionic neurons which initiate the contraction phase. The crustacean heart and other neurogenic hearts of this type usually cease to beat and become quite unresponsive when isolated and empty. In all kinds of hearts, stretching affects contractility; in the neurogenic heart the rhythm may be set by the stretch reflex developed in the filling process.

Thermal effects Temperature has the expected action on cardiac frequency (Fig. 10.2) and Q_{10} values of about 2.0 have often been recorded for both poikilotherms and homeotherms. However, force of contraction as well as frequency is important in cardiac output and, in the frog and dog (heart-lung preparation) increasing frequency is associated with a diminished force. The diastolic filling is less complete, and the reduced tension on the fibers results in a less forceful contraction as discussed above. If in the frog the rate is kept constant, contractions are more forceful at higher temperatures (Mountcastle, 1968). At higher temperatures the optimal fiber length is greater for the development of the maximum tension (Fig. 5.15a). Further comment on temperature effects is reserved for Chapter 10.

Nervous regulation The visceral nervous system frequently plays an important role in cardiac activity. The intrinsic rhythm resides in the pump, and this pump often continues its regular pulsations when isolated from the nervous system. But in the living and intact animal the rate and force of the contraction are modified in accordance with the demands of the organism through a series of well-defined cardiac reflexes. Sensory afferent nerves, stimulated through pressure and chemical receptors, feed information into the cardiac centers. The efferent or motor fibers arising in these centers modify the activities of the cardiac muscle either directly or indirectly, through the pacemakers. The link between motor nerves and cardiac tissues is by way of the accelerating or inhibiting transmitter substances. Thus, as in any reflex system, the cardiac reflexes are based on three components: sensory nerves, cardiac nervous centers, and efferent cardio-regulatory nerves.

Experimental evidence indicates that the most likely transmitter substances are acetylcholine, a variety of amines (catecholamines, 5-hydroxytryptamine) and some of the amino acids (glutamic, aspartic, γ-aminobutyric). These substances were identified as transmitters in Chapter 2. Among vertebrates, the cardiac autonomic effects are mediated through the catecholamines and acetylcholine. Noradrenaline (or adrenaline) excites the heart, increasing both the rate (CHRONOTROPIC EFFECT) and the strength of the beat (INOTROPIC EFFECT); acetylcholine has the reverse effects and brings about inhibition. However, these generalizations are not without exception among the more primitive vertebrates; acetylcholine acts as a cardiac stimulant in the lamprey while adrenaline has little or no effect on the hagfish heart where a different amine, eptatretin, seems to act as the cardiac stimulant (Randall, 1970a).

At present it seems likely that the number of cardiac transmitters is much greater in the invertebrate phyla. Many pharmacological agents which have a marked effect on the heart have been isolated from invertebrate tissues; the evidence of their actions as natural transmitters, however, is still largely circumstantial (Maynard, 1960; J. C. Jones, 1964; Florey, 1966; Hill and Welsh, 1966). Typically, acetylcholine accelerates neurogenic hearts and inhibits the myogenic ones as it does the myogenic vertebrate heart. Catecholamines and 5-HT usually excite the hearts of invertebrates; 5-HT is particularly effective. Sometimes, however, the catecholamines are inhibitory and the effects vary considerably with concentration. Detailed summaries of the effects are given in the reviews already cited.

Special organs for the accumulation and storage of neurohumors or neurosecretory substances are well known in the animal kingdom. The pericardial organs of the Malacostraca are probably structures of this kind (Maynard, 1961). First described in *Squilla*, they have now been identified in many of the Malacostraca and may be a general feature of the anatomy of this group. They differ morphologically and geographically in various genera. Essentially, they are masses of interlacing nerve fibers forming a network through which the blood passes before returning to the pericardium from the gills. The location of the nerve cell bodies has not been determined, but it has been suggested that the neurons arise in the ventral thoracic ganglion (Maynard, 1960). Extracts of the pericardial organs always increase the amplitude of the heart beat, but frequency may be increased in some species and decreased in others. The chemical mediator has an action almost indistinguishable from 5-hydroxytryptamine (Carlisle and Knowles, 1959). Neurohormones are also involved in the regulation of cardiac activity of insects. Accelerating substances have been extracted from the corpora cardiaca of several different species while decelerating effects have been demonstrated with extracts from the corpora allata (Sternberg, 1963; J. C. Jones, 1964).

The CARDIAC REGULATING CENTERS have been localized in several groups of animals. In the vertebrates, both cardio-accelerator and cardio-inhibitor centers are functionally independent although closely associated morphologically in the floor of the fourth ventricle of the brain. Specific ganglia and nerves have been associated with cardio-acceleration or inhibition, or both, in many of the molluscs. In the Malacostraca, cardio-inhibitory fibers arise in the subesophageal ganglion and acceleratory fibers in the region of the third maxilliped and fourth walking leg (Maynard, 1960). Both excitatory and inhibitory effects have recently been examined in *Limulus* (Abbot *et al.*, 1969), in *Squilla* (Pax, 1969; Watanabe *et al.*, 1967) and in a scorpion *Urodacus* (Zwicky, 1968). These workers also discuss the transmitters likely to be involved. It is evident that in many invertebrates cardiac rhythm is modified by nervous reflexes. In some cases only inhibitory fibers have been found and in others only accelerators; Kriebel (1968b) has reported an absence of extrinsic nervous control in the tunicate.

The nervous regulation of cardiac activity is best understood in the vertebrates. The AFFERENT SENSORY LIMB of the reflex is excited by a variety of stimuli. Pressure receptors are located at strategic points in the vascular channels while chemical receptors, responsive to variations in carbon dioxide or pH and perhaps oxygen tensions, may be located in either blood vessels or central nervous tissue; these ensure appropriate changes in pumping action. In addition, many receptor organs associated with protective reflexes alter the activity of the cardiac centers, and in the higher vertebrates cerebral processes, quite divorced from peripheral stimulation, often have a significant effect.

A branchial depressor reflex is well known in teleosts and elasmobranchs (Randall, 1970a). Elevation of the branchial blood pressure brings about a reflex slowing of the heart. This might protect the delicate gill capillaries. The branchial reflex is evidently homologous with the aortic depressor and carotid sinus reflexes in the terrestrial vertebrate. These reflexes have been carefully investigated in the mammal. Stretch receptors in the wall of the aorta, the root of the innominate, and the carotid sinus respond to pressure changes. Under "normal" conditions a flow of tonic impulses along the aortic branches of the vagi and the sinus branches of the glossopharyngeal nerves reaches the cardiac centers in the medulla and maintains a continued vagal restraint on the heart. Increases in the pressure augment the flow of impulses and intensify the vagal inhibition, while a decrease in the pressure has the reverse effect. These reflexes in higher vertebrates are associated with other pressure reflexes originating in the walls of the great veins and adjacent right auricle. Stretching of these tissues which form the receiving chambers of the heart increases the flow of impulses via afferents to the medulla and results in cardio-acceleration through decreased vagal inhibition and increased sympathetic acceleration. These important reflexes provide a nice control of cardiac activity in accordance with variations in blood pressure (Fig. 5.16). Similar reflexes may exist in the lower vertebrates and invertebrates but have not been convincingly demonstrated. In the poikilothermous tetrapods, structures homologous with the carotid sinus are anatomically well known but have not been studied physiologically. The carotid labyrinth of amphibians, for example, appears considerably more complex than the carotid sinus of other groups of land vertebrates (Adams, 1958).

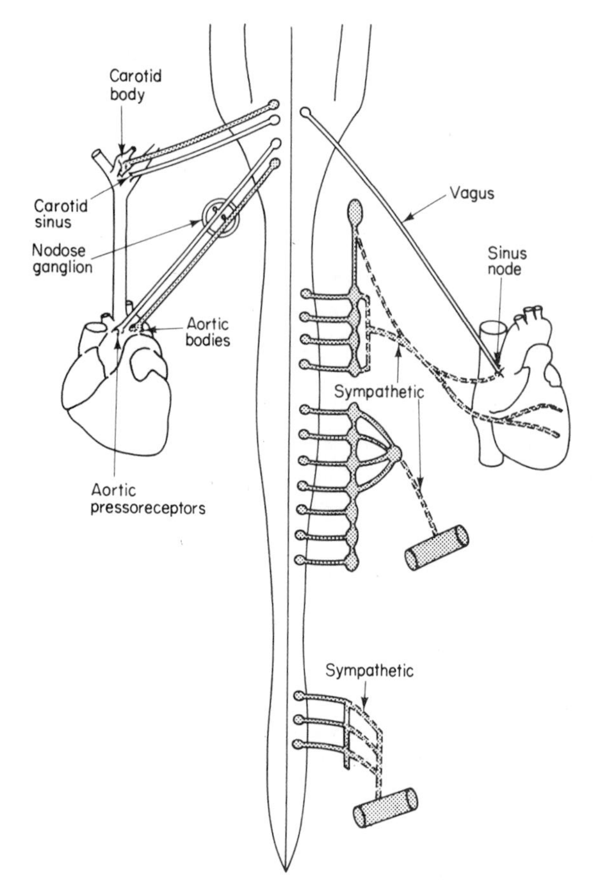

Fig. 5.16 Nervous control of mammalian blood pressure. Sensory elements are shown on the left; effectors are shown on the right. Vagus nerve alters rate through its effect on the S–A node; sympathetic fibers are distributed to both S–A node and ventricular muscle and alter both rate and strength of contraction; sympathetic fibers also alter pressure by widespread innervation of arterioles. [From Ruch and Patton, 1965: Physiology and Biophysics. Saunders, Philadelphia.]

Chemoreceptors, sensitive to changes in carbon dioxide (in reality changes in pH) and oxygen tensions of the blood, modify the cardiac rhythm of higher vertebrates. The carotid and aortic bodies contain such receptors and operate to increase cardiac activity when the blood becomes more acid and anoxic. Alterations in blood pH may also act directly on the cardio-accelerator center of the hindbrain. The aortic and carotid bodies are phylogenetically derived from the branchial pouches of the lower vertebrates; several unsuccessful attempts have been made to demonstrate comparable chemoreceptors in the gill regions of fishes (Hughes and Shelton, 1962; Shelton, 1970).

A decrease in the heart rate has been noted in several different teleosts exposed to water with a reduced oxygen content. There may be little change in the cardiac output since stroke volume tends to increase. However, the important consideration at this point is the reflex slowing of the heart in response to hypoxia. It is not yet known whether these reflexes in the fish are initiated by peripheral receptors or arise centrally through changes in the oxygen tension of the blood. The search for chemoreceptors in the gills and pseudobranch has been fruitless. The experimental work has involved the monitoring of electrical activity in nerves of gills (Sutterlin and Saunders, 1969) and deafferentation of the pseudobranch (Randall and Jones, 1973).

Chemoreceptors involved in cardiac reflexes have not been demonstrated in the invertebrates. However, variations in the carbon dioxide and oxygen tensions of the surroundings alter cardiac activity in *Daphnia* (Maynard, 1960) and probably in other invertebrates, and this action could be a direct one on the heart or its pacemaking system; it could also operate indirectly on these tissues through cardiac reflexes. *Carcinus* develops a progressive bradycardia during hypoxia in water; oxygen receptors are suspected (Taylor *et al.*, 1973).

Sensory stimulation, particularly stimuli eliciting protective reflexes, often activates the cardiac reflexes. In man, an inhalation of irritating vapors or stimulation of the integumentary pain receptors acts in this manner. Mechanical stimulation of the skin and abdominal viscera of the skate (*Raia* sp.) likewise alters the heart rate and raises the blood pressure. In the decapods, removal of a leg or placing foreign materials in the mouth have also been shown to activate cardiac reflexes (Maynard, 1960). This type of protective reflex is widespread among animals.

Blood Pressures and the Control of Peripheral Resistance

Several functions of the circulatory system depend on high and relatively constant pressures. Food and oxygen requirements of dense masses of active tissue require the flow of blood through minute channels, and this creates considerable flow resistance. Adequate blood pressures may also be essential for reasons other than the maintenance of flow through the capillaries. The filtration processes of the kidney and the transfer of nutritive fluids from the capillaries into the tissues are dependent on blood pressures. Locomotion in many invertebrates is contingent upon pressures developed in fluid-filled spaces. The erectile tissue in the copulatory organs of the vertebrates also depends on local accumulations of blood under pressure. This mechanism is found in several places in the animal body where turgidity is required.

Arterial blood pressure is the product of cardiac output and the peripheral resistance which develops when viscous fluids are pumped through a system of small tubes. Factors that modify the output of the heart have already been considered. Under normal conditions, blood viscosity is relatively constant and peripheral resistance is regulated by altering the diameters of the blood vessels.

Among the vertebrates, the smooth muscle of the arterioles controls peripheral resistance. Vasomotor reflexes operate to vary the caliber of these vessels and thus control the blood pressure and the flow to different areas. The finest of the arterioles (metarterioles) constitute precapillary sphincters or stopcocks, and their contractions can effectively stop the circulation into an area of the capillary bed. In many places, arteriovenous anastomoses (thoroughfare channels) can shunt the blood directly from arterioles to venules and thus bypass an area of the capillary bed. The vasomotor reflexes which control the smallest blood vessels are well defined and have been carefully studied in the higher vertebrates.

These reflexes parallel those already described for the heart which, in a sense, is only a specialized region of the vascular channels. The vasomotor centers, like the cardiac centers, are located in the floor of the medulla. Functionally, if not anatomically, a distinct vasoconstrictor and vasodilator region may be distinguished. The vasoconstrictor center, in particular, exhibits a constant tonic action on the arterioles. As in the case of the cardiac centers, a group of reflexes activated by pressure changes in the great veins and right auricle of the heart (vasopressor or McDowall reflex) and in the aortic and carotid bodies leads to appropriate adjustments in blood pressure by a modification of the peripheral resistance at the level of the precapillary sphincters. Pain and temperature receptors as well as receptors in the peritoneum and viscera may also activate the vasopressor reflexes. The chemoreceptors in the carotid and aortic bodies act on the peripheral resistance in a manner comparable to the cardiac reflexes already described. In addition, chemicals, such as carbon dioxide, adrenaline, noradrenaline, histamine, and others can modify the tone of these small vessels directly so that strictly local adjustments in circulation may occur.

Vasomotor control and peripheral resistance of the type found in the vertebrates are not found in most invertebrates. Even the larger representatives have relatively low blood pressures (Table 5.1) which vary greatly with activity. In fact, pressure changes which accompany locomotion in many invertebrates may be considerably greater than those which follow cardiac and other vascular stimulation. The octopus and its allies, however, have an essentially closed circulation and a blood pressure which is comparable to that of some vertebrates (Table 5.1). Changes in the caliber of the blood vessels follow stimulation or sectioning of nerves and indicate the presence of vasomotor reflexes in this animal.

The fluid skeleton Many invertebrates rely on a hydrostatic skeleton. This skeleton, like the more familiar skeleton of the vertebrates, depends on the action of antagonistic groups of muscles; these operate, not on hard skeletal parts, but on a volume of fluid in a fixed space. The rise in pressure of the coelomic fluid of many invertebrates when they become active (Table 5.1) is the reflection of the operation of this fluid skeleton. Chapman (1958) discusses its physiology in many animals ranging from the amoeba to the higher invertebrates. Barrington (1967) also describes many examples. The best known are among the worms, molluscs, and echinoderms.

The body of the annelid operates on a fluid skeleton. Its length and diameter are altered through the alternate contraction of longitudinal and circular muscles acting against a cylindrical tube which contains an almost constant volume of fluid. The molluscs also make extensive use of hydrostatic skeletons. In the bivalves, burrowing and locomotion depend on the turgidity developed in the foot through the movement of hemolymph into extensive connective tissue spaces. In some forms, *Ensis*, for example, burrowing is extremely rapid. The extension of the foot is associated with a relaxation of the pedal muscles and the flow of hemolymph to the foot. The tip of the foot swells into a bulbous anchor; when the pedal muscles contract and the fluid is forced back into the mantle spaces, the animal moves quickly forward. Trueman (1966a,b, 1967, and earlier) has carefully described and compared the digging and burrowing activities of several polychaetes and bivalves. Extensions of the siphon in the lamellibranchs is likewise by the movements of fluid into tissue spaces, while the gastropod foot can be greatly expanded by locking the hemolymph in the pedal sinuses.

In cephalopods the hemal skeleton, as a part of the locomotory machinery, is reduced or completely lost. The activities and high rate of metabolism of these creatures demand a closed circulation, and large volumes of hemolymph can no longer be locked up in sinuses to provide a skeleton. Locomotion is by pallial jet propulsion, and the arms and mantle are operated by an interplay of muscles. The body space is no longer a hemocoel but a true body cavity developed independently in this group of molluscs as an extension of the pericardium around the rest of the viscera.

Erectile tissues of the vertebrates also operate through the accumulation of vascular fluids in blood spaces. The mammalian penis is the most highly developed example. In this case the pressure is generated by the heart and not through muscular action on fluids in a closed space. The penis contains a sponge-like system of irregular vascular spaces between the arteries and veins (the corpora cavernosa penis). These are greatly distended when filled with blood under pressure. Stimulation of the parasympathetic produces dilation of the arterioles and blood flows into these spaces. Venous return is then restricted through pressure on the thin-walled veins, and the organ becomes turgid with pressures equivalent to those in the carotid artery. Valves have been described in the larger veins. Constriction of the arteries (sympathetic stimulation) leads to a gradual escape of blood from the spongy tissue, and during the flaccid condition flow into this tissue is further restricted by longitudinal ridges in the intima of the arteries (Ruch and Patton, 1965).

THE LAWS OF CIRCULATION

In summary, the circulation of the blood and the pressures within the vascular channels depend on three major factors. The first of these is the GENERAL ACTIVITY OF THE ANIMAL. In small and primitive organisms this is the only force involved and it is still an important factor in the maintenance of flow in certain areas of the most highly organized animals—lymphatics and small veins of birds and mammals, for example. The second is the CARDIAC OUTPUT. The amount of blood pumped in a unit time is a function of the blood volume as well as of the force of the heart. Both the blood volume and the force of the heart are subject to considerable variation. The third factor is the PERIPHERAL RESISTANCE which is a measure of the flow resistance of the vascular tubes. This third factor is of major significance only in

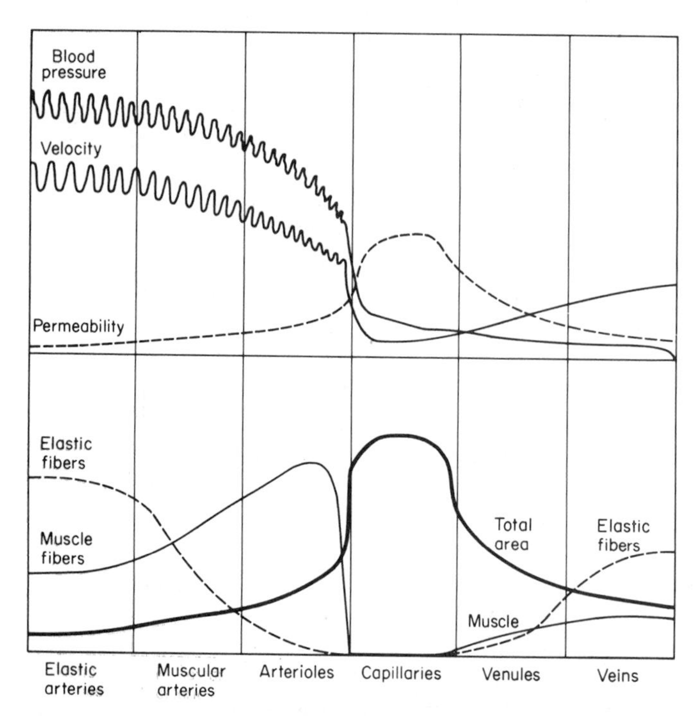

Fig. 5.17 Graph showing some of the changes in structure (below) and function (above) of vessels in the direction of blood flow from ventricle to atrium. [From Finerty and Cowdry, 1960: A Textbook of Histology, 5th ed. Lea and Febiger, Philadelphia.]

the vertebrates and larger invertebrates with closed circulations. In these, the velocity and pressure within different areas of the system follow familiar physical laws. From the arteries to the capillaries and on to the veins there is first a gradual widening of the total cross-sectional area of the system—to a maximum in the capillaries—and thence a narrowing as the veins come together in the heart. The pressure falls throughout the system, and the velocity varies inversely with the cross-sectional area of the stream. These relationships are shown in Fig. 5.17. They are familiar to anyone who has watched a turbulent river widen and discharge its waters into a placid lake; as the water flows from the lake it gains velocity once more while it moves on downstream to the ocean.

6

THE VASCULAR FLUIDS

At the lowest levels of phylogeny the functions of the interstitial fluids or lymph are essentially trophic. At the highest levels the vascular fluids have assumed several additional roles and, as a consequence, have changed both biochemically and morphologically. At all levels the nutritive or trophic functions depend on aqueous solutions of sugars, amino acids and other readily dissolving molecules, together with colloidal droplets of fatty substances. The respiratory gases, however, are insufficiently soluble to be carried in adequate amounts by simple solution. The phylogeny of a gaseous transport system depends on a group of special proteins and becomes one of the major events in the evolution of larger and more specialized animals. Certain other functions which are mainly absent in the simpler forms depend on the phylogeny of the albumins and globulins. The albumins play an important part in the distribution of body water because of their colloidal osmotic pressure. The globulins are active factors in an elaborate defense against disease and foreign materials; blood cells are also involved in this activity. Another group of plasma proteins is concerned with blood clotting mechanisms. The very existence of the higher animal is so dependent on the continuous flow of blood at high pressures that intricate mechanisms become associated with hemostasis and the clotting processes which occur following injury to the vascular channels. These several specialized activities of the vascular fluids (gaseous transport, the regulation of fluid volume, defense mechanisms and hemostasis) are discussed in this chapter.

THE TRANSPORT OF OXYGEN

Water in equilibrium with atmospheric air at 37°C dissolves about 0.46 volumes per cent of oxygen, and this is adequate only for tissues with low rates of metabolism.

Only about 1 per cent of man's total oxygen requirement can be transported in this way. Human blood in equilibrium with alveolar air combines with about 20.0 volumes per cent of oxygen. It is worth noting, however, that the small amounts of oxygen which can be carried in simple solution serve the needs of many invertebrates and some curious vertebrates living a sluggish existence in extremely cold waters. Ruud (1964) first described three fishes (Chaenichthyidae) from Antarctica which lack special transport pigments and have vascular fluids with oxygen-combining capacities only slightly greater than those of water (0.54 to 0.90 vol per cent). The Leptocephalus larvae of the eel also lack transport pigments, and several other fish (goldfish, carp, trout, pike) may, if they are not forced into any activity, live for hours after their transport pigments have been tied up with carbon monoxide (Anthony, 1961; Holeton, 1971, 1972). Tadpoles and young frogs are also known to survive the destruction or loss of their respiratory pigments (Flores and Frieden, 1968). However, even though a specialized gas transport system may be dispensed with under experimental conditions, it is absolutely essential to the larger and active animals with high rates of metabolism; it may also be required for the exploitation of habitats where the oxygen supplies are low or variable (J. D. Jones 1961, 1964 and Fig. 6.8).

Storage and transport of oxygen are achieved by a group of colored proteins capable of forming loose combinations with oxygen when exposed to it at high tensions, and of releasing the gas readily at the lower tensions which prevail in the tissues. They are quite different biochemically in the various phyla. Even in the same phylum there may be several distinct pigments, and more than one pigment may even exist in the same animal (Fox and Vevers, 1960; Manwell, 1960). One can only generalize by saying that they are colored proteins (chromoproteins) which contain a metallic atom in their constitution and have the property of forming loose combinations with oxygen and sometimes with carbon dioxide. Table 6.1 lists the known pigments together with their distribution and some of their properties.

Hemoglobin

Hemoglobin is the most familiar, the most widespread and the most efficient of the respiratory pigments. It is familiar because of its presence in human blood, but it occurs also in the plant world, in some protozoa, and in most of the major animal phyla (Fox and Vevers, 1960; Gratzer and Allison, 1960). As indicated in Table 6.1, the most efficient hemoglobins combine with far greater amounts of oxygen than any of the other pigments.

Hemoglobin is made up of an iron porphyrin compound, HEME, associated with a protein GLOBIN. Heme is a metalloporphyrin. It is composed of four pyrrole rings joined with methene groups to form a super-ring with an atom of ferrous iron in the center attached to the pyrrole nitrogens (Fig. 6.1). This heme component of the molecule is a constant feature of all hemoglobins, but the globin portion varies in different species. In addition, varying numbers of hemoglobin units may unite to form polymers of different size. For example, the muscle hemoglobin of all vertebrates (myoglobin) and blood oxyhemoglobin of cyclostomes correspond to one

Table 6.1 Oxygen Capacities of Some Different Bloods

Pigment	Color	Site	Animal	Oxygen Vol. per cent
Hemoglobin	Red	Corpuscles	Mammals	15–30
			Birds	20–25
			Reptiles	7–12
			Amphibians	3–10
			Fishes	4–20
		Plasma	Annelids	1–10
			Molluscs	1–6
Hemocyanin	Blue	Plasma	Molluscs:	
			Gastropods	1–3
			Cephalopods	3–5
			Crustaceans	1–4
Chlorocruorin	Green	Plasma	Annelids	9
Hemerythrin	Red	Corpuscles	Annelids	2

SOURCE: Values selected from Prosser and Brown (1961) and Nicol (1967).

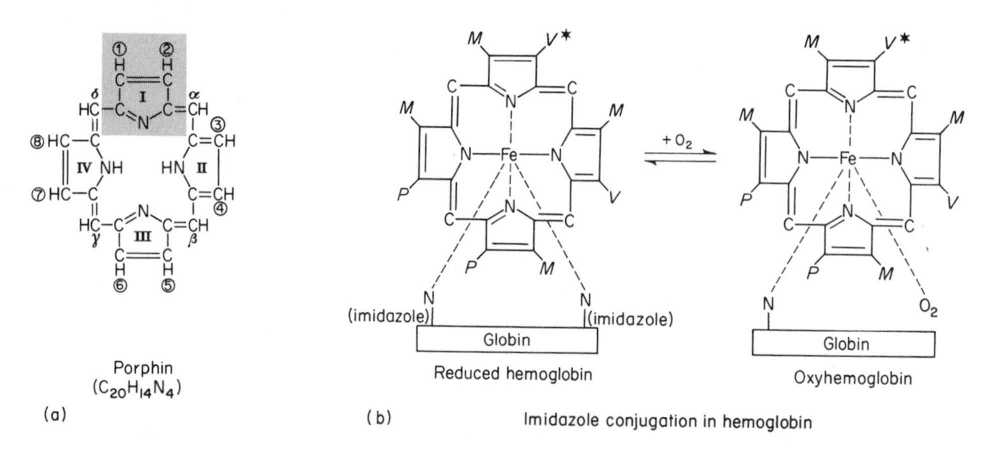

Fig. 6.1 Porphyrins. (a) Skeletal structure of porphyrin molecule (tetrapyrrole ring) with conventional numbering of positions and rings and with shading to show one pyrrole ring. (b) Imidazole conjugation in hemoglobin. *M*, methyl group —CH_3; *V*, vinyl group —CH ═ CH_2; *P*, propionic group —CH_2—CH_2—COOH. In chlorocruorin, position 2 (asterisk) is filled by the formyl group O═CH.

unit with a molecular weight of 16,500 to 17,000. Two basic units (mol wt 34,000) are united in the hemoglobin of the polychaete worms *Glycera* and *Notomastus* and the insect *Gastrophilus*. The molecular weight of the vascular hemoglobin of most vertebrates corresponds to four units (mol wt about 67,000), while in some of the annelids (*Arenicola* and *Lumbricus*) the molecule may correspond to 180 units or a weight of 3,000,000. To repeat, no matter what the variability in molecular weight and protein structure, the heme portion of the molecule is the same. This portion belongs to a phylogenetically ancient group of biochemicals, the metalloporphyrins, and the machinery necessary for its synthesis seems to be universal.

The metalloporphyrins were encountered in Chapter 1. It was pointed out that animal life only became possible when chlorophyll and the cytochromes became a part of the photosynthetic processes which transformed an anaerobic into an aerobic world, and when the cytochrome carriers permitted the release of large amounts of energy in aerobic respiration. All animals possess the biochemical machinery for the manufacture of cytochromes and these materials are structurally very similar to hemoglobin (cf. Figs. 6.1. and 7.7). Cytochrome oxidase is present, whether or not an animal possesses hemoglobin, and thus all animals depend on biochemical combinations of oxygen and a metalloporphyrin compound. In view of the close relation between hemoglobin and the cytochromes and the widespread occurrence of hemoglobin in so many different animal groups, it seems likely that, phylogenetically, it is a very ancient respiratory pigment (Gratzer and Allison, 1960).

The many different hemoglobins vary in oxygen-combining capacities (Table 6.1.) This variation in oxygen capacity is a property of the total molecule and does not depend on structural differences or changes in the metalloporphyrin component. In all cases the atom of ferrous iron (heme unit) is associated with one molecule of oxygen to form oxyhemoglobin. The reaction is readily reversible; the unoxygenated compound is referred to as deoxyhemoglobin or less accurately as reduced hemoglobin. These are not enzymatic reactions; whether or not the heme unit combines with oxygen depends not only on the availability of the oxygen but on the pH and ionic content of the solution as well as on the structure of the total hemoglobin molecule. In contrast, the change of cytochrome from an oxygen-poor to an oxygen-rich state involves a valence change of iron from the ferrous to the ferric state and is governed by specific enzymes (Chapter 7).

Since all proteins are species-specific, it is not surprising that the hemoglobin molecule differs in various animals. The situation, however, is even more complex. Several forms of hemoglobin are regularly found within the same animal. Two or three hemoglobins have been described in molluscs and echinoderms while six to nine were reported in the insect *Chironomus* (Manwell, 1966; K. P. H. Read, 1966, 1968; Terwilliger and Read, 1969). Multiple hemoglobins are characteristic of all of the vertebrates (Riggs, 1970). Further, there may be a succession of hemoglobins during development, with each type adapted to the respiratory needs of the particular stage of development. Details have been studied more carefully in the higher vertebrates (Zuckerkandl, 1965; Perutz, 1969). In man there are four different peptide chains in the vascular hemoglobins with the relative amounts of each characteristic of the

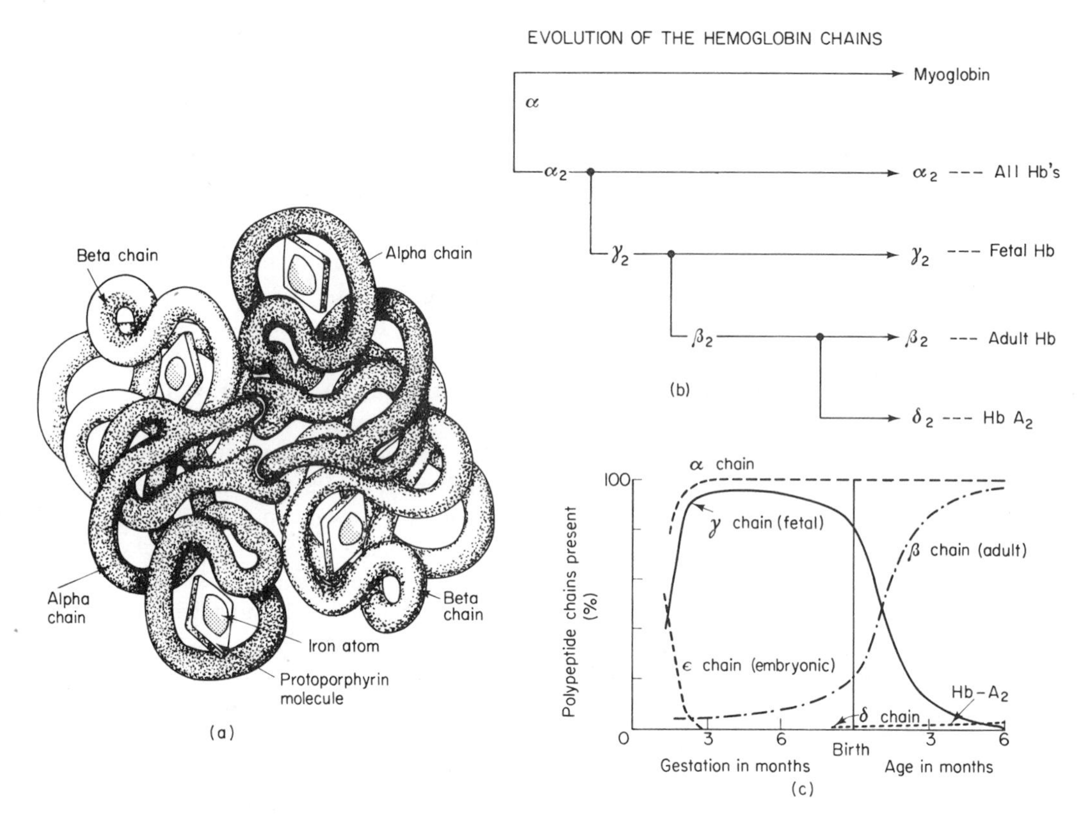

Fig. 6.2 Human hemoglobin. (a) Schematic representation of molecule. [From Stone, 1969: Nutrition Today, *4*(2): 3.] (b) Evolution of hemoglobin chains; the chains which are most different are put farthest apart; the point in time of a gene duplication is indicated by the solid black circles. [From Ingram, 1963: The Hemoglobins in Genetics and Evolution. Columbia U. P. New York.] (c) Peptide chain formation in the human fetus. [From Ingram, 1967: Harvey Lectures, *61*: 58.]

stage of development (Fig. 6.2). In the adult human, 90 per cent consists of two α-chains combined with two β-chains to form single tetrameric molecules (called hemoglobin A). There are 141 amino acids in an α-peptide chain while each β-chain contains 146; when the two chains are placed side by side, 64 of the amino acids are common to both. Adult blood also contains a small amount of hemoglobin A_2 in which the α-chains are replaced by δ-chains. During early embryonic life, somewhat different ε-chains are associated with the α-chains while during later fetal development there is a switch from ε-chains to γ-chains. These molecular changes confer different transport properties which are nicely adapted to the changing respiratory requirements. There is reason to believe that the multiple hemoglobins have evolved from a monomer and that the α-chain is ancestral in man while the other forms have arisen through gene duplication (Fig. 6.2b). In the total structure of the tetramer, a molecule of heme is linked to one of the histidine residues of each chain; the four chains are folded together into an intricate molecule which has now been mapped in considerable detail by methods of X-ray analysis (Perutz, 1964, 1969; Fig. 6.2). Several additional types of human hemoglobin may arise through mutation. Small changes in the protein molecule may greatly alter its physiology. The difference between sickle-cell and normal hemoglobin is a difference in only one of the amino acids of the peptide chain and depends on the presence of a single mutant gene.

Oxygen equilibrium curves An understanding of the functional properties of the different respiratory pigments is greatly facilitated by an analysis of the OXYGEN EQUILIBRIUM CURVES (also called OXYGEN DISSOCIATION CURVES). These curves are developed by determining the amount of oxygen which combines with blood exposed to oxygen at a series of pressures. The amount of gas combined with the blood at equilibrium is expressed as a per cent of the amount when saturated. The equilibrium curve for mammalian blood is sigmoid. It is evident (Fig. 6.3) that almost complete saturation occurs at the tensions of oxygen found in the lungs (about 95 mm Hg) and that the oxygen is very quickly lost at the tensions normally found in the tissues (about 40 mm). In this way, there is an efficient transport and a rapid unloading where required.

The curve for muscle hemoglobin or myoglobin, in marked contrast, is hyperbolic, and the unloading occurs only at very low tensions (about 5 mm). Myoglobin is a storage pigment. The red muscles are thus able to hold considerable amounts of

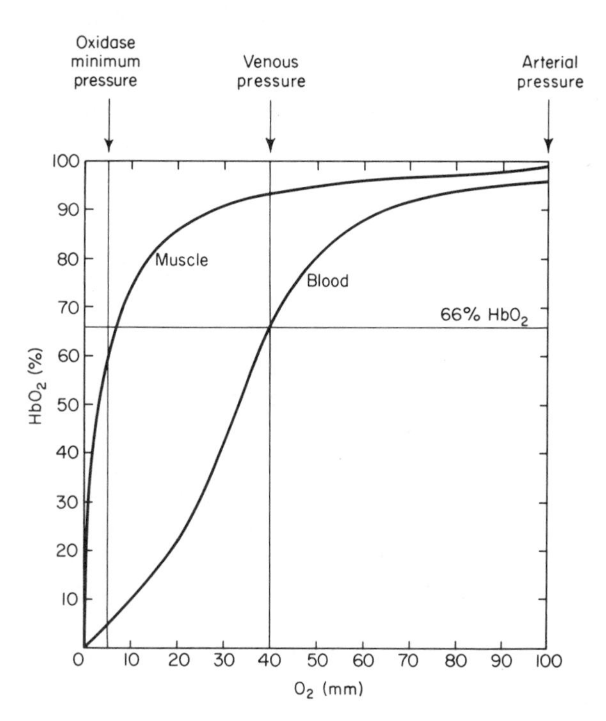

Fig. 6.3 Oxygen equilibrium curves of mammalian myoglobin and hemoglobin measured at body temperature and pH 7.0. The vertical line at the left indicates the oxygen pressure at which the rate of tissue respiration first begins to decline; presumably it represents the pressure at which the cytochrome oxidase system is just saturated with oxygen. The oxygen affinity of myoglobin lies between those of hemoglobin and the cellular oxidase system. The result is that myoglobin readily takes up oxygen from the blood and gives it up to the tissues. [From Wald, 1952: *In* Modern Trends in Physiology and Biochemistry, Barrow, ed. Academic Press, New York.]

oxygen and supply this to the cells when they are thrown into activity which overtaxes the regular delivery service. The oxygen debt is less than it would otherwise be. In the case of myoglobin, the pigment is probably something more than the kind of storehouse which is exploited only in emergencies. As indicated in Fig. 6.3, the oxygen affinity of myoglobin lies between that of the blood hemoglobin and the oxygen tension at which the cytochrome oxidase system is just saturated with oxygen. Consequently, myoglobin will readily pick up oxygen from the blood and deliver it to appropriate enzyme systems of the tissues. This shows that the various hemoglobins may come into the oxygen transfer systems at several different levels.

Many adjustments in the shape of the equilibrium curves have evidently been made during phylogeny. Changes also occur during ontogeny. Developing animals often live in environments of relatively low oxygen tension. The amphibian tadpole in its aquatic habitat has much less oxygen available than the terrestrial adult. Tadpole blood has a more rectangular equilibrium curve than frog blood (Frieden, 1963) and consequently shows a maximum affinity for oxygen at low tensions. Although the waters where the tadpole lives are probably in equilibrium with atmospheric oxygen, the Po_2 at the gill surface may be extremely low because of the slow rates of gas diffusion in water. The rectangular equilibrium curve also indicates that tadpole tissues must operate at low oxygen tensions. In contrast to the tadpole, frog hemoglobin displays a sigmoid equilibrium curve; its hemoglobin is adapted to load oxygen at the higher tensions available to air breathers and unload it readily in the tissues of the more active land animal. Furthermore frog hemoglobin, in contrast to that of the tadpole, is markedly sensitive to CO_2 (Bohr effect) and releases oxygen more rapidly as the tissue CO_2 rises. It has already been noted (Chapter 4) that land animals are much more likely to accumulate CO_2 in their tissues than are the aquatic forms. The equilibrium curves of the avian embryo in its shell and the mammalian fetus in the uterus are likewise adjusted to their environmental limitations. In the case of the viviparous species, this permits an easy unloading of oxygen from maternal to fetal bloods (Fig 6.4).

Chlorocruorin

Chlorocruorin is a beautiful green metalloporphyrin (Fig. 6.1) closely allied to hemoglobin and the cytochromes. Its prosthetic group resembles that of cytochrome *A* (Baldwin, 1967) and, like hemoglobin, suggests a phylogenetic relationship to the cytochromes. Chlorocruorin is restricted in its distribution to four families of polychaete annelids (Manwell, 1960; Florkin, 1969a). It is never found in the cells; but as a plasma chromoprotein it has as great an oxygen-combining power as the comparable hemoglobins (Table 6.1). This is not surprising since the two pigments are so similar biochemically. It is interesting that within the same family of worms (Sabellidae, Serpulidae and Ampharetidae) some species have chlorocruorin while others have hemoglobin. Further, in one genus *Serpula* both of the pigments are present in the blood, and the relative amounts vary with the age; younger individuals

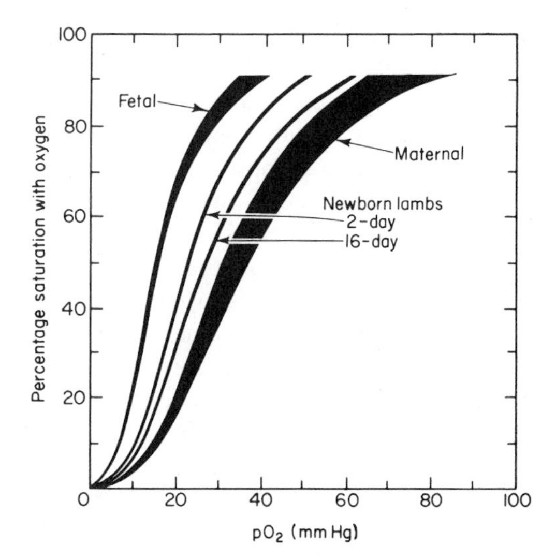

Fig. 6.4 Oxygen equilibrium curves at 38°C and pH 7.4 of fetal, newborn, and maternal sheep blood. For fetuses and adults the area is indicated within which the individual curves fall. [From Meschia *et al.*, 1961: Quart. J. Exp. Physiol., *46*: 98.]

have more of the hemoglobin. In the sabellid *Potamilla,* chlorocruorin is the blood pigment but the muscles contain hemoglobin. These various facts suggest that a genetic mutation produced the chlorocruorin molecule in a world where hemoglobin already existed and that the mutation was, for some reason, preserved (Fox and Vevers, 1960; Jones, 1963).

Other Respiratory Pigments

The remaining respiratory pigments lack the porphyrin nucleus; although they go by the names hemocyanin, hemerythrin, and sometimes hemovanadium, there is, in fact, no heme component. Hemocyanin is of wide occurrence and discharges those functions already discussed for hemoglobin. Hemerythrin occurs only in a few groups of animals and may not be concerned with the transport of oxygen although the storage function is present. Hemovanadium can probably be removed from the list of respiratory pigments (Manwell, 1960; Carlisle, 1968).

HEMOCYANIN is the only one of the non-heme respiratory pigments which is at all abundant in the animal kingdom. Redfield (1934) did much of the pioneer work on this pigment, and his reviews should be consulted for the early literature. More recent literature is summarized in the symposium edited by Ghiretti (1968). Hemocyanin is a copper-containing pigment which occurs in many molluscs and arthropods and is the blood pigment of the largest and most active representatives of these phyla. However, it is not considered to be a phylogenetically primitive pigment; some of the more lowly members of these two phyla possess hemoglobin (the mollusc *Arca* and the crustacean *Daphnia,* for example). Further, the gastropod *Buccinum* and a few other invertebrates have hemoglobin in the muscles and hemocyanin in the blood. Although the hemocyanin molecule seems to be biochemically simpler than hemoglobin, the phylogenetic distribution does not argue for a greater evolutionary antiquity.

Hemocyanin molecules are built up of units consisting of one copper atom associated with a peptide chain of just over 200 amino acids. The molecular weights of these units approximate 25,000 in molluscan hemocyanins and 37,000 in the arthropod pigments. Since two copper atoms are required to hold one molecule of oxygen, it follows that the minimum molecular weights of hemocyanin are 50,000 and 74,000 for molluscs and arthropods, respectively. In deoxyhemocyanin, the copper is in the cuprous form. During oxygenation, the blood changes from almost colorless to blue and it has been claimed that one of the two copper atoms is reversibly oxidized to the cupric form; this, however, is by no means certain (Ghiretti, 1966a). Although the molecule with two copper atoms is theoretically the minimum sized oxygen combining unit, some polymerization is probably universal. Hemocyanin never occurs in blood cells; the quantity of vascular pigment is increased by forming giant molecules. In the spiny lobster (*Palinurus*) the molecular weight is 447,000; in *Octopus vulgaris* it is 2,785,000 and in the snail *Helix pomatia* 6,650,000 (Fox and Vevers, 1960).

The pigment evidently functions in transport as well as in storage. In some cephalopods the oxygen capacity compares with that of the less efficient hemoglobins although it is always much lower than the vertebrate hemoglobins (Table 6.1). The equilibrium curves are rather rectangular, but the shapes depend markedly on the pH (Fig. 6.5) and the temperature (Fig. 6.6). In some species they depend also on

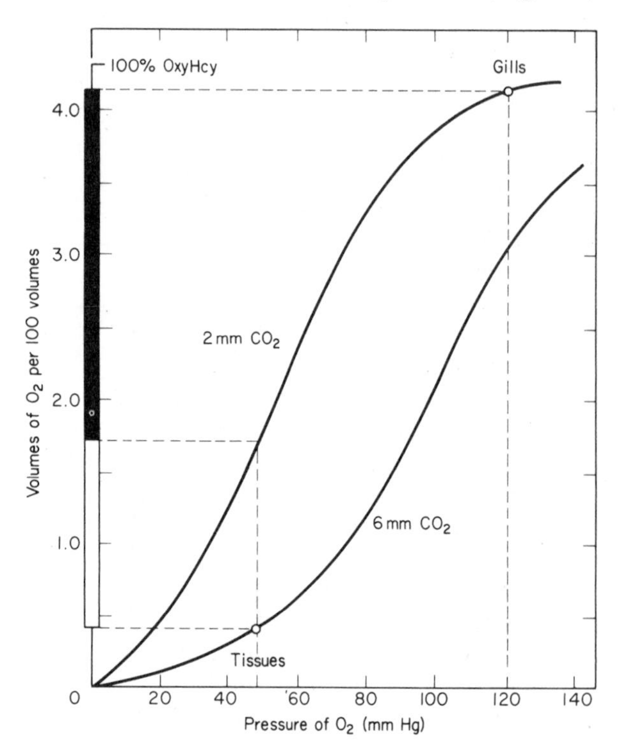

Fig. 6.5 Oxygen transport by the hemocyanin of the squid *Loligo pealeii*. Upper curve, conditions in the gills with 2 mm Hg CO_2 present; lower curve, conditions in body tissues with 6 mm Hg CO_2; vertical dashed lines, average partial pressures of oxygen in blood following passage through gills and tissues; black vertical bar, quantity of oxygen delivered if there was no pH change as blood passed through tissues; white vertical bar, additional oxygen released due to shift in equilibrium curve as CO_2 diffuses into the blood from tissues; change in pH, approximately −0.13 units 23°C. [From Redmond, 1968: *In* Biochemistry and Physiology of Hemocyanins, Ghiretti, ed. Academic Press, New York.]

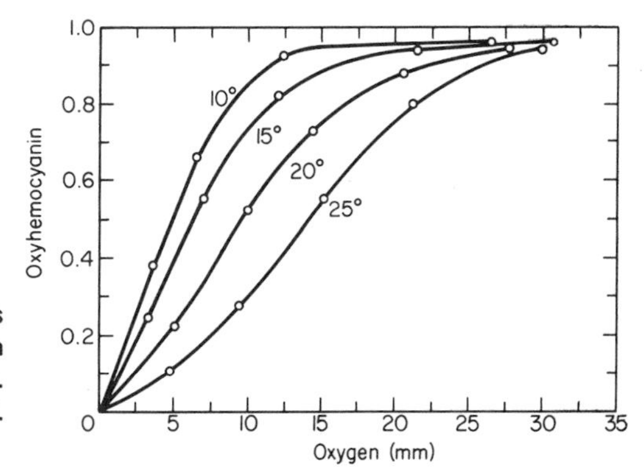

Fig. 6.6 Oxygen equilibrium curves of *Palinurus interruptus* hemocyanin at various temperatures (pH 7.53). [From Redmond, 1955: J. Cell. Comp. Physiol., *46*: 219.]

Ca^{++} (Fig. 6.7), a fact which may be of significance during molting when calcium levels are greatly altered (Chapter 24).

HEMERYTHRIN was first discovered in the ancient brachiopod *Lingula* and has been found in only a few other animals (the sipunculids, the priapulids, and one polychaete annelid, *Magelona*). It may occur in cells (coelomic corpuscles of *Sipunculus*) but is usually in plasma solution (Fox and Vevers, 1960).

Hemerythrin is a reddish violet non-heme iron protein. Molecular weights of 66,000 to 120,000 have been recorded (Manwell, 1960). In the sipunculid worm *Golfingia gouldii* there appear to be two slightly different types of polypeptide chains with molecular weights of 13,000 to 14,000 (Manwell, 1964). The active molecule contains 16 atoms of iron and has a molecular weight of about 107,000; it is thought to be an octamer. The oxygenated violet pink protein develops when one molecule of hemerythrin (16 atoms of iron) combines with eight molecules of oxygen and the

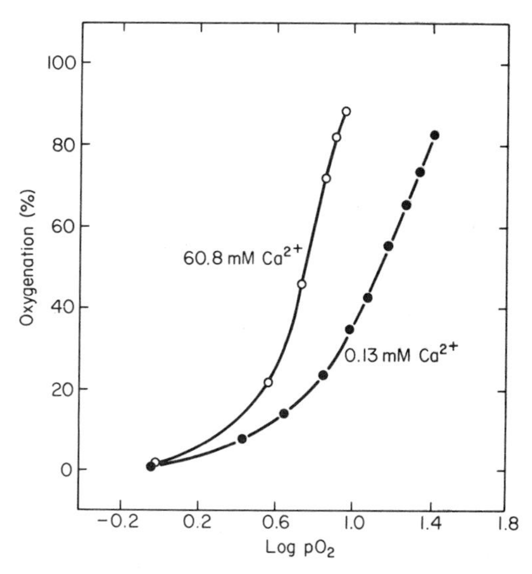

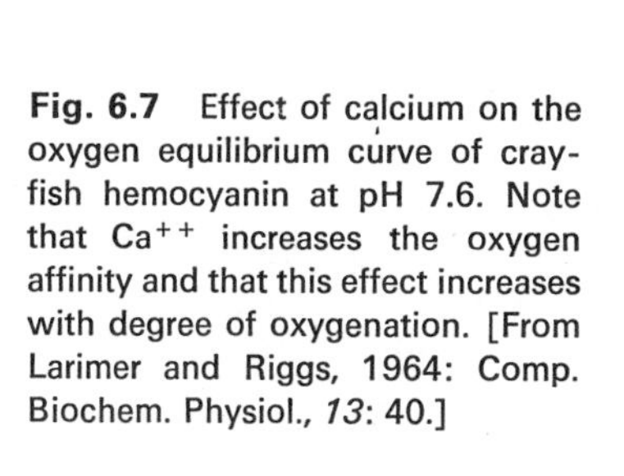
Fig. 6.7 Effect of calcium on the oxygen equilibrium curve of crayfish hemocyanin at pH 7.6. Note that Ca^{++} increases the oxygen affinity and that this effect increases with degree of oxygenation. [From Larimer and Riggs, 1964: Comp. Biochem. Physiol., *13*: 40.]

iron changes from the ferrous to the ferric state (Klotz *et al.,* 1957; Manwell, 1964; Florkin, 1969a). The actual means of coupling is not understood but in both hemerythrin and hemocyanin the oxygen molecule is thought to be held in a bridge between the two metal atoms (Manwell, 1964; Ghiretti, 1966a,b.).

Several other compounds, at one time thought to be respiratory pigments, have now been removed from the list either because more refined techniques have failed to show any capacity for the formation of reversible compounds with oxygen (the vanadium chromogens) or because the identification of the compound as a chromoprotein was probably mistaken (the manganese pigment, pinnaglobin, and the zinc compound, hemosycotypin). Vanadium chromogen (hemovanadium) remains interesting, however, for it is present in the blood cells (vanadocytes) of several families of ascidians and in the plasma of some others (Nicol, 1967). The compound contains pyrrole rings but not the porphyrin complex. It has been suggested that these rings form a chain as in the bilins (Fox and Vevers, 1960). It may be concerned with oxidation-reduction, but its function has not been established.

Phylogeny of Efficient Oxygen Transport

Pioneer investigators regarded respiratory pigments with rectangular equilibrium curves as storage pigments while those which displayed sigmoid curves were obviously concerned with transport. Thus, storage of oxygen for emergencies was thought to be the major function of myoglobin and the invertebrate hemoglobins (also called ERYTHROCRUORINS) while the vascular hemoglobins of vertebrates were concerned with transport. In part, the argument rested on the evident functional capacities of pigments with differently shaped curves. However, it also rested on rather superficial considerations of the environmental oxygen available to the animals. For example, a marine worm such as *Arenicola* might be presumed to obtain sufficient oxygen by diffusion during flood tide but to rely on stored oxygen while ventilation of its burrow is impossible during the intertidal period. The rectangular equilibrium curves of *Tubifex* hemoglobin (Fig. 6.8) were interpreted in the same way; this small animal lives in stagnant muck where oxygen tensions may be very low.

Investigations of the past two decades have shown this concept to be greatly oversimplified. In the first place, there is no clear separation between invertebrate and vertebrate pigments on the basis of shapes of equilibrium curves; marine invertebrates such as the sabellid worm *Schizobranchia* with chlorocruorin or the squid *Loligo* with hemocyanin display sigmoid curves, while the cyclostomes among the vertebrates have highly rectangular ones (Manwell, 1958, 1963; Redmond, 1968). In the second place, it is now realized that transport capacities are not confined to pigments with sigmoid equilibrium curves; pigments with hyperbolic curves like those of the annelids *Arenicola* and *Tubifex,* or the hagfish *Eptatretus* are transport pigments (Manwell, 1963); these animals live under conditions of low environmental oxygen and the very small differences between arterial and venous oxygen levels impose rectangular shapes on the equilibrium curves (cf. Fig. 6.3 and Fig. 6.8). In the third place, the myoglobins, which have been regarded as typically storage pigments,

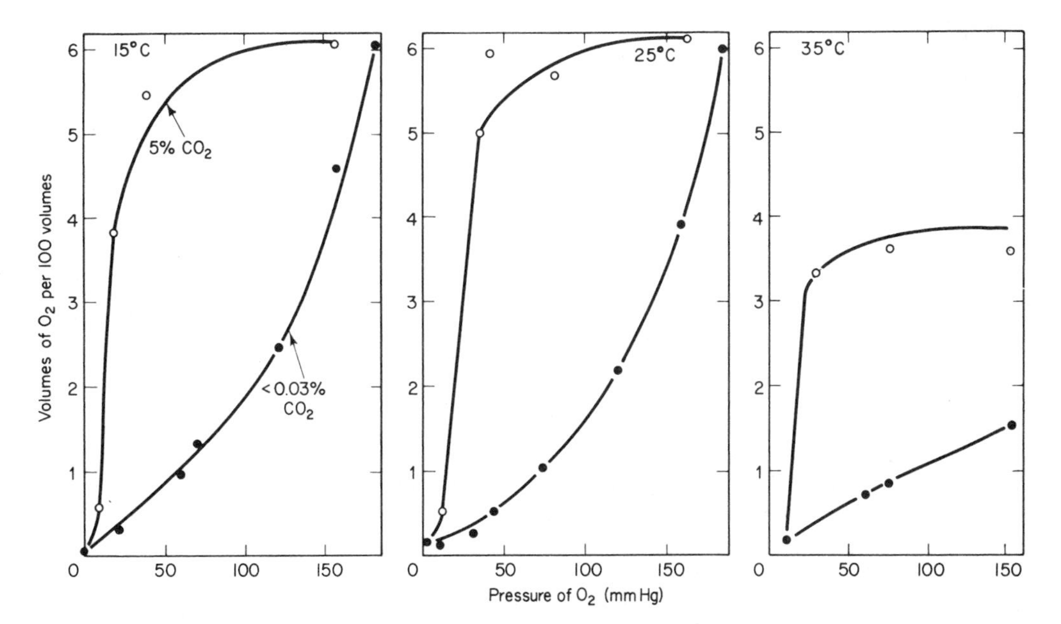

Fig. 6.8 Oxygen equilibrium curves for *Tubifex* hemoglobin. Note that 85 percent of total oxygen carried may be released over the small pressure drop of 40 to 10 mm Hg in the presence of 5 percent CO_2 (= 38 mm Hg P_{CO_2})—conditions which occur in the Thames River where these animals were collected. [From Palmer and Chapman, 1970: J. Zool., *161*: 206.]

may be less concerned with oxygen storage than they are with the facilitation of its transfer at the intracellular level (Jones, 1963; Manwell, 1963; Redmond, 1968); the vascular pigments transfer oxygen from the respiratory surfaces to the tissues while the intracellular hemoglobins may be concerned with its movement within the tissues (Fig. 6.9). Finally, biochemical investigations have failed to show any clear path of phylogeny among the different pigments; even in the single group of heme pigments there are quite different biosynthetic pathways among invertebrate and vertebrate animals (Mangum and Dales, 1965). The phylogeny of the respiratory pigments, like many other evolutionary trails, is not a straightforward path but an intricate maze. The important consideration is the versatility of this group of pigments in meeting the evolutionary demands of particular animals. At many levels in phylogeny, the pigments have capitalized on one or another of their biochemical properties to meet the oxygen transport needs of the animals. Biochemical properties which have been important in phylogenetic adaptation are: (1) an oxygen-combining dependence on pH or CO_2 and (2) on temperature; (3) the pattern of the oxygen equilibrium curve; (4) the concentration of pigment and (5) the effects of certain ions.

Oxygen-combining dependence on pH or CO_2 and on temperature Dependence of blood oxygen tension on pH facilitates the liberation of oxygen to the tissues

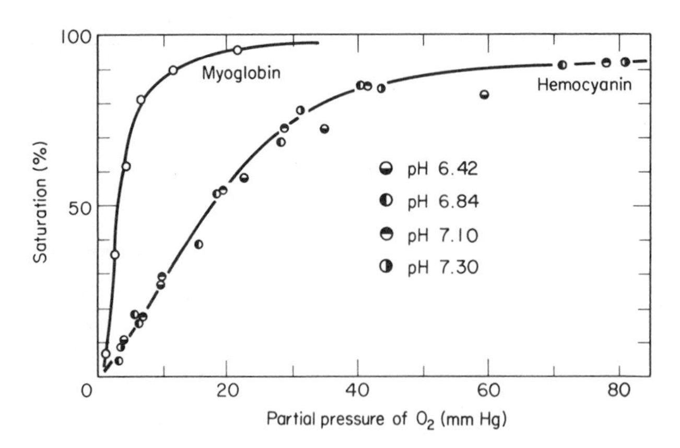

Fig. 6.9 Oxygen equilibrium curves at 10°C and pH 7.3 for myoglobin and hemocyanin for the amphineuran mollusc *Cryptochiton stelleri.* Hemocyanin is concerned with vascular transport and hemoglobin with intracellular or tissue transfer. [From Ghiretti, 1966: *In* Physiology of Mollusca, vol. 2, Wilbur and Yonge, eds. Academic Press, New York. Based on Manwell, 1958: J. Cell. Comp. Physiol., *52*: 345.]

(Fig. 6.5). Increasing acidity, which in life follows the accumulation of carbon dioxide and other metabolites, *reduces* oxygen affinity with a more ready release of oxygen at comparable pressures. The equilibrium curve is shifted to the right, a phenomenon first described by the Danish scientist C. Bohr (1909) and called after him, the BOHR EFFECT. Many invertebrate hemoglobins have little or no Bohr effect while the homeothermic vertebrates have a very definite one. This relationship between oxygen equilibrium and pH varies greatly in the different species of cold-blooded vertebrates (Lenfant *et al.,* 1966) and is clearly related to their ecology and the necessary adjustments of loading and unloading oxygen under different environmental conditions. Since temperature also affects the equilibrium curve (Figs. 6.8 and 12.2), this factor may be balanced against the pH to utilize the oxygen resources of the environment maximally. These problems will be considered in Chapter 12.

In direct contrast to the Bohr effect, several invertebrates (*Limulus, Tubifex, Helix, Busycon*) possesses bloods with an *increased* affinity for oxygen at higher acidities (Redmond, 1968). Thus, at higher levels of CO_2 there is a substantially greater ability to acquire oxygen and deliver it to the tissues (Fig. 6.8). This curious phenomenon, termed a REVERSED BOHR EFFECT, may be an adaptation to permit loading with oxygen under conditions of high environmental CO_2. Redmond (1968) notes that this effect is found in many shelled gastropods which live in well-aerated sea water; he feels that it might permit the utilization of oxygen from water in the mantle cavity when they withdraw into their shells and CO_2 accumulates. The curves for *Tubifex* (Fig. 6.8) show how nicely this capacity is adapted to the stagnant conditions where the animals live (Palmer and Chapman, 1970).

Root (1931) described fish bloods which show a marked loss of *oxygen capacity* at elevated levels of carbon dioxide (Fig. 6.10a). This is now termed the ROOT EFFECT. In the Bohr effect, CO_2 shifts the equilibrium curve to the right and thus unloads oxygen; in the Root effect, CO_2 reduces the maximum level of oxygen saturation. The magnitude of the Root effect has been compared in different animals (Fig. 6.10b); its possible significance in the filling of the swim bladder will be considered in Chapter 13.

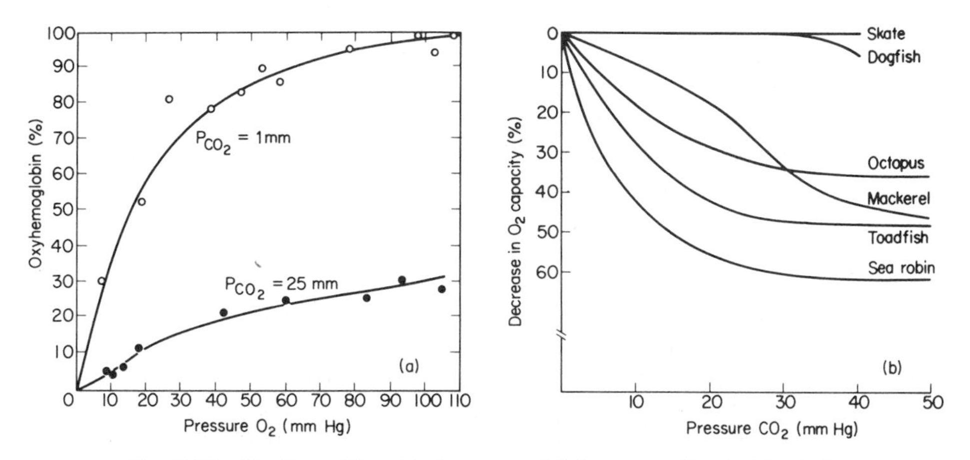

Fig. 6.10 The Root effect. (a) Oxygen equilibrium curves for the blood of the sea robin *Prionotus carolinus* at 20°C. [From Root, 1931: Biol. Bull., *61*: 433. Permission of the Managing Editor.] (b) Root effect compared in various animals. [From Lenfant and Johansen, 1966: Resp. Physiol., *1*: 21. Permission North Holland Publishing Co., Amsterdam.]

It has not yet been possible to identify the particular molecular groups responsible for these CO_2 effects (Manwell, 1964; Antonini, 1965; Riggs, 1965). Biochemical studies of hemoglobin, in which various subgroups of the polypeptide chains have been removed or inactivated, indicate that oxygenation capacity depends on the entire molecule and its configuration, rather than on any specific group such as the sulfhydryl group—at one stage considered significant (Gratzer and Allison, 1960; Riggs, 1965).

Shape of the oxygen equilibrium curve For a given fall in oxygen tension, pigments with sigmoid equilibrium curves release much more of their oxygen load than those with rectangular curves (Figs. 6.6 and 6.8). This must mean that the oxygen binding sites are interdependent so that the presence of oxygen on some of them increases the affinity on others. This phenomenon is usually termed "heme-heme interaction," although "subunit interaction" is probably a more appropriate term (Shulman *et al.*, 1969). The details of the mechanism are not fully understood.

In theory, each of the iron atoms might be expected to combine with one molecule of oxygen but heme alone does not combine with oxygen at all. It requires the associated protein. Each of the four amino acid chains of the tetra-heme unit enfolds one heme group. They work together as a physiological unit so that the first oxygen combinations accelerate additional couplings by several hundred times. Oxygen uptake is thought to alter the conformation of the four subunits—particularly the β-chains—resulting in a separation of the subunits, thus increasing the rate of oxygen combination (Perutz, 1964, 1969). In addition to these intramolecular effects, there may also be interactions between the heme groups of different molecules (Antonini,

1965). These mechanisms have been exploited in different ways during animal phylogeny to gear the loading and unloading of oxygen to particular situations.

Pigment concentration Carrying capacity depends on the actual amount of respiratory pigment as well as on its biochemical characteristics. Thus, any increase in the quantity of pigment will improve the ability of the blood to transport gases. This, however, soon creates a hazard by raising the osmotic content of the circulating fluids. In phylogeny, two quite different physiological mechanisms have appeared in response to this dilemma. In many invertebrates, larger quantities of soluble chromoprotein have been incorporated into the circulating fluids by increasing the size of the molecules. Since osmotic phenomena depend on the actual number of particles (molecules or ions) and not on their size, this is sometimes an adequate solution to the problem. *Lumbricus,* and some of its relatives, with hemoglobin molecular weights in the vicinity of 3,000,000, seem to represent the extreme in this direction of specialization.

The presence of extremely large polymers is at best a partial answer to the increasing oxygen demand. Not only do the very large protein molecules increase viscosity, but they are less active in oxygen transport than the four-heme unit which is common in the most efficient hemoglobins. The vertebrates retained the advantages of the four-heme unit, kept the viscosity at a minimum, and increased the quantity of hemoglobin tremendously by putting it in small packages, the erythrocytes. In this way the pigment was actually removed from the vascular fluids. Subsequent improvements in transport efficiency were achieved through specialization of the erythrocytes—in particular, through the loss of the nucleus and the decrease in the size of the cell to produce a maximum of surface per unit volume (Table 6.2 and Lehmann and Huntsman, 1961). A single human red blood corpuscle contains about 280 million molecules of hemoglobin (Perutz, 1964).

An additional advantage of large blood protein polymers or intercellular vascular pigments should also be noted. Molecules smaller than about 68,000 mol wt are readily filtered through cellular membranes such as the glomerulus of the kidney;

Table 6.2 Size and Number of Erythrocytes in a Series of Representative Vertebrates

Kind of animal	Diameter in microns	Number of cells in millions per cu mm
Proteus anguinus	58.2 × 33.7	0.036
Lacerta agilis	15.9 × 9.9	1.42
Columba domestica	13.7 × 6.8	2.40
Canis familiaris	7–8	6.65
Homo sapiens ♂	6.6–9.2	5.00
Homo sapiens ♀	6.6–9.2	4.50

SOURCE: Herter (1947).

some of the invertebrate excretory organs which depend on filtration may be even more permeable. Thus, if oxygen transport depended on relatively small polypeptides, there might well be a significant wastage of the pigments or a substantial problem of reabsorption (Jones, 1963). A comparative study of the molecular weights of different vascular pigments shows that the smaller molecules (mol wt less than 70,000) are intracellular while the larger ones are extracellular. The only exceptions are found among the insects (*Chironomus* hemoglobin, for example) where filtration pressure is not a factor in urine formation.

The effects of ions Experimentally, the shapes of oxygen equilibrium curves can be readily changed by varying the salt concentration or composition of the medium (Antonini, 1965; Riggs, 1965). The magnitude of the Bohr effect may also depend on the ions present. Increasing the sodium chloride concentrations, for instance, can promote the dissociation of oxyhemoglobin or oxyhemocyanin at low oxygen tensions; an elevation of calcium increases the affinity of hemocyanin for oxygen (Fig. 6.7). Thus, under certain physiological conditions (molting, for example), electrolytes provide another avenue for the adaptation of the chromoproteins to special respiratory demands.

THE TRANSPORT OF CARBON DIOXIDE

Although the solubility of carbon dioxide in water is much greater than that of oxygen, the amounts which can be carried in simple solution are totally inadequate for most animals. A transport based on solubility would take care of less than one tenth of the requirements of a mammal. Carbon dioxide, like oxygen, is carried in a chemical combination of a highly specialized nature as indicated by Fig. 6.11.

The relatively small amounts held in simple solution depend on the pressure

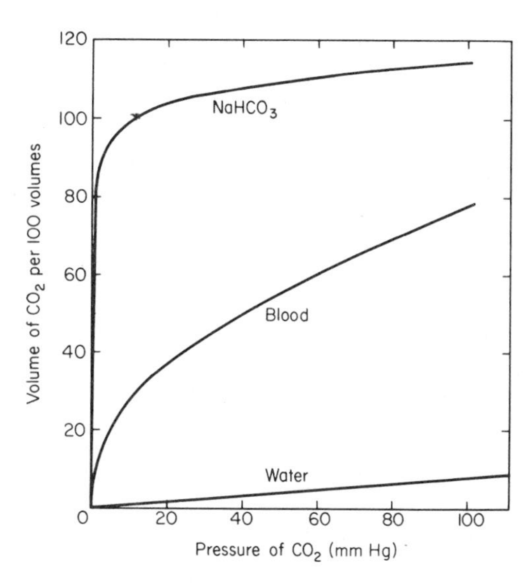

Fig. 6.11 Comparison between the carbon dioxide equilibrium curves of blood and sodium bicarbonate solution of a concentration 0.0484*N*. The amounts of CO_2 held in physical solution in water are shown in the lowest line. [From Evans, 1956: Principles of Human Physiology, 12th ed. Churchill, London.] The relationship for fresh water is curved below $P_{CO_2} = 1$ to 2 mm Hg (Dejours *et al.*, 1968).

(and temperature) but do not exceed about 5 volumes per cent. Sea water, with its excess of strong cations, forms carbonates and hence has a greater capacity than fresh water. A bicarbonate solution contains much CO_2, but in forming this combination the chemical reaction is rapid and complete at relatively low pressures. It is further evident from the equilibrium reaction

$$2NaHCO_3 \rightleftharpoons Na_2CO_3 + CO_2 + H_2O$$

that under vacuum only half of this carbon dioxide of a bicarbonate solution will be discharged. Acid must be added to evolve the other half.

Blood does not behave in this way at all. As indicated in the middle curve of Fig. 6.11, vertebrate blood combines with carbon dioxide rapidly at first and then more slowly; but even at the highest pressures shown there is still a reserve capacity for the transport of this gas. Neither a simple solution nor a bicarbonate compound would satisfy the transport problems of a large active invertebrate or a vertebrate.

Although this is true of many animals, there are some species that do depend on these ordinary chemical mechanisms. This is only possible if they are relatively sedentary or have low rates of metabolism. A group of carbon dioxide dissociation curves is shown in Fig. 6.12. The hemolymph of *Aplysia* or an ascidian has no special transport materials, and the amounts held in the body fluids are essentially the same as those dissolved or in other ways held by the buffers of sea water. The blood of *Urechis,* however, shows an additional capacity for CO_2 (in the form of organic buffers), but this is quickly reached at about 20 mm Hg; thereafter, the curve parallels that of sea water. Finally, the hemolymph of the octopus or the blood of man takes

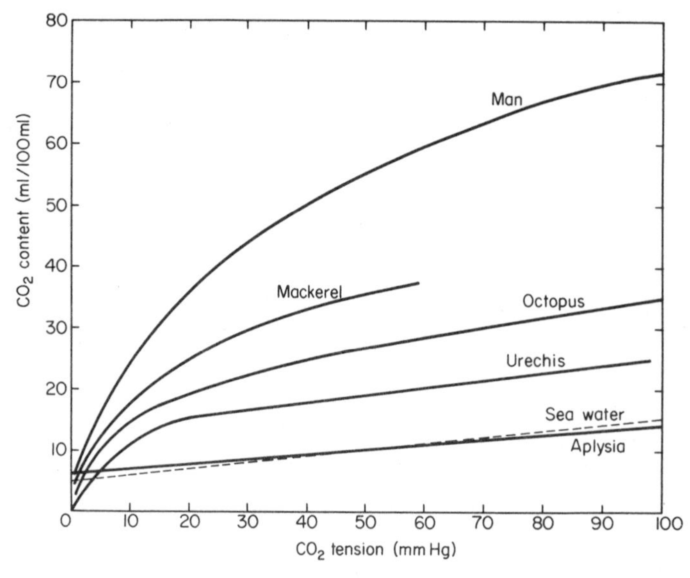

Fig. 6.12 A family of CO_2 equilibrium curves for bloods compared with sea water (broken line). [Data from Parsons and Parsons (1923) and Florkin (1934)].

up more and more CO_2 over the entire range and has evidently some additional capacity at the upper limits shown on the graph. At 100 mm Hg mammalian blood holds about seven times more carbon dioxide than the blood of the sedentary tunicate or the nudibranch. What are the CO_2 transport peculiarities of these more highly specialized bloods?

Three major factors account for the high carbon dioxide capacities of bloods and hemolymphs: the existence of blood buffers, the formation of carbamino compounds, and the presence of the enzyme carbonic anhydrase. First then, several different compounds operate as BLOOD BUFFERS. Of particular importance are chromoproteins involved in oxygen transport. Although other proteins of blood may have a certain buffering capacity, the chromoproteins form the major protein constituent and hence are the main chemicals for the formation of bicarbonate compounds. Hemoglobin is a potassium salt and will combine with carbonic acid as follows:

$$KHb + H_2CO_3 \rightleftharpoons KHCO_3 + HHb$$

Similarly, hemocyanin and the other respiratory pigments can buffer the carbonic acid. In some bloods there are also inorganic buffer systems. In mammalian blood, for example, the phosphate buffers of the plasma (sodium salts) and the corpuscles (potassium salts) will also take up carbonate. Thus, the reaction

$$Na_2HPO_4 + H_2CO_3 \rightleftharpoons NaH_2PO_4 + NaHCO_3$$

accounts for some of the combination, but this is a minor factor in comparison with the blood proteins. As a matter of fact, it has been shown that if hemocyanin is deproteinized it will then combine with no more CO_2 than will sea water.

In the second place, some of the chromoproteins are also able to form DIRECT COMBINATIONS WITH CARBON DIOXIDE. In hemoglobin solutions, carbon dioxide forms direct links with amino groups in the protein portion of the molecule as follows:

$$\text{Hemoglobin}-N\begin{matrix} H \\ H \end{matrix} + CO_2 \rightleftharpoons \text{Hemoglobin}-N\begin{matrix} H \\ COO^- \end{matrix} + H^+$$

The carbamino compounds so formed constitute only a small factor (2 to 10 per cent) in the transport but are nevertheless significant because of the rapidity with which the combination occurs. Other chromoproteins may form similar linkages, but CO_2 transport in the invertebrates is considered to depend primarily on the buffers.

A third important factor in efficient CO_2 transport mechanisms is the presence of a special enzyme CARBONIC ANHYDRASE which promotes the reaction between water and carbon dioxide.

$$H_2O + CO_2 \xrightleftharpoons{\text{carbonic anhydrase}} H_2CO_3 \rightleftharpoons H^+ + HCO_3^-$$

Carbonic anhydrase is a low molecular weight (about 30,000), soluble metalloprotein containing zinc (Maren, 1967a). It is widely distributed among plants and invertebrate animals (Polya and Wirz, 1965; Maren, 1967a) and occurs in all vertebrates where it is known to play a part in the ion movements which occur in several tissues (kidney, gastric mucosa, pancreas) as well as in blood. Its significance in the transport of CO_2 by the invertebrates is speculative, but its presence in their gills is suggestive. It may be concerned with the release of CO_2 from gill epithelia, since the enzyme catalyzes the reaction in both directions. It is also abundant in some muscles and glands of the invertebrates (Wolvekamp and Waterman, 1960) and may serve to accelerate the removal of carbon dioxide from actively metabolizing cells and bring it more quickly into the hemolymph. This is speculation. Its only established functions among invertebrates are concerned with the deposition of calcium carbonate in shells and tests of molluscs, barnacles and corals (Wilbur, 1964; Maren, 1967b) and in the shell-boring organs of some gastropods (Smarsh *et al.*, 1969).

Its role in the transport of carbon dioxide by vertebrate blood is well established. The enzyme is always confined to cells; in vertebrate blood, it occurs only in the erythrocytes. The CO_2 passing from the tissues into the capillaries diffuses quickly into the corpuscles and is held mainly as carbonate. Without carbonic anhydrase the reaction is so slow that only small amounts are directly combined with water in the plasma. This does not mean that the carbonate formed within the red cell is entirely transported in this structure since it can readily diffuse into the plasma. The point emphasized here is that the initial combination occurs almost entirely in the cell (85 to 90 per cent in the mammal).

The details of CO_2 transport in the vertebrate are summarized in Table 6.3.

Table 6.3 Carbon Dioxide Transport in Vertebrates

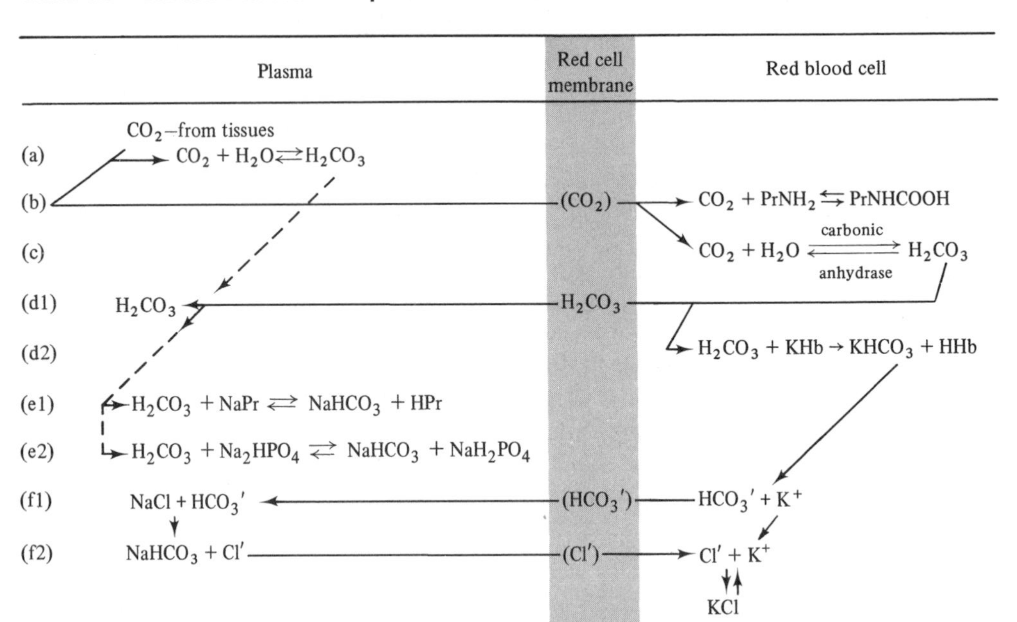

Small amounts of CO_2 entering the plasma dissolve as such or form carbonic acid (line a), but about 90 per cent of it diffuses directly into the corpuscle and forms carbamino groups (line b) or is converted into carbonic acid by the special enzyme system (line c). Some of this carbonic acid may diffuse back into the plasma (line d1), but the bulk of it is tied up by the hemoglobin reaction as bicarbonate (line d2). In the plasma, the carbonic acid (from whatever source) unites with plasma proteins (NaPr) or inorganic buffers (Na_2HPO_4) as shown in lines e1 and e2. In the corpuscle, the ionized potassium bicarbonate provides bicarbonate ions which can exchange with the chloride ions in the plasma in a Donnan equilibrium shift (Davson, 1970) as shown in lines f1 and f2. This is the so-called chloride or Hamburger shift. The net result is that, although most of the CO_2 is first combined with hemoglobin in the corpuscle, a very considerable portion of this is, in reality, transported as carbonate in the plasma.

Usually, the CO_2-combining powers of hemoglobin and hemocyanin (perhaps also other chromoproteins) depend to some extent on their state of oxygenation. Oxyhemoglobin is relatively acid, and deoxyhemoglobin is relatively alkaline so that with the discharge of oxygen in the tissues, the hemoglobin becomes more alkaline and will combine with increasing amounts of carbonic acid. The horizontal distance between the titration curves of Fig. 6.13 shows the pH shift which occurs when hemoglobin is oxygenated or deoxygenated; the vertical distance is a measure of the buffering capacity and shows how much H^+ can be added without a shift in pH. This amounts to 0.7 mole of hydrogen ions taken up by hemoglobin when 1 mole of oxygen is given off. The calculations have been summarized in a lucid manner by Davenport (1958). The mechanisms are reversed in the lungs and the CO_2 is readily discharged into the atmosphere.

Marked species differences in the magnitude of this effect (termed the HALDANE

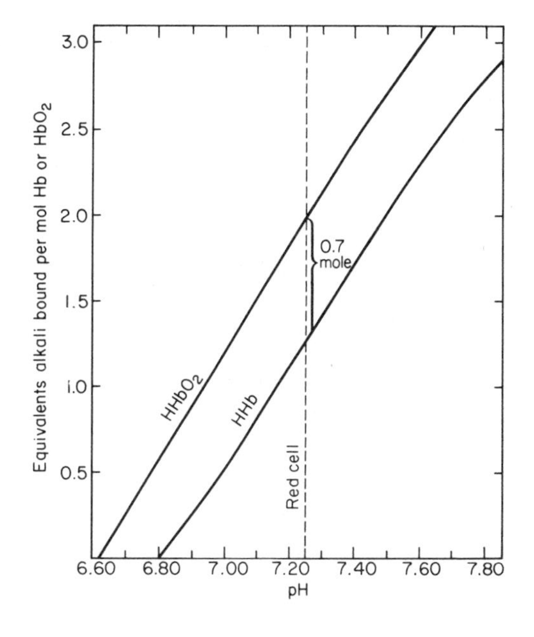

Fig. 6.13 Influence of pH on equivalents of alkali bound per mole of oxygenated ($HHbO_2$) or deoxygenated (HHb) hemoglobin. Note that the titration curves for the two hemoglobins are parallel. At any pH (e.g., the 7.25 pH within the erythrocytes) the oxyhemoglobin binds 0.7 equivalents more base than does the deoxyhemoglobin. Hence, for each mole of O_2 released from oxyhemoglobin in the tissues, 0.7 mole of H^+ formed by release of CO_2 can be neutralized by the deoxyhemoglobin without altering the pH. [From Mountcastle, ed., 1968: Medical Physiology, Vol. 1, 12th ed. Mosby, Saint Louis, Based on Peters and Van Slyke, 1931: Quantitative Clinical Chemistry Methods. Williams and Wilkins, Baltimore.]

EFFECT) have been recognized for many years. Hemocyanins of squid and octopus show a strong Haldane effect while those of the snails *Helix* and *Busycon* are slight and reversed at lower pH values (Wolvekamp, 1961; Lenfant and Johansen, 1965; Burton, 1969). Dogfish hemoglobin lacks a Haldane effect (Lenfant and Johansen, 1966) while that of the lungfish *Neoceratodus* and of some teleosts (*Cyprinus, Opsanus, Prionotus*) exceeds the effect in human blood (Lenfant *et al.*, 1966). The buffering capacities of dogfish and lungfish bloods are compared in Fig. 6.14. It is evident that the buffering capacity of dogfish whole blood differs only slightly from the separated plasma. This must mean that dogfish hemoglobin plays only a minor role in the buffering; there is no Haldane effect and no shift of water and chloride between red cell and plasma when CO_2 is added. Lungfish (also human) blood, in contrast, shows a marked Haldane effect but the plasma by itself has only a slight buffering capacity.

Acid-base balance and temperature The regulation of pH is a universal and essential part of physiological homeostasis. Acid metabolites, of which carbon dioxide is the most significant, are inevitable byproducts of life; unless effectively buffered they would quickly and markedly alter the internal environment. The magnitude of the regulatory problem faced by active animals is emphasized by the observation that a man produces about 13,000 m-moles of carbon dioxide each day; this is equivalent in neutralizing power to over a liter of fuming, concentrated HCl which is about 10 N (Robertson, 1967). Notwithstanding this steady input of acid, the pH of mammalian blood remains slightly alkaline at about pH 7.4.

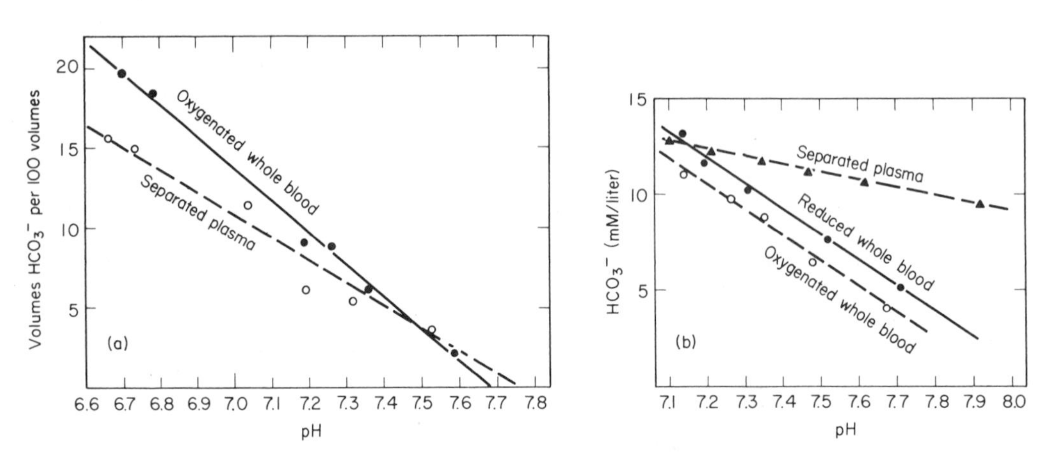

Fig. 6.14 Buffering capacities of whole blood and separated plasma. (a) Dogfish *Squalus suckleyi* at 11°C. The relationship for whole blood does not change with deoxygenation. [From Lenfant and Johansen, 1966: Resp. Physiol., *1*: 17. Permission North Holland Publishing Co., Amsterdam.] (b) Lungfish *Neoceratodus* at 18°C. [From Lenfant, Johansen, and Grigg, 1966: Resp. Physiol., *2*: 9. Permission North Holland Publishing Co., Amsterdam.]

The rate of CO_2 production will vary with the activity of the animal. As previously noted, CO_2 quickly forms carbonic acid, hydrogen, and bicarbonate ions as summarized in the following equation:

$$CO_2 + H_2O \underset{}{\overset{\text{carbonic anhydrase}}{\rightleftharpoons}} H_2CO_3 \rightleftharpoons H^+ + HCO_3^-$$

Table 6.3 shows that most of the CO_2 produced in the tissues goes via the plasma into the erythrocytes, where it either remains as dissolved CO_2 or combines with hemoglobin or is hydrated to carbonic acid which then ionizes and is buffered. The three buffer systems (bicarbonate, phosphate, and proteins) have been described. As buffers, the plasma bicarbonates play a relatively minor role; plasma bicarbonate *and* phosphate carry only about 10 per cent of the CO_2, while the proteins (particularly hemoglobin and oxyhemoglobin) account for 60 per cent of the carrying capacity of mammalian blood (Harper, 1973). However, the bicarbonate system is especially powerful in homeostatic processes since there are two important sites of regulation in addition to the blood buffer system itself; CO_2 is eliminated directly and rapidly in the ventilation of the lungs, while hydrogen ions are excreted more slowly by the renal tubules (Fig. 8.8). Both kidneys and lungs are important organs of acid-base balance in the terrestrial animals. The respiratory center of the mammal is extremely sensitive to changes in hydrogen ion concentration; the rate of ventilation is approximately doubled when the pH falls 0.1 unit and halved if the pH rises 0.1 unit (Gray, 1949). Under physiological conditions, hyperventilation or overbreathing will rapidly reduce the total dissolved CO_2 and cause pH to rise, while an increase in dissolved CO_2 and a fall in pH occurs with hypoventilation or "holding the breath." In summary, acid-base regulation in the land vertebrates occurs at three points: the first and most rapid response is that of the blood buffers themselves, but the respiratory and renal adjustments are responsible for the long-term regulation even though they occur more slowly (1 to 3 min for adjustments in ventilation and 6 to 30 hr for renal adjustments).

The regulation of pH in aquatic animals with gills depends on somewhat different mechanisms. There are three important points to consider when comparing terrestrial homeotherms with the aquatic poikilotherms: (1) the effects of temperature on pH, (2) differences related to the mechanics of ventilation, and (3) ion exchanges which occur through the gills.

Much of the physiological literature emphasizes variability in the blood pH of poikilotherms. In fact, the recorded values for single species are often so different that the lower vertebrates have not been thought to regulate pH at all. The solution to this presumed anomaly has now been traced to a temperature relationship. The theory for interpreting these differences has been known for half a century but the physiological significance and the nature of the regulatory capacities of poikilotherms have only recently been stated (Rahn, 1967; Howell *et al.,* 1970; Rahn and Baumgardner, 1972; Randall and Cameron, 1973). The first point to be noted is that the ionization constant of water, K_w, changes with temperature. The pH of neutrality of water is temperature-dependent and increases by 0.6 pH units when the temperature falls from 37° to 3°C. It is evident from Fig. 6.15 that, over a temperature range of

5° to 35°C, the pH of frog, toad, or turtle blood changes in parallel with the pH or pOH of neutral water (pK_w or pN). In poikilotherms, there is evidently a constant ratio in OH^-/H^+ and the important consideration is the ratio of these two ions rather than the actual number of hydrogen ions. The relative alkalinity of the body fluids remains constant with respect to the neutrality of water at any temperature; the poikilotherms as well as the homeotherms regulate hydrogen ions.

The second point relates to the mechanics of gas exchange discussed in Chapter 4. Terrestrial vertebrates, in contrast to aquatic species, have very high levels of carbon dioxide. The P_{CO_2} values shown in Figure 4.9 for mammals are ten or more times greater than those of such typical fishes as trout and sharks. The aquatic animal lives in an environment low in oxygen, and this requires relatively large ventilation volumes to meet the oxygen demands; moreover, CO_2 is much more soluble than oxygen in water and consequently the CO_2 differences between arterial blood and water are very small. Thus, the aquatic gill breathers are denied the opportunity of *regulating pH by adjustments in ventilation.* Their problems of regulating pH have not been solved by short-term CO_2 adjustments but by long-term ion exchanges.

It is of interest that the pH of the poikilotherm at 37°C (Fig. 6.15) is very close to that of the homeotherm and this might suggest that the blood pH of the hibernator would change when the body temperature is depressed. This, however, does not occur either under natural or experimental conditions and points to the innate differences which occur between the open system of a lung-breather and the closed

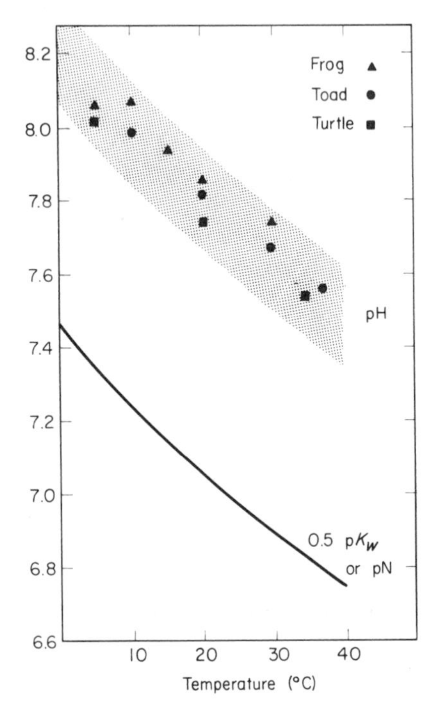

Fig. 6.15 Normal blood pH of frogs, toads, and turtles plotted against body temperature. Mean values for 127 samples. The solid line, pN or 0.5 pK_w, represents the pH or pOH of neutral water. The constant pH–pN difference expresses the constant relative alkalinity. [From Howell, Baumgardner, Bondi, and Rahn, 1970: Amer. J. Physiol., *218*: 602.]

systems of gill-breathers; if mammalian blood is cooled at constant CO_2 content (closed system) then the pH change of the blood closely resembles that observed in poikilotherms. The hibernator (open system) is obviously faced with considerable problems of protein dissociation when the body temperature falls because pH is maintained at 7.4 over a wide temperature range.

The third and final point also relates to the branchial exchange system which serves as a site for ion regulation as well as the exchange of gases. Several freshwater fishes have been shown to take up Cl^- (and Na^+) from the ambient medium in exchange for HCO_3^- (and NH_4^+) in the branchial cells (Cameron and Randall, 1972; Fig. 8.9). This system parallels an acid-base regulating system in the kidney (Chapter 8), where Na^+ in the tubular filtrate is exchanged for H^+ in the renal tubule cells (Fig. 8.8). Since the pK of the bicarbonate buffer system is not affected by temperature change in the same way as the pKs of dissociation of water and proteins, then the necessity for bicarbonate regulation by ion exchanges is obvious when animals are denied the regulation of pH by CO_2 ventilation.

The buffering of acid metabolites by blood has already been described. The efficiency of any buffer system depends on its acid/salt ratio. Since changes in H^+ will alter these ratios, a temperature effect may be anticipated from the relationship shown in Fig. 6.15. If temperature is altered, the relative buffering can only remain constant if temperature acts in an identical manner on the dissociation constants of the buffer and on the ionization constant of water. Some buffers will be more efficient than others in this regard, and the theoretical analysis indicates that hemoglobin with its imidazole groups of histidine plays a dominant role in this regulatory process (Howell *et al.,* 1970; McGilvery, 1970; Reeves, 1972). There are substantial differences in the histidine contents of different hemoglobins, and this is a major factor in the variability found in buffering capacities of blood. The buffer constant for human blood is over 30 while that for dogfish blood is 9–10; in general, blood buffering capacities of terrestrial vertebrates are substantially higher than those of the aquatic species; Prosser (1973) tabulates representative values.

PLASMA PROTEINS AND REGULATION OF FLUID VOLUME

Plasma protein concentrations range from low values of just over 1.0 mg/ml (echinoderms, some molluscs, some annelids) to high values of 100 to 150 mg/ml in the large cephalopods with much hemocyanin in the hemolymph (Engle and Woods, 1960). Values for birds and mammals range from 30 to 75 mg/ml. No strict correlation with phyletic position is evident although, in general, both concentration and number of different protein fractions increase during phylogeny and ontogeny. This lack of strong correlation is not surprising because the plasma proteins as a group function in several different ways (buffer activity, oxygen transport, osmotic pressure, blood coagulation, immune responses). During phylogeny, animals have capitalized on one or another of these capacities in order to meet particular physiological or environmental demands. The role of the plasma proteins in maintenance of fluid volume is considered in this section; other functions will be discussed elsewhere.

The plasma volume of an animal with a closed circulatory system depends on a balance between the rate of filtration from the capillaries into the tissue spaces and the rate of reabsorption of this filtrate (Fig. 5.10). The blood pressure creates the driving force for filtration, while the osmotic pressure of the plasma operates in the reverse direction; the plasma proteins form a large part of the osmotically active material. Osmotic forces are also important in the fluid balance of invertebrate animals with open circulatory systems. Those which operate with a fluid skeleton develop relatively high internal hydrostatic pressures during movement, and these pressures could force fluids from the animals through permeable membranes even though these fluids are isotonic with the environments. Preservation of the body fluids depends on the permeability of the surface membrane and colloidal osmotic pressures of the body fluids.

In higher vertebrates, the albumin fractions amount to about 55 per cent of the total plasma protein (mammals) and are primarily responsible for maintenance of plasma volume. Albumins are readily separated electrophoretically because of the mobility of the relatively highly charged small molecules (mol wt about 69,000 as against 170,000 or more for other plasma proteins). This technique has now been applied to many animals and the albumins have been shown to be characteristic of the bloods of only the higher vertebrates with their more refined homeostatic controls. An electrophoretic analysis of the serum proteins of 26 species of fish (representing 14 families) from the Gulf of Mexico provides an interesting example of the phylogenetic trends (Gunter *et al.*, 1961). The total plasma protein was lower than in the mammals and showed a general increase from primitive to more specialized groups. The elasmobranchs, the gars, and about half of the Clupeidae (relatively primitive teleosts) lacked the albumin fraction. Although the cyclostomes were not included in the above study, both hagfishes and lampreys are known to lack albumin fractions (Rall *et al.*, 1961; Papermaster *et al.*, 1962). The correlation between albumin and phyletic position is not perfect since two of the perciform species (relatively advanced teleosts) also lack albumin. There is, however, a definite increase in complexity of plasma proteins in the more advanced groups; this is reflected in the globulin fractions as well as in the albumins.

The serum proteins may have been of great phylogenetic significance in the evolution of terrestrial life with acute problems of maintaining fluids within the vascular system in the face of desiccation and preserving a balance between the blood and the fluids of the tissue spaces. Several studies of metamorphosing amphibians support this argument (Frieden, 1963). Tadpoles contain little or no serum albumin; virtually all the plasma proteins are globulins. At metamorphosis there is a steady increase both in the amount of serum protein and in the proportion of albumin to a maximum of about 50 per cent of the serum protein. Albumin, because of its large electrical charge, binds smaller organic molecules and salts and is especially important in osmosis.

PHAGOCYTOSIS, THE RETICULO-ENDOTHELIAL SYSTEM, AND IMMUNE REACTIONS

Phagocytic Cells

Amoeboid phagocytic cells are found in the tissues of all multicellular animals (Wagge, 1955). They occur in the circulating fluids, in body spaces, and in the tissues generally. In some primitive invertebrates, amoeboid phagocytes are the only blood cells. Among the invertebrates these cells perform a variety of functions—digestion, excretion, regeneration, and repair. At the lower phylogenetic levels they have two main capacities in the defense mechanisms of the animal: engulfing foreign material and migrating into an injured area to initiate the processes of repair. At the higher phylogenetic levels the system is more elaborate, for these cells have acquired additional capacities in reactions to foreign materials.

In many groups of animals, from nematodes through the phylogenetic series, some phagocytic cells are organized into distinct organs. Huff (1940) describes three different arrangements among the invertebrates. LYMPHOGENOUS ORGANS are made up of masses of primitive blood cells grouped in nodules and held together by connective tissue. They are found in blood sinuses and have been described in annelids, scorpions, cephalopods, and in a few insects and crustaceans. NEPHROPHAGOCYTES, with a dual function of excretion and phagocytosis, occur separately or as masses of cells scattered through the tissues of many animals from annelids through echinoderms. The PHAGOCYTIC ORGANS, consisting of a reticular network filled with phagocytes, are usually situated in the pericardial space where blood must pass over them to reach the heart (Mills and King, 1965). They are best known in the insects (Salt, 1970) but have also been described in nematodes, some annelids, crustaceans and molluscs. In some species of midge, several stages have been described in the ontogeny of phagocytic organs from the phagocytic hemocytes of the blood (Roeder, 1953). A parallel in phylogeny is suggested.

Among the vertebrates the phagocytic system (usually referred to as the reticulo-endothelial system or the macrophage system) is highly organized with circulating leucocytes, amoeboid tissue cells, and extensive reticular nets. Some writers include both wandering and sessile phagocytic cells in this system, while others restrict the term "reticulo-endothelial (R-E) system" to the fixed reticulum of phagocytic organs (Rowley, 1962; Ham, 1969). In any case, wandering and fixed cells operate together to remove the debris from worn-out or injured cells and to maintain the defenses of the body against invading organisms. In the mammal these phagocytic activities depend on the blood neutrophils and monocytes, tissue histocytes or macrophages, the microglial cells of the central nervous system, and the sessile reticulum cells of the liver (Kupffer cells), bone marrow, spleen, lymph nodes, and other sinusoidal tissues of the body. Although all of the jawed vertebrates possess lymphatic tissue, definite lymph nodes, which form compact filters in the lymph drainage, are found only in the birds and mammals (Smith *et al.*, 1967; Good *et al.*, 1967). The phagocytic system, whether composed of wandering or sessile cells, performs the important function of disposing of waste or foreign matter.

Antigen-Antibody Reactions and the Immune Process

Many animals show special capacities for the removal or inactivation of bacteria, viruses, and large organic molecules. This ability depends on the presence of specific proteins which agglutinate, precipitate, neutralize or dissolve the foreign organisms and materials. These substances may either be spontaneously produced (naturally occurring agglutinins or antibodies) or induced by the presence of foreign substances (antigen-induced antibodies). In the latter case, the ANTIGEN is a foreign substance whose presence in the organism evokes the formation of an ANTIBODY. The process is sometimes referred to as the IMMUNE PROCESS OR REACTION, sometimes as an ANTIGEN-ANTIBODY REACTION.

Naturally occurring agglutinins Naturally occurring antigens and antibodies have now been detected at many different levels in phylogeny. The hemerythrocytes of the sipunculid worm *Dendrostomum* has antigens which react with human anti-A and anti-B serum; the serum of the spiny lobster *Palinurus* contains at least ten heteroagglutinins which will agglutinate the sperm or erythrocytes of various animals; the Atlantic lobster has a factor in its serum which clumps the red cells of the herring; seminal fluids of many animals contain natural antibodies which clump sperm or cells of other species. These examples could be multiplied; the literature on blood group reactive substances is extensive and has been carefully reviewed (Boyden, 1965; Cushing, 1970). These reactions depend on highly specific proteins whose syntheses are genetically controlled. In the Crustacea, the blood group reactive substances reside in a small protein fraction of the hemolymph and are not associated with the hemocyanin fraction (Tyler and Scheer, 1945). In some cases the interactions are probably fortuitous and of no particular biological significance; the interactions between sipunculid blood cells and human blood sera would seem to be of this sort. In other cases these reactions are highly important to the species; the chemicals which control the interactions between eggs and sperm must certainly be in this class (Raven, 1966).

The human blood group system is one of the best known of the naturally occurring immune reactions. Landsteiner described the ABO system in 1900; excerpts from his and several other classical papers on human blood groups have been reproduced in Boyer (1963). Human erythrocytes may carry one or the other, both, or neither of two antigens A and B; the blood plasma may carry corresponding antibodies as indicated in Table 6.4. The proportions of the phenotypes (Table 6.4) vary significantly in different races. In Western races the O group forms about 50 per cent; A, 40 per cent; B, 8 per cent; and AB, 2 per cent; but in the pure North

Table 6.4 Human Blood Groups

A. *The blood groups*

Blood group		Antigens in corpuscle	Antibodies in serum
Phenotype	Genotype		
O	OO	–	Anti-A and Anti-B
A	AA AO	A	Anti-B
B	BB BO	B	Anti-A
AB	AB	A and B	–

B. *Determination of human blood groups with two test sera, anti-A and anti-B.*

	Known serum anti-A	Known serum anti-B	Group
Agglutination of unknown blood corpuscles	–	–	O
	+	–	A
	–	+	B
	+	+	AB

SOURCE: Boyd (1950).

American Indians (except for one or two tribes) group AB is unknown, and group B is very rare. In Siam, group B (relatively uncommon in some Western races) occurs to the extent of 35 per cent. Boyd (1950) provides detailed tables of the frequencies of these groups. There are many other naturally occurring antigens in human blood; these were not recognized as early as the AB antigens because there are no corresponding antibodies. They can, however, be readily demonstrated by inducing specific antigens for them in another animal; they are often induced in man through blood transfusions or some other association with the particular antigen (Race and Sanger, 1968). A full discussion of the human blood groups is given in textbooks of medical physiology.

For many years it was tacitly agreed that man was unique in the possession of distinct blood groups. However, first dogs and chickens, then many other groups of animals have been shown to have systems of this nature. Blood groups now seem to be demonstrated wherever a careful search is made (Kovács and Papp, 1972). The dogfish *Squalus acanthias* in the Gulf of Maine has four groups with a clear-cut genetic system; sockeye salmon *Oncorhynchus nerka* are of at least eight different antigenic types or combinations of types, and the frequency of these different types varies in different geographical races (Cushing, 1970). Many other examples will be found in the literature cited above.

Adaptive immunity All vertebrates above the cyclostomes have the ability to form specific antibodies in response to certain foreign substances. This capacity depends on the presence of plasma cells and lymphocytes that elaborate a special group of proteins, the γ-globulins capable of reacting specifically with the foreign substance or antigen. Organisms capable of an adaptive immunity must be able to distinguish "self" from "not self" and possess an "immunologic memory." The phylogeny of this attribute parallels the evolution of the vertebrate lymphoid tissues. Hagfishes lack lymphoid tissue and do not form γ-globulins while the lampreys have a very limited capacity to produce antibodies in response to a few antigens (Papermaster *et al.,* 1963). Lymphoid tissues and plasma cells are characteristics of only the jawed vertebrates. Invertebrate animals appear to be incapable of an adaptive immunity. They usually meet the threat of foreign invaders by cellular mechanisms involving phagocytosis or cyst formation; humoral bactericidal substances occur in some invertebrates but these lack the specificity of antibodies. Although the earthworm is said to reject tissue transplants (Cooper, 1969), the invertebrates usually have a high tolerance of foreign tissue transplants and, in this respect also, contrast sharply with the vertebrates.

Although immune responses are usually evoked by proteins, they may develop with other large molecules (mol wt 10,000 or more) like some of the polysaccharides. The antigenic material is often a part of foreign cells or bacteria, but it may be a protein in solution. A PRIMARY RESPONSE is induced by the first injection. This normally passes unnoticed, but at the tissue level it promotes the elaboration of antibodies. When a second injection is made, the violent reaction of antigen and antibody produces a variety of symptoms in the animal and may lead to death. Basically, the antibody molecules seem to act as bridges between the foreign particles, linking them together in large clumps which may later dissolve or be phagocytized. These reactions are markedly specific. As yet it is not clearly understood how an antigen stimulates cells to elaborate antibodies. Several theories have been proposed, and details may be found in textbooks and monographs devoted to immunology (Nossal, 1964; Haurowitz, 1965).

It is now believed that the higher vertebrates possess a dual system of immunity (Ham, 1969). One part of the system is responsible for the rejection of foreign tissue transplants (skin grafts, etc.) while the other part is concerned with the formation of the circulating antibodies. Until recently, these two reactions to foreign substances were thought to be part of the same immune response; research on the thymus of the mammal and the Bursa of Fabricius of the chicken has provided new bases for studies of immunity and demonstrated functions for these enigmatic organs.

The thymus, present in all jawed vertebrates, arises as a proliferation of gill pouch epithelium which becomes infiltrated with lymphocytes. It grows rapidly to a relatively large size in young animals, only to atrophy in adult life. The organ is absent in the hagfishes and represented by a small accumulation of cells beneath the gill epithelium (a pro-thymus) in the lamprey (Papermaster *et al.,* 1963; Good *et al.,* 1971). In 1961, Miller reported that newborn mice deprived of their thymus failed to develop lymphocytes. The animals survived and grew for several weeks or months but eventually died (see review by Miller and Orsoba, 1967); prior to death these animals were completely tolerant of foreign tissue transplants. The thymus in some way provides for the development of the other lymphoid organs and of the cells responsible for the rejection of skin grafts. Ingenious experiments with baby mice indicate that a humoral factor (hormone) is involved. If a mouse, deprived of its thymus, is implanted with a porous capsule containing the thymus from a second mouse, it will then survive and develop a lymphoid system even though the pores in the capsule are so small that cells cannot pass through them (less than 1 μ or 1/15 the size of a lymphocyte). The thymus from the donor mouse must have provided a substance of molecular dimensions which activated the lymphoid system (Levey, 1964). Such experiments are consistent with the hypothesis that the thymus is primarily concerned with the production of cells whose descendants are responsible for the rejection of transplanted tissues. This line of lymphocytes appears to differ from plasma cells in that they act locally and do not release antibodies into the blood; they have been termed GRAFT REJECTION CELLS (Ham, 1969).

The avian Bursa of Fabricius develops as a dorsal diverticulum of the gut at the junction of the large intestine and the cloaca. The tissue becomes infiltrated with lymphocytes, loses its pouch-like structure and, like the thymus, reaches a maximum size in young individuals and then shrinks during adult life. Experiments have shown that this structure is concerned with the primary production of lymphocytes whose descendants (the plasma cells) are responsible for the *circulating antibodies.* In the mammals, a comparable role is played by lymphatic nodules in the lamina propria of the gut (particularly the lower part of the ileum). These nodules, called PEYER'S PATCHES after their discoverer, are likewise much more conspicuous in young individuals (Ham, 1969). The analogue of the Bursa of Fabricius has not yet been identified in poikilotherms (Good *et al.,* 1968).

Serology as a tool in taxonomy Several of these antigenic differences in blood proteins have been used in taxonomy as adjuncts to gross morphological characters. Boyden (1942, 1963) and his associates have made extensive use of precipitin testing in which antibodies are induced (usually in rabbits) and these antisera then mixed with sera from the same and other species. The strongest reaction (agglutination or protein precipitation) is given with the homologous serum, the intensity of reaction decreasing with distance in phyletic position (heterologous sera). Other workers have used naturally occurring antigens and antibodies such as those which are demonstrated by mixing lobster serum with herring erythrocytes (Sinderman and Mairs, 1959). In these tests, two distinct groups of Atlantic herring were noted, one in which the cells were clumped at dilutions of 1:128 and another in which they clump at 1:4. Other blood tests of a comparable nature include the direct electrophoretic separation of plasma proteins and the rates of hemolysis of erythrocytes suspended in a series of hemolysins (Jacobs *et al.,* 1950). The electrophoretic patterns of the blood proteins are at present singularly popular in taxonomic studies of closely related species. Numerous papers can be readily found in current literature. These different blood tests add a number of useful characters for the taxonomist who must make difficult decisions on animal relationships.

HEMOSTASIS AND THE COAGULATION OF BLOOD

Loss of body fluids becomes a progressively greater hazard as the circulating fluids assume more functions and as specialized physiological mechanisms become dependent on the continuous flow of blood at high pressures. The dangers are probably greater for aquatic organisms in the freshwater habitat (markedly hypotonic) than in the ocean where the invertebrates and some of the vertebrates live in an isosmotic medium. Phylogenetically more advanced animals have a group of safeguards which restrict the loss of fluid in bleeding hazards from traumatized blood vessels. Grégoire and Tagnon (1962) have summarized the available information in tabular form; Needham (1970a) gives a useful taxonomic review of the hemostatic mechanisms of the invertebrates and speculates on the probable phylogeny.

Hemostasis Without Plasma Clotting

The most primitive arrangement depends only on the contractility of body musculature and blood vessels. In many soft-bodied invertebrates this may be the sole hemostatic mechanism. Spasms of blood vessels and contractions of the body wall have been noted among marine worms and sea cucumbers, and in some of the annelids nothing more complex may be required. Such devices, however, would be inadequate for hard-bodied animals (arthropods, many echinoderms and molluscs) or for the soft-bodied animals with even moderate blood pressures. In these forms the vascular fluids contain cellular elements and special jelling proteins which form plugs at the site of injury. Although most groups of animals have not yet been carefully studied, there is now enough information on the echinoderms (Boolootian and Giese, 1959), the arthropods (Florkin, 1960; Grégoire, 1970) and the vertebrates to draw some general conclusions (Grégoire and Tagnon, 1962; Needham, 1970a).

In the more primitive groups of animals only blood cells are involved in clot formation. In several echinoderms (the crinoid, *Heliometra,* the ophiuroid, *Gorgonocephalus,* and the echinoid, *Dendraster*), the blood cells, following injury, show a temporary agglutination but do not lose their identity. This may also happen in some arthropods and molluscs, but usually in these groups, and also in the asteroids among the echinoderms, the cell agglutination is followed by plasmodium formation in which the protoplasm of the agglutinated cells fuses and the cells lose their identity. In the next step, the cells produce a fibrous protein material which serves to entangle other cells and increase the size and strength of the cell coagulum. This phenomenon was carefully studied by Loeb, many years ago, in *Limulus*. The fibrous material produced by the cells was called CELL FIBRIN but is not biochemically similar to the fibrins of the plasma clots. Similar processes of cell coagulation have been described in several groups of spiders and in some crustaceans, such as the edible crab *Cancer* (Grégoire and Tagnon, 1962).

Participation of the Plasma Proteins

In many arthropods, and in all vertebrates, blood clots are primarily composed of tangled protein fibers which develop from a plasma protein—fibrinogen (vertebrate) or coagulogen (invertebrate). The conversion of fibrinogen to fibrin is an enzymatic reaction which is initiated by material released from injured cells. Among invertebrates and lower vertebrates, the thromboplastic activity responsible for the transformation of fibrinogen to fibrin is either completely or almost completely contained in cells and released only when they rupture. In the higher vertebrates, several thromboplastic factors (along with anti-thromboplastic elements) are also found in the plasma.

The crustaceans and insects exemplify the simpler situation. In many species, islands of coagulation develop around agglutinations of blood cells, and then the coagulation extends through the hemolymph. The speed of clotting varies in different species, and several different categories of coagulation are recognized (Florkin, 1960; Grégoire, 1970). Highly specialized fragile blood cells associated with the clotting mechanisms are called HARDY'S EXPLOSIVE CELLS in the crustaceans and COAGULOCYTES in the insects.

In the mammal, no less than thirteen different plasma coagulation factors are recognized. Fibrinogen is the most abundant of these and constitutes 4 to 6 per cent of the total plasma protein. The fibrinogen molecule is large and asymmetric with a mol wt of about 400,000. The following stages have long been recognized in the conversion of fibrinogen into the tangled mass of fibrin threads which are the basis of the blood clot.

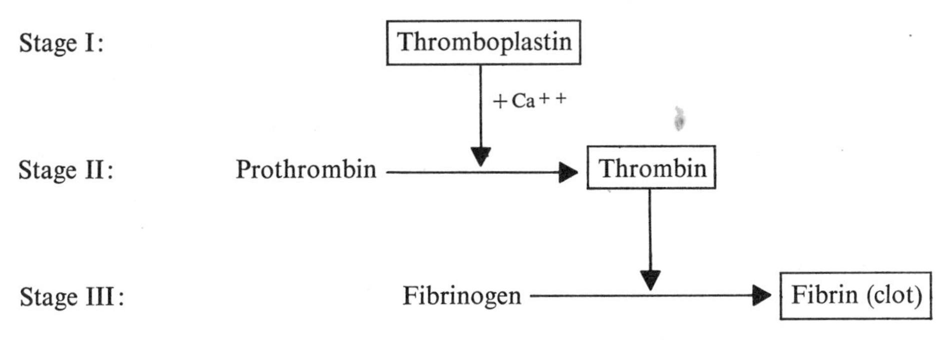

Stages II and III are relatively well understood, but the genesis of thromboplastin remains controversial. Thromboplastin may arise through a series of reactions that follow the rupture of blood platelets (intrinsic prothrombin activator) or it may develop following the injury of body tissues (extrinsic prothrombin activator). These processes, by which intrinsic or extrinsic thromboplastin is formed, involve four to eight different steps or reactions. Macfarlane (1964) suggests that in this series, a succession of factors acts sequentially; each molecule of one factor activates progressively more molecules of the succeeding factors; this exponentially increasing involvement forms a COAGULATION CASCADE. Details will be left to textbooks of medical physiology and the pertinent monographs (Biggs and Macfarlane, 1962; Poller, 1969; Seegers, 1969).

Blood clotting in the lower vertebrates depends much more on cellular factors. The thrombocytes of fishes and amphibians seem to provide all of the thromboplastic factors required in thromboplastin formation (Doolittle and Surgenor, 1962). Reptiles and birds are also deficient in the plasma thromboplastic components; they seem to rely more on vasoconstriction and the rapid liberation of TISSUE THROMBOPLASTIN (Fantl, 1961). The trend in the higher vertebrates has been an extension into the extracellular plasma of mechanisms primitively confined to the intracellular medium. The initial steps always depend on cellular elements, but the presence of abundant clotting factors in the plasma provides for the production of a more permanent type of hemostasis (Grégoire and Tagnon, 1962).

Several anticoagulants in mammalian plasma provide safeguards in this clotting system which is so effective and necessary in trauma, but a potential hazard if activated within the vascular channels. There is enough clotting enzyme in 10 ml of human blood to coagulate all the blood in the human body. An albuminoid material, antithrombin, is believed to neutralize some of the clotting factors, thereby preventing the formation of dangerous intravascular clots. Heparin, a mucopolysaccharide, is a powerful anticoagulant, perhaps acting as a cofactor with antithrombin. Additional safeguards may be present.

In addition to these anticoagulants produced as protective compounds in the animal's blood, there are materials of similar nature which enable predacious species to take vertebrate blood without troublesome clottings. The leech produces an antithrombin (hirudin) which permits the removal and storage of blood in the fluid condition. Blood-sucking insects inject anticoagulants when they puncture vertebrate skin in feeding. Predators have exploited thromboplastins as well as anticoagulants; the venom of the viper contains a powerful proteolytic enzyme which causes disastrous intravascular clots when injected into an unlucky mammal.

7

THE TRANSFORMATION OF ENERGY

Energy necessary for the varied activities of life is generated in a series of chemical reactions, collectively referred to as cellular metabolism or respiration. The term METABOLISM is universally used to describe the many chemical reactions which take place in the living organism whether these are constructive (ANABOLISM) or destructive (CATABOLISM). The term RESPIRATION, however, has several different usages. The Latin word, from which it is derived, means "to breathe" or "exhale" and, in this sense, respiration was originally applied to the exchange of gases between an organism and its environment. It referred to the obvious activities of breathing or their equivalent. As the years went by, it became apparent that the really fundamental exchanges were occurring at the cellular level, and the term INTERNAL RESPIRATION was often applied to this phase of gaseous exchange. At present, the adjective "internal" has been dropped, and respiration is frequently applied to cellular processes. Some writers restrict it to those which involve the uptake of gaseous oxygen, while others use it more generally for all of the energy-yielding reactions of the cell. In this book it will be used in the latter sense. Those activities of an animal which involve an exchange of gases between the organism and its environment will be referred to as VENTILATION. The term INTERMEDIARY METABOLISM is synonymous with cellular metabolism as used here.

ENERGY-PRODUCING REACTIONS

Flow of Energy in the Biological World

The energy transformations involved in tissue respiration form connecting links between the potential energy contained in the animal's food and its utilization in the activities of a living animal. There are three major transfer points along this mainstream in the flow of biological energy (Fig. 7.1). In the first of these, radiant energy is captured by the green plants and transformed into the carbon compounds of protoplasm (Chapter 1). In the second major set of transformations (respiration), the chemical energy contained in carbohydrates, fats, and proteins is transformed into a more labile and directly usable form of chemical energy while, in the third stage, this labile chemical energy (chiefly present in the terminal phosphate bond of ATP) is utilized in such animal activities as movement, biosynthesis, secretion, or the production of light, heat and electricity. The biochemical links between the digested foods and the supplies of directly usable chemical bond energy are outlined in this chapter.

Concept of free energy The free energy of a system is that part of its total potential energy which is available for useful work. Thus, when water flows downhill, the free energy may be used to generate electricity; when a coiled spring unwinds, its free energy can be geared to turn the hands of a clock or drive a mechanical toy; when fossil fuels are burned, heat is liberated and this heat can be used to operate a steam engine or turbine. These are all spontaneous reactions that provide usable free energy. They are "downhill" reactions proceeding with a loss

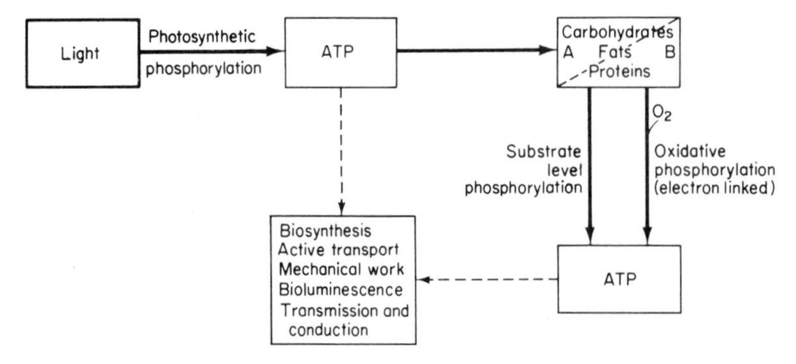

Fig. 7.1 Energy flow through living systems. Photosynthetic phosphorylation provides energy-rich phosphate compounds which are usable energy for carbon fixation and biosynthesis in plants (A). The plant products are ingested by the animals (B) and, through metabolic transformations, their potential energy is reconverted into energy-rich phosphates (ATP) which the animal utilizes to do cellular work.

of free energy; in thermodynamic terms, their standard free energy values are negative and they are said to be EXERGONIC. The reverse reactions do not occur spontaneously; energy must be supplied to pump water uphill or to wind a clock or to synthesize complex carbon compounds from carbon dioxide and water. These reactions require an input of free energy; they are "uphill" and can proceed only when appropriate amounts of energy are supplied; their standard free energy values are positive and they are said to be ENDERGONIC.

The reactions involved in tissue respiration are chemical in nature and can be represented by the following equation:

$$A + B \rightleftharpoons C + D$$

At equilibrium, in accordance with the LAW OF MASS ACTION, the product of the concentrations of the substances formed (C and D) divided by the product of the reactants (A and B) is a constant known as the EQUILIBRIUM CONSTANT K; the relationship is expressed by the equation:

$$K = \frac{[C]\ [D]}{[A]\ [B]}$$

The physico-chemical relationship between the equilibrium constant and the free energy change of the reaction (symbolized by G or F) is as follows:

$$\Delta G^\circ = -RT \ln K$$

where R is the universal gas constant (1.99 calories/mole/degree), T is the absolute temperature, and $\ln K$ the natural logarithm of the equilibrium constant; ΔG° is the standard free-energy change—the gain or loss of energy in calories at molal concentrations of reactants.

It is evident that if $K = 1$ (that is, $\ln K = 0$), no reaction will occur in a system containing molal concentrations of reactants and there is no change in free energy. If the reaction proceeds far to the right (product of concentrations of substances formed is relatively great), then $K > 1$ and the reaction is exergonic with available free energy; but when $K < 1$, the reaction will not proceed from left to right without a supply of free energy.

The first step in the utilization of the blood sugar glucose involves a phosphorylation which will illustrate these formulae. This phosphorylation is an essential preliminary step in glucose metabolism. Although this sugar is a rich potential source of energy (about 3.74 kcal/gm), this energy cannot be released to the organism through direct oxidation to carbon dioxide and water. Actually, glucose in solution (or as a dry substance) is extremely stable and may be kept under sterile conditions for centuries. To perform useful animal work, it must first be changed into a more reactive form and then pass through a series of reactions with gradual release of energy. The initial step in this process is a phosphorylation with ATP (adenosinetriphosphate). Thus,

$$\text{Glucose} + \text{ATP} \rightleftharpoons \text{glucose-6-phosphate} + \text{ADP}$$

At equilibrium, the products of this reaction are some forty times more concentrated than the reactants with the equilibrium constant $K = 1600$ according to the following formula:

$$\frac{[\text{Glucose-6-phosphate}]\,[\text{ADP}]}{[\text{Glucose}]\,[\text{ATP}]} = 1600$$

The reaction proceeds from left to right with a standard free energy change of about -4300 calories/mole calculated at 25°C.

It must be noted that these relationships provide no information concerning the *rate* of the reaction. The standard free energy as calculated above is independent of the rate of the reaction. In fact, some reactions may not occur at all even though they are exergonic with substantial negative free energy values. The situation may be compared to a sledge on the crest of a hill where there is sufficient frictional resistance to prevent its downhill slide; a push may set the reaction in motion with the release of free energy. Biological systems are catalyzed by enzymes that lower the activation energy, detonate the reaction, and permit it to proceed at temperatures compatible with life. The common household sugar sucrose, for example, may be hydrolyzed to the monosaccharides glucose and fructose with a standard free energy change of about -5500 and an equilibrium constant of 10,000. However, sucrose, like glucose, is a stable substance that can be stored almost indefinitely as a dry sugar or in a sterile solution; the addition of a small amount of catalyst such as HCl or the enzyme saccharase lowers the activation energy and accelerates the reaction but does not influence the equilibrium point. In the same way, the enzyme hexokinase is essential for the glucose phosphorylation described above.

Biological systems differ in several important ways from the physical systems on which thermodynamic laws are based. In the first place, living systems are OPEN, constantly exchanging substances with their environment, while thermodynamic principles are only strictly applicable to CLOSED systems which are homogeneous, with the various reactants at unit molarity. Such ideal conditions do not occur in cells; living systems are heterogeneous with many different interconnected reactions taking place in the same milieu. There is a steady flow of reactants into the cell while reaction products are excreted or utilized for subsequent reactions in a complex chain. Even though biological systems are *open*, various homeostatic mechanisms maintain relatively constant levels of reacting substances; living systems are STEADY-STATE systems maintained by the unidirectional flow of metabolites. In addition, the very organization of the living cell confers special properties on cellular reactions. The concentrations of some of the reactants may be extremely low; certain reactants may be locked inside cells or within compartments of cells by semipermeable membranes while others diffuse freely or are readily transported across them. Even though equilibrium constants, calculated under standard thermodynamic conditions, may be small or less than unity, the removal of reaction products from the cell or their utilization in another process may, nevertheless, maintain the reaction in a forward direction. Equilibrium thermodynamics form only a starting point in the analysis of the components of a complex biological system; the concept has been introduced here to explain the coupling of endergonic with exergonic reactions in the transfer of energy from its stable forms in the foods to the labile forms required for animal activities. Detailed discussions are found in textbooks of cell physiology and biochemistry; Lehninger (1971) has given a particularly lucid summary of bioenergetics for the beginner.

Coupling of chemical reactions in the transfer of energy The free energy of an exergonic or downhill reaction is utilized in biological systems to drive endergonic or uphill reactions. In this way, two reactions—one releasing and the other demanding energy—may be COUPLED to do chemical work and to effect syntheses. The glucose phosphorylation discussed in the previous section requires the simultaneous hydrolysis of ATP, an exergonic reaction with a standard free energy of about −7000 cal/mole at pH 7.0 and 25°C. The direct reaction of glucose and phosphate to form glucose-6-phosphate (G + P = G6P) is endergonic with a positive free energy of about +3000 cal/mole. At this point, it should be noted that these free energy values are approximate and will vary with the pH, temperature and concentration of reactants. The important point is that the hydrolysis of ATP provides sufficient free energy to phosphorylate the glucose in a coupled reaction; in fact, there is an excess of free energy which may be spilled out as heat. The overall reactions may be summarized as follows:

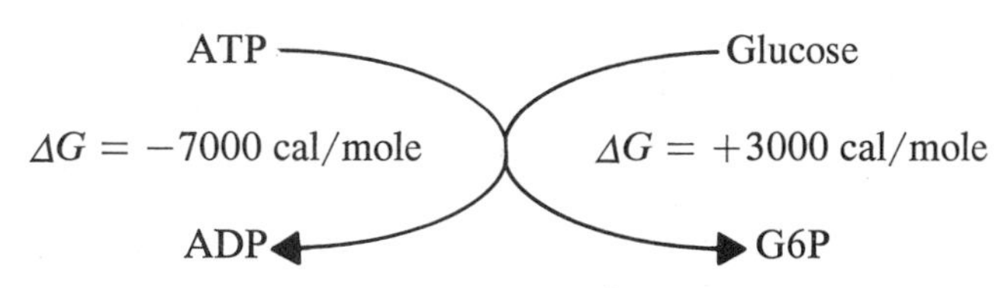

Overall: $\Delta G = -4000$ cal/mole approx.

In tissue respiration, a sequence of energy transfers of this sort is utilized to synthesize HIGH-ENERGY COMPOUNDS which on hydrolysis release relatively large amounts of free energy and perform biological work.

High-Energy Compounds

It is customary for biochemists and physiologists to refer to certain substances as HIGH-ENERGY or ENERGY-RICH COMPOUNDS while others are classed as LOW-ENERGY or ENERGY-POOR. To the physical chemist these terms are misnomers since the bond strengths, in chemical terms, are similar in both groups and not unusual. To the biological chemist, however, the terms have assumed particular meanings and have been found to express useful concepts. They are arbitrarily used to divide compounds into two groups with standard free energies of hydrolysis above or below 5 to 7 kcal/mole respectively. In short, the terms do not describe an energy localized in a particular bond but refer to differences in energy content of reactants and products. Thus, ATP with a free energy of hydrolysis of −7000 cal/mole at pH 7.0 is a high-energy or energy-rich substance while G6P with a free energy of hydrolysis of −3300 at the same pH is a low-energy or energy-poor compound. Low-energy phosphate bonds are ester linkages between phosphoric acid and the alcohol group of a sugar; like other organic esters, they are resistant to hydrolysis and yield comparatively small amounts of energy. High-energy phosphate bonds are normally anhydride linkages between two molecules of phosphoric acid or between a molecule of phosphoric acid and a carboxylic acid. These bonds are relatively unstable and on hydrolysis release large amounts of energy. An understanding of the molecular bases of these differences is left to textbooks of biochemistry. In the present context, the importance of this loose classification rests in the concepts of energy transfer that must always be from the higher- to the lower-energy substances.

Adenosinetriphosphate, or more specifically the ATP-ADP phosphate transfer system, occupies a very special position in these transfers of energy. This is not because it is unique in its high free energy since, in reality, it occupies an intermediate position in a series of phosphate compounds with free energy values ranging downward from about 12.8 kcal/mole. In large part, its importance rests in this intermediate position which permits ready shuttling of phosphate groups between the higher- and lower-energy groups. It seems to be the right size of coin for the rapid exchange of energy (Lehninger, 1964, 1971).

The structure of ATP is shown in Fig. 7.2. The molecule is a combination of three common biochemical materials: a purine base adenine (6-aminopurine), a 5-carbon sugar (ribose), and phosphoric acid. A unit composed of adenine and the pentose forms a combination known as adenosine (a NUCLEOSIDE). This, when associated with one phosphate unit, forms adenosine monophosphate (AMP), also called adenylic acid and known as a NUCLEOTIDE. Two more units of phosphate may also be added as shown in Fig.7.2 to form the diphosphate (ADP) and the triphosphate (ATP).

The energy relationships of the terminal monophosphate bond and those of the diphosphate and triphosphate bonds are quite different. Hydrolyses of the latter are

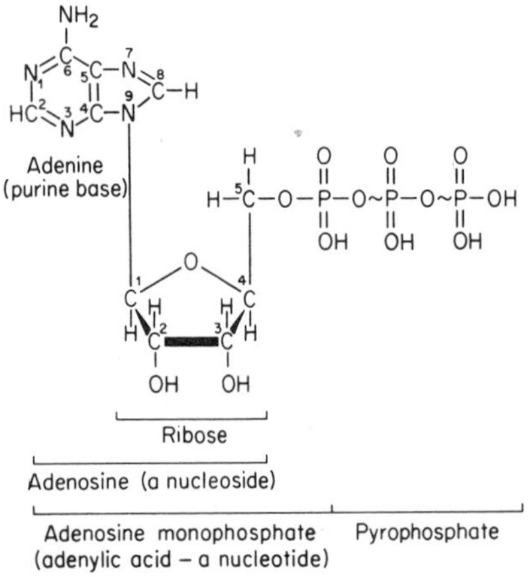

Adenine (purine base)
Ribose
Adenosine (a nucleoside)
Adenosine monophosphate (adenylic acid – a nucleotide)
Pyrophosphate

Fig. 7.2 Structure of adenosine triphosphate.

associated with the release of large amounts of free energy—approximately 8000 cal/mole, while the phosphate bond of adenylic acid is associated with only 1000 or 2000 cal/mole. The high-energy bonds, symbolized by the "swing dash" or "tilde" in Fig. 7.2, are the important energy "accumulators." Most of the specific tasks of the animal cell, such as contraction or secretion, depend directly on the ATP-ADP phosphate transfer system.

Thus, at one level of thinking, cellular metabolism is a matter of transforming the low-energy bonds of many different carbon compounds into the high-energy pyrophosphate bond. In tissue respiration, numerous reactions involving small energy changes lead step by step to a major energy-yielding reaction in which large amounts of free energy are made available for cell work. In these metabolic chains, electrons and hydrogen ions are passed along by a series of compounds, and at certain points ATP molecules appear. In most animals, the final stages require oxygen for the burning or oxidation of hydrogen, and in this series of reactions the greatest amount of ATP is generated (oxidative phosphorylation).

In a sense, this reaction, in which hydrogen combines with oxygen, is the same reaction familiar to every student in general chemistry. As has so often been demonstrated in the high school laboratory, it may occur with explosive violence. In the living cell, however, a series of enzymes operates an extended chain of reactions, and the low-energy bonds are transformed or "stored" in pyrophosphate bonds of ATP. Unlike the mistimed laboratory experiment, the energy never appears suddenly with an explosive production of heat. The energy transformations of life are remarkably efficient. The overall thermodynamic efficiency for the conversion of food energy into pyrophosphate bond energy is 60 to 70 per cent (Krebs and Kornberg, 1957).

Biological classes of energy-rich compounds Although ATP is a member of a select group, it is by no means the only energy-rich compound. Energy-rich compounds as they are known to the biochemist are of two types: derivatives of phosphoric acid and derivatives of carboxylic acid (Huennekens and Whiteley, 1960). In the first group, energy is released in the hydrolytic split of the bond joining phosphoric acid to the rest of the molecule; in the second group, carboxylic acid is hydrolytically split from its derivative. These reactions may be shown as follows:

$$(1)\quad \mathrm{HO{-}}\overset{\displaystyle \mathrm{O}}{\underset{\displaystyle \mathrm{OH}}{\overset{\|}{\underset{|}{\mathrm{P}}}}}\mathrm{{-}}X + \mathrm{H_2O} \rightleftharpoons \mathrm{H}X + \mathrm{HO{-}}\overset{\displaystyle \mathrm{O}}{\underset{\displaystyle \mathrm{OH}}{\overset{\|}{\underset{|}{\mathrm{P}}}}}\mathrm{{-}OH} + \mathrm{Energy}$$

$$(2)\quad R\mathrm{{-}}\overset{\displaystyle \mathrm{O}}{\overset{\|}{\mathrm{C}}}\mathrm{{-}}X' + \mathrm{H_2O} \rightleftharpoons \mathrm{H}X' + R\mathrm{{-}}\overset{\displaystyle \mathrm{O}}{\overset{\|}{\mathrm{C}}}\mathrm{{-}OH} + \mathrm{Energy}$$

In the first reaction, X represents ADP (in ATP) or phosphate (in the polyphosphates) or the guanidino group (in creatine phosphate). In the second reaction, R represents the acetyl group and X' represents coenzyme A in the energy-rich bonds of acetyl CoA. These compounds will be identified in subsequent sections. They are introduced here to exemplify the two biochemical groups of energy-rich bonds.

From a functional standpoint, the energy-rich compounds fall into three major groups. ATP and the other nucleoside polyphosphates, such as the guanosine and uridine polyphosphates, are the PRIMARY PHOSPHORYLATING AGENTS responsible for energy transfers associated with the important animal activities of muscle contraction and protein synthesis. A second group acts as TRANSIENT INTERMEDIATES in phosphorylating reactions and is concerned with the synthesis of the primary phosphorylating agents. Some of these (acetyl CoA, phosphoenolpyruvate and glycerylphosphate) appear to be universal, while others may be more restricted in their distribution. The third group consists of the highly important ENERGY RESERVOIRS. The quantity of primary phosphorylating agent in the tissues is small; where demands are likely to be sudden and of considerable magnitude, the tissues possess a reserve of labile phosphate groups for the rapid regeneration of the primary phosphorylating agents. The primary phosphorylating agent, ATP, is universally distributed in living materials, but the energy reservoirs are quite different in the animal and the plant worlds.

The energy reservoirs The amidine phosphates (also called the guanidine phosphates or the phosphagens) are the important reservoirs of high-energy phosphate in animal tissues. In the plant world the inorganic polyphosphates discharge a similar function. One of the most interesting chapters in physiological chemistry centers around the discovery and recognition of the significance of the phosphagens (Baldwin, 1967; Kalckar, 1969).

In 1907, Fletcher and Hopkins, using improved techniques, had focussed attention on lactic acid as a key substance in muscle biochemistry. Their analyses of isolated and electrically stimulated frog muscle demonstrated the formation of lactic acid in both an anaerobic and an aerobic environment. They showed further that fatigue appeared sooner and more lactic acid accumulated in the anaerobic environment; in an aerobic environment, carbon dioxide was produced as lactic acid disappeared. These studies were followed by those of Meyerhof who proved that glycogen was the source of lactate. Within the next few years a ratio was established between the work done, glycogen used, and lactic acid formed; the chemical events associated with muscle contraction seemed to be fully explained in terms of the metabolism of carbohydrate.

Lundsgaard in 1930 made the significant discovery that muscles poisoned with iodoacetate could no longer form lactic acid but could still contract anaerobically. This quickly shifted biochemical interest from the carbohydrates to the organic phosphates as the primary source of energy. Further work showed that creatine phosphate (phosphagen) was breaking down to form creatine when the iodoacetate-poisoned muscle was activated and that the work done by the muscle was equivalent to the breakdown of phosphagen. Creatine phosphate had been isolated from vertebrate muscle in 1927 and the presence of unidentified organic phosphates had been recognized earlier. However, their true significance was not recognized until after

Lohmann, in 1934, showed that the phosphagen changes in contracting muscle depended on the presence of ATP. In 1937, he suggested a scheme of reactions which has now been generally accepted by physiologists.

In Lohmann's scheme the primary source of energy is provided by the hydrolysis of ATP to ADP and inorganic phosphate (P); the creatine phosphate forms a reservoir of high-energy phosphate which is used to regenerate the stores of ATP. The glycogen-lactic acid changes, studied first during the history of muscle biochemistry, are concerned with the regeneration of creatine phosphate (CP) from creatine (C). Thus:

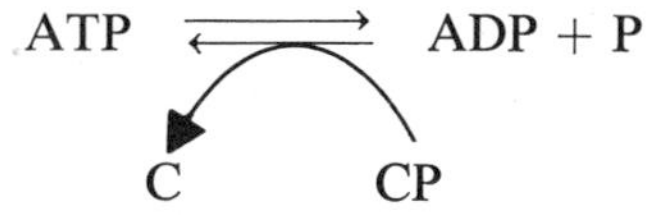

$$\text{ATP} \rightleftarrows \text{ADP} + \text{P}$$

C ← CP

The regeneration of the creatine phosphate is indirect by way of the ATP formed in muscle glycolysis and oxidative phosphorylation.

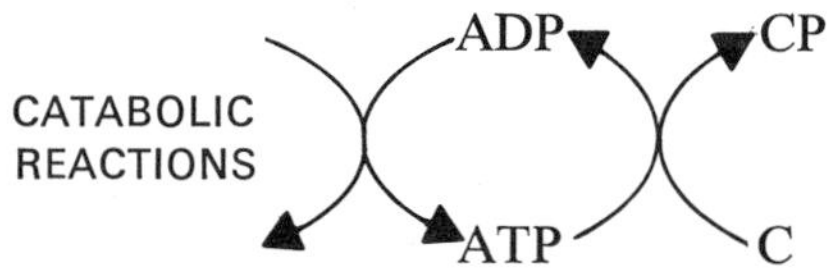

CATABOLIC REACTIONS — ADP → ATP; CP → C

Although the muscles of all vertebrate animals contain creatine phosphate, those of the majority of the invertebrates contain arginine phosphate. When it was discovered that the major exceptions to this general rule are found among the echinoderms and the protochordates, the phosphagens very naturally became prominent in discussions of biochemical evolution. Both arginine phosphate and creatine phosphate are found among the echinoderms and protochordates, and this seemed to be in agreement with speculated lines of evolution which suggest a common ancestry for the echinoderms and the chordates. The formulas of these two common phosphagens are as follows:

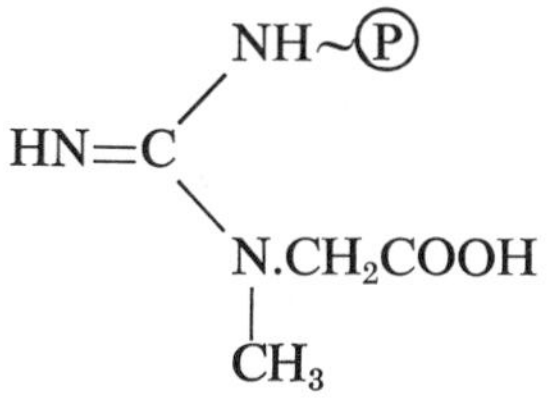

$$\text{HN}{=}\text{C}\begin{cases}\text{NH}\sim\text{Ⓟ}\\ \text{N.CH}_2\text{COOH} \\ \quad|\ \text{CH}_3\end{cases}$$

Creatine phosphate

$$\text{HN}{=}\text{C}\begin{cases}\text{NH}\sim\text{Ⓟ}\\ \text{NH}-(\text{CH}_2)_3-\text{CH.NH}_2-\text{COOH}\end{cases}$$

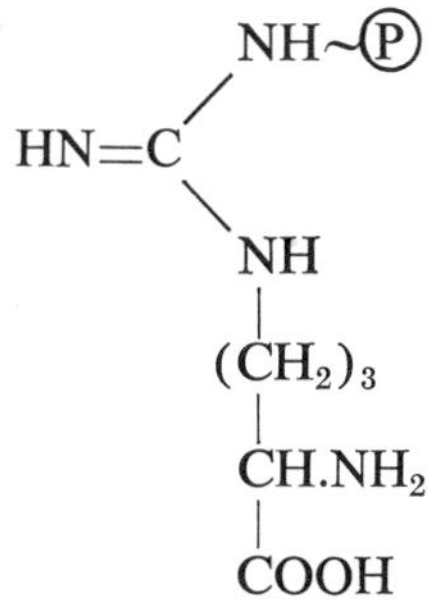

Arginine phosphate

The earlier data have now been greatly amplified; it is apparent that creatine phosphate is not confined to the echinoderms and protochordates among the invertebrates but is also found in at least one sponge (*Thetia lyncurium*), a coelenterate (*Anemonia sulcata*), a sipunculoid (*Sipunculus nudis*) and several of the annelids (Huennekens and Whiteley, 1960). Moreover, five other phosphagens, peculiar to the annelids, nemerteans and sipunculoids have been isolated and studied (Thoai and Robin, 1969; Needham, 1970b). These are also characterized by the presence of the amidine group, shown as a shaded block in the following formulas:

$$HN{=}C(NH_2){-}NH{-}CH_2{-}COOH$$

Glycocyamine

$$HN{=}C(NH_2){-}NH{-}CH_2{-}CH_2SO_3H$$

Taurocyamine

$$HN{=}C(NH_2){-}NH{-}CH_2{-}CH_2O{-}P({=}O)(OH){-}O\cdot CH_2{-}CH\cdot NH_2{-}COOH$$

Lombricine

In spite of these exceptions, the earlier broad generalization remains. The vertebrates do utilize only phosphocreatine; the majority of the invertebrates rely on phosphoarginine, while different members of the echinoderms and protochordates are about equally divided in the nature of their phosphagen. Thinking must, however, be adjusted concerning the phylogenetic antiquity of phosphocreatine. Enzymes and compounds essential for its formation are found in the most primitive multicellular organisms. Moreover, several different, although apparently rare, mutations have given rise to other phosphagens as exemplified by the annelids.

Arginine was probably the starting point in the biochemical evolution of the storage system of amidine phosphates. It is a ubiquitous amino acid and enters into the composition of the protoplasm of animals of all kinds. The guanidino group of the arginine molecule which couples with phosphoric acid to form the high-energy bond of phosphoarginine is common to the other phosphagens and is the functioning portion of all these molecules. Creatine is formed from arginine, glycine and methionine. The enzymes involved in biosynthesis have been studied in mammalian kidney and liver, and ATP is required in some of the steps. In the first step, arginine and glycine react to form glycocyamine (guanidinoacetic acid) and ornithine. This is methylated in a second reaction involving methionine to form creatine. In some of the worms, however, the glycocyamine is directly phosphorylated to form a phosphagen. The taurocyamine molecule is only slightly different; the lombricine molecule contains D-serine linked by phosphoric acid to guanidinoethanol. The guanidino group is common to all these phosphagens.

Animals at almost every stage in phylogeny seem to have experimented with phosphocreatine, and several phyla tried other guanidine phosphates as well before the evolving chordates made the irrevocable decision to use only creatine phosphate in their muscle machinery. This may have been a purely fortuitous "decision"; or phosphocreatine may be superior for the purposes of muscle contraction in the vertebrates. This and many other problems associated with the evolution of the phosphagens await further investigation (Thoai and Robin, 1969; Needham, 1970b; Rockstein, 1971).

Biological Oxidations

An animal's energy budget is mainly derived from dietary carbohydrate and fat with minor contributions from proteins. The energy-generating reactions are oxidations that may be summarized as follows, indicating only the fuel and the end products other than ATP:

$$\underset{\text{Sugar}}{C_6H_{12}O_6} + 6O_2 \longrightarrow 6CO_2 + 6H_2O$$

$$\underset{\text{Fatty acid}}{C_{17}H_{35}COOH} + 26O_2 \longrightarrow 18CO_2 + 18H_2O$$

Now, it is common knowledge that biological oxidations do not involve the direct combination of carbon and oxygen as in burning wood or coal but occur in an extended sequence of enzymatically catalyzed reactions that provide a gradual release of ATP. The direct transfer and incorporation of molecular oxygen into substrate molecules are comparatively rare in biological systems although they do occur in several important reactions catalyzed by the OXYGENASES. Much more frequently, biological oxidations involve the removal of hydrogen (dehydrogenation) in reactions which depend on the DEHYDROGENASES; these latter enzymes are termed OXIDASES if oxygen is utilized as a hydrogen acceptor. A third type of biological oxidation occurs with the direct loss of one or more electrons, as in the oxidation of iron compounds from the ferrous to the ferric state—a type of reaction encountered in the electron transfers of the cytochromes. Of these three classes of oxidations, the dehydrogenations are by far the most numerous.

Two basic facts of elementary chemistry are important in understanding biological oxidations: (1) all oxidation reactions are similar in that they involve the loss of electrons and (2) since free electrons and atoms do not occur in solution, a substance can only be oxidized if its electrons can flow to another substance which is simultaneously reduced. The enzyme lactic dehydrogenase catalyzes the dehydrogenation (oxidation) of lactic acid to form pyruvic acid with the release of two hydrogen ions and two electrons as follows:

$$\underset{\text{Lactic acid}}{\begin{matrix} CH_3 \\ | \\ H\text{—}C\text{—}OH \\ | \\ COOH \end{matrix}} \rightleftharpoons \underset{\text{Pyruvic acid}}{\begin{matrix} CH_3 \\ | \\ C{=}O \\ | \\ COOH \end{matrix}} + 2H^+ + 2e$$

This reaction is only possible in the presence of an electron acceptor which in biological systems is nicotinamide adenine dinucleotide (NAD), a compound which will be identified in the next section. Similarly, the ferrous/ferric oxidation-reduction in cytochrome, which may be indicated thus,

$$Fe^{++} \underset{\text{reduction}}{\overset{\text{oxidation}}{\rightleftharpoons}} Fe^{+++} + e$$

depends on a chain of slightly different cytochromes which are successively reduced and oxidized to permit the flow of electrons. These are oxidation-reduction systems and the two processes go hand in hand (Fig. 7.3).

A reducing agent is an electron donor which, by giving up its electrons, becomes oxidized. An oxidizing agent is an electron acceptor which, by taking on electrons, becomes reduced. Electron donors (reducing agents) may be thought of as having a definite electron pressure while electron acceptors (oxidizing agents) have specific electron affinities. The physical measurement of the tendency of a substance to give up or to accept electrons is its OXIDATION-REDUCTION OR REDOX POTENTIAL. These potentials, measured under standardized conditions, permit the arrangement of substances in accordance with their electron pressures. Values are tabulated in textbooks of cell physiology and biochemistry. In the systems under consideration here, these values range from about -0.015 volts for (succinate $\rightleftharpoons$ fumarate $+ 2H^+ + 2e$) to $+0.815$ volts for ($H_2O \rightleftharpoons \frac{1}{2}O_2 + 2H^+ + 2e$); precise values depend on pH and temperature. Substances with a strong tendency to take up electrons (that is to oxidize other compounds) have large negative redox potentials. The flow of electrons will be from systems with more negative, to systems with more positive redox potentials; transfer of electrons can only take place from donors (reducing agents) with greater electron pressures (negative values) to acceptors with lower electron pressures.

These electron transfers involve changes in free energy and form the basis for the production of ATP. Methods of determining redox potentials and the calculation of the standard free energy ($\Delta G°$) for oxidation-reduction reactions may be found in textbooks of chemistry and cell physiology. Here, it is only important to appreciate that much of the free energy required for the generation of ATP—the driving force of life—comes from the flow of electrons in a series of oxidation-reduction reactions. Several of the specialized molecules universally concerned with this flow of electrons will now be described.

Pyridine nucleotides NAD and NADP These coenzymes are the most usual primary acceptors of electrons from the food metabolites. NICOTINAMIDE ADENINE

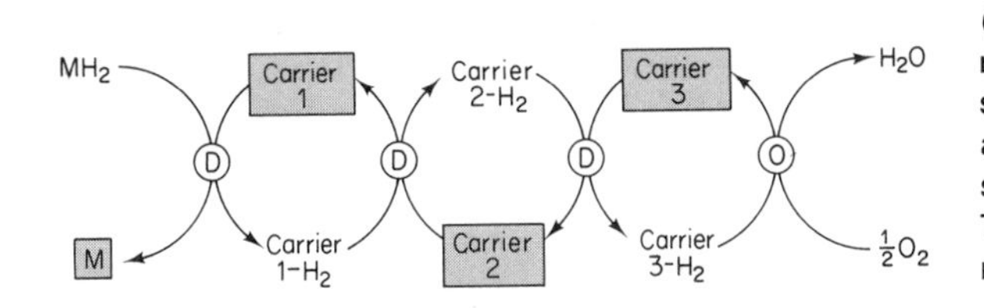

Fig. 7.3 Oxidation of a metabolite (MH_2) in a series of oxidation-reduction reactions catalyzed by a series of dehydrogenases (D) with a terminal oxidase (O) where oxygen serves as the hydrogen acceptor. The shaded blocks are the oxidized reaction products.

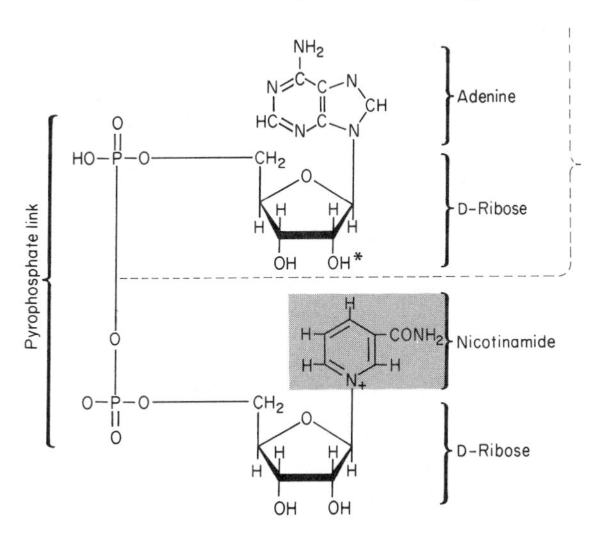

Fig. 7.4 Structure of nicotinamide adenine dinucleotide, NAD. An additional phosphate in the 2 position of the pentose of the nucleoside portion of the molecule* produces nicotinamide adenine dinucleotide phosphate, NADP; shaded block shows group concerned with electron transfer. The Commission on Enzymes of the International Union of Biochemistry has recommended that these names replace the older terms diphosphopyridine nucleotide (DPN) and triphosphopyridine nucleotide (TPN).

DINUCLEOTIDE (NAD), which differs from NADP only in the absence of one phosphate group (Fig. 7.4), transports the bulk of the hydrogen released from the food metabolites to an electron transport system where it is ultimately combined with oxygen to form water in the production of ATP. Biochemically, NAD contains a unit of adenylic acid combined through a pyrophosphate link to pentose; the pentose, in turn, is attached to nicotinamide, a pyridine ring compound which contains a reactive group that can accept two electrons and a proton as follows:

NAD $\rightleftharpoons$ Reduced NAD

$$NAD^+ + 2H^+ + 2e \rightleftharpoons NADH + H^+$$

Nicotinamide comes from nicotinic acid or NIACIN which is the pellagra preventative (P-P) vitamin. Its absence from the diet leads to a serious disease, known as pellagra, which is still one of the most prevalent nutritional deficiencies in the world.

Flavoproteins These important proteins employ riboflavin (vitamin B_2) as the essential cofactor in electron transfer. Riboflavin consists of a sugar alcohol (D-ribitol) attached to a substituted isoalloxazine ring which can accept two hydrogens

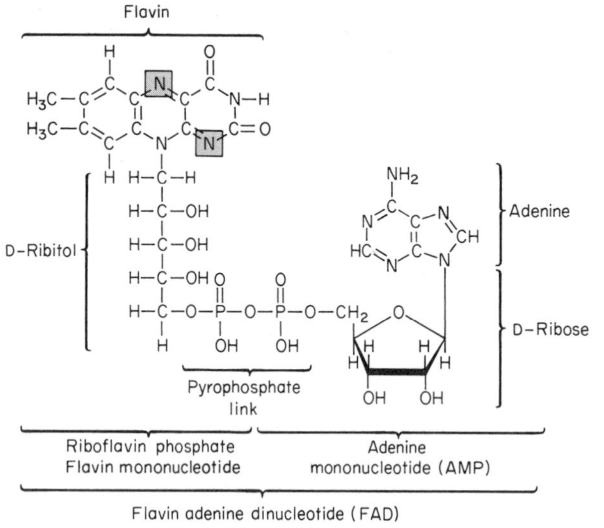

Fig. 7.5 The riboflavin coenzymes. The molecule of flavin adenine dinucleotide (FAD) is formed through the combination of flavin mononucleotide (FMN) and ATP. Each shaded nitrogen in the flavin component takes up a hydrogen from NADH + H^+ to form $FADH_2$.

as indicated in Fig. 7.5. There are two different flavin coenzymes: FLAVIN MONONUCLEOTIDE (FMN) and FLAVIN ADENINE DINUCLEOTIDE (FAD) related biochemically as shown in Fig. 7.5. The prosthetic groups are firmly associated with protein. In the present context, interest centers around FAD which serves as a ready acceptor of electrons from NAD or NADP.

Non-heme iron-proteins Proteins containing iron in a tightly bound form, but lacking an organic prosthetic group such as the iron-flavoproteins, were first recognized in plant chloroplasts and termed FERREDOXIN. Subsequently, comparable iron-proteins have been identified as essential carriers in all major pathways of electron transfer (White *et al.*, 1968). Different members of the group are identified by characteristic absorption spectra. They operate in electron transfer by reversible oxidation-reduction of the iron atom.

Ubiquinone This is a benzoquinone compound with a side chain containing ten isoprenoid units. It is also called COENZYME Q and functions in electron transport as shown in Fig. 7.6. Ubiquinone occurs in the mitochondria of animal tissues. In plant tissues, the chloroplasts contain a comparable substance PLASTOQUINONE which has nine instead of ten isoprenoid residues in the side chain. Some of the

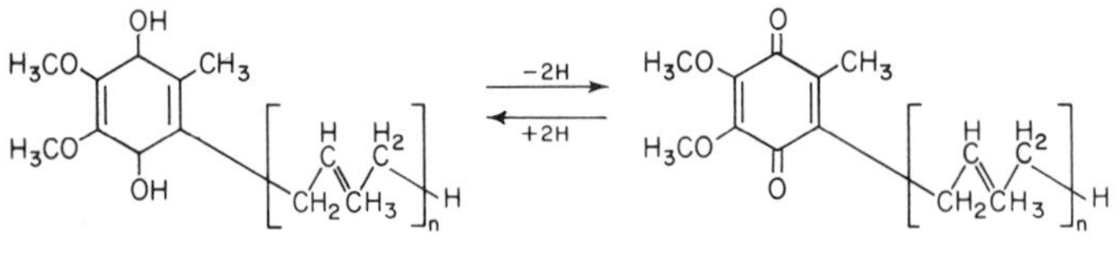

Fig. 7.6 The coenzyme Q oxidation-reduction system. In mitochondria of most mammals, $n = 10$; in some microorganisms, $n = 6$; in plant chloroplasts $n = 9$.

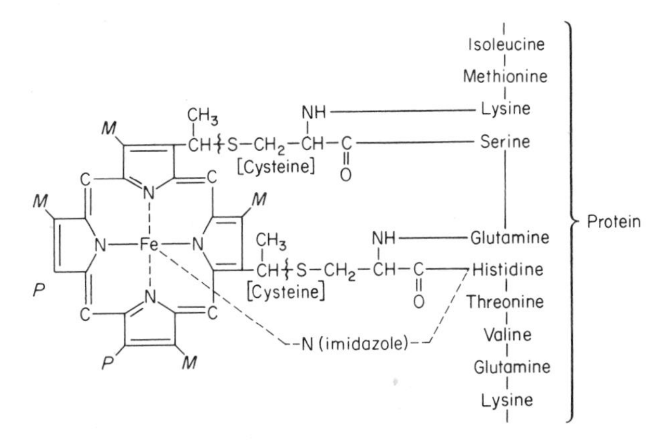

Fig. 7.7 Partial structure of cytochrome *c* (human heart). For species variations, see Lehninger (1970). [From Harper, 1973: Review of Physiological Chemistry, 14th ed. Lange Medical Publications, Los Altos, California.]

yeasts contain six residues; the numbers vary but are less than ten in lower organisms (Crane and Low, 1966; Morton, 1971).

Cytochromes The remaining electron carriers in animal cells are the cytochromes. These are hemoproteins (Chapter 6). Various members of the group differ in the protein portion of the molecule, in the nature of the side chains of the metalloporphyrin, and in the mode of attachment of the heme group to the protein. Cytochrome *c* is depicted in Fig. 7.7; its redox potential is +0.26 at pH 7.0 and 25°C. Cytochrome oxidase in the ferrous form is able to reduce molecular oxygen and in this respect is unique among the molecules operating in the electron transport chain (Lemberg, 1969).

PATHWAYS OF CELLULAR METABOLISM

Animal evolution is founded on the products of photosynthesis. During the phylogeny of the photosynthetic mechanisms protoplasm became associated with most of the organic compounds, enzymes, and reactions involved in the respiration of animal cells. The process of photosynthesis provided the carbon compounds required for the heterotrophic way of life and, at the same time, generated the oxygen atmosphere necessary for high-energy aerobic reactions. Before animal life was possible, the biochemical compounds, the enzymes necessary for their metabolism and the fundamental energy exchanges of life were operating universally in living protoplasm.

It is not then surprising that the production of energy follows the same general pathways in bacteria and plants and in animals at different stages of phylogeny. These universal processes will be outlined first. Energy released in these processes is utilized to do special kinds of work which may be peculiar to certain types of cells and may depend on the way of life of the particular animal. Although these special energy exchanges—whereby a muscle contracts, a gland secretes or a cell produces light—also depend on familiar compounds and reactions, they are, in a sense, more distinctively related to animal life and its phylogeny. They will be considered later.

Only the simple building blocks of the complex foods can serve as fuel for cellular respiration. Carbohydrates are utilized within the cell as monosaccharides, proteins as amino acids and fats as fatty acids and glycerol. Through a series of reactions the monosaccharides, the fatty acids, glycerol, and many of the amino acids are oxidized (with a relatively small release of energy) to the familiar two-carbon compound, acetic acid—actually combined in the cell with coenzyme A in an especially reactive form known as ACETYL CoA or ACTIVE ACETATE. Active acetate (along with two other residues from certain amino acids) may then enter an enzymatic pool (the citric acid or tricarboxylic acid cycle) which generates hydrogen for the reduction of gaseous oxygen and the production of ATP chemical energy. These events are outlined in Table 7.1 and Fig 7.8.

Phase I of Table 7.1 includes the processes of digestion and produces less than 1 per cent of the total energy. The pre-acetate steps of metabolism shown in Phase II are anaerobic and yield about 35 per cent of the energy; this phase is labeled "substrate level phosphorylation" in Fig. 7.1. The third phase, which is aerobic, is responsible for about 65 per cent of the energy liberated (Krebs and Kornberg, 1957). The overall thermodynamic efficiency of these processes is high (of the order of 60 to 70 per cent) and, to a very large degree, this depends on the citric

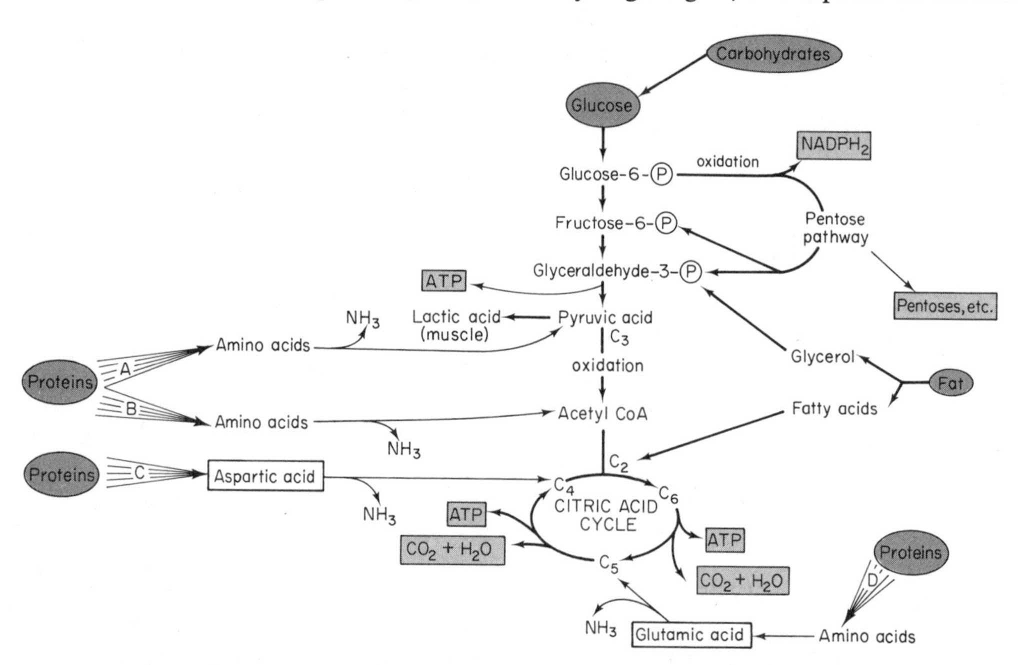

Fig. 7.8 Interrelationships of metabolic pathways for different foods. AMINO ACID GROUPS: A, glucogenic series—alanine, glycine, serine, threonine, methionine, cysteine, valine; B, ketogenic series—leucine, isoleucine, phenylalanine, tyrosine; C, aspartic acid; D, glumatic acid itself or derived from arginine, proline, hydroxyproline, histidine, ornithine.

Table 7.1 The Three Main Phases of Energy Production from Foodstuffs. (Shaded blocks show number of carbon atoms per molecule at different stages in metabolism)

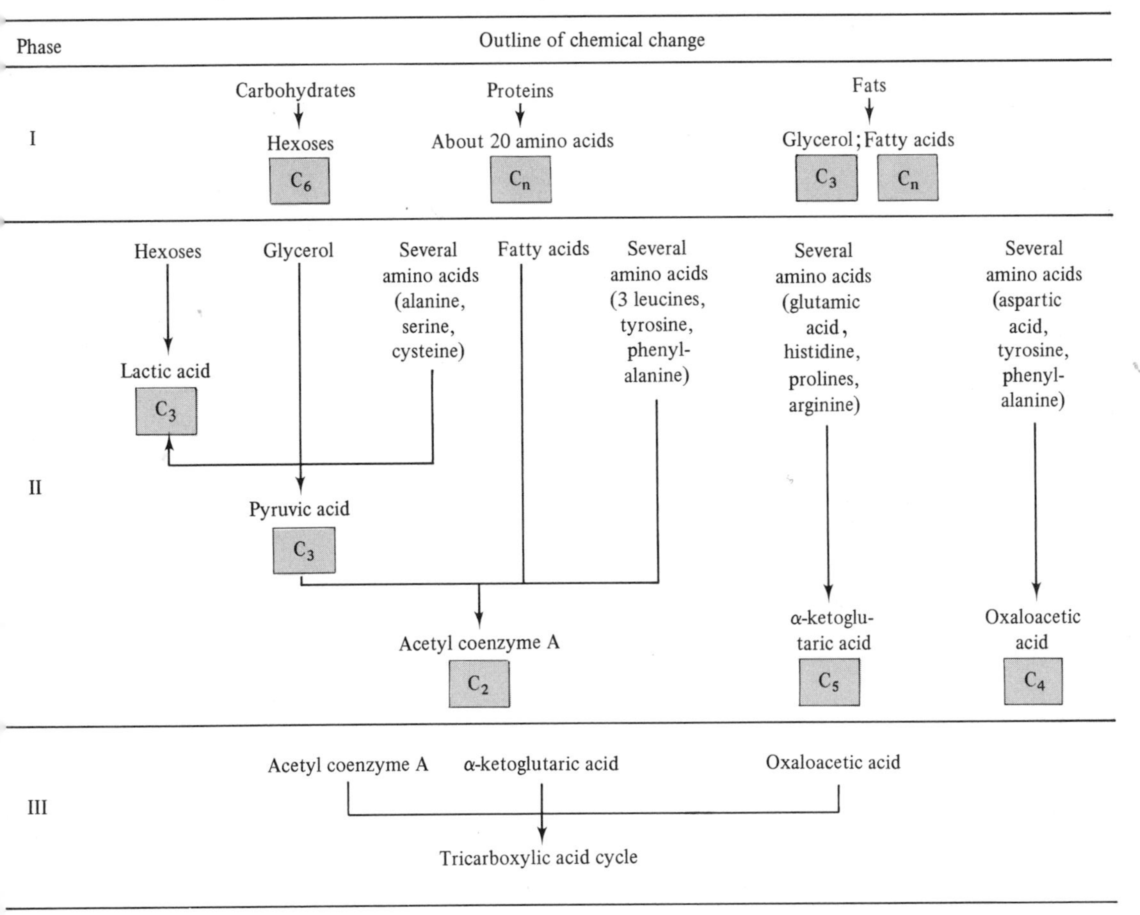

acid cycle and the aerobic processes of respiration. The different steps in energy production will now be described in more detail.

Carbohydrate Metabolism

Monosaccharides to pyruvic acid A fundamental chain of biochemical reactions, known as the "Embden-Meyerhof" sequence was the first series of energy-yielding reactions to be worked out in detail. This is a pathway of glycogen or glucose metabolism. Since animals usually depend on carbohydrate for much of their energy, it forms one of the important sequences of their cellular metabolism. Its elucidation occupied the minds of many other biochemists during the first half of the twentieth century even though the names of Embden and Meyerhof are usually attached to it. This is now known to be the major route in the metabolism of

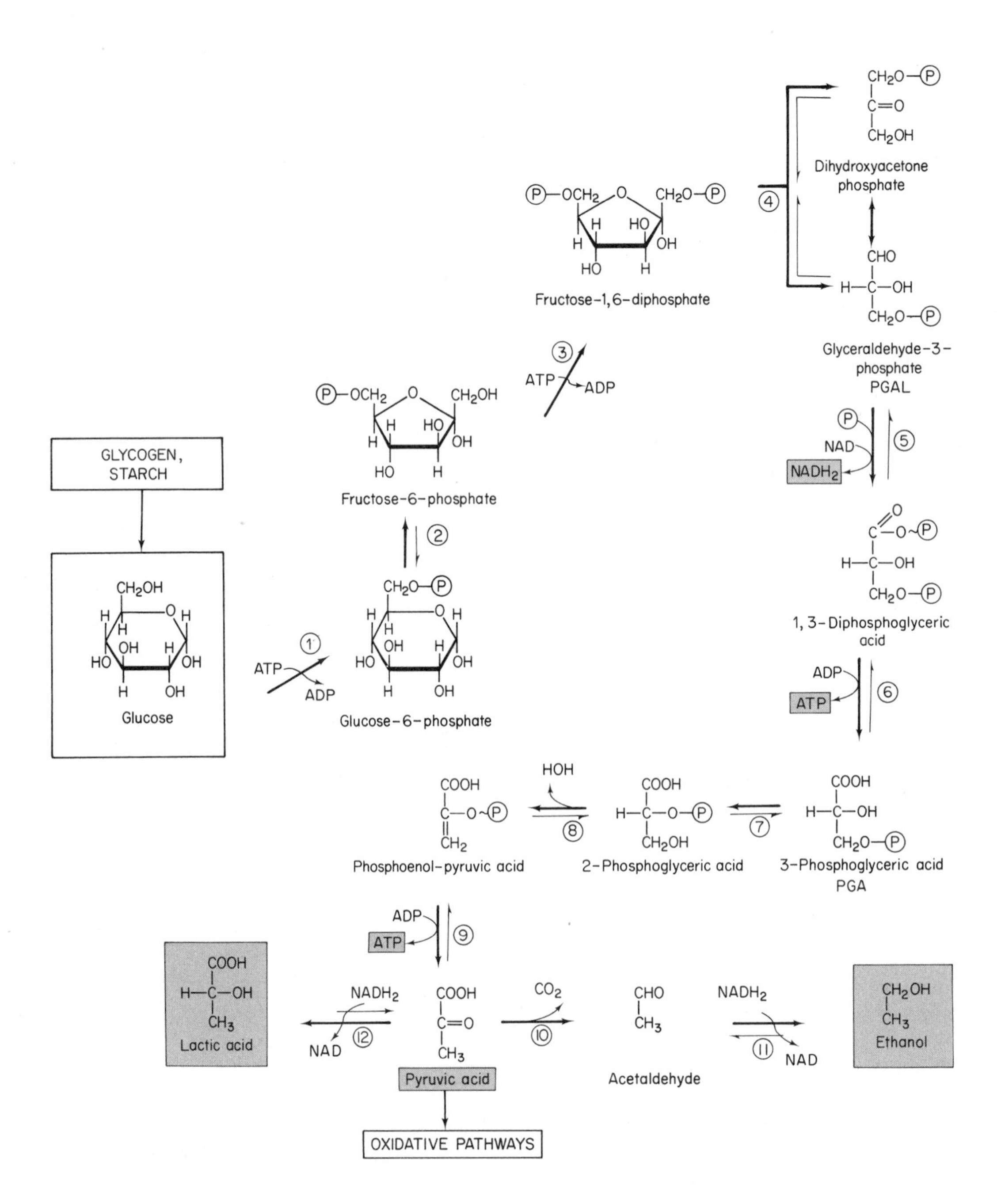

Fig. 7.9 Embden-Meyerhof pathway of glycolysis. ENZYMES: (1), glucokinase; (2), phosphohexoisomerase; (3), phosphofructokinase; (4), aldolase; (5), glyceraldehyde-3-phosphate dehydrogenase; (6), phosphoglycerokinase; (7), phosphoglyceromutase; (8), enolase; (9), pyruvic kinase; (10), pyruvic carboxylase; (11), alcohol dehydrogenase; (12), lactic dehydrogenase.

carbohydrate, whether the sugars are being utilized in the contraction of muscle (GLYCOLYSIS) or whether they are fermented by microorganisms in the production of alcohol. The important steps are outlined in Fig. 7.9.

A comprehensive discussion of this sequence would require many pages. The enzymes, the energetics and the molecular arrangements will be found in textbooks of biochemistry. Here, the general nature of the processs is emphasized and attention is directed to the important energy-transforming compounds, ATP and NAD.

It will be noted that oxidation (dehydrogenation) and the generation of ATP are preceded by several preparatory reactions (Fig. 7.9). Glucose, as previously mentioned, is a stable substance when it enters the cell and must have its energy relations altered through phosphorylation by ATP. The formation of the glucose-6-phosphate in the Embden-Meyerhof sequence is followed by a molecular rearrangement of the sugar to form fructose-6-phosphate and another priming reaction with a second unit of ATP to form a diphosphate sugar (fructose-1, 6-diphosphate). Enzymatic reactions then split this 6-carbon diphosphate compound into two 3-carbon units with no significant change in the energy values of the phosphate bonds. Particular attention is drawn to this 3-carbon unit, D-glyceraldehyde-3-phosphate, also called phosphoglyceraldehyde or PGAL; this substance was encountered in Chapter 1 where it was identified as the primary carbon compound produced by carbon dioxide fixation in photosynthesis. PGAL forms a meeting place in carbohydrate metabolism and photosynthesis as well as in the metabolism of glycerol arising from the lipids (Fig. 7.8).

The formation of the PGAL is followed by a coupled reaction in which it is oxidized in the presence of inorganic phosphate to form a diphosphate, and this then reacts with ADP in a transphosphorylation to form 3-phosphoglyceric acid (PGA) and a unit of ATP. In this way, the two molecules of PGAL which came from the single molecule of glucose pay back the two ATP molecules which were required for the priming reaction. In the next series of enzymatic reactions the two units of PGA release the other phosphate to ADP, and there is thus a net gain of two ATP molecules for each glucose unit metabolized to pyruvic acid. This sequence of reactions has also produced two units of reduced NAD ($NADH + H^+$), but this can only be turned into high-energy phosphate bonds if oxygen becomes available, as described below for the aerobic respiratory chain of oxidative phosphorylation. Otherwise, the $NADH + H^+$ enters into the oxidation-reduction reactions of anaerobic glycolysis to form lactic acid or ethanol (Fig. 7.9). Reactions, in terms of ATP production, may be summarized as follows.

Glycogen to lactic acid:

$$\text{(Glycosyl unit)} + \text{ATP} + 4\text{ADP} + 3\text{P} \longrightarrow 2\ \text{Lactic acid} + \text{ADP} + 4\text{ATP} + 2\text{H}_2\text{O}$$

$$\longrightarrow \text{net gain} = 3\text{ATP}$$

Glucose to lactic acid:

$$\text{Glucose} + 2\text{ATP} + 4\text{ADP} + 2\text{P} \longrightarrow 2\ \text{Lactic acid} + 4\text{ATP} + 2\text{ADP} + 2\text{H}_2\text{O}$$

$$\longrightarrow \text{net gain} = 2\text{ATP}$$

The enzymes and compounds involved in the anaerobic metabolism of glucose by the Embden-Meyerhof sequence have now been identified in both plants and animals at all levels in phylogeny (Lioret and Moyse, 1963).

Other pathways of glucose metabolism Several additional enzyme systems are known to be concerned with the oxidation of glucose. The enzyme glucose oxidase found in the mold *Penicillium notatum* and some other organisms oxidizes this carbohydrate to δ-gluconolactone, while certain bacteria possess pathways involving enzymes and phosphate compounds which bypass many of the steps of glycolysis. In the Entner-Doudoroff pathway, for example, 6-phosphogluconate (Fig. 7.10) is dehydrated to form 2-keto-3-deoxy-6-phosphogluconate which is then split, forming pyruvic acid and PGAL; this pathway is limited to bacteria (Pon, 1964). The

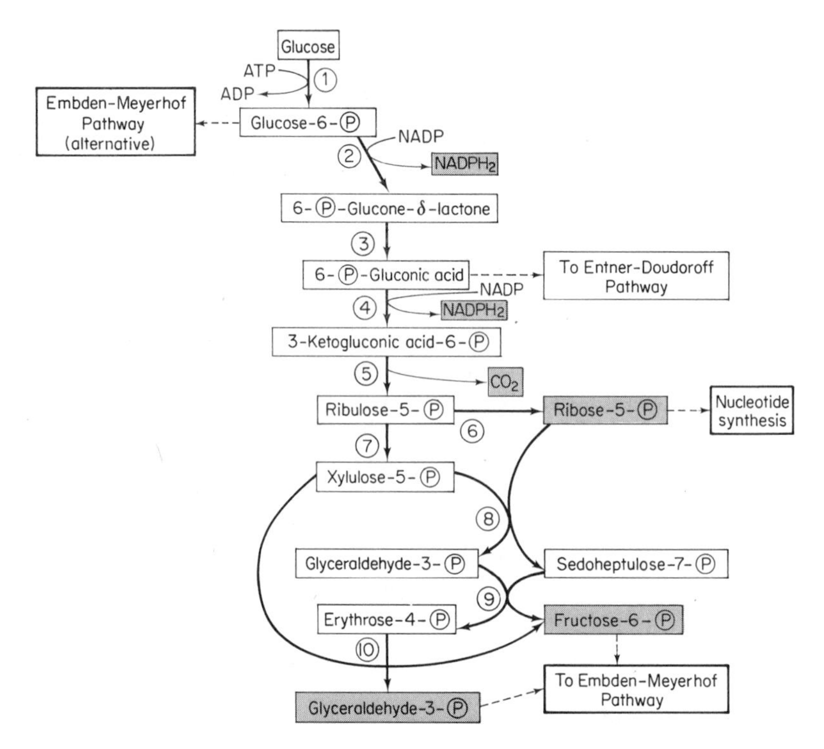

Fig. 7.10 Major steps in the pentose pathway. Since the phosphate group is attached to the terminal carbon, the number that precedes the Ⓟ shows the number of carbon atoms in the molecule. ENZYMES: (1) hexokinase and Mg^{++}; (2) glucose-6-phosphate dehydrogenase; (3) a specific lactonase; (4) 6-phosphogluconic acid dehydrogenase; (5) a β-hydroxyacid oxidative decarboxylase and Mn^{++}; (6) phosphopentose isomerase; (7) phosphopentose epimerase; (8) and (10) transketolase with thiamine phosphate and Mg^{++}; (9) transaldolase. [Based on Bennett and Frieden 1966: Modern Topics in Biochemistry. Macmillan, New York.]

PENTOSE PHOSPHATE PATHWAY, however, is of universal occurrence and forms an essential adjunct to the glycolytic sequence in the metabolism of carbohydrates.

The pentose phosphate pathway, or more properly the oxidative pentose phosphate pathway, is also termed the "Warburg-Dickens pathway," the hexose monophosphate (HMP) oxidation shunt and the phosphogluconate oxidative pathway. Since one of its important functions is to produce the pentoses required for nucleotide synthesis, it will here be termed the "pentose phosphate pathway." The important steps are outlined in Fig. 7.10.

As indicated in this figure, the pentose pathway or cycle may be thought of as an enzymatic pool in which 6-carbon sugars (glucose and fructose) are converted into 5-carbon sugars with the formation of a carbon dioxide. Two 5-carbon units may then be converted into a 7-carbon and a 3-carbon unit (PGAL); the 7-carbon and the 3-carbon unit may be further rearranged into a 6-carbon unit (fructose) and a 4-carbon unit; the 4-carbon sugar and a 5-carbon sugar can then produce a hexose and a triose (PGAL). The reactions are reversible, and this enzyme pool provides the important machinery for rearranging varied monosaccharides. In the overall reaction, every three units of 6-carbon sugar which enter the pool will return two of these units and form one unit of triose phosphate (PGAL) and three molecules of CO_2. In the course of these reactions, the 6-carbon sugars are first condensed to 5-carbon sugars, and the hydrogen transfers are made through NADP. It is noted that this scheme requires NADP instead of NAD, that it generates PGAL which can be added to the pyruvate pool or combined again into hexose, and that pentose compounds appear. The interactions of the carbon fragments are summarized in Fig. 7.11. This summary should be compared with the REDUCTIVE PENTOSE PHOSPHATE CYCLE (Fig. 1.7) which is a part of photosynthesis (Pon, 1964). The action of a second pyridine nucleotide, NADP, as a hydrogen acceptor in these reactions is of biochemical interest. It leads to a clean separation of catabolic from biosynthetic oxidation/reduction reactions. Thus, NADH is used to generate "high energy" compounds such as ATP, whereas NADPH is used in the reductive steps of biosynthesis.

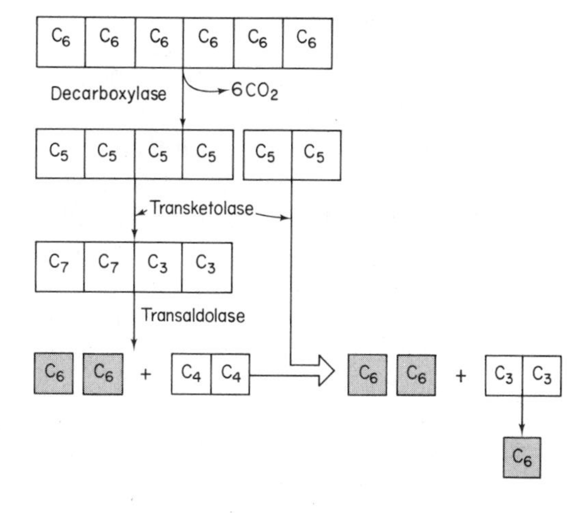

Fig. 7.11 Balance of carbons in pentose phosphate pathway. For every six glucose molecules at the start, one glucose molecule is oxidized while five hexoses are regenerated. In the transketolase reactions, there are transfers of C_2 fragments; in the transaldolase reactions, C_3 fragments are transferred. [Redrawn from Awapara, 1968: Introduction to Biological Chemistry. Prentice-Hall, Englewood Cliffs, N.J.]

The pentose pathway is an essential adjunct to the normal function of aerobic metabolism. It occurs almost universally in living organisms and has been identified in many, but not all bacteria, in plants, and in animals from flatworms to mammals. As a biochemical process, it may be phylogenetically older than photosynthesis (Krebs and Kornberg, 1957). Fermentation processes, as exemplified by the Embden-Meyerhof (EM) pathway, were probably responsible for the release of energy during the evolutionary stages when preformed compounds of many kinds were abundant. Later, however, before the advent of photosynthesis, the dwindling supplies of key substances may have limited energy exchanges, growth and reproduction. The pentose pathway generates the 5-carbon sugars which are essential building blocks in the electron transport system, in genetic material and in synthetic processes. It also provides NADPH which is required for carbon dioxide fixation and for fatty acid and other synthetic reactions; it results in the formation of tetroses and C_7 compounds such as sedoheptulose-7-phosphate whose significance is not yet fully understood. Although this is not an ATP-generating sequence, it is nonetheless important to life and might have provided many important metabolic compounds prior to photosynthesis.

Pon (1964) summarizes criteria for detecting the pentose phosphate cycle in tissues or determining the relative contributions of this cycle and the EM pathway in glucose metabolism. The techniques include the use of metabolic inhibitors, the identification of particular enzymes or intermediates and studies of the metabolism of C^{14}-labeled substrates. These techniques have now been applied to many different organisms. The pentose pathway is more conspicuous in some animals than in others; it is also more active in certain tissues and may change seasonally or in relation to sex, temperature acclimation or other physiological states (Pon, 1964; Chefurka, 1965; Huggins and Munday, 1968; Hochachka and Somero, 1971, 1973). Some variations are to be expected since this pathway provides pentoses and reduced NADP. In particular, syntheses of fatty acids and steroids demand NADPH and this probably accounts for the activity of the pentose pathway in lactating mammary glands, and in the liver, testis and adrenal; mammalian muscle, in contrast to these organs, lacks the pentose cycle. Some insect muscles, however, are active sites of fat metabolism and contain the enzymes and intermediates of the pentose pathway (Sacktor, 1965).

Aerobic steps in carbohydrate metabolism Pyruvic acid or its fermentation products are "dead ends" as far as animal evolution is concerned. They still contain a large part of the potential energy of the original glucose molecule. The evolution of active animals might never have taken place if biochemical evolution had stopped at this point; the many ways of animal life depend on the release of this additional energy. This is effected through a series of aerobic metabolic processes which occur within the mitochondria and involve an additional array of enzymes and electron transport systems and which can be conveniently described as three interlocking sequences. In the FIRST SEQUENCE, an oxidative decarboxylation and molecular activation convert the pyruvic acid into "active acetate" or acetyl CoA. In the

SECOND SEQUENCE, the 2-carbon units of active acetate are fed into an enzymatic cycle where each such fragment unites with a 4-carbon compound (oxaloacetic acid) to form a 6-carbon unit (citric acid); as the cycle revolves through eight major transformations, the 6-carbon citric acid unit reverts to the 4-carbon unit of oxaloacetic acid by releasing two molecules of carbon dioxide and eight hydrogens (Fig. 7.8). The latter are passed to the pyridine nucleotide and other carriers. This fundamental and important cycle is sometimes called the "Krebs cycle" after the biochemist Hans Krebs who did much toward its elucidation (Krebs, 1970). Since the name of Krebs is also associated with other chemical sequences, it is better to call it the "citric acid cycle" or the "tricarboxylic acid (TCA) cycle." In the THIRD SEQUENCE OF REACTIONS, the electrons released in the citric acid cycle are picked up and passed along a series of electron carriers; hydrogen eventually combines with oxygen to form water; and three units of ADP are changed to ATP for each two hydrogens passed down the chain. In this way each turn of the citric acid wheel (which produces eight hydrogens) may be thought of as generating twelve high-energy phosphate bonds in the form of ATP. The first of these sequences is confined to the metabolism of carbohydrates; the second and third are also common to the terminal portions of the metabolism of fats and proteins (Fig. 7.8). It should be remembered that these three sequences are interlocked in cellular metabolism and are separately discussed here only for convenience.

Pyruvic acid to active acetate and acetic acid This is an oxidative decarboxylation in which the 3-C pyruvic acid reacts with water to become a 2-C acetic acid compound, releasing carbon dioxide as a waste product of metabolism and supplying two hydrogens to the NAD carrier system. Arithmetically, these changes can be summarized as follows:

$$\underset{\text{Pyruvic acid}}{C_3H_4O_3} + H_2O \longrightarrow \underset{\text{Acetic acid}}{C_2H_4O_2} + 2H + CO_2$$

The acetic acid, however, does not exist as such. It forms a highly reactive combination with the sulfur atom of coenzyme A. Reference to Fig. 7.12 will show that the complex molecule, known as coenzyme A, or more frequently as CoA, is based on the purine, ribose and phosphate combinations already described as a nucleotide (Fig. 7.2). In CoA there is an attachment to a unit of pantothenic acid and one of β-mercaptoethanolamine with its reactive SH group at one end in Fig. 7.12.

The reaction whereby an acetyl is attached to some other molecule (acetylation) is relatively common; in this case, it constitutes a priming reaction required before further respiration of the carbon units is possible. Acetyl CoA is also an intermediary in the formation of acetic acid from higher fatty acids and forms a meeting place in several metabolic pathways (Fig. 7.8). The activation of the acetyl group is required for its transfer. Just as the phosphorylated group can be transferred from ATP to other compounds, so also can the acetyl group be transferred from acetyl CoA.

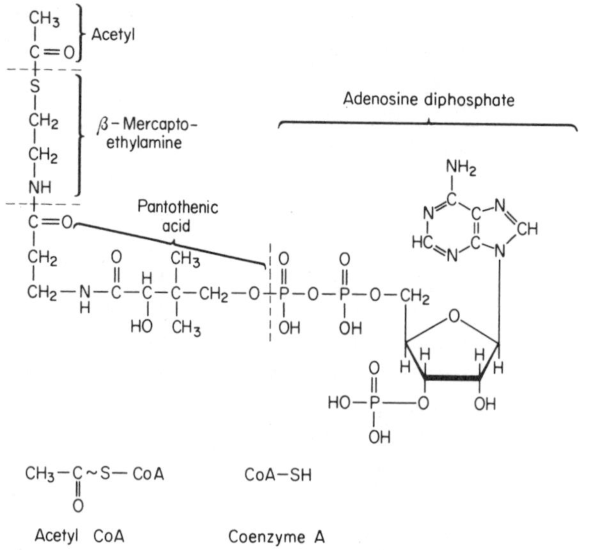

Acetyl

β-Mercapto-ethylamine

Pantothenic acid

Adenosine diphosphate

$CH_3-\underset{\underset{O}{\|}}{C}\sim S-CoA$ Acetyl CoA

$CoA-SH$ Coenzyme A

Fig. 7.12 Structure of acetyl coenzyme A (acetyl CoA). Upper formula, detailed structure of the molecule; lower left, its usual representation showing high-energy bond; lower right, the reactive acetate group replaced by hydrogen to form coenzyme A. The free energy of hydrolysis of acetyl CoA is about −8800 cal/mole; note that this is somewhat higher than the free energy of hydrolysis of ATP.

This particular acetylation is summarized in Fig 7.13. At the end of the sequence, the acetyl CoA may transfer acetate to the citric acid cycle or provide the energy necessary to add inorganic phosphate to ADP. In the first case, C_2 units are provided for the operation of the TCA cycle. In the second case, ATP is generated and acetic acid added to the system. In both cases, CoA has discharged its function and is available for further transfers. In the reversal of the latter sequence (shaded block in Fig. 7.13), acetyl CoA can be synthesized from acetic acid and CoA; other reactions between acetic acid and acetyl CoA have been studied in various organisms (Lioret and Moyse, 1963).

For the animal physiologist, one additional point of interest in this conversion is the array of coenzymes and activator factors involved. Three vitamins—thiamine, lipoic acid (Fig. 7.14) and pantothenic acid (Fig. 7.12), as well as magnesium ions, NAD, and ATP—are required to effect this decarboxylation and acetylation. The majority of animals are unable to synthesize these coenzymes and depend on plants for their manufacture.

The Citric Acid or TCA Cycle

This cycle is the common terminal pathway for the oxidation of foodstuffs. Some animals have modified it to meet the demands of unusual habitats, such as those associated with parasitism, but it seems to have remained dominant in the mainstream of animal phylogeny. In the complete cycle, one acetic acid equivalent is "burned" to CO_2 and H_2O, while a series of di- and tri-carboxylic acids appear as intermediaries (Fig. 7.8). Most writers show the cycle as producing CO_2 and releasing hydrogen which is then picked up by a group of special carriers. These carriers pass the hydrogen to molecular oxygen in a series of oxidation-reduction

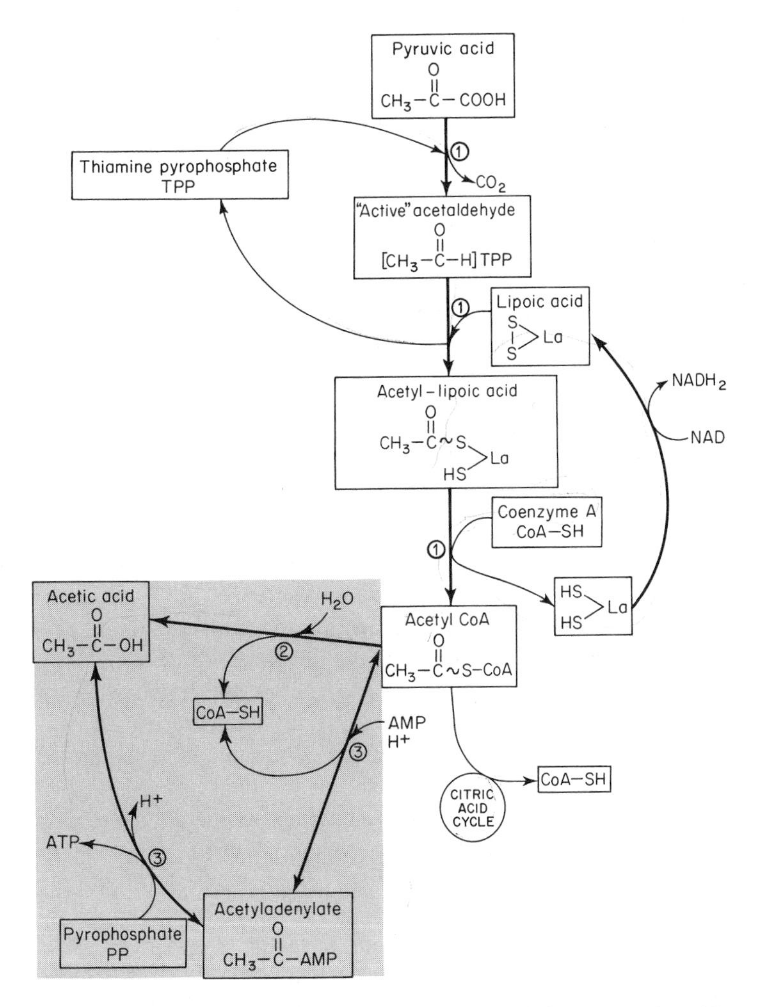

Fig. 7.13 Metabolism of pyruvic acid to acetyl CoA and acetic acid. ENZYMES: (1) pyruvate dehydrogenase complex and Mg^{++}; (2) acetyl CoA deacylase; (3) acetyl thiokinase. Note that dietary acetic acid may be fed into the fuel system by these reactions.

reactions which actually generate the pyrophosphate bonds (OXIDATIVE PHOSPHORYLATION). The many enzymes involved are distributed in an orderly way with the TCA cycle operating within the matrix of the mitochondria while the enzymes concerned with electron transport (oxidative phosphorylation) are arranged in a precise and definite manner on the mitochondrial membranes (Lehninger, 1971).

It should be remembered, however, that the enzymes of the citric acid cycle proper and those which pick up the electrons and shuttle them along to oxygen in oxidative phosphorylation are components of one system. The processes are only described separately as a matter of convenience.

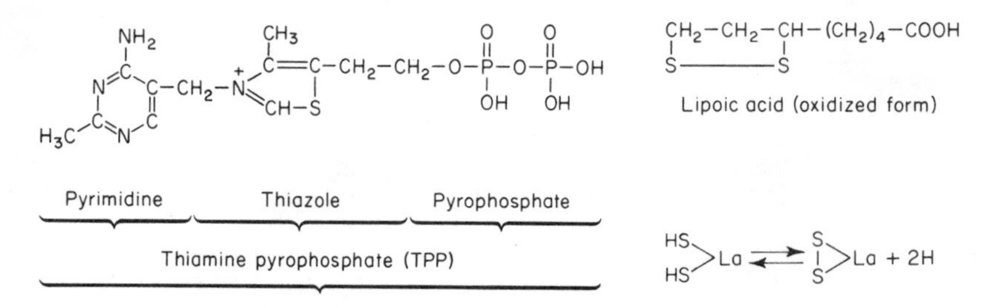

Fig. 7.14 Two factors essential in the metabolism of pyruvic acid. THIAMINE (shown on the left) is also known as vitamin B_1; the oxidation-reduction reaction for LIPOIC ACID is shown diagrammatically below the formula on the right.

The sequences of compounds and reactions in Figs. 7.8 and 7.15 show the points where carbon dioxide and water are released as direct metabolites and where hydrogen is delivered to the special transport system involved in oxidative phosphorylation. Attention is directed to four major points: (1) acetate is not the only carbon compound regularly delivered to and incorporated in the chain; (2) the cycle is not only the prime source of ATP but also supplies carbon skeletons for the manufacture of many of the constituents of protoplasm; (3) both NAD and FAD are concerned with the pickup and transfer of electrons to the cytochrome chain and, finally (4) the energy associated with this portion of tissue respiration forms a major parcel of the total released to the animal.

The process by which glutamate and aspartate are relieved of their amino nitrogen and incorporated in the chain as α-ketoglutarate and oxaloacetate respectively will be considered with the metabolism of the amino acids. It should be emphasized here that the reactions of the citric acid cycle are reversible (Fig. 7.15). This means that, on one hand, several different carbon skeletons maintain its structure; on the other hand, the cycle can supply various carbon fragments for the synthesis of additional compounds. This reversibility of reactions also suggests that carbon dioxide may be used directly to increase the lengths of the carbon chains. This has indeed been shown to occur in protozoa (Ryley, 1967), flatworms (Prichard and Schofield, 1968a), oysters (Hammen, 1966) and several different vertebrate tissues (White *et al.*, 1968; Mounib and Eisan, 1972); carbon dioxide fixation is a recognized cellular activity in animals as well as in the photosynthetic processes of plants. The special enzymes concerned contain the vitamin BIOTIN as a prosthetic group. Other reactions have been studied whereby microorganisms may synthesize C_4 dicarboxylic acids from acetate via the glyoxylate cycle (Fig. 7.18) and thus permit the formation of the necessary carbon compounds of the citric acid sequence and its operation when the organism has only acetate or such highly oxidized compounds as glycolate or oxalate at its disposal. As emphasized by Krebs and Kornberg (1957), the citric acid cycle performs two major functions: the supply of chemical energy in the form of ATP and the formation of C_4, C_5, C_6 skeletons for the synthesis of many different cell constituents.

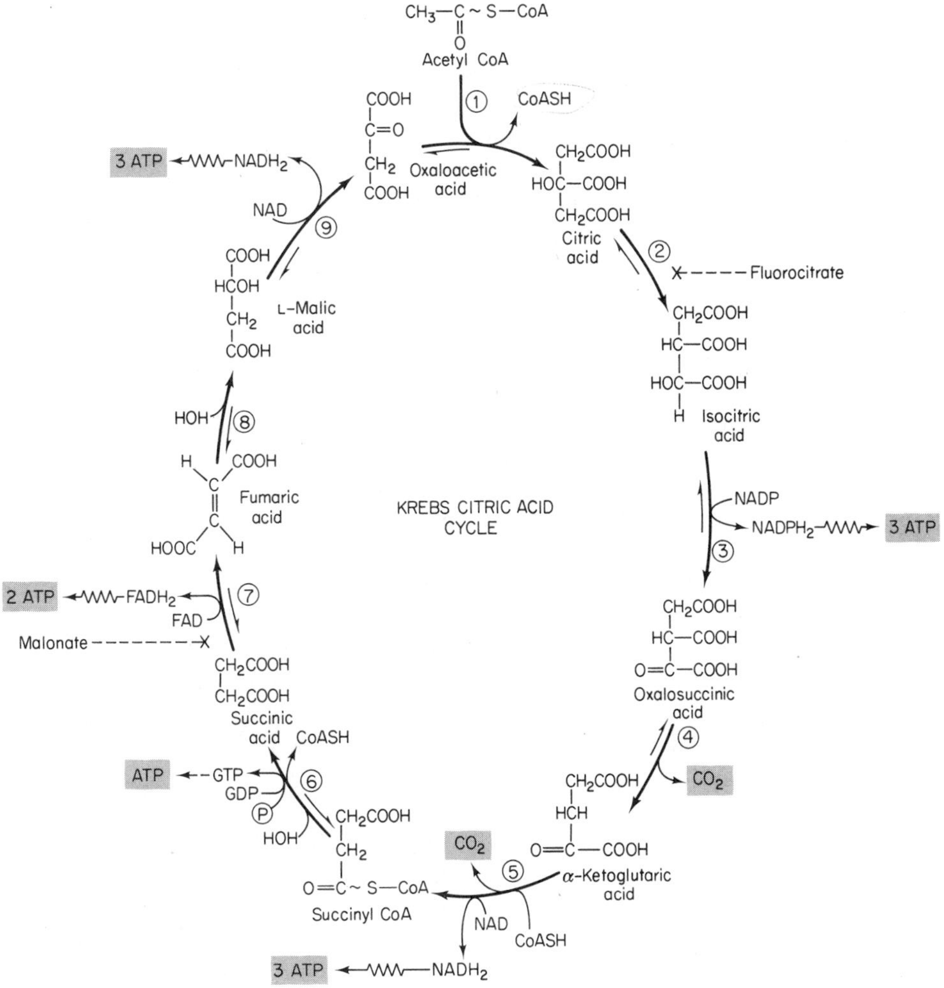

Fig. 7.15 The citric acid cycle. The reactions are stereotyped and only at succinic acid does a randomization occur in the carbon atoms originally in acetyl CoA. Fluorocitrate and malonate block the cycle at the points indicated. ENZYMES: (1) citrate synthase; (2) aconitase; (3) and (4) isocitric enzyme; (5) α-ketoglutarate dehydrogenase; (6) succinyl thiokinase; (7) succinic dehydrogenase; (8) fumarase; (9) malic dehydrogenase. [Based on Bennett and Frieden 1966: Modern Topics in Biochemistry, Macmillan. New York.]

Oxidative Phosphorylation

The release of electrons (hydrogen) at many points in the metabolic pathways has been noted. In the specialized "terminal transport system" of aerobic metabolism, these are passed to a series of carriers which are successively reduced and then oxidized as the electrons move along it until they reach the final cytochrome link (Fig. 7.16); here the enzyme, cytochrome oxidase, catalyzes the reduction of molecular oxygen to form water.

Most of the links in the chain have probably been identified, although there are many puzzling details still to be explained. In the first steps, electrons from the citric acid pool are picked up by the flavoproteins. This seems to be a direct transfer for those coming from succinate (Fig. 7.16), while those released at other points in the cycle are picked up by the pyridine nucleotides and then passed to the flavoproteins. Coenzyme Q forms a link between the flavoproteins and cytochrome *b* while non-heme iron-proteins are probably in some way associated with both the flavoproteins and cytochrome *b*; Cu^{++} is essential in the final cytochrome link.

At three points in the chain there is a release of energy through the coupling of inorganic phosphate with ADP. Probable sites of phosphorylation suggested in Fig. 7.16 indicate the formation of three units of ATP for each two hydrogens passed along the chain and oxidized to water (P/O = 3). To the physiologist, this is the really significant event in oxidative phosphorylation since it provides the major parcel of energy in the aerobic processes of animal life; in this way the mitochondria, through the citric acid enzymes and those involved in oxidative phosphorylations, generate twenty-four units of ATP for each unit of glucose metabolized. Thus, there is a net gain of thirty-eight units of ATP for each unit of glucose completely oxidized and the balance sheet may be written as follows:

$$1 \text{ Glucose unit} \xrightarrow{\text{EM path}} 2 \text{ Pyruvate units} + 2\text{ATP} + 4\text{H}$$

$$2 \text{ Pyruvate units} \xrightarrow{\text{oxidation}} 2 \text{ Acetyl CoA units} + 2\text{CO}_2 + 4\text{H}$$

$$2 \text{ Acetyl CoA units} \xrightarrow{\text{TCA cycle}} 4\text{CO}_2 + 16\text{H}$$

$$24\text{H} \xrightarrow{\text{oxidative phosphorylation}} 36\text{ATP}$$

$$\text{Summary: } C_6H_{12}O_6 + 6O_2 \longrightarrow 6CO_2 + 6H_2O + 38\text{ATP}$$

Alternate terminal pathways The cytochrome system is the predominant terminal pathway in aerobic respiration. It is probably the only one in the free-living animals. However, cytochrome oxidase is not the only enzyme capable of catalyzing reactions involving molecular oxygen; studies of alternative terminal oxidases have been particularly numerous in the field of plant biochemistry. To be of significance, such systems must participate in the oxidation of reduced pyridine nucleotides and flavoproteins. Simultaneously, these oxidations must be coupled with the formation of ATP.

Two possible alternative respiratory pathways are those involving phenol oxidase and those involving ascorbic acid oxidase. These are copper-containing enzymes. The first (also called polyphenol oxidase, phenolase and tyrosinase) occurs widely in both plant and animal tissues; the second is characteristic of plants in which ascorbic acid is often particularly abundant and of vital importance in their respiration. It is also an essential vitamin for many higher animals (Chapter 3).

Insect physiologists have devoted considerable attention to tyrosinase as an

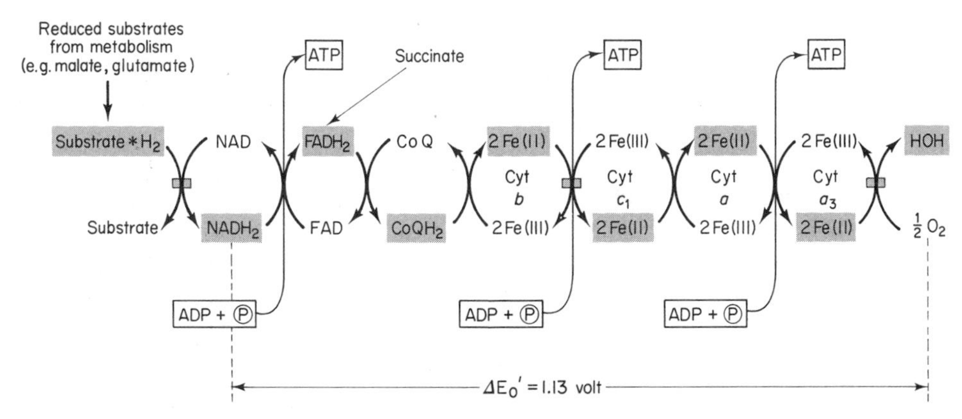

Fig. 7.16 Oxidative phosphorylation and electron transport. Reduced substances are in shaded blocks. Horizontal blocks on crossed arrows indicate, from left to right, points of inhibition by amytal, antimycin A, and cyanide. Cytochromes *a* and a_3 are together called CYTOCHROME OXIDASE or the RESPIRATORY ENZYME.

oxidase of the terminal respiratory transport system. It is especially active at the time of pupation when the cytochrome system may be undergoing changes. However, there is no convincing evidence that the phenolases serve as alternates for the cytochromes in insects or even in plants where these enzymes are very active. The high phenolase activity associated with insect pupation seems to be specifically related to the synthesis of quinones which are active in the tanning of the cuticular protein (Gilmour, 1965).

On the whole, an increasing body of evidence from animals at all levels in phylogeny supports the view that oxidative phosphorylation involving the cytochrome system forms a common denominator in the release of energy in the world of the free-living animals.

Metabolism of Fat

Lipid substances form an essential component of all protoplasm (CONSTANT ELEMENT) and even during extreme starvation considerable amounts can be extracted from the tissues. In addition, however, fat forms an important fuel reserve (VARIABLE ELEMENT) and is often stored in large quantities. The present discussion is confined to a consideration of fats as fuel for the production of ATP.

Many animals live for a long time almost exclusively on the fat reserves. Atlantic salmon enter fresh water, cease feeding and live an active life, sometimes for as long as a year, while they utilize stored fat (Greene, 1926). During this time blood sugars and liver glycogen remain essentially unaltered, indicating that the fat is readily converted into these essential carbohydrates in addition to being used directly as a source of energy.

The neutral fats which form the main source of lipid energy are esters of glycerol and the long chain fatty acids with an even number of carbon atoms. Many cells contain lipolytic enzymes which readily hydrolyze these triglycerides into their constituent fatty acids and glycerol (Fig. 3.5). This is the first preparatory step in their metabolism (Fig. 7.8).

Glycerol metabolism Glycerol, through an enzymatic priming reaction with ATP, is converted into glycerolphosphate. There follows an oxidation (dehydrogenation) in which hydrogen passes to the NAD carrier system and a triosephosphate (PGAL) is formed. This compound has already been identified in the Embden-Meyerhof sequence at the crossroads of several metabolic paths. From PGAL the products of glycerol may be synthesized to carbohydrate or further oxidized through acetate and the citric acid cycle (Fig. 7.8).

Fatty acid oxidation Most of the fatty acids are oxidized by a process referred to as β-oxidation (Greville and Tubbs, 1968). In this, two carbon units are progressively removed from the carboxyl end of the carbon chain to yield acetate equivalents which can then be completely metabolized via the citric acid cycle or, in some organisms, built up into glucose and more complex carbohydrate units if the demands are for these substances. Since the majority of the naturally occurring fats contain even numbers of carbon atoms, oxidation at the β-position is always possible. Several of the important steps in this oxidation are indicated in Fig 7.17.

The reaction sequence commences with an ATP-CoA priming. A fatty acid-CoA derivative results, and this is then desaturated between the α- and β-carbon positions. In the mammal, enzymes capable of performing this desaturation have been found in the liver, and the hydrogen released in this way is first passed to the flavin hydrogen carriers. It may then proceed along the aerobic oxidative phosphorylation chain, yielding ATP. The α, β unsaturated fatty acid CoA compound now undergoes hydration and dehydrogenation with the reduction of NAD, and more fuel is added to the hydrogen carrier chain. At this point a further enzymatic reaction splits off acetyl CoA and leaves a residual fatty acid CoA compound which can likewise cycle the desaturation, hydration and dehydrogenation series. The process is repeated until the long chain fatty acid is completely metabolized (Fig. 7.17).

The acetyl CoA fragments can, of course, enter the citric acid cycle by familiar routes or be synthesized into complex carbon compounds (Fig. 7.8). They may, however, have quite a different fate. Under certain conditions two fragments combine to form acetoacetate. In some animals (man, for example), this seems to happen if large amounts of fat are being metabolized and there is insufficient utilization of carbohydrate, with a deficit of compounds which provide the essential components of the citric acid cycle. Thus, in man, diabetes produces a characteristic ketosis in which ketones (acetone, acetoacetic acid and β-hydroxybutyric acid) accumulate in the blood.

Animals vary considerably in their tendency to form ketone bodies. As already indicated, some fishes may, for prolonged periods, metabolize fats almost to the

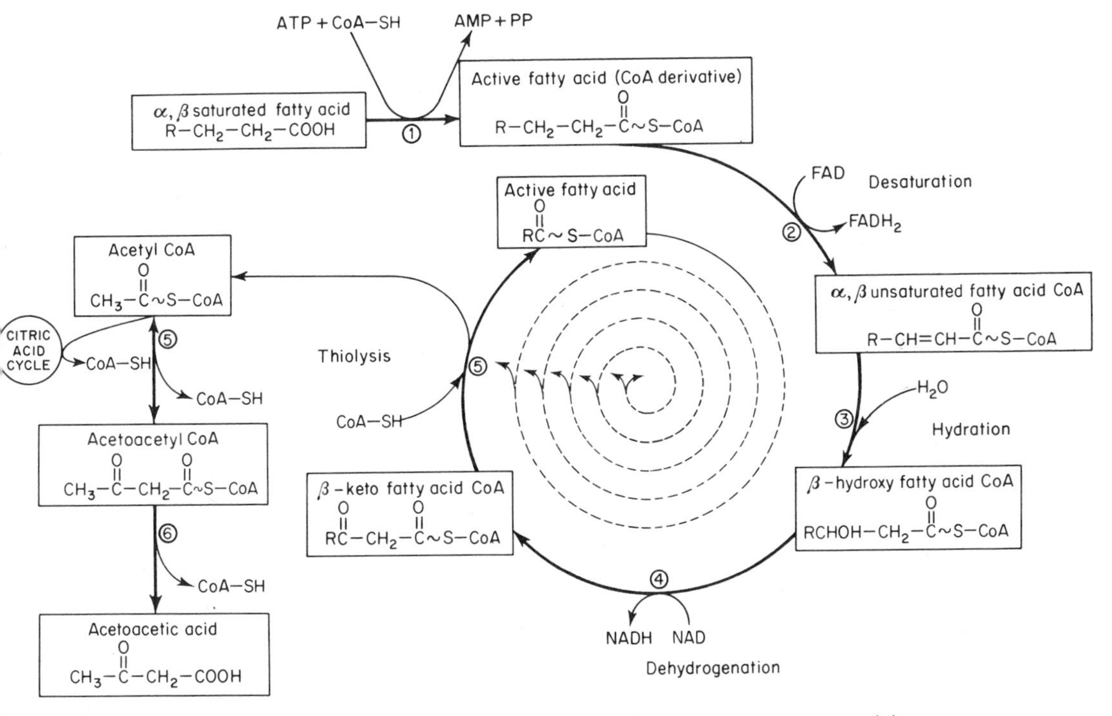

Fig. 7.17 Metabolism of fatty acids. ENZYMES: (1) thiokinase and Mg^{++}; (2) acyl dehydrogenase; (3) enol hydrase (crotonase); (4) β-hydroxy acyl dehydrogenase; (5) thiolase; (6) deacylase.

exclusion of other foods. Similarly, a chick embryo during development depends almost entirely on fats for its energy, and the same is true of many other animals. Variations are common among the mammals and even in the different species of the same genus.

The energetics of fat metabolism Although the glycerol molecule must be primed with one ATP, there is a return of some twenty units of ATP when the glycerol unit is completely oxidized through the citric acid cycle and the associated oxidative phosphorylations. Further, each step in the β-oxidation of fatty acids requires a priming of ATP but yields about eighteen ATP units. Since the common fatty acids of animal cells—palmitic and stearic—contain sixteen and eighteen carbon atoms respectively, it is apparent that fatty acids are rich sources of ATP. Fat, per gram, has more than twice the calorific value of carbohydrate.

The glyoxylate cycle Some organisms are able to use acetate chips from fatty acid metabolism to construct sugars. This capacity depends on the presence of two specific enzymes: *isocitrate lyase* (*isocitrase*) which splits isocitrate to form succinate and glyoxylate, and *malic acid synthetase* which catalyzes the condensation of acetyl CoA and glyoxylic acid to form malic acid. These two special reactions may be represented as follows.

Reaction 1:

$$\underset{\text{Isocitric acid}}{\begin{matrix}\mathrm{COOH}\\ |\\ \mathrm{HCOH}\\ |\\ \mathrm{HC{-}COOH}\\ |\\ \mathrm{CH_2}\\ |\\ \mathrm{COOH}\end{matrix}} \rightleftharpoons \underset{\text{Succinic acid}}{\begin{matrix}\mathrm{COOH}\\ |\\ \mathrm{CH_2}\\ |\\ \mathrm{CH_2}\\ |\\ \mathrm{COOH}\end{matrix}} + \underset{\text{Glyoxylic acid}}{\begin{matrix}\mathrm{CHO}\\ |\\ \mathrm{COOH}\end{matrix}}$$

Reaction 2:

$$\underset{\text{Acetyl-CoA}}{\begin{matrix}\mathrm{CH_3}\\ |\\ \mathrm{C{=}O}\\ |\\ \mathrm{S{-}CoA}\end{matrix}} + \underset{\text{Glyoxylic acid}}{\begin{matrix}\mathrm{CHO}\\ |\\ \mathrm{COOH}\end{matrix}} \rightleftharpoons \underset{\text{Malic acid}}{\begin{matrix}\mathrm{COOH}\\ |\\ \mathrm{CH_2}\\ |\\ \mathrm{HCOH}\\ |\\ \mathrm{COOH}\end{matrix}} + \underset{\text{Coenzyme A}}{\mathrm{CoA{-}SH}}$$

The acetate-combining machine is shown in Fig. 7.18. It is a modification of the citric acid cycle known as the GLYOXYLATE or GLYOXYLIC ACID CYCLE (Baldwin, 1967). The net effect of the reaction cycle is to unite two acetate units to form succinate. This four-carbon compound is then synthesized to sugar via pathways involving fumarate, malate, oxaloacetate, phosphoenolpyruvate and the reverse steps in glycolysis already outlined (Fig. 7.9).

Kornberg and Krebs (1957) first proposed the glyoxylate cycle after the specific enzymes had been identified in bacteria; it was later shown to operate in plants

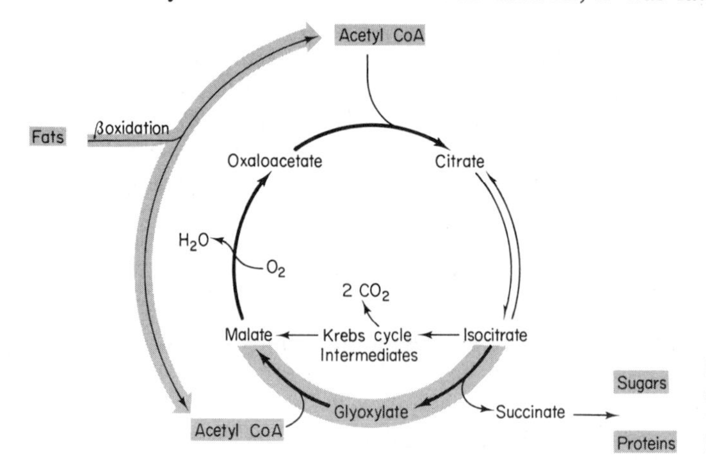

Fig. 7.18 Glyoxylate cycle or by-pass for conversion of fatty acids to sugars. [From Bennett and Frieden, 1966: Modern Topics in Biochemistry. Macmillan, New York.]

(germinating seeds) and in many microorganisms. In animals, however, the two enzymes characteristic of the cycle are very rare. They have not yet been found in the more advanced phyla although they occur in the flagellate *Tetrahymena* and some other protozoa (Dewey, 1967; Ryley, 1967), in some flatworms (Prichard and Schofield, 1969) and in several free-living nematodes including the vinegar eel *Turbatrix aceti* (Rothstein and Mayoh, 1966; Saz, 1969). Contrary to earlier reports, this cycle does not seem to be present in pupating insects (Wyatt, 1967). It seems likely that most animals, including those fasting during migration, can meet the energy demands through direct metabolism of fat; the glycerol component permits synthesis of carbohydrate. However, some organisms (in particular, certain bacteria) may exploit environments rich in C_2-compounds and utilize them directly through the glyoxylate bypass to elaborate more complex carbon units.

Amino Acid Metabolism

Amino acids, derived from the digestion of protein, pass into the fuel system if they are not required for growth, tissue repair or the construction of important secretions and enzymes. Their fate depends on the amount of protein in the diet, the age of the animal and the other sources of fuel available to it. In emergencies, amino acids may be withdrawn from the protoplasm of the cell, and thus during starvation the cellular protein content declines. This is true even though protein is not stored in the same sense that carbohydrate and fat are stored in special organs and depots. The protein reserve of the cell forms an active, although partially expendable, portion of its protoplasm.

Deamination Before entering the fuel system, nitrogen (also sulfur and iodine if present) must be removed from the amino acid molecule. Deamination, the process by which the amino groups are split off, is one of the conspicuous removal processes. During the first half of this century Krebs (1935) showed that slices of kidney and liver from many animals, when incubated with amino acids, took up oxygen and produced ammonia; further, that approximately one mole of oxygen was used for every two moles of ammonia formed. A family of enzymes, the amino acid oxidases, catalyzes these oxidative deamination reactions. Several steps are now known, with flavin compounds acting as intermediary hydrogen acceptors. The overall reaction may be written as follows:

$$\underset{\alpha\text{-amino acid}}{R \cdot CHNH_2 \cdot COOH} + \tfrac{1}{2}O_2 \longrightarrow \underset{\alpha\text{-keto acid}}{R \cdot \overset{\overset{O}{\|}}{C} \cdot COOH} + NH_3$$

Oxidative deamination, although prominent in tissues, is by no means the only way in which amino groups are removed. A series of non-oxidative deamination enzymes are associated with certain specific amino acids, such as the hydroxyamino acids (dehydrases), the sulfur-containing amino acids (desulfhydrases), histidine (histidase), tryptophan (tryptophanase) and others.

In general, these reactions lead to one or the other of two familiar residues, pyruvate or acetate. Dietary experiments with individually labeled amino acids have shown that in mammals the non-essential amino acids usually produce a pyruvate residue (GLUCOGENIC AMINO ACIDS), while most of the essential amino acids form an acetate residue (KETOGENIC AMINO ACIDS). The significance of the terms glucogenic and ketogenic is obvious. The first group of amino acids in large quantities leads to the production and storage of carbohydrate, while fat metabolism is emphasized in the case of the ketogenic group. Through deamination, at least half of the common amino acids are reduced to these two familiar carbon residues, and from this point, as acetyl CoA, they can enter the energy-yielding reactions of the citric acid cycle (Fig. 7.8).

Ammonia metabolism The production of deaminated carbon residues for the citric acid cycle was emphasized in the previous section. This is only a small part of the story of nitrogen metabolism. There are several other ways in which NH_2 groups may be moved about and, although sometimes excreted, they more frequently enter an extremely labile nitrogen pool and become part of the synthetic and regulatory machinery. The importance and lability of the amino nitrogen will now be considered. Its excretion as a waste product is described in Chapter 8.

Several AMINATION reactions couple amino groups with organic acids and other compounds. Two of these are illustrated in Fig. 7.19. At the top of the diagram, glutamic dehydrogenase (in a process of reductive amination requiring energy from reduced pyridine nucleotide) incorporates ammonia into α-ketoglutaric acid to form glutamic acid. In the reversed reaction a deamination occurs. The second amination process is shown where the amides of glutamic and aspartic acids (glutamine and asparagine respectively) are formed through energy derived from ATP.

Through TRANSAMINATION, a family of enzymes (the transaminases associated with the vitamin pyridoxal phosphate as a cofactor) transfers amino groups of various amino acids to keto or other organic acids (Fig. 7.19). At transfer point *A* (Fig. 7.19), the amino group moves from glutamic to pyruvic acid and thus forms the amino acid alanine and α-ketoglutaric acid. The transfer point *B* shows a particularly important molecular rearrangement in which amino acids are linked with the citric acid cycle at two points (the oxaloacetic and α-ketoglutaric links). Through deaminations or transaminations involving glutamic and aspartic acids, α-ketoglutaric and oxaloacetic acids respectively may be added to the citric acid cycle. Conversely, the appropriate carbon skeletons may be picked up from the cycle and coupled with the amino nitrogen for the synthesis of protein. In this way, the metabolism of the proteins is connected with the metabolism of the carbohydrates and the fats. Carbon fragments from the latter two may turn up in the proteins through the oxaloacetic and the α-ketoglutaric links. Alternatively, deaminated amino acids may become a part of the carbohydrates and the fats.

The ammonia used in these various reactions may be derived from deaminations but can also be obtained from inorganic sources. Although animals are unable to

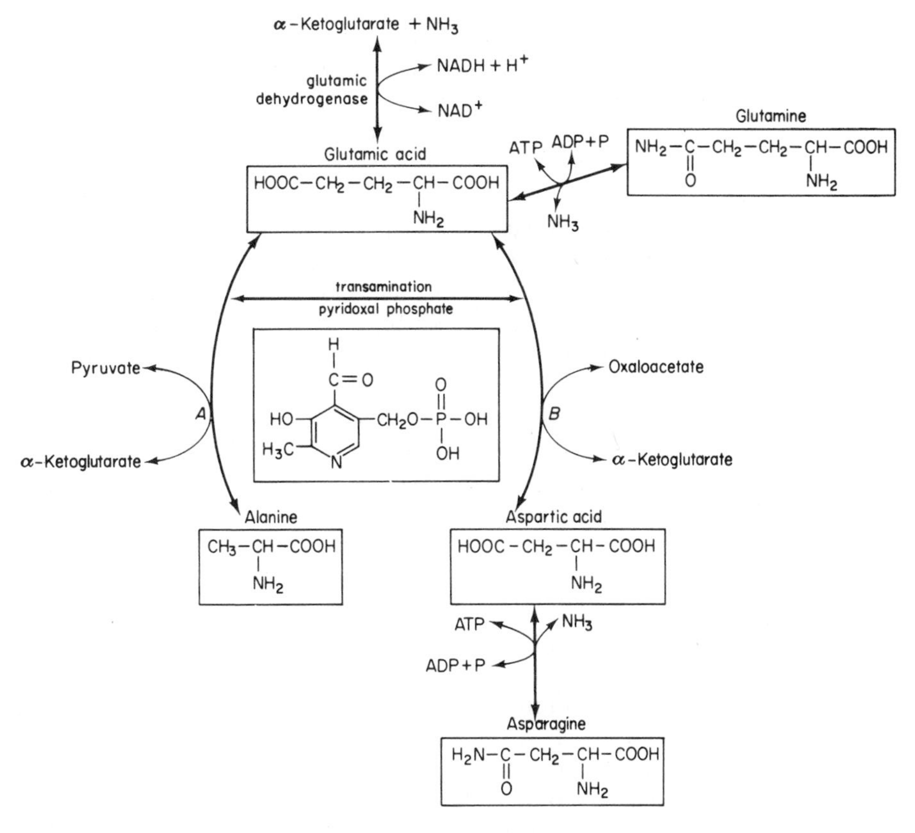

Fig. 7.19 Several amination and transamination reactions. Only the formulae for the amino acids and pyridoxal phosphate are shown; formulae for other compounds in Fig. 7.15. Description in text.

utilize most forms of inorganic nitrogen, they can probably all derive some from ammonia. It has, for example, been shown that rats fed a diet containing N^{15} ammonium citrate will later contain radioactive amino acids (Fruton and Simmonds, 1958). Intestinal bacteria produce ammonia and this, as well as that derived from the dietary ammonium salts, will enter the circulation and be incorporated into metabolically important nitrogen compounds or form nitrogenous wastes such as urea. The ammonium ion itself is rather toxic (Chapter 8) and never occurs as such in significant amounts in animal tissues. However, the nitrogen is indispensable for life and occurs largely as amino (NH_2) or imino (NH) nitrogen, in which forms it is transported and stored.

Energetics of ammonia formation Most deamination reactions produce a certain amount of heat, but there is no gain in ATP. In poikilotherms the production of heat may be thought of as a useless byproduct of metabolism. Useful energy appears in other forms, particularly as in the pyrophosphate bond of ATP. It has been suggested that a coupling of the glutamic acid dehydrogenase reaction with the transaminase systems may be of considerable importance in the production of ATP from protein (Cohen and Brown, 1960). These enzymes are found everywhere in the living world, and glutamic acid dehydrogenase requires NAD or NADP which, when reduced, can enter the chain of oxidative phosphorylation (Fig. 7.20).

Storage of amino groups Glutamine forms an important store of amino groups in the animal body. About one fifth of all the amino acid nitrogen in human blood is contained in glutamine and glutamic acid. In plants, asparagine, as well as glutamine, stores nitrogen. The reactions (Fig. 7.19) whereby these mono-amides of glutamic and aspartic acids are formed require ATP and an appropriate enzyme (glutamine synthetase for glutamine), but the resulting compounds form an indispensable reservoir of amino nitrogen. Specific enzymes remove the amino groups to form glutamic and α-ketoglutaric acids or transfer them in various ways during the synthesis of purines, amino sugars (glucosamine) and numerous amino acids. Thus, glutamine and asparagine (plants) play a central role not only in the storage of amino groups but also in their utilization.

Special Adaptations in the Metabolic Pathways

The steps outlined for the transfer of energy stored in foods to a directly usable form in ATP are found in some organisms at all stages in phylogeny. This general pattern must have evolved in primitive forms and remained the model for the metabolic machinery throughout the history of life. However, the current lively interest in comparative biochemistry is revealing many curious exceptions in the details of energy-yielding reactions and it is now clear that the patterns are not identical in the major animal groups (Awapara and Simpson, 1967). Animals have marked capacities to develop different enzyme systems in response to exceptional biological demands (Chapters 2 and 3); modifications in the regular routes of intermediary metabolism may be expected in organisms which exploit unusual habitats or successfully meet special physiological demands. Some of these special

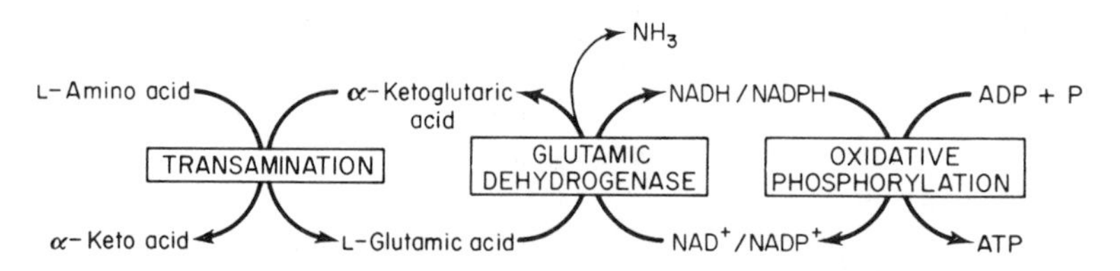

Fig. 7.20 Scheme for deamination of amino acids via coupled transaminase(s) and glutamic dehydrogenase. [Based on Cohen and Brown, 1960: *In* Comparative Biochemistry, Vol. 2, Florkin and Mason, eds. Academic Press, New York.]

adaptations will be described in connection with the physiology of particular organs; only a few general notes are given here. These may be supplemented from the wealth of information compiled in a series of monographs on "Chemical Zoology" edited by Florkin and Scheer (1967 and later) and in current treatises on the comparative physiology of special taxonomic groups.

Knowledge of the intermediary metabolism of invertebrates is still fragmentary. The insects are the best known; they have now been investigated for many years, yet it was not until 1957 that the major blood sugar in the circulating hemolymph was shown to be the disaccharide trehalose and not glucose which is charactertistic of the vertebrates (Gilmour, 1965; Wyatt, 1967). Monosaccharides absorbed in the midgut as well as the energy reserves in the fat body are converted to trehalose and circulate as such in the hemolymph. At the point of carbohydrate utilization, the trehalose is converted to glucose and follows pathways already outlined. Trehalose is also a conspicuous blood sugar in some annelids (Scheer, 1969), molluscs (Goddard and Martin, 1966) and crustaceans (Huggins and Munday, 1968). The nature of the food store varies; most animals rely mainly on glycogen (Awapara and Simpson, 1967) but some specialize in fat storage (Drummond, 1967) or even protein (Zandee, 1966). There are seasonal differences in the utilization of food stores as well as species variations. Many fishes and birds store large amounts of fat prior to migration (Drummond, 1967); frogs depend more on triglycerides as energy reserves during the summer but rely on glycogen in the winter (Jungreis, 1970). Careful studies of the food reserves have not yet been carried out in many of the invertebrates (see Dales, 1969; Scheer, 1969).

Numerous modifications of the conventional pathways have been recorded in the non-segmented worms. Investigations are usually based on *in vitro* examinations of enzyme systems and considerable caution must be exercised in drawing conclusions with respect to functional *in vivo* systems. Quantities of enzymes detected biochemically may be inadequate to operate important pathways to any significant degree. The enzymes of the citric acid cycle have all been found in the liver fluke *Fasciola hepatica* but the importance of the cycle as an energy-producing path is questioned because of the extremely low levels of aconitase and isocitrate dehydrogenase (Prichard and Schofield, 1968b). The situation appears to be the same in the nematode *Ascaris lumbricoides*; the partial citric acid cycle found in these animals is considered important in producing succinate (von Brand, 1966; Saz, 1969). Bennett and Nakada (1968) tabulate values for the glycolytic enzymes in mammals, housefly, earthworm and two molluscs; in extreme cases, the differences are as great as several hundredfold to a thousandfold. Considerable caution is necessary in evaluating the *in vivo* significance of metabolic enzymes.

The vinegar eel *Turbatrix aceti* was the first animal with a predominantly aerobic metabolism shown to lack several of the enzymes in the citric acid cycle (Ells and Read, 1961). It also lacks some of the Embden-Meyerhof enzymes and can only operate the sequence from fructose diphosphate to pyruvate (Ells, 1969); the reasons are not clear but must be related to the high utilization of acetate and the predominant lipid metabolism. Studies of the biochemistry of segmented worms are incomplete but neither the citric acid cycle nor the pentose pathway has been demonstrated in polychaetes, while their importance in the oligochaetes is still uncertain (Dales, 1969).

Anaerobic animals Some invertebrate animals can live indefinitely in an oxygen-free environment. This includes not only many well-known parasitic helminths (von Brand, 1966; Fairbairn, 1970), but also numerous intertidal and benthic organisms. Such parasitic worms as *Ascaris lumbricoides* never utilize oxygen and are OBLIGATE ANAEROBES; intertidal animals such as oysters and polychaete worms metabolize aerobically when the tide is in flood but operate anaerobically when the beach is exposed. They are FACULTATIVE ANAEROBES and representatives of a very large group of animals that live in habitats periodically subject to a deficit of environmental oxygen; at certain points in their metabolic pathways the flow of electrons and protons is switched from molecular oxygen to organic acids. Anaerobic habitats have been recognized for a very long time but the widespread nature of the metabolic adaptations concerned with life under these conditions has been emphasized much more recently (Hochachka and Somero, 1973). It is now apparent that a spectrum of organisms ranges from the strictly aerobic forms with very limited capacities for oxygen debts to the facultative and obligate anaerobes. Indications are that some vertebrates (certain fishes and diving animals) may use pathways comparable to those found in the facultative anaerobic invertebrates.

The scheme outline in Fig. 7.21 accounts for the major end products formed during anaerobiosis. Two points of general interest are emphasized. In the first place, all of the key substances that appear in this scheme have been previously identified (Figs. 7.9, 7.15, and 7.19); they are linked together differently to produce a maximum yield of ATP without oxygen. In the second place, it is evident that the lipids do not serve as a source of energy; ATP is generated only from either carbohydrates or carbohydrates and amino acids. Details of enzyme kinetics have been discussed (Hochachka and Somero, 1973; Hochachka *et al.*, 1973) and will not be considered here. At this point, the general principles noted above are emphasized; this can be most readily done by comparing the metabolism of vertebrate muscle, which has very limited capacities to develop an oxygen debt, with the adductor muscle of the oyster *Crassostera gigas*, which regularly switches from aerobic to anaerobic pathways and back again during a tidal cycle (Hochachka and Mustafa, 1972).

In vertebrate skeletal muscle, glycogen or glucose is converted to pyruvate by the classical glycolytic or EM pathway (Fig. 7.9). Pyruvate is then oxidized to CO_2 and H_2O by the citric acid cycle and oxidative phosphorylation; under conditions of moderate anaerobiosis (for example, vigorous exercise), pyruvate is reduced to lactate in the presence of lactic dehydrogenase. Only a modest oxygen debt is tolerated. Oyster muscle, by contrast, can switch to other pathways and accumulate succinate and alanine as end products (Fig. 7.21). In both vertebrate and bivalve muscle, the sequence of reactions is the same to the level of phosphoenolpyruvate,

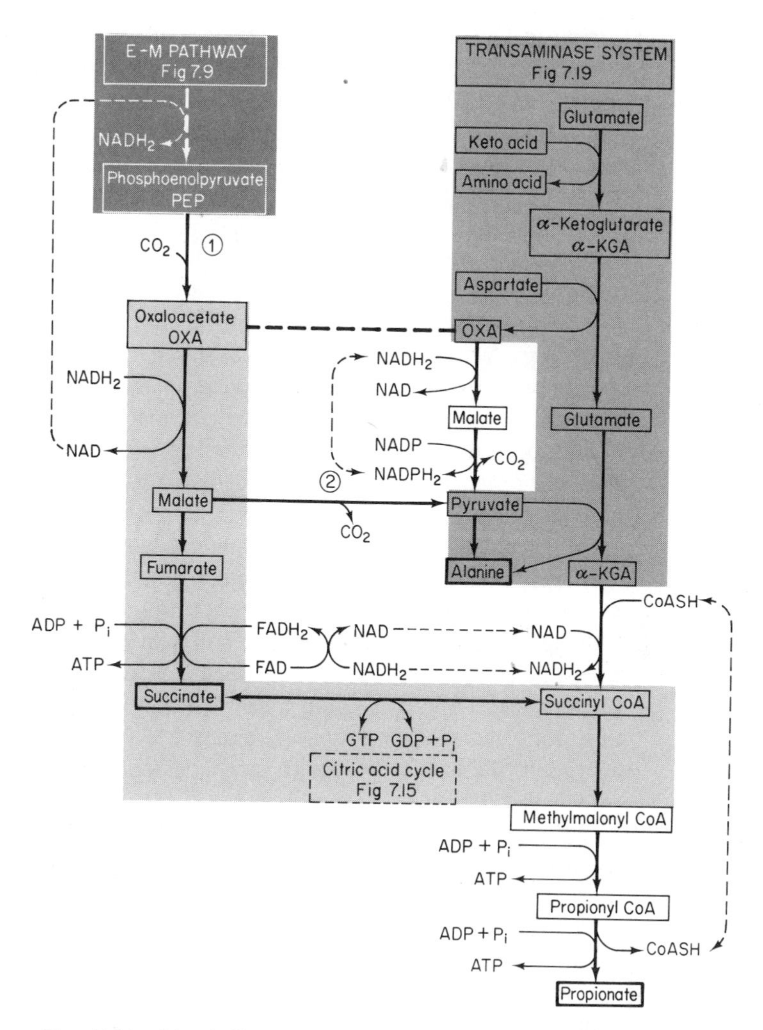

Fig. 7.21 Metabolic pathways accounting for the accumulation of succinate, alanine and propionate as end products of anaerobic carbohydrate and amino acid catabolism in facultative anaerobic animals. ENZYMES: (1) PEP carboxykinase; (2) malic enzyme.

PEP; at this point, however, the anaerobes form oxaloacetate in a carboxylation involving the enzyme PEP carboxykinase. The biochemical kinetics of this switch has been discussed (Hochachka and Somero, 1973); in obligate anaerobes, such as *Ascaris lumbricoides*, this switch is related to an absence of pyruvate kinase and lactic dehydrogenase; in such facultative anaerobes as the oyster, however, other mechanisms must be involved. The fall in pH caused by the accumulation of acidic end products has been suggested.

Succinate forms one of the major end products in anaerobic animals. It may be formed in two ways: (1) from oxaloacetate by way of malate and fumarate in a reversed citric acid sequence and (2) from α-ketoglutarate by way of succinyl CoA (Figs. 7.15 and 7.21). The second important end product is the amino acid alanine. The series of transaminase reactions already outlined (Fig. 7.19) link oxaloacetate and pyruvate (arising by decarboxylation of malate) to the amino acid pool (aspartate and glutamate); the reaction products are alanine and α-ketoglutarate.

Propionate occurs as a third end product in many nematodes, cestodes and trematodes as well as in the swamp-dwelling annelid *Alma emeni* (von Brand, 1966; Coles, 1967). The pathway from succinyl CoA to propionate, as indicated in Fig. 7.21, is an energy-generating one, producing two moles of ATP for each mole of succinyl CoA metabolized. Finally, it should be noted that in addition to these three end products (succinate, alanine and propionate), some of the parasitic helminths can channel pyruvate to acetate; two enzymatic sequences which may operate have already been noted (Figs. 7.13 and 7.18). The proportions of these four different products of anaerobiosis differ greatly in various anaerobes (Hochachka *et al.*, 1973).

Hockachka and Somero (1973) have discussed the energy yields of these various reactions. This may amount to as much as 60 per cent of that possible through the aerobic pathways involving oxidative phosphorylation and the cytochrome system. Although it could not sustain the life of very active animals for prolonged periods, it is sufficient for many sedentary organisms or more active animals during periods of quiescence. Some of these pathways may have figured prominently during the anaerobic stages of evolution (Chapter 1). However, the anaerobic animals considered in this section appear to have evolved from aerobic species.

Evolution of Energy-Transforming Mechanisms

The general processes of energy transformation will not be followed further. Those so far discussed seem to be universal in living tissues. Their relatively small number suggests that they are of great phylogenetic antiquity and that life itself is founded upon them. Krebs and Kornberg (1957), in an outstanding summary of the energy transformations of living matter, suggested the following evolutionary sequence: ANAEROBIC FERMENTATION (glycolytic enzymes, ATP and pyridine nucleotide), PENTOSE PHOSPHATE CYCLE, PHOTOSYNTHESIS (metalloporphyrins), and CELL RESPIRATION (including the citric acid cycle, the cytochrome system and oxidative phosphorylation).

It seems logical, both on biochemical and geochemical grounds, to place fermentation in the basal position. These reactions are the most widespread of biochemical activities and, as discussed in Chapter 1, there is now general agreement that life originated under anaerobic conditions. The evolutionary position of the pentose cycle is more speculative, but pentoses are found in several ubiquitous and indispensable molecules, and a supplementary supply may have been required before photosynthesis provided the conditions essential to the evolution of the plants and animals. If this were the sequence, the development of the pentose cycle could have provided several additional biochemical reactions which were later associated with photosynthesis. Given the pentose path and fermentation, only two more reactions (Fig. 1.6) would have been required in the carbon pathways of photosynthesis itself (Krebs and Kornberg, 1957). Photosynthesis then provided oxygen for aerobic respiration and energy stores for the evolution of an animal world.

Even in the most primitive present-day plants and animals, these fundamental energy-generating reactions are known to operate in orderly sequences only within the architecture of living cells. Cellular morphology probably evolved in association with the biochemical reactions which are part of it. Enzymes and compounds responsible for the release of energy (chiefly in the form of the pyrophosphate bond) are not distributed uniformly throughout the protoplasm but are arranged in an orderly way on a special fabric of cytoplasmic structures (Green, 1962).

The Embden-Meyerhof system and the enzymes required to form pyruvate, acetate and other residues necessary for the citric acid cycle, are located in the cytoplasmic matrix or hyaloplasm. The citric acid cycle itself operates in the matrix of the mitochondria, while the electron transport system of oxidative phosphorylation is spread out in a very definite and organized way (Lehninger, 1971) on the mitochondrial membranes (cristae). Mitochondria are mainly responsible for the liberation of ATP and are integral parts of all the cells of plants and animals. Their evolution would seem to have been associated with the phylogeny of efficient aerobic tissue respiration.

ENERGY-UTILIZING PROCESSES OF ANIMAL LIFE

ATP, liberated through tissue respiration, is transformed in various ways into characteristic animal activities (Fig. 7.1). The catabolic transformations from the food or fuel and the amidine phosphate storage systems have been outlined in previous sections. The transformers which convert the pyrophosphate energy into animal work are less well known. There is, for example, little doubt that ATP provides energy for the operation of the contractile machinery, but the molecular details of the coupling are uncertain. Again, the activation of molecules in such synthesizing processes as the coupling of amino acids to form protein is recognized, but the details are still subjects of intensive investigation.

Active transport of materials across the cell or through its membrane, the maintenance of bioelectric potentials and the excitation processes in nervous tissue are likewise incompletely understood at the molecular level. Heat and light may be regarded as byproducts of metabolism. Natural selection has taken advantage of both these energy byproducts and adapted them in diverse ways.

THE RATE OF METABOLISM

Lavoisier, in the latter part of the eighteenth century, first examined living systems thermodynamically and established relationships between energy liberated in the form of heat and that which entered the animal as food. In 1780, in association with his colleague Laplace, he put a guinea pig in a closed box surrounded by ice and recorded the melting of 341 g of ice in a 10-hour period. These workers also measured the carbon dioxide production of a guinea pig of similar size during a 10-hour period and calculated that the burning of carbon to form this amount of gas would generate enough heat to melt 326.7 g of ice. The agreement was remarkably good; animal respiration was equated to the combustion of carbon compounds. At that time Lavoisier did not have the information necessary to appreciate the differences between the burning of food and the burning of carbon. He concluded that animal heat is produced by the transformation of carbon to carbon dioxide during respiration. In 1781, Cavendish described the oxidation of hydrogen to form water and, in 1785, Lavoisier recognized the probable significance of this reaction in animal respiration and noted that he had been incorrect in assuming that only carbon was oxidized in animal respiration. These pioneer experiments are outlined in several familiar textbooks (Brody, 1945; Gabriel and Fogel, 1955; Kleiber, 1961).

Heat Production

To the scientists of the eighteenth and early nineteenth century, the living machine was a kind of furnace in which foods were burned to liberate heat. The heat was necessary for animal activities in somewhat the same manner that heat is necessary to generate the steam utilized by the steam engine. The scientists of the twentieth century have shown that this concept is entirely incorrect and that the heat produced is really a byproduct of the essential energy-releasing chemical reactions of life. All animals, whether warm-blooded or cold-blooded, produce heat. It is a necessary accompaniment of the transformation of energy. In some cases, the heat produced is lost directly to the environment. In other cases, it is conserved in one way or another to control the temperature of the microclimate or the animal itself. In the course of evolution, animals have utilized this byproduct of their metabolism to permit greater independence of the variable environmental temperature. These adaptations are considered in Chapter 10.

In the thermodynamic sense, the heat produced in metabolism is wasted free energy. The more inefficient the process, the larger is the fraction of free energy which is wasted in heat. In the hexokinase reaction, already described (page 218), the glucose molecule is phosphorylated through a reaction involving the hydrolysis of ATP to ADP. This glucose-priming reaction requires about 3000 cal per mole. The conversion of ATP to ADP is associated with a fall of free energy of the order of 8500 cal. In the thermodynamic sense the difference represents wasted energy or the inefficiency of the reaction, but so far as life processes are concerned this is an almost universal reaction, and glucose cannot be metabolized without it. Heat production can be used as a measure of the rate of metabolism, as Lavoisier discovered almost 200 years ago, but it is related to the fuel of the system in a complex manner through the many enzymatic reactions of intermediary metabolism.

The conversion of food into the active metabolites of the body requires energy. Thus, following the intake of food, there is a heat production in excess of that associated with the operation of the machinery in the post-absorptive condition. Rubner discovered this in his pioneer studies of the metabolism of dogs and named it the SPECIFIC DYNAMIC ACTION (SDA). It is also called the CALORIGENIC EFFECT or the HEAT INCREMENT of the ration. Rubner's early work and many subsequent studies are discussed by Brody (1945). Precise values of the SDA vary with the diet, the plane of nutrition and a number of physiological and environmental conditions. The ingestion of protein may elevate the basal energy expenditure of a bird or mammal by 15 to 40 per cent. The SDA for fat is about 12 per cent and that of carbohydrate about 5 per cent. Warm-blooded animals may utilize this heat increment of the ration to maintain body temperatures, but in the thermoneutral environment or in the poikilotherm it is waste heat. The SDA is not attributed to processes of digestion but to the metabolic interconversions and storage of food molecules. Deamination and the formation of nitrogenous wastes (urea) are thought to account for the high SDA of protein. Several of the amino acids (phenylalanine, tyrosine and leucine) have particularly marked calorigenic effects, and it has been suggested that their metabolism is less efficient than some of the others. Excess of fat or sugar, following a meal, must be stored or otherwise metabolized before the animal machinery returns again to the basal conditions. Explanations of this type are usually advanced for the calorigenic effect, but the matter is still not well understood (Grisolia and Kennedy, 1966).

Direct and Indirect Calorimetry

Lavoisier's crude ice calorimeter has been succeeded by many highly refined instruments which permit the measurement of heat production even in the larger mammals (Benedict, 1938; Brody, 1945; Kleiber, 1961). Energy contents of food and waste products have also been determined. Rubner, about 1895, first compared heat production in animals with the heat liberated when corresponding samples of food were oxidized in a bomb calorimeter (see Brody, 1945 and Rubner, 1968). He found that carbohydrate and fat formed approximately the same amount of heat under the two conditions of "burning" (9.3 kcal per g of fat and 4.1 kcal per g of carbohydrate) but that protein produced less heat in the animal (4.1 in comparison with 5.3 kcal). The protein nitrogen is incompletely oxidized and is excreted in various combinations with hydrogen and carbon. Since Rubner's time, precise relationships have been established between the energy values of the foods consumed, the oxygen used, the nitrogen excreted and the energy released. Living machines, like inanimate ones, are now known to be governed by the first law of thermodynamics or the law of conservation of energy.

Direct measurements of heat production require elaborate equipment. Nowadays, indirect techniques, based on the constants determined in earlier days, are commonly employed. Heat production can be calculated from respiratory exchange and nitrogen excretion using well-established metabolic constants (Brody, 1945; King and Farner, 1961; Kleiber, 1961). These are the methods of INDIRECT CALORIMETRY. They are much more convenient and, under many conditions, more reliable than the direct methods. In actual practice, oxygen consumption alone is commonly used by comparative physiologists. For most purposes it is entirely adequate. Its merits and limitations are discussed in several of the above publications.

It is obvious from the earlier discussion of intermediary metabolism that respiratory exchange and nitrogen excretion will depend on the type of fuel and what is happening to it in the organism. The RESPIRATORY QUOTIENT (R.Q.) or ratio of carbon dioxide produced to oxygen consumed is 1.0 for carbohydrate, about 0.7 for fat and 0.8 for protein. In the carbohydrate molecule, hydrogen and oxygen are present in the proportions to form water, and the ratio of CO_2 formed to O_2 consumed will obviously be 1:1. In the other foods, however, there is relatively much less oxygen and this is reflected in a lower R.Q. The R.Q. will also be affected by metabolic interconversions such as the formation of fat from carbohydrate (lipogenesis) or carbohydrate from protein (gluconeogenesis). Values as high as 1.49 have been recorded for force-fed geese where compounds are being formed with much higher ratios of hydrogen to oxygen (Benedict and Lee, 1937). Thus, respiratory exchange and nitrogen excretion will vary with lipogenesis, gluconeogenesis, growth and starvation, as well as with the composition of the food. These factors must be controlled or evaluated in establishing reliable base lines.

Standard and Active Metabolism

Valid comparisons of the metabolism of animals can only be made under carefully controlled conditions. Most frequently an attempt is made to minimize muscular movement as well as the effects of food ingestion and related metabolic activities by comparing metabolism of animals during periods of fasting and inactivity but not sleep. The term BASAL METABOLISM is used in mammalian studies. It is "the resting energy metabolism in a thermoneutral environment in post-absorptive condition, uncomplicated by heat increments incident to food utilization or to low or high environmental temperatures" (Brody, 1945). Krogh (1914) used STANDARD METABOLIC RATE for the comparable condition of minimal activity in lower animals, and this is the preferred term of comparative physiologists.

Fry and his associates (Chapter 12) recognize three levels of oxygen consumption in fish. The ACTIVE RATE will permit the highest continued level of activity; the ROUTINE RATE is the rate of utilization by fish when all movements are apparently spontaneous; in terrestrial vertebrates, the term TRANSPORT RATE has been used to designate the rate associated with normal minimal locomotion such as walking or slow flight (Brett, 1972). The STANDARD RATE is the nearest attainable approximation to metabolism when all organs are at minimal activity. Many of the recorded measurements of standard metabolism are, in fact, routine metabolism according to these definitions. Some workers have attempted to measure the minimal level by using narcotized fish; others have measured oxygen consumption at several levels of activity and then extrapolated back to zero activity in order to obtain a standard rate (Fig. 7.22). The latter is the approved method, and with refinement of techniques the measurements of standard metabolism have given smaller and smaller values.

Although the minimum rates are the most reliable parameters for comparing the effects of different factors on metabolic demands, the maximum rates (active metabolism) also have physiological significance. The magnitude of the difference between active and standard metabolism varies markedly in different species of animals; in any one species it may change with temperature and other variables. Flying insects show some of the greatest differences between resting and active metabolism. Krogh (1941) reports an almost 170-fold increase for butterflies during flight. The metabolic rate of mammals is increased 10 to 20 times during activity, but the difference is much less in some animals. Fry (1957) found only a fourfold maximum difference in fish. Differences are related to the efficiency of mechanisms which deliver oxygen to the tissues and the ability of the organism to carry on anaerobically and contract oxygen debts. The maximum rate in fish may be primarily limited by the respiratory surface (Fry, 1957) while that of mammals may be limited more by the creatine phosphate stores, the quantity of myoglobin and the ability to accumulate lactic acid. Both groups of factors are involved in fish as well as in mammals; Black's investigations demonstrated the limiting effects of accumulating lactic acid on the activities of fish (Black *et al.*, 1961).

Fry (1947, 1971) argues that the difference between standard and active metabolism represents the extent of respiration available for activity. He terms this "the scope for activity" and finds a good correlation between the maximum steady rate of swimming of a fish and the square root of its scope for activity. Fig. 7.23 shows how the scope for activity or "metabolic scope" (Brett, 1971), as measured by differences between active and standard metabolism, depends on the acclimation temperature of salmon. In several species of fish, maximum differences have been recorded at intermediate temperatures as illustrated in Fig. 7.23. Fry (1957, 1971) and Brett (1963, 1971) have discussed the implications of these relationships in fishes.

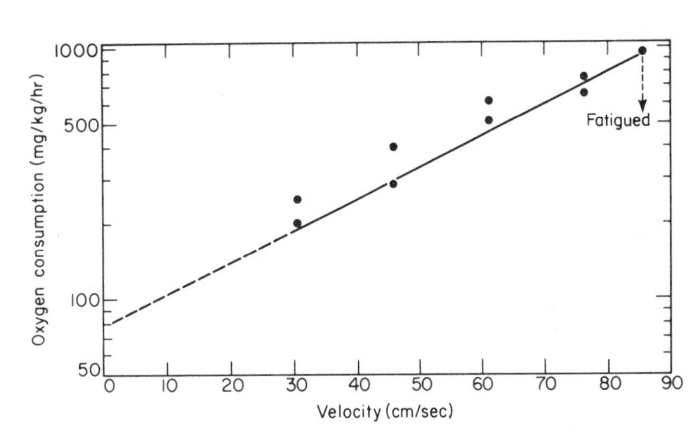

Fig. 7.22 Oxygen consumption in relation to swimming speed of an 88-g sockeye salmon (*Oncorhynchus nerka*) at 15°C (acclimation and test temperature). Projection of line to zero activity gives STANDARD METABOLISM. Line fitted to lower points; higher points due to excitability. Respirometer was a water tunnel through which the flow could be precisely controlled, and measurements were made in duplicate for a series of increasing velocities in which the fish maintained steady swimming for 75 min. [Courtesy Brett, 1963: Trans. Roy. Soc. Canada, *1* (ser. 4): 444.]

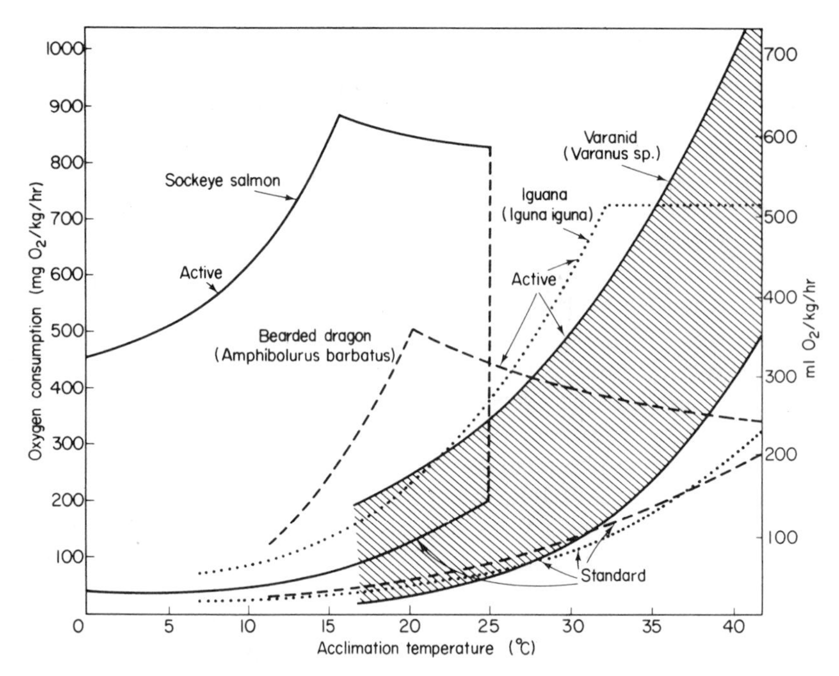

Fig. 7.23 Rate of oxygen consumption of some poikilothermic vertebrates in relation to acclimation temperature and activity. Data compiled by Brett (1972) from several sources. [From Brett, 1972: Resp. Physiol., *14*: 163. Permission North Holland Publishing Co., Amsterdam.]

Variations in the Rate of Metabolism

There are two kinds of factors which modify metabolic rate: those related to an ever-changing environment and those related to the physiological or genetical constitution of the animals. The former category includes oscillating diurnal variations, the cyclical seasonal changes in temperature and photoperiod and the effects of different environmental concentrations of respiratory gases, humidity and salinity. These will be considered in later chapters devoted to environmental relationships. Variations associated more directly with the physiology of the animal are summarized here.

The Relationship to Body Size

The rate of metabolism is proportional to the size of the animal. This is often true of the metabolism of isolated organs and tissues as well as the intact animal, and valid comparisons must usually be based on animals of equal size. Sometimes rates of tissue metabolism are constant over a range of sizes, but until this has been established both the tissue metabolism and that of the whole animal should be measured in a series of individuals of different sizes. Comparisons can then be based on the regression lines as shown in Fig. 7.24, either through the statistical methods for linear regression or by "picking off" equal-sized animals from appropriately fitted lines.

The relationship between size and metabolism is not a direct one. Rather, it seems universally true that the smaller animals have relatively greater rates than the larger ones. About 1883, Max Rubner made the first extensive study of this phenomenon by comparing the metabolism in dogs and other mammals of different sizes, ranging from an 18-g mouse to a 128-kg hog. He found an inverse relationship between metabolism and the body weight but noted a rather constant relation between metabolism and surface area, as expressed by the two-thirds power of the weight. Temperature regulation and heat loss, which are obviously related to surface area, were once thought to be the determinants. However, it is now evident that the situation is more complex, since the same general relationship holds for the poikilotherms (even those of small size) as well as for the homeotherms and often also for isolated tissues from both groups (Fig. 7.24).

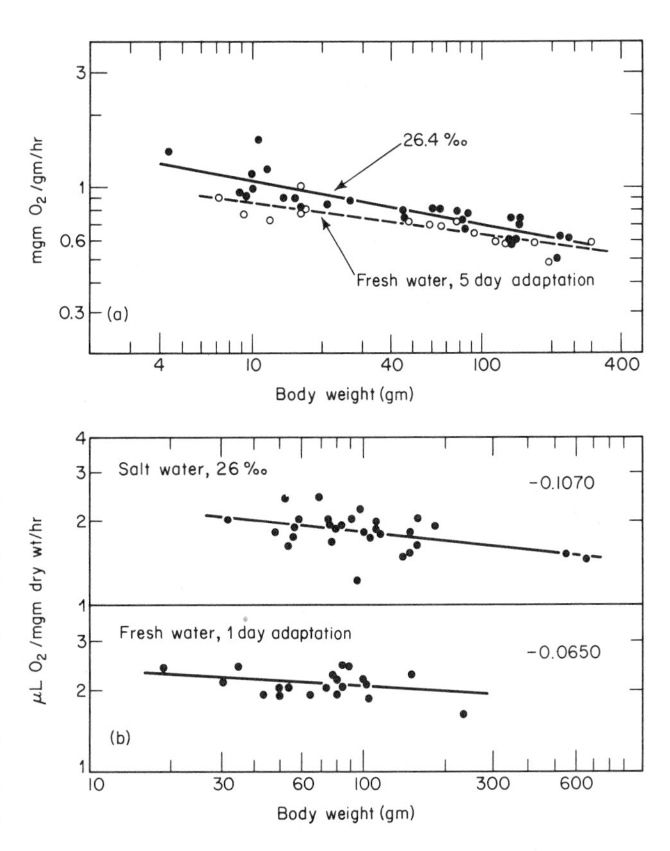

Fig. 7.24 Oxygen consumption of the starry flounder *Platichthys stellatus.* (Top) Standard metabolic rate in fresh water and in sea water measured at 14.8°C; regression lines significantly different at the 1 per cent probability level. [From Hickman, 1959: Can. J. Zool., *37*: 997.] (Bottom) Metabolism of excised gill lamellae measured by standard Warburg manometric methods at 15.0°C (unpublished data of C. P. Hickman, Jr., and W. S. Hoar).

The relationship between size and metabolism is an exponential one of the type which characterizes the relationships between body size and growth rate of many individual body parts (allometric growth of Huxley, 1932). It is expressed as: metabolism – $k \times$ body weight or $M = kW^n$. The logarithmic transformation of the equation is that of a straight line ($\log M = \log k + n \log W$). Thus, in Fig. 7.25 the constant k is the intercept on the y axis of the line with slope n. Rubner equated n to 0.67 (the surface area), but many measurements on all sizes of animals and on some plants show that the best overall value for this constant is about 0.75, with a k which varies for different major groups (Fig. 7.25). It should be noted that this slope, established by Hemmingsen (1960), is based on a smoothed line for a great many values and hides some of the smaller variations to be described in subsequent sections.

The significance of the 0.75 constant has not been established, although many different considerations have been advanced. Several of the arguments are based on the idea that there has been an evolutionary tendency toward increasing size in animal phylogeny and that an impossible situation would develop if certain structures were to grow and if certain processes were to operate in direct relation to body size. Thus, Hemmingsen (1950, 1960) argues that if metabolism increased in proportion to body size from a rat to a rhinoceros, the latter would have to endure

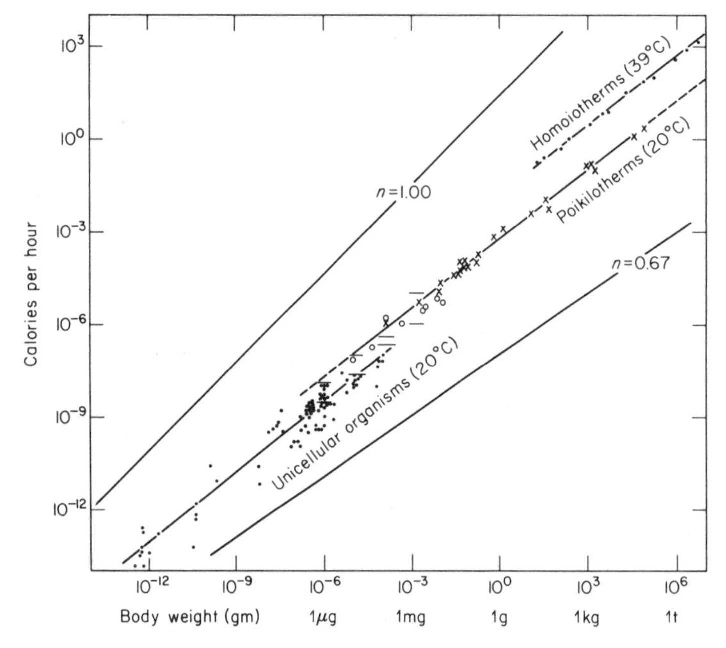

Fig. 7.25 Relation between standard metabolism and body weight in unicellular organisms including marine animal eggs and in many different poikilotherms and homeotherms. Lines fitted by method of least squares assuming n to be identical in the three groups. The slope corresponds to $n = 0.751 \pm 0.015$. [From Hemmingsen, 1960: Rept. Steno Hosp. Copenhagen, *9* (pt. 2): 16.]

surface temperatures of boiling water in order to dissipate the heat. The energy required for such high rates of metabolism would also tax the digesting and absorptive surfaces of the gut. The extent of respiratory epithelium necessary to exchange the gases has also been considered, and similar arguments could be advanced for a number of physiological processes which are directly dependent on the surfaces available for respiration and metabolism. A somewhat different but equally important consideration arises from the variable rates of tissue metabolism. There is relatively more skeletal and connective tissue in a massive organism.

Thus, it is not difficult to argue that the constant n in the weight-metabolism equation should be less than unity. It is also reasonable to argue against any consistent relation to the surface area ($n = 0.67$ in the vertebrates). It has not, however, been possible to explain just why the overall constant should be what it is and why it should be relatively constant for all kinds of organisms so far investigated.

Variations in the Slope of the Weight-Metabolism Line

The foregoing discussion might seem to indicate that differences between metabolic rates of animals of equal size are entirely caused by variations in the constant k. Thus, in Fig. 7.25, unicellular organisms display lower Qo_2 values than multicellular poikilotherms of the same size at the same temperature, although the constant n is the same in both groups. Likewise, active metabolism is much greater than standard metabolism, but the difference may be due to the position of the log weight-log metabolism line and not to its slope. However, it may be quite misleading to emphasize the extreme constancy of the value n. Within limited ranges, values of n may vary from about 0.5 to 1.0, and the points shown in Fig. 7.25 represent means for a whole family of lines of differing slopes. Fry (1957), for example, finds that most species of fish show $n = 0.8$ (approximately), but for some species values of 0.5 to 0.7 have been recorded.

Many different factors can affect the rate of metabolism, and these differences may alter n as well as k. The genetical constitution of the animal may be reflected in the slope of the weight-metabolism line, as indicated by differences in races, closely related species and different sexes. Changes in physiological condition associated with different stages in development and growth, the state of nutrition and a variety of other factors have also been shown to alter the slope or the position of the line. Several reviews of the literature are available (Hemmingsen, 1950, 1960; Zeuthen, 1953, 1955; Prosser and Brown, 1961).

ENDOCRINE REGULATION OF METABOLISM

This chapter has outlined some of the important enzymatic processes which generate metabolic energy. The description started with the monosaccharides, amino acids, and simple fats; it ended with the conversion of their potential energy into

ATP. Many of these processes are now known to be regulated by a group of endocrine glands which produce the METABOLIC HORMONES.

In Chapter 2, it was suggested that the first hormones were probably neurosecretory substances, which timed reproduction and regulated growth. Both reproduction and growth depend on a complex interaction of metabolic events. The maturation of ova dictates the mobilization of food and its storage in yolk; growth accentuates anabolic phases of metabolism with the construction of protein and the storage of fats and carbohydrates. These events are usually timed to take advantage of the most favorable season for survival and it can be assumed that an endocrine regulation of certain aspects of metabolism appeared very early in animal phylogeny. In theory, the hormonal regulation of metabolism might be expected at all levels in phylogeny. However, physiological constants are known to be much more variable among the poikilotherms; in these forms, metabolic rates and processes may be at least partly adjusted by environmental temperatures and the adaptation of enzyme systems (Chapter 2) rather than by hormones. At present this is largely an unexplored field; although there is now a rapidly expanding literature on the endocrine controls of metabolism in lower vertebrates and arthropods, detailed understanding of the mechanisms is still essentially based on studies of mammals.

Carbohydrate Metabolism

Blood glucose levels in man range from 80 mg to 100 mg/100 ml, rising to about 120 mg per cent immediately after a meal and falling to as low as 60 mg per cent with prolonged fasting or severe exercise. The range of values is somewhat greater in the lower vertebrates and invertebrates, but a remarkable degree of constancy prevails in spite of periodic flooding with sugars after feeding and their variable utilization between meals. The nervous system in particular relies on a steady supply of glucose. Vertebrate brain derives almost all its energy from the metabolism of this sugar. In man, this amounts to about 100 g glucose per day and accounts for 20 to 25 per cent of the total basal metabolism. Further, brain tissues—unlike other active tissues like muscle—does not store a reserve of food (glycogen); indeed, brain tissue does not possess the phosphorylase activating system required to convert glycogen to glucose. Thus, it is obvious that this organ, on which an animal's immediate survival often depends, requires a steady perfusion with glucose. If glucose levels fall abruptly, as they do with an overdose of insulin, the brain is rapidly deprived of its primary source of ATP and symptoms very like those of anoxia quickly develop. These peculiarities of brain metabolism are important factors that demand a reliable homeostasis of blood glucose.

The hazards of high blood sugar may be especially great in warm-blooded animals with their periodic feeding habits, their rapid processes of digestion and their steady rates of metabolism. INSULIN is the important hypoglycemic hormone of vertebrates and operates in several different ways to remove excess sugar from the blood and promote its storage in the tissues. One of its functions is to activate the cell membrane systems engaged in the transport of glucose into muscle and fat cells. It seems to be specific for these tissues and does not stimulate glucose absorption from the gut or facilitate its movement into liver cells. Insulin stimulates synthetic processes as well as transport mechanisms. Glycogen storage in muscle and liver and lipogenesis in the fat depots rapidly remove glucose from the blood. A further action of insulin on the hepatic tissues is to depress GLUCONEOGENESIS which is the formation of glucose from noncarbohydrate sources (amino acids and glycerol); this action reduces the blood sugar still further (Clegg and Clegg, 1969).

In the absence of insulin, the blood sugar rises and a condition of DIABETES develops. Although the action of insulin is much more marked in homeotherms, its hypoglycemic effects have been demonstrated in all vertebrates from the fishes (Epple, 1969) to man; insulin or proinsulin has also been extracted from several different invertebrates (Falkmer, 1972). There is, however, relatively little information on hypoglycemic hormones among the invertebrates; this is in contrast to hyperglycemic factors which are now firmly established in several invertebrates; it may be postulated that, among the poikilotherms, the hazards of high blood sugar are less than the problems of mobilizing sugar from reserves.

Several mechanisms come into play as the blood sugar levels decline after the digestion of a meal, while additional factors operate during prolonged fasting (Fig. 7.26). In addition to insulin, vertebrate islet tissues produce GLUCAGON which has

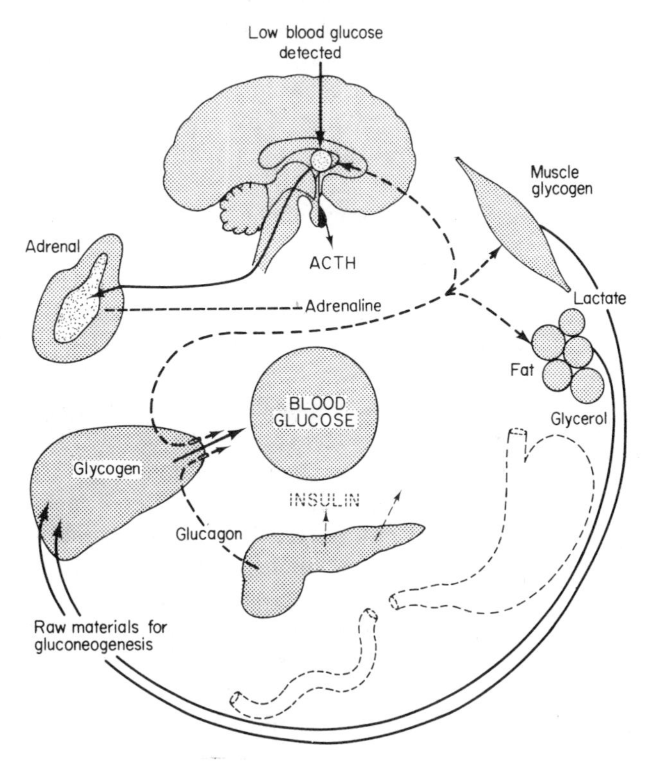

Fig. 7.26 Endocrine control of glucose metabolism in response to a moderate fall in blood sugar after absorption of a meal. With more extreme hypoglycemia (fasting or vigorous exercise), ACTH strongly stimulates glucocorticoid secretion; the glucocorticoids (1) antagonize uptake of glucose by muscle and fat cells; (2) release additional amino acids from muscle for hepatic gluconeogenesis; (3) increase enzymes concerned with hepatic gluconeogenesis. [From Clegg and Clegg, 1969: Hormones, Cells and Organisms. Heinemann, London.]

an opposite effect and raises blood sugar levels. Glucagon operates by activating the phosphorylase system of the liver and thus stimulates glycogenolysis (Chapter 2). This hormone has no effect on the extrahepatic tissues but seems to be specific for the liver phosphorylase system. Its action is much more marked in some vertebrates (lizard, duck, cat) than in others (man). In the cat, 1 μg per kg may raise the blood sugar by about 20 mg per 100 ml, while an infusion of glucagon for about 4 hours can deplete the entire liver glycogen store. It is suggested that some species have an absolute requirement for glucagon while in others it acts as a local hormone associated with the control of insulin secretion in the pancreas (Marks and Samols, 1968). The physiological significance of glucagon in the primitive vertebrates is uncertain (Epple, 1969).

Adrenaline and the glucocorticoids are also involved in the regulation of glucose and in many vertebrates they are probably of more significance than glucagon. A fall in blood sugar (hypoglycemia) stimulates production of adrenaline and this acts at several different points (Fig. 7.26). In the liver, it operates like glucagon, increasing the amount of active phosphorylase in the cells (Chapter 2); it also promotes release of lactate from muscle and glycerol from fat and these provide materials for gluconeogenesis in the liver; finally, during more prolonged fasting or starvation, adrenaline strongly stimulates the secretion of ACTH and then sets in motion another group of homeostatic controls that depend on the adrenal cortex. The glucocorticoids antagonize the uptake of glucose by muscle and fat and promote the transfer of amino acids from muscle for hepatic gluconeogenesis; they also increase the amounts of enzymes concerned with gluconeogenesis (Clegg and Clegg, 1969). These mechanisms operate in emergencies to maintain the levels of blood glucose. The patterns of hyperglycemic control are similar from fishes to mammals although sensitivities to different hormones, drugs and experimental manipulations vary (Grant *et al.*, 1969; Chester Jones *et al.*, 1969).

Several other metabolic hormones also have definite effects on the carbohydrate metabolism of vertebrates. The hypophysectomized dog survives pancreatectomy without the fatal symptoms of diabetes. The dog with both pituitary and pancreas removed is named a "Houssay dog" after the celebrated Argentine physiologist who did the pioneer work with the "diabetogenic hormone" of the pituitary. This factor may be the growth hormone, somatotropin, which has an insulin-like or hypoglycemic action, although its primary effects may be on lipid mobilization and protein synthesis (Snipes, 1968; Trystad, 1968). The gluconeogenic action of the adrenal steroids has been mentioned. The gonadal steroids also have metabolic effects which are reflected in the altered metabolism of carbohydrates (Turner and Bagnara, 1971). Thyroid hormone, in the homeotherms, stimulates many phases of carbohydrate metabolism, from the absorption of sugars to their utilization in the tissues. Although these physiological effects of the metabolic hormones on carbohydrate metabolism are carefully documented, the actions of the hormones at the cellular level are not as well understood as those of the pancreatic hormones and adrenaline.

Steele (1963) described a hormone from the corpora cardiaca of the cockroach which, like glucagon and adrenaline, converts the enzyme phosphorylase from its inactive to its active form (Chapter 2). In this manner, it promotes the synthesis of glucose 6-phosphate from glycogen in the insect fat body. Among the vertebrates, the glucose 6-phosphate thus formed is dephosphorylated and passed into the blood as glucose; among the insects, additional biochemical steps within the fat body are required to synthesize trehalose which is the circulating blood sugar (Gilmour, 1965; Wyatt, 1967). It is now recognized that hyperglycemic factors from the corpora cardiaca also promote trehalose synthesis and transfer to the hemolymph (Friedman, 1967). Action sites of the cockroach hyperglycemic factors are shown in Fig. 7.27. Like glucagon, the cockroach factor acts only in the carbohydrate storage organ (liver or fat body); it has no effect on phosphorylases in other tissues. Biochemically, the insect hyperglycemic factor is a polypeptide like glucagon, but the two factors are not physiologically the same and glucagon does not act on the phosphorylase system of the insect fat body (Steele 1963).

More than a quarter of a century ago it was noted that extracts of the eyestalks would induce a hyperglycemia in crustaceans (Scheer, 1960; Huggins and Munday, 1968). In some species, removal of the eyestalks is followed by hypoglycemia. Extracts of eyestalks vary in their hyperglycemic properties during different stages of the molting cycle. The hepatopancreas is presumably the locus of this hormone's action but details of the mechanism remain uncertain. In part, the eyestalk hormone or hormones may operate by regulating the amount of glucose 6-phosphate which is channeled into either the pentose cycle or the Embden-Meyerhof pathway (Highnam and Hill, 1969).

Fragmentary data imply the presence of hyperglycemic factors in several other invertebrates. Goddard *et al.* (1964) reported a hyperglycemic effect with homogenates of the albumen glands injected into the snail *Helix*. The responsible factor was not characterized. A phosphorylase-activating system has been studied in the liver fluke where glycogenolysis may be stimulated by serotonin. It has been suggested that serotonin among the invertebrates may play a role comparable to adrenaline among the vertebrates (von Brand, 1966; Mansour, 1967).

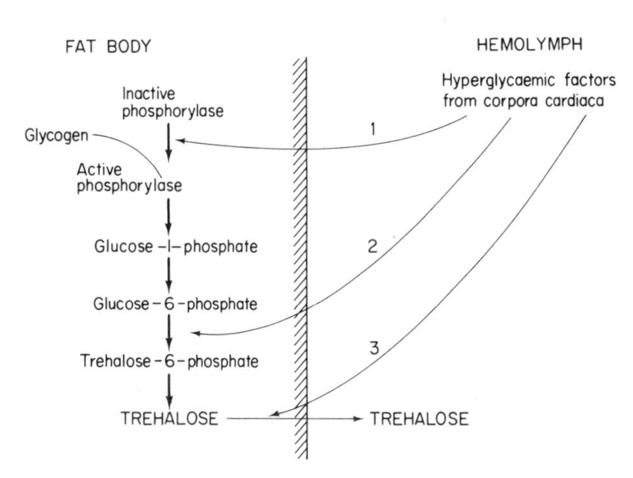

Fig. 7.27 Sites of action of cockroach hyperglycemic factors. Two enzyme reactions are affected (1) and (2) as well as the transfer of trehalose from fat body cells to hemolymph (3). [From Highnam and Hill, 1969: The Comparative Endocrinology of the Invertebrates. Edward Arnold, London, based on Steele, 1963: Gen. Comp. Endocrinol., *3*: 46.]

Other Effects of Hormones on Tissue Metabolism

In the living animal, mechanisms concerned with the metabolism of carbohydrate, fat and protein are interlocked and do not operate as independent units. It is not then surprising that each of the metabolic hormones acts on several different aspects of metabolism. Although insulin and glucagon are primarily concerned with carbohydrate metabolism, the pancreatectomized animal shows increased utilization of fat and protein, an impairment of lipogenesis, and a decreased ability to synthesize protein from amino acids. It follows that the effects of any of the metabolic hormones may be expected to act at several points in the biochemical sequences of tissue metabolism. There is a voluminous literature on the endocrinology of metabolism, but much of it pertains to the mammal, and a large part of it is clinical in nature. Research on the lower vertebrates has already shown some interesting differences between their metabolic controls and the familiar ones in the mammals although several of these apparent differences may disappear when the molecular functions of the hormones are explained.

Many insect physiologists are currently investigating the hormonal regulation of fat and protein metabolism as well as that of carbohydrate (Gilbert, 1967; Highnam and Hill, 1969). Most of these studies relate to the special physiology of reproductive maturation, growth, metamorphosis and molting.

8

EXCRETION

Metabolism produces a variety of byproducts. Some must be removed (excreted); others perform useful metabolic chores. A single substance may sometimes be an excretion product and, at other times, an indispensable metabolite. On one hand, water is a byproduct of metabolism and must often be excreted in large amounts to avoid a serious condition of edema; on the other hand, the only source of water available to some animals is metabolic water and this must be rigorously conserved. Carbon dioxide, although a metabolic byproduct, is also an important component in the synthetic and regulatory machinery of animals and plants. The same is true of ammonia. Similarly, urea, a prominent constituent of the urine in many animals, sometimes discharges useful physiological functions. If the blood urea in man rises above about 0.05 per cent (normal values 0.01 to 0.03) a pathological condition of uremia develops; but the elasmobranch fishes actively retain urea for purposes of osmotic regulation and have normal blood urea values of 2.0 to 2.5 per cent (Smith, 1953). In some mammals too, urea may perform useful functions. In the ruminant it is secreted in the saliva, passes back into the stomach and is a source of nitrogen for the microflora of the rumen. Thus, in one animal or another, almost all of these metabolic products which are commonly considered excretory, have found a use. Conversely, some compounds, not usually classed as waste, must be removed just as regularly as carbon dioxide or ammonia. Many marine vertebrates drink sea water, and by excreting large amounts of salt (usually through gills or special salt glands) they are able to obtain fresh water for their metabolism. Salt in this case is a byproduct of osmotic regulation as ammonia is a product of protein metabolism. No concise definition of excretion can be based solely on the chemical nature of the material removed and it is better to define excretion in a very general way as the separation and ejection of the metabolic wastes, usually in aqueous solution.

In the highly organized animals (both invertebrate and vertebrate) there are three well-defined physiological processes included under excretion: FILTRATION, REABSORPTION and SECRETION. In filtration, non-colloidal solutions are moved by hydrostatic pressure differences through a semi-permeable membrane from the body fluids into a space connected with the outside. This filtered material then passes through a tube or other space lined with cells capable of actively transporting molecules from it back into the body fluids (reabsorption), or of secreting additional substances into the filtered fluid. Obviously, at the cellular level there are but two processes, diffusion and active transport. Understanding of the machinery in the phylogenetically lower invertebrates such as protozoa, hydroids and flatworms is less complete than in many of the higher animals; but it seems likely that the processes are similar at the cellular level. Accordingly, although it is very unsatisfactory to attempt a definition of excretion based on the nature of the componds removed, there is much less ambiguity concerned with the general processes of urine formation.

NITROGENOUS WASTES

Removal of ammonia (the most conspicuous byproduct of the nitrogen-containing compounds) is one of the major tasks of excretion. Whereas other metabolic residues, like carbon dioxide and water, are relatively innocuous and can often be removed as gases, ammonia is highly toxic and is almost always removed in solution —frequently in some detoxified form. Animals, in their phylogeny, have tried several different devices for the removal of excess ammonia and other nitrogenous wastes, such as the purines arising from nucleoprotein. This story has been told in many places but never more concisely and interestingly than in Baldwin's *Dynamic Aspects of Biochemistry* (1967). Baldwin's arguments will be followed in this summary.

Ammonia

The metabolism of ammonia was discussed in Chapter 7, and at that point its lability and importance in biochemical syntheses were emphasized. Its removal as a waste product is considered here.

Free ammonia is extremely toxic and rarely accumulates in living cells or their surrounding media. Simple diffusion experiments show how readily this ion penetrates cell membranes (Heilbrunn, 1952); at the cellular level this may be the basis of its toxicity. Only traces (0.1 to 0.2 mg per 100 ml) are found in human blood. Several simple experiments have demonstrated the toxicity of ammonium ions in mammals and birds. When crystalline urease, which liberates ammonia from urea, was injected into rabbits, the animals died, with ammonia blood levels at about 1 part in 20,000. At this point there was no change in the pH of the blood, and death was attributed to the ammonium ion. Similar injections into a normal hen had no effect, presumably because bird blood normally contains only traces of urea which, in the rabbit, served as a precursor for the free ammonia. When the hen was injected with both urea and urease, it died. The theory of ammonia toxicity, arrived at much earlier through phylogenetic speculations (Needham, 1931), was thus confirmed (Sumner, 1951). Precise levels of toxicity have been evaluated in several vertebrates by intraperitoneal injections of ammonium acetate (Wilson *et al.,* 1968, 1969). Species variations were marked, with LD_{50} values (Chapter 9) ranging from 10.44 m-moles/kg body wt in chicks to 29.34 m-moles/kg body wt in goldfish.

The excretion of ammonia as the primary end product of nitrogen metabolism is usually associated with the presence of abundant water. There are very few exceptions. The best known are the terrestrial crustaceans which are able to withstand levels of tissue ammonia ranging up to 6 mg per cent and liberate the gas directly into the atmosphere (Hartenstein, 1968). In most animals, ammonia is excreted as such only when there is ample water for its rapid removal in solution. This is the situation in marine invertebrates and in all freshwater animals, whether vertebrate or invertebrate. The marine invertebrates are in osmotic equilibrium with their surroundings and, bathed as they are with osmotically equivalent saline, readily lose the freely soluble ammonia by simple diffusion. The freshwater animals, however, live in a decidedly hypotonic solution and are continually flooded with fresh water because of osmotic differences. This is steadily pumped out and the same pumping machinery serves as a flushing system for the soluble wastes, of which the most important is ammonia. Animals which excrete ammonia as the main end product of amino nitrogen metabolism are said to be AMMONIOTELIC. Water, however, is often at a premium, not only in the terrestrial environment but in the environment of many marine fishes which maintain hypoosmotic body fluids and, in consequence, face a constant osmotic desiccation. In these dry environments, ammonia is usually turned into less toxic forms such as urea (UREOTELIC animals) or uric acid (URICOTELIC animals).

These are broad generalizations. Small amounts of urea, uric acid, and other nitrogen-containing compounds appear regularly in the urine of both freshwater animals and marine invertebrates. Amino acids are also secreted as such in the lower forms, but whether this is true excretion or simply leakage is uncertain. The main point is that the excess ammonia arising in metabolism seems to require no further treatment if it can diffuse promptly into the surrounding water, but when water must be conserved, ammonia must be detoxified.

There are two important biological advantages associated with the direct excretion of ammonia (Forster and Goldstein, 1969). One of these is related to the energetics of urea or uric acid biosynthesis; both processes require ATP. While the formation of ammonia from certain sources (glutamine) actually increases the supply of ATP (Fig. 7.19), all forms of detoxification require its expenditure. The second bonus of direct ammonia excretion is associated with the trading of this surplus ion for physiologically indispensable ions, particularly sodium. In this way, important cations of the body may be conserved (Fig. 8.8) or salt may be acquired from the surroundings (Fig. 8.9). These ion exchange processes are discussed in a subsequent section.

Urea

In some animals, urea is a physiologically important compound and occurs in considerable amounts in the tissues. The situation in the elasmobranchs has been mentioned. The aestivating lungfish is another animal whose tissues tolerate very high levels of urea (Smith, 1953; Forster and Goldstein, 1969). Since active lungfishes excrete ammonia directly like most other freshwater fish, Baldwin considers the change to ureotelism during aestivation further evidence for a phylogeny of urea biosynthesis in situations where the water supply is restricted. The evolutionary arguments will be reviewed later. Here, it is noted that the presence of urea at high concentrations in some animals shows that, unlike ammonia, it presents no insurmountable problems in cellular metabolism.

Chemically, urea consists of two molecules of ammonia united to one of carbon dioxide. On paper, the combination can be simply made as follows:

$$CO_2 + 2NH_3 \longrightarrow C(=O)(NH_2)_2 + H_2O$$

Synthesis in living tissues is not so simple.

The history of urea in biochemical literature is an interesting one (Fruton and Simmonds, 1958; Fruton, 1972). Urea was discovered in 1773. In 1828, Wöhler synthesized it and, in so doing, dealt a shattering blow to the proponents of vitalism. It could no longer be argued that only living tissues formed organic compounds. In 1904, the amino acid arginine was identified as a precursor of urea in mammals, and the enzyme arginase was shown to be essential for the reaction which follows:

$$HN{=}C(NH_2)\text{—}NH\text{—}(CH_2)_3\text{—}CHNH_2\text{—}COOH + H_2O \xrightarrow{\text{arginase}} NH_2\text{—}(CH_2)_3\text{—}CHNH_2\text{—}COOH + C(=O)(NH_2)_2$$

Arginine → Ornithine + Urea

The reaction suggests that the building blocks of urea (carbon dioxide and ammonia) are in some way added onto the ornithine molecule and then split off as urea by the enzyme arginase. The precise biochemical steps by which this is accomplished eluded biochemists for more than half a century.

The skeleton of the synthetic machinery was suggested by Krebs, following in-

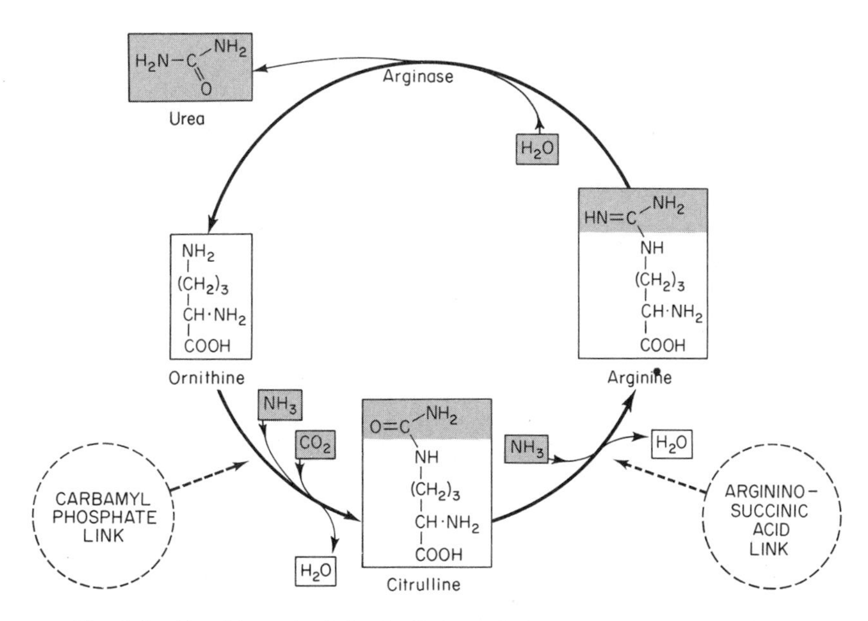

Fig. 8.1 Urea biosynthesis by the Krebs ornithine cycle. Lower circles show links added subsequently and detailed in Figs. 8.2 and 8.3.

tensive studies of the metabolism of rat liver slices and the identification of citrulline as an intermediary between ornithine and arginine (Krebs and Henseleit, 1932). Krebs' "ornithine cycle" (Fig. 8.1) remained the recognizedly incomplete explanation of urea biosynthesis for more than twenty years. Modern tracer and micro-chemical techniques have now added two additional links: (1) the CARBAMYL-PHOSPHATE LINK through which one unit each of CO_2 and NH_3 are added between ornithine and citrulline and (2) the ARGININOSUCCINIC ACID LINK which adds the second amino group to the cycle (Figs. 8.2 and 8.3). A study of the distribution of these compounds and their associated enzymes suggests that the basic components of urea biosynthesis were present in living organisms long before they were organized into the appropriate chains for the detoxification of ammonia in the formation of urea (Cohen and Brown, 1960; White *et al.* 1968). They are widely distributed in bacteria, plants and animals. Details of the enzyme kinetics of urea biosynthesis will be found in books of comparative biochemistry (Hochachka and Somero, 1973). Only a summary of the main reactions is given here.

Carbamyl phosphate This may have been a key compound in biochemical evolution. In its formation there is a primary fixation of both CO_2 and NH_3 with an associated high-energy phosphate bond. The carbamyl group so formed is readily transferred in several organic syntheses; its widespread distribution in bacteria, plants and animals, along with the obvious importance of adding inorganic carbon and nitrogen into organic systems, suggests a key role in evolution (Cohen and Brown,

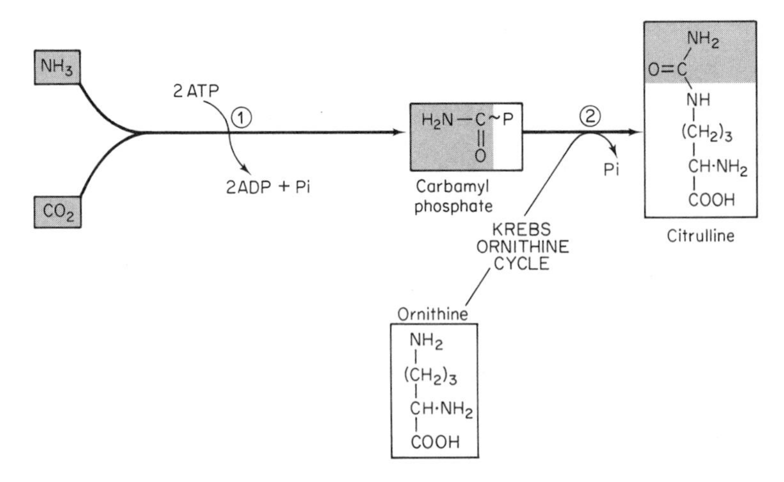

Fig. 8.2 The carbamyl phosphate link in the ornithine cycle. ENZYMES: (1) carbamyl phosphate synthetase, Mg^{++} and ATP; (2) ornithine transcarbamylase (carbamoyl transferase).

1960). In animals, carbamyl phosphate is synthesized in two steps and requires two molecules of ATP along with the enzyme carbamyl phosphate synthetase and certain cofactors, including Mg^{++}. Bacteria can make the combination with one molecule of ATP, using the enzyme carbamate kinase. The overall reaction for animals is represented as follows:

$$NH_3 + CO_2 + H_2O + 2ATP \longrightarrow H_2N{-}\overset{\overset{\displaystyle O}{\|}}{C} \sim P + 2ADP + P_i$$

In urea formation, carbamyl groups are transferred to ornithine, thus forming citrulline in the Kerbs ornithine cycle (Fig. 8.2).

Argininosuccinic acid This link is made through aspartic acid (Fig. 8.3). Aspartic acid (arising through the amination of oxaloacetic acid) couples with citrulline to form argininosuccinic acid which then splits to form arginine for the ornithine cycle and releases fumaric acid for the citric acid cycle and subsequently the formation of more oxaloacetic acid. The link through aspartic acid joins the citric acid cycle with the ornithine cycle by the oxaloacetic acid link of the former. The formation of aspartic from oxaloacetic acid through amination or transamination has been described (Fig. 7.19). These two important metabolic cycles are thus interconnected at the argininosuccinic acid link. One unit of ATP is required for linking of aspartic acid and citrulline. Thus, in terms of metabolic energy, three molecules of ATP are required to make one of urea (Figs. 8.2 and 8.3). This is the cost of detoxifying the ammonia.

In addition to the Krebs ornithine cycle, there are two other recognized routes for the biosynthesis of urea. In the first of these, the amino acid arginine gives rise to urea in the presence of the enzyme arginase (Fig. 8.1). In this manner, dietary arginine is probably the source of urea in the tissues of many animals which lack the key enzymes and compounds of the ornithine cycle. In the second, which will be detailed in the next section, urea is formed as one of the end products of purine metabolism (Fig. 8.6). These two sources of urea, however, are relatively minor and the ornithine cycle remains as the recognized route for the direct synthesis of urea from NH_3 and CO_2. The enzymes and compounds concerned with this cycle are widely distributed from flatworms (Awapara and Simpson, 1967; C. P. Read, 1968b) to the highest vertebrates. Their occurrence, however, is not universal in any of the major groups. At present, it is hazardous to generalize since relatively few animals have been thoroughly investigated, but it appears that the capacity to operate an ornithine cycle is phylogenetically very ancient and has frequently appeared in response to limited supplies of water. There are many examples. The earthworm possesses the biochemistry essential for the ornithine cycle but forms urea only during

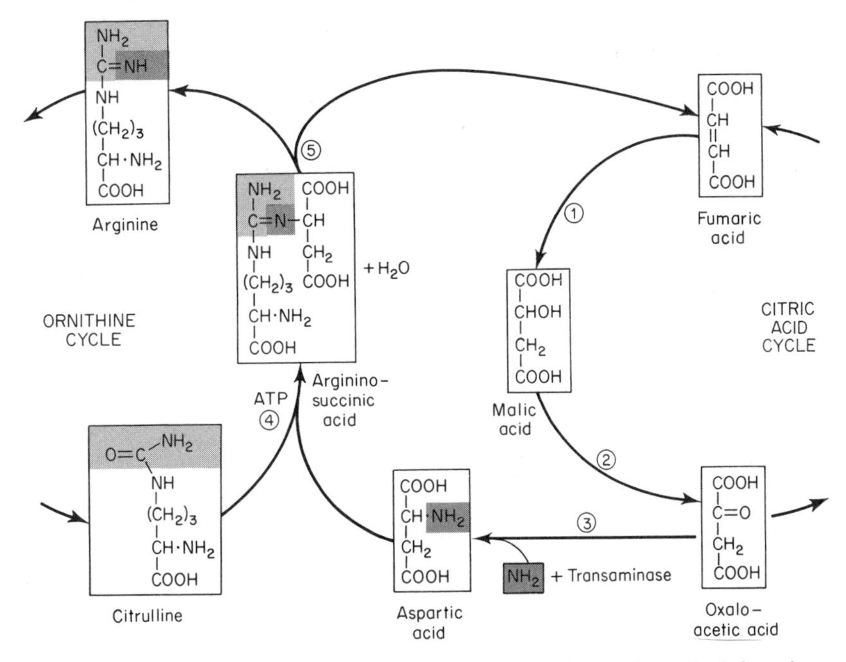

Fig. 8.3 Argininosuccinic acid link with the ornithine cycle at the left and the relation to the citric acid cycle at the right. Shaded blocks, the carbamyl contribution; stippled blocks, the citric acid contribution. ENZYMES: (1) fumarase ($+H_2O$); (2) malic dehydrogenase; (3) transaminase or aminase system (see Fig. 7.20); (4) condensing enzyme + ATP + Mg^{++}; (5) argininosuccinase.

periods of starvation, when increased protein catabolism produces excess ammonia (Bishop and Campbell, 1965; Needham, 1970b). Among the molluscs, the cycle is rare (Duerr, 1968; Speeg and Campbell, 1968) but operates in some of the pulmonates during aestivation (Horne and Boonkoom, 1970) as it does in aestivating lungfishes (Forster and Goldstein, 1969) and in *Xenopus* during periods of desiccation (McBean and Goldstein, 1967). In these examples and many others, water deprivation seems to have been responsible for the phylogeny of urea biosynthesis by the ornithine cycle.

Uric Acid

The third conspicuous nitrogenous waste is uric acid. This too is formed to detoxify ammonia and has the advantage of being highly insoluble and easily precipitated from a supersaturated colloidal solution. With a few minor exceptions, such as the closely related guanine in arachnids (Anderson, 1966; Horne, 1969), uric acid is the only nitrogenous excretory product which can be removed in solid form; it thus permits nitrogen excretion without loss of water. All the successful groups of animals living in arid habitats (the pulmonate snails, insects, and birds and saurian reptiles) are uricotelic. A convergence in biochemical evolution has produced the uricotelic habit at least three times.

PURINE
(general structure)

PYRIMIDINE
(general structure)

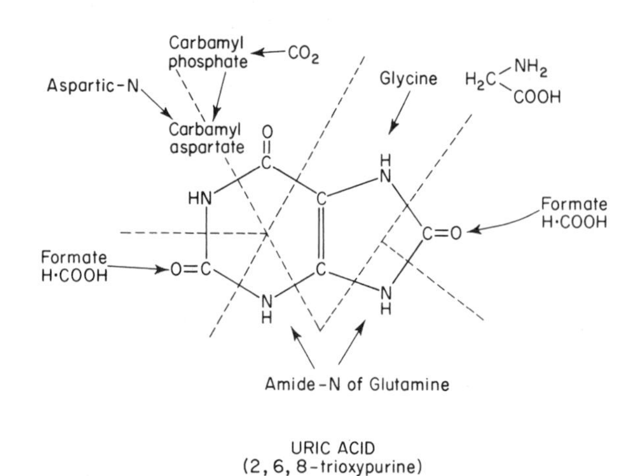

Fig. 8.4 General structure of the purine and the pyrimidine molecules above, with the specific structure of uric acid below. Broken lines divide the uric acid molecule into the building blocks from which its biosynthesis takes place.

Uric acid is a member of the purines. The general structure of the purine molecule and that of several chemical arrangements which appear as excretion products are shown in Figs. 8.4 and 8.7. Studies of liver slices from birds which excrete between 70 and 80 per cent of their nitrogen in the form of uric acid have shown that the hen and the goose form uric acid from added ammonia, while the pigeon forms an intermediate hypoxanthine which is subsequently oxidized to uric acid in the kidney (Baldwin, 1967). For some reason, pigeon liver lacks xanthine oxidase. Experiments such as these established the locus of uric acid formation and the purine intermediates; the compounds involved in the synthesis of the purine molecule itself were only identified after isotopically labeled compounds became available for research.

The different fragments from which the uric acid molecule is assembled are shown in Fig. 8.4. Most of the biosynthetic steps have now been established and are described in recent textbooks of biochemistry. There are indications that the processes are the same in molluscs (Lee and Campbell, 1965), insects (Barrett and Friend (1970), and vertebrates. Purine biosyntheses will probably be found to follow similar patterns throughout the living world.

Products of Nucleoprotein Metabolism

The nucleic acid component of nucleoprotein contains nitrogen in combinations which, during catabolism in some animals, lead to additional nitrogenous wastes. The constituents of the nucleoprotein molecule are shown in Fig. 8.5 and Fig. 22.1. Nucleic acids are polynucleotides. The structure of adenylic acid, one of the nucleotides, is shown in Fig. 7.2. Removal of the phosphoric acid portion of the nucleotide molecule produces a nucleoside which is, in turn, composed of a nitrogenous base and a pentose. There are two major groups of nucleic acids: the ribose nucleic acids (RNA), which appear to have as their primary function the translation of genetic information into protein molecules, and the deoxyribose nucleic acids (DNA) which are the primary carriers of genetic information (some viruses use RNA as their genestuff). The nucleic acids differ in their pentose; RNA pentose has one more oxygen than DNA pentose. Ribose may be associated with either of the purines (adenine or guanine) or with the pyrimidines (cytosine or uracil). Thus, there are four possible kinds of ribose nucleosides or nucleotides. The same number of combinations occur with deoxyribose, but thymine substitutes for uracil (Fig. 22.1). Each type of nucleic acid is, then, built up of multiples of four kinds of nucleotides. The importance of these compounds in the coding of biological information will be discussed elsewhere. In this section only their excretory products are considered. The nitrogen portion of the purine or pyrimidine molecule, like that of the amino acid, is removed in solution.

The pyrimidine molecule, during its catabolism, is completely dismantled into its building blocks of CO_2, H_2O, and NH_3. The further utilization of these, or their excretion, has been discussed. The catabolism of the purines, however, rarely leads to these elemental building blocks. Only a few animal groups (sipunculids, poly-

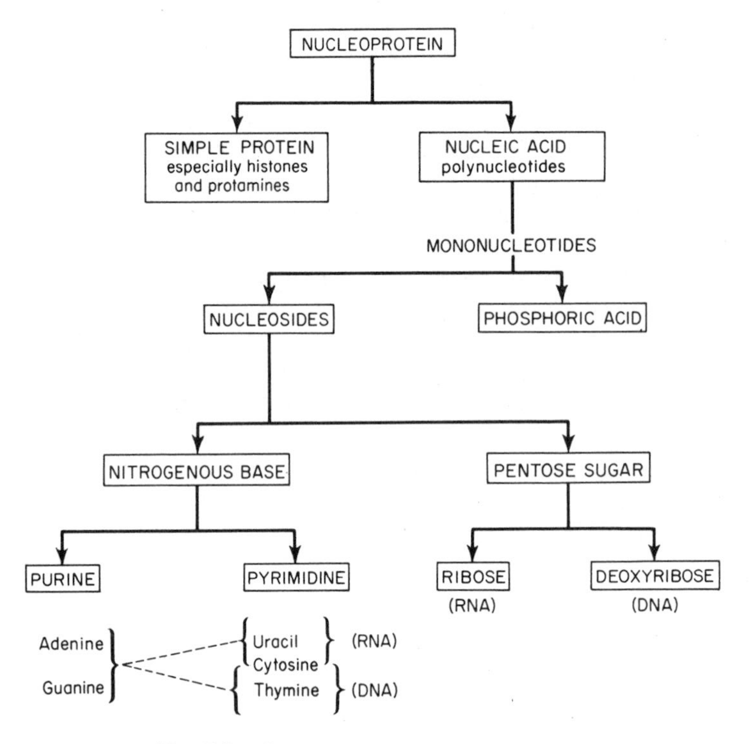

Fig. 8.5 Components of the nucleoproteins.

chaetes, crustaceans, some bivalves) possess all the enzymes for complete deamination and oxidation of purines (Florkin, 1969b, Hult, 1969). In many animals, small amounts of purine are excreted unchanged while, in a few cases, a direct excretion is the rule. This seems to be true in flatworms, some annelids and in the cyclostome *Lampetra fluviatilis* (Prosser and Brown, 1961). In most animals, the excretion product is some intermediate compound between the purine and its simple building blocks (Fig. 8.6).

The biological significance of the particular end product is a matter of speculation. In theory, it would seem advantageous to dispense with as many as possible of the intermediate energy-demanding metabolic steps. This may explain why many groups excrete uric acid or allantoin (Fig. 8.6) rather than degrading these substances to ammonia. However, the nature of the end product must depend mainly on its solubility in relation to the availability of water for excretion and the nature of the excretory apparatus through which the wastes must pass. Solubilities of the end products range from a low of 6 mg/100 ml for uric acid to a high of 120,000 mg/100 ml for urea. The capacities of the excretory tubules differ even in closely related species; the Dalmation dog, for example, is like other dogs in possessing enzymes which oxidize uric acid but, nevertheless, it excretes large amounts of uric acid because of a very low renal threshold for this substance.

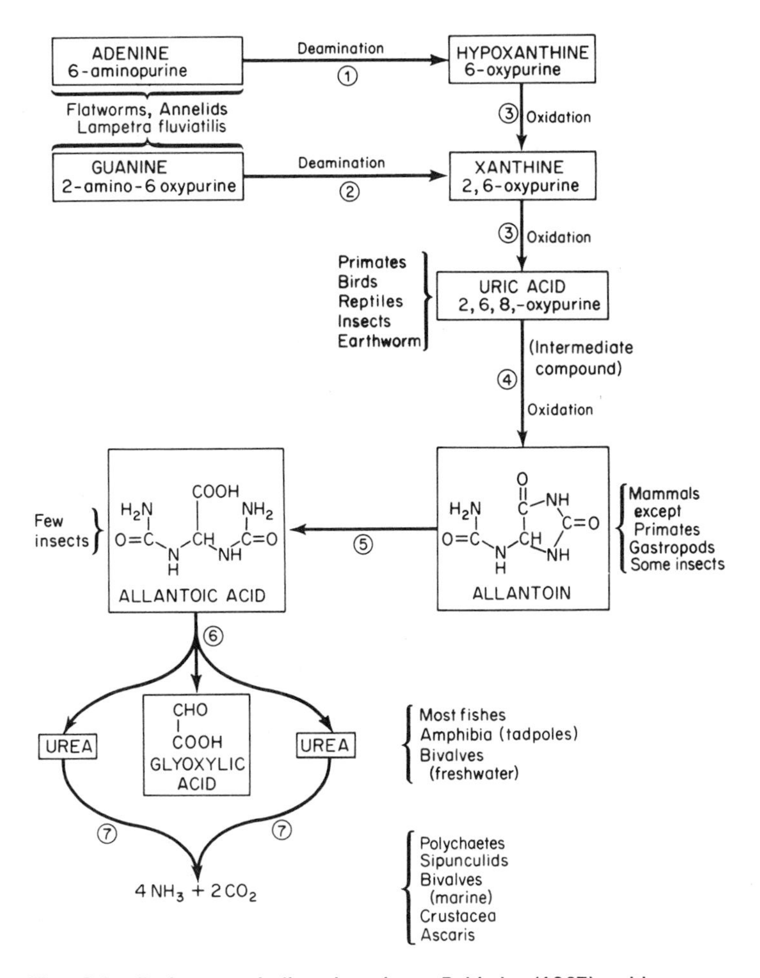

Fig. 8.6 Purine metabolism based on Baldwin (1967) with some revisions from the recent literature. The different animals are grouped with the compounds which they excrete. ENZYMES: (1) adenase; (2) guanase; (3) xanthine oxidase or hydrogenase; (4) uricase (urico-oxidase)—two-step reaction; (5) allantoinase; (6) allantoicase; (7) urease. SOLUBILITIES (mg/100 ml) based on Bursell (1967): hypoxanthine, 70; xanthine, 260; uric acid, 6; allantoin, 60; allantoic acid, slightly; urea, 120,000; ammonia, 89,000.

The excretory products of purine metabolism are shown in Fig. 8.6, together with some scattered data for their distribution in major groups of animals. In this sequence, the purines adenine and guanine are first deaminated and then oxidized as shown in Fig. 8.7. Following the formation of uric acid, first one of the two ring components of the molecule is opened and then the other. From the resulting compound (allantoic acid) two molecules of urea are split from glyoxylic acid, and the

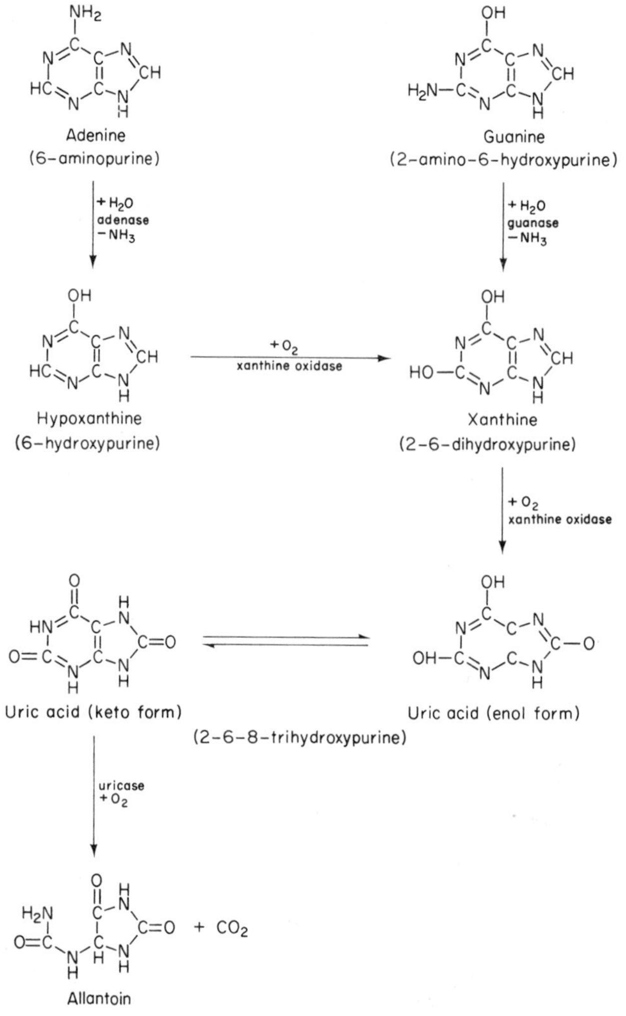

Adenine
(6-aminopurine)

$+H_2O$
adenase
$-NH_3$

Guanine
(2-amino-6-hydroxypurine)

$+H_2O$
guanase
$-NH_3$

Hypoxanthine
(6-hydroxypurine)

$+O_2$
xanthine oxidase

Xanthine
(2-6-dihydroxypurine)

$+O_2$
xanthine oxidase

Uric acid (keto form)

Uric acid (enol form)

(2-6-8-trihydroxypurine)

uricase
$+O_2$

Allantoin $+ CO_2$

Fig. 8.7 Uricogenesis from purine bases. Both the *keto* and *enol* forms of uric acid are shown, but only the *enol* forms of hypoxanthine and xanthine have been depicted. *Keto* forms occur in insects, reptiles, and birds; in these groups, the enzyme xanthine dehydrogenase replaces xanthine oxidase which predominates in mammals.

urea is then separated into CO_2 and NH_3. At each major step some groups of animals have been shown to lack the enzymes for subsequent steps and consequently they excrete the product formed at that stage.

The pathway from uric acid to ammonia is sometimes termed the URICOLYTIC PATHWAY. The presence of urea should be noted in the penultimate step; it indicates one of the routes by which urea may be formed in animals lacking the Krebs ornithine cycle. This biochemical sequence may serve as an important mechanism for the removal of amino nitrogen as well as purine nitrogen. As indicated in Fig. 8.4, uric acid contains nitrogen which arises from aspartate and glutamate. Thus, amino groups may be funneled in by these compounds (Figs. 7.19 and 8.3) and thence pass by urate, allantoin and allantoicate to urea (Fig. 8.6). This route for the synthesis of urea is known as the PURINE PATHWAY. It may be responsible for the synthesis of significant amounts of urea in some animals (Forster and Goldstein, 1969).

Miscellaneous End Products of Nitrogen Metabolism

Several additional nitrogen-containing compounds are normally found in the urine. FREE AMINO ACIDS have frequently been identified. They were long ago shown to form 15 per cent or more of the excreted nitrogen in some invertebrates (Prosser and Brown, 1961; Nicol, 1967a). Through the development of chromatographic techniques and the identification of trace amounts it has been shown that normal adult humans may excrete slightly over one gram of free amino acids per day, or about 1.2 per cent of the total urinary nitrogen (Fruton and Simmonds, 1958). It is now evident that all groups of animals are liable to excrete some free amino acid, but it is questionable whether this is a true excretion of waste material or whether it represents an unavoidable leakage of relatively small molecules; proteins of low molecular weight (less than about 68,000) as well as amino acids are regularly filtered through the mammalian glomerulus but mostly recovered by the tubules (Forster, 1961).

AMINO ACID CONJUGATES formed in several different detoxification processes may also account for a small part of the nitrogen excreted. Benzoic acid, for example, is toxic. It is formed in small amounts during the metabolism of fat and may also occur in foods. In the mammal it is detoxified through conjugation with glycine to form hippuric acid:

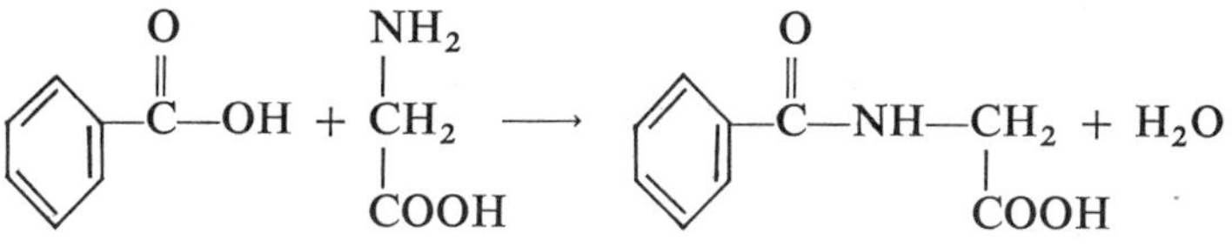

$$C_6H_5\text{—}\overset{O}{\overset{\|}{C}}\text{—}OH + \underset{COOH}{\overset{NH_2}{CH_2}} \longrightarrow C_6H_5\text{—}\overset{O}{\overset{\|}{C}}\text{—}NH\text{—}\underset{COOH}{CH_2} + H_2O$$

Benzoic acid — Glycine — Hippuric acid

In birds the benzoic acid conjugation occurs with ornithine and forms ornithuric acid.

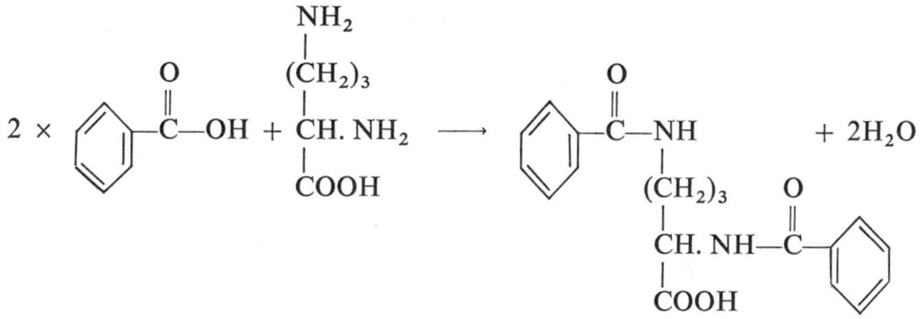

$$2 \times C_6H_5\text{—}\overset{O}{\overset{\|}{C}}\text{—}OH + \underset{COOH}{\overset{NH_2\text{—}(CH_2)_3}{CH.NH_2}} \longrightarrow C_6H_5\text{—}\overset{O}{\overset{\|}{C}}\text{—}NH\text{—}(CH_2)_3\text{—}\underset{COOH}{CH}.NH\text{—}\overset{O}{\overset{\|}{C}}\text{—}C_6H_5 + 2H_2O$$

Benzoic acid — Ornithine — Ornithuric acid

Both of these syntheses require ATP which can be considered the necessary price for safe removal of toxic materials. Cysteine may also be involved in conjugation reactions (with bromobenzene, for example), and there are several additional reactions, involving compounds other than amino acids, which produce detoxification products that appear in the urine (Harper, 1973).

CREATINE and/or CREATININE have been identified in the urine of different classes of vertebrates and in several invertebrates. Creatine plays an important role in the energy transformations of vertebrate muscles. In combination with phosphate (creatine phosphate, one of the phosphagens), it provides a ready store of high-energy phosphate during muscle contraction (Chapter 7). Arginine plays a comparable role among many invertebrates. Creatine is a very special amino acid—not one of the alpha group which enter into the composition of proteins but an end product of the metabolism of glycine, arginine and methionine found predominantly in muscle. If present in amounts in excess of the requirements, it is excreted chiefly in the anhydride form, creatinine.

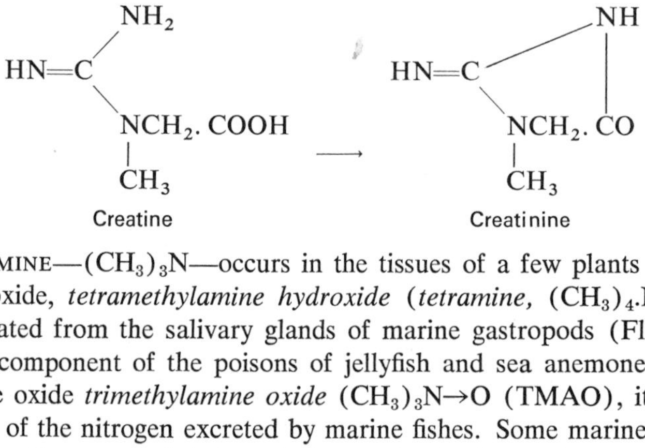

$$HN{=}C(NH_2)\text{—}N(CH_3)CH_2.COOH \longrightarrow HN{=}C\text{—}N(CH_3)CH_2.CO\text{—}NH \text{ (ring)}$$

Creatine Creatinine

TRIMETHYLAMINE—$(CH_3)_3N$—occurs in the tissues of a few plants and animals. As the hydroxide, *tetramethylamine hydroxide* (*tetramine*, $(CH_3)_4.N^+\ ^-OH$), it has been isolated from the salivary glands of marine gastropods (Florkin, 1966) and forms a component of the poisons of jellyfish and sea anemones (Needham, 1965); as the oxide *trimethylamine oxide* $(CH_3)_3N{\rightarrow}O$ (TMAO), it comprises a large fraction of the nitrogen excreted by marine fishes. Some marine teleosts turn out as much as 50 per cent of their total nitrogen in this form. In contrast, the tissue and urine levels of TMAO are very low in freshwater species. Anadromous fishes like salmon show a sharp rise in TMAO levels after migrating to the sea. The kidney tubules of cartilaginous fishes actively absorb TMAO as they do urea and, consequently, the tissue levels are high; the kidney tubules of marine teleosts, on the contrary, actively secrete TMAO so that the tissue levels are only about 20 per cent of those of elasmobranchs (Baldwin, 1967). This substance is thought to participate actively in osmotic regulation (Chapter 11 and Forster and Goldstein, 1969).

The origin of TMAO is obscure. The relatively large amounts excreted and its appearance in elasmobranch embryos (L. J. Read, 1968c) suggest active biosynthesis, but experiments designed to test this theory have given negative results. At present the consensus is that TMAO is mainly of exogenous origin in fishes and arises from the diet and/or intestinal microorganisms. The problem, hovever, is by no means resolved (Forster and Goldstein, 1969).

Phylogenetic Interpretations

The catabolism of protein accounts for 90 per cent or more of the excreted nitrogen. With relatively few exceptions this is predominantly in the form of ammonia or urea or uric acid. The significance of the dominant excretion product was first discussed in relation to animal phylogeny and water economy by Joseph Needham (1931, 1942). From his examination of a voluminous literature (mostly on embryonic tissues and excretion products) he concluded that "the main nitrogenous excretory product of an animal depends on the conditions under which its embryos live, ammonia and urea being associated with aquatic pre-natal life, and uric acid being associated with terrestrial pre-natal life." In particular, his studies of the problems of embryonic life in the three highly successful terrestrial groups of animals, the pulmonates, insects and vertebrates, convinced him that the terrestrial oviparous way of life would have been impossible without the uricotelic metabolism.

The disposal of metabolic wastes is as acute a problem in the embryo as in the adult and, from this angle, Needham sees three major avenues for the evolution of terrestrial forms: VIVIPARITY in which the embryo is, in fact as aquatic as its phylogenetic ancestors; a SEMI-AQUATIC EXISTENCE where the young are incubated in water or extremely moist places, as is true of modern amphibia and some reptiles (Chelonia); and the CLEIDOIC EGG, an impermeable box which must be provided with sufficient stored water to operate the metabolic machinery and dispose of the non-combustible wastes. The success of the pulmonates, the insects, and the birds attests the feasibility of the last arrangement.

The cleidoic egg starts its development with some free water held in the highly retentive colloids of the egg albumen and some metabolic water, primarily in the fats. This meager water supply is used with the greatest economy.

If it were possible to avoid the catabolism of protein, very much less water would be required, but amino acids must be turned over if cells are to live, differentiate and grow. The embryo, like the adult, is highly susceptible to an accumulation of the toxic ammonium ion which is the primary byproduct of amino acid catabolism. Highly insoluble uric acid which precipitates from saturated solution and can be retained in the solid form until hatching, provides the solution to this problem. In a sense, it is not an economical solution since uric acid synthesis depletes the carbon stores and requires ATP. However, it does solve the water problem and is a highly successful compromise, as the egg-laying land animals demonstrate.

Needham's arguments concerning uricotelism are acceptable to zoologists. The hypothesis fits many curious facts into an orderly pattern. One of these is the variety of materials used as sources of energy in different kinds of embryos. The cleidoic egg uses less protein and more fat, thus reducing the ammonia and increasing the water byproducts. A chick during incubation obtains about 6 per cent of its energy from protein and 80 per cent of it from fat, while a frog or fish during this period obtains over 70 per cent of its energy from protein and only 30 per cent from fat

(Needham, 1942). The terrestrial invertebrates follow similar patterns, and in several groups the degree of uricotelism follows closely the availability of water in the environment. The gastropods provide excellent examples since this group, like the vertebrates, has mastered all the habitats from the seashores to the dry hill tops. Patterns established in the embryo, when carried into adult life, permit some of the gastropods, the insects and a few vertebrate species to live in extremely arid habitats. The mammals, which are their only real competitors, are able to perfuse the developing young with maternal water while operating a greatly refined machinery for its conservation.

Needham emphasized the phylogenetic significance of uricotelism in the cleidoic egg. Baldwin (1964; 1967 and earlier) has considered more particularly the advantages of urea as an end product in habitats where water is restricted but not actually absent. In general, ammonia predominates in the excreta of the aquatic invertebrates, whether marine or freshwater; in the marine habitat invertebrate blood is isotonic with the sea water; in fresh water, it is hypertonic. Thus, in neither case is there a restriction in tissue water and, it is argued, the ammonia byproduct of amino acid metabolism diffuses or is carried way in a copious urine. The freshwater vertebrates also excrete large amounts of ammonia, but the marine vertebrates produce more urea (perhaps also trimethylamine oxide). This is attributed to the hypotonicity of their body fluids which creates a constant osmotic desiccation, so that the tissues are really operating in a dry environment. Thus, ammonia predominates as an excretion product when water is abundant throughout life; but when water is restricted, the ammonia is turned into the less toxic form of urea, again at the expense of ATP.

There are many biochemical data supporting this view. Some of the most significant have been found among the amphibians. The wholly aquatic amphibian *Xenopus* excretes large amounts of ammonia throughout its life; the toad *Bufo* turns out 80 per cent ammonia as an aquatic larva but only 15 per cent as a terrestrial adult. Moreover, the nature of the nitrogenous excretion product is sometimes altered in relation to the availability of water. When *Xenopus* is retained out of water, under experimental conditions, for periods of one to three weeks, the nitrogen excretion (ammonia) declines sharply; on return to the aquatic habitat the ammonia excretion is again normal, but there is a large additional urea nitrogen fraction which accounts for the nitrogen retained during the experimental period out of water. Urea is formed but not excreted, and Balinsky *et al.* (1961) argue that in phylogeny the capacity to store urea preceded the ability to excrete it. Urea biosynthesis in the emerging vertebrates was probably an emergency device for the first tentative journeys onto the land. In the more terrestrial of the amphibians, the capacity to excrete large amounts of nitrogen as ammonia is not apparent even when the animals are held in water; there is a change in the patterns of nitrogen excretion as well as in morphology during amphibian metamorphosis (Bennett and Frieden, 1962; Frieden, 1968; Balinsky, 1970a).

Source of Excreted Ammonia

Homer Smith (1953) challenged Baldwin's argument concerning ammoniotelic patterns of excretion and argued that urea is always the main end product of amino acid catabolism (at least in the aquatic vertebrates), and that the ammonia which is so abundantly excreted by freshwater fishes arises peripherally in the regulation of acid-base balance and the conservation of cations.

In this connection it should be noted that the ammonia in human urine arises in this manner—largely from the stored amino groups of glutamine. The mechanisms by which sodium ions in the tubular urine are exchanged for hydrogen ions, which then combine with NH_3 to form NH_4^+ in the regulation of pH, are shown in Fig. 8.8. The exchange of ions in the proximal tubules operates against the $NaHCO_3$ of the tubular filtrate, leading to the exchange of H^+ formed from carbonic acid in the

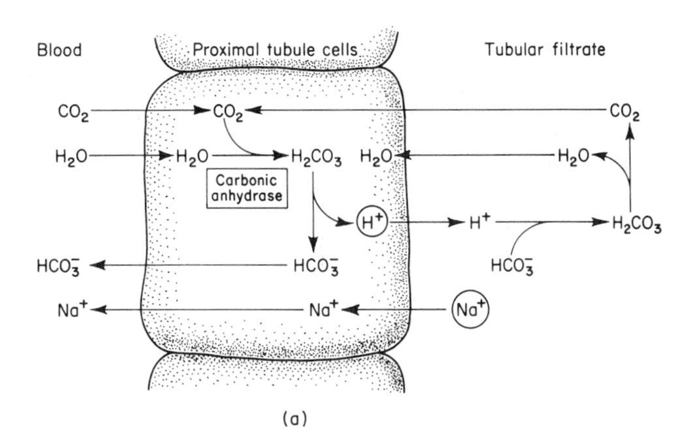

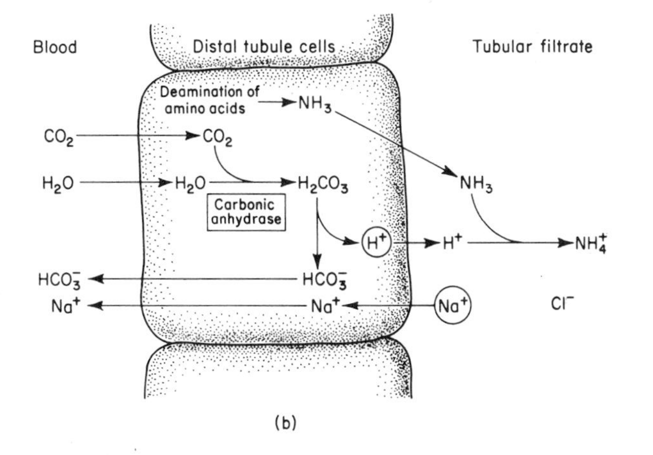

Fig. 8.8 Acid-base regulatory mechanisms in the vertebrate kidney by which hydrogen ions are eliminated and cations are conserved. (a) In the proximal tubule, the exchange first proceeds against sodium bicarbonate, while in the distal tubule a similar exchange (not shown) takes place against Na_2HPO_4 (after all the bicarbonate has been reabsorbed). (b) Ammonia, from the deamination of amino acids, forms ammonium ions which replace sodium of the sodium chloride to conserve this cation. [From Harper, 1973: Review of Physiological Chemistry, 14th ed. Lange Medical Publications, Los Altos, California.]

kidney cells for Na^+ in the urine (Fig. 8.8a). In this manner, Na^+ is returned to the blood and the urine becomes more acid. Not shown in Fig. 8.8, Na_2HPO_4 in the distal tubule may provide the Na^+; thus, the more acidic NaH_2PO_4 is formed for excretion. In the distal and collecting tubules, ammonia arising mostly from glutamine (Fig. 7.19) combines with H^+ in the urine to form NH_4^+ in a process that has been termed DIFFUSION TRAPPING. These several exchanges regulate the pH and conserve the important cations (Fig. 8.8b). Details of the tubular regulation of acid-base balance are discussed in many textbooks of biochemistry and physiology. Here, the important point is that the ammonia excreted in mammalian urine is of peripheral origin; Smith (1953 and earlier) suggested that the ammonia excreted through the gills and kidneys of fishes also arises peripherally. The classical view has also been challenged by Unsworth *et al.* (1969) and others working on amphibians.

A cellular cation exchange mechanism comparable to the diffusion trapping in the kidney has been described in the gills of freshwater fishes (Maetz and García-Romeu, 1964). This leads to the excretion of NH_4^+ and the acquisition of Na^+ as indicated in Fig. 8.9. The point at issue is whether the branchial ammonia is formed in the liver and kidney as in other vertebrates or whether it arises *de novo* in the gills. Branchial ammonia may comprise as much as 60 per cent to 90 per cent of the total nitrogen excreted by teleosts; physiologists have found it difficult to believe that all of this is formed in the gills. Forster and Goldstein (1969) have reviewed the pertinent research. It is currently believed that most of the ammonia excreted by the gills arises centrally through transamination and deamination in the liver (Fig. 7.19), with lesser amounts produced in the kidneys and other tissues, including

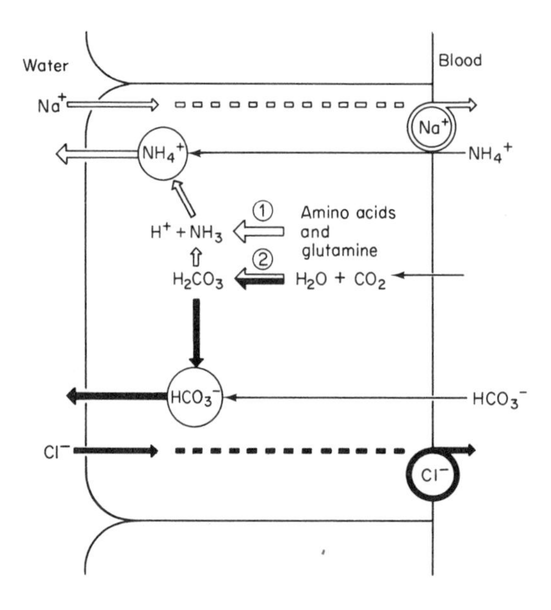

Fig. 8.9 Diagram to show ionic exchanges in a branchial cell of the goldfish. ENZYMES: (1) Deaminase and transaminase systems; (2) carbonic anhydrase. [From Maetz and García Romeu, 1964: J. Gen. Physiol., *47*: 1221. Permission Rockefeller University.]

the gills. The relative importance of branchial and extrabranchial sources of ammonia probably varies with the species and with environmental conditions (Forster and Goldstein, 1969; Goldstein and Forster, 1970).

Evolutionary adaptation follows many curious trails. One of the strangest of these concerns the diffusion trapping of ammonia by some pulmonate snails, followed by the evolution of NH_3 gas. The adaptive significance of this biochemical manipulation appears to be related to $CaCO_3$ deposition in the shell. As already indicated, ammonia nitrogen ($NH_3 - N$) is almost always removed in aqueous solution as NH_4^+ which passes either directly into the ambient water or is excreted in the urine. The extrarenal excretion of NH_3 gas is rare in the animal kingdom, although small amounts are eliminated by direct NONIONIC DIFFUSION from body surfaces and lungs (Hartenstein, 1968; Speeg and Campbell, 1968). In some pulmonate snails, the elimination is far in excess of that which would be formed by nonionic diffusion. Aestivating *Otalia* (*Helix*) *lactea* eliminate 9.6 mg $NH_3 - N$/kg/24 hr while the comparable amount for *Helix aspersa* is 5.5 mg. Speeg and Campbell (1968) have shown that this ammonia is derived from the amino acid arginine by way of urea:

$$\text{Arginine} \xrightarrow{\text{arginase}} \text{Urea} \xrightarrow{\text{urease}} \text{Ammonia} + CO_2$$

This gas is primarily liberated through the shell. The suggested mechanism for volatilization depends on the demonstrated presence of carbonic anhydrase in lung and mantle tissues but its absence from hemolymph. Thus, CO_2 arising in tissue metabolism or from urea will pass down a diffusion gradient into lung and mantle tissues to be trapped as carbonic acid (Fig. 8.10). Snail oxyhemocyanin is highly alkaline and the removal of CO_2 by the carbonic anhydrase mechanism will maintain or increase this pH. The tension of free ammonia gas—formed in the blood and tissues by urease—will rise with the alkalization of the blood; ammonia will also

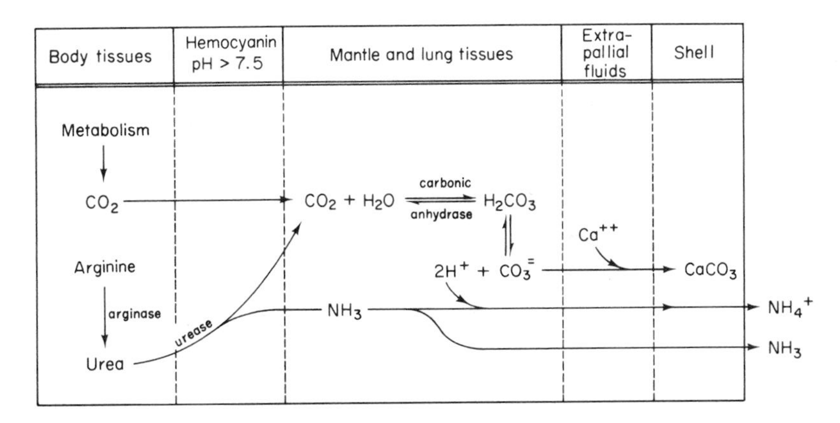

Fig. 8.10 Suggested mechanisms involved in the volatilization of ammonia by the snail *Helix*. [Based on discussion by Speeg and Campbell, 1968: Am. J. Physiol., *214*: 1392.]

diffuse down a concentration gradient into the lung and mantle tissues where it will be trapped by H^+ formed in the carbonic anhydrase reaction. This is another diffusion trapping mechanism like that in the mammalian kidney. The postulated exchanges are shown in Fig. 8.10. The subsequent reactions, as the bicarbonate and ammonium ions pass into the extrapallial space and shell, are assumed to be related to the formation of the shell which is primarily $CaCO_3$. Highly alkaline conditions are required for its deposition and maintenance (Speeg and Campbell, 1968). Ammonia may play a comparable role in the deposition of calcium carbonate in the eggshell of birds (Campbell and Speeg, 1969).

Pulmonate snails are not the only invertebrates which excrete ammonia gas (Hochachka and Somero, 1973). In particular, the terrestrial crustaceans which lack an epicuticular wax, are able to release ammonia directly through the exoskeleton. The isopods have been extensively studied (Hartenstein, 1970). Wieser (1972) terms this capacity "a part of their marine heritage." Several adaptations favor the direct excretion of ammonia by isopods (Wieser and Schweizer, 1970; Wieser, 1972): (a) nitrogen metabolism may be suppressed because of high carbohydrate (leaf mold, rotting wood) consumption; (b) nitrogen metabolism may be reduced during activity and may increase when the animals become quiet in moist places; (c) an alkalinity at the body surfaces due to hemocyanin and the calcified exoskeleton is also important.

ORGANS OF EXCRETION AND THE FORMATION OF URINE

The anatomical unit of the excretory system is most often a minute tubule. Only a few groups of animals, usually of small size, lack excretory tubules and dispose of their wastes through the general body surface or through contractile vacuoles or by phagocytosis. Excreting tubules may occur singly or segmentally in pairs, or they may be collected together into compact organs, often referred to by the non-specific term of KIDNEY. Their function is to move water and dissolved metabolic wastes from a tissue space, body cavity, or vascular fluid to the exterior. The collection of the excretory fluids (urine) is by FILTRATION under hydrostatic and osmotic pressure differences, or by CILIARY MOVEMENT of fluids from body spaces into open ciliated funnels, or by ACTIVE CELLULAR TRANSPORT from vascular fluids into the lumina of the tubules.

Without Special Excretory Tubules

No special organs of excretion have been identified in many of the Protozoa, in the Coelenterata, in most of the Porifera nor in the invertebrate Deuterostomia as a group (Hyman, 1951). Some Protozoa and Porifera have contractile vacuoles concerned primarily with water balance, and many deuterostomes, especially the larger ones, support an extensive system of phagocytic cells for the removal of wastes. However, a large part of the regulation in all these groups is by means of the body surfaces.

Cell membranes The to-and-fro movement of materials between the living cell and its surroundings is regulated by the plasma or cell membrane. This specialized surface layer controls many aspects of excretion whether or not the animal possesses distinct excretory organs. The transfer of substances through a cell membrane depends on both the structural characteristics of the membrane and the forces that drive or transport particular substances through it. Membranes may be freely permeable to certain substances but completely impermeable or semipermeable to others. The functional morphology of the plasma membrane is discussed in textbooks of cell physiology; here, comment is confined to general features of the transfer mechanisms as they relate to excretion.

In discussing food absorption (Chapter 3), a distinction was made between the forces of passive and active transfer. The excretion or removal of waste substances involves processes that are comparable to those of absorption even though they operate in the reverse direction. Historically, PASSIVE TRANSPORT has been attributed entirely to physical forces such as concentration gradients while ACTIVE TRANSPORT depends on forces peculiar to living systems. Although this simple distinction remains useful, critical biophysical analyses have shown its inadequacies and, particularly, the limitations of most definitions proposed for active transport. Relevant discussions will be found in current reviews (Stein, 1967; Curran and Schultz, 1968; Dowben, 1969; Levin, 1969).

One of the most widely quoted definitions of active transport was proposed by Rosenberg (1948) who defined it as the "transfer of chemical matter from a lower to a higher chemical (in case of charged components: electrochemical) potential." It was noted in the preamble to this definition that the transfer of matter was not spontaneous but required the coupling of metabolic processes. The two major concepts emphasized were "uphill" transfer and dependence on intracellular energy-producing processes. Although Rosenberg's definition is not without its difficulties (Curran and Schultz, 1968), these two basic concepts (after a quarter of a century) still remain significant in all definitions.

Molecular mechanisms involved in active transport are speculative. Several different models have been proposed (Bresnick and Schwartz, 1968; Levin 1969). These postulate specific carrier molecules or active binding sites on the cell membrane which combine with the transported material on one side of the cell membrane and then pass through the membrane to be released on the other side. Substances carried through the membrane in this manner then diffuse away from the membrane while the carrier system is reactivated and passes back to transport more molecules. These processes are ATP-dependent and require specific enzymes; the inhibiting effects of respiratory poisons are often cited as evidence of active transport.

Excretion, as well as many other physiological processes, is frequently linked with the flow of water across cell membranes. Although considerable experimental evidence supports the concept of carrier-mediated transfer of ions such as Na^+ and

Cl^- and of non-electrolytes such as sugars and amino acids, no such mechanisms of active transport have been demonstrated for water molecules. At present, the primary transport of water, involving specific energy-utilizing carrier molecules, is considered unlikely; it is believed that water transfer from lower to higher concentrations is always secondary to the transport of solute (Beament, 1964, 1965; Stein, 1967; Phillips, 1970). It must be emphasized, nevertheless, that there are many situations in which water moves "uphill" from lower to higher concentrations and that the mechanisms involved are energy-dependent. This satisfies one of the definitions of active transport. On detailed analysis, however, the mechanisms have been shown to depend on both the architecture of complex cell membranes and the coupling of water movement with the active transport of solute.

One of the classical analyses is based on studies of the gall bladder (Diamond and Tormey, 1966a,b; Davson, 1970). The bile, which is secreted continuously by the hepatic cells, is an isotonic solution containing sodium chloride, cholesterol, bile acids, mucin, pigments, and fat. While stored in the gall bladder, it is concentrated some ten times by rapid absorption of water and sodium chloride; epithelial cells of the gall bladder in the rabbit may absorb as much as 10 ml H_2O/gm/hr. The mechanism, shown diagrammatically in Fig. 8.11, depends on the active transport of salt and the presence of long, tortuous channels between the epithelial cells. These minute channels are closed at the luminal end and gradually widen to a diameter of about 1 μ on the connective tissue side when water is flowing through them; at other times they collapse. On the luminal side, salt is pumped from the cells into the intercellular channels, filling their blind ends with hypertonic fluid; in consequence, water enters osmotically from the gall bladder and the solution becomes diluted as it moves along the gradually widening channel as a result of the inflowing water and diffusion down a concentration gradient. A similar solute-linked flow of water probably occurs in some other vertebrate membranes of the intestine and in the kidney (Schmidt-Nielsen and Davis, 1968; Berridge and Oschman, 1972); another example will be considered in discussing the formation of hypertonic urine by insects (Phillips, 1969).

Cell membranes have been shown to differ markedly in their water permeability. This variable feature of membrane architecture is a significant factor in water balance and excretion at all levels in animal phylogeny. Some are almost or completely watertight. Both marine and freshwater teleost eggs, for example, maintain an osmotic independence after fertilization; this evidently is caused by the impermeability of the vitelline membrane. This membrane, although impermeable to water and solutes, permits the ready diffusion of respiratory gases and ammonia. Varying degrees of membrane permeability have been measured in different tissues and in animals ranging from the Protozoa to the Vertebrata (Prosser, 1973).

Contractile vacuoles These are organs of water balance found in many Protozoa and in some of the freshwater sponges. Because of technical difficulties, there are few direct analyses of their contents. Schmidt-Nielsen and Schrauger (1963) were the first to analyze vacuolar fluids obtained by micropuncture.

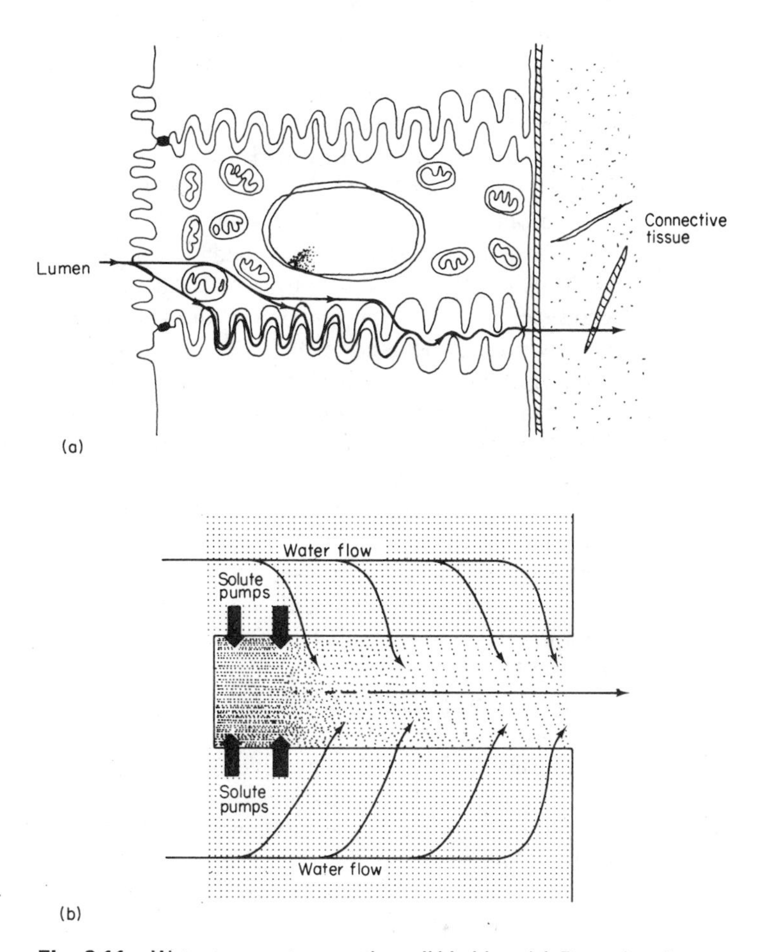

Fig. 8.11 Water transport across the gall bladder. (a) Route from lumen of bladder through intercellular channels; (b) model of fluid transport. Solute is pumped into the closed end of a long narrow intercellular channel, making it hypertonic and pulling in water osmotically; the emerging solution is isotonic because of this gradual dilution. [From Diamond and Tormey, 1966: Fed. Proc., *25*: 1461.]

Measurements of the freezing points showed the osmotic pressures of vacuoles in *Amoeba proteus* to be 32 mOsm/liter in comparison with values of 101 mOsm/liter for the cytoplasm and 6 mOsm/liter for the culture medium. Riddick (1968), using similar techniques, obtained comparable values for the giant amoeba *Pelomyxa carolinensis*. These findings are in agreement with indirect estimates, all of which indicate that the vesicles contain very dilute fluids (Kitching, 1952, 1967). Contractile vacuoles, or water expulsion vesicles, operate as pumps to remove osmotic and metabolic water. The usual arguments for an osmoregulatory function are based on the tonicity of the habitats of the animals which contain vacuoles, on the effects of experimentally changing the tonicity and on the action of metabolic poisons on vacuolar activity. Vacuoles are common in freshwater Protozoa and frequently absent from marine and brackish water species; in freshwater species, vacuolar activity is decreased in proportion to increases in the tonicity of the culture medium (consequently in proportion to the reduction in osmotic flooding); in marine species, vacuolar output increases with dilution of the sea water. Cyanide, which inhibits the cytochrome system and prevents oxidative phosphorylation, stops vacuolar activity and leads to rapid swelling of the cell. These observations are consistent with the hypotheses that the main function of the vacuole is to eliminate osmotic water and that this process demands metabolic energy. Quantitative support for this concept has been obtained by measuring the water permeability of the cell membranes and relating this to the volume output of the vacuole. Satisfactory agreement between the two values has been obtained in several species (for example, *Amoeba* and *Pelomyxa*, Shaw, 1960). Any excretory role which the contracile vacuole may play in the removal of nitrogeneous wastes is probably secondary. Ammonia, the major end product of nitrogen metabolism in the Protozoa, diffuses readily into the surrounding water through the cell membrane. The vacuole is essentially an organ of water balance. In addition, it probably cooperates with the cell membrane in the regulation of electrolytes and may play a specific role with respect to certain ions (Kitching, 1967; Potts, 1968).

There is much less certainty regarding the mechanisms of filling (diastole) and discharge (systole). It is clear that the processes require metabolic energy, probably in the form of ATP. Dilute cyanide suppresses vacuolar activity; cytological studies have revealed rich concentrations of mitochondria or osmiophil substance in close proximity to the vacuolar membrane. Evidently, processes of active transport are involved, but it has thus far been impossible to show just what is being transported. Many theories have been advanced. These include an active transport (secretion) of water into the vacuole, an active secretion of solute with water diffusing osmotically, and the formation of vacuolar membranes around cytoplasmic fluid with subsequent transport of solutes back into the cytoplasm. Many physiologists are not convinced that active transport of water is ever possible and, consequently, a mechanism based on the transport of solute or its enclosure in cytoplasmic membranes with a secondary diffusion of water is favored. Electron microscopy has provided evidence that vacuoles in *Amoeba proteus* are formed by coalescence of minute vesicles whose membranes may be derived from the endoplasmic reticulum. These vesicles (0.03 to 0.1 μ in diameter) contribute their membranes (about 60 Å thick) and their contents to the vacuole (Mercer, 1959).

Riddick (1968) isolated contractile vacuoles from giant amoebae and studied their contents and osmotic reactions in heterotonic solutions. He believes that the minute vesicles when first formed contain fluids isosmotic with the cytoplasm and that these small vesicles fuse, subsequent to the removal of solute (potassium extrusion in excess of sodium accumulation); the action of the cation pump is coupled with an extremely low water permeability of the vacuolar membrane. Similar mechanisms may be involved in other protozoans. Electron micrographs of several genera (*Paramecium, Tetrahymena, Zoothamnium*) reveal complex systems of tubules which are connected with the endoplasmic reticulum and empty into the contractile vacuole through a clearly defined system of canals (Fig. 8.12). Thus, water separation takes place over large areas of intracytoplasmic surface and appears to involve the formation of minute droplets along the tiny tubules; presumably, cation pumps reduce the tonicity of the fluids as they move toward the feeder canals and ampullae (Fig. 8.12). Organ *et al.* (1969) examined this system by high-speed cinematography and described waves of contraction in the coacervate cytoplasmic gel surrounding the ampullae. Fluid is pumped into the major vacuole by these cytoplasmic movements that press around the ampullae from distal to proximal ends.

The discharge mechanism is also uncertain. Theoretically, the contractile force might reside either in the vacuolar membrane or in the cytoplasm of the cell. Earlier workers favored the former view (Kitching, 1954), but recent evidence argues against it. Although electron micrographs of some protozoans have revealed delicate myofibrils winding over parts of the vacuole and its associated ampullae (Fig. 8.12), these fibrils have not been found in all groups. Moreover, some workers feel that fibrils, when present, provide support for the system rather than the contractile force associated with water expulsion. High-speed cinematographic studies of *Amoeba* (Wigg *et al.,* 1967) and *Paramecium* (Organ *et al.,* 1968) indicate that the major contractile force resides in movements of the endoplasmic "gel" which pushes the vesicle against the plasmalemma, bulging it until it ruptures. Fluid then rushes out; the vesicle collapses and inverts into itself; it does not contract. Although the action of contractile fibres may contribute to water expulsion in protozoans such as *Paramecium,* the collapse appears to be mainly related to pressure from the cytoplasm. Jahn and his colleagues (Organ *et al.,* 1968) therefore argue that this structure

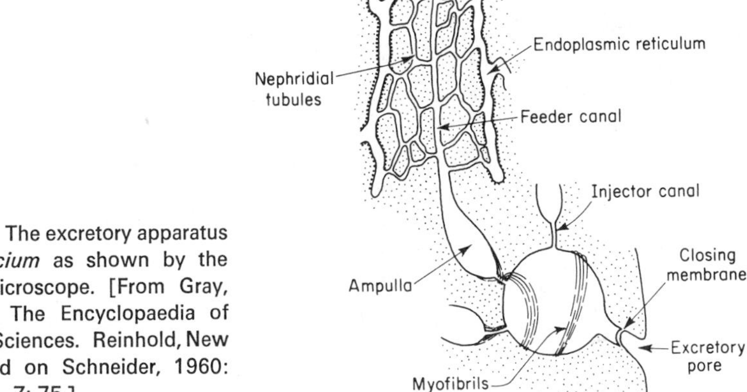

Fig. 8.12 The excretory apparatus of *Paramecium* as shown by the electron microscope. [From Gray, ed., 1961: The Encyclopaedia of Biological Sciences. Reinhold, New York. Based on Schneider, 1960: J. Protozol., 7: 75.]

should be termed the WATER EXPULSION VESICLE since it is neither a "vacuole" nor "contractile"; for similar reasons, they argue that the terms "systole" and "diastole" as applied to this structure should be abandoned.

Finally, it should be noted that some freshwater Protozoa lack contractile vacuoles. In these, as in *Hydra* and a number of other small freshwater animals, the role of water and electrolyte balance must reside in the cell membrane. Small size in itself confers no advantage against osmotic flooding. As linear measurements change, the surface per unit volume alters in the inverse order. Thus, surface areas are relatively greater in smaller forms, and unless appropriate changes in permeability take place, the small freshwater animals have relatively more water to bail out than the large ones. Evidently the plasma membrane alone, either through limited permeability or active transport mechanisms, can cope with this problem; Kitching (1954) suggests that the evolution of the contractile vacuole became necessary with the specialization of the cuticle. The vacuolar membrane probably has essentially the same properties as the plasma membrane. It may be considered an internal plasma membrane which takes care of secretion and possibly excretion. The highly specialized cuticle of many Protozoa made its evolution a necessity.

Storage Excretion

A system of cells for the temporary or permanent storage of wastes is a rather constant feature of all multicellular animals. These either pick up particulate matter phagocytically or elaborate excretory compounds such as guanate and urate granules. The phagocytic (reticulo-endothelial) system was described in Chapter 5; it was noted there that both wandering amoeboid cells and fixed tissues operate to remove foreign material.

Wandering phagocytic cells concerned with the removal of particulate wastes are universal, and in some animals this seems to be the only component of the phagocytic system. This is true of tunicates and echinoderms. These animals are unusual among the more bulky species in that they lack special excretory organs and depend only on the surface epithelium and a colony of wandering phagocytes. Since they have not penetrated fresh waters, their physiology has never demanded a pump to take care of osmotic flooding. Without terrestrial tendencies the problems of salt balance and the disposal of nitrogenous wastes are adequately handled by permeable body surfaces with no specialized areas of excreting epithelia.

Most groups of animals have fixed tissues concerned with storage excretion as well as the wandering amoeboid cells. In a number of the Turbellaria, for example, ATHROCYTES or PARANEPHROCYTES are found singly or in clusters near to, or wrapped around, the nephridial tubes (Hyman, 1951). They take up dyes and are assumed to be "excretory." The "urate cells" of insects are similarly fixed and seem to elaborate the materials which they accumulate. They are often conspicuous in the fat body and, in some insects, the granules gradually accumulate as the animal grows and seem never to be discharged (Wigglesworth, 1972). Malpighian tubules are usually absent in insects which store uric acid in the fat body. Spiders have a well-organized system for storage excretion in addition to the Malpighian tubules. Certain intestinal cells in the region of the hypodermis fill with guanates which are later discharged in liquid form into the lumen of the gut and pass into a cloacal pocket where the water is again removed and the guanates are eliminated as crystals (Kerkut, 1963). Some ascidians possess well-organized storage kidneys that accumulate concretions containing high concentrations of urates; in the renal sacs of the Molgulidae, symbiotic fungi apparently make use of the accumulated excreta (Nicol, 1967a). The small carnivorous land snail *Mesomphix vulgatus* stores guanates and does not excrete any nitrogenous waste during its nine to fifteen month life cycle (Badman, 1971).

Excretory Tubules

Comparative anatomists have encountered some of their most perplexing problems in studies of the structure and phylogeny of the excretory tubules (Goodrich, 1945). Not only are the tubular channels varied, but they are often intimately associated with the genital ducts, both structurally and functionally. The simultaneous evolution of these two discharge systems has produced many anatomical variations and compromises. It seems likely that the functional arrangements differ less than the anatomical details. For present purposes, the many morphological types will be disregarded and the tubules will be discussed in three groups as follows:

1. *Nephridia*—primitive excretory tubules characteristic of most of the Protostomia and subdivided into *protonephridia* and *metanephridia*.
2. *Malpighian tubules*—found in the Myriapoda, Insecta (except Collembola, some Thysanura and Aphidae) and Arachnida other than *Limulus*.
3. *Vertebrate nephron*—phylogenetically distinct from the invertebrate nephridial tubes.

Nephridia The nephridial tubule or the nephridium is the excretory organ of most of the invertebrates. These tubules are found as many separate units throughout the body (as in the flatworms) or serially as a pair of tubes in each segment (as in the annelids) or as a single pair of tubes (as in the crustaceans and molluscs); but they are never gathered together to form compact organs like the vertebrate kidneys. Their walls are single-layered epithelia which often show evidence of being capable of active transport. They may be assumed to secrete materials into the tubular fluids or extract materials from them. Tubules which are closed at the inner end are called PROTONEPHRIDIA; those which open into the coelom by a ciliated funnel, called the nephridiostome or the nephrostome, are termed METANEPHRIDIA. Both types of tubule open to the outside through nephridiopores and frequently enlarge into a storage (urinary) bladder just before they discharge.

Protonephridia are considered more primitive. In the acoelomate and pseudocoelomate groups, FLAME BULBS form the proximal ends of the system of branching tubules in the parenchyma. These bulbs (sometimes single cells, sometimes multinucleate or multicellular structures) are cup-shaped terminations containing tufts of cilia which are assumed to propel fluids through the minute tubules and perhaps also aid in filtering solutions from the surrounding lymph (Fig. 8.13). The ultrastructure has been described (Braun *et al.,* 1966; Clement, 1968; Berridge and Oschman, 1972). In coelomate invertebrates which utilize protonephridia (trochophore and related larvae, some adult polychaetes and *Amphioxus*), the flame bulb is replaced by a SOLENOCYTE. Transitional forms between the two types of termination are found in gastrotrichs, and the solenocyte is evidently a variation of the flame bulbs (Hyman, 1951). The solenocyte cell has a rounded body with a single very long flagellum beating in an exceedingly minute and thin-walled tubule which is attached to the main nephridial system (Fig. 8.13). Solenocytes are usually grouped in packets, sometimes called glomeruli, around the ends of the major nephridial tubes. Further, the bulb-shaped cells are exposed to coelomic fluid and sometimes also applied to the thin wall of a blood vessel. This is obviously an ideal arrangement for filtration or active transport from body fluids, and the possible functional significance of this intimate association of solenocytes and vascular fluids is emphasized. Flame bulbs are similarly bathed in fluid; in the Nemertea they also are closely applied to the blood vessels. Thus, in some of the most primitive animals which possess a distinct system of blood channels there is the possibility of associated excretory filtration. Microanalyses of the nephridial and body fluids of the rotifer *Asplanchna priodonta* have provided direct evidence for protonephridial filtration followed by selective reabsorption (Braun *et al.,* 1966). The findings are consistent with the hypothesis of filtration at the flame bulb, propulsion of fluids by flagella, with reabsorption and secretion by the epithelia of the nephridial tubules. These physiological mechanisms are comparable to those found in the most advanced kidneys (Kirschner, 1967).

Whereas a coelom is sometimes only an adjunct to the protonephridial apparatus, it always forms a physiologically essential partner of the metanephridium. Metanephridial tubes are open at both ends, and the collecting funnel is in the coelom where waste products collect and from which they are wafted by the ciliated nephrostomes into the nephridial tubes. In annelids the coelom is extensive, and there is typically a pair of nephridial tubes in each of its segmental compartments. In molluscs the coelom is represented only by the pericardial cavity and the cavities of the kidneys and gonads. The pericardial coelom communicates with the renal coelom (renal tube) through the renopericardial canal, and the associated structure forms the kidney of the mollusc (Fig. 8.14). In those arthropods which depend on nephridial tubes (segmental tubes in the Onychophora, antennary and maxillary glands in the Crustacea and coxal glands in the Arachnida), the coelom is reduced to a small, thin-walled sac attached to the proximal end of the excreting tubule (Fig. 11.6). In all cases, solutions first formed in the coelom move slowly through a long tubule to the point of discharge, and both coelom and tubule are involved in regulating the composition of the urine.

These are broad generalizations, and some adult animals have a highly modified

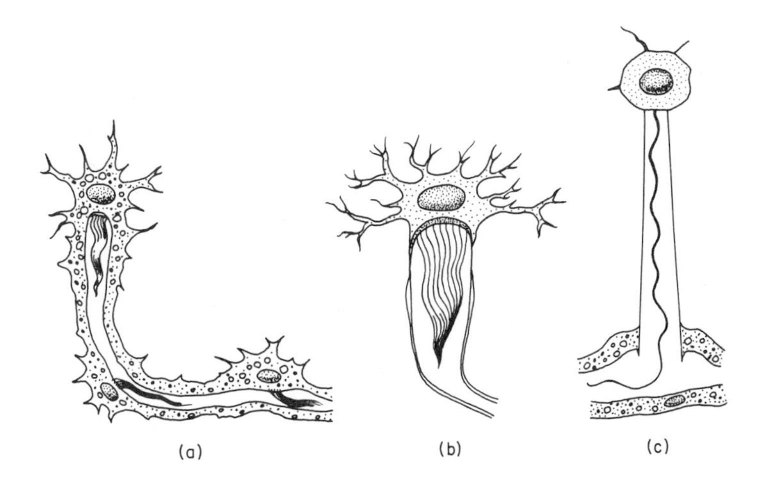

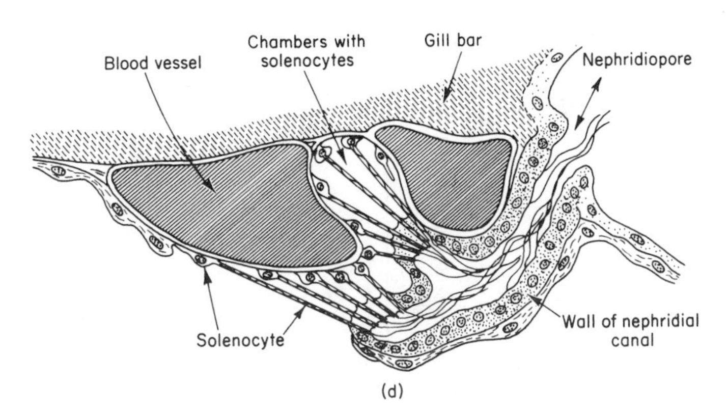

Fig. 8.13 Protonephridial apparatus. (a) Flame bulb of a polyclad turbellarian. (b) Flame bulb of a cestode. (c) Solenocyte of a polychaete. (d) Transverse section through the nephridium of *Amphioxus* showing solenocytes in close proximity to blood vessels and the glandular wall of the nephridial canal. Further description in text. [From Goodrich, 1945: Quart. J. Micr. Sci., *86*: 121, 363.]

metanephridial apparatus with secondary closure of nephrostomes and internal openings of the nephridiopores. In the Oligochaeta, for example, Bahl (1947) describes a series of modifications associated with dry terrestrial habitats. Nephrostomes are lost, and the tubules must operate only by secretion and osmotic differences like Malpighian tubules of insects and aglomerular kidneys of teleosts. In some oligochaetes, secondary openings drain the tubules into the gut (enteronephric) where water is reabsorbed. Annelids with intestinal enteronephric systems live in hot, dry climates, and the arrangement forms one of their adaptations for water economy (Bahl, 1947). The enteronephric tubules of annelids find their parallel in the cryp-

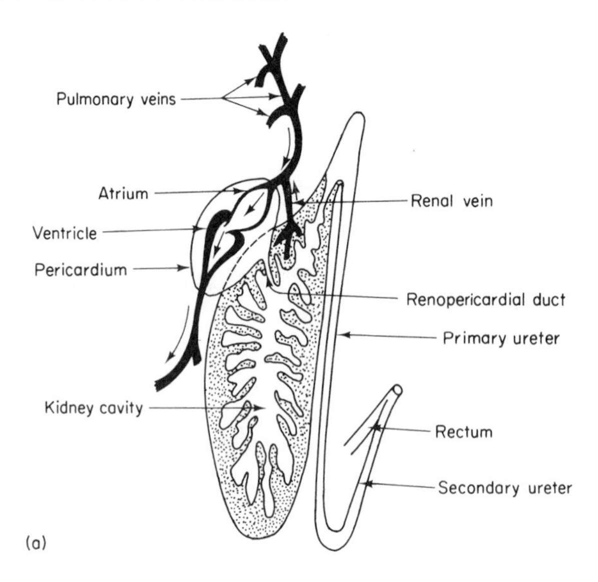

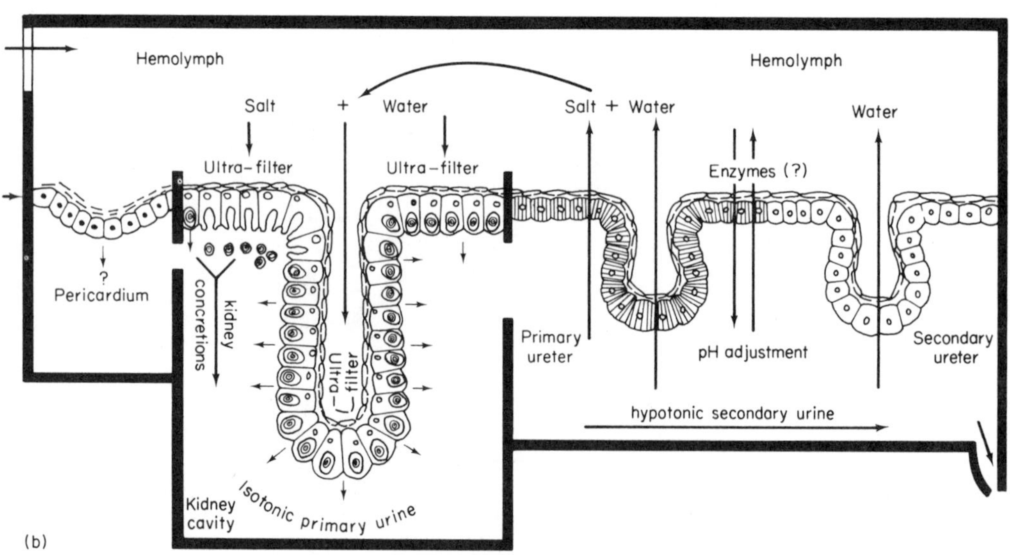

Fig. 8.14 Excretion in a land snail such as *Helix pomatia*. (a) Anatomical relationships based on several sources (see Plate, 1891; Gostan, 1965). (b) Suggested physiological mechanisms. [Based on Vorwohl, 1961: Z. Vergl. Physiol., *45*: 44.]

tonephridial tubes of certain insects (page 304) and the two appear to be an evolutionary convergence.

Urine formation in the metanephridial apparatus has been investigated in a number of the larger invertebrates. Both filtration and active transport are involved (Martin, 1957; Kirschner, 1967). The physiological techniques are those first used with success in studies of vertebrate urine formation. With delicate cannulae, fluids are removed from different areas of the tubular apparatus, and the composition and osmotic pressure are compared with blood or other body fluids. Sometimes the hydrostatic pressures in these different compartments can also be measured by micropuncture.

Considerable information about the nature of the kidney processes can be adduced from the relative concentrations of different substances in urine and blood. If the ratio of the concentrations (U/B ratio) is unity, then the substance is being excreted in direct proportion to its concentration in the body fluids, and the kidney shows no evidence of actively regulating the output either by secretion (U/B ratio greater than unity) or by reabsorption (U/B ratio less than unity). This is well illustrated with U/B ratios for several electrolytes in a small shore crab *Hemigrapsus oregonensis,* maintained in waters of different salinities (Fig. 8.15). Sodium, potassium and calcium values are almost the same in urine and blood, indicating a lack of regulatory activity with respect to these ions. Magnesium, however, becomes more and more concentrated in the urine as the salinity of the environment (hence also the hemolymph concentration) increases (Dehnel and Carefoot, 1965). Clearly, the kidney of *Hemigrapsus* excretes magnesium actively to regulate the blood level of this ion but has little control over the levels of the other three electrolytes.

Such techniques can demonstrate the nature of the excreting mechanisms but do not provide quantitative data for the various rates of filtration, reabsorption and secretion. For this, renal physiologists often use the "plasma clearance" test. A non-toxic chemical is injected into the blood and subsequently the amount of this substance is determined simultaneously in blood and urine.

Plasma clearance may be defined as the volume of plasma cleared of a given substance per minute *or* the volume of blood or plasma which contains the amount of a particular substance excreted by the kidney in a minute. If P is the plasma concentration, U the urine concentration, and V the rate of urine flow in ml/min, then

$$\text{Clearance (ml/min)} = \frac{U \times V}{P}$$

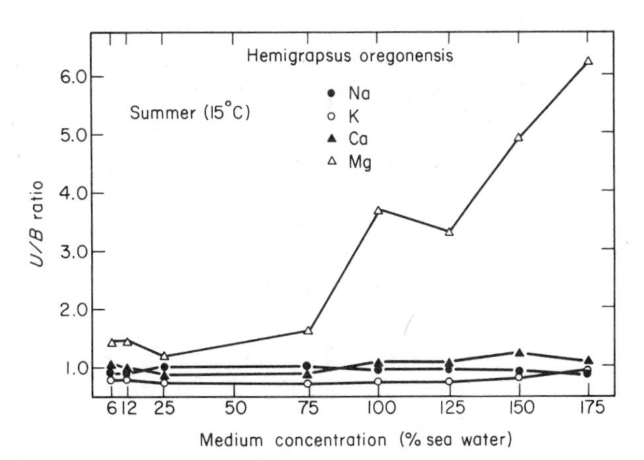

Fig. 8.15 Relationship between the environmental salinity and the urine-blood (U/B) cation ratios in the shore crab *Hemigrapsus*. [From Dehnel and Carefoot, 1965: Comp. Physiol. Biochem., *15*: 390.]

Thus, if urine is being formed at the rate of 100 ml/hr and found to contain 5 mg/ml of the injected material while the plasma contained 0.2 mg/ml, then the clearance is 42 ml/min, and this means that 42 ml of plasma would have to be filtered or cleared of the substance every minute to provide the quantity of the substance found in the urine. Since the kidney is seldom, if ever, capable of truly cleansing the plasma of any given substance which passes through it, "clearance" is actually the amount of plasma needed to provide the quantity of substance found in the urine.

The usefulness of clearance studies was long ago demonstrated in renal physiology. Vertebrate physiologists found that inulin, a polysaccharide of molecular weight about 5000 was excreted only by filtration. The evidence was based on comparisons of its concentration in the plasma and glomerular filtrate of the frog, its inability to pass into the urine of aglomerular fishes and, more directly, from experiments where tubules were blocked with oil droplets or where tagged inulin was used. In the vertebrates, at any rate, inulin clearance is a good measure of glomerular filtration and substances which are secreted as well as filtered will have correspondingly higher clearance rates, while substances which are reabsorbed will have lower ones. Different rates can thus be quantitatively evaluated.

The techniques of micropuncture and clearance have demonstrated both filtration and active transport in the metanephridial tubule as well as in the vertebrate nephron. For example, inulin has been successfully used to demonstrate filtration in the end sac of the antennal gland of the crayfish (Riegel and Kirschner, 1960), while microanalyses of fluids from different parts of the tubule of the gland show that chloride is reabsorbed as the urine passes along it, resulting in a hypotonic urine and the conservation of salt (Fig, 11.6). Again, Martin (1957) and his coworkers perfused the closed vascular system of the octopus with inulin and later found urine-to-blood ratios of approximately one, indicating a filtration of this material in the octopus. This theory was confirmed when poisons (such as phlorizin and dinitrophenol), known to inhibit active transport, had no effect on the inulin clearance, although they did alter the clearance of other materials. Reabsorption of materials such as glucose and the secretion of compounds such as phenol red were also shown in the octopus by the classical methods of renal physiology; these same techniques have proved fruitful in studies of other molluscs and the larger crustaceans (Martin and Harrison, 1966; Potts, 1967).

The main source of the filtration pressure in the molluscs and arthropods is the hydrostatic pressure of the blood. This will be counteracted to a small degree by osmotic pressure differences between the blood and the coelomic fluids as well as by the hydrostatic pressure in the nephridial apparatus. The wall of the coelomic sac is highly vascular in the crustacean (Fig. 11.6), while in the mollusc the heart actually passes through the filtration cavity or pericardial sac (Fig. 8.14). The filtering membranes differ in various groups but, in general, blood pressure forms the driving force, and the coelomic space is located near the heart or in a region of high blood pressure. Active transport is dependent on the tubules which drain the coelomic space. The mechanisms have been less satisfactorily demonstrated in the annelids because of their small size and the problem of collecting samples of blood and urine. However, filtration seems possible through the nephridial capillary network or through the vascular wall of the coelom; the tubules are certainly capable of active transport (Martin, 1957; Laverack, 1963; Kirschner, 1967).

Malpighian tubules In the most successful terrestrial arthropods, the duties of excretion have been taken over by the gut and the tubular glands of Malpighi which discharge into the posterior portion of it. Usually, no vestige remains of the nephridial system, characteristic of the aquatic invertebrates (Goodrich, 1945). Rather, it is replaced by a system which requires neither the forces of blood pressure for filtration nor the coelom for the collection of fluid wastes. The relegation of excretion to the alimentary tract is associated with the novel mode of gas exchange in the terrestrial arthropods; the evolution of the Malpighian tubules may have been a response to the reduced importance of the circulatory system. When gas exchange depends on efficient fluid transport, then the excretory machinery seems always to be coupled to it and to operate, at least in part, by the pressures set up in the contraction of the heart. In tracheates, the pressure of the hemolymph is extremely low and thus one of the major forces for urine production in many other groups of animals is absent. The glandular lining of the hind-gut takes over the responsibilities for removal of metabolic wastes and the conservation of essential electrolytes. Tubular glands collect and transport solutions (isosmotic) from the hemolymph into the hind-gut where water and physiologically important compounds are absorbed by the specialized hindgut epithelium. Malpighian tubules and hindgut operate together to remove the wastes and, sometimes, to conserve water. The dynamics of this system were beautifully demonstrated in mosquito larvae and in the stick insect, *Dixippus,* by Ramsay (1958 and earlier) who developed novel micromethods for collecting tubular and gut fluids and for determining their osmotic pressures and chemical constituents (Craig, 1960).

Ramsay found that the tubules accumulate potassium from the hemolymph with which they are bathed (Fig. 8.16). This active cellular transport of potassium (against an electro-chemical gradient) is probably a fundamental feature of tubular activity and represents the major metabolic work of their epithelia. In the stick insect, the concentration of potassium in the tubular liquid may be ten times greater than in the hemolymph (Ramsay, 1955). This secretory process seems to be the prime mover in generating the flow of urine. It leads to the diffusion of water and low molecular weight solutes, such as inorganic salts, sugar and urea, into the tubules (Ramsay, 1956). By means of tracer techniques Ramsay followed the movements of ions and organic molecules into isolated tubules from drops of fluid which bathed them. Several other insects have now been examined and shown to depend on mechanisms comparable to those described for the stick insect; Phillips (1970) has summarized the fluxes of ions and water for the desert locust in Fig. 8.17.

In general, then, the mechanism of excretion by Malpighian tubules depends on the steady movement of an isosmotic solution of soluble low molecular weight substances from the hemolymph into the hindgut where there is a recovery of essential compounds and an elimination of the wastes. This is an active circulation of fluids;

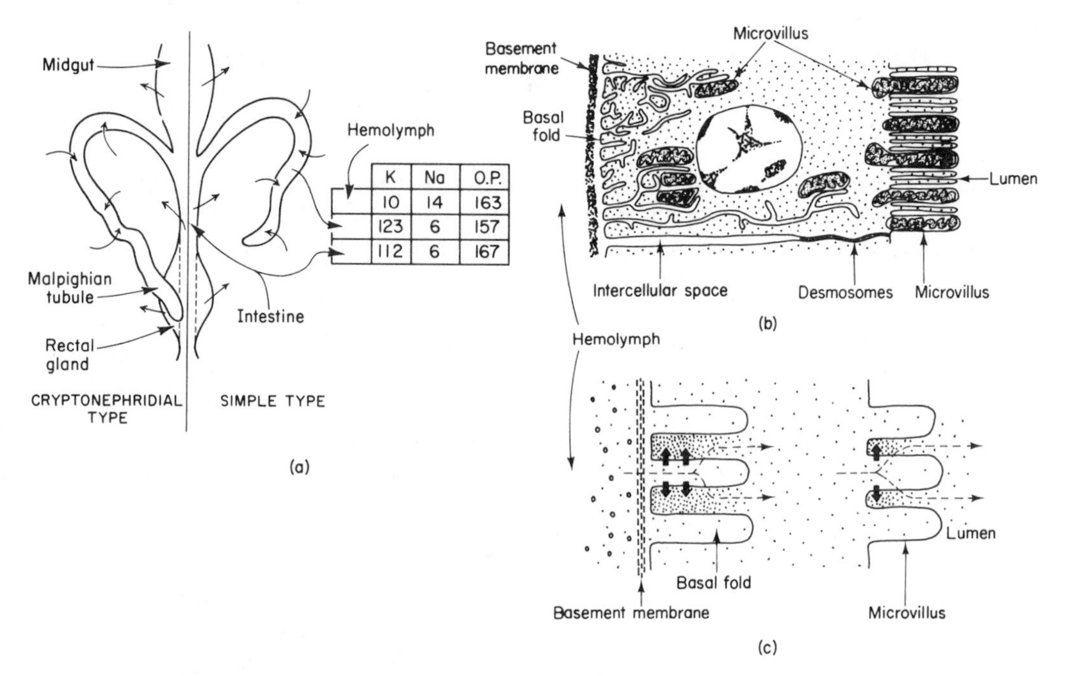

K	Na	O.P.
10	14	163
123	6	157
112	6	167

Fig. 8.16 Malpighian apparatus. (a) Gross morphology with small arrows to show direction of liquid flow; values for potassium and sodium in m equiv/liter and osmotic content (O.P.) in mM. liter; data are for *Dixippus* from Ramsay (1955). (b) Ultrastructure of a primary cell in the wall of the Malpighian tubule. (c) Mechanism responsible for flow of liquid into the tubule; osmotic gradients (heavy stippling) created by ion uptake (solid arrows) draw water (broken arrows) through the basement membrane and cell by osmotic filtration. [From Berridge and Oschman, 1972: Transporting Epithelia. Academic Press, New York.]

the tubular contents are steadily flushed into the hindgut where the appropriate amount of water and physiologically important substances are reabsorbed before discharge. Minute intrinsic muscles are often present in the tubules, and these agitate them in the hemolymph to improve the collection of solutes and perhaps also to facilitate circulation of the solutions within them (Roeder, 1953). The reabsorption process has a parallel in the vertebrate nephron, where filtration under hydrostatic blood pressure at the glomerulus produces an isotonic filtrate which is appropriately altered as the filtrate flows along the nephron. In the Malpighian tubule, filtration under hydrostatic pressure is absent, and active transport takes its place.

The hindgut or rectum has remarkable capacities for regulating fluid loss in accordance with water availability and the ecology of the insect. At one extreme, freshwater species and terrestrial forms living on wet foods produce isotonic or hypotonic urines; at the other extreme, saltwater species and most terrestrial insects produce distinctly hypertonic urines with excreta which may be dust dry. In the

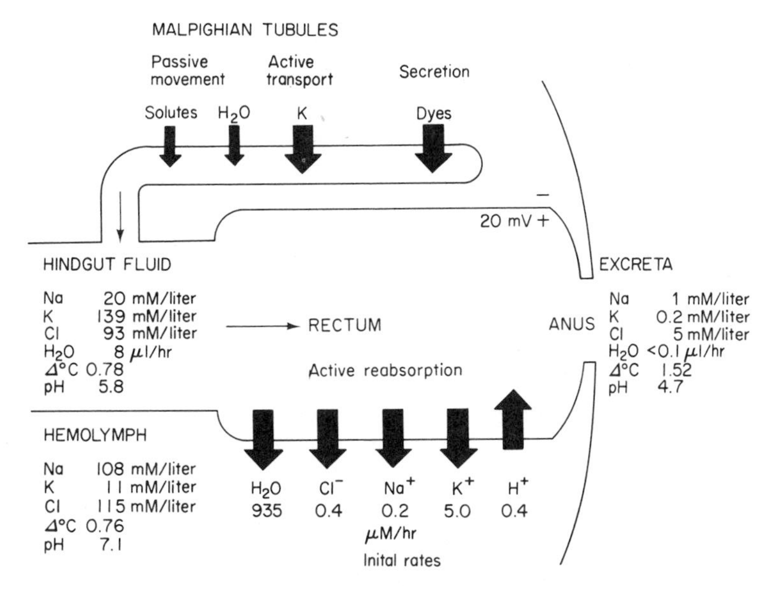

Fig. 8.17 Diagrammatic summary of the various secretory and reabsorptive processes occurring in the excretory system of the desert locust. The rates and concentrations of body fluids are for hydrated animals; the reabsorption rates for individual ions are those observed in the isolated rectum when the initial concentration of the rectal fluid is the same as normal hindgut fluid. [From Phillips, 1970: Am. Zool., *10*: 417.]

desert locust and cockroach, the urine may be two to four times more concentrated than the hemolymph while mealworms living on dry foods are even more resourceful (excreta up to 10 times more concentrated than blood). These capacities to concentrate the urine are comparable to those of specialized mammalian kidneys.

The most perplexing physiological questions concern the mechanisms of water reabsorption. In mealworms, living in dry environments and depending only on metabolic water, this absorption is so efficient that Ramsay (1964) believes it must depend on the extraction of water from the feces as water vapor. Spaces between the fecal pellets in the posterior part of the rectum are filled with air and Ramsay suggests that the absorption of water vapor by the rectum creates an atmosphere of lowered humidity into which the feces lose progressively more water by evaporation until they are voided as dry, hard pellets in equilibrium with an atmosphere of 75 per cent to 90 per cent relative humidity (Ramsay, 1964; Grimstone *et al.*, 1968).

There are several well-known biological systems capable of concentrating solutions by the removal of water. The gall bladder has already been described (Fig. 8.11) and the vertebrate nephron will be considered in the next section (Fig. 8.20). In these systems, solvent removal is clearly linked with the active transport of salt. Factors that interfere with solute transport have corresponding effects on water movement. If, for example, the physiological saline solutions surrounding the experimental preparations are replaced with a non-penetrating solute such as sucrose, water transfer ceases. The insect rectum appears to be more complex. In the desert locust, for example, water absorption from the rectum continues for some time when the organ is filled with a non-penetrating solution such as trehalose; this movement is against increasing gradients without a measurable net transfer of solute in either direction (Phillips, 1964, 1970). By itself, this experimental evidence argues for an active transport of water, a concept which physiologists find hard to accept. A true understanding of the mechanisms depends on an appreciation of the complexity of the rectal wall (Berridge and Oschman, 1972).

The wall of the insect rectum is not a simple membrane but an organ of considerable complexity—particularly in those species capable of producing highly concentrated urines (Wigglesworth, 1972). In some insects (Dermaptera, Orthoptera, Carabidae), the large columnar cells lining the rectum are grouped in six specialized cushions which may be extended to form distinct papillae; these cushions are termed "rectal glands" or "pads" or "papillae." Although structural details vary, electron microscopy has revealed a similarity in that they all contain sinuses, intricate intercellular channels, and spaces with complex infoldings of membranes; mitochondria are numerous and there is a rich tracheal supply and conspicuous neurosecretory terminals. This is evidently a cellular system of active metabolism that is probably regulated by neurosecretions; the complexity of channels and sinuses is important in the development of local ion differences and in the recycling of ions which couple water flow with solute transport. In lepidopteran larvae, Coleoptera, and several other insects, the distal ends of the Malpighian tubules are intimately attached to the wall of the gut, so that the lumina of the gut and tubules are separated by only thin membranes and the hindgut is clothed with a plexus of tubules, the cryptonephridial tubes (Grimstone *et al.*, 1968; Wigglesworth, 1972).

The sinuses and channels in the rectal pads provide spaces for the recycling of ions on which water absorption depends. Over relatively short periods, water transport appears to be independent of the presence of ions. Over prolonged periods, however, there is a depletion of ions in the pads; absorption of water by this organ is secondary to the transport of ions, as it is in the gall bladder and the kidney. Although water is being absorbed from the rectal lumen without a demonstrable net movement of solute across the wall, ion pumps are in fact operating within the tissue spaces and the water flow depends on concentration differences. J. E. Phillips and his associates (see Goh, 1971), using everted rectal sacs of the desert locust, measured transfer of ions for extended periods. In these preparations, a steady transport of water will occur for as long as 4 hours. This, however, is only possible if certain monovalent ions (Na^+, K^+, Cl^-) are present on the lumen side. The principle of moving water uphill by the use of ion pumps has been exploited by animals at several levels in phylogeny. However, this is not to argue that water absorption is *always* secondary to the transport of ions; similar solutions to biological problems have sometimes been achieved by quite different evolutionary paths (see Chapter 11).

The vertebrate nephron The vertebrate nephron begins with a PRESSURE FILTER, the renal corpuscle, and extends as a tube of variable length and structure which carries the filtrate to the outside while REABSORBING useful substances from it and SECRETING additional wastes into it. These three basic mechanisms of urine formation are the same as those of the metanephridium, although the anatomy and phylogeny of vertebrate nephron and invertebrate nephridium are quite different.

Comparative anatomists have agreed that there is no phylogenetic connection between these two types of excretory tubule (Goodrich, 1945), even though the more primitive nephric units of both show several physiological parallels. In many embryos and some adult lower vertebrates, the tubules open into the coelom by a ciliated funnel (coelomostome); it has often been suggested that this represents an ancestral condition. In the most primitive situation, a tangle of blood vessels, the glomus, is found in the neighboring wall of the coelom, and this serves to increase filtration of fluids into the body cavity, whence they can be wafted into the ciliated funnels (Fig. 8.18a). In the larval *Petromyzon,* for example, there are four segmental funnels; close to each there is a glomus which disappears in the adult *Petromyzon* but persists in the closely related hagfish *Myxine.* In most vertebrates this primitive vascular arrangement is superseded by one in which the glomus is pushed into the wall of the tubule as a glomerulus and thus forms the renal corpuscle (Fig. 8.18c). The coelomostome usually disappears (or is not even formed embryologically), although it is found in some adult sharks, in *Amia* and in a few amphibians (Smith, 1953). These physiological associations between excretory organs and coelom are similar to but not phylogenetically related to those of the annelids. Excretion in both cases is originally a drainage of coelomic fluids. In vertebrate evolution the nephric tubules are grafted onto the closed circulatory system which, through its very high blood pressures, provides a potent force for filtration. The happy marriage of these two systems required several rearrangements by both partners.

The nephron is a highly adaptable structure. With the exception of the excretory pore, there is really only one portion of it—the proximal convoluted or "brush border" segment—which is present in all vertebrates. Two major variations occur: the suppression of the glomerulus and the elaboration of the median and the distal ends of the tube. Both of these modifications are associated with life in dry habitats and the problems of water conservation. Since water balance and osmoregulation are discussed in Chapter 11, only an outline is given here.

The aglomerular condition (found in some teleosts, a few amphibians and reptiles) reduces the unit to a secretory tubule which is comparable to the Malpighian tubule of many terrestrial arthropods. Marine teleosts are subjected to constant osmotic desiccation because of the hypotonic condition of their body fluids, and thus they experience a water shortage comparable to that of amphibians and reptiles in arid habitats. Reduction or elimination of glomerular filtration was a part of the evolutionary response.

The difficulty was met in quite a different manner by the mammals and some of

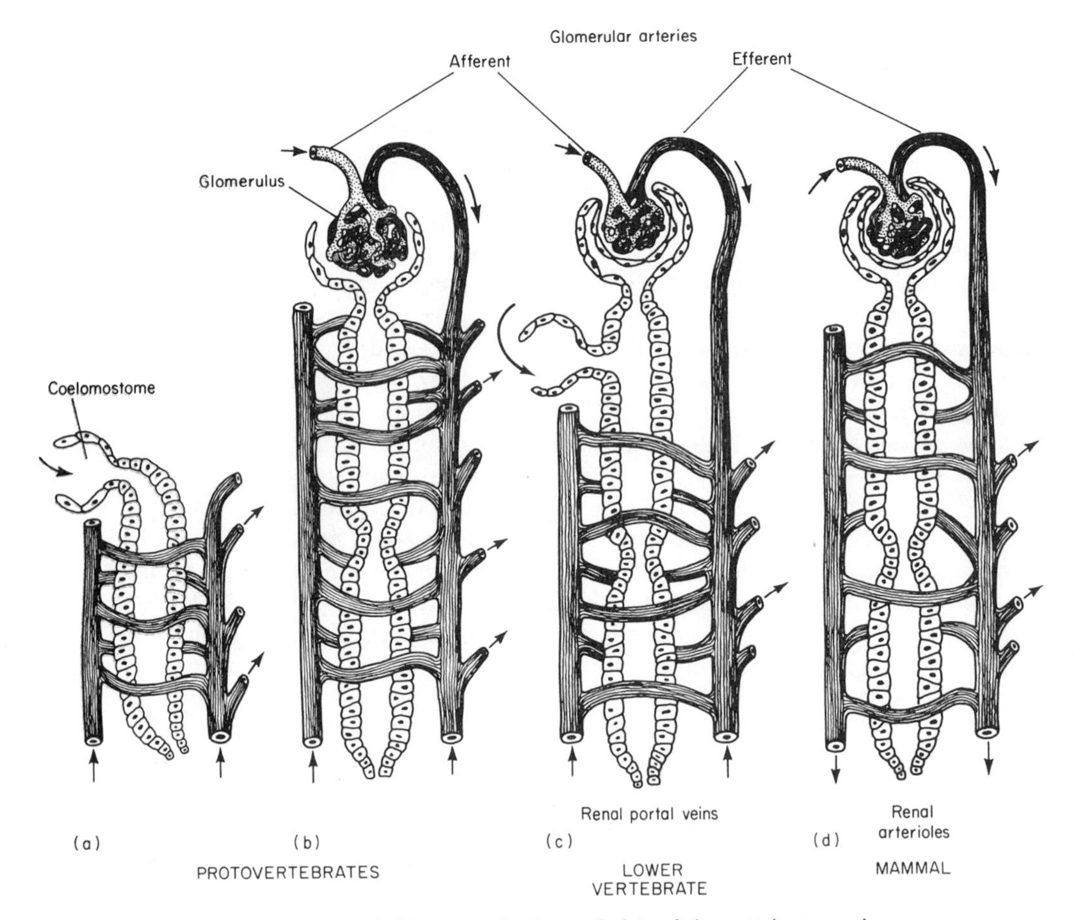

Fig. 8.18 Four probable stages in the evolution of the vertebrate nephron. (a) In the protovertebrate, the renal tubule drained the coelom by means of an open coelomostome. (b) The glomerulus was first only loosely related to the coelomostome. (c) Later, the glomerulus became sealed within the end of the tubule, the coelomostome persisting in many species. (d) In the higher vertebrates, the coelomostome disappears entirely. The blood supply to the tubules is by way of the renal portal in the lower vertebrates; in some reptiles and birds a part of the renal portal bypasses the kidney; in the mammals the renal portal disappears. [From Smith, 1953: From Fish to Philosopher. Little, Brown and Company, Boston.]

the birds. In these, the active glomerular filter remains, and a special unit is added for the recovery of the water used in filtration. The loop of Henle and the distal convoluted tubule are so effective in water conservation that mammals like the kangaroo rats (*Dipodomys* has been intensively studied) can meet all their water demands without drinking and rely entirely on metabolic water (Smith, 1953; Riegel, 1972).

The physiology of the tubule cannot be divorced from the circulatory system at any point along its length. An active tubule requires a good blood supply; when tubules are numerous, closely packed, and no longer bathed in coelomic fluid, a rich peritubular capillary network is essential. This was first obtained from the renal portal vein (gnathostomous poikilotherms) while the efferent glomerular artery had no direct connection with it. In birds and mammals the renal portal disappears, and the efferent glomerular artery is spliced into the proximal portion of the peritubular capillary network. Blood filtered in the glomerulus then passes directly through the capillaries around the tubules where its constituents are brought to proper physiological levels before passing into the general circulation.

These two patterns of blood supply (Fig. 8.18) are associated with two quite different lines of physiological specialization in the tubule. The renal portal vein drains a large area of active tissue in the tail and posterior appendages and, as the returning blood circulates through the peritubular capillaries, wastes can be actively transported or can readily diffuse into the tubules. If water is limited, glomerular filtration can be reduced or eliminated without restricting the secretory activities of the nephric units because they receive an independent venous supply; this trend reached an extreme with the development of aglomerular kidneys in some species. Purely secretory tubules, however, are no longer possible when the tubular capillaries receive their blood directly from the glomerular artery. The kidney is then committed to continuous glomerular activity, and mechanisms for water conservation are transferred to the distal end of the tube. Constant high pressure filtration at the glomerulus is imperative. The efficiency of this arrangement is dependent upon the relatively high blood pressure of birds and mammals, and this in turn on the evolution of the double heart which supplies constant high pressure filtration at the glomerulus.

Urine formation in the vertebrates In the middle of the seventeenth century, Marcello Malpighi first saw renal corpuscles which he described as hanging from the small arterioles "like apples" on the branch of a tree (Fulton, 1966). He rightly surmised that they were concerned with urine formation, but 200 years elapsed before Bowman in 1842 published his classical description and demonstrated their anatomical relationships with the rest of the tubule. Precise functions have gradually been assigned to the different regions of the nephron in some of the most impressive physiological research of the past century. There are still unanswered questions, but the basic mechanisms are now evident. The history is described in several places, and detailed citations for the following summary can be readily checked in standard texts (Smith, 1951; Mountcastle, 1968).

Bowman noted the glandular nature of the tubular epithelium and thought that the glomerulus produced fluid to wash the secretions of the tubules into the ureter. Ludwig, about the same time, correlated urine flow with changes in the blood pressure and proposed a mechanical theory of urine formation by hydrostatic filtration from the glomerular arteries. Such mechanistic concepts were violently opposed in the latter part of the nineteenth and early twentieth century. The proof of filtration

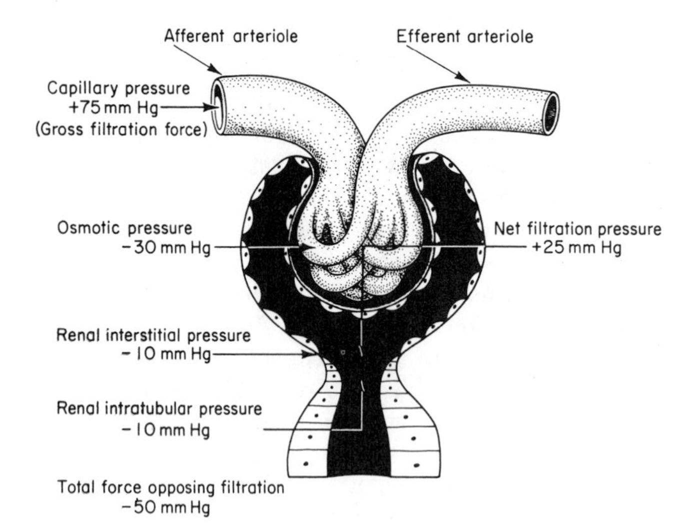

Fig. 8.19 The renal corpuscle showing the pressures that operate to produce a net filtration pressure of about 25 mm Hg. [From Sharp and Dohme, 1947: Seminar, Vol. *9* (3).]

came only in 1924 when Richards and his associates published the results of their micropuncture experiments. With delicate needles Richards withdrew fluid from the renal capsule of the frog and mudpuppy and showed that it contained glucose and chloride in about the same concentration as the plasma but was normally free of protein. In succeeding years, pressures were measured and the dynamics of the corpuscle were established in physical terms of blood pressure, osmotic pressure and pressures of the capsular fluid (Fig. 8.19).

A few years later Bayliss determined the limits of the permeability of the mammalian glomerular membrane at about 68,000 mol wt dimensions by following the excretion of a series of proteins ranging in size from gelatin (mol wt 35,000) to *Helix* hemocyanin (mol wt 5,000,000). Much more precise values of glomerular permeability have now been determined using dextrans of different molecular sizes; dextrans with mol wt >50,000 are cleared very slowly (Arturson, 1970). The pioneer studies, however, clearly demonstrated that glomerular filtration is the first step in urine formation. The fluid formed contains all the low molecular weight solutes of the plasma and is isosmotic with it. Large volumes of fluid are continually pushed through these small filters. The human kidney with its million or more corpuscles, each containing about 50 capillary loops, filters approximately 120 ml per min or some 40 gal of fluid per day. But this is only part of the story, for it is actually the tubular epithelium which does the real work in the kidney. In the human, a cellular surface of about 6 sq m is assigned to this job.

One of the functions of the tubular epithelium is the reabsorption of physiologically important solutes such as glucose, low molecular weight proteins (Forster, 1961) and chloride. This became apparent when glomerular filtration was demonstrated and concomitant analyses were made on the plasma and urine. The mechanisms, however, whether active transport or passive diffusion, were not immediately clear. Moreover, the quantities of certain substances such as urea and creatinine in the urine could be explained equally well by tubular secretion (excretion) or water absorption, leading to their concentration in the bladder urine.

The first clear answer to some of these questions came in studies of renal function in the algomerular fishes (*Lophius, Opsanus*) where only the tubular epithelium can be involved in urine formation. In a series of studies the aglomerular kidney was shown to excrete water, creatine, creatinine, urea, uric acid, magnesium sulphate, potassium, chloride and a variety of foreign substances such as thiosulphate and phenol red; but glucose, inulin and a number of other materials never appeared in the urine. Active secretion was thus demonstrated in a vertebrate nephron which is considered homologous with the proximal tubule of higher forms. Many analyses of fluids obtained in micropuncture of renal tubules of amphibians and mammals have proved that the tubules secrete as well as reabsorb. The *Necturus* tubule has been particularly useful for studies of this sort.

Another extremely productive technique for the study of tubular activity was also developed in studies of comparative physiology. Forster (1948 and subsequently, Hickman and Trump, 1969) first demonstrated active accumulation of phenol red in isolated pieces of flounder mesonephros. Tubules were observed microscopically, and the dye was seen to concentrate gradually within the tubules. Photometric measurements of the changing dye concentration in the suspension media also showed that it was actually passed into the tubules. These tests have also been performed on tadpole and fetal human kidney tubules. The transport of phenol red evidently depends on active cellular transport since it is readily inhibited by such enzyme inhibitors and poisons as iodoacetate, cyanide and dinitrophenol.

Tracer techniques have also provided valuable evidence for the active transport abilities of the nephron (Solomon, 1962). Perfusion fluids containing radioactive materials have been trapped between two oil droplets in the relatively large tubules from *Necturus* by first injecting a drop of oil into the capsular space and then following this with perfusion fluid and another drop of oil. As the droplets move along the tubules, changes in their composition leave little doubt that substances such as sodium are being actively transported. Quantitative estimations of filtration, reabsorption and secretion are based on the renal clearance studies already discussed (page 299 and Smith, 1956).

Mammals produce a hypertonic urine; the locus and mechanism of the water recovery have only recently been satisfactorily explained. The nephron of mammals, and of some birds, is unique among the vertebrates in having a long "hairpin" tube, the loop of Henle, which extends deep into the medulla of the kidney. The association of this anatomical feature with the production of hypertonic urine suggests a locus for water recovery but provides no evidence for the mechanism. Analyses of fluids obtained through microcatheter and puncture and the direct cryoscopy of kidney slices have shown that the urine remains isotonic in the proximal tubule but becomes progressively more hypertonic as it slowly descends the loop of Henle. Within the ascending limb of this loop it becomes gradually less hypertonic. Within the distal tubule it is either hypertonic or once more isotonic to the surrounding tissue fluids (Fig. 8.20). As the urine passes through the collecting tubule in the medulla,

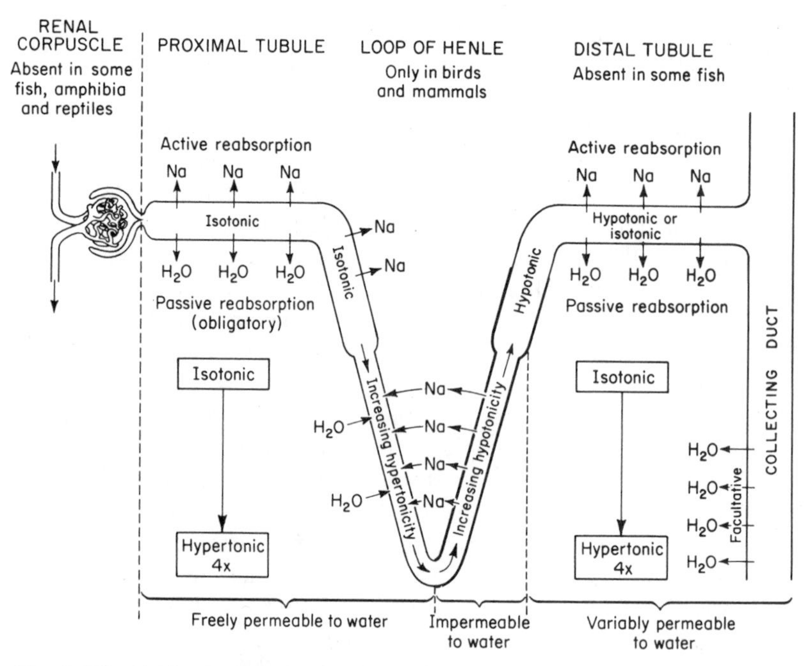

Fig. 8.20 Mechanism of urine formation in the mammal. [After Pitts, 1959: The Physiological Basis of Diuretic Therapy. C. C. Thomas, Springfield, Ill.]

it again becomes hypertonic. It is suggested that the medullary tubules constitute a counter-current multiplier (Gottschalk, 1960; Ullrich, 1960; Wirz, 1961) and that the active mechanism is a cellular transport of sodium from the urine in the thick portion of the ascending limb of Henle's loop. The thin portions of the loop are freely permeable to sodium (or may actively absorb it), and thus an increasing hypertonicity develops in the lower part of the loop and in the surrounding tissue spaces (Fig. 8.20). Fluids move very slowly through these passages. This active transport of sodium from the ascending limb produces a concentrated brine bath (about four times as concentrated as the fluids of the cortex) which withdraws water osmotically as it flows through the delicate thin-walled collecting tubules. The microscopic anatomy of the medulla, with its thousands of closely packed parallel tubules, supports such a concept (Smith, 1956; Mountcastle, 1968). The actual length of the loop of Henle shows considerable variation in different mammalian species and is correlated with their ability to concentrate the urine (O'Dell and Schmidt-Nielsen, 1960).

The arrangement of the blood vessels in the medulla (vasa recta) is also important to the operation of this system (Fig. 8.21). They form a plexus of hairpin loops parallel to the loops of Henle and operate as another countercurrent system. The vascular loops do not form a multiplier system like the loops of Henle but are passive exchangers which irrigate the renal medulla without interfering with its osmotic stratification (Wirz, 1961).

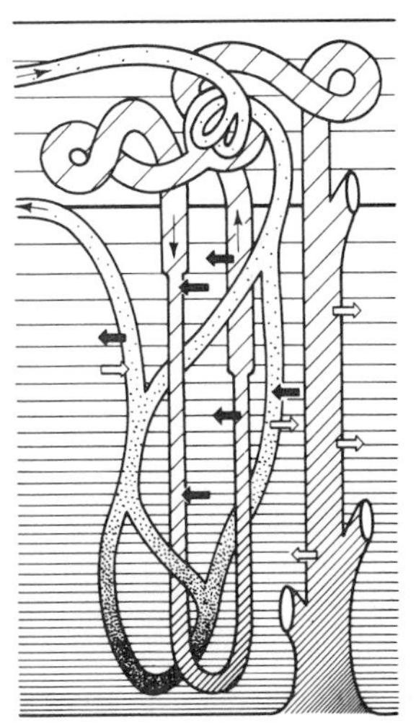

Fig. 8.21 The blood supply (stippled) to the mammalian nephron. Angled lines, renal tubule; horizontal lines, interstitial fluids; intensity of shading indicates stratification of osmotic concentration; white arrows, water transfer (passive); black arrows, transfer of crystalloids (active and passive). [From Wirz, 1953: Helv. Physiol. Acta, *11*: 27.]

Regulatory Mechanisms

Physiological machines such as nephridial and renal tubules that filter, secrete, and partition solutions must be geared to meet variable environmental stresses and metabolic demands. Theoretically, regulation might occur either at the filter by altering the blood pressure or in the tubules by controlling the movements of substances through the cells. Among the vertebrates, it has often been experimentally demonstrated that variations in blood pressure alter the filtration rate and that urine production is subject to the autonomic nervous system. Under normal conditions, however, this factor is of minor significance in the regulation of filtration and apparently without any effect on tubular activities. The adjustments among both invertebrates and vertebrates are primarily hormonal. In this chapter, the factors that regulate the responses of the excretory organs will be identified; in Chapter 11, the broader questions of water balance and ionic regulation will be examined.

The water balance hormones of invertebrates Strong indications that osmoregulation is under hormonal control have been found in many invertebrate animals, ranging from flatworms and starfish to snails and crustaceans (Ralph, 1967). Much of this evidence comes from changes in neurosecretory activity which follow manipulations of water balance. Convincing experimental proof of specific hormonal factors was first demonstrated by Maddrell (1963, 1964) in the blood-sucking bug *Rhodnius*. This animal makes the most of its limited opportunities to feed by ingesting an extremely large volume of fluid. The extra water is rapidly eliminated to bring the body back toward normal size. Maddrell found that isolated Malpighian tubules showed markedly increased activity when exposed to the hemolymph of bugs which had recently gorged themselves on blood (Fig. 8.22); hemolymph from fasting bugs lacked this property. The source of the activity was traced to the mesothoracic ganglionic complex, which was shown to be activated by stretch receptors in the

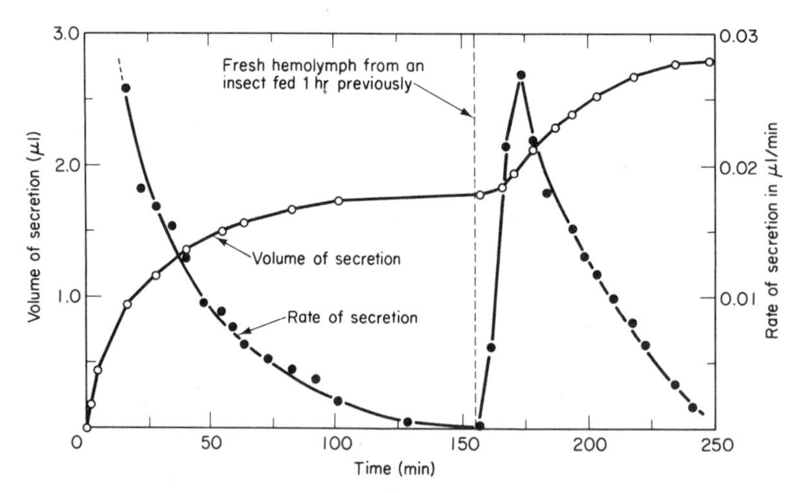

Fig. 8.22 Urine production by the isolated Malpighian tubules of *Rhodnius*. Vertical broken line shows time when the hemolymph which bathes the tubules was changed from that of an "unfed" insect to that of a "fed" (diuretic) insect. [From Maddrell, 1963: J. Exp. Biol., *40*: 249.]

abdomen. This neurosecretory substance is termed the DIURETIC HORMONE; it has now been found in several other insects (Highnam and Hill, 1969). In the desert locust, it inhibits water resorption from the rectum and also stimulates water uptake by the tubules (Mordue, 1969). In the cockroach, locust and some other insects, the primary source of diuretic hormone is in the cerebral neurosecretory cells rather than the mesothoracic ganglionic mass—the center of its formation in *Rhodnius*.

In some insects, there is also evidence of an ANTIDIURETIC HORMONE. Again, using *in vitro* preparations of Malpighian tubules and rectum, it has been found that hemolymph from dehydrated cockroaches contains a factor of neurosecretory origin that *lowers* the rate of water output by the tubules and *increases* water absorption by the rectum (Wall and Ralph, 1964; Wall, 1967). Cockroaches seem to have a dual hormonal control of tubules and rectum with diuretic as well as antidiuretic factors. The diuretic factor was demonstrated in cockroaches that had been led into a bout of overdrinking by depriving them of water for several days (Mills, 1967). Thus, in at least some insects, there appears to be a dual control of water balance through hormones which arise in neurosecretory cells and act at the level of the Malpighian tubules and rectum. It seems likely that comparable factors are present in other groups of invertebrates but, at present, the evidence is less complete—particularly with respect to the locus of action. Highnam and Hill (1969) have summarized the literature.

Hormones and vertebrate kidney function Three groups of tissues are intimately concerned with the regulation of the activities of the renal tubules. One of these, the JUXTAGLOMERULAR APPARATUS, is located within the kidney; the other two—adrenal cortex and certain constituents of the neurosecretory system—operate from a distance.

There are two cellular components of the juxtaglomerular apparatus (Ham, 1969). The granular juxtaglomerular cells (JG CELLS) develop as a cuff or band of modified smooth muscle cells in the median layer of the afferent arteriole at its junction with the glomerulus; cells of the MACULA DENSA occur as a plaque in the wall of the nearby distal convoluted tubule. The granular JG cells are the most likely source of the enzyme renin; the evidence is based not only on cytological and histochemical studies of the granules but also on actual sampling and assay of the granules obtained by micropuncture of the JG cells in the mouse (Cook, 1968). The function of the macula densa cells is uncertain but is thought to be related to that of the JG cells since there are parallel cytological changes in the two groups. Granular JG cells have been found in all groups of vertebrates from the phylogenetic level of the bony fishes (Capréol and Sutherland, 1968). Functions of the corpuscles of Stannius, found only in the bony fishes, have been recently linked with those of the JG cells (Ogawa, 1968; Chester Jones *et al.,* 1969).

The long, long trail of research that led to our current understanding of the functions of the JG cells began with the studies of high blood pressure (hypertension) which were carried out by Goldblatt (1947 and earlier) during the fourth decade of this century. He showed, in several different mammals, that experimental procedures which resulted in deficient oxygen in the renal tissues, were followed by a persistent rise in blood pressure. The usual physiological manipulation was by clamping the renal artery, or in some other manner, to restrict the flow of blood to the organ. Investigations of this phenomenon focused attention on an enzyme RENIN which is formed in the kidney and acts on a circulating plasma protein (synthesized in the liver) to convert it into a decapeptide called Angiotensin I. This is followed by a second plasma reaction in which a converting enzyme splits off two amino acids to form the octapeptide Angiotensin II (also called hypertensin or angiotonin). Angiotensin II is a powerful constrictor of arterioles, one of the most potent vasopressor substances known. The sequence of biochemical steps leading to its release is as follows:

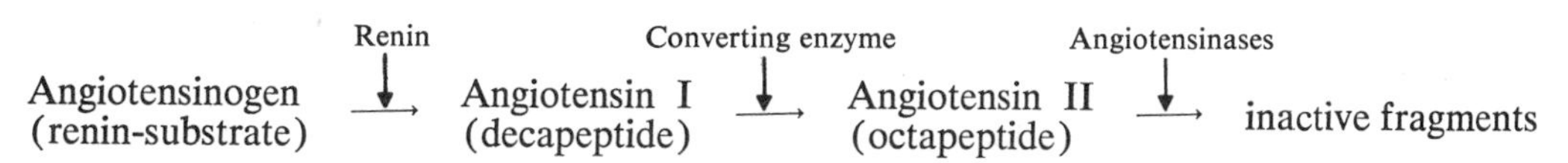

At one time, the renin-angiotensin system was thought to be primarily responsible for the regulation of renal blood flow by regulating the systemic circulation. It was believed that the pressure-sensing JG cells in the walls of the afferent arterioles safeguarded the renal tissues from a sluggish circulation which would reduce urine formation. This action is now considered secondary to the intrarenal effects and particularly those concerned with the regulation of water and electrolyte balance through the adrenocortical steroids.

The renin-angiotensin system seems to regulate water and electrolyte balance through three interrelated effects (Brown *et al.*, 1968). The systemic action is probably minor. In addition, purely local actions have been noted whereby the renin-angiotensin mechanism may alter intrarenal circulation, glomerular filtration, and tubular functions. In the mammal, however, the function of major importance is the regulation of aldosterone output by the adrenal cortex. Although other functions have been suggested, the regulation of electrolyte balance by the adrenal steroids is considered to be the major one. The main links in these controls of the mammalian system are shown in Fig. 8.23.

The renin-hypertensin system has been found in the bony fishes and higher groups of vertebrates (Brown *et al.*, 1968; Chester Jones *et al.*, 1969; Ogawa *et al.*, 1972; Sokabe and Nakajima, 1972). It has been suggested that, phylogenetically, renin was first concerned with purely local functions affecting water and electrolyte balance within the kidney; during this early phase of evolutionary history, the renin-angiotensin biochemical system may have been confined to the renal lymph. The control of adrenocortical functions may have been acquired as a simple extension of these local hormone activities while the systemic effects produced through the action of these agents in the general circulation were probably subsidiary and supplementary to the local actions (Brown *et al.*, 1968).

Several of the adrenal steroids modify electrolyte balance in mammals, but aldosterone is the most potent and seems to be the physiologically important mineralocorticoid. In mammals, its action is on the distal tubule where it depresses the Na: K ratio by promoting the absorption of sodium and favoring the excretion of potassium. It has a similar action on some other mammalian tissues like salivary

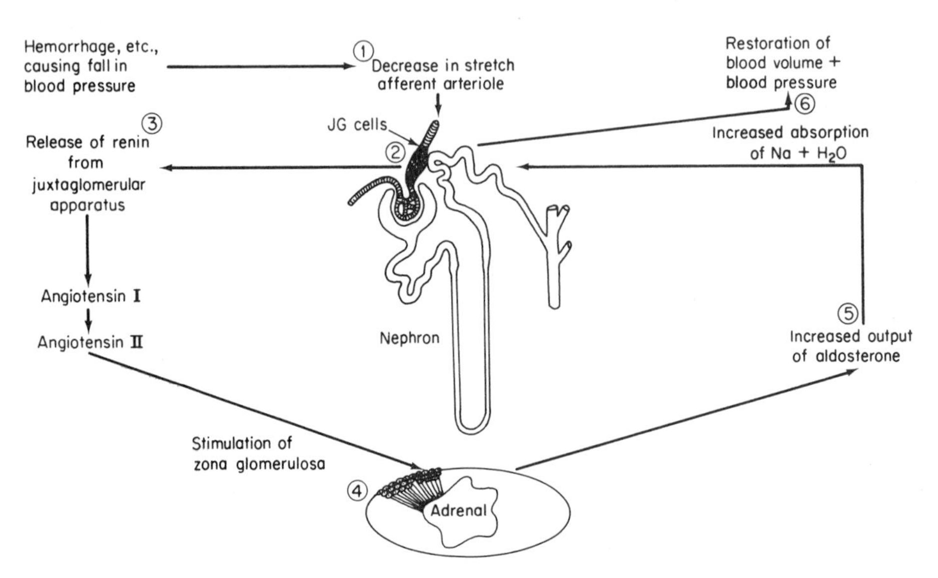

Fig. 8.23 Regulation of renal function in the mammal by the JG cells and aldosterone. [Based on Clegg and Clegg, 1969: Hormones, Cells and Organisms. Heinemann, London.]

gland cells and the epithelium of the gut, and this may indicate a more generalized action on the electrolytes of cells. The regulation of aldosterone secretion by the changing levels of blood electrolytes was described in Chapter 2.

It seems likely that the kidney tubules of the lower vertebrates are also regulated in some manner by the adrenal steroids but experimental evidence of this remains inconclusive (Maetz, 1968; Chester Jones *et al.,* 1969). Although it has been repeatedly shown that injections of certain adrenal steroids into fishes and amphibians are followed by alterations in water and electrolyte balance, the precise localization of these effects is uncertain. There are two hurdles facing investigators of these problems. First, one of the classical techniques of endocrinology cannot be used because surgical removal of the adrenal cortex from the majority of fishes and amphibians is impossible. Second, excretion in the lower vertebrates depends on several organs in addition to the kidney (gills, skin, urinary bladder, nasal and rectal glands); each of these organs must be separately evaluated in establishing the role of the cortical steroids in excretion. This topic will be discussed further when osmotic and ionic regulation are considered in Chapter 11.

Antidiuretic hormone (ADH) is the second major factor regulating renal tubular activity. Several slightly different octapeptides from the vertebrate hypothalamus are stored in the neurohypophysis and released on demand to perform different functions, one of which is the absorption of water by the kidney tubule (Chapter 2). Arginine vasotocin is the antidiuretic hormone of amphibians, reptiles and birds, while the mammals rely on vasopressin. In amphibians, ADH increases water permeability of the skin and bladder, decreases glomerular filtration rate and reduces tubular reabsorption (Maetz, 1968). Thus, water may be taken into the body through the skin or bladder wall and its leakage through the kidney reduced. In the other terrestrial vertebrates, the only effect of ADH appears to be on the kidney tubule. Osmoreceptors in the hypothalamus control ADH release in response to changes in the osmotic pressure of the plasma circulating through it. In mammals, ADH acts by increasing the water permeability of the distal convoluted tubules and collecting ducts (perhaps by enlarging pores in the membranes) and hence will promote the removal of water from the tubular urine (Fig. 8.21). Thus, it is an important factor in the production of hypertonic urine by mammals and is always found as a part of the water-regulating machinery of terrestrial vertebrates. Among the fishes, the physiological effects of ADH are primarily extrarenal and concerned with electrolytes; there is scant evidence of direct effects on water movement such as those found in terrestrial forms (Maetz, 1963, 1968; Perks, 1969). The regulation of water permeability and the control of renal tissues may have come gradually under the control of the neurohypophysial hormones during the evolution of terrestrial life.

TWO

ENVIRONMENTAL RELATIONS

9

PHYSIOLOGICAL COMPENSATION FOR ENVIRONMENTAL VARIATION

An animal rarely lives under constant conditions. Some habitats, such as the depths of the ocean or the interior of a warm-blooded animal which houses its parasites, provide relatively constant conditions. Yet even in the depths of the ocean supplies of food may be uncertain, while the habitat of the parasite can change with the health and nutrition of its host. Most animals face not only nutritional uncertainties but also marked diurnal and seasonal oscillations which alter rates of metabolism and activity; sudden extremes may tax the physiological machinery to the limit. An animal does not exist apart from its environment; the comparative physiologist recognizes this association and attempts to describe and explain the varied mechanisms by which animals compensate for all sorts of environmental alterations and stresses.

It follows that some comparative physiologists study the habitat as well as the animals which live there. In this they share common interests with ecologists. Both measure the environmental parameters and record their normal ranges and extremes; but they study the animals in rather different ways. Both ecologist and physiologist may record the magnitude of the environmental variations found in a tidal estuary with its wide range of salinity and its shores which are sometimes stagnant and always prone to marked temperature alterations. The ecologist examines these factors as they affect the distribution of animals and regulate their population dynamics; he is also interested in the behavioral responses of the animals and their orientations to temperature and salinity gradients or any of the other features of the estuarial habitat.

The physiologist examines the same environmental variables, and he may describe them in the same terms; but his interest is primarily in the physiological machinery and the manner in which this compensates for the environmental alterations. He is much less interested in the way in which salinity affects the distribution, abundance and success of the animals than he is in processes which enable the animal to acquire salts from low salinity waters or to remove excesses of electrolytes when the water is excessively salty. He describes the respiratory problems of the aquatic organism stranded in the air by the ebbing tide, and he attempts to understand the acclimatization which occurs seasonally and the lethal processes associated with extremes.

There is, of course, much common ground, and it is pointless to attempt a sharp division of interests. Ecologists have become increasingly interested in the physiological mechanisms basic to an understanding of the animal-habitat interrelationships; the physiologist has become much more aware of the curious adaptive mechanisms associated with different and peculiar habitats. These studies of the physiological compensations for environmental oscillations and stresses are often considered together as a subdivision of comparative physiology—ENVIRONMENTAL OR ECOLOGICAL PHYSIOLOGY. This subdivision is largely one of convenience. It does, however, recognize a changing emphasis in comparative physiology (Hoar, 1967; Prosser, 1969). The pioneers of this science were usually taxonomically oriented and excited by the variations in physiological processes found at successive levels in phylogeny; present-day comparative physiologists are more often environmentally oriented and find their problems in the curious adaptations to unusual habitats and in the physiological capacities which permit life under changing and adverse circumstances. Environmental physiology focuses on this latter group of problems but inevitably merges with other areas of comparative physiology, particularly with studies of physiological adaptation in the evolutionary sense. In this chapter, principles concerned with physiological compensations for environmental change will be discussed; in succeeding chapters, compensations for specific environmental variables will be considered. Valuable and extensive reviews of the literature on environmental relations have been compiled in Section 4 of the *Handbook of Physiology* (American Physiological Society, 1964), while several recent books have emphasized different segments of the field (Folk, 1966; Vernberg and Vernberg, 1970).

NATURE OF THE INTERACTION WITH THE ENVIRONMENT

Tolerance and Resistance

As a part of its genetic endowment, every animal has a capacity to compensate for environmental change. It can live within a certain range of variations, whether the variable is temperature, humidity, oxygen supply or any other environmental factor. This is its TOLERANCE, and it will not be killed or damaged by any particular environmental factor, provided this does not exceed the tolerance limits. Beyond these limits, however, the organism is damaged. Although it may RESIST the change for a longer or shorter period, it will eventually succumb as a result of the change. Thus, an organism has a certain capacity for both TOLERANCE and RESISTANCE. Under appropriate conditions a catfish may live at temperatures ranging from 1°C to 35°C. This is its RANGE OF TOLERANCE, with a lower and an upper INCIPIENT LETHAL LEVEL (Fry, 1971); exposure to temperatures less than the lower or greater than the upper lethal level will kill the animal after a resistance time which depends on the magnitude of the temperature differences.

Acclimation and Acclimatization

An animal has not only a capacity for tolerance and resistance but also one for ACCLIMATION and ACCLIMATIZATION. This means that its previous history with respect to any factor may modify its subsequent tolerance and resistance to changing conditions of this factor. Again, explanations are easiest in terms of familiar temperature effects, but it should be noted that the same principles apply to many of the other variables. If a catfish is maintained for a week or more at 30°C instead of 25°C, then its upper and lower incipient lethal levels are elevated by 2 to 3°C as shown in Fig. 9.1. In short, there is a whole family of upper and lower incipient lethal levels, and the range of tolerance is really a ZONE OF TOLERANCE bounded by a ZONE OF REALM OF RESISTANCE (Fig. 9.2).

Many of the vital functions change in response to altered environmental conditions, whether the environmental changes fall within the zone of tolerance or the zone of resistance. These zones should not be thought of as static physiological areas. Temperature acclimation has, for example, been shown to alter the nature of the body fats of goldfish, the oxygen-binding of the blood of frogs, the heat resistance

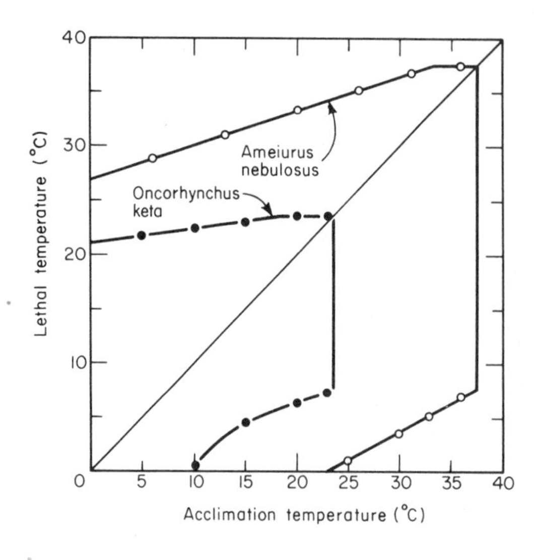

Fig. 9.1 Relation between the acclimation temperature and the upper and lower lethal temperatures for the catfish, *Ameiurus nebulosus* and for the chum salmon, *Oncorhynchus keta*. [From Brett, 1956: Quart. Rev. Biol., *31*: 76.]

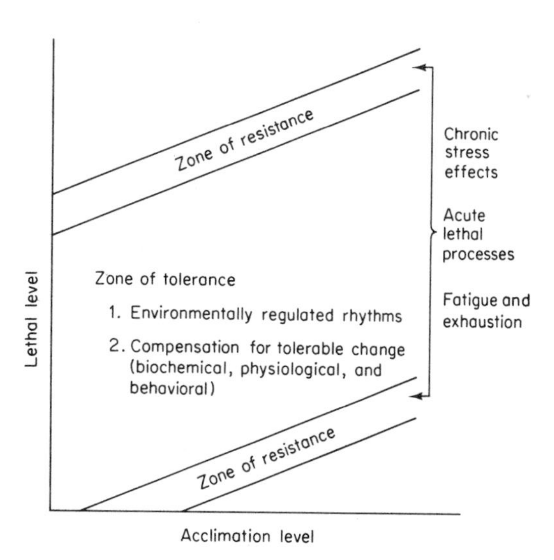

Fig. 9.2 Summary of factors involved in physiological compensation for environmental variation. Further description in text.

of the proteolytic enzymes in the stomach juices of snails (*Helix*), the excitability of the isolated foot of the gastropod (*Lymnaea*), and the pathways and kinetics of several different metabolic enzymes in fish. Precht (1958) and Hochachka and Somero (1971, 1973) have documented these as well as many other physiological and biochemical responses to temperature change. Similar examples of responses to other environmental factors will be discussed in subsequent sections. An animal makes a dynamic physiological response, whether the environmental change is within its normal (zone of tolerance) or more extreme range (zone of resistance). In temperature work, Precht (1958) uses the term CAPACITY ADAPTATION for the compensations which take place within the range of normal temperatures (Fig. 10.3) and RESISTANCE ADAPTATION (cold and heat resistance) for compensations to extremes which alter the lethal level of the environmental factor (Prosser, 1969). Alderdice (1972) defines adaptive responses in terms of response maxima.

Although the terms ACCLIMATION and ACCLIMATIZATION have essentially the same meaning in the English language and are frequently used interchangeably in biological literature, there is a tendency on the part of environmental physiologists to restrict their usage to somewhat different compensatory changes. In the current terminology, "acclimation" is the descriptive term applied to compensatory changes which occur in the laboratory where the animals are maintained under controlled conditions of the factor in question, while "acclimatization" refers to the more complex situation in nature. With reference to temperature, the catfishes described in Fig. 9.1 were acclimated in the laboratory by holding them in aquaria at different constant temperatures. In nature, these animals also show seasonal changes in their temperature tolerance. These are partly due to the seasonal temperature cycle but may also be associated with photoperiod and other seasonally changing conditions. The term acclimatization is reserved for compensatory changes occurring under such natural conditions. It is the sum of the adjustments which follow repeated and prolonged exposure to natural environmental change.

There is also a third level (the species level) on which temperature compensation is possible. Zones of tolerance may be altered by natural or artificial selection through changes in genotypes. This is not the particular concern of the comparative physiologist.

Interaction of Environmental Factors

The best descriptive term in the English language for the environmental pressures which require physiological compensation is STRESS. The word is used here in this general sense and without any connotation of its specialized medical usage in the "stress syndrome" popularized by Selye (1949). Unlike the single stress laboratory tests by which the zones of tolerance are established, the natural environment is liable to create several different stresses simultaneously. Adverse humidity and temperature often occur at the same time; the salinity and temperature in the tide pool may rise, and the oxygen may be depleted during the intertidal period. There is thus an interaction of stresses. The experimental analysis of such an interacting set of stresses has not often been attempted (Alderdice, 1972). McLeese (1956), however, studied the interaction of temperature, salinity, and oxygen in the survival of the American lobster. He found that an acclimation to any two of these factors produced a marked alteration in the incipient lethal levels of the third and that acclimation could be readily demonstrated for each factor singly.

At the physiological level these different stresses are operating in several different ways. Low oxygen, for example, restricts metabolism because of the reduction in an essential metabolite required for the production of ATP. Lack of nutrients, vitamins or trace elements would act in the same manner. These factors have been described as LIMITING. Poisons, narcotics and extremely low temperature act in a different way by suppressing the rates of metabolism. They have been called INHIBITING. Factors such as increased salinity, high temperature, or excessive muscular exercise may produce a stress through their excessive demands on metabolism. There may be little or no surplus energy for normal physiological processes. Brett (1958) refers to these factors as LOADING. It is thus evident that the interaction which McLeese described in his studies of the lobster is exceedingly complex physiologically and, further, that the natural environment may produce even more complicated interacting stresses. Alderdice (1972), in a critical review, considers the techniques and importance of multiple factor studies.

Measurement of the Lethal Level

The precise boundaries of the apparent zone of tolerance will depend on the method of determining the lethal levels. Two techniques have been commonly used. In one, the level of the environmental variable is gradually altered until the animal succumbs; in the other, separate animals (in practice, groups of animals) are placed in a series of constant but lethal environments and the time to death is noted. Certain arbitrary decisions are required in each case. In the first, it is the RATE OF CHANGE which must be standardized. Huntsman and Sparks (1924), for example raised the temperature of the sea water by 1 °C every five minutes until the animals died; Tsukuda (1960) raises or lowers the temperature at a steady rate of 0.5 °C per minute until temperature coma is observed. The second type of test involving an exposure to one lethal level presents no problem when the majority of the animals are soon killed by the lethal agent. However, if by definition the lethal level is the level which kills after an indefinite exposure, then, obviously, the experimenter must decide how long the lethal tests are to last. In practice, it is usually possible to determine from the course of the mortality curve whether the lethal agent is continuing to operate and to adjust the experimental procedures accordingly. In lethal temperature work, tests have been continued for 12 to 14 hours or even as long as seven days because various species react so differently (Brett, 1956; Fry, 1971). In general, tests involving sudden exposure to the lethal environment are preferred since there is less chance of acclimation during the test and since the data are more readily susceptible to standard statistical methods of analysis. Most determinations of lethal levels follow this technique.

The analytical techniques have been carefully studied by toxicologists and are described in many places (Finney, 1964; Bliss, 1967). They will not be detailed here, but several points are mentioned to facilitate understanding of the physiological literature. The analyses are based on the normal variability which every population shows with respect to its morphological and physiological characteristics. There are giants and dwarfs; there are also individuals which are extremely resistant and others which are particularly susceptible. In between there are the average individuals that make up most of the population. In short, we are usually dealing with a normal distribution curve, and its pattern is the same whether the measurements are sizes of animals or their incipient lethal levels.

The pioneer work was carried out by pharmacologists in their attempts to standardize drugs by bioassay techniques (Burn *et al.,* 1950). In one of the earliest studies the lethal dose of digitalis was determined for each of 573 cats; the lethal doses were distributed around a mean value in the pattern of the normal distribution curve. In another very precise early investigation, 146 frogs were slowly infused with *k*-strophanthin until they died. When the frequency of deaths at different lethal doses was plotted, a distribution of the same type was obtained. These pioneer data are shown in Fig. 9.3.

The average individual is represented by the peak of the curve, and this dose (the median lethal dose or LD_{50}) is the best representation of the lethal dose for the sample. In actual practice it is not determined from data such as those shown in Fig. 9.3. On the contrary, relatively small samples are used; different samples are exposed to a single lethal dose, and a graded series of doses is used; the one which kills 50 per cent of the animals in the test period is recorded as the LD_{50}. Usually five or six dose levels are sufficient to estimate the 50 per cent level when these are appropriately analyzed (Fig. 9.4). The mathematics has been carefully

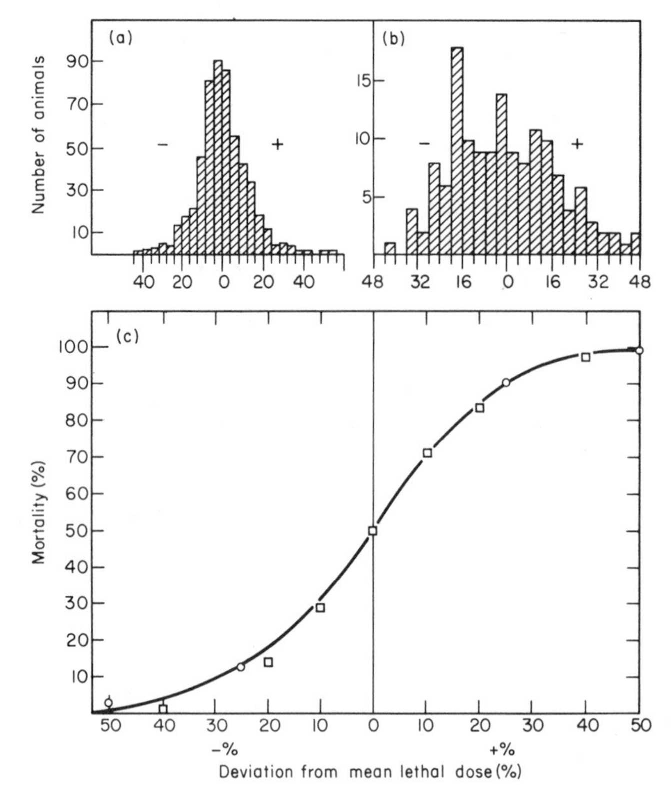

Fig. 9.3 Dose-mortality relationships. (a) Distribution about mean value of the lethal dose of digitalis for different cats. The abscissa 0 represents the mean with lethal dose given in percentages below and above the mean. (b) Same for lethal doses of strophanthin for frogs. (c) Data plotted as per cent mortality (ordinate) with circles for digitalis and squares for strophanthin. [After Burn, Finney, and Goodwin, 1950: Biological Standardization, 2nd ed. University Press, Oxford.]

studied, and easy graphical methods are now available for the ready determination of lethal levels. These methods are based on the fact that the bell-shaped dose mortality curve becomes a sigmoid when cumulated deaths (or per cent dead) are plotted against the dose or the log dose, as was evident in the very early studies of cats and frogs (Fig. 9.3). A close approximation to the LD_{50} can often be obtained directly from such a sigmoid; for precise comparisons it is rectified through the probit transformation which converts it into a straight line (Fig. 9.4). The LD_{50} values can be obtained from these curves, or the lines can be compared by standard methods of linear regression analysis.

The zone of tolerance illustrated in Figs. 9.1 and 9.2 is bounded by points representing temperatures at which 50 per cent of the sample died in tests lasting 14 hours. This provides the best representation of the way in which the population may be expected to respond to a single lethal factor. It is, in a sense, the mean reaction for a sample. However, there are times when the experimentalist may wish to evaluate the level which will kill all of the population or, alternatively, permit them all to survive. The trapezoid could just as well have been drawn for 5 per cent deaths and 95 per cent survival or any other level. The area would vary accordingly.

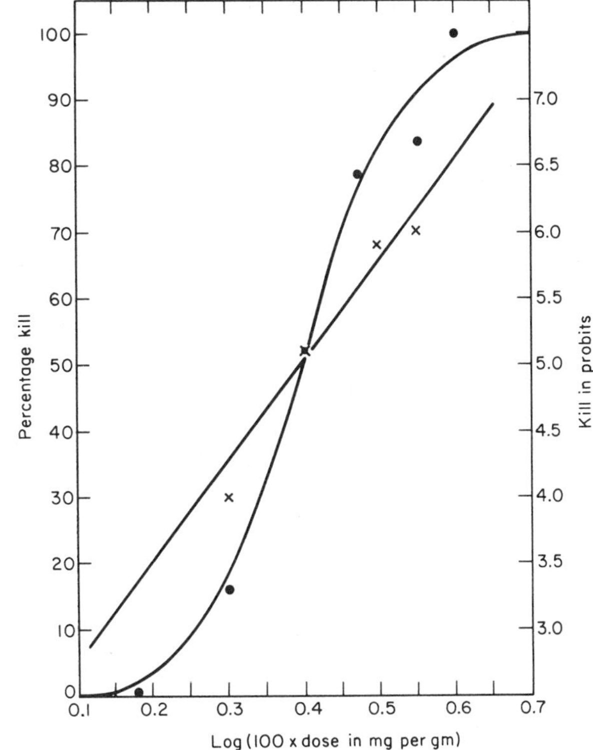

Fig. 9.4 Toxicity of cocaine hydrochloride to mice plotted as per cent kill (sigmoid curve) and as probits (straight line) against log dose. [From Burn, Finney, and Goodwin, 1950: Biological Standardization, 2nd ed. University Press, Oxford.]

It should also be emphasized that the trapezoid (Fig. 9.1) is a zone of survival in the face of environmental change. It does not describe, in any way, the effects of this change on various vital processes such as growth, activity or reproduction. It is known, for example, that young salmon fail to grow at temperatures slightly above the lower lethal level and slightly below the upper lethal level. Thus, a trapezoid representing tolerance levels for growth would be considerably smaller. Environmental limits for reproduction may be even narrower. Additional areas could be marked off to show the environmental variation compatible with each of several different activities.

Physiological Compensation in the Zone of Tolerance

There are two different kinds of physiological compensation for successful living in altered environments. Many animals have tissues with the capacity to operate over wide ranges. In these the internal environment reflects the external environment, and changes in the latter are followed by corresponding alterations in the former. The temperature of a spider crab fluctuates with that of its surroundings; it is said to be POIKILOTHERMIC (poikilos = manifold). The osmotic content of the body fluids of the polychaete worm *Arenicola* almost matches that of the sea water over a range of dilutions down to about 12 per cent; it is said to be POIKILOSMOTIC. The physiological processes of these animals operate well at a series of different temperatures or under varying osmotic conditions. They exemplify an environmental compensation referred to as CONFORMITY OR ADJUSTMENT (Fig. 9.5).

In contrast to the conformers, many animals preserve relatively constant conditions in their tissues. They control or regulate their internal environment and are killed if this fluctuates beyond rather narrow limits. Temperature variations or fluctuations in osmotic and other environmental conditions activate the regulatory homeostatic machinery which preserves the constancy of the internal environment. Excess heat is dissipated or heat is generated to make good the losses; water taken in osmotically from a dilute solution is excreted, and ions are absorbed to compensate for those lost through excessive water removal. The homeothermic or the homeoosmotic condition is maintained. The animal is said to show REGULATION as opposed to the CONFORMITY (or adjustment) displayed by the first examples (Fig. 9.5).

At least two different types of regulation are frequently observed. The type already mentioned depends on the homeostatic machinery of the animal—primarily the integrated responses of the autonomic nervous system and the hormones. In addition, many animals show behavioral responses which likewise compensate for environmental change. Terrestrial isopods, for example, became active when the humidity falls and crawl at random (kinesis) until they once more find a humid habitat where they become quiet. Many animals in a gradient of an environmental variable such as temperature, humidity or salinity will move about while they are in less favorable areas and remain quiet when they arrive in the more favorable ones.

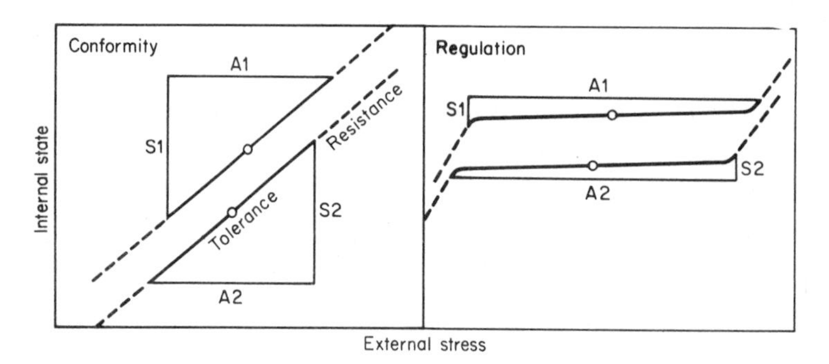

Fig. 9.5 Diagram illustrating the relation between external stress and internal state for the physiological adjustors or conformers (left) and the physiological regulators (right) at two levels of acclimation (A1 and A2). Note the difference in change of internal states (S1 and S2) for the two types. [Based on Brett, 1958: H. R. MacMillan Lectures (Fisheries), Univ. Brit. Col., Vancouver, after Prosser, 1955: Biol. Rev., *30*: 233.]

The environment in this case is acting as a DIRECTIVE FACTOR, and the preferendum response results in a behavioral regulation as opposed to the physiological regulation associated with homeostasis.

Prosser (1955, 1969) refers to these functional properties of animals which favor their continued successful living in altered environments as PHYSIOLOGICAL ADAPTATION. However, the term adaptation has several different usages in biology, and for this reason the term COMPENSATION will be used here.

Resistance to Extreme Conditions

Within the zone of resistance, the physiological mechanisms are taxed so heavily that they eventually collapse (Fig. 9.2). If the lethal factor is not far beyond its incipient level, the deterioration will be slow; if conditions are extreme, death will be sudden. Consequently, the changes which might be traced in an organism while it continues to exist in this zone will vary, and no description can satisfy all the cases. The ultimate failure may occur at any level, from the enzyme system to the morphological or chemical organization of the cells, from the particularly susceptible organs or tissues—such as the mammalian cerebrum in anoxia—to the complex homeostatic machinery of the higher vertebrate. All levels may be affected simultaneously, but there is probably one particularly vulnerable link which finally breaks. Extreme heat will quickly coagulate proteins; but less excessive temperatures may merely alter adversely the kinetics of enzyme activity (Hochachka and Somero, 1973). Either change can kill the cell, although the temporal relations and cytological alterations will be very different.

The mammalian physiologists have accumulated an impressive body of literature on the harmful effects of extreme conditions on the autonomic nervous and endocrine systems. In many cases the collapse of the pituitary-adrenal system seems to be ultimately responsible for death. Selye (1949, 1961) has emphasized this in his concept of "stress." Any one of a variety of damaging agents, such as trauma, extreme temperatures, poisons, infections or social stimulation, can produce a series of stereotyped physiological responses in a mammal. These are essentially the same for all "stressors." When, for example, rats or other laboratory animals have been exposed to prolonged low temperatures, or subjected to trauma, or injected many times with small amounts of formalin, or repeatedly damaged in any one of a variety of ways, there are characteristic changes in the blood constants, in the circulatory machinery and in the lymphatic tissues. In particular, there is an excessive activation of the adrenal cortex; this seems to be a very fundamental response to "stress." Stimulation of the hypothalamus, directly by way of neural pathways or indirectly through the sympathetic nervous release of adrenaline, increases the production of ACTH (often at the expense of other pituitary hormones such as the gonadotropins or growth hormone) and hence the activation of the adrenal cortex. The corticoids mobilize glucose and in other ways meet the demands produced when homeostasis is drastically altered.

Resistance is maintained for a time, but eventually the pituitary-adrenal mechanisms fail, and a condition of adrenal insufficiency develops which produces circulatory collapse and other systemic changes found in "stress." The medical physiologist has accumulated convincing evidence for this sequence of changes through his studies of adrenalectomized animals and the effects of injected steroids and ACTH, as well as by comparisons of animals maintained under many kinds of altered environments. The higher vertebrates, with their more precise homeostatic controls, are most susceptible; but a similar, although less extreme, picture also develops in many of the lower vertebrates under comparable conditions (Hoar, 1966; Donaldson and McBride, 1967; Fagerlund, 1967). It is conceivable that the more complex invertebrates respond to environmental extremes in a comparable way.

10

TEMPERATURE

Within an animal's zone of tolerance, temperature change frequently produces a prompt, direct, and proportional alteration in the rate of its physiological processes, although the range of temperature for which these relationships hold is usually not more than 10° to 20°C. At temperatures not far outside the normal range biological processes are retarded or completely inhibited. Thus, for many phenomena there is a TEMPERATURE OPTIMUM; this has been demonstrated many times in such diverse processes as the growth rates of fish, the luminescence of bacteria and the movements of animals (Fig. 10.1). Enzyme-catalyzed reactions

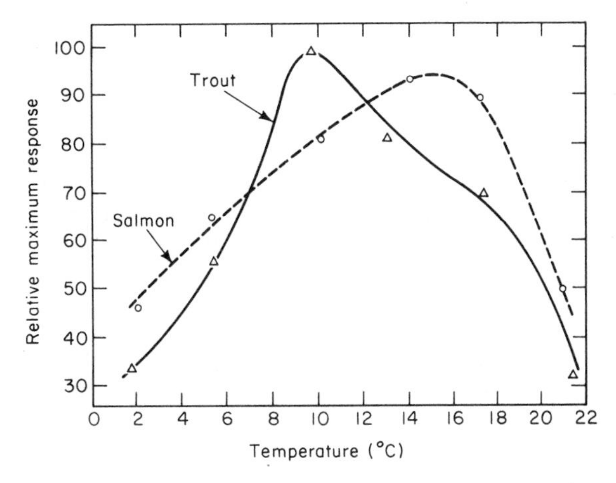

Fig. 10.1 Maximum distance moved by salmon and trout in response to an electrical stimulus at different temperatures. [From Fisher and Elson, 1950: Physiol. Zool., *23*: 32.]

follow the same pattern, and the obvious explanations are in terms of the chemical reactants.

This line of argument was at one time extended to the concept of a MASTER OR CONTROLLING REACTION for each of the many different physiological processes. It was postulated that the dominant enzyme reaction for complex activities like heartbeat or breathing might be identified by studying the temperature relations and comparing them with different enzyme reactions. This concept was never fully satisfactory and has now been largely discarded, even though it is recognized that some physiological processes may be dominated by one important enzymatic activity or another. Biological phenomena are exceedingly complex and involve temperature dependent physical (diffusion, absorption) as well as chemical phenomena. Whether chemical or physical, they depend on molecular activities, and their bases should be sought in physicochemical terms. The fact that there is no simple or single description for temperature effects on physiological activities should not discredit the approach nor lead to its discard in favor of purely empirical relationships such as those proposed by Bělehrádek and others (Bělehrádek, 1930; Johnson *et al.*, 1954).

TEMPERATURE AND THE RATES OF BIOLOGICAL ACTIVITIES

Arrhenius, in the latter part of the nineteenth and early twentieth century, first deduced the relationships which are used by both chemists and biologists to describe the effects of temperature on reaction velocity. From a study of the rate of hydrolysis of sucrose, he formulated the empirical relationship:

$$\frac{d \ln k}{dT} = \frac{A}{RT^2}$$

where k is the reaction velocity constant, T is the absolute temperature, R is the gas constant (1.987 cal per degree per mole) and A is a constant, the significance of which remained to be demonstrated. The integrated form of the equation between temperatures T_1 and T_2 corresponding to reaction velocities k_1 and k_2 is:

$$\ln \frac{k_2}{k_1} = \frac{A}{R}\left[\frac{1}{T_1} - \frac{1}{T_2}\right]$$

Hence, a straight line relationship may be expected when the $\ln k$ is plotted against the reciprocal of the absolute temperature, $1/T$ as shown in Fig. 10.2. The slope of the line is equal to A/R or approximately $A/2$ if natural logarithms are used ($A/2.303R$ if the rate was plotted as $\log_{10}$ reaction velocity). The value of the constant A for a great many enzymatic reactions and biological processes falls between 1000 and 25,000 cal. In some of his later studies of complicated biological reactions, Arrhenius used the term μ instead of A for the constant, and this symbol is now universally used by biologists and referred to as the "critical thermal increment," the "apparent activation energy" or the "temperature characteristic." Its

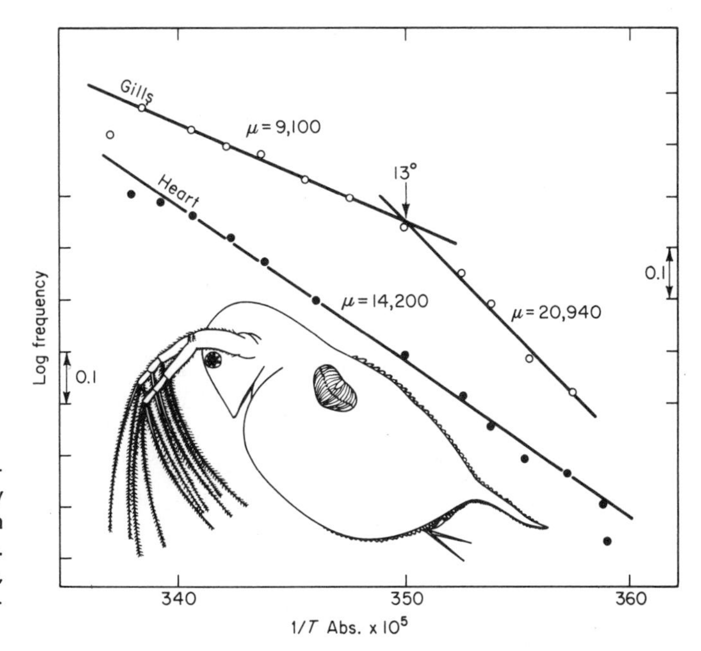

Fig. 10.2 Temperature characteristics for heart rate and respiratory movements in *Daphnia*. [Data from Stier and Wolf, 1932: J. Gen. Physiol., *16*: 367, after Barnes, 1937: Textbook of General Physiology. Blakiston, Philadelphia.

significance was not apparent when the formula was first deduced, and a quarter of a century or more passed before the constant A or μ secured a firm theoretical and experimental basis. A historical account of physical chemistry with theoretical discussions of the constant will be found in texts of cell physiology and molecular biology (Johnson *et al.*, 1954; Giese, 1973). Only a brief comment is given here.

Chemical reactions depend on molecular collisions and interactions at the electronic level. Thus, temperature change may be expected to alter reaction rates through its effects on molecular activity and the kinetic energy of the reaction system. The physical chemists of the nineteenth century noted, however, that the temperature effect was much greater than would be expected on this basis, and thus Arrhenius introduced the hypothesis of an activated state. According to this theory all elementary rate processes (diffusion, solubility, oxidation, hydrolysis) can be considered unstable equilibria between reactants or molecules in the normal state and those in an activated condition (Johnson *et al.*, 1954). In short, of the numerous molecular collisions which occur in a reaction system, only that small fraction which involves active molecules will result in chemical change. These active complexes are very rapidly formed with rising temperatures, so that reaction rates go up by about 12 per cent per degree, in contrast to less than 1 per cent which would be the case if only the increase in average kinetic energy of the molecules in the system were considered.

Experimental evidence in physical chemistry is in full agreement with this theory. Subsequent developments have provided formulae for the calculation of the changes in activation energy associated with temperature alterations (Maxwell-Boltzmann distribution law). The precise meaning of this activated state can only be obtained from an understanding of advanced quantum and statistical mechanics; but the Arrhenius constant A or μ can be appreciated from these brief comments. For our purposes, this constant represents the energy which molecules in their initial state must acquire before they can participate in a chemical reaction. It is the energy of activation for the particular chemical reaction or biological process and remains constant at least over a limited temperature range. Thus, the hydrolysis of β-glycerophosphate by bone phosphatase has a μ value of 9940 cal over the range of 10° to 40°C; the rate of creeping of ants, the chirping of crickets and the flashing of fireflies have μ values in the neighborhood of 12,200 cal; respiratory and cardiac rhythms show higher μ values of about 16,700 cal (Fruton and Simmonds, 1958).

In many cases the value of μ does not remain constant over the temperature range compatible with the particular activity. Marked changes are common in simple enzyme-catalyzed reactions as well as in complex physiological processes (Fig. 10.2). Crozier (1924–25 and later) interpreted these changes in terms of MASTER OF CONTROLLING REACTIONS which governed the overall rate of complex physiological processes. According to the Crozier theory, a physiological process depends on a catenary series of reactions, each with its characteristic critical thermal increment; the rate of the entire process is governed by the slowest reaction in the series, and this is the master reaction. The μ value for a complex physiological activity is the μ value of the slowest step.

In a general way, this much of the theory is probably true since temperature may have a much more significant influence on one type of process than it has on another. Photochemical reactions, for example, have much lower μ values than thermochemical ones. A process such as photosynthesis which involves both photochemical and thermochemical processes, may show very different μ values over a range of temperatures. At lower temperatures, the limiting factor is the rate at which light quanta can be absorbed (a photochemical process); this is characterized by a much lower μ value than the thermochemical or enzymatic reactions which control the rate of photosynthesis at higher temperatures, where the chlorophyll machinery is saturated with radiant energy (Giese, 1973). This concept of the limiting or master reaction which was developed earlier (Blackman–Pütter principle) has remained a part of the thinking of the general physiologist and has found a place in the discussions of the environmental physiologist (Chapter 9: Johnson *et al.*, 1954; Fry, 1971).

Crozier went further, however, and argued that the sharp changes in the slope of the Arrhenius plot (Fig. 10.2) represented a change from one master reaction to another and that it should be possible to identify the master reactions from their μ values. This is an appealing hypothesis and was the basis of considerable research during the early years of this century. The hope of identifying precise master reactions, however, was never realized, and the expectation of this now seems somewhat illogical. Activities like the creeping of ants, the breathing of frogs, or any one of the many other phenomena examined are as liable to be limited by such physical phenomena as diffusion, viscosity, or protein denaturation as they are by the enzyme-catalyzed processes emphasized by Crozier. There must often be an interaction of several potentially limiting processes, physical as well as chemical, and it is not surprising that the relationship between velocity and temperature can sometimes be better represented by a curve than by a series of straight lines with sharp breaks (Fig. 10.2). Excellent discussions of these theories are available.

The critical thermal increment, even though its significance in terms of specific physiological reactions is uncertain, remains the most realistic of the several temperature characteristics which biologists have used to describe temperature effects on reaction rates. It is not the simplest, however; the familiar Q_{10} value (calculated readily from the same data used in determining μ) is frequently as useful as the critical thermal increment for descriptive purposes. Q_{10} is the increase in reaction velocity caused by a 10°C rise in temperature. Thus,

$$Q_{10} = \frac{k_{t+10}}{k_t}$$

where k_t is the velocity constant at temperature t and k_{t+10} the velocity constant at 10°C higher. The value is easily calculated with data obtained over any temperature range from the general formula

$$Q_{10} = \left(\frac{k_1}{k_2}\right)^{10/(t_1 - t_2)} \quad \text{or} \quad \log Q_{10} = \frac{10\,(\log k_1 - \log k_2)}{t_1 - t_2}$$

where k_1 and k_2 are the velocities at t_1 and t_2 respectively.

Numerous Q_{10} values have been tabulated. Like μ values, they vary somewhat with temperature range and the conditions of the material. In general, however, Q_{10} values associated with physical processes such as diffusion or conductivity and those associated with photochemical reactions are less than 1.5, while thermochemical (enzymatic) reactions range from 2 to 3. Values for protein coagulation and heat death are much higher; the Q_{10} for coagulation of egg albumen is 635 and of hemoglobin, 13.8 (Giese, 1973). Similar ranges have been recorded for heat death of protozoans (891–1000) and of rabbit leucocytes (28.8).

Physiological Compensation for Rate-Limiting Temperature Effects

These, then are the rules which govern the effects of temperature on reaction rates. At the molecular level, the varied life processes are governed by a logarithmic law relating velocity of reaction to the temperature.

Free life has escaped the rigidity of this Arrhenius relationship, and some of the most significant steps in organic evolution were made in exploring these escape routes. Among the poikilothermic animals, physiological compensations adjust rates and activities to seasonally changing temperatures and to latitudes. A study of cardiac or respiratory rhythm at a series of temperatures during the summer may indicate that these processes would be almost suspended during the winter. This is not the case. On the contrary, acclimatization may lead to similar rates at both seasons. Likewise, the same animal species in different latitudes may show similar or identical rates even though the temperatures vary greatly. The species has escaped "from the tyranny of a simple application of the Arrhenius equation" (Barcroft, 1934). Among homeothermic animals, temperature compensation, which in the poikilotherm occurs at the cellular and biochemical level, depends more particularly on a specialized thermostat in the brain; this controls the body's regulating machinery so that the internal environment is temperature-constant.

Homeothermism was one of the most progressive steps in animal phylogeny. It took place in the evolution of the brain; perhaps its greatest significance has to do with the constant rates and the continuous high level of activity provided for nervous coordination. Barcroft (1934) was the first to discuss these matters in a lucid way. In his words, "Here then is a very fine issue—the cold-blooded animal successfully adopting ingenious mechanisms, first biochemical, then physiological, in order to adapt its heart to the variations of its environment; the warm-blooded animal discarding what its cold-blooded predecessor has laboriously beaten out, invoking the nervous system to reverse the normal biochemical relationship and gaining a new freedom by adapting, not itself to the internal environment, but the internal environment to itself." The mechanisms associated with temperature compensation are so radically different in these two animal groups that their physiology will be discussed separately in the following sections.

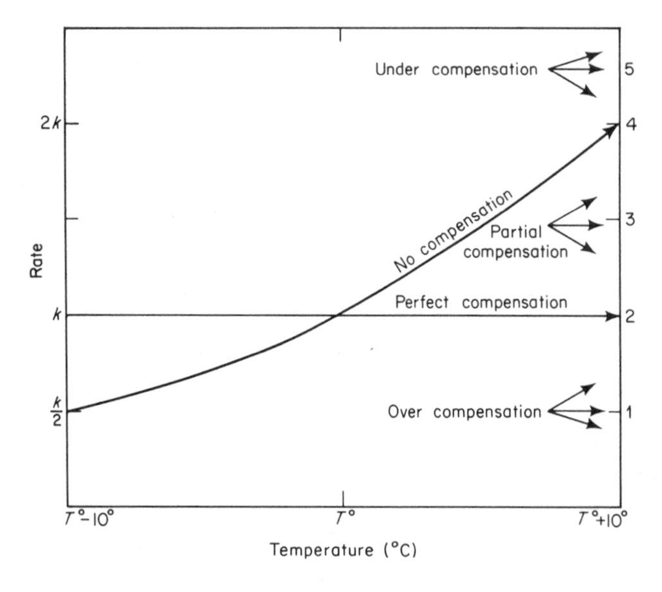

Fig. 10.3 Effect of temperature change on the rate of a biological process. The "no compensation" line shows the immediate effect, based on a Q_{10} of 2.0. Other lines and points show various degrees of compensation that have been recognized. Numbers at the right are the numerical types described by Precht (see text).

TEMPERATURE COMPENSATION IN POIKILOTHERMS

The theoretical consequences of the temperature-variable environment are shown in Fig. 10.3. As indicated by the solid line, raising or lowering the temperature by 10°C will alter the rates of the physiological processes according to the Q_{10} coefficient of about 2. In the laboratory such effects are readily demonstrated. It is obvious that a strict adherence to the van't Hoff–Arrhenius relationships in life processes would result in markedly different seasonal rates and latitudinal differences. As a matter of fact, this is the exception rather than the rule. The immediate consequences of a temperature change such as that shown in Fig. 10.3 are usually followed by gradual compensations (capacity adaptation) which often bring the altered rate back to the original rate or somewhere near it. Following the terminology of Precht (1958) and Prosser (1958), the compensation is referred to as "perfect" if the original rate is attained, or "partial" if an intermediate condition occurs. Examples of "over" and "under" compensation have also been described (Prosser, 1973).

This type of description can be extended by measuring the velocity of a physiological process at a series of temperatures. The velocity-temperature curves thus obtained may show rather different patterns in relation to the thermal history of the organism. If, for example, metabolic rates of "winter" and "summer" animals are compared at a series of temperatures, the velocity-temperature curves will be expected to fall along one line if no temperature compensation has taken place between winter and summer (Fig. 10.4) but show different positions if the basic biochemical controls have been altered. Several different patterns have been described. The curves are said to show "translation" if the winter and summer lines are parallel and "rotation" if they are crossed (Fig. 10.5). Prosser (1973) gives many examples. These changes in position indicate alterations at the molecular level during the seasonal adjustments. Several probable physiological mechanisms are suggested below.

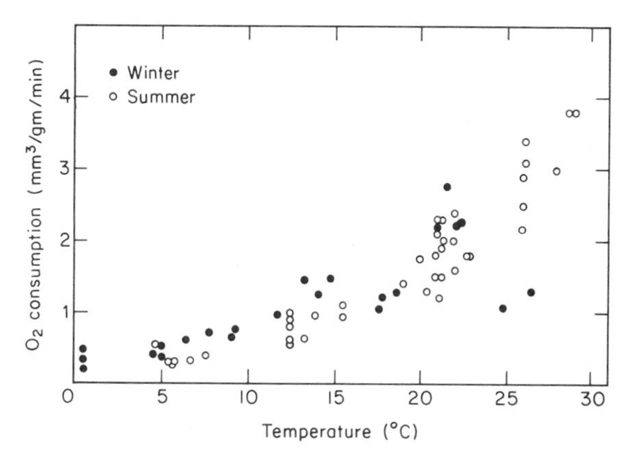

Fig. 10.4 The metabolism of a teleost fish *Tautogolabrus adspersus* during the winter and summer showing little or no seasonal temperature compensation in the rate of function. [Data from Haugaard and Irving, 1943: J. Cell. Comp. Physiol., *21*: 23.]

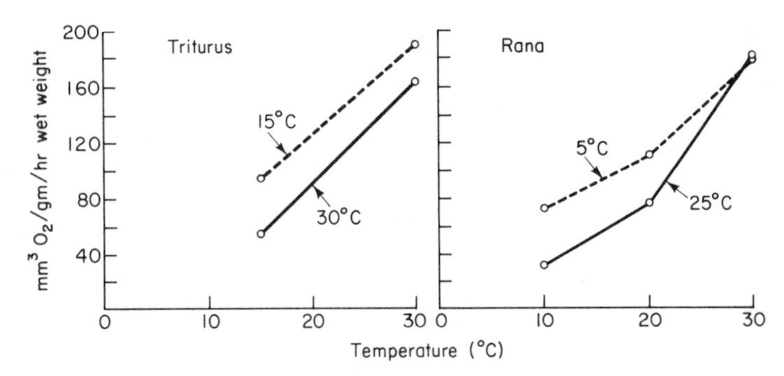

Fig. 10.5 Total oxygen consumption for the amphibians *Triturus viridescens* and *Rana pipiens* acclimated to the two temperatures shown on the graph and measured at a series of ambient temperatures as shown on the horizontal axis. The lines for *Triturus* show TRANSLATION and those for *Rana* show a clockwise ROTATION combined with translation. [Based on Rieck, Belli, and Blaskovics, 1960: Proc. Soc. Exp. Biol. Med., *103*: 437.]

Effects of acclimation on the temperature responses of poikilotherms are more completely described in the analysis popularized by Fry (1947, 1967, 1971). In this analysis, the effects of both thermal history and activity are depicted in a single graph (Figs. 7.23 and 10.6); the curves for active metabolism usually rise to a peak near the optimum temperature while the curves for standard metabolism rise progressively with acclimation temperature until the lethal level is reached. These relationships have now been depicted for many different poikilotherms (Newell, 1970; Brett, 1971; Prosser, 1973).

Mechanisms for Temperature Compensation

The various patterns of temperature compensation have been more satisfactorily described than the physiological mechanisms responsible for the differences. At the ecological level, it is customary to recognize three types of compensation: acclimatization, acclimation and phylogenetic adaptation (Chapter 9). However, it is unlikely that these processes are physiologically distinct, and comparative physiologists have sought for compensatory and lethal mechanisms in the operation of enzyme systems, cellular processes, the functioning of the neuroendocrine systems and the operation of the total animal.

At the most primitive level, temperature compensation is a BIOCHEMICAL and CELLULAR PROCESS. Exposure to a different temperature alters enzyme kinetics in organisms at all levels in phylogeny. The rapidly growing literature testifies to the importance of temperature-induced alterations in enzymes, their substrates and reaction kinetics during the compensatory response (Chapter 2; Hochachka and

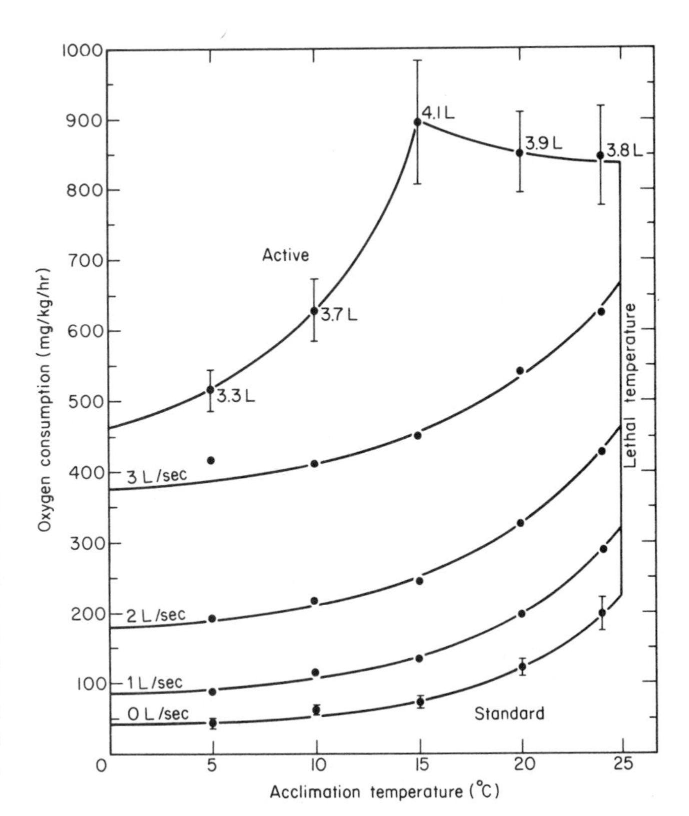

Fig. 10.6 Metabolism of yearling sockeye salmon in relation to acclimation temperature and activity expressed as swimming speeds in lengths per sec (L/sec). Limits include ±2 S.E. Active metabolism shows maximum sustained speeds for 60 min. See also legend for Fig. 7.22. Difference between ACTIVE and STANDARD is the SCOPE FOR ACTIVITY. [Courtesy J. R. Brett, 1963.]

Somero, 1973; Prosser, 1973). At the organ-system level of the multicellular animal, temperature effects may be more critical in one tissue or organ system than in another, even though all physiological processes are based on enzyme-catalyzed reactions. In the higher vertebrates the NEUROENDOCRINE system attains the ultimate control by regulating the temperature of the entire organism; even among lower forms a measure of neuroendocrine adjustment of physiological processes occurs. In addition, BEHAVIORAL responses may assist in the adaptation to changing thermal environments. The mechanisms for thermal compensation in the poikilotherms can thus be conveniently considered as biochemical, neuroendocrine and behavioral.

Biochemical and cellular adjustments Here the adjustments involve enzymes, substrates, and reaction media. Extensive research on the biochemistry of temperature compensation and thermal death has focused attention on adjustments in the water content of the tissues, the temperature relations of the enzyme reactions, the structure of the proteins, the organization of the lipids (especially those of the protoplasmic membranes), and the formation of toxic byproducts.

Metabolism takes place in an aqueous solution, and consequently life is only possible while water is in the liquid state. The actual biokinetic range is narrower than the temperature range for liquid water (0° to 100°C) and rarely exceeds 10° to 45°C. Temperatures outside these limits usually inhibit vital enzyme reactions or bring about irreversible changes in proteins and lipids. It is important to note that low-temperature death may occur at temperatures well above freezing (10° to 15°C in warm-acclimated fishes, Fig. 9.1) and that heat death occurs well below the boiling point of water. It is likewise known that a gradual temperature change is much less damaging than a sudden one. Bacteria (*E. coli,* for example) are killed if cooled suddenly from 37°C to 0°C, but they may be gradually cooled to this same temperature without damage even though the cooling period is only 30 minutes (Smith, 1954). There are many similar examples in species ranging from the protozoans to the vertebrates. The hazards of EXTREME temperatures (ice formation, for example) cannot be denied, but thermal death may be expected at temperatures well above the freezing and far below the vaporization of water.

In very cold environments, several adjustments are known to prevent water crystallization, which damages cells either through a mechanical disruption of their organization or merely by removing the solvent (water) from the reaction medium, leading to excessively high concentrations of solutes. The latter effect is probably the more damaging since it usually occurs first in the interstitial spaces and leads to osmotic desiccation of the cells. The tissues of some intertidal animals have been shown to tolerate considerable ice formation. Mussels and periwinkles have been found to survive at −22°C for months in the Arctic, with about 75 per cent of their body water frozen and their tissue fluids four times the normal concentration (A. U. Smith, 1958). This tolerance of high salinities has not been fully explained. It appears to be related to a high content of "bound" intracellular water; while the cells of most organisms contain about 4.5 per cent bound water (Webb, 1965), those of *Mytilus* in very cold environments have 20 per cent of the cellular water in the bound state and consequently incapable of being osmotically withdrawn from the cell. Williams (1970) postulates the presence of an unidentified "antifreeze" (see below). Several writers have reported changes in the proportions of bound and free water during thermal adjustments (Smith, 1954: Fry, 1958).

In a test tube, crystal formation at temperatures below zero can be avoided by rapid cooling which either leads to the formation of minute crystals or to an amorphorus solidification (vitrification); a number of lower organisms have been experimentally taken to very low temperatures in this way and shown to survive. Ice formation can also be avoided by supercooling, a phenomenon easily demonstrated by cooling water or a solution while avoiding agitation. In this way, temperatures considerably below zero develop without ice formation; crystallization occurs rapidly when these supercooled fluids are agitated or "seeded" with an ice crystal or other particulate matter.

This phenomenon has been recognized in nature where fish living in the deep waters off Labrador were shown to have body fluids with freezing points of −0.95°C in waters at −1.75°C. Thus, they are swimming about in a supercooled state with body temepratures about 0.8°C below the freezing point of their blood; they freeze when their bodies are seeded with ice, even though they are surrounded by fluid (DeVries, 1971). Insects may also be supercooled to temperatures as low as −30° to −40°C. Unlike the Arctic fishes studied by Scholander, some insects (the European corn borer, for example) can survive freezing, subsequent to supercooling (A. U. Smith, 1958; Asahina, 1969). Studies of the European corn borer showed that the physiology was subject to seasonal acclimatization, since summer and autumn insects, in contrast to the winter and spring ones, could not tolerate freezing.

Finally, the damaging effects of low temperature may be avoided by altering the freezing points. The freezing point of any solution is lower than that of the pure solvent; a molal solution of any non-electrolyte has a freezing point of −1.86°C. Thus, an increase in the osmotic content of the body fluids will lower the freezing point and, other things being equal, protect the organism from the damage of ice formation. This phenomenon is well recognized both in insects and in fishes. There are many records of such changes in the physical properties of solutions during compensation to freezing temperatures. Only a few will be noted here.

"Winter-hardened" insects may show a fantastic increase in the osmotic content of the hemolymph and this, coupled with their capacities for supercooling (down to −47.2°C in overwintering larvae of *Bracon cephi,* a parasite of the wheat stem sawfly), produces the lowest temperature tolerances yet recorded in multicellular animals. Salt (1959, 1964) showed that glycerol and sorbitol were the solutes mainly responsible for this remarkable phenomenon. In *Bracon,* concentrations of glycerol may rise to 5 molal at the time of hibernation. This depresses the freezing point of the hemolymph to −17.5°C. Glycerol has several properties which make it a particularly useful metabolite in low-temperature resistance (Smith, 1954). It has a strong tendency to supercool, partly because of its high viscosity. Although its melting point is + 18.0°C, it is seldom seen in the crystalline state. Small amounts of water depress its freezing point and, conversely, it depresses the freezing point of aqueous solutions Its utility as an "antifreeze" is well known to the layman, while biologists find it a most useful medium for the low-temperature storage of sperm and blood cells. The literature is discussed in textbooks of cell physiology.

Scholander and associates (1957) found that shallow-water fishes in the fjords of northern Labrador are swimming about in ice water at −1.7° to −1.8°C protected by a solute which develops with the onset of winter and doubles the osmoconcentration of their body fluids. It is now known that several changes in osmotically active substances may occur in the plasma of fishes exposed to low temperatures. In some temperate zone fishes, low-temperature acclimation induces an actual decline in the plasma electrolytes (Houston and Madden, 1968) although the total osmolality rises because of an increase in the small weight organic molecules (Umminger, 1971). In *Fundulus,* glucose has been shown to play this role; several aspects of carbohydrate metabolism have been studied in supercooled fish (Umminger, 1972 and elsewhere). In polar fishes, proteins and/or glycoproteins rise as the water temperatures fall (DeVries, 1971; Hargens, 1972). The glycoproteins in Antarctic fishes of the genus

Trematomus have been most carefully investigated. These molecules have a molecular weight of about 21,500 and, like glycerol, are characterized by the presence of many hydroxyl groups; the action of these substances as "antifreezes" is destroyed by chemically blocking these groups.

The mechanisms, by which such substances as glycerol and the glycoproteins depress the freezing point and prevent ice formation, are not fully understood (DeVries, 1971; Hochachka and Somero, 1973). In part, their action depends on the numerous hydroxyl groups which may be expected to bond with the polar groups of water and thus reduce water-water interactions and the formation or growth of ice crystals. In addition, glycerol and similar substances may exert a protective influence on cellular proteins (enzymes); these cryoprotective materials appear to prevent the changes in tertiary and quaternary protein structure, which frequently occur at low temperatures; in this way they may maintain important enzymes in the active state at low temperatures (Hochachka and Somero, 1973).

Many years ago it was suggested that changes in the organization of the lipids, especially those of cell membranes, might be important in temperature compensation. It was noted that animals of the same species from different latitudes often show characteristic differences in tissue fats. Salmon from northern Pacific waters, for example, have fats with lower melting points than those from more southern areas, and similar differences were recorded in several other groups of animals (Heilbrunn, 1952). The character of the body lipids seems to be correlated with environmental temperatures. Experimental evidence is in agreement. At the very beginning of the present century. Swedish scientists dressed pigs in sheepskin coats and, after two months, compared their body fats with those of controls maintained at about 0°C and about 35°C; the findings were consistent with the theory that an elevated temperature (through insulation) will raise the melting point of the body fats. Similar results were obtained by acclimating fishes and tadpoles to different temperatures. Heilbrunn (1952) and Prosser (1973) cite many references.

There are several places where lipid organization is critical to life. The most obvious spots are the cell membranes and nervous tissues, both of which contain much fatty substance. Older theories stressed the probable importance of fat in the molecular organization and physical state of cells and tissues; changes in lipids during thermal compensation might be critical in regulating permeability and ion exchanges at altered temperatures. Alternatively, or in addition, lipids may play their special role in enzyme reactions; several recent investigations have stressed this possibility (Hochachka and Somero, 1973). Many of the enzymes concerned with mitochondrial activity, for example, contain lipid as an essential part of their structures. Hazel (1972) investigated one of these enzymes, succinic dehydrogenase, from goldfish acclimated to 5°C and 25°C. Significant differences were found in the activities of his two extracts. However, when these were purified and fractionated into the lipid and protein components, the two enzymes (proteins) no longer showed a difference and appeared to be biochemically the same. The significance of the lipid component was shown by recombining lipid and protein (Fig. 10.7); further tests with different

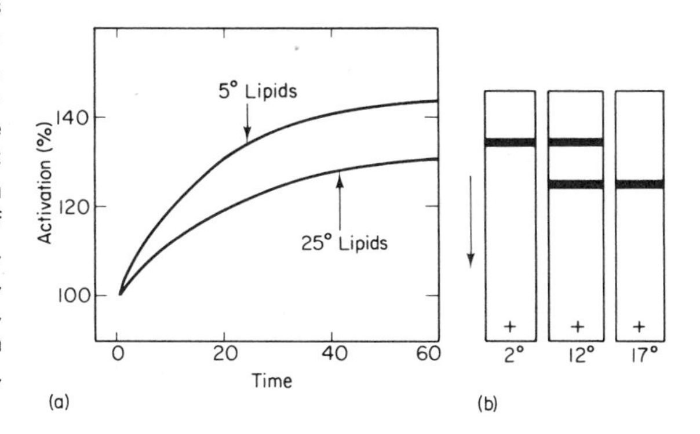

Fig. 10.7 Effects of acclimation on two enzyme systems in fish. (a) Activation of purified succinic dehydrogenase (protein) by purified lipids isolated from mitochondria of goldfish acclimated to 5°C or 25°C. Note that the time-dependent activation process leads to much higher enzyme activities when the lipids from cold-acclimated mitochondria are present. (b) Electrophoretic patterns of rainbow trout brain acetylcholinesterase isozymes from 2°, 12°, and 17°C acclimated fish. [From Hochachka and Somero, 1973: Strategies of Biochemical Adaptation. Saunders, Philadelphia; (a) based on Hazel, 1972: Comp. Biochem. Physiol., *43B*: 837; (b) based on Baldwin and Hochachka, 1970: Biochem. J., *116*: 884.]

classes of lipids support the hypothesis that the important change during temperature compensation relates to the lipid (Hazel, 1972).

It must be stressed, however, that the lipoprotein enzymes are not the only ones which may change during thermal compensations. Characteristic isozyme patterns are recognized in poikilotherms living in different latitudes while distinct changes have been charted during acclimation (Fig. 10.7). There may also be adjustments in the actual quantities of some enzymes (Somero, 1969; Hochachka and Somero, 1973); different rates of synthesis for both proteins and nucleic acids have been listed among the biochemical changes associated with thermal compensation (Somero and Hochachka, 1971).

An adaptive response in the enzymic systems appears to be fundamental to thermal compensation. These adjustments occur at more than one point in the reaction series. Not only the types of enzymes and their modulators (Chapter 2) but also the reaction media (electrolytes, lipids and other organic molecules) have been shown to change and to affect rates of reaction. The biological significance of these alterations in enzyme-substrate affinity (Chapter 2) can be appreciated from the comparative studies (for example, Fig. 10.8), even though the detailed mechanisms are not yet fully explained (Hochachka and Somero, 1973).

Finally, various metabolites may appear in excess at extreme temperatures, and these may alter protoplasm and contribute to the lethal process. Toxic substances, such as histamine which increases cell permeability, or thromboplastic materials which coagulate protein, have often been identified in injured tissues. Within the ZONE OF TOLERANCE adjustments occur in the enzymes, the substrates and the reaction media;

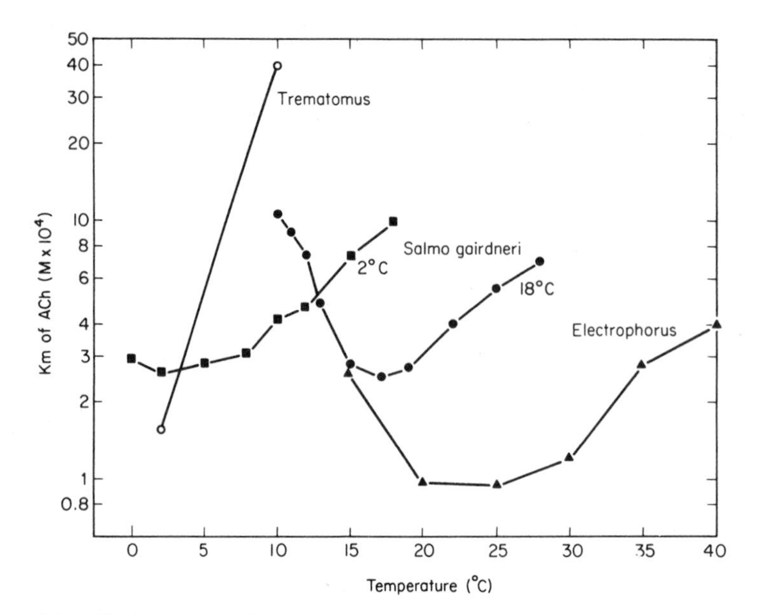

Fig. 10.8 Enzyme (brain acetylcholinesterase) activity in relation to temperature for trout acclimated to two temperatures and for an Antarctic fish *Trematomus* and an Amazon river fish *Electrophorus*. [From Baldwin, 1970: Temperature Adaptation in Enzymes of Poikilotherms: Acetylcholinesterase in the Nervous System of Fishes. Ph.D. thesis, Univ. Brit. Col., Vancouver.]

within the ZONE OF RESISTANCE denaturation of the proteins, disorganization of the lipids, changes in the distribution of water and electrolytes and the development of toxic materials are some of the factors which lead to cellular death.

Neuroendocrine mechanisms It has been suggested that heat death in larger organisms is ultimately caused by some failure in the nervous system, with a loss of indispensable reflexes such as cardiac or respiratory rhythm. Many years ago Battle (1926, 1929) tested the responsiveness of skate tissues at a series of progressively rising temperatures and found that such processes as the automaticity of the heart, myoneural junctions, and peristaltic movements in the gut were much more sensitive than tissues such as somatic muscle. Synapses were found to fail at or below the lethal temperature, while some other tissues remained responsive after the animal's death. There are some confirmatory data for other species (Fisher, 1958). Battle also found distinct differences between the resistances of synaptic processes in closely related species of *Raja* and even between small and large individuals of a single species of flounder (*Pseudopleuronectes americanus*). These data suggested a specific locus for failure of vital processes in thermal death.

More recently, Roots and Prosser (1962) have examined the effects of low temperature on caudal reflexes, peripheral nerve conduction, conditioned responses and activity of goldfish (*Carassius auratus*) and bluegills (*Lepomis macrochirus*) acclimated to a series of temperatures. Peripheral nerve conduction continued at temperatures lower than those required to block the caudal reflexes; the reflexes were, in turn, still evident at temperatures lower than those required to block conditioned reflexes. These data indicate that the locus most sensitive to cold is in the central nervous system. Konishi and Hickman (1964) have recorded midbrain potentials of trout at different temperatures in response to optic nerve stimulation. The response pattern is substantially prolonged when fish acclimated to 10°C are transferred to 4°C. With acclimation to the lower temperature, however, there is a progressive decrease in the potential duration at 4°C showing significant compensation. A comparable change was not observed with acclimation to higher temperatures (16°C) and these workers argue that acclimation to low temperatures involves changes in biophysical properties of nervous tissues while high temperature acclimation depends more on metabolic compensation. Prosser (1967, 1973) discusses several other attempts to identify temperature-sensitive loci in the nervous system.

The endocrine system is also involved in thermal compensation. Seasonal cycles in the endocrine activity of poikilotherms are well documented and may be basic to the temperature acclimatization processes. Goldfish maintained under constant laboratory conditions of temperature show relatively greater resistance to heat during the summer and a relatively greater resistance to cold during the winter (Hoar, 1955a). Temperature acclimation is not responsible for the compensation. Since the phenomenon can be partially regulated by altering the photoperiods, it is assumed that the neuroendocrine system (the usual link between the vertebrate photoreceptors and metabolism) is responsible for this compensation (Hoar, 1965). However, many of the details necessary for unequivocal evidence of a hormonal involvement in thermal compensation or temperature death are still lacking (Hoar and Eales, 1963; Johansen, 1967, 1968; Umminger, 1972). It seems clear, nevertheless, that in the more complex poikilotherms compensatory changes in the neuroendocrine system play a significant role in temperature compensation. They may have marked the first phylogenetic steps in the evolution of the neuroendocrine controls of the homeothermic animals, but there is at present little pertinent research (Fry, 1958).

Behavioral regulation This line of argument may be extended to the behavioral regulation shown by several poikilotherms, both invertebrate and vertebrate (Gordon *et al.*, 1972). Some make use of solar energy, while others utilize metabolic heat to raise their body temperatures. Many different species of animals, when placed in a temperature gradient, have been shown to "prefer" one particular temperature area. They are less active at the PREFERRED TEMPERATURE and thus arrive there by a THERMOKINESIS (Chapter 21).

The terrestrial environment is more prone than the aquatic to sudden temperature changes. The most successful terrestrial poikilotherms (insects and reptiles) have made good use of behavioral responses to avoid temperature extremes or to elevate temperatures sufficiently for certain activities. Some reptiles have particularly well-developed sensory organs for this purpose. The infrared sense organs in the facial pit of the rattlesnake can detect a temperature difference of the order of 0.001 to 0.005°C—a sensitivity which is highly valuable in detecting warm-blooded or cool (moist) prey, as well as in orienting the animals to warm or cool environments (Bullock and Diecke, 1956; Barrett *et al.*, 1970; Gamow and Harris, 1973).

The entomological literature gives many examples of insects which use solar or metabolic energy to warm up before flight (Heath and Adams, 1967). Furthermore, such social insects as ants, termites and bees may regulate temperatures in their nests or hives through varied activities (Scherba, 1962; Southwick and Mugaas, 1971). Prosser (1973) cites numerous references to the original literature.

TEMPERATURE COMPENSATION IN HOMEOTHERMS

In eutherian mammals the body temperature lies somewhere between 36° and 38°C; in birds it is slightly higher, between 39° and 42°C while in the monotremes and marsupials it is somewhat lower—30° to 35°C (Prosser, 1973). This stabilization of body temperature removes one of the variables of the internal environment and permits a steady high level of activity, both metabolic and locomotory. The advantages (at the metabolic level) are somewhat speculative but very obvious in behavioral, social and cultural evolution, demanding continuous associations of individuals. Increasing complexity of organization (especially behavioral organization) makes homeothermy a necessity; or conversely, one may argue that complexity, both physiological and behavioral, becomes progressively more feasible as the internal environmental temperature, especially that of the nervous system, is stabilized. It is none the less curious that all species of mammals or birds should have fixed on about the same body temperature and that this should be of the order of 35° to 40°C rather than, let us say, 10° higher or lower. It may represent a particularly economical level for cellular activity in the environment where the homeotherms first evolved (Davson, 1970), but again this is speculative (Burton and Edholm, 1955; Young, 1957).

Maintenance of a constant body temperature is a neat balance between heat production and heat loss. It demands a sensitive thermostat in the brain, a capacity not only to use the heat formed as a byproduct of metabolism but also to increase the output of metabolic energy in accordance with demands. In addition, it requires several anatomical correlates such as appropriate insulation and special heat exchangers. In extreme conditions the metabolic price of a regulated body temperature may be impossibly high, so that some species temporarily suspend temperature control (torpidity and hibernation) or migrate to more favorable climates. In man, there is a behavioral evasion of extremes with the development of clothing, air conditioning and other technological devices.

Phylogeny and Ontogeny

Among the homeotherms there is a good general correlation between the precision of temperature control and the complexity of behavioral organization. This is true both phylogenetically and ontogenetically. The variations in body temperature of a series of mammals measured after a two-hour exposure to temperatures ranging from 5° to 35°C are shown in Fig. 10.9. There is a graded series from the poikilothermic reptiles through the primitive mammals (Monotremata and Marsupialia) to the Eutheria.

Likewise, there is variation within the Eutheria; many of them are virtually poikilothermic at birth. Full temperature control is not attained in the newborn rat for 73 days (Adolph, 1957). A similar situation is found in many birds (Kendeigh, 1939). Species of homeotherms which are born or hatch in a helpless condition depend for some time on their parents for heat as well as food and shelter; but there are other species which are quite independent at birth and well able to regulate body

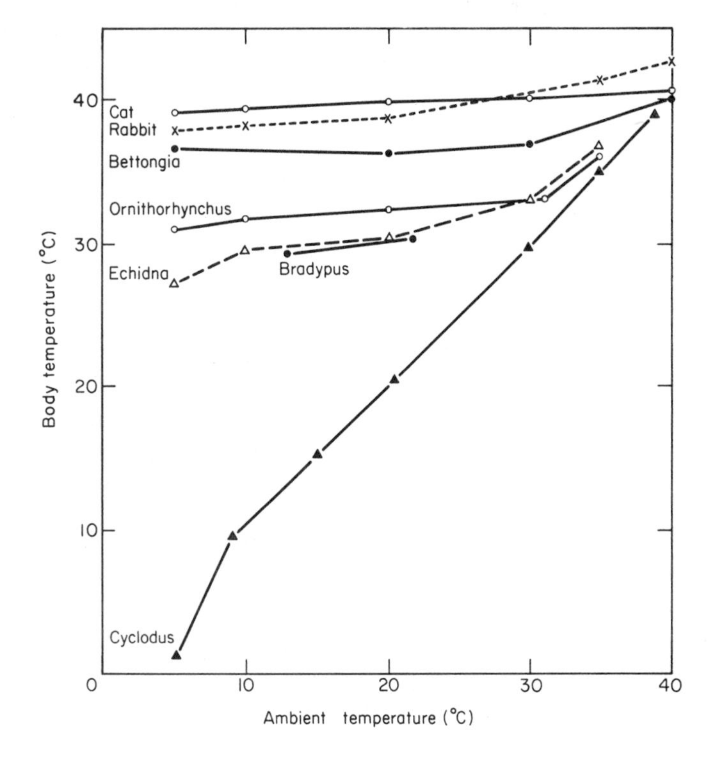

Fig. 10.9 Relationship between body temperature and the ambient temperature for the lizard (*Cyclodus*), the monotremes (*Ornithorhynchus* and *Echidna*), the marsupial (*Bettongia*), the sloth (*Bradypus*, an imperfect homeotherm), and the true homeotherms (cat and rabbit). [From Johansen, 1962: *In* Comparative Physiology of Temperature Regulation, Part 1, Hannon and Viereck, eds. Arctic Aeromedical Laboratory, Alaska.]

temperature. The caribou calf is a good example; it can take care of its own heat production even though it is born into an extremely rigorous environment (Hart *et al.*, 1961).

Different species of birds and mammals vary greatly in their temperature-regulating capacity. Some breeds of dogs (also ducks) can withstand an environmental temperature of −100°C *for one hour* before there is a depression in the "core" temperature of the body. This LOWER CRITICAL TEMPERATURE, as it has been called, is by contrast −1°C in naked man, −15° in a porpoise, −30° in a sparrow, −45° in a rabbit and −90° in a goose (Precht *et al.*, 1955). Several other comparative measurements of temperature resistance are on record. Scholander (1955) defined the CRITICAL TEMPERATURE as the lowest air temperature at which the animal can maintain a resting or basal metabolic rate without losing body temperature. By this criterion, most tropical mammals, including naked man, have critical temperatures between 25° and 27°C while larger arctic mammals such as the fox range from −30° to −40°C. In smaller species like the lemming it is considerably higher (about +15°C). The variability in metabolic rate, which appears long before alterations in the core temperature, is shown in Fig. 10.10. The critical temperature is a most useful index of the temperature-regulating and insulating properties of an animal. In Fig. 10.10 it is shown as the point where metabolism diverges from the standard value.

The ability of birds and mammals to face elevated temperatures is just as variable and again reflects a genetic capacity related to the thermal conditions in the animal's natural environment. The fur seal may actually die of overheating on a warm day in the Arctic at 10°C while an essentially tropical animal like man can tolerate four times this temperature without damage, even though his cooling machinery may be strongly activated under such conditions. Within the same species

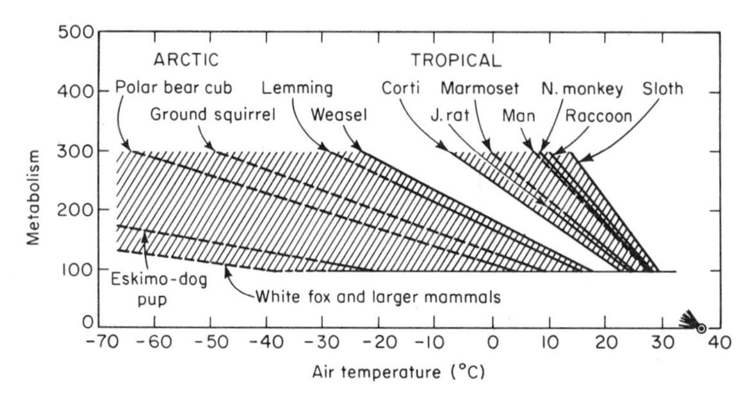

Fig. 10.10 Metabolism of resting mammals in relation to environmental temperature. The base line (100) is the standard or basal metabolism. Solid lines, observed values. Broken lines, extrapolated values. [Based on Scholander, Hock, Walters, Johnson, and Irving, 1950: Biol. Bull., *99*: 254.]

(men or cattle, for example), marked racial differences in thermal tolerance are known. Prosser (1973) cites these and many other interesting examples.

Phylogeny of homeothermy The relationship shown in Fig. 10.9 for the reptile *Cyclodus* is misleading because it suggests that reptiles are unable to regulate their body temperatures. Although this is the case in an acute experiment such as that shown in the figure, it is by no means true in nature. Many reptiles maintain steady body temperatures for prolonged periods through BEHAVIORAL THERMOREGULATION (Gordon *et al.*, 1972; Prosser, 1973). The precision and complexity of the thermal reactions of the horned toad *Phrynosoma* are shown in Fig. 10.11; comparable compensatory reactions have been observed in many reptiles. These animals regulate their temperatures by sun basking or moving into the shade and by making contact with a thermally favorable substratum.

Four important physiological reactions are associated with these behavioral responses: metabolic heat production, cardiovascular reactions, skin color, and

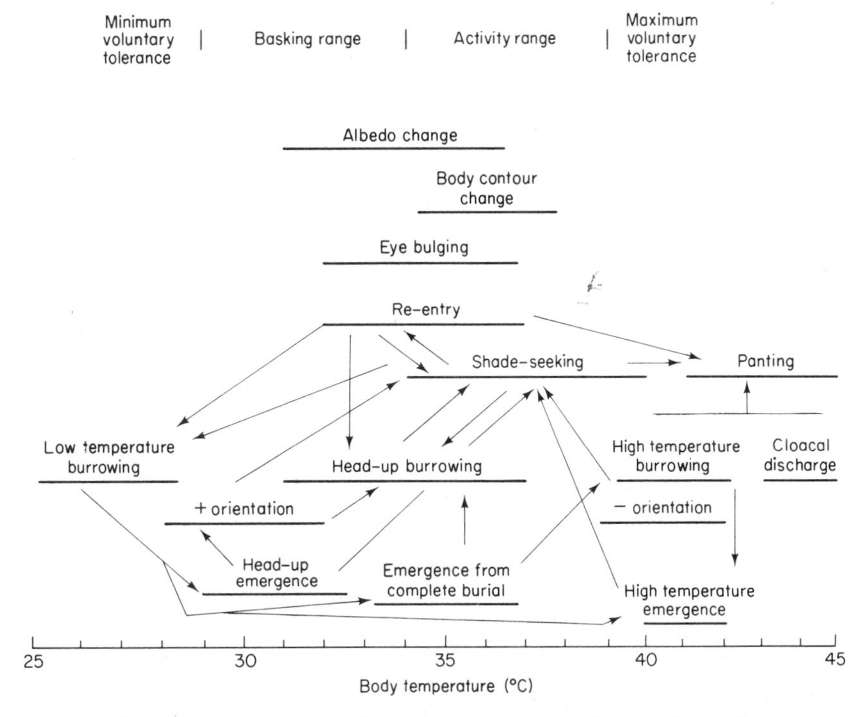

Fig. 10.11 Behavioral thermoregulation in the horned lizard *Phrynosoma*. The diagram shows the range of body temperature for each behavior pattern. [From Heath, 1965: Univ. Calif. Publ. Zool., *64*: 124.]

reflectivity with evaporative cooling through skin or respiratory surfaces (Cloudsley-Thompson, 1972; Gordon *et al.*, 1972). The degree to which each is developed depends on the species, its size and environmental pressures. Thus, the python, while incubating its eggs, maintains a constant temperature of about 33°C by spasmodic muscular contractions; body temperatures may be elevated by as much as 7.3°C above ambient temperatures by means of metabolic heat production analogous to the homeotherms (Hutchinson *et al.*, 1966). By contrast, the lizards do not appear to utilize metabolic heat when temperatures fall. One further example, however, shows how effectively a lizard can use cardiovascular responses to regulate its temperature. The marine iguana *Amblyrhynchus* of the Galapagos spends much of its time on rocky shores under the intense tropical sun regulating its body temperature during the day to about 37°C by behavioral means. It feeds on algae from waters which are about 10° to 15°C cooler and, consequently, it faces sharp temperature changes at frequent intervals. It is to the iguana's advantage to cool slowly on entering the water and to heat rapidly on emerging and it has been shown that the cooling rates are about one-half of the heating rates (Bartholomew and Lasiewski, 1965). Further, the heart rate is more rapid during heating than cooling and this is coupled with an increased cutaneous blood flow. Thus, the slowing down of the general circulation and the restricted flow of blood to the skin will reduce heat loss in water while the more rapid circulation and flow of blood through surface layers will increase the warming from the environment. A more complete discussion of thermoregulation in the reptiles will be found in the literature already cited. The point emphasized here is that this group of animals exploits many of the same physiological mechanisms as do birds and mammals; clearly, homeothermy was preceded by a physiology that permitted the ectothermic animal to maintain a constant body temperature. It should be noted that different reptiles regulate their body temperatures at somewhat different levels; that the extent of regulation varies seasonally and diurnally; that their important behavioral responses depend on a precise temperature sense; and that metabolic heat usually plays little or no part in the regulatory processes. The reptiles differ from the homeotherms in lacking the central autonomic (hypothalamic) controls, the continuously high body temperatures, the emphasis on metabolic heat and the insulation, in form of feathers and fur. However, the reptilian ancestors exploited all the important physiological mechanisms which were later refined in the evolution of warm-blooded birds and mammals. In this way they set the stage for the true homeotherms.

Regulation of Body Temperature

The "DU BOIS TEMPERATURE BALANCE" has been shown in many textbooks of physiology during the past thirty-five years and is still worth careful study (Fig. 10.12). It shows graphically how the physiological and metabolic reactions which produce heat must be matched against those which radiate or conduct it away in

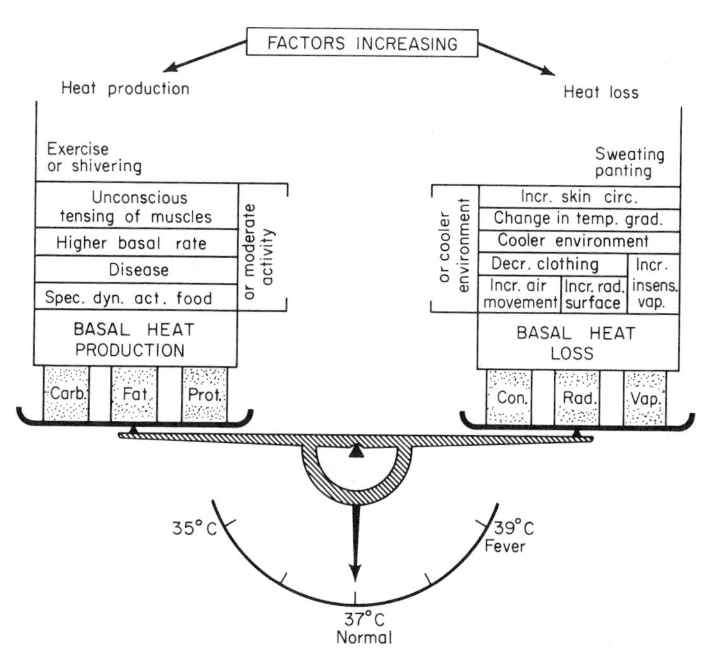

Fig. 10.12 The Du Bois temperature balance. [From Du Bois, 1937: Lane Medical Lectures. Stanford University Press. Stanford, Calif.]

order to provide a constant body temperature. Except in a very narrow THERMALLY NEUTRAL ZONE, the maintenance of a constant body temperature makes a steady demand either on the chemical processes of heat production or on the physical devices for heat loss.

Heat production A man, when appropriately acclimated, may show an eighteen-fold increase in heat production within twelve minutes if suddenly immersed in a bath at 4°C. In contrast, moderate exercise such as walking at the rate of three miles per hour increases the total heat production by only three times. The eighteen-fold increase is comparable to that observed in maximum physical activity. This phenomenal heat production (of the order of 16,000 kcal/m^2/24 hr) is associated with violent shivering; even under much less extreme conditions this is the familiar response to the cooling of the body surfaces.

In some animals an increased muscle tone can be measured before overt shivering or piloerection is evident. These muscular responses, referred to as SHIVERING THERMOGENESIS, generate a large amount of heat but are by no means responsible for the total. All metabolizing tissues produce some heat, and in the maintenance of homeothermy the fat and viscera may be as greatly involved as the muscles. The contributions of the various organs have been measured in several animals (Jansky,

1971); during cold stress, acclimated rats produce as much as 65 per cent of their heat in fat and visceral organs (Jansky and Hart, 1968). This heat produced from extramuscular sources is termed NONSHIVERING THERMOGENESIS.

Animals vary greatly in their capacity to generate heat. Measured on the basis of calories per kilogram per day, the energy output of a canary is 1000 times that of a sturgeon. One of the greatest contrasts between poikilotherm and homeotherm is in the heat production curves over a series of environmental temperatures (Fig. 10.13). Even among the different species of homeotherms the capacity shows a twenty-fold variation in terms of energy output per kilogram per day (Table 10.1).

Table 10.1 Energy Production Under Basal or Standard Conditions for Representative Vertebrates

Animal	Body weight (kg)	Body surface (m^2)	Energy output per day (kcal/kg)	(kcal/m^2)
Man	56–65	1.65–1.83	23.2–25.5	790–910
Beef cattle	400–500	3.20–4.7	15.2	1635
Sheep	42.7–49.5	0.95–1.1	25.7–26.3	1160–1180
Dog	11.7–15.5	0.58–0.65	33.5–38.5	770–800
Cat	3.0	0.2	50	750
Rat	0.2	0.03	130	830
Marmot	2.6	0.18	28.3	420
Elephant	3670	23.8	13.3	2060
Duck	0.93	0.1	90	855
Pigeon	0.28	0.04	100	670
Canary	0.016	0.006	310	760
Lizard	1.2	0.11	2.5	29
Frog	0.05	—	—	130
Sturgeon	1400	11.8	0.3	31

SOURCE: Data from Spector, 1956: Handbook of Biological Data. Saunders, Philadelphia.

Such temperature-metabolism comparisons might suggest that adjustments in heat production would provide important avenues for climatic adaptation to temperature. This is not the case. Metabolism is roughly proportional to body surface, and when comparisons are made on this basis, the differences between various species of true homeotherms are less evident. There is no consistent difference between the heat generating capacities of tropical and arctic species (Table 10.1). The evolution of homeothermic living in the extreme environments was primarily by way of physical regulation of heat dissipation rather than by chemical processes involving heat production.

Brown fat—a special thermogenic tissue in mammals All organs produce heat as a byproduct of metabolism but the mammals, and only the mammals, have a

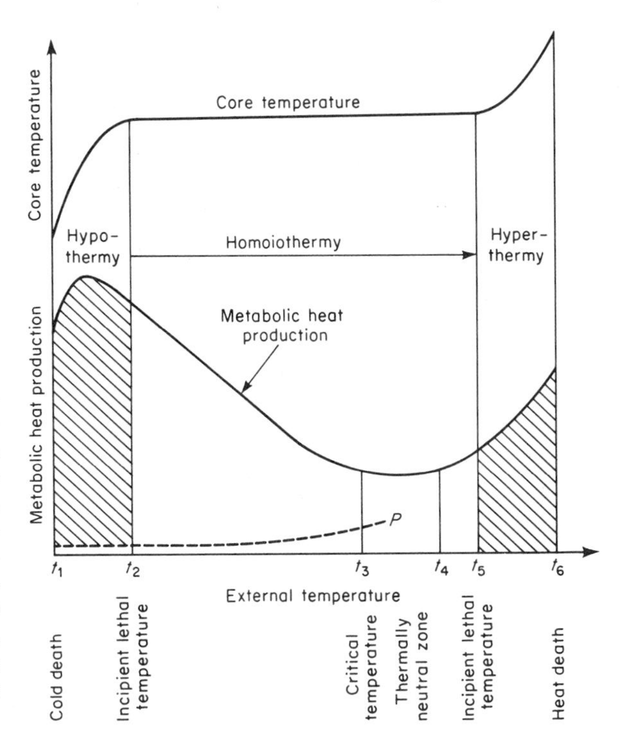

Fig. 10.13 Pattern of heat production and variations in body temperature of homeotherms exposed to various external temperatures. The broken line *P* shows, for comparison, the heat production of a poikilotherm. [From Allen, 1960: *In* Comparative Biochemistry, Vol. 1, Florkin and Mason, eds. Academic Press, New York; after Precht, Christophersen, and Hensel, 1955: Temperatur und Leben. Springer Verlag, Berlin.]

special thermogenic organ or tissue known as BROWN FAT or BROWN ADIPOSE TISSUE (Dawkins and Hall, 1965; Smith and Horwitz, 1969; Hochachka and Somero, 1973). This tissue is most conspicuously developed and active in many newborn mammals (including the human), in hibernating animals at the time of arousal and in cold-acclimated mammals; like a smoothly operating automatic furnace, it responds to cold stress with a large burst of heat. The tissue is strategically localized in the neck and thoracic regions in relation to major blood vessels so that its heat is quickly transported to those organs (brain and heart) whose continuous high temperatures are vital to the operation of a homeotherm. The physiological mediator for the stimulation of brown fat appears to be noradrenaline; sympathetic stimulation activates the adenyl cyclase system with rapid lipase hydrolysis and oxidation of fats.

Brown fat was first described over four centuries ago in hibernating marmots, *Muris alpinus*. It was called the HIBERNATING GLAND and assumed to play an active role in hibernation. This role was not established until the second half of this century; there are still many unsolved problems concerned with its intermediary metabolism. Histologically, brown fat is an extremely vascular tissue with relatively high concentrations of myoglobin, cytochromes and flavin compounds; these give it a pale buff to dark reddish brown color which contrasts with ordinary (white) adipose tissue. It also differs cytologically from white fat in that the cells are smaller, polygonal rather than spherical, with central nuclei and numerous fat droplets, which are dispersed rather than concentrated in a single vacuole; mitochondria are small, numerous and closely associated with the fat droplets. Brown fat also differs biochemically from white fat, with higher concentrations of water, protein, phospholipid, cholesterol and certain mitochondrial enzymes; its *in vitro* rate of respiration is much higher (Smith and Horwitz, 1969). Details of mitochondrial respiration in brown fat are speculative but heat, rather than ATP, appears to be the main end product; an uncoupling of oxidative phosphorylation has been suggested (Hochachka and Somero, 1973). This tissue can be thought of as a rapidly firing furnace that warms some mammals up quickly in an emergency, without muscular activity or shivering.

Heat loss Temperature regulation would be extremely uneconomical if it depended solely on variations in metabolism. The rapid generation of heat can mean survival in a crisis, such as that which man would face if suddenly immersed in water at 4°C. However, the energy requirements in such a situation are comparable to those of vigorous muscular exercise; they could only be met by increased fuel consumption, and this during seasons and in environments where food is liable to be minimal. Among the warm-blooded animals, temperature adaptation in the genetic sense has been a refinement of the mechanisms for heat conservation or dissipation. Many different anatomical and behavioral correlates have appeared —particularly in connection with body insulation, vascular heat exchangers and economical body forms (Scholander, 1955).

Three weights (conduction, radiation and vaporization) are shown on the right side of the Du Bois temperature balance (Fig 10.12). Convection need not be listed as a separate factor since it only contributes to conduction or radiation. The relative significance of the three different weights varies with the environmental conditions and the structure of the integument. In an aquatic environment, conduction will account for the entire transfer. In the terrestrial habitat, however, only relatively small amounts of heat are exchanged in this way. Most of the homeotherms are terrestrial. In man, radiation usually accounts for 55 per cent or more of the heat lost and evaporation for 40 per cent or less—the amounts depending on the environmental temperature and humidity (Hardy, 1961; Ruch and Patton, 1965). Loss of heat by radiation and conduction is only effective in a cool environment. It is obvious that, at high temperatures, the animal will actually take on heat by these routes. Evaporation, however, is always a negative factor. About 0.6 kcal is required to vaporize 1 g of water from the moist surfaces of the skin or respiratory epithelia. This technique for cooling has been exploited in quite different ways by the birds and mammals.

Birds have a dry skin with no special integumentary organs to increase cooling by vaporization. They do, nevertheless, show excellent temperature tolerance and heat regulation. Relatively high body temperatures are often coupled with a capacity to tolerate a hyperthermia of 2°C or more (Bartholomew, 1964); these high temperatures provide a distinct biological advantage since they reduce the environmental temperature differential and heat load. This is particularly important in small birds which have fewer escape routes from the heat than mammals or large birds; small mammals often burrow under ground or/and are nocturnal while large birds may soar into the cooler upper atmosphere. Even though the skin is dry and insulated, evaporative cooling is still the major factor in lowering body temperatures. Buccal and respiratory surfaces are the important routes for evaporation. Two different movements have been distinguished: PANTING which refers to rapid breathing movements (thorax and abdominal muscles) and GULAR FLUTTER which involves a quick fluttering of the floor of the mouth and upper region of the throat; the latter activity is metabolically less expensive. At elevated temperatures, 100 per cent of the heat loss may be by evaporation. A pigeon's rate of breathing at 41.7°C is 46 per min with a tidal volume of 4 ml and a minute volume of 185 ml. At a body temperature of 43.6°C the corresponding values were 51 per min for respiratory rate, 1.2 ml for tidal volume and 610 ml for minute volume (Salt and Zeuthen, 1960). The air sacs are of particular significance. Inactivation of the abdominal and thoracic sacs in the pigeon reduces the water loss by 50 to 65 per cent; at high air temperatures such an animal is no longer able to maintain its normal temperature. Several anatomical adaptations regulate air sac evaporation under particular conditions (Salt and Zeuthen, 1960). If plenty of water is available, some birds may increase their cooling by urinating on their legs. The wood stork will excrete as often as once per minute when placed in a very warm environment, and the evaporation of water on the vascular appendages has a significant effect in lowering body temperature (Kahl, 1963).

A mammal, such as a panting dog, may also dissipate considerable heat through evaporation from respiratory surfaces. However, the integument forms the specialized route for water vaporization in most mammals. Mammalian skin is provided with sweat glands, and these seem to be a direct evolutionary response to the pressures of temperature regulation. At any rate, there are many obvious correlations between the cooling problems of the species and the development of these skin glands. Aquatic mammals, for example, lack skin glands and do not depend on vaporization for cooling; man, however, has sweat glands all over his body and can maintain a steady temperature in a warm environment.

It should be noted that considerable evaporation takes place from the skin even in the absence of sweat glands. As a matter of fact, the integument probably played an important part in temperature compensation long before the development of homeothermy. Cowles (1958) has speculated on the origin of dermal temperature regulation and has emphasized the phylogenetic significance of the highly vascular dermal layers of the skin in many amphibians and reptiles. In the poikilothermous amphibian it often serves as a supplementary respiratory organ; in some reptiles it is a valuable collector and dispenser of heat (Cloudsley-Thompson, 1972); and finally, in the mammals, it becomes a major organ of temperature regulation. In each case there is a system of dermal blood vessels which dilate in the heat and constrict in the cold to alter the peripheral blood flow. This physiological response in the poikilotherms is a nice preadaptation for the refined temperature regulation of the homeotherm.

Scholander (1955) discusses two major lines of adaptation in the control of heat dissipation: "one is the increase in body insulation as we go toward colder climates; another is an adaptation of extremities and other peripheral parts to tolerate, and remain functional at, low tissue temperatures, sometimes even approaching zero degrees." A third might be an adaptation of the body form (Bergmann's and Allen's rules), but Scholander considers this of little or no significance. Several good discussions of the last point will be found in the literature (Wilber, 1957; King and Farner, 1961; Dobzhansky, 1962).

Insulation by fur and feathers has been compared in many species of mammals and birds, and a significant relationship has been established between this capacity of the integument and the natural environment of the animal. In general, tropical species are less well insulated than the arctic species (Irving, 1972). In addition, there is usually a well-marked seasonal change in the insulative quality (Fig. 10.14).

Animals which lack fur can achieve insulation with a layer of cold superficial tissue. Pigs as well as aquatic mammals have been studied. This protection is somewhat akin to that afforded the skin diver inside his neoprene suit. It differs, however, in that the cold skin is vascular and can be quickly changed from an insulator to a radiator or a conductor through circulatory adjustments. In the seal, sharp temperature gradients have been measured from the skin surface at about 0°C (with animals in ice water) to a skin depth of about 5 cm where the normal mammalian body temperature is found (Irving and Hart, 1957). These cold tissues are sensitive and vascular, although about 35°C cooler than the body proper. Even more astonishing, they can warm quickly if the animal exercises or comes into warm air. Such abrupt thermal changes cannot be tolerated in most animals, even in aquatic poikilotherms. Presumably, the surface tissues of the cold-skinned mammal have achieved a rather unique biochemical organization.

Special insulative qualities have been claimed for fat. The fatty layer tends to

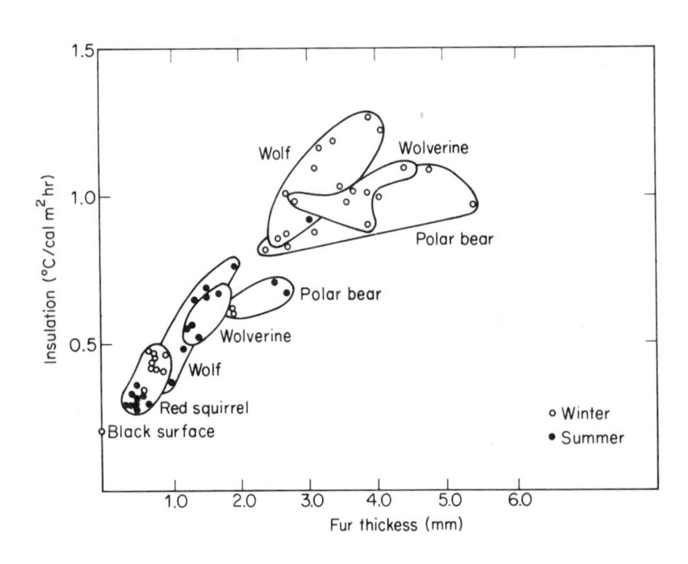

Fig. 10.14 Comparison of the insulative effectiveness of fur in summer and winter. [From Hart, 1956: Can. J. Zool., *34*: 55.]

be thicker in animals acclimatized to cold and particularly so in bare-skinned animals. A comparison has shown that fatter men (body fat about 20 per cent of the body weight) are 30 per cent better insulated than lean individuals with fat about 2 per cent of the body weight (Fry, 1958). The increased thickness provides better insulation, but it is questionable whether white adipose tissue has peculiar qualities in this regard. Dead fat is admittedly a better insulator than dead skin or muscle (the ratios of thermal conductance for the three are of the order of 3:2:1), but living fat is reasonably vascular and an active tissue.

There are, however, other properties of fat which may be as significant as its insulative capacities for homeotherms in the cold. The selective deposition of more fluid fats (lower melting point) in colder environments and in more exposed tissues such as those in the extremities of the Arctic animal (Heilbrunn, 1952; Irving *et al.*, 1957) may be important in preserving the flexibility of the tissues at low temperatures. Again, white fat has significant advantages as a metabolic source of energy (Chapter 7). The heat combustion of animal fats is approximately 9.3 kcal per g in comparison with about 4.1 for the carbohydrates and proteins in the human body (Ruch and Patton, 1965). Finally, adipose tissue may be highly specialized in the form of brown fat to yield quick bursts of heat in emergencies. Thus, the fat diet or the storage and metabolism of fats is particularly advantageous for the homeotherm in a cold climate (Pagé, 1957).

The Arctic fox can sleep in snow at −40°C and preserve his resting metabolic rate because of his heavy body insulation. The skin surface remains warm like that of a well-clothed man. This same insulation, however, poses a problem when the animal becomes active and generates excess heat associated with the rise in metabolism which may be twenty to thirty times the resting level. The compromise has been an uneven covering with poorly insulated extremities which serve as avenues for heat dissipation. To be useful, these extremities must be resistant to cold and capable of rapid changes in temperature. Although they can be kept warm by fur (or feathers) when the animal curls up to sleep, they are fully exposed during activity. The web in a gull's foot may be at 0°C and its leg at 7°C while the "drum stick" is 30°C and the core temperature of the body 41°C. The reindeer's leg may be at 8° to 10°C with a core body temperature of about 37°C. Nerves remain sensitive and cells uninjured in these cold tissues.

The anatomical correlate for this interesting specialization is the counter-current heat exchanger (Scholander, 1955). The principle has already been discussed (Chapter 4). The major arteries to the appendage are located centrally and surrounded by numerous thin-walled veins. When heat must be conserved, the peripheral circulation is restricted, and the warm blood flowing into the appendage passes heat to the cool blood of the nearby veins; there may be little more than 1°C between the temperatures of the arterial and venous blood in the proximal areas. When the animal faces the reverse problem of heat dissipation, more blood returns through the peripheral vessels, and cool blood rather than warmed blood returns to the body. In many birds and mammals (both from the tropics and the frigid regions) this vascular arrangement is elaborated into a multichanneled RETE made up of bundles of hundreds of intermingled arteries and veins. The tail, as well as the legs, often forms the site for these specialized vascular bundles.

Many different animals use heat exchangers to meet environmental stresses or to exploit a particular way of life (Schmidt-Nielsen, 1972; Taylor, 1972). Gazelles and antelopes may tolerate body temperatures as high as 46.5°C while the temperature of the brain remains about 2.9° cooler because of an efficient rete in the carotid artery; this brings the warm blood of the body in close contact with cooler venous blood from the nasal passages, which are kept cool by evaporation from their moist surfaces. Likewise, the temperature of the arterial blood to the testes of many mammals is lowered as it flows in close contact with an intricate plexus of veins returning blood from the testes and scrotum, which may also be cooled by evaporation; core body temperatures are incompatible with the development of fertile sperm. Some poikilotherms have also made good use of heat exchangers. The most carefully studied examples are the tunas and some of the sharks (family Lamnidae) which can maintain internal body temperatures as much as 10° to 15°C higher than their environment (Carey and Teal, 1966; Carey, 1973). The cooled arterial blood coming from the gills flows through cutaneous arteries near the surface; a tangled net of arterioles arising from these cutaneous arteries carries the blood medially toward the spinal column; this tangled net is in intimate contact with a venous plexus returning blood from the inside muscles to the superficial cutaneous veins. This efficient system provides these warm-blooded fishes with many of the advantages of the homeotherm; muscles that are 10°C warmer can generate three times as much power; the advantages in high-speed swimming are obvious. Some of the large flying insects have comparable specializations; further examples will be found in the literature already cited.

Regulatory mechanisms The immediate responses to acute temperature change are mediated through the nervous system. The physiological thermostat has been localized in the hypothalamus. Through appropriate surgical procedures and electrical and temperature stimulation it has been shown that the anterior hypothalamus of the mammal is responsible for protection against heat while the posterior hypothalamus confers resistance to cold (Burton and Edholm, 1955; Ruch and Patton, 1965).

These centers are reflexly activated by the temperature receptors of the skin or mucous membranes and directly through changes in the temperature of the hypothalamus or the blood circulating through it. The efferent nerve fibers are both somatic to the muscles controlling respiration (panting) and voluntary activities (shivering) and visceral to the autonomic system, which regulates the cutaneous blood vessels, the sweat glands and the piloerector muscles. Destruction of the autonomic fibers does not entirely eliminate vascular responses to temperature change, indicating, in addition, a direct effect of temperature on the blood vessels.

These are the acute responses to temperature change. If the exposure is more prolonged, the endocrine system enters the picture and metabolism is altered, particularly by way of the thyroid and the adrenal glands. These adjustments are considered below.

Homeotherms in Extreme Environments

The homeotherm has the same general avenues for temperature compensation as does the poikilotherm. Several well-marked changes have been described during laboratory ACCLIMATION; the seasonal and latitudinal differences due to ACCLIMATIZATION are recognized in both tropical and frigid regions. The homeotherms, even more than the poikilotherms, have achieved BEHAVIORAL COMPENSATION through migration, social cooperation and the development of favorable microhabitats. Finally, at the genetic level a specialized physiology permits some homeotherms to retreat periodically to the poikilothermic state through TORPIDITY or HIBERNATION.

Compensation through acclimation and acclimatization Both birds and mammals have refined capacities for compensation to heat and cold. There are excellent reviews of the extensive literature (Hart, 1964a, b; Chaffee and Roberts, 1971); detailed citations will not be given here. Many species variations have been noted and it is unwise to generalize from studies of any one animal, for example, the white rat, which has so often been used in laboratory work.

In birds and mammals, critical temperatures are adjusted both seasonally and during laboratory acclimation. The compensatory responses include changes in insulation, alterations in the storage and utilization of fat, adjustments in general metabolism and changes in the activities of the thyroid and adrenal (Fig. 10.15). Although the metabolic rates of tropic, temperate and polar homeotherms are similar, cold-acclimated and winter animals may maintain higher rates of heat production for long periods of time. Thus, the basal metabolic rate of the cold-acclimated rat is 20 per cent higher than the controls while Korean women divers (unlike men acclimated to the Antarctic) are found to show a 17 per cent increase in BMR during the winter; several species of Arctic birds respond to lower temperatures by an increase in metabolism.

Changes in body insulation have been carefully studied. In some homeotherms,

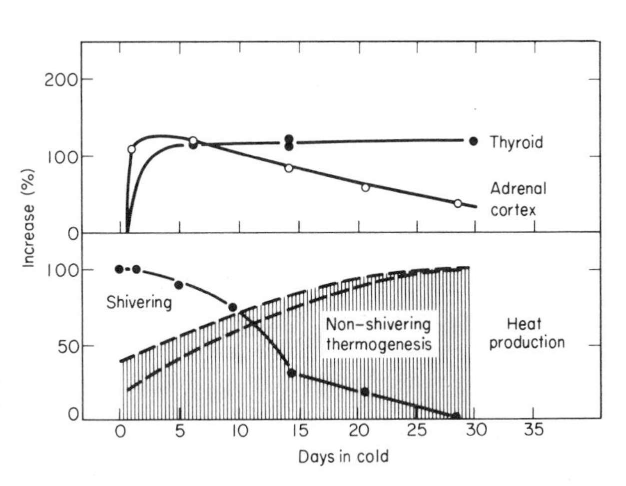

Fig. 10.15 Cold acclimation in rats. (Top) Summary of endocrinological changes; (bottom) nature of the thermogenic response. See original for sources of data. [From Hart, 1963: *In* Temperature—Its Measurement and Control in Science and Industry, Vol. *3* (3): 393. Reinhold, New York.]

insulation actually decreases (by as much as 75 per cent) during cold acclimation, evidently because of greater peripheral circulation and heat flow (Hart, 1957). However, in the acclimatization which occurs seasonally (under the influence of photoperiod as well as temperature) there is a significant increase in insulation and this reduces the demands for thermogenesis. The thickening of fur or feathers, which may increase insulation by 10 to 50 per cent, is associated with changes in the biochemistry of some of the peripheral tissues permitting them to operate when very cold.

Birds respond to the cold by shivering. Muscular activity appears to be their only means of increasing heat production; nonshivering thermogenesis has not been demonstrated in these animals; they lack brown fat and fail to show a thermogenic response to catecholamines. Although brown fat is absent, birds rely heavily on white adipose tissue in compensation to the cold and winter conditions; there is an increase in the degree of unsaturation of the fats, while their storage and metabolism are accelerated.

The metabolic significance of brown fat in cold-acclimated and winter mammals is still not well understood. It responds to noradrenaline with a sharp burst in metabolism, but the total contribution of heat in nonshivering thermogenesis is modest (only about 8 per cent in the cold-stressed rat). It has been suggested that the heat from brown fat is important in warming thermosensitive centers in the cervical spinal cord, thus suppressing shivering, or that it produces a hormonal factor. This tissue which has excited scientific curiosity for so many centuries may yet hold some scientific surprises.

The capacity for temperature compensation varies greatly in different animals. The red fox may extend its low temperature limits by at least 38°C, a porcupine by 33°C, a house sparrow by 25°C and a lemming by 17°C (Hart, 1957; Barnett, 1970). Man's capacities are much more limited but nevertheless real (Scholander, 1958). Although the body temperature remains the same, a lowered perspiration threshold, increased production of dilute sweat, and improved vascular responses have been measured when man is exposed to heat for prolonged periods. White men moving to a tropical desert climate may show a slight initial rise in body temperature with an elevated pulse rate; but within a few weeks the physiological constants are back to normal, and there is less and less discomfort with increased moisture production and improved cardiovascular responses (Adolph, 1947). Likewise, temperature tolerance of certain parts of the body may change with exposure. The hands of fishermen living in cold climates develop marked tolerance to cold and resistance to frost damage, primarily through adjustments in peripheral circulation (Prosser, 1973).

Cold narcosis and hibernation If the body temperature of a typical homeothermic animal such as man is appropriately lowered, cold narcosis develops. This form of anaesthesia has been extremely valuable in certain types of surgery. In clinical work, temperatures down to about 28°C can be readily produced in several ways—for example, by immersion in a water bath at 6° to 10°C while lightly anaesthetized to prevent shivering (McMillan and Machell, 1961). Temperatures considerably below this (15° to 18°C) can also be produced with impunity if an artificial heart and respiration are combined with the hypothermia (Drew, 1961). Experimental animals, such as dogs and rats, have been cooled to about 0°C, but recovery from this deep hypothermia requires artificial respiration and an external source of heat during re-warming. The hibernator does not require any such assistance while resuming the homeothermic state, and this points to major differences between cold narcosis and physiological hibernation.

These differences are evident at every stage in the process (Lyman and Dawe, 1960). Cold narcosis develops only as a response to a rather sharp lowering of the body temperature. In contrast, although the onset of hibernation comes during cold weather, it is seasonally controlled by internal (endocrine) factors and regularly appears in species such as *Citellus* at ambient temperatures of 30° to 35°C. Heat production falls precipitously with the onset of hibernation. In the marmot, at very low temperatures, it is reduced to less than 2 per cent of the resting value prior to hibernation (Davson, 1970).

Just as striking are the differences during the resumption of the homeothermic condition. The cold narcotized homeotherm requires an external source of heat; the hibernator warms spontaneously with an explosive burst of energy which amounts to as much as 3000 kcal/m^2/24 hr in the marmot and a temperature rise of 20°C in an hour.

A third important difference is found in the functioning of the nervous tissues. The true homeotherm shows respiratory arrest at about 19°C while the hibernator continues to breathe at 5°C or less. Isolated nerves of the rat (a non-hibernator) fail to conduct at 9°C while the hamster (a hibernator) shows nerve conduction down to 3.4°C. Likewise, the isolated auricle of the hedgehog maintains its spontaneous rhythm down to 5°C while that of the rat ceases at 16° to 18°C (Lyman and Chatfield, 1955). These observations attest a special adaptation of rhythmic and nervous tissues for activity at low temperatures. This is also evident in the physiological vigilance which the hibernator retains at low temperatures. Even though the animal is, at that stage, almost poikilothermic, its thermoregulatory machinery is not completely suppressed; the body temperature does not follow the ambient temperature when the latter comes dangerously close to freezing. At this point the hibernator awakes and becomes homeothermic.

Hibernation is then a specialized physiological state associated with the evolution of homeotherms in certain extreme habitats. Several steps between the true homeotherm and the hibernator may be expected. Pearson (1960) groups the birds and mammals into three categories: OBLIGATE HOMEOTHERMS, STUBBORN HOMEOTHERMS, and INDIFFERENT HOMEOTHERMS. Most of them belong to the first group and mobilize all their metabolic resources when the body temperature falls; they hold the temperature line just as long as possible and then die. Less common are the stubborn homeotherms which maintain a warm body over a wide range but become torpid when cold, especially if short of food. Deer mice (*Peromyscus*) resort to this expediency. Likewise, swifts, poor-wills and nighthawks, which feed on flying insects, are able to survive in a torpid state when the weather is cold and there are no airborne insects. Normally they do not show this torpidity if there is adequate food, even though their resting body temperatures may be somewhat below normal. The hummingbird, however, may maintain its body temperature for a period without food but quickly becomes torpid, even at mild temperatures, after dark. It is an indifferent homeotherm, like most of the North Temperate bats, whose body temperatures, even in moderately warm environments drop almost to those of the surroundings whenever they fall asleep. Some members of either the stubborn or the indifferent homeotherms will truly hibernate—that is, show a seasonally regulated fall in temperature associated with winter conditions.

The biological significance of overnight or unfavorable weather torpidity is just as great as that of hibernation. Pearson calculates that a hummingbird, which has a high energy demand because of its small size (Chapter 7), would use 10.3 kcal per 24 hr if it sleeps at night without becoming torpid but only 7.6 if it lowers the thermostat in the evening. This daily reduction in metabolism is only economical in small animals which can quickly cool and warm. A hummingbird may warm at the rate of 1°C per min. A massive animal such as a bear could not do this; the equivalent of the 24-hr heat budget is required to warm a 200-kg bear from 10° to 37°C (Pearson, 1960). It is much more profitable for the bear to carry a reserve of high energy foods and keep the metabolic fires burning during the night. However, even the bear may sometimes find it economical to lower the temperature, and this has become a part of the normal physiology of species living in northern latitudes. It should be noted that the body temperature of the bear is lowered only a few degrees (3° to 6°C) during periods of "winter sleep"; this is in marked contrast to a true hibernator, such as an Arctic ground squirrel, whose body temperature approaches that of the environment, down to about 2°C (Hock, 1960).

A low environmental temperature is only one of the unfavorable conditions which can be met by a period of hypothermia. The California pocket mouse *Perognathus californicus* and the kangaroo mouse *Microdipodops pallidus*, for example, become torpid when food is in short supply; even brief periods of torpidity will substantially reduce metabolic demands (Bartholomew, 1972). Likewise, white-footed mice sometimes escape the hazards of moisture shortage by becoming torpid. It is evident that many homeotherms have exploited hypothermia as a means of escaping from high metabolic demands during unfavorable conditions. The physiological differences between the short-term torpidity and the seasonal torpidity of animals that hibernate or aestivate are probably differences in degree rather than kind.

Physiological parameters have been recorded for many hibernating animals. Numerous reviews of the literature are available (Lyman and Dawe, 1960; Hoffman, 1964; Suomalainen, 1964; Fisher, 1967). The hibernator is essentially poikilothermic down to almost freezing. At low environmental temperatures the body cools rapidly on entering hibernation (in the ground squirrel from about 32°C to 8°C in 10 hours) and can warm even more rapidly on arousal (4° to 35°C in 4 hours).

The metabolic rate declines to between 1/30 and 1/100 of the "resting" level with a very low respiratory quotient which indicates fat utilization. The R.Q. is greater than 1 during the preparation for winter sleep when fat is being stored from the carbohydrate or other foods; but it is only about 0.3 to 0.4 during hibernation while energy is being drawn from the stored fats. The rate of breathing declines (in the ground squirrel from 100–200/min to about 4/min) and is markedly periodic (Cheyne–Stokes respiration).

Circulatory and vascular changes are equally spectacular. Again, in the ground squirrel the normal cardiac rate of 200–300/min falls to 10–20/min; there may also be an initial fall in blood pressure, but later this rises because of vasoconstriction and increased blood viscosity. In some species, a pooling of the blood has been noted in the spleen, with this organ several times its normal size during hibernation. An associated prolonged clotting time avoids the hazards of thrombosis during this period of stasis. Increases in the erythrocyte count and hemoglobin levels have also been reported.

The changes in fat metabolism, electrolyte balance, and endocrinology have most frequently been examined for causative roles. There may be as much as a 100 per cent weight increase caused by rapid lipogenesis just prior to hibernation —at the rate of about 2 g per day in the meadow jumping mouse *Zapus hudsonius* (Morrison and Ryser, 1962). An unsaturation of the body fats occurs with localized developments of "brown fat." The role of this tissue in the rapid thermogenesis associated with arousal is now well established (Hayward *et al.*, 1965; Lindberg, 1970).

The hibernator shows marked changes in electrolytes but the significance of these is not understood. The major alteration is an elevation of magnesium. This increases by as much as 25 per cent in the golden hamster, 65 per cent in the ground squirrel, some bats and the woodchuck, to a maximum of 92 per cent recorded by Suomalainen in the hedgehog (Riedesel, 1960). High magnesium levels are evidently universal during hibernation and have been of particular interest because of the well-known anaesthetic action of this ion (Heilbrunn, 1952). However, it has never been shown that magnesium actually causes the winter sleep; it has been noted in some species that the levels rise somewhat later than the first depression of the temperature and do not return to normal until after arousal.

The endocrinology of the thyroid, the islets of Langerhans, the adrenals and the pituitary-hypothalamic neurosecretory system have been frequently investigated. In general, thyroid activity is low during hibernation; there may be an increase in noradrenaline; the cortical tissues fail to show the usual response to low temperature "stress"; there may be hypertrophy of the pancreatic islet tissues, and well-marked changes have been followed in the hypothalamic neurosecretory system and in the pituitary.

The seasonal nature of hibernation makes it tempting to assign a regulatory role to the endocrine system. Many physiological processes are triggered seasonally by hormone action which is in turn regulated through the hypothalamus by the seasonal cycles of photoperiod and temperature. This may also be true of hibernation, but in some species, at least, the phenomenon seems to depend on an "internal seasonal clock." Pengelley and Fisher (1963) kept ground squirrels (*Citellus*) for two years under constant conditions of low environmental temperature and twelve-hour photoperiods and found seasonal periods of activity and hibernation which corresponded to the normal cycles in nature. However, the fact remains that hibernation is associated with a specialized endocrine physiology, whatever may be the role of hormones in the onset or the arousal from it. Its evolution has probably included adjustments in all the endocrine tissues, most frequently resulting in their involution prior to the winter sleep (Lyman and Chatfield, 1955). As some writers intimated, the physiology of hibernation may have been, in an evolutionary sense, built upon the "general adaptation syndrome" (Selye, 1949), which is the characteristic response of most mammals to extremes of temperature.

Behavioral regulation and migration Many of the homeotherms avoid extreme temperatures by seeking or constructing more favorable microclimates; others migrate seasonally to less rigorous climates. The physiology of migration has been intensively studied, and some of the more pertinent data will be considered in a later chapter.

11

IONIC AND OSMOTIC BALANCE

The tissue fluids are dilute saline solutions with sodium chloride as the predominant electrolyte. Small changes in composition are always permissible, but in most cases the variation compatible with life is extremely limited. The maintenance of this constancy is a major physiological task.

The marked similarity between these dilute salt solutions in tissues and the saline waters of the ocean has long intrigued physiologists. All evidence—both geological and biological—points to the marine habitat as the ancestral home of primitive animal cells, and some phylogenetic relationship between sea water and body fluids is a seemingly logical postulate. Macallum (1926), one of the first scientists to think seriously about this, saw in it evidence of a gradually changing composition of the oceans during animal evolution. He assumed that the oceans were much less salty when life made its appearance and that some portion of this ancestral sea became enclosed within the cells and tissues of primitive organisms; the body fluids of present-day animals would thus reflect the composition of the sea during the period when their ancestral lines were established. This hypothesis may be traced through the literature for almost half a century, but neither biologists nor geologists have ever found a shred of evidence for it. The palaeochemistry of sea water has now been carefully studied, and the indications are that the waters of the ocean have changed little in composition during the long period of animal evolution (Rubey, 1951; Riley and Chester, 1971). The salinity of the Palaeozoic seas was probably not unlike that of the present oceans.

However, the fact remains that life originated in sea water and that the body fluids are like sea water in their general composition. These facts may well be connected, even though it is not possible to adduce from them any evidence as to the nature of the ancient oceans. It seems likely that the marine habitat, in contrast to all others, provided conditions which were most suitable for the organization of protoplasm and the metabolism of cells. In all the animal phyla, from protozoa to vertebrates, there are representatives with body fluids of the same osmotic content as sea water; further, many more of the species in primitive phylogenetic groups are isosmotic with sea water. Perhaps the problem of water balance did not exist at the beginning of animal phylogeny. It might first have appeared when animals spread from the oceans into the estuaries, up the rivers, into the ponds and marshes and onto the land.

The problem of electrolyte balance, on the contrary, has certainly existed since the organization of the first cell. Unless primitive cells were very differently organized from those of today, their lives depended on the maintenance of a transmembrane potential brought about by an unequal distribution of ions (particularly sodium and potassium) across the cell membrane (Chapter 15). This bioelectric potential, characteristic of all living cells, depends on the active transport of ions; the organization of the first cell demanded efficient machinery for the regulation of its electrolytes and the transport of ions. IONIC REGULATION BY CELLS IN ISOSMOTIC SOLUTIONS is a part of the life of all cells in a multicellular animal as well as in the single-celled organism living in sea water; it is the first physiological topic for consideration in this chapter.

Most of the major phyla met the challenge of adapting their physiology to life in brackish and fresh waters. Some groups were more successful than others; the coelenterates and the echinoderms have been notably unsuccessful. The brackish water environment created a new water balance problem and accentuated the original difficulties of maintaining the balance of electrolytes. The osmotic flooding of water, which is inevitable in a less saline medium, requires a good water pump unless extreme dilution can be tolerated. The removal of this osmotic water leaches out the soluble salts. One of the first physiological challenges in spreading from the marine habitat was the PROBLEM OF MAINTAINING A HYPEROSMOTIC STATE. The opposite difficulty was encountered when phylogenetic lines, which had stabilized their body fluids at osmotically lower levels than the sea water, reinvaded the marine habitat. There are several examples of this; the teleost fishes provide a particularly successful one. Here the difficulty is that of MAINTAINING A STABLE HYPOOSMOTIC STATE in the face of continuous osmotic desiccation. Organisms which live in highly saline ponds and lakes must face the same stress.

In TERRESTRIAL LIVING, neither water nor electrolytes can be obtained directly from the ambient medium but are acquired through food and drink. In the most arid of terrestrial environments there may be absolutely no drinking water for long periods, and the animal must produce all of its body water through metabolism. Similarly, the land animal, with its relatively dilute body fluids, faces essentially desert conditions when it returns more or less permanently to the ocean and lives in a hyperosmotic environment.

Thus, depending on its habitat, every animal faces one of four major problems in the regulation of the electrolytes and water of the tissues (Fig. 11.1): (1) ionic regulation in isosmotic media; (2) maintenance of a hyperosmotic state; (3) maintenance

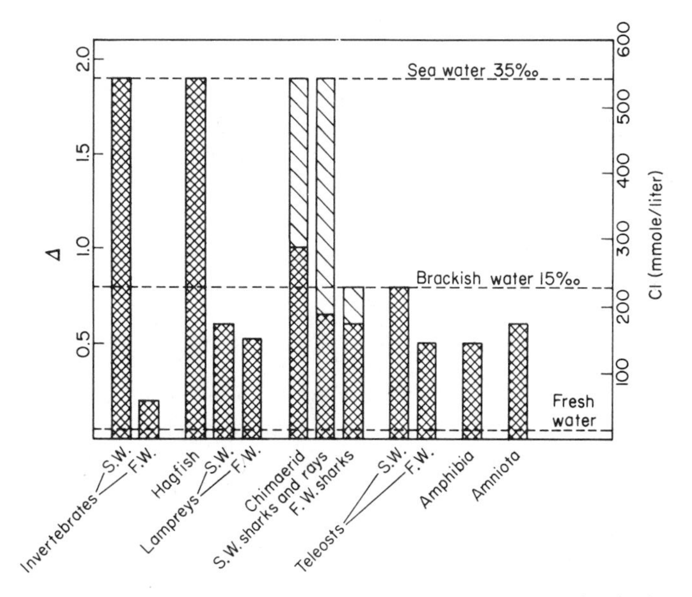

Fig. 11.1 Osmolality of body fluids in different groups of animals in relation to the tonicity of their environments. F.W., fresh water; S.W., sea water; Δ, depression of the freezing point. [Data from Robertson (1957), Schlieper (1958), Fänge and Fugelli (1962).]

of a hypoosmotic state; and (4) terrestrial living. Before discussing these, the range in electrolyte and water content of the various habitats will be considered.

THE ENVIRONMENTAL PARAMETERS

The concentration of sea water varies slightly in different geographic regions, but the percentage composition of its different electrolytes is remarkably constant. Surface waters of the open oceans contain about 3.4 to 3.6 per cent salt, with the highest concentrations at the equator where evaporation is greatest. The salinity of the deep waters is almost constant between 3.48 and 3.49 per cent. The major constituents are chloride (about 55 per cent) of which sodium is the most abundant (about 31 per cent). Because of the chemical constancy in proportions of the inorganic compounds, the total salt content can be readily determined by measuring any one of the constituents. Chloride, because of its abundance and ease of determination, is the element of choice. The classical technique is a silver nitrate titration which in reality determines the total halides; chlorinity by definition is

> *The total amount of chlorine, bromine, and iodine in grams contained in one kilogram of sea water, assuming that the bromine and iodine had been replaced by chlorine* (Sverdrup *et al.*, 1942).

To avoid apparent changes in chlorinity which can be brought about by more precise determinations of atomic weights of the elements, an International Commission in 1902 prepared a primary standard of "normal water" at the Hydrographical Laboratories in Copenhagen. Chlorinities are now related to this standard or subsequent standards based on it. A new primary standard prepared in 1937 has a chlorinity of 19.381 parts per thousand (‰). The CHLOROSITY of a sample is the chlorinity expressed as grams per 20°C-liter; it is obtained by multiplying the chlorinity of the sample by its density at 20°C. A chlorinity of 15.00‰ corresponds to a chlorosity of 15.28 g/liter.

Sea water is also described in terms of the total dissolved solids or the SALINITY defined as

> *The total amount of solid material in grams contained in one kilogram of sea water when all the carbonate has been converted to oxide, the bromine and iodine replaced by chlorine, and all organic matter completely oxidized* (Sverdrup *et al.*, 1942).

It is readily calculated from the empirical formula

$$\text{SALINITY} = 0.03 + 1.805 \times \text{CHLORINITY}$$

These relationships are shown in Table 11.1.

Silver nitrate titrations are tedious and technically difficult under conditions which often prevail in oceanographic vessels. Electrical conductivity is roughly proportional to the salinity and reliable instruments have now been developed to provide rapid and automatic records of ocean salinities (Riley and Chester, 1971). The density and refractive index also vary with the salinity but methods based on these properties are less accurate (density) or less readily adapted to oceanographic work. Textbooks of oceanography should be consulted for details.

Table 11.1 Physicochemical Values for Brackish Water and Sea Water. (Specific gravity measured at 0°C in relation to distilled water at 4°C. Salt content based on standard of 293 mmole Cl/liter = Δ 1.00°C)

Salinity (‰)	Chlorinity (‰)	Specific gravity	Freezing point-Δ°C	mmole/liter
5.0	2.76	1.0040	0.266	79
10.0	5.56	1.0080	0.533	156
15.0	8.38	1.0121	0.795	233
20.0	11.21	1.0161	1.077	317
25.0	14.08	1.0201	1.350	396
30.0	16.95	1.0241	1.628	477
35.0	19.86	1.0281	1.907	559

SOURCE: From Schlieper, 1958: Physiologie des Brackwassers. Die Binnengewasser, *22*:217–230.

Physiologists frequently describe the salinity of the environment as a percentage of sea water. This is quite satisfactory, but the implication must always be that it is a percentage of "normal water" (chlorinity 19.381‰) unless otherwise defined.

Near the shores, where the fresh water from the land meets and mixes with the waters of the ocean, there is every gradation in salinity of habitats. Lakes and rivers are even more variable. Unlike the ocean, they differ not only in salinity but also in the composition of the dissolved salts. Clarke (1924) has summarized a wealth of information on the analyses of waters from all over the world. Waters considered "fresh" range from approximately 15 to 500 parts per million (ppm) of dissolved solids, corresponding to salinity values of 0.015 and 0.5‰. Hutchinson's (1957) mean salinity values for lakes and rivers the world over is 100 ppm. The actual composition depends on the geological formation and the amount of precipitation. It is not unusual to find saline rivers in the interiors of the continents with salinities over 1000 ppm while the salt lakes may be more than 100 times this (10 to 30 per cent).

There are two very different kinds of salt lakes. Both develop in isolated basins from which the drainage never reaches the sea; but in one group (represented by the Great Salt Lake and the Bonneville Basin in North America) the salts have been leached from sedimentary rocks or salt deposits of ancient seas, while in the other they come from volcanic rocks. Lakes of the first group are chloride waters derived directly as remnants or indirectly through leaching from the ancient oceans; they often have the same general composition as sea water. The waters of Great Salt Lake, for example, are like sea water, although the salinity is four to seven times greater, depending on the precipitation. Lakes of the second group, such as the Lahontan Basin in northwestern Nevada, are alkaline and more variable in composition (Clarke, 1924).

The saltern or saltworks provides another interesting habitat of high salinity. Sea water is admitted into a series of ponds through which it slowly circulates as it evaporates. In this process the water not only becomes progressively more saline but shows a changing salt composition with, first, a precipitation of calcium carbonate and traces of iron oxide at a specific gravity of 1.050, followed by the calcium sulfate with further evaporation at a specific gravity of 1.264. As long as calcium is present, such ponds contain a characteristic bacterial and algal flora and a sparse fauna of protozoans (for example, the autotroph *Dunaliella*), crustaceans (particularly *Artemia salina*) and insects (especially *Ephydra millbrae*); this ecological community disappears with the precipitation of the calcium and the rising magnesium concentration (Baas Becking, 1928; Boone and Baas Becking, 1931).

In studies of osmotic regulation the environmental physiologist is often more concerned with the relative osmotic properties of the internal with respect to the external environment than he is in the precise chemical constitution of the two media. When the environment displays the same osmotic pressure as the body fluids, it may be either ISOSMOTIC or ISOTONIC, but the two terms do not mean the same thing since the latter is a function of the cell membrane while the former is not.

This can be most easily explained with an example. A sea urchin egg in a 0.53 molar solution of sodium chloride does not swell or shrink; the osmotic content is evidently the same on both sides of the cell membrane. Now, on the basis of theoretical calculations, a 0.37 molar solution of calcium chloride should have the same osmotic pressure, and the size of an egg immersed in it should likewise show no change. Experimentally, however, the 0.37 molar calcium chloride will shrink the egg; it remains unchanged in a more dilute solution of 0.30 molar concentration. It is evident that these salt solutions have in some way modified the permeability properties of the sea urchin egg membrane so that it is not responding like an ideal semipermeable membrane. Determinations of osmotic pressure by physical techniques depend on the number of solute particles and are calculated for an ideal membrane permeable only to the solvent (water). The different response of the sea urchin egg to isosmotic solutions of sodium and calcium chloride illustrates one of the differences between an isotonic and an isosmotic solution. Solutions which produce no osmotic stress are isotonic; those which are theoretically of the same osmotic pressure are isosmotic. Solutions may or may not be both isotonic and isosmotic at the same time. This argument indicates that the terms HYPEROSMOTIC and HYPOOSMOTIC are preferable to hypertonic and hypotonic for the osmotically more concentrated and dilute solutions in discussions of environmental physiology (Potts and Parry, 1964a).

The term OSMOLALITY is used to describe concentrations of biological solutions in terms of *osmotically active particles*. An OSMOLE is the amount of solute which, if dissolved in 1000 g water, will exert the same osmotic pressure as one mole of an ideal non-electrolyte. Thus, for substances which dissociate, the osmole equals the number of ions *and* undissociated molecules which give the same osmotic pressure as one mole of an ideal non-electrolyte. In biological work it is more convenient to use the MILLIOSMOLE; mammalian body fluids are about 300 milliosmolar. See Hoar and Hickman (1975) for methods of determining osmotic pressures and osmolality of solutions.

IONIC REGULATION IN ISOSMOTIC MEDIA

It has been argued that the evolution of an active ion transport system was one of the prime requisites for the organization of primitive cells (Chapter 15). According to this hypothesis, primordial cells—evolving in a marine environment—encountered osmotic problems which were solved by a regulated permeability of the plasma membrane and the active transport of cations.

The inevitable organic metabolites of protoplasm are assumed to be the basis of the transmembrane osmotic gradient. An impermeability of the membrane to these organic molecules is a necessity since they are the valuable coinage of life; however, their formation inside the cell is a potential source of osmotically active materials; the organic anions (amino acids and metabolites such as pyruvate, lactate and acetate) are especially significant. If the osmotic content were to increase during metabolism, water would flow into the cell; thus, even in an isosmotic environment, osmoregulation becomes essential for the preservation of a constant cell volume. If this concept is correct, the osmotic problem existed even in the ancestral (isosmotic) cell environment.

There are several theoretical ways in which primitive cells might have adapted their physiology to counteract this flooding. The permeability of the cell membrane might have been reduced until the organism was watertight. Alternatively a water pump might constantly discharge the excess osmotic water, or a salt pump might be used to reduce the electrolytes and maintain an osmotic balance. The latter mechanism seems to be universal, although it is often assisted by the other two. Ion pumps are a part of the physiology of all cells, and they may have evolved first to maintain a constant cell volume in an isosmotic medium.

It follows that the structure and physiology of the plasma membrane should be considered first in any discussion of osmotic and ionic regulation. The metabolic processes of active transport, the physical forces of diffusion and the dynamics of the living membrane which modifies these forces are discussed in many recent reviews (Hendler, 1971; Lockwood, 1971) and in several excellent textbooks of cell physiology (Dowben, 1969a; Davson, 1970; Giese, 1973); they will not be detailed here. It is, however, important for the comparative physiologist to remember that cell membranes may be organized so that they are freely permeable to water, to other small molecules, or to ions; but they may also be extremely tight and practically impermeable. Thus, the penetration of materials, which involve the physical forces of diffusion, electric potential gradient and solvent drag, depends very greatly on the character of the plasma membrane. It is also noted that the environment can markedly alter the permeability of cells; divalent ions, such as calcium, tend to reduce permeability while the monovalent ones, such as sodium, increase it. A proper balance is essential, and the lack of cations, such as calcium, in the environment may greatly reduce the survival time in otherwise innocuous habitats (Cuthbert, 1970).

Many marine (and parasitic) animals at all levels in phylogeny live in isosmotic environments. Although their total osmotic content matches that of their environment, the ionic composition of the body fluids is never the same as that of the ambient medium. Often the potassium is much higher and the sulfate lower, but no sweeping generalizations are possible; the significant point is the continuous regulation of electrolytes which is an essential part of life in all habitats. In complex multicellular animals, many mechanisms for the regulation of water and electrolytes come into play, but active ion transport is the key to them all. Surface areas permeable to water and ions are regularly reduced to a minimum, water pumps in the form of contracile vacuoles, nephridial tubes and kidneys are often present, but the active transport of salts remains a part of the machinery of every cell. In certain organs (the gills of fishes or crustaceans), cells exploit this capacity by forming highly specialized tissues for the excretion of large amounts of salt.

The transfer of regulatory responsibilities from relatively undifferentiated cells to specialized tissues is nicely shown during the ontogeny of such organisms as the marine teleosts (Blaxter and Holliday, 1963; Parry, 1966; Holliday, 1969). Herring (*Clupea*) eggs tolerate a wide range of salinities and are essentially isosmotic from 5 to 50‰ during the early stages of development. They swell and shrink in relation to changing tonicity of the environment, with only the chorion of the egg mechanically restricting the size increase in salinities lower than 5‰. After gastrulation, however, when the egg is completely enclosed in the expanding cellular layers of endoderm and ectoderm, the egg contents are regulated at a tonicity equivalent to a salinity of about 12‰ ($\Delta = 0.72$). The relatively undifferentiated surface cells are apparently responsible for this regulatory capacity, and this is probably also the case throughout the larval stages. The newly hatched larvae tolerate salinities ranging from 2.5 to 52.5‰, with the most likely site of regulation in the epidermis. In older fish, however, osmotic and ionic regulation become the responsibilities of other tissues (kidney and gills) as the skin takes on its protective functions and develops its scaly layers. The mesonephros of the herring has a high glomerular count and serves as a good water pump in dilute habitats; the gills are the locus of salt transfer and regulation. This sequence of physiological changes during ontogeny probably parallels changes during phylogeny from the first single-celled organisms to the complex multicellular forms.

LIVING IN A HYPOOSMOTIC ENVIRONMENT

Aquatic animals which are able to live within only a narrow range of outside salinities are said to be STENOHALINE; there are many examples both in fresh waters and in the ocean. Such familiar freshwater invertebrates as planarians, mussels and crayfish are in this class, as well as the majority of the marine invertebrates, ranging from protozoans to protochordates. Both freshwater and marine fishes provide numerous examples. In contrast to stenohaline animals, the EURYHALINE forms are able to withstand a wide range of salinities by either conformity or regulation. Again, there are many examples in most phyla, although coelenterates and echinoderms are notably inconspicuous in the estuarial habitats where euryhaline species abound.

This section deals with the physiological problems of animals which maintain a hyperosmotic condition in their body fluids. These are the animals of fresh and brackish waters; they may be either stenohaline or euryhaline. They share the characteristic of living in environments which are osmotically less concentrated (hypoosmotic) than their body fluids and are consequently faced with continual flooding of water and a leaching of their salts. During phylogeny the estuary must have been their ancestral home, with the stenohaline freshwater species making their way to rivers and lakes through progressively less brackish habitats.

The electrolyte and water problems of these ancestral lines can be appreciated from simple experiments in which marine or estuarial animals are placed in heterotonic seawater solutions and subsequent measurements made of the body volume and the composition of the fluids. The two major hazards of living in a hypoosmotic environment are indicated in Fig. 11.2. These data were obtained by simply measuring the weight or volume of a marine polychaete (*Eudistylia vancouveri*) in dilute sea water and in isosmotic sugar solutions. The amount of swelling is proportional to the dilu-

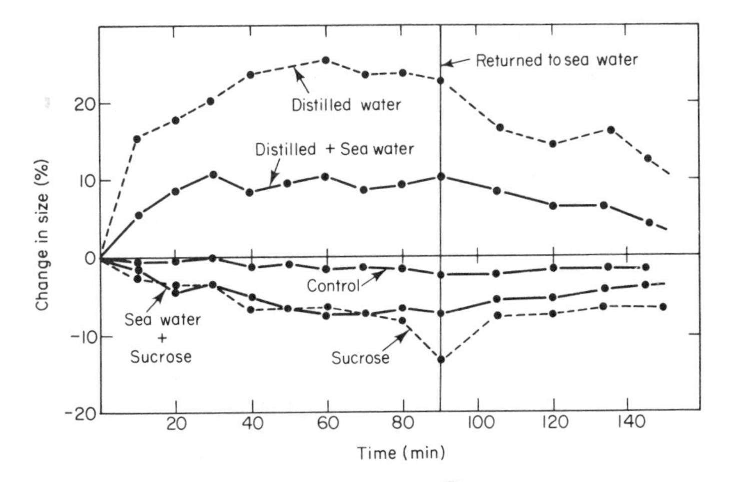

Fig. 11.2 Change in weight or volume of *Eudistylia vancouveri* as per cent difference from the initial values determined in sea water at about 28‰. Sucrose was isosmotic with sea water; mixtures were 50:50 by volume. Each value is an average for 50 worms measured individually in an introductory physiology class.

tion of the medium: it is roughly twice as great in fresh water as it is in 50 per cent sea water. Although the swelling reached a peak in 60 minutes, recovery did not occur so quickly on return to sea water. This could be because of a leaching of the salts with a resultant decrease in the amount of osmotically active material; the sucrose experiments support this suggestion. Loss of weight in isosmotic sucrose must be due to the loss of ions and electrolytes. A similar experiment is summarized in Fig. 11.3, but in this case recovery was followed for a much longer period. It was then evident that the weight, on return to the natural habitat, actually exceeded the original—indicating that the surfaces were not quite impermeable to sucrose but that some of this sugar entered to give an added osmotic component in the body fluids.

Figure 11.2 also illustrates the presence of regulatory mechanisms in *Eudistylia.*

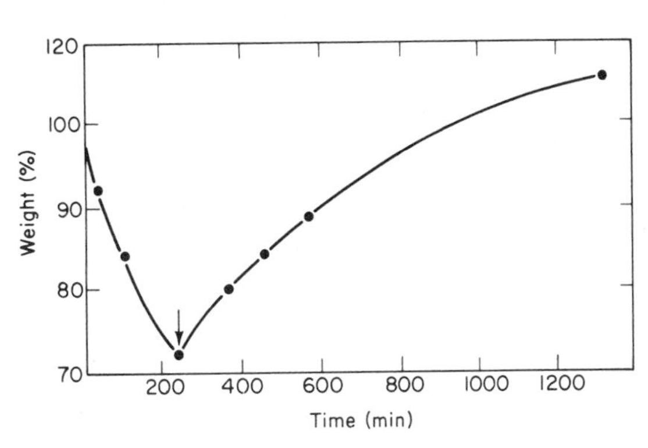

Fig. 11.3 Loss of weight of *Aplysia* in one part of isosmotic cane sugar and three parts of sea water. Vertical arrow, return to full strength sea water. [From Krogh, 1939: Osmotic Regulation in Aquatic Animals. University Press, Cambridge.]

The rapid initial swelling is followed by a period of more gradual weight increase; then a static condition or a slight decrease in size occurs even before the animals are returned to their natural environment. The initial flooding is apparently stemmed by regulatory mechanisms which not only prevent further flooding but pump out some of the osmotic water.

Data for another annelid, *Nereis diversicolor,* shown in Fig. 11.4, show a much greater regulatory response and also emphasize the importance of calcium. Cell physiologists have often measured plasma membrane permeability in relation to electrolyte changes in the environment; the necessity of an appropriate balance of monovalent and divalent ions has been stressed repeatedly. In general, calcium and certain other divalent ions tend to decrease membrane permeability while the monovalents such as sodium and potassium have the reverse effect. In this particular experiment (Fig. 11.4) the impaired ability to regulate might be traced to several specific changes, but at the cellular level it is probably due to the changed permeability of plasma membranes.

These simple experiments underline the major problems of life in a hypoosmotic habitat—the flooding with osmotic water and the loss of the electrolytes. There are three obvious solutions to the osmotic flooding. The body of the animal might be encased in a rigid box which would balance the osmotic pressure and prevent the flux of water into the cell. The cellulose wall plays this role for plant cells in hypoosmotic environments. Animals may also counteract some swelling in dilute solutions through their elastic or rigid body coverings, but this mechanism cannot provide a completely satisfactory answer since it is incompatible with animal mobility and the holozoic methods of nutrition which demand a constant flow of the external materials into the body.

Animals living in hypoosmotic environments have then two major avenues for specialization. One of these is to tolerate a dilution and the other is to pump out the water. Limitations on the first of these possibilities are obvious, and the strictly freshwater habitats could never have been attained without water pumps. Only such relatively unspecialized membranes as the egg membranes of freshwater teleost eggs, and perhaps some protozoans, are sufficiently impermeable to maintain a water balance without some special mechanism to expel water. The animals which live in

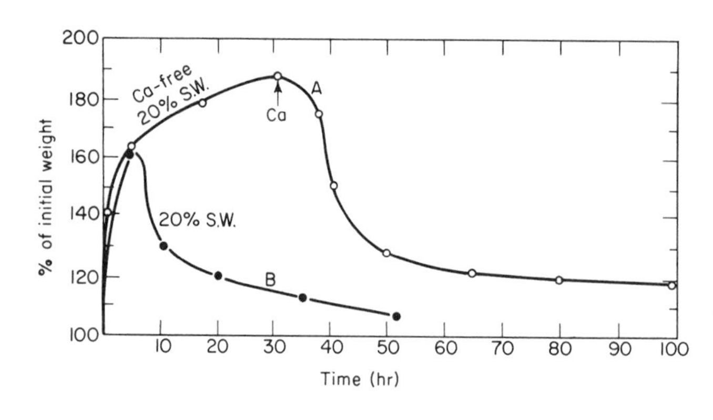

Fig. 11.4 Weight change as per cent of initial weight of *Nereis diversicolor* when transferred from 100 per cent sea water to 20 per cent sea water (curve B) or to 20 per cent sea water lacking calcium (curve A). Calcium added to 20 per cent sea water at the arrow. [From Prosser, 1950: Comparative Animal Physiology. Saunders, Philadelphia.

brackish and fresh waters vary remarkably in their abilities to tolerate dilute body fluids. Some representative values for animals living in media of different concentrations are shown in Fig. 11.5. The freshwater mussel *Anodonta* lives successfully with body fluids equivalent to a freezing point of −0.078°C while a marine bivalve has a corresponding value of almost −2.0°C.

Some osmotic conformers can live comfortably over a wide range of osmotic concentrations. In at least one species of flatworm, *Procerodes* (=*Gunda*) *ulvae* which has been carefully studied at Plymouth, England, some of the osmotic water is temporarily stored in special tissues (Beadle, 1934; Potts and Parry, 1964a). This animal lives in estuaries where it is periodically exposed to fresh water, and by experiment it has been shown that considerable osmotic flooding, with volume increase, occurs after the ebbing of the tide. Provided, however, that there is sufficient calcium in the environment (down to 0.5 mg/liter) an equilibrium is soon attained with some subsequent decline in weight. Evidently, the animal has a capacity to remove or exclude water but, in addition, there is the interesting fact that large amounts of this osmotic water pass through the tissues and are stored in large vesicles of the endoderm. When the animals return to sea water, these vesicles shrink and the volume of the worm decreases.

Water pumps are almost always present in organisms which live in hypoosmotic environments. Contractile vacuoles, flame cells and other protonephridial tubules, metanephridia and the nephron of the vertebrate are all capable of removing large amounts of fluid (usually by filtration); it may be argued that their primary function was water balance rather than excretion of nitrogenous wastes. Even the closed units such as Malpighian tubules (in mosquito larvae) or the aglomerular kidney (in a freshwater pipefish, *Microphis boaja*) may discharge large volumes of water (Chapter 8). In a few animals there are no special organs for the removal of water. *Hydra* is the most familiar example. Its ability to regulate both water and salt depends on the active transport of sodium; the whole process breaks down if sodium and calcium are absent from the external environment (Macklin, 1967). Sodium is pumped into the

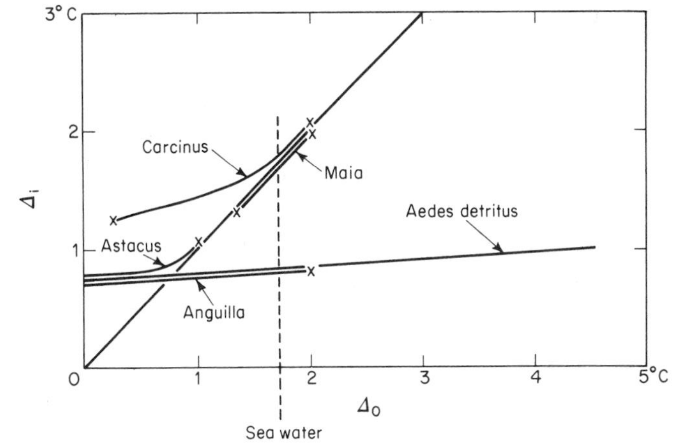

Fig. 11.5 Osmotic regulation in various animals shown by plotting the freezing point depression Δ_i of the blood (vertical axis) against the freezing point depression Δ_o of the external medium (horizontal axis). [From Ramsay, 1968: A Physiological Approach to the Lower Animals, 2nd ed. University Press, Cambridge.]

gut and water follows passively along the osmotic gradient. The mesoglea acts like an extracellular fluid space. In the freshwater medusa, *Craspedacusta sowerbyi,* two pumps seem to be involved; the first of these transports sodium into the mesoglea while the second moves it into the gut (Hazelwood *et al.,* 1970). Similar mechanisms probably occur in *Hydra* (Macklin *et al.,* 1973). Water taken in osmotically is expelled through the mouth; vigorous contractions of the body wall have been observed when the pressure rises in the gastrovascular cavity. Thus, both osmotic and volume regulation depend on the active transport of sodium; the gut serves as an organ of both digestion and excretion. Structurally, these transporting epithelia of coelenterates appear to be similar to frog skin and toad bladder; the trick of transporting ions through epithelial cells was evidently discovered very early in phylogeny.

The Absorption of Salts

Ion regulation is an essential physiological process in all aquatic habitats. In general, the marine and estuarial invertebrates produce an isosmotic urine. In the marine environment where these species—with the exception of some of the crustaceans—are in osmotic equilibrium with their environment, urine formation might not be expected to create any ionic imbalance. However, even in the isosmotic habitat the electrolyte composition of the tissues differs from that of sea water; ionic fluxes always occur at the permeable surfaces, and mechanisms are essential to preserve the equilibrium. In brackish water these problems are greatly magnified; the body fluids are hyperosmotic, and the production of an isosmotic urine may soon deplete the electrolytes. The loss will be particularly rapid during periods of fasting. Nevertheless, many invertebrates have not only penetrated the dilute waters of estuaries but have moved on into the rivers and met the hazards of osmotic flooding by pumping out large amounts of isosmotic urine (Lockwood, 1964). A classical example is the mitten crab, *Eriocheir sinensis.* This small crab is indigenous to Asia but has now spread widely throughout Europe; it has been carefully studied because of its marked euryhalinity. Spawning takes place in the sea, but the growing individuals make their way into rivers where they spend most of their lives until it is time to return to the sea and spawn. Body fluids of this crab are considerably more dilute while in fresh water, but the blood and urine are isosmotic to each other in both the sea and the river (Krogh, 1939). In fresh water, the animal is continually losing its salts, and it has many times been demonstrated that the loss is made good by the active absorption of ions from the ambient water by the epithelial cells of the gills. Even isolated gills take up salts, and such preparations have been used to study some problems of active ion transport. *Eriocheir* has such an efficient mechanism for the accumulation of ions that its blood is actually osmotically more concentrated than that of the crayfish, a strictly stenohaline freshwater species ($\Delta = 1.2°C$ in contrast to 0.8°C, Schlieper, 1958).

Rates of ion uptake by *Eriocheir* have been carefully studied in relation to both external and internal salt concentrations (Lockwood, 1962, 1964; Potts and Parry, 1964a). It is known that *Eriocheir* actively absorbs both the sodium and the chloride ions; although both ions may be absorbed simultaneously to preserve an electrical balance, they may be independently absorbed with an exchange of sodium for ammonium, or possibly hydrogen, and an exchange of chloride for bicarbonate. Sodium trapping is better understood than chloride uptake but there are still unanswered questions with respect to both of them (Schoffeniels and Gilles, 1970).

The brackish water prawn, *Palaemonetes varians,* is another crustacean with remarkable capacities for exploiting a great range of saline habitats, in spite of a urine isosmotic with the blood. Potts and Parry (1964b) have studied its salt-regulating machinery and found the animal to be isosmotic with somewhat dilute sea water (about 65 per cent). At this point of osmotic neutrality, there is no electrical potential difference (EMF) across the body surfaces and the exchange of sodium and chloride ions takes place by passive diffusion. However, when in more dilute or in concentrated sea water, the interior of the prawn's body became negative with an EMF difference up to 30 or 40 mv. By the use of tagged ions the animals were shown to be actively accumulating bromide (*chloride*) when in dilute solutions and actively extruding *sodium* in concentrated ones. Although the anion was being actively transported in one case and the cation in the other, the effects on the electrical potential were similar, and in both cases there was a rapid flux of sodium chloride; the oppositely charged ion must closely follow the actively transported one. The body surfaces of *Palaemonetes* are extremely permeable, and this animal has proved most valuable for the analysis of ionic regulation. The studies of ionic fluxes in teleost fishes, as well as in crustaceans, have clearly shown that either the chloride *or* the sodium may be the actively transported ion (Potts, 1968; Kirschner, 1970). The movement of a single ion, however, cannot take place from a sodium chloride solution unless an appropriate exchange of other ions occurs to preserve electrical neutrality. In many places, ion exchange mechanisms (Fig. 8.9), rather than active transport, may be involved.

Ion absorption at the body surfaces can, then, make good the entire loss of electrolyte even in fresh waters. However, the majority of the more highly organized freshwater aquatics have additional machinery for ion absorption located in the excretory tubules; ion absorption from the environment is associated with ion absorption from the excretory tubule. In this way the tonicity of the urine is lowered and dependence on ion trapping from the environment is greatly reduced. It is never entirely eliminated, since a hypoosmotic urine still contains electrolytes (Fig. 11.6) and some leaching of salts may also be expected at such permeable surfaces as the gills.

Formation of a Dilute Urine

Production of a dilute urine has now been demonstrated in some representatives of all the more advanced phyla. Findings, summarized in Fig. 11.6 for the crayfish, show the magnitude of this ion-concentrating capacity. Comparable data have been obtained for several other arthropods, earthworms, freshwater molluscs, teleosts, and amphibians. The experimental proof of such mechanisms can be visualized from Fig. 11.6. Isotonicity of blood and glomerular urine is readily demonstrated in *Nec-*

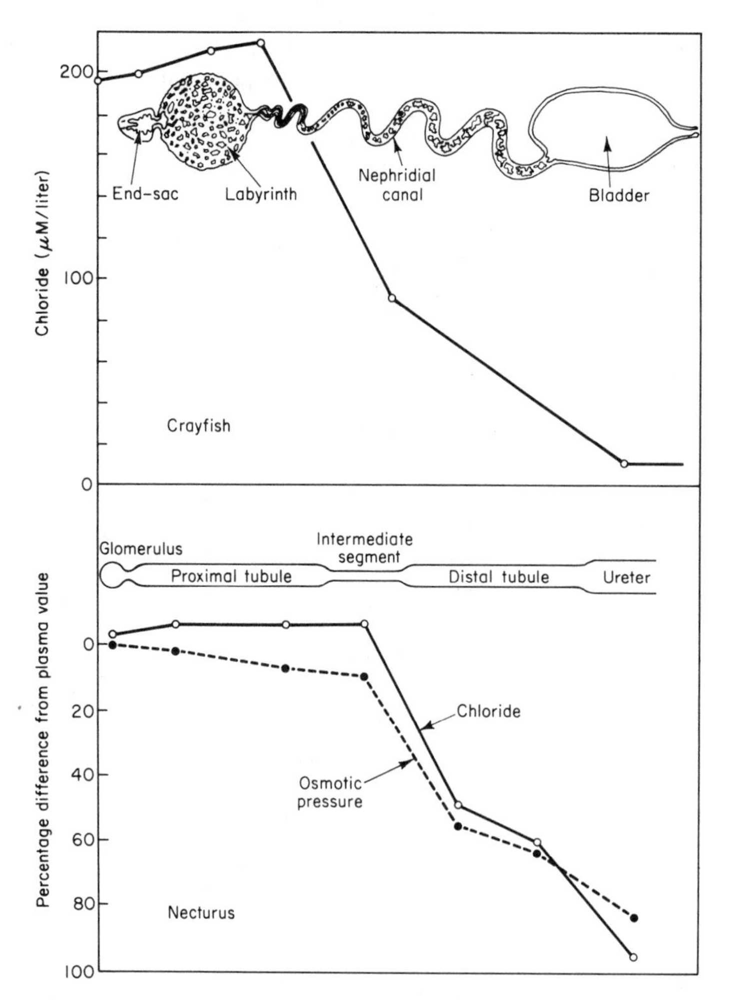

Fig. 11.6 The formation of a hypoosmotic urine in the crayfish and in the mudpuppy (*Necturus*). [Upper, from Parry, 1960: *In* The Physiology of the Crustacea, Vol. 1, Waterman, ed. Academic Press, New York. Lower, from Shaw, 1960: *In* Comparative Biochemistry, Vol. 2, Florkin and Mason, eds. Academic Press, New York.]

turus by analysis of fluids obtained by micropuncture; since these tubules are relatively large, the tonicity can readily be followed from segment to segment. The observed dilution of the urine in the distal tubule and ureter might be theoretically attributed to the addition of water or to the extraction of the salts. However, the techniques of renal clearance (described in Chapter 8) as well as the use of metabolic poisons (which prevent active transport) have clearly demonstrated that it is because of an absorption of materials rather than the addition of water.

In summary, the colonization of fresh waters depended on the gradual phylogeny of less permeable cuticles (Fig. 11.7) and the restriction of semipermeable boundary membranes to relatively small areas; concurrently, excretory organs acquired the capacity to form hypoosmotic urines and special cells appeared for the trapping of ions from the ambient fluid. The metabolic price of this regulation is considerable. It requires energy to transport ions and perhaps also to operate filters; tolerance of dilute body fluids thus becomes metabolically advantageous for animals living in hypoosmotic environments, and this has also been a vital factor in the colonization of fresh waters (Lockwood, 1962). *Mytilus edulis* provides one of the most extreme examples, with body cells which can operate isotonically with salinities ranging from full sea water to as low as 15 per cent sea water in parts of the Baltic Sea (Lockwood, 1964).

LIVING IN HYPEROSMOTIC ENVIRONMENTS

Marine fishes provide the outstanding example of aquatic animals living in hyperosmotic habitats (Fig. 11.1). Only the myxinoid cyclostomes among the vertebrates have an electrolyte concentration comparable to that of sea water; their osmotic prob-

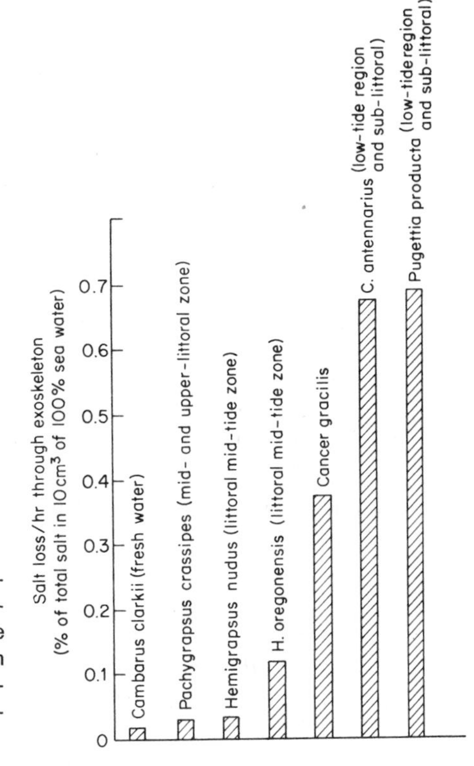

Fig. 11.7 Permeability of the exoskeleton of crustaceans from different environments, showing the tendency for lower permeability in animals from more dilute or semi-terrestrial habitats. [From Lockwood, 1962: Biol. Rev.. *37*: 266.]

lems are thus minimal or absent, although ionic regulation is essential. Elasmobranchs through the retention of osmotically active nitrogenous wastes (particularly urea), also have minimal osmotic gradients. In contrast, the marine lampreys and teleosts are continually dehydrated through the osmotic loss of water at their permeable surfaces. A few arthropods live under comparable conditions of osmotic stress. Some grapsoid crabs and palaemonid shrimps are hypoosmotic (about 86 per cent of sea water), although the osmotic gradient is considerably less than that of the marine teleosts, where body fluids are less than half the concentration of sea water (Fig. 11.1). *Pachygrapsus crassipes* and *Hemigrapsus oregonensis* thrive in hypersaline lagoons where the salinities reach 50 to 55‰ (Gross, 1961). The best invertebrate examples are found among insects (larvae of the mosquito *Aedes detritus*) and the primitive crustacean *Artemia salina* living in salt lakes and brine ponds.

Drinking Salt Water and Excreting Excess Electrolytes

Pioneer investigations of the physiology of regulating in a hyperosmotic habitat were carried out on the marine teleosts. These early experiments (Krogh, 1939) showed that water lost osmotically was recovered by drinking sea water and excreting the salts extrarenally. Drinking was first demonstrated by placing fish in sea water containing phenol red and later measuring the amount of dye in the gut. In other experiments, balloons were inflated in the esophagus of the eel to block water ingestion. Such animals were able to regulate in fresh water, indicating that the drinking of fresh water was not a necessary part of their osmoregulatory response; but in sea water they lost weight continually and could never achieve a steady state. Measurements of drinking rates indicate that most marine teleosts swallow a volume of water equivalent to 0.3 to 1.5 per cent of their body weight per hour; much higher rates have been recorded for some species (Conte, 1969).

Extrarenal excretion of salt was first shown in the classical experiments of Keys (1931) who perfused a heart-gill preparation from the eel. When analyses of the balanced saline solutions used to perfuse the gills were compared with analyses of the solutions bathing the branchial cavities, it was evident that chloride was actively passed through the gills from the internal to the external medium. Acidophilic cells, the "chloride secreting cells," at the base of the gills were considered responsible for the salt transfer, although these studies did not actually demonstrate which cells were responsible or whether the active processes involved the chloride or the sodium ion or both. The structure of the "chloride cells" has now been studied with both light and electron microscopes (Conte, 1969).

Keys' experiments, performed many years ago, have now been substantiated by the use of much more refined techniques. The marine bony fishes and cyclostomes (Petromyzontia) compensate for the loss of osmotic water by drinking sea water and eliminating the salts; in fact, they live in a desert and distill sea water to make good the water lost through osmotic desiccation. Most of the monovalent ions and a part of the water are absorbed from the gut, while about 80 per cent of the divalent ions remain behind. Their elimination from the gut requires a part of the ingested water since the intestinal residue is essentially isotonic with blood (Smith, 1953). The absorbed ions are either used metabolically or excreted by kidneys and gills. Almost all the magnesium and sulfate are removed renally while most of the monovalent ions, together with the nitrogenous wastes, are excreted through the gills. Urine production is scanty in marine teleosts (only about 1 to 2 per cent of that in corresponding fresh-water species) but continuous (Parry, 1966; Hickman and Trump, 1969).

Several anatomical adaptations are associated with the scanty urine production of the marine teleosts. With reduced demands for filtration, the glomeruli are often partially or entirely eliminated. All mesonephroi possess glomeruli during early stages of development, but the adults of some species are actually aglomerular. Even when glomeruli are present, the constriction or closure of the neck segment of the tubule (Edwards, 1928–35) indicates a minor role for filtration in urine formation. In aglomerular species like the toadfish, *Opsanus tau,* the nephron is essentially a proximal (brush border) convoluted unit. These kidneys have been of unusual interest to physiologists since they obviously function only by secretion; some of the first evidence for secretion by the vertebrate nephron was adduced from the aglomerular kidneys of teleosts.

In general, the physiological mechanisms for hypoosmotic regulation among arthropods are comparable to those described for marine teleosts. They have been beautifully demonstrated by Croghan (1958a, b) in *Artemia salina,* an animal which weighs only about 8 mg as a large adult and hence requires the nicest of micro-techniques for its study. *Artemia* is able to regulate over an unusually wide range of salinities. It can survive as long as 24 hours in glass-distilled water but requires an environment of at least 10 per cent sea water for normal life and activities. Its body fluids are hyperosmotic in media more dilute than 25 per cent sea water; they remain relatively constant (equivalent to 1 to 2 per cent NaCl) in habitats of increasing salinity up to about 10 times sea water. In nature, the animals may be found in crystallizing brine.

Like the marine teleosts, *Artemia* is able to accomplish this remarkable feat by constantly drinking water, absorbing it together with NaCl through the gut epithelium and actively excreting the salt. The first ten branchiae are the organs of active sodium excretion; the transfer has been shown to be extremely fast. Sodium efflux figures for *Artemia* may be in excess of 150 mmole/liter hemolymph/hr in contrast to 11 mmole for *Daphnia* and 1.4 mmole for the guppy *Lebistes* (Croghan, 1958c). During larval life, before the development of the branchial glands, *Artemia* extracts salt with a remarkable organ (NECK ORGAN) located on the top of the cephalothorax (Conte *et al.*, 1972). *Artemia* is permeable both to water and to salt and has achieved an environmental independence by improving its abilities to transfer salts rapidly away from the tissues.

Adjustment of Osmolarity with Amino Acids and Urea

A different mechanism for osmotic compensation depends on the adjustment of the osmolarity of the body fluids to match that of the environment by the formation, or retention, of low molecular weight organic molecules. This adaptation was first studied in elasmobranch fishes where the retention of urea and trimethylamine oxide (TMAO) raises the osmotic content of the blood to that of sea water (Fig. 11.1). This phenomenon, however, is part of a widespread capacity to compensate for osmotic differences with small size organic molecules. Among the invertebrates, the tissues of worms, molluscs, arthropods and echinoderms often contain free amino acids or products of their metabolism, such as taurine (probably derived from cysteine) and glyoxylic acid (from glycine). A part of the intracellular osmotic regulation which some members of these phyla show is caused by the presence of these substances (Schoffeniels and Gilles, 1970, 1972). Figure 11.8a shows the correlation between osmotic content of the environment and the amount of intracellular amino acids and other ninhydrin-positive substances (NPS) in the mussel *Mytilus*. Relatively greater concentrations of taurine at the higher salinities are thought to be important in sparing essential amino acids (Lange, 1963).

Thus, low molecular weight nitrogenous substances may be used to adjust either extracellular or intracellular osmolalities. Examples of both situations occur among the fishes. The ancient marine crossopterygian *Latimeria,* like the elasmobranchs, maintains high blood urea concentrations to match the osmolality of sea water (Pickford and Grant, 1967), while some of the cyclostomes and teleosts build up the osmolality of the intracellular fluids in a manner comparable to *Mytilus* and many

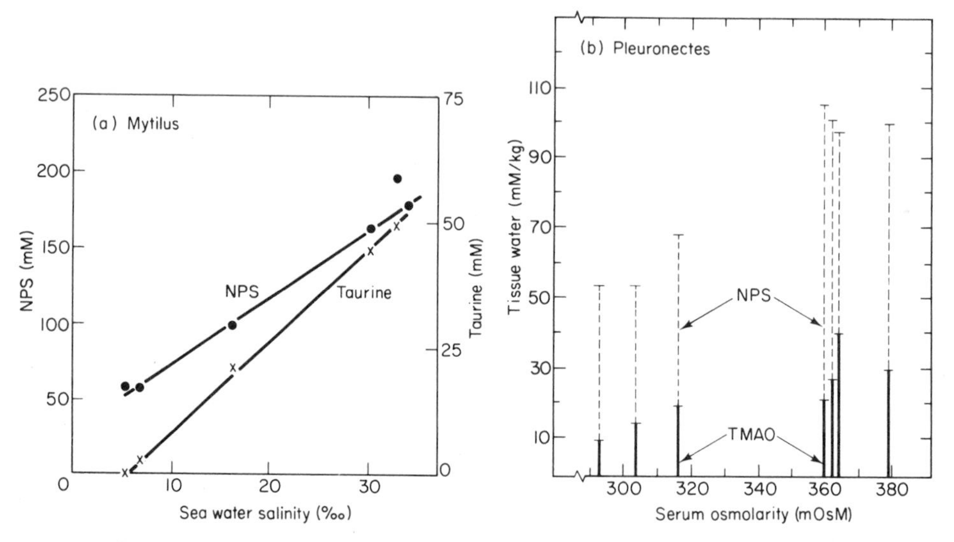

Fig. 11.8 Osmotic adjustments with low molecular weight nitrogenous substances. (a) Effect of environmental salinity on tissues of the blue mussel. (b) Serum osmolality and NPS/TMAO content of muscle in a flounder. Total free ninhydrin positive substances (NPS), determined as taurine equivalents. [(a) From Lange, 1963: Comp. Biochem. Physiol., *10* (2): 175. (b) From Lange and Fugelli, 1965: Comp. Biochem. Physiol., *15*: 291.]

other invertebrates. In the myxinoid cyclostomes, the osmotic equivalence of plasma and sea water is due to the high salt concentration of their blood, while that of the muscles depends on low molecular weight organic compounds—particularly amino acids (Robertson, 1960; Morris, 1965). Likewise, trimethylamine oxide and ninhydrin-positive substances are important in intracellular osmotic regulation of euryhaline teleosts (Forster and Goldstein, 1969; Fig. 11.8b); through adjustments in the amounts of these organic substances, cell volumes remain constant in both fresh water and sea water, even though substantial changes occur in the extracellular fluids.

Again, it is emphasized that water and ion regulation are two distinct problems. Thus, the sharks and their relatives achieve an osmotic neutrality with their high blood urea and TMAO, but they still face problems of ion regulation; the RECTAL GLAND of the elasmobranch serves as a salt pump for the extrarenal elimination of ions which would otherwise accumulate because of the high external salinities (Burger, 1962; Conte, 1969).

Some amphibians have osmoregulatory mechanisms comparable to those of the elasmobranchs (Gordon *et al.,* 1961; Gordon, 1962; Gordon and Tucker, 1965, 1968). This group of vertebrates, unlike the fishes, has not exploited the marine habitat, but there are a few representatives with marked euryhalinity and the ability to live in brackish or marine coastal areas. The green toad (*Bufo viridis*) of Europe and the Middle East can tolerate environmental salinities as high as 19‰ by matching its plasma to that of its surroundings, primarily by increasing its NaCl content (about 84 per cent) but to a small extent by accumulating urea (about 5 to 10 per cent).

The crab-eating frog (*Rana cancrivora*) of Southeast Asia is also able to exploit saline habitats and can tolerate salinities as high as 28‰ at 30°C. Tadpoles are even more tolerant (39‰ at the same temperature). Unlike the green toad, urea retention plays a major role (about 60 per cent) in the crab-eating frog and together with sodium chloride raises the plasma to slightly hyperosmotic levels. The urea concentrations may reach 2.9 per cent.

The skin permeability of these amphibians is the same as that of other anurans and does not account for any part of their exceptional abilities to live in salt water; further, they neither drink water nor put out large amounts of extrarenal salt. Their significant physiological achievement is that of tolerating a uremia, or a very high salt content, in their tissues. They share the former in common with the elasmobranchs and the lungfish and the latter with the marine invertebrates and the hagfishes. Adjustments of tissues to high concentrations of salt or urea may have been a major biochemical achievement; urea concentrations comparable to those found in the crab-eating frog would denature certain enzymes and affect oxygen transport in many vertebrates (Gordon *et al.,* 1961). In terms of evolutionary potentialities, the distillation of sea water may have been a simpler solution to the problem.

Phylogeny Animals which live in a hyperosmotic environment have probably evolved from freshwater ancestors. Biologists generally agree that this has been the case with the marine teleosts, brine shrimps, and mosquito larvae, all of which drink salt water and excrete salts extrarenally. The main controversy centers around the home of the ancestral vertebrates and the position of the myxinoid cyclostomes. Homer Smith (1953), on the basis of kidney structure, and Romer (1946) from palaeontological evidence, argued for the freshwater origin; the opposing view has been just as vigorously maintained by both biologists (Robertson, 1957) and palaeontologists (Denison, 1956). In some of the more recent discussions the hagfish with its isosmotic plasma has been used as evidence for the marine ancestry because of the primitive position of the Cyclostomata and the similarity of their osmotic mechanisms to those of the marine invertebrates which, everyone agrees, evolved in the sea. The fact that certain amphibians—a group which in phylogeny probably crawled onto the land from stagnant freshwater pools—may live in concentrated sea water by physiological adaptations comparable to those of the myxinoids and the elasmobranchs, has reduced the force of some of these arguments. At present, neither the physiological nor the palaeontological evidence seems adequate to settle the matter.

THE WATER AND ELECTROLYTE PROBLEMS OF TERRESTRIAL LIVING

The land animals have made good the water deficits of their environment through special modifications of many of the mechanisms discovered by their aquatic ancestors and by evolving a few new ones of their own. In the first category, there is the tightening up of water-permeable coverings, the drinking of water, the extrarenal excretion of salts, the reduction of glomerular filtration and the ability of protoplasm to function efficiently with different relative amounts of electrolytes and water. In the second category, that of evolutionary innovations, are the capacity to recover large amounts of water from the urine (hypertonic urine), uricotelism (Chapter 8), the capacity to absorb significant amounts of water at the surfaces, the ability to depend largely on metabolic water, and the behavioral responses of avoiding desiccating microhabitats. Some representatives of each of the major groups of animals (oligochaetes, snails, insects, amphibians, reptiles, birds and mammals) are found in extremely arid habitats where drinking water is rarely or never available. The special mechanisms exploited in xeric environments may be conveniently divided into anatomical, physiological and behavioral. There are several excellent reviews (Edney, 1957; Chew, 1961; Berridge 1970; *Am. Zool.*, 8 (3), 1968); recent studies of the biology of desert animals forms one of the most fascinating chapters in environmental physiology (Schmidt-Nielsen, 1964; Yousef *et al.*, 1972; Maloiy, 1972).

Anatomical Specializations

The waxy chitin of insects or the impervious layers of epidermal keratin which cover the terrestrial vertebrates may reduce evaporation to a minimum. There are as many degrees of water tightness between the emerging terrestrial arthropods (land isopods) and certain insects which live entirely on metabolic water (*Tenebrio*) as there are between the semiterrestrial amphibians and the desert-dwelling amniotes (Fig. 11.9). Detailed comment would be superfluous.

Water conservation in the face of excretory demands for the removal of soluble wastes is met in one of two ways. The phylogenetically lower classes of terrestrial vertebrates followed the pattern of the marine teleosts and reduced filtration. Unlike the teleosts, they seem never to have dispensed entirely with the glomeruli—according to Smith (1953)—because their ancestors had no other route for the elimination of chloride. A reduction of the filtering machinery, however, is found in some amphibians, reptiles and birds. The desert-living frog, *Chiroleptes (Cyclorana)*, from Australia is the classical example among amphibians (Dawson, 1951), while this anatomical trend is quite general among reptiles and birds (Marshall, 1934). In the latter groups there are relatively few glomeruli, and those present show reduced capillary development with two or three short loops sometimes associated with a syncytial core of non-vascular tissue. The glomerular surface is considerably smaller

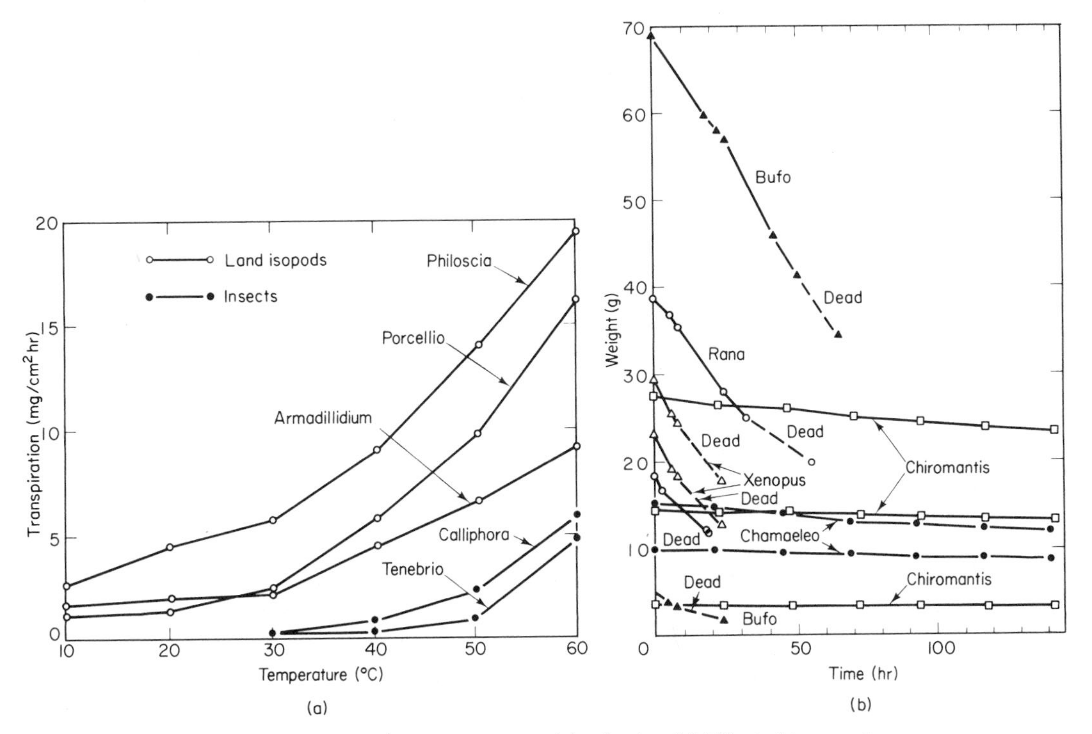

Fig. 11.9 Water loss in some terrestrial animals. (a) Effect of temperature on rates of transpiration into dry air for several arthropods. [From Edney, 1960: *In* The Physiology of the Crustacea, Vol. 1, Waterman, ed. Academic Press, New York.] (b) Weight changes in frogs and lizards at 25°C and 20 to 28 percent relative humidity. *Chiromantis* is an unusual amphibian which, like many reptiles, seems to excrete uric acid rather than ammonia. [From Loveridge, 1970: Arnoldia (Rhodesia), *5*: 4.]

among the arid-dwelling snakes and lizards than it is among semi-aquatic reptiles such as turtles and crocodiles; the glomeruli of the birds' kidneys are the smallest known (Marshall and Smith, 1930). However, minimal filtration seems to be unavoidable in these groups, perhaps because of the absence of alternative routes for the removal of chlorides, as Smith (1953) suggested, or because of the reduction of the renal portal system (Chapter 8). It is also possible that glomeruli have been retained because of the flexibility of a filtration-reabsorption sequence which permits endless modification of urine to meet changing excretory requirements; such a system also simplifies the task of excreting unusual wastes (pollutants, toxins) for which specific excretory pumps have not evolved.

The other technique for the conservation of urinary water depends on an innovation in renal anatomy, the appearance of the loop of Henle. This operates by actively pumping salt into the intercellular spaces about the tubules of the medulla and recovering water osmotically from the collecting tubules into the saline fluid thus formed (Chapter 8). There is a telling association between the length of the loop of Henle and the water available to the particular species. The beaver has short-looped nephrons which produce a maximum urinary concentration of 600 milliosmolal (mOSM), the rabbit with a mixture of short and long-looped tubules can produce a urine which is two and one-half times this concentration, while the African desert rodent *Psammomys obesus* with only long-looped units produces urine of 6000 mOSM. Some birds show the reptilian pattern of renal structure with reduced glomerular development while others show the mammalian type with loops of Henle. Several anatomical specializations in the excretory tubules of the arid-dwelling invertebrates were noted in Chapter 8.

Schmidt-Nielsen and his associates first described a counter-current heat exchanger in the nasal passages of the kangaroo rat. This device cools the expired air and thus reduces the moisture dissipated in breathing. Comparable adaptations in the morphology of the vascular and respiratory system have now been described in some lizards and birds as well as in several other mammals (Schmidt-Nielsen *et al.,* 1970; Murrish and Schmidt-Nielsen, 1970). Inhaled air is warmed and humidified as it passes over the extensive vascular surfaces; this cools the respiratory passages so that the exhaled air moving over them gives up some of its heat and water. Large amounts of water are saved in this way; at 15°C and 25 per cent relative humidity, a cactus wren recovers 74 per cent of the water added to the inhaled air while the kangaroo rat does even better with 83 per cent recovery (Fig. 11.10). A kangaroo rat breathing air at 28°C may actually exhale air at a lower temperature (24°C).

A few land vertebrates have taken to the marine habitat and thus excluded themselves from fresh water for drinking and from foods which contain isosmotic or hypoosmotic fluids. Reptiles such as the sea snakes and turtles and the marine iguana, as well as many aquatic birds, possess orbital glands which secrete watery fluids containing sodium and potassium chlorides. Thus, like the marine teleosts, they can drink sea water and "distill" it. The extrarenal saline excretions may be more concentrated than sea water, and the glands which produce them show a morphological development which is well correlated with the demands for salt elimination; there is now an extensive literature (Potts and Parry, 1964a; Dunson

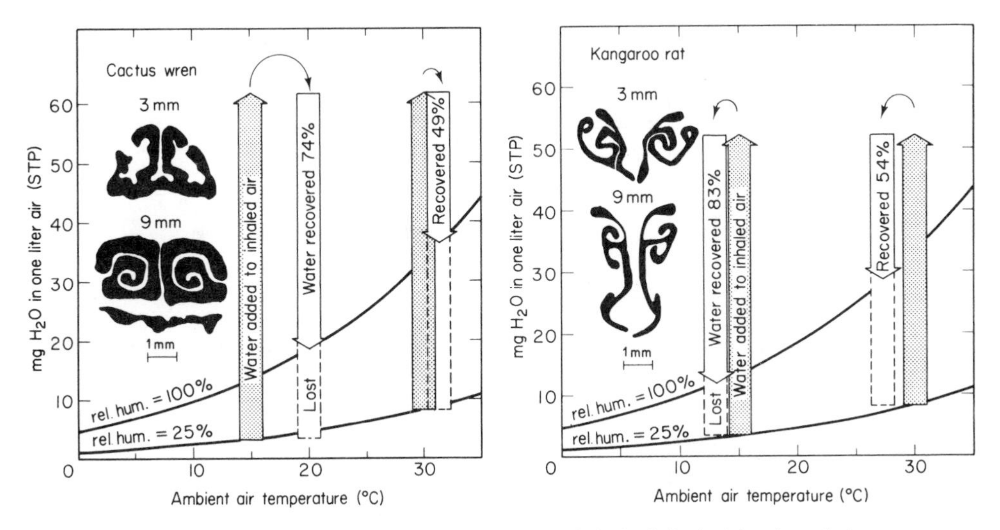

Fig. 11.10 Water vapor added to the inhaled air (stippled bars) and the amount recovered on exhalation (open bars) by the cactus wren (left) and the kangaroo rat (right). Temperatures of the inhaled air and the exhaled air are indicated by the positions of the bars. Cross sections of the nasal passages shown diagrammatically at the left of each diagram for depths of 3 mm and 9 mm from the external opening. [From Schmidt-Nielsen, Hainsworth, and Murrish, 1970: Resp. Physiol., *9*: 272, 273.]

and Taub, 1967; Bentley, 1971). The marine mammals have followed the same phylogenetic trail as their relatives in the desert, with kidneys which produce strongly hyperosmotic urines; their water problems are not nearly as great as those of the desert mammals since they do not experience high temperatures with marked evaporative water loss. Mammals, like the fish-eating bat *Pizonyx vivesi* which feeds on marine fish, can concentrate salt sufficiently to drink sea water (Carpenter, 1968). Techniques for the regulation of salt and water balance are compared in Fig. 11.11.

Biochemical and Physiological Adaptations

Some animals can safely tolerate a significant reduction in their body water. This is well illustrated by the group of amphibians listed in Table 11.2. Although each of them contains about 80 per cent water when in a moist environment, the amount of evaporation which can be safely tolerated varies greatly and is associated with the habitat conditions. The Florida spadefoot, a terrestrial species, can lose an amount of water equivalent to 60 per cent of its body weight while an aquatic frog, such as *Rana grylio,* is killed with a loss of less than 40 per cent.

Similar comparisons have been made with several other terrestrial groups. Man's capacity to tolerate dehydration is relatively limited. In the desert he cannot tolerate a loss of body water equivalent to more than 12 per cent of his weight, but the camel

	Liability to osmotic water loss	Liability to evaporative water loss	Blood conc. relative to medium conc.	Urine conc. relative to blood conc.	
Elasmo-branchs	○	○	Isotonic	Isotonic	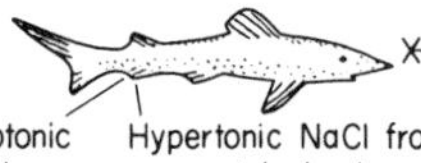 Does not drink sea water; Isotonic urine; Hypertonic NaCl from rectal gland
Teleosts	●	○	Hypotonic	Isotonic	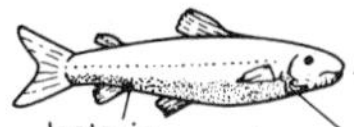 Drinks sea water; Isotonic urine; Secretes salt from gills
Reptiles	○	●	Hypotonic	Isotonic	 Drinks sea water; Isotonic urine; Hypertonic tears
Mammals	○	●	Hypotonic	Hypertonic	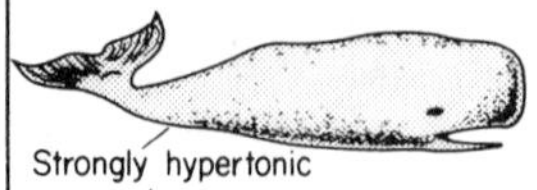 Does not drink sea water; Strongly hypertonic urine
Birds	○	●●●	Hypotonic	Hypertonic	 Drinks sea water; Hypertonic nasal secretion; Weakly hypertonic urine

○ = No liability ● = Liability

Fig. 11.11 Methods used by marine vertebrates to regulate their water and salt. [From Lockwood, 1964: *In* Animal Body Fluids and Their Regulation. Harvard, U.P., Cambridge, Mass.]

may lose twice this amount (about one-third of the water in its system) without being seriously weakened (Schmidt-Nielsen, 1959). The blood volume of the camel shows no serious reduction under these conditions, indicating a withdrawal of tissue water. It is evident that the cells of vertebrate animals may function at very different tissue water levels. However, one of the major differences between the water demands of man and the camel is the remarkable contrast in their temperature-regulating mechanisms. The camel does not sweat freely until its body temperature rises to about

Table 11.2 Correlations of the Habitats of Different Anurans with Their Ability to Survive the Loss of Body Water

	Habitat	Body water per cent weight	Vital limits of water loss as per cent	
			Body weight	Body water
Scaphiopus holbrookii	Terrestro-fossorial	79.5	47.9	60.2
Scaphiopus hammondii		80.0	47.6	59.5
Bufo boreas	Terrestrial	79.8	44.6	55.8
Bufo terrestris		78.8	43.3	54.9
Hyla regilla		79.4	40.0	50.3
Hyla cinerea	Terrestro-arboreal	80.1	39.3	49.0
Rana pipiens	Terrestro-semi-aquatic	78.9	35.5	44.9
Rana aurora	Semi-aquatic	79.7	34.3	43.0
Rana grylio	Aquatic	77.5	29.5	38.0

SOURCE: Thorson and Svihla, 1943: Ecology, *24*: 374.

40.5°C (Schmidt-Nielsen, 1959). The camel can tolerate a temperature fluctuation of 6.5°C (34 to 40.5°C) without taxing its regulatory machinery while man reacts to an elevation of 1°C. Like the camel, several other desert homeotherms economize on water by tolerating an elevation in body temperature (Maloiy, 1972).

This capacity for tissue dehydration may be associated with the ability to absorb large amounts of water rapidly from the surroundings. The Australian desert frog *Cyclorana* can be desiccated until lean and dry but within two minutes it will take up enough water to be as round as a "knobly tennis-ball" (Buxton, 1923). Its absorptive capacity is so great that the aborigines are said to use water-loaded animals as a source of drinking water. The desert frog is by no means the only amphibian capable of absorbing water from its environment. As a matter of fact, this may be the only route open to them. Water absorption through the anuran skin has been frequently measured, and since these animals are not known to drink water, all environmental water must be acquired in this way (Chew, 1961). Immersion is not essential; contact with moist filter paper or moss is sufficient.

Many amphibians can store large volumes of urine in the bladder and reabsorb the water as required (Warburg, 1972). There is a nice correlation between the available water and the capacity of the urinary bladder to hold urine (Bentley, 1966, 1971). *Cyclorana* and some of its near relatives can store a volume of fluid equivalent to 50 per cent of their body weight, while the capacity of the bladder of the aquatic toad *Xenopus* is only 1 per cent of its body weight; both anurans and urodeles show variable bladder capacities (Bentley, 1971).

Many reptiles (snakes, crocodiles, some lizards) and all adult birds (except the ostrich) lack a urinary bladder (Romer, 1970). Urine drains directly into the cloaca. By adding dyes and radio-opaque substances to the urethral urine it has been shown that urine quickly passes anteriorly in the intestine to about the level of the cecum. Thus, the cloaca and large intestine serve as an extensive reservoir for the retention of fluid (Bentley, 1971; Skadhauge, 1972). The importance of this fluid in ion-osmoregulation has been clearly demonstrated in several birds. In some experiments, ureters were surgically transferred to the exterior of the body in order to bypass the cloaca; these experimental animals require more water and salt. Other experiments have shown that the large intestine is the site of absorption and that relatively little uptake occurs from the cloaca itself. Dehydrated chickens may absorb as much as 50 per cent of the water and salt that pass into the cloaca from the ureters. This becomes a major factor in water economy of birds in dry habitats (Skadhauge, 1972).

Some terrestrial arthropods can also withstand desiccation and rapidly absorb water when conditions are suitable (Berridge, 1970; Shaw and Stobbart, 1972). Dry environment insects, such as the firebrat *Thermobia* and the prepupa of the flea *Xenopsylla,* can absorb moisture from subsaturated atmospheres as dry as 45 to 50 per cent relative humidity (Beament, 1965). The capacities of the terrestrial arthropods as well as most insects are much more limited; the wood louse *Armadillidium* can only take up water in still air at 98 per cent relative humidity while other species of land isopods can only absorb water when the air is fully saturated (Edney, 1954). The absorption of water from subsaturated air has long been one of the most perplexing problems in insect physiology. The thick waxy cuticle is a truly marvellous organ for regulating water loss (Locke, 1964; Beament, 1964), but theories of water absorption based on its pores have not been entirely satisfactory. It now appears that the rectum is the site of absorption for atmospheric moisture as well as fluids passing into it from the gut. Noble-Nesbitt (1970) found that firebrats could no longer take up water after the anus had been occluded with wax; the ability to absorb water was not impaired by blocking the mouth in a similar manner. If this situation is general in insects, it follows that the long-debated mechanism of moisture absorption will become another example of water movement which is secondary to the active transport of salt. The capacities of the insect rectum to absorb water were considered in Chapter 8.

Water is a byproduct of tissue respiration (Chapter 7) and, in some species, this metabolic water takes care of the animal's entire requirements. Foods vary in their water potential. One gram of carbohydrate, when metabolized, produces 0.6 g of water, in contrast to 0.4 g from an equivalent amount of protein and 1.07 g from fat (Schmidt-Nielsen and Schmidt-Nielsen, 1952). Consequently, an emphasis on fat metabolism will go far toward alleviating water shortage; unlike fat, the metabolism of protein produces only relatively small amounts of metabolic water and forms nitrogenous wastes which must be removed in solution. There are many examples of dependence on metabolic water. Some of the best are found in chemical embryology; an embryonic chick, with its cleidoic egg developing in a dry environment, obtains 90 per cent of its energy from fat, while a fish or amphibian in contrast metabolizes 90 per cent protein. In this way, the chick forms much more water and much less of the nitrogenous wastes which demand energy or water for their removal (Chapter 8). Some of the desert rodents find the same metabolic advantages in eating large amounts of fatty seeds; when this is coupled with an efficient renal mechanism for the reabsorption of water and a behavior of avoiding the extreme desert heat by seeking moister microclimates, they can dispense entirely with the drinking of water. The urine of the kangaroo rat is about twice as concentrated as that of the laboratory rat and three times as concentrated as that of the dog. This remarkable ability permits the animal to actually drink seawater under experimental conditions. The kangaroo rat is only one of several animals now known to be independent of exogenous water; examples will be found in the literature already cited.

It has often been pointed out that the potential benefit of metabolic water to an active land animal is much less than might be expected since the oxidation of the hydrogen requires ventilation with the attendant evaporation of additional water from the respiratory surfaces. This is indeed true, but such animals as the kangaroo rat, *Dipodomys,* which rely on metabolic water, avoid this hazard (during the heat of the day) by remaining underground where the air is cooler (30°C or less) and the humidity higher (30 per cent relative humidity or greater). Again, one of the most significant differences between the water demands of the desert mammal and that of its near-relatives, living in moister environments, is associated with their different demands for temperature regulation (Chew, 1961).

A certain amount of water is always required for the excretion of nitrogen. One of the major events in biochemical evolution was the phylogeny of metabolic pathways which reduce these demands to a minimum when water is in short supply. The varied end products of nitrogen metabolism have already been described, with the phylogenetic implications of the conversion of the primary end product, ammonia, into less toxic nitrogen compounds (Chapter 8). "The nature of the predominant end product in any particular case seems to be conditioned by the nature of the habitual environment of the particular organisms, and the known facts are best explained on the supposition that the conversion of ammonia to other products is an indispensable adaptation to limitation of the availability of water" (Baldwin, 1967).

Behavioral Adaptations

The first experiments with terrestrial living were probably based on behavioral rather than physiological adaptions. Primitive land animals, both invertebrate and vertebrate, are limited in their capacities to acquire water or to resist dehydration. Their major achievement is an ability to select an appropriate microclimate. The shore crab *Pachygrapsus* prefers 100 per cent sea water when given a choice between this and 50, 75, 125, or 150 per cent sea water (Gross, 1957); the littoral isopod, *Ligia baudiniana,* prefers distilled water to normal sea water (Barnes, 1940); *Birgus,* an air-breathing land crab, selects drinking water of an appropriate salinity to maintain its ionic balance (Gross, 1955); terrestrial isopods (wood lice) are active in dry air and show random movement until they find themselves in a moist place where they become quiet again (Fraenkel and Gunn, 1940). Each species thus relates itself to an appropriate habitat and a suitable supply of water and electrolytes.

Similar examples are found among the vertebrates. Local distributions of the plethodontid salamander (*Aeneides lugubris*) are nicely correlated with the moisture conditions of the habitat (Rosenthal, 1957); a freshwater race of the snake *Natrix sipedon* is killed in sea water because it drinks the water, but a race which lives in salt marshes, although still preferring fresh waters, tolerates salinities up to 73 per cent of sea water because it avoids drinking (Pettus, 1958). The green turtle *Chelonia mydas mydas* drinks sea water, probably to obtain the necessary sodium to balance the excretion of large amounts of potassium obtained in its natural foods (Holmes and McBean, 1964). Some constancy in the ratios of these two ions is maintained in their metabolism and excretion; if the intake of potassium is exceptionally high, additional sodium is required in the diet. Some herbivorous mammals may travel long distances to "salt licks" for this reason.

Water loss is greatly reduced when animals aestivate or become dormant. Pulmonate snails living in deserts provide one of the most extreme examples. Several different species may be found dormant and exposed to the heat of the sun in environments where water is rarely available. Schmidt-Nielsen and his associates (1972) recorded very high lethal temperatures (about 55°C) for desert dwelling pulmonates (genera: *Sphincterochila* and *Helicella*) and estimated that dormant animals could go without water for several years. Water stored during the brief rainy season is lost very gradually through the shell and operculum (estimated rate: 0.5 mg/day).

REGULATORY MECHANISMS

Since the permissible variation in tissue water and electrolytes is usually small, it would be surprising if animals did not possess a refined machinery for osmotic and ionic regulation. A chemical rather than a nervous integration might also be anticipated, since the necessary adjustments often follow gradually modified environments associated with seasonal periods of rainfall, tidal cycles or migration.

Neuroendocrine factors concerned with hydromineral regulation have been convincingly demonstrated in several more advanced invertebrates. The freshwater pulmonate snail *Lymnaea stagnalis* seems to produce two neurosecretory factors; one of these hastens the elimination of water while the other stimulates the uptake of salt (Wendelaar Bonga, 1972). Martoja (1972) summarizes the rather sketchy evidence of hormonal regulation of water or ion balance in a few other molluscs. Among the crustaceans, extirpation of the eyestalks or sinus glands has been shown to disturb hydromineral balance. It seems likely that there is a hormone concerned with the absorption of salt in freshwater species and one regulating the influx of ions in marine species (Kamemoto and Tullis, 1972). The most convincing evidence of osmoregulatory hormones among the invertebrates has been found in studies of insects. Both diuretic and antidiuretic factors are now recognized (Mordue, 1972; Chapter 8). These substances are protein in nature and appear to be of relatively low molecular weight (Goldbard *et al.,* 1970). There is only meager evidence of hormones concerned with the regulation of water and electrolytes in the other invertebrate groups.

In the vertebrates, several different hormones collaborate to control the balance of water and ions. The principal agents regulating the balance of water and salt are prolactin, the neurohypophysial hormones, and the cortical steroids. The hormones of the urophysis and corpuscles of Stannius are also involved in some of the aquatic vertebrates. The levels of calcium and phosphate depend on the ultimobranchial and parathyroid glands; the physiology of these organs was summarized in Chapter 2 and will not be considered further.

The target organs of the principal actors in this osmo-ion regulatory troupe are skin, gills, kidney, urinary bladder, gut, chloride cells and salt glands. The literature concerned with the endocrinology of hydromineral regulation is now voluminous and a large component of it emphasizes the compensatory mechanisms concerned with environmental change. It is these environmental relations which are emphasized in this chapter; the hormones and structures involved have already been identified (Chapters 2 and 8). Only a few examples can be selected from a vast literature. In making this selection the attempt has been to summarize the different components involved and emphasize that a group of hormones and several target organs operate as a system, to regulate osmotic phenomena in the habitats and environments which frequently show drastic variations in available water and salt. The extensive literature is summarized in a general monograph by Bentley (1971) and in several special reviews (Maetz, 1968; Benson and Phillips, 1970; Maloiy, 1972; Motais and García-Romeu, 1972).

Prolactin and Hydromineral Regulation

Pickford's classical studies of the hypophysectomized *Fundulus heteroclitus* first focused attention on the osmoregulatory effects of prolactin (Pickford and Atz, 1957; Ball, 1969). Unless the pituitary is intact, *Fundulus* soon dies in fresh waters with a low content of dissolved solids although it can survive for prolonged periods in dilute sea water. Death is associated with a marked loss of chloride and can be prevented by injections of prolactin; persistent attempts to implicate other hormones in this reaction have failed. Several other euryhaline teleosts also require prolactin for freshwater survival. Prolactin is also concerned with hydromineral regulation in some of the tetrapods. Effects in amphibians include the skin mucus gland secretion in newts and the transport of sodium across the bladder wall. Prolactin effects on the plasma levels have been recorded in the lizard *Dipsosaurus dorsalis;* nasal salt glands are stimulated in some birds, while a renotropic, salt-retaining effect occurs in several mammals including man (Buckman and Peake, 1973).

The migratory physiology of the anadromous stickleback *Gasterosteus aculeatus* has been selected to illustrate an involvement of prolactin in environmental relations. Lam (1972) summarizes the literature. Marine races of this small fish migrate into fresh water during the spring, prior to spawning. At this time, there are changes in osmoregulatory capacities, which involve the morphology of the kidneys and the ability of the animal to excrete water and trap, or retain, ions. During the autumn and winter, while the animals live in the sea, a sudden transfer to fresh water of low ion content (especially low calcium) is lethal; death is associated with the loss of salt and an inability to produce a copious urine of low ion content. Prolactin injections prior to the experimental immersion of "winter" animals in fresh water increase their survival time and prevent the excessive loss of salt. Thus, it appears that the marine stickleback is stenohaline during the winter but becomes euryhaline in the spring; further, it is indicated that prolactin secretion promotes the physiological change.

There is further evidence of prolactin involvement. Seasonal changes in behavior, as well as physiology associated with the spring migration, are triggered by increasing day lengths and may be induced during the fall and winter by experimentally exposing animals to long (spring and summer) photoperiods (Baggerman, 1972; Lam, 1972). Again, prolactin injected into "winter" fish induces the osmoregulatory physiology of the spring animal. Although it has not been possible to perform the crucial tests of hypophysectomy, histophysiological changes in the pituitary support the hypothesis that prolactin secretion and output are higher during spring and summer than during autumn and winter. This implies that the increasing length of days in spring stimulates pituitary activity in the stickleback and that this is a major factor in the physiological changes which occur prior to migration. Studies of the ion and water fluxes argue for an action of prolactin on both gills and kidney, and probably on gut and urinary bladder as well (Lam, 1972 and earlier).

The stickleback is only one of several animals in which prolactin plays a part in environmental compensations. All the target organs concerned with ion and water balance have been implicated in one or another of the vertebrates studied; the mechanisms are uncertain and may involve more than one cellular process (Nicoll and Bern, 1971; Lam, 1972).

Water Balance and the Neurohypophysis

The neurohypophysial hormones were described in Chapter 2. They form a closely related family of octapeptides which serve two distinct functions; one group acts on smooth muscle (uterine, vascular and lactating glands) while the other group affects semipermeable membranes such as those of the kidney tubules, frog skin, cloaca and urinary bladder. The smooth muscle effects are due to the neutral peptides (oxytocin and its allies) while the permeability effects are caused by the basic peptides (vasotocin and vasopressin). It is the latter two basic peptides which are of interest here; historically, they are usually referred to as the ANTIDIURETIC HORMONES (ADH). One or the other of these factors has now been identified in all the vertebrates (Table 2.3, p. 55) and it is apparent that they have played a major role in the adaptation of the vertebrates to habitats of varying salinity and aridity. Their versatility in evolution is shown both in a variable capacity for hormone production and in changing relationships with important organs of salt and water balance. In this chapter, comment is restricted to several examples which stress these adaptive responses in the output of ADH and the changing relationships to target organs.

There is a good correlation between the size of the neurohypophysis and the demands for conservation of water (Fig. 11.12). Moreover, the bioassays reveal markedly higher levels of ADH in the pituitaries of animals which experience a shortage of fresh water; for example, the following biological activities of vasopressin in microunits per gram have been reported (Vizsolyi, 1972): guinea pig, 600; red kangaroo, 10,000–15,000; sheep, 16,000; Indian elephant, 19,600; camel, 132,000; harbor seal, 55,600. In part, these differences relate to gland size, but the relatively high values for animals such as the camel and seal suggest greater hormone production associated with the lack of drinking water.

A study of the renal effects of the antidiuretic hormones in different vertebrates reflects the changing functions of the hormone during phylogeny (Fig. 11.13) In freshwater fishes, arginine vasotocin is diuretic and adaptive to a physiology which demands steady elimination of large volumes of water taken in osmotically; in the tetrapods it is antidiuretic and plays its part in the production of hypertonic urine; however, in non-mammalian forms it acts on the glomerulus, as it does in the fishes, while in mammals its target site is restricted to the tubules (Sawyer, 1972). These changing relationships demonstrate the versatility of ADH in the evolutionary pro-

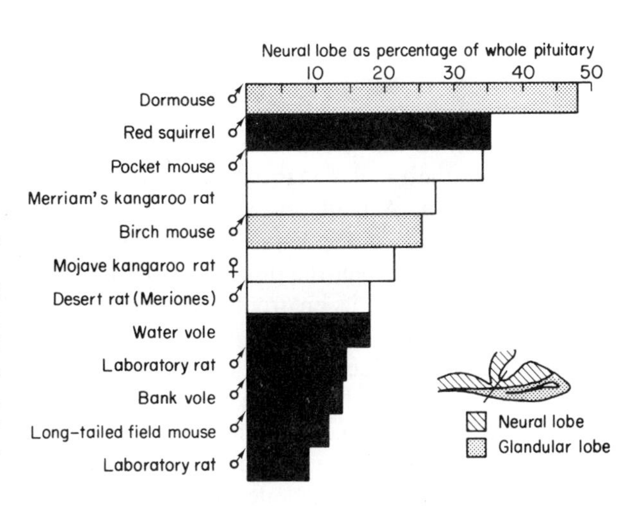

Fig. 11.12 Relative sizes of the neural lobe of the hypophysis in several different rodents. Stippled bars, hibernating species; open bars, desert-dwelling species; solid bars, non-hibernating rodents from temperate zone. [From Bentley, 1971: Endocrines and Osmoregulation. Springer-Verlag, Heidelberg. By permission.]

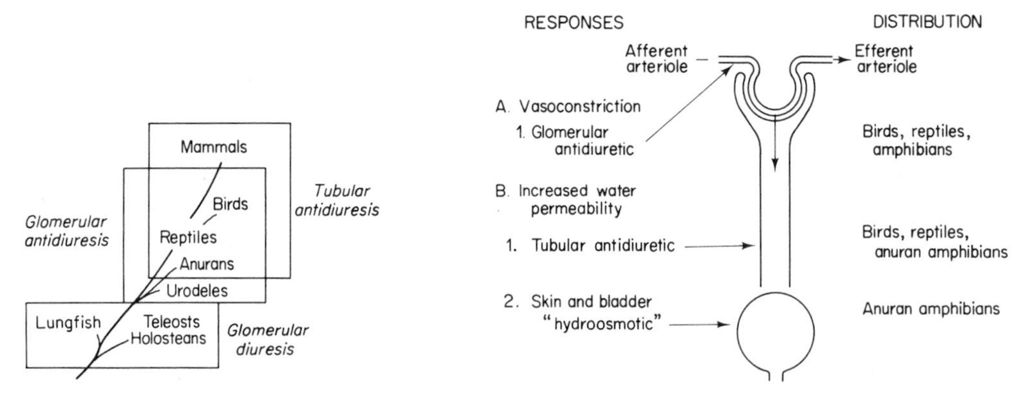

Fig. 11.13 Renal responses to antidiuretic hormones in different groups of vertebrates: (left) Phyletic distribution of responses; (right) types of action of arginine vasotocin in nonmammalian tetrapods. [From Sawyer, 1972: Gen. Comp. Endocrinol., Suppl. *3*: 346, 348.]

cess and support the old aphorism that it is not the hormones that change during phylogeny but the uses to which they are put.

The Adrenocortical Steroids

The adrenal cortex, or its homologue in the lower vertebrates, is the dominant chemical regulator of electrolytes in the vertebrates. This gland is relatively easy to remove from laboratory mammals and the sequelae have often been recorded: a loss of sodium and chloride through the kidneys, a lowered serum sodium, an elevated serum potassium, a decreased urinary potassium and a decreased plasma volume. Death is inevitable if the animals are not provided with the appropriate cortical steroids (Chapter 2).

In mammals, aldosterone is recognized as the most potent of the adrenal steroids concerned with electrolyte balance. Its action at the level of the organism is to increase the tubular reabsorption of sodium and to promote the renal excretion of potassium. Actually, it has a general regulatory action on all cells of the body, increasing the intracellular sodium and decreasing the concentration of potassium; but in this chapter the primary concern is with the regulation of electrolytes between the animal and its environment, and the general action of aldosterone on cell membranes is not discussed. Although several factors have been shown to alter the secretion of aldosterone, the renin-angiotensin (JG) system provides the dominant mechanism regulating mineralocorticoids in mammals. The mammalian JG system is activated by changes in fluid volume or blood electrolytes (Chapter 2; Fig. 8.23). Although there is evidence of an involvement of the JG system in all groups of submammalian vertebrates, its dominance in the regulatory controls of lower forms is by no means certain (Maetz, 1968; Bentley, 1971; Krishnamurthy and Bern, 1973). An evolutionary change in controls, from direct effects of plasma sodium and potassium concentrations on the adrenal to an indirect action by way of the JG system, has been suggested (Denton, 1965).

In mammals, the kidney is the organ of significance in a balance of electrolytes. In the lower vertebrates, on the contrary, several extrarenal organs may play a part or even the dominant role in maintaining the salt balance (Bentley, 1971). In fishes, the gills are often active in the absorption of salt (freshwater teleosts) or in its elimination (marine forms). The gastrointestinal epithelium may also be involved in fish which drink sea water to keep up their water supply; in the elasmobranchs there is a special rectal gland concerned with extrarenal salt excretion. Amphibian skin controls the passage of electrolytes as well as water. Marine reptiles and birds can discharge large amounts of salt in a mucous secretion from specialized NASAL or SALT GLANDS located in the supraorbital region. These specialized tissues for the extrarenal removal of salt permit their possessors to drink sea water, a feat which only a very few mammals have been able to achieve through the refinement of water reabsorption in the kidney (for example, the desert rat).

All these extrarenal salt-regulating tissues have now been shown to respond to cortical steroids. The effects of adrenalectomy and salt-loading experiments have clearly demonstrated that both extrarenal and renal tissues are under comparable controls. Several recent reviews are available (Holmes *et al.,* 1963; Bentley, 1971). It is only possible to generalize in a very broad way because the direction of the facilitated ion flux is not constant with respect to corresponding organs in different species; nor is the mammalian distinction between the "glucocorticoids" and the "mineralocorticoids" faithfully maintained in the other phyla.

At this stage in comparative endocrinology it seems logical to follow the arguments of Romer (1970) and Chester Jones (Chapter 2) and to look upon the adrenal cortex as a regulating tissue which evolved along with the primitive kidney or holonephros. The latter has probably been concerned, from the beginning, with ionic as well as osmotic balance, and the cortical tissues which arise from adjacent regions of embryonic mesoderm produce a variety of regulatory steroids; the dominant hormone is aldosterone in the mammal but may be something quite different in some of the lower forms. The reviews cited give many examples of the recorded variations. At present the most useful generalization is that the cortical steroids are associated throughout vertebrate phylogeny with the regulation of water and electrolyte balance and that they probably act at both renal and extrarenal sites in a direction dictated by the needs of the body to maintain homeostasis.

Other Factors Regulating Water and Electrolyte Balance

The hormonal regulation of water and electrolytes is an integrated process dependent on the cooperative action of several chemicals. Although the neurohypophysial and cortical hormones were considered separately, their action is frequently a cooperative one. Thus, the formation of hypertonic urine by the mammalian kidney depends on the control of membrane permeability by ADH *and* the regulation of sodium transport by aldosterone (Chapter 8). Likewise, the interaction of these hormones may be required for the regulated transport of sodium by anuran skin; both vasotocin and aldosterone have been shown to promote sodium transport, even though experimentally the effects of the latter are much more prolonged. Further, in addition to these two factors, which are mainly concerned with the movement of water and electrolytes, several metabolic hormones are known to increase sodium transport by amphibian skin; adrenaline, thyroid hormones and insulin have all been tested with positive results. Optimal rates of sodium transport *in vivo* may depend on an interaction of several hormones (Bentley, 1971).

Many fishes also depend on a concert of hormones to regulate the balance of water and electrolytes (Maetz, 1968; Bentley, 1971). Several metabolic hormones, as well as those of the neurohypophysis and adrenal cortex, have been shown to have effects; the importance of prolactin has already been stressed. Moreover, the teleost fishes possess two unique endocrine structures—the urophysis and the corpuscles of Stannius—which may also play a part in hydromineral regulation. There is now an extensive literature on these puzzling glands and a considerable portion of it relates to their possible roles in electrolyte balance. However, neither the urophysis nor the corpuscle of Stannius has yet been shown to play an essential part in the maintenance of water and salt balance; their *in vivo* functions have still not been satisfactorily explained (Maetz, 1968; Lederis, 1970; Bentley, 1971; Chan, 1972).

12

THE GASEOUS ENVIRONMENT

Joseph Priestley, in the mid-eighteenth century, seems to have been the first to demonstrate scientifically a similarity between life and fire (Kleiber, 1961). By simple experiments he showed that either a mouse or a flame in an enclosed space would change the air so that neither the life of the mouse nor the flame was any longer possible, and that the conditions in the enclosed space were about the same when either of these processes came to an end. Priestley's explanations, in terms of PHLOGISTON, have now been relegated to the curiosities of scientific history; but the similarity between life and fire remains. Both are combustion processes which require oxygen as Lavoisier, a contemporary of Priestley, recognized. Priestley prepared oxygen in 1777 by heating mercuric oxide and showed that it would support the life of animals; but he fitted his facts into the wrong theory, and it remained for Lavoisier to break new ground with the assumption that both fire and animals make the air unfit for their existence, not by producing phlogiston but by removing oxygen to form carbonic acid.

The call for oxygen is almost continuous throughout the life of active animals. As Lavoisier realized, life is a combustion, but the similarities between life and a fire do not really go any deeper than the science of Lavoisier's century. Our century has shown the far greater complexity of metabolic fires and established in some detail the molecular changes through which potential energy in the fuel is channeled into the high-energy phosphate bonds of ATP; these, in final analysis, are the source of power for animal life (Chapter 7). The production of large amounts of ATP requires a continuous supply of oxygen. The rate of supply can be limited by certain anatomical and physiological characteristics of the organs of respiration and the transport pigments. These were considered in Chapters 4–6. In addition, the actual oxygen content of the environment may be a limiting factor while other environmental variables such as temperature, carbon dioxide or salinity may impose extra demands for oxygen or affect the rates of exchange. These environmental restrictions in the availability of oxygen, and some of the interrelated effects of the environment on metabolism, will now be considered.

OXYGEN RESOURCES OF THE ENVIRONMENT

From sea level to the tops of the highest mountains, the earth's atmosphere is everywhere about 21 per cent oxygen. Thus, each liter of air which circulates over the respiratory epithelia contains 210 ml of oxygen. The exchanges of gas between the animal and its environment, however, depend on concentration gradients, and it is the actual number of oxygen molecules in any volume of gas rather than its volume which is important. As Robert Boyle discovered many years ago, the volume occupied by any mass of gas varies inversely with the pressure. At sea level and 0°C a mole of oxygen occupies a volume of 22.4 liters, but at a height of about 18,000 ft where the pressure is reduced to half an atmosphere the volume occupied by the same amount of gas is doubled. The respiratory epithelia of an air breathing animal living at this altitude will then be exposed to only half the number of oxygen molecules in each volume of gas circulated over them. Pressure, rather than volume, is the meaningful parameter for an air breather.

In a mixture of gases, such as the atmosphere (Table 12.1), the pressure exerted by each gas depends on its percentage in the mixture (Dalton's Law of Partial Pressure). Thus, at sea level, where the atmospheric pressure is 760 mm mercury, the oxygen partial pressure (also referred to as the TENSION) is 20.948 per cent

Table 12.1 Composition of the Atmosphere

	Per cent	Partial pressure (mm Hg)
Oxygen	20.948	159.20
Carbon dioxide	0.030[1]	0.23
Nitrogen	78.00	592.8
Argon	0.94	7.15
Other	0.082	0.62

SOURCE: Krogh, 1941: The Comparative Physiology of Respiratory Mechanisms. U. Penn. Press, Philadelphia.

[1] May rise to about 0.04 per cent in streets of large cities and is increasing significantly during this century because of the burning of fossil fuels and greater agricultural activities (Plass, 1959; Newell, 1971).

of 760 mm or 159.20 mm Hg. On the top of Mt. Everest, although the percentage composition of the air is the same as at sea level, the oxygen partial pressure is only about 49 mm Hg, and this is much too low to maintain the necessary gradients for oxygen exchange.

Temperature also changes the volume of oxygen in a unit volume of air at any pressure (Gay-Lussac's or Charles' Law), but this effect is of little physiological significance; at 0°C a gram-mole of gas occupies 22.4 liters under atmospheric pressure, and this becomes 25.4 liters at the mammalian body temperature of 37°C (an increase of only about 13 per cent). As indicated below, the temperature effects for aquatic organisms are much greater.

Each of the atmospheric gases dissolves in water according to its partial pressure, its solubility coefficient and the temperature. The solubility coefficient is characteristic for each gas. Oxygen is a little more than twice as soluble as nitrogen, and carbon dioxide is about thirty times more soluble than oxygen. Thus, the proportions of the various atmospheric gases are very different in air and in water (Tables 12.1 and 12.2).

Rising temperature reduces the solubility of gases, and the magnitude of the effect is also characteristic for each different gas. This may be readily calculated from the solubility coefficient, defined as the volume of gas dissolved in one volume of water exposed to the gas at 1 atmosphere pressure. At 0°C, the coefficient for oxygen is 0.0486, for nitrogen 0.0235, for carbon dioxide 1.704; the corresponding values at 20°C are 0.0326, 0.0163, and 0.921. Consequently, at sea level and 0°C, water exposed to the atmosphere will dissolve 20.948 per cent of 0.0486 (= 0.0102) ml O_2 per ml water or about 10.2 ml/liter; at 20°C, the value becomes about 6.6, a reduction of about 40 per cent. The temperature effect on oxygen availability is thus much greater in water than in air.

The solubility of gases is also reduced by the presence of dissolved solids, and in consequence sea water contains considerably less oxygen than the fresh waters (Table 12.2). The importance of dissolved solids and the magnitude of the temperature effect on the solubility of gases are so great that a measure of partial pressure alone has little meaning as a useful parameter for the availability of oxygen

Table 12.2 Oxygen (ml/liter) in Water at Different Temperatures and Chlorinities When Saturated with Atmospheric Air

Temperature	Chlorinity (‰)		
	0	10	20
0°C	10.29	9.13	7.97
10°C	8.02	7.19	6.35
15°C	7.22	6.50	5.79
20°C	6.57	5.95	5.31
30°C	5.57	5.01	4.46

SOURCE: Krogh, 1941: The Comparative Physiology of Respiratory Mechanisms. U. Penn. Press, Philadelphia.

to the aquatic animal. It is more useful and meaningful to record the quantities of dissolved respiratory gases in milligrams or milliliters per liter of water; the former is preferred for the expression of oxygen supply and consumption of aquatic animals.

Still another physical limitation on the availability of oxygen to aquatic organisms is the rate of diffusion. This is very much slower in water than in air and, in many habitats, does not keep pace with the rate at which oxygen is being used. It would require thousands of years for the waters of deep lakes to become saturated with oxygen if they were calm and if only diffusion were involved in gas transport.

Textbooks of oceanography and limnology provide details and generalizations concerning the oxygen-rich and the oxygen-poor habitats of the fresh waters and the ocean. They range from the anaerobic to the supersaturated. The former condition is not uncommon in deep lakes at certain seasons and is characteristic of certain sea waters (Black Sea, Gulf of California). Supersaturation may develop in ponds where there is active photosynthesis, and it is also found in torrents at the base of waterfalls where atmospheric air is carried deep enough to be dissolved in large amounts under the increased pressures. Supersaturation may be sufficiently great to kill fish (Harvey and Smith, 1961). Bishai (1960) summarizes the literature.

OXYGEN AS A LIMITING FACTOR IN THE ENVIRONMENT

Life commenced in an oxygen-less world (Chapter 1); but animal evolution was built on the abundance of oxygen required for bulk production of ATP in the processes of oxidative phosphorylation (Chapter 7). Secondarily, however, many animals have acquired the capacity for life with little or no oxygen. In a few habitats (bottom of water basins, certain soils and the intestines of larger animals) the oxygen levels are continuously low; in others, this condition develops occasionally or seasonally (von Brand, 1946). Animals which exist in these habitats are able to live without oxygen for a long time; some of them are FACULTATIVE ANAEROBES which switch to aerobic metabolism when oxygen is available; a few are OBLIGATE ANAEROBES which never require oxygen. The subject of SUSTAINED ANAEROBIOSIS was introduced in Chapter 7 where the probable biochemical pathways are outlined (Fig. 7.21).

Many animals resort to temporary anaerobiosis during short bursts of muscular activity. The immediate demands for oxygen exceed the rates of delivery and the animal incurs an OXYGEN DEBT which is paid for after the emergency. A comparable condition will be considered later in this section when the physiology of diving animals is discussed. Metabolically, temporary anaerobiosis is quite different from sustained anaerobiosis. In the former, such acid metabolites as lactate accumulate, while in the latter, the end products are succinate, alanine and propionate (Fig. 7.21). Hochachka and Somero (1973) find that ATP production in true anaerobiosis may be as high as 60 per cent of the ATP yield of oxidative phosphorylation.

The line between the temporary and sustained states may not be the sharp one implied above. The early investigations were mostly confined to the parasitic worms, while later studies have focused on the intertidal invertebrates (see symposia in *Am. Zool.*, 11 (1), 1971 and *Am. Zool.*, 13 (2), 1973). Interest is now shifting to some of the vertebrates. In fact, there has been good evidence of facultative anaerobiosis among the fishes for almost two decades. In 1958, Blažka reported that the Crucian carp *Carassius carassius* can live in an oxygen-free environment for two to three months at temperatures of about 5°C. During this period, carbon dioxide is eliminated and the animals actually accumulate fat. Unlike salmon and trout, there is no accumulation of lactate and no oxygen debt. During the winter, Crucian carp regularly become "ice-locked" in small ponds which gradually become anaerobic and remain free of oxygen until the spring thaw. Blažka reviewed the literature and found evidence of similar adaptations in some developing fish eggs during periods of anoxia. Since that time, several species of fish have been noted in environments which appear to be essentially anaerobic (Hochachka and Somero, 1971).

It now seems likely that the Crucian carp exploits pathways similar to those described for other facultative anaerobes (Fig. 7.21). It is unlikely that the teleost fishes are alone among the vertebrates in their abilities to switch metabolic pathways to anaerobic channels under conditions of restricted oxygen. Prolonged diving is characteristic of some species in all classes of terrestrial vertebrates; newborn mammals are notoriously resistant to lack of oxygen (Adolph, 1973); there is still a wealth of material for those interested in studies of life under anoxic conditions.

Oxygen Consumption in Relation to Oxygen Availability

The important relationships are shown diagrammatically in Fig. 12.1. At minimum levels of metabolism (standard or basal metabolism shown by the lower

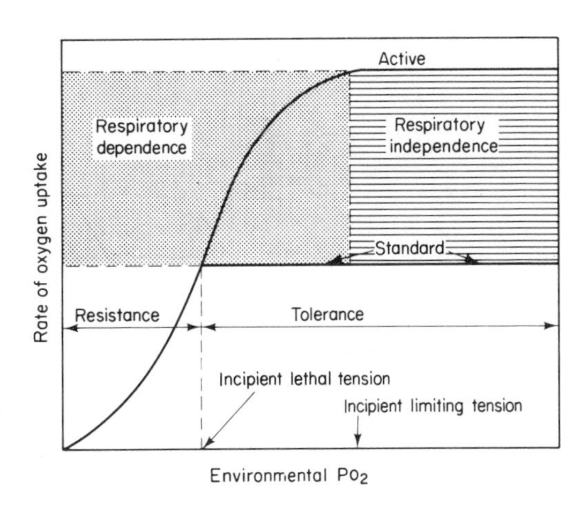

Fig. 12.1 Relation between standard and active (maximum) rates of oxygen uptake at different environmental oxygen concentrations.

horizontal line in this figure), the animal's oxygen requirements are at their lowest point. An environment which supplies this much oxygen is adequate (INCIPIENT LETHAL LEVEL), but at any lower level a condition of anoxia may be expected, and sooner or later the animal will die (below the incipient lethal level an animal is living in a ZONE OF RESISTANCE). The upper horizontal level (active metabolic rate) shown in Fig. 12.1 occurs at the INCIPIENT LIMITING LEVEL or the critical pressure (*Pc*) of Prosser (1973). This is the demand level for active metabolism. At higher oxygen tensions, most animals (regulators) will maintain a steady, independent rate (RESPIRATORY INDEPENDENCE or REGULATION) until toxic levels develop. Between these incipient lethal and limiting levels, metabolism or oxygen consumption is dependent on the availability of oxygen (RESPIRATORY DEPENDENCE or CONFORMITY); the organism can tolerate the situation, but the operation of its machinery must be adjusted in accordance with the supply. In a poikilotherm this dependence is markedly affected by temperature (Fig. 12.2).

Species differences in oxygen requirements The actual values of the lethal and limiting levels are characteristic of the animal species; they have become established in its phylogeny in response to the oxygen conditions of its environment and its way of life. In general, the warm-blooded species have greater oxygen demands than the

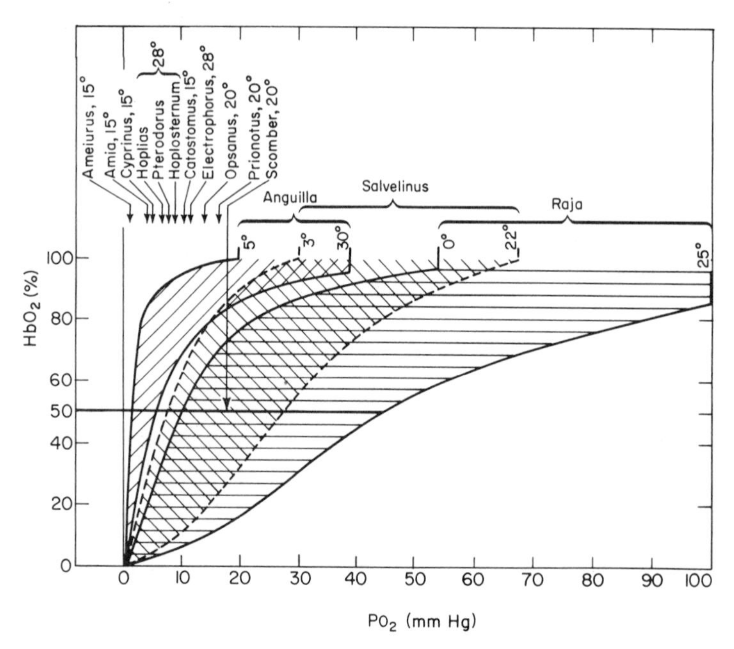

Fig. 12.2 Oxygen equilibrium curves for the bloods of three species of fish in relation to temperature, together with the positions of the half-saturation points of the bloods of various other species determined at single temperatures. [From Fry, 1957: *In* The Physiology of Fishes, Vol. 1, Brown, ed. Academic Press, New York.]

cold-blooded ones; there are, however, substantial variations even within a single class of vertebrates. This is readily apparent in the rates shown for several animals of equal weight in Fig. 12.3. This genetic or species level of compensation depends on variable capacities of the ventilating system, the transport pigments and the enzymatic machinery of the individual cells. In fishes, for example, the number of gill lamellae show a good relationship to the metabolic demands of the animal and the kind of environment where it lives. Active fish such as mackerel have about 31 lamellae per mm of gill filament and a gill area of 1158 mm^2 per gram of body weight while a toadfish, at the other extreme, shows corresponding values of 11 and 197 (Hughes, 1966; Steen, 1971). Similar measurements have been made for the respiratory surfaces of crabs (Gray, 1957). The magnitude of the species difference in hemoglobin transport capacity is indicated by Fig. 12.2.

Acclimation and acclimatization Compensation, through acclimation and acclimatization, may also alter an animal's critical oxygen demands. An improved

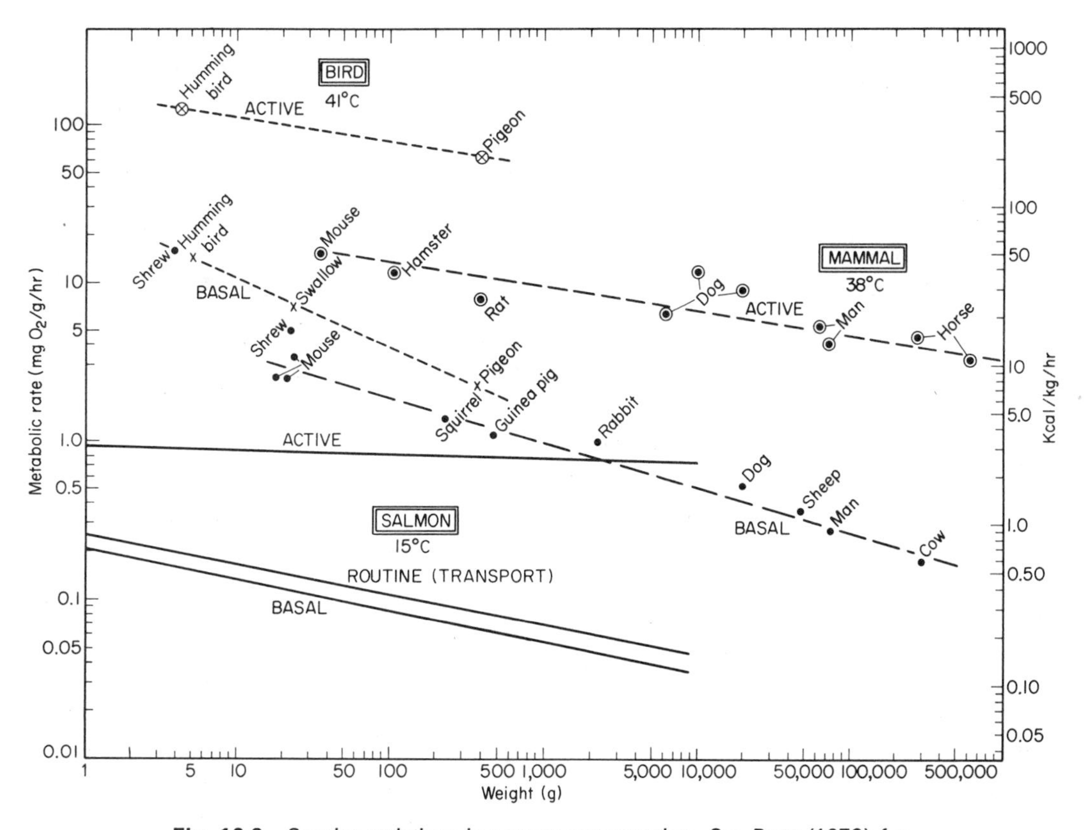

Fig. 12.3 Species variations in oxygen consumption. See Brett (1972) for sources of original data. Mean body temperatures indicated for all cases. [From Brett, 1972: Respir. Physiol., *14*: 164.]

capacity to tolerate hypoxic environments has been recorded for several different vertebrates, including both poikilothermic and homeothermic species. The ability to extract oxygen may improve while, in some species, both active and standard metabolic rates change during acclimation (Fig. 12.4). Although the higher vertebrates, particularly man, have been most intensively studied under hypoxic (high altitude) conditions (Hock, 1970), there are also several critical investigations of aquatic species (Fry, 1971). In general, there are two different levels on which compensation may occur; there are adjustments that enhance the supply and adjustments that restrain the utilization. Improved gas transport through circulatory adjustments, changes in the circulating hemoglobin, and, possibly, adaptations in enzyme systems may act to improve the supply of oxygen in environments where it is deficient. In addition, an animal may decrease the demand for oxygen by modifying its behavior. As is true in other avenues of evolution, it seems likely that species which live for extended periods under hypoxic conditions will differ in their particular adaptive mechanism.

Several invertebrates are also known to respond to partial anoxia by synthesizing hemoglobin (Fox, 1955; Steen, 1971). *Daphnia, Artemia, Chironomus* larvae and *Planorbis* are the most familiar examples. The phenomenon can be easily demonstrated by comparing *Daphnia* cultured in air-saturated water with those maintained in water at about 20 per cent air saturation. The animals become conspicuously red in the latter environment and have been shown to survive much better than the pale ones when tested at low oxygen tensions; they can also take up more oxygen. It can be shown that the hemoglobin synthesized under these conditions may be actually necessary for their survival; for example, carbon monoxide treatment is lethal to these red individuals in their partially anoxic habitat but has no effect on either the pale or the red animals when living in oxygen-saturated waters (Fox and Phear, 1953). The brine shrimp *Artemia salina* shows a similar response (Bowen *et al.*, 1969). Under experimental conditions, the gain or loss of hemoglobin may be

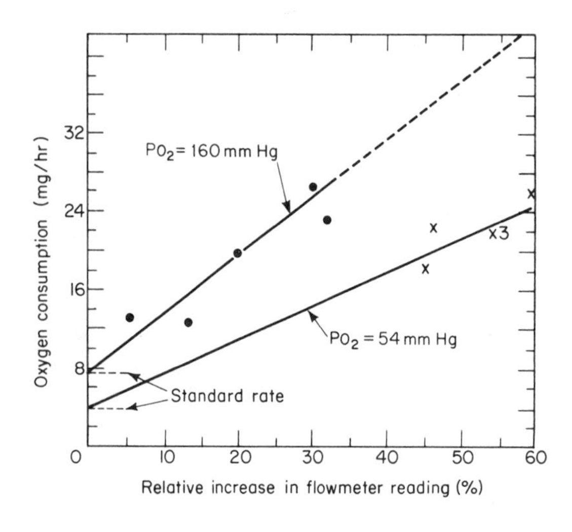

Fig. 12.4 Effect of oxygen acclimation on the active and standard rate of metabolism of a fish. Flow meter readings indicate degrees of activity in respirometer. Tests performed at P_{O_2} = 54 mm Hg. Upper line, fish acclimated to P_{O_2} = 160 mm Hg. Lower line, fish acclimated to P_{O_2} = 54 mm Hg. [From Beamish, 1964: Can. J. Zool., *42*: 364.]

demonstrated in a conspicuous way within two or three weeks. Natural populations of *Artemia* become redder as the salinity of their habitat increases since, of course, the oxygen content of the water gradually declines.

Compensation in metabolism through acclimatization is also well documented. Several species of fish have been shown to use more oxygen in the summer than they do in the winter, even though they are maintained at a constant temperature (Wells, 1935; Beamish, 1964). The photoperiod is probably the dominant environmental regulator of the seasonally changing metabolism of the poikilotherm (Newell, 1971; Songdahl and Hutchinson, 1972).

Living at high altitudes The highest human settlement is a mining camp in the Chilean Andes at 17,500 ft. The Indian miners who live there work in a sulfur mine at 18,800 ft but return to their homes each night to sleep and recover. They know, and it has been proven scientifically (Hock, 1970), that man gradually deteriorates at altitudes of 19,000 ft even when acclimatized as completely as possible. Unacclimatized man is seriously incapacitated at much lower altitudes (9000 to 10,000 ft); fully acclimatized man can spend short periods at much higher levels without supplements of oxygen. Hillary and Tenzing, in the greatest of all mountain climbing epics, removed their oxygen masks on the summit of Everest at 29,028 ft where the partial pressure of oxygen is about 47 mm Hg. They slept for a time without oxygen masks at 27,900 ft (Po_2 about 50 mm Hg), but when three French scientists in 1875 ascended directly in a balloon to over 26,000 ft, two of them died and the other lost consciousness at about 26,000 ft. These experiences underline two major problems: the physiological limitations at high altitudes and the acclimatization processes which enable men to extend their range of high-altitude living in such a spectacular way. In fact, more than ten million people live permanently at altitudes above 12,000 ft where they face a continuous deficit in oxygen (Fig. 12.5). How is this possible?

The limitations of life at high altitudes are not caused by the reduced pressures, unless the ascent has been very rapid (Chapter 13), but rather by the lack of oxygen. Figure 12.5 shows atmospheric pressure with corresponding oxygen partial pressures at different altitudes. An oxygen tension of about 80 mm is required to load human hemoglobin (Fig. 6.3), and it can be seen that this is just possible at 18,000 ft. At 36,000 to 37,000 ft even pure oxygen delivered at the pressure of the atmosphere would barely suffice for human needs since there is always a vapor pressure of about 47 mm Hg in the alveoli of the lungs; when this is added to the alveolar CO_2 pressure (ranging from 40 mm to 24 mm, depending on the extent of acclimatization) and the necessary oxygen tension of 80 mm, the total $(47 + 40 + 80)$ just about matches the atmospheric pressure (Fig. 12.5). At 63,000 ft the atmospheric pressure is about 47 mm Hg and, theoretically, beyond this altitude human blood would boil. It has been calculated that a man suddenly decompressed to an elevation of 70,000 ft would boil away 4 lb of water from his lungs before dying in about 3 mins (Guyton, 1971).

Many terrestrial animals probably experience similar limitations at high altitudes,

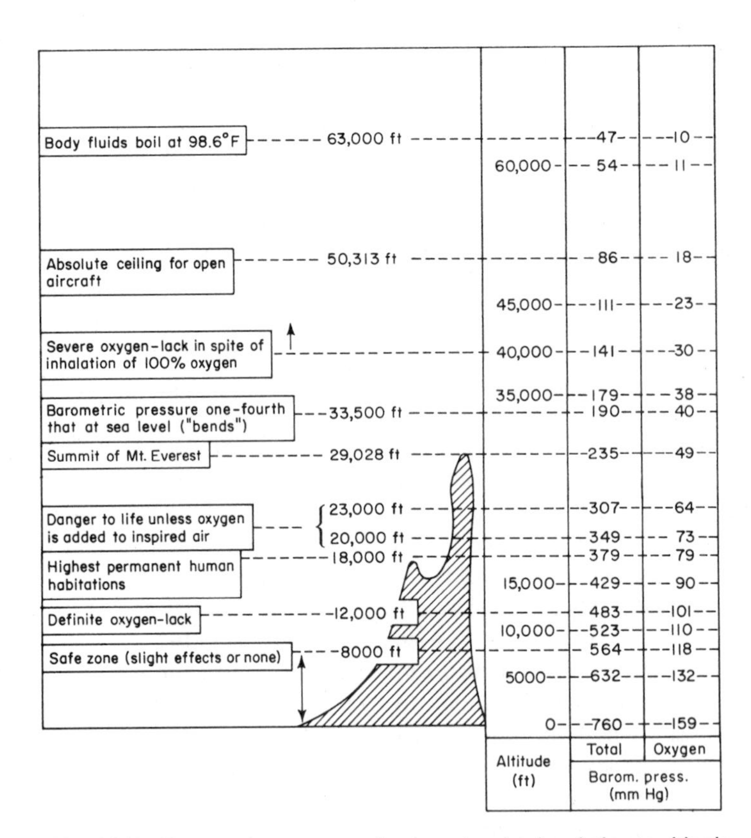

Fig. 12.5 Barometric pressure and oxygen tension in relation to altitude. [Data from Bard, ed., 1961: Medical Physiology, 11th ed. Mosby, St. Louis, Mo.]

and life is sparse beyond 18,000 ft (Swan, 1961). However, some primitive insects and spiders live permanently above 19,500 ft, and many birds travel through these altitudes. The bar-headed geese are known to fly from sea level in India on a non-stop flight over the Himalayas to the lakes of Tibet.

Barcroft, who led one of the most famous of the early scientific mountain climbing expeditions into the Peruvian Andes in 1921–22, recorded all the symptoms of progressive anoxia in his men—an increased rate of breathing (hyperventilation), rapid heart beat (tachycardia), cyanosis, fatigue, insomnia. At altitudes where the supply of oxygen to the cortical layers of the brain is inadequate, the symptoms become those of anaesthesia or alcoholism: lack of discipline, quarreling and laughing, poor judgment and loss of memory, nausea and vomiting. The manifestations of anoxia recede after two to four days and gradually disappear with acclimatization. There are several excellent monographs and symposia (*Handbook of Physiology*, Sect. 4, 1964; Hart, 1969; Porter and Knight, 1971); only a summary is appropriate here.

The acclimatization process is largely one of improving the efficiency of ventilation and accelerating the delivery of oxygen to the tissues. The first response to an oxygen deficit is an intensified pulmonary ventilation. This appears when the oxygen falls to about 15 per cent of the sea level pressure; first the depth and then the frequency of breathing increase. The response is a direct one resulting from stimulation of the chemoreceptors of the aortic and carotid bodies (Chapter 4). There is an improved flow of oxygen through the lungs but, at the same time, there is an excessive elimination of carbon dioxide with a resultant increase in blood pH. This disturbed acid-base equilibrium is later adjusted by the kidneys, but, before this takes place, oxygen transport by the hemoglobin is somewhat decreased since oxyhemoglobin dissociates less readily in alkaline solutions (Chapter 6). Further, the respiratory center, which is first strongly activated by carbon dioxide, is inhibited with the elevated pH in opposition to its stimulation by way of the O_2-sensitive chemoreceptors. Alveolar tensions of CO_2 are about 40 mm at sea level and gradually adjust to about 26 mm after 35 days at 14,000 ft; at this altitude the respiratory minute volume is about 50 per cent above the sea level value. The thoracic dimensions actually change if residence at high altitudes is prolonged for many years; Barcroft found the high Andes inhabitants to have chests of larger volumes with the ribs placed more horizontally.

It is worth noting that the metabolic costs of the increased ventilation are considerable. The oxygen demands of the respiratory muscles and heart are greater since they work harder. In addition, the loss of heat caused by evaporation of water in the breath creates an added metabolic demand for a homeothermic animal. Beyond a certain altitude the operation of the ventilating machinery requires more oxygen than the air can supply.

The transport of oxygen also becomes more and more efficient during acclimatization. Anoxia produces a prompt elevation in the number of circulating erythrocytes due to contraction of the spleen. This is followed by an accelerated multiplication of red cells and a stimulation of hemoglobin synthesis as the bone marrow is activated by the anoxia. At sea level, human blood contains 4 to 5 million erythrocytes per mm^3, but the number rises to about 8.3 million at 18,000 ft. In rats, studied experimentally, the maximum changes occurred at pressures equivalent to 6000 ft when the hemoglobin had increased to 2.2 gm/100 gm body weight from a sea level value of 0.75 gm; the hematocrit, at the same oxygen tension, had altered by 85 per cent (Tribukait, 1963). When these changes are coupled with an increase in blood volume, the circulating hemoglobin may rise by as much as 90 per cent. The oxygen affinity of the hemoglobin also changes during acclimatization (Hurtado, 1964). The oxygen curves of individuals living at high altitudes are displaced significantly to the right; this means that oxygen will be released more readily to their tissues.

The blood circulating machinery also shows its capacity to respond to prolonged stresses of anoxia. Greater cardiac output (by as much as 20 to 50 per cent) is transient, and the heart rate of the acclimatized person is near normal. There is, however, a very real improvement in the vascularity of tissues such as muscle, heart and brain. The number of capillaries is actually greater in men living at high altitudes and this, together with the enrichment of the blood and the improved ventilation, permits man to make maximum use of the scanty oxygen in these environments.

Finally, adaptive changes at the cellular and biochemical levels have also been well documented. The actual quantity of myoglobin in the muscles increases while the mitochondrial enzyme systems concerned with electron transport appear to be adapted to make maximum use of the reduced oxygen. Ou and Tenney (1970), for example, found a 40 per cent increase in the number of mitochondria in the heart muscle of cattle acclimatized to an altitude of 4250 meters (about 14,000 ft). Measurements of enzyme activity, particularly the NADH-oxidase system, indicate an adjustment in the mitochondrial system to permit higher rates of oxygen utilization. Although any one of these adaptive changes may provide only a slight advantage, the combined action of all of them permits abundant life in an otherwise inhospitable habitat.

Birds in flight If a sparrow and a mouse, two animals of comparable size, are exposed to a simulated altitude of 20,000 ft, the mouse is soon incapacitated and scarcely able to crawl while the sparrow is as perky as ever. Moreover, if stimulated, the sparrow will fly, an activity which creates an additional ten-fold increase in its demands for oxygen (Tucker, 1968; Schmidt-Nielsen, 1972). The oxygen tension at an altitude of 20,000 ft is only about 73 mm Hg (Fig. 12.5) and this is entirely inadequate to meet the respiratory demands of a mammal. Birds and mammals, including the bats which are excellent aerial performers, have comparable metabolic demands (Fig. 12.3: Schmidt-Nielsen, 1972) but the bird has an ability to fly at high altitudes while the mammal is incapacitated unless highly acclimatized and shielded by complex technical devices. The migrations of geese which fly non-stop over the Himalayas from sea level in India to the lakes in Tibet at altitudes of 20,000 to 30,000 ft, have already been mentioned.

These achievements are caused by the pattern of air flow through the lung (Chapter 4). Tucker calculates that with a counter-current flow of air and blood, the arterial blood leaving the bird's lung is still 80 per cent oxygenated while the pulmonary blood of the mouse, with its tidal exchange of air, is only 24 per cent saturated. The disadvantages of the dead air spaces in mammalian lungs are emphasized by these calculations, as are the pressures that have probably shaped the evolution of the highly specialized system of air tubes and sacs in birds.

The insects, which vie with birds as aerial travelers, also have a unique mechanical design for the delivery of oxygen to their tissues. The almost direct exposure of cells to atmospheric oxygen in the thin-walled tracheal capillaries permits the high levels of aerobic metabolism which occur in their flight muscles (Miller, 1966a; Steen, 1971).

Air-breathing animals under water All the major groups of terrestrial animals have representatives capable of living under water. Insects, spiders, water mites and members of each class of land vertebrates are familiar residents of the aquatic habitat. Life in water is easily possible for a tiny animal like a water mite which can get sufficient oxygen by diffusion alone, or for larger animals, such as earthworms and amphibians, with highly vascular and permeable skins. Simple diffusion and cutaneous respiration, however, are not possible for most terrestrial animals under water, since they retain their highly impervious coverings of chitin or keratin. Usually, they depend on oxygen which is obtained periodically at the surface and used most economically while under water. Their capacity to make efficient use of these oxygen stores is astonishing; sperm whales are said to go as deep as 1000 meters and remain under water for one or more hours without showing evidence of anoxia. The ability to remain under water varies greatly in different animals (Andersen, 1966; Kooyman, 1972).

In general, the major hazard of living under water is the danger of anoxia—essentially the same difficulty encountered at high altitudes. It is obvious, however, that the problem required quite a different solution in the two habitats. In the water, specialization has been in the direction of temporary oxygen storage, with very economical usage and the capacity to deal with oxygen debts, while the adaptations for life at high altitudes have been improvements in the rates of delivery of oxygen to the tissues.

The tracheates show the most varied arrangements. During submergence, a mosquito larva depends entirely on the oxygen in its tracheal system and, to replenish the oxygen supply, must come to the surface and hang there by hydrophobic hairs grouped around the functional spiracles at the posterior end of the body. Replacement depends entirely on diffusion; a *Culex* larva with a tracheal volume of about 1.5 mm^3 obtains enough oxygen in this way to stay under water for 5 to 10 minutes.

With a relatively larger tracheal volume and by actively ventilating the system when at the surface, some of the larger insects have extended their periods of submergence for as long as 30 minutes—for example, the water scorpion, *Nepa* (Hemiptera). The period of submergence can be further increased by trapping atmospheric air on the outside of the body beneath the modified forewings (elytra) or in a pubescence of hydrofuge hairs. In this way, the predacious diving beetle, *Dytiscus*, extends the period of submergence up to 36 hours (Roeder, 1953). The water bugs (Hemiptera) provide other good examples. Air stores act in a curious way as a gill, since the oxygen dissolved in the water gradually replaces that which the insect uses in its respiration and the respiratory carbon dioxide diffuses away readily. It has been calculated that the air supply lasts 10 to 30 times as long because of this replacement of oxygen in the trapped bubbles (Krogh, 1941).

Aquatic plants may also supply oxygen to insects. Some species can capture the bubbles released in photosynthesis; others (some of the beetles and Diptera larvae) have especially developed hard cutting edges on their spiracles which enable them to puncture into intercellular spaces of the plants.

In all the examples so far mentioned the peculiar adaptations are for the acquisition of atmospheric air. In addition, however, many insects (unlike the air-breathing vertebrates) have adapted their respiratory organs for the extraction of oxygen from water. These tracheal gills, which are characteristic of many aquatic insect larvae, permit the animal to live continuously under water and avoid the hazards of journeying to the surface to replenish the air supply. Dragonfly and damselfly nymphs (Odonata) are good examples, with their feathery rectal gills or abdominal gill plates consisting of a rich system of tracheal tubes beneath thin cuticle (Roeder, 1953; Steen, 1971). The rigidity of the tubes prevents their collapse when the gases dissolve into the water. Exchange of gas depends entirely on diffusion; this is improved by moving the gills in the water or by creating water currents over them.

On the whole, terrestrial vertebrates have been more conservative than insects in the variety of their underwater breathing methods. Some turtles ventilate the mouth and richly vascularized pharynx when submerged; in other species, the cloaca and associated bursae serve as organs of respiration. However, absorption of oxygen from water is exceptional among the diving amniotes and even turtles only obtain relatively small amounts of oxygen in this manner (Steen, 1971). Reptiles, birds, and mammals under water use stores of atmospheric oxygen which must be regularly replenished by surfacing. The physiology of the submerged alligator illustrates several of the basic adaptations (Andersen, 1961). The resting oxygen consumption of a 3 kg alligator is about 4 ml/min at laboratory temperatures of 22° to 27°C. This animal has a lung volume of 250 to 300 ml and this, under optimal conditions, will contain 51 ml of oxygen. The blood volume of the alligator is just over 5 per cent of its body weight or about 150 ml for this animal. A fair estimate of the amount of oxygen in this volume of alligator blood is 8 ml (Andersen, 1961) and this means that the total supply of oxygen on submergence is not more than 60 ml. At a consumption rate of 4 ml/min this store would be completely exhausted in 15 minutes, and yet these animals may stay under water for as long as 2 hours without difficulty. Physiologists have shown that the alligator and other diving amniotes accomplish such remarkable feats not through anaerobiosis but by special oxygen-saving mechanisms.

The most important of these mechanisms is the slowing of the heart (BRADYCARDIA) and the restriction of the circulation to the most essential of the vascular beds—especially the brain and heart. In the alligator, the heart rate may decline from a pre-diving value of 41/min to 2–3/min after 10 minutes of submergence. Lactic acid increases only slightly in the arterial blood during the dive but rises ten times or more on emerging; this indicates that the circulation to the muscles is shut off during the dive. Reflexes, which in most mammals lead to vasodilation with anoxia in these tissues, no longer operate but the details have not yet been completely explained. The carbon dioxide rises only slightly because of the small amounts of stored oxygen for its production and the efficient buffering in the tissues (Andersen, 1961).

The diving abilities of birds and mammals are based on adaptations similar to those found in the alligator. The classical work was reviewed by Irving (1939), Scholander (1940), and Krogh (1941); the more recent literature is discussed by

Andersen (1966) and Angell James and Daly (1972). The diving bradycardia has been repeatedly confirmed; careful measurements of blood pressure and blood flow have shown that circulation to the brain and heart is well maintained while the blood flow to muscle, skin and viscera is virtually closed down because of the increased peripheral resistance in these organs. Blood volumes and oxygen capacities are somewhat greater in the good divers but this is a relatively minor factor in the marked tolerance to asphyxia which a good diver shows. Likewise, and contrary to earlier views, the relatively high myoglobin is not a significant factor in the total oxygen reserves although it may be important in providing oxygen for ischemic muscles during diving activity. The important factors are the reduced sensitivity to asphyxia, the bradycardia and a restriction of the circulation to vital organs, particularly the brain and heart.

It should be noted that these circulatory adjustments are refinements of mechanisms which also occur in the non-diving vertebrates. It has long been known that submergence triggers bradycardia in many terrestrial animals, including man. This is characteristic not only of the amniotes but also of anuran amphibians (Jones and Shelton, 1964; Jones, 1972). It appears to be a very basic reflex in terrestrial vertebrates. Likewise, muscle ischemia during anoxia is in no way peculiar to the amniotes.

Some fishes have also discovered this trick. The grunion (*Leuresthes tenuis*) is a teleost with the peculiar habit of spawning on beaches. The animals show considerable activity while out of water; physiological responses include bradycardia and a sharp rise in the muscle lactate with little change in the lactate of the blood until the animal returns to water. During the early phases of recovery, lactic acid in the blood rises acutely as it is released into the circulation from the muscles (Scholander *et al.*, 1962). During phylogeny, the diving vertebrates have capitalized on certain basic reflexes to permit longer and longer exploits under water.

Several interesting morphological specializations have been described in diving animals (Slijper, 1962; Elsner, 1969; Kooyman, 1972). From the point of view of the physiologist, those associated with the circulatory channels and the respiratory system are the most interesting. The peculiarities of the vascular system have been known since the days of John Hunter (1728–93). In contrast to terrestrial mammals, seals and whales have large venous reservoirs (particularly in the abdominal cavity), a specialized sphincter to regulate the blood flowing to the heart (most pinnipeds but not cetaceans) and numerous *retia mirabilia.* The venous pools are contained in a much enlarged thin-walled posterior vena cava which is often duplicated, and in the expanded hepatic veins which may form a conspicuous sinus; these pools are associated with the large blood volume characteristic of diving vertebrates. The blood which they contain is remarkably well oxygenated prior to the dive and provides a substantial reserve of oxygen during the period of anoxia. The sphincter of striated muscle which surrounds the vena cava just anterior to the diaphragm is thought to protect the heart from venous engorgement during bradycardia. The functions of the retia mirabilia are most speculative. These complex and extensive arterial nets are most extensive in the cetaceans; they occur in many places but especially in the thorax, between the ribs and vertebral column, at the base of the brain, in relation to the eyeball, and in the flippers. The retia of the appendages are concerned with temperature regulation (Chapter 10) but the functions of the other vascular nets are uncertain; they may provide reservoirs of oxygenated blood or serve to equalize pressure differences or create a resilience when organs are put under great pressures during deep dives.

The peculiarities of the respiratory system may be partly related to problems of buoyancy and oxygen supply (Kooyman, 1973), but they are mainly safeguards against the hazards associated with compressed air dissolving in tissues under high pressures.

The human diver is liable to experience decompression sickness (the "bends") if he surfaces too suddenly from a great depth (Couteau and Corriol, 1971; Lambertson, 1971). In this case, the lung gases, dissolved under increased pressures, come out of solution rapidly enough to form bubbles which block small blood vessels with disastrous results (Chapter 13). The problem in human diving is created by the necessity of breathing compressed air, maintaining a full lung volume, continuously oxygenating the blood and perfusing all of the tissues. In contrast to this, some of the pinnipeds (grey seals and sea elephants) actually exhale before submerging. Cetaceans, however, fill their lungs to capacity before the dive; but since air is not renewed during the dive, the actual volume of oxygen or nitrogen which might dissolve in the tissues is relatively small. In addition, there are several anatomical specializations which greatly reduce the potential hazard. The cartilaginous supporting rings of the upper respiratory passages extend much farther down into the lung than they do in terrestrial mammals, thus preventing the collapse of these passages under pressure. Consequently, air will be forced from the thin-walled alveoli, which collapse under pressure, into the thick rigid air passages. Besides this, some whales have a special system of sphincters or valves around the respiratory bronchioles. The function of these valves is unknown; in theory, they could trap air in the alveoli and promote absorption of oxygen, but alternatively they could cut off the air from the alveoli to restrict absorption of gases which would be likely to occur in these thin-walled and highly vascular sacs.

The thorax of the diving mammal is very flexible because of the presence of numerous "floating ribs" (ribs not attached to the sternum); the great whales lack a sternum and have only one true rib (Kooyman, 1972). Whatever may have been the evolutionary pressures which reduce the importance of a rigid thoracic cage, the end result is a thorax which transmits the pressure experienced during deep diving to the highly compressible air spaces of the lungs; these will collapse and restrict the capilliary circulation so that excessive amounts of gas are not dissolved. Only a slight supersaturation of the body fluids has been found even after dives to great depths.

These structural and functional modifications are now well understood. By contrast, the reflex mechanisms which elicit, control and integrate them are still only partly explained and subjects of active research (Angell James and Daly, 1972; Jones and Johansen, 1972). A part of the problem can be understood by comparing the human heart rate while breath-holding in air with the rate during breath-holding with the face under water; the first test produces a negligible or variable effect but the second causes an immediate bradycardia. It is evident that contact of the face with water reflexly slows the heart. Comparable experiments have been carried out many times with diving animals. Another test, which was performed long ago with ducks, shows that asphyxiation in air and asphyxiation in water are quite different experiences for a diving animal. Ducks which are asphyxiated by occlusion of the trachea while under water live three times longer than those asphyxiated under similar conditions in air. Water immersion evidently initiates responses which are important to survival during long dives; unlike the terrestrial animal during asphyxiation, the diving animal does not waste energy trying to breathe while submerged.

Many attempts have been made to explain the mechanisms responsible for diving bradycardia (Angell James and Daly, 1972; Jones and Johansen, 1972). Diving mammals appear to rely on rather different cardiovascular reflexes during submergence from those relied upon by diving birds. In a variety of mammals, water receptive reflexes arising in the nasal region trigger the cardiac slowing. Although specific receptor cells have not been identified, similar reflex pathways have now been found in several different mammals (Angell James and Daly, 1972). In sharp contrast, ducks do not appear to have these nasal reflexes. Butler and Taylor (1973) found that neither the chemoreceptor sensitivities nor the cardioinhibitory neurons in the medulla are influenced when the mallard duck's head is immersed in water; submergence in water causes cessation of breathing (APNEA) and this appears to be the direct cause of the bradycardia. There are still several puzzling questions for physiologists interested in these controls.

Oxygen toxicity Paul Bert (1878) recorded the toxicity of molecular oxygen in one of the early classics of environmental physiology. In our century, both medical and cellular physiologists have intensively studied these effects in aviation and space medicine and in deep-sea diving (Haugaard, 1968; Lambertsen, 1971).

At atmospheric pressure, man cannot breathe pure oxygen safely for longer than 12 hours. Under increased barometric pressure of diving the hazards are much greater, but at high altitudes equivalent to 35,000 ft with P_{O_2} of only 179 mm Hg, man has been exposed to pure oxygen for longer than two weeks without damage. Lower organisms are less sensitive but, as Bert noted in his early study, invertebrates as well as vertebrates, and plants as well as animals are prone to oxygen poisoning. The phylogenetically older groups seem to be more resistant; ferns stand molecular oxygen better than the angiosperms; the poikilotherms better than the homeotherms. Fundamentally, this seems to be related to the rates of tissue metabolism since cells with higher rates of metabolism are evidently more sensitive. Nervous tissue, such as the retina of the newborn, is extremely sensitive; convulsions and paralysis are common symptoms of oxygen poisoning in both vertebrates and invertebrates. These early findings have been frequently confirmed in our century (Clark and Cristofalo, 1961; Walker, 1970). The mechanisms are less well understood than the symptoms. It is evident, however, that rates of cellular metabolism are altered and, in particular, oxygen consumption is reduced at high oxygen tensions. There is probably a blocking of metabolic pathways concerned with oxidative phosphorylation. The steps concerned with the early stages of electron transfer seem to be involved; cytochrome oxidase is not affected but dehydrogenases containing sulfhydryl (SH) groups may be. It is almost certainly true that there are several enzyme sites at which excess oxygen poisons cellular function (Haugaard, 1968; Lambertsen, 1971). For this reason, there may be no single and simple explanation of oxygen poisoning.

EFFECTS OF THE ENVIRONMENT ON OXYGEN DEMAND

Available oxygen as a factor which regulates oxygen consumption or metabolic rate has just been discussed. This, in Fry's (1971) terminology, is a LIMITING FACTOR or one that acts by virtue of its operating in the metabolic chain. Oxygen is indispensable to oxidative phosphorylation (Chapter 7), and below certain minimal levels the production of ATP is curtailed; the metabolism and activities of the animal are restricted or limited in accordance.

There are other ways in which metabolism is environmentally regulated. The environment may, for example, control oxygen consumption by altering the medium in which the enzymatic processes of metabolism operate. Environmental factors which operate in this way are CONTROLLING FACTORS (Fry, 1971); they govern both the maximum and minimum rates of metabolism while the limiting factor acts only on the active metabolism.

Temperature, Salinity, and Photoperiod

Temperature is a good example of a controlling factor in metabolism; salinity and photoperiod also probably operate in this manner. The action of temperature on cellular metabolism is direct, with Q_{10} values of about 2 to 3; its effect on the total metabolism of both poikilotherms and homeotherms has been considered (Chapter 10).

It is more difficult to generalize concerning salinity effects. There are many contradictions in the recorded literature. At least in part, this is caused by a failure to acclimate animals prior to testing and to control activity during the experiments. Newell (1970) summarizes some of the important experiments on the intertidal invertebrates, while Fry (1971) discusses more recent literature on the teleost fishes. The most consistent pattern shows minimum metabolic rates at salinities which are closest to the osmotic content of the body fluids; in short, it seems that the metabolic costs of ion-osmoregulation are major factors in determining the effects of salinity on oxygen consumption.

K. P. Rao (1958) compared the oxygen consumption of prawns, *Metapenaeus monoceros*, from marine and brackish-water habitats. The brackish-water prawns had minimum oxygen consumptions in 50 per cent sea water which was equivalent to their natural habitat and probably closest to the osmotic content of their body fluids; metabolic rates were elevated in both lower and higher salinities. In contrast, prawns from marine habitats had minimal rates in 100 per cent sea water—again their natural habitat and isotonic with their body fluids. The most critical of the experiments on teleost fishes provide data in line with K. P. Rao's findings. Thus, G. G. M. Rao (1968) acclimated rainbow trout, *Salmo gairdneri*, to a series of salinities and measured metabolic rates at the salinity of acclimation; minimum rates were recorded in isosmotic dilutions of sea water. In comparable studies, Farmer and Beamish (1969) studied *Tilapia nilotica* and again obtained the lowest oxygen utilization in isosmotic salinities. Although there are contradictory data in the literature, many of the experiments could be interpreted within this framework.

Circadian and seasonal rhythms in oxygen consumption are often well marked. They are usually correlated with the animal's normal activity rhythm. Experimentally, photoperiod has been shown to modify oxygen consumption in certain animals; it can be assumed that light triggers these changes through neurosecretory centers and changing levels of hormones.

Carbon Dioxide

Carbon dioxide is a vital physiological constituent and acts in several different ways to modify oxygen consumption. Although it is formed as a waste product in the internal environment, it performs several well-marked regulatory activities. It modifies the rate of ventilation through its direct action on the centers of respiratory control; it alters the oxygen-combining properties of hemoglobin through Bohr or Root effects (Chapter 6); it acts directly on the vasomotor centers and is thus a factor in the regulation of blood pressure. In general, carbon dioxide, along with other metabolites, serves as a delicate signal for the oxygen demands of the tissues; as the pH falls, the ventilation, circulation and delivery of oxygen are often improved. At very high environmental levels, carbon dioxide becomes toxic through a depressing action on the nervous tissues; it is a useful anaesthetic in insect physiology.

Under natural conditions, low oxygen is likely to become limiting long before CO_2 levels are elevated sufficiently to affect metabolism. The CO_2 tension in sea water is only about 0.25 mm Hg and, in fresh waters, varies up to a usual maximum of about 5 mm. A significant depression in respiratory metabolism is unlikely until the CO_2 of natural waters rises some ten-fold (Brett, 1962), although the effect varies considerably with the temperature. Carbon dioxide effects, however, are not infrequent under anaerobic conditions which often develop during experimental studies with aquatic species. The action of elevated environmental CO_2 on the respiration of teleost fishes has been studied many times; Fry (1971) cites the earlier literature. In two of the more recent studies, Basu (1959) and Beamish (1964) found that unless carbon dioxide levels were extremely high, the STANDARD metabolic rate was unaffected by ambient CO_2; the ACTIVE metabolic rate, however, declined exponentially with any increase in carbon dioxide (Fig. 12.6). Fry (1971) notes

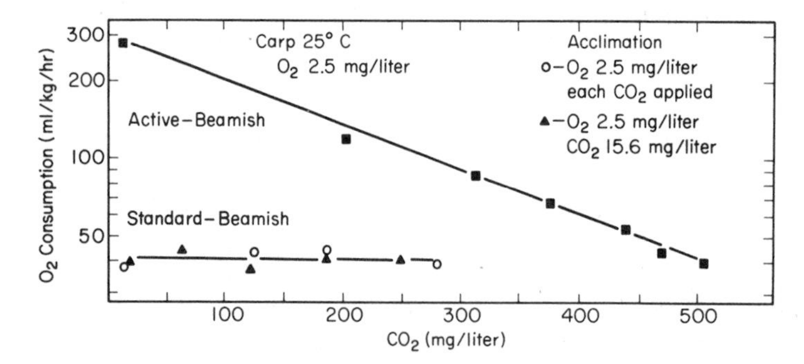

Fig. 12.6 Active and standard metabolic rates of carp in relation to various levels of carbon dioxide at 25°C and P_{O_2} 64 mm Hg (2.5 mg/liter). Measurements of active metabolism made only on fish acclimated to air-saturation water. [From Fry, 1971: *In* Fish Physiology, Vol. 6, Hoar and Randall, eds. Academic Press, New York; based on Beamish, 1964: Can. J. Zool., *42*: 847.]

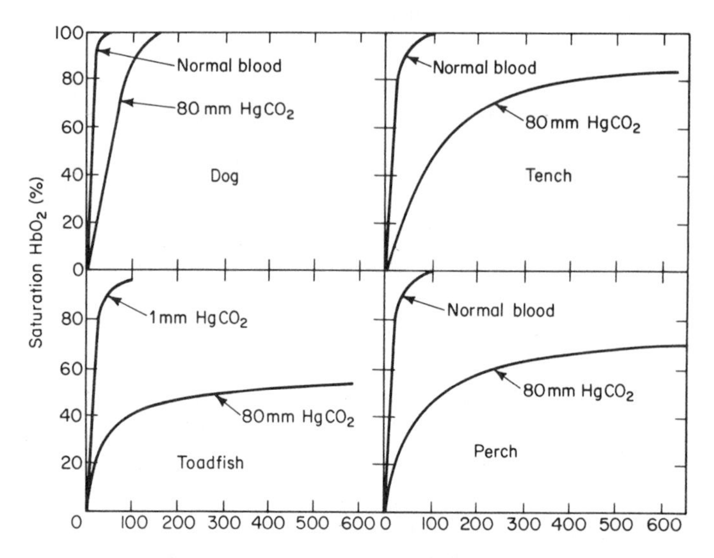

Fig. 12.7 The effect of carbon dioxide on the oxygen equilibrium curves of mammalian and fish bloods. In each case the left curve is for blood exposed to minimal amounts of carbon dioxide (1 mm Hg or less) and the right curve for blood at 80 mm Hg carbon dioxide. Dog blood shows a typical Bohr effect; toadfish blood shows the typical Root effect; in the presence of CO_2, it cannot be saturated with oxygen even at the highest of oxygen pressures. [Based on Jones and Marshall, 1953: Biol. Rev., *28*: 33.]

that the fish used for studies of active metabolism were not acclimated to different levels of CO_2 and that acclimation might significantly affect the findings.

The physiology of these reactions is probably complex. It seems likely that CO_2 will affect both ventilation and transport systems. In fish, as in other animals, elevated carbon dioxide accelerates the rate of ventilation. This in turn, through increased activity of the respiratory muscles, creates greater demands for oxygen. The augmented ventilation volume might be thought to satisfy these demands by increased utilization of oxygen. This, however, does not happen, since hyperventilation results in less intimate contact between water and blood; in final analysis, this contact is the important factor because the exchange is a diffusion phenomenon. As already noted (Chapter 4), the tips of the gill lamellae move apart in hyperventilation; the more rapidly flowing water is in less intimate contact with the lamellae and in contact for a briefer period. Both of these factors operate against the efficiency of exchange.

The transport pigments are also modified by carbon dioxide and usually carry progressively less oxygen as the CO_2 tension rises (Figs. 6.5 and 12.7). However, there does not seem to be any simple relationship between the loading capacity of the blood and the ability of fish to utilize oxygen in the presence of carbon dioxide. There are marked species differences which are not understood. Thus, partial pressures of CO_2 must be well over 200 mm Hg to produce respiratory stress in the bullhead, *Ameiurus nebulosus*, but salmonids, bass and many other species are limited at tensions of 50 to 80 mm Hg (Black *et al.*, 1954). Explanations must apparently be sought in terms of cardiac output, circulation efficiency and utilization of oxygen in the tissues as well as in the ability of hemoglobin to capture oxygen from the environment (Black, 1940; Basu, 1959).

13

PRESSURE AND BUOYANCY

The aquatic habitat, because of its density, creates several special physiological problems. Two of these will now be discussed: the forces of hydrostatic pressure and the maintenance of a neutral buoyancy.

At sea level, a terrestrial animal experiences a maximum pressure of 14.7 pounds per square inch (760 mm Hg). Under natural conditions this can increase only if the animal descends into a deep cave or a mine shaft; the decrease in barometric pressures at higher altitudes has been considered (Fig. 12.5). These pressures of the terrestrial habitat are, in themselves, of no physiological significance, although they do exert profound effects on the exchange of gases (Chapter 12).

The aquatic animal, however, must withstand not only the atmospheric pressure prevailing at sea level but also the additional weight of water at that level. This hydrostatic pressure increases by about one atmosphere for every 10 meters or 33 ft, and yet life exists in the very deepest marine trenches (over 11,000 meters) where pressures exceed 1000 atmospheres. From these extreme depths, the *Galathea* brought up sea anemones attached to stones and caught fish at 7000 meters, more than three-and-a-half miles beneath the surface of the sea (Marshall, 1954). The pressures which exist at these depths, when applied experimentally to protoplasm at the earth surface, will alter its constituents (particularly the proteins) and disturb the normal organization and physiology of cells; we now know that barophilic organisms have, in their evolution, acquired special adaptations for life under these great hydrostatic pressures (*Am. Zool.*, 11 (3), 1971).

The different species of aquatic animals are ecologically adapted to particular depths. Some are narrowly restricted; others move freely up and down, but none is able to survive and flourish outside a normal range which is a small part of the vertical distribution of life (Table 13.1). The specific gravity of protoplasm (exclusive of such dense materials as mollusc shells or echinoderm tests) lies between 1.02 and 1.10, while sea water has a maximum value of about 1.028 (Marshall, 1954; Alexander, 1972). Hence, without specializations to counteract gravity, aquatic organisms will sink or continually expend energy to maintain their normal depth. In terms of energy requirements, there are obvious advantages in maintaining the same density as the aquatic habitat or a NEUTRAL BUOYANCY as it is called. The metabolic economy is considerable (Alexander, 1966, 1972); the mechanisms are curious and varied (Denton, 1963).

THE EFFECTS OF HYDROSTATIC PRESSURE

The *Talisman* dredging expedition of 1882–83 sparked the first scientific research on the effects of hydrostatic pressure. The French scientist Regnard, in particular, was stimulated by the discovery of animals living beneath 11,000 meters of sea water at pressures of about 1000 atm. He designed an apparatus which permitted him to test, and sometimes observe, the effects of hydrostatic pressures up to 1000 atm on many different organisms. His findings were described in a classical monograph which appeared in 1891; they have been summarized in several places (Cattell, 1936; Johnson *et al.*, 1954).

Perhaps the most surprising of the early observations was that modest pressures, up to at least 100 atm, had little or no effect on many organisms and that the changes observed at pressures up to 1000 atm were often reversible. At pressures of about 500 atm, bacteria, yeasts, algal cells, salmon eggs, tadpoles, skeletal muscle and many other tissues or processes often showed depressed activity or retarded growth but no permanent injury following moderate exposure times. Some larger organisms such as echinoderms and coelenterates slowly recovered after an hour at pressures as high as 1000 atm while others (for example, molluscs, crustaceans and fishes) were much more sensitive and could be killed at pressures in the 500 atm range. Precise effects are a function of time; when Regnard watched small aquatic organisms such as *Daphnia*, *Cyclops* or *Gammarus* through the quartz window of his experimental chamber, he observed their agitation at pressures of about 100 atm. At somewhat higher pressures they stopped swimming and fell to the bottom. If soon released from the pressure they recovered, but longer application led to progressive paralysis or coma and then death. The nervous system seemed to be particularly sensitive.

Technical advances since Regnard's day now permit the use of experimental pressures up to 10,000 atm (in some experiments even 100,000 atm), and it is apparent that a sharp physiological distinction is to be drawn between the *moderate pressure* effects of 100 to 1000 atm and the *excessive pressures* between 1000 and 5000 atm or more. A highly diversified group of animals has exploited the lower of these two ranges, which spans the hydrostatic pressures encountered under natural conditions (Table 13.1 and Fig. 13.1). Changes observed under these moderate

Table 13.1 **The Range of Biological Pressure Phenomena. Lake Baikal is the deepest body of fresh water in the world (about 1750 meters deep); the deepest ocean abyss is the Marianas trench in the Western Pacific near the Philippines (11,034 meters or almost 7 miles); ATA, atmospheres absolute**

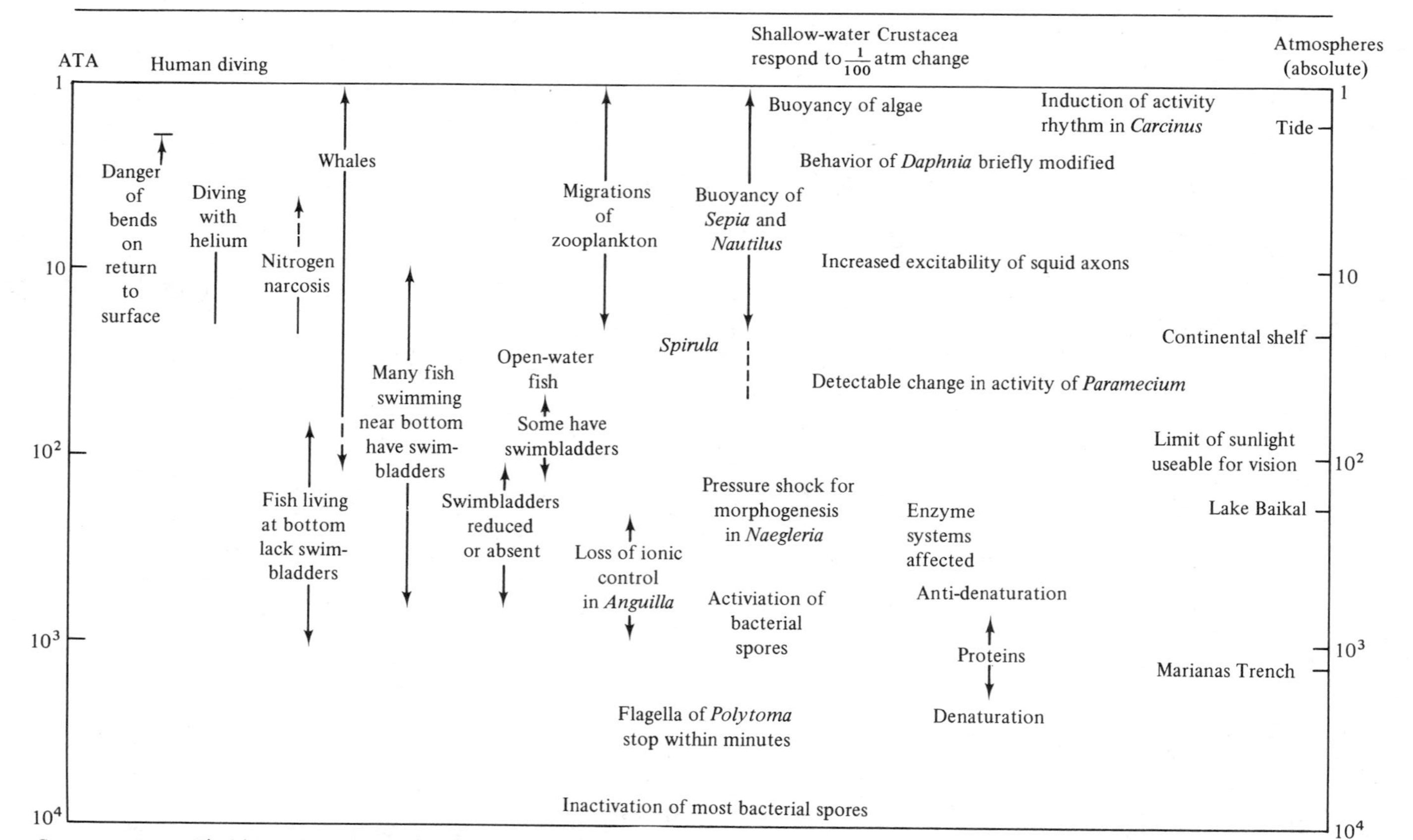

SOURCE: From Kitching, 1972: Symp. Soc. Exp. Biol. *26*: 474

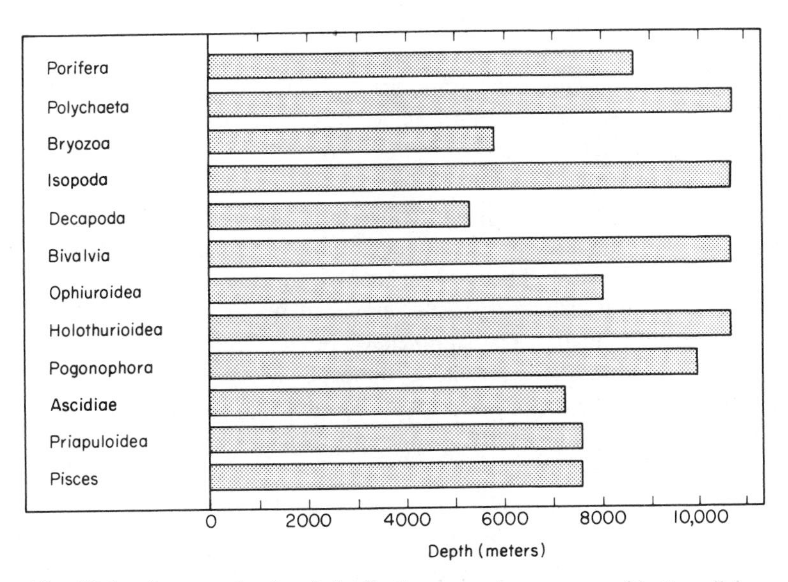

Fig. 13.1 Greatest depths of distribution on various groups of bottom-living animals. [From Flügel and Schlieper, 1970: *In* High Pressure Effects on Cellular Processes, Zimmerman, ed. Academic Press, New York; based on a table in Zenkevitch (1963).]

pressures are often reversible. While excessive pressures accelerate protein denaturation in an irreversible manner, moderate pressures retard denaturation and stabilize proteins in situations which usually denature them, for example, exposure to high temperatures.

There is a substantive recent literature on the effects of moderate pressures on primitive organisms. Regnard's early observations have been confirmed and extended to the cytological, ultrastructural and biochemical levels. Many of the effects relate to changes in the structure of proteins. Marsland (1956, 1958, and earlier) carried out some of the pioneering investigations with *Amoeba proteus*. As early as 1936, he and his colleagues focused on several well-defined alterations in the sol-gel relationships of protoplasm. They noted the withdrawal of pseudopodia and rounding up of amoebae under pressure, with a reduction in the gelatinous characteristics of the plasma gel (ectoplasm) and a sharp decline in the viscosity of the endoplasm. Since that time, many workers have emphasized the changes in sol-gel states of cells with solation of protoplasm at high pressures. Frequently observed changes include disorganization of cilia, flagella and microtubules, alterations in the shape and rigidity of cells, disappearance or disorder of the mitotic apparatus with failure of normal cell division, changes in RNA and protein synthesis, suppression of growth and changes in enzyme kinetics. Bacteria, protozoa and the dividing eggs of marine invertebrates have been favored objects of study. The experimental work is documented in several symposia [*Am. Zool.*, 11 (3), 1971; *Symp. Soc. Exp.*

Biol., 26, 1972], a review paper (Zimmerman, 1971), and a monograph edited by Zimmerman (1970).

These descriptive studies underline many physiological problems which living systems must encounter in exploiting deep-water habitats. They do not, however, suggest just how barophilic organisms may have solved the problems during phylogeny (Hochachka and Somero, 1973). There is now ample evidence that many organisms have solved these problems; ultimately, the solution must have depended on adaptations in proteins, both structural and metabolic. Johnson and Eyring (1970) and Hochachka and Somero (1973) have discussed the biochemical kinetics of pressure effects. Hochachka and his associates have directed their attention to several enzyme systems known to be adapted to high-pressure conditions in deep-sea animals. The significance of the adaptive process is apparent in Fig. 13.2. In an abyssal fish, fructose diphosphatase, a key enzyme in the flow of carbon during gluconeogenesis (Chapter 7), is clearly adapted to "work better" under pressure. At physiologically significant concentrations of enzyme and substrate, reaction velocities for enzymes extracted from livers of trout (a surface-dwelling fish) are sharply depressed by pressure while those extracted from the livers of rat-tails, *Coryphaenoides,* which live at depths of about 3000 meters, are unaffected. It is evident that pressure has little effect on the maximum velocities of the enzyme preparations from these two animals; the unique properties of the rat-tail enzyme are revealed only at physiological enzyme-substrate concentrations. Hochachka and Somero (1973) note that comparative studies of the effects of pH, temperature, and so forth, when carried out in isolation, may reveal little of significance concerning the adaptive process; the physiologically important adaptive "strategies" are the

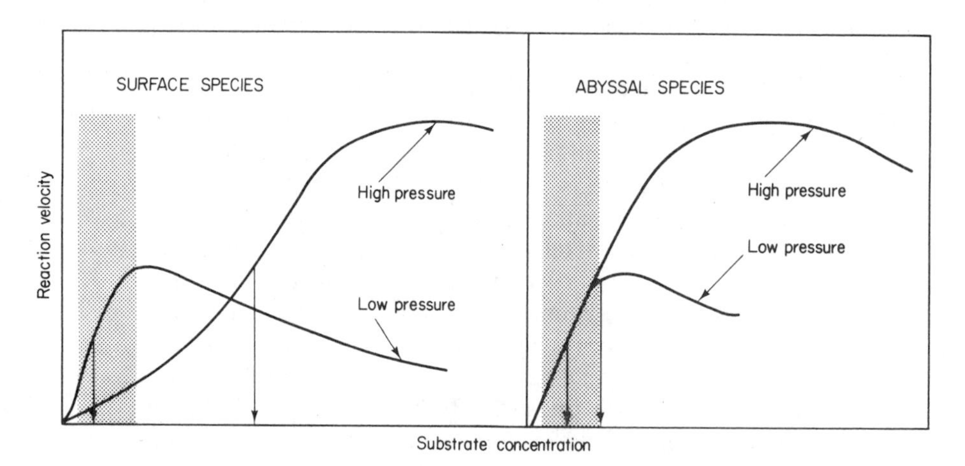

Fig. 13.2 Substrate saturation curves for liver fructose diphosphatase from the trout *Salmo gairdneri,* a surface-dwelling fish, and from an abyssal fish, the rat-tail *Coryphaenoides*. Shaded bar, approximate physiological range of fructose diphosphate levels. [From Hochachka and Somero, 1973: Strategies of Biochemical Adaptation. Saunders, Philadelphia.]

net result of several opposing changes in the total catalytic system. The comparative physiologist can readily appreciate the importance of the adaptive process from a study such as that depicted in Fig. 13.2; substantial further biochemical research will be required to clarify the molecular details.

Man as a Deep-Sea Diver

Man has no specialized physiological machinery for life under water and must continue to ventilate his lungs and perfuse all his tissues with oxygenated blood. His capacity to incur an oxygen debt is very limited. Thus, when he goes beneath water for more than a few minutes, some device must be used to deliver air to the lungs under the pressures which exist at that level. Unless this is done, the lungs collapse as the pressure increases in accordance with Boyle's Law. The relationship is shown in Fig. 13.3. At a depth of 33 ft, the lung volume would be halved with a doubling of the pressure from 1 atm at sea level. Caisson workers in deep tunnels, men in diving bells and the thousands of underwater explorers with their SCUBA diving equipment (SELF-CONTAINED UNDERWATER BREATHING APPARATUS) are maintaining a full lung volume with air at whatever pressure exists in their immediate environment. This creates two serious hazards. The first of these is caused by the toxicity of the excessive quantities of gases which dissolve in the tissues under pressure and the second develops if the diver ascends so rapidly that these gases come quickly out of solution or the air expands too rapidly in the lungs.

Oxygen and nitrogen are the dangerous gases under these conditions. Although carbon dioxide is toxic in high quantities (above 10 per cent), it occurs in only small amounts in atmospheric air and does not usually accumulate dangerously in most of the devices used for diving. Oxygen toxicity was mentioned in the previous chapter. Dangerous amounts can dissolve in the tissues within half an hour at 33 ft

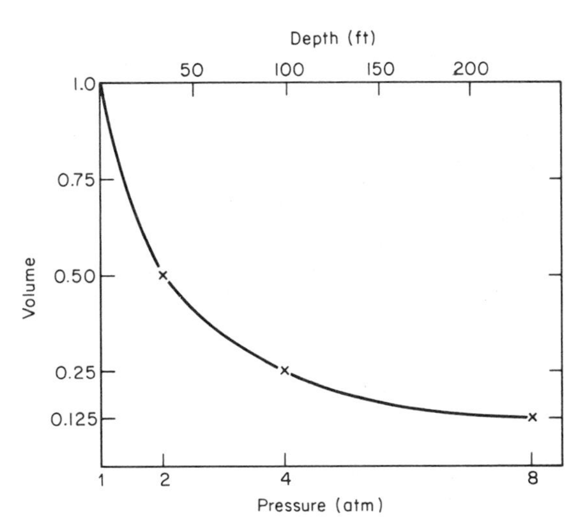

Fig. 13.3 The effect of hydrostatic pressure (depth) on the volume of a gas in an enclosed space.

(increased pressure of 1 atm) if the breathing apparatus provides pure oxygen and the individual is active.

In breathing compressed air the hazards are primarily from nitrogen which accounts for about four-fifths of the volume of the air and is particularly soluble in tissues. At sea level, a man has about 1 liter of nitrogen dissolved in his tissues, with less than half of this in the body water and somewhat more than half of it in the lipids (Folk, 1966; Guyton, 1971). Although only about 15 per cent of the body is fat, the dissolved nitrogen is largely in this fatty tissue. At sea level, this is innocuous but, in deep water, large amounts in solution under pressure are extremely toxic and produce a nitrogen narcosis ("raptures of the depths"). The nitrogen dissolves slowly, and it requires an hour or more at any pressure to produce an equilibrium. If, however, nitrogen saturation is permitted, a mild narcosis may be expected at 130 ft where the pressure is about 5 atm; the diver will become quite helpless at 300 ft with about 10 atm pressure.

The physiology of nitrogen narcosis is not well understood; superficially, it is similar to that of other anaesthetics. Nitrogen solubility in plasma membranes, particularly in the neurons, depresses the excitability of the cells in general and the nervous system in particular. Theories and discussions of these matters will be found in textbooks of pharmacology and medical physiology. Because of its highly toxic nature, nitrogen is often replaced by helium in the gas mixtures prepared for divers. Helium is less soluble than nitrogen, lacks the narcotic effect, and, because of its low molecular weight, diffuses much more quickly from the tissues.

The second major hazard is the expansion of the gas and its release from solution if the pressure is too quickly reduced. If the diver ascends rapidly, air can expand suddenly enough in the lungs to rupture membranes and capillaries and permit gas to enter the circulatory system (air embolism). Gases dissolved throughout the tissues may also be released as air bubbles with symptoms and damage which vary with the location of the gas. The dangers depend both on the depth and the length of the dive: a diver can stay 5 hours or more at a depth of 35 ft without fear of injury upon sudden decompression, while little more than 10 minutes at 130 ft make a sudden decompression hazardous. Clinical symptoms seen in aviation medicine and deep-sea diving, safe limits of decompression, types of diving gear and many other details are discussed in texts of medical physiology and specialized symposia (Lambertsen, 1971).

Man is not the only animal that runs the risk of sudden decompression. Similar damage has been described in fish which have come through the turbines of a power dam and were suddenly released into shallow water after their residence in a deep reservoir behind the power dam (Hamilton and Andrew, 1954).

BUOYANCY

In theory, an organism might adjust its weight and counteract gravity either by excluding some of the heavier elements or by including lighter materials which operate as floats or buoyancy tanks. There are many examples (both in the plant and animal kingdom) of organisms which have solved the buoyancy problem in one or the other of these ways. In addition, it is possible to improve flotation by altering the surface-to-volume relationships. Some of the radiolarians (*Acanthometra*), for example, possess a system of symmetrically radiating spicules on which contractile threads of protoplasm (the myomeres) are anchored. The myomeres arise in the extracapsular protoplasm of the cell and are inserted on the tips of the spicules; when they contract, the gelatinous surface layer is greatly expanded and the animal can take advantage of water currents to rise in the sea; with relaxation of the myomeres and a shrinking of the protoplasmic mass, the animal sinks once more toward the bottom. The *Nautilus*, an immensely larger and more complex animal, also improves flotation by greatly extending the soft parts of the body. In this case, a large part of the animal can be extended from or withdrawn into the beautifully coiled shell. Other curious examples are described in books by Jacobs (1954) and Marshall (1954).

Exclusion of Heavier Elements

Sea water has a specific gravity of about 1.026. A hypothetical animal from one of the lower phyla, containing body fluids isosmotic with sea water, would obtain a lift of 26 mg per ml of fluid if its salts were replaced by fresh water. This is, of course, biologically impossible, but a measure of hypotonicity is sometimes permissible, and the pelagic egg of the marine teleost owes a portion of its buoyancy to its lowered tonicity (Denton, 1963). Fertilized eggs become impermeable to sea water and retain the tonicity characteristic of adult fish; unfertilized eggs take on sea water rapidly and sink. It has been suggested that one of the major advantages of hypotonicity in the marine teleost is its contribution to the formation of pelagic eggs.

A lift can also be obtained by replacing the heavier ions with lighter ones. Our hypothetical animal could retain its isotonicity and achieve a lift of 3.5 mg per ml if pure sodium chloride replaced the other sea salts (Denton, 1961). Again, a complete replacement is physiologically impossible, but some of the algae and protozoans, the gelatinous planktonic ctenophores, medusae, nudibranchs and tunicates have reduced amounts of the heavier ions such as Ca^{++}, Mg^{++} and So_4^{--}. Exclusion of the latter ion seems to be particularly important in these gelatinous forms where the proportions of water are relatively great (about 95 per cent in a jellyfish). A partial replacement of sodium chloride with ammonium chloride is said to account for the buoyancy of the luminescent protozoan *Noctiluca miliaris* (Denton, 1963). The feeble ossification and reduction in the caudal and trunk musculature of many bathypelagic fishes contribute markedly to their buoyancy (protein has a density of about 1.33). The density of some of these forms may be within 0.5 per cent of sea water, with a body containing less than 5 per cent protein in comparison with 17 per cent in coastal fishes (Denton, 1961). They have become floating traps through the loss of the swimming muscles and the retention of heavily muscled jaws.

One of the most elegant buoyancy machines based on the exclusion of salt is the cuttlebone of the cuttlefish, *Sepia officinalis*. This operates as a buoyancy tank in which the amount of liquid can be varied by a "desalting apparatus"; removal of the salt from the liquid in the tank lowers its tonicity and, in consequence, water moves out osmotically. This buoyancy tank (the cuttlebone) contains only a small amount of gas, mostly nitrogen; the effective changes are the result of the osmotic flow of water.

The cuttlefish may range to depths of more than 600 ft where the pressures are 20 atm. Its cuttlebone is beautifully built to withstand these pressures. About 100 plates or lamellae of calcified chitin are placed one above the other and held apart by sturdy vertical pillars. The chambers thus formed are further divided by thin membranes parallel to the lamellae (Fig. 13.4). The whole structure is sealed off along the dorsal, lateral and anterior-ventral surfaces by a tough calcified membrane, but the posterior-ventral or siphuncular surface is covered with a vascular layer of

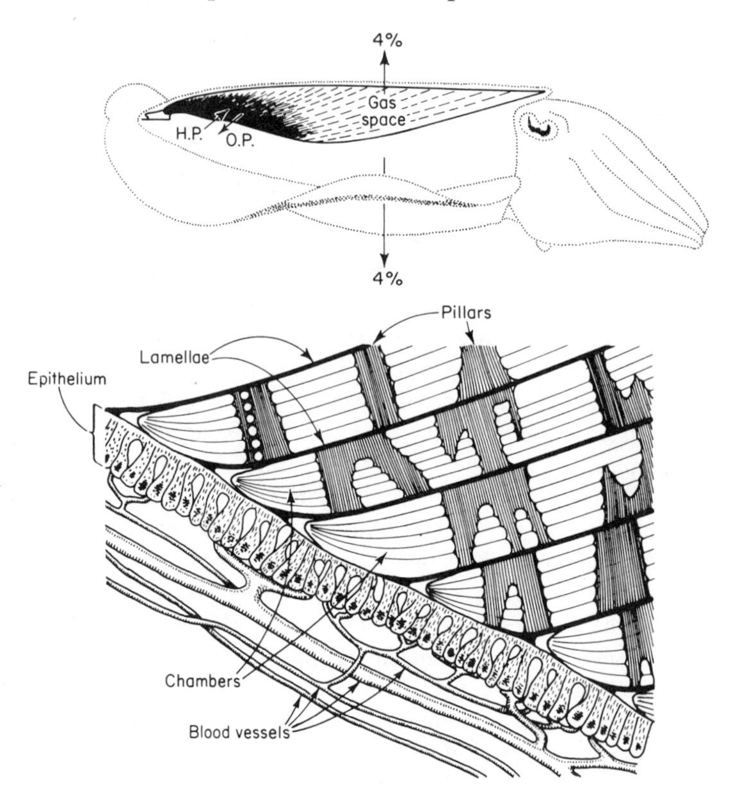

Fig. 13.4 The cuttlebone of the cuttlefish, *Sepia officinalis* (outlined above) has a gas space and a liquid-filled area (shown in black). Below, section of cuttlebone showing the epithelium lying along the posterior ventral surface. H.P., hydrostatic pressure of the sea is balanced by an osmotic pressure, O.P., between the cuttlebone liquid and the blood. In sea water, the cuttlebone gives a net lift of 4 per cent of the animal's weight in air and thus balances the excess weight of the rest of the animal. [After Denton, 1961: Progr. Biophys., *11*: 223.]

epithelial cells. This is the ion pump or salt extractor. Denton (1961) and his associates have shown that the animal can use it to alter buoyancy rather quickly in accordance with behavior. Cuttlefish lie on the bottom during the day and hunt in the surface waters at night (Fig. 13.5). The density of the cuttlebone varies from about 0.5 (containing 10 per cent liquid) to almost 0.7 (containing 30 per cent liquid). At a density of 0.6, an animal of 1,000 gm would achieve an upthrust of 40 gm, and this will just about balance the excess weight of the animal in sea water (Denton, 1961).

Inclusion of Lighter Materials

In addition to exchanging heavy for lighter ions, animals may actually accumulate or secrete low specific gravity materials such as fats, ammonium salts or gases and thus attain a neutral buoyancy.

Oil droplets and fat depots The specific gravity of fish oil is about 0.93. Oil droplets in pelagic fish eggs and diatoms add to the lift achieved by excluding heavy ions. Some of the larger vertebrates, such as sharks and whales, have exploited fats as their major buoyancy device. Sharks (Family Squalidae) have extremely fatty livers making up about 25 per cent of their body volume and containing a special fatty hydrocarbon, squalene, with a specific gravity of 0.86 (Corner *et al.*, 1969). Per gram, squalene has 70 per cent greater lifting power than fish oils. Whales also have generous amounts of low-density fats—in this case, esters of long chain aliphatic alcohols with fatty acids. Many different aquatic vertebrates have utilized fatty and watery tissues to improve their buoyancy (Brawn, 1969; Roberts, 1969; Bone, 1972).

Accumulations of ammonium salts Like Piccard's bathyscaphe with its gasoline-filled buoyancy tank, the deep-sea squids (Cranchiidae) attain a neutral buoyancy

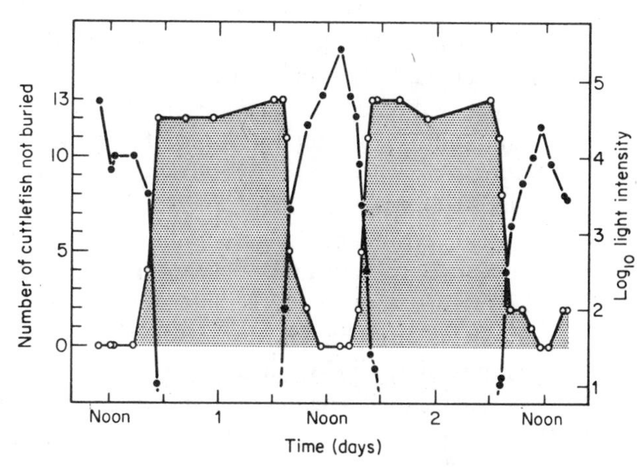

Fig. 13.5 Diurnal changes in the behavior of cuttlefishes. Open circles, number of animals not buried (total number of animals, 13). Solid circles, light intensity on a logarithmic scale. Light corresponds approximately to that found in the sea at about 125 ft. [From Denton and Gilpin-Brown, 1961: J. Marine Biol. Assoc., *41*: 345.]

with a float of low-density liquid—in this case, a solution of ammonium chloride. Again, the cephalopods have shown their originality since, like the cuttlebone, this large buoyancy tank is without parallel in the animal world.

Although bulky, the buoyancy tank has a definite advantage over cuttlebone for an animal exploiting the depths of the ocean. If the water in the cuttlebone were completely desalted, the osmotic difference could not act against hydrostatic pressures of more than 20 to 25 atm (about 800 ft); cuttlefish are most frequently found between 100 and 250 ft. Since water is virtually incompressible, the major limitation on the use of a fluid filled buoyancy tank is one of bulk.

The ammonium chloride solution is accumulated in the large coelomic cavity which forms about two-thirds of the total volume of the squid. Denton (1961) has measured its density at 1.010 to 1.012 and found that it contains 480 mM of ammonium and only about 90 mM of sodium. In contrast, the body fluids of *Sepia* contain approximately 465 mM of sodium (Nicol, 1967a). The pH of the fluid in *Cranchia* is around 5; this acidity accounts for the retention of the ammonia in the form of ammonium salts and avoids the hazards of free ammonia toxicity. No special structures are required to generate ammonia since it is the end product of nitrogen metabolism in cephalopods (Gilpin-Brown, 1972). The efficiency of this fluid as a buoyancy device is easily demonstrated by draining the coelomic cavity and watching the animal sink. Some squids improve their buoyancy by accumulating large amounts of ammonia in vesicles throughout the tissues of the mantle and arms. The success of the squids as bathypelagic animals is attested by their range of form and size. Some of them are the largest of the invertebrates; *Architeuthis* exceeds 50 ft in total body length (Morton, 1958).

Floats containing gas If the technical problems can be solved, a tank of compressed air (density 0.00125 at sea level) is by far the most efficient device for flotation. The air bladder of a teleost fish forms about 5 per cent of its body volume, the cuttlebone about 10 per cent of *Sepia,* and the coelomic fluid of the cranchid squid approximately 65 per cent.

Gas-filled spaces are common in floating plants and in animals at many levels in phylogeny. Active processes are often involved in the accumulation of this gas, as evidenced by unique constituents and the tremendous pressures which can be maintained. For example, the pneumatophores of some giant kelps may contain 5 to 10 per cent carbon monoxide (Rigg and Swain, 1941), and this may also contribute significantly (0.5 to 13 per cent) to the float of the Portuguese man-of-war (*Physalia*) where it is evidently formed from a substrate of L-serine (Wittenberg, 1960; Hahn and Copeland, 1966). The gas pressures are just as astonishing as the strange constituents; teleosts with air bladders have been caught at 4500 meters where the pressures in the bladder must be in the region of 450 atm (Denton, 1961). This is an extreme example, however, and most of the animals with gas-filled spaces are above 1000 meters. In deeper waters the swim bladder is either lost or its gas is replaced by oil (Marshall, 1960).

Jacobs (1954) and Marshall (1954) describe many curious examples of animals with gas floats. The minute rhizopod protozoan *Arcella*, not more than 1.1 mm in diameter, adjusts its buoyancy with bubbles of oxygen (Bles, 1929); the giant coelenterate *Physalia* with trailing tentacles that may be 30 ft in length, operates a float or pneumatophore of up to one liter capacity with a gas-generating layer of cells and a pore through which gas can be emitted to adjust the volume (Copeland, 1968). This gas mixture is about 20 per cent oxygen, with the remainder carbon dioxide, nitrogen, argon and carbon monoxide (Wittenberg, 1960).

The cephalopods (nautiloids and fossil ammonoids and belemnoids of the Palaeozoic and Mesozoic seas) also experimented successfully with gas-filled buoyancy tanks. Two living representatives, the tiny *Spirula* and the familiar pearly nautilus, have been carefully investigated by Denton and Gilpin-Brown (1966, 1971). Both these cephalopods are deep-water animals of tropical seas; *Spirula* has a small coiled internal shell while that of the nautilus serves as a protective external armor. Both shells contain gas which is mostly nitrogen under subatmospheric pressures. The gas-producing mechanisms are similar to those described for the cuttlefish. In short, gas accumulates incidental to the active transport of salt; there is, in fact, *no* gas-generating system comparable to that of the fishes. The system is not yet fully understood (Gilpin-Brown, 1972), but it appears to require a waterproof lining of the tank with an active ion-transporting epithelium (perhaps also an ion-concentrating system) to create hypoosmotic fluids within the shells; the gas gradually diffuses in from the tissue fluids to occupy the vacuum created by the osmotic extrusion of fluid. A highly pressure-resistant wall is essential to the operation of this system; the main walls of the nautilus shell can withstand pressures equivalent to depths of 450 meters or more. In comparison with the teleost swim bladder, the cephalopod gas tank has the advantage of being incompressible; it has the disadvantage of requiring a heavy weight tank. Eighty per cent of the gas space in the shell of the pearly nautilus is required to float the shell itself.

In phylogeny, the air or swim bladder of fishes probably served first as an accessory organ for the exchange of gases (Chapter 4). It has assumed several different functions in the evolution of the teleosts but its role as a hydrostatic organ dominates. Because of pressure-volume relationships (Fig. 13.3), a gas float is only useful over a range of depths if the quantity of gas can be readily altered. In short, when the fish descends 33 ft, the volume of its swim-bladder gas will be halved; this must be recovered by acquiring gas and retaining it at a pressure of 2 atm to achieve the same buoyancy as it had at sea level. The formation of gas and its removal are the major physiological problems concerned with the function of this structure as a hydrostatic organ. Only these aspects of its physiology are summarized here. There are many excellent reviews of the literature on swim bladders (Jones and Marshall, 1953; Marshall, 1960; Denton, 1961; Alexander, 1966; Steen, 1970).

The teleostean swim bladder develops from the anterior region of the digestive tract. The connection with the gut (PNEUMATIC DUCT) is retained in the more generalized groups (Physostomi), and air can be readily passed in either direction through it. In the least specialized examples (salmon, for example), there are no other arrangements for gassing or degassing. The pneumatic duct is under the autonomic nervous control characteristic of this region of the gut, and a well-defined series of reflexes operates to adjust the volume, but the organ can only be filled if the animal comes to the surface and gulps air. In the more specialized teleosts, and particularly those that live in deep water, the pneumatic duct is lost during ontogeny (Physoclisti) and the gas supply depends on the blood transport system. Intermediate conditions also occur, and many fish (for example, the Atlantic eel) have both a pneumatic duct and a specialized gas gland.

It is the closed or physoclistous bladder which presents the major physiological problems; there are still many questions concerning the details of gas flow through it. In general, however, the organ is a gas-tight sac in the physoclist, with thick elastic walls through which gases dissolve slowly or not at all. Exchanges depend on two specialized areas: the RESORBENT PART (oval or posterior chamber), concerned with the removal of gas, and the SECRETORY PART, an organ of gas production, consisting of a gas gland with adjacent rete mirabile.

The oval is a vascular pouch on the dorsal wall of the bladder. It is surrounded by a ring of muscle which can be relaxed to expose its vascular surface to the gases in the main bladder or constricted to separate the two areas. The removal of gas from the physoclist bladder depends on its resorption by the blood in the vascular net of the oval. The significant factors are pressure differences, blood flow and the transport capacities of the hemoglobin; the principles are not different from those which operate at other places in the animal. Instead of an oval, some of the physoclists have a special resorptive posterior chamber separated by a partial diaphragm, but the physiology of gas removal is the same.

The filling of the physoclist swim bladder has excited the curiosity of physiologists for more than a century. The task of generating gases and maintaining them under pressures which exist at depths of 200 to 7000 meters appears to be a formidable one. The many theories that have been proposed and discarded are discussed in several of the reviews previously cited. The physiology is now understood in terms of *gas solubility* and *diffusion pressures* operating in a highly specialized organ called the GAS GLAND. The process is often termed SECRETION, but it should be clearly understood that this "gland" does not "secrete" in the usual sense of the word but "concentrates" gases delivered to it by the arterial blood. Tracer techniques have clearly shown that the ambient water is the source of the gas (Wittenberg, 1961) and that the gas is not elaborated or manufactured by the secreting cells of the gland.

There are two anatomical parts to the gas generating apparatus: the epithelium of the gas gland and an associated rete mirabile through which blood circulates to this epithelium (Fig. 13.6). The significant physiological activity of the secreting epithelium is an intense glycolysis; lactic acid is produced at a great rate even under aerobic conditions (D'Aoust, 1970). The formation of lactate has been demonstrated experimentally and is evidently a key factor in oxygen release. The epithelium is also characterized by the presence of carbonic anhydrase, and a release of CO_2 is probably important in lowering pH or in promoting Root and Bohr effects (Chapter 6; Fig. 12.7). The significant feature of the rete mirabile is

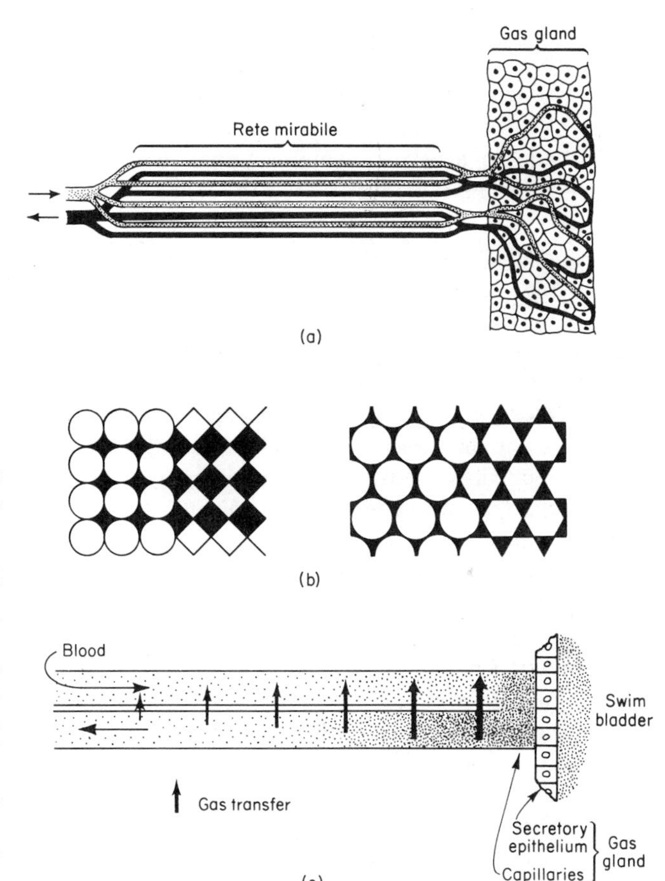

Fig. 13.6 Diagrams of parts of the gas gland and rete mirabile of the teleost swim bladder. (a) Arterio-venous loop to gas gland. [From Marshall, 1954: Aspects of Deep Sea Biology. Hutchinson, London.] (b) Cross section through the loops showing some possible arrangements of venous (open) and arterial (black) capillaries. [From Scholander, 1954: Biol. Bull., *107*: 263. Permission of the Managing Editor.] (c) To show how the counter-current diffusion system could reduce the loss of gas once this gas had been secreted.

the densely packed arrangement of very long, thin-walled blood vessels; in cross section they look much like a checkerboard (Fig. 13.6b). These capillaries are the longest in the animal kingdom, measuring 4 mm to 25 mm in different species; in comparison, the very long capillaries in skeletal muscle are only about 0.5 mm. Almost half a century ago, Krogh (1929) made several impressive calculations of the capillary surface in the retia of the eel; he counted 88,000 venous and 116,000 arterial capillaries in the two retia and estimated aggregated lengths of 352 and 464 meters. Steen (1970) reviews his findings and those of more recent workers. The tremendous surface area for exchanges between the venous and arterial vessels is emphasized here.

How do these two structures, the secreting epithelium and the rete mirabile, operate to concentrate gases in the swim bladder at high pressures? Haldane (1927) seems to have been the first to note the probable role of acid in the release of gas from the blood and the function of the rete mirabile in its retention within the bladder. It remained for Kuhn and his associates (1963 and earlier) to demonstrate that a vascular supply of this sort could generate gases at high pressures within the bladder; these workers built on concepts of COUNTER-CURRENT MULTIPLICATION which had been worked out earlier for the concentration of urine by the mammalian kidney (Chapter 8). Steen (1970) and his associates have provided substantial additional experimental support for the hypothesis and explained some of the problems initially raised by the concept.

In brief, the lactic acid (and possibly other metabolites) produced in the epithelium, lowers the pH of the venous blood returning from the gland. The lowering of the pH releases oxygen from the hemoglobin of the venous blood; in consequence, oxygen tension (pressure) rises in the venous channel and then diffuses across to the arterial capillaries. At the same time there is also a diffusion of lactate from the venous to the arterial channel. Thus, blood that enters the arterioles of the rete with oxygen tensions and pH characteristic of the general arterial circulation is enriched with O_2 and acid as it flows towards the epithelium of the gas gland. As it circulates through the epithelium, it receives still more acid which further increases the O_2 tension resulting in the diffusion of oxygen into the bladder. It should be remembered that the blood returning to the rete from the epithelium of the gland has a lower pH and a reduced content of oxyhemoglobin but a *higher tension* of O_2; gas solubility is probably also reduced by the addition of ions (lactic acid, carbonic acid) and this is important in the "salting out effects"; it should also be noted that nitrogen as well as oxygen may be concentrated in the bladder. Moreover, Root and Bohr effects are very marked in some species of fish (Fig. 12.7) and may be coupled with these other factors in release of oxygen. However, the really important trick appears to be the formation of lactic acid with the *rapid* release of gas from the blood into the counter-current multiplier system. Berg and Steen (1968) have shown that acid very quickly releases O_2 from hemoglobin (half time, about 50 msec), while binding of O_2 to hemoglobin occurs much more slowly (half time, 10 to 20 sec). Hence, the released oxygen does not have time to recombine during its short passage along the venous capillary. Further details of the current understanding of the physiology of this system are discussed by Steen (1970, 1971).

There are several interesting morphological correlates. Thus, the wall of the bladder is highly impermeable to gases; except in the region of the oval and gas gland, a thick layer of connective tissue impregnated with guanine reduces the O_2 and CO_2 permeability to about 10 per cent of that of other connective tissues (Kutchai and Steen, 1971). Further, a good correlation has been established between the lengths of the rete capillaries and the depths at which fish live; thus, fish ranging in depths down to 2500 meters have capillaries ranging in lengths from 4 mm to 12 mm, while teleosts found at depths of 5000 to 7000 meters have capillaries varying from 15 to 25 mm in length (Marshall, 1972). The efficiency of counter-current multiplication can be expected to vary directly with the length of the rete capillaries and inversely with the flow of blood through them. Another curious phylogenetic adaptation has been noted in several species of mesopelagic fishes in which the capillaries are not particularly long but the blood cells have become highly specialized. The erythrocytes in these fishes have followed the same phylogenetic trail as those of the mammal and dispensed with the nucleus; this increases the surface/volume ratio and will presumably aid in rapid exchange of gases and increase the oxygen carrying capacity of the cell. The capillaries in some of these fishes are also extremely narrow so that the erythrocytes seem to squeeze through them in single file. Marshall (1972) discussed these and other techniques that improve the exchange between the venous and arterial capillaries.

A system designed to adjust buoyancy in accordance with depth may be expected to have some means of automatic control. Well-defined reflexes concerned with inflation and deflation have been investigated in several species (Fänge, 1953, 1966; Steen, 1970). The autonomic supply is well developed, with parasympathetics of the intestinal vagi and sympathetics from the coeliac ganglia. Fibers pass both to the smooth muscle of the bladder wall and to the gas gland. Tension receptors in the wall of the bladder initiate reflexes appropriate to the volume changes required for an efficient buoyancy device (Fänge, 1953; Qutob, 1962; Nilsson, 1971).

14

LIGHT

Solar radiation is the ultimate source of energy for all life. When a photon (the packet or quantum of radiant energy) strikes and interacts with particles of matter, it sends an electron into a higher energy level or an excited state. These electrons drop back to the ground state after little more than 10^{-7} seconds, but during this brief interval they provide the electronic energy which powers the machinery of life. Highly efficient biochemical substances capture the excited electrons, uncouple them from their partners and permit them to return to the ground state through a sequence of energy-yielding reactions. In photosynthesis this electronic energy is channeled into the pyrophosphate bonds of ATP and the reduced pyridine nucleotides; in turn, these are used to synthesize the complex organic molecules which are the fuels of metabolism. These events have been sketched in Chapter 1. Emphasis, in the present chapter, is not on radiant energy as the source of power for life but on light as an environmental factor which limits, controls and orients animal processes.

Twentieth-century theories of quantum mechanics provide a unified concept for the properties of light and matter. Details are beyond the scope of this book, but the principles are pertinent to the present discussion. According to theory, light can be accurately characterized in two very different ways. On the one hand, it has the properties of particles or corpuscles as Newton argued in the seventeenth century while, on the other hand, it exhibits wavelike properties just as his contemporary Huygens so vigorously insisted. Modern theory not only shows that both particle and wave concepts are to be used in certain contexts but also gives the precise quantitative relationships between them. Each packet of light or photon has an energy content which can be calculated from its wavelength by Planck's formula. $E = hc/\lambda$

where h is Planck's universal constant (1.58×10^{-34} calorie seconds), c is the velocity of light (3×10^{10} cm per second) and λ is the wavelength. Einstein's law of photochemical equivalence teaches that photochemical reactions occur only when a molecule absorbs a photon. Hence a mole of substance (containing 6.02×10^{23} particles) can be expected to absorb a mole of photons in a photochemical reaction. The energy of this number of photons is the EINSTEIN, equivalent to 2.854×10^{7} gram-calories. Thus, the physicist is able to calculate the energy associated with light of different wavelengths or, conversely, wavelengths or radiation which correspond to energies of activation for chemical reactions (Chapter 10). Figure 14.1 shows broad bands for the energies of activation of ordinary chemical reactions and for photochemistry, with much narrower bands for several important photobiological processes.

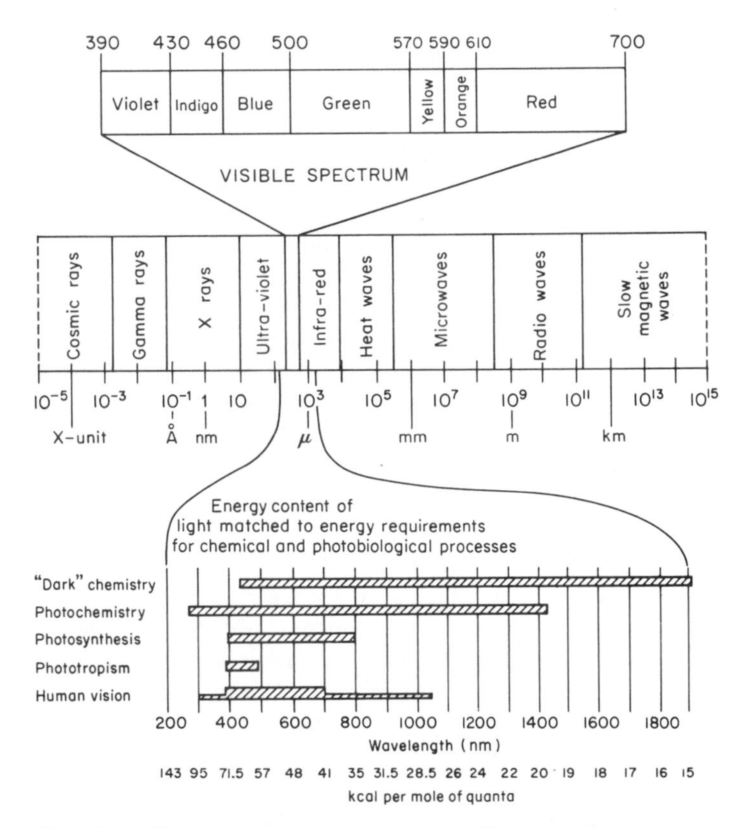

Fig. 14.1 The electromagnetic spectrum with wavelengths in nanometers at center. Visual spectrum enlarged above and (at the bottom) the energy content of light matched to the energy requirements of chemistry and photobiological processes and to the absorption spectra of photoreactive substances. [Based on values from Ditchburn, 1952: Light. Blackie, London, and from Wald, 1959: Life and Light. Copyright 1959 by Scientific American, Inc. All rights reserved.]

PHOTOBIOLOGICAL PROCESSES

Significance of the Term "Light"

Life exists in and operates on a relatively narrow band of the electromagnetic spectrum (Fig. 14.1). The entire spectrum extends from the cosmic and gamma rays with wavelengths of only a ten billionth of a centimeter to the radio waves which may be miles in length. Within this broad expanse of radiant energy, there is a narrow band which we call LIGHT because of the sensation which it creates when it falls on the retina of the human eye. Its wavelengths extend from 380 to 760 nanometers (nm), with extreme limits of 310 to 1050 nm in very intense artificial sources. This is our range of conscious vision, and we justly attach great significance to it. However, the comparative physiologist could scarcely confine attention to this particular band unless its action had a much broader basis in animal physiology. This is indeed the case, and the particular band of radiant energy, which we recognize as LIGHT, controls all of the important photobiological processes; the action spectrum for photosynthesis extends from about 400 to 760 nm, photoperiodism in plants from about 500 to 800 nm, while photoreception in all animals is almost covered by the extreme human range. For this reason alone the phenomena discussed here are more appropriately grouped under the heading of LIGHT rather than under the broader term of RADIANT ENERGY. There are however, more cogent reasons for this distinction in photophysiology.

The environmental physiologist might reasonably restrict his attention to this band of the spectrum because it encompasses almost all the solar energy which actually reaches the earth's surface (Fig. 14.2). The ultraviolet is cut off sharply at wavelengths shorter than about 300 nm by the blanket of ozone which surrounds the earth in its upper atmosphere; in the aquatic habitat the spectral band is still further restricted as the sunlight penetrates deeper and deeper (Fig. 14.2). The biologist's attention can also be focused on this band because only this portion of the spectrum is primarily active in photochemical reactions. Wald, in many stimulating papers, has argued that biological processes, no matter where they exist (on this planet or elsewhere), must be confined to the same range of wavelengths because of the physical nature of the action of radiant energy in chemical processes.

Because of the complex and delicate nature of many organic compounds, the radiant energy compatible with photobiological reactions is considerably more restricted than that which is concerned with many other photochemical processes. The delicate secondary and tertiary bondings which give the proteins and nucleic acids their highly specific properties are destroyed by radiation shorter than 300 nm (95 kilocalories per mole). Proteins are denatured and nucleic acids are depolymerized; the living cell is destroyed. This sets a lower limit on the radiation compatible with life. There are other limits associated with particular reactions. Energies required to break single covalent bonds fall between about 40 and 90 kcal per mole (710 to 310 nm); those which excite valence electrons to higher orbital

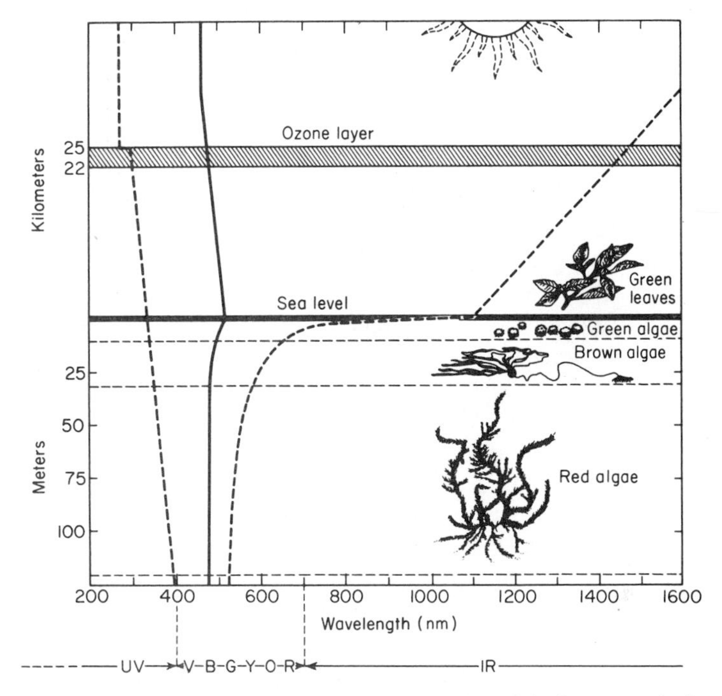

Fig. 14.2 The narrowing of the spectrum of sunlight by atmospheric absorption and by absorption in sea water. The heavy solid line from top to bottom locates the wavelengths of maximum intensity. Broken lines from top to bottom limit the wavelength boundaries within which 90 per cent of the solar energy is concentrated at each level in the atmosphere and ocean. UV, ultraviolet. V, violet. B, blue. G, green. Y, yellow. O, orange. R, red. IR, infrared. [After Wald, 1959: Life and Light.]

levels involve energies of 20 to 100 kcal per mole (1430 to 280 nm). These are the important photochemical reactions of life, and they all fall approximately in the solar spectrum which reaches the earth's surface and drives the indispensable photobiological reactions. On these bases, Wald (1959) argues that only the radiant energy which we call LIGHT is a suitable source of power for living machines.

Photobiological Effects

Three different effects follow the irradiation of protoplasm by various components of the electromagnetic spectrum. The more energetic wavelengths associated with gamma rays, X rays and the extreme ultraviolet (shorter than about 300 nm) shatter the molecules by displacing electrons and producing IONIZATION;

disintegration of the atomic nucleus may also occur with very high-energy radiation. Inorganic as well as organic cellular constituents are altered; the ionized products of water are particularly reactive. It is the chemical changes which follow ionization which are even more destructive than the ionization itself. The literature on ionizing radiation is extensive, for it is now a subject of great medical and biological importance (Lea, 1962; Arena, 1971; Giese, 1973). Strictly speaking, however, it is not part of the environmental physiology of animals, for the curtain of ozone in our upper atmosphere screens life from most of the ionizing effects.

In the visible portion of the spectrum, from the ultraviolet to the infrared, radiant energy has a PHOTOCHEMICAL and a THERMAL ACTION. The thermal effects extend beyond the visible, but as the wavelengths become longer and longer, protoplasm gradually becomes transparent to the radiant energy and is quite unaffected by long waves such as radio waves.

Both the photochemical and thermal effects are caused by the activation of molecules. The necessary energy of activation for a chemical reaction may be acquired through thermal agitation or collision with a photon. In both cases, chemical reactivity is increased through the activation of molecules. Electrons may be raised to higher energy levels, but this is a much less drastic effect than the ejection of an electron which takes place in ionizing radiation. However, even though the molecular alterations are less drastic in photochemical and thermal reactions, the effects can still be damaging or lethal.

The significant photochemical processes of life depend on a few different pigmented molecules which absorb radiant energy and thus initiate indispensable biochemical and physiological processes. These colored molecules may provide the mechanisms which generate electronic power for the operation of photosynthesis (as in the case of chlorophyll), or they may trigger the release of energy in some entirely different biochemical system; the visual pigments do this through a molecular rearrangement which initiates a nerve impulse.

Most, and perhaps all, of these special pigmented molecules are synthesized only by plants from which the animals must acquire them in order to capitalize on this process. In plants, the chlorophylls operate photosynthesis; the carotenoids are involved in phototropism; the phytochromes regulate photoperiodism. In animals, one set of pigments, the visual pigments, is concerned not only with reactions to light but, indirectly, also with phototaxis and photoperiodism. These are the carotenoid pigments and the animal must obtain them or their important building blocks from the plants.

Absorption and Action Spectra

When white light is passed through a colored solution, the colored solute molecules absorb photons of particular energy values in accordance with the atomic structure of the molecules. A solution of chlorophyll appears green to us because only light of wavelengths in the vicinity of 520 nm (green) is transmitted through it. The chlorophyll molecules absorb photons associated with the other wavelengths, but the band of green light is transmitted and gives rise to green sensations when it, in turn, is absorbed by the photosensitive pigments of our eyes.

A characteristic curve is obtained when the percentage of light transmitted through a solution is graphed as a function of different wavelengths (Fig. 14.3). Such a curve is called an absorption curve if the "peaks" correspond to high absorption, or a transmittancy curve if the "valleys" correspond to high absorption. The ABSORPTION MAXIMUM of the former corresponds to the energy required to boost an electron into a higher orbit or to *excite* a molecule of the particular solute. The absorption spectra for chlorophyll and the visual pigments of the human eye are compared in Fig. 14.3; it is evident that the point of maximum transmission for the former is the point of maximum absorption for the latter.

One might logically postulate some correlation between the absorption spectrum of a substance such as chlorophyll or visual purple and the rate of its biochemical or physiological action in light of different wavelengths. The rates of photobiological processes have often been measured under light of different wavelengths, and the data have been recorded in the form of ACTION SPECTRA. The postulated correlation between absorption and action spectra is evident (Fig. 14.3). Thus, the action spectrum has become a useful tool for the identification of compounds associated with different photobiological processes. The spectral sensitivity curve for human twilight vision (scotopic vision) has a peak at 500 nm which corresponds to the absorption spectrum of rhodopsin or visual purple, the rod pigment concerned with vision in dim light. The spectral sensitivity for daylight or photopic (cone) vision is centered around 550 nm and it has been possible to identify three pigments in the fovea or color-sensitive area of the human retina with absorption maxima at 450 nm, 525 nm, and 555 nm (Chapter 18).

INJURIOUS EFFECTS OF SUNLIGHT

Sunburning is a familiar experience which leaves little doubt that the radiant energy at the earth's surface can be damaging to protoplasm. From 1 to 5 per cent of the total solar radiation which reaches the earth is in the ultraviolet (between 300

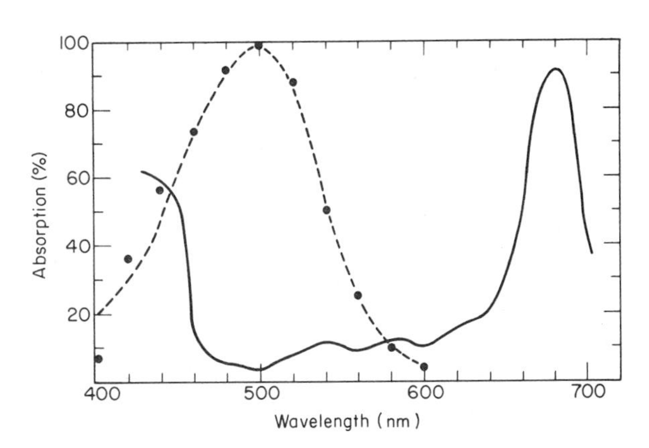

Fig. 14.3 Absorption spectra for human rhodopsin (broken line) and for a group of chlorophylls (solid line). The rhodopsin curve represents the absorption spectrum while the plotted points represent human scotopic sensitivity plotted as percentage of maxima. [Rhodopsin based on data from Crescitelli and Dartnall, 1953: Nature, *172*: 196. Chlorophyll based on data from Emerson and Lewis, 1942: J. Gen. Physiol., *25*: 587.]

to 390 nm). This not only creates sunburn in the lightly pigmented human skin but can also injure or kill other naked animals such as earthworms, planarians or protozoans in shallow water. This band produces rather non-specific photochemical effects on the cell proteins in contrast to the shorter ultraviolet band with its distinct nuclear and genetic effects. The shorter ultraviolet range, particularly between 200 and 300 nm, has been intensively used in cell physiology since Müller's discovery of the mutagenic action of X rays on *Drosophila*; induction of *Drosophila* mutants with ultraviolet came a few years later. The literature is reviewed by Giese (1950), Hollaender (1955), and texts already cited but is not considered here since animals do not experience the effects of the short or far ultraviolet in their natural habitats. There are, however, several interesting physiological effects of sunlight on exposed organisms which are not mediated through the specialized photoreceptor pigments; these will now be discussed.

Photodynamic Action

Not only ultraviolet but visible radiation up into the red region of the spectrum may be damaging under certain conditions. Raab discovered this accidentally, near the end of the nineteenth century, while studying the toxic effects of acridine on paramecia. Cultures were killed in 6 minutes with acridine concentrations at 1 : 20,000 in bright sunlight but survived 1 hour in diffuse sunlight; identical cultures in the dark were uninjured. In 1900, he performed another experiment which demonstrated that effects of this sort were not peculiar to protozoan cultures. Mice, which had been injected with eosin, were obviously uncomfortable and irritated when in the light. They scratched their skins and sought the shade. In a short time, sores appeared on exposed areas, the tissues become necrotic and the animals died. Another scientist, Meyer-Betz, in 1913 boldly injected himself with 0.2 gm of hematoporphyrin and carefully described the symptoms which followed irradiation of a small area of his arm: erythema (red coloration), edema, pain, hemorrhagic sore, followed by a scab in about three weeks and eventually a deep scar. Like the mice, he experienced intense discomfort in the light and was forced to seek total darkness during a period of photosensitization which lasted several weeks. All of the symptoms did not disappear for about six months. These pioneer experiments are summarized in several places (Laurens 1933; Blum, 1941; Spikes and Livingston, 1969).

Several plant pigments are now known to cause photosensitization under natural conditions (Giese, 1971). White pigs which feed on the roots of *Lachnanthes* or white sheep eating the St. John's-wort, *Hypericum crispum,* may become ill and die while the black members of the species are uninjured. Buckwheat (*Fagopyrum esculentum*) may poison cattle, swine, and sheep in the same way. Photosensitization can be readily demonstrated by feeding fresh buckwheat plants to guinea pigs for about 4 days. Giese (1971) has studied a photosensitizing pigment which develops in the protozoan, *Blepharisma*, when cultured in darkness; only the pigmented individuals are killed when exposed to strong light.

This PHOTODYNAMIC ACTION, as it is called, is the result of a photosensitized oxidation (Spikes, 1968). Anaerobic photosensitized photobiological reactions are rare; oxygen is almost always required, and the reactants of the best-known photodynamic systems consist, therefore, of the dye or sensitizer, the substrate and oxygen. The molecules of the photodynamic dye, which are more or less fluorescent, hold or trap quanta of absorbed radiant energy for a brief period of about 10^{-7} or 10^{-8} seconds before passing it on to *excite* one of the other reactants. Light of wavelengths up to 800 nm may be effective. Both the substrate and oxygen have been postulated as the secondarily excited substance, and indeed they both may be (Spikes, 1968). In any case, a photooxidation of the substrate follows. Proteins appear to be the most sensitive to injury and it is probably their oxidation which forms toxic byproducts that damage cells. The presence of a reducing agent, or the exclusion of oxygen, interferes with the reaction. Photodynamic action may be demonstrated in the laboratory by adding small amounts of Rose Bengal or eosin to suspensions of erythrocytes, or cultures of protozoans, and exposing them to the light, with adequate controls in darkness.

Sunburn and Suntan

Most land animals possess special coverings which protect them from the action of ultraviolet rays and at the same time provide moisture-proofing and, in some cases, temperature regulation. Arthropods achieve this protection with a heavily chitinized, noncellular, waxy cuticle produced by the underlying epidermis; in the land vertebrates, superficial epidermal cells are modified into scaly plates through the loss of their nuclei and the development of keratohyalin granules of highly insoluble protein. Even thin layers of keratin absorb or reflect a large part of the ultraviolet, while specialized epidermal appendages (scales, feathers, fur) increase this screening and cut off all ultraviolet from the actively dividing layers of the epidermis (stratum germinativum) and the nutritive layers of the dermis. For the most part, man lacks these specialized appendages and depends on thickening of the stratum corneum and on epidermal pigmentation for his protection.

The action of sunlight on human skin has been studied many times since the pioneer experiments of Finsen at the end of the nineteenth century (Blum, 1945, 1961; Johnson *et al.*, 1968; Wiskemann, 1969). The reactions may be systemic as well as cutaneous and are caused by both heat and ultraviolet radiation; only the effects of the ultraviolet on the skin are considered here.

The marked erythema or red coloration of sunburn appears an hour or more after exposure, is usually confined to the irradiated area, and persists for several days. The action spectrum for erythema is a narrow one with a peak at about 300 nm; studies of the transmission of this spectral band through pieces of human skin show that it is mostly reflected or absorbed before it penetrates to the dermal layers where all of the blood vessels are found (Fig. 14.4). Although it is now evident that penetration is greater than originally thought, and of possible importance in directly stimulating blood vessels, it is still agreed that the action of ultraviolet is primarily

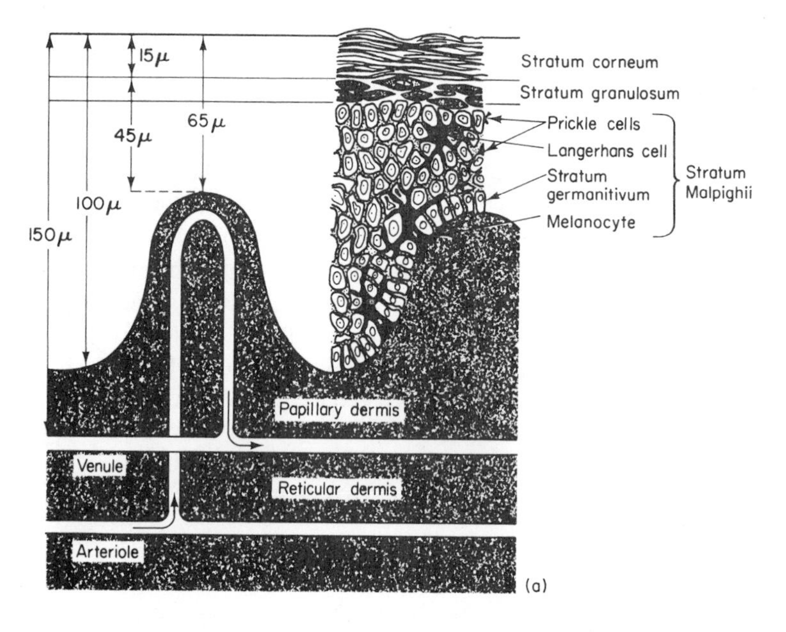

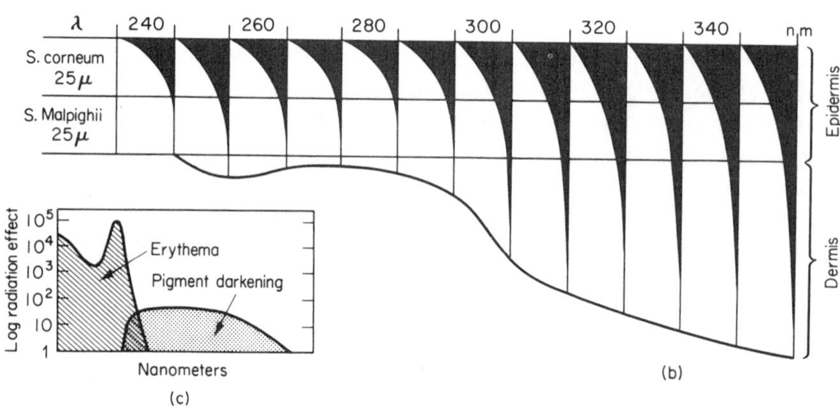

Fig. 14.4 Human skin and the penetration of ultraviolet through its surface layers. (a) Transverse section through the skin from the back of a young adult. [From Johnson *et al.*, 1968: *In* Photophysiology, Giese, ed. Academic Press, New York.] (b) Depths of penetration of radiant energy into skin on inside of forearm. [From Wiskemann, 1969: *In* An Introduction to Photobiology, Swanson, ed. Prentice-Hall, Englewood Cliffs, N.J.] (c) Action spectra for erythema and pigment darkening in human skin. [From Henschke and Schulze, 1939: Strahlentherapie, *64*: 36.]

in the epidermis (Johnson *et al.*, 1968). A photochemical decomposition of amino acids and proteins produces an active vasodilator which diffuses from the non-vascular epidermis and produces an enlargement and engorgement of the minute vessels of the dermis, with an accompanying intracellular edema and migration of leucocytes into the surrounding tissues. The vasodilator has not been identified. In the pioneer studies, these ultraviolet effects were compared with reactions induced by pricking histamine under the skin or with mild cutaneous damage, such as scratching, which elicits a TRIPLE RESPONSE in the form of a local RED REACTION, a spreading FLUSH or FLARE and a local EDEMA or WHEAL (Davson and Eggleton, 1968). At present, it is by no means clear that these parallels are justified or that histamine is the active vasodilator; other biogenic amines such as 5HT have been considered (Blum, 1955; Johnson *et al.*, 1968; Wiskemann, 1969).

In lighter colored skins, increased pigmentation (suntan) follows the erythema with a very gradual change from red to brown. This does not prove, however, that the erythema and pigmentation are causally related. In fact, there are actually two processes involved in the suntanning itself; one of these, directly related to the erythemal spectrum, depends on the formation of additional melanin in the deeper layers of the epidermis while the other, produced by a different spectral band, is a photooxidation of bleached or leuko-melanin to a black form. This latter reaction is called "pigment darkening" and has a much wider action spectrum (maximum near 340 nm) than the erythemal band (Fig. 14.4c). Common window glass cuts out all wavelengths below 320 nm, and this excludes the entire erythemal band and much of the pigment darkening band (Blum, 1955).

Many of the links in melanogenesis have not yet been fully described. In general, however, it is clear that melanin is formed from the amino acid tyrosine through a series of intermediates (Fig. 20.17) and that its production depends on a special dendritic cell, the melanocyte. During ontogeny, melanocytes arise from the neural crest and migrate into the epidermis where they occur in varying numbers up to about 4000 per mm^2 among the basal cells of the Malpighian layer. In some manner, not yet clearly explained, the melanin is transferred from the dendrites of the melanocytes to the epidermal cells (Montagna, 1961). In blond individuals, this black or brown pigment is confined to the deep epidermal layers, but in dark-skinned people it extends into the outer cornified layers.

Biochemical significance of pigmentation The physiological advantages of pigmentation have been debated for a very long time (Blum, 1961). In the pioneer studies, Finsen recorded protection from sunburning on areas of his arms which were coated heavily with India ink. It was assumed that melanin, like India ink, was opaque to ultraviolet and in this way protected the living cells of the Malpighian layers. Indeed, the histology of the skin supports this hypothesis since the black granules normally form supranuclear caps in this layer and are thus advantageously placed for screening the sensitive nucleoproteins. Further support for this theory comes from critical comparative measurements of the transmission of solar ultraviolet through detached outer layers of skin from Africans and Europeans; black skin transmits significantly less light (Thomson, 1955).

This, however, is now known to be only a partial explanation of increasing resistance to sunburning. Albino skins and areas of vitiliginous (non-pigmented) skin also show acclimatization. Further, it is recognized that protection from sunburn is more transient than the pigmentation of suntanning. Exposure to ultraviolet produces a prompt and marked increase in the thickness of the stratum corneum, and this probably provides more protection than the increased pigmentation (Johnson *et al.,* 1968). Racial as well as acclimatization differences in the thickness of the stratum corneum have been recorded.

Blum (1961) and Dobzhansky (1962) summarize the hypotheses and rather speculative evidence for an adaptive significance of skin color in racial evolution. Since the darkest races usually live in the hottest climates, an association between skin color and the amount of sunlight seems obvious. There is, however, no very satisfactory evidence to support the hypothesis. Physiological studies indicate that the thickening of the corneum is more critical than melanization in excluding the ultraviolet; further, a white skin actually reflects about 30 per cent more sunlight than a black one (Blum, 1945). Thus, the black body has an increased heat load; this could be an advantage if the cooling machinery were activated at a lower temperature or if it operated more efficiently (Cowles *et al.,* 1967). Other suggested advantages of a dark skin in the sunnier lands have been protection from skin cancer (Blum, 1961) and camouflage (Cowles, 1959). The statistics support the cancer theory, but there is no evidence of its significance in evolution.

The possible advantages of the non-pigmented skin have also been emphasized. One line of argument is based on the importance of vitamin D in the development of bones and the prevention of rickets. Ultraviolet wavelengths shorter than 320 nm are responsible for the photochemical production of this vitamin from sterols in the outer layers of the epidermis. There is no question of the importance of vitamin D during growth and of the limited amount of effective sunlight for this purpose in the more frigid regions of the earth. It should be noted, however, that the requirements depend on the type of diet and that rickets may be quite unknown among the Eskimo who are at the greatest disadvantage as far as sunlight is concerned. A second line of argument stresses the hazards of hypervitaminosis (Loomis, 1967). Ingestion of more than 2.5 mg vitamin D per day may produce excessive calcification of tissues along with high blood calcium and phosphorus. Calculations indicate that sufficient ultraviolet could reach the deeper layers of skin to create these high levels. However, the counter arguments note that this requires constant and complete exposure of the body to intense sunlight, while under natural conditions, tropical man seeks shade and tends to be less active during the heat of the day (Blos *et al.,* 1968). Thus, the antirachitic argument is no more satisfactory than the others and, at present, there is a general lack of convincing evidence that any of these factors is sufficiently critical to provide a selection pressure in evolution.

PHOTORECEPTOR PIGMENTS

The carotenoid pigments are the most widespread of all the colored compounds in the living world; their importance in animal evolution seems to rank with that of the chlorophylls. It is pointless to argue that any one of life's indispensable compounds is more important than another; but it seems safe to maintain that the chlorophylls and the carotenoids are members of a family of very special molecules which were required to launch animal evolution. The chlorophyll molecule is the primary link in trapping radiant energy which indirectly operates animal machinery; the carotenoid pigments play their indispensable role in the orienting and directing mechanisms which coordinate feeding activities, as well as much of the animal's general behavior and social organization. In both cases there is a complete dependence on the plant world. Animals, with the exception of a few of the protozoans (Protista), are unable to synthesize either chlorophyll or carotenoids. The latter are universal dietary requirements for animals. Even the beautifully colored sea anemone becomes colorless if fed on white fish muscle devoid of carotene (Fox and Vevers, 1960).

Chemistry of the Carotenoids

The carotenoids take their name from carotene, the yellow pigment of carrots, which was first isolated in 1831 (Fruton and Simmonds, 1958). It is one of a group of fat-soluble substances characterized by a long skeletal chain of carbon and hydrogen with alternating single and double bonds between the carbon atoms. The color which is attributed to this alternating arrangement of bonds varies from yellow through orange and red to violet, depending on the increasing number of double bonds and the presence of certain radicals.

The formula at the top of Fig. 14.5 represents a molecule of β-carotene, one of

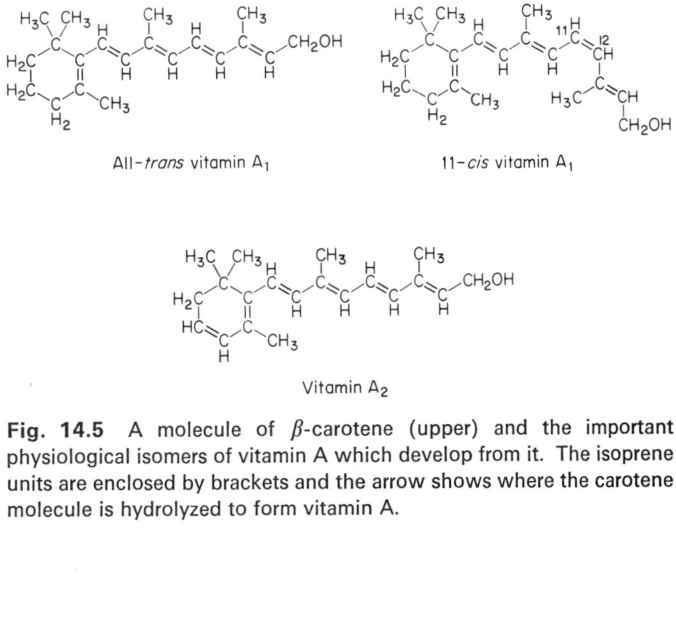

$$+\ 2H_2O$$

All-*trans* vitamin A_1

11-*cis* vitamin A_1

Vitamin A_2

Fig. 14.5 A molecule of β-carotene (upper) and the important physiological isomers of vitamin A which develop from it. The isoprene units are enclosed by brackets and the arrow shows where the carotene molecule is hydrolyzed to form vitamin A.

the most important plant pigments in animal nutrition. The molecule is built of eight 5-carbon isoprenoid units, linked to form a long chain of 40 carbon atoms with an ionone ring at each end. Halfway along its length the molecule is turned on itself and, at this point, may be broken hydrolytically to yield two molecules of vitamin A. The latter exists in two forms which differ only in a pair of hydrogens (presence or absence of a double bond) in the terminal ring (Fig. 14.5). This formula typifies the carotenoids; the ends of the different kinds of carotenoid pigments differ with incomplete terminal rings or a ring at only one end (Goodwin, 1962; Needham, 1974).

***Cis-trans* isomerism** A family of differing molecular configurations is possible. By changing the position of either the —H or the —CH_3 attached to the carbons of the double bond, these groups may appear on the same (CIS) or on opposite (TRANS) sides of the chain. The result is that the *cis*-isomers are kinked molecules while the all-*trans* forms are straight (Fig. 14.5). Most of the naturally occurring carotenoid molecules have their double bonds in the all-*trans* configuration. Irradiation of the natural pigments with ultraviolet will produce a variety of *cis*-isomers. There are also some naturally occurring *cis*-isomers. One of these, the 11-*cis* or *neo*-B isomer of vitamin A, which has the two hydrogens on the same side of the double bond between carbons 11 and 12, is particularly significant in visual processes. The excitation of the photoreceptor seems to depend on the straightening out of the kinked molecule (*cis* to *trans* isomer). This, in some way, triggers an impulse in the optic nerve (Wald, 1961). Further details are considered in Chapter 18.

Carotenoproteins The carotenoid pigments of animals are either dissolved in the tissue fats or chemically combined with specific proteins; in either form, they play a part in animal coloration (Fox and Vevers, 1960). As prosthetic groups of proteins, two of them can form the highly specialized molecular machines which are the basis of photoreception.

Visual pigments have now been investigated in representatives of each of the major phyla with highly specialized eyes. In all cases, the active pigment is an aldehyde of vitamin A, known as RETINENE, combined with a protein called OPSIN. Retinene, or retinaldehyde, is also called RETINALD. Relationships between the two forms of vitamin A and their associated pigments are as follows:

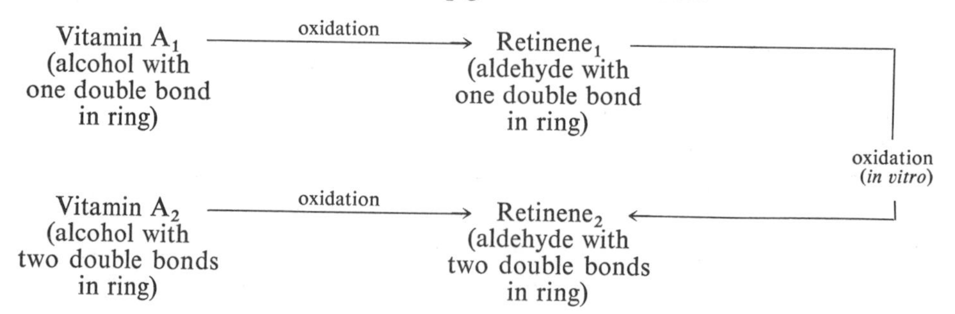

Although there are only two known retinenes, many different visual pigments have been categorized on the basis of their absorption and action spectra; the absorption maxima of the known pigments range from 430 to 562 nm in the A_1 series and from 510 to 620 in the A_2 series (Dartnall and Tansley, 1963). These differences are presumably caused by the opsin component of the molecule (either the species-specific nature of the proteins or differences in the manner of coupling the prosthetic groups). A completely satisfactory explanation of the differences has not yet been found (Wald, 1960a; Dartnall and Tansley, 1963).

Visual pigments in biochemical evolution The phylogeny of the visual pigments is still one of the interesting problems of biochemical evolution.

The first of these pigments to be examined came from a frog. It was called RHODOPSIN or visual purple, and its bleaching to a yellow color in the light was noted as early as 1876 (Fox and Vevers, 1960). Its relationship to vitamin A_1 and retinene$_1$ was subsequently established. Somewhat later it became apparent that some freshwater fishes possessed visual pigments based on vitamin A_2 and retinene$_2$. The first of these was named PORPHYROPSIN. These two terms (rhodopsin and porphyropsin) are still useful, although it must be appreciated that they refer to broad classes of pigments and not to single substances with distinct absorption maxima.

Interest in the phylogenetic implications of these early findings was sparked by the discovery of an apparent correlation between the habitat of an animal and its retinene. Thus, the mammals, birds, and almost all of the marine fishes were found to have rhodopsin, while the first freshwater fishes studied possessed porphyropsin. Of even greater interest was the observation that several groups of vertebrates which make a profound change in habitat during their life cycle show associated changes in their rhodopsin-porphyropsin system. Comparative studies of the eyes of tadpoles and frogs showed a predominantly porphyropsin system before metamorphosis with a rhodopsin system during adult life; a newt such as a *Diemyctilus,* which shows a second metamorphosis before returning to water at spawning time, alters its pigment in the reverse direction at this stage in life. Euryhaline fishes have a mixture of porphyropsin and rhodopsin; this is also true of the anadromous and catadromous species, but in these the dominant pigment is characteristic of the habitat where the fish spawns. All this seemed to fit neatly into a phylogenetic pattern with rhodopsin in habitats where water is in short supply (land and sea water), while porphyropsin dominates in strictly freshwater environments. In 1960, Wald summarized the work of three decades and emphasized probable evolutionary implications.

Further studies, using more precise analytical techniques, have thrown considerable doubt on these evolutionary speculations. To begin with, the invertebrates do not follow any simple pattern of pigments based on retinene (Wolken, 1971). Whether terrestrial, marine or freshwater, invertebrate animals possess visual pigments of the A_1 or rhodopsin type; yet, within this framework, their pigments show a wide range of spectral characteristics and have been adapted to the major problems of animal vision, including color vision. Further, it is now apparent that the freshwater fishes—at one time thought to be characterized by porphyropsin type eyes—more frequently possess mixtures of porphyropsin and rhodopsin (Schwanzara, 1967; Munz, 1971); in fact, pure porphyropsin eyes are rare among the teleosts (mostly freshwater catfishes and centrarchids), while some cyprinids and poeciliids possess pure rhodopsin type eyes. It is also pertinent that the proportions of the two retinenes may change seasonally in some species or be altered experimentally by illumination (Beatty, 1969a, b; Munz, 1971).

All this does not argue that visual pigments have not been adapted to the photic environment. On the contrary, the evidence suggests that they are highly adaptive but that there are several avenues of adaptation, and some of these may be more important than the type of retinene. The opsin (protein) portion of the molecule is far more variable than the prosthetic group; in teleost fishes alone, the range of absorption spectra in rhodopsin type pigments extends from about 465 nm to 540 nm (Munz, 1971). Varying the proportions of the pigments seems a likely third avenue of adaptation; again, suggestive evidence for this is found among the teleosts where absorption spectra in visual pigments from eyes with mixed retinenes extend from about 485 nm to 540 nm. Even though the rhodopsin pigments are dominant in land vertebrates and marine fishes, the facts seem to argue against an evolutionary pressure toward the phylogeny of porphyropsin for problems of freshwater life—even though its evolution may have occurred in fresh water as Wald (1960a) argued. The entire pigment molecule and varying mixtures of pigments have been adapted to the visual needs of animals. This, like several other speculative trends in biochemical evolution, is more complex than it first appeared to be.

Visual Cycles

Little is known about carotene metabolism in the lower animals. In the mammal, dietary carotene is converted to vitamin A in the wall of the gut or the liver; in some animals (rats, pigs, goats and others) this occurs only, or almost entirely, in the intestinal wall while in man it may be confined to the liver (Harper, 1973). In all animals, the vitamin is stored in the liver which is the major organ concerned with its regeneration from the *trans* form to the *cis* form during the visual cycle. Livers of some mammals (polar bear, for example) store such tremendous amounts of vitamin A under natural conditions that they are toxic in human diets (Lewis and Lentfer, 1967). From these liver stores, or directly from the wall of the gut, this vitamin passes by way of the blood to replenish the wastage which occurs through the visual processes in the retina. It circulates only in the all-*trans* form and must be converted to the 11-*cis* isomer in the retina before it is available for rhodopsin synthesis (Fig. 14.6).

Most of the biochemical studies of photoreception have been confined to rhodopsin (Wald, 1961). A series of chemical transformations follows exposure to light, but only one of these is entirely dependent on the radiant energy. When rhodopsin absorbs a quantum of light, the 11-*cis* retinene of the molecule is isomerized to all-*trans* retinene, and in some way the photoreceptor is stimulated to fire the nerve impulse (Wald, 1968). The succeeding events, outlined in Fig. 14.6, are enzymatic thermal reactions which do not require radiant energy. Small amounts (30 per cent or less) of the 11-*cis* retinene may be regenerated directly from the

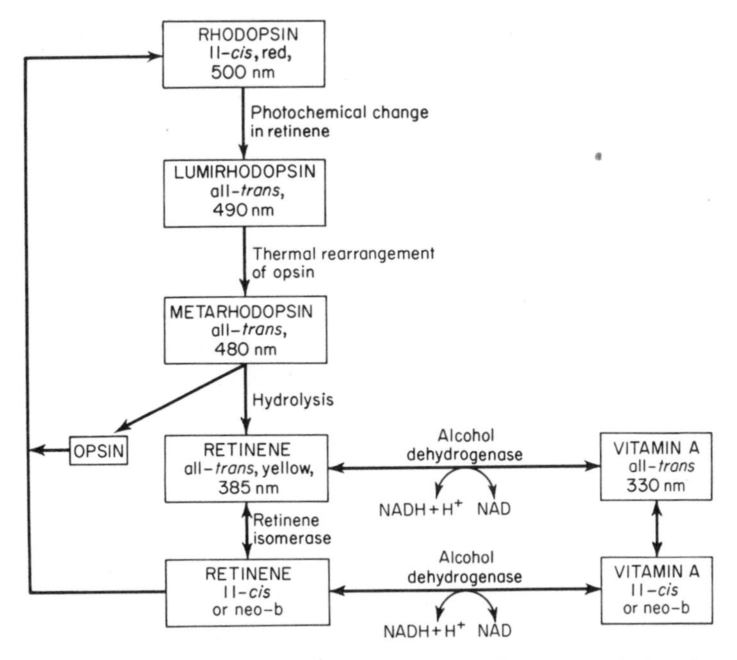

Fig. 14.6 Chemical events in the visual cycle of rhodopsin. Values for absorption maxima are approximate.

all-*trans* retinene with the enzyme retinene isomerase; this reaction is stimulated by light of short wavelengths (blue) but can occur in darkness. Most of the 11-*cis* retinene is resynthesized (probably in the liver) by the oxidation of 11-*cis* vitamin A; this, in turn, comes from the all-*trans* form of the vitamin by steps which are not yet established. Retinal tissue has a high content of NAD and this, with the enzyme retinene reductase, is responsible for the oxidation-reduction reactions which take place between retinene and vitamin A (White *et al.,* 1968).

PHOTOTACTIC RESPONSES

Practically all living organisms are light-sensitive. A variety of orientation responses are characteristic of both primitive plants and animals. In phylogeny, these simple adaptive movements must have preceded the complex activities associated with the visual systems just considered. Both are based on the carotenoids, but unfortunately there is not nearly enough precise knowledge of the biochemical mechanisms responsible for phototaxis in the primitive animals to reveal any definite trail of biochemical evolution. Only a brief comment on phototaxis is appended here.

Over a period of almost 300 years scientists have frequently recorded their observations on the turning of plants toward the light. The phenomenon was first called

PHOTOTROPISM in the middle of the nineteenth century. This term is now reserved for fungi, higher plants and sessile animals such as hydroids, which are anchored at one end and develop a curvature in response to light; the comparable reaction in motile organisms is called PHOTOTAXIS. In this, there is a free movement of the entire organism, with the light acting either directly or indirectly on organs of locomotion such as muscles and flagella. There are several current reviews of the literature (Thimann and Curry, 1960; Clayton, 1964, 1971; Giese, 1964; Page, 1968).

Although the analysis of phototropism has been somewhat more successful than that of phototaxis, there is still no satisfactory explanation for the molecular mechanisms involved in either process. Growing shoots of oats or corn, and the fungus *Phycomyces*, have been most frequently studied by investigators of phototropism. More than half a century ago it was suggested that the plant hormone AUXIN (indoleacetic acid) was responsible for the bending of oat seedlings toward light. Auxin promotes growth, and the light was assumed to cause an asymmetry in its synthesis or distribution which caused bending of the seedling because of the unequal growth of cells on the two sides of the shoot. This theory has been presented in several different forms (Giese, 1964), but the role of auxin remains uncertain. Investigators of *Phycomyces* have also looked without success for a specific growth substance (possibly an enzyme) which might be localized in such a manner as to cause bending (Page, 1968). It is obvious that the light must be acting through some photosensitive pigment to trigger these growth effects. The most likely pigments are the carotenoids and the flavinoids (riboflavin); current evidence suggests that both may be involved. However, identity of the pigments, their mechanisms of action, and the precise location of the photoreceptor areas or structures are still speculative.

The carotenoids appear to be involved in most phototactic responses. There are exceptions among the photosynthetic bacteria and some algae where changes in the photosynthetic rate seem to govern the response (Clayton, 1964, 1971); however, the carotenoids are clearly involved in primitive animals such as the flagellates. Two pigments sometimes appear to be active, but these may be different carotenoids, or a carotenoid and a flavinoid. The pigments may act directly to trigger motile organs such as flagella, but the mechanisms remain to be demonstrated. The responses of higher animals which depend on special photoreceptors and the coordinated activities of nervous system and effector organs are discussed in Chapters 18 and 21.

BIOLOGICAL CLOCKS

Life is strongly rhythmical. In our own species there are rapid oscillations in the muscles of the heart and in the neurons of the breathing centers, diurnal rhythms of sleep and activity, lunar periodicity in reproduction and seasonal changes in physiology as well as in social activities (Mills, 1966; Luce, 1971). The longer oscillations measured in periods of days, months, or years are environmentally determined. Even though they sometimes persist stubbornly under constant experimental conditions, the oscillations usually change or disappear sooner or later unless reinforced or "corrected" by environmental change.

The biological advantages of activities geared in an adaptive way to changes in the surroundings are obviously great. Light is the most likely force in regulating such cycles with its persistent and constant differences from day to night and from summer to winter. Tidal oscillations, temperature cycles and the seasonal variations in the quality and quantity of food are additional but indirect forces which trace their rhythms back to the rotation of the earth and its changing exposure to the sun. The literature is now voluminous; it may be readily traced in volume 25 of the "Cold Spring Harbor Symposia on Quantitative Biology" (1960) and through many papers in "Photophysiology" (A. C. Giese, ed.).

Circadian Rhythms

Phylogenetically, the more primitive of these light-induced rhythms were probably diurnal and tied to the cycles in energy production through photosynthesis. In the existing plants, not only photosynthesis and respiration but many other physiological processes such as growth, spore discharge and flowering are markedly diurnal (Bünning, 1967).

The photosynthetic dinoflagellate, *Gonyaulax polyedra,* displays an interesting daily rhythm (Sweeney, 1960; 1969). This organism, which can be claimed by both botanists and zoologists, shows regular oscillations in cell division and luminescence as well as photosynthesis (Fig. 14.7). Its rhythm of luminescence has been carefully studied. When stimulated by agitation, *Gonyaulax* produces a brief flash of light (about 90 milliseconds duration) which is much more intense during the night. The major point of interest is that the rhythm of flash intensity is not rigidly tied to the cyclical supply of photosynthetic energy but will persist for long periods when the organisms are grown in darkness or in dim light; it gradually becomes less intense as food resources are exhausted because of the absence of photosynthesis, but the rhythm is maintained. In short, the rhythm appears to be ENDOGENOUS; there are many such rhythms in both plants and animals (Enright and Hamner, 1967; Hawking, 1970; reviews cited and Fig. 13.5). They were called CIRCADIAN by Halberg in 1959 to indicate that the period length is about (circa) one day (diem). Frequently the rhythm is short of 24 hours; it can be altered (ENTRAINED) by light exposure or certain other persistent environmental changes; it is, however, remark-

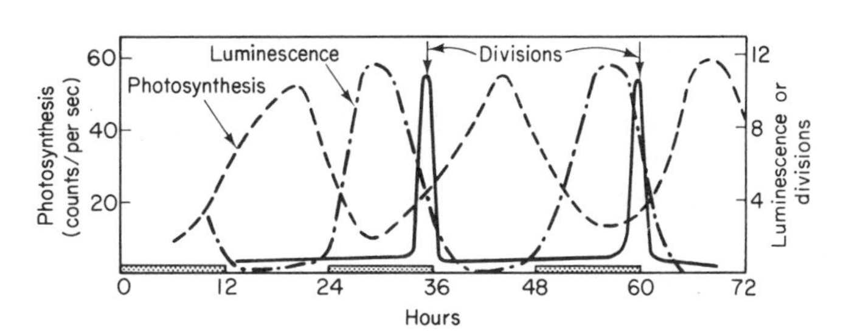

Fig. 14.7 Diurnal rhythms of luminescence, photosynthesis, and cell division in *Gonyaulax polyedra* kept under alternating light and dark periods of 12 hr each at 25°C. Dark periods are indicated by gray rectangles. Rhythms persist when cells are transferred to constant light conditions. [From Hastings, 1959: Ann. Rev. Microbiol., *13*: 299.]

ably stable to irregular temperature variation and sublethal doses of chemical inhibitors and narcotizing agents. Further, many of these rhythms are innate and require no learning, imprinting, or entertainment to initiate them (Brown *et al.,* 1970). In short, they have many of the characteristics of a chronometer or clock. The timing mechanisms are not yet fully understood. All workers recognize an endogenous mechanism, but while most argue that the "clock" will eventually be understood in terms of a cellular timing device, F. A. Brown, Jr. and his associates maintain that subtle "geophysical cues" provide additional information necessary to regulate the precise timing.

Lunar Periodicity

As the moon travels around the earth on a cycle of about 29½ days, it reflects the maximum amount of solar illumination to the earth at the time of the "full moon"; its "size" progressively increases before the "full moon" while it gradually decreases from night to night until it "disappears" thereafter. This lunar cycle is also of potential significance in the evolution of cyclical natural phenomena. Many half-lunar or full-lunar cycles have been recorded (Korringa, 1957), as well as physiological events which occur only once or a few times during the year but always in association with some particular phase of the moon (Korringa, 1957; Bünning, 1967; Sweeney, 1969; Brown *et al.,* 1970). One of the classical examples is the Palolo worm (*Leodice*) which is a tube-dweller in coral reefs. The West Indian species, *L. fucata,* spawns only during the third quarter of the June-July moon. The anterior part of the animal remains in its burrow beneath the sea, but the posterior ends, distended with ripe gametes, break off and wriggle to the surface in such vast numbers that the waters are milky for an hour or two with the eggs and sperms. This swarming occurs at dawn and thus is precisely timed both to sun and moon. Another spectacular example is the beach spawning of the Pacific grunion, *Leuresthes tenuis,* which swims up on the beaches, spawns, and then flips back into the water just after the turn of the tide on the second, third, and fourth nights after the full moon in the months of March, April, May and June. Precision of this sort greatly increases the chances of successful reproduction and may also synchronize the hatching of larvae with suitable plankton blooms.

The type of lunar rhythm which occurs throughout the year in association with the tidal cycle has also been carefully documented. F. A. Brown (1957) analyzed the daily changes in chromatophore appearance and oxygen consumption of the fiddler crab, *Uca,* and found a tidal as well as a diurnal cycle. Both are endogenous and persist for long periods under constant light conditions; they also show a measure of temperature independence. The tidal cycle could be altered by appropriate manipulation of the illumination, thus indicating that the light from the moon played a direct part in its establishment. This could, of course, be reinforced by the changing tidal amplitude, but the principle remains that the moon may serve as a potential cue for the establishment of biologically useful rhythms.

Photoperiodism (Circannual Rhythms)

The success of a species often depends on a life cycle which is precisely timed to take full advantage of the changing seasonal conditions. In temperate and frigid regions, young animals are most likely to survive if their birth coincides with the onset of warm weather and unlimited supplies of food; in the tropics, the rainy season may be equally important for both aquatic and terrestrial life. Animals may prepare for the rigors of winter by storing fat, by hibernating or by migrating; diadromous migrations are often timed to take advantage of the flooding rivers in the spring or autumn. The examples could be multiplied. Plants as well as animals show seasonal rhythms which maintain advantageous relations between the major physiological demands and the seasonal environmental cycles. They are examples of a different type of physiological chronometer. In this case, the mechanism has its parallel in the interval marker or the hourglass rather than our familiar timepiece; something starts the mechanism and it terminates after a definite interval.

The length of the day (PHOTOPERIOD) is the most dependable cue for this seasonal timing; it most frequently SETS the biological clock. At any point on the earth's surface, the length of the day depends only on the latitude and the time of the year. Temperature, annual rainfall, and the availability of food also show orderly seasonal changes, but these tend to fluctuate more from year to year than the photoperiod. In many animals, seasonal physiological rhythms have apparently become independent of the environmental controls and are endogenously regulated; they persist under constant conditions or may depend on the environment only for the "correction" of the endogenous clock. This type of rhythm has been mentioned in the hibernating ground squirrel (Chapter 10 and Pengelley and Asmundson, 1971).

Photoperiodism was first recognized about 1920 as an environmental regulator of flower development in many plants. It was shown experimentally that certain plants failed to flower unless the days were of the "right" length; in some cases flowering was induced by short days, and in other cases long days provided the effective stimulus. The extensive studies on animal photoperiodism were initiated by Rowan's studies of bird migration in 1925 and Kogure's experiments with silkworms in 1933. The voluminous literature is reviewed in several places (Withrow, 1959; Beck, 1963; Lees, 1968; Lofts, 1970; Farner and Lewis, 1971), and there are popular accounts of the classical experiments mentioned above (Beck, 1960; Butler and Downes, 1960).

What is the nature of this clock which "predicts" the oncoming seasons? Since it appears to be based on the changing lengths of days and nights, there are really two questions: (1) What is the photosensitive pigment which detects the light change? (2) What is the photoperiodic timing mechanism on which the light (photochemical reaction) operates?

Unlike photoreceptors which are all based on the carotenoids, photoperiodism is regulated by colored molecules that are biochemically different in plants and in animals. In 1959, plant physiologists isolated a pigment, PHYTOCHROME, which is the active transducer of the light effects in plants: this is a chromoprotein whose prosthetic group is a phycobilin, probably an open tetrapyrrole related to the bile pigments (Briggs and Rice, 1972; Shropshire, 1972). It exists in two forms, an inactive form (P_{660}) with an absorption maximum at 660 nm and an active form (P_{735}) with a maximum at 735 nm. During the photoperiod response the pigment is changed from the inactive to the active form, and this then stimulates the growth phenomena. The reaction is reversible, with P_{735} changing to P_{660} in darkness at a rate and to a degree which depends primarily on the genetics of the plant; thus, there are "short-day" plants, such as chrysanthemums, which flower in the autumn and "long-day" plants, such as petunias, which require long summer days and short nights to maintain the necessary supplies of P_{735}. Not only flowering, but fruiting and many growth processes are regulated in this manner. The molecular mechanisms by which phytochrome operates are still uncertain; research has centered on gene action, enzyme reactions, and membrane permeability.

Animal photoperiodism is more complex. It has not yet been possible to identify the photosensitive pigment with certainty, and indeed there may be more than one. More significantly, the regulated processes in animals are not controlled by single events like permeability or an enzyme reaction—as they appear to be in plants—but through a complex hormonal sequence involving the hypothalamic-pituitary pathway and one or more target organs and tissues. Finally, the seasonally triggered events appear to be considerably more varied and include, in addition to diversified growth processes, such phenomena as coat color in the snowshoe hare, salinity preference and temperature resistance in certain fishes, and the behavior associated with migration and reproduction in several different groups of animals.

Some of the first analyses of photoperiodism were carried out on the ferret, an animal whose reproductive cycle is strongly dependent on the seasonal changes in day-length (Hammond, 1954). The essential links in hormonal regulation were established by removal of the eyes, the pituitary, or the gonads. Removal of any one of these three organs eliminates the response. The photoreceptors were logically assumed to be in the retina and the photosensitive pigment a carotenoid. This conclusion has stood the test of time for the ferret but subsequent research has shown that the extension of it to certain other groups of animals is unwarranted. First in insects and then in birds, photoperiodism has been shown to be independent of the eyes. In some of the lower vertebrates the pineal may be involved, but in several birds and arthropods the photosensitive receptors have been shown to be confined to the brain. The experimental evidence for non-visual light reception is now overwhelming (Lees, 1960, 1968; Farner and Lewis, 1971; Menaker, 1972) and will not be repeated here. The precise location of the receptor cells within the brain has not yet been determined and neither has the identity of the photosensitive pigment. The pigment, however, has been shown to have a broad absorption spectrum which varies in different animals but spans the visual range and extends beyond it in certain species (farther into the blue in insects and into the red in some birds). Although the carotenoids appear to be the most likely candidates, it is quite possible that more than one pigment may be involved in animals as appears to be the case in plants.

This now brings us to the second major question raised at the beginning of this section; what is the mechanism for measuring time? Two different theories receive attention in the current literature. The first of these is the HOURGLASS HYPOTHESIS. This assumes that some critical physiological process (possibly the elaboration of a hormone) requires a specific length of time. The theory has not been very precisely stated; it finds its best support in certain insects where the length of the dark period provides the regulatory cue. Thus, in the aphid *Megoura viciae* the short days of winter induce the development of egg-laying females while the long days of summer induce formation of parthenogenetic females. In this case, it is the duration of the dark period (long night) which determines the change; provided that the darkness lasts more than 9.5 hr the change to egg laying occurs even though the light periods may be extended to over 30 hr (Lees, 1968; Lofts, 1970). This suggests that some process requires 9.5 hr or more and that it is not the light-dark oscillation which matters.

The second hypothesis, termed the BÜNNING MODEL, assumes that the circadian periodicity serves as a chronometer to "count" the time. This model was proposed many years ago for plant photoperiodism (Bünning, 1967 and earlier); it was later extended to several groups of animals and has been stated in various modified forms (Pittendrigh and Minis, 1964 and reviews cited). It can be most readily visualized by postulating a basic physiological rhythm of almost 24 hr, made up of altering "dark-requiring" (scotophil) and "light-requiring" (photophil) periods of equal length. If the photoperiod response is normally triggered by increasing day length (spring and early summer), then light interrupts the dark period (night) earlier and earlier and the animal interprets this as the onset of summer conditions (longer days); an opposite situation develops in the autumn with night lasting longer and longer and prolonging the onset of subjective day. This theory has now been tested experimentally with many different animals. Fig. 14.8 shows the response of Japanese quail, subjected to 6 hr light periods followed by dark periods of varying lengths from 6 to 66 hr. A sharp increase in testicular size occurs only when the light comes in the

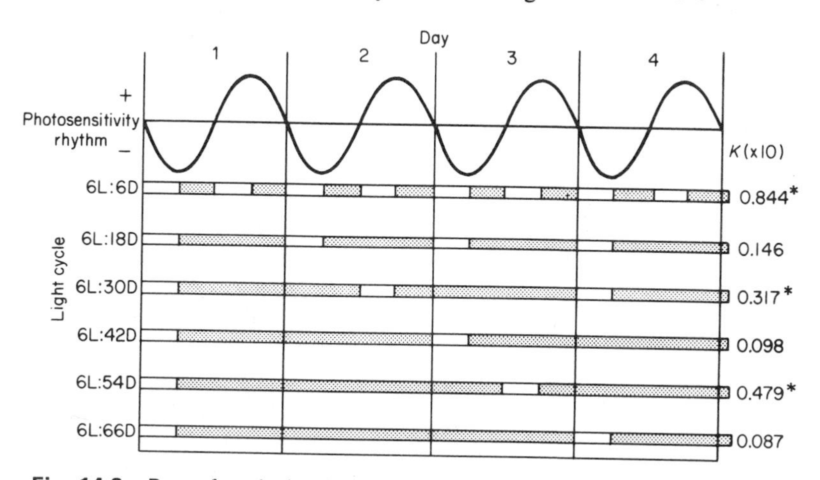

Fig. 14.8 Rate of testicular development (K) in groups of Japanese quail maintained under six different experimental light regimes. Asterisks, significant stimulation of gonadal growth. [From Lofts, Follett, and Murton, 1970: Mem. Soc. Endocrinol., *18*: 554.]

period of normally increasing photosensitivity (night); in other words, it is not the quantity of light that is important but when it occurs in relation to the PHOTOINDUCIBLE PHASE.

This is emphasized in another kind of experiment in which brief light flashes during the dark period initiate the photoperiodic response. The data illustrated (Fig. 14.9) are also for the Japanese quail where the flash was only 15 min during the darkness of a 6 hr light 18 hr dark experimental regime; the only light pulses which were effective fell within the period of 11.5 and 16 hr after the onset of the main photoperiod; in other words, the photoinducible phase has a duration of about 4 hr. Comparable data are available for several other animals (Farner and Lewis, 1971; Baggerman, 1972).

Many annoying problems have been glossed over in this brief summary. The examples, however, illustrate certain basic points. For one thing, it appears that the fundamental circadian oscillations, which appear to be a part of life at all levels in phylogeny, have been used to measure time on a seasonal basis; it should be emphasized once more that many of these circannual rhythms are endogenous but require a chronometer to adjust the time accurately. The second important point is that some animals use the dark period to measure time, while the majority use the light period. In theory, at any rate, there is no reason why any one of the seasonally changing components of daily light duration (dawn, dusk, lengths or relative lengths of night and day) might not serve as a reliable cue for the prediction of seasonal changes. A measure of expediency seems common in the workings of organic evolution; future experiments may be expected to show that different animals have selected the component of the light cycle which best fits their physiology, behavior, and habitat.

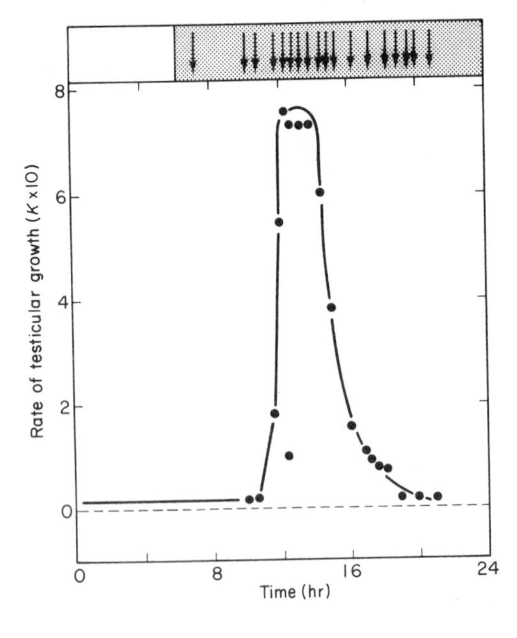

Fig. 14.9 Effect of brief daily pulses of light on testicular growth of Japanese quail. All groups of animals maintained under daily photoperiods of 6 hr light followed by 18 hr of darkness; pulses of light (15 min duration) were given at times indicated by the series of arrows; note that there are 18 experiments and the birds in any experiment always received the pulse at, and only at, the time indicated. [Based on Follett and Sharp, 1969: Nature, *223;* 968.]

THREE

NERVOUS INTEGRATION AND ANIMAL ACTIVITY

15

IRRITABILITY

Animal life usually depends on a capacity to move about and encounter food, to respond in a purposive way to the food and to react in an adaptive manner to a varied habitat as well as to the different living organisms encountered in the search for food. Activity is one of the most obviously distinctive features which separate the animals from the plants. It has been a dominant force in phylogeny and the basis for many of the refinements in neurosensory structures and in the feeding and digesting machinery. This capacity for the active and effective exploitation of the varied resources of the environment is expressed in many different ways but stems from a single fundamental property of all living cells—their IRRITABILITY or EXCITABILITY.

The concept of irritability as a general attribute of living matter can be traced back to Francis Glisson in the seventeenth century (Heilbrunn, 1952). Verworn (1899 and earlier) in his classical textbooks of general physiology defined it clearly in a modern form and discussed it at length as one of the distinctive properties of the living in contrast to lifeless substances. According to Verworn's definition, the irritability of protoplasm is its "capacity of reacting to changes in its environment by changes in the equilibrium of its matter and its energy." Claude Bernard also discussed irritability as a universal property of protoplasm, stressing the excitation or activity which follows environmental change (Bayliss, 1960). It is interesting that Verworn's definition covers inhibition as well as excitation. He emphasizes in several places that the significant feature of irritability is the CHANGE IN VITAL PHENOMENON—sometimes excitation, sometimes depression or inhibition. The effective environmental change is the STIMULUS; the resulting protoplasmic reaction is the RESPONSE. That irritability need not necessarily be expressed in activity but that inhibition might be the normal response to stimulation has been recognized for more than 100 years; the brothers Weber demonstrated cardiac inhibition following stimulation of the vagus nerve of the frog in 1845 (Bayliss, 1960).

Pioneer physiologists also recognized CONDUCTION (the progressive spread of excitation) as a fundamental property of living cells and one which is inseparably connected with irritability. Although the most familiar examples of irritability and conduction are found in nervous and muscular tissues, these properties can be readily demonstrated in many other types of cells, both plant and animal. Verworn (1913) and Heilbrunn (1952) cite many examples of protoplasmic conduction in the rhizopod protozoa, in elongated algal cells like *Nitella,* and in higher plants such as *Elodea.*

MOLECULAR BASIS OF CELLULAR IRRITABILITY

At the molecular or ionic level, cellular irritability is related to minute differences in electrical potential across the surface membrane of every cell. In the so-called "resting cell," this transmembrane potential is of the order of 20 to 100 millivolts (mv), with the inside of the membrane negative to the outside. The classical materials for the demonstration of these bioelectric potentials were the giant axon of the squid, with a diameter of about 1 mm, and large algal cells such as *Halicystis.* The RESTING POTENTIAL in the former is about 50 to 75 mv and in the latter 70 to 80 mv. Today it is possible to explore some of the smaller cells with microelectrodes and highly sensitive recorders. The principle is the same, however: when one electrode is placed inside the cell and the other on its surface, a steady transmembrane potential is evident and the inside of the membrane is always negative with respect to the outside. Moreover, any stimulus that evokes a response in the cell is associated with a change in this membrane potential (Fig. 15.1); it is this potential change which is the most fundamental cellular attribute of irritability and

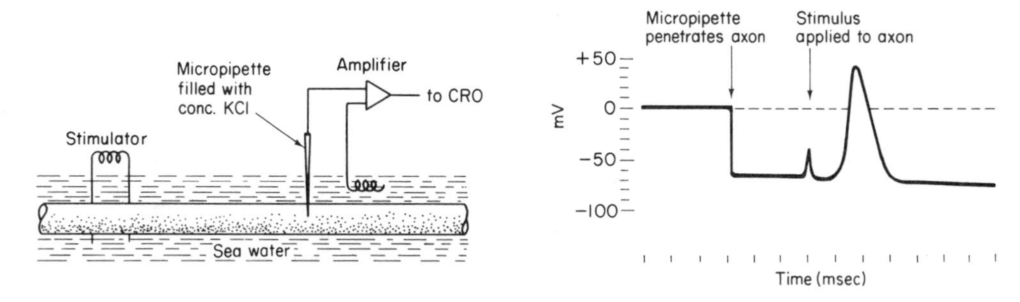

Fig. 15.1 Intracellular recording from a squid giant axon. As the electrode (micropipette of diameter less than 1 micron) penetrates the axon membrane, a shift in the baseline of the beam on the cathode-ray oscillograph (CRO) indicates a membrane potential of about −70 mv. Electrical stimulation generates an action potential during which the membrane potential is briefly reversed. [From Miles, 1969: Excitable Cells. Heinemann, London. Based on Hodgkin, 1958: Proc. Roy. Soc. London, *B148*: 1.]

the one on which the excitability of tissues and the phylogeny of animal activity has been founded. The living cell is, in reality, a small voltaic battery of potential 20 to 100 mv. When the terminals of the battery (the inside and outside of the plasma membrane) are connected through a suitable circuit, some of the energy is discharged and may produce a signal.

The Ionic Basis of the Resting Potential

As long as a cell is alive, its surface membrane separates solutions of different chemical composition. The osmolarity of the external solution equals that of the internal solution (except for the surface cells of freshwater and some marine organisms), but the ionic species are very different; it is this difference which produces the resting membrane potential. In a vertebrate, for example, more than 90 per cent of the osmotic content of the extracellular medium is made up of sodium and chloride ions while these ions account for less than 10 per cent of the solutes inside the cell. In the intracellular fluid, potassium takes the place of sodium, and the organic anions produced during metabolism (aspartate, glutamate, phosphate esters) replace most of the chloride to produce *an electrically balanced solution.*

The latter point should be emphasized: the bioelectric potentials under discussion are membrane potentials. At the plasma membrane, where the internal and external fluids would tend to mix by diffusion, there are localized accumulations of positive and negative charges separated by this bimolecular layer. This is a purely localized phenomenon; in most cells ionic diffusion fronts of K^+ press toward the outside while Na^+ presses toward the inside (Fig. 15.2). Exchanges of ions between the cell and its surroundings are restricted but not entirely prevented by the permeability properties of the cell membrane; the balance is maintained by cellular processes of active ionic transport (ion exchange pump). Thus, electrical neutrality is preserved both in the extra- and the intracellular fluids, but the diffusion fronts result in sufficient separation to create the transmembrane bioelectric potential.

Physiological evidence, based primarily on studies of squid axons, vertebrate

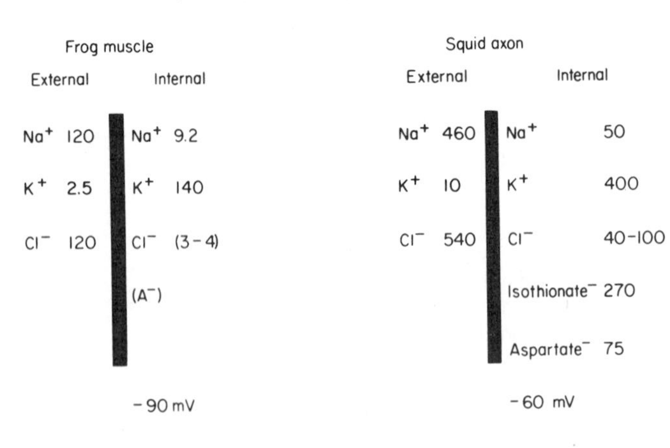

Fig. 15.2 Electrolyte concentrations (mM/liter) and potential differences across cell membranes. Note that the potassium ion is 40 to 50 times more concentrated inside the membrane while the sodium ion is 10 to 15 times more concentrated outside. [Values from Katz, 1966: Nerve, Muscle and Synapse. McGraw-Hill, New York.]

muscles and certain algal cells, points to the K^+ as the principal source of the "resting" potential. For more than half a century it has been realized that cell membranes behave as though they are freely permeable to K^+ but relatively impermeable to Na^+. Hodgkin and Katz (1949) estimate that the resting membrane of the squid axon is 25 times more permeable to the K^+ than to Na^+. Because of this difference in free movement of cations through the cell membrane, the "passive" diffusion of potassium toward the outside will be much greater than the "passive" diffusion of sodium toward the inside (Fig. 15.2). Thus, there is a net loss of cations (K ions) that charges up the membrane so that the inside is negative with respect to the outside. It should be emphasized once more that these outwardly diffusing K^+ do not drift away from the membrane but are held at the membrane by equal numbers of negatively charged particles such as Cl^- to create the transmembrane potential. In support of this hypothesis, it has been found that increasing the concentration of potassium in the extracellular fluid (hence reducing the K^+ gradient between intra- and extracellular fluids) reduces the resting potential in a directly dependent way (Fig. 15.3). In comparable experiments, the resting potential has been found to be independent of the external Na^+ concentration (Katz, 1966). In such experiments, many cell membranes behave as though impermeable to all cations other than potassium, and calculations show that the resting potential is approximately what might be expected from the potassium concentrations (Bayliss, 1960). Textbooks of cellular physiology and biophysics discuss the theoretical background and quantitative details (Cole, 1968; Tasaki, 1968; Davson, 1970; Giese, 1973). Beginners will find the brief monographs by Katz (1966) and Miles (1969) particularly helpful.

Based on these considerations, Bernstein (1902) developed a hypothesis for nervous excitation which was, for many years, the cornerstone of physiological thinking on cellular irritability. He postulated a membrane which was selectively permeable to potassium ions and impermeable to sodium, chloride and the cellular organic anions. Excitation, according to Bernstein, was caused by a sharp increase

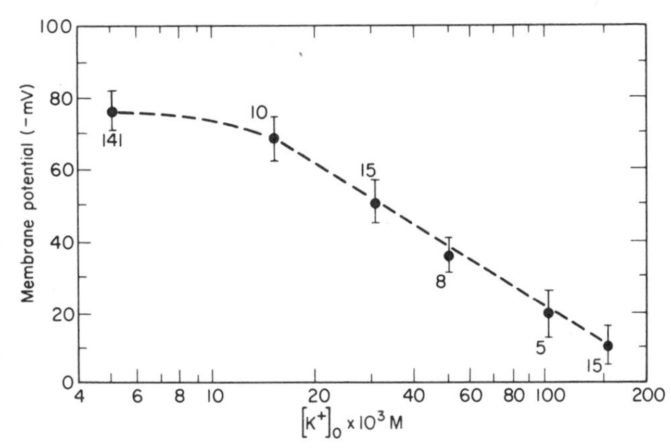

Fig. 15.3 Effect of external K^+ on resting membrane potential of the isolated electroplax of *Electrophorus* (Fig. 20.1). Steady potential across the innervated membrane as a function of potassium ion concentration $[K^+]_o$ in the solution facing the innervated membrane; solution facing the non-innervated membrane was unmodified Ringer's. Temperature, 23°C; vertical bars, one standard deviation on either side of the mean for numbers of observations indicated. [From Higman, Podleski, and Bartels, 1964: Biochem. Biophys. Acta, *79*: 140.]

in permeability to the extracellular ions so that the membrane was rapidly depolarized and the potential dropped toward zero. This concept served as a workable hypothesis for about fifty years until modern electronic recording instruments and radioactive tracer techniques demonstrated two major deficiencies. First, the membrane is usually not simply depolarized during excitation; the potential is often substantially reversed (40 to 50 mv with the inside positive in nerve) and, moreover, there are many cases where membranes are actually hyperpolarized on stimulation. Second, the use of radioactive isotopes has clearly shown that the membrane is permeable to sodium and chloride as well as to potassium.

Consequently, it has been necessary to modify Bernstein's hypothesis; the present theories place a "cation pump" in the membrane to do the job which Bernstein attributed merely to membrane properties of selective permeability. Although Bernstein's detailed postulates are no longer tenable, his generalized concepts of the importance of the plasma membrane, its selectivity and variable permeability, remain basic to all modern theories (Katz, 1959, 1966). The membrane is considered to be differentially permeable, with potassium entering much more readily than sodium; the slow leakage of sodium into the cell is controlled by the "sodium pump."

There are several lines of evidence for active ion transport mechanisms in the maintenance of the membrane potential. Resting potentials in *Halicystis* remain

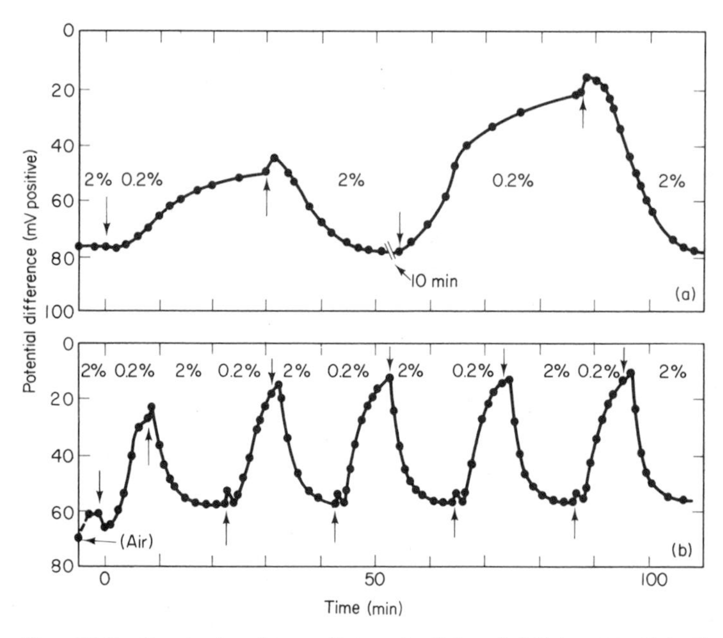

Fig. 15.4 Graph showing resting potential and fall in potential of impaled cells of *Halicystis* when 0.2 per cent O_2 is substituted for 2 per cent O_2 (in nitrogen) and recovery when 2 per cent O_2 is again bubbled. [From Blinks, 1955: *In* Electrochemistry in Biology and Medicine, Shedlovsky, ed. Wiley, New York.]

steady for hours but fall off sharply if the environmental oxygen is reduced; recovery quickly follows restoration of the oxygen supply (Fig. 15.4). The resting potentials are clearly oxygen-dependent. Similar evidence has been obtained by the use of metabolic inhibitors. When squid axons are treated with dinitrophenol, azide or cyanide, the active efflux of tracer sodium ceases but not its leakage into the cell; significantly also, the "pumping" is temporarily resumed when ATP or arginine phosphate is injected into the axoplasm of the cyanide-inhibited squid fiber (Katz, 1959). Thus, it is evident that the pyrophosphate bond supplies the energy for the cation pump and that the "resting" cell is not really resting but continually expending energy to maintain a balance. In fact, cellular metabolism is involved in two ways: (1) in the synthesis of the organic anions which cannot diffuse through the plasma membrane and (2) in operating the ion exchange mechanism. A substantial portion of the energy budget of the cell is required for the operation of the ion pumps (Keynes and Maisel, 1954).

The selectivity of the plasma membrane remains a significant part of the Bernstein hypothesis, both in its impermeability to the organic anions and in its differential permeability with respect to sodium and potassium. Sodium leaks through very slowly, potassium much more rapidly so that the potential difference across the resting membrane approximates that of a potassium-concentration cell. Moreover, potassium seems to be actively transported into the cell as well as diffusing in "passively." The cation pump evidently accumulates potassium within the cell at the same time that it eliminates sodium (Fig. 15.5). Some of the most convincing evidence comes from experiments with squid axons which show (1) that metabolic poisons which *inhibit* the efflux of sodium *reduce* the inward movement of potassium and (2) that the rate of tracer sodium efflux declines when potassium is withdrawn from the bath solution. Both observations support the concept of a coupled transport mechanism in which the rates of sodium efflux and potassium influx change simultaneously and in a parallel manner. This ion exchange pump is evidently located in the cell membrane; the axoplasm of the squid neuron can be almost completely replaced by isotonic solutions without destroying the pump, provided that the cell membrane is not damaged (Baker *et al.*, 1962; Baker, 1966). The molecular mechanisms responsible for the operation of the pump remain obscure (Katz, 1966) although it is assumed that the two ions share common carrier molecules.

Phylogeny in Protoplasmic Irritability

It can scarcely be argued that the membrane potential evolved to meet the demands for animal activity. It must certainly have been one of those pre-adaptations which served as a springboard for protoplasmic specializations ranging from simple cellular movements to the integrated and complex behavior of modern man. A seemingly logical argument for this delicately balanced condition of ions at the plasma membrane has already been presented in a discussion of ionic regulation (Chapter 11). The hypothesis enunciated was that VOLUME REGULATION formed

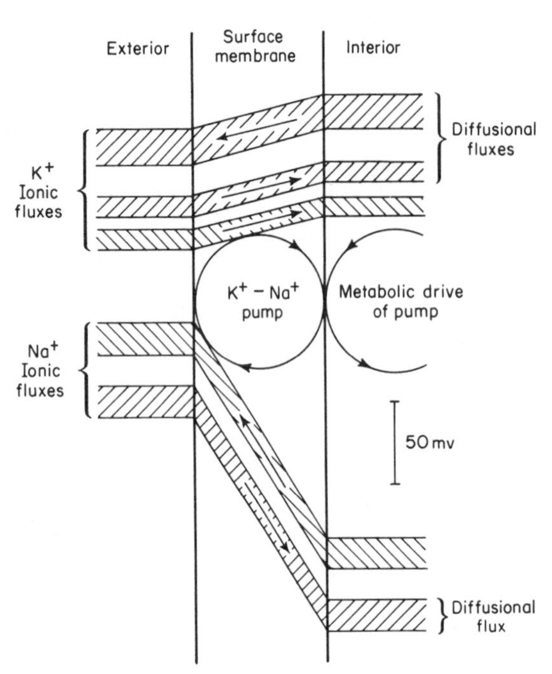

Fig. 15.5 Diagrammatic representation of K^+ and Na^+ fluxes through the surface membrane in the resting state. The slopes in the flux channels across the membrane represent the respective electrochemical gradients. At the resting membrane potential (−70 mv) the electrochemical gradients, as drawn for the K^+ and Na^+ ions, correspond respectively to potentials which are 20 mv more positive and about 130 mv more negative than the equilibrium potentials (note the potential scale). The fluxes caused by diffusion and the operation of the pump are distinguished by the direction of hatching. The outward diffusional flux of Na^+ ions would be less than 1 per cent of the inward and so is too insignificant to be indicated as a separate channel in this diagram, because the magnitudes of the fluxes are indicated by the widths of the respective channels. [From Eccles, 1957: The Physiology of Nerve Cells. The Johns Hopkins Press, Baltimore.]

the first step in the evolution of the ion pumps characteristic of living cells (Brown and Stein, 1960; Grundfest, 1966).

The organic anions (aspartate, phosphate esters) produced during the metabolism of the cell are essential byproducts. They must be retained since they are links in the production of energy. Cell membranes are impermeable to them and their presence creates a net excess of osmotically active material within the cells which could result in osmotic flooding. A primitive cell in its marine environment might maintain its volume in the face of osmotic flooding by pumping out the water or by pumping out some of the diffusible ions. The cation pump is the most economical measure since membranes, in general, are far more permeable to water than to ions. Thus, it is suggested that volume regulation was the first evolutionary pressure for the development of cation pumps and the creation of unequal distributions of ion species on the two sides of cell membranes (Table 15.1). Plant cells, under similar conditions, avoid the hazards of volume increase by living in rigid cellulose boxes. There will still, however, be an unequal distribution of ion species between the extra- and intracellular fluids because of the presence of the organic anions which impose a Donnan distribution on the diffusible ions. This distribution, known as the DONNAN EQUILIBRIUM, gives the concentrations of ions on either side of a membrane permeable to some but not all of the ions. For example, a membrane

Table 15.1 Ionic Gradients Across Cell Membranes Expressed as Ratio of Intracellular to Extracellular Concentrations for the Cations and the Reciprocal of this for Cl^-. External fluids are plasma except for sea urchin egg and squid nerve which were in sea water.

Organism	Cell	Na^+	K^+	Ca^{++}	Mg^{++}	Cl^-
Sea urchin—*Paracentrotus*	Egg:					
	Unfertilized	0.11	21	0.36	0.2	7
Squid—*Loligo*	Nerve:					
	Resting	0.09	31			7.9
	Stimulated	0.3	23			6.4
Crab—*Carcinus*	Muscle	0.09	9.2		0.7	9.9
Frog	Muscle	0.03	58	2.5	12	50
Turtle	Erythrocyte	0.15	13.6			
Rat	Muscle	0.02	24	0.61	11	24
	Erythrocyte	0.18	23			1.44

SOURCE: Values selected from Brown and Stein (1960).

bag could be prepared that would be freely permeable to Na^+ and Cl^- but impermeable to proteinate (Pr^-). If such a bag were partially filled with a sodium proteinate solution and immersed in a solution of sodium chloride, Donnan's relationship states that at equilibrium, the product of the concentrations of the diffusible ions inside (i) will equal the product of the concentrations of the diffusible ions on the outside (o). Thus, $[Na^+]_i \times [Cl^-]_i = [Na^+]_o \times [Cl^-]_o$. Electrical neutrality, however, can only be preserved within the bag or in the external solution if the sum of the cations remains equal to the sum of the anions. In short, $[Na^+]_i + [Pr^+] = [Cl^-]_i$ while $[Na^+]_o = [Cl^-]_o$. This relationship could, for example, be satisfied by inside (2×8) and outside (4×4) ratios for the diffusible ions. Donnan's theory is supported by both theoretical and experimental studies; its importance in understanding ion distributions in biological systems is obvious (Donnan, 1927). More complete statements of the theory with discussions of the relevant physical chemistry will be found in many recent texts (Davson, 1970; Giese, 1973; Harper, 1973).

If the above arguments are sound, the organization of the very first cell produced a transmembrane potential due to an unequal distribution of ion species; the steady state or resting potential depended on the properties of the cell membrane and the presence of the cation pump. This has remained a universal attribute of living cells at all stages in their phylogeny. In addition, it can be assumed that the most primitive cell encountered a variety of alterations in its environment (stimuli) which were of sufficient magnitude to disturb this delicate balance of ions and change the value of the resting potential, in other words, to cause ELECTROGENESIS. This constitutes a stimulus-response situation and, at the risk of some oversimplification, the capacity for electrogenesis may also be listed with the universal attributes of cells.

The discharge of electricity as a terminal physiological activity occurs only in the electric organs of some fishes. Usually, cellular electrogenesis is coupled with some other process to effect such characteristic responses as locomotion or color change. It follows that the primitive response of electrogenesis required the evolution of a battery of specialized effector organs to achieve the complexities of animal physiology and behavior. Likewise, at the point of the stimulus, the evolutionary process has produced delicate transducers with specialized sensitivity for particular categories of environmental change; electrogenesis is the fundamental response, but it may be triggered in many different ways and it, in turn, may trigger several different kinds of activities. Both receptors and effectors have been built onto the primitive electrogenic activity of cells.

PHYSIOLOGICAL PROPERTIES OF EXCITABLE TISSUES

Electrogenesis

The excitation of receptor organs, transmitting cells and effector tissues can be most satisfactorily explained in terms of the ionic permeability theory of electrogenesis. This has been stated in a modern form by Grundfest (1961, 1966), Katz (1966) and others; only a brief summary is given here. Comprehensive discussions may be found in textbooks of neurophysiology.

The ionic permeability theory of electrogenesis can be most easily visualized by reference to Eccles' (1957) diagram of the cell membrane shown in Fig. 15.5, with the addition of channels for the flux of chloride as well as Na^+ and K^+. The cation pump functions in the maintenance of stability and in the recovery processes which follow excitation; it may be neglected for the moment. The membrane itself appears to be a barrier perforated with pores which are capable of passing specific ions. Many of the facts fit a concept of "valved" openings (gates or channels) which operate on an all-or-none principle to admit or exclude specific ions. In familiar preparations like the squid axon or frog muscle, three types of valves have been distinguished (Na^+, K^+ and Cl^-), but in some other tissues there are also Ca^{++} and/or Mg^{++} valves (Hagiwara and Naka, 1964). Excitation, according to this hypothesis, involves a change in the number of open valves rather than an increased flow of ions through channels which remain continuously open. In some membranes, the valves are operated only electrically (ELECTRICALLY EXCITABLE VALVES), while in others specific chemicals or a mechanical stretching of the membrane provides the stimulus (ELECTRICALLY INEXCITABLE VALVES); frequently both types of valves occur in the same membrane.

Nerve axons (particularly squid giant axons) and muscle fibers have most often been used to study the electrical properties of cell membranes. These cells are relatively large, readily available, easily excited and of a suitable shape for stimulation and recording. For experimental convenience, electrical currents are almost always used since they can be precisely varied in duration, intensity and rate of rise to peak value. Although the following discussion is based on nerve axons and muscle fibers investigated with electric stimuli, the principles described apply to other tissues excited by different stimuli. Examples of electrogenesis triggered by mechanical and chemical stimuli will be considered in subsequent sections devoted to receptor organs.

The plasma membrane of a resting cell is said to be POLARIZED. It displays a steady potential; a front of positive charges pressing toward the external fluid is held at the boundary membrane by the negative charges just inside the cell. Passage of an electric current through the membrane (stimulus) will inevitably alter the distribution of these charges. When the redistribution of ions increases the potential difference (i.e., the current flows inward and raises the charge separation), the membrane is said to be HYPERPOLARIZED; on the contrary, when a reduction in membrane potential occurs (i.e., current flows outward), the membrane is DEPOLARIZED (Fig. 15.6).

With a squid axon, a family of curves similar to those depicted in Fig. 15.6 can be obtained by applying brief stimuli of fixed duration but increasing strength and different polarity. The three hyperpolarizing stimuli (membrane potential changing from a resting value of −80 mv to −120 mv) produce a series of potential changes that speedily return to the resting value. These potential differences are LOCAL (confined to the area stimulated) and GRADED in relation to the stimulus strength. Likewise, the first two or three depolarizing stimuli produce similar potential differences in the opposite direction. With stronger stimuli, however, recovery is slower and the degree of depolarization ceases to be directly proportional to the increase in stimulus. At a critical level (−40 mv in Fig. 15.6), the membrane is suddenly depolarized completely and the potential momentarily reversed. At this THRESHOLD

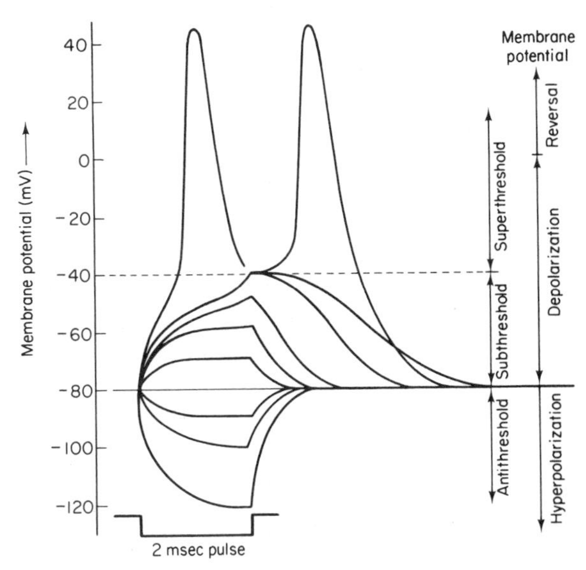

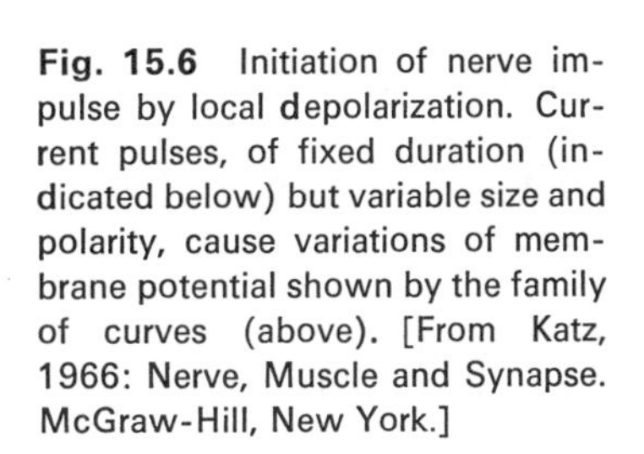

Fig. 15.6 Initiation of nerve impulse by local depolarization. Current pulses, of fixed duration (indicated below) but variable size and polarity, cause variations of membrane potential shown by the family of curves (above). [From Katz, 1966: Nerve, Muscle and Synapse. McGraw-Hill, New York.]

or FIRING LEVEL the graded responses suddenly disappear and the stimulus triggers a SPIKE DISCHARGE (POTENTIAL)—usually referred to simply as the SPIKE.

Although these events are characteristic of the excitation process in many tissues, the specific patterns of the potentials vary in different cells. Spikes may terminate in long-lasting, low amplitude after potentials; in some cells these are depolarizing; in others they are hyperpolarizing; in still others they are absent. Although spikes usually take the form of brief pulses, they are sometimes stretched out to form a plateau. Many variations have been recorded (Grundfest, 1966). They are manifestations of the specialized functions of the cells and depend on differences in the kinetics of electrogenesis. Spikes, graded potentials and the various patterns of electrogenesis are adapted to different aspects of the stimulus-response machinery. Some examples will be given in chapters devoted to receptor and effector organs. Considerations at this point are confined to comparisons of the major types of response and the ionic changes associated with them.

Spike potentials Sodium ions have been found to play the key role in initiating the spike potential of squid axons and vertebrate muscle fibers. The importance of sodium in muscle irritability has been recognized for more than half a century; Overton (1902) reported that the contractility of frog muscle which was suppressed in sucrose solution could be restored by the addition of sodium. Modern microanalytical methods applied to single cells show that the briskly rising spike is associated with a rush of Na^+ into the cell. In resting cells, Na^+ valves appear to be mainly closed while the K^+ and Cl^- valves are open only in sufficient number to balance the electrochemical forces and establish the resting potential. An electric stimulus opens the sodium gates and permits rapid influx of Na^+ which decreases and then reverses the transmembrane potential; the sodium valves are only momentarily open (0.3 msec in squid axon); as they close, there is a rapid loss of K^+ associated with the opening of the K-valves (Fig. 15.7a). When the K-efflux exceeds the Na-influx, the membrane again becomes polarized. For a few milliseconds, while the gates are returning to their resting positions, the high potassium conductance may hyperpolarize the membrane. The ion exchange pump, which restores the normal balance of cations, comes into operation only after these events. Thus, it is believed that the spike potential depends entirely on the passive Na^+ and K^+ fluxes associated with the permeability changes and that the pump operates during the recovery period.

Katz (1966) stresses the small magnitude of these ion fluxes. During a single nerve impulse, the squid axon acquires Na^+ and loses K^+ to the amount of approximately 3 to 4 $\times$ 10^{-12} mole/cm^2 of cell surface; this amounts to a decline of about 1/1,000,000 of the cell potassium. Even in the absence of the cation pump, the squid axon transmits thousands of impulses without exhausting its store of ions (Hodgkin, 1964). Under normal conditions, a brief acceleration of the cation pump maintains the axon in the fully charged state. Many tissues are capable of brief bursts of high level activity followed by longer periods of recovery.

Several lines of experimental evidence support these hypotheses. Overton's

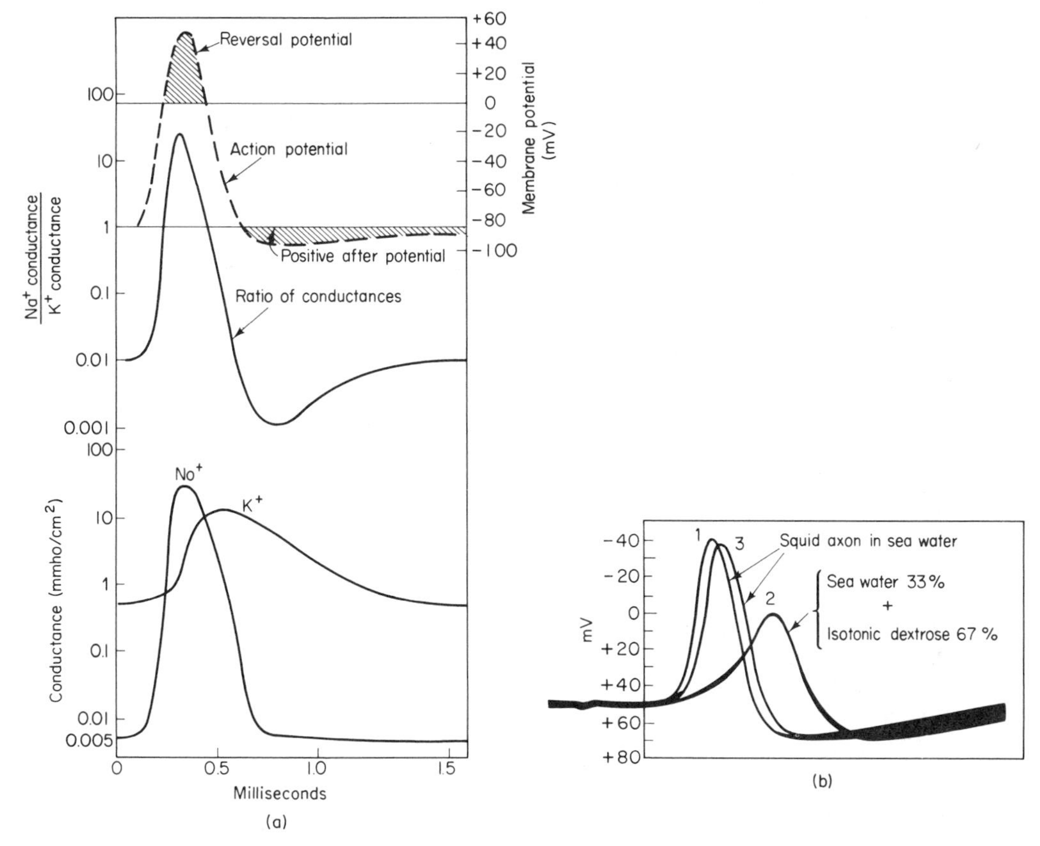

Fig. 15.7 Genesis of the spike potential. Left series of curves (a), sodium and potassium ion fluxes (conductances) in relation to the spike; right series of curves (b), reduction in magnitude of potential due to lowering of external Na^+ for 16 min (curve 2) with recovery after 13 min when returned to sea-water (curve 3). Note that Na^+ conductance increases several thousand-fold during the early phases, while K^+ conductance increases only about thirty-fold during the latter part of the action potential and for a short time thereafter. [(a) From Guyton, 1971: Medical Physiology, 4th ed. Saunders, Philadelphia, based on Hodgkin and Huxley (1952) but transposed from squid axon to large mammalian nerve fibers. (b) Based on Hodgkin and Katz, 1949: J. Physiol., *108*: 48.]

(1902) classical studies of the excitability of whole muscles soaked in different solutions have been followed by sophisticated tracer measurements of ion fluxes in individual cells (Hinke, 1961; Katz, 1966). These fluxes are in accordance with the theory as are also the effects produced by changing the ionic content of the fluids surrounding the axon. Thus, the spike potential (unlike the resting potential) is abolished by the removal of sodium from the extracellular medium (Fig. 15.7b). Several pharmacological agents have proved valuable in sorting out the ion fluxes. Metabolic poisons such as cyanide will eliminate the pump without altering the valves. Likewise, there are chemical agents which alter ion permeability in a selective manner. For example, in *Onchidium* neurons the entrance of Na^+ (Na-activation) can be diminished by poisoning with urethane. This eliminates the spike but leaves a small graded response indicating that the spike associated with all-or-none conduction is primarily caused by flooding with Na^+. The small graded response, which has been shown to be associated with the outward flux of K^+ (K-activation), is unaffected by urethane but diminished by tetramethylammonium ion (Grundfest, 1961). Another poison useful in studies of excitation is tetrodotoxin (tarichatoxin) which occurs in some tropical marine fishes and in newts of the family Salamandridae (Fuhrman, 1967; Russell, 1969). This substance interferes with neuromuscular transmission by specifically blocking the sodium channels. The increase in Na-conductance which normally follows stimulation is reduced or abolished without affecting potassium permeability (Hille, 1968; Russell, 1969). Tetrodotoxin (TTX) seems to have a very specific blocking action on the Na-channels of the cell membrane. Experiments such as these demonstrate the existence of different types of valves and the significance of the movements of specific ions in the electrogenic response.

Some of the most convincing evidence of differential ion movements and permeability changes comes from VOLTAGE CLAMP studies. This technique was introduced in 1949. Cole (1968), one of the pioneers in its development, summarizes the history of its use and provides technical details. Hodgkin *et al.* (1952) applied the technique to the squid axon and in a series of classical papers published in the *Journal of Physiology* (1952) established the basic principles on which all discussions of this topic are based; these papers were summarized by Hodgkin and Huxley (1952) and reference to them can be found in the monographs of nerve physiology already cited. The technique is relatively simple in principle but technically demanding in its operation (Baker, 1966; Miles, 1969). Two elongated silver electrodes are introduced longitudinally into a segment of axon; one of these is connected to an external membrane electrode and serves to measure membrane potentials; the other long internal electrode is a current-carrying electrode, connected through a feedback amplifier into this system. This arrangement permits the membrane potential to be quickly altered and fixed or "clamped" at any desired level by the electronic feedback. In addition, a pair of current-sensing electrodes, attached to the membrane of the section of axon under study, permit the recording of membrane potentials. Many different experiments are possible. When, for example, the intact membrane of an axon is depolarized at some fixed value near zero (as happens in a stimulated nerve), the current that flows is first directed inward and then rapidly reversed in direction in accordance with the theory of differential ion movements. These currents can be predictably altered by ionic changes in the fluids surrounding the membrane; the poison TTX first reduces the inward (Na-dependent) current and then abolishes it while the outward (K-dependent) current is unaffected (Fuhrman, 1967). A host of voltage clamp experiments supports the theories of differential ion conductance in excitation.

Na-conductance (proportional to membrane permeability—Miles, 1969) triggers the action potential in squid axons and vertebrate muscle cells. Although this is also characteristic of many electrically excitable cells, it is by no means universal (Grundfest, 1961, 1966); the electrically excitable plant cell, *Chara,* produces a spike as a result of Cl-activation; divalent ions (especially Ca^{++}) rather than Na ions play the major role in production of action potentials in crustacean muscle fibers (Hagiwara and Naka, 1964). In some crayfish and in giant barnacles, the muscle spike potential is caused by an increase of membrane conductance to Ca^{++} which is then followed by K^+-conductance as in the preparations already described. As might be expected, these "Ca-spikes" are not suppressed by TTX while Mg^{++} inhibits them competitively (Hagiwara and Nakajima, 1966). Grundfest (1966) gives further examples of the participation of ions other than sodium in spike electrogenesis.

Propagated action potentials and the spread of excitation A spike discharge at any point on an electrically excitable membrane will disturb the delicate balance of ions in neighboring regions and create a flow of electric current between the resting and active areas. This flow of current is the basis of nerve impulses, muscle excitation and conduction found in several different types of cells. In fact, it seems likely that the most primitive conducting tissues were sheets of epithelial cells similar to those found in some present-day coelenterates (Mackie, 1965). Although the basic mechanisms concerned with the spread of excitation were first investigated with squid axons, the findings also apply to other kinds of electrically excitable membranes.

A segment of an axon is shown diagrammatically in Fig. 15.8. The influx of Na^+ which occurs at the point of stimulation reverses the membrane potential to inside positive; electric current then flows in a local circuit between the resting and active regions. The first point of importance to be emphasized is the *reversal* of membrane charge. The local circuit thus established reduces the membrane potential just ahead of it sufficiently to activate the adjoining region. Permeability rises with the opening of sodium valves and the "inside-positive" region moves along the membrane. Thus, there is a progressive reversal of charge that extends from the point of stimulation as though an electrode were moving along the membrane at the rate of conduction (Brazier, 1968, 1969). The second important feature of membrane conduction is the *explosive nature of the spike* which occurs when the threshold or firing-level is reached (Fig. 15.6); in the squid axon, the threshold charge may be amplified by a factor of 5. "This automatically boosts the signal to full strength all along the line and so makes up for imperfections of the cable structure" (Katz, 1966). As a consequence, the wave of excitation moves along WITHOUT DECREMENT

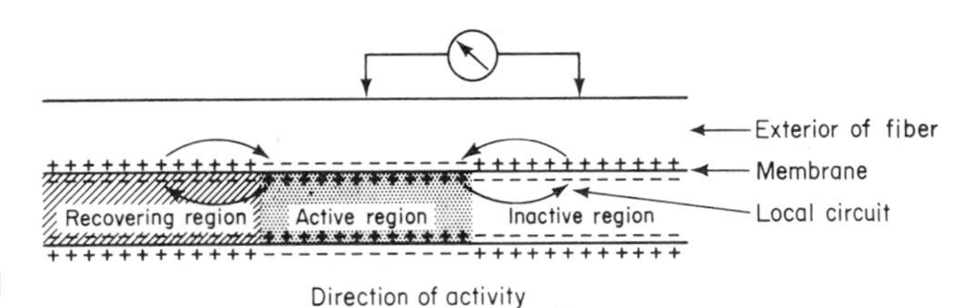

Fig. 15.8 Diagram illustrating local circuit theory in squid axon.

or attenuation in all-or-none manner. The third important point is associated with the *rapid closing of the sodium gates* after a brief open period (0.3 msec in squid axon); it follows that recovery occurs speedily but after a brief REFRACTORY PERIOD. Thus, excitable membranes can be repeatedly activated to full intensity and may be specialized to transmit volleys of impulses. In summary, the LOCAL CIRCUIT THEORY for the spread of excitation rests on the reversal of membrane potential during stimulation, the amplification of threshold potentials to provide a large margin of safety for the stimulation of neighboring regions and the brief period of sodium influx followed quickly by potassium efflux.

The local circuit theory has been tested in different ways (Hodgkin, 1964; Katz, 1966; Miles, 1969). Conduction velocities in squid axons have been experimentally changed by altering the electrical conductivity at the external surface of the axon membrane; placing a section of the axon in oil reduced rates of impulse conduction while placing a region of the fiber on a grid of platinum strips increased it (Hodgkin, 1964). Many of the findings already described for spike potentials and those from experiments with local anaesthetics (cold, pressure, procaine) are satisfied by the local circuit theory.

Graded responses The propagated action potential is a highly specialized cellular attribute. In many membranes the transducer action which follows stimulation produces only local responses in which the electrogenesis varies in amplitude with the stimulus strength and is decrementally propagated (GRADED RESPONSES). This type of response is phylogenetically older and considered to be the basis for the evolution of the all-or-none propagated activity required for long-distance transmission of excitation in multicellular animals (Grundfest, 1959). Whereas propagated action potentials characterize the spread of excitation within cell boundaries, non-propagated or graded local potentials are usually responsible for the activation of the cells. The spread of excitation from one cell to another generally depends on graded potentials established at the point of cellular contact by the release of a chemical. Examples will be considered in the discussions of nerve synapses and myoneural junctions. In both cases, the gap between the membranes of the two cells is too great to permit the development of local circuits between them with the electrotonic spread of excitation; among vertebrates, the space between the two membranes of the nerve synapse is usually 150 to 200 Å while the gap at the myoneural junction is of the order of 500 to 1000 Å (Katz, 1966). The chemical released by one cell bridges the gap to excite the other. Receptor cells provide other examples of membranes capable of graded responses (GENERATOR POTENTIALS); these may be triggered by mechanical or chemical stimuli and lead to excitation in the electrically excitable nerve membranes associated with them. The generator potentials in the Pacinian corpuscle, a good example of a mechanical receptor, will be discussed in Chapter 16.

Membranes capable of graded responses are electrically inexcitable. Studies of the electric eel provided the first evidence for membranes of this sort. At a time when the electrical stimulus was almost universally used to excite tissues experimentally, the existence of tissues which would not respond to it was clearly a novelty. It is now known that these particular electric organs are modified myoneural junctions; they are chemically excitable but unresponsive to electric current (Chapter 20).

The myoneural junction (of the frog sartorius, for example) is a particularly useful preparation for the study of graded responses. Like the electric organs just mentioned, these junctions (Fig. 19.22) are electrically inexcitable although the neighboring parts of the muscle membrane contain electrically excitable valves. If one sticks a microelectrode into the muscle fiber at any point, a potential difference of about −90 mv will be recorded. Over almost all of the fiber this potential remains constant for prolonged periods. Only at the end plate (Fig. 19.22) is there a difference and here, intermittent potential differences of the order of 0.5 mv occur at random and last for about 20 msec (Fig. 15.9). These tiny spontaneous potentials result from leakage of small amounts of the transmitter (acetylcholine) from the nerve ending. Acetylcholine is quickly destroyed by the enzyme cholinesterase and insufficient in amount to excite the neighboring muscle membrane. Experimentally, it has been possible to imitate potentials of this sort by squirting minute quantities of acetylcholine close to the end plate region (Katz, 1966, 1969).

The application of different drugs is most useful for sorting out the events associated with the end plate or JUNCTION POTENTIAL. Curare (an arrow poison used by South American Indians), seems to compete with acetylcholine for the receptor substance in the membrane. When applied to the junction, it reduces the potentials in accordance with its concentration. When the junction potential is reduced to about 30 per cent of the maximum, the neighboring electrically excitable valves are no

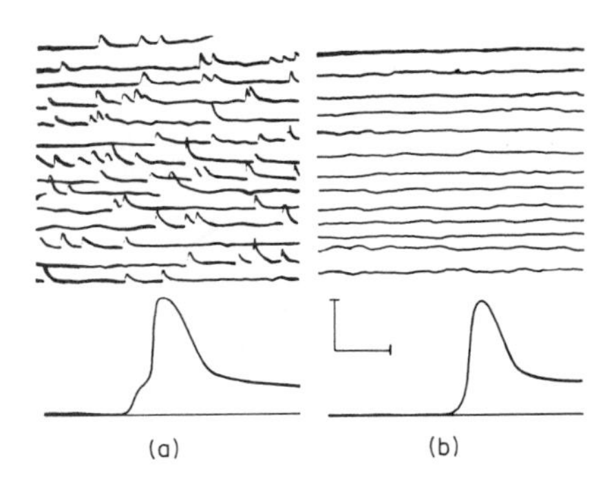

Fig. 15.9 Intracellular recordings of electrical potentials in frog skeletal muscle: (a) was recorded at the end plate and (b) at a distance of 2 mm from the end plate in the same muscle fiber. The upper recordings, taken at low speed and high amplification, show the spontaneous activity confined to the end plate; the lower recordings, made at high speed and low amplification, show the response to the nerve stimulus (shock applied at the beginning of the sweep); (a) shows the step-like initial end plate potential which leads to the propagated wave; (b) shows only the propagated action potential delayed by conduction over a distance of 2 mm. Voltage and time scales: 3.6 mv and 47 msec for the upper part; 50 mv and 2 msec for the lower part. [From Fatt and Katz, 1952: J. Physiol., *117*: 110.]

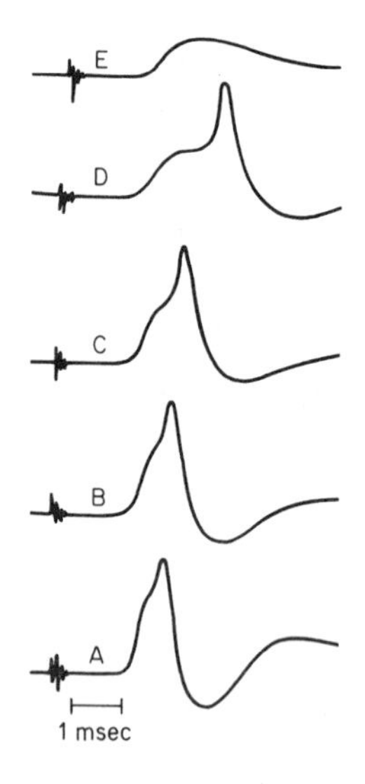

Fig. 15.10 Action potentials from the region of the end plate in frog sartorius. A, before curarization; B to D, increasing degrees of curarization; E, curarization complete. Propagated spike arises from junction potential in A to D. [Based on Kuffler, 1942: J. Neurophysiol., *5*: 23.]

longer activated, and the spike is not propagated along the muscle (Fig. 15.10). Physostigmine (a plant alkaloid used by native people of West Africa in ordeal trials) or its synthetic counterpart neostigmine, however, competes with acetylcholine for the enzyme cholinesterase. Hence, acetylcholine is not rapidly destroyed in the presence of neostigmine and the junctional potential is greatly prolonged (Fig. 15.11).

Only the myoneural junction of nerve-muscle preparation from adults shows this sensitivity to acetylcholine. Embryonic frog muscles or the fetal muscles of rats show a general chemosensitivity; the entire cell membrane can be excited by applying acetylcholine at any point. After the innervation has been established, only the end plate responds to acetylcholine; chemosensitivity has been lost over the surface of the fiber (Thesleff, 1961). If the nerves to the adult muscle are cut, the general chemosensitivity gradually returns. In some way the motor nerve can control the size of the chemoreceptor surface.

Spikes and graded potentials are contrasted in Table 15.2. Most of the differences have already been noted. Three important points, however, remain to be

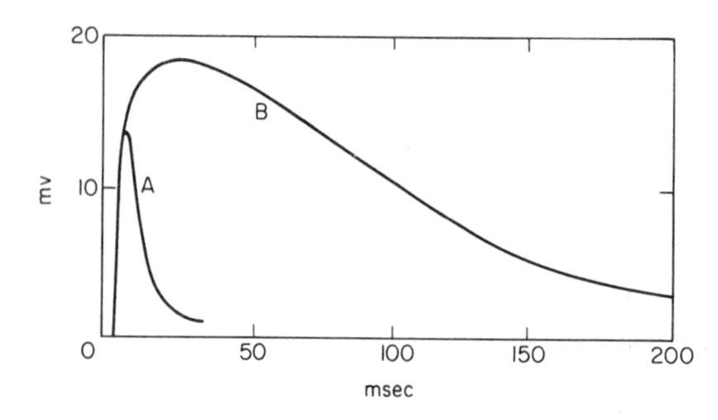

Fig. 15.11 Effects of an anticholinesterase drug on end plate potential of single muscle fiber produced by stimulation of motor nerve. A, when neuromuscular transmission is blocked after reduction of sodium concentration in bathing medium. B, from same fiber after addition of neostigmine to the sodium-deficient bathing medium. [From Fatt and Katz, 1951: J. Physiol., *115*: 338.]

Table 15.2 Comparison of the Properties of Action Potentials (Propagated Spikes) and Local Graded Potentials. (Numerical values refer to cat motor neuron.)

Property or condition	Action potential	Local graded potential
1. Initiated by:	Local current of graded potential	Chemical transmitter substance or mechanical stimulus
Stimulus sensitivity:	Electrically excitable membrane	Chemically or mechanically excitable membrane
2. Polarity change:	Reversal of potential	Depolarization or hyperpolarization
Size:	Up to 130 mv	E.P.S.P. = 10–50 mv I.P.S.P. < 15 mv
Rate of rise:	Rapid (670V/sec)	Slower (220 v/sec)
3. Conduction:	Regenerative, i.e. non-decremental, propagated	By local electrotonus only, decremental
4. Response to varying stimulus intensity:	Threshold, all-or-none	No threshold, graded response
5. Response to repeated stimulation:	Refractory period	No refractory period, but summation and facilitation
6. Nature of membrane permeability change:	Increased initially to Na^+ and during recovery to K^+.	Increased to all ions (excitatory) or to K^+ and/or Cl^- only (inhibitory)

SOURCE: Courtesy of J. E. Phillips.

considered: membrane permeability changes, effects of repeated stimuli and the phenomenon of inhibition.

Spike potentials in axons and vertebrate muscle fibers have been ascribed to the sudden brief rise in Na-conductance followed by a sharp increase in K-conductance (Fig. 15.7a). The graded potentials, caused by release of acetylcholine, are associated with the *simultaneous* increase in permeability to both Na^+ and K^+ (Katz, 1966). The transmitter reacts with the postsynaptic membrane for a very brief period (5 msec or less) to open the channels for all small cations at the same time. The number of channels opened is directly related to the amount of chemical released. The simultaneous opening of these valves creates a "short circuit" across the membrane and if this reaches threshold value at the nearby electrically excitable areas of the membrane, a spike potential will ensue.

The effects of repeated stimuli are quite different when applied to the electrically excitable and the electrically inexcitable membranes. With the former, any stimulus of threshold or greater value triggers a spike which is maximal. Moreover, the spike is followed by a refractory period during which the membrane must be repolarized before it can again respond. Excitation of these membranes is characteristically "all-or-nothing" and appears as a volley of discharges. In contrast, the electrically inexcitable membranes even while "at rest" frequently show small spontaneous potentials (Fig. 15.9) and when excited, display a characteristically graded response in accordance with the amount of stimulus (Figs. 15.12 and 16.3). There is no refractory period; potential changes of subthreshold value add to one another and produce greater and greater responses (SUMMATION); moreover, the amounts of transmitter released with successive stimuli increase during a repetitive series and each successive potential is greater than the preceding one (FACILITATION). In a number of preparations (actinian sphincters, slow crustacean muscle fibers), the first stimulus, no matter how strong, has no obvious effect whatever. Finally, it is to be noted that local graded potentials may be inhibitory (hyperpolarizing) as well as excitatory (depolarizing).

Inhibition An integrating nervous system could never have been perfected with excitatory responses only. Inhibitory activity is an essential component of both the central nervous system and the peripheral control.

The nerve-muscle preparations of crustaceans have been most useful in the study of nervous inhibition (Chapter 19). Their skeletal muscles receive separate nerve fibers for excitation and inhibition. Recordings with intracellular electrodes show greatly reduced end plate potentials following a stimulation of the inhibitory fibers. When only the inhibitory fibers are stimulated, there will be little change in the membrane potential if this is at or near the resting level at the moment of stimulation; if, however, it be displaced somewhat (as can be done experimentally), then muscle potentials develop as transient hyperpolarizations or depolarizations, depending on the direction of displacement. In short, the membrane potential tends to be stabilized by the stimulation of the inhibitory fiber; these fibers counteract changes in the

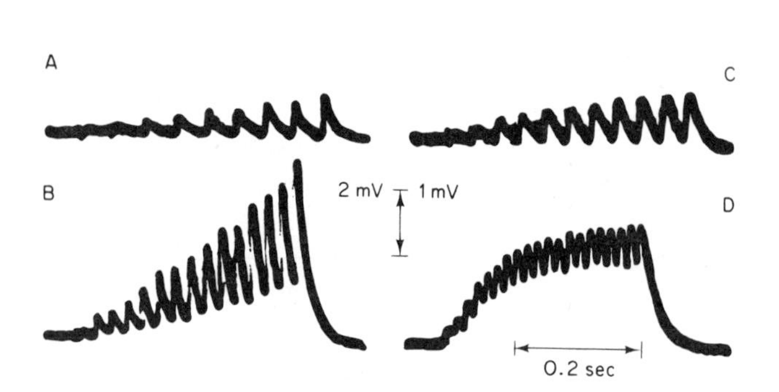

Fig. 15.12 Summation and facilitation of excitatory (A and B) and inhibitory (C and D) potentials at the crayfish neuromuscular junction. Intracellular recording of resting potential is near 80 mv. Stimulation rates for A and C, 20 and 23/sec; for B and D, 40 and 58/sec. Note that with repetitive stimulation, each junctional potential adds to the remnant of the previous one (lower traces) to increase the overall depolarization (SUMMATION) *and* that the individual junctional potentials become progressively larger with successive impulses—due to augmented transmitter release (FACILITATION). [From Dudel and Kuffler, 1961: J. Physiol., *155*: 531.]

resting potential. This is at least in part due to increased Cl^- permeability (Cl-conductance) or influx of Cl^- which tends to neutralize any depolarizing effects. This, however, is not the entire story. The transmitter, which is very likely gamma-aminobutyric acid (Kravitz *et al.*, 1968), has two distinct effects. It not only increases Cl-conductance and stabilizes or repolarizes the membrane potential of the muscle cell but it also acts on the excitatory nerve fibers to reduce their output of transmitter and hence their excitatory action (Katz, 1966; Kennedy, 1966; Miles, 1969).

Kennedy (1966) summarizes the literature for several other invertebrate preparations which exhibit hyperpolarizing potentials. These are usually attributed to transmitters which increase permeability to K^+ and/or Cl^-. It is becoming apparent, however, that inhibition in some of the invertebrates may also depend on the action of the cation pump. If the pump is speeded and adjusted so that sodium is removed more rapidly than potassium is acquired, it is obvious that hyperpolarization will occur. Certain inhibitory neurons in ganglia of the snail *Helix aspersa* suggest that their hyperpolarizing potentials may depend on quite different mechanisms; various components of the membrane physiology have evidently been modified during phylogeny to effect inhibition. In some *Aplysia* neurons, acetylcholine produces hyperpolarization by increasing Cl-conductance (Kerkut *et al.*, 1969a); in others, a different transmitter—perhaps dopamine—increases potassium permeability (Kerkut *et al.*, 1969b). Of particular interest, however, is the evidence that inhibition in certain neurons of *Aplysia* may also depend on the action of the cation pump (Kerkut *et al.*, 1969a; Pinsker and Kandel, 1969).

Peripheral inhibition is confined to the visceral nervous system among the vertebrates. The cardiac arrest which follows stimulation of the vagus is the most familiar example in the elementary physiology laboratory. This was shown to depend on the release of a chemical substance (since shown to be acetylcholine) in classical experiments by Otto Loewi, published in 1921 (Heilbrunn, 1952). It is now known that acetylcholine produces hyperpolarization or stabilization of the resting potentials of heart muscle, an activity which is entirely opposite to its effect on the myoneural junction. In the membrane of the cardiac muscle cell it increases only K^+ ion conductance, and the membrane tends to move toward, or be held at, the potassium equilibrium potential (Katz, 1966). In the myoneural junction, on the other hand, Na-conductance is also increased and general depolarization is initiated.

Inhibition has also been keenly investigated in the central nervous system of the vertebrates. Eccles (1964, 1968) and his associates believe that the inhibitory transmitter opens K^+ and/or Cl^- valves. Like the inhibitory machinery of crustacean muscle, central inhibition in vertebrates also seems to operate by reducing the effect of the corresponding excitatory system. Further comment will be found in Chapter 21.

Pacemaker potentials Before leaving this section on bioelectric potentials, it should be recalled that in addition to spike potentials and graded potentials, some cells possess membranes which generate regularly re-occurring potentials called PACEMAKER POTENTIALS. These have already been described in connection with the pacemaking activity of the heart (Fig. 5.13).

Steps in phylogeny In summary, Grundfest, (1959, 1965a) argues that the generalized excitable cell exhibits three distinct physiological capacities. These appeared at different times in evolution and have been exploited to different degrees by various types of excitable cells to produce highly integrative animal systems. Receptor cells which are not electrically excitable and which show a graded electrogenesis (either depolarizing or hyperpolarizing) in accordance with the strength of the stimulus must be closest to the more primitive excitable cells. The spike-generating membrane structure which transmits coded messages from the receptor to the effector is phylogenetically more recent. The coded messages flow from receptors to effectors in all-or-none pulses which vary only in number and frequency. Finally, the hyperpolarizing, inhibitory type of synaptic electrogenesis provides a control on excitatory activities; the interaction of excitation and inhibition is the physiological basis of integrated animal activity.

General Characteristics of Stimulus and Response

Many classical exercises in the physiology laboratory have been built around quantitative descriptions of tissue responses to carefully measured stimuli. The electric current has been the favored type of stimulus because its intensity, duration, and rate of rise to peak value can be precisely controlled and because of its effectiveness in eliciting responses in readily available tissues, such as the heart and nerve-muscle preparations from the frog. These relationships are detailed in textbooks of vertebrate physiology and will not be described here. It is, however, important to note that most of them find a ready explanation in terms of the ionic permeability theory of electrogenesis.

An effective stimulus must obviously be of a certain THRESHOLD or minimal value if it is going to open sufficient numbers of valves to produce the transmitted type of electrogenic response. The subthreshold, subminimal, or subliminal stimulus, however, may be expected to produce some electrogenesis (LOCAL EXCITATORY STATE) which, under certain conditions, can be added to and results in the SUMMATION OF INADEQUATE STIMULI. As noted in the last section, any change in the membrane potential will initiate recovery processes which involve both the action of the cation pump and the valves; it follows then that the stimulus must rise to peak value within a certain period of time, or the recovery processes will counteract those which tend toward depolarization. A stimulus must also last long enough (MINIMAL EFFECTIVE DURATION) to open a sufficient number of valves to trigger a response. Thus, the effective stimulus is characterized by its STRENGTH, DURATION and RATE OF RISE to peak value. Strength-duration relationships for several activated organs are shown in Fig. 15.13.

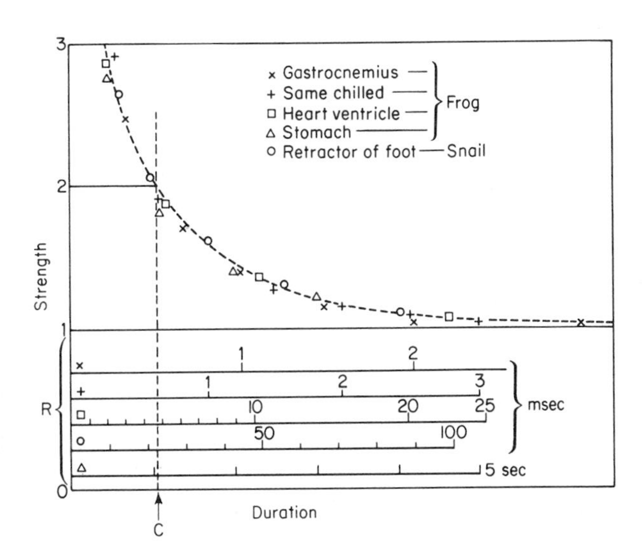

Fig. 15.13 Five strength-duration curves superimposed by adjusting the time scale. Stimulus strength 1 is the strength that just fails to excite in an infinitely long time; it is called the RHEOBASE. The duration required to stimulate when the strength is twice the rheobase (stimulus 2 on the ordinate) was termed the CHRONAXIE by Lapicque and is shown at the broken line C. [From Heilbrunn, 1952: An Outline of General Physiology, 3rd ed. Saunders, Philadelphia. Based on Lapicque.]

Responses measured in the effector organs are characterized by time relationships which also find a ready explanation in terms of membrane electrogenesis. The time required for an action current to develop at a myoneural junction or to move along a muscle or nerve fiber creates a measurable LATENT PERIOD between stimulus and observed response; the depolarization of the membrane leaves it in a refractory state from which it gradually recovers through stages which are first ABSOLUTE REFRACTORY PERIODS and then RELATIVE REFRACTORY PERIODS.

Intercellular transmission often requires more than one stimulus. Several stimuli in rapid succession may be necessary to activate certain synapses or the myoneural junctions of crustaceans, or vertebrate junctions which have been treated with curare. SUMMATION and FACILITATION depend on releasing sufficient transmitter to produce a depolarization at these junctions. FATIGUE and SENSORY ADAPTATION or ACCOMMODATION, in which repeated stimulation leads to reduced responsiveness, may also be explained by the membrane hypothesis; cation pumps require a steady supply of metabolites, neurohumors gradually become exhausted, and the responsive membrane can only tolerate limited changes in the distribution of ions. The present membrane theories of cellular irritability provide a satisfactory explanation for many of the classical demonstrations in the physiology laboratory.

16

RECEPTOR MECHANISMS

"Man doth not live by bread only." The free life of an animal depends on a constant flow of information from its environment, the computation and integration of this into meaningful and purposive instructions for the effectors and the relay of the instructions to the appropriate glands, muscles, chromatophores and other organs concerned with animal activities. This flow of information pertaining to the environment is every whit as important as the energy-yielding foods which must also be acquired from the outside. The first link which unites an animal with its habitat is the sensory receptor system, specialized to transform chemical, radiant, electrical or mechanical energy into a train of membrane potentials which are the only data transmitted to the computing centers of the brain or coordinating ganglia.

It has already been argued that the ancestral animal cells were endowed with protoplasmic irritability. It can be assumed that these simple cells took advantage of this capacity to avoid some of the environmental hazards. Like the familiar *Amoeba*, they probably showed protoplasmic responses to a shift in environmental temperature, to changing light conditions, to vibrations and to foreign chemicals. With nothing more than simple avoiding reactions, a measure of contact with the more suitable parts of the habitat is possible. With the evolution of specialized receptors and the associated integrating machinery, animals have been able to exploit the world in a much more positive manner: to locate and utilize particular foods, to find mates and care for their young, to establish territories and to orient their travels.

In the higher animals, these capacities were further enlarged with the specialization of cells and tissues for the production of distinctive animal signals and the gradual incorporation of these into systems for communication. These "languages" are as often based on odors as they are on sounds and sometimes incorporate light or electrical fields. As the range and variety of environmental stimuli have gradually changed during animal evolution, the receptor organs have made use of more and more of the potential information. Primitive animal reactions are positive or negative movements with respect to broad bands of the stimulus spectrum; the delicate transducers of the more advanced phyla discriminate among different wavelengths or the planes of polarization of light; they permit responses to specific frequencies and amplitudes of vibratory stimuli and detect complex organic chemicals which may have meaning for only one sex or one species out of a multitude of animal species.

The receptor system of a complex multicellular animal must collect information from the internal as well as the external environment. Co-ordinated activities depend on a delicate balance between the movements of the different parts. Receptors in muscles, tendons and joints (PROPRIOCEPTORS) provide the basic information for these adjustments. Signals from the visceral receptors (INTEROCEPTORS) are concerned with much of the internal regulation of visceral functions (for example, the cardiovascular pressoreceptors or the osmoreceptors of the hypothalamus); they may also modify the overt activity of an animal (visceral pain, for example). These parts of the receptor system which signal changes from within the organism, although superficially inconspicuous, would rank in bulk with the distance receptors if they were collected together in one place.

Physiologists interested in the sense organs have acquired most of their information from two different lines of investigation—behavioral and electrophysiological. In the behavioral type of analysis, some obvious activity of the animal is related to a measured change in environmental energy. A change in light intensity or sound, the exposure to a chemical or the change in an electric field may modify an animal's activities in a real and obvious manner. Through appropriate conditioning experiments it is possible to evaluate an animal's ability to discriminate extremely small differences. Bull (1957), for example, found that some species of fish can perceive differences of only 0.06‰ salinity or 0.03°C temperature. In the electrophysiological studies, the generator currents of the receptor cells or the action potentials of the associated neurons are measured during exposure of the cells to controlled environmental stimuli. The capacity of the transducer can be accurately evaluated; infrared detectors in rattlesnakes, for example, respond to temperature changes of the order of 0.001°C (Bullock, 1959a).

The recorded information on the structure and function of the receptor organs forms a large segment of the physiological literature. The present discussion will be restricted to a brief comment on the transducer mechanisms, an attempt to discover some phylogenetic trends in the functional anatomy, and an emphasis on the receptors as links between the different animals and between the animals and their habitats. This seems to be the logical course in an endeavor to trace a phylogeny in physiological mechanisms. Comprehensive surveys of the literature can be found in the *Handbook of Physiology* (American Physiological Society), the *Handbook of Sensory Physiology* (Springer-Verlag), and in physiological monographs devoted to the various animal groups. Receptors concerned with light, chemicals, vibrations and temperature are discussed in the two subsequent chapters, while comment on reception of electrical stimuli is reserved for the description of electrical organs in Chapter 20.

General Physiology

Information is collected from the environment by the RECEPTOR CELLS. Particular areas of these cells, known as the RECEPTOR MEMBRANES, are exposed to the environment and serve as transducers to change the stimulus energy of the environment into the electrical energy of the cell membrane. Any particular receptor membrane is usually sensitive to only one kind of stimulus—mechanical, photic, chemical and other (RECEPTOR SPECIFICITY). Although the stimuli vary both qualitatively and quantitatively, the electrical responses of the receptor membranes are always similar and generate trains of spike potentials in associated transmitting fibers; the conductile or transmitting portion of the system may be a part of the same cell or an associated nerve (Fig. 16.1).

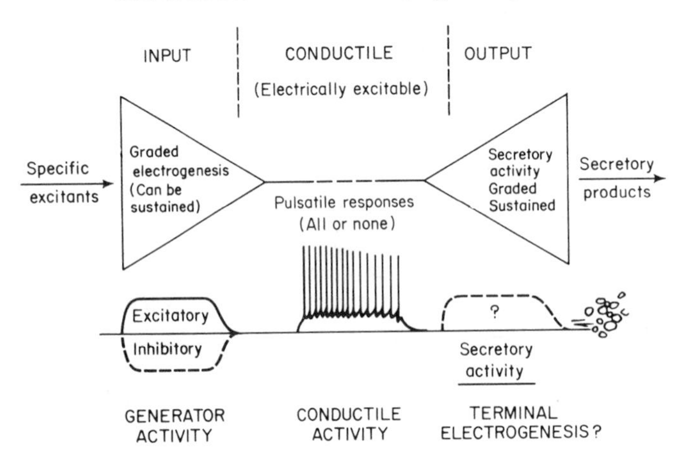

Fig. 16.1 The different responsive components of an excitable cell. The conductile portion is often absent (glands, muscle fibers, electroplax). The types of electrogenesis are shown on the lower line of the diagram; the possibility of inhibitory, hyperpolarizing responses is indicated by the broken lines. [From Grundfest, 1957: Physiol. Rev., *37*: 343.]

In the previous chapter, it was noted that nerve impulse conduction is an "all-or-nothing" process. The spike potentials are constant in amplitude but variable in frequency. Thus, an animal's reactions are based on *pulse-coded messages*. Its ability to respond appropriately depends on the capacities of the central nervous system to interpret these sequences of action potentials and to transmit similar pulse codes to the effector organs. The nature of pulse-coded messages is shown in Fig.

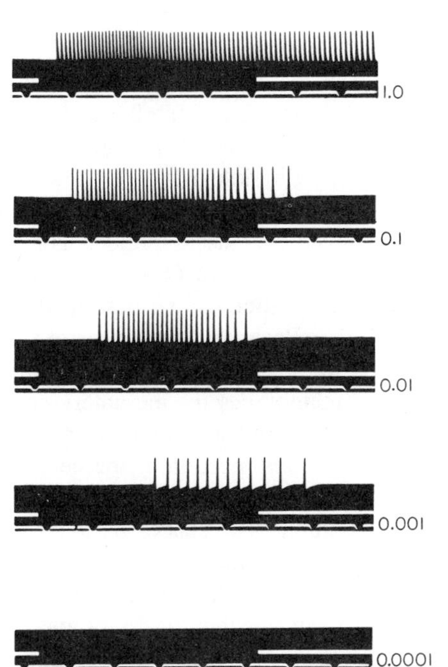

Fig. 16.2 Coding of impulses in a single optic nerve fiber of *Limulus* by a one-second flash of light of relative intensities shown at the right. Oscillograph records of action potentials. The lower white line marks 0.2 sec intervals and the gap in the upper white line gives the period for which the eye was illuminated. [From Hodgkin, 1964: The Conduction of the Nerve Impulse. Univ. Press, Liverpool, after Hartline, 1934: J. Cell. Comp. Physiol., *5*: 237.]

16.2; light of different intensities generates impulses which vary in number and frequency but not in amplitude.

Basic problems in the general physiology of sense organs concern (1) the mechanisms by which receptor membranes change stimulus energy into action potentials and (2) the quantitative relationships between the magnitude of the applied stimulus and the timing of these action potentials. In considering these general problems, examples will be drawn from several different receptors which seem best to illustrate the points.

Origin of Generator Potentials

In 1950, Bernard Katz made the remarkable discovery that the electrogenic properties of the muscle spindle were quite different from those of the nerve associated with it. The muscle spindle is a mechanoreceptor concerned with the adjustment of tensions in skeletal muscle. Katz (1950) found that a modest stretching generates a very weak and purely localized electric current while more vigorous stretching gives a greater response until a point is reached where an impulse is triggered in the associated nerve. These localized potential changes, referred to as GENERATOR POTENTIALS, are graded responses in direct relation to the energy of stimulation; the nerve impulse which they generate is an all-or-none propagated wave of depolarization. Within the muscle spindle, the relationship between input and output is like that found in a carbon microphone where the mechanical deformations of the disk of carbon by the sound waves reduce the electrical resistance so that an electrical current flows through it, and the strength is proportional to the sound (Loewenstein, 1960; Matthews, 1972).

Physiologists have now investigated several different mechanoreceptors, and it is apparent that Katz discovered a fundamental transducing mechanism in his work with the muscle spindle. The crustacean stretch receptor (Fig. 16.10) and the Pacinian corpuscle (Fig. 16.3) have been most useful in developing the current ideas. The latter will be described to illustrate several of these.

Pacinian corpuscles are pressure receptors found in the deeper layers of the skin, in the connective tissues around the tendons, muscles and joints, and in the serous membranes and mesenteries of the viscera. They are extremely large, reaching almost 1 mm in length and 0.6 mm in diameter. These are primary sense cells in which the terminal portion of the nerve fiber is surrounded by a relatively thin granular mass and covered by concentric layers of connective tissue like the many leaves of an onion. The spaces between these layers contain fluid which may be important in transmitting the superficial pressure changes to the delicate nerve ending within. However, the coverings are not essential to the generation of electric potentials; it has been possible to dissect away 99.9 per cent of this structure without destroying its capacity to transduce mechanical stimuli. The myelin sheath of the associated nerve extends for about one-third of its distance within the corpuscle, but the terminal part is unmyelinated. As indicated in Fig. 16.3, there is an important node of Ranvier inside the corpuscle.

In studies of the Pacinian corpuscle, most of the connective tissue lamellae are removed and the almost naked nerve ending is stimulated with a minute glass stylus. This is made to vibrate very precisely at one point by a controlled electric current operating through a piezoelectric crystal which converts the electrical energy into

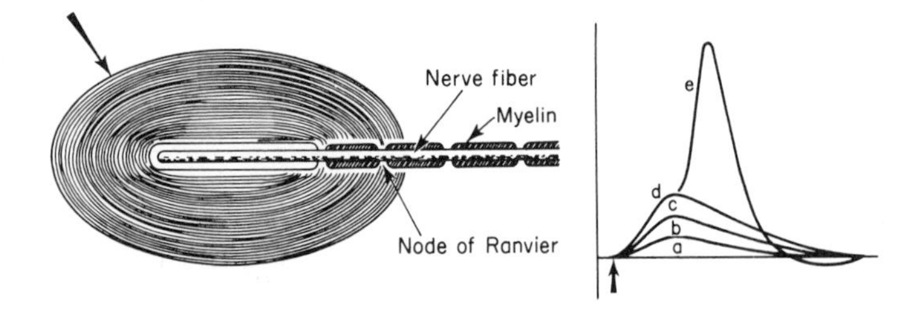

Fig. 16.3 Diagram of the Pacinian corpuscle (left) and the electrical potentials (right) which follow progressively stronger stimulation applied at the point of the arrow. A weak stimulus produces a weak generator current (a). Progressively stronger stimuli produce correspondingly stronger generator currents (b) and (c). The threshold stimulus (d) fires an all-or-none nerve impulse (e). If the first node of Ranvier is blocked, the all-or-none impulse cannot be induced. [From Biological Transducers, by W. R. Loewenstein. Copyright © 1960 by Scientific American, Inc. All rights reserved.]

Fig. 16.4 Spread of excitation from the point of stimulus (S) along the nerve ending of a Pacinian corpuscle. *Upper,* diagram of apparatus. The lamellae of the corpuscle have been removed. The mechanical pulses of a piezoelectric crystal are applied by a glass stylus (S). The resulting generator potentials are recorded with a microelectrode (R) from the surface of the receptor membrane or, alternatively, between the first (I) and second (II) nodes of Ranvier by electrotonic spread. *Lower,* a small region of the receptor membrane (E) is stimulated with equal mechanical pulses while the surface of the receptor membrane is scanned with a microelectrode (R). Four selected samples of generator potential recorded at the distances indicated (d) have been superimposed. [From Loewenstein, 1961: Ann. N.Y. Acad. Sci., *94*: 511.]

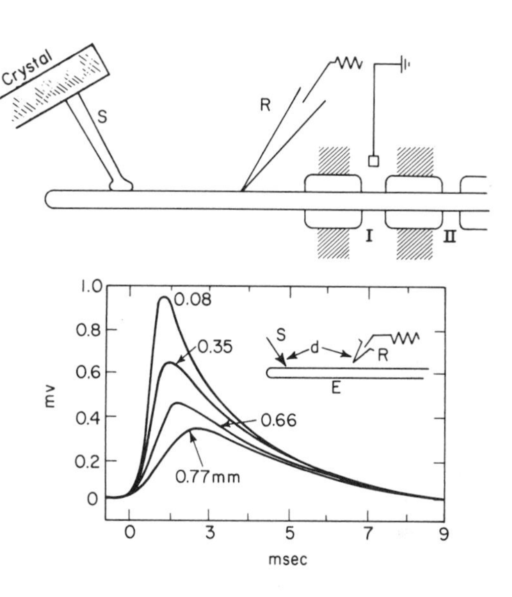

mechanical vibrations of the stylus (Loewenstein, 1960, 1961). At rest the membrane potential is steady. When a mechanical pulse is applied at one point and the generator current measured at varying distances from the point of stimulation, the potentials are found to decrease with distance (Fig. 16.4). Clearly, the excitation is restricted to the region of stimulation, and the signals fade rapidly in the surrounding regions of the membrane.

When the almost naked fiber is stimulated at two regions separated by about 0.5 mm, the generator currents are added to produce a single large generator potential (Fig. 16.5). It is evident that the flow of current increases in proportion to the area of the membrane deformed (spatial summation). In the terminology of Chapter 15, wrinkling of the cell membrane opens valves (mechanically operated valves) which permit the flow of ions and produce electrogenesis. The facts are consistent with the view that the mechanical operation of valves or the opening of holes is in direct relation to the amount of stretching and that, when this reaches a point where the excitation passes the first node of Ranvier, an all-or-none depolarization of the nerve fiber occurs. The local generator currents increase until they initiate a "spike" (Fig. 16.3).

The Pacinian corpuscle has been described to illustrate the nature of a receptor membrane. Each receptor membrane is specialized to generate electric potentials which are graded in accordance with the amount of stimulus. The processes that occur at the molecular level are not fully understood. In the Pacinian corpuscle, it is suggested that valves are opened mechanically by stretching the membrane; in

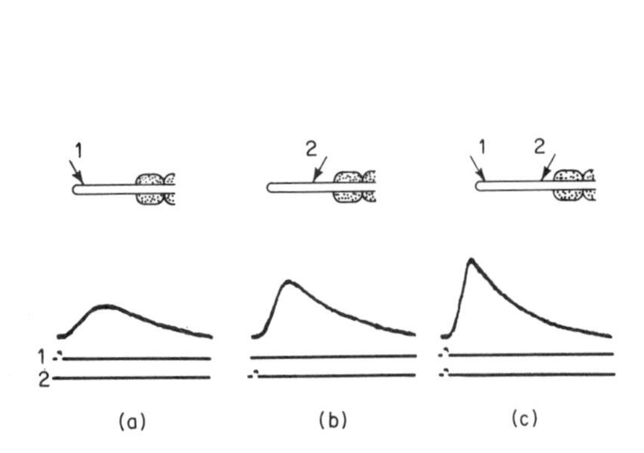

Fig. 16.5 Spatial summation in the receptor membrane. Styli 1 (about 30 μ diameter) and 2 (about 20 μ diameter) belong to independent crystals that stimulate two membrane spots about 400 μ apart. Generator potentials (*upper beam*) in response to a mechanical pulse applied in (a) to spot 1; in (b), to spot 2; and in (c) to both spots simultaneously. *Lower beams* signal pulses of styli. Calibration: 1 msec.; 50 μv. [From Loewenstein, 1961: Ann. N.Y. Acad. Sci. *94*: 513.]

chemoreceptors, the exciting chemical is thought to combine with a receptor substance in the membrane; in photic receptors, light presumably alters a photosensitive pigment. In all cases, increased permeability occurs; it seems likely that both Na^+ and K^+ (perhaps also other ions) are involved. With sufficient depolarization, a spike potential is triggered in the associated electrically excitable and conducting portion of the system.

Each receptor membrane has a limited capacity to trigger nerve impulses. At the lower level, the membrane has a definite threshold for excitation; at the upper level, it shows SATURATION—an inability to respond when the stimulus exceeds a certain maximum. In a complex sense organ, such as the eye or the ear, the monitoring of a broad range of environmental stimuli depends on the interacting populations of receptor cells with units of differing thresholds and saturation levels (Mellon, 1968; Miles, 1969).

Primary and Secondary Sense Organs

PRIMARY RECEPTOR CELLS are neurons in which a portion of the dendrite or afferent fiber forms a receptor membrane (Figs. 16.3, 16.6, 16.10). SECONDARY RECEPTOR CELLS are not nerve cells but specialized epithelial cells which form synapses with transmitting neurons (Fig. 16.7). In both types, the receptor membrane frequently develops from or is associated with cilia (Vinnikov, 1965).

Primary receptor cells are usually considered to be more primitive since most of the invertebrate sense organs are of this type while both kinds are found among the vertebrates. This view, however, has been challenged (Grundfest, 1965a). It is now recognized that at least one invertebrate receptor (the eccentric cell of the *Limulus* eye) is a secondary receptor; moreover, relatively few invertebrate receptors have been carefully studied and other examples of secondary receptors may yet be found among more primitive animals. Grundfest (1965a) argues that both sensory receptors and neurons have developed from ancestral epithelial cells which

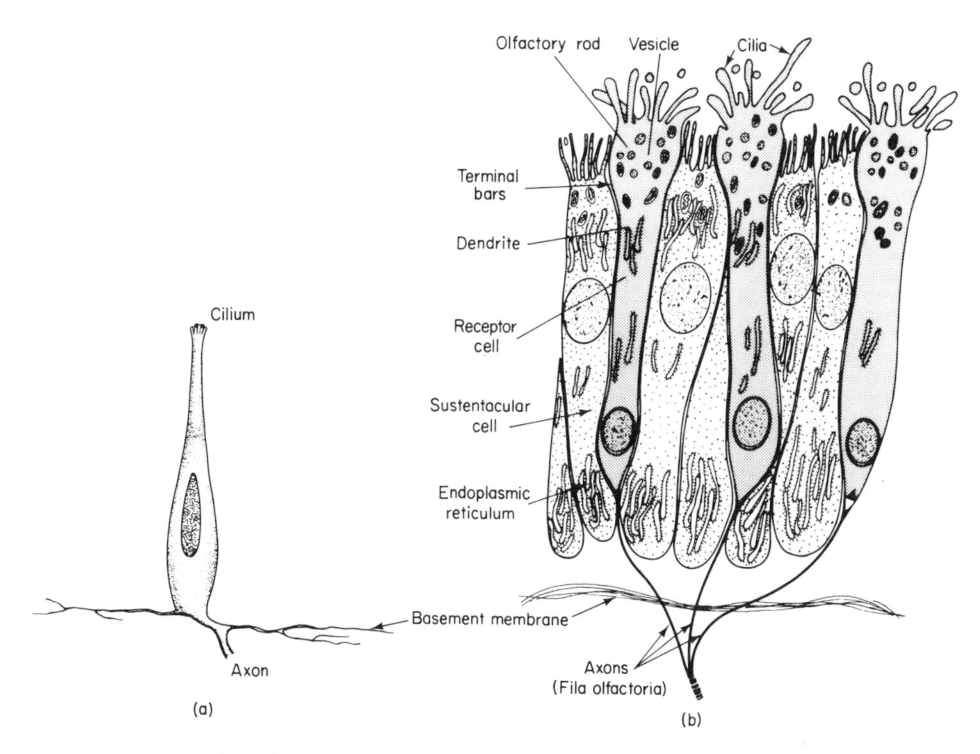

Fig. 16.6 Primary receptor cells. (a) From the sucker of *Octopus vulgaris*. [From Graziadei, 1964: Z. Zellforsch., *64*: 511.] (b) Olfactory mucosa of man. [Based on de Lorenzo, ed., 1963: *In* Olfaction and Taste, Zotterman, ed. Pergamon, Oxford.]

were secretory in nature. These ancestral cells specialized in two directions: (1) the development of stimulus-specific receptor surfaces which are highly sensitive to some particular stimulus and (2) the differentiation of an elongated cellular process which resulted in neurosecretory cells on the one hand and discrete neurons on the other. Thus, one line of phylogeny led to receptor cells while the other produced conducting neurons (Chapter 21).

Physiologically, sensory neurons or primary receptors usually monitor a narrower range of stimulus intensity than do the secondary sense organs. It has been suggested that the interposition of a receptor cell between the stimulus and the neuron, as found in secondary receptors, may be physiologically important and account for the broader range of stimulus energies detected by many secondary receptors (Grundfest, 1965b). However, only very broad generalizations can be drawn with respect to differing sensitivities of these two types of receptors. Some insect chemoreceptors respond to an extremely broad range of intensities while, among the vertebrates, the taste buds, which are secondary sense cells, are intermediate in receptor capacity between the free nerve terminals which subserve the relatively

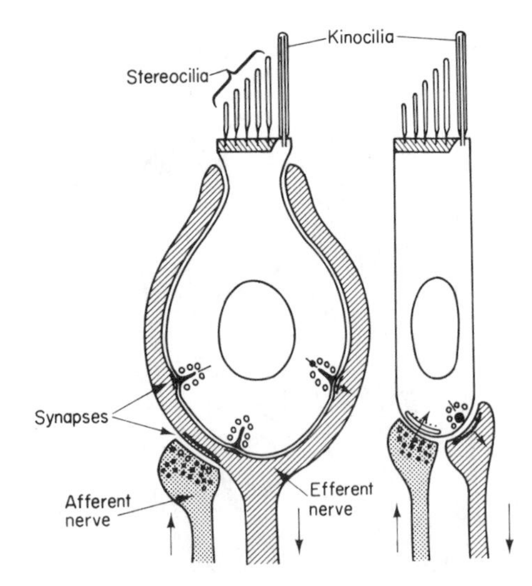

Fig. 16.7 Schematic drawing of the relation between afferent and efferent nerve fibers and the vestibular sensory cells in the mammalian labyrinth. [From Wersall, Gleisner, and Lundquist, 1967: *In* Myotactic, Kinesthetic and Vestibular Mechanisms, A. V. S. de Reuck and J. Knight, eds. J. and A. Churchill, London.]

crude common chemical sense and the olfactory cells which detect extremely complex organic molecules in trace quantities.

Relationships Between Stimulus and Response

The adequate stimulus Receptor specificity is well illustrated in chemoreception. The two morphologically similar dendrites in the labellar hairs of the blowfly provide an excellent example; one of these responds to sugar and the other to salts. Moncrieff (1967) cites numerous examples based on human responses. For instance, in the following series of three nitrotoluidines, the first one is very slightly bitter, the second is tasteless and the third sweet.

CH_3 / NO_2 / NH_2 $\qquad$ CH_3 / NH_2 / NO_2 $\qquad$ CH_3 / NH_2 / NO_2

This display of differential sensitivity is akin to the concept of the ADEQUATE STIMULUS in human sensory physiology. Each receptor is especially sensitive to one form of energy. The classical example is the human retina, the cells of which normally respond to radiant energy of wavelengths between about 400 and 650 nm. Although they can also be stimulated mechanically, they are much more sensitive to radiant energy; light is their adequate stimulus.

Müller introduced the concept of SPECIFIC NERVE ENERGIES to human physiologists many years ago (Heilbrunn, 1952). He pointed out that a sense organ could be stimulated by other than the adequate stimulus, but the subjective response is always the same and not influenced by the *kind* of energy. Stimulation of the retina gives the sensation of light, whether the cells are stimulated by radiant energy or mechanically. Physiologists now understand that this must be the case since the only functional activity of a nerve fiber is the membrane potential. The same general principle can be illustrated from comparative physiology. The chemically sensitive hair fibers of the blowfly can also be activated by bending or mechanical stimulation, but the resulting movement of the proboscis is the same whether the hair fibers are excited chemically or mechanically (Hodgson and Roeder, 1956). It should, however, be added that receptor specificity is somewhat less rigid than these early concepts suggest. The human touch receptor, for example, responds not only to tactile stimuli but also to temperature changes (Eyzaguirre, 1969).

Phasic and tonic receptors Spontaneous activity in receptor nerve fibers was first noted by Hoagland (1933) while recording from the lateral line nerves of the catfish *Ameiurus nebulosus*. Instead of the usual smooth base line characteristic of the recordings from nerves of many unstimulated receptors, the lateral line nerves were found to emit continuous high-frequency pulses. This curious activity changed in frequency when the skin was stimulated by pressure, jets of water and movements of the fish or when the temperature was altered; it was only silenced, however, by the destruction of a section of the nerve connected to the receptor organs (Fig. 16.8).

Since Hoagland's time, physiologists have recognized spontaneous activity in

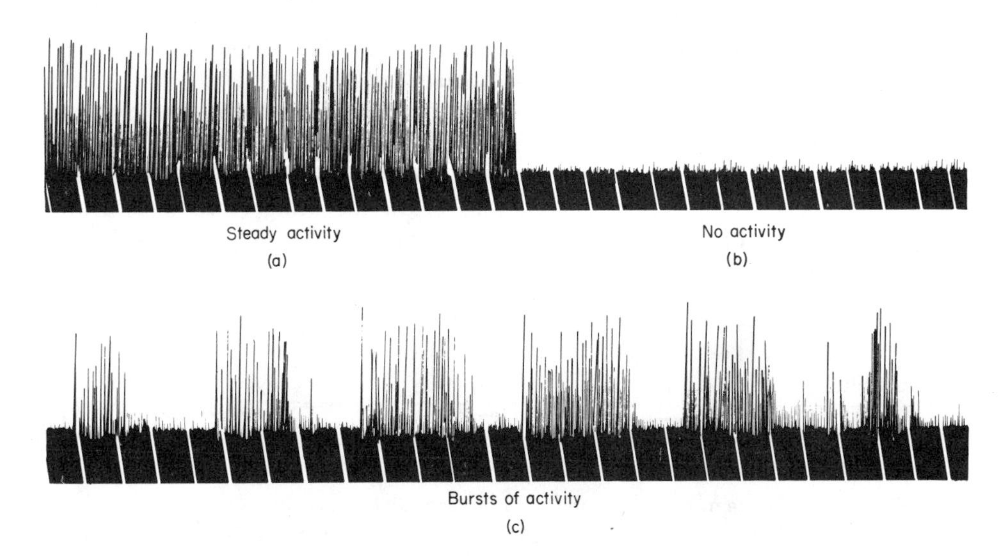

Fig. 16.8 Electrical potentials recorded from the lateral line nerves of the catfish. Time marked in tenths of seconds; (a) spontaneous firing; (b) effect of cutting nerve between its peripheral region and the recording electrodes; (c) response of a branch of the facial nerve to the lower lip when the lip is stroked with a feather. [From Hoagland, 1933: J. Gen. Physiol., *16*: 698.]

many mechanoreceptors (statocyst of the lobster, slowly adapting stretch receptors in crustaceans, receptors in the vertebrate semicircular canals), in some photoreceptors (eyes of *Limulus* and certain molluscs) and in a few chemoreceptors such as those of the carotid sinus in mammals (Florey, 1966). Receptors which are spontaneously active when apparently unstimulated are called TONIC RECEPTORS while those which remain silent unless stimulated are termed PHASIC RECEPTORS. Contrasts between the electrical activity of the two types are shown in Figs 16.2 and 16.8. The basic physiological difference between the two seems to be related to the magnitude of the resting potential (Florey, 1966). This is very close to the firing level in tonic receptors so that firing is continuous, but in phasic receptors considerable generator potential must develop before firing occurs.

Adaptation Even though a receptor is steadily stimulated, its excitability gradually declines (Fig. 16.9). This phenomenon is called ADAPTATION. In tonic receptors the discharge frequency falls very gradually (SLOWLY ADAPTING RECEPTORS), but in the phasic type it is extremely rapid (RAPIDLY ADAPTING RECEPTORS). The two patterns of adaptation have been beautifully demonstrated in recordings from the stretch receptors in crustacean muscles. These tension detectors, first described in Alexandrowicz (1951), occur in pairs (two on each side) in the abdominal segments (Fig. 16.10). They are primary sense organs whose dendrites end in brush-like terminations on special muscle fibers. Each receptor neuron is associated with a single specialized muscle fiber; their axons pass to the ventral nerve cord. One member of this paired system is phasic and rapidly adapting while the other is tonic and slowly adapting. Stretching of the muscles excites the receptors, which are easily accessible and of ample size to permit recording intracellularly either near the sensory terminals or from the associated axons.

Effects of stretching, recorded intracellularly near the sensory terminals, are shown diagrammatically in Fig. 16.9. The rapidly adapting (phasic) receptor responds with only a short burst of electrical activity; even under steady stimulation, the generator potential quickly falls below the firing (threshold) level for spike initiation. In contrast, the slowly adapting (tonic) organ continues to generate

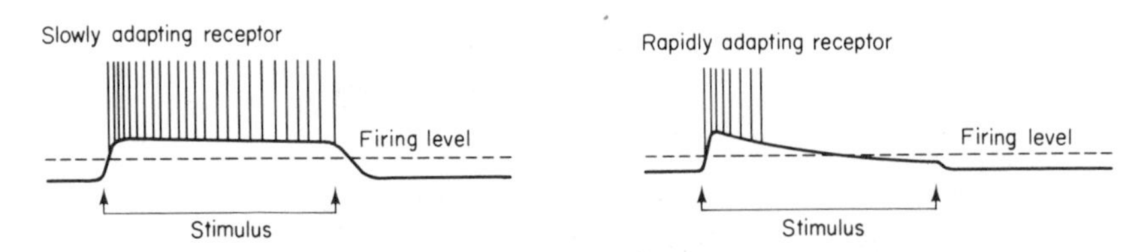

Fig. 16.9 Diagram of events recorded intracellularly near the sensory terminals of the stretch receptors of the crayfish. Note the spike activity (vertical lines) and generator potentials (solid line from extreme left to right). Distance between arrows shows period of stimulation. [From Eyzaguirre, 1969: Physiology of the Nervous System. Copyright © 1969, Year Book Medical Publishers, Inc. Used by permission. Based on Eyzaguirre and Kuffler, 1955: J. Gen. Physiol., *39*: 87.]

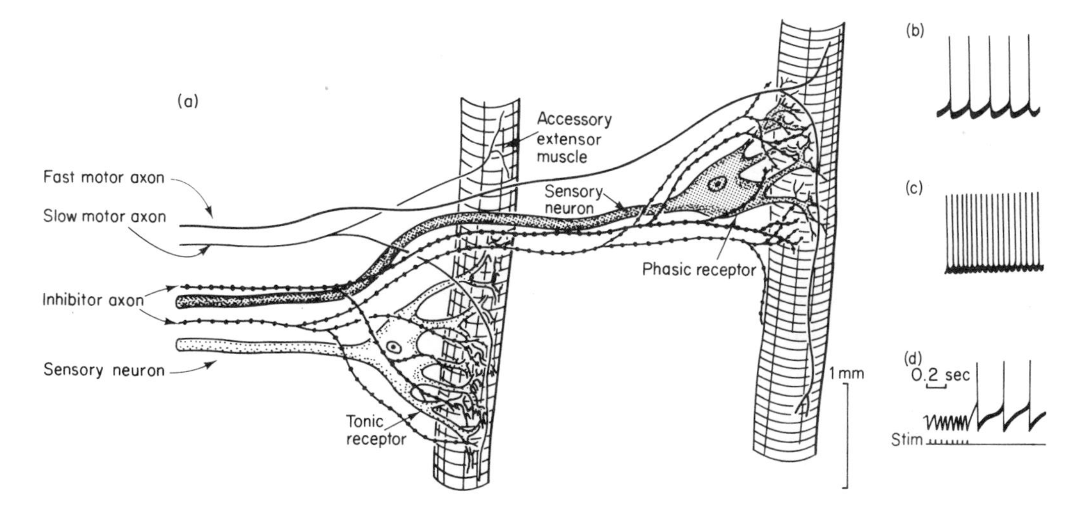

Fig. 16.10 Diagram of the abdominal stretch receptors between the segments in the tail of the crayfish or lobster. (a) Two modified slips of accessory extensor muscle from one side of one intersegmental joint, showing the separate nerves; (b) intracellular response of the slow (tonic) receptor neuron to a small degree of stretch; (c) the same, to greater stretch; (d) suppression of activity by stimulation of the thick inhibitor axon at 20/sec. [From Horridge, 1968b: Interneurons. W. H. Freeman, London, and from Kuffler and Eyzaguirre, 1955: J. Gen. Physiol., *39*: 155.]

potentials throughout the period of stimulation with only a gradual decline in spike frequency and a slight fall in generator potential.

Adrian and Zotterman (1926) did the classical work on sensory adaptation using the stretch receptors in the sterno-cutaneous muscle of the frog. Systematic dissection of this delicate muscle, which contains only three or four stretch receptors, gradually removed all but one of the sense organs; with progressive stretching of the muscle an increased frequency of electrical discharge was quite evident. Although recording instruments have been vastly improved during the last half century, these pioneer investigators were able to establish many of the basic principles of sensory adaptation. The adaptation curve shown in Fig. 16.11a shows not only the gradual decline in activity with steady stimulation but also the momentary inhibition of electrical activity which often follows the sudden withdrawal of the stimulus.

Another important feature of receptor activity is shown in Fig. 16.11b. The magnitude of the response depends on the rate of rise of the stimulus to peak value. This phenomenon is familiar in human sensory experience; gradual changes in temperature or the slow development of odors in a closed room may pass unnoticed while the change can be readily perceived if suddenly encountered by entering the room.

The rapidly adapting receptor plays its biological role by constantly informing the animal of "new" developments in the environment. These receptors soon fail

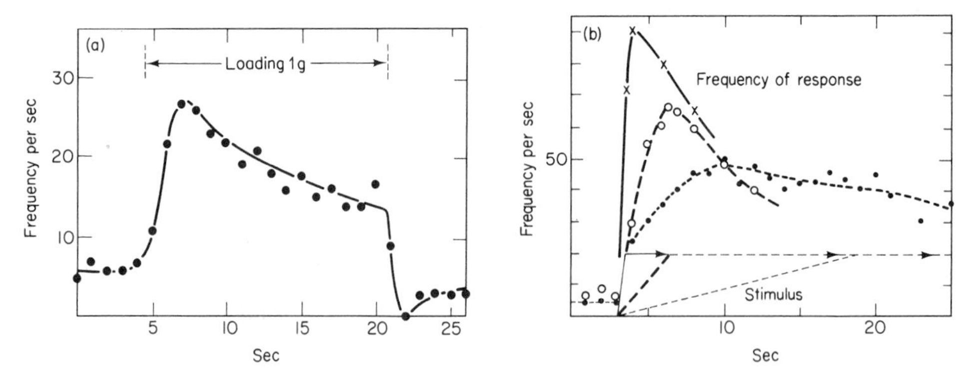

Fig. 16.11 Responses of single sensory end organs in the sternocutaneous muscle of the frog. (a) Frequency of impulses before, during, and after loading with a constant weight at 14°C; (b) effect of various rates of loading. [From Adrian and Zotterman, 1926: J. Physiol., *61*: 168.]

to respond to stimuli which are of minor or transitory importance. The tactile stimulation of clothing quickly ceases to be noticed after dressing; a delicate odor is detected only briefly in spite of repeated attempts to recapture the experience. In part, our alertness to environmental change depends on rapidly adapting receptor organs. The slowly adapting receptor is equally important biologically for it continuously monitors where a failure of input information would lead to disastrous results. The balance in groups of contracting muscles must be constantly regulated if equilibrium is to be maintained; if tension receptors adapted quickly, they would be quite useless. Different receptors are nicely specialized to meet these varied demands. Adrian's (1928) classical diagram showing responses to constant stimulation is reproduced in Fig. 16.12.

Quantitative relationships The frequency of impulses in a receptor neuron is directly related to the magnitude of the generator potential. This direct, arithmetic relationship has been studied in several different sense organs (Mellon, 1968); it

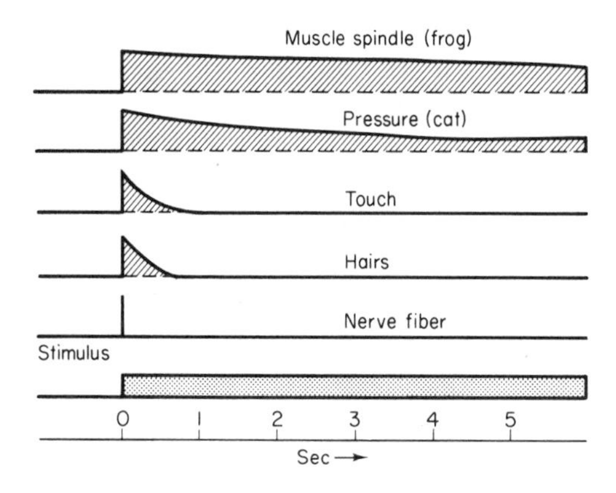

Fig. 16.12 Duration of firing of impulses by various types of end organ, and a single nerve fiber, to a continuous stimulus. [From Young, 1957: The Life of Mammals. Clarendon Press, Oxford, based on Adrian, 1928: The Basis of Sensation.]

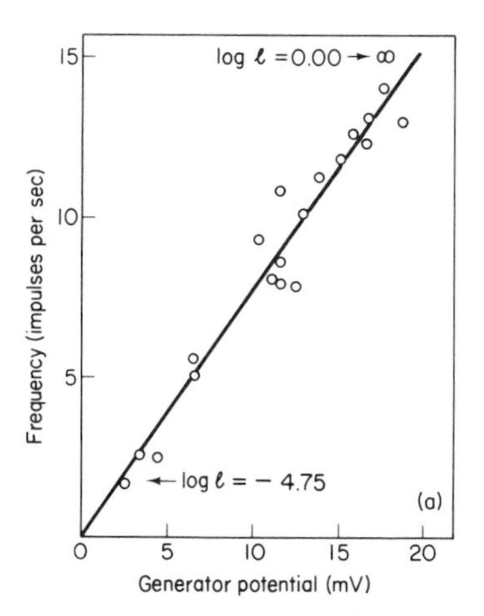

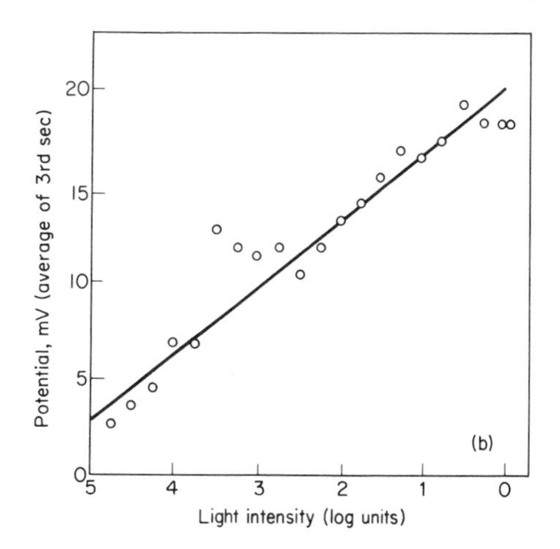

Fig. 16.13 Relation between illumination and electrical potentials in the *Limulus* eye. (a) Generator potentials and frequency of spike discharge; (b) logarithms of illumination and generator potentials. [From MacNichol, 1958: Exp. Cell Res., Suppl. *5*: 416.]

is illustrated by data from a photoreceptor in Fig. 16.13a. In contrast, the quantitative relationships between stimulus intensity and impulse frequency in receptor axons is quite different and usually found to be semilogarithmic—sometimes logarithmic (Werner and Mountcastle, 1965). The semilogarithmic relationship was first studied by B. Matthews (1931) who added progressively heavier weights to one of the delicate muscles in the toe of a frog while recording from a single stretch-sensitive neuron (Fig. 16.14). In some receptors it has been possible to record both generator and spike potentials simultaneously while varying the intensity of stimulus. MacNichol's (1958) data for the *Limulus* eye are shown in Fig. 16.13. An approximately linear relationship between generator potentials and the logarithm of light

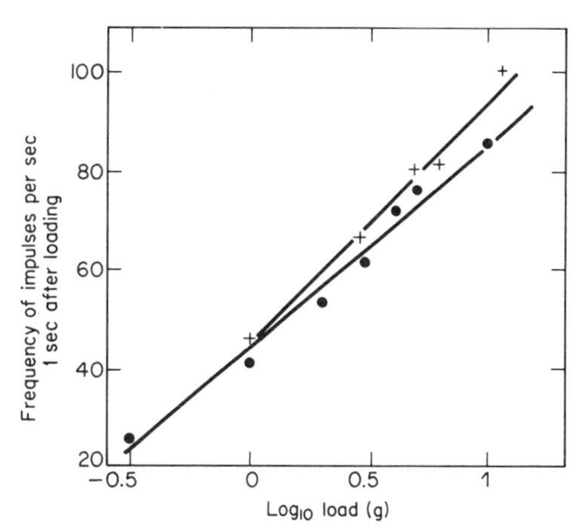

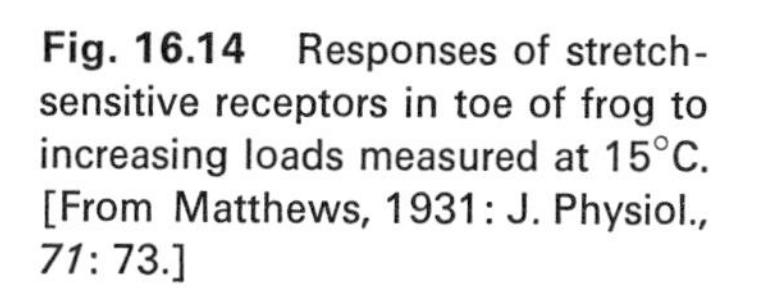
Fig. 16.14 Responses of stretch-sensitive receptors in toe of frog to increasing loads measured at 15°C. [From Matthews, 1931: J. Physiol., *71*: 73.]

intensity was found over a 60,000-fold change in illumination (4.75 log units in Fig. 16.13b); by contrast, the frequency of spike discharge was directly related to the slow generator potentials.

Although the molecular basis of this logarithmic relationship is uncertain, its physiological importance in extending the range of stimulus response is readily apparent. The intensity of sunlight is about 30,000 times greater than that of moonlight, and yet we can see moderately well in both. The range in capacity of nerve fibers to conduct waves of excitation is not more than several hundredfold; the logarithmic relation permits responsiveness of receptors over vastly greater ranges. This phenomenon is discussed in books of general and sensory physiology and is sometimes given the dignity of a law ("Weber-Fechner Law") named after the men who first emphasized its importance. It is not an invariable law but frequently applies over certain ranges of intensities (Schmitt, 1959).

The logarithmic relationship is only one of the physiological adaptations concerned with the monitoring of broad ranges of stimulus intensity. Sense organs consist of batteries of receptor units which differ in excitation thresholds, saturation levels, and rates of adaptation. Differences in these capacities, as well as the modulating influences of the central nervous system, broaden the ranges of stimulus perception. While conduction rates in the individual nerve fibers are fixed at an upper level of about 1000 per sec by the duration of action potentials and refractory periods, the range of stimulus perception is enormously greater—as great as 10^{10} for changes in light intensity. Alert animals must maintain high sensitivity to very small variations in the environment but must still respond to changes over very broad ranges of intensities.

17

RECEPTION OF CHEMICAL, MECHANICAL, AND THERMAL STIMULI

CHEMORECEPTION

Excitation in response to chemicals is fundamental to orientation and communication at all stages in biological organization. At the molecular and ionic levels, specific chemicals alter membrane permeability, trigger the flow of ions and create electrogenesis. The evolutionary process has capitalized on this basic response of cell membranes and produced a system of highly sensitive and specialized chemical receptors. An oriented response toward or away from chemicals (CHEMOTAXIS) has been carefully studied in many single cells such as bacteria, protozoa, leucocytes and spermatozoa (Adler, 1966; Miller, 1966b; Keller and Sorkin, 1968). Many multicellular organisms respond differentially to hundreds of odors and tastes, while their individual cells communicate by way of several well-known nerve transmitters and their organs function in concert through the interplay of chemicals secreted by the endocrine glands. The development of sensitivity to particular chemicals occurs among the most primitive organisms; it assumes a commanding role in all physiological processes and becomes the dominant sense in some of the most specialized animals.

Present understanding of the physiology of chemoreception is based almost exclusively on studies of the arthropods and vertebrates. The insects in particular, because of the superficial location of their receptors and, more especially, because of their suitability for electrophysiological studies, have contributed basic information in this field. Although other invertebrates have been a rich source of interesting data concerning chemical-dependent behavior, they have played a very minor role in physiological studies of chemoreception. The vertebrate literature is extensive, but a large component of it relates to human cultural activities and has been generated by the pharmaceutical and food industries.

Human experience draws a sharp line between the sensations of taste and smell. Taste is a contact chemical sense related to food sampling while the olfactory receptors monitor airborne chemicals from more distant sources. Moreover, the olfactory cells are primary receptors with axons running directly to the olfactory bulb of the forebrain; gustatory receptors, on the contrary, are secondary sense cells which "synapse" with the afferent fibers of three (four in fishes) different cranial nerves passing to the hindbrain. There are several further differences between the organs of taste and smell in higher vertebrates. Taste bud cells, like other epithelial cells, have a limited lifespan (about 10 days in mammals); they are constantly renewed from less differentiated cells at the margin of the bud (Murray, 1969; Beidler, 1970); olfactory cells, in contrast, are neurons with limited capacities for regeneration even under experimental conditions (Guth, 1971; Takagi, 1971). Finally, the relative sensitivities of taste and smell often differ with characteristically higher thresholds for gustatory excitation. This difference is probably a function of neural connections and central processes rather than diversity of receptor membrane physiology. There are millions of olfactory receptor cells in man with as many as a thousand of them connected to a single central neuron (Case, 1966); the possibilities of summation are very great. In comparison, the number of taste receptors is usually small (about 10,000 in man). It should be added, however, that the gustatory machinery may be much more extensive in some lower vertebrates since taste buds are widely distributed on the body surfaces of many fish (Bardach and Atema, 1971; Hara, 1971).

Insects, like vertebrates, have a battery of chemoreceptors on their mouth parts for sampling food, while a different set of sensory cells on the antennae respond to air- or water-borne chemicals from more distant sources. Like the vertebrates also, the receptors on the mouth parts are usually less numerous and more limited in the range of chemicals detected. Antennary chemoreceptors of some insects are incredibly specialized; a single feathery antenna of the male moth *Telea polymorphys* bears about 150,000 receptor cells of which several discrete groups are specialized to detect specific chemicals in molecular amounts. This distinction between the organs of taste and smell in insects is supported by both electrophysiological and ultrastructural evidence (Schneider, 1969). Whether the chemoreceptor systems of other invertebrates have modalities analogous to olfaction and taste seems doubtful; at present, there is insufficient information to extend these comparisons beyond the vertebrates and the insects.

Functional Morphology of Chemoreceptors

Insect chemoreceptors vary in structural details but are all composed of (1) a small group of SENSORY NEURONS, (2) protecting CUTICULAR PARTS that cover the delicate sensory dendrites, and (3) SHEATH CELLS that form the cuticular processes. There are many excellent anatomical descriptions (Dethier, 1963; Schneider and

Steinbrecht, 1968; Steinbrecht, 1969; Slifer, 1970) and only a general summary will be given here; Fig. 17.1 is a diagram of an olfactory sensillum.

There are usually from two to six sensory neurons, but a few organs with single neurons have been identified while some are known to have as many as fifty (Slifer, 1970). These are grouped at the base of the sensillum and send their delicate axons directly to the brain. The dendritic portion of the cell extends peripherally and narrows first into a typical ciliary segment with the characteristic arrangement of nine peripheral fibers. The OUTER SEGMENT of the dendrite, which extends beyond the ciliary portion, may remain unbranched but usually divides into numerous delicate sensory filaments; these fine branches are actually microtubules of 0.02 μ diameter or less.

The chitin-protected space, which surrounds the dendrites, may take the form of a pit, peg, or hairlike papilla (Fig. 17.1). This cuticular process and its base or socket are formed by the sheath cells (TRICHOGEN and TORMOGEN). Liquid of unknown composition surrounds the dendrites within the process. There is either a small pore at the tip of the cuticular process (thick-walled contact receptors) or the wall is pierced by numerous, extremely minute holes; these holes admit the chemical stimulants but restrict or prevent the loss of moisture. Some workers believe that

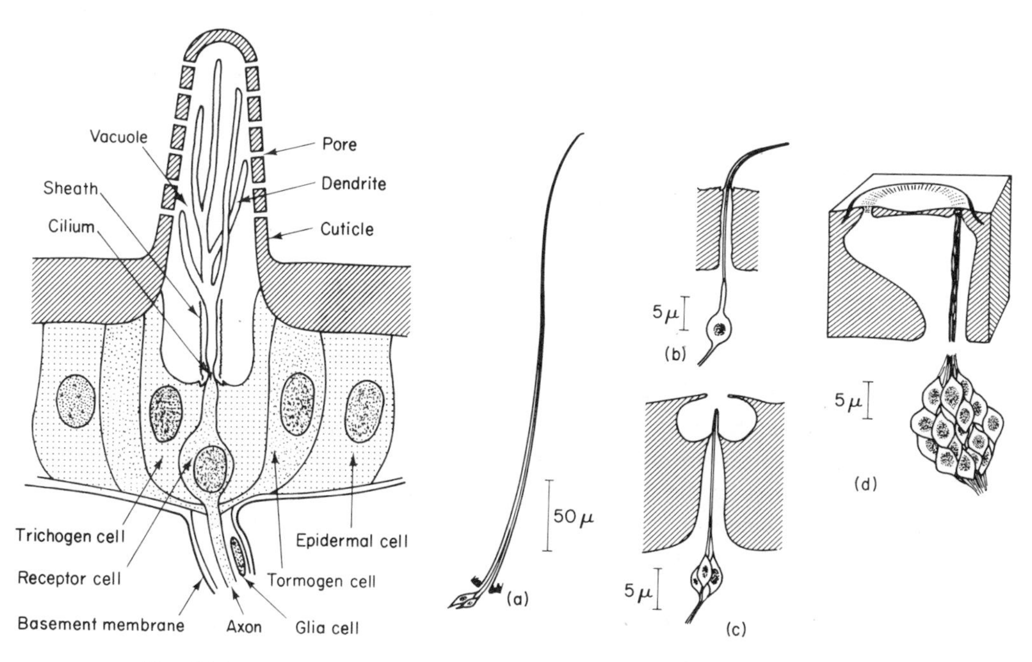

Fig. 17.1 Diagram of insect olfactory receptor. Left, receptor cell and associated structures; right, several types of olfactory sensilla: (a) long hair with dendrites of two sensory cells; (b) short hair; (c) pit organ; (d) pore plate. [From Schneider and Steinbrecht, 1968: Symp. Zool. Soc. London, No. *23*: 281, 282.]

excessively fine microtubular extensions of the dendrites extend in groups into these pores so that the receptor membranes are in direct contact with the outside world; others describe pore-tubule systems which lead as channels through the chitin into the sensillum liquor. In any case, there is an ultrastructural arrangement to permit ready passage of chemicals from the environment to the dendrites.

The different neurons in a sensillum may be highly specific. The chemosensory hairs on the labium of the blowfly have been particularly useful for the analysis of the transducer mechanism, since action currents can be recorded directly from the dendrites. The evidence indicates that one of the neurons responds only to sugars while another is activated by monovalent salts; the neuron which terminates at the base of the hair is a mechanoreceptor while a fourth neuron is excited by water (Dethier, 1962, 1963). This preparation has intriguing possibilities for the student of animal behavior. The stimulation of a single sensory neuron will initiate complete behavior responses of proboscis extension in response to sugar or proboscis retraction or inhibition when the non-sugar receptor is activated.

Several cells from the mammalian olfactory mucosa are shown in Fig. 16.6b. The cells of the taste bud are similar in appearance. However, taste cells are secondary receptors and not specialized nerve endings; consequently, they synapse at their bases with the dendrites of the gustatory nerves. Both taste and olfactory receptors are slender, rod-shaped epithelial cells which bear microvilli or sensory hairs on an exposed surface and are associated with sustentacular cells which lack associations with sensory nerves. In addition to the receptors concerned with taste and smell, there are secondary chemosensors in the aortic and carotid bodies which monitor oxygen and carbon dioxide (Eyzaguirre and Koyano, 1965), while unspecialized dendrites in various mucous membranes (also the skin of aquatic forms) respond to a wide range of chemical irritants. This latter sensory system was termed the COMMON CHEMICAL SENSE by Parker (1922) who considered it phylogenetically primitive and the evolutionary basis of the more specialized senses of taste and smell. A comparable receptor system has been claimed for insects (Roeder, 1953), fishes (Hara, 1971), and many primitive organisms. However, these naked nerve endings are usually stimulated by high concentrations of irritants (particularly acids and alkalis) and the reactions seem to lack the specificity normally associated with organs of special sense; they may be nonspecific manifestations of tissue damage.

Taste receptors are usually confined to the oral cavity while olfactory organs occur in pouches on the snout (aquatic forms) or within the olfactory chambers of the nasal air channels (terrestrial forms). There are many curious differences in the details of functional morphology. Gustatory receptors occur on the barbels, fins, and body surfaces of some teleosts—mostly freshwater, bottom-dwelling species; many nice correlations between dermal taste cells and the ecology and feeding habits have been recorded (Bardach and Atema, 1971); olfactory sacs are likewise variously adapted to the ecological requirements of fishes (Kleerekoper, 1969; Hara, 1971). The VOMERONASAL (JACOBSON'S) ORGAN is a highly specialized olfactory chamber in many terrestrial forms (Romer, 1970). This organ is vestigial in birds but found in the majority of other tetrapods (Tucker, 1971); it reaches its climax in some of the reptiles (Parsons, 1970). In the more primitive situations, grooved channels or pockets of olfactory epithelium develop along each side of the nasal sac and open into the roof of the mouth; in lizards and snakes, these pouches become completely separated from the nasal passages and occur as blind pouches opening into the roof of the mouth. They are served by distinct branches of the olfactory nerves and appear ideally located to pick up odors from food after it enters the mouth.

Transducer Mechanisms in Chemoreception

There is no lack of speculation concerning the mechanisms of chemoreception. "Olfactory theories are as numerous as pebbles on the beach" (Davies, 1971). Many of the older theories can now be discarded and attention focused on several which postulate a physical interaction between the stimulating chemical and a receptor site of molecular dimensions on the cell membrane (Beets, 1971); this interaction is highly specific and reversible. The most convincing evidence comes from studies of the behavior of intact animals and the electrophysiological responses of their receptor cells to specific stimulants. Insects have been especially valuable experimental subjects since large numbers can be inexpensively tested; their responses are stereotyped and easily quantified; and their receptor cells may be stimulated individually with satisfactory monitoring of the electric potentials. The vertebrates have also played their part in both types of investigations. The relationships most frequently studied have been concerned with the chemical constitution of the stimulating molecule, molecular concentrations, enzyme inhibitors and temperature.

Transducing mechanisms are probably similar in all chemoreceptors. The most likely sites of transduction are the microvilli and cilia of the sensory cells (Ottoson, 1970; Davies, 1971). In some manner, not yet understood, the interaction between stimulant and receptor molecule alters membrane permeability and leads to its depolarization or hyperpolarization. The process may be basically similar to that which occurs at synaptic and myoneural junctions where the complexing or adsorption of the transmitter is thought to open ion valves and trigger electrogenesis. The receptor molecules are probably proteins. Although the evidence for this is still largely circumstantial and based on arguments of specificity of interaction, recent chemical work indicates that highly specific receptor proteins can be extracted from chemoreceptors of both insects and vertebrates (Amoore, 1970).

The nature of the interaction between stimulant and receptor molecule remains speculative. Davies (1971) describes the process as "penetration and puncturing" and postulates the adsorption of a stimulating molecule to a specific receptor molecule on the cell membrane; the diffusion of this complex through the lipid membrane leaves a hole by which ions penetrate. Amoore (1970) postulates a lock and key transducing mechanism with a series of molecular traps or holes of differing but specific shapes in the cell membrane; the shape of the stimulating chemical corresponds to that of the hole; in some manner, this fitting together of molecules creates generator potentials. Wright (1964) has attempted to correlate the far infrared vibrational frequencies of certain molecules with odor quality (Wright and Burgess, 1970). The active molecules might initiate nerve impulses through their vibrations or they might excite pigments which are conspicuous in some chemoreceptors. There is currently more support for hypotheses involving an interaction between stimulant and receptor molecule than there is for vibrational theories (Beets, 1971; Blum *et al.,* 1971; Moulton, 1971), but the details of the complexing remain uncertain.

The Chemical Sense and Animal Orientation

The more primitive animals possess only a CONTACT CHEMICAL SENSE while the more advanced ones have specialized DISTANCE RECEPTORS as well. The contact receptors of lower forms have relatively high thresholds for stimulation and the resultant behavior is largely trial and error. Specialized distance receptors are excited by minute quantities of chemicals which often arise at great distances and initiate very complex behavior such as aggression, migration or copulation. This distinction between contact chemical and distance receptors is a useful one in comparative physiology. Contact receptors are usually localized in the mouth region and their excitation elicits feeding or avoidance reactions; distance receptors have much lower thresholds and generate very complex activities. During evolution, simple trial-and-error behavior has differentiated into highly specific oriented behavior based on separate anatomical parts for testing foods and sampling distance odors.

The contact chemical sense Fraenkel and Gunn (1940) depict trails of protozoans, flatworms, insects and other invertebrates in the presence of food attractants or repellent chemicals. The animals always reach their goals by indirect (trial and error) routes. The reactions are classed as kineses and klinotaxes. At low concentrations only the speed or frequency of locomotion may be altered (ORTHOKINESIS); at somewhat higher concentrations there may be an increase in the amount or frequency of turning per unit time (KLINOKINESIS). Orientation becomes more precise in close proximity to the chemical and is then a DIRECTED REACTION either toward the chemical or away from it (TAXIS). In chemotaxis, however, movements with respect to the stimulus are not straightline orientations (TROPOTAXIS), but the goal is attained as the animal moves its body from side to side testing the environment during its progression (KLINOTAXIS).

This sort of classification is adequate to describe the movements of planarians, leeches or mites toward bait or their avoidance of acids but does not exhaust the capacities of animals to exploit the contact chemical senses. The honeybee which informs its companions of a source of nectar by the form of its dance also tells them how sweet the nectar is by the vigor of the dance (von Frisch, 1971). Spiders sample the insects caught in their webs and treat a fly very differently from a wasp. The wasp is bundled up in a web without being "tasted," and a fly will be treated in the same way if it is first soaked in turpentine (Carthy, 1958). The starfish *Asterias rubens* responds positively to live mussels and, under experimental conditions, can locate them in a simple maze; dead mussels are avoided, as are mussel extracts, certain amino acids and some starfish predators (Castilla and Crisp, 1970; Castilla, 1972). This simple behavior involves highly adaptive preferences, avoidances, and the recognition of complex chemicals. Juvenile Pacific salmon of the genus *Oncorhynchus* show a preference for water of a specific salinity; this preference changes seasonally in a progressive manner and may lead the fish through estuaries into the ocean (McInerney, 1964). Adult Pacific salmon recognize "home stream" waters and respond positively to them; they avoid water containing molecular amounts of L-serine, a substance often found in the skin of their predators (Hara, 1971).

Sensing of distant chemicals Several evolutionary advances were possible as animals extended their range for detection of elaborate organic molecules. One of the most significant is the production of species-specific odors—PHEROMONES (Karlson and Lüscher, 1959). This may be thought of as the positive approach to the development of a language based on the chemical senses. These species-peculiar scents are elaborated by different epithelial glands in the epidermis, the oro-anal and the urinogenital regions. Chemically, they belong to diverse groups of compounds (amino acids, alcohols, organic acids, lipids). Two functional categories are conveniently recognized. RELEASER PHEROMONES initiate specific patterns of behavior; they serve as powerful sex attractants, mark territories or trails, initiate alarm reactions, or bring about aggregations of individuals. PRIMER PHEROMONES trigger physiological changes in endocrine activity or metabolism—usually related to sexual maturation, growth or metamorphosis. A summary of the literature and excellent bibliographies will be found in a volume edited by Pfaffmann (1969).

The full exploitation of the chemical sense in animal integration depends on a storage of information pertaining to the particular molecule. In the lower forms, both invertebrate and vertebrate, there is a considerable capacity for the inheritance of odor-dependent reactions. Complex interactions between the sexes, between predator and prey or between different members of a species often depend on innate behavior patterns which are released by odors. In the higher groups these relationships are more frequently built up through learning and experience. With increasing encephalization there is a progressively later differentiation of reactions in accordance with the experiences of the animal.

A second evolutionary trend is also evident in odor-dependent behavior. Students of animal behavior recognize *species odors* which signal relations with other members of the same species, *individual odors* which permit mutual recognition of particular individuals, and *community odors* which are distributed throughout the colony or territory to mark its boundaries; in this line of phylogeny, specific chemicals have acquired discrete signaling functions and form the basis of complex social behavior. The ontogeny of odor communication is correlated with the development of the endocrine glands as well as the differentiation of the nervous system (Schultze-Westrum, 1969; Pfaffmann, 1971).

Some examples of chemoreceptor-dependent behavior Only a few examples have been selected to illustrate the range of behavior which has become odor-dependent. The extensive literature is reviewed in many places (Carthy, 1958; Busnel, 1963; Wilson, 1965; Sebeok, 1968).

Chemoreception associated with feeding has sometimes been elaborated into complex food-finding behavior and communication of information concerning the source or nature of the food. Von Frisch's (1971) study of the honeybee provides a classical example. Through simple conditioning experiments he first showed the capacity of the bee to detect and to discriminate many different odors. He then demonstrated that bees which had located a rich source of scented nectar or sugar water carried this odor back to the hive and passed it on to their companions both directly from the outside of their bodies and by regurgitating some of the scented material. In addition, special scent glands are used to mark flowers which are rich in nectar and these odors also guide the searchers in their hunt for the food. The odors of both the plants and the bees are woven into a language of communication associated with "food finding" (Wenner *et al*, 1969). There are many similar examples among the insects. Male bark beetles feed on certain species of *Pinus* and produce a substance in their feces which attracts beetles of both sexes. This aggregating substance has now been identified (Vité and Renwick, 1971); it brings the males and females together where there is suitable food.

Many social insects produce alarm substances which excite other members of the species to attack invaders. These chemicals are among the best known of pheromones—both biologically and chemically (Blum, 1969). The honeybee, for example, injects 2-heptanone into the victim it stings; this marks the intruder so that its chances of getting more stings increase (Free and Simpson, 1968).

Many animals depend on their olfactory organs for the location and selection of a mate. The attractions which females of certain species of moth show for their males have long excited the wonder of naturalists. Females of the common silk moth *Bombyx mori* produce a scent which is attractive to males in concentrations of 10 ng. This material (a complex alcohol) has been isolated and its chemical structure established in one of the great feats of microchemistry (Karlson, 1960). The literature on insect sex attractants is now extensive (Jacobson, 1972). This is also a well-known means of sexual recognition in some fishes and many mammals (Gandolfi, 1969; Rossi, 1969; Pfaffmann, 1969, 1971).

Parasitism and commensalism often depend on chemoreception. The larvae of the ichneumon wasp, *Ephialtes ruficollis*, parasitize the caterpillars of the pine shoot moth, *Rhyacionia buoliana*. The adult parasite is repelled by pine oil when it emerges from its host in the pine woods. It leaves the pines for three or four weeks, but as it becomes sexually mature the pine oil becomes strongly attractive, and it then returns to the woods to parasitize the caterpillars of the pine shoot moth (Carthy, 1958).

A curious story in the evolution of epidermal glands in one group of teleost fishes (Ostariophysi) has been selected for the final example (Pfeiffer, 1962, 1963). Specialized epidermal cells (CLUB CELLS) elaborate a chemical which is another of the pheromones; this substance, when released from the fish, initiates a fright reaction in the same or closely related species. These special glandular cells have no ducts or openings to the exterior and, in histological sections, look like tiny sacs or bubbles of homogeneous staining material. When the skin is injured, their contents are liberated into the water. Although the skin contains active cells at hatching, the fright reaction cannot be elicted until the fish are considerably older (50 days in *Phoxinus*). Thus, the very young fish when injured will frighten the adults but will not themselves be alarmed; this reaction provides important insurance against cannibalism. The behavior is innate but is intensified through experience.

Like other groups of fish, the Ostariophysi can be conditioned rapidly to odors, and the innate fright reaction (*Schreckreaktion*) can be elaborated and intensified through conditioning. For example, when a pike attacks and alarms a school of minnows, the odor of the pike also becomes associated with the fright reaction; pike odor then takes on a meaning which it did not have before the encounter. Reactions to pike odor are not innate although the *Schreckreaktion* is. Likewise, visual stimuli may become associated with the fright reaction to give it added emphasis in certain situations. This fright reaction is confined to the Ostariophysi among fishes and the Bufonidae among amphibians (Pfeiffer, 1966); it has considerable biological significance for schooling animals. Although it does not protect the individuals under attack, it operates to the benefit of the group.

MECHANORECEPTION

Many cells are excited by contact stimuli which stretch or wrinkle their surface membranes. This is easily demonstrated with protozoa and some other cells not particularly specialized as receptor organs. Avoiding responses to contact stimuli were probably the phylogenetic bases of this highly important system of receptors which, in more advanced groups, collects information from distant sources as well as local contacts and informs the central nervous system of both the internal body pressures and those which arise in the external environment.

The proprioceptors in the organs of locomotion transmit steadily to the central nervous centers so that tensions are properly adjusted and body equilibrium is maintained. There is also a constant monitoring of the position of the body in relation to gravity and angular acceleration. Pressure receptors in the viscera may function with respect to such diverse phenomena as the sensing of hydrostatic pressure or visceral disturbances which produce pain. Distant vibrations often provide information concerning dangers, mates and foods while the actual production of vibrations becomes the basis of a special language and orientation in a manner parallel to that already considered for the chemical receptor system.

The Pacinian corpuscle, one of the better known mechanoreceptors, was selected to illustrate several general properties of receptors (Chapter 16). The transduction of mechanical energy to electrical activity was related to the stretching or deformation of the receptor membrane; stretching was assumed to open ion gates or valves and trigger electrogenesis. Although this is a satisfactory working hypothesis, there is little experimental evidence for it (Catton, 1970; Aidley, 1971) and the actual physiology may be quite different. Current concepts of the mechanism are speculative and will not be pursued further. This section is restricted to brief descriptions of the functional morphology of several types of mechanoreceptors and to comments on mechanoreception in relation to animal behavior and communication.

Statocysts and Tactile Hairs

Superficial receptors that monitor vibrations and gravity in the invertebrates usually take the form of statocysts and tactile hairs. The receptor membranes are specialized dendrites (primary receptors) in both types of organ and probably function in a similar manner. The actual organs in which these dendrites operate, however, occur in many curious models.

Statocysts The statocyst of the scallop *Pecten*, illustrated in Fig. 17.2a, is a simple sac lined with hair cells and supporting cells. The sac contains a central STATOLITH; this is a heavy calcareous structure formed by some of the lining cells. The delicate hairlike filaments of the sensory cells are activated by changes in the position of the statolith. As this falls on different groups of hairs, impulses are relayed to the nervous system to provide information concerning the position of the animal with respect to gravity. Barber and Dilly (1969) have described the electron microscopy of this organ.

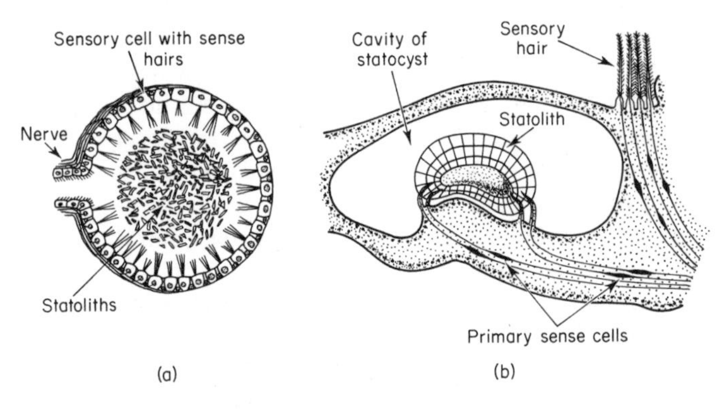

Fig. 17.2 Statocysts of invertebrate animals. (a) *Pecten* with a free statolith. (b) *Leptomysis* with dendritic hairs attached to the statolith. Free sensory hairs with attached dendrites are also shown in (b). [From Heidermanns, 1957: Grundzüge der Tierphysiologie. Gustav Fischer Verlag, Stuttgart.]

Similar statocysts are found in many different invertebrates at all levels in phylogeny from the coelenterates and flatworms to the molluscs and arthropods. Such organs are very ancient evolutionary acquisitions. The sensory cells are all built on the same principle; their receptor membranes appear to have evolved from motile cilia or flagella which characteristically respond to bending and behave as simple mechanoreceptors (Horridge, 1971a). Statocysts, although all built on similar principles, have probably evolved many times. Non-motile cilia associated with high-density material have succeeded motile cilia. Many statocysts have openings or ducts to the exterior, and this suggests that surface vesicles preceded closed sacs during evolution. The cephalopod statocyst is the most specialized of these invertebrate organs, with several functional parts and a capacity to respond to angular acceleration and low-frequency vibrations as well as changes in gravity (Katsuki, 1965; Barber, 1968).

Statoliths may be compact homogeneous concretions or masses of small crystals (STATOCONIA) imbedded in gelatinous material. In the type illustrated for *Leptomysis,* a crustacean, the dendrites are anchored in a sac containing the statolith (Fig. 17.2b). Although statoliths are usually secreted by the lining cells, the decapod crustaceans rely on the haphazard accumulation of sand grains through a special opening into the sac; these particles must be renewed after each molt (Cohen and Dijkgraaf, 1961; Case, 1966).

Tactile hairs and chordotonal sensilla Non-motile cilia, sensory hairs and bristles are common on the exposed surfaces of many different invertebrates. They occur in great variety and serve as tactile and sound receptors, detectors of water currents (RHEORECEPTORS) and monitors of air movements.

The most widely distributed of these are non-motile cilia which serve as delicate detectors of vibrations at all levels in phylogeny. They appear to be the evolutionary basis of the most specialized distance receptors, the auditory organs of higher vertebrates. There are now several carefully studied examples from the lower metazoa. The finger-like processes of the ctenophore *Leucothea* bear several types of epithelial receptor cells with non-motile cilia (Fig. 17.3). The "fingers" which remain flaccid when the animal is undisturbed suddenly erect to about twice their resting length if there is the slightest disturbance in the water; clearly, they are involved in locating and capturing the prey of this voracious carnivore (Horridge, 1965b). The rheoreceptors of the turbellarian *Mesostoma* are large neurons with long filamentous dendrites which extend beyond the ordinary surface cilia (Hyman, 1951). The chaetognath *Spadella* feeds in total darkness, localizing small copepods and other planktonic forms with highly sensitive vibration detectors (Horridge, 1966a). The receptors are fan-shaped groups of ciliated neuroepithelial cells (Fig. 17.4) which respond to the swimming vibrations of plankters; *Spadella* will grab the end of a vibrating glass needle (frequency, about 10/sec; amplitude, about 300 μ). In the internal ear of the vertebrate, sensory cells bearing non-motile cilia not only detect vibrations but also monitor frequency and loudness of sound as well as rotational movements, acceleration, and gravity.

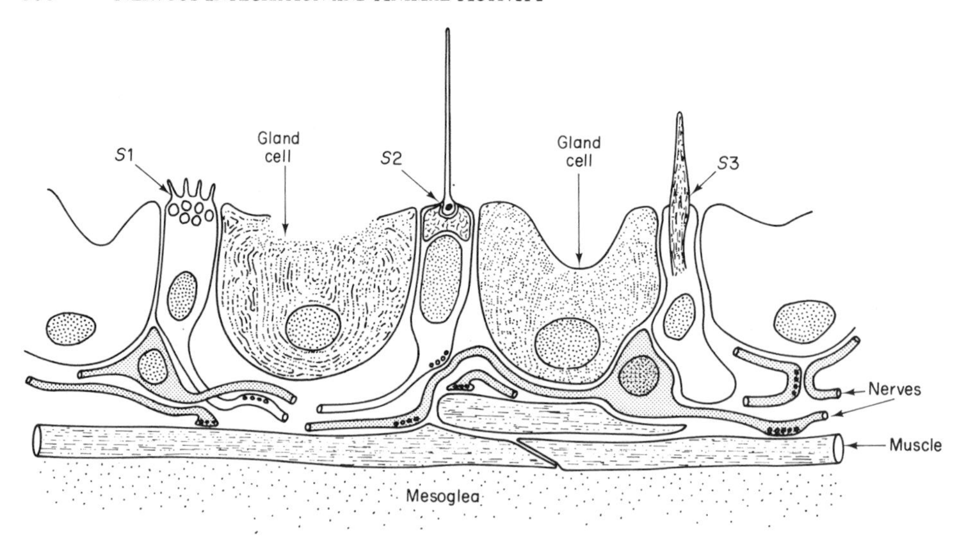

Fig. 17.3 Tactile hairs on the epithelium of the "finger" of the ctenophore *Leucothea*. Three types of sensory cell are shown from left to right; *S*1 bears a group of short sensory cilia; *S*2 bears a solitary long cilium while *S*3 is in the form of a sensory peg. Rows of small vesicles at the bases of the sense cells show synapses with underlying nerves and muscle; large gland cells occur between the sense cells. [From Horridge, 1965b: Proc. Roy. Soc., *B162*: 339.]

Two forms of tactile hair from insects are shown in Fig. 17.5 The presence of a chitinous exoskeleton requires a special articulation between the dendrite proper and the stiff tactile bristle. Simple hair sensilla may be sufficiently specialized to serve as a primitive acoustic organ when they vibrate in their tiny sockets and pull on the nerves (Schwartzkopff, 1963). However, some insects have very acute hearing and this requires batteries of receptor units or scolopidia arranged in specialized hearing organs.

The SCOLOPIDIUM or scolophorous sense cell (Fig. 17.5c) is similar to a hair cell but does not protrude from a body surface. The dendritic filament is surrounded by an envelope and attached to a cap cell; both the cap cell and the axonal end of the sense cell are anchored, either directly or indirectly, to membranes of pliable cuticle while the receptor itself is stretched across a fluid- or air-filled space. There are many varieties of scolopod sensilla; they have been utilized by insects to form delicate organs for sensing tactile stimuli and vibrations from both near and far.

A group of scolopidia forms a CHORDOTONAL ORGAN. These organs may stretch across a fluid-filled space as in the subgenual organs in the legs or Johnston's organ in the second antennal segment. The former serves as a tactile organ or pickup for low-frequency vibrations (200 to 6000 cycles per sec) from the ground or substratum; the latter appear to be "statical organs" used to register antennal movements or positions and maintain flight patterns in variable air currents, or normal swimming positions in water. Johnston's organ is most highly developed in the Culicidae and Chironomidae where it contains several thousand closely crowded sensilla. It has an accessory auditory function in mosquitoes (Culicidae).

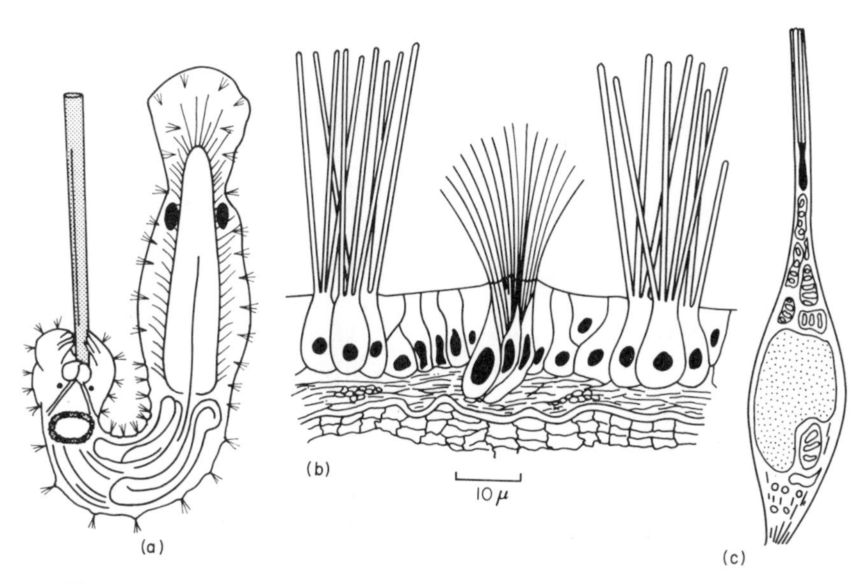

Fig. 17.4 Setae and non-motile cilia of the arrow-worm *Spadella cephaloptera.* (a) Whole animal turning to grab a vibrating probe; (b) two tufts of setae and a fan of ciliated neurons (center); these structures alternate all around the margin of the body; (c) a single sensory neuron showing the base of the non-motile cilium. [From Horridge, 1966a: *In* Some Contemporary Studies in Marine Science, Harold Barnes, Ed. Reprinted with permission of Macmillan Publishing Co., Inc. © George Allen & Unwin Ltd. 1966.]

The TYMPANAL ORGANS of insects are the most specialized of invertebrate sound receptors. An area of thin cuticular membrane (TYMPANUM) on the surface of the body separates the ambient air from an internal air space. When the membrane vibrates, in response to sound, it activates a group of scolopidia which are directly or indirectly attached to it and stretch across an air space. Numbers of scolopidia vary from two in some moths to as many as 1,500 or more in cicadas (Dethier, 1963; Wigglesworth, 1972). Tympanal organs are found only in certain families of Orthoptera, Homoptera, Hemiptera and Lepidoptera.

The insects as a group detect a very wide range of sound. The capacities of some of the night-flying moths are much greater than those of the human ear and in a class with the ears of bats, which are among the most versatile sound receptors of vertebrates. Electrophysiological and behavioral studies show that tympanal organs are sensitive to changes in rhythm (an analysis of temporal patterns) but do not discriminate differences in frequency of the sound waves themselves, i.e., pitch and harmonics (see below).

Insects with tympanal organs are capable of sound production. The noctuid moths, which respond to the ultrasonic cries of bats, are capable of producing ultrasonic sounds, probably warning the bats that they are loaded with nasty odors or tastes (Dunning, 1968). Communication between night-flying moths and insect-

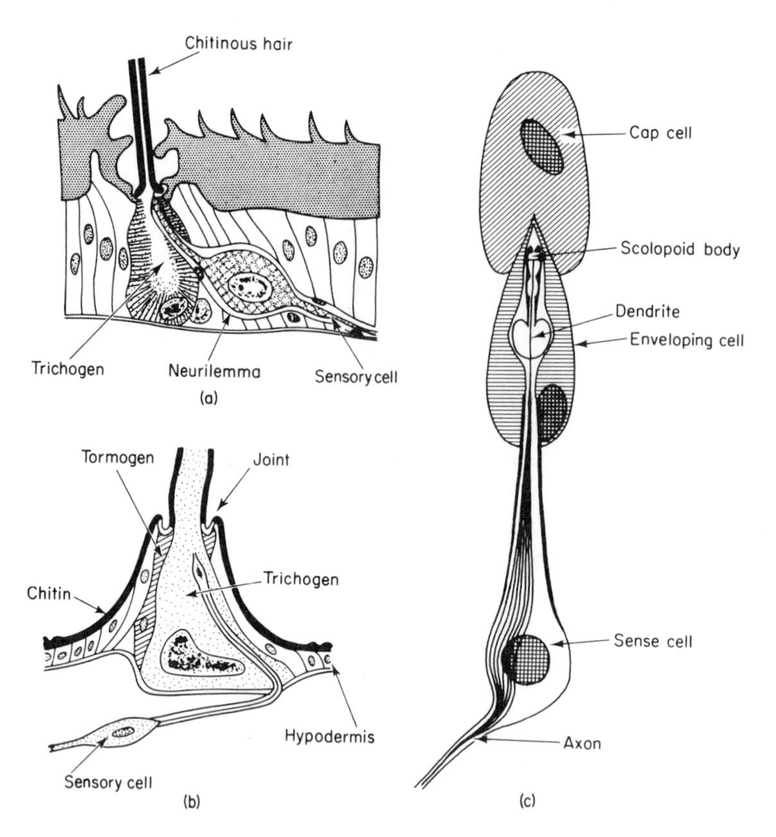

Fig. 17.5 Mechanoreceptors of insects. (a) Hair sensillum from the cercus of the cricket *Liogryllus*; (b) hair sensillum from the caterpillar *Pieris*; (c) diagram of a chordotonal sensillum. Note that the trichogen is the hair-forming cell while the tormogen secretes the chitinous joint membrane. [(a and b) After Weber, 1933: Lehrbuch der Entomologie. Gustav Fischer Verlag, Jena. (c) From Dethier, 1963: The Physiology of Insect Senses. Methuen, London after Schwabe, 1906: Zoologica, Stuttgart, *20*: 1.]

eating bats, animals which are very distantly related phylogenetically, is based on organs of transmission and reception which are constructed on entirely different, but equally effective plans.

The Acoustico-Lateralis System

The neuromasts or sensory hillocks of the aquatic vertebrates proved to be the most versatile of all mechanoreceptors in the evolution of specialized transducers for vibratory stimuli. Basically, they are organs of touch and monitor water currents at short distances. In phylogeny they have been turned to the detection of gravitational forces, accelerations, and the precise analysis of sounds from distant sources. A review by Dijkgraaf (1963) and the symposium volume edited by Cahn (1967) summarize earlier theories, pioneer investigations, and our current understanding based on sophisticated techniques of electron microscopy and physiology.

A neuromast or sensory hillock of a fish consists of a cluster of pear-shaped secondary sense cells supported in a basket-like arrangement of tall epithelial cells (Fig. 17.6). Elongated sense hairs project from these receptors into the gelatinous material forming the cupula, a product of the neuromasts. Movements of the cupula activate these delicate sense hairs. In the aquatic amphibians, the cyclostomes and some of the more advanced groups of fishes, all of the sensory hillocks are free and exposed on the body surface, but in many of the fishes some or all of them are located in canals or tunnels beneath the epidermis. The canal organs have evidently evolved from surface neuromasts through the development of pits and grooves, and there is a good correlation between the behavior of the fish and the degree of protection of its neuromasts; the canals are better developed in the more active forms. Lateral line canals usually open to the surface through regularly spaced pores and contain a fluid, the canal endolymph. A cupula is a constant feature of the neuromast, whether it is exposed or hidden in a canal. The number of hair cells beneath a cupula varies from 6 to 10 in an epidermal organ such as that of the mudpuppy *Necturus* to as many as several hundred in a fish such as *Lota* (Flock, 1971). Efferent (inhibitory) as well as afferent nerve fibers terminate near the bases of these cells (Russell, 1971). The supporting cells seem to be both sustentacular and secretory in function.

Internal ears of vertebrates The evolution of the membranous labyrinth or internal ear from an anterior portion of this canal system is a familiar story in com-

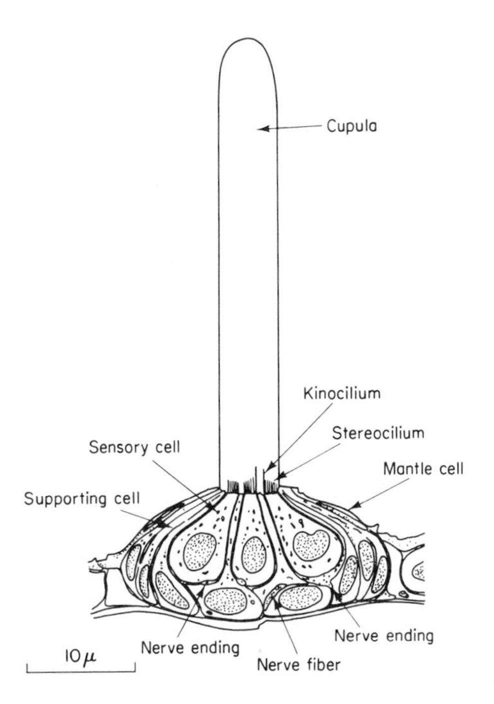

Fig. 17.6 Diagram of a naked neuromast from a 4.5 mm larva of the teleost fish *Oryzias latipes*. [From Iwai, 1967: *In* Lateral Line Detectors, Cahn, ed. Indiana U.P., Bloomington.]

parative anatomy and will not be detailed here. In brief, the embryonic ectodermal sac (OTIC VESICLE) can be thought of as an enlarged anterior portion of the lateral canal system. In the gnathostomes, this differentiates into two sacs, a ventral SACCULUS (the more primitive portion) and a dorsal UTRICULUS. Three semicircular canals, set at right angles to one another, open into the vestibule, an expanded portion of the utriculus; one end of each canal is expanded into a bulb-like AMPULLA. A ventral lobe of the sacculus is extended into a structure which varies from a tiny flap (the LAGENA) in the fishes and amphibians to an elongated coiled duct (the COCHLEA) in mammals. This system of sacs and ducts (Fig. 17.7) is filled with fluid, the ENDOLYMPH, and housed in a bony or connective tissue space filled with PERILYMPH.

The flask-shaped receptor cells (Fig. 16.7) are grouped in several distinct patches within the membranous labyrinth (Fig. 17.7). There is a transverse dike of them, referred to as the CRISTA AMPULLARIS, in each of the three ampullae as well as a distinct patch (a MACULA) in the utriculus, the sacculus and the lagena; the cochlea bears a long ridge known as the ORGAN OF CORTI. The phylogenetic antecedent of the organ of Corti is the BASILAR PAPILLA, which occurs first in some of the amphib-

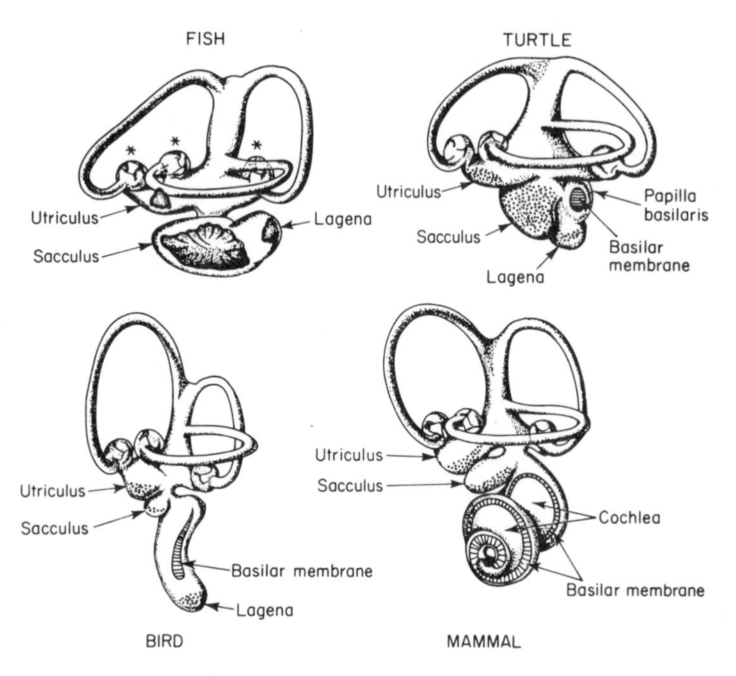

Fig. 17.7 Inner ear (membranous labyrinth) of some representative vertebrates. Cyclostomes possess only one (myxinoids) or two (petromyzontids) semicircular canals; not shown is the MACULA NEGLECTA—a small sensory spot found in the utriculus of some fish and lower tetrapods; asterisks, just above ampullae in fish. [From von Frisch, 1936: Biol. Rev., *11*: 236.]

ians as a small pad of receptor cells in the posterior wall of the sacculus; from a small papilla it appears to have extended during reptilian phylogeny to form a BASILAR MEMBRANE in the elongated lagena, and is synonymous with the organ of Corti in the crocodiles and birds (Baird, 1970; Romer, 1970).

Anatomically, the receptors of the ampullae are most like the sensory hillocks on the skin of a fish (Figs. 17.6 and 17.8a). The cupulae which arise from the cristae extend completely across the ampullae like flap valves or swinging doors. Thus, the cupulae, because of their positions in fluid-filled canals set in three different planes, are subject to displacement during angular acceleration or rotation of the head. Rotational movements provide the adequate stimuli; the cristae are probably not influenced by linear acceleration which is monitored by the maculae (Gernandt, 1959; Aidley, 1971).

The sense cells in the utriculus, sacculus and lagena are excited by the movements of heavy mineralized concretions, or granules (OTOLITHS or OTOCONIA). Hence, these three maculae are sometimes grouped as the otolith organs (Fig. 17.8b). The utricular macula (PARS SUPERIOR) is concerned with gravitational stimuli and linear acceleration; throughout the vertebrates it is an organ of major importance in postural reflexes (Lowenstein, 1971). The PARS INFERIOR, including both the sacculus and its appendage, the lagena, has had a much more varied history; the role of the sacculus in the higher vertebrates is still somewhat obscure (Gernandt, 1959). In the lower vertebrates the pars inferior (both sacculus and lagena) is concerned with hearing as well as positional changes. As the lagena elongated, the cochlear duct, which developed as a consequence, became a delicate organ for the analysis of sound (organ of Corti). In the mammals the lagena (which is at the end of this duct in the reptiles and birds) does not occur.

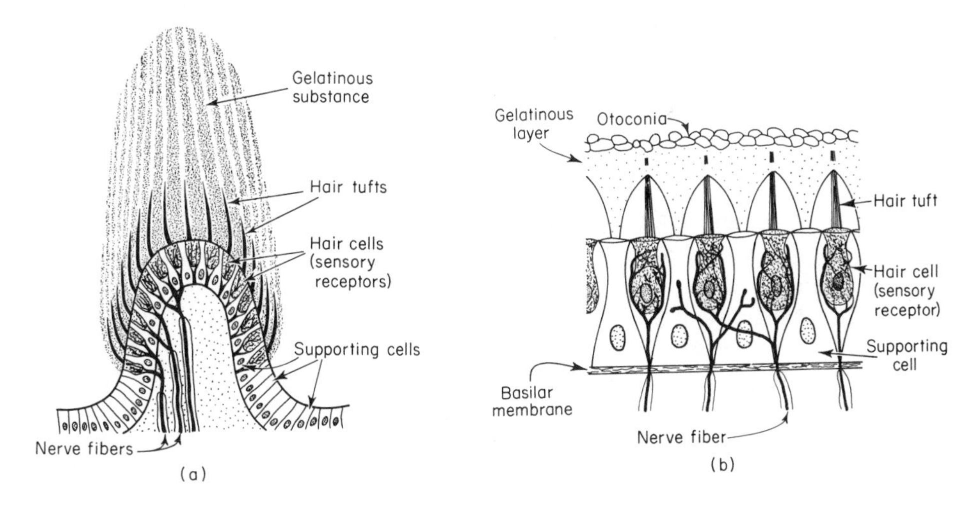

Fig. 17.8 Vestibular organs. (a) Mammalian crista; (b) mammalian macula. [From Netter, 1953: The Ciba Collection of Medical Illustrations, Vol. 1. Ciba, Montreal.]

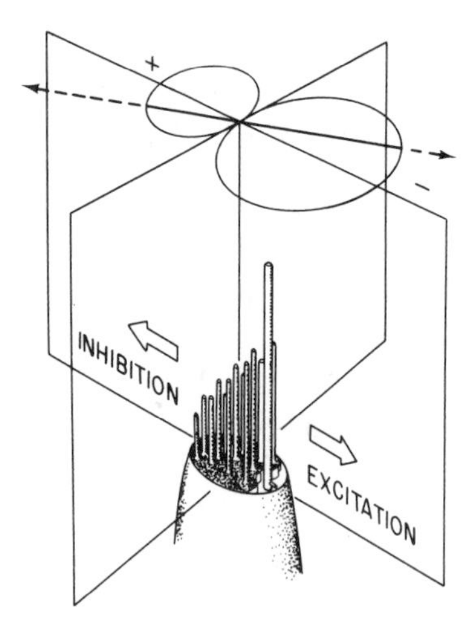

Fig. 17.9 Transducing mechanism in hair cells. The amplitude of the receptor potential is proportional to the component vector of displacement parallel to the rows of stereocilia. The receptor potential generated by equal amplitudes of cupular displacement in opposite directions is proportional to the distances cut out by the two circles. The amount of depolarization is larger than that of hyperpolarization. [From Flock, 1965: Cold Spring Harbor Symp., *30*: 143.]

The monitoring of these different motions and vibrations depends on the orderly positions of the receptor cells and the directional arrangements of their sensory hairs (Fig. 16.7). Ultrastructural studies have revealed diametrically opposed hair cell orientations in different regions of the labyrinth (Wersäll *et al.*, 1965; Flock, 1965; Lowenstein, 1971). This morphological polarization of the sensory cells has been associated with electrophysiological studies; the consensus is that depolarization with excitation occurs when the sensory hairs bend in one direction while hyperpolarization with inhibition takes place when they bend in the opposite direction (Fig. 17.9).

Thus, the membranous labyrinth consists of three receptor systems which analyze three different kinds of movements—those produced by gravitational forces, rotational stimuli and distant vibrations. The phylogenetic sequence in the differentiation of these systems is debatable (Pumphrey, 1950; Dijkgraaf, 1963; Bergeijk, 1966). It seems clear, however, that the lateral line from which they have evolved was basically concerned with the detection of shearing forces induced by movements of fluids at close quarters and that otolith and auditory organs which pick up distant vibrations came later. Rheoreceptors evidently preceded organs of hearing.

Physiology of the ordinary lateral line organs The role of the neuromasts of fishes has remained controversial since the pioneer studies at the beginning of the twentieth century (Dijkgraaf, 1963; Lowenstein, 1967). Claims have been made for a short-distance auditory function, a temperature receptor and a chemical sense but the evidence is not at all convincing. From both classical behavioral studies and modern electrophysiological recording there is strong evidence that local water displacements provide the normal stimulus. It is obvious that disturbances of many kinds in the surroundings of a fish can activate the neuromasts, but the adequate stimulus is moving water on the surface of the body.

The patterns of electrogenesis are interesting. Hoagland (1932) was the first to record a continuous "resting" discharge of action potentials in the lateral line nerves of fish. This has been many times confirmed, and it is now recognized that these

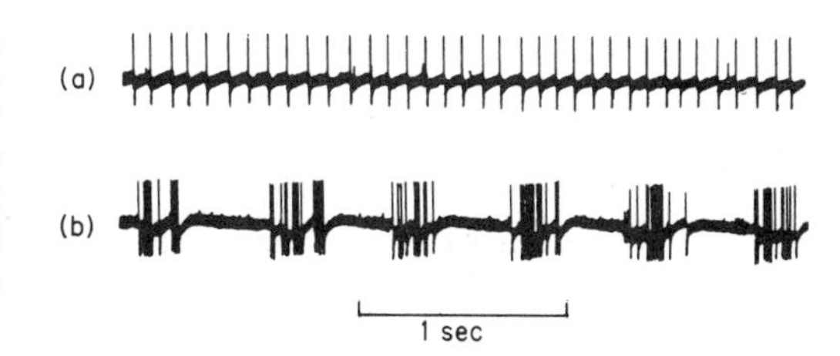

Fig. 17.10 Impulses in the lateral nerve fibers of *Xenopus*. (a) Spontaneous activity. (b) Responses of sensory unit when surrounding body of water was made to swing to and fro. Periods of activity correspond to headward current. [From Dijkgraaf, 1963: Biol. Rev., *38*:63.]

organs, unlike many other receptors, send a steady spontaneous train of bioelectric potentials along their nerves and that the rhythm of the pulses changes during stimulation (Fig. 17.10).

The message of the neuromast is not merely "disturbance in the water nearby" but "water moving from head to tail" or "water moving from tail to head." This discrimination is based on an asymmetry in the distribution of the filaments within the sense hairs (Dijkgraaf, 1963; Flock, 1971). Each receptor cell bears a tuft of sensory hairs consisting of a single kinocilium and 15 to 35 stereocilia arranged in step-like fashion leading up to the kinocilium (Figs. 16.7 and 17.11). The kinocilium is characterized internally by the 9 + 2 microfilaments typical of cilia and flagella; microfilaments are absent from the stereocilia. Neighboring cells are oriented in opposite directions, with the kinocilium facing forward or backward in alternating cells. The asymmetry in the sensory hairs and the directional polarization of the cells have been beautifully shown by the scanning electron microscope (Flock, 1971). Electric potentials have been recorded both intracellularly (receptor potentials) and in the associated nerves of the lateral line. The findings are consistent with the hypothesis that generator potentials are triggered by the bending of the sensory hairs

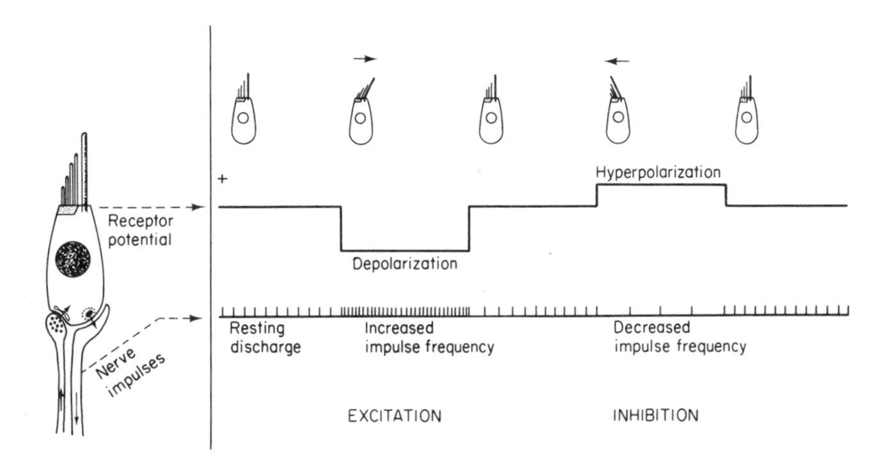

Fig. 17.11 Proposed theory for hair cell function showing the relation between receptor potential and frequency of nerve impulses when the sensory hairs are inclined toward or away from the kinocilium. [From Flock, 1965: Cold Spring Harbor Symp., *30*: 142.]

and lead to the release of transmitter substance which effects synaptic transmission between the secondary sense cell and its sensory neuron (Flock, 1971).

The information received by an animal with such a system might, at least in theory, be of use in rheotaxis; in reality, this is probably of minor importance since vision seems to play the dominant role in rheotaxis of fish (Dijkgraaf, 1967). More important functions of the lateral line organs are probably concerned with detecting and locating moving animals (predators, prey, and social partners).

Analysis of distant vibrations (sound) Sound has two main properties: its frequency, which is perceived as the *pitch,* and its intensity, or *loudness,* which is determined by both frequency and amplitude. Most sound sources in the natural environment produce compound vibrations with *overtones,* or *harmonics,* in addition to the simple wave, or *fundamental;* a tuning fork or an electronic oscillator generates a simple wave, but the human voice or a musical instrument produces a fundamental, which determines the pitch, and a series of overtones, which give the sound its characteristic *timbre,* or *quality.* The human ear is said to be able to resolve several hundred thousand sounds, although the audible frequency range is from about 20 to 20,000 cycles/sec. In mammals, these complex sounds are sorted out by the organ of Corti and its associated nervous centers in the brain; the lagena of the bird performs functions similar to those of the organ of Corti. In lower forms, the ability to analyze sound is more restricted; insects depend mainly on the rhythm of the sound pulses; the lower vertebrates have capacities intermediate between simple frequency and amplitude perception, and the capacity to separate compound vibrations (Lanyon and Tavolga, 1960; Grinnell, 1969).

During life the cochlea, like a tiny beautifully spiraled snail, is housed in a bony case. When uncoiled and examined in cross section (Fig. 17.12), it is found to consist of three fluid-filled canals, the central COCHLEAR canal containing the organ of Corti and the two peripheral TYMPANIC and VESTIBULAR canals which communicate through a fine hole, the HELICOTREMA, at the apex and have movable membranes or WINDOWS at their base. The base of the cochlear canal, in contrast, is solid bone, and this is important in transducer action.

The long ridge of receptor cells which make up the organ of Corti sits on a dense mat of connective tissue, the BASILAR MEMBRANE. These cells, like those of other regions of the labyrinth, are regularly arranged with a precise organization of stereocilia on their surfaces. There are three or four rows of outer hair cells and a single row of inner cells (Fig. 17.12); about 12,000 of the outer cells and 3,400 of the inner ones are concerned with the analysis of sound in the human cochlea. Each of the outer cells is equipped with about 100 stereocilia precisely arranged in the form of a W; the inner cells bear about half as many sensory hairs, organized in two parallel straight rows. The tips of the hairs are evidently embedded in the gelatinous under-surface of the tectorial membrane but not in the membrane itself. Kinocilia are absent although their basal bodies remain and are thought to be important in the transducer action (Gulick, 1971).

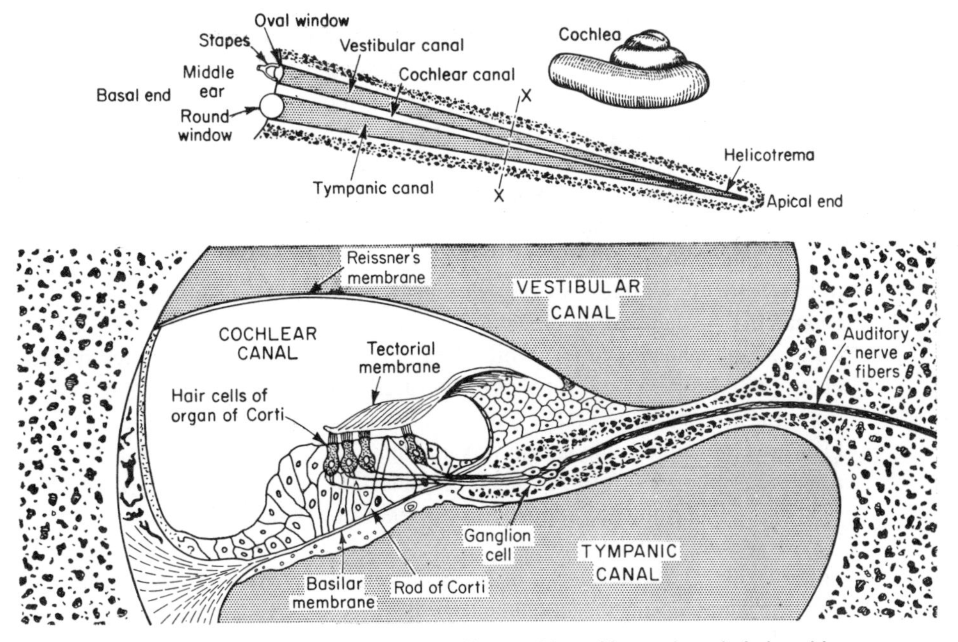

Fig. 17.12 Diagrams of the mammalian cochlea. Above, the spiraled cochlea has been uncoiled and split open to show the relationships of its three canals. Below, a cross-section through X–X to show the organ of Corti sitting on the basilar membrane. The receptor cells for hearing (hair cells) are secondary sense cells innervated by fibers of the auditory nerve. The cochlea is embedded in bone.

Helmholtz, in the mid-nineteenth century, was the first scientist to develop a workable theory of hearing. He noted that the basilar membrane became gradually wider toward the apex of the spiral and argued that fibres of different lengths resonated in accordance with the frequency of the sound; this membrane with its fibres under some tension and of different lengths (like the strings of a piano) was assumed to be the basis of hearing (Gulick, 1971, and texts of medical physiology). The RESONATOR THEORY has now been disproved, although his PLACE THEORY, which was a component of it, remains, since the varying width of the basilar membrane is still considered important.

Current thinking is based on the careful studies of von Békésy (1956 and earlier) who explored the physical properties of the fluids and membranes and made direct microscopic observations of their movements in preparations from fresh cadavers. A TRAVELING WAVE THEORY seems to fit the facts best. This wave is first generated in the vestibular fluids by the movement of the stapes on the oval window. It travels varying distances along the canals in accordance with the diameter of the canal and the width of the basilar membrane. Above about 60 cps the basilar membrane begins to vibrate equally over its length. Short wavelengths die out more quickly than the longer ones. In other words, maximal vibrations of different areas of the membrane are related to the different frequencies, with the higher frequencies operating closest to the oval window. Finer aspects of discrimination depend on the central nervous system; the analysis at the level of the basilar membrane is a rough, mechanical frequency analysis. The nervous system sharpens the analysis in some manner not yet understood (von Békésy, 1957).

It is still impossible to explain just how sound is converted into action potentials. The seat of transduction must certainly reside in the superficial sensory hairs, where movements in the surrounding fluids affect the ion gates. Two resting potentials and a generator potential are considered to be important and have been recorded repeatedly. An electrode placed within a receptor cell indicates a resting potential of 20 to 70 mv inside negative, while the fluids within the cochlear duct are 50 to 80 mv electrically positive with respect to the perilymph in the surrounding canals. Excitation creates oscillating COCHLEAR POTENTIALS or MICROPHONICS which reflect rather faithfully the stimulating sound waves (Fig. 17.13). These potentials may be recorded at various places within the cochlear duct (scala media) and surrounding tissues (Davis, 1965); they are believed to be generator potentials set up in some way by the traveling mechanical waves which bend the stereocilia (Gulick, 1971). Generator potentials lead to the release of transmitter at the synapse between the receptor cell and the dendrite of the cochlear nerve (Fig. 16.7); the nerve then transmits the

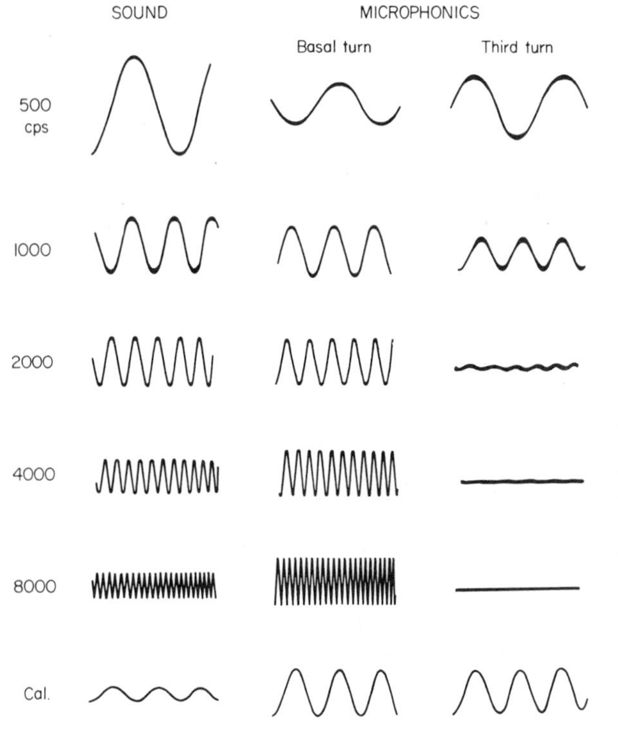

Fig. 17.13 Relation between frequency of sound (left column) and microphonic responses recorded from basal turn (middle) and third turn (right) of cochlea. Compare relative amplitudes and phase differences. Calibration at bottom indicates sound pressure of approximately 57 db above 0.0002 microbar at 1000 cps (left) and 0.33 mv peak to peak (middle and right). [From Tasaki, 1954: J. Neurophysiol., *17*: 97–122, 1954, Fig. 2]

information in the form of action potentials to the central nervous system. It should be noted that the pattern of the action potentials *does not* parallel that of the microphonics which are true generator potentials (Tasaki, 1954; Aidley, 1971; Chapter 16). The operation of the ion-passing valves, the nature of the transmitter and many other molecular events remain speculative.

Communication Based on Mechanoreceptors

Twentieth-century recording instruments have revealed a much "noisier" world than man's receptor capacities would lead him to suspect (Lanyon and Tavolga, 1960). The human ear is limited to an extreme range of about 20 to 20,000 cps, with a maximum sensitivity at about 2000 cps. Night-flying moths, in contrast, make much use of supersonics up to more than 80,000 cps while the frequency range of hearing in some bats extends out to 200,000 cps (200 kc). Thus, there are many sounds of which we are unaware; these are, however, accessible to analysis by appropriate instruments, and careful studies have shown that the majority of the invertebrates and many of the lower vertebrates make no use of this modality in communication. Hearing is much more widespread than the ability to communicate. Some animals may show alarm or avoidance reactions to distant mechanical vibrations but lack the capacity to produce sounds. Communication by sound involves the reception of meaningful vibrations produced by another animal at some distance. A wealth of information will be found in the volumes edited by Busnel (1963) and Sebeok (1968).

Among the invertebrates, only relatively few insects (particularly the Orthoptera, Hymenoptera and the Cicadidae) have well-developed systems of communication based on sound. Although many of the decapods produce sound, it has not been conclusively shown that they have mechanisms for its reception and analysis (Carthy, 1958; Cohen and Dijkgraaf, 1961). Phonoreception is universal among the major groups of vertebrates but communication based on it is far less common. Some of the teleosts, the anurans and a few representatives of each of the surviving orders of reptiles produce meaningful sounds. The remaining groups of poikilotherms seem to be silent (Lanyon and Tavolga, 1960).

Insect sounds are usually produced by rubbing together two different parts of the hardened integument. Frequently, a ribbed or toothed file is moved rapidly across a fixed surface to produce a shrill stridulation. This may be called the file-and-peg technique and has numerous variations in detail. The cicadas use quite a different method; they produce sounds by buckling modified areas of the exoskeleton. Insect sounds are often of very high frequency, perhaps because of the small surfaces concerned with their reception (Haskell, 1964). The evidence is that insects with tympanal organs are able to localize sounds but that their capacities for analysis are extremely limited. Many different patterns of vibration are produced, and these often form meaningful phrases in the language of one species (Faber distinguished

12 in the grasshopper), but it seems that only one of the characteristics of the sound, viz., its pulse rhythm, has significance (Pumphrey, 1940, 1950); insects cannot analyze for pitch and tone quality but rely only on the pattern of pulses (Roeder, 1953).

Many of the teleost fishes also produce sounds by drumming or stridulation of hard body parts. In the black bass (*Centropristes striatus*), for example, the pounding of the opercula against the cleithra and other supporting pectoral bones creates a series of drumming thumps; in the sea catfish (*Galeichthys felis*), there is a high-pitched creaking sound when a tuberosity at the base of the first dorsal is rubbed against the inside of its socket formed by the cleithrum. There are many variations of this drumming, rasping and creaking (Tavolga, 1971).

Some teleosts also make use of the air bladder, either as an auxiliary to the skeletal machinery or as an independent sound producer. Specializations of this organ have been incorporated into refinements of sound reception as well as sound production. Members of the Ostariophysi, which show the greatest acuity among fishes, have a series of modified vertebral elements between the swim bladder and the internal ear (Weberian apparatus); these transmit vibrations from the wall of the swim bladder to the auditory apparatus. Sound transmission is essentially the same in fish protoplasm as it is in water, and the detection of distant vibrations only becomes feasible through the incorporation of materials of different density into the system—the ear stones and sometimes sacs of air as in the Ostariophysi. Fish sounds are of low frequency, generally below about 800 cps, although some whistles and grunts have been recorded in the 2000 to 8000 cps range. The communicative value of many of the recorded sounds is as yet unknown. It seems clear, however, that fish do use sound as "warnings" and in reproductive behavior and intraspecific communication; this remains a potentially productive field for research in both physiology and ethology (Tavolga, 1971).

The organs of ventilation provide the basic machinery for sound production in the terrestrial vertebrates. With a few exceptions, such as the warning "rattle" of the rattlesnake, the sounds of tetrapods are produced by the controlled movements of air through respiratory passages. There are many anatomical modifications: the larynx and vocal cords, the resonating vocal sacs of the anurans, the syrinx and certain air sacs of birds and the lips of man. The most primitive communicating noise may have been the hiss which is common in tetrapods with and without a specialized voice. In any case, during evolution, this system has expressed many of the potentialities of the woodwind instruments of an orchestra. With a refined motor control for the organs of breathing, the development of the organ of Corti and the associated neural centers for the discrimination of pitch and tone, a vastly more elaborate system of communication was available to the birds and the mammals.

Sound reception as well as sound production are based on quite different principles in the insects and the terrestrial vertebrates. Pumphrey (1940) classes the insect receptors as DISPLACEMENT RECEIVERS and points out that although the tympanum looks superficially like an ear drum, and also has air at atmospheric pressure on both sides, yet as far as the receptor cells are concerned, the action of its movement is the same as that of the moving bristle. The inner ear of the vertebrate, however, is a PRESSURE RECEIVER with the distant vibrations creating a system of waves in a fluid-filled space.

Echo orientation Several groups of animals make use of reflected sound in orientation. The bat is the best known example. The series of researches which led to the discovery of this highly specialized capacity is one of the really great classics in zoological research. Griffin, who wrote one of the later chapters in this classic, tells the history in his books (Griffin, 1958, 1959). Lazzaro Spallanzani of Pavia, in 1793, became much interested in animals which can navigate in the dark. He soon showed that the owl could not really find its way about in complete darkness and was evidently using its eyes at low light intensities. The bat, however, posed a different problem. Spallanzani carried out many experiments. Blinded bats easily found their way back to the bell tower of the cathedral where he had captured them earlier. What was even more surprising, their stomachs were filled with freshly caught insects, proving that they could capture insects as well as navigate in the dark. He and others who became interested in the problem during the eighteenth century showed that the senses of vision, touch, smell and taste were of no significance but that the ears were indispensable. The problem was insoluble since everyone agreed that bats produced no sounds whatever, and by 1800 Spallanzani's findings were dismissed as ridiculous and soon were almost forgotten.

Approximately a century and a half elapsed before Griffin brought the first bats into range of equipment capable of recording ultrasonic vibrations and started a chain of research which has shown that bats and several other animals are operating in a world of sounds which is completely inaccessible to the human ear. Griffin and his associates showed that covering the mouth was just as effective as excluding the ears in disorienting the bat.

It is now known that the bats (Chiroptera) vary in their capacity for echo orientation with many differences in the organs for the transmission and reception of sound (Griffin, 1958; 1962; Henson, 1970). The old-world fruit-eating bats (Megachiroptera) all have large eyes and orient visually as would seem to be necessary in animals which feed on fruit, pollen and nectar. With the exception of one genus, *Rousettus*, the Megachiroptera evidently rely solely on vision. *Rousettus* also has good eyes and can orient visually in the light; but when flying in dark caves, it emits clicks which are clearly audible to human ears. These animals will start clicking as soon as the lights in the laboratory are extinguished, and the sounds become much more pronounced as the problems of orientation are increased. Many experiments show that *Rousettus* echo-orients in the dark but not in the light.

There are two suborders of bats, and the Microchiroptera, in contrast to the Megachiroptera, all emit high-frequency and short wavelength sounds which are used in orientation. Most of these bats feed on flying insects and, as Spallanzani discovered, these are regularly captured in the darkness. The sounds are characteristic of the different groups and are emitted in short pulses ranging in length from a tenth of a second (100 msc) to less than one millisecond. Most of the frequencies are beyond the range of human ears; they vary in different species from about 10 kc/sec (wavelength 34 mm) to over 130 kc/sec (wavelength 2.5 mm). The intensity of the sound is considerably higher in species which hunt insects on the wing, and the pulse repetition rate increases with the difficulties of orientation.

In comparing the sounds of four families of Microchiroptera, Griffin (1962) found some representatives which emit sounds of almost constant frequency throughout the duration of the pulse (Rhinolophidae). But in most the pulse is frequency-modulated (Vespertilionidae, Mollossidae, and Noctilionidae) with a downward sweep by about one octave during the pulse. The Rhinolophidae or horseshoe bats also differ from the other groups in their techniques of sound transmission; their faces are modified into trumpet-like structures (hence the name "horseshoe bats") which confine their calls to narrow beams (Moehres, 1960). The animals direct these in accordance with their problems of navigation. In other groups the sound spreads around the animal in all directions. There are no sharp lines between the vocabularies of the four families mentioned; the species form a continuum in frequency patterns. The pulse patterns have been adapted to the navigation problems of the particular species.

Since some insects are known to create and respond to high-frequency vibrations, it is not surprising to find a behavioral interaction between the predatory bats and their flying insect prey. Roeder and Treat (1957) recorded responses in the tympanic nerves of several species of noctuid moths; these responses were closely related to the cries of bats. Electrogenesis followed the bat's cry immediately and persisted for some time after it ceased. These moths are known to respond to sound in the range of 3 to 240 kc, with peak sensitivity between 15 and 60 kc. The reactions of some of these moths in the presence of bats substantiates the theory that they can actively evade their predators (Griffin, 1958; Roeder, 1966).

Bats are not alone in the use of echo orientation. In 1953, Griffin showed by experiments similar to some of the classical ones carried out earlier on bats that *Steatornis,* the oilbird of Caripe (Venezuela), flies swiftly and safely in completely dark caves emitting clicks (6000 to 10,000 cps) which are well within the range of human ears. Swifts of the genus *Collocalia,* famous for the production of the salivary glue used in bird's nest soup, likewise orient themselves in caves with low-frequency clicks (Novick, 1959). Moreover, it now appears that Spallanzani was not as correct about owls as he was about bats; the barn owl *Tyto alba* can locate prey in total darkness with an error of less than 1° in both horizontal and vertical directions; this ability also depends on listening to echoes (Payne, 1971). Nor is the capacity confined to bats among the mammals. The abilities of marine species to use sonar are now well known (Kellogg, 1961; Grinnell, 1969). At the opposite end of the animal size range, tiny shrews use a variety of sounds (frequency 30 to 60 kc) to explore strange places or unfamiliar objects (Gould *et al.*, 1964), while baby mice and certain other rodents call to their mother—particularly when cold—using ultrasonic calls of 60 to 90 kc/sec (Pye, 1968; Okon, 1970). Sound ranging may also be utilized by such insects as the noctuid moths and certain other groups of animals. It seems to appear in habitats or situations where it may be presumed to confer a biological advantage. A rich literature on the biological uses of echoes is now available (Griffin, 1958, 1959; Milne and Milne, 1962).

TEMPERATURE RECEPTORS

A potentiality for response mechanisms based on temperature change is to be anticipated from the universal action of temperature on both physical and chemical processes. This maxim may be coupled with another; the lethal effects of extreme temperatures and the advantages of constant body temperatures place a high premium on behavior which relates an animal reliably to suitable temperatures. This potentiality and the strong biological advantage associated with its expression have been basic to the phylogeny of a delicate system of thermal reception as well as to many curious and intricate animal activities. Several examples have already been described (Chapter 10). Both poikilotherms and homeotherms avoid extremes and "select" a favorable temperature when placed in a gradient; warm-blooded animals regulate their body temperature through metabolic adjustments which are controlled by temperature receptors in the hypothalamus; many cold-blooded animals regulate their temperature through behavioral responses which are triggered by the cutaneous receptors; blood-sucking arthropods (mosquitos, lice) often locate their warm-blooded prey by delicate receptors sensitive to a gradient as narrow as 0.5°C; the rattlesnake, with the most acute temperature receptors known, is able to detect a rat-sized object 10°C warmer than its environment at a distance of 40 cm after only 0.5 sec (corresponding to a threshold of 0.001° to 0.002°C). Behavior based on temperature reception may be extremely intricate. This is well illustrated by the curious activities of a family of birds, the Megapodidae in Australia, which incubate their eggs in mounds of rotting vegetation and adjust the temperature of the nest almost continually throughout the day by changing the amount of insulating or fermenting material which covers the eggs. *Leipoa* males may be busy for three to seven hours each day from September to March. They evidently obtain their cues from thermal receptors on the face or inside the mouth; the animals show a characteristic probing behavior, repeatedly thrusting their heads into the mound beside the eggs with mouth open and then proceeding to adjust the blanket which covers the eggs. These and other interesting examples of behavior which depend on temperature receptors are reviewed by Murray (1962).

Although human experience, as well as these observations of temperature-dependent behavior at all levels in phylogeny, leave no doubt of the existence of temperature receptors, there are still many uncertainties about the detailed mechanisms. Murray (1962) has reviewed the literature and emphasized the unsatisfactory state of present knowledge concerning the nature of the receptor organs and their mode of excitation.

Morphology

At one time textbooks of mammalian physiology showed two distinct nerve endings associated respectively with the sensations of heat and cold. Among the lower vertebrates, the lateral line was considered to play a dominant part in temperature reception because of the marked effect of temperature on the electric potentials which can always be monitored in its nerves; evidence for thermal reception by the ampullae of Lorenzini in elasmobranchs seemed even more convincing. Among the arthropods, thermosensitive hairs have been described on the antennae or legs of several different species. In each case investigators attributed temperature reception to morphologically distinct receptor organs. The evidence, however, was circumstantial and is now considered inadequate (R. W. Murray, 1971). On the contrary, temperature reception probably depends on the branching processes of free nerve endings. This at least is true of the two situations most carefully analyzed by electrophysiological methods, viz., the thermal receptors in the cat's tongue (Zotterman, 1959; Hensel, 1966) and the facial pits of the crotalid snakes (Barrett *et al.,* 1970). In the latter, nerve fibers end freely in palmate expansions with numerous fine branching processes; from 500 to 1500 of these expanded endings are found in one square millimeter of the pit-membrane of the rattlesnake.

Transducer Mechanism

Concepts of the physiology of thermal reception have had a history which is somewhat parallel to that of the morphology. Earlier hypotheses assumed some indirect action of temperature on the dendrites concerned with electrogenesis. Temperature-sensitive mechanical or thermal processes were postulated as steps in the excitation of nerves. Thermal expansion of materials in the ampullae of Lorenzini, changes in volume of gases or liquids in thermal sensilla of insects and changes in tension of elastic elements were some of the suggested physical means of exciting cell membranes; chemical theories were based on temperature coefficients of reactants involved in the syntheses of transmitter substances (acetylcholine) or of the enzymes concerned with their destruction. These hypotheses seem to be disappearing with those which attribute the receptor action to special morphological cell types.

It now seems likely that the temperature effect is directly on the processes concerned with electrogenesis in the free nerve endings which serve as receptors. After a careful review of the available information, Murray (1962, 1966) concludes that "enough is now known or will soon be known of the physical and chemical basis of nerve activity and of the way in which the various processes are affected by temperature, for a plausible explanation of thermoreceptor transduction to be made without invoking any other special mechanisms." An understanding of the details is still very inadequate, but there are temperature-sensitive processes in electrogenesis which are associated with both membrane permeability and the operation of the cation pump; understanding of temperature reception will probably be in terms of its action on these processes and not by indirect mechanical or chemical transducer mechanisms.

18

PHOTORECEPTION

Many of the complex organic molecules of protoplasm are easily altered by radiant energy. Proteins are denatured, nucleotides are depolymerized and isomeric arrangements in carotenoid pigments are changed. A reaction of protoplasm to radiant energy of wavelengths between about 400 to 800 nm is the basis of a receptor system which permits the animal to exploit the advantages and avoid the disadvantages of radiant energy and especially to relate its activities to the objects in its environment through the light which is reflected from them.

Light as an environmental factor was considered in Chapter 14. The effects described there were generalized photochemical reactions and the long-range physiological consequences of a diurnal and seasonal nature. The photochemical mechanisms considered in this chapter relate the animal quickly and adaptively to its environment. This is a chemical sense as far as the transducer action is concerned. It is, however, a highly specialized one in which the active chemicals are produced endogenously and altered photochemically to generate nerve action potentials. Like all functioning receptor systems, this one becomes progressively more useful with increasing capacities of the integrating nervous system and specialization of the effector organs.

During phylogeny more and more of the potentialities of the radiant energy were utilized. Responses which first depended only on the presence or absence of light became directional and permitted precise orientations when the receptor cells were grouped and the light only reached them from certain angles. Modifications in the morphology of the receptor cells and in their sensitive pigments permitted utilization of different wavelengths and the exploitation of the plane of polarization; the associated evolution of central nervous mechanisms opened the possibilities for analysis of images, detection of movement or the appreciation of distances and form.

THE DERMAL LIGHT SENSE

Steven (1963), in a helpful review of this subject, has compiled the evidence for a dermal light sense in all the major phyla of animals. Diffuse photosensitivity over a large part of the body is extremely common; it is found in many animals with localized photoreceptors as well as in those without them. It is rare in the terrestrial arthropods and may not occur in cephalopods and amniote vertebrates with the possible exception of direct chromatophore responses. The soft, moist coverings of aquatic animals are more likely to contain dermal light receptors than the dry integuments of the terrestrial animals.

The Dermal Photoreceptors

In many cases, the cells actually responsible for the dermal light reactions have not been satisfactorily localized. A direct action of light on protoplasm is recognized, and this can be demonstrated not only in many protozoans but also in certain cells and tissues from multicellular animals, such as the isolated mesenteries of *Metridium* or the isolated iris of the eel where the muscle fibers are thought to be stimulated directly. Light-induced responses have also been established in several other cells and tissues which are not recognized as receptors. Chromatophores sometimes operate as independent effectors and show changes in pigment distribution associated with light and darkness. It has also been shown that some parts of the nervous system can be activated directly by light and, as a consequence, initiate adaptive behavioral reactions. In studies with blinded minnows, von Frisch (1911) found evidence for photosensitivity of the diencephalon associated with color changes; Young (1935) obtained photokinetic responses in ammocoetes by illuminating the spinal cord; Prosser (1933) and Welsh (1934) monitored changes in the action potentials of the ventral nerve of the crayfish when the last abdominal ganglion was illuminated and noted that these animals continued to avoid light after both eyes were destroyed but only so long as the ganglion was intact; extraretinal photoreceptors are presumed to occur in ducks and sparrows (probably also other vertebrates), since blinded animals continue to show photoperiodic and circadian responses (Menaker, 1972). Steven (1963) gives other examples.

The evidence for specialized photoreceptor cells associated with the diffuse sensitivity of the integument is less satisfactory. Millott (1968) has evaluated the literature and concludes that the dermal light sense probably depends on photosensitive nerves within or just below the translucent skin. These nerves contain photosensitive pigment but in other respects are structurally unspecialized.

Dermal Light Reactions

Reactions which follow stimulation of the dermal receptors are varied, sometimes elaborate and complex. In addition to the direct responses of the stimulated cells, such as the chromatophores already mentioned, there are three major types of activity initiated through this sense: (1) the photokinetic locomotion of free moving animals, (2) the bending of the body of a sessile animal such as a hydroid or the local movements of tentacles, tube feet and spines, and (3) the shadow reflexes or withdrawals of exposed parts in response to sudden illumination.

When light operates through the dermal photoreceptor system to control the direction of locomotion, the response is a KINESIS. A TAXIS, the other major category of orienting response (Fraenkel and Gunn, 1940), is precluded in the absence of eyes which are localized receptors organized to admit light only from certain angles. Most of these photoorientations are orthokineses, although the movements of blinded planarians and ammocoetes are klinokinetic (Steven, 1963 and Chapter 17).

Shadow reflexes are common in many of the invertebrates and lower vertebrates. The adaptive value is obvious. Usually, the response is most marked when the light intensity decreases, but sometimes the reverse is the case. The retraction of the siphon of the bivalve *Mya* is a carefully investigated example of the shadow reaction in response to suddenly increased illumination.

One example will serve to illustrate the complexity of behavior which may depend on dermal light sensitivity. Millott (1960; 1968) and others have studied the responses of the spines of the sea urchin, *Diadema*. The reaction to a change in illumination is a sharp swinging movement of the spines which may be repeated several times depending on the intensity of the stimulus. The shadow reaction, following a decrease in light ("off" reaction), is much easier to elicit than the "on" reaction which follows sudden illumination. The urchins are eyeless, and the responses depend on the dermal sense and associated nervous system. The radial nerve is indispensable, indicating the reflex nature of the reaction. Like the crayfish ganglion mentioned above, the radial nerve of the sea urchin can be stimulated directly by light; shading a spot of radial nerve not more than 10 μ in diameter will elicit a movement of spines, but no reaction is obtained from shading a spot of equal size just outside the margin of the nerve. No receptors have been located, and Millott (1960) concludes that the nerve elements themselves must be excited by the light. The photosensitive pigment has not been identified, but its action spectrum shows a peak between 455 and 460 nm.

These shadow reflexes of sea urchins have been built into rather complex behavior. When *Diadema* is placed under a checkerboard of electric lamps, it will rapidly adjust the angle of its spines and point them at any lamp which is suddenly turned off. Another urchin (*Lytechinus*), which inhabits coral reefs in the Caribbean, responds to a narrow beam of light directed at its aboral surface by picking up opaque objects such as seaweed with its tube feet and pedicellariae and placing them in the beam of light, like a parasol (Milne and Milne, 1959). Thus, it is evident that complex behavior can be built onto responses mediated through the dermal light sense and that they probably depend on reflexes set up in "naked" nerves. The extent of specialization shown by the dermal light sense suggests that "it is no mere evolutionary relic" (Millott, 1968).

Some Physiological Parameters

Several of the general properties of receptor systems can be very nicely demonstrated with the dermal light response. Reaction times are slow in contrast to comparable periods for responses mediated through eyes, and instructive measurements are possible with simple techniques. For example, the minimum time for the retraction of the siphon of *Mya* is about 1 second and for movements of *Myxine* about 20 seconds. The strength-duration relationships (Fig. 18.1) show clearly that similar reactions may be expected with a series of flashes of increasing intensity but constant duration, or with a series of progressively longer flashes at constant intensity. A minimum dose of light is necessary to elicit a response. It has also been found unnecessary to illuminate the receptors for the entire latent period; the minimal exposure time is only a part of the total reaction time (Fig. 18.1).

Extensive experiments of this type were first carried out by Hecht almost half a century ago. Steven (1963) summarizes his findings. From studies of the retraction of the siphon in *Mya*, Hecht (1937 and earlier) proposed a photochemically controlled mechanism which proved basic in the development of physiological thinking on photoreception. This represents one of the many fundamental contributions which comparative studies have made in the history of physiology. Hecht found that the first part of the reaction which depended on the light was independent of temperature, but the "dark" process was temperature-dependent. He assumed that the retraction of the siphon was a coupled reaction with an initial photochemical phase followed by a thermochemical reaction. With *Mya*, his facts fitted the theory remarkably well; the fit of the equations is less satisfactory with several other animals such as *Myxine*, and later workers have pointed out that the two phases or reactions probably overlap or proceed simultaneously (Newth and Ross, 1955).

Rates of adaptation, action spectra and the Weber-Fechner Law (intensity discrimination) can also be effectively demonstrated with the dermal light responses

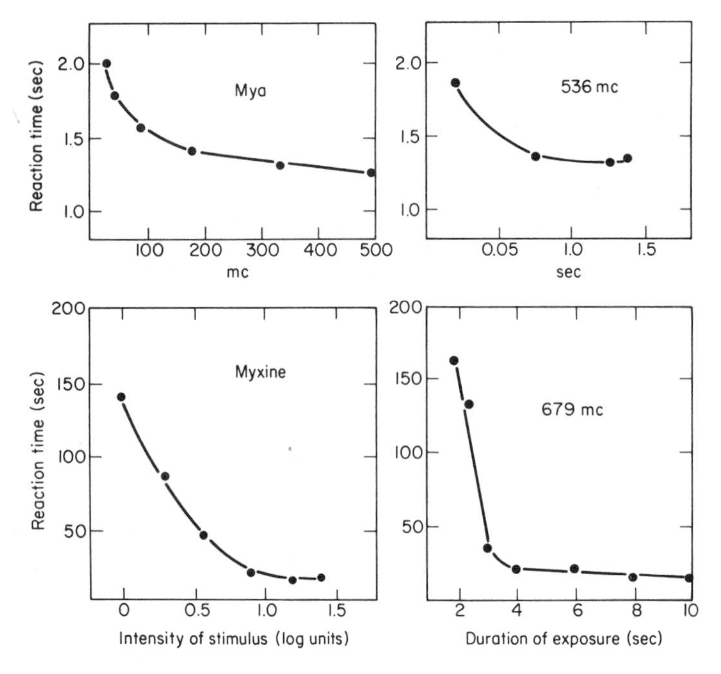

Fig. 18.1 The reaction times of the mud clam *Mya* and the hagfish *Myxine*. Left, intensity measured in metric candles (mc) was varied while duration remained constant. Right, duration was varied while intensity was constant. [From Steven, 1963: Biol. Rev., *38*: 215.]

of many invertebrates and lower vertebrates. Steven (1963) cites several examples, and many others will be found in Hecht's original papers.

LOCALIZED PHOTORECEPTORS OR EYES

The advantages of precise orientation probably provided the evolutionary force toward a localization of photoreceptors. When light can only strike the receptor cells from certain definite angles, movements with respect to it become directional. It can be assumed that the change from a photokinesis to a phototaxis would often confer a sizable advantage on the ancestral species. The flat retina, which is presumably the first step, is found in primitive representatives of all the major animal phyla (Fig. 18.2). The stigma or eyespot of the protozoan *Euglena* exemplifies this step within a single mass of protoplasm. This light-sensitive bit of protoplasm is a swelling at one side of the flagellum and is shaded from one direction by a curtain of orange-red pigment (Wolken, 1971).

The directions from which light reaches the receptors can be limited through the formation of cups and vesicles with progressively narrower apertures or with

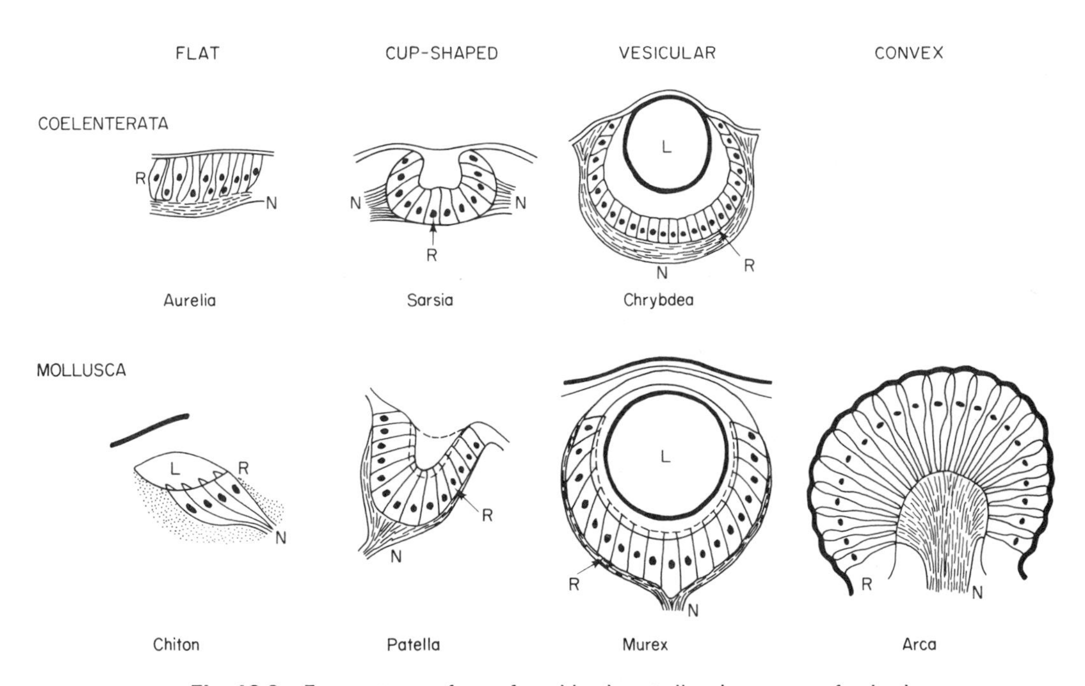

Fig. 18.2 Four patterns of eyes found in almost all major groups of animals. Examples selected from Coelenterata and Mollusca; see Novikoff (1953) for examples from other major phyla. *R*, retina; *N*, nerve; *L*, lens. [Based on Novikoff, 1953: Systematic Zool., *2*: 59.]

lenses in the apertures to focus the light on the retina. Cup-shaped and vesicular retinae are also found in each of the major phyletic groups. A lens is usually, but not always, present. The frequency with which these basic structural patterns recur in distantly related groups of animals is suggestive of both the evolutionary trail and the limited directions possible with the biological materials available (Novikoff, 1953).

The fourth structural pattern shown in Fig. 18.2 is the convex retina found in many of the annelids, molluscs and arthropods. In this, many tiny eyes called OMMATIDIA are arranged in a radiating pattern on some eminence or knob-like development of the body surface. In some ways this is a radical departure from the vesicular type of eye. However, it should be noted that the sensory cells are still located in a vesicle. The photoreceptor cells of each ommatidium are surrounded by a curtain of pigment so that they are actually at the base of a deep cone (Fig. 18.8). Just as in the vesicular eye, light enters the unit from only one direction (at least when the pigment is dispersed) and thus, the unit has capacities for directional orientation which are similar to those of the vesicular eye. The convex retina, like the other patterns, has been tried many times in evolution (Milne and Milne, 1959). In its most specialized form, as found in the insects, it has proved capable of almost

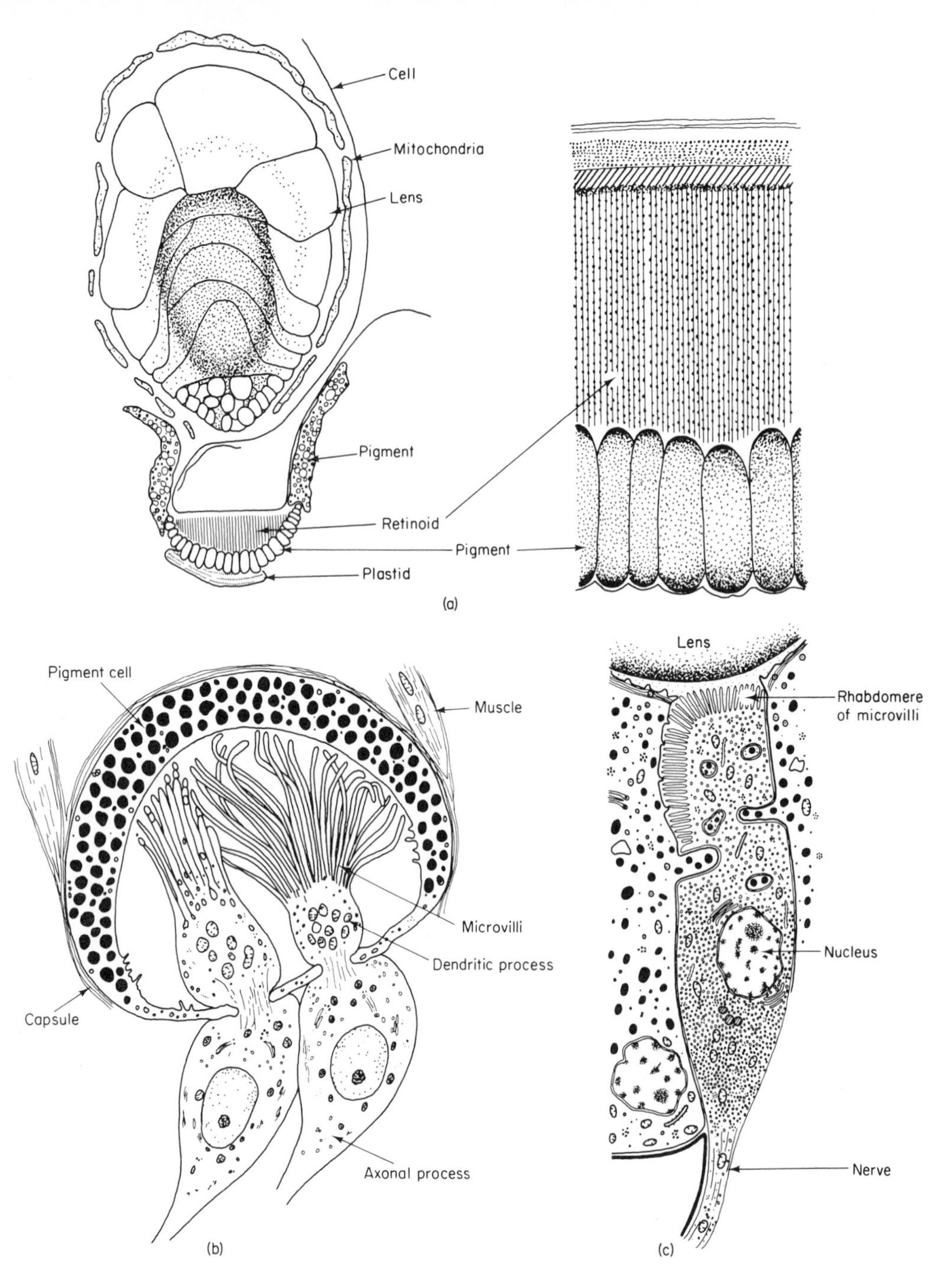

Fig. 18.3 The structure of three primitive photoreceptors as shown by the electron microscope. (a) Longitudinal section of the eyespot of the dinoflagellate *Nematodinium*. [From Francis, 1967: J. Exp. Biol., *47*: 496; and Morin and Francis, 1967: J. Microscopie, *6*: 768.] (b) Cross section of the ocellus of the marine polyclad flatworm *Notoplana*. [From MacRae, 1966: Z. Zellforsch, *75*: 472.] (c) Photoreceptor cell in the nudibranch mollusc *Hermissenda*. [From Eakin *et al*., 1967: J. Cell. Sci., *2*: 351.]

as many capacities as the camera eye. Differences between the visual abilities of insects and vertebrates are often differences in central nervous mechanisms rather than in the capacities of the photoreceptor cells.

Functional Anatomy of Photoreceptor Cells

A photoreceptor is a light trap that converts radiant energy into nerve impulses. The metabolic machinery, in the form of mitochondria and associated cell organelles, is always a prominent feature (Figs. 18.3 and 18.7). This machinery must look after the routine operation of the cell, assemble or generate the chromoproteins and produce transmitters which effect the synapse. These cells are very active metabolically, for chromoproteins are subject to destruction by light (Young, 1970) and transmitters must be steadily passed into synaptic vesicles. The fuel system will not be considered further; the following description is confined to the light traps.

Photoreceptor cells show a unity of plan both in the nature of their photochemical pigments (Chapter 14) and in the organization of their light-trapping apparatus. Light traps, whether in photoreceptors or in chloroplasts of plants (Chapter 1), are layered or laminated structures. There are two somewhat different patterns in the photoreceptors: one of these, exemplified by the vertebrate rod or cone, is a pile of discs stacked up like poker chips (Figs. 18.6 and 18.7); the other, exemplified by the rhabdomeric receptor of arthropods is made up of neat piles of microtubules (Figs. 18.3 and 18.8). Both kinds arise from cell membranes which are either folded into coin-like lamellae or rolled into rods (Fig. 18.4). It follows that all light traps are made up of orderly arrangements of lipoprotein molecules,

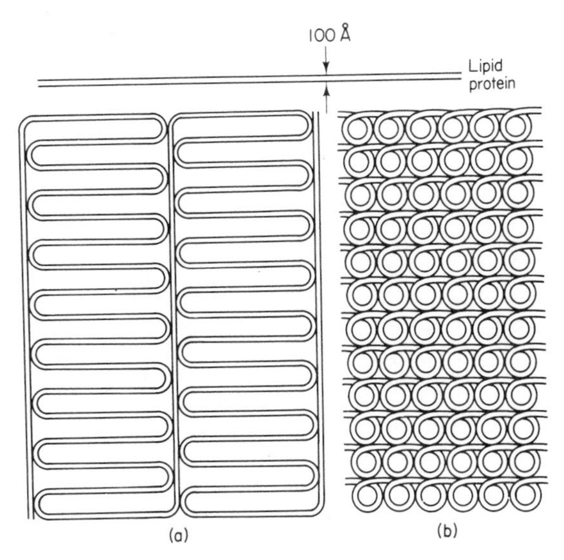

Fig. 18.4 Diagram to show how cell membranes can form either: (a) lamellae or (b) microtubules. [From Wolken, 1971: Invertebrate Photoreceptors. Academic Press, New York.]

as is true of all cell membranes. Wolken (1971 and earlier) suggests that all photoreceptor membranes have evolved from cilia or flagella. Ciliary or flagellar processes are clearly associated with photoreceptors in several groups of animals and it is assumed that photosensitive pigments became associated with these structures in primitive motile cells; such a system, seen in the protozoan flagellates, provides for rapid, light-oriented responses in an aqueous environment. Eakin (1965, 1968) postulates two main lines of evolution: (1) the ciliary line which is characteristic of the deuterostomes and culminates in the vertebrate rods and (2) the rhabdomeric line, which arose as an offshoot of the ciliary line, and is characterized by a remarkable folding of the cell membrane to form microtubules (Fig.

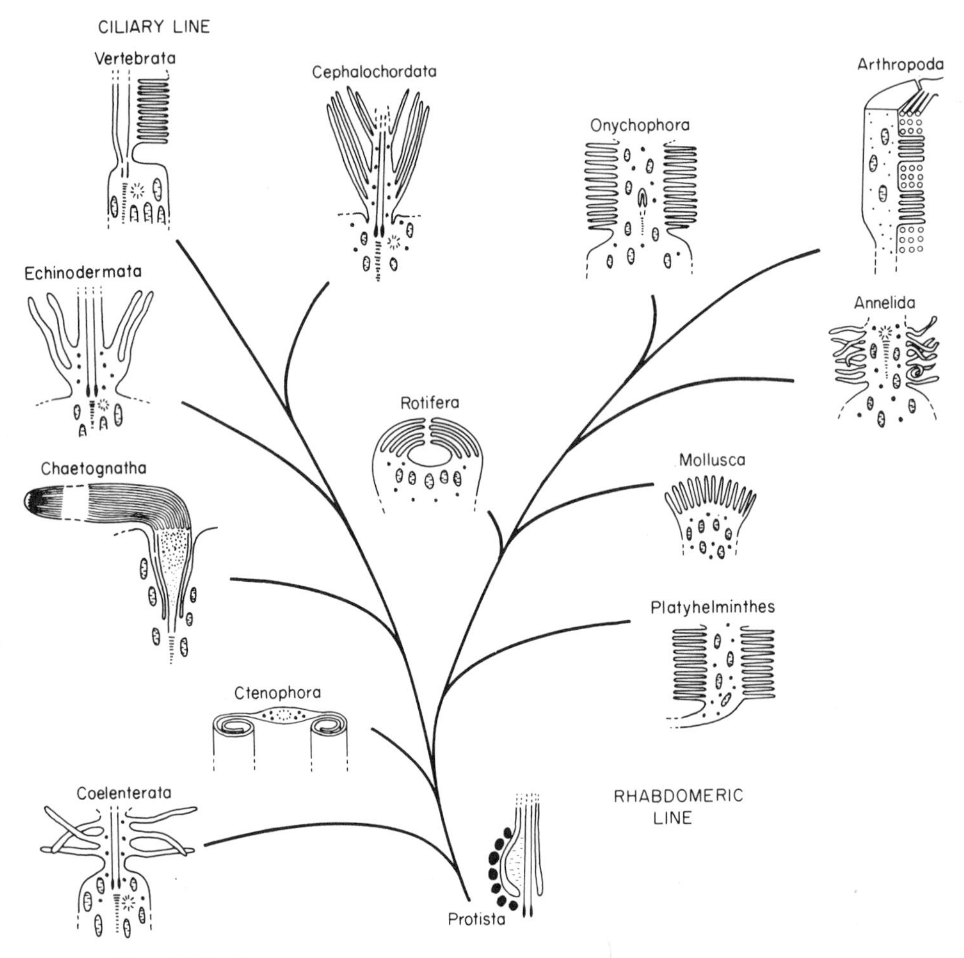

Fig. 18.5 Suggested phylogeny of light receptors along either ciliary or rhabdomeric lines of evolution. [From Eakin, 1968: Evolutionary Biology, Vol. 2. Appleton-Century-Crofts, New York.]

18.5). Such speculation is valuable in emphasizing once more the unity of plan in biology since both types are clearly elaborations of cell membranes.

Rods and cones More than a century ago, two morphologically different types of receptor cell were described in the vertebrate retina (Pedler, 1969). One of these appears as a long, thin bristle or rod, while the other is short, thicker and conical in shape. These cells, the RODS and CONES, have now been examined in many different vertebrates and categorized in terms of their light sensitivity, photochemical pigments, and synaptic connections. As indicated by the examples depicted in Fig. 18.6, there are fat rods and thin cones; shape alone is an unreliable guide to the receptor capacities of the cell. Several other criteria, sometimes used to distinguish rods and cones, have also proved unreliable and it is now apparent that the distinction should be based on relative light sensitivity (Underwood, 1968; Pedler, 1969). The rod is a low-threshold cell that is highly sensitive to radiant energy; the cone is a high-threshold receptor that operates in bright light. The low-threshold receptor (rod) has a much longer light trap (OUTER SEGMENT) which, at least in the frog, is being continually renewed as more and more discs are formed at its base to be lost at the outer tip; the cone has a relatively small outer segment

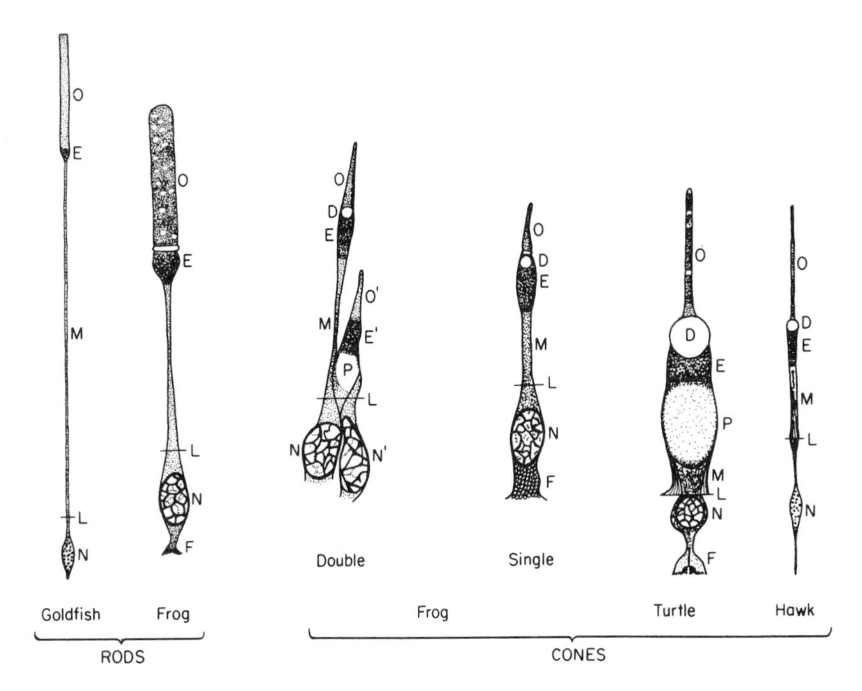

Fig. 18.6 Representative types of rods and cones (O, outer segment; D, droplet of oil; E, ellipsoid; M, myoid; L, external limiting membrane; P, paraboloid; N, nucleus; F, foot-piece). [From Walls, 1942: The Vertebrate Eye. The Cranbrook Institute of Science, Bloomfield Hills, Michigan.]

with discs which are not replaced although new protein is continually synthesized for them (Cohen, 1963; Young, 1970). One rarely finds a phylogenetic trail without many branches and alternate paths; the opportunism of natural selection, as well as the restrictions imposed by general cell morphology, are beautifully shown in a detailed study of the adaptations of vertebrate photoreceptors (Pedler, 1969).

A model of the vertebrate retinal rod is shown in Fig. 18.7. Plates of protein alternate with bimolecular layers of lipoprotein, the visual pigment, like stacks of coins. In detail, the lipoprotein layers are closely packed macromolecules arranged vertically in the plates. This outer segment of the vertebrate rod containing the visual pigment is joined to the inner metabolic segment by a narrow connecting piece which is a modified cilium, as shown by its embryology and submicroscopic structure (Sjöstrand, 1959; Cohen, 1963). During the histogenesis of the vertebrate photoreceptor, a bulge containing the structural elements of a cilium differentiates into the outer segment of a rod or a cone. Synaptic pedicles of three

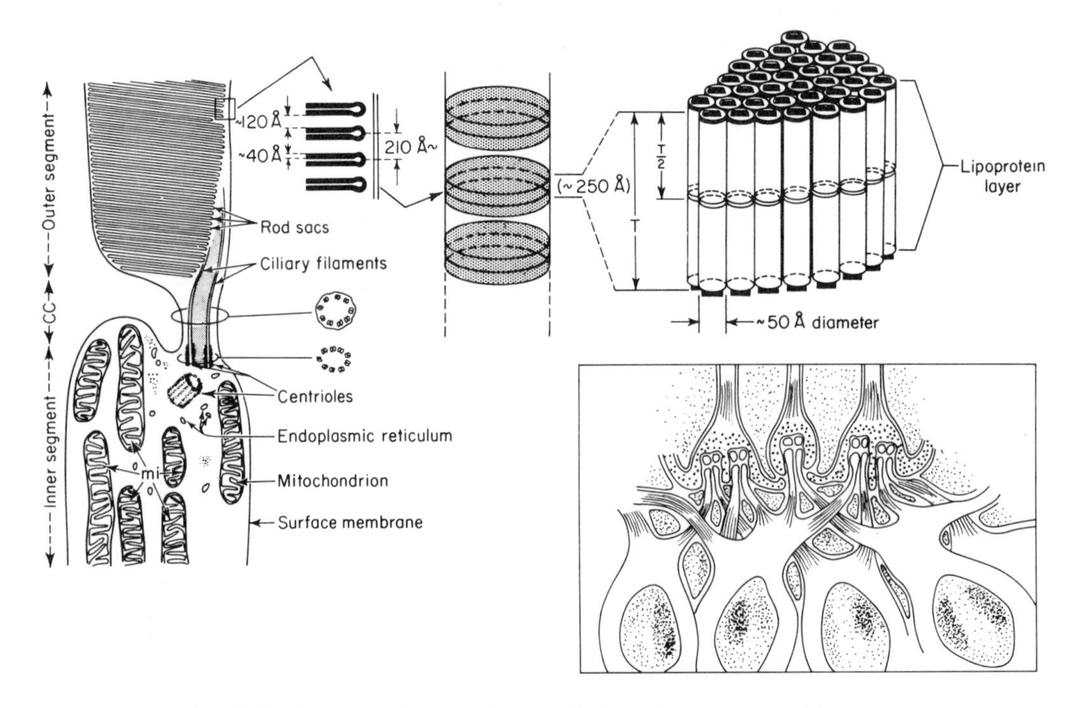

Fig. 18.7 Structure of mammalian rod. Left, portion of outer and inner segments with connecting cilium (*CC*); upper right, detailed structure of outer segment; lower right, synaptic connections of three rods in guinea pig. [Left, from Giese, 1973: Cell Physiology, 4th ed. Saunders, Philadelphia, after DeRobertis *et al.*, 1970: Cell Biology, 5th ed. Saunders, Philadelphia. Upper right, from Wolken, 1956: Trans. N.Y. Acad. Sci., *19*: 317. Lower right, from Sjöstrand, 1961: *In* The Structure of the Eye, Smelser, ed. Academic Press, New York.]

guinea pig rods are also shown in Fig. 18.7. Visual capacities of the eye depend on these synaptic elements and the intricate ganglionic connections of the retina as well as on the light trap. The electron microscope has greatly extended our understanding of the cellular physiology of both the transmitting and the receiving ends of the visual receptors (Davson, 1972). Further comment is reserved for the section on the physiology of the retina.

Retinular cells of compound eyes The image-forming compound eyes of arthropods and molluscs are made up of optical units called OMMATIDIA. The number of ommatidia varies from only a few in certain ants to 2000 in the dragonfly and up to about two million in the octopus; the latter number is roughly equivalent to the number of retinal rods in a vertebrate (Wolken, 1971). Each ommatidium is, in fact, a small eye (Figs. 18.8 and 18.11). The photoreceptors are the RETINULA CELLS, which stand vertically as a cylinder beneath the lens system, with their highly specialized border of microtubules directed centrally in line with the lenses and in the path of light.

The number of retinulae in any one ommatidium varies from as few as two in the dorsal ocellus of the cockroach to eight in the bees (Ruck, 1962; Dethier,

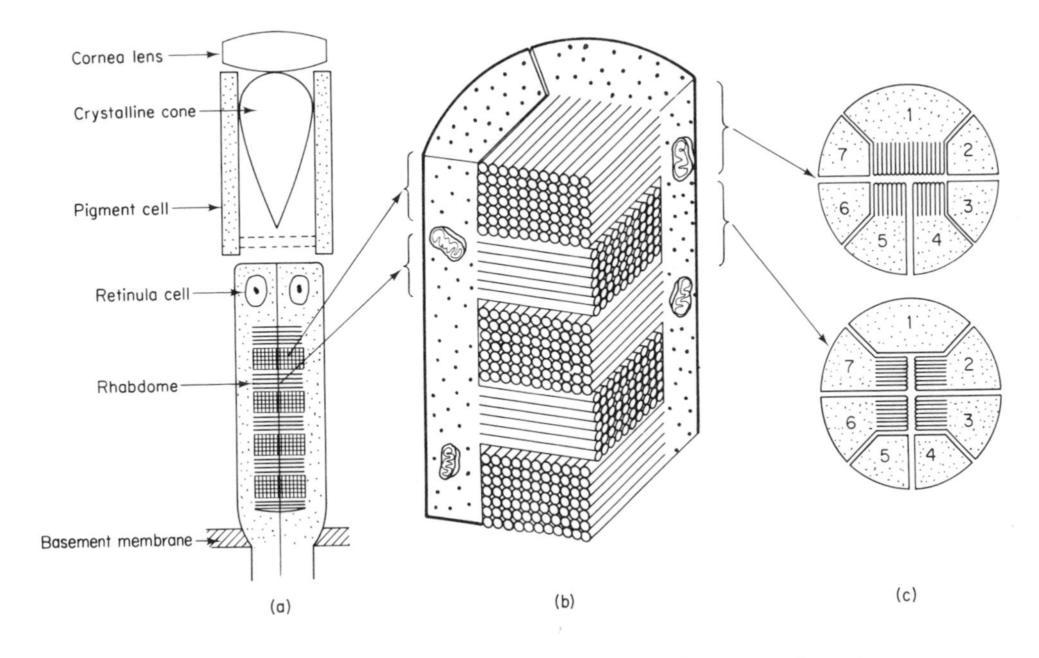

Fig. 18.8 An ommatidium and three-dimensional diagram to show the rhabdom in the compound eye of the crayfish, *Procambarus clarkii.* (a) Ommatidium shortened by the omission of the elongated cone stalk (see Fig. 18.11b); (b) stereogram to show patterns of packed microtubules; (c) cross sections through the seven retinular cells at the levels indicated. [From Eguchi, 1965: J. Cell. Comp. Physiol., *66*: 413.]

1963). The compound eye of *Drosophila* consists of about 700 ommatidia, each with seven retinula cells which measure about 17μ in diameter and 70 to 125μ in length (Wolken, 1958).

The inner border (RHABDOMERE) of the elongated retinula cell forms the light trap. It is made up of closely packed rods or tubes of lipoprotein which are actually microvilli (400–500 Å diameter) differentiated from this region of the retinular cell membrane. The several rhabdomeres of an ommatidium jointly form a central rod, the RHABDOM (Fig. 18.8). The ommatidia of *Limulus* differ from all others by the presence of a special ECCENTRIC CELL. This cell is actually a second-order neuron or ganglion cell which sends an elongated, sturdy dendrite up the central canal of a cluster of retinular cells (Miller *et al.*, 1961; Wolbarsht and Yeandle, 1967). There is no rhabdomere; the eccentric cell appears to be the site of integration of excitation and inhibition for all the cells of the ommatidium (Purple and Dodge; 1965). The arrangement is unique among the arthropods and thought to be primitive. In addition, it should be noted that some insect eyes contain one or two rudimentary retinulae per ommatidium and that these are sometimes called "eccentric cells" (Goldsmith, 1964; Scholes, 1965).

Photoreceptor cells, no matter how they may differ morphologically, operate as transducers of light energy into membrane potentials or nervous action. Many facts concerning the chemistry of vision have now been established, but it is still not known exactly how the photochemical change induces generator potentials (Chapter 14 and Arden, 1969). The *Limulus* eye and some of the simple eyes of insects have been particularly valuable in studies of the membrane potentials, but the connection between photochemistry and electrogenesis remains theoretical (Ruck, 1962; Hartline, 1969).

SPECIALIZED VISUAL FUNCTIONS

Just what does an animal see? Shadow reactions, kineses and taxes are found at all levels in phylogeny. The light-dependent behavior becomes more varied in the advanced animal phyla as the visual apparatus acquires the capacity to form images, discriminate color and brightness, detect movements, appreciate distances and operate over a wide range of intensities. These refinements confer expanding potentialities in relating the animal to its habitat and to other forms of life. Again, as with the other sense organs, the understanding of visual capacities is based on observations of behavior in relation to light or on the recording of action potentials in the visual apparatus.

Perception of Form and Movement

A few of the molluscs, some of the arthropods and most of the vertebrates have the necessary visual machinery to distinguish shapes and to utilize, in an adaptive manner, the different patterns of light which are reflected to their eyes from various objects in their surroundings. This is a matter of observation in some of the cephalopods, crustaceans, insects and many of the lower vertebrates. It has also been often demonstrated with conditioning experiments. There are many summaries of the pertinent literature (Carthy, 1958; Thorpe, 1963); only a few examples will be given here.

An octopus, for instance, will learn to attack a crab when it sees a square of one particular size if this has been associated with a food reward; but it will avoid or fail to attack a somewhat different-sized square if this presentation has been associated with an electric shock. Experiments of this sort show that the octopus can discriminate a variety of different figures: a horizontal stripe from a vertical stripe, a square from a diamond (Young, 1961, 1971). In comparable experiments, bees may be trained to locate sugar-water in a cardboard box with a certain pattern pasted around the entrance (von Frisch, 1950). A sphere is easily separated from a cross, but other shapes, such as a sphere and a square or a triangle (which are quite distinct to us) are not distinguished. Bees also show spontaneous preferences for certain patterns and use the markings on flowers to guide them to sources of nectar. Although these findings clearly indicate abilities to utilize light patterns in orientation, they do not require any precise analysis of the details, and most of the evidence indicates that the insect eye is highly adapted for the detection of motion but has limited capacities for the analysis of pictorial details (Burtt and Catton, 1966; Horridge, 1966b).

Fish also recognize the spatial relations of their environment and can be trained to discriminate among a variety of patterns. The three-spine stickleback, *Gasterosteus aculeatus*, builds a nest in a particular part of its environment and relates its activities to this through vision; other male sticklebacks are attacked while females are courted; the differences between the two sexes are recognized visually by body form, its position and characteristic movements (Tinbergen, 1951). Movement becomes an integral part of the visual stimulus in many of these more complex behavioral responses.

The ability to detect changing light patterns or movement in a light source is a more primitive and widespread capacity than form perception. It probably stems from the shadow reaction through temporal specializations of response and movement in relation to changing light. Buddenbrock and Moller-Racke (1953), for example, showed that the scallop, *Pecten*, reacted in a predictable manner to certain speeds of movement. This animal has a row of rather complex eyes along the mantle; when white stripes on a black background are moved past its eyes, responses of the tentacles and shell valves are often evident. If the movements are of the order of 0.66 mm/sec, the animal shows no reaction, but at about 1.7 mm/sec the tentacles are markedly extended from the shell. At 7.7 mm/sec the tentacles are extended and then the shell partially closes; at speeds of movement between 11.6 and 29.4 mm/sec the tentacles are withdrawn and the valves of the shell close. These visual reactions are associated with chemosensory responses, and the extended tentacles are sampling the water for odors from starfish and whelks

—the natural predators of the scallop. The significant point for the present discussion is that the speed of movement controls the reaction and, in nature, this reaction time is adapted to the speed of the enemies (Carthy, 1958).

Physiological comparisons of the capacities of different animals to detect movements are based on tests such as those just described for *Pecten* with a series of moving stripes (or flashing lights). As the stripes pass the eye at slow speeds, an optomotor reaction (for example, turning the head or moving the eyes) can often be observed; at higher speeds the reaction disappears (CRITICAL FLICKER FUSION FREQUENCY); as our eyes see it, the pattern disappears or the light ceases to flicker. The reaction can also be followed by recording action potentials in the retina or optic nerves. Eyes vary greatly in their ability to resolve the temporal features of the stimulus; some values for maximum flicker fusion frequencies in flashes per sec, tabulated by Waterman (1961), follow: frog 5, pigeon 143, cricket 5, bee, hornet, and fly 200, crab and crayfish about 55. The critical flicker fusion frequency depends on the intensity of illumination and the temperature (Hanyu and Ali, 1963).

These few examples, selected from the many now recorded, attest the capacities of different types of eyes to form images and detect movements. Two lines of specialization in the functional morphology of the receptor organs have been particularly significant: the increasing number of interconnected sensory neurons and the dioptric apparatus or the lens system. A good photograph depends on both the emulsion of the film and the lens system of the camera; pictures will be poor if either of these is inferior.

The vertebrate retina By analogy with the photographic camera, the retina is often likened to the emulsion and the photosensitive cells to the "grain" of the film. The likeness, however, is only superficial and the comparison rather deceptive. A picture is never fixed in the retina but sets up a continuous series of impulse patterns more akin to the television camera than the photographic camera. Only a few types of receptor cells (proprioceptors in muscles, for example) fire impulses steadily in response to a constant stimulus. Most sense cells adapt rather quickly, and this is the case with the photoreceptors. When constantly stimulated by a fixed light they cease to respond. The picture which is apparently stable to the human eye is really dancing over a multitude of photosensitive cells; the "steady" eye is rarely motionless. This dynamic feature, related to the excitatory processes of sensory cells, is one of the two factors which preclude the functioning of individual retinal units in isolation to form steady or fixed images (Rushton, 1962). The second factor is static and depends on the cytological arrangements of the neurosensory fibers and the interconnecting neurons between them and the visual centers in the brain.

The human retina has been calculated to contain over 100 million rods, six million cones, and only about one million optic nerve fibers (Ham, 1969). Thus, there are over 100 times as many rods and six times as many cones as there are conducting fibers to the visual centers. Unless the rods and cones were divided evenly into clusters associated with different optic fibers, the concept of a photographic "grain" in the retina would largely disappear. As indicated in Fig. 18.9, there is no such orderly arrangement. On the contrary, the interconnections are varied and elaborate. Neither are the receptors connected in uniform bundles to conducting fibers nor are there any receptors with direct transmitting lines to the brain, as was at one time considered essential for the resolution of details. On the contrary, both the receptors and the ganglionic interconnections are variously adapted to the visual requirements of different animals.

For more than 100 years the two morphological types of vertebrate photoreceptor have been credited with different visual capacities. Rods are the cells of twilight vision while the cones operate in bright light. The physiology of this distinction is still not fully understood. It is known that the cells contain different pigments, and this is one factor in the explanation. In addition, the axons from

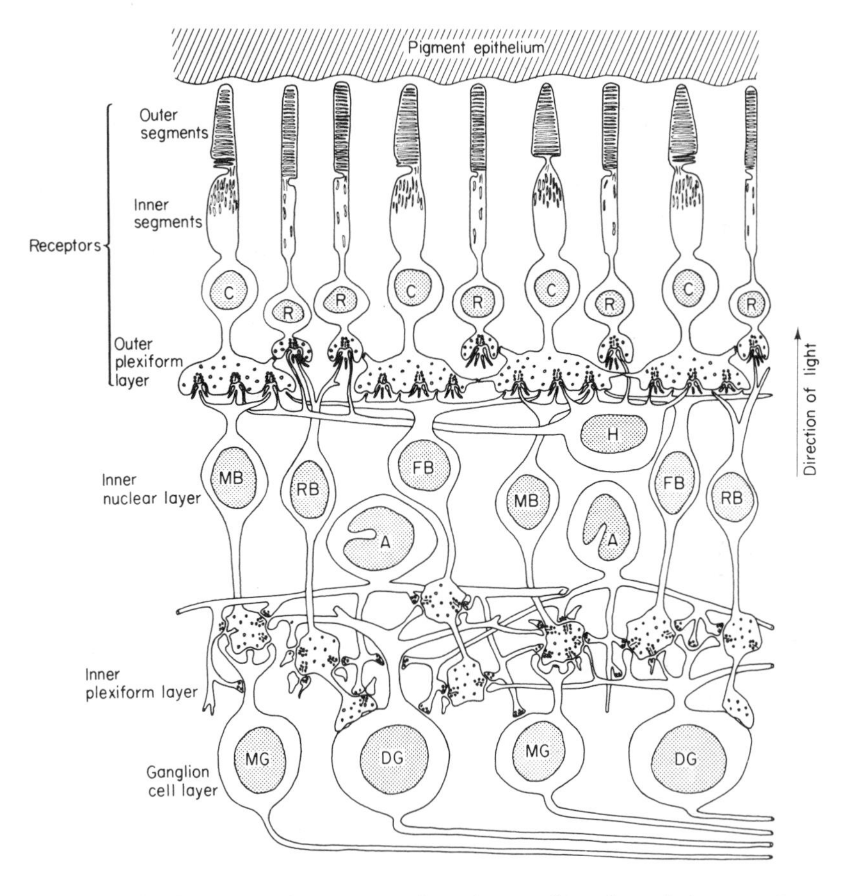

Fig. 18.9 Diagrammatic reconstruction of a small portion of the central part of the primate retina, based on electron microscopy. C, cone; R, rod; H, horizontal cell; MB, midget bipolar; RB, rod bipolar; FB, flat bipolar; A, amacrine cell; MG, midget ganglion; DG, diffuse ganglion. [From Dowling and Boycott, 1966: Proc. Roy. Soc. London, *B166*: 104.]

the rods are elaborately interconnected in the ganglionic layer so that summation becomes particularly significant; minimal amounts of light falling on many rods can excite an optic nerve fiber. A single quantum of radiant energy can evidently produce electrogenesis in one human rod, but six rods must be excited to produce a detectable visual response (Hecht *et al.*, 1942). This however, does not explain the significance of the two cell types—one long and thin, the other short and thick. It is thought that the thicker cone with its refractile ELLIPSOID (sometimes also an oil droplet) may be important in converging the rays on the photosensitive pigments of the outer segment of the cone (Fig. 18.6). It seems as though the cones are designed for efficient capture of radiant energy coming from a particular direction—along the focal plane. However, the rhodopsin in the rod captures minimal amounts of light falling on it diffusely from different angles.

Rods and cones are not uniformly distributed. There are pure cone retinas in a few diurnal animals and pure rod retinas in some strictly nocturnal species (Tansley, 1965); there may be varied mosaics of cones which have been correlated with different feeding habits of fishes (Ahlbert, 1969); and there are highly specialized patches of cones in areas requiring high visual acuity and sharp focus of detail.

The most constant of these patches is the FOVEA CENTRALIS, found as a small depression in the center of the retina where the light passing through the lens comes to sharp focus (Fig. 18.10). This is our area of clear focal vision where the details of the picture are clearly resolved; more peripheral areas—consisting predominantly of rods—are concerned with the detection of movement and dim light vision. Most vertebrates have only a central fovea but some reptiles, birds, and teleost fishes have temporal foveas as well (Prince, 1956). The human fovea centralis is small, measuring only about 1500μ in diameter; in its densest portion

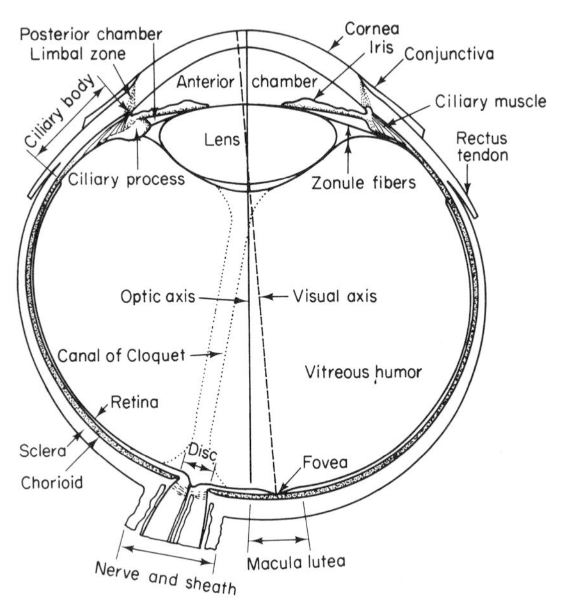

Fig. 18.10 Horizontal section through the right human eye. [From Walls, 1942: The Vertebrate Eye. The Cranbrook Institute of Science, Bloomfield Hills, Michigan.]

it contains approximately 150,000 closely packed cones per square millimeter and no rods (Davson, 1972). There are no blood vessels in this tiny spot, and the ganglionic nerve fibers diverge to admit the light directly to the cones. (Fig. 18.10).

Finally, the visual capacities of an animal depend on the synaptic organization of the plexiform layers of the retina as well as the distribution of the rods and cones. The electron microscope has greatly extended the earlier understanding which was based on Golgi silver staining. The main lines of transmission are vertical, with a flow of information from receptors through bipolar nerves to the ganglion cells which form the optic nerve (Figs. 18.7 and 18.9). Contrary to earlier views, there are no bipolars forming direct links between receptors and ganglion cells without any cross connections between them; at all levels, horizontal and amacrine cells provide for a horizontal flow of information. However, although one-to-one connections are no longer recognized (Pedler, 1969; Davson, 1972), the bipolar links are much more direct with fewer horizontal synapses in the central retinal areas (cones of the fovea) than in peripheral areas, where excitation from many rods impinges on particular ganglion cells to permit a high degree of summation. Details of synaptic organization vary considerably in different vertebrates. In general, the interconnections are more intricate in lower vertebrates like the frog and pigeon than they are in more advanced ones such as the primates and cats (Dowling, 1968). This may seem strange but must be interpreted in relation to the ontogeny of the eye and the phylogeny of the brain. The eye develops as an outpocketing of the forebrain and the plexiform layers of the retina are really extensions of the neuronal network of the central nervous system. With increasing encephalization in higher vertebrates, there has been a transfer of integrative activity from the lower to the higher centers, and the retina has been simplified accordingly. Thus, synaptic organization is simplest in the retinas of animals where the cerebral cortex is best developed (Dowling, 1968).

The arthropod retina Theories of arthropod vision have been dominated for almost a century by the classical studies of Müller (1826) and Exner (1891). The background literature will not be detailed; it has been summarized many times (Dethier, 1963; Goldsmith, 1964; Horridge, 1971b). In brief, Müller noted that the long, tubular ommatidia, which are the visual units of the compound eye, appear to be ideal machines for the detection of separate points of a picture and their direct translation to the central nervous system. The many points of light, projected separately by the ommatidia, would form a pattern, or mosaic, of luminous dots varying in brightness—hence the MOSAIC THEORY of insect vision. Exner's contribution was based on comparative studies in which he distinguished two main types of compound eyes. The eyes of diurnal insects such as flies, bees and butterflies were termed APPOSITION EYES. In these, the light-sensitive retinulae abut directly onto the cone of the lens and receive light parallel to the ommatidial axis. Moreover, each separate ommatidium is surrounded by a curtain of pigment, the distribution of which is not altered greatly by light intensity; the whole arrangement seems ideal for gathering light from narrow sectors of the visual field as required by the mosaic theory. The second type of eye, characteristic of nocturnal insects and many crustaceans, was termed a SUPERPOSITION EYE. In this the ommatidia are much elongated and the retinulae separated from the cone of the lens by a considerable space filled with a non-refractile transparent medium; pigment cells are confined to the outer clear-zone areas with granules distributed throughout the cells in bright light but accumulated up around the cones when dark-adapted. Exner considered these units ideal for vision in dim light since the retinulae could be excited by light from neighbouring lenses as well as from their own. Such a system, like the rods of the vertebrate retina, would be ideal for dim light and summation phenomena but unsuited for good resolution of the picture. The two types of eye and their operation, as explained by Exner, are shown in Fig. 18.11.

Although it is still not possible to say precisely how or what an insect sees, these pioneering concepts remain basic to all discussions of the visual processes of insects. After almost a century of general acceptance, both the mosaic theory and the concept of a superposition eye were critically tested with electrophysiological methods and precise measurements of the capacities of the dioptric system. At first there appeared to be little support for the mosaic theory (Burtt and Catton, 1962, 1966; Kuiper, 1962), but more recent studies show that it is substantially correct, at least for the diurnal insects (Goldsmith and Bernard, 1973). The rays of light are brought to focus part way down the crystalline cone (Fig. 18.11c); beyond this point they diverge but the lateral rays are lost in the surrounding cytoplasm. The refractive indices of the cytoplasm are probably more important than the pigment granules. Both crystalline cone and rhabdom have higher refractive indices than the neighboring cytoplasm of the pigment cells or the outer cytoplasm of the retinula cells. Thus, the rhabdom serves as a wave guide or light pipe for the central rays (Fig. 18.11c). Analyses of the physical properties of the system indicate that there is very little overlap in the fields of view of the ommatidia (Varela and Wiitanen, 1970); the electrophysiology shows that the retinula cells of an ommatidium are electrically coupled through tight junctions and behave as a unit (Shaw, 1969b). Thus, both optical measurements and electrophysiological data support the mosaic theory of Johannes Müller, proposed so long ago on the basis of behavioral responses.

Exner's sharp distinction between superposition and apposition eyes was seriously questioned after the eyes of several different nocturnal insects had been examined by modern techniques, particularly electron microscopy. Some superposition eyes can form erect images of the type usually associated with the apposition eye, while the eyes of some diurnal insects have superposition characteristics. As oft demonstrated in biology, the full range of adaptation is so great that functional phenomena rarely fall into tidy categories. It now appears that Exner was essentially correct but did not fully appreciate the visual capacities of the nocturnal eyes (Shaw, 1969a,b; Horridge, 1971b).

Horridge (1971b) prefers the term CLEAR-ZONE EYES for those of the nocturnal insects since this does not prejudge the character of their operation. He describes

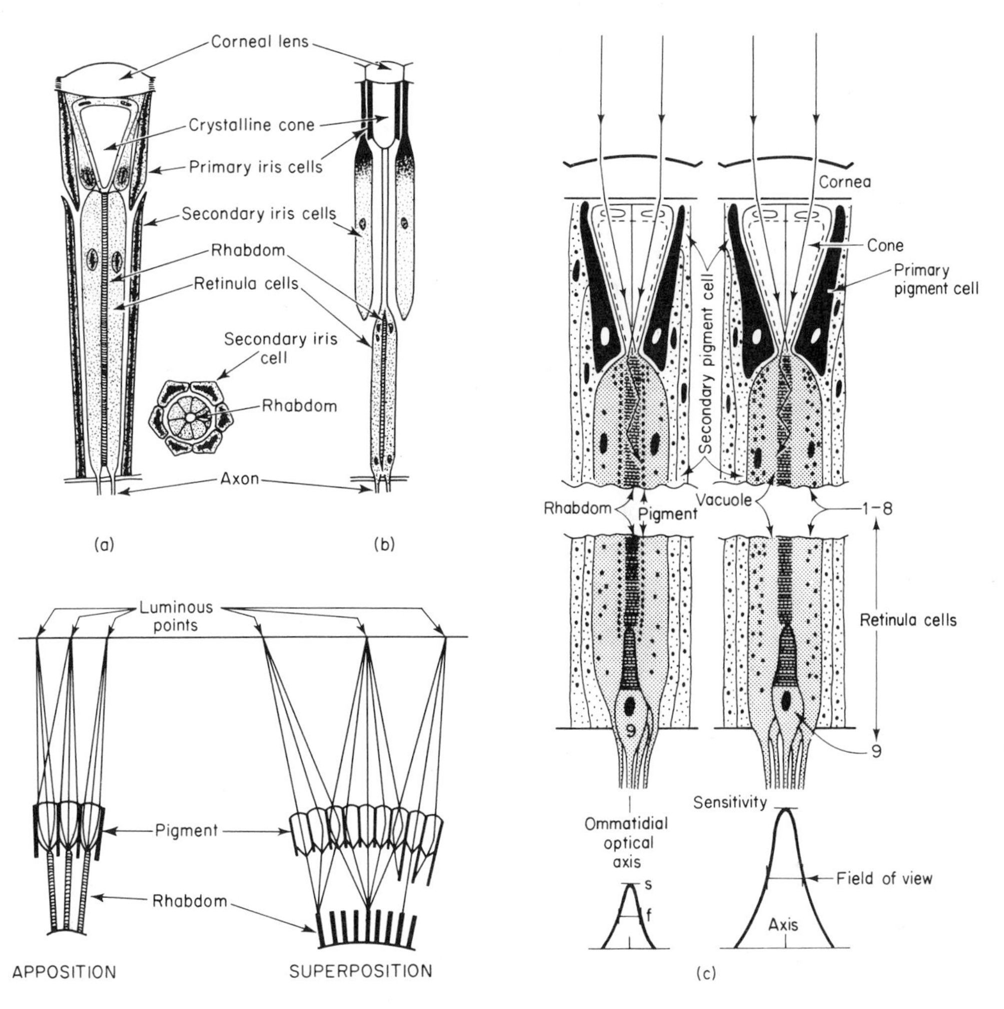

Fig. 18.11 Functional morphology of insect ommatidia. (a) and (b) Structure (above) and optics (below) of apposition and superposition eyes as described by Exner in 1891. [After Wigglesworth, 1972: Principles of Insect Physiology, 7th ed. Chapman & Hall London.] (c) Structural details of ommatidia characteristic of most diurnal insects. The difference between light-adapted (left) and dark-adapted (right) lies in the distribution of pigment and refractive index around the rhabdom; this, by changing its optical properties as a light guide through a critical range can control its sensitivity (s) and to some extent the size of the field of view (f) as shown at bottom of diagram. The 9th retinula cell, lying behind the filter formed by the other eight can show enhanced detection of the plane of polarization and is the receptor for navigation by this means in the honeybee. [From Horridge (1974); figure courtesy G. A. Horridge.]

two patterns of "clear-zone"; those in which the clear-zone or light guide is formed by elongated corneal or crystalline cones, and those in which the distal portions of the retinula cells are elongated and clear while the rhabdoms are confined to the proximal part of the cell. In all cases, light may enter from many facets in the dark-adapted eye to sum upon individual receptors on the far side of the clear zone. However, in addition, a light-guide mechanism, provided by the tracts or columns in some clear-zone eyes, provides a separate optical pathway for the formation of erect images.

The dioptric apparatus Some of the cup-like eyes are broadly open with only a transparent lining over the photosensitive cells (the limpet *Patella*) and some of the vesicular eyes (the *Nautilus*) operate as pinhole cameras. Most eyes, however, have a lens or dioptric apparatus of transparent tissue which serves to concentrate the light on the receptor cells.

A lens is found in the eyespot of some dinoflagellates and in certain hydroid medusae such as *Sarsia* (Fig. 18.2). It is evidently a phylogenetically ancient device for the convergence of light rays on the photoreceptors. In most of the cup-like or vesicular eyes of invertebrates the lens almost or completely fills the cavity and is quite immobile. The focus is fixed like that of an inexpensive camera. The same is true of the aggregate or compound eyes (Figs. 18.8 and 18.11). On the other hand, the eyes of all the vertebrates and a few of the invertebrates (cephalopod and heteropod molluscs and alciopid polychaete worms) possess a dioptric apparatus which can be changed to focus the eye. This capacity is referred to as ACCOMMODATION and, like a fine camera, permits the lens to form a sharp image on the retina for objects both near and far.

In focusing a camera, the position of the lens is altered with respect to the photographic film by moving the lens forward for near objects. Invertebrates and lower vertebrates also accommodate by altering the lens/retinal distance. Several devices have been described. In the polychaete worm, *Alciopa*, the lens/retinal distance seems to be adjusted by changing the fluid volume in one of the two regions of the optic cup. In the heteropod and cephalopod molluscs the distance is altered by squeezing the optic cup or moving the lens forward (Carthy, 1958; Nicol 1967a). The lower vertebrates also accommodate by changing the position of the lens. In lampreys and teleosts the lens is said to be moved backward to focus on distant objects; Munz (1971) notes several exceptions and reviews recent developments. Elasmobranchs, amphibians, and snakes move the lens forward to adjust for near objects; these adjustments are like those of the camera or microscope.

The mammals, birds and reptiles other than snakes went one step further. In their visual systems a soft, pliable lens in a transparent capsule can be squeezed by altering the tension on the tough ligaments attached to it. The "camera" is focused by changing the shape of the lens. The action is rapid and under autonomic control. Parasympathetic stimulation contracts the circular fibers of the ciliary muscle, and this relaxes the ligaments of the lens which then bulges and increases its refractive power to permit focusing on objects near at hand. Contraction of the radial fibers innervated by the sympathetic system as well as a relaxation of the circular fibers takes place during accommodation for far vision. Thus, in many amniotes, accommodation is quick and precise; it is probably rather crude in the invertebrates and lower vertebrates.

There are several curious devices which make accommodation unnecessary. Certain bats, for example, have a corrugated or folded retina which brings many rods into focus at different distances; in the ray, the horse, and probably some other vertebrates the retina is arranged on a "ramp," with the axial length of the eyeball changing continuously in the vertical meridian (Walls, 1942). Such devices serve well when visual acuity is less important than sensitivity.

Photopic and Scotopic Eyes

The world is alive with animal life both night and day, but quite different species are often abroad and active at these times. Both the arthropods and the vertebrates show several distinctive retinal specializations associated with daily activity rhythms. Some of the differences are morphological; others are physiological. The monographs by Walls (1942), Prince (1956) and Tansley (1965) contain a wealth of details; only a brief summary is given here.

Several adaptations for PHOTOPIC (light) and SCOTOPIC (dark) vision have already been noted. The clear-zone eyes of arthropods and the retinal rods of vertebrates are low-threshold units, arranged to accumulate light from wide angles and permit maximum summation in the optic nerves. The complete catalogue of photopic/scotopic adaptations is much longer and includes additional items associated with the eyes themselves as well as the visual units.

The AREA CENTRALIS is a specialization of the diurnal eye of vertebrates. A well-marked series of changes during its evolution have greatly improved the resolving power of this tiny retinal spot which lies in the focal plane of the optic lens system (Fig. 18.10). In the more primitive situation, the area centralis is a retinal THICKENING caused by a concentration of cones. In animals with more precise diurnal vision, the area becomes very THIN since the tangled nerve fibers diverge from the cones on all sides to form a FOVEA where light falls directly on the photoreceptors. Some foveae (higher primates) contain an abundance of yellow pigment which corrects for chromatic aberration. The area centralis is then referred to as a MACULA LUTEA.

The retina reaches its climax as a diurnal photoreceptor in some of the birds. In water birds and species which live in open plains the fovea may be greatly extended into a horizontal band, and cone density may reach 1 million/mm^2 in contrast to about 145,000/mm^2 in the human fovea (Pumphrey, 1961; Shaler, 1972). A temporal as well as a central fovea is common in birds which depend on accurate distance judgment (hawks, swallows, hummingbirds or kingfishers). These eyes are capable of resolution over a large part of the visual field. Increasing cone density is associated with a decrease in the general vascularity and the development of a special organ, the

PECTEN, which protrudes from the back of the optic cup and assists in the diffusion of metabolites into the vitreous humor. The pecten is a sizeable comb-like structure of folded vascular tissue which, in its phylogeny, has acquired an optic function in addition to its trophic one. The early evidence (reviewed by Pumphrey, 1961) came from morphological correlations between the habits of different species of birds and the development of their pectines. More recent studies support the earlier conclusions (Wingstrand and Munk, 1965; Barlow and Ostwald, 1972).

Scotopic eyes are specialized for maximum dim light sensitivity at the expense of the resolving power of the retina. Batteries of elongated rods connect in groups to single bipolar nerve cells and thus effect a maximum summation. Bats may have as many as 1000 rods connected to one nerve cell (Young, 1957). Moreover, the retinae of many nocturnal animals contain reflecting tissue (the TAPETUM LUCIDUM) which acts as a mirror to reflect the light back through the photosensitive layers and double its effect on the rods. A tapetum lucidum is characteristic of many nocturnal arthropods as well as vertebrates from fish to mammals. Both the nature and distribution of reflecting materials vary in different species. Guanine crystals are common, but some insects rely on shiny tracheal tubes, certain fishes utilize minute spheres of triglyceride located in the pigment epithelial cells, and some mammals (the musk ox, for example) achieve the same advantages by glistening white collagenous fibers. A layer of golden yellow riboflavin crystals creates a very effective mirror in the eyes of the bush baby (lemuroid *Galago*) and the garpike (Walls, 1942; Nicol and Arnott, 1972, 1973).

There are also good correlations between the shape of the optic cup and the daily activity rhythms. In general, the lens/retinal distance is greater in the diurnal forms,

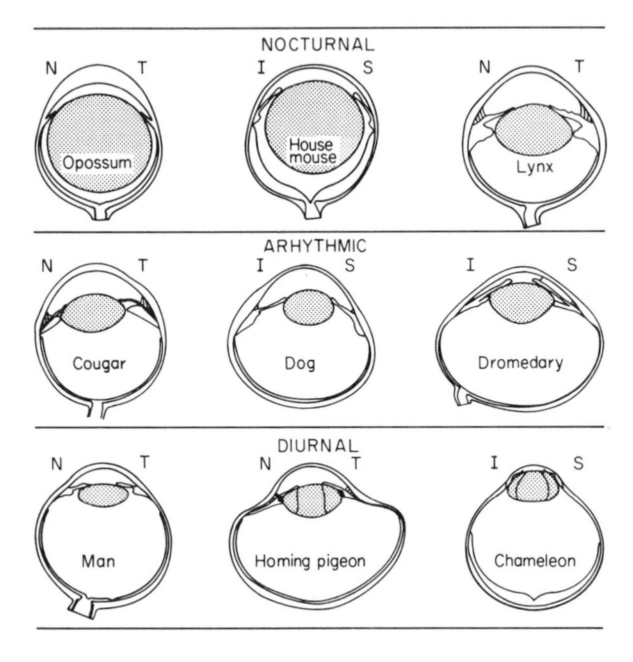

Fig. 18.12 Intraocular proportions in relation to daily activity habits. I, inferior side of eyeball; N, nasal side; S, superior side; T, temporal side. [From Walls, 1942: The Vertebrate Eye. The Cranbrook Institute of Science, Bloomfield Hills, Michigan.]

permitting the projection of larger images on the retina and the possibility of greater resolution of details (Fig. 18.12). The use of color filters may also increase visual acuity as well as correct for chromatic aberration. The macula lutea has already been mentioned. In addition, many diurnal animals have a yellowish lens or cornea, and in some amphibians, reptiles and birds the retina is studded with a beautiful mosaic of oil droplets (usually yellow), which occur just beneath the light trap in the cone (Fig. 18.6).

Color vision Wavelength discrimination is a special capacity of the diurnal eye and its associated neural centers. The human eye is capable of distinguishing about 1,500 different hues between 400 nm in the blue and 800 nm in the red. This means the separation of wavelengths not more than 0.2 to 0.3 nm apart (Tansley, 1965). Although relatively few groups of animals possess color vision, this capacity seems to be potentially present in all of the more highly organized optical systems and is capable of expression under evolutionary pressures for high visual acuity. Von Frisch, more than half a century ago, demonstrated color vision in the honeybee (Burkhardt, 1962; von Frisch, 1971), and since that time it has been found in several orders of insects (Coleoptera, Diptera, Hymenoptera, Lepidoptera). It may also occur in other invertebrate groups (crustaceans, cephalopods), but the evidence is largely circumstantial.

Among the vertebrates, hue discrimination has been demonstrated with certainty in primates, birds, lizards, turtles, frogs, and teleost fishes. It is associated with bright-light vision, foveae with rich areas of cones, and eyes with good mechanisms for accommodation. These associations are obviously linked with high visual acuity and pictorial analysis rather than with taxonomic position. Electrophysiological studies indicate that the retinal factors essential for color perception are sometimes present even when the animal often behaves as though it were color blind (cats, rats, and guinea pigs, for example). The importance of the analytical centers in the brain is apparent (DeReuck and Knight, 1965; Tansley, 1965).

The first acceptable theory of color vision was formulated by Thomas Young in 1801 (literature reviews by Walls, 1942; Granit, 1955). This is based on the well-known fact that a proper mixing of three "primary colors" (blue, yellow and red for pigments or blue, green and red for lights), will produce white or any of the colors recognized by the human eye. Young made three fundamental assumptions: that color reception is organized by the retina, that the number of color-sensitive elements in the retina is limited and that they represent widely different regions of the spectrum. These basic assumptions remain fundamentally correct, although the detailed nineteenth-century theory proposed by Helmholtz has been extensively modified by later investigations (Davson, 1972). The YOUNG-HELMHOLTZ OR TRICHROMATIC THEORY ascribed color perception to the interaction of three specific types of retinal element (sensitive to red, green and violet respectively) with zonal representation of the colors in the cortex. It has now been possible to test the theory by studying the retinal action potentials (electroretinograms) associated with the stimulation of areas of the retina with monochromatic light and also to measure the absorption spectra of pigments in individual cones of the vertebrate fovea. In addition, the examination of the insect retina has been particularly valuable because electrodes can be inserted into individual retinula cells.

Burkhardt (1962) summarizes studies of the spectral sensitivity of the fly *Calliphora erythrocephala.* Recordings were made from individual retinula cells while the eye was stimulated with flashes of monochromatic light (wavelengths separated by about 20 nm). Burkhardt's review should be consulted for details. The findings show that *Calliphora* has three visual substances with peak sensitivities at 350 nm, 470 nm and 520 nm; these are distributed in a rather precise manner in the ommatidia. *Calliphora's* information about wavelength appears to depend on the distribution of excitation among these three cell types. The peak at 350 nm is associated with the ultraviolet sensitivity which marks one of the striking differences between insect and vertebrate eyes. In comparison with the vertebrates, the insect spectral sensitivity is shifted about 100 nm toward the shorter wavelengths.

It is also possible to record from individual rods and cones. Fain and Dowling (1973), working with the mudpuppy retina, first achieved this difficult feat. Although the attempts to penetrate single receptor cells failed for more than a quarter of a century, substantial support for the trichromatic theory was obtained by recording from cells in the ganglionic layers. Granit (1955 and earlier) pioneered the electrophysiology of the vertebrate retina and his book should be consulted for these important investigations.

Crucial support for the trichromatic theory of color vision in vertebrates was first obtained by Wald and his associates, who devised techniques which permitted measurements of the absorption characteristics of individual cones (Brown and Wald, 1964). Human retinae were removed and mounted on the stage of a microscope, with the visual cells pointing upward and the light passing through them axially in the direction of incidence in the living eye. Absorption maxima of the rods occurred at 505 nm. Three types of cones were convincingly separated with maxima at 450 nm (blue receptors), 525 nm (green receptors) and 555 nm (red receptors), as shown in Fig. 18.13. These findings have now been confirmed in studies of the retinas of several different animals known to have good color vision, including the goldfish (Munz, 1971). There is, undoubtedly, a firm biochemical basis for the kind of theory which Helmholtz developed in the nineteenth century.

Several different color receptor systems are probably present in the animal kingdom, and some may be much simpler than that of the human retina. The retina of the frog evidently contains three visual pigments, two in the rods and one in the cones (Muntz, 1964). The peak sensitivity of the green rods is at 440 nm and of the red rods at 502 nm; the cone pigment shows peak sensitivity at about 560 nm. Both behavior and action potential recordings from the optic nerves show that frogs are especially sensitive to blue, although this is not associated with any single pigment in the retina. The frog's preference for blue and the tendency to jump toward it or toward the light when startled seems to be adaptive and favors escape toward the aquatic habitat.

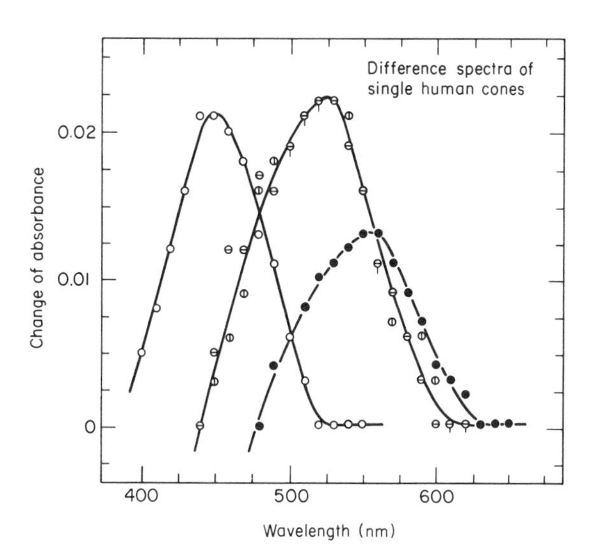

Fig. 18.13 Difference spectra of visual pigments in single cones of the parafoveal region of the human retina. In each case the absorption spectrum was recorded in the dark from 650 to 380 nm, then again after bleaching with a flash of yellow light. The differences between these spectra are shown. One of these cones, apparently a blue receptor, has λ max about 450 nm; two cones, apparently green receptors, have λ max about 525 nm; and one, apparently a red receptor, has λ max about 555 nm. In making these measurements, light passed through the cones axially in the direction of incidence normal in the living eye. [From Brown and Wald, 1964: Science, *144*: 51.]

Polarized light sensitivity In addition to its intensity and wavelength, the light which reaches the earth has an important quality associated with the DIRECTION OF ITS WAVE MOTION. The wave motion of white light radiated from the sun is considered to be in all directions perpendicular to the direction of transmission. If, for any reason, this vibration is restricted to one plane, the light is said to be PLANE POLARIZED. Certain types of glass, plastic or crystals will polarize light passing through them because their orderly molecular arrangement permits vibrations only in one plane. Light may also be polarized by reflection and, in nature, this occurs in the upper atmosphere when the short wavelengths (blue) are scattered by molecules of air (especially the inert gases and nitrogen) to produce the blue sky in what is often spoken of as "Tyndall scattering." The "azure vault of heaven" is in reality a layer of reflected short wavelengths of light about twelve miles above the earth.

Both the extent and the angle of polarization in any patch of sky depend on the relative position of the patch of sky with respect to the sun (Fig. 18.14). The effect is most pronounced at right angles to the sun's rays; if one looks straight up at sunrise or sunset with a suitable detecting instrument (called an ANALYZER or POLARIZER), the blue sky becomes a distinct and predictable map.

Von Frisch (1950, 1971) was the first to demonstrate clearly that the blue sky is being used in animal navigation. Since that time it has been shown that many different terrestrial arthropods can orient themselves with respect to polarized light, that light penetrating the aquatic environment often becomes polarized, and that several different crustaceans, cephalopods and teleost fishes are capable of orienting themselves to it (Carthy, 1958; Jander *et al.*, 1963; Dill, 1971; Waterman and Forward, 1972). The occurrence of this ability, in separate phyla with different kinds of eyes, indicates that the basic visual machinery (probably the orderly arrangement of lipoprotein

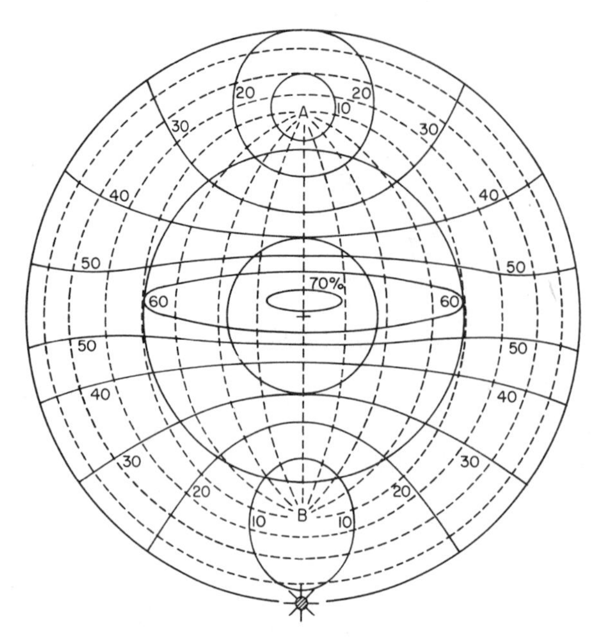

Fig. 18.14 Distribution of polarized light in the sky when the sun is on the horizon. Solid lines indicate the points of equal percentage of polarization and the broken lines indicate points where the angle of polarization is the same. *A* and *B* are the two points where the light is unpolarized. The diagram represents the hemisphere of the sky, + being the highest point and the two rings concentric with + being the lines subtending angles of 30° and 60° with the horizontal. [From Carthy, 1958: An Introduction to the Behaviour of Invertebrates. George Allen and Unwin, London.]

molecules in rods or plates) has the capacity to acquire sensitivity to the plane of polarization if this confers an evolutionary advantage.

Von Frisch, in his pioneer work, constructed a polaroid model which produced precise light patterns when aimed at specific patches of sky at particular times in the day. His analyzer, like the ommatidium of the honeybee, consisted of eight parts (retinular cells); the eight triangular plates were fitted together to give four different planes of polarization as shown in Fig. 18.15. Waterman and his colleagues, working mostly with decapod crustaceans, have now shown that retinular cells do respond to light polarized in various directions, that light sensitivity (including detection of polarized light) is localized in the rhabdomeres and that, unlike the four-channel model proposed by von Frisch (Fig. 18.15), the analyzer is a two-channel system (Fig. 18.8). At the molecular level, analysis depends on the alignment of the molecules of visual pigment in the parallel stacks of microvilli (Waterman and Horch, 1966). Cephalopods depend on a similar type of analyzer, but the basis of polarized light orientation in the teleost fishes is not yet known (Waterman and Forward, 1972).

Photomechanical Responses and the Arhythmic Eye

The arhythmic eye, which performs well over a wide range of light intensities, has several adaptive mechanisms associated with activity during both night and day. These are primarily devices for controlling the amount of illumination which reaches the photosensitive cells. The maximum available light should impinge on the receptors during the night, but during the day, when illumination is adequate, the significant

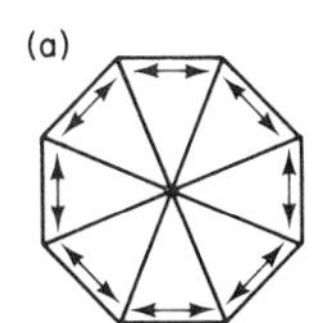

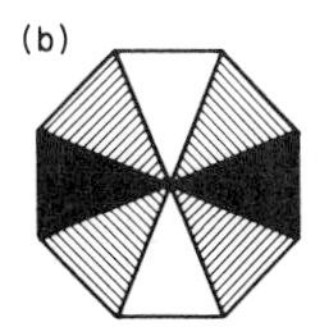

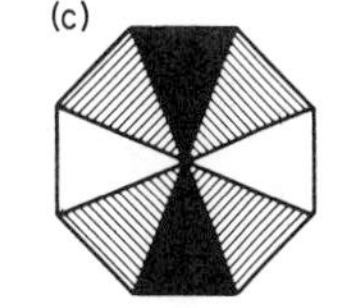

Fig. 18.15 Polaroid model of the honeybee eye suggested by von Frisch. (a) Analyzer consists of eight triangular pieces of Polaroid, fitted together so that the plane of vibration of the light transmitted by opposite pieces is the same. (b and c) Different patterns of light seen when the analyzer is directed toward different areas of blue sky. [From von Frisch, 1971: Bees, Their Vision, Chemical Senses and Language. Cornell U.P., Ithaca, N.Y.]

problem is the resolution of pictures. In other words, the retinal elements or small groups of the cells must be excited separately by different points from the picture during the day; but at night acuity is sacrificed for sensitivity, and light is collected from many angles to excite the receptor cells. The human eye works well over a 30,000-fold intensity range of illumination (sunlight to moonlight), and moderately well over about three times this intensity range; it can detect changes in intensity of the order of 2 to 5 per cent. Comparable capacities are found in several groups of animals. This twenty-four-hour habit depends on several different photomechanical responses involving rapid changes in the distribution of pigment and the action of contractile elements in the iris and/or the retina.

Pigment migration Photomechanical responses in the arthropods are confined to pigment migration. The apposition eye has been described as photopic and the superposition eye as scotopic. Pigment migrations occur in both but are more striking in the superpositional organ, and this is the visual system which is best adapted to arhythmic behavior. Pigment occurs both in the retinulae and in the envelope or iris cells surrounding each ommatidium (Fig. 18.11). Among the crustaceans, pigment migration occurs in both kinds of cells (Kleinholz, 1961); among the insects it is confined to the iris cells (Wigglesworth, 1972). It should be noted that while pigment distribution may be altered, the cells in which it is found do not change their shapes. In many species of arthropods, marked diurnal rhythms of changing pigment distribution have been observed, and these may persist for a long time under constant light conditions. The phenomena appear to be under nervous regulation in insects but under hormonal control in the crustaceans (Highnam and Hill, 1969; Wigglesworth, 1972).

Retinal pigment migration also occurs in many vertebrates. It is said to be rapid and extensive in the teleosts, anurans and birds; it is slow and less marked, or slight, in turtles and crocodilians; it is absent in snakes and mammals (Walls, 1942). In darkness, the pigment granules within the epithelial cells surrounding the rods and cones move to the back of the retina; in the light they are dispersed through the receptor layer and the outer layers of the retina (Fig. 18.16).

A most curious example of pigment migration was described many years ago in

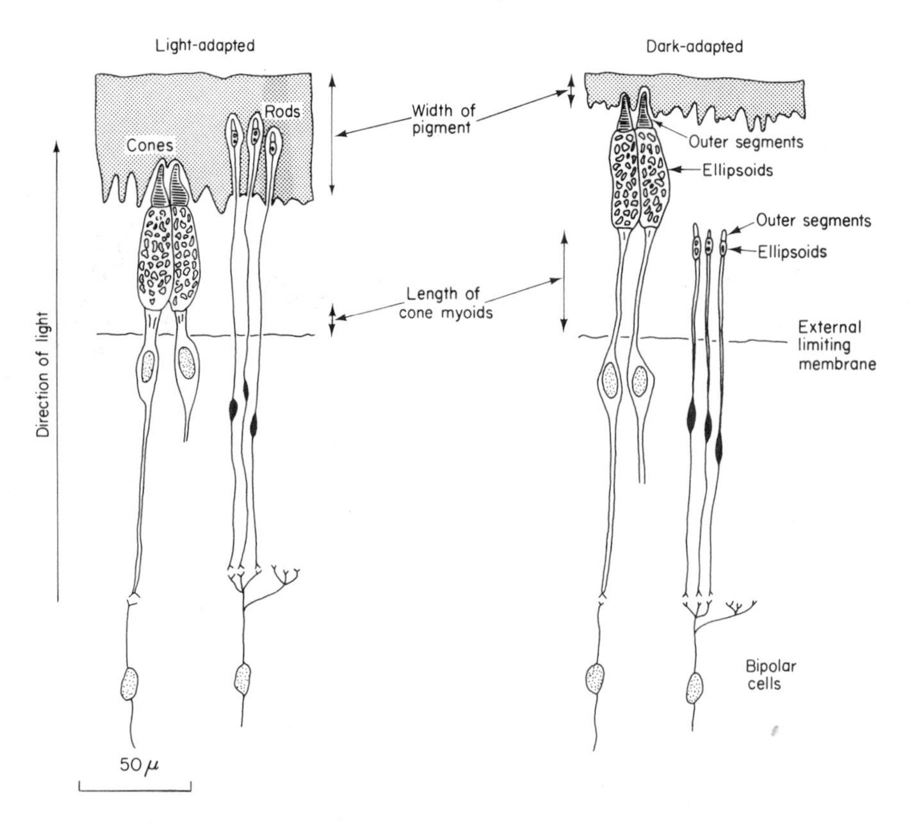

Fig. 18.16 Photomechanical responses in the retina of the adult herring [From Blaxter and Jones, 1967: J. Marine Biol. Assoc. U.K., *47*: 679.]

the elasmobranch eye. The epithelial cells associated with the rods and cones are devoid of pigment in these fishes, but the choroid contains chromatophores which are said to spread like curtains over a series of slanted silvery plates of the choroidal tapetum (Fig. 18.17). The original investigation was carried out by Franz on *Mustelus laevis* (reviews by Walls, 1942, and Nicol, 1963).

Nicol (1961, 1963) has extended the observations to several species of elasmobranchs and finds support for Franz's arguments in some of the pelagic sharks such as *Mustelus* and *Squalus;* in benthic species like *Raja* and *Scyliorhinus,* however, the tapetum is fixed and non-occlusible. Species with a non-occlusible tapetum lucidum show rapid and extensive pupillary movements in contrast to forms with a fixed tapetum. Tapetal chromatophores are independent effectors; pigment granules continue to move normally in excised eyes (Nicol, 1965).

Careful studies of the optics of the elasmobranch eye have shown that the silvery plates, composed of guanine and/or hypoxanthine, are set at angles so that the light impinging on the retina is not scattered as it would otherwise be from a concave surface. The points of light are reflected directly back through the photoreceptors, thus

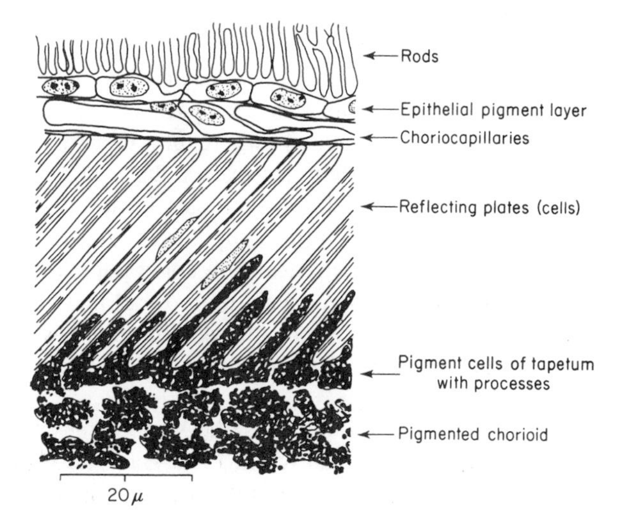

Fig. 18.17 Section through the tapetum lucidum of the dorsal region of the eye of *Scyliorhinus canicula*. [From Denton and Nicol, 1964: J. Marine Biol. Assoc. U.K., *44*: 220.]

avoiding a blurring of the image (Denton, 1970); the tapetum so greatly increases the efficiency of the eye that the quantity of visual pigment appears to be reduced to about half that found in fishes living in the same environment without such reflectors in their eyes. The movement of pigment over the plates probably reduces the risk of eyeshine which might betray the fish to predators (Denton, 1971).

Visual cell movements In those vertebrates which show retinal pigment migration there is usually an associated photomechanical change in the position of the receptor cells. The few exceptions include some of the fishes where only the pigment moves and others where the activity is largely confined to the rods or cones (Walls, 1942; Nicol, 1963).

Movement of the photoreceptor depends on the MYOID, the highly contractile inner stalk-like portion of the rod or cone between the nucleus and the ellipsoid (Figs. 18.6 and 18.16). These associated movements of the pigment granules and receptor cells are not rapid and can be easily traced by histological techniques in which eyes are fixed after intervals of exposure to different intensities of illumination (Ali, 1959; Nicol, 1963). In teleost fishes, where this type of retinomotor response is conspicuous, as much as an hour may be required for complete change from light-adaptation to dark-adaptation (Ali, 1959). However, this is quite adequate to keep pace with the changing daily light conditions and provides an effective means of adjusting the retina to the external illumination.

Pupillary responses A mobile iris which, like the diaphragm of a microscope or camera, rapidly alters the aperture of the camera-like eye is the most efficient of the light-regulating adaptations for the arthythmic habit. It reaches its climax in the mammals where its smooth muscle is under rapid autonomic control. The shape of the aperture varies from a circle in some species to a slit in others, depending on the demands of the animal for light at low intensities. The cat, with its reasonably good twenty-four-hour vision, provides a familiar example of the slit pupil. Like a pair of curtains, this can be drawn apart to permit a full view of the stage.

The mammals have no particular monopoly on pupillary responses. They occur in the cephalopod eye where they are under nervous control (Nicol, 1967a) and in all classes of vertebrates except the cyclostomes (Walls, 1942). Only a few of the teleosts (the eel, for example) have a mobile iris, but the elasmobranchs show extensive, although slow, pupillary responses. Other fishes (ganoids, dipnoans) and the terrestrial vertebrates have some pupillary control. It is slight in the lower forms, becoming progressively more rapid and extensive as the retinomotor responses decrease in significance through the amniotes. The birds are unique in preserving rapid and extensive retinomotor responses and at the same time possessing a highly mobile iris.

In the lower forms the smooth muscle of the iris is directly but very slowly activated by light (Walls, 1942; Barr and Alpern, 1963). A nervous control is absent; the pupil may require two or three minutes to close in bright light and as long as an hour to reopen in the darkness. A measure of autonomic control is achieved in the amphibians; the pupillary responses become much more rapid and extensive in some reptiles, birds and mammals where there is not only improved nervous regulation but also an increased development of the iris musculature. In general, phylogeny among the vertebrates has seen a gradual reduction and disappearance of retinomotor responses with a progressive development of iris mobility and pupillary control.

Adaptations to Space, Motion, and Unusual Habitats

The basic physiology of the photoreceptor system has been adapted in many different ways to meet varied habits and habitats. The morphology of the eye may be curiously altered in association with the habitat, as in deep-sea fishes. Through learning and the specialization of visual centers in the brain, depth and distance may be accurately evaluated. The position of the eye and its mobility may greatly enlarge the fields of vision. Special coverings of tissue may form "spectacles" or "goggles" for protection in swift swimming or burrowing. These and many other curious devices are based on the machinery already considered; the visual system is one of the most versatile in the variety of evolutionary adaptations.

Like many phylogenetic trails, that which has led to the most specialized of photoreceptors shows numerous diverging paths. Sometimes a more primitive mechanism is preserved by a more advanced group, as the birds, which retain the photomechanical activities of the visual units along with a mobile iris. The reverse may also be true; lower groups may acquire certain devices which are usually associated with the higher forms. Pupillary responses occur among some of the invertebrates and lower vertebrates, although they are most characteristic of the birds and mammals. Darwinian evolution works with a strictly limited number of building materials and weaves these into patterns which frequently show clearly marked trends of specialization but, at the same time, display random elements of the potential variation.

19

EFFECTOR ORGANS AND THE PHYSIOLOGY OF MOVEMENT

It is by the effector organs that animals express their unique characteristics. Specialized structures for locomotion, the production of electricity, bioluminescence and rapid changes in color are virtually absent from the plant world. It is true that plants may show movements, but these are usually based on growth or turgor and not on a specialization of contractile proteins as is the case with animals. Only the amoeba-like slime molds, the flagellated algal cells or the sex cells of a few higher plants possess this kind of motility. All of the multicellular animals, on the other hand, depend on muscular movements for the activity of visceral organs and usually for locomotion as well. Glands also are universal effector organs among multicellular animals. Electric organs, luminescent structures and the activities of chromatophores occur in many different animals, although they do not play as general a role as the organs of locomotion and secretion. There are also several unusual effectors of very limited distribution, such as the nematocysts or stinging hairs of coelenterates and the colloblasts or adhesive cells of the ctenophores; their physiology is not considered here since it has little bearing on the story of animal phylogeny. Furthermore, secretion will not be separately treated but mentioned only with the physiology of the organs and systems which depend on this process. The present discussion is confined to the mechanisms which produce movement, to the electric organs, the luminescent structures and the chromatophores. Motility is considered in this chapter; the other three processes are dealt with in Chapter 20.

MECHANISMS PRODUCING MOVEMENT

Animal locomotion and complex visceral activities, such as the propulsion of materials through ciliated or muscled tubes, may have evolved from the changing shapes shown by some of the enzyme and protein molecules which participate in the fundamental energy exchanges of life. Mitochondria, both *in vitro* and *in vivo,* may very slowly swell and contract, sometimes with oscillating regularity (Lehninger, 1962, 1964; Harris *et al.*, 1969). These changes, which are caused by the *non-osmotic* uptake and extrusion of water, are directly linked to cellular respiration. Lehninger postulates that the enzymes of the respiratory chain are "mechano-enzymes." They convert the oxidation-reduction energy not only into phosphate-bond energy of ATP but also into mechanical energy by changing their molecular arrangement in the mitochondrial membranes. It is of interest that a number of other important protein molecules change their configuration with their functional state; the change from oxygenated to deoxygenated hemoglobin and the change of cytochrome *c* from the oxidized to the reduced form are characterized by significant alterations in protein structure (Lehninger, 1962; Perutz, 1964). In the present context, the most significant observations are the striking similarities between the enzymatic and mechanochemical properties of the muscle protein actomyosin and the energy-coupling mechanisms in the mitochondrial membranes. Lehninger finds similarities in pH optima, in the effects of heavy metals and azide, in stimulation of ATPase by DNP, involvement of divalent cations and other biochemical characteristics as well as the changes in shape. Substantial amounts of mitochondrial actomyosin are known to be present in the membranes of mitochondria (Lehninger, 1964).

Mitosis is another universal cellular activity which may be based on a physiology similar to that of the motile proteins of cilia and muscle. The slow movements of the elements in the mitotic spindle display several similarities to the more rapid ones of cilia, pigment granules and muscle fibrils (Inoué and Sato, 1967; Jahn and Bovee, 1969). Fundamental principles common to all motile phenomena were postulated almost one hundred years ago (see Weber, 1958), and it now seems that the molecular biologists of the twentieth century may provide the evidence to support such a hypothesis.

The Basic Mechanisms Involved in the Movements of Cells

Movement depends on special protein structures which actively change their form or position to produce an elongation or a contraction. The ancestral unicellular animals probably experimented with several different systems before muscles and cilia were selected as the most satisfactory structures for locomotion. Present-day protozoans support such a hypothesis. An *Amoeba* rounds up when stimulated and then lengthens markedly (without any pull from the outside) as it spreads out and "crawls away." Some ciliates possess TRICHOCYSTS and others have MYONEMES in addition to their rhythmically beating cilia. Trichocysts are small bodies in the outer layers of the cytoplasm which may explode either spontaneously or when stimulated. In exploding they discharge elongated threads some 6 to 7 times their original length; myonemes are contractile fibrils in the pellicle which contract to alter the body of the animal. The myoneme in the stalk of *Vorticella* has attracted considerable physiological interest because of its accessibility and superficial resemblance to a muscle fibril. *Vorticella* is shaped like an inverted bell with a long stalk resembling the handle on the bell. The stalk which attaches the animal to the substrate consists of a central contractile myoneme within a flexible sheath. When stimulated, the contraction of the myoneme bends the sheath into a tight spiral and the animal rapidly withdraws. As the myoneme relaxes, the stalk straightens out because of the elasticity of the sheath.

Thus, the protozoans have realized three different kinds of movement: elongation, contraction and rhythmic vibration (cilia). Studies of the biochemistry and cellular physiology of these systems, however, have shown the superficiality of such a classification of fundamental mechanisms. At the structural level, motility seems always to depend on the action of ultrafine fibrils. The electron microscope has revealed a fibrillar differentiation in almost all kinds of cells (Pitelka, 1969) and animals at all levels in phylogeny seem to have adapted these fibrils to effect movement. At the biochemical level, the contractile machinery depends on comparable proteins whose changes in shape or position are dependent on ATP. Details may differ in superficially similar systems like the muscle fiber and the stalk of *Vorticella* but an ATP-driven movement of an actomyosinoid protein is probably universal.

Contractile models Dead cell models have proved most valuable in analyzing different processes involved in movement. The first of these was prepared from vertebrate skeletal muscle by Szent-Györgyi (1949) who used 50 per cent aqueous glycerol to extract pieces of the psoas muscle from rabbits. This extraction removes the soluble constituents of the sarcoplasm (proteins, enzymes, substrates) and leaves little more than the fibrous contractile elements. Cellular metabolism is no longer possible, but the models still contract when ATP is added under appropriate conditions, and in doing so they exhibit many of the characteristics of intact contracting muscle. This isolation of the power plant from the contractile machinery has permitted a much more precise analysis of the activities of both. Hoffman-Berling (1960) has applied the same technique to other motile structures such as flagella, trichocysts, *Vorticella* stalks and pseudopodia.

By means of strong salt solutions it is possible to dissolve the contractile proteins of muscle. When these solutions are squirted through a fine orifice, well-organized threads of actomyosin are formed, and these also will shorten and perform work under conditions comparable to those which initiate muscle contraction. Such threads were first produced by Weber in 1934 (see reviews by Weber, 1958, 1960).

Contractile protein systems of this kind have also been prepared from slime molds and sarcoma cells, but several other motile systems have not yet been successfully dissolved and reorganized (Hoffman-Berling, 1960).

ATP-actomyosin systems The major protein components of vertebrate muscle are the MYOSINS (mol wt about 450,000) and the ACTINS (mol wt about 60,000). These may be readily extracted from muscle; when mixed in salt solutions of suitable composition and ionic strength, they combine in the ratio of about three molecules of myosin to one of actin and form the complex contractile protein ACTOMYOSIN. *In vitro* preparations of actomyosin threads shorten in the presence of ATP. One of the central questions in the comparative physiology of motile systems concerns the distribution of actomyosinoid proteins and their behavior in the presence of ATP. Jahn and Bovee (1967, 1969) review the studies on motile systems of Protozoa and tabulate data for organisms ranging from bacteriophage "tails" and bacteria to higher plants and Metazoa. It is apparent that actomyosinoid proteins which contract in the presence of ATP and possess ATPase activity are universal; the actomyosinoid biomechanical system was probably a very early evolutionary development.

Studies of cell models and contractile protein threads reveal four physiological types of motile systems (Weber, 1958): (1) the stretching and (2) contraction of organelles by Ca^{++} (these are prevented or reversed by ATP), (3) the stretching movements induced by ATP and other polyphosphates and (4) the contraction produced exclusively by ATP and related nucleoside triphosphates. All of these movements are either initiated by ATP or reversed and inhibited by ATP, and in each case this is evidently the ultimate source of energy.

Movements inhibited or reversed by ATP All motile systems which have been examined in multicellular animals are driven by ATP. Among the protozoans, however, two are known to be induced by ATP-free reactions (Hoffman-Berling, 1960). One of these, the trichocyst, is an elongation; the other, *Vorticella's* myoneme, is a contraction.

Although the myoneme of the *Vorticella* stalk is like a minute muscle, models of the stalk do not contract on the addition of ATP. On the contrary, their contraction seems to be physiologically triggered by calcium ions. Several other substances may produce a similar effect. Sr^{++} and the ions of the quaternary ammonium bases are very effective; Ba^{++} has a weak action while Mg^{++} and Be^{++} are inactive. Removal of the calcium by the chelating agent EDTA (ethylenediaminetetraacetate) reverses the contraction and relaxes the stalk. Under natural conditions calcium is assumed to be the active agent. Since the preparations used in these studies are cell models and lack the capacities to generate energy, it can be argued that the calcium is acting directly on the contractile proteins.

Even though ATP does not initiate contraction in the stalk, it is assumed that the energy of the system is ultimately derived from its high-energy bonds. The details are not yet clear, but if ATP is added to the contracted model, the system relaxes; under certain conditions ATP will induce a rhythm of spontaneous contractions and relaxations. The relaxation is quite different from that produced by EDTA which is a permanent condition caused by the removal of Ca^{++}. The evidence summarized by Hoffman-Berling (1960) indicates that the ATP effect is on the contractile proteins rather than on the calcium. In the *Vorticella* stalk, the chemical energy of ATP is apparently utilized in the relaxation phase and stored in the contractile system.

Isolated trichocysts elongate explosively in a manner comparable to living trichocysts when Ca^{++}, Sr^{++}, or certain other ions (but not Mg^{++}) are added. Again, Ca^{++} is probably the physiologically active agent. This system is also inhibited by ATP; in the presence of ATP the isolated trichocysts fail to elongate when Ca^{++} is added. In this case, also, the ATP inhibition is not due to Ca^{++} binding but seems to depend on a complexing with the contractile proteins. These interesting models are not yet completely understood, but they show very clearly that there are motile systems in the animal world which operate very differently from those present in muscles and cilia.

ATP-driven movements At all levels in animal phylogeny some motile cells are activated by ATP (Jahn and Bovee, 1969). The two ATP-inhibited systems just described are exceptional even among the protozoans.

Active contractions are readily induced by ATP in glycerol-extracted amoebae and amniotic fibroblasts as well as in muscle cells. This activity also depends on the presence of Mg^{++}, a suitable pH and specific ionic concentrations. The other nucleoside triphosphates may replace ATP but are neither as powerful nor as effective; adenosine monophosphate (AMP), creatine phosphate, other organic phosphates and the inorganic polyphosphates are inactive. Conditions required for contraction of the amoeboid and muscle cell models are similar in every respect; differences are quantitative, and the indications are that comparable machinery is responsible in both cases.

Hoffman-Berling (1960) summarizes the biochemical studies on the movements of the mitotic apparatus; these involve active elongations as well as contractions. Both are ATP-driven systems with biochemical requirements similar to those of the muscle and fibroblastic models. The inescapable conclusion seems to be that mechanisms responsible for the motility of such highly specialized structures as muscle cells are both phylogenetically and ontogenetically older than the muscle cells themselves.

Protoplasmic movements These ATP-powered biomechanical systems probably operate in several different ways to move protoplasm and consequently to propel animals from one part of their environment to another. Many of the earlier theories, including the coiling or folding and unfolding of long protein molecules, have now been discarded. Jahn and Bovee (1967, 1969) argue that there are only two major applications of the actomyosinoid protein mechanism: (1) a CONTRACTION-HYDRAULIC SYSTEM in which an outer layer of gel contracts and simultaneously liquefies to a sol which flows under hydraulic pressure as in *Amoeba* and (2) an ACTIVE-SHEARING OR SLIDING SYSTEM in which active tangential movements occur between two surfaces, either both of gel, or one sol and one gel. The contraction-hydraulic system has been studied in the slime molds (Mycetozoa) as well as in *Amoeba* and its allies; the active-shearing or sliding system is best known in striated muscle but also seems to operate in cyclosis, in the movements of the filipods of Foramenifera and the axopods of Heliozoa, in the migration of chromosomes on spindle fibers and perhaps also in the movements of the fibrils in cilia and flagella. Both systems are presumably phylogenetically ancient. In fact, a taxonomic split in the protozoan class Sarcodina has been proposed on the basis of these fundamentally different motile systems (Jahn and Bovee, 1964). Although there are still facts which do not seem to fit readily into the hypothesis proposed by Jahn and Bovee (1969 and earlier), current evidence suggests that the actomyosinoid proteins have been utilized in relatively few different ways to operate the varied organs of locomotion.

AMOEBOID MOVEMENT

The elusive *Amoeba* has provided generations of students with their first great challenge in zoology. Ever since these minute masses of protoplasm were first described by Rösel von Rosenhof in 1755 (Allen, 1961) microscopists have attempted to explain their gentle agitations and delicate gliding movements, for it is felt that these activities may be fundamentally primitive and might provide clues to the general physiology of locomotion by cells.

The functional morphology of amoeboid cells is detailed in many familiar zoological texts and will not be repeated here. Excellent bibliographies are available (Allen, 1961; Jahn and Bovee, 1967, 1969). The present discussion is confined to current hypotheses of the mechanics of the movement.

The first half of the twentieth century saw the formulation, experimental examination and eventual discard of hypotheses which attempted to localize the motive forces in changing surface tensions or in sol-gel transformations of colloidal cytoplasm (De Bruyn, 1947). At present there is a return to the much earlier theory of protoplasmic contraction. In detail, the current theories of contraction bear little similarity to the early ones proposed by Dujardin (Allen, 1961), but the general concept is the same, since both postulate forces which depend on the contraction of the cellular proteins.

Allen (1961, 1962) advocates the terminology shown in Fig. 19.1; this involves an abandonment of the terms "plasmagel" and "plasmasol" and the theories associated with these terms. Instead, an active contraction of the cytoplasm is postulated either in the fountain zone or in the tail process (uroid). If this contraction is in the fountain zone, then the animal is being pulled along; but if it occurs in the uroid, the animal is being pushed forward. Allen regards the former theory as the more likely while Jahn and Bovee (1964, 1967, 1969) support tail-end contraction just as vigorously. The action of the contractile proteins is accepted as the dynamic propulsive force; the controversy centers around the locus of contraction. Is the *Amoeba pulled* along or *pushed* forward?

There may be no single explanation. Even protozoans of the class Sarcodina utilize extremely diverse motile organs ranging from the short, blunt, lobose processes

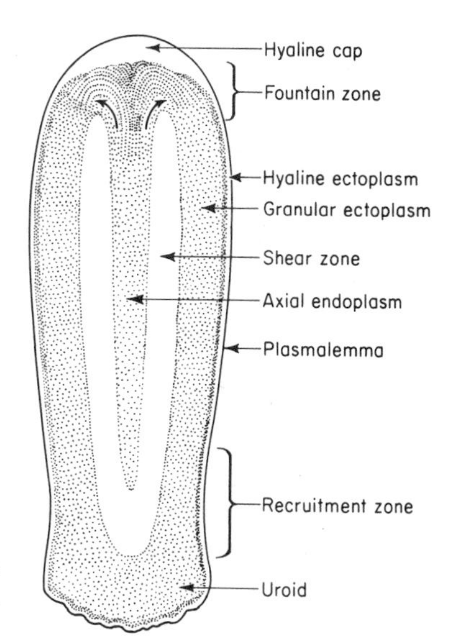

Fig. 19.1 Structure of an *Amoeba* as described by Allen, 1960: J. Biophys. Biochem. Cytol., *8*: 395.

which arise from the naked protoplasm of the Amoebina or the Mycetozoa to the long, slender, branched, anastomosing and sticky reticulopods which emerge through numerous pores in the delicate shells of the Foramenifera. The contractile proteins may well be differently adapted in organs so different in appearance. Jahn and Bovee (1967, 1969) recognize two fundamental systems as previously noted; Allen (1968) suggests that the mechanisms may be even more variable. Current theories are detailed and evaluated in the reviews already cited. In addition, there are many pertinent papers in the symposium volume edited by Allen and Kamiya (1964).

According to Allen's "fountain zone" theory, an amoeba pulls itself forward by contraction of the fibrous anterior endoplasm. He maintains that the endoplasm is not homogeneous but contains a central core or plug of more rigid material separated from the ectoplasmic tube by a fluid mantle, the shear zone. A shortening or movement of the molecules in the front end pulls this axial endoplasm toward the forward end of the ectoplasmic tube. The hyaline cap or watery area at the front end of an advancing *Amoeba* is presumably caused by SYNERESIS or the squeezing of liquid from the contracting gel. The fluid of the hyaline cap appears to move posteriorly between the plasmalemma and the ectoplasmic tube, to be reabsorbed in the recruitment zone as the ectoplasmic material softens and relaxes to join the endoplasmic stream. Allen suggests that this theory may also account for a two-way streaming of the cytoplasm sometimes observed in intact amoebae and regularly seen in the fine pseudopodia of foraminiferans (Jahn and Rinaldi, 1959). A localized pressure caused by posterior contraction could not create a two-way stream, but a shortening of proteins at a sharp bend in the cytoplasm such as the fountain zone or the tip of a reticulopodium might do just this.

Jahn (1964) notes that both SOL and GEL refer to a spongy, contractile protein network which enmeshes various granular inclusions. The sol and gel states in

Amoeba differ only in the density of the protein network or the number of protein fibers per unit volume of framework. In the CONTRACTION-HYDRAULIC theory of motility, the contracting gel at the posterior end causes the less fibrous endoplasm or sol to flow forward while at the anterior end the sol is diverted peripherally and becomes denser adding to the length of the tube. Sol forms at an equivalent rate posteriorly as the available ATP is used up in the contracting and compacting gel. Bhowmick (1967) finds that electron microscopy supports this theory of amoeboid movement.

Jahn and Bovee (1969) argue that an entirely different system operates in the pseudopodial filaments of a forameniferan. Their earlier conclusions were based on observations of the active bidirectional movement of granules on the threads of plasmagel which form the filopods. These delicate protoplasmic structures seem to lack outer hyaline layer, gel tube, and central core. Instead, the actomyosin is organized in two parallel gel filaments and movement is caused by the active shearing or parallel displacement forces located between their adjacent surfaces. Further studies will be necessary to establish the precise mechanism or mechanisms. However, whether the true explanation is in terms of "a pull" or "a push" and a system of sliding filaments, it seems to be based on a system of contractile actomyosinoid proteins activated by ATP.

MOVEMENTS OF CILIA AND FLAGELLA

Cilia and flagella are vibratile extensions of the cell surface which permit mechanical work without any marked change in the form of the effector cell. All the major groups of animals, except the nematodes (Andreassen, 1966), make use of them to propel either individual cells, such as protozoans and spermatozoa, or entire multicellular animals such as planarians and ctenophores, or to move fluids and adherent materials over the lining surfaces of tubular organs. Among the arthropods, motility based on cilia or flagella occurs very rarely and is confined to the spermatozoa of certain groups. In some animals, such as the hemoflagellate *Trypanosoma,* cilia are modified to form undulating membranes; in others they may be grouped to form cirri and macrocilia. Each macrocilium of the common ctenophore *Beroë* is composed of 2,000 to 3,000 cilia of typical internal filament structure, arranged within a single membrane to form vibratile organs about 6 to 10 nm thick and 50 to 60 nm long (Horridge, 1965c). In all groups of animals, including the nematode worms, parts of certain sensory receptor cells develop from modified cilia (Chapter 18).

Several ingenious theories for ciliary locomotion were proposed by earlier workers (Gray, 1928). However, with only the light microscope to reveal their micromorphology, these hypotheses were scarcely more satisfactory than Leeuwenhoek's concept of "diverse incredibly thin feet, or little legs, which were moved very nimbly" (Satir, 1961). Here again the electron microscope has provided a clue which will probably lead to a true understanding of these delicate machines. Although the mechanism is still not understood, the revelation of a constant fibrillar pattern suggests a common basis for ciliary movement and muscular contraction.

Cilia may range in length from several microns to several hundred microns, but their diameter is surprisingly constant—ranging from about 0.1 to 0.5 μ. This uniformity in diameter is a reflection of the regular pattern of internal fibrils (Fig. 19.2). The covering membrane of a cilium is continuous with the plasma membrane. Just inside the plasma membrane, around the periphery, are nine double fibrils about 0.028 μ in diameter with two additional fibrils of smaller size in the center of the organelle. The fibrils appear hollow so that the peripheral doublets form figures-of-eight in cross section. Along the shaft of the organ, these doublets possess arms pointing in one direction as shown in Fig. 19.2. These arms, which are probably fundamental to the movement of the filaments (Sleigh, 1968), contain the protein DYNEIN; this protein has ATPase activity (Gibbons, 1965). The nine peripheral fibrils seem to merge at the base into a hollow tube which forms the basal body; they terminate

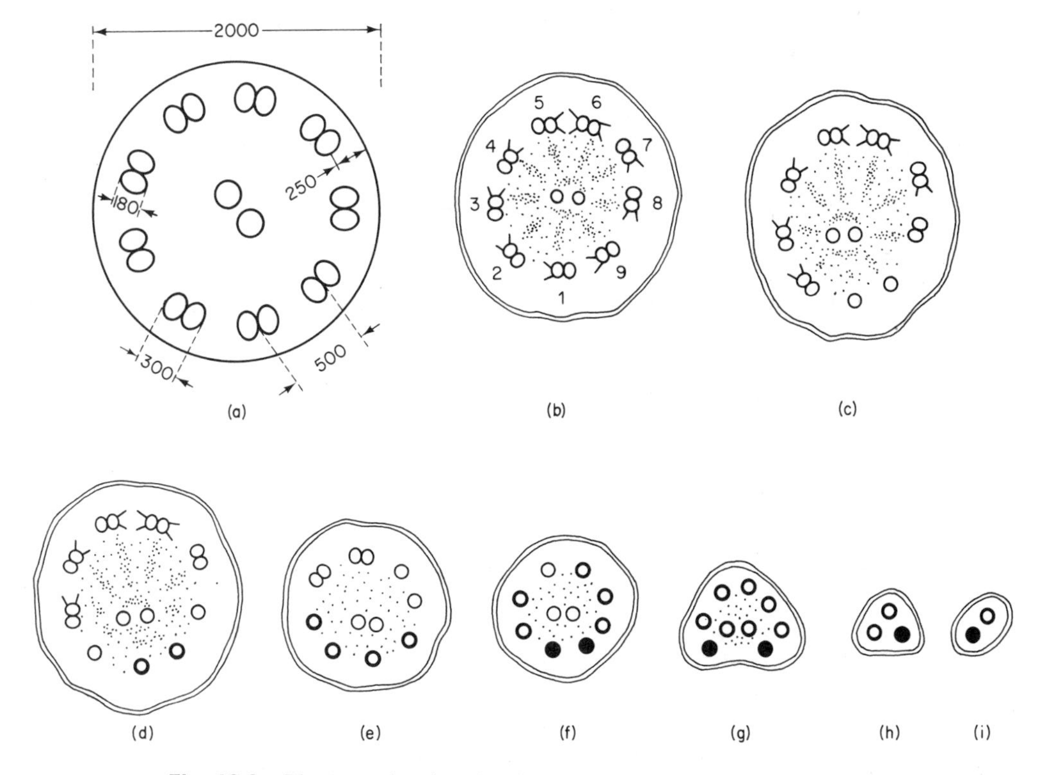

Fig. 19.2 Diagrams showing the disposition of fibrils in cilia. (a) and (b) Cross sections through shaft of cilium in gill of a freshwater mussel giving approximate dimensions in angstrom units. Note interlocking of arms between filaments 5 and 6 and the orientation of the arms which project in a clockwise direction when viewed along the cilium from base to tip. (c) to (i) Sections made at progressively more distal levels through the tip of a bent cilium showing first the disappearance of the doublets on one side (c), (d), and (e) and then on both sides (f) to (i). [(a), After Fawcett, 1961: *In* The Cell, Vol. 2, Brachet and Mirsky, eds. Academic Press, New York. Others, after Satir, 1965: J. Cell. Biol., *26*: 806.]

separately at the outer tip of the cilium; the two central fibrils do not reach the basal body. Further details of ultrastructure are summarized by Holwill (1966). Exceptions to this standard arrangement of fibrils, whether in organs of motility or receptor cells, are extremely rare and are confined to the sperm flagella of certain insects (Phillips, 1968). This uniformity suggests very strongly that the motility depends on these fibrils and that the arrangement is significant for their action.

Current discussions of the bending mechanism center around two different hypotheses. One of these assumes that the filaments are contractile and shorten alternately on opposite sides to bend the cilium or flagellum from side to side; according to the other hypothesis, the fibrils are "sliding filaments" which do not change shape but move past one another to produce a curvature of the cilium. The weight of evidence, derived largely from studies of electron microscopy, favors the latter theory (Sleigh, 1968). In part, this evidence depends on studies of the arrangements of the tips of the peripheral filaments (Fig. 19.2). Satir (1965), using a technique which instantaneously fixed the gill cilia of freshwater mussels, was able to study the relative positions of the fibril tips in cilia during effective and recovery strokes. Fibrils on the *concave side* of the cilium were 0.5 to 1.0 μ longer than those on the convex side; if bending were caused by contraction, the fibrils should be shorter on the concave side (Fig. 19.3).

Horridge (1965c) has also provided significant support for the sliding filament theory in studies of the macrocilia around the inside of the lip of the ctenophore *Beroë*. Hundreds of cilia operate together in these giant vibratile organs to bend them as units. The individual cilia of one of these structures are interconnected by relatively

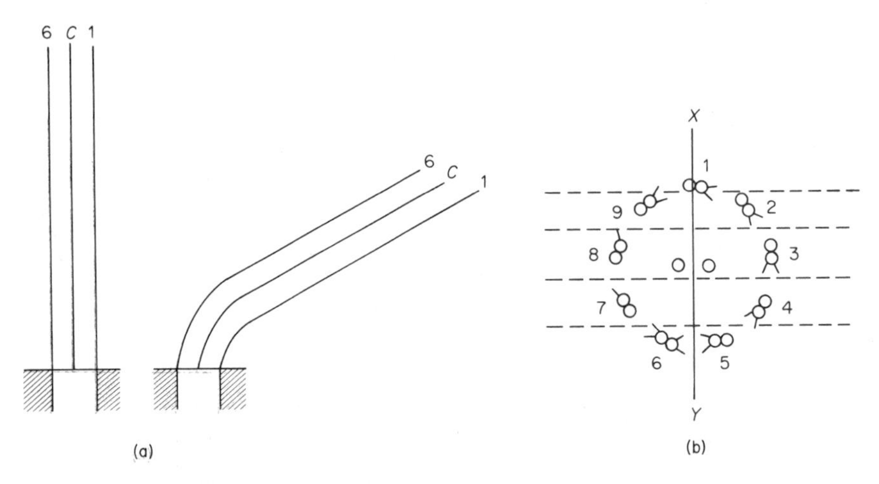

Fig. 19.3 Sliding filament theory of ciliary movement. (a) The change in position of the tips of peripheral fibrils 1 and 6 relative to a central fibril when a cilium bends during its effective stroke. (b) Transverse section of a ciliary shaft, with broken lines showing possible lines of sliding when the cilium bends in the plane *X–Y*. [From Sleigh, 1968: Symp. Soc. Exp. Biol., *22*: 145.]

rigid bridges set at right angles to the line of beat. No evidence was found for changes in the diameter of the fibril or for buckling during bending and relaxation—as might be expected if an actual contraction were taking place. Moreover, the basal bodies remain flat even at the height of the power stroke. These observations lend strong support to the sliding filament hypothesis. Calculations indicate that the required shortening of the fibrils on their concave sides would be of the order of 40 to 50 per cent to produce the degree of bending observed in ctenophore macrocilia; this degree of contraction should produce some visible differences in the convex and concave sides of the fibrils.

Patterns of movement Both flagella and cilia show movement patterns which are frequently constant and characteristic of the particular organism and tissue (Holwill, 1966; Sleigh, 1968). The beat of the velar locomotor cilia in the larva of the nudibranch *Jorunna* is analyzed in Fig. 19.4. These cilia beat slowly, show a large amplitude, and rest at the start of each effective stroke. Reviews cited contain graphical and mechanical analyses for cilia and flagella from many other organisms.

A nice coordination of movements is present when cilia are associated in rows, tracts, or sheets. Ciliated surfaces, when examined under the light microscope, show rhythmic waves constantly moving over them. This rhythm is said to be ISOCHRONAL when all of the cilia beat together and METACHRONAL when the successive cilia in each row beat in sequence to form a regular series of waves (Fig. 19.5). The latter is characteristic of most ciliated surfaces and assumes several forms in different organisms (Sleigh, 1962; Parducz, 1967).

Coordination of rhythm There appear to be several ways in which groups of cilia are entrained to beat in a harmonious fashion. In the simplest situation, a mechanical interaction through the fluids between the cilia may be the only requirement. Many years ago, Gray (1928) noted that when the heads of individual spermatozoa are in intimate contact, their tails beat synchronously and that the

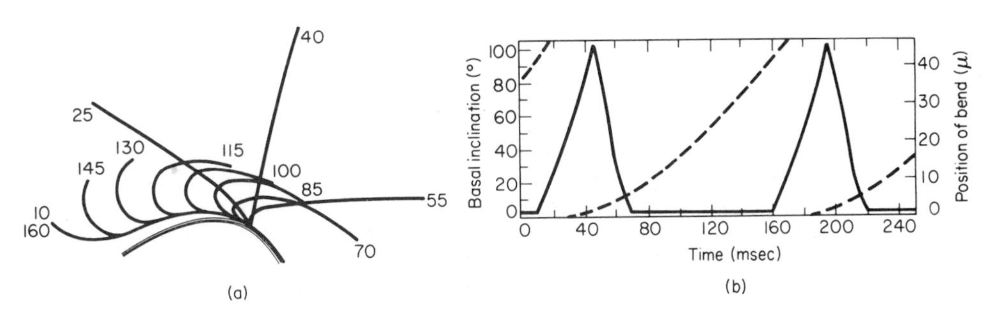

Fig. 19.4 The beat of a cilium on the velum of a veliger larva of *Jorunna tomentosa* filmed at 340 frames/sec at 22°C. (a) Position of cilium at times shown in msec. (b) Change with time of the angular position of the basal region of the ciliary shaft (continuous line) and the position of the bend along the ciliary axis (discontinuous lines rising from left to right). [From Sleigh, 1968: Symp. Soc. Exp. Biol., *22*: 135.]

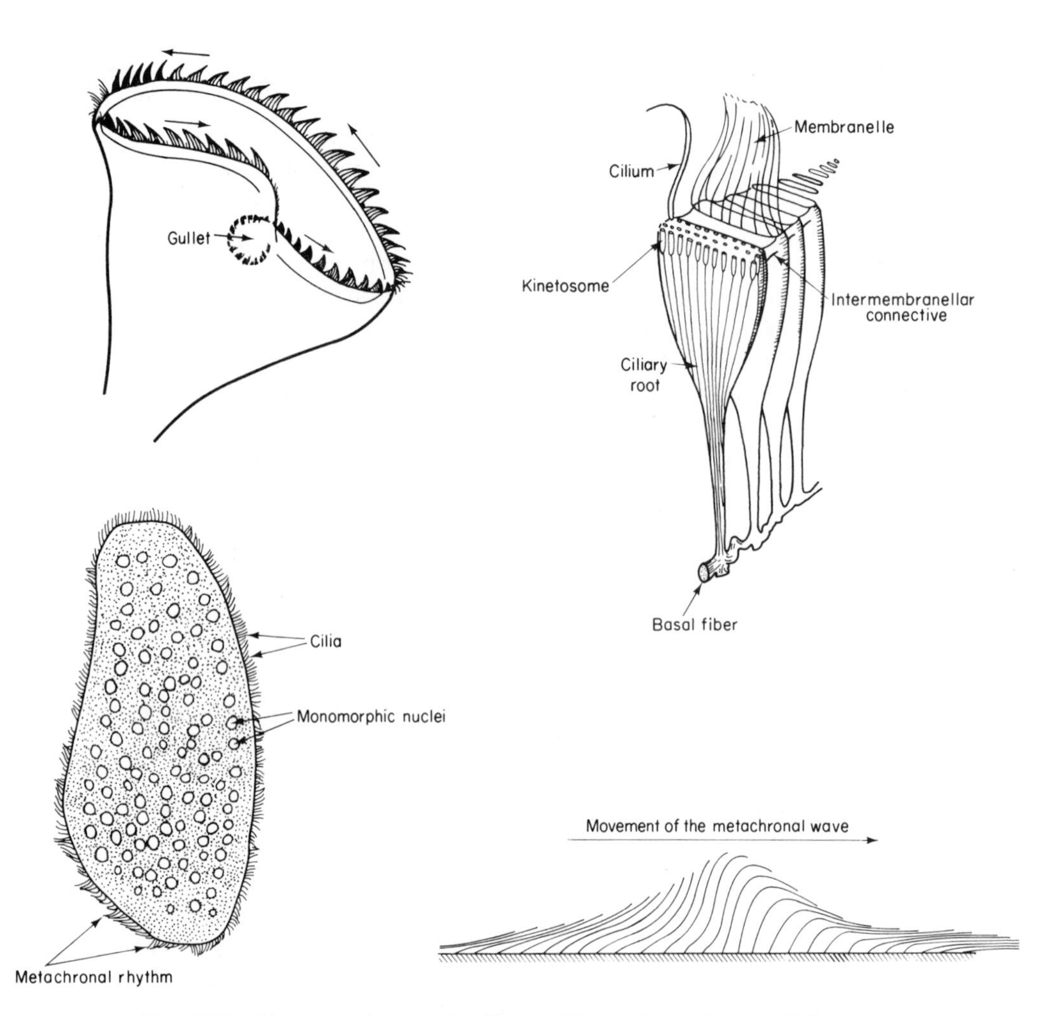

Fig. 19.5 Metachronal waves in ciliates. Above, the peristome of *Stentor* with arrows to show the direction of wave transmission on left, and diagram of ciliary rootlets and basal fibers as revealed by electron microscopy on right; below, *Opalina* showing the component cilia and metachronal waves. The plane of beat is approximately in the plane of the paper and successive movements of a cilium can be followed by moving along the row from right to left. *Opalina* is moving toward the left in the lower right diagram. [Ciliary waves, from Sleigh, 1962: The Biology of Cilia and Flagella. Pergamon Press, Oxford. Rootlet system in *Stentor* based on Randall and Jackson, 1958: J. Biophys. Biochem. Cytol., *4*: 820. Diagram of *Opalina* from Freeman and Bracegirdle, 1971: An Atlas of Invertebrate Structure. Heinemann, London.]

undulations of spirochaetes tend to come into phase with one another when they are swimming close together. Automatically contractile cells seem to coordinate their movements without the existence of any permanent connection between them, and Sleigh (1968, 1969) finds this hypothesis of MECHANICAL INTERFERENCE satisfactory to explain the rhythmic waves of the ciliates *Opalina* and *Paramecium* or the gills of *Mytilus*. The cilia of these forms are spontaneously excited from their basal bodies, but the beat is synchronized mechanically so that there is a minimum of interference in their movements. If this hypothesis is correct, changes in the viscosity of the bathing fluid should have less of an effect on the speed of the beat than it has on the metachronism. This has been shown experimentally; manipulations of the viscosity alter the beat in accordance with the presumed viscous drag. In *Opalina, Paramecium,* and the gills of *Mytilus,* the main action of an increased viscosity is on the form of the wave while the rate of beat declines only slightly.

The situation is entirely different in the membranelles of the ciliate *Stentor* (Fig. 19.5) or the ciliated combs of a ctenophore. Here, increased viscosity reduces the rate somewhat but has no effect whatever on the metachronism which is assumed to depend on a NEUROID MECHANISM.

The ctenophore *Beroë* is a beautiful transparent animal about the size of a small plum. It has a mouth at the more pointed end and a sensory APICAL ORGAN containing a statolith at the opposite pole. Its locomotion depends on eight ciliated combs arranged in four pairs; each pair shares one of the four nerves which radiate from the apical organ. *Beroë* may suspend, reverse or accelerate the movements of the cilia on the combs in response to vibrations, increased hydrostatic pressure or tilting of its body. Simple experiments readily demonstrate complex behavior which depends on inhibition as well as excitation of different groups of cilia. A nervous regulation, based on the apical organ and a nerve net, has been established in many classical studies of anatomy and physiology (Hyman, 1940; Horridge, 1968b). In electron micrographs, Horridge and Mackay (1964) identified synapses between the nerve fibers and the bases of the giant ciliated cells. Long-lasting depolarizations are associated with the beat of the cilia. The physiology of this elaborate system is still speculative, but the evidence of nervous regulation of ciliated combs from the apical sense organ strongly supports the concept of a NEUROID TRIGGER THEORY of control.

Sleigh's (1969) analysis provides for several coordinating mechanisms. He avoids many of the difficulties posed by hypotheses which attempt to explain the synchrony in extremely divergent ciliated organs by either a mechanical interference or a nervous coordination. Studies of the membranelles of *Stentor,* for example, suggest another possibility. Electron microscopy has revealed bundles of root fibrils arising from the ciliary cells (Fig. 19.5). Viscosity effects, behavioral responses and microsurgery indicate that these root fibrils maintain the phase between constantly beating cilia. In *Stentor* as in *Paramecium* the beat is triggered in the basal body; the phase, however, seems to be internally initiated and maintained by a conducted "impulse" in *Stentor* while only a mechanical interaction is responsible in *Paramecium.*

Cilioregulatory nerves Activity, in many different ciliated epithelia, may be altered by stimulating nerves (Kinosita and Murakami, 1967). Responses to nerve stimulation have been clearly demonstrated in ciliary systems of ctenophores, turbellarians, annelids, molluscs, and frogs; both excitatory and inhibitory effects have been noted. Cilia on the pharyngeal epithelium of the frog beat more rapidly with waves of greater amplitude when certain cranial nerves are stimulated; they appear to be under autonomic control. Both adrenaline and acetylcholine have been tested and, although some recorded data are debatable, the acetylcholine response seems incontestable in a number of preparations. Cilioregulatory effects of the branchial nerve have been studied in the gill epithelium of many bivalve molluscs (Aiello, 1970). Inhibition as well as excitation sometimes follow stimulation of this nerve. The excitatory effect is probably mediated through the transmitter 5-hydroxytryptamine (serotonin).

In summary, the movements of cilia and flagella are basically autonomous. Synchronization of movements in groups of cilia sometimes depends on a simple interaction of closely packed elements, but in many places a more elaborate regulation is present and depends on fibrils or nerves associated with the bases of the cilia-bearing cells. Complex behavior and locomotory responses may depend on regulating nerves and sensory organs.

MUSCULAR MOVEMENT

Studies of muscle have formed a large segment of physiology—particularly in the laboratory—ever since the beginning of experimental work in animal biology. Almost two centuries ago, Galvani (1737–98) was fascinated by the twitchings of skinned frog legs in response to electrical stimuli; these preparations still pose some unanswered questions (Wilkie, 1956; Bendall, 1969) and never fail to provide interest for the beginner in zoology.

An appreciation of the structural basis of contractile tissues stems from Bowman's classical description of its fibrillar nature in 1840. Details, insofar as they could be revealed by the light microscope, were gradually established during the latter half of the nineteenth century, and the mechanical properties were carefully described within the limits of the recording instruments available. Familiar experiments with the frog gastrocnemius, loops of intestine and the hearts of turtles and frogs were introduced into physiology during this period (Verworn, 1899; Bayliss, 1920).

The basic biochemistry of muscle was established during the first half of the twentieth century. This period opened with Fletcher and Hopkins' description of lactic acid production in 1907—to be followed by an elucidation of the glycolytic changes during the next 25 years. Lohmann's scheme for the ATP-creatine phosphate source of triggering energy was postulated in the middle thirties (Chapter 7). This was largely the biochemistry of the sarcoplasm; the studies of the muscle proteins (although initiated by Kühne in 1859) came somewhat later. Some of the most exciting discoveries in muscle biochemistry took place between 1940 and 1950 with the recognition that myosin was an ATPase, with its crystallization and with the descriptions of actomyosin, tropomyosin, and paramyosin.

The second half of this century has been marked by an extension of the morphological analysis with the electron microscope and persistent attempts to associate the biochemical with the mechanical events. Many of the pertinent details have probably now been recorded, but there is still some uncertainty regarding the linking of the chemical reactions which surround the binding and splitting of ATP with the contraction, relaxation or movement of the fibrillar proteins.

Although the pioneer physiologists devoted considerable attention to a variety of contractile systems such as cilia, flagella, protoplasmic steaming, amoeba and *Vorticella* (Verworn, 1899), interest in the comparative physiology of muscular tissues is recent. Especially since 1950, details of micromorphology in many different species have been recorded, and an intensive examination has been made of the neuromuscular physiology of animals from all phyletic levels. The functioning of intact muscles depends on the manner in which they are activated and also on such passive elements as connective tissue. Studies of neuromuscular physiology are basic to an understanding of muscular movement.

There is an extensive literature on the physiology of muscle. The monographs edited by Bourne (1960), Volume 37 of the *Cold Spring Harbor Symposia on Quantitative Biology* (1972) and a short book by Carlson and Wilkie (1974) cover most aspects of it; Hoyle (1957, 1962a, 1969) has carefully reviewed the comparative work.

Energy for Muscle Contraction

Like other types of animal cells, the muscle cell contains an elaborate system of enzymes for the production of ATP (Chapter 7). Power production and cell movement operate as a unified system. The glycolytic enzymes concerned with the formation of lactic acid from glycogen or glucose are found in the structureless areas of the cytoplasm (sarcoplasm in muscle), while the carbon fragments which they form are utilized to produce large amounts of ATP within the mitochondria (sometimes called sarcosomes in muscle).

These enzyme systems probably show many functional correlations with muscle physiology, although there is still relatively little information on the comparative aspects (Hoyle, 1969). It has, however, been noted that vertebrate skeletal muscle, which can accumulate a considerable oxygen debt, has relatively small mitochondria in comparison with cardiac muscle where the mitochondria are both larger and more numerous. The sarcosomes of insect flight muscle are the most numerous and the largest observed in any animal cells and attain as much as 40 per cent of the weight of the muscle in *Phormia*. Likewise, the very active mantle muscle fibers of the cephalopods have a much more conspicuous mitochondrial system than the tonic fibers of the lamellibranch adductors. Similar correlations between activity and the development of the mitochondrial system have been made in some other groups, but relatively few forms have been examined (Hanson and Lowy, 1960). Adaptive differences in several specific enzyme systems have also been recorded. For example, in comparative studies of invertebrates, as well as vertebrates, ATPase is always higher in more rapidly contracting systems; there is a 200-fold difference between cat striated muscle and dogfish mesenteric smooth muscle (Hoyle, 1969).

The Contractile Proteins

The ubiquity of actin and myosin, two fibrous proteins associated with contractile tissues, has already been noted. Intensive biochemical and biophysical work has revealed many aspects of their organization and contractile behavior. Questions still remain, but some of the mysteries of the system have been solved and the beauty of its organization clearly exposed. In addition, several other fibrous proteins associated with contraction have been identified.

Myosin Just over half of the contractile protein in vertebrate skeletal muscle is MYOSIN. Estimates of molecular weights of myosin from different muscles range from 220,000 to 550,000 (Gergely, 1966). Purified solutions, examined with the electron microscope, show masses of elongated thin rods with an expansion at one end; although smaller in size, these rods resemble spermatoza with distinct tails and heads. Digestion with trypsin or chymotrypsin splits these molecules into two fragments which have been found to possess quite different biochemical and physiological properties. The tail fragments are termed LIGHT MEROMYOSIN (LMM); while the heads are HEAVY MEROMYOSIN (HMM); their dimensions are shown in Fig. 19.6.

On precipitation these molecules fall into a very orderly arrangement with tails parallel and staggered so that the heads point in one direction along half of the filament and in the opposite direction along the other half as shown in Fig. 19.6. The tail portions consist of two (possibly three) polypeptide chains of light meromyosin coiled around each other in an α-helix. In the discussion of fibrillar structure, it will become clear that the rod-like tails form the backbone structure of the thick filaments of the A-band in skeletal muscle while the heads release the power required for movement *and* are also the sites of interaction with the actin filaments.

Organized myosin filaments have not been identified in vertebrate smooth muscle (Elliott, 1967). In this respect it appears to be exceptional. Although the protein myosin is present and presumably discharges its function as an ATPase, the indications are that the thick filaments of LMM are absent and that the myofibrils are primarily actin. There are still many unsettled questions with respect to the molecular organization of smooth muscle (Mountcastle, 1968).

Actin Spherical molecules of molecular weight about 60,000 and diameter 55 Å may also be extracted from muscle. These are known as GLOBULAR (or G-) ACTIN and polymerize in salt solutions of certain ionic strengths to form double-stranded helixes known as FIBROUS (or F-) ACTIN. F-actin looks like two chains of beads

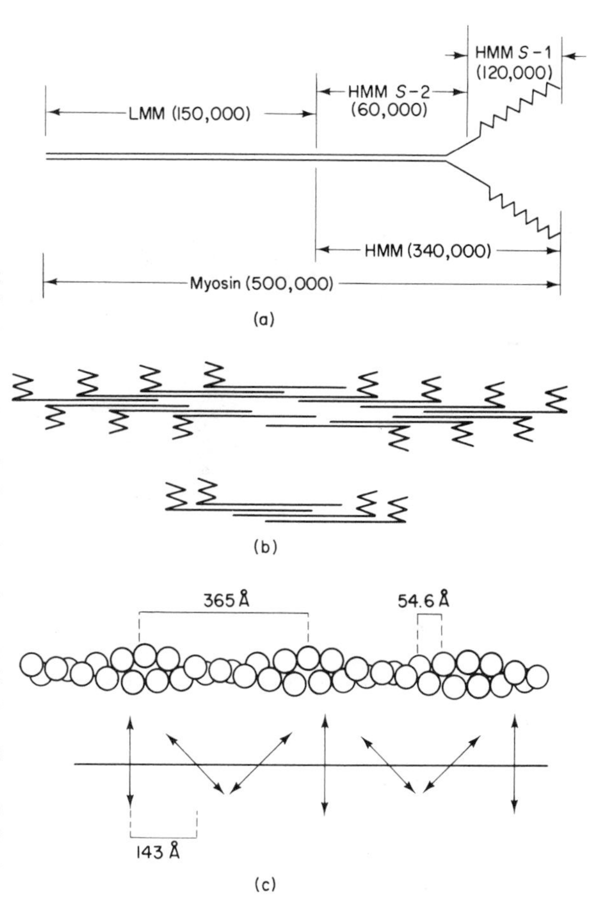

Fig. 19.6 Myosin and actin molecules. (a) Schematic representation to show dimensions of light meromyosin (LMM) and heavy meromyosin (HMM); HMM may be split into two fragments: *S*-1 and *S*-2. [From Slater and Lowy, 1967: Proc. Nat. Acad. Sci., *58*: 1617.] (b) Aggregation of myosin molecules to form filaments whose structural polarity reverses at the midpoint; the LMM parts of the molecules form the backbone of the filament while the hooked ends of the HMM components form the projecting cross-bridges. (c) Arrangement of G-actin monomers in actin filaments, with pitch of helix and subunit repeat in actin related to those of myosin, indicated schematically at the bottom. Note that cross-bridges would act asynchronously, and a sequence of them would develop a steady force as the filaments moved. [From Huxley, 1969: Science, *164*: 1356.]

twisted together (Figs. 19.6 and 19.7); ATP is required for its formation from G-actin.

Actomyosin Myosin and actin form moderately viscous solutions. When mixed in certain proportions (about 3 to 1 for vertebrate muscle extracts), there is a marked increase in viscosity with the formation of another protein ACTOMYOSIN which displays many distinct and unique properties. Actomyosin may be formed into threads, films or variously shaped masses and these have been repeatedly studied as models of contractile protein activity for more than a quarter of a century.

In the presence of ATP with Ca^{++} and Mg^{++} at specific ionic strengths these systems shrink or shorten—a process which is termed "superprecipitation." Since ADP is formed in this reaction, it is evident that myosin retains its ATPase activity when it associates with actin in the formation of actomyosin. These reactions, which lead to the formation of actomyosin in the absence of ATP, with contraction or dis-

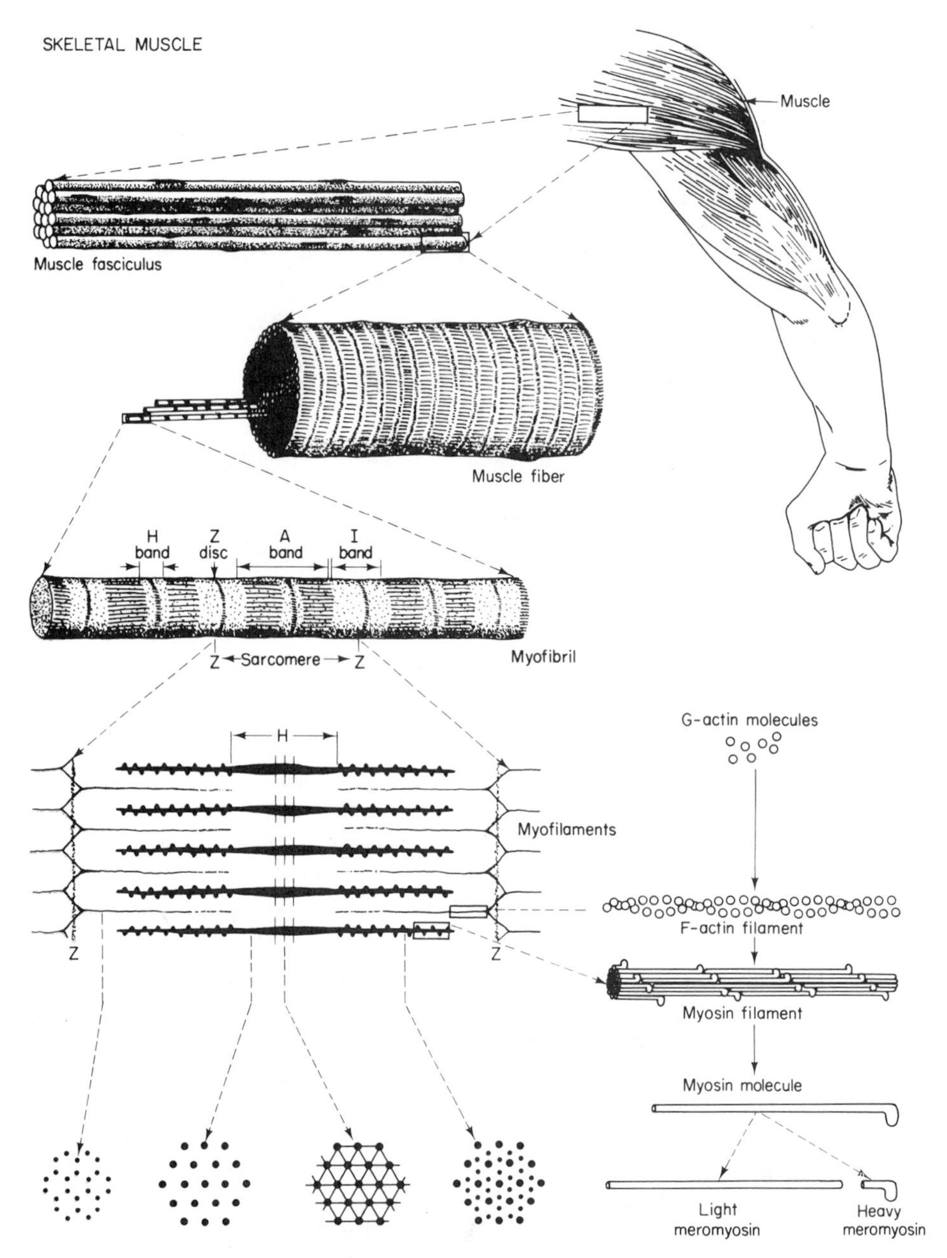

Fig. 19.7 Organization of vertebrate skeletal muscle. Groupings of dots at bottom left are cross sections at the levels indicated. [From Bloom and Fawcett, eds., 1968: A Textbook of Histology, 9th ed. Saunders, Philadelphia.]

sociation at certain levels of Ca^{++} in the presence of ATP, have their counterparts in muscle rigor, muscle contraction and relaxation (Wilkie, 1968). However, the actual contractile process differs in living muscle and depends on the linear and regular arrangement of the fibrous protein molecules in muscle tissue. Structural as well as biochemical organization is essential in the physiology of movement. The properties of *in vitro* actomyosin systems are still subject to intensive biochemical investigation.

Tropomyosin Small amounts of another fibrous protein known as TROPOMYOSIN can also be extracted from all muscles. This substance has a lower molecular weight than myosin and differs also in solubility and amino acid composition; its rod-like molecules lack ATPase and correspond to the LMM or tail portions of the myosin molecule.

Two different tropomyosins are recognized. One of these, termed TROPOMYOSIN A (TMA) or PARAMYOSIN was first identified in the adductors of bivalves where it may account for as much as 50 per cent of the extracted protein (Rüegg, 1961). Its molecular weight is about 135,000. Paramyosin seems to be very limited in distribution and bivalve muscles are the major known source of it; smaller amounts have been found in some annelids (Millman, 1967) and in the notochord of *Amphioxus* (Guthrie and Banks, 1970a). In the scallop *Pecten*, the TMA content varies from less than 1 per cent in the striated adductor to about 55 per cent in the smooth opaque adductor (Rüegg, 1961). The distribution seems to be correlated with certain mechanical properties of these muscles and supports the theory of long standing that paramyosin plays some special role in the prolonged maintenance of tension or the "catch" mechanism which forms one of the special features of these muscles.

TROPOMYOSIN B (TMB) is a universal component of muscle and appears to be of uniform character wherever found. Molecular weights of 50,000 to 70,000 have been recorded (Young, 1969; Katz, 1970). Its functions are still speculative but associations with the thin (actin) filaments and with the Z discs are indicated. As a partner of the actin filament, tropomyosin seems to be related to calcium ions whose binding to the myofibrils is now a recognized part of the contractile mechanism; TMB may be actually coiled with actin in the I band filaments. It has also been thought to give mechanical strength to the Z discs, but its association with the Z line has not been conclusively demonstrated (Katz, 1970).

Other contractile proteins Several other myofibrillar proteins occur in lesser amounts but little is known of their physiology. The substance first described as tropomyosin (now termed "native" tropomyosin) is actually composed of two substances: tropomyosin B and TROPONIN (Young, 1969; Katz, 1970). Muscle biochemists are very busy unraveling the puzzles of the troponin-tropomyosin system. Two proteins—α-ACTININ and β-ACTININ are also under active investigation. These substances are biochemically similar to actin. They are found to participate in some of the *in vitro* reactions of actomyosin but their roles remain obscure (Peachey, 1968; Young, 1969).

In addition, there are numerous proteins other than the fibrous proteins in muscle cells (myoglobins, enzymes, for example); the emphasis of uniformity in the actomyosin system should not obscure the fact that these other proteins vary markedly in different organs and different species of animals. Muscle proteins, like the plasma proteins discussed in Chapter 6, show distinct species specificity and have been used in systematic studies to investigate phylogenetic relationships (Tsuyuki *et al.,* 1962).

The Fibrillar System

Muscle cells are called fibers because of their elongated form. They all contain longitudinally arranged, submicroscopic elements, the MYOFILAMENTS, composed of long protein molecules concerned with the motility of the cells. These myofilaments may be organized into MYOFIBRILS (Fig. 19.7).

It is customary to distinguish two histological types of muscle on the basis of the presence or absence of distinct transverse striations under the light microscope. These were first described in vertebrate muscles. The striated fibers are characterized by alternating light ISOTROPIC (weakly birefringent) and dark ANISOTROPIC (strongly birefringent) transverse bands. Two types of striated fiber are found in the vertebrates. In the skeletal muscles, large multinucleate cells are arranged in parallel bundles; in cardiac muscles, smaller uninucleate cells are separated by INTERCALATED DISCS and connected in a sheet to form a net with an irregular arrangement of longitudinal slits between the fibers. The smooth or non-striated fiber of vertebrates is uninucleate and spindle-shaped and varies in length from about 20 μ in small blood vessels to about 500 μ in the pregnant uterus.

Although both smooth and striated types are also characteristic of invertebrate animals, there are several variations of the smooth type and such a range in size and form, with heterogeneity in the same muscle, that it is more logical to think of a broad spectrum rather than of two types. No clear phylogenetic trends have been detected. The range among primitive multicellular animals is almost as great as it is in the entire animal kingdom. Some coelenterates contain classic smooth fibers, and others have evenly striated fibers (swimming muscles of some of the medusae); there are, moreover, many curious types of the musculo-epithelial cells which combine the functions of motility with those of digestion or protection. Coelenterate contractile elements range from epithelial cells with mere tails containing contractile threads to highly specialized fibers with cross-striations superficially similar to those of vertebrate skeletal muscle (Hyman, 1940; Horridge, 1954; Hoyle, 1957). The latter specialization is evidently associated with the rapid rhythmic swimming movements of the medusae; there seems to be a similar trend toward cross-striations associated with more rapid movements throughout the animal world, but the correlation is by no means perfect.

Perhaps the most peculiar muscle cells in the entire animal kingdom have been found in the nematode *Ascaris lumbricoides*. In these cells the contractile fibrils are concentrated in a thick ribbon-like band from which the nuclear-bag or belly of the muscle rises as a distinct balloon with basal extensions which interconnect with other muscle cells and also with the central nerve cord (del Castillo *et al.*, 1967). The notochord of *Amphioxus* provides an example of a curious chordate muscular organ; notochordal contractions increase the stiffness of this rod which serves as a hydrostatic skeleton (Guthrie and Banks, 1970a).

The largest muscle cells have been found among the invertebrates. The somewhat atypical "smooth" muscle cells from *Ascaris* (Rosenbluth, 1967) measure as much as 1 mm $\times$ 0.2 mm in cross section while other smooth muscles, whether invertebrate or vertebrate, rarely exceed 15 μ in diameter. Record sizes for striated fibers are found among the arthropods where flight muscle cells of some diptera may reach 2 mm in diameter and the giant cells in a few crustaceans, such as large barnacles or the Alaska king crab and spider crabs, may be even larger; some of these fibers exceed 3 mm in diameter and 6 cm in length. These cells have been extremely valuable in physiological studies since direct cannulation is readily feasible (Hoyle, 1969). There is no evidence that size *per se* confers any particular advantage; it appears to be a consequence of general growth in a system with a fixed number of fibers.

Ultrastructure of muscle There are few places where electron microscopy has contributed more to physiological understanding than in studies of muscle. The histological structure, described long ago by light microscopists, is now largely understood in terms of the distribution and arrangements of individual molecules.

STRIATED MUSCLE owes its appearance to the orderly arrangement of the myofilaments within the myofibrils. High magnifications provided by the electron microscope, together with X-ray diffraction and phase contrast or interference microscope studies of fresh preparations, indicate that the alternating light and dark bands (Fig. 19.8) are produced by the arrangement of two types of protein filaments (Fig. 19.7). The thick filaments (about 110 Å in fixed vertebrate muscles) lie parallel to one another about 450 Å apart and form the anisotropic band; they probably consist of the protein myosin. The thin filaments of actin (about 50 Å in vertebrate muscle) are disposed in an orderly array between the thick ones as shown in Figs. 19.9 and 19.11. They are bisected by the Z discs or lines (delimiting the myofibril units or sarcomeres) and almost meet in the middle of the anisotropic band when the muscle

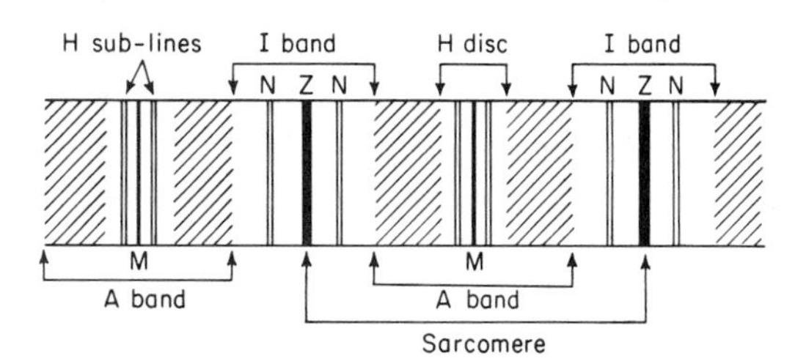

Fig. 19.8 Diagrammatic representation of cross bands and lines of the striated myofibril. The M, N, and H sublines are not always evident. [From Perry, 1960: *In* Comparative Biochemistry, Vol. 2. Florkin and Mason, eds. Academic Press, New York.]

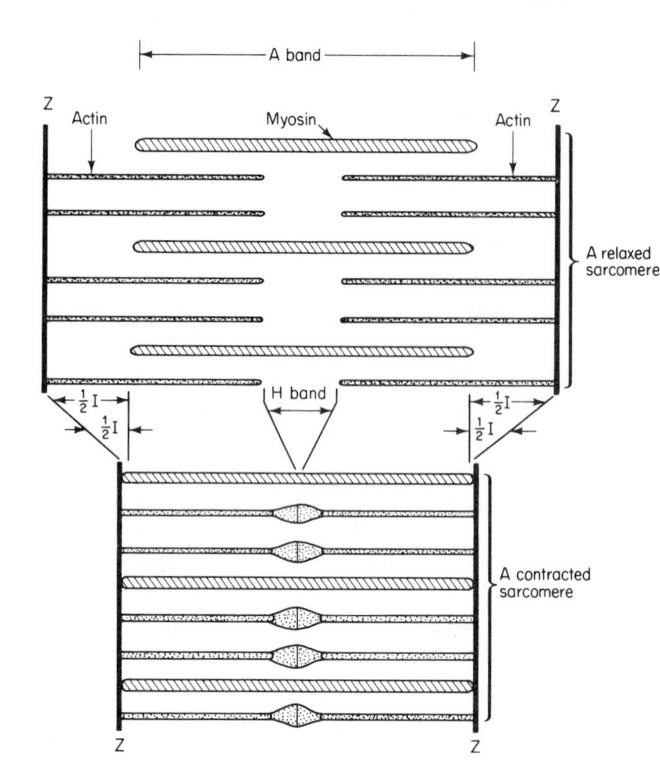

Fig. 19.9 Diagram to show the distribution of thick and thin filaments in a sarcomere of vertebrate skeletal muscle when relaxed and when contracted. The thin filaments of actin are not thought to crumple at the ends during contraction but slide over one another as shown in Fig. 19.11. [From Ham, 1969: Histology, 6th ed. Lippincott, Philadelphia.]

is relaxed. The space between the ends of these thin actin filaments forms the H zone of the fiber (Fig. 19.8). A pattern of large and small dots (shown in the lower part of Fig. 19.7) is evident in electron micrographs of muscle cross sections and has been important in establishing the arrangement of the myofilaments. Regularly spaced lateral projections from the thick filaments appear to touch the thin filaments and are thought to be significant in the sliding or ratchet-like action of the filaments which, according to a popular theory of contraction, is the significant event in the shortening of the muscle (Figs. 19.6 and 19.11).

The banding of striated muscle sometimes shows an M line (where there is a slight thickening in the middle of the myosin filaments) in the central area of the H disc. Less frequently, N lines of unknown significance appear in the I band on either side of the Z membrane (Perry, 1960).

Structural diversity has become increasingly evident as the electron microscope has been focused on more and more different kinds of striated muscle (Hoyle, 1967). Interdigitating thick and thin filaments running parallel to the long axis are probably universal, but there is little uniformity in other structural features. This is particularly true of invertebrate muscles where recorded major differences include: length and number of both thick and thin filaments, sarcomere length, extent and form of sarcoplasmic reticulum, locations of SR and T tubules (Fig. 19.12), nature of the Z region and the number of subunits in the filament. In short, invertebrate muscles vary markedly in all details examined and, once again, comparative studies may be expected to provide additional clues to the contractile mechanism and thus eliminate certain ill-founded hypotheses. Striated muscles from the vertebrates are more uniform in structure. There is, for example, a consistent ratio of 2:1 in thin to thick filaments which results in the familiar hexagonal lattice shown in cross sections (Fig. 19.7); invertebrate muscles by contrast often have relatively more thin filaments. Some insect skeletal muscles have ratios of 4:1, 5:1 or 6:1 and one kind of slow muscle from crab has a 7:1 ratio (Hoyle, 1969). It may be phylogenetically significant that some vertebrate embryonic muscles have high ratios—7:1 in the chick (Fischman, 1967). Again, sarcomere lengths are relatively uniform among the vertebrates (1.8 to 2.8 μ) while in some invertebrates a 20-fold range in length of both thick and thin filaments has been found with a good correlation between length and contraction speeds (Hoyle, 1969).

OBLIQUELY STRIATED MUSCLES occur in many invertebrates (locomotor muscles of nematodes, annelids, cephalopods, intervertebral muscles of ophiuroids). They were first described as HELICAL SMOOTH MUSCLES (Hanson and Lowy, 1960) and the oblique striation attributed to a spiral winding of ribbon-like myofibrils (bundles of myofilaments) around a central core of sarcoplasm. Investigations of the ultrastructure have not supported this interpretation (literature review in Mill and Knapp, 1970). In the nematode *Ascaris* and in several annelids, double arrays of thick and thin myofilaments have been found as in transversely striated muscle. The oblique striation is caused by a regular displacement of each row of myofilaments with respect to the next row along the long axis of the muscle; this produces oblique rather than transverse striations (Fig. 19.10).

Oblique striations seem to be characteristic of invertebrate muscles which permit whip-like or crawling movements. Rosenbluth's (1967) analysis indicates that the oblique organization confers greater extensibility and other plastic properties on these

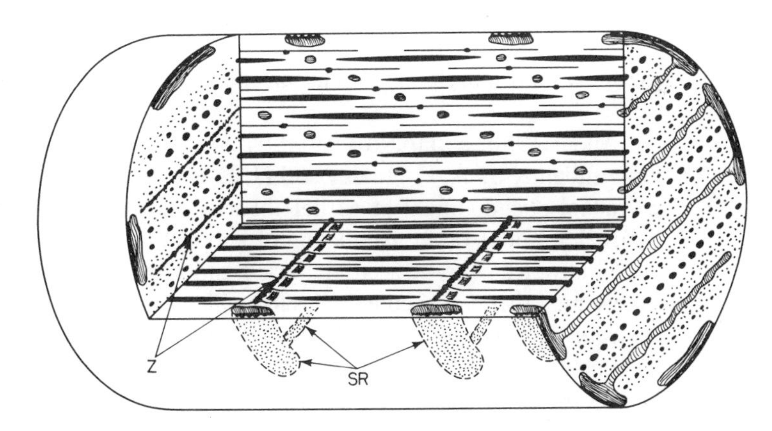

Fig. 19.10 Diagram to show arrangement of myofilaments in the obliquely striated muscle of the polychaete *Glycera*. [From Rosenbluth, 1968: J. Cell Biol., *36*: 258.]

muscles which are not attached to a rigid skeleton but depend on an intracellular connective tissue "skeleton." The obliquely striated muscle may combine some of the functional properties of smooth muscle with those of the transversely striated types. In some CLASSIC SMOOTH MUSCLES (retractors of the pharynx and penis of *Helix*, retractors of *Phascolosoma*, vertebrate smooth muscle) the myofilaments are not grouped into myofibrils but scattered through the cell parallel to its long axis. Thus, with the light microscope, the sarcoplasm appears homogeneous and displays little or no evidence of its fibrillar nature.

The PARAMYOSIN SMOOTH MUSCLE is best known in the lamellibranch molluscs where it apparently serves a tonic function associated with the prolonged and continuous contractions of adductor muscles. The tenacity with which a bivalve can maintain tightly closed shells is evidence of this capacity. Structurally, these fibers contain two types of filaments: (1) thick myosin-paramyosin filaments which range from 500 Å to 1500 Å in diameter and appear to have a central core of paramyosin with myosin molecules on the outside; (2) thin filaments (50 Å diameter) which, like the actin filaments of other muscles, probably contain tropomyosin B in association with actin. Six to ten thin filaments are normally grouped about each thick filament (Lowy *et al.,* 1964; Millman, 1967). These muscles differ from skeletal or striated muscle both in the massive ribbon-like filaments which are about ten times thicker, and in the presence of a unique protein paramyosin.

Mechanisms of Contraction

For more than a century, muscle physiologists have wrestled with two major problems. One is the mystery of the actual movement or change in length of the muscle, while the other, which has proved no less baffling, concerns the mechanism which excites and energizes thousands of myofilaments to work simultaneously. The first analysis has now firmly focused on sliding molecules of fibrous protein while the second analysis has revealed a unique telegraphic communication within and between the individual muscle cells. The mechanics of fibril movement will be considered first.

About 1950, a SLIDING FILAMENT MODEL of muscle contraction was proposed independently by A. F. Huxley and H. E. Huxley. The two subsequent decades saw a gradual abandonment of earlier theories, such as those based on folding protein molecules, and a rapid marshaling of support for the sliding of myofilaments. This model now seems to be securely based and applicable to all types of muscle as well as to some other contractile tissues like cilia and flagella; at present, scientists are concentrating on the molecular details of the sliding mechanism. The classical studies on which the model is based have been detailed by A. F. Huxley (1957) and H. E. Huxley (1960); the development of the hypothesis can be traced through recent reviews and monographs (Davies, 1963; Bendall, 1969; Huxley, 1969); several more popular accounts are also available (Huxley, 1958, 1965).

In brief, the model assumes a slipping or sliding of the interdigitating filaments of actin and myosin; these filaments remain at almost constant length while temporarily interacting at specific contact points. The force of contraction is developed in the overlap region of the thick and thin filaments where the hook- or bridge-like globular ends of heavy meromyosin crawl along the beads which make up the twisted chain of the F-actin filaments (Figs. 19.6 and 19.11). A movement or tilting of the bridges occurs through a flexible connection between LMM and HMM *S*-2 (Fig. 19.6). Calculations indicate that the cross bridges must hook on at a succession of points during a cycle since mammalian muscle can shorten by as much as 30 per cent while the distance between cross bridges is only about 5 per cent of the length of half a sarcomere. In some muscles, where the maximum length changes are much smaller (only about 5 per cent in some insect flight muscles), a succession of attachments may be unnecessary and a simple swinging of the bridge may account for the movement. Further, the quantitative analysis indicates that the actin filaments of vertebrate skeletal muscle will slide, or be pulled, about 100 Å along the myosin filament and then return to the original position ready for the next pull (Huxley, 1965); in

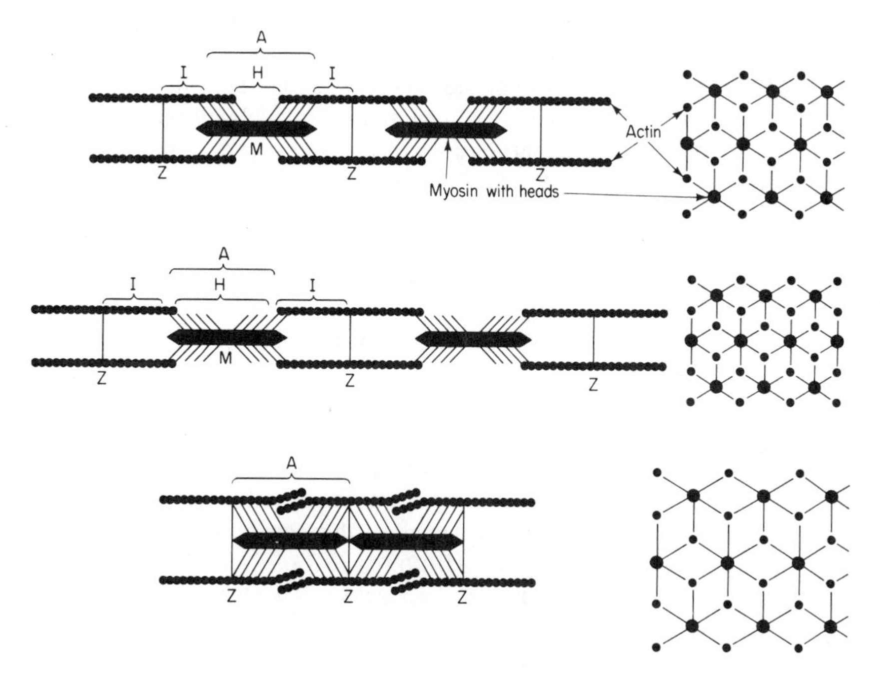

Fig. 19.11 Fine structure of a sarcomere in longitudinal and cross sections. The length of the sarcomere has been reduced tenfold in relation to the thickness of the filaments, for ease of drawing. Top, sarcomeres at rest length ($= 2.4\ \mu$ per sarcomere); middle, sarcomeres after stretch to 1.3 times rest length; bottom, sarcomeres contracted to 0.6 times rest length. [From Bendall, 1969: Muscles, Molecules and Movement. Heinemann, London.]

mammalian skeletal muscle, 50 to 60 cycles of this activity may occur each second. The theory requires the rapid but temporary coupling of actin and myosin at the expense of ATP (probably one molecule for each bridge connected); this aspect is considered in the next section.

The sliding filament hypothesis was first suggested by electron micrographs of skeletal muscle fixed at different stages of contraction or stretching. The pictures clearly showed that the length of the A band remains constant while the I band changes in relation to the length of the muscle (Figs. 19.9 and 19.11). Further, the length of the H zone (lighter region in the center of the A band) increases and decreases with the length of the I band; the distance from the end of one H zone through the Z region to the beginning of the next H zone is approximately constant (Huxley, 1958). The obvious conclusion seemed to be that the two sets of filaments slip past each other while remaining approximately the same length under varying degrees of stretch or contraction. At certain stages in contraction the ends of the filaments appeared to be crumpled or overlapped, but this was interpreted as a result of shortening and not the cause of it.

Although the sliding filament model was first advanced to explain the contractions of vertebrate skeletal muscle, it is considered adequate to account for the movements of smooth muscles as well as other structural types (Lowy and Millman, 1962, 1963; Millman, 1967). Many of the details of myofibrillar movement are still unresolved, but it seems likely that the sliding filament model will remain one of the most significant physiological contributions of the past decade. Alternate theories of contraction are reviewed by Peachey (1968).

Excitation–Contraction (E-C) Coupling

The cytoplasm of both plant and animal cells, with very few exceptions (erythrocytes, bacteria), is pervaded by a complex system of membranes which form tubules, vacuoles and other variously shaped compartments. Although known to light microscopists, this system was frequently regarded as an artifact of fixation. The electron microscope revealed its reality and cell biologists are now acquainted with many details of its complexity and significance. The ENDOPLASMIC RETICULUM (ER) forms the major component of this vacuolar system; within muscle cells it is termed the SARCOPLASMIC RETICULUM (SR).

During the latter part of the nineteenth century, light microscopists, using gold or silver impregnations, described a delicate reticulum which was intimately associated with the fibrils in skeletal and cardiac muscles. The true significance of this reticulum was not suspected until the electron microscope showed it to be a highly intricate and specialized endoplasmic reticulum (Fig. 19.12) with morphological relationships suggesting some special functional relationship to the myofibrils (Porter and Palade, 1967).

Sacroplasmic reticulum The organization and distribution of the sarcoplasmic reticulum varies considerably in different muscles and in various animals (Smith, 1966; Peachy, 1968). Its salient features are apparent in the three-dimensional diagram shown in Fig. 19.12. Many tubules, often standing like columns parallel with the myofilaments, are connected by transverse channels in the Z and H regions of the fiber. This system of tubules and cisternae forms a loose network which ensheathes the greater part of the surfaces of the myofibrils. The tubules are dilated and particularly extensive in the Z region. Such a system is obviously ideal for the

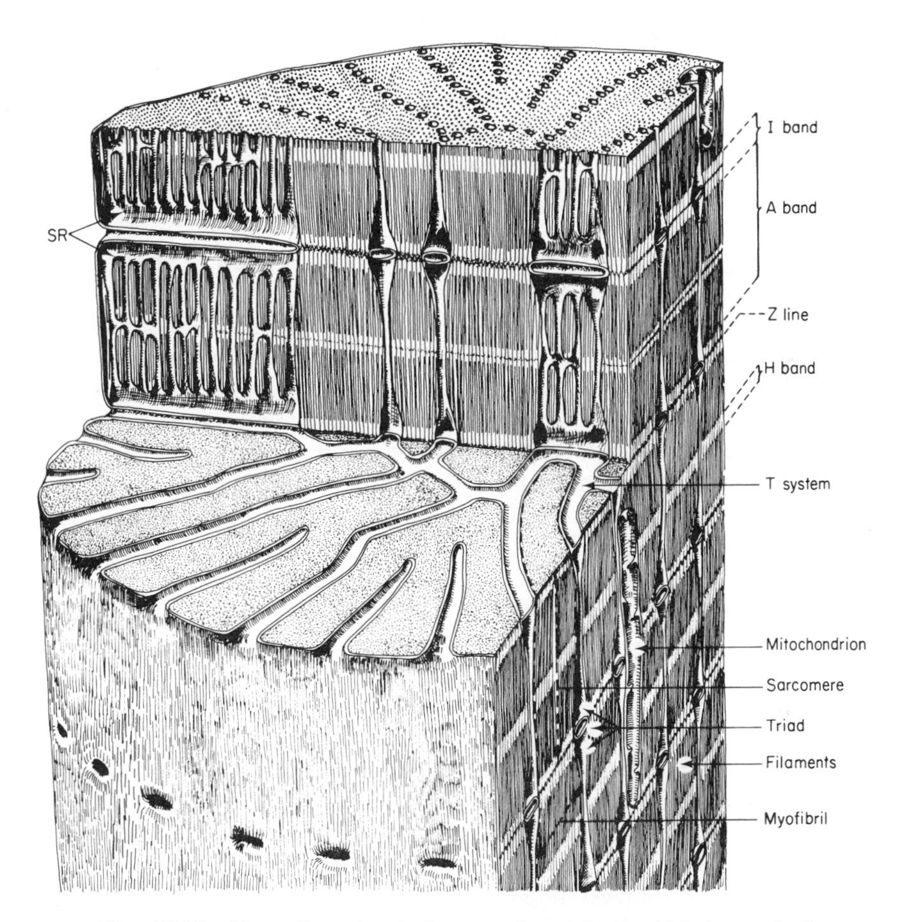

Fig. 19.12 Three-dimensional diagram of vertebrate striated muscle to show sarcoplasmic reticulum and muscle filaments in longitudinal sections with the transverse T system of tubules in cross sections and their interconnections (triads) at the Z regions. [From Porter and Franzini-Armstrong, 1965: Sci. Am., *212* (3): 75. Copyright 1965 by Scientific American, Inc. All rights reserved.]

speedy distribution of energy-rich substances to thousands of closely packed myofibrils.

The T system and the triads Electrophysiologists, exploring the surfaces of muscle fibers with tiny electrodes, have found them to be excitable only at certain specific points (Huxley and Taylor, 1958). Electron microscopists, following this lead, identified another system of tubules which penetrate the muscle fibers transversely from pores in the plasmalemma at the Z region (or A-I region in some animals) and which interconnect with one another in an elaborate manner. This transverse or T SYSTEM, as it is now called, has been shown to provide for telegraphic communication from the surface of the fiber to excite multitudes of filaments in a synchronous way. As is evident in Fig. 19.12, the T tubules and the vesicles or sacs of the sarcoplasmic reticulum come together in the Z region to form a TRIAD. In some muscles, T tubules are also present at the junctions of the A and I bands (Smith, 1966).

Although SR and T tubules make intimate contact and operate together as integral components of a distinct physiological machine, there is no direct connection between the channels of the two. Indeed, they develop from quite different membranes; the T tubules are invaginations from the surface of the myoblast while the SR develops from evaginations of the rough-surfaced endoplasmic reticulum. The intimacy of the two systems of tubules in the region of the triads varies with the type of muscle; it is particularly extensive in rapidly contracting and relaxing fibers. In some rapidly contracting muscles, triads also occur at the A-I junctions and in this manner provide a double system for excitation (Peachey, 1968; Bendall, 1969). Smith (1966) tabulates the positions of the T system in different muscles from many animal species.

The SR and T systems provide the complex machinery required for speedy communication, energy supply and plumbing for thousands of closely packed fibrils. The understanding of further details of EXCITATION-CONTRACTION (E-C) COUPLING have come from studies of the long-suspected importance of calcium ions in muscular activity.

Calcium ions and muscle contraction More than thirty years ago, Heilbrunn (1940, 1952), one of the pioneer cell physiologists, performed simple experiments which demonstrated the indispensable and highly specific role of calcium in muscle contraction. In some of the most critical of his experiments, the lengths of isolated muscle fibers were observed microscopically and measured before and after intracellular injections of various ions. Of the ions normally present in muscle, only calcium caused contraction (Heilbrunn and Wiercinski, 1947). Although Heilbrunn's (1952) interpretations in terms of colloid chemistry and his calcium release theory of stimulation have now been discarded in favor of more soundly based molecular concepts, his early experiments are noteworthy landmarks in the long trail of research which has found calcium at the very center of E-C coupling.

In brief, current evidence indicates that the electrical depolarization of the sarcolemma, which follows the arrival of the nerve impulse at the myoneural junction, spreads into the muscle fiber by the T system. It will be recalled that in most cells, excitation involves a momentary influx of Na^+ and efflux of K^+; this also happens at the plasma membrane of most muscles. Within the T system, however, the spread of excitation triggers (probably by depolarizing the membranes) the release of Ca^{++} from binding sites or pools within the sarcoplasmic reticulum—especially the terminal cisternae. This flood of calcium into the myoplasm throws two pieces of machinery into instant action: (1) it "releases a brake" which prevents the sliding of the myofilaments and (2) it starts the chemical plant which generates phosphate bond energy. The first of these operations seems to depend on the protein troponin which is closely associated with the thin actin filaments and, in its calcium-free form, prevents the interaction of actin and myosin; when troponin combines with Ca^{++}, the sliding of thick and thin filaments becomes possible. The second of the activities depends on ATPase, an enzyme which requires a specific balance of Mg^{++} and Ca^{++}. The dephosphorylation of ATP provides immediate energy for the operation of the sliding filaments. Thus, the free calcium ions couple the excitation of the plasma membrane with both the contractile system and the metabolic system. Its effects on the metabolic system probably go beyond the activation of ATPase since Ca^{++} seems also to stimulate glycogenolysis.

Relaxation of muscle is just as vital an operation as its contraction. For many years, biochemists have extracted and studied "relaxing factors" from muscle cells. Relaxation is now known to be an active process associated with the return of calcium ions from the myoplasm to binding sites in the sarcoplasmic reticulum; a CALCIUM PUMP in the membranes of the SR is assumed to perform this operation speedily and efficiently. Within a matter of milliseconds, the free calcium is pumped back into the SR against steep concentration gradients; the concentrations of bound calcium in the terminal cisternae may be 500 times or more greater than the free Ca^{++} in the myoplasm of the relaxed muscle. Processes leading to contraction are initiated by the depolarization of the plasma membrane but terminated by the action of the calcium pump in the sarcoplasmic reticulum; both contraction and relaxation are active operations. Like the Na^+/K^+ pump of the plasma membrane, the calcium pump is an energy-requiring machine which depends on some of the phosphate bond energy released by the flooding of calcium. The mechanics of the calcium pump are as obscure as those of the sodium pump.

The actual shortening of a skeletal muscle has been found to last much longer than the action potential and the flood of free Ca^{++} (Fig. 19.13); the twitch of a frog muscle is from 10 to 1000 times longer than the associated action potential (about 1 msec). Thus, this calcium-operated machine can be reactivated long before the end of the muscle's contraction and, consequently, a tetanus of the muscle is possible. This lag in the operation of the machinery is related to the viscoelastic properties of muscle which will be considered in the next section.

The evidence of calcium fluxes associated with E-C coupling has been ably reviewed in many places (Sandow, 1965; Ebashi and Endo, 1968; Peachey, 1968; Bendall, 1969; Hoyle, 1970; Katz, 1970). Although the experimental work is too extensive to be summarized here, it seems worth noting the contributions of comparative physiologists which once more underline the rich rewards from studies of comparable mechanisms in diverse organisms. Darwinian evolution has perfected many different models of contractile machinery and some valuable clues to the essential components have been found by comparing their manifold workings.

The giant muscle cells of arthropods have been singled out to emphasize this point. Whereas muscle fibers rarely exceed a diameter of 100 μ, the fibers of the giant barnacle *Balanus nubilus*, the Alaska king crab *Paralithoides kamchatka* and certain other arthropods measure 2 mm to 3 mm in diameter. These large cells may be cannulated and samples withdrawn or injected; chemical analyses of component parts are possible and substances may be precisely localized in different parts of the cell by methods of radioautography. Gillis (1969), for example, was able to apply Ca^{++} at several discrete places on single sarcomeres of crab muscle and thereby localized the point of calcium action in the region of overlap of the myofilaments (A-I junctional area). Several physiologists have drawn on their knowledge of curious biological materials to support the hypothesis of calcium involvement in E-C coupling. Two further examples will be noted here.

Tyrian purple (Fox, 1966), one of the valued commodities of antiquity, is an indole pigment which can be extracted from several species of marine snails (*Murex, Purpura, Mitra*). Its biological function is uncertain; its commercial value as the regal dye or the "purple of the ancients" is recorded in ancient and biblical literature; its interest in muscle physiology is relatively recent but significant. This pigment changes color in the presence of calcium and Jöbsis and O'Connor (1966) were able to demonstrate that toad muscles stained with murexide showed a rapid color change following excitation and then quickly reverted to the original murexide color even before the tension in the muscle reached its peak.

In comparable experiments, Hoyle and his associates obtained additional support for calcium fluxes using the luminescent protein aequorin (Ridgway and Ashley, 1967). This substance, which is present in the jellyfish *Aequorea aequorea* (Johnson and Shimomura, 1968), will be described in Chapter 20. In the presence of Ca^{++}, it emits a beautiful blue light which can be measured photometrically. Barnacle muscle fibers injected with aequorin were found to emit light when stimulated electrically. The burst of light which appeared just after stimulation disappeared before the muscle developed maximum tension (Fig. 19.13). The findings with murexide and aequorin were comparable, but the latter permitted a more precise study of the quantitative relationships because of the injection into individual cells (Hoyle, 1970).

These experiments have been selected to illustrate the importance of comparative studies. They add substantially to the mounting evidence of a central role for calcium in E-C coupling. The quantitative relationships in terms of ionic fluxes and mechanical responses have been established in a number of different preparations; details can be found in the reviews cited.

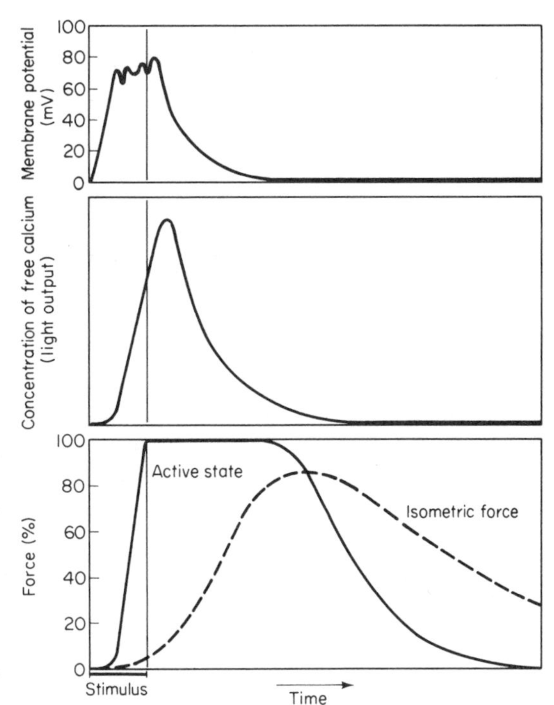

Fig. 19.13 Factors in muscle contraction. Top, changes in membrane potential; middle, concentration of free calcium; bottom, active state (or true force developed by the contractile elements) and isometric force recorded externally for the entire muscle fiber. Note slow development of isometric force due to viscoelastic properties of muscle. [From Hoyle, 1970: Sci. Am., *222* (4): 93. Copyright 1970 by Scientific American, Inc. All rights reserved.]

Mechanical Properties of Muscle

An emphasis on the homogeneity of the contractile proteins throughout the animal kingdom should not obscure the fact that the muscular organs are among the most versatile physiological machines (Pantin, 1956b; Prosser, 1960). Contraction times range from those of the flight muscles of certain insects which have been clocked at less than a millisecond to those of the circular muscles in the column of sea anemones where a single contraction-relaxation sequence may require five or six minutes. Insect flight muscles are striated while those of the sea anemone are smooth, and, in general, striations are usually associated with the more rapidly contracting muscles. Speed of relaxation is just as variable (see Table 50 in Prosser and Brown, 1961). There may also be significant differences between contraction and relaxation times of a single muscle; both slow and fast fibers are often present within the same muscular organ.

Extent as well as speed of contraction is variable. In visceral organs such as the urinary bladder, which shows regular and remarkable changes in size, the smooth muscle may be passively extended to many different resting lengths. Striated muscles usually operate through the production of tension and do not change their length by more than about 20 to 30 per cent (Morales, 1959); the smooth muscle of the retractor of the snail *Helix* can shorten by 80 per cent of its extended length.

The responses of muscles also depend on the manner in which they are activated.

Some, such as the skeletal muscles of vertebrates, respond to a single stimulus with a maximal twitch while others, such as crustacean skeletal muscles, fail to respond to a single stimulus no matter how intense; they require a rapid sequence of stimuli for excitation.

This versatility in the action of the muscular organs depends on three factors: (1) the organization of the myofilaments within the fibers as previously discussed, (2) the arrangement of the muscle cells and the non-contractile connective tissues and (3) the nervous control. Some of the mechanical properties of muscle, which depend on the first two of these, will now be considered. Nervous control is discussed in the final section on the contractile tissues.

Types of muscle contraction If an isolated skeletal muscle such as a frog gastrocnemius (also a single muscle fiber) is momentarily stimulated, it changes from a flaccid organ into a much firmer structure which then more slowly becomes soft as it relaxes. If allowed to do so, it will shorten and then lengthen during the process. This is the familiar muscle TWITCH, a characteristic response to a single stimulus of threshold or greater intensity. By appropriate recording devices a graphic representation of the phenomenon (Fig. 19.14) is readily obtained; the time relationships and form of the contraction curve vary with the load.

If a muscle is stimulated a second time before it has relaxed, it is again activated and responds to a somewhat greater degree, the precise response depending on the interval between successive stimuli. With a rapid series of stimuli (about 30/sec for frog muscles at 0°C) there is a complete fusion, individual responses are no longer evident, and the muscle shows the smooth maintained contraction called a TETANUS (Fig. 19.14).

Muscles may do work by changing their lengths and moving weights for variable distances as in lifting or moving a heavy object. They may, however, develop tension and exert a force without any perceptible shortening as is the case when one attempts to lift an object which is impossibly heavy. Physiologists refer to the first kind of muscular activity as ISOTONIC (equal tension or constant load) and the second as

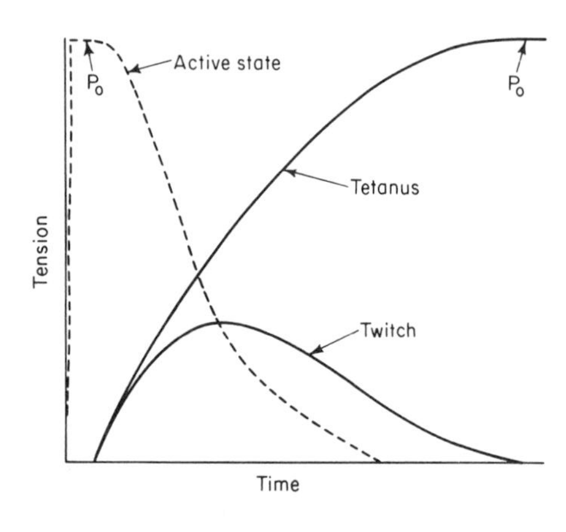

Fig. 19.14 Diagram showing the relation between tension and time for a twitch, a tetanus, and the active state (page 588) in vertebrate striated muscle. P_0 is the maximum tension of which the fiber is capable. Stimulus begins at zero time. [From Mountcastle, 1974: Medical Physiology, 13th ed, Vol. 1. Mosby. St. Louis. Mo.]

ISOMETRIC (equal or fixed length). In either case, the time sequence of changes depends on the load lifted or the tension exerted as well as on the temperature, the physiological condition of the muscle, the species of animal and the type of muscle.

In the older techniques, isometric contractions were recorded mechanically with muscles lifting weights or operating against almost, but not quite, immovable springs or metal strips to produce movements of levers which recorded on moving papers. Suitable transducers are now available, and recordings obtained in this way permit a much more precise analysis of the contraction and relaxation phenomena.

The elastic elements in muscle If the length of an isolated muscle is recorded at rest in relation to a series of weights, the LENGTH TENSION or stress-strain curve for the resting muscle is curvilinear as shown in the lower right area of Fig. 19.15a. Tension begins to develop when the muscle is slightly longer than its normal *in situ* length and is a function of length up to the breaking point. This elasticity is due to the VISCOUS-ELASTIC ELEMENTS attributed not only to the elasticity of the muscle fibers themselves but also to the sarcolemma and the connective tissues. These non-contractile elements are important in smoothing out rapid changes in tension. Length-tension relationships of resting muscles vary considerably (Fig. 19.15b; Hanson and Lowy, 1960).

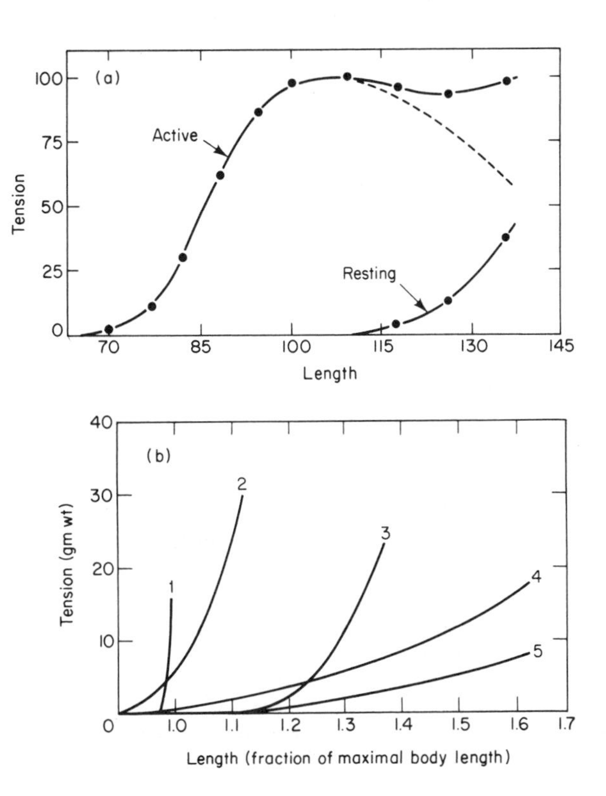

Fig. 19.15 Tension-length relations in muscles. (a) Vertebrate striated muscle. Broken line, difference between stress-strain and tension-length curves; length 100, the resting *in situ* length and length of maximum tension. [From Mountcastle, 1974: Medical Physiology, 13th ed., Vol. 1. Mosby, St. Louis, Mo.] (b) Several invertebrate muscles compared with frog muscle (resting). (1) Bumblebee flight muscle; (2) locust flight muscle at 11°C; (3) frog sartorius at 0°C; (4) anterior byssal retractor of *Mytilus* at 14°C; (5) pharynx retractor of *Helix* at 14°C. [From Hanson and Lowy, 1960: *In* The Structure and Function of Muscle, Vol. 1, Bourne, ed. Academic Press, New York.]

The viscous-elastic system is made up of three components. Studies of length-tension relations in isolated muscles working against different loads, and stimulated to contract at (brief) intervals following a sharp stretch, indicate the presence of SERIES ELASTIC ELEMENTS as well as PARALLEL ELASTIC COMPONENTS; the third source of elasticity resides in the viscous proteins of the muscle fiber. Wilkie (1956, 1968) discusses the experimental analysis. This will not be described here; the important principles can be readily appreciated by reference to Fig. 19.16. The potential contractile force of a muscle is substantially greater than the force measured in the presence of the elastic elements. Wilkie (1956) defines ACTIVE STATE as "the isometric tension which the contractile component can develop (or just bear without lengthening) at that instant." It is the tension which depends on the contractile elements alone without the intervening series elasticity. The importance of the elastic elements in the movements of intact muscles is obvious; they provide a damping effect on a system which is suddenly activated by nervous impulses. The morphology of the elastic elements is somewhat speculative. The tendons and Z bands probably operate as series elastic elements while sarcolemma sheaths and connective tissues serve as parallel elements (Mountcastle, 1968).

Properties of isolated muscles The physiological activities of isolated muscles have most often been examined in relation to (1) the latency, contraction and relaxation phases of isotonic twitches, (2) summation and the development of tetanus, (3) length of the muscle in relation to the tension which it can develop (LENGTH-TENSION CURVES), (4) the speed of shortening (FORCE-VELOCITY CURVES and LENGTH-VELOCITY CURVES) and (5) duration and magnitude of the "active state" (ACTIVE-STATE CURVES). These various parameters are described for vertebrate skeletal muscle in many textbooks of medical physiology (see also Wilkie, 1956,

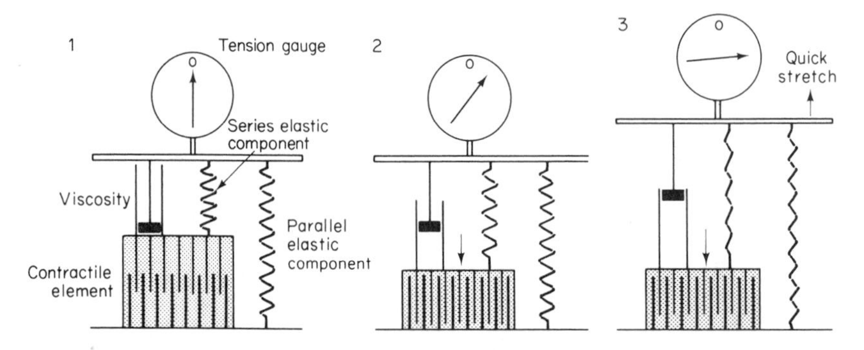

Fig. 19.16 Diagram shows effects of elasticity on muscle action. Stimulated muscle must first stretch the elastic elements and this creates a damping effect so that the externally measured peak tension is considerably less than the force actually developed in the fiber. If at the time of stimulation, the fiber is given a quick stretch that pulls out the elastic elements (far right diagram), then the full contractile force can be measured. [From Hoyle, 1970: Sci. Am., *222* (4): 91. Copyright 1970 by Scientific American, Inc. All rights reserved.]

1968; Bendall, 1969), for vertebrate smooth muscle by Csapo (1960) and for several invertebrate muscles by Hanson and Lowy (1960) and Lowy and Millman (1962, 1963).

In general, the response patterns are similar for all kinds of muscles, but the quantitative relationships vary considerably. Selected examples such as those illustrated for the length-tension relationships of resting muscle (Fig. 19.15b) show a spectrum of responses varying in magnitude and time relationships but all basically alike. In the present context, the similarities are more significant than the differences since they argue for essentially similar contractile mechanisms in all muscles. Detailed comparisons will not be attempted here.

The Versatility of Muscle

Survival and success frequently depend on the versatility of the muscular system. During evolution, the contractile machinery has been adapted to an almost infinite range of movements: the sudden withdrawal of the startled tube-dwelling worm, the jet propulsion of the squid, the gentle breathing and fin movements of a resting fish, the rapid oscillations of wings, the finely modulated and precise movements of the skilled craftsman or artist. In part, this precision of movement depends on the architecture of the muscle cells; in part it is a function of their nervous control. Some examples of specialized muscle types will be considered before discussing neuromuscular control.

Twitch (fast) and tonic (slow) muscle fibers The popularity of the frog gastrocnemius in elementary physiology laboratories has sometimes given the false impression that this preparation is representative of most muscles. In the broad spectrum of animals, the TWITCH or FAST fibers, characteristic of frog leg muscles, are much less frequently encountered than the TONIC or SLOW ones. Fast fibers are sometimes also termed PHASIC.

Frog legs are "white meat" and it is a matter of everyday experience that meats like fish and chicken may be white or red. The distinction is evident in color, taste and texture. Physiologists have long associated the white muscles with capacities for rapid, vigorous movements such as jumping or flying while the red muscles are concerned with more leisurely and steadily maintained activities. It was, for example, suggested almost a century ago that only the red muscle is active when fishes cruise along slowly while the white muscles come into play when they are startled or otherwise stimulated to swim vigorously. Critical physiological work has confirmed this hypothesis; there appear to be two separate motor systems in fish and these operate independently and use different metabolites (Bone, 1966; Gordon, 1968).

The characteristic color of red muscle is caused by the presence of myoglobin and mitochondrial cytochromes. This tissue is also more vascular, has smaller fibers with more granular cytoplasm and contains many mitochondria and much fat in contrast to the white muscle; the large white muscle cells are closely packed with myofibrils and contain glycogen which is utilized during short bursts of activity. These characteristics, as well as the many enzyme systems investigated (George and Berger, 1966), accord well with the distinctive functions ascribed to red and white muscle. However, physiologists now know that a critical separation of these fundamentally different muscle types cannot be made on the basis of color and other gross features of the tissue. Pigeon breast muscle, for example, is both fast and red and performs sustained work during flight while the very white pectoral muscles of chickens contain some red fibers. Nevertheless, at the cellular level, it is clear that at least two types of skeletal muscle fibers operate in all classes of vertebrates and in many invertebrates.

Hess (1970) has reviewed the literature on fast and slow fibers in the vertebrates and finds several distinctive features. There is, first of all, a difference in innervation and the physiological implications of this will be considered in a later section; here, it is noted that fast or twitch fibers are supplied by a single robust motor nerve with a specialized end plate while the slow fibers have multiple, delicate nerve terminals (Figs. 19.20 and 19.21). This distinction can be made only in the vertebrates since the motor endings of the invertebrates are all of the multiple type.

In addition to this difference in innervation, Hess (1970) finds that vertebrate slow fibers have a reduced sarcoplasmic reticulum, an irregular T system, myofibrils of varying size and irregular distinction, and zigzag Z lines. In many places, fast and slow responses are directly related to these basic differences in fibers rather than to the innervation *per se*. Direct applications of Ca^{++} to frog slow fibers with the sarcolemma removed give characteristic slow responses indicating that slowness is a property of the myofibrils and not the E-C coupling (review by Peachey, 1968). It is still not clear whether or not slow muscle fibers conduct action potentials. In any case, action potentials cannot be readily elicted in slow fibers and the response to stimulation is a slow contraction and graded response rather than a twitch and tetanus (Fig. 19.17).

Slow and fast fiber systems have been investigated in several different arthropods and a number of differences in structure and physiology have been recorded (Cohen and Hess, 1967; Fahrenbach, 1967; Hoyle and McNeill, 1968). In general, the differences are similar to those found among the vertebrates; additional features have been noted, for example, higher ratios of thin to thick filaments in slow muscles from some crabs. Several investigators have found a spectrum of fiber types ranging from fast to slow with no sharp lines of demarcation between the two. Hoyle and McNeill (1968) describe an interesting example in the levator muscles of the eyestalk of the portunid crab *Podophthalmus vigil*. This animal spends much of its time buried in the mud with its eyes above the substratum on extremely long stalks (up to 7 cm) which are moved about by minute and delicate muscles. A variety of movements is obviously important if the eyes are to "follow" moving objects, maintain steady positions, or be withdrawn suddenly to avoid hazards. There are rapidly contracting fibers which are white, and very slowly contracting ones which are pink because of cytochromes; between these extremes are other fibers of intermediate colors and rates of movement. It is possible that many contractile systems have more than two

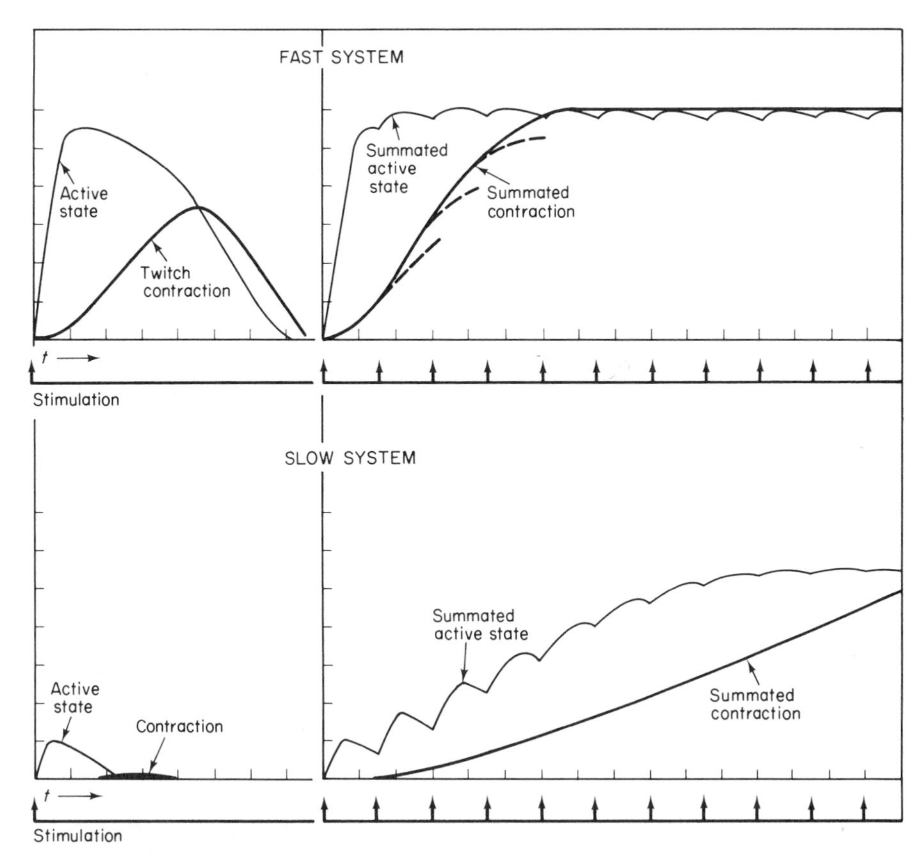

Fig. 19.17 Active state and tension development during contraction in a fast (e.g., frog gastrocnemius) and a slow (e.g., opener of crayfish claw) muscle. In both, the active state can be evoked repeatedly, but in the fast system a single nerve impulse leads to nearly maximum development of active state while in the slow system the active state develops only partially. [From Florey, 1966: General and Comparative Animal Physiology. Saunders, Philadelphia.]

types of fibers. Bone (1966) described four types in elasmobranch fishes; two of them were intermediate, in position and contractile capacities, between the red superficial fibers just under the skin and the very fast fibers located in the inner part of the myotome.

Lamellibranch "catch" muscles The adductor or holding muscles of lamellibranchs (paramyosin smooth muscles) are able to maintain high levels of tension for long periods without signs of fatigue. Two hypotheses have been considered. The older theories postulated a "set" or "catch" of certain elements while the muscle is in the shortened state and a prolonged maintenance of tension without expenditure of energy. However, it is now known from electrical recordings that these muscles are not continuously passive during prolonged contraction; they must be periodically activated or the tension will decline to zero (Abbott and Lowy, 1958).

According to the other hypothesis, tension is maintained by a continuous activity comparable to tetanus in vertebrate skeletal muscle. Evidence of "spontaneous" bursts of activity has been found in some of these muscles, suggesting an INTERMITTENT ACTIVATION MECHANISM (Lowy and Millman, 1963). According to this concept, the physiological differences in the contractile mechanisms of the holding muscles and familiar skeletal muscles are quantitative rather than qualitative (Hanson and Lowy, 1960; Millman, 1967).

Recent work suggests that the "catch" may be caused by persistence of free intracellular calcium following E-C coupling. The protein paramyosin, which is always present in these muscles, is assumed to play some part in the special "catch" mechanism, but it should be noted that paramyosin is also found in some molluscan muscles which lack "catch" properties; the molecular details of relaxation and contraction in these intensively investigated and interesting muscles remain speculative; Hoyle (1969) has summarized the recent experimental work.

Insect fibrillar muscle The flight machinery of the vertebrates and of the insects is patterned on two entirely different principles. The wings of birds and bats are expanded sheets of tissue developed from the anterior limb and flapped by the modified muscles of this appendage. Insect wings, by contrast, are dorsolateral expansions of flat folds of the body wall and bear no direct relationship to the legs. During phylogeny, they were probably first used to glide about from trees to the ground and lacked articulations with the exoskeleton. In all present-day insects, articulations are present between the wing on the one hand and the tergum and pleuron on the other (Fig. 19.18). There are two different arrangements of the body wall muscles which furnish the power for insect flight. In the SYNCHRONOUS or DIRECT type, the contractions of an inner pair of muscles raise the wings by pulling on a lever-like structure at their bases; the heavier outer pair of muscles then contract to depress the opposite ends of the levers and provide the downward power strokes of the wings. The action of the muscles of the ASYNCHRONOUS or INDIRECT type is not directly on the wing articulations but on the thoracic exoskeleton to which they are attached; heavy vertical and longitudinal thoracic muscles contract alternately to deform the thorax and move the wings indirectly. Contraction of the vertical column pulls the tergum toward the sternum and raises the wings; the longitudinal muscles operate as antagonists and their contractions, assisted by the elasticity of the tergum, provide the powerful downstroke of the wing. It is these indirect flight muscles which are of particular interest for they have been found to have unique properties of rapid oscillation. In some small flies, the indirect flight muscles may operate at frequencies greater than 1000/sec (Smith 1965). Because of peculiarities in structure, these rapid oscillating muscles are termed FIBRILLAR MUSCLES; their physiology has been a subject of intensive research since Pringle (1949) first described some of their specialized activities. The early work may be traced in reviews by Pringle (1957, 1965a, 1967) and Boettiger (1957, 1961).

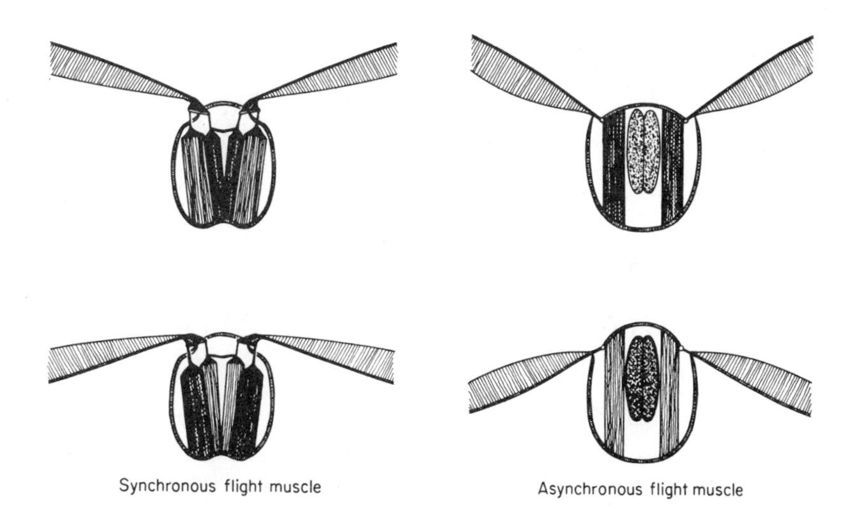

Fig. 19.18 Wings and flight muscles of insects. Synchronous flight muscles occur in all but four orders of flying insects and operate a direct up-and-down movement of a lever at the base of the wing. Asynchronous flight muscles, found in wasps, flies, mosquitoes, beetles, and some bugs, operate indirectly on the wings by deforming the thorax in a rapidly oscillating manner. [From Smith, 1965: Sci. Am., *212* (6): 78. Copyright 1965 by Scientific American, Inc. All rights reserved.]

What Pringle (1949) demonstrated in the first instance was a lack of synchrony between the muscle potentials and the vibrations of the thorax associated with the wing beat in the blue bottle fly *Calliphora erythrocephala.* During steady flight, muscle potentials occur at about 3/sec while wing beat frequency is 120/sec. It is now known that this capacity for rapid myogenic oscillations is characteristic of some groups of flying insects (Diptera, Hymenoptera, Coleoptera, Hemiptera) while in others (Orthoptera, Odonata, Lepidoptera), there is a strict 1:1 relationship between the neuromuscular electrical events and the mechanical action, as in vertebrate skeletal muscle. Roeder (1951) first emphasized this difference; the two types of insect muscle are termed ASYNCHRONOUS and SYNCHRONOUS (Fig. 19.19).

The asynchronous type of activity is also present in the tymbal muscles which produce the pulses of sound in cicadas of the genus *Platypleura* (Pringle, 1954). These tymbal muscles can be experimentally isolated with the associated skeletal attachments and a sufficient length of motor nerve to permit stimulation and recording from the component parts of the system. This preparation has played an important role in the studies of fibrillar muscle. Pringle found that a single stimulus excited the muscle to contract almost isometrically against the elastic resistance of the tymbal cuticle; at a certain tension the cuticle buckles and clicks to the IN position thus producing the sound. The tension is reduced, and the myofibrils are deactivated by this sudden release in tension. The exoskeleton then rebounds to its former OUT position; the original condition is restored in the system and the muscle

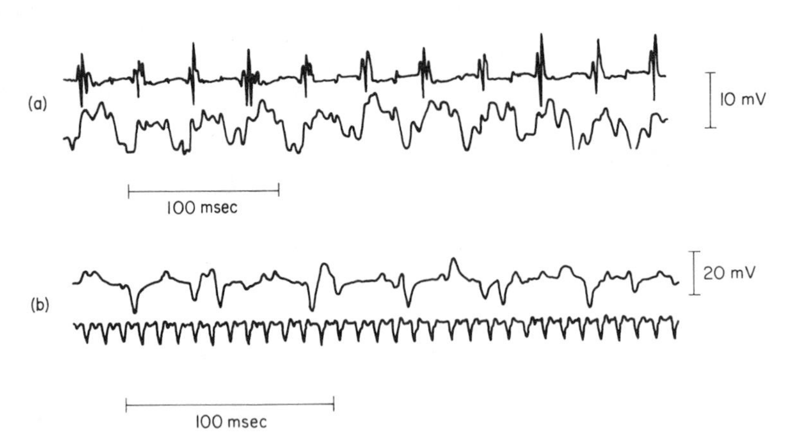

Fig. 19.19 Spike potentials (upper traces) and the thoracic movements (lower traces) during steady flight. (a) The cockroach *Periplaneta* which shows synchronous wing beating; (b) the wasp *Vespa* which shows asynchronous pattern. [From Roeder, 1951: Biol. Bull., *100*: 97. Permission of the Managing Editor.]

contracts once more. A single stimulus applied to the motor nerve at 30°C elicits four sound pulses; at stimulus frequency of 50 per sec there is a rapid oscillation of 320 per sec (Pringle, 1957).

The various histological features of fibrillar muscle are detailed in a monograph by Tiegs (1955) and summarized by Pringle (1965b) and Wigglesworth (1972). There is no sharp line of separation between the structural features of the direct and indirect flight muscles. The differences are quantitative rather than qualitative; a range in structural details is found even within the different orders and species of insects capable of asynchronous wing action (Pringle, 1957). Physiological capacities differ in a like manner. Although fibrillar muscles provide the extreme examples in several physiological features, the differences are again in degree of specialization rather than in radically different mechanisms. This is nicely shown in the length-tension curves depicted in Fig. 19.15b; the fibrillar muscle of the bumblebee shows scarcely any change in length with rising tension while lamellibranch smooth muscle, at the opposite extreme, steadily stretches over the entire experimental range of tensions. Limited extensibility is one of the most important mechanical properties of fibrillar muscle and is evident in muscles both at rest and under conditions of maximum isometric tension. Fibrillar muscles cannot be stretched by more than 5 to 10 per cent of their resting length without damage. In comparison, frog muscle can be stretched by 30 per cent or more of its rest length before injury occurs (Fig. 19.15). Again, however, fibrillar muscle provides the extreme example and the synchronous muscle of the locust, which is also very resistant to stretch, stands between it and frog leg muscle (Fig. 19.15b).

The oscillations of fibrillar muscle depend on the elasticity of the thorax, the limited extensibility of the muscle itself (non-compliant series elastic elements) and its capacity to develop tension *only after a delay*. Wings, thorax, and muscles operate as a resonant system and the muscles pull in synchrony with the resonant frequency of the system (Tregear, 1967). Nervous impulses are not required for each wing beat but must occur at intervals or the oscillations gradually die away. The delayed rise in tension is unique to fibrillar muscle and an essential factor for the maintenance of oscillation. Detailed experimental analysis of insect fibrillar muscle is beyond the scope of this discussion, but it should now be apparent that the contractile system is similar to that of other animals, but specialized to oscillate rapidly under suitable loading conditions. The action of the muscles on the flexible exoskeleton stores energy; the elastic force then contributes to the action of the wings. Limited extensibility and delayed tension operating on the elastic thorax produces the rebound with rapid oscillations in a system which receives a periodic reinforcement from the nervous system. The unique features seem to be the nature of the E-C coupling (reduced SR) and the delayed development of tension (Pringle, 1965a). The sarcoplasmic reticulum is markedly reduced while an extensive ramification of intracytoplasmic intermediary tubules (IT) are present as invaginations of the sarcolemma. By contrast, the SR is well developed in synchronous flight muscles and comparable to that of the vertebrates. Structural comparisons of the SR and T tubules in the synchronous and asynchronous muscles might suggest that nervous excitation occurs in a similar manner in both types but that the sarcoplasmic reticulum is not required for the release and recapture of Ca^{++} (calcium pump) in the asynchronous type (see review by Peachey, 1968).

Nervous Control of Muscles

The pioneer studies of fast and slow responses were carried out on coelenterates. These primitive animals have obviously experimented with a variety of neuromuscular controls. The jellyfish swims slowly and gracefully by gently and rhythmically pulsating its bell; the extended sea anemone quickly withdraws when touched and then, if further disturbed, contracts more and more until it shrinks into a rounded mass tenaciously sticking to its rocky holdfast. The physiology of these responses was carefully investigated for many years by Pantin (1935 to 1965). He and his followers, using isolated strips of jellyfishes and sea anemones, demonstrated a spectrum of neuromuscular controls ranging from the direct "all-or-nothing" (nonfacilitated) twitches to very highly facilitated contractions. The facilitated types include quick contractions which are set in action only by two or more stimuli and slow responses requiring several stimuli and showing no direct relationship to the stimulus sequence. Facilitation is markedly characteristic of these systems and is developed to varying degrees in different forms (reviews by Hoyle, 1957, 1962a,b).

The physiological basis of this versatility is still rather obscure. In the coelenterates it has not proved possible to isolate nerve, muscle and synapse in order to study separately the properties of each. Consequently, the locus of control has remained speculative. Although these technical difficulties are gradually being resolved (Josephson, 1966; Robson and Josephson, 1969), most of the solid information concerning fast and slow responses comes from studies of arthropod and vertebrate muscles. In these muscles it is now apparent that all three components—muscle, nerve and synapse—may be variously adapted to effect the precision of control.

Myoneural junctions Anatomical adaptations for the transmission of information from the central nervous system to the muscle cells are extremely diverse. At one end of the spectrum, the muscle fibers themselves stretch out to contact the central nervous system; the so-called ventral roots of the spinal cord of *Amphioxus* are, in fact, stretched out lamellae (muscle fibers) of the V-shaped myotomal muscles; the thinner dorsal fibers probably operate as a slow system while the thicker ventral ones provide a fast system (Flood, 1968; Guthrie and Banks, 1970a,b). At the other end of the spectrum, impulses are rapidly relayed for long distances over myelinated motor nerves extending from the spinal cord to distant parts of the appendages. The presence of distinct motor nerves is the rule; the situation described in *Amphioxus*—also in nematodes and some echinoderms—provides an exception (references in Hoyle, 1969, and Guthrie and Banks, 1970a,b).

Naked nerve endings may be applied to the sarcolemma, or the endings may be specialized to form MOTOR END PLATES of varying complexity. Muscle cells may be activated by single nerves with end plates or there may be many points of nerve contact scattered over the surface of the muscle cell; the range in complexity and terminology is shown diagrammatically in Fig. 19.20.

Most motor nerves are MULTITERMINAL with minute branching ramifications which come into close contact or actually fuse with the sarcolemma; the nerves, however, never actually penetrate the muscle cells (Bullock and Horridge, 1965; Hoyle, 1965). Among the invertebrates, specialized motor end plates occur in only a few insect skeletal muscles where they form multiterminal claw or plate-like ex-

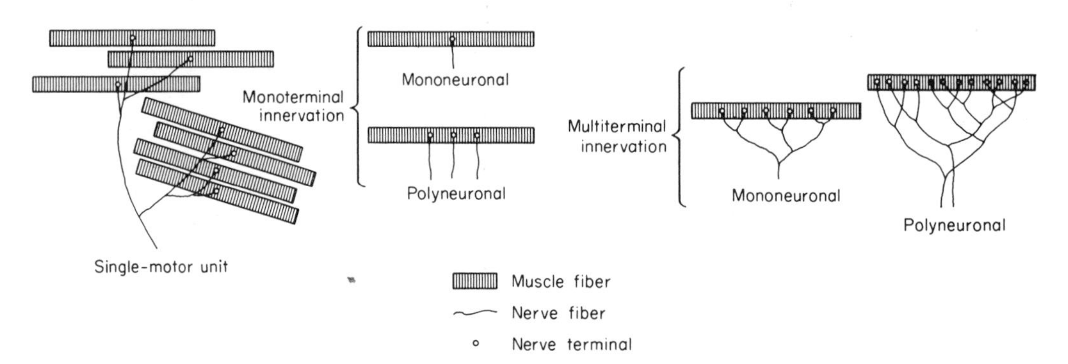

Fig. 19.20 Patterns of efferent innervation of muscle. [From Florey, 1966: General and Comparative Animal Physiology. Saunders, Philadelphia.]

pansions without subsynaptic folds (Hoyle, 1965). Characteristically, invertebrate muscle cells are served by more than one motor nerve (POLYNEURONAL) whether or not specialized nerve endings are present.

Specialized motor end plates have long been known in vertebrate skeletal muscle of the twitch type but are not recognized in vertebrate smooth or cardiac muscle nor in the slow fiber junctions. The motor endings on fast fibers of the lower vertebrates are comparable to those described in some insect skeletal muscles (Fig. 19.21).

The motor end plates of higher vertebrates vary in detail (Coërs, 1967) but are similar in that the myelin sheath of the axon ends near the muscle while the nerve ending, covered only by its plasma membrane, expands into an irregular area in contact with the sarcolemma. The postsynaptic membrane is specialized among the higher vertebrates and the nerve ending sits in synaptic gutters or troughs formed as depressions in the muscle fiber (Fig. 19.22). The sarcolemma lining the troughs or clefts is thrown into a series of folds which, in vertical sections, look like tiny perpendicular rods (the PALISADES). The plasma membranes of muscle and nerve are separated by a very short gap (0.02 to 0.03 μ) termed the SYNAPTIC CLEFT. Both the axoplasm and the sarcoplasm of the end plate contain numerous mitochondria, indicating a region of high metabolic activity.

The subneural apparatus seems to have evolved in association with the quick responses of higher vertebrates (Coërs, 1967). The evolutionary trend has been from polyneuronal and multiterminal endings to a mononeuronal innervation with single specialized end plates. Invertebrate motor nerve endings and those on slow vertebrate fibers never show postjunctional foldings of the sarcolemma (Bullock and Horridge, 1965; Hess, 1970). The twitch fibers of fishes are innervated by scattered multiterminal endings or small groups of endings with claw-like expansions which grasp a restricted area of the muscle fiber, but single plate-like junctions with subneural foldings of the sarcolemma are absent (Jansen *et al*, 1964; Bone, 1966;

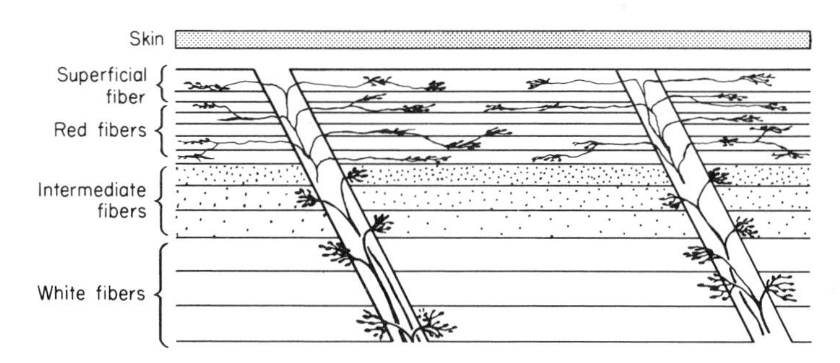

Fig. 19.21 Innervation of different muscle types in a dogfish myotome. White fibers are innervated only at their myoseptal ends by a characteristic basket-like arrangement which clasps the end of the muscle; the red fibers possess several motor terminations which curve around and along the muscle fiber with numerous small swellings distributed at short intervals. [From Bone, 1966: J. Marine Biol. Assoc. U.K., *46*: 329.]

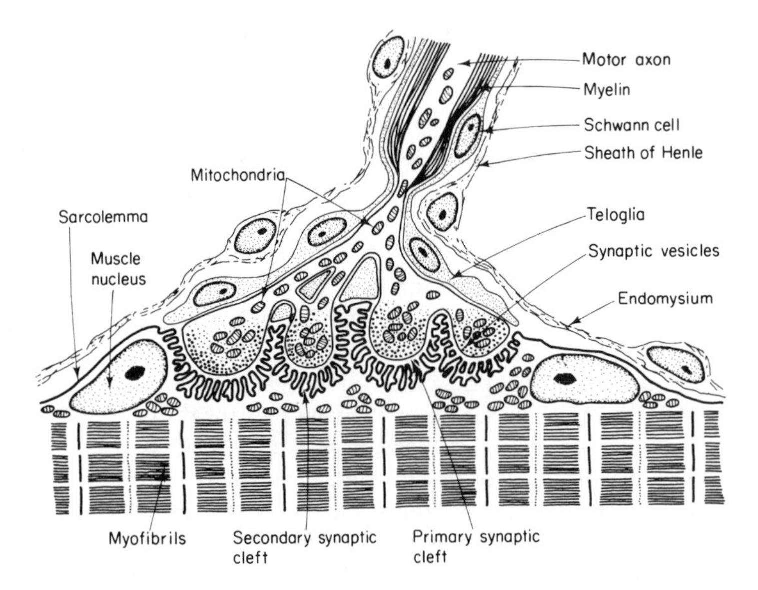

Fig. 19.22 Diagram of the structure of the motor end plate of mammalian skeletal muscle. [From Coërs, 1967: *In* International Review of Cytology, Vol. 22. Academic Press, New York.]

Coërs, 1967). Typical motor end plates localized in the middle of the muscle fiber become established in the reptiles; the extensive folding of the sarcolemma reaches an extreme in some of the muscles of mammals (Coërs, 1967; Hess, 1970).

Motor units in vertebrate muscle The functional unit of skeletal muscle in higher vertebrates is not the muscle itself nor the individual muscle cell but the MOTOR UNIT (Fig. 19.20); this consists of a single motor neuron and the group of muscle cells which its single axon innervates (Buller, 1970). The number of muscle cells comprising the unit varies from as few as two to six in some of the eye muscles to as many as several hundred in the limb muscles of the mammal. The units are smaller where the movements are delicate and finely graded. It has been estimated that the mammal may contain about a quarter of a billion individual muscle cells but about 450,000 myelinated nerve fibers, and it is thus evident that most of the motor axons are associated with a great many muscle cells. The individual fibers belonging to a motor unit are scattered throughout the muscle, and hence the action of the unit is on the muscle as a whole and not in a localized muscle area. A muscle receives a large number of axons while the individual muscle cells are innervated by very few motor end plates (usually only one).

In contrast, arthropod skeletal muscles are innervated by a small number of axons (two to five usually), and each of these divides and subdivides to supply a significant percentage of all the cells which comprise the muscle (Fig. 19.20).

Motor units are absent. Moreover, each of the axonal branches to a muscle fiber is subdivided into numerous fine twiglets which contact the muscle cell at many points.

Fast and slow contractions Some muscle preparations respond promptly with a quick contraction or TWITCH following a single threshold stimulus applied to the motor nerve. This is true of individual cells as well as organs. The twitch response is familiar in the frog gastrocnemius-sciatic nerve preparation so often demonstrated in physiology laboratories. It is also characteristic of several other muscles, such as the closer muscles in crayfish claws, the pharynx retractor of sipunculids, the jumping muscles of locusts and the jet propulsion muscles of squids. In these preparations, the active state develops rapidly to peak value and the muscle tension follows directly and appears as the twitch contraction. Summation occurs with repeated stimuli leading to a TETANUS or sustained contraction; these relationships are shown diagrammatically in Fig. 19.17.

The twitch is by no means the most common reaction to stimulation of motor nerves. In most muscles, the first stimulus leads to only a partial development of the active state and little or no change in muscle tension; the full development of the active state and the mechanical response follow progressively with repeated stimuli (Fig. 19.17). Thus, FACILITATION is marked and the full contraction may be significantly greater than the tetanus of the twitch response (Lang *et al.*, 1970). Fast or phasic and slow or tonic contractions in the closer muscle of the king crab *Paralithodes* are compared in Fig. 19.23.

It has already been noted that fast and slow responses are frequently associated with structural differences in the muscle cells. Fast and slow cells or fibers may occur in the same muscle or they may be segregated into different muscles (Atwood, 1967). However, fast and slow responses cannot always be explained in terms of muscle cytology or membrane properties. In some muscles, morphologically uniform populations of cells contract rapidly or slowly depending on the nature of the inner-

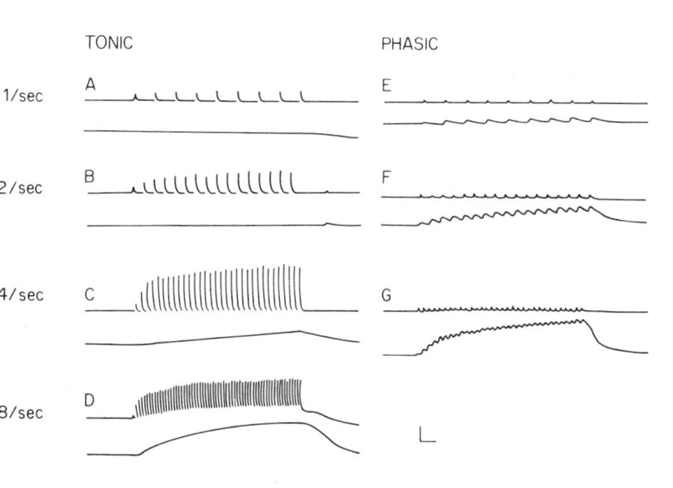

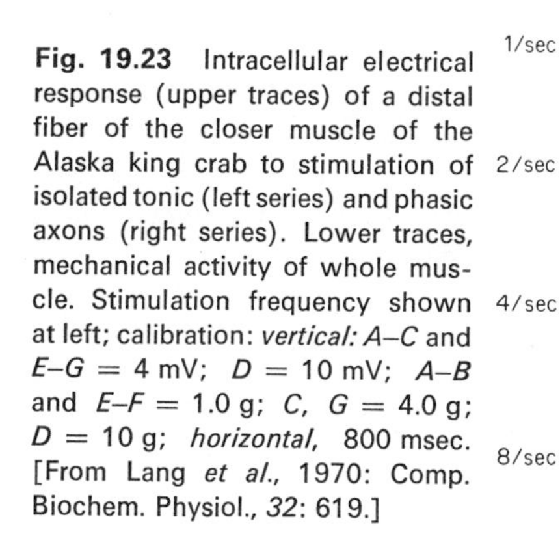
Fig. 19.23 Intracellular electrical response (upper traces) of a distal fiber of the closer muscle of the Alaska king crab to stimulation of isolated tonic (left series) and phasic axons (right series). Lower traces, mechanical activity of whole muscle. Stimulation frequency shown at left; calibration: *vertical: A–C* and *E–G* = 4 mV; *D* = 10 mV; *A–B* and *E–F* = 1.0 g; *C*, *G* = 4.0 g; *D* = 10 g; *horizontal*, 800 msec. [From Lang *et al.*, 1970: Comp. Biochem. Physiol., *32*: 619.]

vation. Phasic and tonic contractions have been extensively studied in the arthropods where distinct axons may be associated with different speeds of contraction in their associated muscles. The innervation is frequently polyneuronal (Fig. 19.20) and the stimulation of one axon leads to a rapid contraction while stimulation of another produces a slow contraction; some of these muscles may also receive axons which initiate responses of intermediate speed (Wiersma, 1961; Atwood, 1967; Usherwood, 1967). The axons are sometimes termed "fast" and "slow" but this is misleading; even though the nerves may differ in size and have different conduction velocities, control of muscle contraction speed does not seem to reside in the axon but in the myoneural junction. Some muscles with a single axon may also show fast and slow contractions in accordance with the stimulus frequency applied to the motor nerve. The opener muscle of the crayfish *Procambarus,* for example, has a single axon; the muscle responds differently to high and low frequencies of stimulation—because different groups of nerve fibers within the axon possess endings specialized to excite the muscle to respond rapidly or slowly. Stimulation of phasic axons or fast components give large junction potentials and spikes while the tonic axons or fibers give small junction potentials which show strong facilitation and produce smooth tonic contractions at high frequencies. In these muscles, the differences in speed of contraction reside neither in muscle fibers themselves nor in the nerve fibers; they seem to reside in the nerve endings (Bittner, 1968).

Peripheral inhibition Almost a century ago, stimulation of certain nerves to a crayfish claw were found to produce an inhibition rather than an excitation of the associated muscles. This phenomenon has now been investigated repeatedly in crabs and crayfishes in which both presynaptic and postsynaptic inhibition are recognized (Chapter 21 and Atwood, 1967). Peripheral inhibition is restricted to certain groups of invertebrate animals. It occurs in many muscles of the decapod crustaceans but has not been found in barnacles; it seems to be present in the closer muscle of *Limulus* but is absent in the spiders; postsynaptic but not presynaptic inhibition occurs in at least some of the insects (Usherwood, 1967; Atwood, 1968).

In crustacean muscles, the inhibitory fibers are usually the smallest of the three types (fast, slow, and inhibitory). When stimulated, they either produce no change in the muscle or reduce its tonus, usually after a definite latent period. If inhibitory and excitatory neurons are stimulated together, the excitatory activity will be reduced or completely suppressed; the slow system is more sensitive to inhibition than the fast. Membrane potentials likewise vary in accordance with the state of the muscle at the time of stimulus. If the potential is near its normal resting value, the effect of inhibitory stimulation is either hyperpolarizing or inconspicuous. If, however, the potential has been displaced from the resting value, then inhibitory stimulation tends to return the potential to the resting value. It is assumed, although the evidence is far from complete, that different transmitter substances are involved in excitation and inhibition. It has already been noted that inhibition seems to be associated with increased permeability to chloride ions (Chapter 15).

In vertebrate animals, inhibition following the stimulation of peripheral nerves is found only in the autonomic nervous control (the vagal inhibition to the heart, for example). It does not occur in the somatic musculature; inhibition of skeletal muscles is centrally controlled. This marks another interesting contrast between the nervous control of skeletal muscles in the decapod crustaceans and the vertebrates. In the former group versatility depends on the interaction of three types of neurons which may produce different degrees of excitation at many points in the muscle fiber. In the vertebrates, motor units are activated by varied patterns and frequencies of impulses from the central nervous system. Somatic motor nerves producing peripheral inhibition are unknown among the vertebrates. Inhibition is also a central phenomenon in some groups of invertebrates, the echinoderms, for example (Cobb and Laverack, 1967).

Transmitter substances The spread of excitation from nerve to muscle at the vertebrate motor end plate is effected through the release of acetylcholine (Chapter 15). In some manner not clearly understood this chemical leads to an increase in the ionic permeability of the plasma membrane; a wave of depolarization spreads from the end plate along the surface of the muscle cell. Although the transmitter substance is certainly not the same in all groups of animals nor in all muscles from the same animal, the principle of a chemical mediation of the events at the neuromuscular junction is generally accepted as applying to most of them.

At rest, the permeability properties of muscle cells are such that an electrical potential of 30 to 100 mv is maintained across their plasma membranes. In nervous excitation, transmitter substances, released at the nerve endings, increase ionic permeability and lead to junction potentials which then spread along the surface of the muscle fiber. Current understanding of excitation-contraction coupling was discussed in an earlier section.

Acetylcholine is also the transmitter at the junctions of the fast fibers in fishes (Bone, 1966) and may operate at many other neuromyal junctions. It is still uncertain whether fast and slow fibers rely on the same or different transmitters; the weight of the evidence, however, favors a single transmitter with more chemical per impulse released to effect the faster responses (Aidley, 1967). There is considerable evidence for the action of glutamate as an excitatory transmitter in arthropods (Atwood, 1967; Usherwood, 1967). However, it may act to enhance the release of transmitter substance rather than serve as a transmitter (Florey and Woodcock, 1968).

Gamma-aminobutyric acid (GABA) mimics the action of the chemical transmitter at inhibitory synapses of crustaceans and insects. It is considered the most likely candidate for the role of an inhibitory transmitter (Aidley, 1967; Atwood, 1967, 1968; Usherwood, 1967). The properties of these transmitters were considered in Chapter 2.

Retrospect Animal activity is based on ATP-driven systems of protein molecules which change their positions to effect the movements of cells. This system seems to be a conservative one in the homogeneity of the contractile proteins and in the activation by ATP. It has expressed its versatility in the many cell types concerned with movement and in the varied ways in which they are governed by nerves. Cell types range from amoeboid masses of protoplasm and cells with ciliated surfaces to many kinds of muscle fibers. Variety in myofibrillar organization ranges from simple myonemes in the epitheliomuscular cells of coelenterates to irregularly arranged masses of myofilaments in classical smooth muscle types and to the orderly arrangement of myofilaments in the cross-banded striated types. The trend toward orderliness in the myofilaments seems to be associated with speed of movement. Precision and versatility in muscular control depend largely on the manner in which the muscles are activated by the nerves. Separate neuron types are associated with quick and slow contractions in all multicellular phyla while some groups have inhibitor fibers as well. A precision of control is possible through a grading of activities at the myoneural junctions. Among the vertebrates, however, this control is largely taken over by the central nervous system.

Thus, these ATP-driven systems of contractile proteins show adaptation at all levels from the molecular to the cellular and from the organ (muscle) to the neuromuscular system. At the cellular level both the contractile machinery and the fuel systems have been variously adapted to the needs of the organism; at the level of the organism itself, several different systems operate together to provide the versatility of movement characteristic of a living animal. Contractile systems still offer many challenging areas for investigation. Out of the great variety of animal forms, only a very few have been carefully studied; although the picture may now be essentially correct in its outline, many of the details may be quite different when fully understood.

20

ELECTRICAL DISCHARGE, LIGHT PRODUCTION, AND COLOR CHANGES

ELECTRIC ORGANS

Although electrogenesis occurs in all receptor-effector systems, the discharge of electricity as an effector action has been described only in some of the fishes. There are well-known representatives among both elasmobranchs and teleosts; the elasmobranch species are marine while the teleost species, with one exception (the stargazer *Astroscopus*) live in tropical fresh waters (Keynes, 1957; Grundfest, 1960; Bennett, 1970, 1971a). Several of the families of electric fishes are quite unrelated; apparently, these organs evolved independently in at least six different groups.

Electric fishes have long excited man's curiosity. The writings of the ancients record astonishing experiences with these strange animals while Charles Darwin, in his classic *On the Origin of Species,* wrote that "the electric organs of fishes offer another case of special difficulty." His dilemma in this instance was to explain the evolution of specialized and unique structures, such as the powerful electric organs, by natural selection when there was apparently no conceivable function for them in ancestral species lacking an ability to deliver paralyzing electrical discharges. Darwin's faith, however, that "we are far too ignorant to argue that no transition of any kind is possible" was vindicated almost exactly a century later when Lissmann (1958) demonstrated the usefulness of low voltages in some species of fish which depend on regular pulsating discharges for their orientation; these fishes are able to sense changes in their electrical fields and thus avoid objects or react to other animals. Since most electric organs develop from the myoneural apparatus during ontogeny, it can be assumed that their physiological phylogeny has been built onto the electrogenesis associated with the activities of skeletal muscle. Electrogenesis was thus a preadaptation which became useful in the evolution of electric organs for orientation, communication and interaction between certain aquatic animals. The extent to which electrocommunication enters into the behavior of fishes is still under active investigation.

Morphology

Morphologically, electric organs are composed of units known as ELECTROPLAQUES or ELECTROPLATES or ELECTROCYTES (Bennett, 1970). With one recognized exception, these units are modifications of the myoneural apparatus—usually transformed muscle fibers (cells), sometimes specialized motor end plates and, in one family of gymnotid eels (Sternarchidae), the modified terminals of the motor axons themselves. The known exception to a myoneural involvement occurs in the accessory electrical organs of the chin region of the sternarchid *Adontosternarchus* where sensory neurons are specialized to emit high-frequency electric pulses. The electrical properties of these differently evolved electrocytes are detailed by Bennett (1971a).

Typical electrocytes are thin, wafer-like, flattened units arranged in an orderly manner with one surface (relatively smooth) a specialized nervous layer and the other a papilliform nutritive layer; like surfaces all face in the same direction. In some species, the nervous layer is directly supplied with a dense net of nerve fibers; in others, the nerve supply is indirect to one or several sturdy stalks emerging from this surface (Fig. 20.1). Electrocytes are embedded in a jelly-like material and individually housed in a series of connective tissue compartments.

The arrangement and number of the electroplates are as variable as their cytology but show many obvious adaptations to the special requirements of the species. In the giant ray *Torpedo nobiliana* they are horizontal, piled up from ventral to dorsal surface in columns like stacks of coins; there are over 1000 "coins" in a column and approximately 2000 columns. In the electric eel *Electrophorus electricus* the electrocytes are vertical in longitudinal columns parallel to the spinal cord. This animal has from 6000 to 10,000 units in each column with about 60 columns on each side of the body (Grundfest, 1960; Bennett, 1971a). About 40 per cent of the eel's total bulk is devoted to this specialized tissue which forms the most powerful bioelectric generator known, with the capacity to discharge something more than 500 volts. The electric eel inhabits fresh waters and thus requires a large number of electroplaxes in series to overcome the resistance; *Torpedo,* in contrast, generates a lower voltage in salt water with lower resistance, but by having 2000 columns in parallel it achieves an extraordinary amperage. The electric catfish *Malapterurus* exhibits still another arrangement with a jacket of electric tissue which surrounds the body in a thick layer extending from the gills to the tail. This organ was at one time thought to be exceptional in arising from skin glands rather than muscles; but it is now known to arise from myoblasts in a manner comparable to other electric organs (Johnels, 1956).

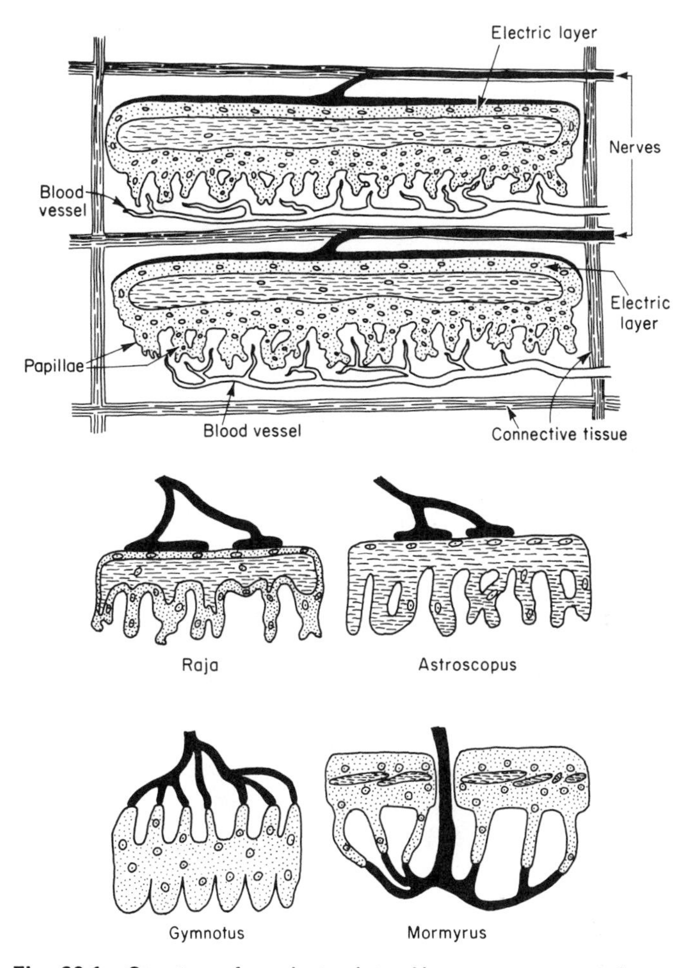

Fig. 20.1 Structure of an electroplate. Above, arrangement of two neighboring plates; below, four kinds of electroplates: cross-hatching, modified muscle. For detailed references to anatomy, see Bennett (1970, 1971a). [Above, based on Ihle *et al.*, 1927: Vergleichende Anatomie der Wirbeltiere. Springer, Berlin. Below, based on Dahlgren and Kepner, 1908: Principles of Animal Histology. Macmillan, New York.]

Physiology

The scientists of the nineteenth century appreciated the significance of the arrangement of the electroplates. Their hook-up in series adds the output of the units to build up the voltage; the arrangement of the columns in parallel builds up the amperage. Pioneer workers noted that the nervous face of the electroplax became negative to the non-nervous layer during the discharge, but they had neither the theoretical knowledge nor the delicate instruments necessary to carry the analysis further. The first significant theoretical discussions of the mechanism responsible for the potential differences are those of Bernstein and his associates (Bennett, 1970, 1971a).

Bernstein's classical concepts of cell permeability and ion fluxes during excitation were described in Chapter 15. These theories, applied to the electric organs, postulated a selective cell membrane permeability for potassium with an impermeability to sodium (as in muscle and nerve) but, in addition, assumed a difference in the response of the two faces of the electroplate during activity. Bernstein argued that the nervous face became depolarized during stimulation while the non-nervous face retained its resting potential and thus the non-nervous membranes, lying parallel to one another with their resting potentials of about 85 mv (in *Electrophorus*) are suddenly connected in series like an electric battery. This hypothesis could not be experimentally tested until microelectrodes and suitable amplifiers became available almost fifty years later. The recordings with intracellular electrodes quickly demonstrated that Bernstein's basic assumption of differences in the permeability of the two membranes was correct but incomplete. During excitation the nervous layer is not merely depolarized but shows a reversal in polarity (to about 67 mv in *Electrophorus*) as described earlier for muscle (Chapter 15). In this way the nervous face develops a potential difference in the same direction as that of the non-nervous face, and when these become connected in series, both faces of the electrocyte contribute to the charge of "the battery" (Fig. 20.2).

The electric eel uses this discharge system to deliver high-voltage paralyzing electric shocks. Subsequent studies show that this is only one of several discharge systems used by fishes and emphasize once more the remarkable opportunism in the evolutionary process (Bennett, 1971a). The greatest diversity occurs among the

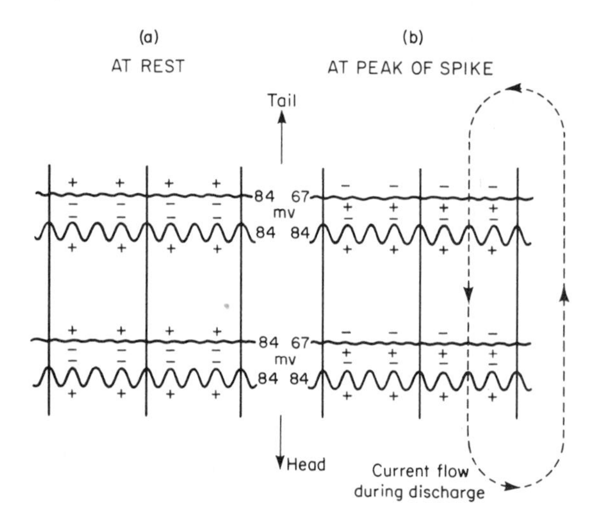

Fig. 20.2 Mechanism of additive discharge in the electroplates of the organ of Sachs in *Electrophorus*. (a) At rest, there is no net potential across the electroplates. (b) At the peak of the spike, all the potentials are in series, and the head of the eel becomes positive with respect to its tail. [From Keynes, 1957: *In* The Physiology of Fishes, Vol. 2, Brown, ed. Academic Press, New York.]

weakly electric fishes which use pulses of electricity for electro-echo orientation. Some of the knife fishes (*Gymnotus*) and some of the mormyrids (*Gnathonemus*) produce diphasic or triphasic rather than monophasic pulses. Both faces of the electroplate show changing permeability (potentials) but not simultaneously, so that the alternating activity of nervous and non-nervous faces produces very rapid diphasic pulses (0.3 msec in *Gnathonemus*). The duration of the spikes may vary so greatly that the resulting pulse can be almost monophasic; in the mormyrid *Mormyrus rume* (also the strongly electric African catfish *Malapterurus*) the rostral spike is very long and the caudal spike extremely short. Representative examples of the spikes from the two surfaces and the resulting external discharge are shown in Fig. 20.3.

Vertebrate skeletal muscle cells are normally activated through motor nerves by way of the motor end plate. It will be recalled (Chapters 15 and 19) that the motor nerve initiates junctional potentials in the end plate through release of acetylcholine; these potentials rise to a certain peak value and then trigger the all-or-none action potential in the muscle cell. Thus, in vertebrate skeletal muscle there are two major bioelectric events—the end plate potentials and the action potentials of muscle. During the phylogeny of electric organs from muscle, some fishes have capitalized on the end plate potentials (elasmobranchs) and others have utilized muscle action

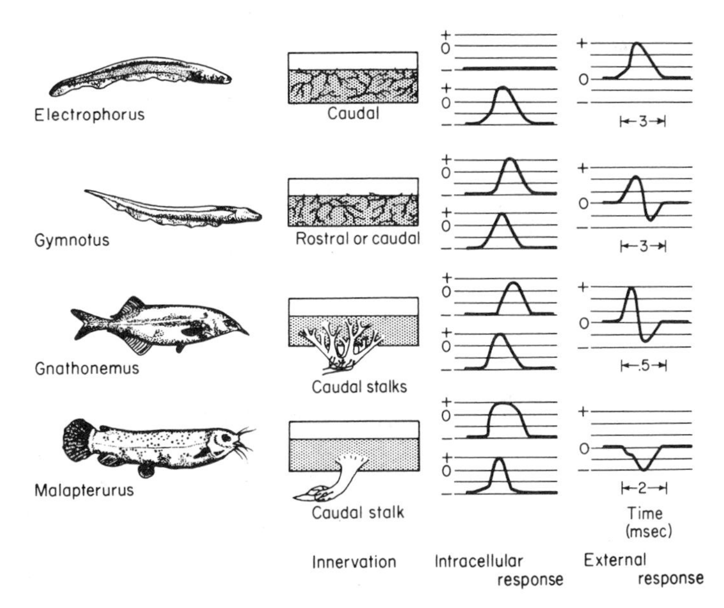

Fig. 20.3 Discharges of different electric fishes. Shaded area of the electroplates (second series of diagrams) shows the innervated membranes. Top curve of each pair (third column) shows intracellular potential across uninnervated membrane during a response. Bottom curve, response of innervated membrane. External response (right curves) is algebraic summation of oppositely directed potentials of the two membranes. [From Grundfest, 1960: Sci. Am., *203* (4): 122. Copyright 1960 by Scientific American, Inc. All rights reserved.]

potentials (freshwater teleosts). This interesting variation in physiological phylogeny was also discovered after the advent of electrophysiological recording instruments, although the findings were based on observations made about 1880 (Grundfest and Bennett, 1961). At that time it was noted that the electric organs of the ray *Torpedo* could not be stimulated electrically while those of the eel *Electrophorus* could. When *Electrophorus* has been forced to discharge its electric organs to the point of exhaustion, the organs can still be excited by direct electric or mechanical stimuli; *Torpedo,* in a similar test, is completely unresponsive to direct stimulation. The fatigued organs of *Torpedo* will, however, respond to a direct application of acetylcholine.

The solution to this puzzle was evident when it was found that the myoneural junction is an electrically inexcitable membrane, while the muscle cell itself can be directly excited by electrical stimuli (electrically excitable membrane) as well as indirectly through its nerves and, further, that the end plate potentials develop at the myoneural junction in accordance with the release and accumulation of acetylcholine. Thus, the electric organs of the rays are specialized synaptic membranes (electrically inexcitable) while those of the freshwater teleosts are specialized muscle membranes and are electrically excitable—although normally activated through nerves. The marine teleost *Astroscopus,* in contrast to all freshwater teleosts, has an electrically inexcitable system, suggesting opportunism rather than sequential phylogeny.

Nervous Control

Like all effectors, the electric organs are only useful physiological machines when they respond appropriately to meet the exigencies of a complex environment. As a means of stunning prey or discouraging predators these organs act in response to information gathered by the visual and tactile receptors; as a system for detecting objects in the environment and orienting in dark habitats they require a battery of specialized electroreceptors. In any case, sensory information must be received and coordinated in the computing centers of the brain, and these neural centers must discharge the appropriate motor impulses to the electric organs. Certain areas of the brain (particularly the cerebellum and parts associated with the lateral line system) are almost as highly specialized as are the muscles which give rise to the electric organs; the phylogeny of a specialized electric discharge system has modified these fishes in many curious ways (Grundfest, 1960; Lissmann, 1963; Bennett, 1971b).

Some of the problems of coordination can be readily appreciated in *Electrophorus.* An electric eel may reach a length of six to eight feet, and yet the electric organs which stretch all along its sides discharge in a matter of about 3.0 msec; this is only slightly longer than the discharge time of a single electroplate. Clearly, there is some very precise system for synchronizing the discharges from several thousand electroplates.

Another interesting problem in physiological control is exemplified by those electric fishes which produce continuous streams of electric pulses for the purpose of sensing their environment. The different species show characteristic discharge frequencies ranging from 50 to as high as 2000 per sec; the rate may be astonishingly steady to within 0.5 per cent in some of the knife fishes even when stimulated. In sensing unfamiliar objects, some species vary only the amplitude of the pulses (*Gymnarchus*) while others can vary frequency of discharge as well as amplitude (Mormyridae). These two systems have evidently evolved in parallel in unrelated families both in Africa and South America (Lissmann, 1961, 1963; Bennett, 1970).

Lissmann did the pioneer work with *Gymnarchus niloticus,* an African fish which lives in turbid waters of very low visibility and orients by sensing changes in the pattern of discharges from the small electric organs at the end of its long pointed tail. This animal sends out a steady stream of pulses at the rate of about 300 per sec; these vary only in amplitude, waxing and waning as the fish approaches an unfamiliar object. The animal swims with its beautifully undulating fins but maintains a relatively rigid vertebral column. At each electric discharge the tip of the tail becomes momentarily negative to the head so that the animal is surrounded by an electric field, the configuration of which depends on the conductivity of the water and the presence of objects of variable electric conductivity. The characteristic swimming with rigid body is no accident, for the configuration of the field is the main sensing device in dark and turbid waters; the animal can operate as easily in reverse as in forward swimming.

The ability of *Gymnarchus* to use this system in electro-orientation has been carefully established through classical conditioning experiments (Lissmann and Machin, 1958). *Gymnarchus* is sensitive to potential differences as small as 0.03 mv/cm. The reactions of the fish to electric currents, magnets, conductors and non-conductors when introduced into their surroundings prove their sensitivity to any change in electric fields; they detect the presence of magnets and can distinguish conductors and non-conductors. They can detect the presence of another fish by this system, but it is evident that their prey (species lacking this system) are unaffected by the discharges.

Electric organ command systems Electric organs are controlled by small groups of neurons (nuclei) situated in the medulla and neighboring regions of the spinal cord; in some species the control system may also extend into the back part of the midbrain (Bennett, 1971a). The number of neurons which make up the nucleus varies from two in the electric catfishes to half a hundred in some of the gymnotids. In species which operate the system intermittently to stun enemies or prey, the neurons of the command system are "turned on" by the reception of pertinent information from the environment. In species which utilize the system for electro-orientation, pacemaker neurons within the command system maintain a rhythmic output of pulses which can be appropriately varied in amplitude or frequency in response to environmental cues.

The neurons of the command system are interconnected through electrotonic synapses. This important feature ensures autoactivation and synchronous discharge of all the neurons in the nucleus—a property which could not be achieved in a system of polarized chemically operated synapses with their relatively long latent periods. Synchronous activation of electrocytes results in larger outputs in terms of voltage and power and becomes particularly important when the discharge is diphasic or triphasic, where slightly out-of-phase activity would lead to cancellation (Bennett, 1971a). The phylogenetic trend in command systems appears to have been toward a reduction in the number of neurons involved. Bennett (1970, 1971a) has reviewed the rapidly growing literature on the circuitry of these systems.

Synchronization of electrocyte activity Millions of electrocytes located at varying distances from the command center must be simultaneously activated. Electrotonic synapses ensure a synchronous output from the brain. Within the electric organs themselves, each of the individual units has its separate nerve supply; thus, it is apparent that the simultaneous discharge of all the electrocytes in an organ requires modifications in conduction rates which connect command centers with electrocytes distributed over variable distances (two to three meters in *Electrophorus*). In part, this has been achieved by an adjustment of path lengths, with the nerves to the electrocytes becoming progressively shorter from the cephalic to the caudal segments; in detail, the nerves to the more anteriorly located electrocytes are longer and take a more devious route to the organ so that total distances from the command centers are equalized. In part, also, the conduction velocities have been varied by modifications in the axon diameters (stalks) to the electroplates or in the spinal relays. In some species, synaptic delay may be varied at the chemically transmitting synapses on the electrocytes or elsewhere along the connecting paths. Thus, the several different components of the transmission system have been adapted to ensure simultaneous discharge of the millions of units which may make up an electric organ (Bennett, 1970, 1971a).

Electroreception

All living cells respond to electrical stimuli. The reasons for this should be evident from the discussions of stimulus-response mechanisms (Chapters 15 and 16). There are, however, certain receptor cells in some fishes which sense extremely small voltage gradients in their natural habitats and use this information adaptively. The lower limit of detection is in the region of 0.01 μV/cm (behavioral studies) which exceeds that of other receptors of fish by a factor of 10^4 or more. The sensory organs concerned are true ELECTRORECEPTORS; they monitor distortions in an electric field which they generate for electrolocation; they also detect signals of external origin such as those produced by electric organs (or possibly skeletal muscles) of another fish. Systems involved in electrolocation have been termed ACTIVE ELECTROSENSORY SYSTEMS in contrast to those which detect only external signals and are called PASSIVE (Bennett, 1971b). Dijkgraaf (1968), working with the catfish

Ameiurus nebulosus, was the first to demonstrate electroreceptors in a fish without electric organs. It is now recognized that most elasmobranchs, as well as several catfishes, are similar to *Ameiurus* in being able to detect electric currents, even though they lack special organs to generate them. Although many questions about electroreceptors remain to be answered, the microanatomy and many of the physiological properties are now well known; the literature can be readily traced through the symposium volume edited by Cahn (1967) and comprehensive reviews by Waltman (1966) and Bennett (1970, 1971b).

Electroreceptors have been found only among fishes. There are two quite different lines of speculation concerning their phylogeny. In earlier discussions, the evolution of electrosensitivity was considered *a consequence of* the appearance of electric organs; the more recent arguments assume the evolution of receptor organs *before* the corresponding effectors (Bennett, 1971b). This reversal of the argument followed the recognition of highly developed electroreceptors in fishes which do not possess electric organs. It now seems entirely possible that these receptors first evolved as monitors of electric signals arising external to the fish or as a consequence of muscle contractions in the fish itself. At any rate, it is apparent that electroreception is important in many present-day fishes which lack electric organs; in those which do possess electric organs, it has evolved into a remarkable system for orientation and probably also for intraspecific communication.

Kinds of electroreceptors Electroreceptors are specialized lateral line detectors and show some of the morphological characteristics of a neuromast. The sensory

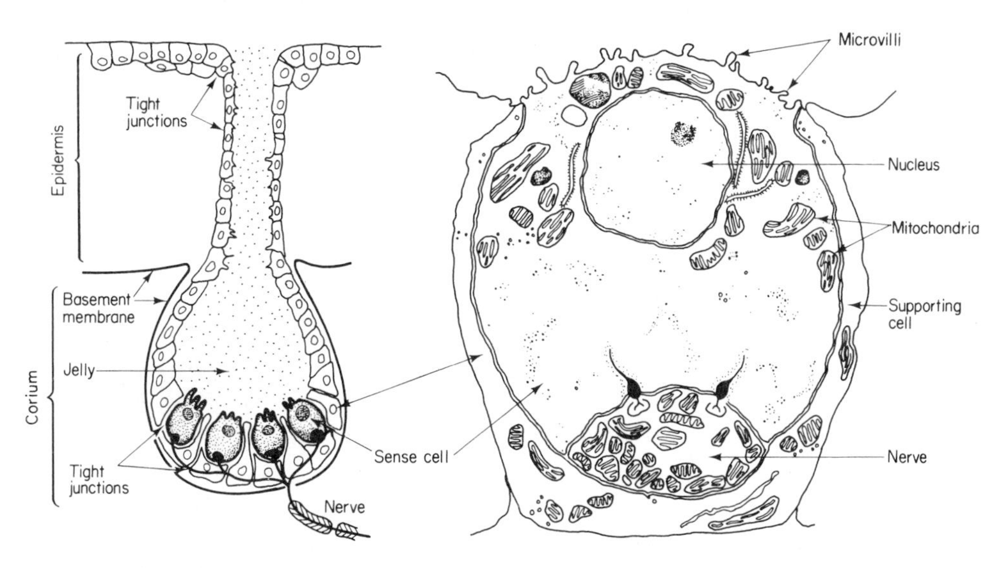

Fig. 20.4 Diagram of an ampullary (tonic) electroreceptor with the ultrastructure of a single sense cell as revealed by electron microscopy. [Sense cell after Szamier and Wachtel, 1969: J. Morph., *128*: 263.]

cells are cuboidal or columnar; each cell synapses basally with an afferent neuron while its opposite surface is extended in numerous microvilli directed toward the outer world. Kinocilia, which are present in ordinary lateral line hair cells, are absent, except for the ampullae of Lorenzini.

Early anatomists described several epidermal structures now known to be electroreceptors: AMPULLAE OF LORENZINI (heads of elasmobranchs), AMPULLARY ORGANS (gymnotids), PIT-ORGANS (silurids) and MORMYROMASTS (mormyrids). Bennett (1970, 1971b) recognizes only two main types which differ both in anatomy and physiology. The AMPULLARY RECEPTORS are long-stemmed, flask-like structures or ampullae, filled with a gelatinous substance; several sensory cells are embedded in the wall at the bottom of the flask so that only a small portion of the cell is exposed to the lumen and in contact with the jelly (Fig. 20.4). TUBEROUS RECEPTORS lack the jelly-filled canal; instead, the connection between the cavity of the organ and the body surface is filled with loosely packed epithelial cells. As a further contrast, the receptor cells of the tuberous organ protrude far into the space beneath the epithelial plug so that most of the cell surface, covered with microvilli, is exposed (Fig. 20.5).

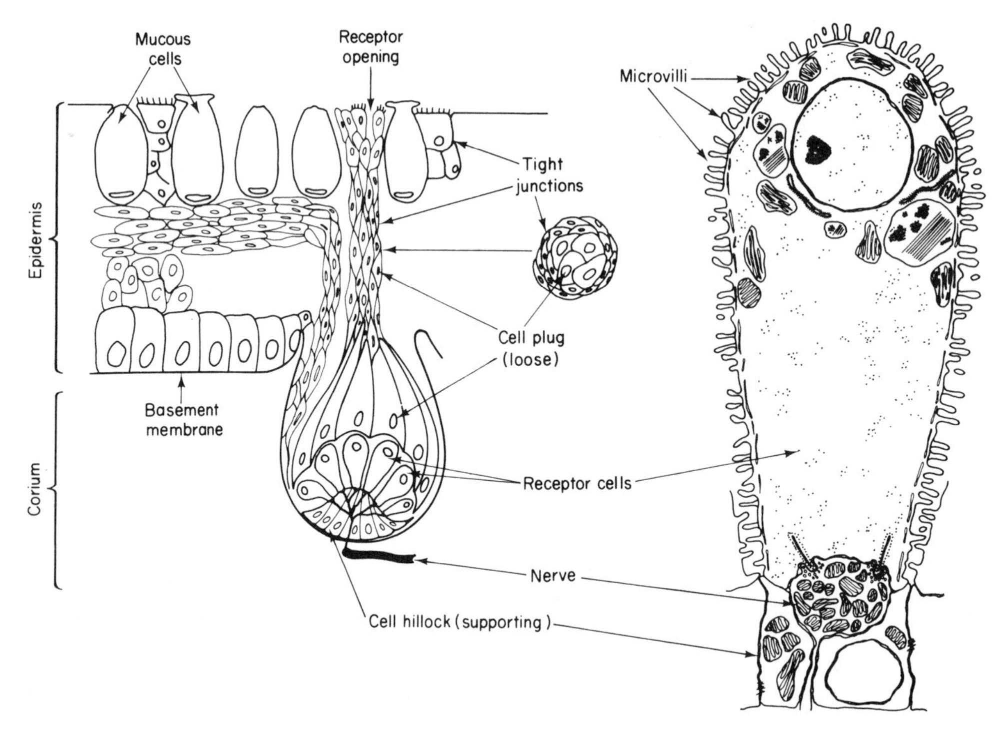

Fig. 20.5 Diagram of a tuberous (phasic) receptor. [From Szabo, 1965: J. Morph., *117*: 231.] With ultrastructure of a single sense cell based on electron microscopy. [From Szamier and Wachtel, 1969: J. Morph., *128*: 264.]

Gymnotids, mormyrids, and probably *Gymnarchus* possess both types of electroreceptors while elasmobranchs and catfishes have only ampullary organs (Bennett, 1971b).

Physiological differences between the two kinds of receptors are even more striking. Ampullary organs are tonic receptors (Chapter 16) with a steady resting discharge which may decrease or increase on stimulation (Fig. 20.6). They are constantly vigilant and monitor for low frequency or dc stimuli. Tuberous organs, by contrast, are phasic and remain silent or almost silent unless stimulated by relatively high-frequency currents; they are insensitive to maintained or dc stimuli. In short, they monitor for changes in the electric field and are the basis for electrolocation; to do this, it is obvious that they must fail to maintain a response to steady stimuli.

Transducer mechanisms Even though many of the electrical properties of these sense cells are known, details of the ionic and molecular events, which transduce the environmental changes in electric fields into the action potentials of the sensory nerves,

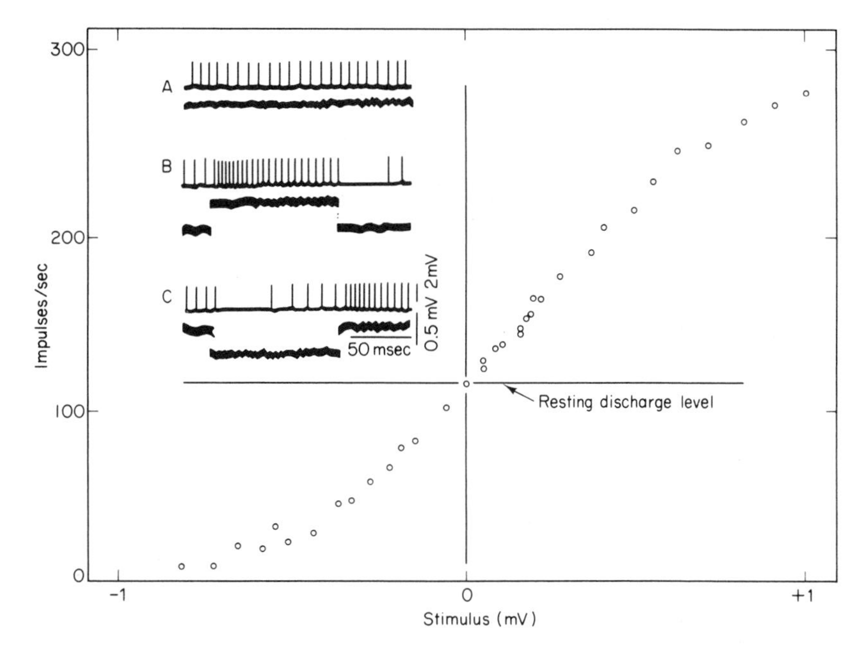

Fig. 20.6 Responses of a tonic receptor in *Gymnotus*. *Inset:* A, Spontaneous discharge; B, excitation by an anodal stimulus; C, inhibition by a cathodal stimulus; in each case, the upper traces are impulses in the afferent nerve, while lower traces show stimulating potentials applied externally; positive potentials and anodal currents indicated as upward deflections. *Graph:* Average impulse frequency during a stimulus of about 100 msec plotted against stimulating voltage; anodal stimuli to the right; cathodal to the left. [From Bennett, 1968: *In* Physiological and Biochemical Aspects of Nervous Integration, Carlson, ed. Prentice-Hall, Englewood Cliffs, N.J.]

remain speculative. At this level of understanding, biologists are still asking questions comparable to those which remain in other areas of sensory physiology.

It may be profitable to compare cellular events in electroreceptors with those already noted for ordinary lateral line hair cells (Chapter 17). In both, an external stimulus alters the electrical properties of the cell and leads to release of transmitter at a synapse; the transmitter, in turn, triggers or alters action potentials in the sensory nerve. In ordinary hair cells, movement of the kinocilium is thought to open ion gates and alter the transmembrane potential; in electroreceptors, an external electrical stimulus serves this function. Bennett (1971b) suggests that the electrical properties of the electrosensing cell membrane are probably similar to those of many receptors, and that sensitivity is mainly the result of the special arrangements of the external structures.

The histology of the epidermis has been carefully studied. In brief, the epidermal structures which surround the electroreceptors are arranged to channel the stimulating currents directly through the receptor cells and prevent leakage through the intercellular spaces. Many layers of closely packed, flattened epithelial cells cover much of the body and provide an electrical resistance to potentials applied tangentially to the surface. Thus, the receptors normally respond only to potentials applied *across* the skin. Moreover, tight junctions unite the closely packed cells which surround the flask-shaped organs to shield them electrically from the surrounding tissues; zonal tight junctions between the receptor and supporting cells must prevent leakage of current between the cells and channel it through them. Bennett (1970, 1971b) provides many additional details and discusses current concepts of receptor physiology; only a bald summary is given here.

Perhaps the simplest situation to describe is that which appears to operate in the tonic receptors of gymnotids, mormyrids and catfishes. In these, both the jelly in the canal and the microvillar surface of the receptor (Fig. 20.4) are good conductors and create only a low resistance to the stimulating current. Thus, there is an almost direct reflection of the environmental voltage change to the synaptic membrane, where depolarization or hyperpolarization alters transmitter release and subsequent activity in the afferent nerve (Fig. 20.6). The ampullae of Lorenzini in elasmobranchs are also tonic receptors but differ both in electrical properties and sensitivity. The transducing mechanisms are more speculative, but it seems likely that the first step occurs at the outer (microvillar) surface of the sense cell with a change in transmembrane potential; in short, these receptor cells seem to have an electrically excitable outer membrane.

The phasic receptors, which detect distortions in the electric field (Fig. 20.7), differ in at least one important electrical property. The very extensive outer (microvillar) face of the cell (Fig. 20.5) behaves as a series capacitor. The immediate effect of a long-lasting stimulus pulse is to depolarize or hyperpolarize the inner (synaptic) membrane as the stimulus spreads through the low-resistance tissues to the base of the sense cells. Gradually, however, the capacitor is charged and the effect of the maintained voltage on the synaptic membrane becomes attenuated and ceases; the opposite

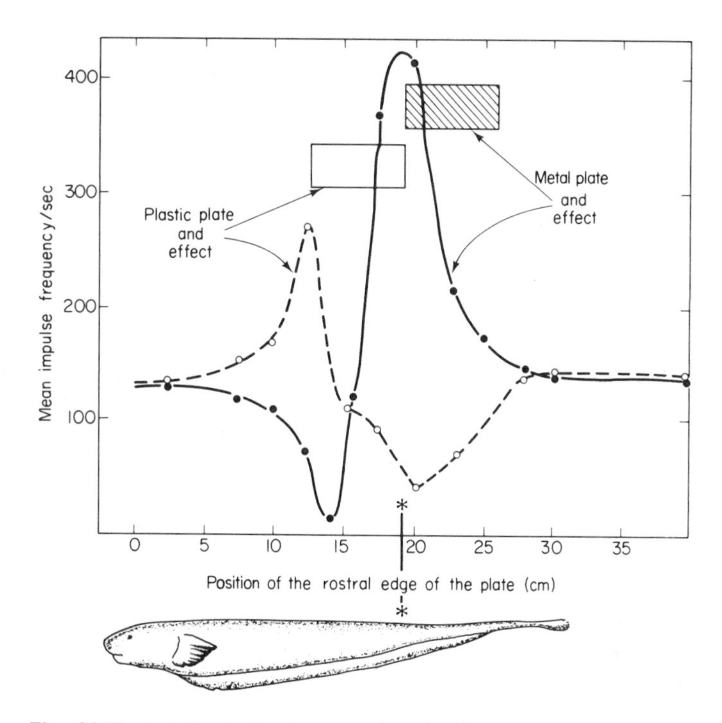

Fig. 20.7 Relation between nerve impulse frequency in a particular electroreceptor nerve fiber of *Sternarchus* and the position of metal and plastic plates placed close to the fish. Arrow indicates the location of the electroreceptor ending; plates drawn to scale in the positions where they elicited maximal response. [From Hagiwara, Szabo, and Enger, 1965: J. Neurophysiol., *28*: 788.]

sequence occurs when the stimulus pulse ends (Fig. 20.8). This transient action of a maintained stimulus leads to an "on and off" effect; converse sequences occur with anodal and cathodal stimuli (Fig. 20.8). With a single known exception (Bennett, 1971b), transmission at the synapse is chemically mediated. A burst of impulses in the afferent nerve follows depolarization of the synaptic membrane; a hyperpolarization probably stabilizes the membrane. Various fishes have different response patterns and the whole receptor system is nicely adapted to the particular discharge frequency of the "radar" system of the fish.

Central nervous system Adaptive responses to electrical pulses and fields in the environment depend on an intricate network of neural centers. The electroreceptors provide the input while the cerebellum seems to be the major center for the processing of information and the initiation of appropriate commands. The cerebellum is conspicuously large in all electric fishes and reaches a climax in the mormyrids, where other parts of the brain are completely covered by its extensive lobes (Bennett,

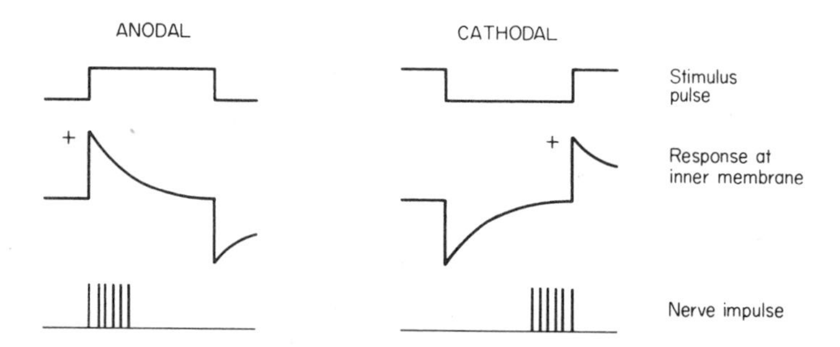

Fig. 20.8 Diagram showing the probable response at the synaptic membrane and in the afferent nerve of a phasic receptor when a voltage pulse is applied at the external opening. Positive intracellular potentials and anodal currents indicated by upward deflections.

1971b)—a development which is superficially similar to the cerebral hemispheres of mammals.

Retrospect At some point in the phylogeny of fishes, electrosensitivity evidently appeared in certain mechanoreceptors of the lateral line. This initiated a long and intricate evolutionary trail which culminated in tonic and phasic electroreceptors of great sensitivity, and the development of massive integrating centers in the brain; here, the sensory impulses are received and appropriate messages are relayed to the effector organs—particularly the muscles and electrocytes. The latter are effectors unique to the world of fishes; they almost always arise from the myoneural apparatus and seem to have evolved as a consequence of the electrosensitivity in the lateral line. This phylogenetic trail is marked by many curious modifications of external anatomy as well as neuromuscular and sensory organs. As a basis for a special system of communication and a unique "radar" system it has permitted added dimensions in the ecological and behavioral adaptations of certain fishes.

LUMINESCENT ORGANS

In some chemical reactions the changes in free energy lead to an emission of light. Electrons of reactant molecules are hoisted to a higher energy level (excited) and, in returning to the ground state, give off their excess energy as quanta of light or photons. Many examples of chemiluminescense have now been studied in both physical and biological systems. The release of radiant energy as a consequence of metabolism occurs in some of the bacteria and fungi and in certain representatives of all the major phyla. Although widespread, the distribution is sporadic and does not seem to follow any definite evolutionary pattern. In some organisms the role of bioluminescence is clearly defined; in others its function is obscure, and its presence may be a fortuitous consequence of tissue metabolism (Harvey, 1952, 1960; McElroy and Glass, 1961; Johnson and Haneda, 1966).

Bioluminescence, like bioelectricity and animal heat, seems to have evolved as a byproduct of tissue metabolism. Its scattered distribution among the lower forms, the varied and curious functions of luminescent organs and the emission of light without obvious function suggest a phylogeny based on widespread biological processes. Some of the best evidence for this concept comes from studies of the luminescent bacteria. Many different species are recognized, and several of these (particularly representatives of the genera *Photobacterium* and *Achromobacter*) have been isolated and grown in cultures for many years. They are common on dead fish or spoiling meat. The importance of oxygen to their luminescence was recognized by Robert Boyle three centuries ago, long before the metabolic basis of living light was appreciated. Boyle noted that very small amounts of "air" were required to maintain the glow of "shining meat," and we now know that the important constituent of the air is oxygen and the reaction an oxidative one, requiring in addition flavin mono-

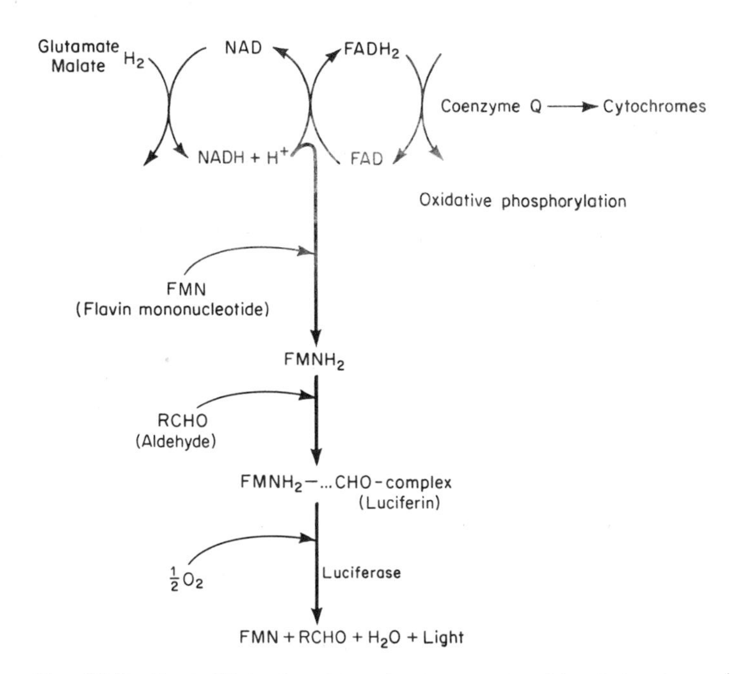

Fig. 20.9 Bacterial luminescence is shown as a side chain of the electron transport system of oxidative phosphorylation. Luciferin is a complex of flavin mononucleotide and an aldehyde; this is oxidized in the presence of the enzyme luciferase; the suggested sequence is still speculative (Cormier and Totter, 1968). [Based on McElroy and Seliger, 1962: *In* A Symposium on Light and Life, McElroy and Glass, eds. Johns Hopkins, Baltimore.]

nucleotide, a long-chain aldehyde and a specific enzyme or luciferase. The biochemical steps in bacterial luminescence are thought to be linked to the electron transport chain of oxidative phosphorylation as indicated in Fig. 20.9.

The metabolic significance of these reactions is unknown, and the value of the light to the bacteria is questionable. It has been suggested, however, that such a reaction may have been important in the removal of free oxygen during the early stages of biochemical evolution (McElroy and Seliger, 1962). It will be recalled (Chapters 1 and 7) that life evolved under anaerobic conditions and that the first energy-yielding pathways were anaerobic. Oxygen, liberated in small amounts as a metabolic byproduct, may have been toxic during these stages, and reactions such as those illustrated in Fig. 20.9 may have effectively removed it. The biochemical sequence involves several of the universal components of respiratory metabolism (reduced pyridine nucleotide and riboflavin phosphate) together with a specific enzyme and an aldehyde. If this speculation is valid, the light emission of present-day bacteria is probably a vestigial process since oxygen removal now serves no useful purpose. In support of this concept, mutant strains of some of these bacteria have been shown to lack one or the other of the constituents required for luminescence without obvious disadvantages to their growth and metabolism (Harvey, 1960). There are distinct "bright," "dim" and "dark" mutants. One of the dim strains becomes bright on the addition of a flavin to the culture fluid; another particularly interesting dark mutant seems to lack only the long-chain aldehyde and glows brightly when minute amounts of dodecaldehyde are supplied. The enzyme (luciferase) is evidently present, but a part of its substrate is absent. These bacterial studies support the theory of an evolution associated with oxidative cellular processes but quite divorced from any significance with respect to the emission of light.

It would be bold to argue that the phylogeny of animal bioluminescence had any direct connection with the processes of bacterial metabolism just described. The biochemistry is different in its specific details, and it seems more logical to assume an independent evolution in many different phyla, with the initial stages derived from the byproducts of different metabolic processes. Luminescent earthworms of the genus *Eisenia* provide suggestive evidence for this hypothesis (Harvey, 1960). When disturbed, *E. submontana* exudes a yellowish slime from dorsal pores which open from its body cavity; on contact with the air this glows with a yellowish-green light. The pigment appears to be riboflavin. The significant point is that a closely allied species of worm (*E. foetida*) releases a yellowish fluid of similar biochemical nature when disturbed, but this is non-luminous. It seems likely that the fluid plays an adaptive role in the biology of these animals but that the light-emitting reaction is coincidental.

Distribution of Bioluminescence

Biologists owe a great debt to E. Newton Harvey and his associates who, for more than half a century, accumulated information on the distribution of luminescent organisms, studied the biochemistry of light emission, and gathered evidence concerning its biological significance and probable phylogeny. His monograph, published in 1952, contains a wealth of information on the distribution and biology of animals which emit light. In summary, most of the phyla of free-living animals have luminescent representatives, and these are particularly numerous among the larger phyla—protozoa, coelenterates, annelids, molluscs, arthropods and vertebrates. In all, some forty to fifty different groups emit light (Harvey, 1960). The majority are marine animals, ranging in habitat from the surface to the depths where only animal light flashes in the abyssal blackness. In contrast, the fresh waters are devoid of bioluminescence except for certain bacteria and a limpet, *Latia neritoides,* in the lakes and streams of New Zealand. In some parts of the world the terrestrial habitat is brightened at night by the flashing of fireflies, glowworms and lightning bugs (all members of the Coleoptera—beetles); in addition, there are several other orders of insects, centipedes, millipedes, earthworms and one land snail capable of producing light. However, this is a relatively small array in contrast to the marine species; there are no luminescent organs among the terrestrial vertebrates or plants (other than bacteria and fungi).

Biochemistry of Luminescence

The pioneer experiments were performed by the French physiologist Raphaël Dubois (1885, 1887). He used the photogenic organs of the West Indian elaterid beetle *Pyrophorus* for the first of these. When brightly luminescent organs were plunged into boiling water, the lights were extinguished, but the water, if quickly cooled, contained something which produced light when mixed with an extract prepared from similar organs by triturating them in cold water and letting the mixture stand until the luminescence had disappeared. These classical experiments were followed by comparable and much more detailed studies of the bivalve mollusc *Pholas dactylus.* This animal has the habit of boring in soft rock and thus creating the tunnels where it lives; its luminescence has excited the interest of naturalists since the days of Pliny (Harvey, 1952). Dubois' investigations of *Pholas* extended over a period of more than forty years. They have been summarized many times; Harvey (1952) has provided an extended bibliography.

Dubois coined the term LUCIFERIN for the principle in his hot water extracts and LUCIFERASE for that in the cold extracts. Very early in these studies he concluded that light production took place when luciferin was oxidized by molecular oxygen in the presence of a catalyst (luciferase). Subsequent research has shown that he was correct. Bioluminescence is an instance of chemiluminescence in which complex organic molecules (luciferins) are oxidized in the presence of specific enzymes (luciferases); molecular oxygen is almost always required for the reactions (Cormier and Totter, 1968). More than sixty years of work since these early studies of Dubois has demonstrated the varied nature of the organic molecules involved but confirmed the universality of a generalized luciferin-luciferase reaction. The three groups of or-

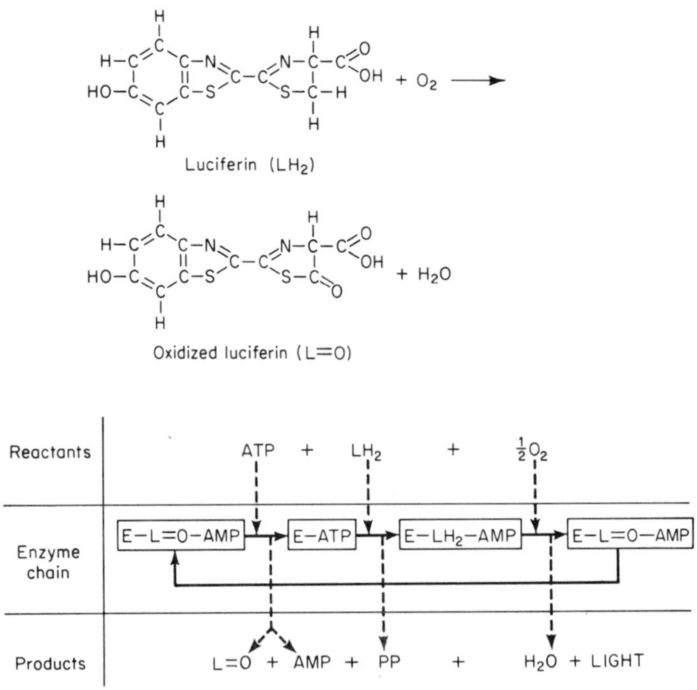

$+ O_2 \longrightarrow$

Luciferin (LH_2)

$+ H_2O$

Oxidized luciferin (L=O)

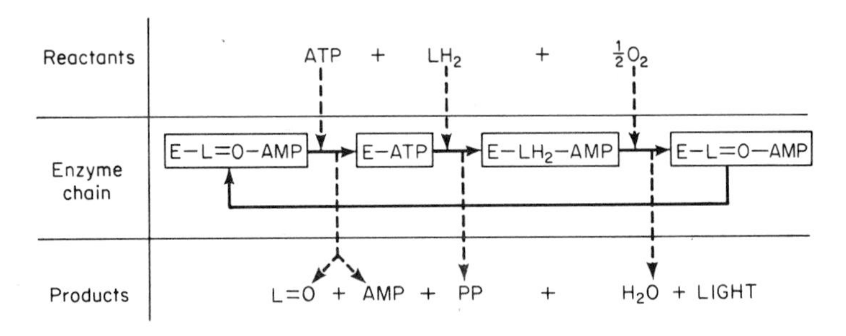

Fig. 20.10 Structural formula of firefly luciferin (above) with the suggested chemical sequence of luminescence below. *E*, luciferase enzyme; reactions also require Mg^{++}. [Based on McElroy and Seliger (1962).]

ganisms most carefully investigated biochemically are the luminescent bacteria (described above), the fireflies and the marine crustacean *Cypridina*.

The structural formula of firefly luciferin (Fig. 20.10) was established after almost half a century of continuous research (Harvey, 1914; McElroy and Seliger, 1961, 1962). Its chemical structure has now been confirmed by synthesis, and the associated luciferase has been purified and shown to contain about 1000 amino acid residues. The reaction requires magnesium ions and phosphate bond energy; ATP is split to form adenosine monophosphate and inorganic pyrophosphate as summarized at the bottom of Fig. 20.10. The reaction occurs in an aqueous medium; dried lanterns from fireflies which have been dead for many years will luminesce brightly when broken and moistened. This is also true of many other bioluminescent tissues. During wartime, the Japanese soldiers turned the reaction to practical use by pulverizing dried bodies of tiny crustaceans (*Cypridina*) in their hands and moistening them to provide sufficient light for map reading (Harvey, 1952).

Cypridina hilgendorfii, a small (2 to 3 mm long) ostracod from Japan, has long been a favored object for the investigation of the luciferin-luciferase reaction. Harvey, who first introduced it to research workers in 1916, describes the biology of this interesting little animal together with a history of the biochemical studies (Harvey, 1952, 1960). Structurally, ostracod luciferin has been shown to be composed of residues of tryptamine, arginine and isoleucine linked together in an unusual manner as follows (Shimomura *et al.*, 1969):

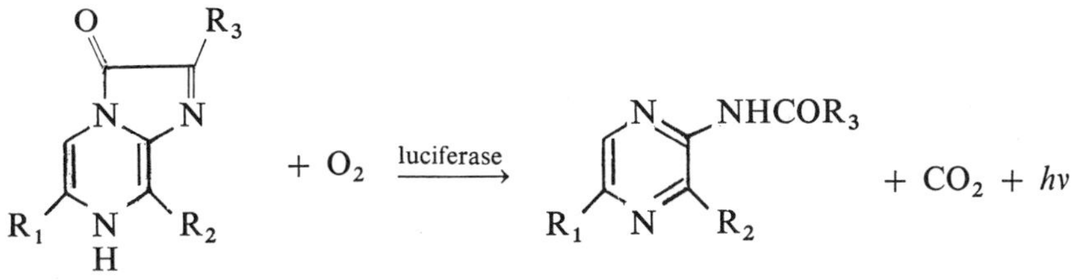

Luciferin $+ O_2 \xrightarrow{\text{luciferase}}$ Oxyluciferin $+ CO_2 + h\nu$

R_1 = 3-indolyl
R_2 = $-(CH_2)_3NHC(=NH)NH_2$
R_3 = $-CH(CH_3)CH_2CH_3$

The oxidation of this material also requires molecular oxygen but does not depend on the presence of Mg^{++} and ATP. Many other luciferins are now under active investigation, and it is apparent that the oxidizable compounds and reactants are as different as the groups of animals which produce them (Cormier and Totter, 1968).

Light emission in the coelenterates depends on a photoprotein instead of the usual ("luciferin-luciferase") system. This protein luminesces with Ca^{++}; the reaction is independent of the oxygen tension. Energy transfer mechanisms vary in different coelenterates (Morin and Hastings, 1971) but the presence of a calcium-activated photoprotein is constant. This system was first identified in the hydromedusa *Aequorea*, where the photoprotein was identified as AEQUORIN; more recently about two tons of jellyfish yielded 125 mg of electrophoretically pure protein from which 1 mg of the light-emitting moiety was obtained and tentatively identified as follows (Shimomura and Johnson, 1972):

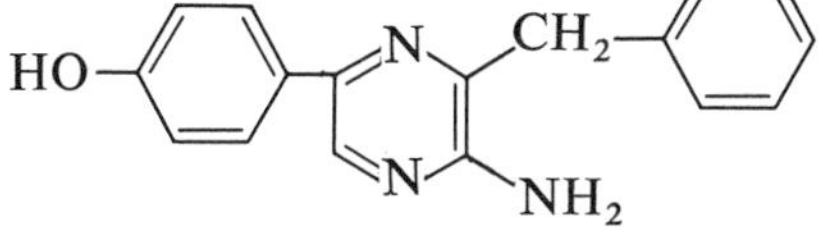

HO–C_6H_4–(pyrazine)–CH_2–C_6H_5, NH_2

Other unusual systems have also been identified. *Balanoglossus* luciferase seems to be a peroxidase, with peroxide rather than oxygen required for the emission of light. These curious exceptions to the usual type of reaction, first described by Dubois, emphasize an opportunism in the evolution of luminescent processes and suggest that there may still be many surprises for the biochemist who studies them. Recent work is summarized by Johnson and Shimomura (1972).

Physical Properties

The colors of living lights vary all the way from blues, ranging from 410 nm in some bacteria, polychaetes and teleosts, to the reds of some beetles which extend from 650 nm through orange and yellow to yellow-greens. Bioluminescence is most frequently blue, blue-green or white; green and yellow are less common; orange and red are comparatively rare (Nicol, 1962). A few animals possess two kinds of luminous organs, each emitting its particular color of light. The "railroad worm" (larval stage of *Phixothrix*) from Central and South America emits a greenish-yellow light from 11 pairs of luminous spots on the posterior lateral margins of the second thoracic to the ninth abdominal segments of the body, and a bright red glow from paired organs on the head. The two sets of lamps operate independently, and when only the headlights are on, the organism looks like a glowing cigarette; when disturbed and crawling, the green lights also flash on to suggest a moving train with red head lamps (Harvey, 1962; McElroy and Seliger, 1962).

Spectral emission curves are usually rather narrow and sharply peaked; occasionally they are broad and sometimes bimodal in form (Fig. 20.11). The light is always of low intensity, although the brilliant flashes of the ctenophore *Mnemiopsis* may be one to two million times brighter than the dim flashes of the dinoflagellate *Noctiluca,* when comparisons are made on the basis of comparable areas of receptor surface and distance. For the multicellular animals this range can be reduced by a factor of 10 to 100 times. Nicol (1960, 1962, 1969) has summarized data on spectral quality, brightness of illumination and various temporal characteristics of the flashes.

Morphological Correlates

"There are always two aspects to any evolutionary problem: (1) the first beginnings of a new organ, in this case the appearance of chemical reactions which emit

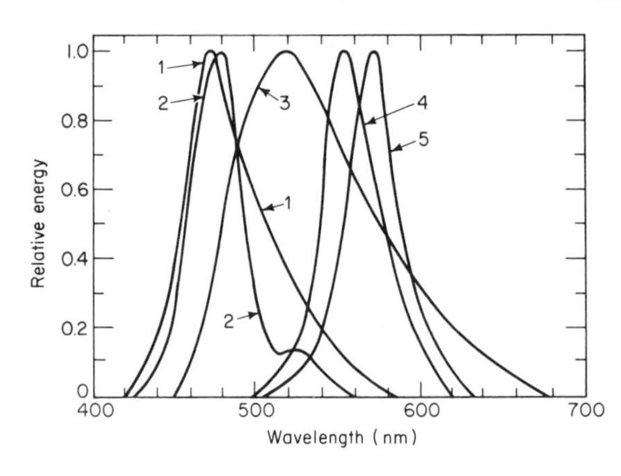

Fig. 20.11 Some relative spectral emission curves: (1) *Noctiluca miliaris*; (2) *Euphausia pacifica*; (3) polynoid worms; (4) *Photuris pennsylvanica*; (5) *Photinus pyralis*. [From Nicol, 1962: Adv. Comp. Physiol. Biochem., *1*: 260.]

light; (2) the further evolution of accessory structures, in this case the development of lenses, reflectors, and pigment screens" (Harvey, 1960). In no system is this more completely illustrated than in the bioluminescent organs which have achieved considerable variety and great complexity in most of the phyla where luminescent organs occur. Many of these beautiful lanterns have been described in detail and are well illustrated in books by Harvey (1952) and Nicol (1967a). There are three fundamentally different patterns: SPECIAL TISSUES OR ORGANS UTILIZING SYMBIOTIC BACTERIA, the discharge of luminous secretions (EXTRACELLULAR LUMINESCENCE) and the presence of light-emitting cells (INTRACELLULAR LUMINESCENCE). A few animals possess both types of intrinsic luminescent tissue.

Use of symbiotic bacteria Some of the myopsid squids and a few families of deep-sea teleosts depend on symbiotic bacteria which grow in special tissue sacs; these sacs may be provided with reflectors, lenses and pigment curtains to control the emission of the light. Fishes of the family Anomalopidae have a conspicuous elongated organ just below the eye. This is formed of long glandular tubes filled with luminous bacteria. Numerous blood vessels supply the tubes. The organ has pores to the outside and a reflector layer at the rear. The bacteria luminesce continuously, but the light can be intermittently concealed in some species by turning the organ downward until its light surface is covered by a pocket of black pigmented tissue, while in others a fold of black tissue can be drawn up over the light surface like an eyelid.

Extracellular luminescence Luminous secretion has been described in representatives of all the major groups of light-emitting animals, from coelenterates and nemerteans to the balanoglossids and fishes. It is common among marine invertebrates but rare among the terrestrial invertebrates and fishes. Exceptions among the latter groups include some myriapods and oligochaetes and a single group of fishes, the alepocephalid genus *Searsia.* Secretions may appear as a luminous slime over the surface of the body as in balanoglossids, or as discrete scintillating points of light in the beautiful little transparent nudibranch *Phyllirrhoe bucephala;* often these exudates are suddenly released in a brilliant luminescent cloud as in some of the pelagic shrimps, in the boring bivalve *Pholas dactylus,* in the polychaete *Chaetopterus* or in the fireworm *Odontosyllis.*

The photocytes are elongated flask or club-shaped cells. They usually occur as patches of unicellular glands interspersed with mucus cells (Fig. 20.12). Sometimes they are grouped into compact masses which open through a common duct like an alveolar or tubulo-alveolar gland. The epithelial area concerned may be epidermal, as in *Chaetopterus* and *Pholas,* or it may be coelomic, as in the earthworms or pelagic shrimps (green gland secretion).

Most of the details of extracellular luminescence remain unsolved. The biochemistry and morphology have been described in relatively few forms; physiological regulation is understood in only general terms. Extrusion could be dependent on associate muscle fibers as illustrated in Fig. 20.12, but the photocyte itself may be

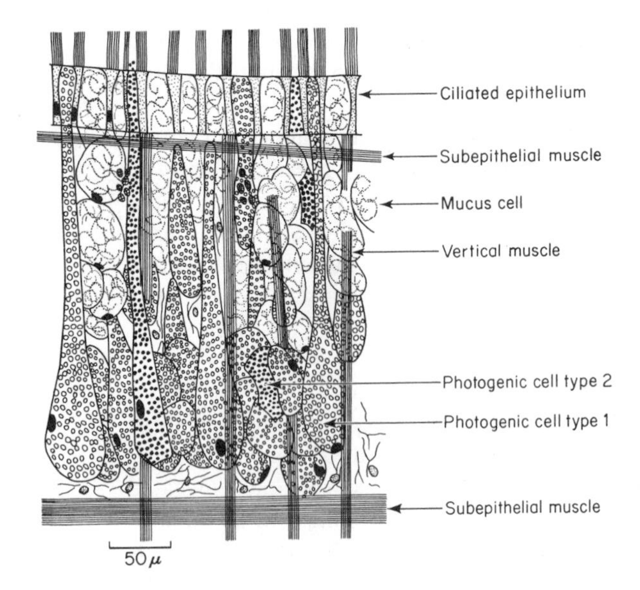

Fig. 20.12 Cross section through tissue responsible for extracellular luminescence in *Pholas dactylus*. [From Nicol, 1962: Adv. Comp. Physiol. Biochem. *1*: 252.]

responsible since nerve endings have been traced directly to it in some forms (Nicol, 1962). Extracellular luminescence is usually intermittent, and nervous control has been satisfactorily demonstrated in certain forms, from coelenterates throughout the phylogenetic series (Nicol, 1960, 1962); but details of receptors, pathways and transmitter substances remain to be elucidated. In some animals, such as *Cypridina,* two types of granular cells are evident, and it is presumed that luciferin and luciferase are separately discharged and mixed in water; in many forms no such morphological separation of cell types is apparent, and the manner in which enzyme and substrate are brought together remains obscure.

Intracellular luminescence The greatest structural complexities occur in photophores where light is produced intracellularly. In many cases the light from groups of photocytes is concentrated by an arrangement of mirrors and lenses to sparkle like a brilliantly lighted jewel in the darkness. Some of these organs (photophores) look very much like eyes in which the receptor surface or retina has been replaced by a photogenic light-transmitting surface (Fig. 20.13).

Luminescence is almost always intracellular in the more advanced groups such as fishes, terrestrial arthropods and cephalopods. It also occurs in some more primitive groups where there are no specialized photophores (Nicol, 1962). Several species of protozoans luminesce when stimulated. In the dinoflagellate *Noctiluca miliaris* photogenic granules of two sizes are arranged along strands of protoplasm which radiate from a concentrated area of granules near the oral groove; these join together around the periphery so that the cell, in surface view, looks like a shimmering fan (Harvey, 1952). The granules seem to be permanently located, but the flashing is intermittent and passes over the cell in waves which originate near the oral groove.

The luminous polynoid worm *Acholoë astericola* is another carefully described

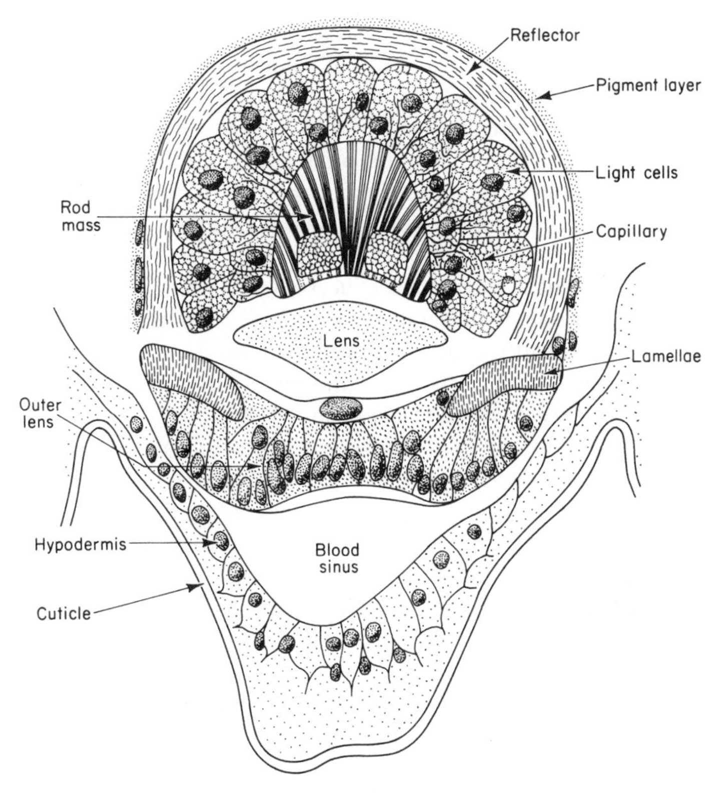

Fig. 20.13 Photophore of the euphausid *Nyctiphanes norvegica.* Light-emitting surface at bottom. [Based on Harvey, 1952: Bioluminescence. Academic Press, New York; after Dahlgren, 1916: J. Franklin Inst., *181*: 823.]

example in which intracellular photogenic tissues are diffuse rather than concentrated in photophores. Luminescence is confined to elytra or scale-like plates which cover the body segmentally. The photocytes occur in a single layer just above the cuticle of the lower surface of the scale and are well supplied by branches of the elytral nerve. When the animal is irritated, a flash of light runs along the scales, passing from segment to segment; sometimes elytra are detached, and the animal swims away leaving its glowing scales behind it. Characteristically, the luminous polynoids have easily detached scales and swim with rapid movements (Harvey, 1952).

Intracellular luminescence is no better understood than extracellular luminescence; problems associated with the physiological control are similar in both. Intracellular light is always intermittent. In the multicellular forms, control by nerves is well documented at all levels in phylogeny, and there is no evidence of hormonal influence (Nicol, 1960, 1969).

The physiological regulation has been most carefully studied in the fireflies where flashing lights form a complex signaling system between the males and females. The precise timing of the flashes and their relation to an intricate mating sequence indicate a nervous control, but the details of its organization are still somewhat speculative. Two anatomical systems have long figured prominently in the theories. One of these is the tracheolar system which in most species provides the photocytes with numerous closely applied air capillaries (Buck, 1948); the other is the innervation which is well developed and often parallels the tracheolar tubules. Two different theories have been current for a long time (Buck, 1948; Carlson, 1969). According to one of these, nervous excitation leads to a sudden release of air, and this puff of oxygen to the chemical reactants produces a flash; according to the other theory, adequate oxygen is normally present, but release of the transmitter substance acetylcholine triggers the chemical reaction. Current evidence has decided in favor of the transmitter theory although uncertainties still remain about both the identity of the transmitter and the details of its action (Carlson, 1969).

Physiological controls of other complex photophores have not been at all well investigated. Among fishes, only the midshipman *Porichthys* has been carefully studied. Nicol (1969) reviews the older literature. There is morphological as well as physiological evidence for a nervous control from centers in the spinal cord—probably mediated through sympathetic nerves. Although a hormonal regulation by adrenaline has been suggested, the evidence is largely against it.

Functional Significance of Bioluminescence

Bioluminescence has been woven into the lives of many animals in ways that affect the survival of individuals or the reproductive success of the species. Sometimes it seems as though there may be no functional significance but, although this was suggested in the earlier section on phylogeny, the concept is never completely acceptable; the lack of apparent functional significance for the light may reflect limitations in knowledge concerning the species. The literature already cited contains abundant evidence for the biological importance of living light. A few examples have been selected to indicate the various ways in which it is used.

One of the most striking examples of the use of BIOLUMINESCENCE IN FOOD GATHERING is provided by the glowworms of the famous caves at Waitomo about 200 miles north of Wellington, New Zealand (Harvey, 1952). These glowworms are dipteran larvae (Bolitophilidae) which live in vast numbers on the ceilings of the caves. Each spins a long, sticky, glistening thread which hangs down 15 to 60 cm and serves as a trap for other insects. The glowworms are carnivorous, and this unique method of trapping their food probably depends on attraction of prey by the lights as well as the air currents which move through the caves. The lights are rapidly extinguished if there is any unusual disturbance in the cave.

A number of marine invertebrates emit light in ways which suggest a significant role in FACILITATING ESCAPE FROM PREDATORS. Some, such as the deep-sea shrimp *Systellaspis* or the squid *Heteroteuthis*, suddenly discharge a cloud of luminous secretion when irritated, and this might confuse a predator and permit an escape into the darkness. Easily detachable luminous scales of the polynoid worms might serve in a similar manner to distract a predator while the remainder of the animal swims away. These possibilities have never been tested experimentally.

There are several well-known instances of bioluminescence during sexual behavior; in some cases, the light plays a part in the TIMING OF REPRODUCTION and in synchronizing the activities of the males and females; in other cases, well-defined photophore patterns or specific timing of flashes seem to serve as INTRASPECIFIC RECOGNITION SIGNALS. The swarming of fireworms, *Odontosyllis* (Polychaeta), is timed by lunar and diurnal light rhythms (Chapter 14), but the mating sequence seems to be controlled by the flashing of living lights. The much larger females commence swimming in the surface waters at a precise time in the evening; they suddenly become brilliantly phosphorescent as they swim rapidly in small circles two to three inches in diameter, discharging eggs and luminous secretions to form halos into which the males, emerging from deeper waters, dart to discharge their sperms (Harvey, 1952).

The mating of fireflies often depends on a very accurate signaling system (Buck, 1948; Lloyd, 1966; Buck and Buck, 1968, 1972; Papi, 1969). Females of the common Eastern North American species (*Photinus pyralis*) do not fly about but crawl onto a blade of grass in the evening and wait in the darkness for a proper signal. The males flash their lanterns as they fly approximately 50 cm above the ground; if a female sees one of these flashes within three to four meters, she may be expected to flash back after an exact interval (2 sec at 25°C). This attracts the male in her direction. After four or five exchanges of signals the male reaches the female and mating occurs. In this species, recognition evidently depends on the flash interval. Other species have somewhat different signaling systems; females of the European glowworm *Lampyris noctiluca* are flightless and emit a long-lasting light which attracts the much less brilliantly luminous males; in some species the males lack luminescent organs. These signaling systems of the fireflies have reached a high level of specialization. Other groups of animals may have equally complex light-emitting behavior, but there is at present no accurate information. The teleost fishes, which are assumed to make use of their highly complex lanterns, are deep-sea animals rarely seen by man. Sexual differences in the patterns of the photophores and in the colors of their lights are suggestive of sex recognition structures, but no experimental work has been done, nor are there any satisfactory observations of mating behavior. The significance of the light to most of the luminescent animals is still an unexplored field of animal behavior.

PIGMENT EFFECTOR CELLS

Pigmented compounds are found in animals at all levels in phylogeny. In the photoreceptors, the carotenoids capture radiant energy and through photochemical changes excite nerves; metalloporphyrins and metalloproteins as blood pigments combine with oxygen and store or transport it to serve aerobic respiration; the body coverings of multicellular animals are usually pigmented, and it is a matter of general observation as well as careful experiment that such colorations are frequently cryptic and often provide recognition marks of importance in behavior. In contrast to the visual and respiratory pigments, the integumentary pigments are extremely varied biochemically, ranging from the almost ubiquitous melanins and carotenoids to the less common quinones, pterins and flavins (Fox and Vevers, 1960; Needham, 1974).

Most multicellular animals are characterized by a distinct color and often by a particular color pattern. These pigmented patterns are static or change only slowly from juvenile to adult stage, or with sexual maturation, or seasonally in accordance with the formation or destruction of pigment (MORPHOLOGICAL COLOR CHANGES). They may also change gradually under selection pressures, as in the slow replacement of the typical light-colored pepper moth *Biston betularia* by a black melanotic variety in the industrial areas of Britain (Kettlewell, 1961).

In addition to these fixed or very slowly changing colors, several groups of animals have a specialized system of effector organs which can rapidly alter the amount of exposed pigment in the integument so that the animal appears darker or lighter or assumes a matching background pattern. These are PHYSIOLOGICAL COLOR CHANGES and depend on specialized effector cells (chromatophores) in which the pigment is moved about in response to specific stimuli. This capacity is most highly developed among cephalopods, crustaceans, and the poikilothermous vertebrates—especially teleost fishes and lizards. Some annelids and echinoderms become lighter during the day and darker at night, while a few insects adapt to the color of their background. But these groups are not characterized by species which habitually alter their colors to match their surroundings. Chromatophores as active effectors are not found among birds and mammals.

Morphology of Chromatophores

Two types of effector are concerned with physiological color change. One of these, evidently confined to the molluscs, is a minute organ consisting of a sac-like cell surrounded by a stellate series of radial muscle fibers (Fig. 20.14). The central cell contains a highly elastic sacculus filled with pigment granules. Around the equator of the cell, the elastic sacculus is attached to the cell membrane which is in turn anchored to the radial muscles. The plasma membrane is intricately folded in the retracted state. Contraction of the radial muscles stretches the whole cell, including the elastic sacculus, to form a sheet of highly pigmented cytoplasm; on muscle relaxation, the elasticity of the sacculus quickly changes the shape from a flat disc to a small sphere. Thus, the rapid contractions and relaxations of these obliquely striated radial muscles, by changing the shape of an elastic sac of pigment, quickly alter the color of the animal (Florey, 1969). The scintillating colors which characterize some of the cephalopods depend on these small organs. A similar but much more slowly operating structure has been described in some of the nudibranchs (Nicol, 1964).

The other type of pigment effector is a single, irregularly shaped cell containing

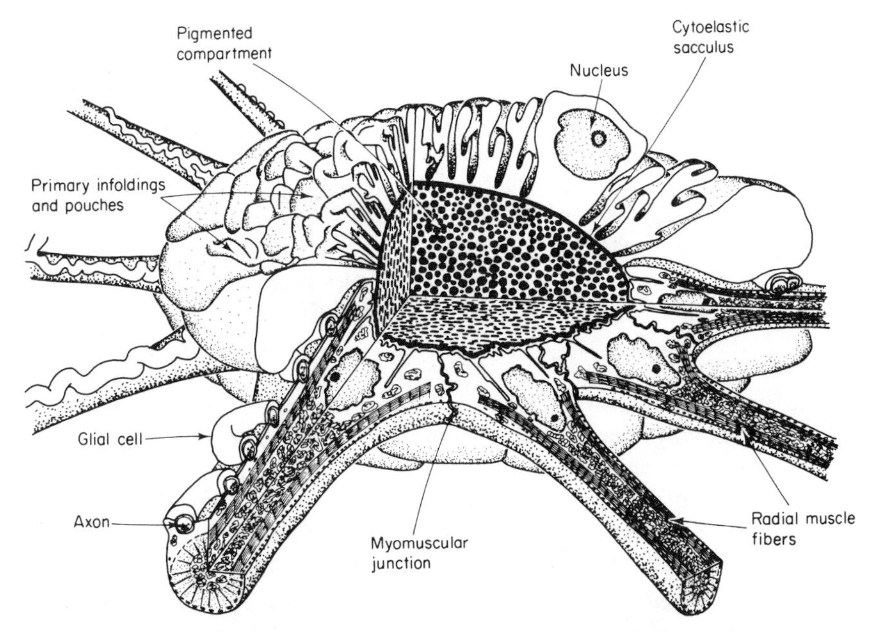

Fig. 20.14 The retracted chromatophore organ of the squid *Loligo opalescens*. [From Cloney and Florey, 1968: Z. Zellforsch., *89*: 254.]

pigment granules, which can be concentrated in a small area or dispersed throughout the protoplasm. Only one kind of tissue is present in the second type of effector, and movement of the pigment does not depend on extracellular contractile elements. The cell boundaries are indistinct, and special techniques are required to make them visible. Shapes of chromatophores vary from lenticular or plate-like forms to intricate arborizing and stellate structures. As indicated in Figs. 20.15 and 20.16, pigment granules move out of the processes or away from the periphery to concentrate in a small area in the central part of the cytoplasm. Rapidity of movement and degree of concentration depend on the type of chromatophore and the nature of its physiological controls. Much of the quantitative work has been based on a five-point MELANOPHORE INDEX (Fig. 20.15), first used by Hogben and Slome (1931) for amphibian chromatophores but subsequently adapted to other groups. Several different photometric methods are discussed by Fujii (1969).

Four kinds of chromatophores have long been distinguished on the basis of color and the biochemical nature of their pigments. The distinctions were first made when it was thought that all black and brown pigments were melanins while the reds and yellows were carotenoids and the silvery deposits were guanine. On this basis, brown and black pigment cells were called MELANOPHORES; the red chromatophores were called ERYTHROPHORES; the yellow ones XANTHOPHORES; the silvery ones GUANOPHORES or IRIDOPHORES. These terms are still satisfactory, although is it now clear that chromatophore color may depend on several pigments other than the melanins and the carotenoids. The beautiful colors of the cephalopod, for example, are neither

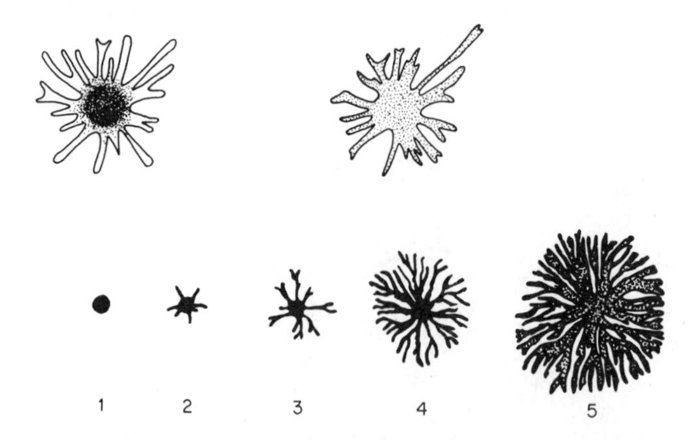

Fig. 20.15 Diagrams of fish chromatophores. Upper, melanophores of *Fundulus* with pigment aggregated on the left and dispersed on the right. Lower, the five-point melanophore index: (1) punctate; (2) punctostellate; (3) stellate; (4) reticulostellate; (5) reticulate. [Upper after Matthews, 1931: J. Exp. Zool., *58*: 475; lower after Healey, 1951: J. Exp. Biol., *28*: 300.]

melanins nor carotenoids but ommochromes. This is a varied group of pigments which may appear black, brown, red, orange or yellow. The iris pigments of insects and crustaceans are ommochromes, as are also some of their body colors. Pterins form another group of pigmented substances which may be responsible for the yellow, orange and red tints. Pterins are common in fishes and amphibians; they have also been found in crustaceans and insects. The semantics of chromatophore terminology

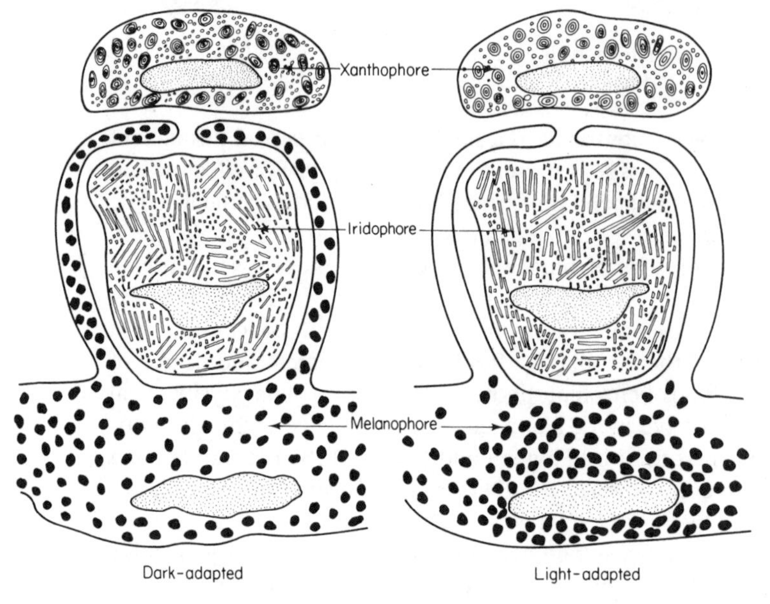

Fig. 20.16 The frog dermal chromatophore unit consisting of three cells: a xanthophore (just under the epidermis), an iridophore, and an innermost melanophore with finger-like processes extending around the iridophore. Note the lamellated pterinosomes and the carotenoid-containing vesicles in the xanthophores. [From Breathnach, 1971: Melanin Pigmentation of the Skin. Oxford U.P., London; based on Bagnara, Taylor, and Hadley, 1968: J. Cell Biol., *38*: 67.]

is discussed in several recent reviews (Fujii, 1969; Fingerman, 1970; Bagnara and Hadley, 1973). The familiar terms—melanophore, erythrophore, iridophore and xanthophore—will be used here without any implication as to the biochemical nature of their pigments.

Studies of ultrastructure have removed many longstanding uncertainties about the internal organization of chromatophores. The pigment is contained in discrete cellular organelles bounded by definite membranes. The MELANOSOME is a spherical electron-dense package of melanin which varies in diameter but averages about 0.5 μ; PTERINOSOMES are spherical or elliptical granules of pteridine, of about the same size as melanosomes but showing definite concentric lamellae (Fig. 20.16). The organelles of the iridophores are usually called REFLECTING PLATES (or LAMELLAR RIBBONS in cephalopods); they are purine in nature and vary in size from small motile granules to large square or octagonal plates, which appear to be stationary within the cell and may reach almost 2.0 μ in length and 0.2 μ in thickness. The plates are precisely oriented to reflect the light in definite directions (Fig. 20.16). The fat-soluble carotenoids are contained in VESICLES. It should be noted that more than one kind of organelle may occur within a pigment cell although a single type usually dominates; melanophores may contain pterinosomes; xanthophores may house melanosomes; other combinations occur and pigment cells containing all three types of organelle are recognized. Finally, it is apparent that different types of chromatophores may be grouped in a definite manner within the skin to form CHROMATOPHORE UNITS or COMPOUND CHROMATOPHORES; Parker (1948) called these multicolored structures CHROMATOSOMES. The chromatophore units of amphibians and reptiles are beautifully illustrated in the ultrastructural studies of Bagnara and his associates (Bagnara and Hadley, 1973).

Chromatophore Pigments

The chromatophore pigments are the melanins, the ommochromes, the carotenoids, the pteridines (pterins) and the purines. Several other groups of chemicals are also involved in the varied color displays of animals (Fox and Vevers, 1960; Needham, 1974), but the present description is confined to those which are common in chromatophores.

Melanins Melanin is formed through the oxidative metabolism of the aromatic amino acid tyrosine. A copper-containing enzyme, tyrosinase (also called phenol or polyphenol oxidase) catalyzes the initial oxidation to 3, 4-dihydroxyphenylalanine (DOPA). Several of the subsequent steps are outlined in Fig. 20.17, but some details are still uncertain. It should be realized that melanin is not a single compound but a group of polymers of different sizes (Thomson, 1962). Their colors are as variable as those of human hair. In chromatophores, the melanin granules may be linked to protein.

The metabolism of tyrosine and phenylalanine was already mentioned in con-

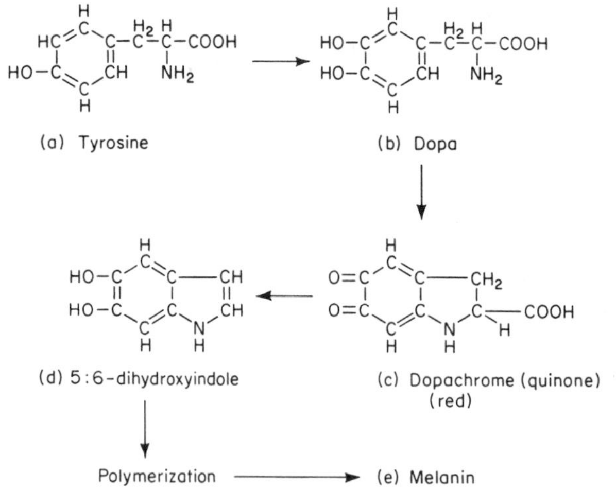

(a) Tyrosine → (b) Dopa

↓

(d) 5:6-dihydroxyindole ← (c) Dopachrome (quinone) (red)

↓

Polymerization → (e) Melanin

Fig. 20.17 Several steps in melanogenesis. See also Fig. 2.8.

nection with the biosynthesis of noradrenaline and adrenaline (Chapter 2, Fig. 2.8). These hormones or transmitter substances, like melanin, arise from tyrosine by way of "DOPA"; it is of interest that the cells concerned with melanogenesis in vertebrates, as well as the tissues of the adrenal medulla, develop from the neural crests. Embryonic cells with the capacity to synthesize melanin are called MELANOBLASTS; both the MELANOPHORES (pigment effector cells) and the MELANOCYTES are their direct descendants. The latter are also responsible for melanogenesis but do not show pigment movement (Gordon, 1959). They are found in various tissues throughout the vertebrates and were identified in Chapter 14 as responsible for melanogenesis in human skin. The melanocytes of mammalian skin occur at the dermo-epidermal junction, with dendritic processes extending into the stratum germinativum to which they pass melanin as described in Chapter 14. They are usually colorless cells, but their biosynthetic characteristics may be easily revealed histochemically by the "DOPA" reaction (Ham, 1969).

Melanogenesis is not confined to the descendants of melanoblasts from the neural crests of vertebrates. It probably occurs in plants (some mushrooms and bacteria) and in at least a few animals at all levels in phylogeny (the anemone *Metridium senile,* the echinoderm *Diadema*, the polychaete *Chaetopterus* and the gastropod *Lymnaea*). However, among invertebrates melanin probably is less common than some other black or brown pigments (Fox and Vevers, 1960; Thomson, 1962). One of the most active melanin-producing tissues is the ink gland of cephalopods. With the exception of *Nautilus*, these molluscs secrete melanin from a glandular mass in the region of the digestive tract and store it in the ink sac. When the animal is excited, ink can be squirted through the anus in a black cloud to serve as a smoke screen behind which the animal escapes. As evidence of the chemical stability of melanin, fossil ink sacs have yielded melanin which was stored in the fossil form for as long as 150 million years.

Ommochromes Until about 1940 all the black or brown pigments were considered to be melanin. First, the brown tanned protein SCLEROTIN was identified in the cuticle of insects and, at almost the same time, ommochromes, and not melanins, were shown to form the iris pigments of insects and crustaceans. Ommochromes can be readily separated from melanins on the basis of solubility and colors in certain specific reagents (Needham, 1974). They are common in molluscs and arthropods and responsible for fixed colors as well as the pigments of chromatophores. Ommochromes have also been identified in the eggs of *Urechis*; they are unknown in the vertebrates and appear to be confined to the protostomes. Ommochromes occur in organelles resembling melanosomes.

There are two classes of ommochromes: the ommatins and the ommins. The former are less complex biochemically and better known (Bagnara and Hadley, 1973). Both are derived from the heterocyclic amino acid tryptophan. The metabolic breakdown of this essential amino acid involves an oxidation with the opening of the pyrrole ring (enzyme, tryptophan pyrrolase) followed by a hydrolysis to form kynurenine. Several of the subsequent steps in the biosynthesis of the yellow pigment xanthommatin are shown in Fig. 20.18. In the higher animals, the products of tryptophan metabolism are excreted as kynurenine, 3-hydroxykynurenine or derivatives (Henderson *et al.,* 1962); in some of the lower forms (particularly the molluscs and arthropods) these substances are converted into integumentary and chromatophore pigments. Information on the comparative biochemistry is still sketchy. The basis of the color differences, ranging from yellow to black, is unknown; the genetics of the enzyme systems has been investigated only in some of the fruit flies (*Drosophila*).

Carotenoids Knowledge of the carotenoids is much more extensive and of much longer standing than that of the ommatins and the ommins. Perhaps because of their

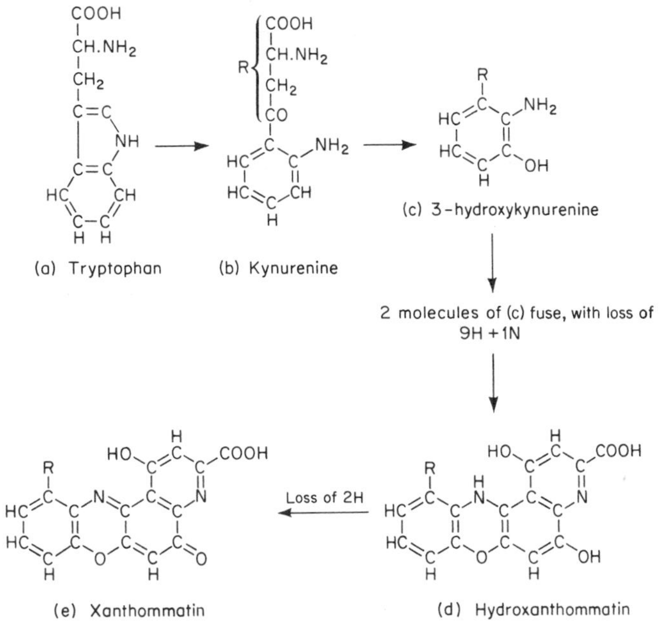

(a) Tryptophan

(b) Kynurenine

(c) 3-hydroxykynurenine

2 molecules of (c) fuse, with loss of 9H + 1N

(d) Hydroxanthommatin

Loss of 2H

(e) Xanthommatin

Fig. 20.18 Biosynthesis of xanthommatin.

ubiquitous distribution and their prominence in plants they have been studied by chemists for well over a hundred years (Chapter 14). No animal is known to synthesize carotenoids; they are obtained directly or indirectly from the plants, and yet at all levels in animal phylogeny they play an essential role in visual processes and in integumentary pigmentation; in some animals they also function in color changes and probably in other processes. The orange, brown and red colors of coelenterates and sponges are often carotenoids, as are also the bright colors in the exoskeleton of crustaceans like the lobster and the crayfish (astaxanthin), or the dermis of the goldfish or the feathers of the canary. In crustaceans the carotenoids share with the ommatins the responsibility for a range of colors from yellow to black; in the vertebrates they function in a similar manner with the melanins and the pterins. The diversity of carotenoid colors may depend on differences in the proteins with which they combine to form carotenoproteins as well as to variations in the carotenoid part of the molecule. Carotenoids may not be present in mollusc chromatophores (Goodwin, 1962). The chemistry of carotenoid pigments was considered in Chapter 14.

Pteridines The pteridines or pterins were discovered during the early part of the present century in the wings of butterflies. The first of these was a yellow pigment XANTHOPTERIN. Its name comes from the Greek roots for yellow and wing. Thus, it records the history of the first biochemical isolation of these substances from lepidopteran wings, almost a quarter of a million of which were used in one of the pioneer studies (Fox and Vevers, 1960).

The pteridines became much more than a biochemical curiosity when it was discovered in 1945 that an important vitamin, folic acid (pteroyl-L-glutamic acid) is a pterin-containing compound. Another essential vitamin, riboflavin (Fig. 7.5) seems to be closely related to the pteridines in its biosynthesis (Forrest, 1962). These substances are described here because of their presence in chromatophores. They vary in color from the white of leucopterin, to the yellow of sepiapterin and the red of drosopterin. They are often associated with carotenoids in determining the yellow, orange and red colors of fishes and amphibians. They have been isolated from several different tissues of crustaceans and may also occur in their chromatophores (Goodwin, 1971).

The pterins are related to the purines from which they may be derived in their biosynthesis. F. G. Hopkins, who discovered them in the latter part of the nineteenth century, thought that he was dealing with a derivative of uric acid (2, 6, 8-trioxypurine) which is present along with isoguanine (2-oxy-6-aminopurine) in butterfly wings. However, the biosynthetic pathways of the pterins are still uncertain (Forrest, 1962). The structures of xanthopterin, biopterin (a coenzyme in several aromatic hydroxylations), sepiapterin from insect eyes, and the vitamin folic acid are shown in Fig. 20.19. The pteridine nucleus (a combination of a pyrimidine and a pyrazine ring), which is characteristic of this group of compounds, appears to be of universal occurrence in living organisms. It is a constituent of certain vital enzyme systems and some animals, including the mammals, are unable to synthesize it (Chapter 3). The microorganisms of the intestinal tract provide a ready source.

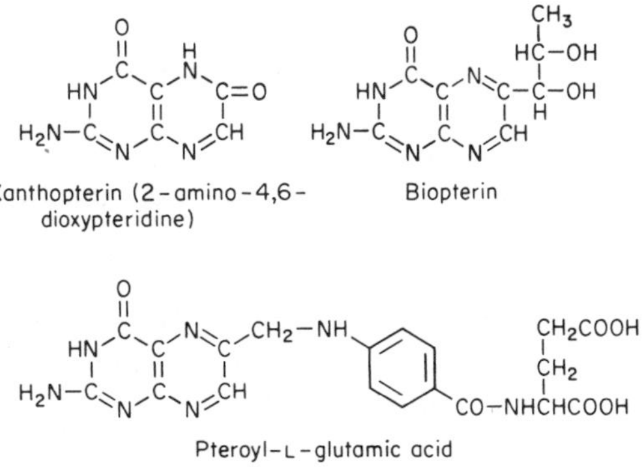

Xanthopterin (2-amino-4,6-dioxypteridine)

Biopterin

Pteroyl-L-glutamic acid

Fig. 20.19 Some pteridine compounds.

Purines As constituents of the nucleic acids and several compounds concerned with energy transformation, the purines are ubiquitous in both plants and animals. The comparative physiologist also knows several of them as excretion products and as the basis of some of the most beautiful animal colors. Guanine is widely distributed in vertebrate iridophores; uric acid, adenine and especially hypoxanthine also occur. The white pigments, reflecting platelets and silvery ribbons in crustaceans and cephalopods may also be purine but this has not yet been established (Bagnara and Hadley, 1973).

The role of purines in animal coloration may be a fortuitous adaptation of highly insoluble nitrogen excretion products. The purines tend to form relatively insoluble deposits, especially in habitats where water is in short supply; their use in integumentary colors may have been built on a process of storage excretion. The storage excretion of urates and guanates in the terrestrial arthropods was noted in Chapter 8. Special cells take up these highly insoluble nitrogenous wastes and store them either permanently or temporarily, and it seems reasonable to assume that such cells might be adapted to other purposes. It may be significant that guanine occurs abundantly in the connective tissues of internal organs like the swim bladder as well as in the integument of some marine teleosts, a group of animals in which the supplies of water for removal of soluble nitrogenous wastes is very limited.

Glitter and iridescence, as well as many of the blue and green colors, are caused by light interference, diffraction and Tyndall scattering. This STRUCTURAL COLORATION depends on the orderly arrangement of dermal chromatophores or their organelles. Reflecting platelets may be associated with light filters provided by yellow, red, brown and black pigments. In some cephalopods, for example, myriads of precisely arranged crystals form a mirror-like background for melanophores and xanthophores whose pigments shimmer over this shiny surface in a spectacular fashion. In some teleosts the guanine particles are so minute that the incident light undergoes a Tyndall scattering which, against a background of melanophores, appears blue. The visual effects may be silvery or white, depending on the size and distribution of the particles, while associated chromatophores may produce a varied iridescence and different metallic hues (Monroe and Monroe, 1968; Denton, 1971).

Mechanism of Pigment Migration

Chromatophores, except for the contractile pigment cells of molluscs, are of fixed form. This was not appreciated by early workers who attempted to explain their apparent changes in shape by theories of amoeboid movement and muscular contraction. There is still no satisfactory explanation for the movement of the cellular organelles. Factors currently thought to be involved are: sol-gel transformations, electrophoretic effects and the action of microtubules and microfilaments (Bagnara and Hadley, 1973).

S. A. Matthews (1931) made the first significant contribution when he demonstrated conclusively that the melanophores of *Fundulus* have an almost constant size during all phases of activity. He observed the cells in tissue culture and proved that the pigment changes position, but the shape of the cell is not altered; these pigment movements may be observed in isolated cell processes as well as in intact melanophores. Micromanipulations led Matthews to conclude that granule dispersion was associated with decreased protoplasmic viscosity while the protoplasm became more viscous when the granules were clumping. Marsland's (1944) investigations of the action of hydrostatic pressure on *Fundulus* melanophores also linked granule movement with changes in protoplasmic viscosity. High pressure, which is known to increase fluidity of protoplasm (sol condition), inhibits the clumping of granules; the effect is proportional to the pressure applied up to 7000 pounds per square inch. Low temperature (6°C) which has a similar action to high pressure on protoplasmic viscosity likewise inhibits clumping; high temperature (30°C) has the reverse effect. These responses are independent of nerves. Marsland was convinced that the movement of granules was a physical rather than a chemical process and depended on the capacity of protoplasm to undergo sol-gel transformations.

Kinosita (1963) postulates an electrophoretic migration of pigment granules in the melanophores of *Oryzias*. His studies, like those of Matthews, were made on isolated melanophore processes as well as on intact cells. Evidence is based on measurements of melanophore potentials; these vary, in a predictable manner, with cellular activity and in accordance with the position of the inserted electrode. The granules appear to be negatively charged since they move toward the anode. In the dispersed state the central part of the pigment cell is electrically more negative than the cell processes; consequently the granules move out into the processes. Kinosita also finds a change from the sol condition in the dispersed phase to the gel condition in the clumped phase; these colloidal changes are coupled with the electrophoretic movement. Agents such as KCl and adrenaline, which clump melanin granules, alter the cell potentials in accordance with a theory of electrophoretic migration. Kinosita has also investigated *Oryzias* iridophores. The pigment granules are likewise negative, but the chromatophore potential is reversed, with the processes electrically more negative than the center of the cell. In accordance with theory, KCl has an opposite effect on the movements of melanophore and guanophore pigment granules.

Birkle *et al.* (1966) first demonstrated microtubules in *Fundulus* melanophores.

These workers pointed out that melanosomes move along relatively fixed channels and that the cytoplasm around these channels contains numerous microtubules (about 225 Å in diameter), aligned parallel to the direction of pigment movement. The microtubules appear to be cytoskeletal elements and not directly responsible for the propulsive force. There might, however, be an important interaction between the melanosomes and the surface of the stationary microtubules (Fujii and Novales, 1969). Microtubules have not yet been found in other fishes and do not appear to be universal in melanophores; they have been described in some amphibians and reptiles (Bagnara and Hadley, 1973).

Intracellular microfilaments occur in the melanophores of the guppy, the dogfish (Fujii and Novales, 1969) and the frog (McGuire and Moellmann, 1972); they also form a conspicuous lattice arrangement in the iridophores of the lizard *Anolis* (Rohrlich and Porter, 1972). Pigmented organelles are closely associated with the microfilaments and may be attached to them. There is a distinct possibility that these filaments are involved in the changing distribution of the melanosomes or the maintenance of the complex array of reflecting plates in iridophores. There may be no single mechanism responsible for granule movement in all kinds of chromatophores.

Factors Regulating the Movement of Pigment

Chromatophore responses have long been described as PRIMARY if they are evoked by non-visual stimuli, and SECONDARY if the visual pathways are involved. This distinction was first drawn in studies of larval salamanders where the earliest responses of the pigment effector cells are independent of the eyes (Parker, 1948). Larval *Ambystoma* of several different species, ranging in length from about 1.5 cm to 5.0 cm, are light colored when in complete darkness but become dark when in bright light; animals which are larger than about 5 cm show almost the reverse reaction. A similar sequence of primary, followed by a secondary response has been noted in many other fishes and amphibians. The secondary color response, characteristic of late larval and adult life, depends on the nature of the background, more particularly on the ALBEDO, which is the ratio of the incident to the reflected light reaching the retina. The primary response is better defined without reference to the possible mechanisms involved; it occurs in larvae before they acquire the ability to show a background response (Bagnara and Hadley, 1973).

The eyes are not always involved in the chromatophore responses of adults. Temperature and humidity may alter pigment distribution in blinded animals or in animals under constant light conditions. In a few cases, chromatophores seem to operate as independent effectors, responding directly to changes in illumination. Waring (1963) lists only *Xenopus* and *Phrynosoma* as unquestionably showing chromatophore changes after both denervation of the skin and removal of the pituitary; usually, pigment movement is controlled either by hormones or through nerves. The hormonally regulated changes are relatively slow; those mediated by direct innervation may be exceedingly rapid. The latter appears to be more specialized and more recent phylogenetically. Waring (1963) tabulates data suggesting that the more ancient animals had a hormonal control of chromatophores while those of recent origin may be controlled only by reflexes. The more ancient fishes (cyclostomes and elasmobranchs) and the primitive tetrapods (amphibians) show relatively slow chromatophore responses which are hormonally regulated; some of the teleosts and reptiles depend only on nerves. The teleost fishes as a group present the most convincing picture, with a completely humoral control in the Anguillidae, a group which can be traced back into the early Eocene, a mixed nervous and endocrine coordination in families such as the Siluridae and Pleuronectidae which seem to have evolved in the late Eocene, and an entirely nervous regulation in families of very recent origin such as the Gasterosteidae and Cyprinodontidae—evolving sometime in the Pleistocene. Waring presents the picture tentatively, for there is much that is speculative and many blanks still exist. The concept, however, finds further support in the two groups of invertebrates which show marked physiological color changes. Chromatophores of the decapod crustaceans, with a probable origin in the Triassic, are entirely controlled by hormones while the highly specialized reflexly controlled pigment cells of the dibranchiate cephalopods belong to animals of Pleistocene or Recent origin.

Nervous control in cephalopods The chromatophores of cephalopods are the most highly specialized structures concerned with color change. They are in reality pigment effector *organs* rather than cells. The movement of the pigment depends on the stretching of an elastic sac by radiating extrinsic muscle cells. Contractions may be extremely rapid (0.14 to 0.5 sec in *Loligo* and about 1 sec in *Sepia*) and can produce waves of varied hues which sweep swiftly over the animal to assist in camouflage by breaking up the body outline. These oscillating colors may produce definite patterns which are precisely associated with different aspects of the animal's behavior. The innervation of the chromatophore muscles has been described and the coordination has been shown to depend on color centers in the subesophageal ganglia of the brain. The pathways from the neurons of these ganglia are direct to the muscles of the chromatophores. The early literature has been summarized several times (Parker, 1948; Fingerman, 1963; Nicol, 1964). Current understanding is based on the beautiful ultrastructural studies of Cloney and Florey (1968) and the electrophysiological investigations of Florey and Kriebel (1969). Florey (1969) has summarized their findings.

In brief, the darkening reaction is produced by the contraction of radiating muscles which stretch an elastic sacculus filled with pigment (Fig. 20.14); muscle relaxation is followed by an elastic recoil of the sacculus with concentration of the pigment. The numerous, obliquely striated, radiating muscles are innervated in groups by several different motor nerves; inhibitory fibers have not been found. Although many drugs have been tested and several are known to alter pigment distribution, the nature of the true transmitter of the motor nerves remains to be established. The most spectacular color displays in the animal world depend on these minute effector organs. The selective activation of different groups of muscles permits varied degrees of stretching of the elastic sacculus; the flashing of black, brown, red and yellow pigments against a silvery mirror of iridophores creates a shimmer of colors, which excited man's aesthetic interests long before the mechanisms could be understood in prosaic physiological terms.

Hormonal control in crustaceans Physiological color changes in arthropods depend on the hormonally regulated movements of pigment granules with chromatophores. There is no innervation of the effector cells; there are no extrinsic contractile elements. Thus, the system is built on mechanisms which seem to be phylogenetically most ancient. The extent of specialization within this framework, however, is such that some of the arthropods are capable of remarkably varied and beautiful color patterns which change adaptively in relation to environmental factors, both visual and non-visual. It is not known when pigment effector cells first appeared in phylogeny, but the present-day crustaceans with a fossil record extending back almost 300 million years show some of the most notable color reactions in the animal world. This ability seems to have been achieved through a variety of different pigments in complex associations of chromatophores and through the specialization of different hormones to regulate the independent movements of these varied pigments.

The arthropods as a group show a considerable range in chromatic abilities. Although the capacity for morphological change in color is widespread, physiological change is restricted to a few of the insects and to several orders of the Malacostraca among the crustaceans (Isopoda, Stomatopoda, Decapoda). The least specialized condition is shown by some of the isopods which change from darker to lighter phases under environmental conditions but have no real capacity to alter their colors or color patterns because of the limited variety of chromatophores present. From this simple situation, some of the brachyuran (true) crabs (*Uca, Hemigrapsus, Callinectes, Eriocheir*) represent an intermediate condition between the isopods and the natantian decapods (shrimps). The brachyurans have three (occasionally four) kinds of monochromatic chromatophores (black, red and white) scattered throughout the hypodermis but not arranged so as to produce striking color patterns; many of the shrimps (Natantia) have an elaborate system of intimately associated chromatophores; as many as four different pigments may occur within one complex with as many as eight differently responding chromatophore patterns in the genus *Crangon*. These are often arranged in such a way that changes in color patterns as well as shade and tint of color are possible.

The literature of the nineteenth and early twentieth century records many attempts to associate the nervous system with the activities of the pigment effector cells. It is now agreed that the control is entirely hormonal. The classical experiments were performed by Koller (1929) who adapted shrimps (*Crangon*) to black or to white or to yellow backgrounds and showed that the blood from black-background animals would disperse the black chromatophore pigments when injected into shrimps on a white background, while the blood from animals adapted to a yellow background would disperse the pigment of the xanthophores in similar white-adapted animals. Perkins (1928), at about the same time, showed that although cutting of nerves had no effect, the occlusion of the blood supply to certain regions prevented color changes which were restored when the circulation returned to these areas. Perkins went further and localized the source of the hormones in the eyestalks by injecting seawater extracts of different tissues. Parker (1948) summarizes these significant pioneer experiments.

The sinus gland is the immediate source of chromatophorotropins in the crustaceans. As indicated in Fig. 20.20, the secretions which are stored in this neurohemal organ may arise from the neurons of the *X*-organ, the brain or certain ventral ganglia. Although the *X*-organ is known to be a major source of chromeactivating factors, it is probable that, at least in some species, these substances are also secreted by neurons arising in other areas. At any rate, it has been experimentally demonstrated that extracts of ganglia other than those of the *X*-organ sometimes activate chromatophores and, further, that distinct color changes may follow electrical stimulation of the central surfaces exposed by removal of the *X*-organ-sinus-gland complex. The primary source of these neurosecretory materials may vary in different groups. In the few insects which show physiological color change (the stick insect *Carausius mormosus,* for example), the brain and the corpora cardiaca probably function in a manner comparable to the *X*-organ and sinus gland in the crustaceans (Fingerman, 1959, 1963).

The chromeactivating hormones of arthropods (CHROMATOPHOROTROPINS) are relatively small polypeptides with molecular weights probably in the range of 1000 to 2000 (Fingerman, 1970). The blanching hormone of the prawn *Pandalurus borealis* contains eight amino acids and has been shown to be pGlu-Leu-Asn-Phe-Ser-Pro-

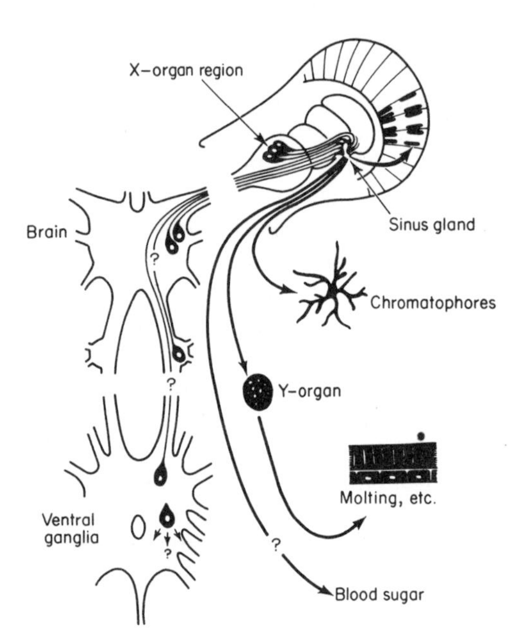

Fig. 20.20 Diagram of brachyuran sinus gland neurosecretory system. [From Welsh, 1961: *In* The Physiology of the Crustacea, Vol. 2, Waterman, ed. Academic Press, New York.]

Gly-Trp-NH_2; this peptide is active in picogram amounts when tested in shrimps (Fernlund and Josefsson, 1972).

Although, in theory, a single hormone might either concentrate (or disperse) pigment while the reverse effect might take place in its absence, current evidence points toward a polyhumoral control (Fingerman, 1970; Bagnara and Hadley, 1973). There appear to be different hormones for the different types of pigment cells as well as pigment-dispersing and pigment-concentrating factors. Considerable biochemical work will be required to settle this long-standing problem. There may be wide variations in different species. The mode of action of hormones on the colored organelles is also uncertain, although several possible mechanisms have been proposed (Fingerman, 1969).

Pigment movements in arthropod chromatophores are environmentally regulated by several different factors and may show persistent rhythmical changes under "constant" conditions. Greater TOTAL ILLUMINATION often increases the dispersal of chromatophore pigment; at higher TEMPERATURES the dark pigments are sometimes concentrated and the white pigments dispersed so that reflection is greater, and a measure of body temperature regulation may be achieved (Fingerman, 1970). In the stick insect *Carausius*, changes in the HUMIDITY, as well as light, alter the distribution of the chromatophore pigments. The animals become darker at higher humidities; the reaction is mediated through the neurosecretory system (Bagnara and Hadley, 1973). Background or ALBEDO RESPONSES are also characteristic of crustaceans. These depend on the ratio of incident to reflected light, so that the dark pigments disperse and the light pigments concentrate on dark backgrounds while the reverse happens on white backgrounds. The principles are similar to those found among the vertebrates and will be discussed in the next section. Finally, some crustaceans show rhythmic diurnal changes in the pigments of the chromatophores and the eyes, which may be in phase with solar and lunar changes (Chapter 14) but persist for a considerable period when animals are maintained under constant conditions of temperature and illumination. These PERSISTENT RHYTHMS have been extensively studied by Brown and his associates (Brown, 1972).

Neurohumoral regulation in vertebrates Hormones and nervous reflexes, either separately or in combination, provide the lower vertebrates with a highly varied and rapid control of their pigment effector cells. Although there are still many puzzling details, the major components of the regulatory machinery are now known. Some of the most distinguished biologists of the past century have been associated with its investigation. Parker (1948) provides a comprehensive bibliography, and the historical notes which follow can be amplified from his monograph and those of Waring (1963) and Bagnara and Hadley (1973).

Joseph Lister, of antisepsis fame, reported the first experimental work on vertebrate color changes in 1858. He noted that blinded frogs were light in color and failed to show pigmentary responses to altered background. This he interpreted in terms of a nervous regulation and concluded that the cerebrospinal axis was essential to color changes and that the regulation of chromatophores was "chiefly, if not exclusively" dependent on the central nervous system.

A neural control was further emphasized by the distinguished French scientist Pouchet (1871–76) who found that sectioning the peripheral nerves of the flatfish resulted in a darkening of the denervated areas while electrical stimulation resulted in a pallor. It seemed obvious that autonomic nervous stimulation excited the chromatophores and led to a concentration of pigment while the pigment dispersed in the absence of nervous excitation. Von Frisch (1910–12) confirmed Pouchet's findings. Using the minnow *Phoxinus,* he was the first to trace the distribution of the chromatophore nerves from their origin in the medulla. Although these early studies focused attention on the autonomic nervous system, there were, at the same time, indications of regulatory factors apart from the nerves. Injections of extracts of the adrenal gland or adrenaline were shown to cause blanching in frogs; application of adrenaline to the melanophores in isolated fish scales concentrated their melanin granules; decapitated amphibian larvae became light in color and failed to respond to changes in background; hypophysectomized tadpoles and adult frogs behaved in the same way.

The full significance of these many experimental observations was not appreciated until the publication of a series of papers by Hogben and Winton in 1922 and 1923 (see Waring, 1963). These workers demonstrated that hypophysectomized frogs were permanently pale even when on a dark background and that the frog pituitary contains a factor which, when injected into the hypophysectomized or normal animal, will produce a jet black frog. A single pituitary gland was shown to be sufficiently potent to cause melanin dispersal in 50 hypophysectomized animals. Hormone production was localized in the pars intermedia and the factor was termed INTERMEDIN. This term is still used, although MELANOPHORE STIMULATING HORMONE MSH is now more common. Hogben and Winton demonstrated further that stimulation of nerves or sectioning of nerves to the skin did not produce color changes. Thus, the frog was shown to be unlike the turbot or the minnow in its chromatic responses, and although Lister's experimental results with blinded frogs were sound, his interpretation was not supported by the later work.

Somewhat later, Hogben and Slome (1936) and Hogben (1942), using the African clawed toad *Xenopus,* explained the ocular responses in a series of classical experiments which have been summarized by Waring (1963). In brief, the position and anatomy of the toad's eye is such that the central portions of the retina are only stimulated by direct, transmitted light from above while the dorsal segment of the retina is stimulated by light reflected from the background (Fig. 20.21). When only the ventral segment of the retina is excited (black background and absence of reflected light) the animal becomes dark, and for this reason the ventral part of the retina is called the *B* (black) area. Likewise, when the dorsal area of the retina is strongly stimulated (white background and reflected light) the animal assumes a lighter color, and the associated area of the retina is called *W* (white). This, in essence, is the basis of the secondary or background responses of the lower vertebrates (also called albedo responses).

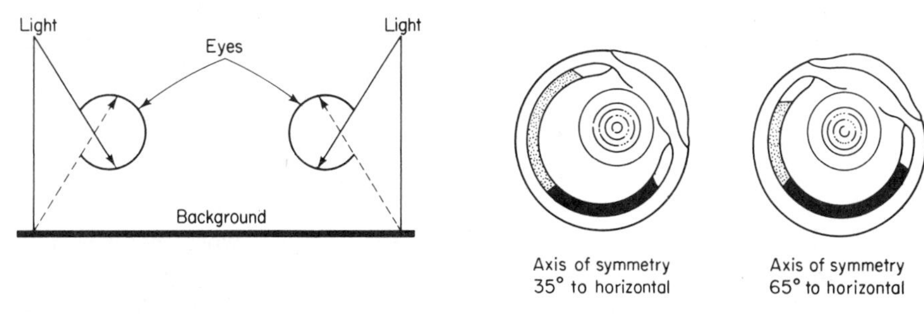

Fig. 20.21 The eyes and albedo responses of lower vertebrates. The left diagram shows how the dorsal, or *W*, areas of the retina are stimulated only by reflected light from the background while the ventral, or *B*, areas receive rays transmitted directly from above. Right diagrams for the eye of *Xenopus* show the extent of the *B* (black) and *W* (stippled) areas as shown by precise measurements. [From Hogben, 1942: Proc. Roy. Soc. London, *B131*: 117. Permission of the Royal Society.]

These experiments, and numerous others carried out during this period, created a unified picture of chromatophore responses in the lower vertebrates. Lancelot Hogben and his associates described a MELANOPHORE CONCENTRATING HORMONE (MCH), and a lively controversy developed between his followers and those of G. H. Parker on the opposite side of the Atlantic. In addition, many workers recognized both adrenergic and cholinergic fibers in the nervous regulation. Vertebrates were known to differ in the relative strengths of hormonal and neural controls, but the generalized model of the machinery concerned was based on two pituitary factors with antagonistic actions and an autonomic regulation; the autonomic sympathetics caused melanosome concentration and blanching while the parasympathetics were thought to disperse melanosomes and darken the animal. The peripheral receptors were appropriately located in the ventral (black) and dorsal (white) areas of the retina.

This classical picture prevailed for about two decades following the publication of Parker's (1948) monograph. It is now regarded as a useful model which has been substantially modified. There appears to be neither a melanophore concentrating hormone nor a cholinergic autonomic supply. Both blanching and darkening can be caused by the sympathetic, because chromatophores possess two kinds of adrenergic receptors (Chapter 2). Stimulation of the *beta* receptors leads to melanosome dispersion while stimulation of the *alpha* receptors concentrates the pigment. In addition, the pineal gland now occupies a firm position in the team of chromatophore regulators, while the nature of the hypothalmic controls of MSH secretion have been described, and cyclic AMP has been established as a second messenger (Chapter 2) in the action of MSH. Details are summarized in the monograph by Bagnara and Hadley (1973); only a few of the highlights are noted here.

The general physiology of the two types of catecholamine receptors was outlined in Chapter 2. In brief, the effector may possess only one type of receptor substance, or both may occur on a single cell; responses to *alpha* and *beta* receptor stimulation are opposite in character; *alpha* receptors usually dominate *beta* receptors; either circulating catecholamines of adrenomedullary origin or the adrenergic nerve transmitters can act on the receptor substances. Teleosts, amphibians and reptiles are known to have exploited these possibilities in various ways to achieve a physiological regulation of their chromatophores; not only species variations but also differences between geographic races of the same species (*Rana pipiens*) have been recorded (Hadley and Goldman, 1970). The experimental work involves both *in vivo* and *in vitro* preparations with the use of pharmacological agents, which mimic sympathetic effects or selectively block the two receptor sites (Bagnara and Hadley, 1973). The most complete analysis has been carried out on the American lizard *Anolis carolinensis,* whose melanophores lack an autonomic innervation but respond strongly to MSH (albedo responses) and to the hormones of the adrenal medulla. Catecholamines from the medulla are responsible for the "excitement pallor" and "excitement darkening" which appear in *Anolis* under conditions of stress, such as electrical stimulation (Kleinholz, 1938; Hadley and Goldman, 1969). This lizard has a mosaic of melanophores, some possessing both kinds of receptors and others having only *beta* sites. A remarkable range of color control has been achieved through albedo responses involving MSH, reactions to the catecholamines of the adrenal medulla, and direct responses to temperature changes and light.

Iridophores as well as melanophores may carry the two types of catecholamine receptor substance. In the dermal chromatophore unit of the frog (Fig. 20.16), the responses of the two chromatophores are complementary; *alpha* receptor stimulation aggregates melanosomes and disperses reflecting plates while the reverse movements of the pigment follow *beta* receptor stimulation (Bagnara and Hadley, 1973).

Evidence of pineal involvement in color changes goes back more than half a century. Karl von Frisch did the pioneer work using the minnow *Phoxinus* (reviews by Healey, 1957 and Fenwick, 1970). He described the innervation of the melanophores and showed that the pineal organ and neighboring parts of the diencephalon as well as the lateral eyes contain photoreceptors which regulate their responses. Much more recently it has been shown that the pineal organ (and lateral eyes) secrete a substance melatonin (N-acetyl-5-methoxytryptamine) which concentrates melanosomes and produces a marked blanching. This appears to be a hormone concerned with adaptation to darkness and acts only during early (larval) life of amphibians and fishes. Larval amphibians, as well as embryonic and very young *Fundulus,* become light in color when in the dark; this blanching may be partially caused by a reduced secretion of MSH but the experimental evidence indicates that melatonin of pineal origin is primarily responsible. Although melanosomes are somewhat aggregated in hypophysectomized animals, they become much more concentrated when the animals are placed in the dark; an even more convincing experiment may be carried out with hypophysectomized tadpoles immersed in MSH solution which disperses the melanosomes; tadpoles treated in this way remain dark in the light but blanch if placed in darkness. The lateral eyes are not essential to these reactions. During later stages of development and in adult life, the melanophores are unresponsive to melatonin. Melatonin was first extracted from mammalian sources and may be a hormone related to coat color and reproductive functions in higher forms (Chapter 2). However, its most clearly established function is that associated with the blanching of amphibian and teleost larvae in darkness (Turner and Bagnara, 1971; Bagnara and Hadley, 1973).

Etkin (1941, 1962) was the first to show that lesions in the hypothalamus of the frog tadpole cause a darkening of the skin which is presumably due to the release of MSH. At first it was assumed that neurosecretory fibers from the hypothalamus produced an inhibitory hormone which regulated the activities of the pars intermedia. Later evidence of both a physiological and morphological nature indicated a control by ordinary nerves rather than neurosecretions; the findings pointed toward an inhibitory adrenergic supply and possibly inhibitory cholinergic fibers as well. Bagnara and Hadley (1973) summarize the work; it is still uncertain just how significant the hypothalamic innervation is, but it seems likely that there are species variations, that the relative importance of nervous and endocrine regulation may vary with age, and that the nervous regulation is less significant in phylogenetically more advanced groups.

Although uncertainties remain concerning the nervous supply, it is now clear that the hypothalamus does secrete one and possibly two hormonal factors which regulate the pars intermedia. Two MSH-release inhibiting factors (MIF) have been isolated from beef hypothalami (Schally *et al.,* 1973); the more active principle is a tripeptide Pro-Leu-Gly-NH_2 while the second factor contains five amino acids (Pro-His-Phe-Arg-Gly-NH_2). The first of these is the carboxyl terminal tripeptide chain of oxytocin and can be formed *in vitro* by incubating oxytocin with an enzyme found in the hypothalamus; a biosynthetic relationship between these two substances is suggested. Although the biochemical work has not progressed nearly so far, there appears also to be a MSH-releasing hormone (MRH); the existence of two factors having opposite effects seems theoretically likely since no feedback of information is expected in the regulation of color changes. The comparative endocrinologist can anticipate exciting developments in understanding of pars intermedia regulation, since the research is currently very lively.

In summary, the classical explanations which developed and prevailed during the first half of the century have now been substantially modified. There is no convincing evidence for either a pituitary MCH or a cholinergic autonomic; the opposing autonomic effects are achieved through *alpha* and *beta* receptor sites. Besides these two corrections, several additional mechanisms are now recognized. The pineal substance melatonin acts on the chromatophores of some vertebrates at certain stages; the hypothalamic factors which regulate MSH secretion have been demonstrated both biochemically and physiologically. In addition, cyclic AMP is known to be a "second messenger" concerned with the movements of pigment in melanophores and iridophores; adrenergic substances, as well as MSH, probably act by way of this mechanism (Novales and Fujii, 1970). Different vertebrates have adaptively exploited hormonal and nervous controls, or some combination of the two, in varying ways during their evolution. The cyclostomes, most of the elasmobranchs and the amphibians probably depend entirely on hormonal regulation. The teleosts depend on an interplay of endocrine and nervous control; in some, such as the eel *Anguilla*, hormones dominate, while in others, such as *Fundulus*, the autonomic nervous system can override the hormones. A parallel situation is found among reptiles. In the Iguanidae (*Anolis*), chromatophore control is predominantly humoral while the autonomic system holds sway among the Chamaeleontidae.

Finally, there are a few cases where vertebrate chromatophores make direct responses to changes in illumination (UNCOORDINATED NON-VISUAL RESPONSES) and must behave as independent effectors (Waring, 1963). There are also a few cases in which photoreceptors occur in the integument as well as in the eyes (COORDINATED NON-VISUAL RESPONSES). Hogben and Slome (1931), for example, found that blinded *Xenopus* showed dispersion of melanin in the light and its aggregation in darkness; the response persisted after complete destruction of the spinal cord. Confirmation of this direct photosensitivity of dermal chromatophores in *Xenopus* has been obtained by stimulating the cells directly with sharp beams of light and by investigating them in tissue culture (review by Fingerman, 1970).

Coordinated non-visual responses have also been studied in *Xenopus*, but these are more completely analyzed in some of the reptiles. The presence of dermal photoreceptors was shown by Zoond and Eyre (1934) in an ingenious series of experiments on the chameleon *Chamaeleo pumilus*. Blinded animals usually become darker in bright light. If an area of the skin of such an animal is covered with an opaque object, the precise pattern of the shaded area becomes lighter than the exposed regions in about two minutes. This does not happen if the skin is denervated or removed, although the melanophores may still be active as evidenced by their responses to electrical stimulation. These experiments show that the chameleon melanophores do not behave as independent effectors. Reflex control through integumentary receptors was demonstrated by decapitating animals and eviscerating them, so that the spinal cord and peripheral nerve trunks were intact but the circulatory system (hence the possibility of endocrine regulation) was removed. The light and dark reactions remained in such preparations. When, however, the nerves were cut on one side of the body, the color of that side was permanently dark while the opposite (intact) side responded to light and darkness as before. These experiments demonstrate a coordinated non-visual response which can be activated by dermal photoreceptors. High temperature (37° to 40°C) overrides the darkening reactions in response to light and induces a pallor. Zoond and Eyre (1934) discuss these changes in relation to temperature regulation. Waring (1963) reviews this and other pertinent literature on the non-visual responses to background illumination.

21

NERVOUS INTEGRATION

The diversified activities of a complex animal depend on a steady flow of information from the receptors to the effectors. Pertinent environmental stimuli are coded by the receptor cells as a series of electrogenic pulses; these are transmitted through the nervous system to activate muscles, glands, chromatophores, and other effector structures. The nervous system is the indispensable link between the receptors and effectors and serves the dual functions of transmission and integration. Physiologically, it is much more than a telegraphic or telephonic system of communication where there is a one-to-one correspondence between input and output. No such simple relationships are usually found in animal responses. Activities such as nest building, fighting, or courtship may follow a relatively simple visual stimulus as an extended sequence of highly varied muscular movements. In the lower forms, vertebrate as well as invertebrate, such behavior may be innate and expressed in full measure without prior experience. In the phylogenetically higher vertebrates the relationships between input and output are regularly modified through learning. Thus, in its most specialized form, the nervous system permits the complexity of action which a musician displays when he appears with his instrument before an audience or the organizational abilities of the construction engineer who directs an intricate operation from a sheaf of blueprints.

The phylogeny of this integrating system depended on the evolution of highly specialized transmitting cells (the NEURONS) and their associations in coordinating ganglia and in a central nervous system. The neuron is an evolutionary product of three lines of cellular specialization: an INPUT or receptor surface, a conducting fiber or TRANSMITTING area of cell membrane, and an OUTPUT region usually concerned with the release of a specific secretion or transmitter substance (Figs. 16.1 and 21.1).

Excitability as a general property of cells was considered in Chapter 15. Primitive

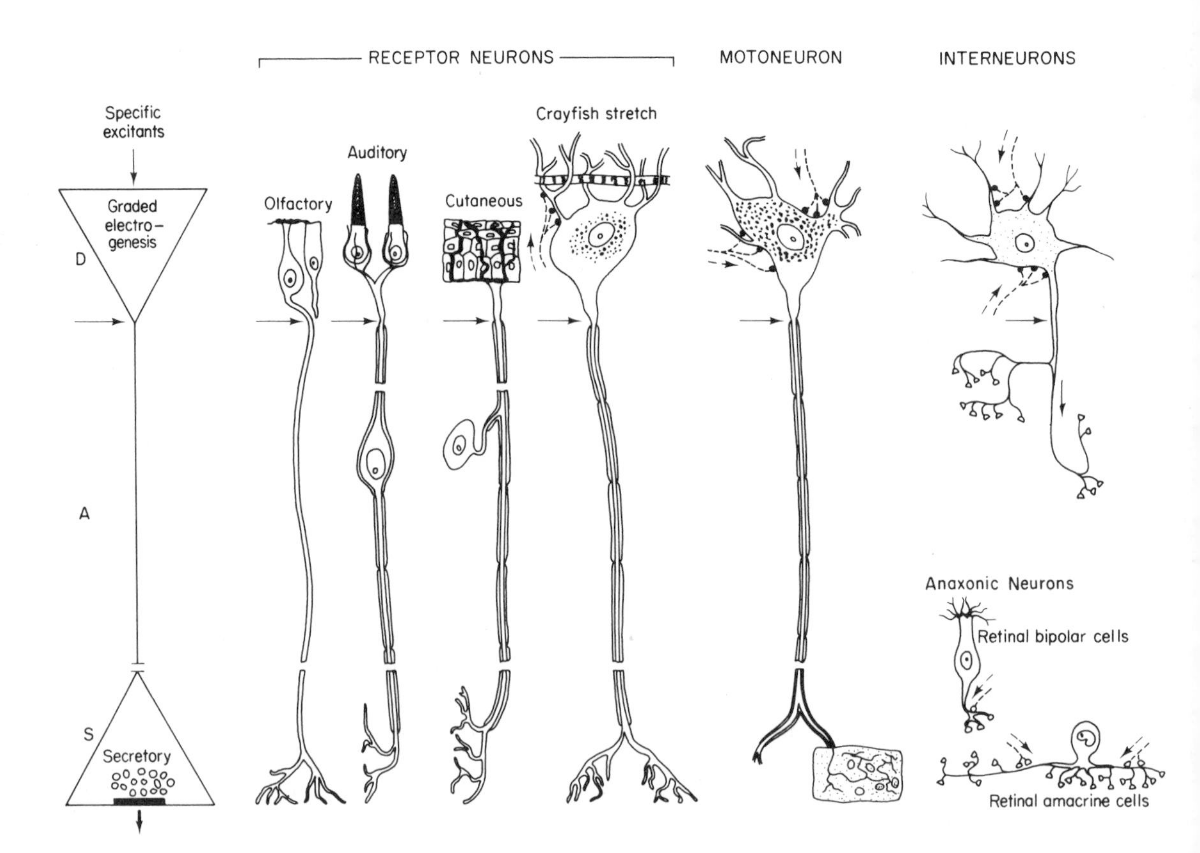

Fig. 21.1 Comparative morphology of several different neurons. Arrow shows the origin of the axon in each neuron related to the diagram at the left; diagram at left (based on Grundfest, 1957) shows the three essential components of a neuron: D, dendritic zone with its input or receptor surface; A, the transmitting axon; S, the synaptic output surface. [From Bloom and Fawcett, 1968: A Textbook of Histology, 9th ed. Saunders, Philadelphia, based on Bodian, 1967: *In* Neurons, Circuits and Neuroglia, Quarton, Melnechuk, and Schmitt, eds. Permission of Rockefeller University Press, New York.]

organisms and cells were probably responsive to a wider variety of stimuli, but in receptor cells and neurons the input areas are only excited by specific environmental changes, such as a wrinkling of the cell membrane or the application of some particular chemical. An electrogenesis is the characteristic consequence of this excitation. Phylogenetically, the input area may be considered the oldest portion of the excitable cell; its many specializations in the receptor organs have been discussed in previous chapters. Whatever the nature of the specific excitant, the consequence at the input is a series of electrogenic pulses, graded in frequency according to the strength of stimulus (Chapter 16).

Electrogenesis, initiated at one point on the cell membrane, spreads as a wave of depolarization (sometimes hyperpolarization) over the cell. The extension of nerve cell membranes as slender fibers or filaments provides for the distant transmission of this electrical activity. These nerve fibers are electrically excitable, spike-generating structures which transmit faithfully and rapidly the information coded at the receptor surface. They become progressively more numerous and extend much farther in the larger multicellular animals where the problems of coordination are greater and the distances between receptors and effectors are longer. A single afferent nerve fiber in man may be as long as a meter. Increasing speed of transmission is the evident phylogenetic trend in the physiology of this conductile component of excitable cells.

The terminal part of the excitable cell is concerned with the release of a specific transmitter. This, in turn, excites another nerve cell or an effector organ. The concept of chemical transmission has been considered (Chapter 15), and the recognized transmitter substances were described (Chapter 2).

Explanations of the coding of stimuli and the transmission of this information to another cell are based on these three lines of cell specialization (Fig. 16.1). The integrative properties of the nervous system, however, depend on the intricate synaptic associations of these units into nervous tissues. Architecturally, these arrangements vary from diffuse nerve nets to massive aggregations of cells in centralized nervous systems with distinct areas devoted to particular functions. Explanations of the higher capacities of the nervous system, such as instinctive behavior patterns, learning and memory, are almost completely speculative, but many of the elements of neurophysiology can be understood in terms of established properties of neurons and synapses and their anatomical arrangements. The present discussion is confined to neurons, synapses, the physiological properties of several different arrangements of nerve cells, and some of the simple functional units of animal behavior.

NERVE CELLS

Although the cell theory was clearly enunciated during the first half of the nineteenth century, its application to the central nervous system remained controversial for another fifty years (Bullock, 1959b). Routine histological preparations often reveal only a tangle of delicate fibers with scattered cells and provide scant basis for understanding the physiological complexity of this system. Many of the scientists of the past century believed in a continuity between the processes of neighboring neurons, and this concept was staunchly supported by Camillo Golgi of the University of Pavia who, in 1873, developed the classical silver staining methods for the differentation of nerve fibers. His beautiful preparations served only to strengthen his faith in the reticular organization of nervous tissues. It was the Spanish investigator Ramon y Cajal, using improved Golgi techniques, who was primarily responsible for demonstrating that neurons are the functional units of the nervous system, that there is no direct continuity between the protoplasm of one cell and the next, and that definite points of contact between nerve fibers can be recognized. This concept, sometimes known as the NEURON DOCTRINE, is applicable to the nervous tissues of all animals; a syncytial organization occurs in only a very few places, such as the giant nerve fibers and at certain points in nerve nets of invertebrates (Bullock and Horridge, 1965). In 1906, Golgi and Cajal shared the Nobel prize for their contributions to our knowledge of the structure of nervous tissues.

Phylogeny of Nerve Cells

For more than a century biologists have speculated about the phylogeny of neurons and the nervous system (Horridge, 1968a; Lentz, 1968). During almost half of this period Parker's (1919) views, as expressed in his classic book *The Elementary Nervous System*, were dominant. Parker argued that the nervous system, in particular the neuromuscular system, evolved from independent effectors and he selected the porocyte of the asconid sponge as his archetype (INDEPENDENT EFFECTOR STAGE). A porocyte is a conical cell pierced by a tube which actually forms one of the numerous pores or ostia through which water streams into the body of the animal. The water currents are created by the flagella of the choanocytes which line the cavities of the sponge, but the flow can be interrupted in unfavorable situations through the closure of the ostia by a slow contraction of the porocyte; receptor, conductor, and effector are combined in a single cell, the porocyte. The *second step* in Parker's phylogenetic series was represented by a simple RECEPTOR-EFFECTOR SYSTEM thought to be present in the tentacles of actinians. Through modifications of their exposed (receptor) surfaces, epithelial cells in close proximity to underlying muscle fibers provided this receptor-effector combination. Parker noted that the connections of the sensory epithelial cells with the muscle fibers were usually not simple but arborized in a communicating web of primitive nerves (PROTONEURONS); he considered these protoneurons to be a *third type* of cell, which formed the phylogenetic basis of conducting and integrating nerves and ganglia. Systems exemplifying each of these theoretical steps appeared to be present in the coelenterates.

These views have been substantially modified during the past 25 years (Pantin, 1965; Lentz, 1968). Recent evidence now indicates that they should be wholly abandoned (Horridge, 1968a). Tight junctions between cells, with the possibilities of intercellular communication, are common in many epithelia—both invertebrate and vertebrate (Loewenstein, 1966, 1970; Horridge, 1968a); it is felt that non-nervous conduction in epithelial tissues may well have preceded nerve cells phylogenetically. The long-distance transmission of impulses with the release of chemical transmitters probably evolved subsequent to a simple direct spread of excitation through tissues which may have been provided with pacemakers for primitive coordination.

Non-nervous or NEUROID CONDUCTION was described in several kinds of cells by the pioneer physiologists (Verworn, 1899; Heilbrunn, 1952). More complex situations, involving many cells, were recognized in several plants. The rapid closing responses to touch of the leaves of the sensitive plant *Mimosa pudica* are well known;

spread of excitation from the point of stimulation is associated with characteristic spike potentials. These action potentials can be propagated over the whole length or the entire area of certain regions of the *Mimosa* leaf, indicating a non-decremental and unpolarized spread of excitation. Similar observations have been made on certain other plants such as Venus flytrap *Dionaea muscipula* (Sibaoka, 1966; Mackie, 1970).

Of more direct pertinence to the present discussion is Mackie's (1965) demonstration of conduction across the epithelium of the swimming bells of siphonophores. In preparations, entirely devoid of nerves, conduction occurs at velocities of 20 to 50 cm/sec with refractory periods of 2 to 3 msec; it is non-decremental and unpolarized. Non-nervous electrical conduction seems also to occur in the rows of ciliated cells along the combs of ctenophores (Horridge, 1965a, 1968a) and will probably be found in other places as more tissues from different animals are investigated. Cell to cell conduction of electrical impulses across tight cell junctions is probably widespread and may be phylogenetically basic to the evolution of the nervous system. These recent findings indicate that spikes preceded the evolution of axons and suggest that electrical transmission is more ancient than the chemical transmission of the sort found at typical nerve endings.

If these arguments are sound, the neuron must have evolved from quite a different basis than Parker's receptor-effector system or the epithelio-muscle which figured prominently in even earlier discussions of neuron phylogeny. Several workers have recently focused on neurosecretory cells as the probable direct forbears of neurons (Grundfest, 1965a, b; Horridge, 1968a; Lentz, 1968). These writers emphasize the capacities of some sheets of non-nervous epithelia to secrete, to respond to stimulation, and to conduct impulses; they suggest a phylogeny of neurosecretory cells from such epithelial cells through the specialization of an outer receptor surface and an elongated inner conducting fiber; they also stress that many neurosecretory cells are capable of propagated action potentials (Bern, 1966) and that many neurons—both invertebrate and vertebrate—secrete substances which have powerful effects on growth and especially on regeneration (Chapter 24). The unique feature of a neuron resides in its capactiy to transmit impulses over long distances without exciting all the intervening cells en route. Horridge (1968a) suggests that neurons first appeared as neurosecretory or growth regulating cells; their elongated processes were later adapted to rapid conduction and chemical transmission by release of transmitter at their endings. Other phylogenetic sequences have been proposed (Novák, 1964; Lentz, 1968; Mackie, 1970); the important point is that simple reflexes and two-neuron arcs were probably absent from the most primitive coordinating systems.

Morphological Types of Neurons

The vertebrate motor neuron from the ventral horn of the spinal cord is most often selected to illustrate a typical nerve cell (Fig. 21.2). This is a multipolar cell with numerous short fibers (DENDRITES) associated with the input or reception, and one

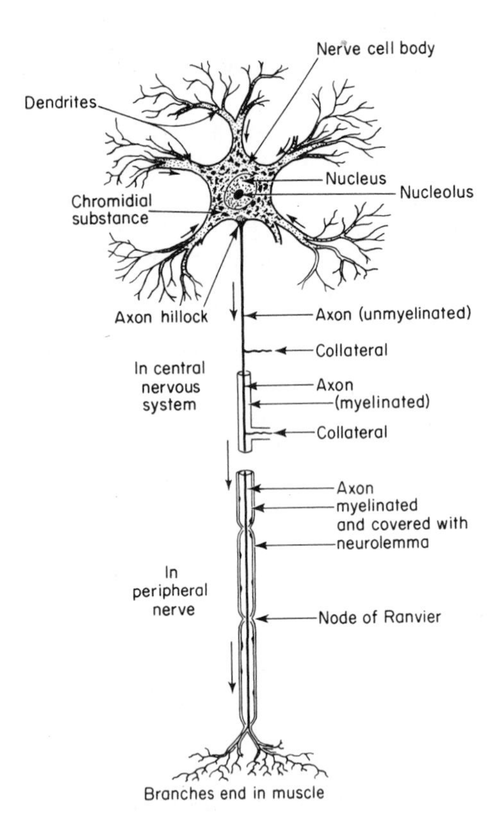

Fig. 21.2 Diagram of a multipolar neuron. [From Ham, 1965: Histology, 5th ed. Lippincott, Philadelphia.]

long fiber (AXON) which transmits to the muscle. Characteristically, the soma or cell body contains a central spherical nucleus with prominent nucleolus and fine chromatin granules while the cytoplasm is filled with prominent granules, flakes, or clumps of chromidial material, the Nissl bodies. These are composed of ribose nucleoproteins and are thought to be responsible for the synthesis of proteins—possibly the enzymes concerned with acetylcholine synthesis. With appropriate techniques, fine neurofibrils can be shown to course through the cytoplasm and to extend into the dendrites and axon. Neurons of all animals, from hydra and jellyfishes to chickadees and men, show common staining reactions based on the presence of similar cytological elements (Bullock and Horridge, 1965).

This multipolar cell is only one of the many forms of neurons. Its use as the prototype of a nerve cell in many elementary accounts of nervous tissue quite inaccurately suggests that the cell body is the focal point for nervous conduction. It has already been noted (Chapters 15 and 16) that impulse transmission is a property of cell membranes; the cytoplasm of the neuron body is trophic since it is concerned with the outgrowth and maintenance of the processes rather than the direct transmission of impulses (Bodian, 1962). Some of the variations in location of the cell body with respect to processes as well as the extent of fiber development and the site or origin of the impulse are shown in Fig. 21.1. Most of the neurons of invertebrates lack

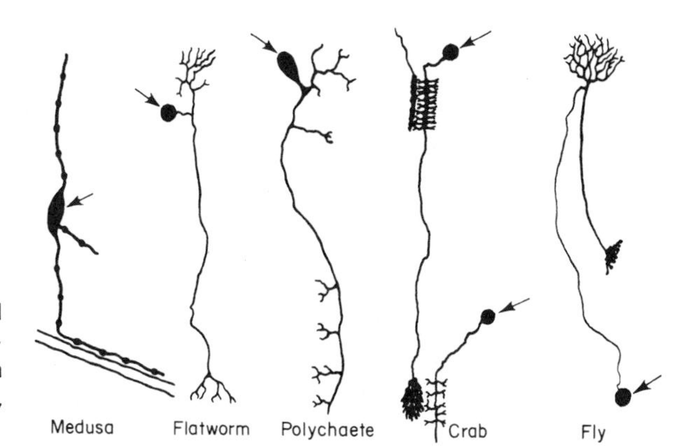

Fig. 21.3 Different morphological types of invertebrate neurons. Arrows point to the cell bodies. [From Bullock, 1952: Symp. Quant. Biol., *17*: 268.]

dendrites. They are monopolar with axo-axonal synapses. Frequently, the nerve cell body or soma is spatially distant from the conducting fiber (Fig. 21.3).

Neuroglia

Nerve cells are supported by a special connective tissue whose cells are called the NEUROGLIA or simply the GLIA. Neuroglia is present in the central nervous system and in ganglia; in most animals it also extends peripherally to surround the nerve fibers where the enveloping cells are often termed SCHWANN CELLS.

Glial cells are rare or absent among the coelenterates and ctenophores; they become progressively more conspicuous as the phylogenetic complexity of animals increases (Bullock and Horridge, 1965; Bunge, 1968; Horridge, 1968a). Among the vertebrates, the neuroglia ("nerve glue") includes the ependyma which lines the central nervous system, the neuroglia proper which mingles with the neurons of the central nervous system and the retina, the satellite or capsule cells of the peripheral ganglia, and the cells of Schwann which surround the nerve fibers.

Although the insulating functions of the peripheral neuroglia have long been recognized, the physiological importance of the central neuroglia and the ependyma have been less systematically studied. The central neuroglia has been assumed to be a supporting, insulating, or trophic tissue but much further work is indicated; it may contribute to the synaptic and storage ("memory") properties of the neurons (Galambos, 1961; Hydén, 1967).

Nerve Fibers

The membrane theory of nerve conduction was discussed in connection with the general physiology of excitable cells in Chapter 15. Some of the special properties of nerves as transmitters operating over long distances will now be examined.

The transmitting properties of a nerve fiber depend not only on the nature of its semifluid protoplasm but also on its diameter and its insulation. Although the electrical conductivity of protoplasm is high, the small diameter of the fiber creates an extremely great longitudinal resistance. The axis cylinders of human nerves vary in diameter from about 0.1 μ to 10 μ; Hodgkin (1964) calculates a resistance of about 10^{10} ohms per cm for a nerve fiber of about 1 μ diameter containing axoplasm with resistivity of approximately 100 ohm/cm. The resistance in a meter length of such a small nerve is comparable to that in 10^{10} miles of 22-gauge copper wire—a distance roughly ten times that between the earth and the planet Saturn. This formidable problem was solved during animal evolution by enlarging the diameter of nerves (other things being equal, conduction is proportional to the cross-sectional area) and/or by increasing the insulation. The former solution is most characteristic of the invertebrates and the latter of the vertebrates. Representative conduction rates are given in Table 21.1, and more detailed comparative tables may be found in Prosser (1973) and in Nicol (1967a).

Table 21.1 Conduction Velocities of Representative Nerves. (*A* fibers are myelinated somatic; *B* fibers, myelinated autonomic; *C* fibers, non-myelinated; temperatures for poikilotherms, about 21°C; homeotherms, body temperature)

Nerve	Conduction velocity (m/sec)	Fiber diameter (μ)
Calliactis nerve net:		
Mesentery, longitudinal	1.2	
Column, longitudinal	0.1	
Circular	0.15	
Radial	0.04	
Myxicola giant fiber	6–20	100–1000
Lumbricus:		
lateral giant	7.5–15	40–60
median giant	15–45	50–90
Loligo giant fibers	18	260
	35	520
Homarus leg nerves	2–10	35–70
Ameiurus Mauthner's fiber	50–60	22–43
Esox olfactory nerve	16–24	
Rana motor nerve	20–30	10–15
Dog: *A* fibers	100	
B fibers	4.5	3
C fibers	0.6	1–5
Man, sciatic nerve	65	1–30

SOURCE: Data from Spector, 1956: Handbook of biological data. Saunders, Philadelphia.

Giant nerve fibers Giant fibers occur in many of the more complex invertebrate groups (particularly the annelids, the crustaceans, and the decapod cephalopods) and in many of the anamniote vertebrates (Stefanelli, 1951; Nicol, 1967a). They are characterized anatomically by their large size which is usually greater than that of ordinary fibers and, physiologically, by their role in the coordination of quick withdrawal movements of a startled animal.

Two types of giant fiber have been distinguished on the basis of their ontogeny. One of these is clearly the enlarged axon of a single cell; the other develops through the fusion of several or many neurons (Fig. 21.4). The multicellular type has been considered more primitive although there is no clear phylogenetic trend of decreasing multicellularity. Within a single closely related group of animals such as the polychaete worms, some of the species have a few long unicellular fibers while others have complex multicellular types formed through the fusion of hundreds of individual neurons. Both types of giant fiber also occur in the more advanced phyla. Among the aquatic vertebrates, the giant fibers (called MAUTHNER FIBERS) are clearly unicellular and morphologically distinct. There are usually two of them with cell bodies in the medulla and giant axons which pass caudad in the spinal cord.

In contrast, a true fusion of many separate axons has long been recognized in the cephalopod molluscs. Hundreds of axons, arising in cells concentrated in a ganglion,

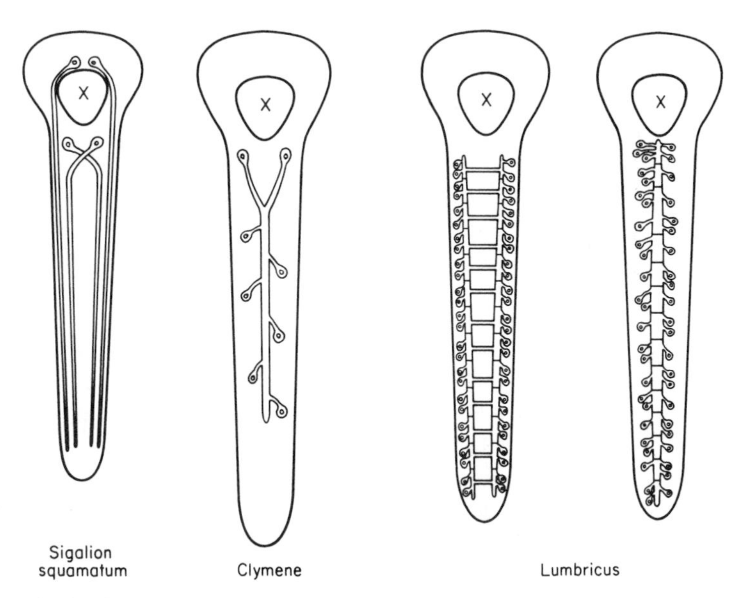

Fig. 21.4 Representative giant axons from annelids. X, esophagus. From left: Unicellular giant axons arising in the supraesophageal ganglion (upper) and the anterior nerve cord of *Sigalion squamatum*. Multicellular giant axons of *Clymene*. Paired lateral and single median giant axons of *Lumbricus* illustrating the multicellular septate variety; the lateral giants are connected by cross anastomoses. See Wilson (1961): Comp. Biochem, Physiol., *3*, 274. [Based on Nicol, 1948: Quart. Rev. Biol., *23*: 295, 301, 304.]

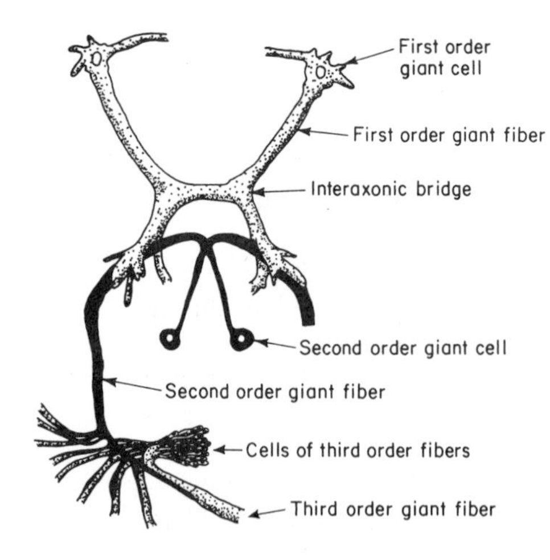

Fig. 21.5 Giant nerve system of the squid *Loligo pealii.* Cells of third-order fibers form stellate ganglion. [From McLennan, 1963: Synaptic Transmission. Saunders, Philadelphia; based on Young, 1939: Phil. Trans. Roy. Soc. London, *B229*: 465.]

fuse to form giant conducting cables (Fig. 21.5). The third-order giant fibers of the squid *Loligo* represent one of the most highly organized examples of this type with axons which receive contributions from as many as 1500 nerve cells (Young, 1939). In some of the segmental animals like annelids and arthropods, a multicellular septate type of giant fiber has been described. Oblique septa or partitions (synapses), indicative of a segmental origin, divide the fibers segmentally (Fig. 21.4).

Some of these descriptions are based on work carried out many years ago using routine techniques of anatomy and light microscopy. Mulloney's (1970) work on the earthworm indicates that these multicellular systems, particularly those of the annelids, should be reexamined with modern methods. Mulloney failed to find any evidence of fiber fusion when he injected a fluorescent dye into giant axons of the earthworm (Fig. 21.4). Septa form the anterior and posterior boundaries of the axons which are connected longitudinally through tight electronic junctions to form the median and lateral giants. These fibers, which have long been described as multicellular, appear to be unicellular. Although the evidence for fiber fusion seems satisfactory in cephalopods, older morphological descriptions of multicellularity should now be reexamined.

Nerves which are classed as "giant" vary in diameter from about 20 μ to something over 1500 μ in the very large multicellular fibers of the squid *Loligo* or the tubicolous polychaete *Myxicola*. Some ordinary nerves reach this lower limit. Lateral line nerves of a teleost fish often measure 20 to 60 μ in diameter; the Mauthner fibers in the same group also fall within this range. Although the mammals do not have giant fibers, some of their nerves may attain a diameter of 20 μ.

The distinction between giant and ordinary nerves is only partially based on size. There is a fundamental physiological difference. The giant fiber system serves the rapid coordination of large groups of muscles in quick escape or withdrawal responses. Three examples will illustrate this. *Myxicola* dwells in a mucoid tube from which it extends the anterior part of its body to feed and carry on other activities; it can withdraw into the tube with lightning rapidity when disturbed. This capacity rests on a highly organized giant fiber system where a single giant axon forms a final common path to all the longitudinal muscles concerned with the withdrawal reflex, and excitation can occur at all levels (Nicol, 1967a). The squid represents the climax in the evolution of a swimming mollusc. The quick and simultaneous contraction of a number of different mantle muscles operates a jet-propulsion system. These muscles are supplied by third-order giant axons arising in the stellate ganglion (Fig. 21.5). Since the axons to the different groups of muscles vary in diameter with their length and since transmission velocity is proportional to the size of the fiber, the many muscles can thus be operated in a unified manner. Finally, in the lower vertebrates, two large neurons (Mauthner's cells) with enormous cell bodies in the floor of the medulla send out axons which extend to the tip of the spinal cord. Fibers from neurons in the cerebellum, the sensory nucleus of the trigeminal, the vestibular nerve, and the optic tectum form synapses with Mauthner's cells. The system serves as a final common path for the speedy coordination of complex swimming movements. It is well developed in fishes and amphibians with functional lateral line systems and a pronounced tail musculature but reduced or absent in some forms which live on the bottom and in tailless species (Stefanelli, 1951). Diamond (1970) suggests that it may be particularly important in escape from diving birds.

Medullated nerves Although the velocity of the nerve impulse increases with the size of the conducting fiber, the most rapidly conducting nerves are not the giant axons but the much smaller medullated fibers of vertebrates. Some mammalian nerves (cat *B* fibers) conduct at rates of 80 to 100 m/sec while the median axon of the earthworm *Lumbricus*, one of the most rapidly conducting giant axons of invertebrates, transmits at about 30 m/sec. Mauthner's fibers in teleosts are also rapidly conducting giant fibers, but the conduction rates are only about half those recorded for cat *B* fibers. Bullock (1952) emphasizes the unsatisfactory correlation between size *per se* and conduction velocity in giant axons of invertebrates. Species differences in axoplasm and the nature of the insulation are presumed to be responsible for this variation.

Nerve fibers, invertebrate as well as vertebrate, are often covered by cells which provide insulation (in an electrical sense) and support or protection (in a mechanical sense). The development of these sheaths varies, with a definite trend toward increasing insulation in the phylogenetically higher animals. Glial cells appear to be absent from the neurons of coelenterates and ctenophores, but elsewhere naked neurons are rare except in the very small axons (0.5 μ). Single Schwann cells may provide the only covering in many fibers of both the invertebrates and the vertebrates (NON-MEDULLATED NERVES); large numbers of naked axons may be bundled together within a single sheath or Schwann cell. In the larger and more complex animals, where transmission distances are greater, the Schwann cells often wind around the nerves to form loosely or densely organized layers of myelin (MYELINATED or MEDULLATED NERVES). In mammals, all nerves greater than 1 μ in diameter are covered with myelin which may reach a thickness of 2.5 μ on larger fibers. In some crustacean nerves the organization of the myelin sheath may be just as complex as in the higher vertebrates (McAlear *et al.*, 1958; Bullock and Horridge, 1965). This

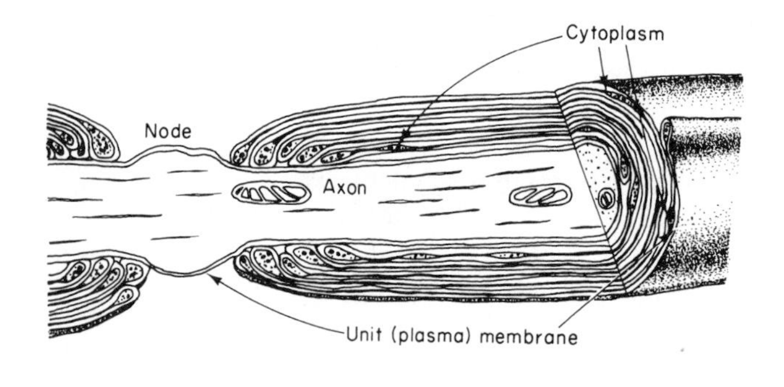

(a)

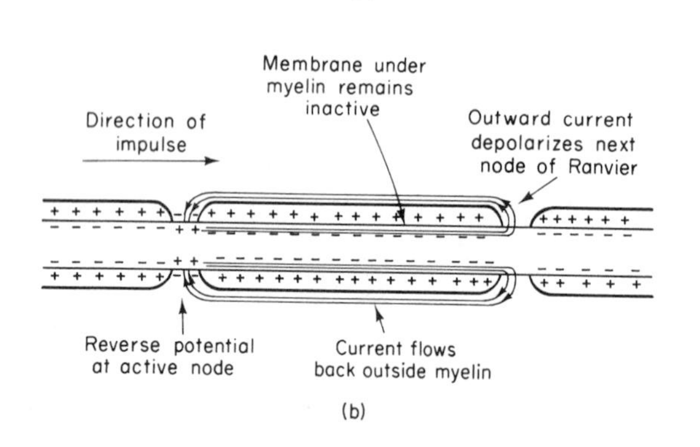

(b)

Fig. 21.6 Myelinated nerve fibers. (a) Three-dimensional diagram showing microanatomy of myelinated mammalian nerve. [From Bunge and Ris, 1961: J. Biochem. Biophys. Cytol., *10*: 79.] (b) Saltatory conduction in myelinated nerve fiber. Only the membrane in the nodal region becomes active; current flows from the next nearest resting node back over the outside of the myelin and returns in the core of the fiber. [Based on Hoyle, 1962b: Am. Zoologist, *2*: 13.]

substance is mainly composed of lipids (cholesterol, phospholipids, glycolipids) with smaller amounts of protein; it is similar biochemically to the cell membranes from which it arises. A few exceptions to the general trend in phylogeny of myelin development have been noted; this substance seems to be absent from the spinal cord of the cyclostome *Entosphenus* while some nerves in fishes are covered with loose rather than dense layers of myelin (Bullock and Horridge, 1965).

Seen with the ordinary light microscope, the myelin sheath of a vertebrate medullated nerve is a homogeneous osmophilic layer closely surrounding the axoplasm and neatly covered with the cytoplasm of Schwann cells (Fig. 21.6). Outside this again are the connective tissues which will be omitted from further discussion. Deep constrictions in the myelin sheath (nodes of Ranvier) occur at intervals of about 1 mm. At these points the myelin is interrupted, and the axoplasm is covered only by the neighboring Schwann cell. A single Schwann cell nucleus is found in each internodal segment and this is, in fact, the nucleus of the cell responsible for the myelin of this region. One of the very remarkable achievements of electron microscopy has been the demonstration that myelin is a laminated material with regularly arranged layers of lipid molecules alternating with thinner layers of protein; these lipid-protein "sandwiches" measure about 18 mμ perpendicular to the planes of the layers. The Schwann cells produce these layers, wrapping themselves around the axon by continuous infolding of the outer Schwann cell membrane as indicated in (Fig. 21.7).

The physiological importance of the insulating myelin sheath is well established.

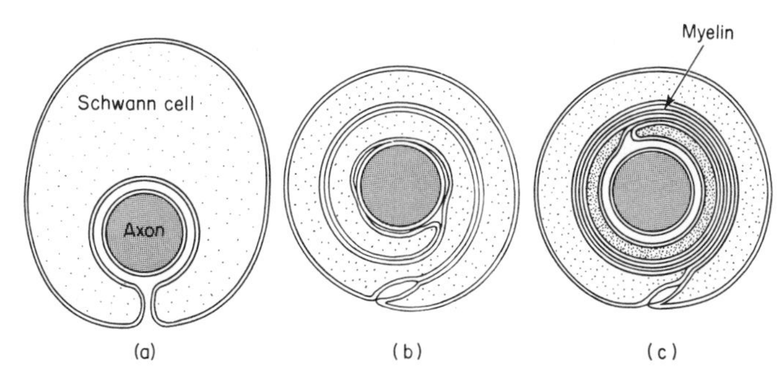

Fig. 21.7 Diagram of the structure of the medullated nerve fiber based on Geren's (1954) theory for the origin of myelin sheath. [From Robertson, 1961: Ann. N.Y. Acad. Sci., *94*: 344.]

Its significance in transmission might be argued from comparative measurements of conduction velocities. Thus, in a shrimp, a giant axon measuring 50 μ diameter, with the myelin sheath forming about 20 per cent of the fiber thickness, conducts with the same velocity (about 25 m/sec) as a squid giant axon of 650 μ lacking this insulation (Nicol, 1967a). The physiological studies of vertebrate medullated nerves have, however, given much more concrete evidence concerning the insulating capacities of myelin and the part which it plays in the conduction of the nerve impulse. Hodgkin (1964) summarizes the evidence. Much of it is based on the application of microphysiological techniques to single myelinated nerve fibers of frogs. Internodal distances may be as great as 1 mm, and this permits precise application of stimuli with respect to nodes and internodal regions.

The threshold of an electrical stimulus, applied at various points along such a fiber, is lowest when the cathode is opposite the node and highest when it is at the middle of the internode. When the current is confined to the internode, the threshold is virtually infinite, but if the cathode and anode are at different internodes, the spread of current along the fiber will act at an intermediate node. Many different experiments have shown that the effective stimulating current acts at the node. Moreover, temperature changes or blocking agents such as cocaine and urethane act at the nodes but have no effect on the internodes. These experiments are in good agreement with the theory that, during excitation, depolarizing permeability changes associated with the flood of Na^+ and loss of K^+ occur only at the nodes and that a "cable conduction" takes place through the internodal areas. If this is true, measurements of action currents over different areas of nerve during the passage of an impulse should show the typical diphasic potential change at the node but only a leakage or outward current along the internodes. This again has been convincingly supported by several ingenious experiments. One of these is illustrated in Fig. 21.8; further details are summarized by Hodgkin (1964) and many textbooks of medical physiology.

In summary, myelin provides electrical insulation; the generation of the membrane potentials associated with the passage of the nerve impulse takes place only at the nodes of Ranvier (Fig. 21.6). In non-medullated nerve and muscle, the spread of depolarization is continuous along the plasma membrane and conduction is a uniform

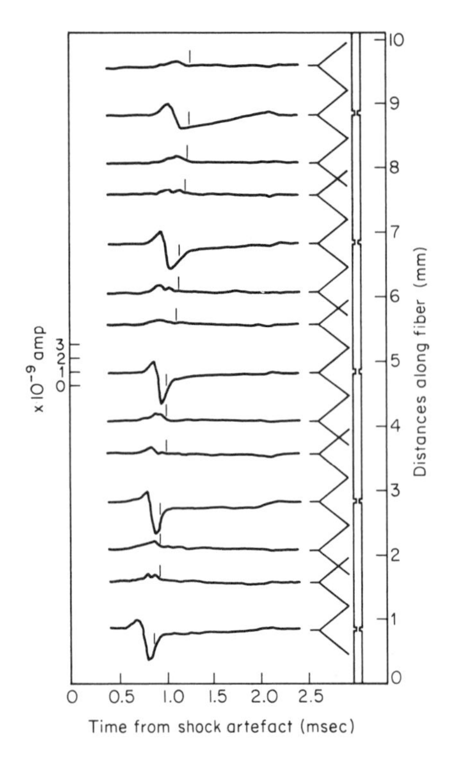

Fig. 21.8 Distribution and time course of membrane current in frog's myelinated nerve fiber. Each curve shows the difference between the longitudinal currents at two points 0.75 mm apart; the position of these two points relative to the nodes is indicated on the right. Vertical marks show the time of peak membrane potential. Outward current is plotted upward. [From Huxley and Stämpfli, 1949: J. Physiol., *108*: 327.]

process; in medullated nerves, the active generation of current is confined to the nodes of Ranvier, and the impulse behaves as though it jumped from one node to the next (SALTATORY CONDUCTION). Actually, it does not "skip from node to node" but spreads along the internodal regions at a finite rate by cable conduction from amplifier to amplifier. Saltatory conduction is an evolutionary achievement of the vertebrates and possibly of some higher invertebrates. Although nodes are present in myelin sheaths of a few of the higher invertebrates (crustaceans), these are irregularly arranged and conduction has not yet been shown to have the saltatory features which have been demonstrated in vertebrate nerves. This type of transmission confers two distinct advantages. The first of these is high conduction velocity; a frog's medullated fiber of 20 μ diameter conducts at the same rate (20 m/sec) as a squid's giant axon of 500 μ diameter. The second advantage is one of metabolic economy. If only the nodes are depolarized, there is relatively less movement of ions across the membranes and appreciably fewer demands on the "ion pumps."

INTERNEURONAL TRANSMISSION

Sir Charles Sherrington (1861–1954) who spent a long and productive life studying the integrative action of the nervous system first applied the term "synapsis" to the point of contact between two nerve cells. He and the pioneer physiologists of the late nineteenth century established several of the important properties of synaptic trans-

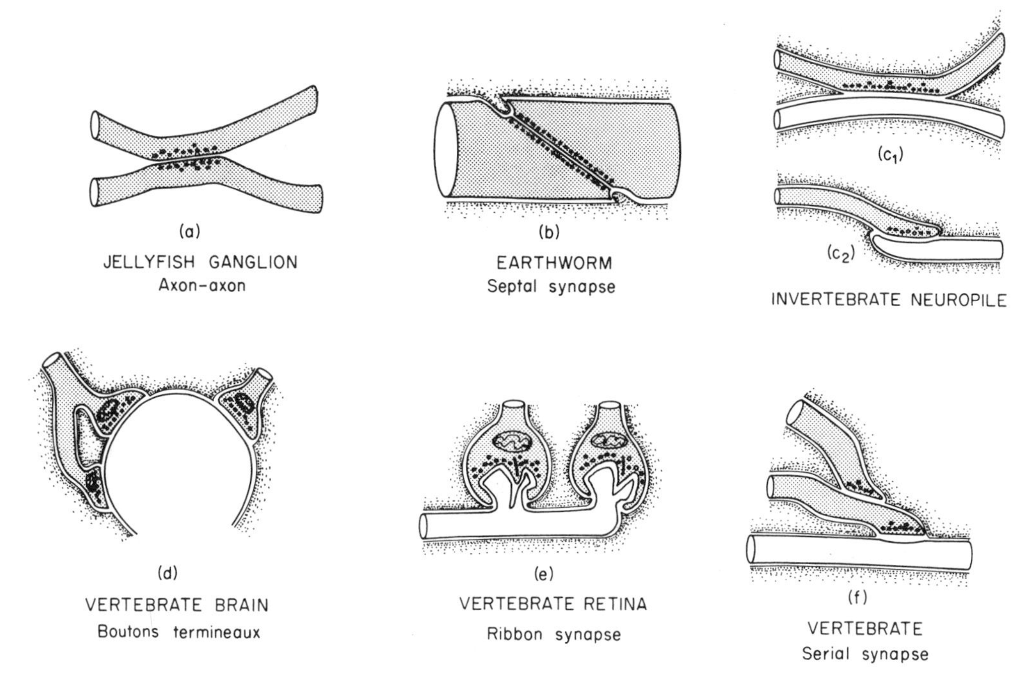

Fig. 21.9 Some types of synapse revealed by electron microscopy. Stipple, presynaptic fiber; clear, postsynaptic fiber, fine surrounding line and stipple, the innermost extent of sheath. (a) *Cyanea* ganglion with vesicles on both sides and without adhering sheath; (b) earthworm with close apposition of membranes and sparse vesicles; (c_1) axon-axon synapse *en passant* typical of neuropile in many invertebrates; (c_2) axon to fine dendrite; (d–f) typical vertebrate synapses; the serial synapse offers many possibilities for presynaptic inhibition and other complex interactions. Further examples and details in original. [From Bullock and Horridge, 1965: Structure and Function in the Nervous Systems of Invertebrates. W. H. Freeman and Company. Copyright 1965. San Francisco.]

mission long before there was any concrete supporting morphological evidence for the phenomena which they described (Eccles, 1964). In particular they noted that there was a certain delay in the passage of an impulse over a synapse, a susceptibility to fatigue, and a definite polarity or one-way transmission not found in nerve fibers. Moreover, synaptic phenomena were found to be less stereotyped than those in nerve fibers and permitted a SUMMATION and AFTER DISCHARGE which was very different from the all-or-nothing action found in nerves. The first electron microscope studies of the synapse, made about 1953, revealed a very real morphological basis for the delay, polarity, fatigue, and potentialities for summation and after discharge.

It has already been argued that, phylogenetically, an electrotonic transmission through tight junctions between epithelial cells probably preceded chemical transmission between neurons. Thus, it is not surprising that tight junctions with electrotonic conduction unite some neurons and that this situation is more frequently encountered among the invertebrates. The rich variety in synaptic connections (Fig. 21.9) was not fully appreciated until the advent of the electron microscope. Physiologically, they can be classed as either TIGHT JUNCTIONS characterized by direct electrical or ephaptic transmission or SYNAPSES proper where transmission is chemically mediated.

Tight Junctions (Ephaptic Transmission)

Tight electrotonic junctions are recognized at many axo-axonic connections in the giant fibers of invertebrates and in certain fishes at the synapses of giant (Mauthner neuron) fibers with motor neurons. In contrast to the chemically mediated synapses, these are usually characterized by an absence of presynaptic vesicles and by extremely narrow spaces between the synaptic membranes; clefts usually measure 200 Å or less in contrast to spaces of 250 to 500 Å for other synapses (Bullock and Horridge, 1965). Regions where the membranes actually fuse occur in the electrotonic junctions of several different fishes (Bennett *et al.*, 1967). The delay in transmission at a tight junction (0.12 msec in crayfish, 0.2 to 0.5 msec in fishes) is about half or less than half of that which occurs at chemically mediated synapses (cf. Figs. 21.10 and 21.20).

In principle, ephaptic transmission is the same as impulse propagation in an axon (Grundfest, 1969). The septa are regions of low resistance; the action current generated on one side flows into the adjacent segment and excites it to produce a spike potential. One might anticipate a lack of polarity and this is usually found. When an electrode is placed on either side of the septum of a lateral giant fiber in the crayfish, the electrical response is essentially the same whether the stimulus is delivered to the rostral or the caudal part of the nerve cord (Fig. 21.10). In a few cases, however, transmission at the electrotonic junction is definitely polarized and synaptic in the strict sense of the term (Furshpan and Potter, 1959; Grundfest, 1969). The electrical

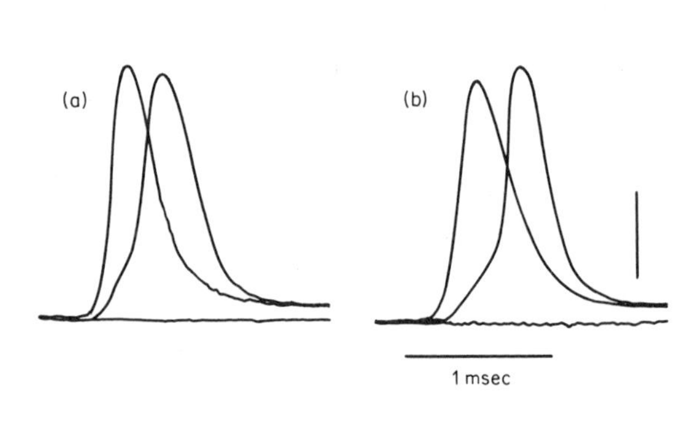

Fig. 21.10 Septal transmission in lateral giant fibers of crayfish. Two microelectrodes are inserted, less than 50 μ apart but on either side of the septum. In (a), a stimulus was delivered to the rostral side of the nerve cord and to the caudal side in (b). The spike of the rostrad segment was slightly larger. Transmission across the septum in either direction involved a septal potential. Caudorostral delay was somewhat longer than in the reverse direction. [From Watanabe and Grundfest, 1961: J. Gen. Physiol., *45*: 276. Permission of Rockefeller University.]

properties responsible for this one-way transmission have been discussed (Bennett, 1966; Katz, 1966; Auerbach and Bennett, 1969).

Inhibition, as well as excitation, is known to occur at electrotonic synapses (Katz, 1966; Horridge, 1968b; McLennan, 1970). It was first investigated in the Mauthner neuron of the goldfish (Furukawa and Furshpan, 1963). This particular system shows two inhibitory mechanisms: the familiar type which presumably depends on a chemical transmitter, and an electrical type associated with positive extracellular potential changes which hyperpolarize the Mauthner cell. The morphological site of this activity is the axon hillock which is surrounded by finely coiled fibers called the SPIRAL SYNAPSE (Fig. 21.11a).

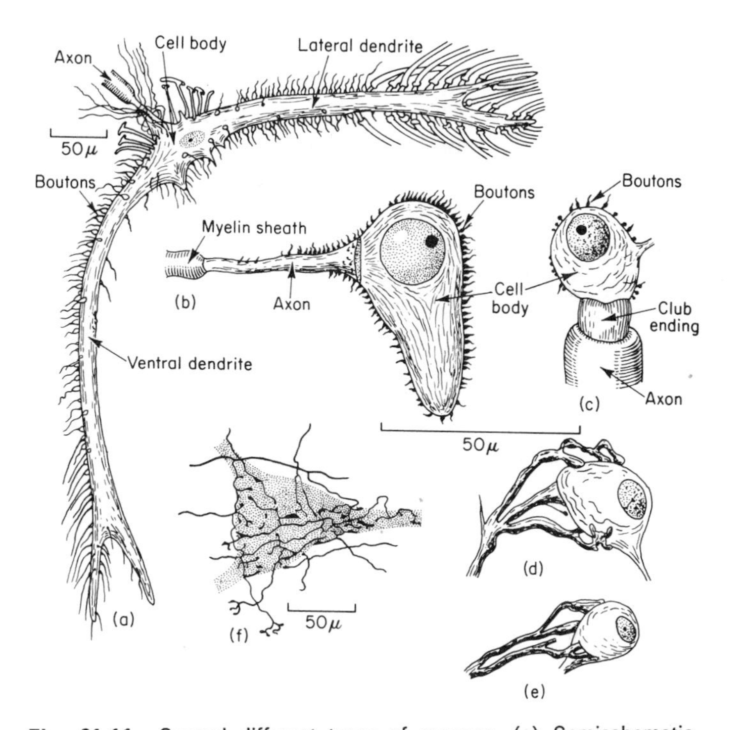

Fig. 21.11 Several different types of synapse. (a) Semischematic representation of the synaptic apparatus found on Mauthner's cell in the goldfish. The endings on the lateral dendrite are all of vestibular origin while those on the ventral dendrite and cell body come from other sources. (b) Large motor cell from the reticular formation of the goldfish showing relatively uniform distribution of homogeneous boutons on the cell body and proximal part of the axon. (c) Cell from the reticular formation of the goldfish showing a single large club ending as well as small boutons. (d) and (e) Two cells from the oculomotor nucleus of the goldfish showing a basket-like system of club endings derived from a single large branching axon. (f) Large interneuron from the spinal cord of a young cat. [From Young, 1957: The Life of Mammals. Clarendon Press, Oxford; after Bodian, 1942: Physiol. Rev., *22*: 148.]

Ephaptic transmission seems to be a feature of sudden activity like the flip of the crayfish tail or the "startle response" of a fish. Reductions by as much as 50 per cent in synaptic delay must have made its specialization biologically advantageous. It appears to be significant in synchronizing the activities of many nerves, as in the neuron pools which regulate electric organs of fishes and certain muscle groups (Grundfest, 1969; Kriebel *et al.*, 1969). There may be an added functional advantage in that the electrotonic synapse is immune to pharmacological and metabolic substances which might interfere with chemical transmission (McLennan, 1970).

Synaptic Transmission

The details of classical synapses, as revealed by the light microscope, are extremely variable. Several curious arrangements described in vertebrates are shown in Fig. 21.11. Axons may be applied to dendrites, to other axons, or to cell bodies; they may branch profusely, ending in numerous tiny swellings (*boutons terminaux,* END BULBS, END FEET), or they may divide only a few times to form a basket-like system which "holds" the postsynaptic cell. The molecular organization responsible for transmission is probably less variable.

A typical motoneuronal bouton is shown in Fig. 21.12. There may be as many as 2000 of these small bulbs (1 to 5 μ diameter) applied to a single motor neuron of the mammalian spinal cord, sufficient to cover about four-fifths of its surface area. Each terminal swelling contains many mitochondria and numerous transparent spheres or tubes, the SYNAPTIC VESICLES (diameter 200 to 600 Å), which presumably contain the transmitter substance. Vesicles are often clustered near the PRESYNAPTIC MEMBRANE which covers the bouton in the region of the postsynaptic cell. The delicate neurofibrils of the nerve fiber may be evident where the axon swells out into

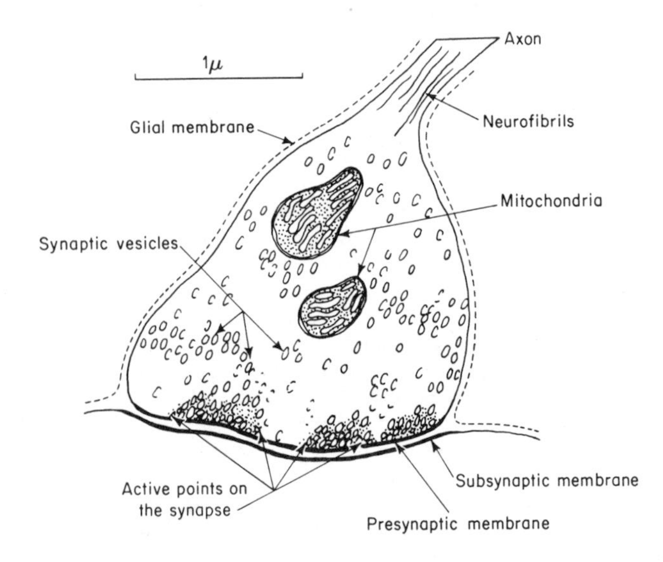

Fig. 21.12 Diagram of the fine structure of a typical motoneuronal bouton as revealed by the electron microscope. [From De Robertis, 1960: General Cytology, 3rd ed. Saunders, Philadelphia.]

the bouton. Presynaptic and subsynaptic membranes (about 60 Å thick) are separated by a narrow SYNAPTIC CLEFT (average width about 200 Å) which is continuous with the extracellular spaces and fluids. In the cortex of the brain, where synaptic clefts are wider, a system of intersynaptic filaments about 50 Å thick, appears to connect the two synaptic membranes while a web of filaments or fine canaliculi—the subsynaptic web—is attached to the subsynaptic membrane and penetrates the cell for varying distances (De Robertis, 1959; 1971).

There is an obvious anatomical similarity between the synaptic junction of the bouton and the end plate of mammalian skeletal muscle (Fig. 19.22). The physiological mechanisms are likewise comparable; these have already been discussed with the general properties of excitable tissues in Chapter 15 and with particular reference to muscle stimulation in Chapter 19. Evidence of the synaptic events has been frequently reviewed (Eccles, 1965; Miles, 1969; McLennan, 1970). In summary, the release of synaptic transmitter through the presynaptic membrane alters the ionic permeability of the subsynaptic membrane (chemically excitable valves). The resulting changes in electrical potential can be recorded with intracellular microelectrodes. During excitation there is an increased permeability to all ions, resulting particularly in a flood of Na^+ and a depolarization of the membrane, the EXCITATORY POSTSYNAPTIC POTENTIAL (EPSP). During inhibition the increased permeability involves only Cl^- and/or K^+, and stimulation is followed by hyperpolarization (INHIBITORY POSTSYNAPTIC POTENTIALS—IPSP). These potentials rise to a peak in from 1 to 2 msec and then decline exponentially with a time constant of about 5 msec. Like the potentials at a receptor membrane, synaptic potentials are graded in relation to the intensity of stimulus (quantity of transmitter released) and are thus very different from the all-or-nothing action currents of the axon or skeletal muscle cell.

Inhibition may be PRESYNAPTIC as well as POSTSYNAPTIC. In the presynaptic situation, activity is in some way blocked before the presynaptic spikes reach the synapse to trigger the release of the transmitter so that there is no change in permeability or conductance of the postsynaptic membrane. This form of inhibition has now been studied in several different preparations (neuromuscular systems of crustaceans, vertebrate central nervous system, abdominal ganglia of snails). The degree of the depression of the postsynaptic potential, as well as the recovery time from inhibition, depend on the intensity of priming stimulation (Tauc, 1966, 1967). Presynaptic and postsynaptic inhibition are contrasted diagrammatically in Fig. 21.13. Postsynaptic inhibition operates by modifying the postsynaptic membrane while the presynaptic type depends on effects which act at the level of the presynaptic membrane or excitatory terminal (Miles, 1969; McLennan, 1970).

In theory, excitation and inhibition might be due to either different synaptic transmitters or to differences in the synaptic membranes. Both possibilities seem to have been exploited by animals during their evolution. Acetylcholine, although only one of several transmitters, is the most firmly established of these substances (Chapter 2). Among the vertebrates, it is excitatory at the motor end plates of somatic muscles but inhibitory at the vagal endings in cardiac muscle. Similarly, in the gastropod *Aplysia* acetylcholine excites one membrane while inhibiting another. Particular cells

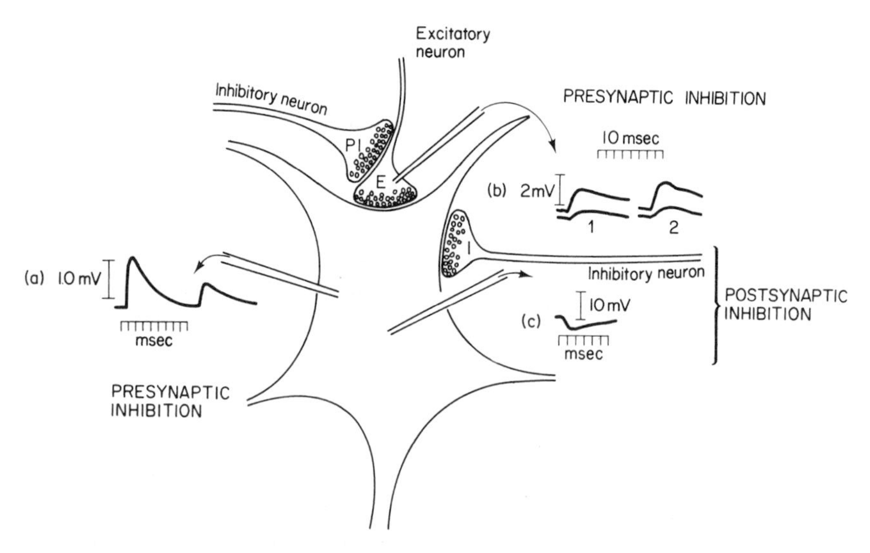

Fig. 21.13 Presynaptic and postsynaptic inhibition. An imaginary neuron bearing an inhibitory terminal (I), an excitatory one (E), and a presynaptic inhibitory one (PI). The following typical responses are shown: (a) normal and reduced EPSP before and during presynaptic inhibition; (b) intracellular (upper trace) and extracellular (lower trace) from an excitatory synaptic terminal (the difference between these traces shows the *depolarization* of the terminal itself as a result of presynaptic inhibition; this depolarization is thought to reduce the output of transmitter at each impulse); (c) typical hyperpolarizing IPSP. [After Eccles, 1964: The Physiology of Synapses. Permission, Springer Verlag, Heidelberg.]

in the abdominal ganglia respond to Ach by producing EPSP; these cells are termed D CELLS because of the depolarizing effect of Ach. Certain other cells in the same ganglia are hyperpolarized (H CELLS) in response to Ach and show IPSP. Both potentials may be blocked with curare (Fig. 21.14) and show other pharmacological responses typical of Ach mediation (McLennan, 1970). Furthermore, studies of these interesting ganglia have shown that specific individual nerve fibers may have opposite effects on different postsynaptic membranes (Tauc and Gerschenfeld, 1961). The results demonstrate that the action of the transmitter may depend on the nature of the synaptic membrane.

These different mechanisms of interneuronal transmission (inhibitory, excitatory, electrotonic, chemical), provide for many integrative features of nervous activity. In addition, however, specific spatial and temporal relationships are fundamentally basic to the functional organization of neurons. It will be recalled that the motoneuron may have as much as four-fifths of its surface covered with terminal boutons; the potentialities for SPATIAL SUMMATION, through the release of transmitter substance from many different boutons, are considerable, and the electrical events recorded with microelectrodes are fully in accord with such a concept. TEMPORAL SUMMATION is

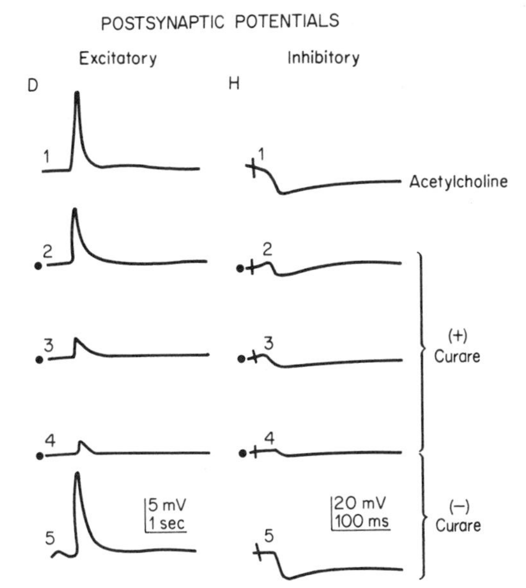

Fig. 21.14 Effects of *d*-tubocurarine (10^{-4} gm/ml) on the evoked EPSP of a D cell (at left) and on the IPSP of an H cell (at right) of *Aplysia*. 1, Control; 2, 3, 4, during continued contact with curare; 5, recovery after washing. [From Tauc and Gerschenfeld, 1961: Nature, *192*: 366.]

also recognized. Successive stimuli from the same presynaptic terminal add to the amounts of transmitter already there. A second subliminal stimulus, applied before the complete restoration of the membrane (up to about 15 msec), will produce a further permeability change; two or more such stimuli may thus summate to raise the potential to the critical value (about 20 mv for vertebrate motoneurons) where excitation occurs in the postsynaptic neuron.

In retrospect, this outline of synaptic transmission has described a rich variety of mechanisms at interneuronal junctions. The examples show that many of these properties are present in phylogenetically lower as well as in the more advanced animals; they suggest further that the greater integrative properties of the vertebrate brain will be found in the enormously greater number of neurons involved and not in the capacities of individual cells. In fact, at the level of the single neuron some of the invertebrates may be physiologically more versatile than the vertebrates; this, combined with the anatomical simplicity of their ganglia and nerve cords, accounts for the rewarding investigations of crayfish and snails (Kennedy, 1966).

INTEGRATING SYSTEMS OF NEURONS

Animal behavior depends on the integrative properties of the nervous system as a whole. The complexity increases phylogenetically from such very simple movements as the feeding reactions of a hydroid or the phototactic responses of a planarian to the intricate social behavior of the higher vertebrates. Even at the very primitive multicellular level of the coelenterates, which depend mainly on nerve nets, curious and prolonged sequences of precisely coordinated movements may follow rather simple stimuli. The swimming anemone *Stomphia* rocks to and fro, tugs itself loose, and swims away when stimulated by small amounts of secretion from the starfish *Dermasterias*; the anemone *Calliactis* hoists itself onto the shell of the hermit crab *Pagurus* in a series of seemingly deliberate movements; these are unrelated to any stimuli provided by the hermit crab but depend on a factor present in the periostracum of the empty mollusc shell occupied by the crab (Ross, 1965). This is integrative action at a high level and, it must be admitted, at a level which has defied satisfactory descriptions in physiological terms. The physiologist has, however, had some success in describing the integrative properties of smaller systems of neurons, and some examples of integration at this level will now be considered.

Integration is said to occur when the output from a cell, a group of cells, or an organism bears no direct relation to the input (Bullock, 1957; Horridge, 1963). Understanding of the physiological mechanisms is largely restricted to those properties of cell membranes which have already been described. Neurons are regularly excited to fire their own messages when stimulated by several neighboring neurons or by temporally separated stimuli from one neuron; such properties as facilitation and spatial and temporal summation are integrative properties at this level of organization. The descriptions which follow are in terms of cellular excitation and synaptic transmission as found in several different anatomical arrangements of neurons. The examples are not selected to show a phylogenetic sequence, but they do illustrate several different integrative properties of groups of neurons which are involved in animal behavior. Neurophysiologists in many places are searching for additional properties of neurons in groups and masses, for there is a conviction that complex behavior, memory, and mind can be explained in such terms (Bullock, 1958).

Coelenterate Nerve Nets

The nerve net represents the most primitive organization of neurons into a system for the integration of the behavior of an entire animal. It is the nervous system of coelenterates and, with modest ganglionic additions, remains the basis of the nervous system in several other primitive invertebrate phyla. Further, in phylogenetically more advanced groups the integration of autonomic and vegetative processes often depend on nerve nets. Hence, it is not surprising that zoologists have looked so carefully at the coelenterate nervous system for clues to the evolution of neurological mechanisms.

Morphological arrangement of neurons The coelenterate nervous system is basically a plexus of relatively short-fibered bipolar and multipolar nerve cells, together with neurites from sensory cells. The simple colonial hydroid *Cordylophora* seems to exemplify a very primitive situation with only one diffuse nerve net and no evidence of centralization (Jha, 1965). This pattern is relatively rare; most coelenterates, including *Hydra*, have at least two nets: a main plexus between the epidermis and the musculature, and a second less highly developed network associated with the gastrodermis and connected at various points with the epidermal plexus. Ever since the classical work of Schafer and the Hertwigs in the 1870's, the nerve net has been recognized as a synaptic association of neurons in which fibers make contact but do not fuse (Hyman, 1940; Pantin, 1952). Schafer was much impressed by the histological resemblances between this system and the autonomic plexuses of higher forms.

The coelenterates are a highly successful and extremely diversified phylum of animals, and it is not surprising that this basic pattern of the nerve net (as seen in a simple form like *Hydra*) is considerably modified in the larger representatives with their complex reactions. In *Hydra* the neurons are only slightly more concentrated in the hypostome and the pedal disc, where the plexus has a circular arrangement suggesting a nerve ring, but in the scyphozoans (jellyfishes) the neurons are concentrated and the fibers aligned to form a thick "through conducting" nerve ring at the margin of the bell. Special transmission lines often occur also along the radial canals, and in some groups there are radially arranged marginal sensory bodies (RHOPALIA) which contain ganglionic concentrations of neurons. In the mesenteries of an actinozoan such as the sea anemone *Metridium* some of the fibers are greatly elongated (up to 7 or 8 mm), and the net is stretched out to form an irregular ladder or lattice. These long bipolar nerve cells form a THROUGH CONDUCTION SYSTEM for rapid transmission and are comparable to the giant fiber systems of the higher invertebrates (Pantin, 1952). Such specializations may permit an independence of different functions. In the jellyfishes, a strongly developed two-net system provides for the separate regulation of feeding and swimming activities (Horridge, 1968b). In *Aurelia* the rapid coordination of the swimming beat depends on a network of very large bipolar cells or giant fibers while the feeding responses are controlled by a diffuse nerve net; these two nets interact in the marginal ganglia where a concentration of sensory cells presumably responds to environmental stimuli and where the pacemaker of the swimming beat can be modulated.

Mackie (1960) describes two distinct components in the ectodermal nervous system of *Velella*, a colonial, free-swimming hydrozoan (Siphonophora). The fibers of THE CLOSED SYSTEM are relatively large (diameters 1 μ to 5 μ and much more conspicuous than those of THE OPEN SYSTEM (fiber diameters 0.25 μ to 0.5 μ). The closed system is characterized by a continuity between neighboring fibers so that the neurons form a syncytium of the type visualized for all coelenterate nerve nets by the early anatomists. Mackie suggests that these connections between neurons are secondarily established through adhesion bridges formed during development and that this is, in fact, a giant fiber system comparable to that of the annelids. In the open system the fibers are not continuous but run independently, frequently coming close together to form synapses of the type described as *en passant*. Clearly, the coelenterates have adapted the nerve net to the processes of rapid transmission in their through conducting or giant fiber types of nerves and have achieved some measure of centralized control in sensory ganglionic masses.

Physiological properties of nerve nets In many classical studies, nerve nets have been regarded as unpolarized systems in which transmission could occur with equal ease in many directions. Although this may be true in some places, studies of the ultrastructure indicate a complexity of synaptic contacts which is comparable with that found in the more advanced nervous system (Horridge, 1968a; Lentz, 1968). Asymmetrical as well as symmetrical synapses have been identified (Jha and Mackie, 1967). Very large synaptic vesicles may occur on only one side of a wide synaptic cleft; these vesicles measure 100 to 150 nm in hydromedusae and 50 to 100 nm in jellyfishes compared to 20 to 40 nm in the vertebrates or 25 to 50 nm in arthropods. In the scyphomedusae, symmetrical synapses appear to be the rule; the synaptic clefts are narrow (about 20 nm) and small vesicles cluster inside both synaptic membranes. Symmetrical synapses of this type are also present in the marginal ring of the hydromedusae and in some echinoderms; it is not known if the vesicles are the same on the two sides of the membrane. It seems likely that symmetrical junctions devoid of synaptic vesicles occur also, and in some coelenterates (*Cordylophora,* for example), nothing comparable to a synapse has been identified. Thus, on the basis of ultrastructure alone it seems safe to assume that the coelenterates as a group possess both electrotonic and chemically mediated synapses. Histochemical studies are providing support for the hypothesis of chemical mediation (Lentz, 1968), and it seems likely that these animals utilize the same range of synaptic mechanisms found in higher forms.

Several of the physiological properties of nerve nets are understandable in terms of the large number of synapses. Transmission in the nerve net, by comparison with other nervous systems, is slow (Table 21.1); the nerve processes are relatively short and the large number of junctions decreases the rate of conduction. The actual synaptic delay, as calculated by Pantin (1952), is of the order of a few milliseconds (2.5 msec in the mesenteries of *Metridium*) and comparable to that of synapses in higher animals (squid stellate ganglion, 1 to 2 msec at 9°C and 0.5 msec at 20°C; cat spinal cord, 0.5 msec). Thus, the slow rate of conduction is caused by the number of synapses rather than by differences in the rate of synaptic transmission.

Even where morphological polarity is absent at the synapses, impulses may pass more easily across some parts of the nerve net than others. This is in part structural, due to varying lengths of nerve fibers and their distribution; it is in part also due to greater interneural facilitation in certain directions. In the coelenterate nervous system, facilitation is marked, and repeated stimulation is normally required to elicit a response. Examples of neuromuscular facilitation in the coelenterates have already been discussed (Chapter 19). The interneuronal facilitation has similar properties.

Interneuronal facilitation is also responsible for a "decrement" evident in the nervous conduction of some coelenterates, (Pantin, 1935; Carter, 1961). If the disc of a sea anemone is touched lightly, the edge of the disc will frequently bend inward; the bending is marked near the point of stimulation and gradually fades around the disc but extends progressively farther with stronger stimuli. The stimulus evidently becomes weaker as it passes over the net, seemingly in contradiction to the principle of an all-or-nothing response in nerves. Actually, this decrement is observed only

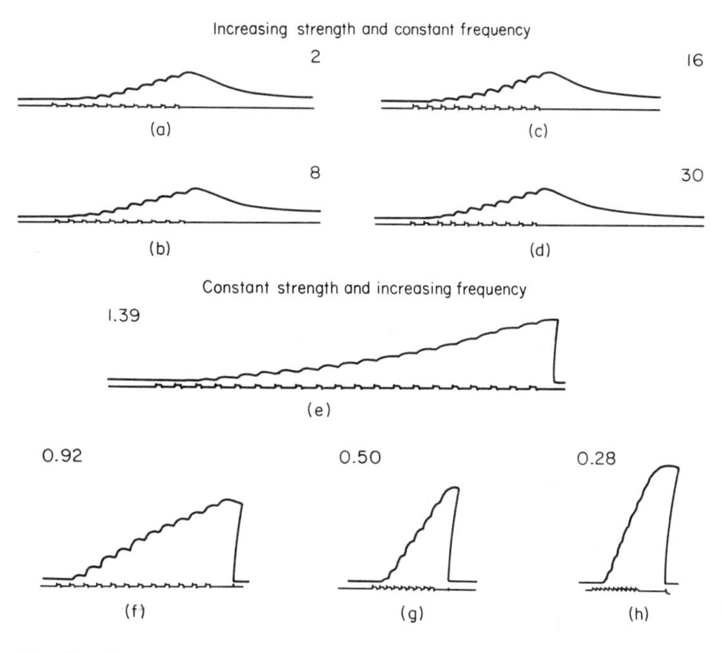

Fig. 21.15 Response of the sphincter muscle of *Calliactis*. (a), (b), (c), and (d) Responses of the sphincter to stimuli of different strengths but constant frequency (1 stimulus/sec). The threshold was 1.8 and the intensity was increased as shown by the numbers on the right of each trace. (e), (f), (g), and (h) Responses to a series of increasing frequencies of stimulation. Numbers on these traces are intervals (in seconds) between the stimuli. [From Pantin, 1935: J. Exp. Biol., *12*: 126.]

in response to repeated stimuli such as a tactile (pressure) stimulus or a series of electric shocks; a single electric shock either sets up no disturbance or one that disappears suddenly. In a strongly facilitated system of neurons the first of a series of repetitive stimuli will only facilitate the receiving neuron while the second stimulus may excite it to transmit to the next neuron which will in turn be facilitated. The third stimulus will thus excite the second neuron and so on. The response will depend on the number of stimuli rather than on their intensity (Fig. 21.15). Studies of different regions of the nerve net of the sea anemone show variations from areas which show little or no interneuronal facilitation to the highly facilitated condition just described.

Coelenterate nerve nets are also characterized by long refractory periods. In the *Metridium,* net refractory periods are 50 to 100 times longer than those of crustacean nerves which are, in turn, about three times longer than those of mammalian nerves (Pantin, 1935). These, however, like other properties, differ from those of other animals in degree rather than in kind. The same is true of inhibitory processes and spontaneous (pacemaker) activity. Inhibition among coelenterates was at one time denied, but it has now been studied in a particular system of

Tubularia (Josephson and Uhrich, 1969) as well as in several intact hydroids (Ross and Sutton, 1964; Rushforth, 1965); pacemaker activity was demonstrated many years ago in well-controlled experiments (Pantin, 1952).

One of the most significant facts revealed by studies of coelenterate nets is their dependence on the same basic physiology of excitation and transmission known in higher forms. In other words, the molecular basis of excitation and transmission was established in the primitive multicellular forms. It has been refined, but the mechanisms are similar at all levels in phylogeny.

Pacemaker Neurons

An unexpected feature to emerge from some of Pantin's later work on the sea anemone was the discovery of spontaneous and periodically active pacemaker systems (Pantin, 1952). A slow, rhythmic activity may be recorded in an apparently resting actinian such as *Metridium* (Fig. 21.16). A coordinated series of rhythmic movements continues for hours under constant environmental conditions and in the absence of vibrations. The conclusion is that these movements are inherent, spontaneous, and timed by some, as yet unidentified, cell or group of cells within the animal. More complex movements may also arise spontaneously; a starved anemone sometimes goes through the entire series of feeding activities "as though starvation has so lowered the threshold that the complex machinery started to operate spontaneously." The rhythmic pulsations of a swimming jellyfish are much more obvious.

The behavior of primitive animals such as hydroids is basically dependent on systems of pacemakers. Electrophysiological recordings from the surface or subsurface of *Hydra* reveal two important pacemaking systems—one giving rise to bursts of contraction and the other generating a regular rhythmic series of potentials; additional pacemakers may also be present in this animal (Passano and McCullough, 1965). The spontaneous behavior of *Tubularia* depends on three interacting pacemakers: (1) a major system in the neck region, (2) another in the hydranth and (3) additional pacemakers in the tentacles (Josephson and Mackie, 1965). The cellular

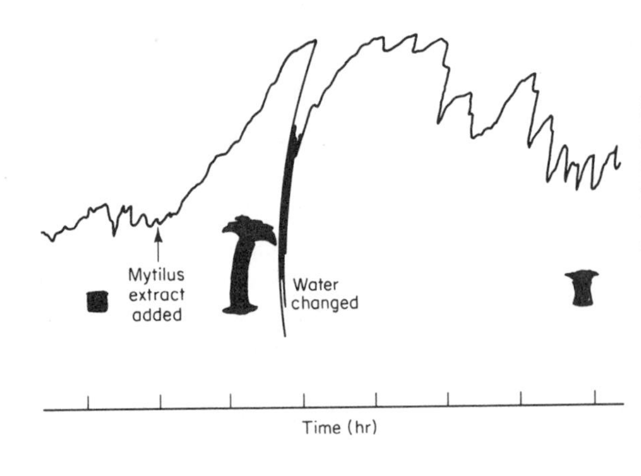

Fig. 21.16 Inherent activity of *Metridium* shown at the left and right sides of the figure with the feeding movements (induced by the addition of *Mytilus* extract) in the central part of the trace. The partially rhythmic contractions (registering downward) and extensions are recorded isotonically with the record running from the left. [From Pantin, 1950: Symp. Soc. Exp. Biol., *4*: 186.]

site of origin of these rhythms is uncertain. At present, it is by no means proven that this activity originates in neurons (Josephson and Mackie, 1965). Horridge (1968a) postulates pacemaking cells in the primitive non-nervous transmitting systems which preceded nerves in phylogeny. The most ancient rhythm-generating cells may not have been neurons.

Pacemaker systems have now been identified in several different tissues and in many animals at all levels in phylogeny. Embryonic limbs may show coordinated movements before the reflex arcs are formed; a pacemaker of myogenic origin sets the rhythm in the vertebrate heart; ciliated epithelia often show a metachronal rhythm; tube-dwelling, polychaete worms show spontaneous activity cycles associated with life in a burrow. Bullock (1961) cites these and other examples. The point of interest in the present discussion is that certain neurons may fire spontaneously and in a rhythmic manner. This capacity of nerve cells to generate activity independent of external stimulation evolved early in animal phylogeny and plays an important part in regulating behavior as well as in visceral functions. Many of these systems have not yet been systematically investigated. Two which have been analyzed in considerable detail will be described briefly.

The first of these is a FIXED-FREQUENCY ALARM CLOCK concerned with the production of characteristic sounds in the cicada *Graptosaltria nigrofuscata* (Hagiwara and Watanabe, 1956; Bullock, 1961). The duration of the sound varies with external stimulation, but the frequency of the buzzer is constant at about 100/sec. The sound depends on the rapid oscillation of certain muscles which are controlled on either side by a motor nerve arising in the mesothoracic ganglion. The triggering of the sound depends on sensory nerves from hair sensilla. The pacemaker is a single interneuron which fires impulses at a fixed frequency of 200/sec. This pacemaker interneuron excites the motoneuron controlling the sound-producing muscles. These muscles, left and right, are alternatively active; there is evidently a reciprocal inhibition of the two sides. Every other interneuron impulse fires the motoneurons to produce the muscle activity of 100/sec. Sensory stimulation changes only the duration of the burst; its oscillation frequency is fixed. This interesting system can be isolated experimentally; recordings have been made with microelectrodes placed on sensory or motor nerves or penetrating the pacemaker neuron in the ganglionic center. In this preparation, nervous integration may be studied in a nervous pathway involving only a few neurons.

The second example is the pacemaker system of the crustacean heart. In this case a group of nine neurons (the cardiac ganglion) maintains rhythmic activity but permits a variable rate. Lobster hearts (*Homarus americanus* and *Palinurus interruptus*) were used in many of the classical experiments (Maynard, 1955, 1960; Bullock, 1957, 1961). There is now convincing experimental evidence that the cardiac ganglion is the pacemaker of the decapod heart (Maynard, 1960). The heart (or any isolated part of it) displays spontaneous contractions only as long as the ganglion or some part of it is present. Moreover, the ganglion may continue to show electrical activity even after it has been isolated from the heart. In the intact heart, electrical activity of the ganglion precedes by about 10 to 14 msec both the electrical and mechanical events in the cardiac muscle. The rhythm of the decapod heart depends on a neurogenic pacemaker; the timing of the cardiac cycle can be readily modified experimentally and varies under physiological conditions in response to a variety of factors such as stretching, inorganic environment or drugs, as well as special cardiac-accelerating and inhibiting nerves (Maynard, 1960; Chapter 5).

The arrangement of neurons in the cardiac ganglion of the spiny lobster is shown diagrammatically in Fig. 21.17. The five anterior cells are considerably larger than the four posterior ones. Their arrangement is such that the individual neurons can be successfully penetrated with microelectrodes and their activities recorded. The ganglion is clearly a spontaneously active integrating system in which the neurons fire in a coordinated burst at the beginning of each systole and then remain quiet for the remainder of the cycle; this discharge pattern may remain constant for long periods.

The organization of the system is intricate, and the details of the interaction of

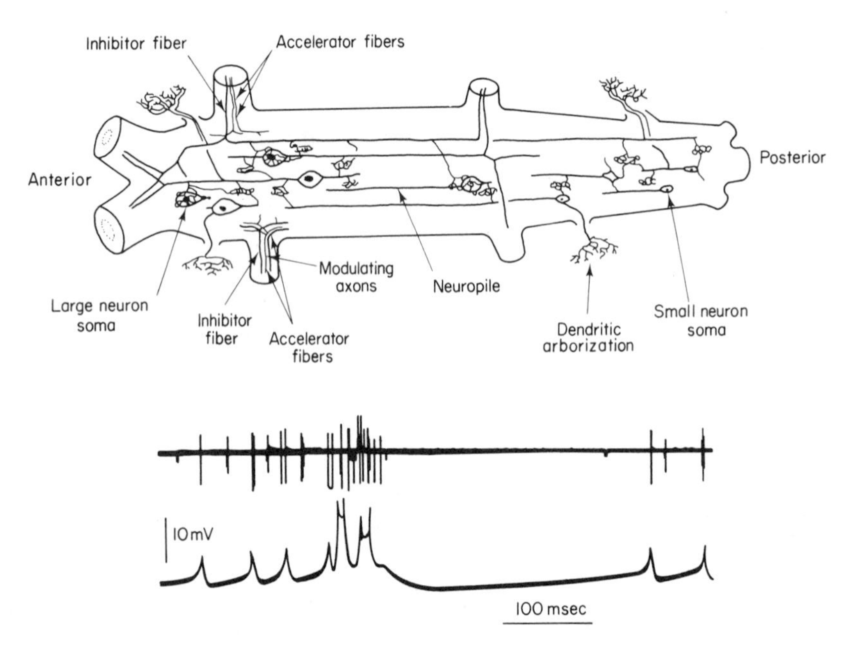

Fig. 21.17 Cardiac rhythm in the spiny lobster *Palinurus interruptus*. Above, diagram of the heart showing cardiac ganglion cells. The ganglion is about 12 mm long and the largest cells are about 50 μ; much of the ganglion is occupied by the neuropile. Below, typical burst of impulses in which all neurons of the ganglion participate (upper trace) and an intracellular record from a single follower cell (lower trace). [Cardiac ganglion based on Bullock, 1957: *In* Recent Advances in Invertebrate Physiology. Oregon State U.P., Corvallis, Ore.; records from Hagiwara and Bullock, 1957: J. Cell. Comp. Physiol., *50*: 39.]

its nine neurons have not yet been established. It has, however, been demonstrated that any particular burst of activity involves several impulses from each neuron and that no individual neuron fires at the burst frequency. Evidently, this is an integrating system in which the output bears no direct relation to the input.

The primary pacemaker resides in the posterior group of smaller cells and is probably a single cell—at least insofar as an individual patterned burst is concerned (Bullock, 1961). The five larger anterior nerve cells are the major motoneurons to the cardiac muscle; they are sometimes called the follower neurons. Not only does the excitability and responsiveness vary in the different neurons, but there is feedback of information among them. Several possible connections of neurons have been suggested (Bullock, 1957). The morphological arrangement of the neuropile is intricate and could satisfy several different theoretical organizations.

Ganglia and Central Nervous Systems

The nervous system of the platyhelminth has been likened to a coelenterate nerve net with a concentration of neurons in the head region. This is a misleading concept. Although the anatomical similarities in neuron arrangements are obvious (Fig. 21.18), a novel and significant phylogenetic step in integrative mechanisms was taken during the evolution of the primitive worms.

Planocera gilchresti is a large pelagic flatworm occurring intertidally on the coast of Africa. It crawls about and swims, feeds carnivorously and reproduces sexually by copulating with other individuals. Thus, it is capable of a wide range of behavior involving complex movements and reactions to many different environmental situations. The nervous system of *Planocera* consists of a brain from which seven

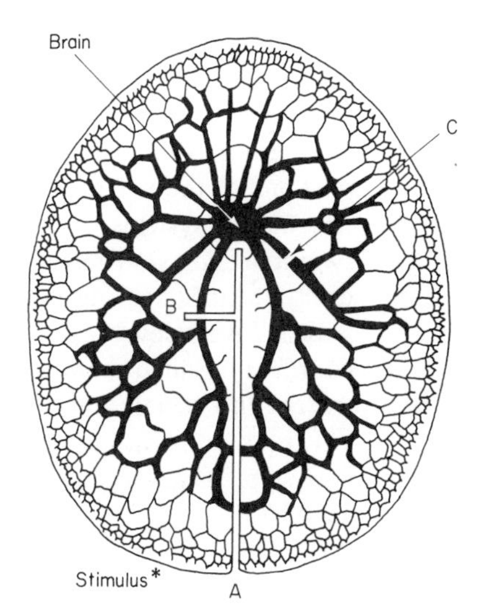

Fig. 21.18 Diagram of the nerve reticulum of *Planocera gilchresti* showing positions of incisions and point of stimulation as detailed in the text. [From Ewer, 1965: Am. Zoologist, *5*: 564.]

pairs of major nerve trunks pass peripherally and are interconnected to form a complex net (Fig. 21.18). Gruber and Ewer (1962) studied the effects of destroying various components of this system. They argued that if *Planocera's* behavior depends on an unpolarized nerve net of the coelenterate type, the destruction of parts of the reticulum should not unduly interfere with the flow of impulses, since alternate pathways should be available. This hypothesis was first tested by making a longitudinal incision as indicated at *A* in Fig. 21.18 and stimulating at the point marked by the asterisk. Contraction of the muscles on the opposite side occurred as in the intact worm and indicated that excitation spread from the point of stimulation through the brain. However, contractions were no longer evident in these muscles following the transverse incision shown at *B* on the left. A vital sensory pathway must have been destroyed. Clearly, something more than the reticular formation peripheral to the cut is required for the integration of sensory stimuli from the left side. In addition, definite motor pathways were demonstrated by making the incision shown at *C* on the upper right (Fig 21.18). The muscles peripheral to this cut no longer contract when the opposite side is stimulated or when the animal swims; swimming must be coordinated through a flow of impulses from the brain.

The brain is indispensable for coordinated locomotory behavior. Decerebrate worms move by gliding on their ciliated ventral surfaces but do not make the complex coordinated movements of crawling and swimming. Although feeding and egg laying continue, these activities also lack the coordination of the normal movements. Decerebrate worms persist in feeding behavior even though it becomes mechanically impossible for food to enter the pharynx; intact worms suspend feeding when satiated. Complex behavior such as copulation is never seen in the decerebrate animals.

Planocera and its allies possess separate sensory and motor pathways with a complex system of interneurons in the brain or cerebral ganglia. In brief, *Planocera* has a CENTRAL NERVOUS SYSTEM. The coelenterates, on the contrary, lack these features and depend primarily on unpolarized nerve nets. It has already been noted that considerable specialization is possible within the framework of the nerve net. Different nets may be associated with particular functions like feeding or swimming; portions of the net may concentrate to form rapidly conducting bundles of fibers; pacemaker systems are well developed in all forms; rapid withdrawal, rhythmic swimming and slow feeding activities are characteristic of many coelenterates. However, a central nervous system is absent and a new dimension in neural organization appears at the level of the flatworms with separate sensory and motor pathways and coordinating interneurons. Nerve nets and pacemakers are also present in the flatworms and remain important components of the nervous organization of the most advanced animals, but the appearance of ganglia forms a new element and marks a first step in the march toward the integrating systems of higher forms (Horridge, 1968b).

Invertebrate ganglia The ganglia of invertebrates show many structural similarities (Tauc, 1967; Kandel and Kupfermann, 1970). The somata or cell bodies of the neurons form an outer coat or rind which covers an inner core, the NEUROPILE. The neuropile is a thick network of nerve fibers with many ramifications and collaterals. Glia cells are present and discrete fiber tracts may often be traced through the ganglia (Fig. 21.19). Covering the exterior is a connective tissue capsule which may contain smooth muscle. The neurons themselves are usually unipolar and lack dendrites (Fig. 21.3); in the ganglia of invertebrates, dendrites occur only in the cardiac ganglia of crustaceans (Fig. 21.17) and in the first-order giant cells of the squid (Fig. 21.5). Ganglionic synapses are confined to the neuropile and are axo-axonic (Fig. 21.19).

In several of these features, invertebrate ganglia are in strong contrast to those of the vertebrates. Vertebrate ganglia lack zonation; neuron bodies are scattered throughout the ganglia and mingled with the fibers. Again, vertebrate ganglion cells often possess dendrites and the synapses are axodendritic or axosomatic in contrast to the invertebrates which have zones of synaptic impingement spread along the cell processes in the neuropile.

Total numbers of neurons, varieties of cell types, and complexities of the ganglia vary with both the level in phylogeny and the integrating responsibilities of the organ. The brain of the octopus is estimated to have 170×10^6 cells while that of the crayfish has 9×10^4 and the sea hare *Aplysia* only 1×10^4 (Tauc, 1967). In comparison, the human brain contains 1 to 2×10^{10} neurons (Young, 1964). With increasing complexity, greater numbers of individual cells are devoted to more elaborate functions; thus, 75 per cent of the cells in the octopus brain are concerned with the highly developed visual system (Young, 1963) while 95 per cent of the cells in the human brain are found in the cerebral cortex (Young, 1964). The varieties of cells also increase with the problems of integration. The brain of the flatworm *Notoplana* (Fig. 21.19) contains five named groups of cells; the brain of the octopus has about 50 named groups and the human brain several hundred.

Many of the current concepts of neurophysiology stem from studies of invertebrate ganglia (Kennedy, 1967). Although the complexity of neural organization in the vertebrate brain defies repeated investigations of precisely identifiable small

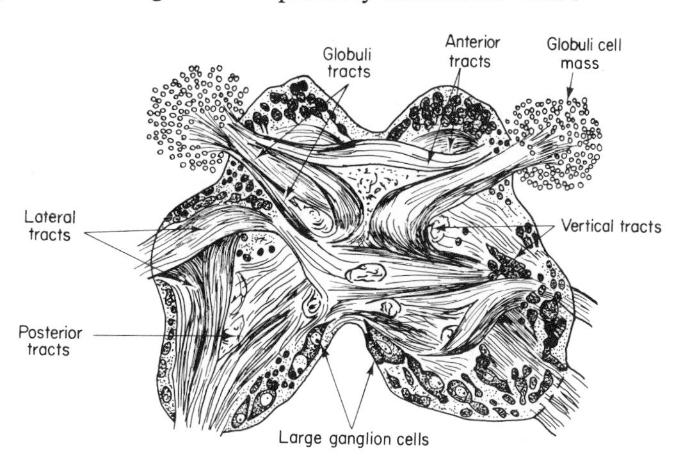

Fig. 21.19 Transverse section through the brain of *Notoplana atomata*. This is one of the most complex platyhelminth brains and shows features found in many ganglia of higher invertebrates; there are five distinct cell masses and ten named fiber tracts; further description in Bullock and Horridge (1965). [Based on Hadenfeldt, 1929: Zeitschr. Wissenschaft Zool., *133*: 607.]

groups of neurons, several invertebrate preparations are easily accessible for physiological manipulations and do permit such investigations. The giant cells of the sea hare *Aplysia,* the giant axons of the squid, and the segmental ganglia of crustaceans have provided classical material for the neurophysiologist (Hughes, 1967b; Kennedy *et al.*, 1969). Studies of these preparations have been used to illustrate several synaptic events already discussed (Figs. 21.10, 21.13, 21.14, 21.15). They also provide simple preparations for the study of behavior. Complex activities such as turning and swimming can be elicited by stimulating particular identifiable motor units in the opisthobranch *Tritonia* (Willows, 1967); small groups of neurons such as those found in *Aplysia* have been used to investigate the cellular mechanisms involved in simple learning and conditioning (Tauc, 1966; Kennedy, 1967; Horridge, 1968b).

Synaptic Transmission in the Stellate Ganglion of the Squid

The synaptic junction in the stellate ganglion of the squid is the only known chemically transmitting synapse which permits the insertion of microelectrodes into the synaptic regions of both presynaptic and postsynaptic fibers (Tauc, 1966). This unique preparation has provided classical information about the evolution and time sequence of synaptic events.

Young (1939) has detailed the morphology of this giant fiber system (Fig. 21.5). The nine to eleven third-order giant fibers innervate the mantle musculature which is concerned with swimming and ventilation of the mantle cavity. All the third-order giant fibers are excited simultaneously at the synapse. However, because they are graded in size, with the largest going to the most distant muscles, and because the rate of transmission is proportional to size, the different parts of the musculature are also excited simultaneously to produce the coordinated movements involved in the jet propulsion. Microelectrodes can be inserted into the presynaptic and postsynaptic elements to permit recordings of the electrical activity, following stimulation of the various nerves. The following important principles of synaptic transmission have been demonstrated with this preparation.

One-way conduction This is a typical polarized synapse and conducts in only one direction. Stimulation of the second-order giant fibers elicits activity in the third-order fibers, but the reverse is not true. A direct stimulation of the third-order fibers excites a conducted impulse in them, but this does not cross the synapse to activate the second-order giants.

Obligatory transmission The unfatigued preparation is a non-integrating system. Threshold or greater stimuli to the presynaptic fiber invariably evoke conducted impulses in the postsynaptic fiber; the transfer is a one-to-one phenomenon, and the synapse can be driven at rates up to 400/sec for short periods. This is a direct axo-axonic connection, and there is no evidence of facilitation.

Synaptic delay Recordings of the postsynaptic activity show an initial small decrease in membrane potential followed directly by the conducted action potential. The SYNAPTIC DELAY between the arrival of the presynaptic impulse and the local postsynaptic response is of the order of 0.5 to 1 msec (Fig. 21.20). The time interval is somewhat longer when measured to the point of the conducted action potential in the postsynaptic (POSTSYNAPTIC RESPONSE TIME).

Fatigue and temporal summation With repeated stimulation of the presynaptic fiber, the activity develops more and more slowly in the postsynaptic fiber and eventually fails to appear (Fig. 21.20). At this point of failure, local postsynaptic responses spread electrotonically for a short distance around the synapse but do not result in a conducted impulse. The system may now show integrative capacities since these local postsynaptic responses can be summed, and two presynaptic impulses arriving at a short interval will excite a postsynaptic. Under these conditions the synapse is not obligatory and shows TEMPORAL SUMMATION, a characteristic of integrative action.

These synaptic events are best understood in terms of the release of transmitter substance at the presynaptic membrane, and although the chemical has not yet been identified, the evidence for its existence is strong. Extracellular calcium is involved in release of the transmitter from squid presynaptic fibers (Miledi and Slater, 1966); the action of calcium ions in interneuronal transmission becomes more and more evident (McLennan, 1970).

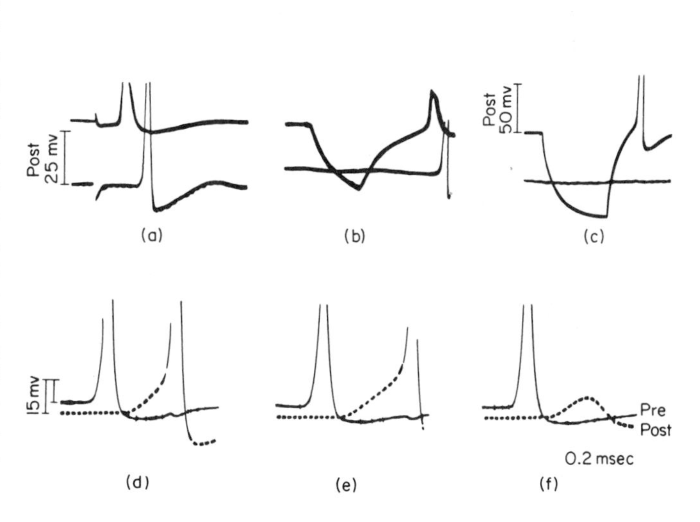

Fig. 21.20 Simultaneous recording of the action potentials in the presynaptic and postsynaptic fibers of the synapse of giant fibers in the stellate ganglion of the squid *Loligo pealii.* (a) Stimulus to the presynaptic (upper trace) excites the postsynaptic (lower trace) after a synaptic delay. (b) A hyperpolarizing pulse was applied to the presynaptic without effect on the postsynaptic (traces in same position. (c) Hyperpolarizing pulse applied to the postsynaptic (upper trace) without effect on the presynaptic (lower trace). (d), (e), and (f) Simultaneous records in the presynaptic (continuous line) and postsynaptic fibers (broken lines) taken just as transmission begins to fail, during prolonged high-frequency presynaptic stimulation. [From Hagiwara and Tasaki, 1958: J. Physiol., *143*: 118, 120, 121.]

Reflex Action

A reflex is a stereotyped response to a stimulus. The term implies a REFLECTION of stimulus excitation by the central nervous system to the effectors. The spinal frog has long been the classical preparation for its demonstration. Generations of students have now watched the suspended pithed frog respond to a stimulus of dilute acid applied to one of its toes. It is not difficult to devise experiments which will show that this simple FLEXION REFLEX depends on sensory nerves from the pain receptors of the toe and motor nerves to the muscles of the leg, with an area of spinal cord connecting them in between. The simplest of spinal reflexes (for example, the knee jerk) involves only two neurons, the dorsal sensory and ventral motor neuron with a direct synapse in the spinal cord (Fig. 21.21). Most reflexes, however, are considerably more complex, and this may also be demonstrated with the spinal frog by placing the acid on a different part of the animal. If the acid is placed under the arms, for example, the animal will wipe the spot with its leg or legs and "attempt" to remove the irritation. This is a complex reaction involving several different groups of muscles, some of which must be inhibited while others are excited. Obviously, many different connector neurons must be concerned with the integration since it requires responses on both sides of the body and in several segments of the spinal cord. Many familiar reflexes also depend on the special sense organs as receptors and the higher centers of the brain as parts of the connector system. Detailed discussions are given in texts of medical and vertebrate physiology. The term reflex is

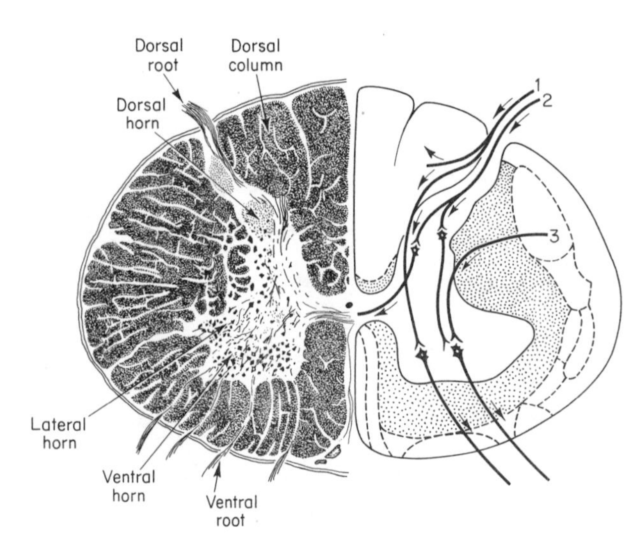

Fig. 21.21 Transverse section of the human spinal cord in the cervical region showing on the left the histological appearance with the medullated nerve fibers and cells stained. On the right, a diagrammatic representation of some fiber connections and pathways. (1) A monosynaptic pathway involving only two neurons as in the stretch or myotactic reflex. (2) Flexor and crossed extensor reflexes involving connector neurons and a crossing of fibers from one side of the cord to the other. (3) The corticospinal pathway by which impulses from the higher centers of the brain may modify the response. Areas surrounded by broken lines on right are nerve tracts running from brain through spinal cord. [From Young, 1957: The Life of Mammals. Oxford U.P., London.]

also applied to those homeostatic mechanisms which depend on the autonomic nervous system. Several of these—such as the cardiac and the vasomotor reflexes —have already been described.

Physiologists have sometimes regarded the reflex as the functional unit of the nervous system and looked upon an animal's behavior as a complex chain of reflexes. It is postulated that, during phylogeny, increasingly complex reflexes became organized into a hierarchy of movements through the integrating nervous system and, further, that an animal's repertoire of reflexes can often be modified through experience. An obvious distinction is made between the innate or unlearned reflexes and those which are learned through conditioning or other processes.

There are many unsatisfactory features in these descriptions of behavior in terms of reflexes. The most obvious is the failure to recognize the spontaneous activities which depend on pacemaker systems of the sort already described. A sea anemone displays a characteristic sequence of movements in response to food (Fig. 21.16), but the starved animal may spontaneously show a similar pattern of activity without apparent stimulation from the outside world. Bullock (1961) lists several rhythmic behavior movements which may occur in the absence of peripheral stimulation: the motoneurons to a swimmeret of a crayfish continue to discharge rhythmically after isolation of the abdominal cord from the periphery; the swimming movements of a medusa, the peristaltic creeping of the earthworm, the swimming of a leech, and the rhythmic flying movements of the locust depend on a neurogenic rhythm rather than a phasic input from the periphery. There is evidently a central automatism which persists in the absence of peripheral stimulation, although feedback from the periphery may alter or modify it. Bullock (1961) considers several theoretical possibilities and emphasizes the importance of pacemakers, neurogenic rhythms, and the feedback of information in the organization of animal behavior.

The student of animal behavior, as well as the physiologist, finds the reflex a very inadequate unit of animal behavior (Tinbergen, 1951). Classical reflexes occur as relatively direct responses to stimulation, while many of the instinctive movements or innate behavior patterns identified by the ethologist follow the releaser situation by relatively long intervals and continue as intricate patterns of movements for extended periods. Several examples are given in the final section of this chapter. Here, it is emphasized that the reflex, either innate or conditioned, is only one of the components of animal behavior. Unless the definition is substantially altered, the reflex should not be termed *the* functional unit of the nervous system.

Principles of Convergence and Divergence

Many of the properties of the integrating nervous system are due to the multiplicity of presynaptic endings applied to a postsynaptic cell. Several motoneurons, beset with numerous terminal boutons, are illustrated in Fig. 21.11. The dendrites and basal portions of the axon, as well as the soma of the cell, may serve as synaptic points. It is estimated that the soma of a motoneuron in the vertebrate spinal cord may receive several hundred to several thousand boutons and have up to 40 per cent or more of its surface covered with these synapses (Haggar and Barr, 1950; McLennan, 1970).

Sherrington (1929) described several of the physiological consequences of this architecture in his pioneer studies of the spinal reflexes. He noted that the result of a simultaneous stimulation of two or more afferents might be very different from the total effect of their separate stimulation. If the stimulation is of high intensity, the simultaneous effect is less (OCCLUSION); if the stimulation is of low intensity, the simultaneous effect may be greater (FACILITATION or SPATIAL SUMMATION). These integrative properties of neurons in groups find their explanation in terms of the arborizing presynaptic terminals. Any particular presynaptic makes contact at many points with several postsynaptic cells (PRINCIPLE OF DIVERGENCE). It follows that any postsynaptic serves as a FINAL COMMON PATH for several presynaptics (PRINCIPLE OF CONVERGENCE) as shown in Fig. 21.22.

Suppose that two boutons of either *A* or *B* must be activated to excite any one of the fibers on the right of Fig. 21.22. If *A* is separately stimulated, it will only excite fibers 1 and 2; likewise, *B* will excite 5 and 6, but neither 3 nor 4 will be excited under these conditions. If, however, *A* and *B* are simultaneously stimulated, then all six fibers will be excited because of the overlap or convergence on the central fibers leading to a summation or facilitation with respect to fibers 3 and 4. However, if only one active bouton is required to excite the fibers on the right, then either *A* or *B* stimulated by itself will excite four fibers but, stimulated simultaneously, they excite only six fibers, and there is an OCCLUSION of two fibers because of the overlap. The diagram is oversimplified and, in point of fact, several presynaptics are probably always required to excite a postsynaptic, but the principle is the same.

This divergence of presynaptic terminals to several postsynaptic neurons leads to the facilitation and occlusion which Sherrington noted when afferents were separately or simultaneously stimulated at different intensities. Each presynaptic may be expected to have a large number of terminals on some neurons and a smaller number on others. A weak stimulus to an afferent may excite only those neurons on which it makes many synapses but may set up a local excitatory state on the neighboring ones where it has relatively few junctions. However, if two neighboring afferents are simultaneously stimulated with relatively weak stimuli, then the neurons in their overlap areas or subliminal fringes will be excited, and the total effect will be greater

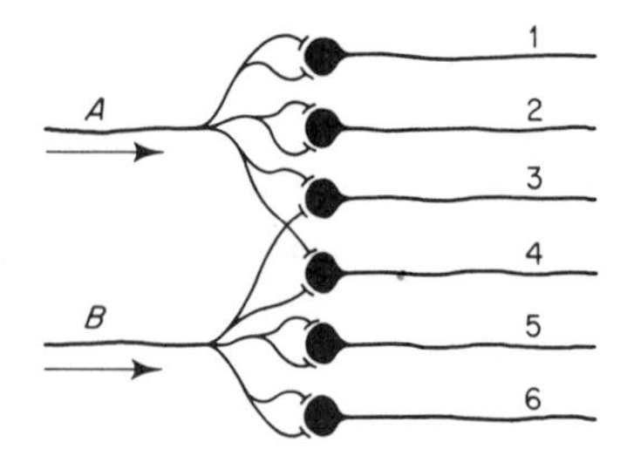

Fig. 21.22 Diagram to show occlusion and spatial summation. Explanation in text.

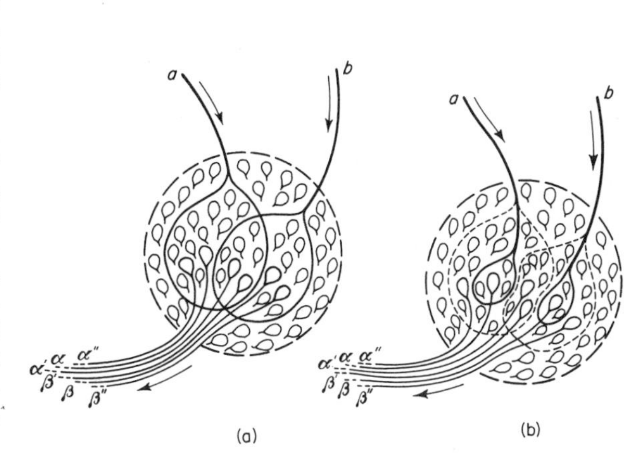

Fig. 21.23 Sherrington's classical diagrams of (a) occlusion and (b) summation. (a) Two excitatory afferents, *a* and *b*, with their fields of supraliminal effect in the motoneurone pool of a muscle. *a* activates by itself four units (α', α, α'', and β'); *b* by itself, four (β', β, β'', and α'). Concurrently they activate not eight, but six, i.e., give a deficit by occlusion of α' and β'. (b) Weaker stimulation of *a* and *b* restricting their supraliminal fields of effect in the pool as shown by the continuous-line limit. *a* by itself activates one unit; *b* similarly; concurrently they activate four units (α', α, β', and β) owing to summation of subliminal fields outlined by dots. (Subliminal fields of effect are not indicated in diagram (a). [From Sherrington, 1929: Proc. Roy. Soc. London, *B105*: 338.]

than the separate effects (facilitation or spatial summation). With intense stimulation the neurons in the overlap area are excited in either case, and the simultaneous stimulation leads to an occlusion. Sherrington's classical diagrams of these phenomena are shown in Fig. 21.23.

Action of Interneurons

Sometimes, as in the myotactic or stretch reflex, there is a direct linking of sensory input to motor output with only two neurons involved. These monosynaptic pathways, however, are rare. More frequently, masses of interconnecting neurons provide integrating connections between receptors and effectors. In the centralized ganglia of invertebrates and in the brains of vertebrates these internuncial cells have short fibers which appear to form a disorganized "felt" in routine preparations. However, it is usually assumed that there are definite patterns of organization and that these provide an anatomical basis for precise and purposeful behavior. What often appears to be a tangle of nerve fibers invites speculation and many different neuron patterns have been suggested (Bullock and Horridge, 1965).

One of the earlier studies, based on Golgi silver preparations and careful physiological experiments, grouped the many arrangements of interneurons in the mammalian brain into two basic patterns: the MULTIPLE CHAIN and the CLOSED CHAIN (Fig. 21.24). These patterns offer considerable scope for explanation of several integrative features of nervous action. In either pattern an excitation of the main

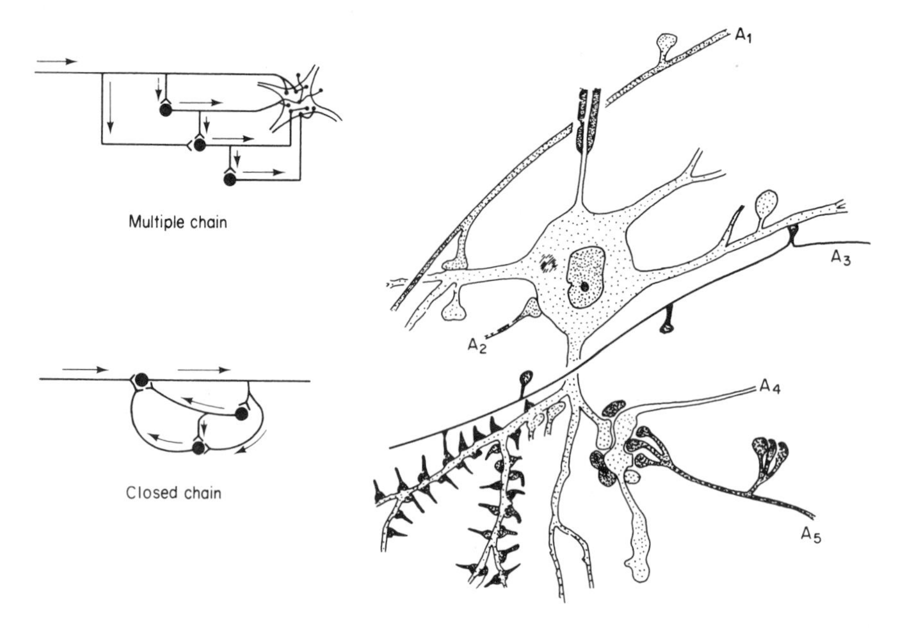

Fig. 21.24 Interneuronal connections in the mammalian brain. Left, two types of chains formed by internuncial cells in mammalian brain. Right, five different types of axons (A1-A5) which form synapses on a cell in the lateral geniculate nucleus of the cat; note the serial synapse A3 suggesting presynaptic inhibition. [Left, as described by Lorente de Nó (1938); right, from Peters and Palay, 1966: J. Anat., *100*: 480.]

transmitter will generate extended activity in its postsynaptic area. In the multiple chain a parallel series of neurons is connected through collateral branches, and the postsynaptic bombardment is prolonged (perhaps also rhythmic) in accordance with the number of synapses along the connecting route. In the closed chain the collaterals are interconnected to reexcite neurons and set up a reverberation of impulses with continued prolonged postsynaptic effects. Either of these architectural arrangements provides for a sustained action or after-discharge.

Even though multiple and closed chains may exist in the central nervous system, current understanding of neurophysiology argues strongly for models which incorporate additional features and are considerably more versatile. Inhibitory as well as excitatory synapses must be present and electron microscopy suggests that more than one type of each often occurs on a single cell (Fig. 21.24). At present, the discussions of organization of interneurons are almost completely speculative. Horridge (1968b) lists not only those features which seem to be general in all animals but also the important differences between the interneurons of invertebrates and vertebrates. Much more precise knowledge will be required before it is possible to understand the interaction of these small neurons, which are neither sensory nor motor, but filter multitudes of incoming action potentials and channel the coded information into meaningful directions to the effector organs. Bullock (1958) speculates that "the main factor in evolutionary advance is not just numbers of cells and connections" and that a satisfactory explanation of complex behavioral sequences, learning, and memory must probably await "the discovery of new parameters of neuronal systems."

Cephalization and a Hierarchy of Nervous Centers

A single bilobed ganglion forms the central nervous system of the platyhelminths and several other groups of primitive Metazoa. In the larger and more complex invertebrates, however, several ganglia are strategically located to take care of the affairs of particular geographic areas. Thus, the supraesophageal or cerebral ganglia at the anterior end dorsal to the gut, and the subesophageal ganglia below the gut, appear early in phylogeny with the differentation of a head, its complex feeding structures, and sensory instruments; the anus, genital organs, and other anatomical parts may have separate ganglia for their control. Annelids and arthropods possess a segmental nervous system which permits relatively complex segmental reflex activities independent of the cerebral ganglion or brain. In molluscs there are several ganglia but in the gastropods and cephalopods, which are the more active members of the phylum, nervous control is dominated by the brain and complex reflexes are not possible in decerebrate animals; pedal, pleural, and visceral ganglia act as local reflex centers.

This trend toward a dominating neural mass is associated with complex behavior and the increasing range of special sense organs. It reaches a climax in the vertebrate brain which is fundamentally different, both embryologically and structurally, from the ganglionic masses of the invertebrates.

At present, descriptions of the phylogeny of higher nervous systems are largely morphological; experimental work has been essentially an exercise in the localization of functions. An excellent recent example on a little-known nervous system is a study of the brain and behavior of the *Octopus* (Wells, 1962, 1966). The capacities of these animals for sensory perception and learning were first established through many careful conditioning experiments; lesions were then placed in different areas of the brain and their effects on behavior were noted. Comparable studies have also been made on the effects of brain lesions on the instinctive behavior of fishes (Segaar and Nieuwenhuys, 1963; Bernstein, 1970). Various experiments of a similar nature, extending over many years, have localized motor and sensory areas of higher forms while numerous rather fruitless attempts have been made to find the "memory trace" in mammals (Lashley, 1950). Valuable information of this type has come also from the association of pathological disturbances with morphological changes and from electrical stimulation of different areas of the brain. The fact remains, however, that these are *localizations of functions* and do not provide a satisfactory explanation of how the nervous system controls behavior.

A hierarchical organization of controls is apparent, both from morphological studies of the interconnections between nerve centers and from the experimental

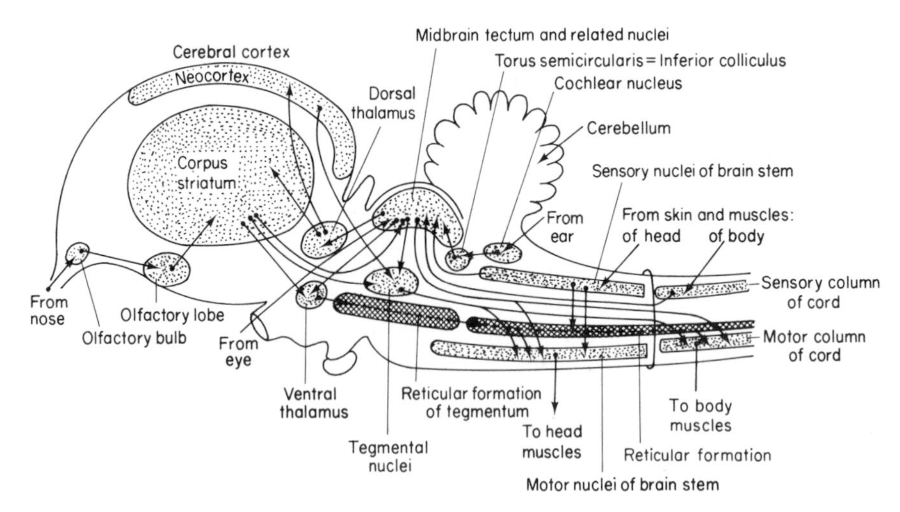

Fig. 21.25 Main centers and major nerve connections in a bird brain. The reticular formation (cross hatched) carries impulses to nuclei of the stem and cord. Many pathways, as well as the cerebellar connections, have been omitted. [From Romer, 1970: The Vertebrate Body, 4th ed. Saunders, Philadelphia.]

localization of functions. Purposeful behavior demands an integrating system of controls for the many separate ganglia or nerve centers, and this develops in the head region which receives the main input of sensory information from the environment. This process of encephalization is illustrated in the morphology of several different groups of animals. The vertebrates provide the most familiar example (Fig. 21.25). In fishes and amphibians the tectum (dorsal part of the midbrain) receives information from all other areas of the brain and serves as the highest level for overall integration of behavior. In amniotes the forebrain which develops first as an olfactory area becomes progressively more dominant. Among reptiles forebrain and midbrain are about equally significant as the dominant regulating centers. In birds the corpus striatum or basal ganglia assume this function, while in mammals the cerebral cortex (neopallium) receives sensory information from all parts of the body and transmits impulses to the motor centers which control the effector organs (Diamond and Hall, 1969). These relationships are detailed in textbooks of comparative anatomy (Romer, 1970).

PHYSIOLOGY OF BEHAVIOR

There is now a rapidly growing interest in the physiology of behavior. Physiologists, zoologists and comparative psychologists have been active in developing this exciting field. Physiologists, for their part, have extended the analysis of movements from nerve muscle preparations to simple reflexes, complex reflexes, and conditioned reflexes; in doing so, they have become more and more convinced of the inadequacy of the reflex as the unit of behavior and increasingly aware of the existence of spontaneous movements and rhythmic activities arising in the central nervous system. At the same time, students of behavior have extended their investigations from the descriptive to the analytical and, in attempts to understand the immediate causation of behavior, have adopted many of the techniques of the physiologist. Physiologists and behaviorists are meeting more and more frequently on common ground.

Many components of animal behavior have now been described in physiological terms. The sensory physiologists have been particularly successful in demonstrating the environmental cues which orient animal activities. Von Frisch (1971), in one of the outstanding classics of modern zoology, has shown that honeybees depend on their eyes and a keen sense of smell to locate nectar and that they use a stereotyped dance pattern associated with characteristic sounds to inform their fellow colony members of the location and direction of rich sources of food. Again Creutzberg (1961) finds that the shoreward movement of young eels, *Anguilla anguilla,* is determined by a tidal mechanism coupled with positive reaction to odors in the waters from the land. Numerous interesting demonstrations of this sort have now been made; several were described in Chapters 17 and 18, devoted to sensory physiology.

Sensory physiology is only one aspect of the physiology of behavior. Neurophysiologists are actively seeking areas of the brain and integrative mechanisms responsible for complex behavior patterns. Current work on brain stimulation has been particularly successful in providing an insight into the neural basis of intricate activities (von Holst and von Saint-Paul, 1962; Demski, 1973). Some of the earliest work was carried out on cats. By stimulating appropriate areas of the brain through implanted electrodes the animals could be induced to eat, attack, or flee. Von Holst and his associates used chickens. These animals continue in good health while the implanted electrodes remain "permanently" in place; the leads may be attached to the electrical circuit for any particular experiment. Progressively increased stimulation of one particular area leads to a gradual arousal of escape behavior; in another area the stimulation provokes an attack on an object which would normally arouse no such response. Many clearly distinct behavior patterns can be elicited in this way. Curiously, the simultaneous stimulation of two areas may provoke quite a different response. One particularly interesting one is the THREAT which may appear with the simultaneous arousal of ESCAPE and ATTACK. This neurophysiological demonstration provided an interesting confirmation of a relationship which ethologists had predicted from observational data.

A further line of study in the neurophysiological field is based on chemical rather than electrical stimulation of areas of the brain (Fisher, 1964; Lisk *et al.*, 1972). Implantation of gonadal steroids in localized areas of the rat's brain may elicit specific patterns of sexual behavior. With appropriate hormones male animals display female behavior and vice versa. The work again demonstrates the localized nature of the brain centers controlling complex motor acts but goes further in showing that specific chemicals applied to groups of neurons facilitate a particular behavior pattern in a predictable manner. Both the electrical and chemical stimulation experiments are pushing the analysis beyond a mere localization of neurons and showing something of their interactions and the significance of the internal chemical environment in controlling their activities. Endocrine studies are by no means confined to the stimulation of brain centers with hormones. The role of hormones in social behavior (pheromones) and in the regulation of seasonally changing patterns of behavior are subjects of lively research (Liley, 1969; Hinde, 1970).

Physiological understanding of behavior must be in terms of recognizable movements or coordinated groups of movements which combine to form the animal's complete repertoire. The biologists, rather than the physiologists, have been responsible for the dissection of behavior into discrete sequences of movements, and for demonstrating that these are predictable and stereotyped elements of extremely complex activities. This anatomy of behavior must obviously precede the physiology.

Jacques Loeb (1859–1924) made the first attempt to describe the total behavior of lower forms in purely mechanistic terms. His thinking was strongly influenced by the plant physiologists of the late nineteenth century who had introduced the concepts of TROPISM to describe the growth movements of plants. Loeb (1918), whose work in this field started about 1888, did a great service to biology by so forcefully rejecting the prevailing anthropomorphic and subjective explanations of animal activities and by substituting hypotheses which were susceptible to experimental analysis and explicable in terms of known physical and chemical laws. According to the thinking of the nineteenth century, a moth flies toward the light because it is curious about this bright and shining object; according to Loeb there are certain energy forces which compel the oriented flight of the moth to the light; the light is controlling the response through the animal's eyes. It is a forced movement of the whole organism toward or away from the stimulus and determined by internal and external forces. The term tropism is now used for the growth movements of plants and sessile animals while the locomotory orientations of mobile animals are called TAXES. Loeb, in his analysis, emphasized a physiological as well as a morphological bilateral symmetry in animals and argued that the processes inducing forward locomotion are equal in both halves of the central nervous system, in the symmetrical muscles, and in the peripheral receptors. The animal normally moves in a straight line. If, however, the velocities of reactions are changed in one half of the animal as happens when a light falls on one eye, then the physiological symmetry is disturbed; the animal changes direction and turns until the two sides are again in equilibrium. A unilaterally blinded moth flies in circles. This is a forced response of the whole organism in accordance with the strength of stimulus.

Fraenkel and Gunn (1940), Tinbergen (1951), Thorpe (1958), and others have described the opposition of biologists to Loeb's extreme views. Some of the earliest and most telling criticisms came from work on protozoans which might be presumed, from their lowly phylogenetic position, to support these theories best. Jennings (1923), who made many classical observations, found that numerous activities of the invertebrates were trial-and-error orientations; during the early part of the twentieth century more and more examples were found which did not fit well into the concept of tropisms or taxes. In 1919, Kühn published a classification of behavior which incorporated the concepts of undirected locomotory reactions as well as the directed orientation reaction or taxis. A modification and elaboration of this classification is given by Fraenkel and Gunn (1940). The term KINESIS is introduced for the undirected locomotory reactions, in which the speed of movement or the frequency of turning depends on the intensity of stimulation.

This period during which biologists attempted to categorize behavior in terms of taxes and kineses of several different varieties was the period in which physiologists saw the intact animal as a reflex machine. Pavlov was born in 1849, ten years earlier than Loeb, and was awarded the Nobel prize in 1904 for his concepts of the conditioned reflex. According to the Pavlovian analysis, the instinctive act was an inherited reflex, and learning was synonymous with conditioning.

The mid-twentieth century witnessed the resurgence of a branch of biology devoted to the causal analysis of animal behavior and known as ETHOLOGY (Tinbergen, 1963; Jaynes, 1969). It stems from the observations of naturalists with a genuine interest in what animals are doing and from the penetrating analysis of innate behavior patterns by Konrad Lorenz. Changes in thinking have taken place since Lorenz published his classical paper in 1935, but the broad principles on which his analysis was based have remained the same (Eibl-Eibesfeldt, 1970, Chapter 1; Baerends, 1971); there are many detailed accounts and current evaluations by physiologists (Bullock and Horridge, 1965) as well as by ethologists (Marler and Hamilton, 1966; Manning 1967; Hinde, 1970).

Behavior as a Series of Stereotyped Movements

Lorenz saw dissectable components in the behavior of the birds and fishes which he studied. He regarded these elements or patterns of behavior as ORGANS or attributes with special functions. Tinbergen (1963) considers this emphasis on the FIXED ACTION PATTERNS of the species as the first of Lorenz's three important original concepts. The second was that these patterns are vastly more complex than the reflexes usually studied by physiologists, and the third was that internal factors played a much larger part than neurophysiologists of that period believed.

These patterns of movement are innate, stereotyped, and performed under standard conditions by all members of the species in essentially the same way; they are as characteristic and fixed in their form as are the anatomical parts of the animal. Any particular behavior pattern occurs in relation to a specific internal physiological state (food requirement, sexual maturity, rhythmic activity in neural centers) and usually in response to definite RELEASING SITUATIONS in the external environment (presence of food, a potential mate). Both the internal and the external contributions to behavior are susceptible to analysis.

The releasing situations are as predictable as are the behavior patterns with which they are associated. The relationship is comparable to the key and the lock. A male stickleback (*Gasterosteus aculeatus*) fights in response to a visual releaser in the form of a red stripe on the underside of another male or in response to a crude model of another fish with the appropriate color contrast. A female stickleback spawns in response to a mechanical releaser created when the male nudges her body while she is in the nest; the same reaction may be induced by probing with a glass rod. The male sheds his sperm partly in response to a chemical releaser diffusing from the recently laid eggs, and he fans these eggs during incubation at an intensity which is related to the carbon dioxide content of the environment. Thus, a small part of the total sensory information which the animal receives sets in motion a complex group of stereotyped muscular movements. This information may be received through any of the sense organs, and the activity, unlike a simple reflex, often continues for some time after the presentation of the releaser. Further, the intensity of the response is usually related to the amount of information received. The effectiveness of many different releasing situations varies with the intensity of color or the structural relationships of different parts in a complex visual releaser. Moreover, the releasing information for a particular behavioral pattern is sometimes received through more than one sensory channel. Spawning of the pike is triggered by a rising temperature (thermal receptors) when the fish are in the presence of a particular type of vegetation (visual receptors). What actually happens in such a case seems to depend on the way in which these stimuli are summed in the central nervous system (HETEROGENEOUS SUMMATION). Under normal circumstances a certain amount of information comes through each sensory channel, but the response may also occur when one of the sensory channels receives little or no information, provided that the other is strongly stimulated. Pike may spawn if the temperature is high even though the vegetation is sparse (Fabricius, 1950).

One additional feature is characteristic of simple instinctive behavior. There is not only the fixed pattern of movement, but there is normally also a steering or orienting component to the movement. The male stickleback, in fanning his eggs, uses definite groups of muscles and performs stereotyped activities; but the location of the activity is governed by the entrance of the nest and certain other features in the geography of its environment. This orienting factor may be quite distinct from the releasing situation.

This then is the ethologist's general concept of the anatomy of behavior. A series of innate movements are set in motion by specific releasing situations and guided by conditions or factors in the environment. They may be preceded by considerable locomotion (APPETITIVE BEHAVIOR) before the appropriate releasing situation is encountered. Complex activities such as reproduction are organized as a rigid (hierarchical) series of such movements. The immediate causation of the varied components of this framework is susceptible to physiological analysis.

Complexity and Versatility in Behavior

The implicit suggestion of rigidity in the foregoing statements is at variance with the obvious versatility of animals and their abilities to adjust their activities to the problems of living. The ethologist is well aware of this, and his analysis has shown several of the factors involved. The spontaneous or pacemaker type of activity has already been mentioned several times in this book; the effect of varying amounts of releasing-information on the intensity of some behavioral movements (heterogeneous summation) is another source of diversity. Additional factors are discussed in books and reviews already cited. Of particular interest to the physiologist are the effects of the variable internal environment and the adaptive changes in behavior which result from experience or learning. Developmental and seasonal changes in the activity of the endocrine glands and alterations in the levels of blood sugar and body water may modify an animal's readiness to perform certain behavior patterns; in the terminology of behavioral science, these factors alter the INTERNAL MOTIVATION. Levels of hormones, blood sugar and tissue moisture can be precisely measured and related to specific behavior patterns.

Again, all groups of animals show an ability to learn. The investigation of learning has been traditionally the concern of the psychologist; while ethologists have investigated behavior by observing animals under natural conditions, comparative psychologists have concentrated on laboratory analyses. At one time there were lively debates between the two groups of scientists; much of the heat has now been dissipated with the recognition of both innate and learned components in an animal's behavior and the validity of both methods of investigation. Thorpe (1950, 1963) has provided a useful classification of learning processes for zoologists. Physiological interest centers on the effects of experience on neural mechanisms; biochemists as well as neurophysiologists are active in this analysis (Rosenzweig *et al.*, 1972).

Analysis of the physiology of behavior provides some of the most challenging problems in biology. Although students of behavior have dissected many aspects of animal behavior into describable components and recorded the environmental situations which spark their expression, the intervening links are frequently treated as a "black box." It is in this "black box" that the physiologist finds his problems as he attempts to expose the links which join the cues or releasers with the behavioral movements. Although only a start has been made, the current work shows great promise at all three important levels of organization: nerves or isolated groups of nerves (Willows, 1967; Mendelson, 1971; Eisenstein, 1972), complex nervous systems (Rosenzweig *et al.*, 1972), and whole animals (Hölldobler, 1971; Dingle, 1972).

FOUR

REPRODUCTION AND DEVELOPMENT

22

REPRODUCTION

The structures and processes described in this chapter are concerned with the survival of the species rather than the individual. The physiology of reproduction depends on the same metabolic machinery and is served by the same sensory, motor and neural mechanisms as the other organ systems but is not essential to the functioning of any of them. Animals, both primitive and advanced, may develop, grow and live a normal life span without reproducing. However, the continuance of the species and all evolutionary changes would quickly come to an end without this constant replacement of the old with similar but somewhat variable offspring. Depending on one's philosophy, the events concerned with reproduction may be considered the ultimate objective of all other life processes or they may be considered "a privilege which may or may not be indulged" (Hisaw, 1959).

THE GENETIC MATERIAL

Even the layman with only a casual interest in living phenomena is now aware that his biological inheritance depends on genes which are arranged on chromosomes and that the actual code of the hereditary information is in the deoxyribonucleic acids (DNA) of the genes. The evidence linking genes and DNA with the genetic code will not be repeated here. Excellent summaries are available (Crick, 1962; Barry, 1964; Sullivan 1967; Watson, 1970). There are, however, several considerations which are pertinent to the concepts of a phylogeny of living processes, and these will be summarized before discussing the physiology of reproductive organs and systems.

Deoxyribonucleic Acid (DNA)

The probable significance of DNA in the coding of genetic information was recognized many years ago, and this is its most frequently emphasized role in introductory courses and in popular writings. Much more recently, the broad outlines of its activities in directing the synthesis of proteins have been established. These two interrelated functions must have evolved together since the replication of entire animals is probably an extension of the capacities to replicate their most characteristic building blocks (the proteins) and the enzymes which control metabolism. Different forms of life owe their peculiar characteristics to their proteins; the continuation of life requires a faithful duplication of these units and of the organism which houses them; the evolution of new species depends on the chance appearance of small variations (mutations) which are in some way superior to the existing arrangements. The point to be emphasized here is that the DNA machinery responsible for these processes seems to have remained basically the same throughout evolution. Only minor variations (mentioned below) have been noted in the base composition of the nucleic acids. Moreover, it appears that identical "code letters" are used to assemble particular amino acids in the synthesis of proteins from bacterial systems to men (Nirenberg, 1963).

Nucleic acids, like proteins, are long-chain polymers of high molecular weight which can be split chemically into repeating units. About two dozen amino acids form the repeating units of proteins; only four different NUCLEOTIDES are joined in varied sequence to form the nucleic acids. A nucleotide was defined in Chapter 7 as a substance composed of a nitrogenous base (either a purine or a pyrimidine), a pentose and phosphoric acid. Adenosine monophosphate (Fig. 7.2) is a mononucleotide which, together with two additional phosphate units, forms the important high-energy phosphate compound ATP. The same nucleotide is found in several vitamins and enzymes (riboflavin, NAD, coenzyme A). The other nucleotides also play a part in metabolism as well as in the coding of genetic information. Uridine triphosphate, for example, is a high-energy compound involved in polysaccharide biosynthesis, while the syntheses of porphyrins and some proteins depend on guanidine triphosphate, and cytosine triphosphate plays a part in lipid formation (Lehninger, 1971). Thus, the essential building blocks of the nucleic acids are by no means unique to them but are also constituents of important energy-transferring compounds of metabolism. Again, the ubiquity of a relatively few indispensable biochemical combinations is apparent.

There are two kinds of nucleic acids (Fig. 22.1). In animal cells, *deoxyribonucleic acid* (DNA) is found primarily in the chromosomes of the nucleus; *ribonucleic acid* (RNA) is also found in the nucleus, but the amount there is relatively small and much larger quantities are present in the cytoplasm (particularly in the ribosomes). The chemical differences between these two compounds are shown in Fig. 22.1. In DNA the sugar is 2-deoxy-D-ribose; in RNA the sugar is D-ribose.

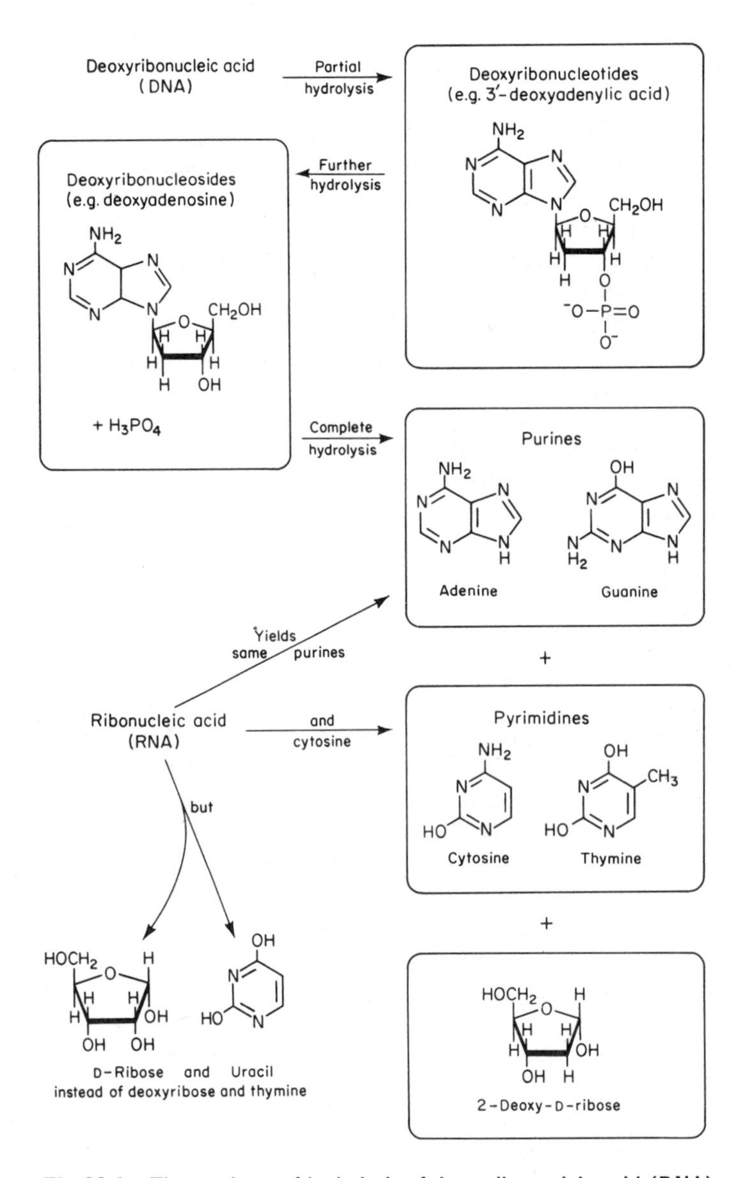

Fig 22.1 The products of hydrolysis of deoxyribonucleic acid (DNA) and ribonucleic acid (RNA). [Based on Anfinsen, 1959: The Molecular Basis of Evolution. Wiley, New York.]

There is also a difference in one of the pyrimidines; the irregular member of the pair is thymine in DNA but uracil in RNA. Small variations occur in the base composition of DNA in some plants and microorganisms (Anfinsen, 1959); methylcytosine, for example, bears the same relation to cytosine that thymine does to uracil (McGilvery, 1970). Further, the DNA of some organisms contains a minor, but easily identifiable, portion with highly variant DNA sequences (Watson, 1970). This SATELLITE DNA, as it is called, was first studied in crustaceans where it is particularly abundant and extremely rich in adenine-thymine sequences (Laskowski, 1972). The significance of these variations is not understood.

The quantity of DNA per cell has increased sharply during evolution (Fig. 22.2); this appears to have been one of the major phylogenetic requirements for the expansion of information necessary to operate progressively more complex organisms (Ohno *et al.*, 1968; Britten and Davidson, 1969). Amphibians provide one of the notable and unexplained exceptions to this generalization (Watson, 1970).

One of the most remarkable biochemical achievements of our century has been the demonstration of the order and significance of the arrangement of nucleotides in the DNA molecule. The Watson-Crick model, proposed in 1953, shows two chains coiled together in a double helix (Fig. 22.3) The two ribbons of the helix are composed of a series of sugar-phosphate units joined through 3′–5′ phosphate diester bridges. The two ribbons are cross-linked through a hydrogen bonding formed between the oxygen and nitrogen atoms on the adjacent bases of paired nucleotide molecules. This is only possible between specific base pairs as shown in Fig. 22.4, so that thymine is *always* joined to adenine and cytosine to guanine. It is also to be noted that the molecules in each pair run in opposite directions and that the four nucleotides can be arranged in almost infinite numbers of sequences, thus providing for the existence of many different kinds of DNA (Fig. 22.3).

The evidence for this model was based on both chemical and physical data. The chemical analyses showed that the sum of the purines always equals the sum of the pyrimidines and, further, that the adenine content equals the thymine content, the cytosine equals the guanine, but the total amount of adenine and thymine rarely or never equals the total amount of guanine and cytosine. Hydrogen bonding was also evident from the chemical analyses while X-ray diffraction studies clearly showed a regular and uniform molecular pattern. The historical events leading up to the

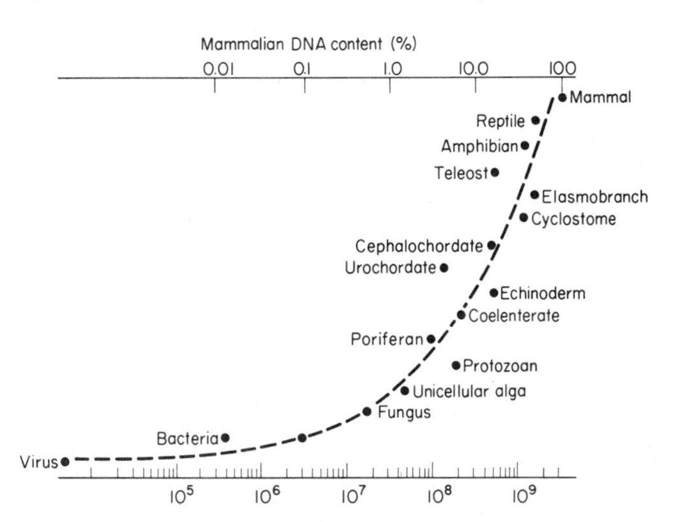

Fig. 22.2 Minimum haploid amounts of DNA recorded for species at different levels in phylogeny. The ordinate is not numerical and the shape of the curve has little significance. [After Britten and Davidson, 1969: Science, *165*: 352.]

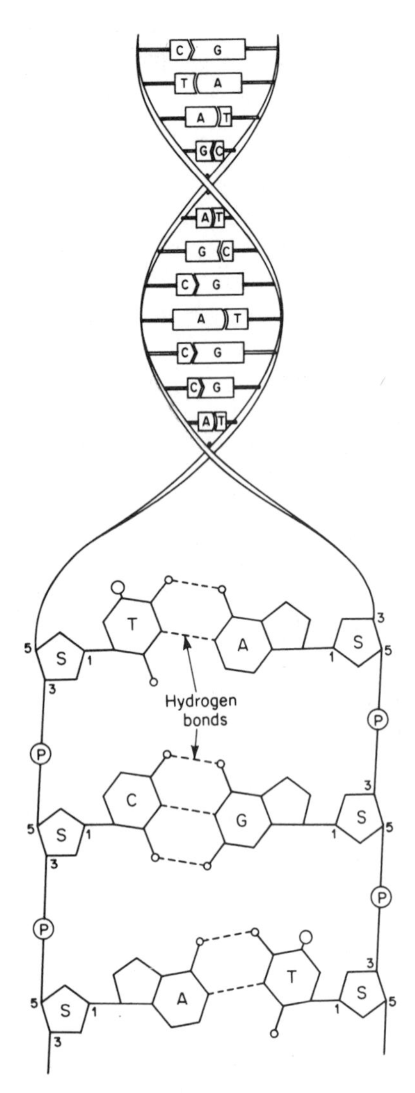

Fig. 22.3 Schematic diagram of the DNA molecule showing the base pairing and the helical structure. A, adenine; C. cytosine; G, guanine; P, phosphate; S, pentose; T, thymine. For a discussion of single-stranded DNA, see Sinsheimer, 1962, and Watson, 1970. [From McElroy, 1971: Cell Physiology and Biochemistry, 3rd ed. Prentice-Hall, Englewood Cliffs, N. J.]

Watson-Crick model of DNA are summarized in reviews already cited; Watson (1968) has vividly described the drama of the discovery. This model is now supported by convincing evidence from many sources and has provided a most fruitful impetus to the studies of inheritance and protein synthesis. Watson, Crick and Wilkins were awarded the Nobel prize in 1962.

Many elegant experiments have demonstrated that DNA can make copies of itself. According to the Watson-Crick hypothesis, the strands of the double helix separate and serve as templates for the assembling of matching nucleotide chains. Kornberg was awarded a Nobel prize in 1959 for his convincing *in vitro* experiments

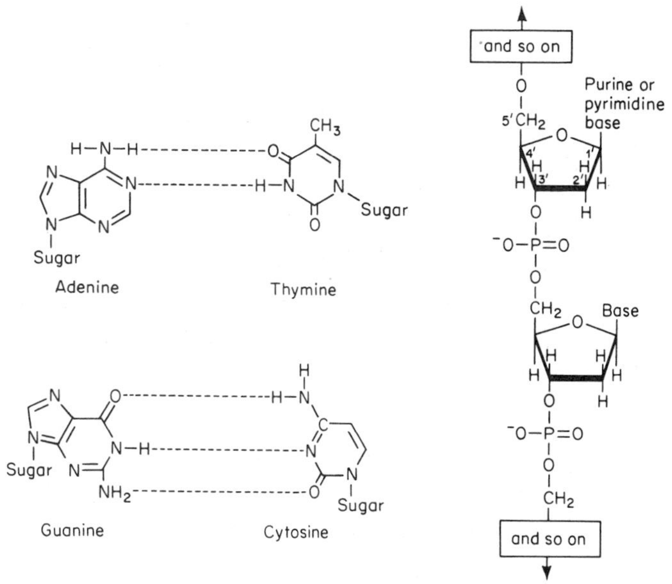

Fig. 22.4 Linkages in the DNA molecule. Left, hydrogen bonding between adenine and thymine and between guanine and cytosine. Right, linkage of nucleotides through 3′–5′ phosphate diester bridges. [Based on McElroy, 1971: Cell Physiology and Biochemistry, 3rd ed. Prentice-Hall, Englewood Cliffs, N. J. and on Anfinsen 1959: The Molecular Basis of Evolution. Wiley, New York.]

demonstrating this. Recent work goes much further and has shown how the DNA of the nucleus transfers its code to RNA which passes into the cytoplasm and how the RNA system of the cytoplasm assembles amino acids into specific proteins. It is now accepted that the genetic alphabet, consisting of four letters (adenine, guanine, uracil and cytosine) is used to spell three-letter words which are quite specific for the different amino acids. The first of these to be recognized was the code for phenylalanine which is uracil-uracil-uracil or UUU. Once the techniques were developed, the genetic code was almost completely deciphered in one decade (Bennett and Frieden, 1966).

Diploidy and Sex

The units of heredity were named "genes" by Johanssen in 1911. The presence was adduced from genetical studies starting with those of Gregor Mendel (1822–84) on garden peas. Only in recent years has it been generally recognized that all genes are DNA, with the exception of some of the viral genes which are RNA (Temin, 1972). A gene is a functional unit, an area of DNA in a chromosome which contains directions for the synthesis of messenger RNA (mRNA) and hence of the proteins (STRUCTURAL GENES), or the information required to regulate the function of the structural genes (REGULATOR and OPERATOR GENES). A gene is, in short, a center of specific enzymatic activity.

It was also apparent from breeding experiments, carried out long before the significance of DNA was appreciated, that almost every gene in animal cells has an identical partner and that the expression of any unit character depends on the combined action of two genes. Animals are diploid organisms, and this is a point of particular significance in their evolution and in the development of sexually reproducing mechanisms. Since their observable characteristics (phenotypic) depend on the existence and interaction of two partners, a stability can be expected in the processes of vegetative reproduction. In ordinary cell division (MITOSIS) the genes are duplicated so that the diploid number appears in each of the daughter cells. Unless the same alterations (mutations) occur simultaneously in both members of a genic pair, the phenotype will not be altered during vegetative reproduction or mitosis. In haploid organisms such as molds and bacteria with one set of genes, mutations can lead to a rapid phenotypic expression of the mutant allele. In diploid organisms there is only a remote chance that two members of a gene pair will mutate in the same way at the same time. The significance of diploidy in conferring genetic stability during growth and vegetative reproduction is emphasized. Animals which reproduce by budding or parthenogenetically, self-fertilizing hermaphrodites or gynogenetic clones may be highly successful in the environment where they are found but lack the plasticity necessary to meet gradual changes in their environment or to penetrate different environments; they may be quickly eliminated.

While diploidy confers stability, sex provides for diversity. The two fundamental processes of sexual reproduction are MEIOSIS and FERTILIZATION. In meiosis or reduction division the members of the homologous pairs of chromosomes separate at some stage, and each of the two resulting haploid germ cells contains half of the DNA of the diploid cells. Actually, meiosis is not a single nuclear division but a double division with only one duplication of the chromosomes. The significant point for present arguments is that the diploid cells containing two of each kind of chromosome give rise to haploid gametes containing one of each kind of chromosome. In fertilization, two haploid cells combine to form a new diploid which then divides mitotically during the growth and development of the individual. Now, the geneticist has shown that the paired chromosomes of the diploid cell are RANDOMLY ASSORTED into the haploid cells or gametes. Moreover, exchanges of the homologous portions of double chromosomes (CROSSING-OVER) frequently occurs during maturation division. These processes occur in all kinds of organisms from bacteria to men and are the major sources of genetic variability.

Meiosis followed by fertilization is a process which maintains a constancy in the chromosome content of the species and, at the same time, continually generates new individuals with random assortments of parental characters. Organic evolution is based on this steady flow of new varieties into the population. Within the limitations of the DNA of the species, there is thus the possibility that the population will contain individuals which can take advantage of new environments or survive in the face of environmental changes. "Sex arose in organic evolution as a master adaptation which makes all other evolutionary adaptations more readily accessible. Perhaps no other biological function appears in such a bewildering diversity of forms in different organisms as does sexual reproduction. Yet, despite its diversity, sex everywhere serves the same basic function—production of genetic variability by gene segregation and recombination" (Dobzhansky, 1955).

The evolution of sex itself has played a key role in the diversification of animals. Many curious morphological parts, intricate physiological controls and complex behavioral adaptations are the direct consequences of the functional importance of sex in organic evolution. Presumably, the first sexual processes involved the combination of similar cells of different mating types. In some primitive algae and protozoans the partners are morphologically indistinguishable; if either is grown by itself, it will produce only asexual progenies (CLONES) without any fusion of individuals, but when cells from different clones are mixed, a pairing of individuals takes place. Even among the Protozoa, however, a morphological differentiation of male and female individuals is not uncommon. This can range from slight differences in cell size, as in some species of *Paramecium,* to the development of distinctly different male and female forms. In extreme cases, sexual union takes place between a large, passive, non-motile, food-filled female macrogamete and a minute, motile, flagellated male microgamete (Hyman, 1940). These are, in fact, eggs and sperm. Sporozoans, such as the malarial parasite *Plasmodium,* provide the most familiar examples of eggs and sperms in the protozoans. The formation of two types of gametes is general throughout all the multicellular groups. In many primitive representatives and in some of the most advanced, the two types of gamete-producing structures are housed in the same individual (monoecious or hermaphroditic condition), but even in the most primitive groups male and female cells frequently develop in separate animals (dioecious condition). Many complex features of physiology and behavior are associated with this separation of male and female individuals and the necessity of bringing their sex products together at the time of reproduction.

REPRODUCTIVE MECHANISMS

Some Alternatives to Sex

Protozoans frequently divide mitotically for many generations to form clones of like individuals. At certain times, however, most species have been shown to enjoy the biological advantages of sex, with a typical meiosis and mating of either similar or dissimilar gametes (Kudo, 1966; Curtis, 1968).

Sexual reproduction is the rule in all multicellular phyla—including the sponges. Many of the more primitive groups, however, have marked capacities for extensive regeneration and often reproduce asexually for longer or shorter periods in their lives. Some sponges will regenerate after being reduced to small masses of cells by squeezing them through a filter; many of the sponges reproduce by budding; all freshwater and some marine species form special asexual bodies called GEMMULES; these consist of hard, round balls containing masses of food-laden cells which are released when the sponge disintegrates. AGAMIC REPRODUCTION by budding and fragmentation occurs in several different phyla—particularly among the coelenterates, the platyhelminths, the nemertines, a few of the annelids and tunicates. Sometimes lateral buds are formed as in *Hydra;* sometimes the animal multiplies by a series of transverse divisions as in the scyphistoma of *Aurelia,* in many of the flatworms, both free-living and parasitic, and in some nemertines and oligochaetes; sometimes the fragmentation takes a less regular course and results in variously shaped pieces as in some platyhelminths. Many examples are given in Hyman's monographs and in the book by Vorontsova and Liosner (1960). In general, however, these are *not* alternatives to sexual reproduction but an adjunct to it. The asexual multiplication alternates with a sexual stage and provides for a rapid increase in numbers of individuals through vegetative processes without a loss of the evolutionary advantages of sexual reproduction.

Strictly parthenogenetic species, gynogenetic forms, and self-fertilizing hermaphrodites, in contrast, form clones of genetically similar individuals. These are closed genetic systems which have exchanged the advantages of sex for the biological security of unisexual proliferation.

PARTHENOGENESIS is the development of a new individual from an egg or a spermatozoon without the participation of a germ cell from the opposite sex. Among animals only the maternal germ cells are known to give rise to parthenogenetic individuals, but certain algae arise in this manner from paternal cells. Although many examples of artificial and occasional parthenogenesis have been recorded, natural populations of parthenogenetic animals are restricted to some groups of invertebrates (particularly the insects) and to several lizards (Olsen and Marsden, 1954; Austin and Walton, 1960; Cuellar, 1971; Maslin, 1971). In some platyhelminths and rotifers and certain wasps and sawflies (Hymenoptera), parthenogenesis is the only known method of reproduction. In many parthenogenetic invertebrates there is a cyclical alternation of asexual with bisexual reproduction. Parthenogenesis may be seasonal and related to temperature or food supply, or it may appear at irregular intervals. In several insects (honeybees, some wasps and sawflies), unfertilized eggs develop into haploid males while fertilized eggs give rise to diploid females. Suomalainen (1962) discusses the evolutionary implications. The nuclear changes associated with the formation of gametes vary considerably in different parthenogenetic species (White, 1973). Several useful discussions and extensive bibliographies will be found in a series of symposium papers edited by Oliver (1971).

In GYNOGENESIS, a spermatozoon activates an egg but does not contribute any genetic material to it. The resulting embryo carries only maternal chromosomes. The reverse condition, where the chromosome contribution comes exclusively from the male, is called ANDROGENESIS, but in animals this is known only experimentally. From the standpoint of genetics, gynogenesis, androgenesis and parthenogenesis are the same, since, in each case, the offspring possess chromosomes from only one parent. Naturally occurring gynogenesis is recognized in a few non-segmented worms, a ptinid beetle, three genera of teleost fishes and an amphibian (Austin and Walton, 1960; Asher and Nace, 1971; Schultz, 1971). The natural history and genetics have now been studied in several different species. In some flatworms, gynogenesis is of the racial type, with eggs of the gynogenetic race activated by sperm of a bisexual race. In contrast, *Poecilia* (*Mollienesia*) *formosa*, a teleost fish, borrows sperm from a different species (*P. sphenops or P. latipinna*) to effect fertilization. The offspring are exclusively females. The homozygosity of different clones was demonstrated by tissue transplant techniques; tissue grafts among members of the same clone always survive whereas grafts between members of different clones are rejected (Kallman, 1962a, b).

In HERMAPHRODITISM, or the monoecious condition, both male and female gametes are formed in the same individual. It is common among primitive animals, and all the major groups, including the vertebrates, have their hermaphroditic representatives. Cross-fertilization is usual, and under these conditions the significance of reproduction in evolutionary processes is the same as that of dioecious species. Self-fertilizing hermaphrodites, on the other hand, form clones of identical individuals. The teleost fishes provide interesting examples. Some species of Sparidae and Serranidae are invariably hermaphroditic, producing eggs and sperm in different areas of the same gonad; they may be self-fertilizing (D'Ancona, 1949). However, self-fertilization is not obligatory. In *Serranellus subligarius* a single individual may shed both eggs and sperm in an aquarium, and these produce normal offspring. In nature, however, there is an elaborate sexual play, with paired individuals in different color phases playing distinct roles. Curiously enough, an animal may show a change of color phase and behavior during any mating sequence (Clark, 1959). Cross-fertilization is probably the usual outcome.

The cyprinodont *Rivulus marmoratus,* in contrast, is a self-fertilizing hermaphrodite (Harrington, 1963). The homozygous nature of the genotypes in clones of *Rivulus* has been established by tissue transplantation (Harrington and Kallman, 1968). Thus, the teleost fishes show many curious reproductive specializations; parthenogenesis seems to occur only sporadically, but gynogenesis and obligatory self-fertilization are established beyond doubt. Both evidently represent a secondary loss of the potential for genetic variability through sexual processes.

According to traditional views, parthenogenesis, gynogenesis and hermaphroditism are evolutionary "dead-ends." The advantages of producing new varieties by shuffling and recombining the genes from two parents have been lost. Evolutionary adaptation to environmental pressures depends on the regular occurrence of new and slightly different combinations; according to the argument, animals which have lost these advantages of sex will be more liable to extinction when conditions change.

This view has now been challenged by extensive studies of natural history and genetics (Davey, 1965; Oliver, 1971) and by several considerations of a more theoretical nature (Tomlinson, 1966; Ghiselin, 1969). It seems more reasonable to regard these unisexual alternatives to sex as specialized biological ways of life which have evolved in response to particular pressures. In short, the disadvantages of "selfing," which have already been noted, must be outweighed by distinct advantages to the parthenogenetic or hermaphroditic organism. This problem has been discussed for over a century; Darwin (1877) argued that it was more advantageous to self-fertilize than not to reproduce at all. In theory, at any rate, parthenogenetic and hermaphroditic organisms show increasing biological advantages as populations become sparse, or where habitats become marginal, or when animals lack mobility. Thus, a *low-density model* predicts self-fertilization among sessile organisms, deep-sea animals, parasites and species with similar life styles. There may also be biological pressures toward the *reduction* of variability in small isolated populations where genetic drift might be deleterious. However, the major factor, which was missed in the early discussions, is the extent of genetic variability which is still present in unisexual species. The triploid populations of unisexual lizards and salamanders probably possess considerable genetic flexibility; higher reproductive potentials may offset deleterious mutations; mutation itself, and "mitotic" crossing-over may produce sufficient genotypic diversity for evolutionary change in rapidly expanding populations.

Morphology of the Reproductive Organs

Some of the most heated discussions of classical embryology have centered around the origin, segregation and lineage of the germ cells, the source of non-germinal gonadal tissues and the differentiation of the gonad. These topics are reviewed in textbooks of embryology and in monographs devoted to the reproductive organs (Brambell, 1956; Balinsky, 1970a). The present discussion will start with the mature gonads containing a stock of germ cells, from whatever source, and will consider them as organs for the production of gametes and the integration of reproductive processes.

In some primitive animals the gamete-producing tissues are diffuse, consisting of numerous scattered loci for the proliferation of the sex cells. In all the more advanced animals the gonads are localized, and in the bilaterally symmetrical forms they arise as paired structures. Sometimes one of the gonads degenerates secondarily. The birds provide a familiar example with paired testes in the male and a single left ovary in most females.

Primordial germ cells which have set out on the path of female development are called OOGONIA; those which are on the male path are SPERMATOGONIA. Oogonia and spermatogonia, like the somatic cells, are diploid and may divide many times mitotically. However, when they commence their maturation divisions and undergo meiosis to form haploid cells, they are called OOCYTES and SPERMATOCYTES, respectively. A PRIMARY OOCYTE or a PRIMARY SPERMATOCYTE is undergoing its first reduction division; the stage referred to as SECONDARY is associated with the second maturation division, and the various transformations beyond this produce the mature eggs and sperms. Vitellogenesis or the storage of yolk is usually initiated after the early stages of the first meiotic division. The number of mitotic divisions and the timing of meiosis varies in different groups. In certain representatives of advanced phyla the processes of vitellogenesis and the initiation of meiosis have been shown to be hormonally regulated.

The ovary The primary and universal function of the ovary is to generate the female cells or eggs. A second function which is an almost universal adjunct to this is the elaboration of a store of nutritive materials (yolk) for the early stages of embryonic development. The synthesis of hormones for the chemical coordination of reproductive functions is also an ovarian responsibility in vertebrates and in some of the more advanced invertebrates; information on this point is lacking for many lower forms. A fourth ovarian function, present in some viviparous animals, is the housing and nourishment of developing young (FOLLICULAR GESTATION). There are many radical differences in ovarian morphology (Raven, 1961; Zuckerman, 1962). The examples which follow have been selected to illustrate these four functions.

The first examples, drawn from the turbellarians, illustrate some of the most unusual and highly specialized methods for the storage of reserve food (Hyman, 1951). In the Acoela, Polycladida and some of the Rhabdocoela, there are no special follicle or nurse cells, but the developing ovum draws its nourishment from surrounding maternal tissues (SOLITARY EGG FORMATION). In other rhabdocoels

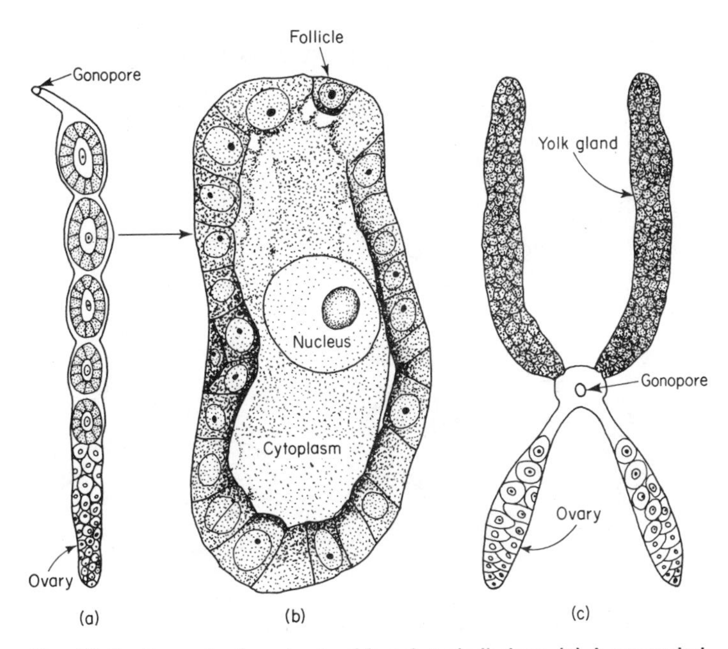

Fig. 22.5 Reproductive glands of female turbellarians. (a) A germovitellarium composed of maturing follicles in the Prorhynchidae. (b) A follicle of *Prorhynchus*. (c) The condition found in most rhabdocoels and alloeocoels with two yolk glands and one or two ovaries. [From Hyman, 1951: The Invertebrates, Vol. 2. McGraw-Hill, New York.]

and in the Alloeocoela, however, egg formation is ALIMENTARY as in most animals (Raven, 1961) with groups of specialized yolk-synthesizing cells associated with the developing eggs. The follicle of *Prorhynchus*, illustrated in Fig. 22.5b, shows the alimentary condition where one of the ovocytes becomes an egg and others are arranged around it as follicle cells which supply it with food. In the elongated ovary these follicles become progressively more differentiated toward the gonopore (Fig. 22.5a). This follicular organization of a developing ovum within a ball of epitheloid cells is characteristic of most multicellular animals, from the primitive to the advanced. In a more specialized condition of many turbellarians, groups of cells are set aside as a special yolk-producing gland or vitellarium (Fig. 22.5c). These animals are hermaphroditic but cross-fertilizing, and after copulation the sperm fertilize the ova in the oviduct; eggs then receive the products of the yolk glands as they move toward the genital atrium where a cocoon or shell is added. The details vary; the important point is that nutritive functions have become separated and that these primitive animals show highly specialized methods of provisioning their eggs.

An example of solitary egg formation is illustrated in the section of the gonad of the bivalve *Sphaerium* (Fig. 22.6). This is a hollow structure formed of columnar epithelial cells, some of which give rise to oocytes. The developing oocytes bulge into the lumen but remain attached by a stalk with elongated germinal epithelial cells arranged around it. These elongated cells are assumed to supply nourishment to the developing egg; sometimes they seem to be ingested by the ovum (Woods, 1931; Raven, 1961).

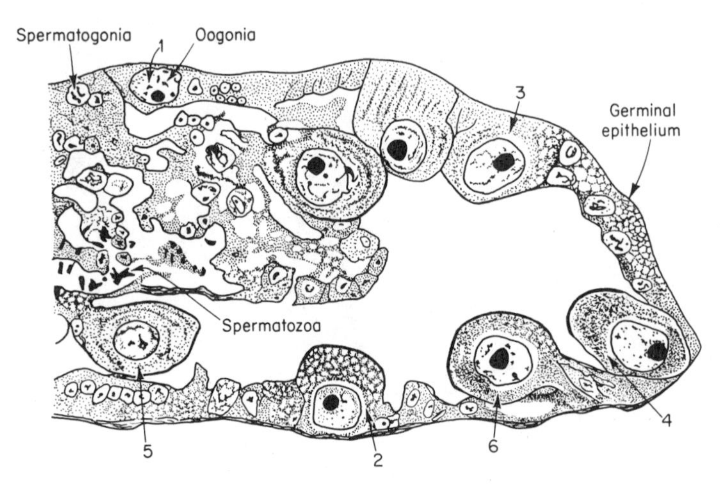

Fig. 22.6 Section through a maturing gonad of the bivalve *Sphaerium striatinum*. Numerals indicate successive stages in the growth of the oocytes. Note the absence of typical follicle cells. Solitary egg formation has been noted in several bivalves (Lammens, 1967), but it does not appear to occur in other classes of Mollusca (Selwood, 1968). [Based on Woods, 1931: J. Morph., *51*: 552.]

The insects show several types of alimentary egg formation, one of them comparable to the vitellarium of the turbellarians (Bonhag, 1958; Davey, 1965). The ovary consists of a series of egg tubes or ovarioles varying in number from one in the viviparous dipteran *Glossina* to more than 2000 in certain termites (Wigglesworth, 1972). Three different situations are illustrated in Fig. 22.7. In each case the elongated ovariole gradually widens from its terminal filament to the broad base which opens into the oviduct, and the developing eggs become progressively larger and more mature from pointed apex to base. The apex consists of densely packed cells and is called the germarium. Some of its cells are primordial germ cells and differentiate into oocytes while others take on a nutritive function. In the most primitive type (PANOISTIC) the nurse cells become arranged around the oocyte in a follicle (Ephemeroptera, Orthoptera); in the most specialized form (TELOTROPHIC) the nutritive cells form specialized TROPHOCYTES and are localized in the apex and connected to the developing ova by long nutritive cords (Hemiptera, some Coleoptera). In the Diptera, Hymenoptera, Lepidoptera and some others there is an intermediate condition (POLYTROPHIC) in which each oocyte has a number of trophocytes enclosed with it in its follicle. Further details with descriptions of the lineage of these cells will be found in textbooks of insect anatomy and physiology. Endocrine production by the insect ovary is questionable (Highnam and Hill, 1969). It is well known, however, that gestation may take place in the ovaries, although

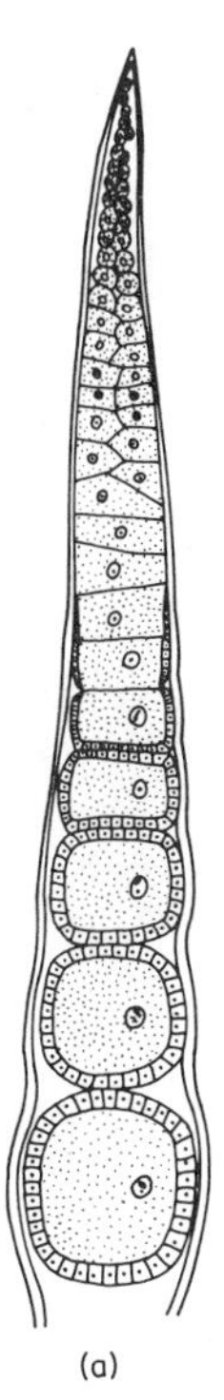

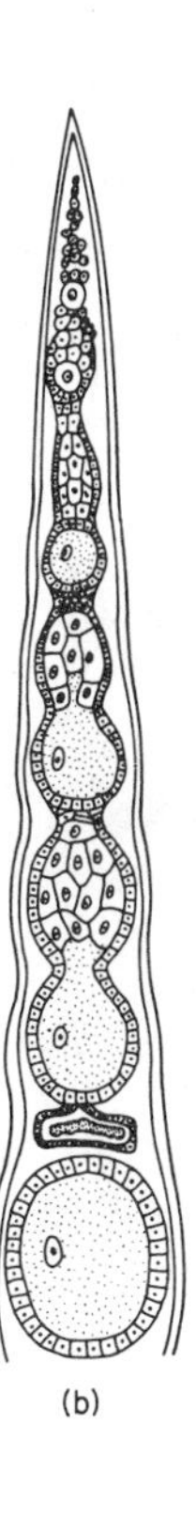

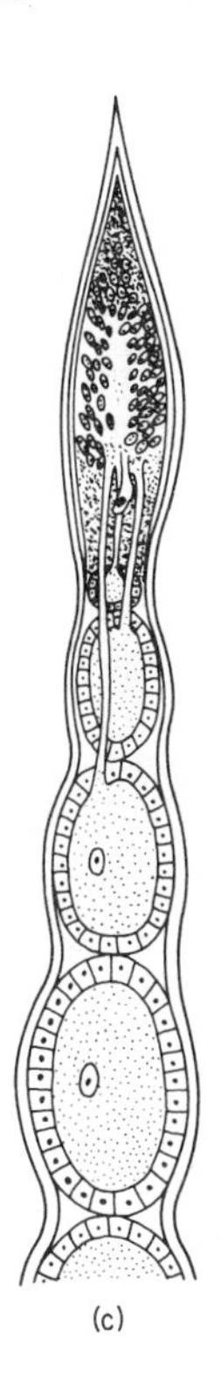

Fig. 22.7 Histology of the insect ovary. (a) Panoistic type represented by the firebrat *Thermobia domestica*. (b) Polytrophic type represented by the queen honeybee. (c) Telotrophic type represented by the milkweed bug *Oncopeltus fasciatus*. [From De Wilde, 1964: *In* The Physiology of Insecta, Vol. 1, Rockstein, ed. Academic Press, New York.]

larval development in the more specialized viviparous insects occurs in a uterus or in the hemocoel (Davey, 1965; Wigglesworth, 1972).

The physiologically important features of the vertebrate ovary are illustrated in Figs. 22.8 and 22.9. Within the cortex of the gland, single oocytes differentiate inside a ball of follicular cells which are epithelial in nature and vary in number from a single layer in some forms to numerous layers in others. This is the FOLLICULAR EPITHELIUM OF GRANULOSA, and its primary function was probably that of providing materials for the synthesis of yolk and the growth of the ovum. The stroma of the ovary surrounding the granulosa becomes organized into connective tissue layers

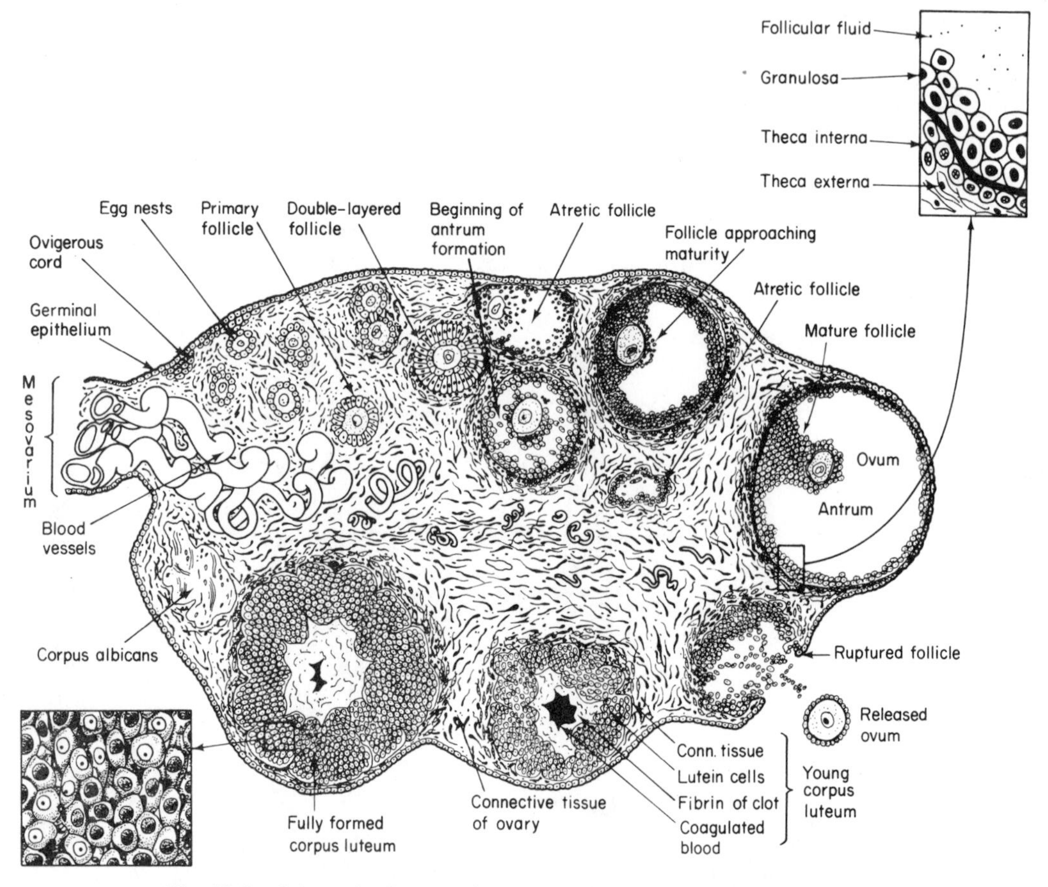

Fig. 22.8 Schematic diagram of ovary showing sequence of events in origin, growth, and rupture of ovarian (Graafian) follicle and formation and retrogression of corpus luteum. Follow clockwise around ovary, starting at mesovarium. [From Patten, 1953: Human Embryology, 2nd ed. McGraw-Hill, New York.]

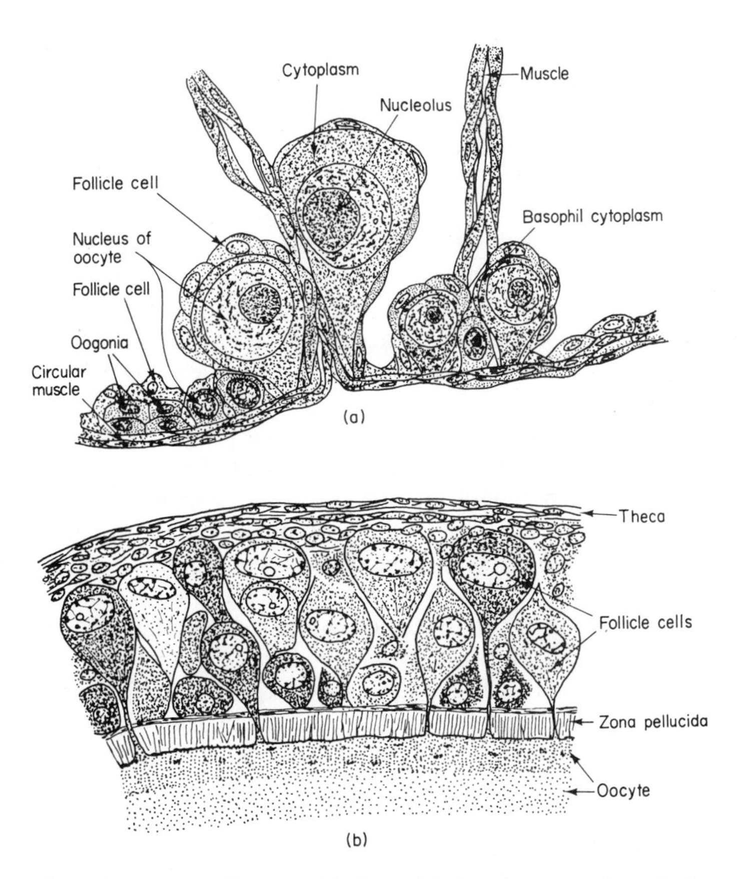

Fig. 22.9 The follicular epithelium. (a) Development of previtellogenetic eggs in the chiton *Sypharochiton septentriones*. [From Selwood, 1968: J. Morph., *125*: 78.] (b) Part of the follicular wall in the lizard *Lacerta muralis*. Note the canalicular prolongations of the large follicular cells which traverse the radially striated zona pellucida into the cytoplasm of the oocyte. [Based on Loyez, 1906: Arch. Anat. Micr., *8*: 69.]

called the theca. In higher forms, the THECA INTERNA becomes a glandular and vascular estrogen-synthesizing layer which is distinct from the theca externa. In vertebrates the mature eggs are usually shed into the peritoneal cavity where they pass into the open ends of the oviducts (most species) or from the body through abdominal pores (some fishes). In some teleost fishes the mature eggs are discharged into the ovarian lumen which is continuous with a short oviduct (Hoar, 1969). It is within this type of ovary that follicular gestation occurs in viviparous species.

The follicular epithelium or granulosa discharges several different functions in vertebrate animals. As previously mentioned, its primary role seems to be nutritive, and the cells are sometimes highly specialized with canalicular prolongations associated with the transfer of food to the differentiating ovum (Fig. 22.9). In addition, the granulosa plays a phagocytic role after the ovum is discharged or if it becomes atretic. The ovary, like many other organs, operates with a considerable margin of safety, and many follicles never produce a mature egg. Variable numbers, depending on physiological conditions, degenerate, and the granulosa cells are concerned with the removal of the yolk and other substances. They perform the same function in removing blood clots and other debris following ovulation. This phagocytic activity is often evident in ovaries and can be demonstrated experimentally in most vertebrates by removing the pituitary gland; vitellogenesis ceases without the gonadotropic hormones and the developing follicles are cleared of yolk by the granulosa.

In addition to its nutritive and phagocytic activities, the granulosa may take on the functions of an endocrine gland. Following ovulation in the mammal there is a proliferation of the granulosa and theca interna to form the CORPUS LUTEUM concerned with the synthesis of progesterone. Comparable morphological bodies are formed in the ovaries of many vertebrates, ranging from the cyclostomes to the mammals; they may arise during follicular atresia (pre-ovulatory corpora lutea) or after ovulation (post-ovulatory corpora lutea). After many years of controversy (Hoar, 1969), it now seems likely that these bodies produce hormones in some of the fishes and reptiles as well as in the birds and mammals (Chapter 23). Thus, during the evolution of the vertebrates, the ovarian follicle has assumed the duties of providing two different hormones (estrogen and progesterone) as well as the generation of female gametes.

The testis The gonad of the male plays a less varied role and for that reason is structurally simpler. The complexity of the male reproductive system is associated with its accessory structures rather than the testes since the male animal is usually more active in bringing the two sexes together and in the transfer of gametes.

When the spermatogenetic tissue is organized into compact testes, as it is in the more advanced phyla, the organ is made up of a series of elongated follicles or seminiferous tubules. The form of these units varies greatly as does also the organization of the seminiferous epithelium within them. Often all stages in spermatogenesis are evident at any level in cross sections of the tubules (Fig. 22.10); spermatogonia are located just inside the basement membrane of the tubule, and the series of maturing stages occur toward the lumen where the mature sperm are released. This arrangement is characteristic of most vertebrates. In the elasmobranchs, however, the testes are made up of ampullae which contain only one stage of developing germ cells (Hoar, 1969). This feature proved most useful in localizing the source and action of the pituitary gonadotropins of these fishes, since the removal of the ventral lobe of the pituitary and no other part affected one particular stage of development (Dodd *et al.*, 1960). Testes of insects are also zonated structures and classical organs for the study of spermatogenesis. Testicular follicles contain a suc-

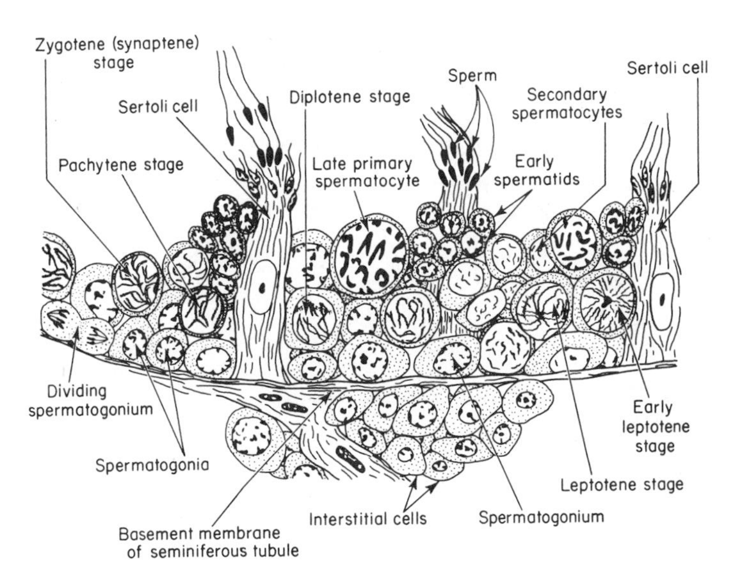

Fig. 22.10 Diagram of a part of the seminiferous tubule of the cat testis. [From Nelsen, 1953: Comparative Embryology of the Vertebrates. McGraw-Hill, New York.]

cession of zones, each composed of cells in one particular stage of development; progressively more mature stages in differentiation occur from the apical area of spermatogonia to the basal area of spermatozoa (Wigglesworth, 1972).

Androgen-producing endocrine tissue is recognized within the testes of all vertebrate animals. It may be identified histochemically by its staining reactions. Small groups of relatively large polyhedral cells with granular cytoplasm and large nuclei are situated in the spaces between the seminiferous tubules (CELLS OF LEYDIG, INTERSTITIAL TISSUE) or are arranged as LOBULE BOUNDARY CELLS in the walls of the seminiferous tubules. The lobule boundary-cell arrangement is known only in some of the teleost fishes and may be a secondary development (Hoar, 1965b). The staining properties of these two types of androgen-producing tissue are similar, and they probably elaborate similar hormones. Endocrine-producing tissues have not been identified in the gonads of most invertebrates (Chapter 23).

SERTOLI CELLS are believed to regulate the differentiation of spermatozoa and to nourish them. Clusters of spermatids (haploid cells which arise from spermatocytes) become engulfed in or attached to the Sertoli cells while developing into mature sperm (spermiogenesis). The differentiating sperm are obviously supported by the Sertoli cells and presumably draw nourishment from them (Figs. 22.10 and 22.11). In addition, there is now abundant evidence that these cells elaborate androgens in the vertebrates (Bell *et al.*, 1971; Lofts, 1972). It is suggested that the androgens produced by Sertoli cells act locally to regulate spermiogenesis while those formed by the interstitial cells enter the circulation and operate peripherally on such structures as secondary sex characters and the brain (Lofts, 1972). Thus, the Sertoli cells in

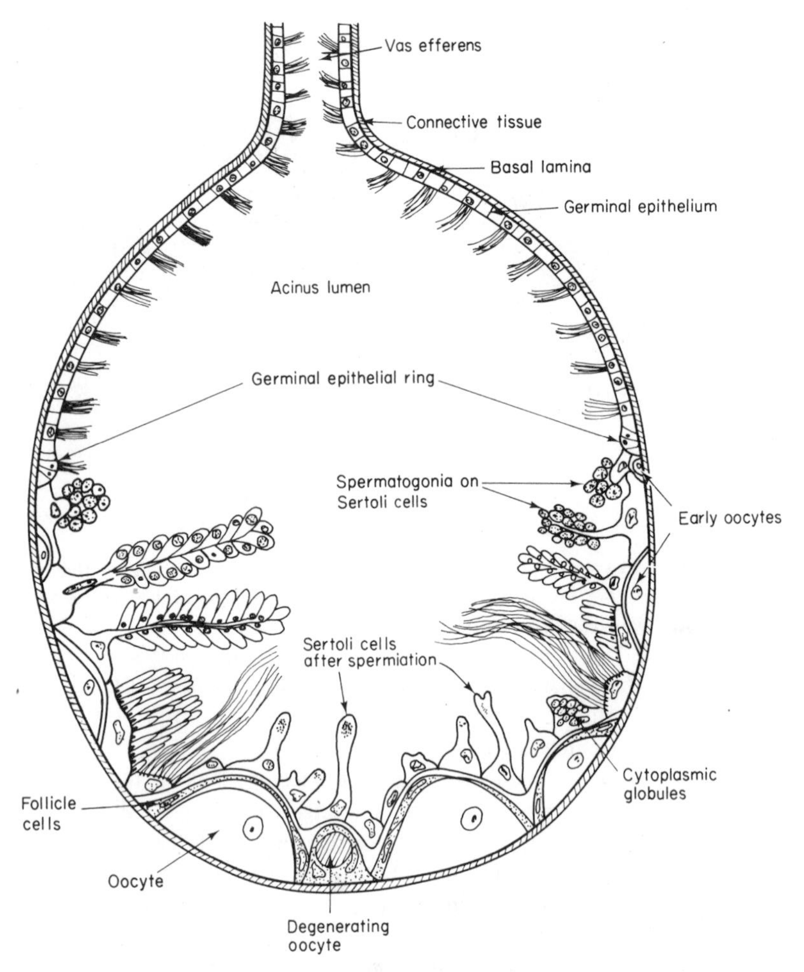

Fig. 22.11 Section through the wall of the ovotestis of the snail *Lymnaea stagnalis*. Note oogonia, follicle cells, and Sertoli cells arising from the germinal epithelium. [From Joosse and Reitz, 1969: Malacologia *9*: 104.]

the testis and the follicle cells in the ovary play comparable roles. Embryologically, both probably arise from the germinal ridge (Balinsky, 1970b); their close morphological relationships are shown suggestively in the ovotestis of the gastropod *Lymnaea* (Fig. 22.11).

Intersexual Relationships

The mechanisms which bring sperm into contact with eggs and the subsequent processes of fertilization must be absolutely reliable or the species will perish. This requirement of life has been satisfied through a host of special adaptations; some of these are cellular, others occur at the tissue or organ level, and still others at the species and interspecies level of social organization.

At the cellular level there are the egg secretions which influence sperm (FERTILIZIN) and sperm secretions which influence eggs (ANTIFERTILIZIN). These interacting chemicals produced by the gametes form the chemical basis of fertilization. The bisexual condition of higher forms in which ovaries and testes are housed in separate individuals eliminates the possibility of self-fertilization but increases the importance of the chemical system of attraction between eggs and sperm. Animals, from the most primitive to the most advanced, display a range of specializations which, on the one hand, ensure cross-fertilization through a separation of the sexes and, on the other, magnify the opportunities for union between the eggs and sperm from different individuals. Cross-fertilizing hermaphrodites with internal fertilization have many curious morphological arrangements to reduce the chances of self-fertilization; both hermaphrodites and bisexual species which broadcast their gametes into the ambient water have timing devices (some chemical and based on egg and sperm substances and others neuroendocrine) which often induce mass spawning and thus increase the chances of cross-fertilization by the sheer concentration of cells. Internal fertilization of bisexual individuals is also based on chemical reactions between the gametes as well as on a host of morphological specializations and behavioral adaptations to ensure that sperm are deposited in the right place at an appropriate time.

Fertilization Research extending over more than half a century has now established the presence of interacting egg and sperm chemicals which hold the cells together once contact has been made and, in some cases, stimulate mass spawning (Austin, 1965; Monroy, 1965; Raven, 1966; Metz and Monroy, 1967–69). Many of the pioneer studies were carried out with sea urchins. A specific chemotaxis whereby an egg attracts sperm to its surface is unlikely among animals (Rothschild, 1956; Balinsky, 1970b). However, once the sperm, through its motility or otherwise, makes contact with the egg, a definite reaction occurs. If sperm are added to "egg-water" (water in which ripe eggs of the same species were retained for a period of time), they are first activated and then agglutinated or clumped. The sperm-activating and agglutinating material was named FERTILIZIN by F. R. Lillie who worked on sea urchin eggs. Comparable materials have been demonstrated in many animals, vertebrate as well as invertebrate; they are usually species-specific and probably form a complex of materials in any one species rather than a single substance. In sea urchin eggs, fertilizin is localized in the gelatinous coat; it is a glycoprotein in which the carbohydrate component varies among different genera. Sperm contain a chemical or chemicals called ANTIFERTILIZIN which unites with the fertilizin of the egg so that the spermatozoon is trapped on its surface. Antifertilizins are small molecular weight acidic proteins which are located on the sperm head. The fertilizing reaction also depends on the presence of sperm lysins, which are present in the acrosome and dissolve a path for the entrance of the sperm through the egg membranes. In mammals, HYALURONIDASE, another enzyme produced by sperm, brings about the disintegration of the follicular cells associated with the mammalian egg at ovulation. Reactions between eggs and sperm are species-specific and bear many resemblances to immunological reactions. Comparable reactions occur on the surfaces of conjugating protozoans. The different mating types "stick together"; the reacting substances seem to be associated with the cilia in *Paramecium*, but this is not true of some other ciliated forms (Bonner, 1958; Kudo, 1966).

Reciprocal induction of spawning Egg and sperm secretions have been shown to induce spawning in some of the marine invertebrates (Nelsen, 1953; Rothschild, 1956). This phenomenon was observed in the spawning of oysters *Ostrea virginica* many years ago (Galtsoff, 1938, 1961). The presence of oyster sperm stimulates spawning of female oysters; the presence of oyster eggs stimulates the males to release sperm. The reactions are specific, and foreign eggs or sperm are quite ineffective. A mass spawning is triggered in this way, and the chances of successful fertilization are increased. Similar reactions have been demonstrated in some other invertebrate species and are probably of wide occurrence—particularly in sessile or sedentary forms. They may also be important in some more advanced aquatic animals such as the fishes, but the coordination at this level is more apt to depend on neuroendocrine and behavioral mechanisms.

The associations of males with females The evolution of the bisexual or dioecious condition was a most significant phylogenetic advance. The localization of ovaries and testes in different individuals removed all chances of self-fertilization and insured the advantages which come through the shuffling of two sets of genes. Moreover, the bisexual species has freedom, both to specialize in gamete production and to divide the labors of protecting and nourishing the young. It is obvious that at the top of the phylogenetic tree man's peculiar evolution, which has attained a certain freedom from the germ plasm, owes much of its rich variety and interest to the existence of separate male and female personalities.

Many biologists have noted the marked diversity in reproductive structures, physiology and behavior. It seems to exceed that of any other system. In part, this may be attributed to the existence of two kinds of individuals, the males and females. In part, however, it is due to the freedom of reproductive structures and processes from the demands of other organs and systems. There is a definite autonomy of the reproductive system; it exists for the future of the species and not for the individual which houses it. This autonomy seems to have been responsible for the frequent appearance of highly specialized processes, such as viviparity, in groups widely separated phylogenetically.

The array of mechanisms concerned with sexual reproduction ranges from simple interactions between morphologically similar males and females to complex interactions between highly dimorphic sexual partners. There is a vast and fascinating descriptive literature; only a few examples are recorded here. Equally curious mechanisms occur in most major groups of animals.

Many marine invertebrates and some teleost fishes reproduce by mass spawnings of apparently similar individuals. By contrast, the mother sea turtle lumbers up a sandy tropical beach, after a long life of extensive ocean travel, to deposit previously fertilized eggs in a deep pit and hasten back to the deep waters; the female salmon migrates hundreds of miles to scoop out a crude nest for her eggs and to mate with a highly dimorphic partner; the male stickleback *Gasterosteus* selects an appropriate territory, builds a nest by gluing weeds together with secretions from his kidneys, courts his partner in a graceful dance and leads her into the nest where he fertilizes and guards the eggs which she deposits.

Fertilized eggs may be broadcast into the vastness of the ocean with no insurance for the future except the safety of numbers. Very often, however, the species depends on a relatively small number of eggs which are carefully protected during their development. Gobiid and cottid fishes glue their eggs to the underside of a protecting ledge where the male fertilizes, aerates (fans) and guards them until hatching; many birds are skilled architects, using varied nesting materials to construct intricate and beautiful homes. Most of the successful groups of animals, at all levels in phylogeny, have exploited the body of the parents as a nesting site. Often the body provides only shelter during the development of yolky eggs; many of the cichlid fishes incubate their eggs in the oral cavity; the male sea horse has a brood pouch on the under side of the abdomen; the marsupial toads carry their eggs under a flap of dorsal skin; the female bitterling *Rhodeus* uses an elongated ovipositor to deposit her eggs in the body of another animal (the freshwater mussel). In all these cases, the animal body provides only a protecting microhabitat and is comparable to an external nest.

When, however, the eggs are retained within the reproductive organs during development, there is often a reduction in the amount of yolk and an elaboration of mechanisms for the provision of the developing young. Although protection of the young must have been a major factor in the evolution of viviparity, it may not have been the only pressure operating in the direction of internal fertilization. Parker (1970) suggests, in a discussion of sperm competition, that there may be real advantages in bringing fresh sperm into close contact with eggs and that internal fertilization itself, as well as mating plugs and extended copulation, reduce sperm competition.

The oviduct, or some modified part of it (a uterus), is the usual place for viviparous development. In some insects and teleost fishes, however, fertilization occurs prior to ovulation and the egg undergoes considerable development within the ovary (Davey, 1965; Hoar, 1969; Wigglesworth, 1972). Gestation among the teleosts may occur either in the cavity of the ovary (OVARIAN GESTATION) or inside the ovarian follicle (FOLLICULAR GESTATION). In some insects, the young develop within the hemocoel. The maternal-embryonic connections are extremely varied, ranging from the secretion of a "uterine milk" to the development of specialized vascular absorptive folds, villi or ribbons from the external surface of the young; in our own species, which represents one of the most specialized situations, there is an elaborate PLACENTA with vascular, finger-like fetal processes bathed in pools of maternal blood.

Maturation of gametes is often a seasonal phenomenon. It is uneconomical to produce sperm and food-laden eggs continuously when the chances of survival and growth are much greater at certain seasons. Physiological mechanisms, that synchronize the development of gametes and regulate sexual and parental behavior, provide many of the interesting problems in the biology of reproduction. The endocrinology, which is primarily responsible for the timing, will be discussed in the next chapter; several environmental pressures, that seem to have forced a precise timing of events, are noted here.

Gray seals *Halichoerus grypus* of the North Atlantic give birth to their young and mate on the dense ice packs of the Arctic seas. A "platform" for these activities may exist for only about one month of the year. During this period, females are concentrated and available; at other times, they are scattered on their feeding journeys throughout the northern seas. Mating follows closely after parturition but the ovum, after undergoing early cleavage, remains dormant in the oviduct for about three months before implantation in the uterus and the subsequent development of the blastocyst. Nine months later, at the end of gestation, another ice platform has formed and the seals "haul out" on it to start a new cycle of reproductive activity (Fisher, 1963). DELAYED IMPLANTATION occurs in many mammals; in various species, it has been found to last for periods as short as four weeks (weasels) to as long as ten months (European badger). Periods of four to six months are not uncommon (van Tienhoven, 1968).

This curious phenomenon seems to have been noted first by the famous seventeenth-century physician William Harvey on his deer-hunting trips with King Charles I. It permits mating at a time when females are available while the time of implantation is adjusted to the gestation period. The delay of implantation is not always related to seasonal conditions but may depend on lactation. Mice, rats and guinea pigs become pregnant but delay implantation while the mother is busy nursing the babies—more than two babies in the case of mice but more than five in mother rats. There are many other curious biological adjustments; useful reviews and an entry to the literature will be found in Enders (1963), Sharman and Berger (1969) and Lanman (1970).

Some oviparous animals are also able to arrest development and delay hatching to avoid unfavorable environmental situations. Many aedine mosquitos lay their eggs in moist soil, water puddles or other places subject to temporary flooding. The eggs must remain moist for a few hours until the shell becomes impermeable to water; the embryo continues to develop within the water-tight shell and remains viable for periods ranging from several weeks in some species to as long as several years in others (Clements, 1963; Chapter 24). This ability to arrest development in the egg stage permits some species to hibernate and others to aestivate or to colonize habitats which are only occasionally flooded.

A comparable condition occurs in certain cyprinodont fishes of tropical Africa and South America (Greenwood, 1963). Spawning takes place before the dry season and the partially developed embryos remain for extended periods in the bottom debris after the ponds dry up. Young hatch shortly after the next annual rains, grow to adult size and spawn within a year. Since the adults die after one spawning whether or not the ponds dry up, these fishes have been called ANNUAL FISHES. Wourms (1967) has described a number of their developmental peculiarities.

There are many ways to ensure an adequate number of contacts between the two sexes. These range all the way from complex reproductive behavior, involving long, precise migrations with an intricate courtship, to bizarre morphological adaptations in the parasitic males of the deep-sea ceratoid angler fishes (Greenwood, 1963). Storage of sperm in the female genital tract is often an important adjunct to internal fertilization; a single insemination may permit steady production of fertile eggs for months or years and provide reliable insurance against the hazards of infrequent mating (Parkes, 1960b). The best-known examples are found among terrestrial poikilotherms—particularly, the insects and reptiles. Van Tienhoven's (1968) tabulation shows that the reptiles hold the record for prolonged sperm storage with periods ranging up to six years recorded for one of the colubrid snakes. In contrast, homeotherms, which tend to be much more active, usually show a close synchronization between insemination and ovulation. In fact, numerous mammals like the domestic cat, the rabbit and the ferret require the stimulation of copulation before ovulation will take place. These animals are said to be INDUCED OVULATORS in contrast to the SPONTANEOUS OVULATORS.

In some viviparous teleosts of the family Poeciliidae, sperm storage is associated with simultaneous development of several broods of young (SUPERFETATION). *Heterandria formosa* seems to exemplify the climax in this specialization with as many as nine broods developing simultaneously in one female (Turner, 1937). The necessity for frequent sexual contacts has thus been reduced and a few young are produced at regular intervals in a hospitable environment where the conditions for development and growth are suitable throughout the year.

In summary, sex as the major device for evolutionary adaptation has been curiously exploited in an amazing number of different ways. The seasons of spermiogenesis, oogenesis, copulation and fertilization are nicely adjusted to the best seasons for the survival and growth of the young. High concentrations of sperm are maintained in the vicinity of freshly produced eggs by precisely triggered mass spawnings, or through localized nesting sites, or by intricate sexual behavior and internal fertilization. Sperm may remain viable for long periods in the female or the development of fertilized eggs may be arrested in anticipation of the best season for the appearance of the young. The young are offered varying degrees of protection through carefully hidden and guarded nests, viviparity and parental care. Adjustments occur at each stage in reproduction, so that the maximum biological advantages of sex are realized in the evolutionary or adaptive sense.

23

ENDOCRINE REGULATION OF REPRODUCTION

Reproduction is usually a seasonal or cyclical activity. A sockeye salmon *Oncorhynchus nerka* lives for about four years, attains a weight of five to ten pounds, travels many hundreds of miles into the ocean and returns to the headwaters of a river where it spawns and dies. At the same time, one of its internal nematode parasites *Philonema oncorhynchi*, acquired from the food (*Cyclops*, a copepod) during the salmon's freshwater juvenile existence, rides along during these migrations and matures and reproduces in the peritoneal cavity of the female salmon. When the salmon spawns, the nematodes, now fllled with thousands of larvae, pass with the eggs into the water where they swell and burst, releasing the larvae to infect the copepods (Platzer and Adams, 1967). In this example both the salmon and its parasite have a single period of reproduction at the end of their lives, and this is timed seasonally in accordance with the requirements for the success of the next generation.

Sometimes the most advantageous period occurs rarely, and reproduction only takes place at irregular intervals. This is the case with some Australian birds whose gonads may remain quiescent for several seasons of drought but respond quickly to rainfall or its effects (Serventy, 1971). Often reproduction is cyclical as in many female mammals with regular estrous cycles; occasionally sexual activity may be continuous as in males of some higher vertebrates. However, in the animal kingdom as a whole, continuous reproduction is the exception and seasonal or cyclical reproduction the rule.

If reproduction is seasonal or occurs only once in the animal's life cycle, it must be precisely timed so that young appear when food is abundant and other conditions are optimal for survival. Seasonal changes in temperature, photoperiod, moisture, food or chemicals may serve as triggers for the timing of the associated physiological changes. In each case it seems that the most reliable environmental change has been utilized. In the stickleback *Gasterosteus aculeatus,* the physiological changes can be experimentally manipulated by photoperiod, with temperature playing a subsidiary role (Baggerman, 1972). The sexual cycle in the minnow *Couesius plumbeus,* however, shows only a slight photoperiod effect under experimental conditions but a marked response to temperature change (Ahsan, 1966). Sexual maturation of the nematode parasite of the sockeye salmon, mentioned previously, is probably timed by the gonadal hormones of the female salmon and this in turn by photoperiod and/or temperature.

Hormones, rather than nervous reflexes, control the development of sexual structures and time the reproductive events. Differentiation and growth of gametes are necessary preludes to reproduction. In many species this involves the gradual mobilization of massive amounts of food in the form of yolk; sometimes both males and females store abundant food in subcutaneous tissues, muscles, or liver, in anticipation of extended migrations or prolonged fasting. Secondary sex characteristics are frequently present and may require the accumulation of quantities of inorganic material (as in the antlers of the deer) or the synthesis of brilliantly colored substances or odoriferous pheromones. Brood pouches may develop; periodic preparatory changes frequently occur in reproductive passages to facilitate the movements of gametes or the implantation of the developing ovum. Finally, the actual breeding behavior synchronizes the activities of males and females; this may include nest building, intricate courtship and prolonged parental care as well as the act of mating. In at least some animals, each of these events has been shown to depend on specific hormones. Only in the behavior of reproduction, which is obviously a neuromuscular function, are the controls somewhat divorced from chemical regulation—and here, only among the higher mammals where increasing encephalization leads to greater components of learned behavior (Beach, 1958).

Neurosecretory cells have now been identified in all groups of enterozoic multicellular animals and usually provide the direct link between the environment and the physiological machinery for reproduction (Highnam and Hill, 1969; Tombes, 1970). The evidence associating neurosecretion with reproduction is tenuous but suggestive among the lower invertebrates. However, at the annelid and higher levels of phylogenetic organization, the regulation of reproduction has now been clearly associated with neurosecretion. In most invertebrates, the neurosecretory substances act directly on the effectors, but in some of the molluscs, the arthropods and the vertebrates there are often intermediate links; the neurosecretory cells pass their more specific responsibilities on to other parts of a complex endocrine system. In many animals the gonads also attain a glandular status and secrete substances which act directly on the effectors; the interrelationships between trigger, neurosecretory center, intermediate endocrine gland and effector may be complex (Scharrer and Scharrer, 1963) and very nicely balanced through a feedback of information (Chapter 2).

INVERTEBRATE HORMONES OF REPRODUCTION

Evidence of hormonal regulation among the coelenterates rests largely on studies of the freshwater hydrozoa (Burnett and Diehl, 1964; Highnam and Hill, 1969). The hypostome of an adult *Hydra* possesses concentrations of cells with distinct neurosecretory characters. This is the region of new growth. Buds fail to develop if separated from the parent body prior to the appearance of neurosecretory cells in the budding hypostome. Moreover, extracts of fully formed hypostomes show growth-promoting effects on transverse blocks cut from hydroid cylinders. With the onset of sexual maturity, neurosecretory cells disappear and extracts of the hypostome no longer stimulate growth. Thus, growth and reproduction are antagonistic processes. The neurosecretory growth hormone appears to have two functions: (1) it stimulates cell proliferation and (2) it causes interstitial cells to develop into somatic structures such as nematocysts. In the absence of this hormone, growth ceases and the interstitial cells form gametes. Factors inhibiting growth (low temperature, short photoperiods, starvation or stagnant water) stimulate the development of gametes and sometimes the formation of overwintering zygotes. The biological significance is obvious; an endocrine control is now reasonably well established (Highnam and Hill, 1969).

Evidence of hormonal controls is less satisfactory in the flatworms and the other acoelomate and pseudocoelomate animals. Suggestive correlations between numbers of neurosecretory cells and the state of reproductive development have been noted in the flatworm *Dendrocoelum* (Ude, 1964; Tombes, 1970), but current evidence is all indirect and rests primarily on studies of growth, regeneration and molting. These processes will be considered in the next chapter.

Studies of the endocrinology of the coelomate invertebrates have been confined to the annelids, molluscs, arthropods and echinoderms. In all cases, the dominant controls are neurosecretory. The notes which follow are arranged phylogenetically since generalizations, other than the neurosecretory nature of the controls, seem to be absent.

Annelids

The annelid worms have, inconspicuously but successfully, exploited a wide variety of marine, freshwater and terrestrial habitats. Several of their peculiar biological adaptations are obviously well-suited to the type of control provided by the endocrine system: temperate, frigid and arid environments have imposed seasonal constraints on growth and reproduction—two processes which are frequently antagonistic because of the excessive metabolic demands of each; their soft bodies and habits of living in the substratum expose them to hazards of predation and damage, so that regeneration and prolonged growth become precious biological assets; their sedentary ways of life favor such specializations as asexual divisions, seasonal swarming of morphologically distinct sexual forms (heteronereis), hermaphroditism (either self-fertilizing or copulating) and viviparity (R. I. Smith, 1958). Numerous investigations of the past two decades make it clear that neurosecretions play a commanding role in the regulation of these processes (Highnam and Hill, 1969; Tombes, 1970; Golding, 1972).

The evidence is both morphological and physiological. A series of investigations, commencing with Berta Scharrer's classical paper on *Nereis* in 1936, has localized several types of neurosecretory cells in the brain and other ganglia; these studies have not, however, shown whether the different animals depend on one or several hormones. Likewise, the physiological work has demonstrated a well-established endocrinology of growth, metamorphosis, sexual development and reproductive behavior but has not yet identified the biochemical factor or factors involved. Surgical manipulations, including transections of the nerve chain, removal of various ganglia and implantation of cerebral ganglia into decerebrate individuals, have predictable effects on reproductive processes. In general, the facts argue for the existence of one or more hormones of neural origin.

Marked species differences are recognized. Among the polychaetes, brain hormones inhibit gametogenesis and reproductive activities (Fig. 23.1); in the lugworms, earthworms and leeches, by contrast, the gonadotropic hormones are stimulatory.

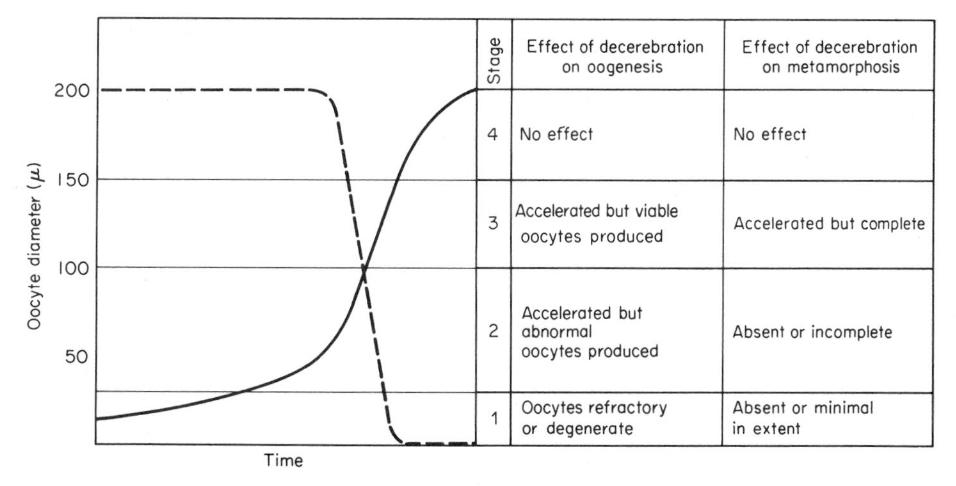

Fig. 23.1 The sigmoid growth pattern of nereid oocytes (solid line) with a summary of the consequences of decerebration at different stages in the life cycle. Broken line indicates how the titer of brain hormone may change during development. The final oocyte diameter and that corresponding to the onset of each stage vary from species to species. [Based on several sources, from Golding, 1972: Gen. Comp. Endocrinol. Suppl., *3*: 581.]

The nereid worms have been most carefully studied but it is by no means clear that they exemplify the most primitive patterns of endocrine control (Golding, 1972).

Molluscs

Links in the endocrine control of cephalopod reproduction are recognized and shown to have certain parallels with the more familiar vertebrate patterns. The octopus has been the main target of investigation (Wells and Wells, 1959, 1969). In this animal, which is one of the greatest of evolutionary achievements, the brain (subpeduncular lobes) regulates a pair of OPTIC GLANDS through the action of inhibitory nerves. The brain, in turn, receives environmental cues through the eyes and probably elsewhere as well. The optic glands, in a manner somewhat comparable to the vertebrate pituitary, dominate the reproductive endocrinology. These glands are located on the optic stalks, one on either side of the central part of the supraesophageal brain (Fig. 23.2). The direct inhibitory action of nerves on these glands differs from the vertebrate situation where hypothalamic neurosecretions link the nervous system with the pituitary.

Optic gland secretions promote vitellogenesis in females and spermatogenesis in males. They also regulate the differentiation of the female genital tract; in contrast, male sex characteristics and behavior are under direct control of testicular hormones (Wells and Wells, 1969). Convincing experimental evidence for these findings

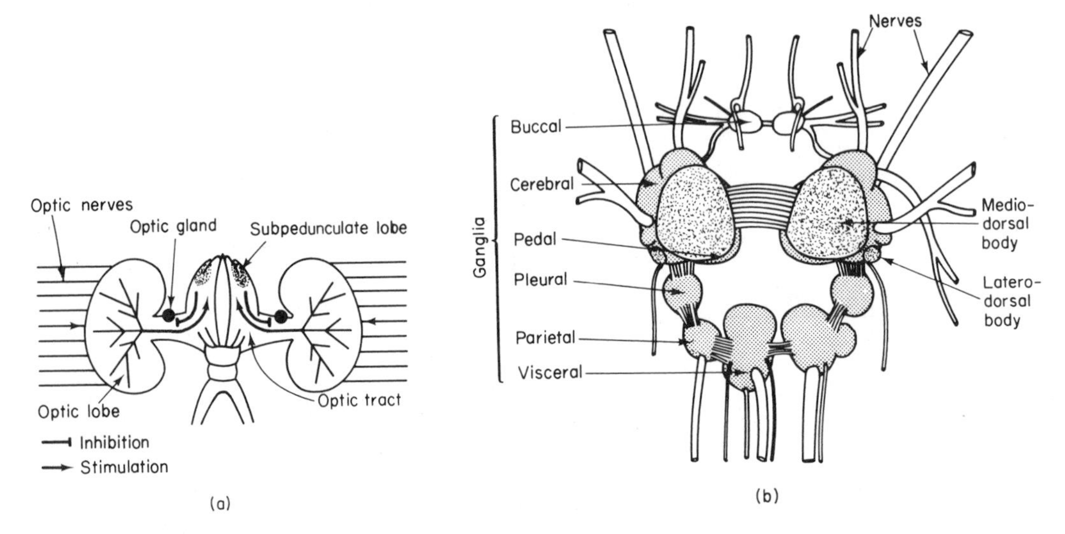

Fig. 23.2 Some endocrine glands of molluscs. (a) Position of the optic glands in the octopus and their relationships to the eyes and brain. [From Wells and Wells, 1959: J. Exp. Biol., *36*: 28.] (b) Central nervous system of *Lymnaea stagnalis*. [From Lever *et al.*, 1961: Proc. Kon. Ned. Akad. Wetensch., C, *64*: 532. Permission North Holland Publishing Co., Amsterdam, Netherlands.]

comes from detailed anatomical studies and physiological research involving gland extirpation, replacement therapy, organ culture and observations of behavior. In broad outline, the controls in the octopus parallel the vertebrate scheme but are simpler in that neurosecretions and ovarian hormones seem to be absent.

Only the gastropods, among other molluscan groups, have as yet provided evidence of a special endocrinology of reproduction. This statement is made while recognizing numerous correlations between neurosecretion and reproduction. Attempts to confirm these have proved so disappointing—particularly among the bivalves—that conclusions based on them must be examined with great skepticism (Simpson *et al.*, 1966; Joosse, 1972).

The best-known endocrine organs of gastropods are the DORSAL BODIES of the freshwater pulmonate snails. Their location varies somewhat but in *Lymnaea stagnalis,* the species most carefully studied, they are small, discrete bodies on the cerebral ganglia—conveniently exposed for surgical work (Fig. 23.2b). Their functions are more varied than those of the cephalopod optic glands and include osmo-ion regulation (Wendelaar Bonga, 1972). Dorsal bodies develop from glial cells and the perineurium of the cerebral ganglia. Their secretions stimulate vitellogenesis and the differentiation of the female reproductive tract. Beyond this, the entire field of gastropod reproductive endocrinology bristles with problems: oogenesis may be autonomous; spermatogenesis may be regulated by the cerebral ganglia; neurosecretions are probably involved in some species; ovotestes and optic tentacles may produce reproductive hormones and show strong evidence of steroid biosynthesis

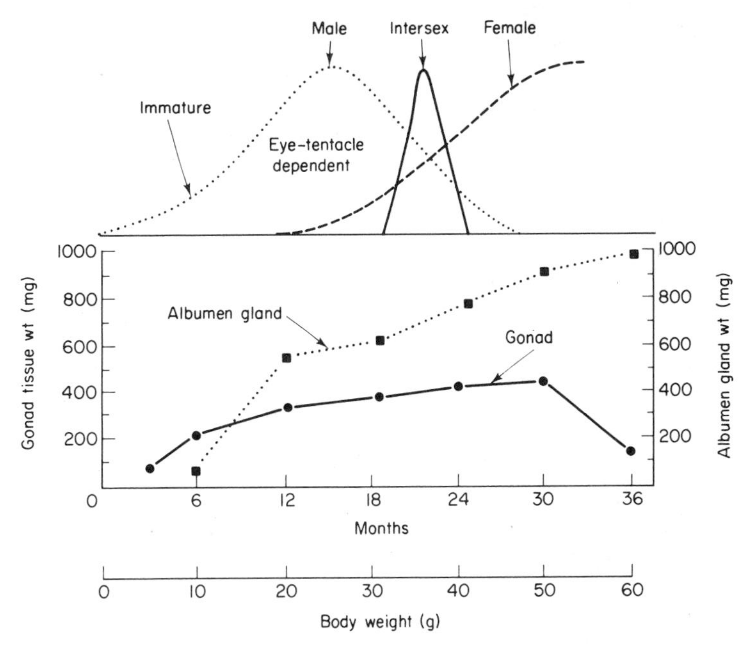

Fig. 23.3 The gonad cycle in a protandrous hermaphroditic mollusc. [From Gottfried and Dorfman, 1970: Gen. Comp. Endocrinol., *15*; 105.]

(Gottfried and Dorfman, 1970). A complex endocrinology with marked species variations can be expected in animals which possess curious secondary sex characteristics and lead a protandrous hermaphroditic life that may include the experiences of a male for a year or more, to be followed by a typical female existence (Fig. 23.3). The molluscs offer many inviting avenues for research.

Crustaceans

The major components of the crustacean endocrine system are shown in Fig. 20.20. Removal of the eyestalks in juvenile or non-breeding females leads to a rapid increase in ovarian weight through the stimulation of vitellogenesis (Fig. 23.4). The neurosecretory cells in the eyestalk complex secrete a gonad-inhibiting hormone, and when the eyestalks are removed surgically or when the activities of the cells are suppressed, ovarian development with yolk production is initiated in females while testicular development occurs in males. Under natural conditions the changing photoperiod seems to regulate the activity. A high titer of sinus gland extract also inhibits molting in egg-bearing females. In crayfishes, for example, the spring molt is postponed in females while they are carrying developing eggs attached to their pleopods. If their sinus glands are removed, they molt at the same time as the males or non-breeding females. Thus, the sinus gland hormone first inhibits vitellogenesis; then, as its secretion is gradually suppressed under changing environmental conditions (photoperiod), the ovary increases rapidly in weight and accumulates the food supply for the future embryos; following mating and egg-laying, the eyestalk hormone(s) is again secreted in sufficient amounts to inhibit vitellogenesis and to prevent molting while the eggs are attached to the female.

The complete explanation of these findings may be more complicated than sug-

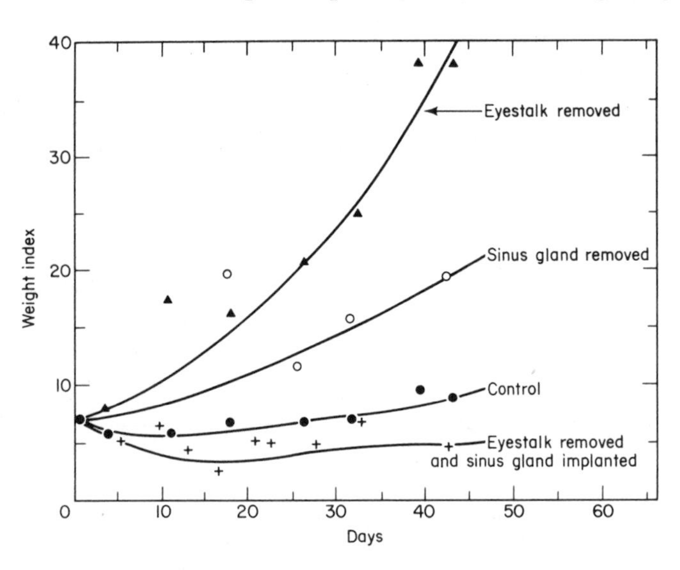

Fig. 23.4 Rate of increase in ovarian weight of the shrimp *Leander*. Weight index determined by dividing the wet weight of the ovaries by the body length cubed. [Data from Panouse, 1944: C. R. Acad. Sci., *218*: 293.]

gested above. In several species of decapod crustaceans, implantation of brain or thoracic ganglion has been found to accelerate gonadal development. Thus, in addition to the gonad-inhibiting hormone (GIH) of the eyestalks, there may also be a gonad-stimulating hormone (GSH) of neurosecretory origin (Adiyodi and Adiyodi, 1970). According to this bi-hormonal hypothesis, reproduction is inhibited by the secretion of GIH from the eyestalks and stimulated by GSH from the brain and/or thoracic ganglion.

Most of the crustaceans are bisexual with distinct sex dimorphism and behavior. Sex characters and mating behavior depend on a second group of hormones which are comparable to the gonadal hormones of the vertebrates. Once vitellogenesis has been initiated, the ovary takes over several of the endocrine functions associated with reproduction. It is particularly responsible for certain secondary sex characters (ovigerous hairs, brood pouches) and brooding behavior. The dominant endocrine gland of the male is the ANDROGENIC GLAND. This is a small mass of cells attached to the distal end of the vas deferens. Its removal is followed by failure of testicular development and the appearance of female characters in the operated males. Further, transplantation of vas deferens glands into females will convert their ovaries into testes and cause them to assume male sexual behavior. The androgenic gland is clearly the source of a potent male hormone and seems to be the major endocrine gland concerned with reproduction in male crustaceans. The hormone may be a steroid (Gilgan and Idler, 1967). In some species of isopods the testis itself, rather than the androgenic gland, is the source of the male hormone (Carlisle and Knowles, 1959; Scheer, 1960). The androgenic gland may have arisen phylogenetically from interstitial cells of the testis.

In addition to neurosecretory and gonadal hormones, there is some evidence for an involvement of the Y-organ in reproductive controls. Removal of this gland before sexual maturity is followed by degeneration of the gonads and sex failure in both sexes. Since a similar operation on adults has little or no effect on sex functions, its action may be on the general maturation of juveniles rather than specifically on the reproductive processes.

Insects

The corpora allata (Chapter 2 and Fig. 24.2) link the neural centers with the reproductive glands. Wigglesworth (1936) performed the pioneer experiments on the bug *Rhodnius,* a bloodsucker of mammals. If the corpora allata are removed from this animal, egg development ceases at the point where yolk deposition should begin. Testicular development is unaffected by the operation, but the secretory activity of the accessory glands of the male (as well as the female) *Rhodnius* is impaired. Implantation of corpora allata (from either sex or from larvae) restores oocyte development and permits development of accessory sex structures. It is evident from these experiments that the corpora allata are essential for vitellogenesis and sexual development. The hormone is neither sex specific nor confined to adult life. Neurosecretions do not seem to be involved since normally fed animals, when decapitated in front of the corpora allata, produce eggs. It has been concluded that the corpora allata of *Rhodnius* produce a gonadotropic hormone which is identical with the juvenile hormone (Chapter 24). Peripheral stimuli arising from the distention of the gut with blood trigger activity through nerves from the brain.

Although the corpora allata hold a central position in the reproductive controls of most insects, the links have been shown to be more varied in many other species. There are excellent reviews (Davey, 1965; Highnam and Hill, 1969; Engelmann, 1970; Wigglesworth, 1970). The experimental evidence will not be detailed here; the major links are shown for a generalized insect in Fig. 23.5 and may be summarized as follows:

1. In the stick insects (Phasmidae) and the butterflies and moths (Lepidoptera), egg development proceeds normally in the absence of corpora allata. These insects are unusual; stick insects are considered neotonous (reproduce as juveniles); lepidopterans have a very brief adult life with eggs almost fully developed before the final molt.
2. Neurosecretions from the brain may inhibit the corpora allata (locusts).
3. Cerebral neurosecretions (via the corpora cardiaca) act in association with the allatal hormone in such insects as blowflies, locusts and cockroaches. Hormones from the corpora cardiaca stimulate the synthesis of protein which is then incorporated into the oocyte under the stimulus of the gonadotropic hormone.

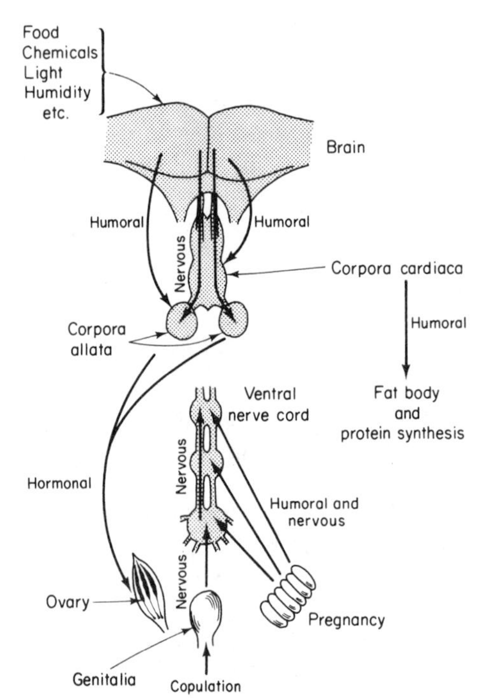

Fig. 23.5 Hormonal control of reproduction in a "generalized" female insect. Only some of the links are present in any given species. [Based on Engelmann, 1970: The Physiology of Insect Reproduction. Pergamon Press, Oxford.]

4. In addition to food, which triggers reproduction in many species, several additional stimuli have been shown to act by nervous pathways: copulation stimulates egg formation (locust); pregnancy inhibits it (cockroach); male pheromones accelerate maturation in other males and females to synchronize maturation in gregarious insects (desert locust); female pheromones may inhibit maturation of other females (social insects).
5. In various insects, neurosecretions and/or allatal hormones seem to be involved in ovulation, oviposition, spermiogenesis, spermatophore formation and reproductive behavior.

A diversity of endocrine controls is not surprising in a class of animals which has successfully explored so many different avenues of evolution.

Bisexuality with internal fertilization is the rule among insects. Organs of copulation are present and, in addition, there may be color and morphological differences. Early experiments indicated an absence of sex hormones with a genetic regulation of male and female characters. It is now known, however, that the testis of at least one insect, the firefly *Lampyris noctiluca*, elaborates a potent hormone—referred to as the androgenic hormone. This factor, first described by Naisse (1966), is formed in the larval testis and causes the development of maleness. Larval testes induce a sex reversal when transplanted to females; males implanted with ovaries, however, show no change in sex characters. The secretion of androgenic hormone appears to be regulated by neurosecretions (Englemann, 1970).

Many of the insects live in a world dominated by odors, and some of the best known examples of pheromones are associated with insect reproductive behavior. The female gypsy moth *Porthetria dispar* cannot fly and depends on a sex attractant to lure the winged males. Scientists who isolated and identified the pure substances used 500,000 female gypsy moths to obtain 20 milligrams; they found the male gypsy moths would curve their bodies and make copulatory motions when exposed to as little as a trillionth of a microgram (1×10^{-12} μg) of this chemical. Only molecular amounts are required to attract males from some distance; traps baited with this chemical are proving an effective weapon in controlling the moth (Beroza and Knipling, 1972). Several other insect pests are now being controlled by the use of their specific sex attractants; the organic chemist has been able to synthesize many of them so that the applied entomologist is not dependent on the natural source of supply.

The "queen substance" provides another interesting example of a specific sex substance used in the social life of insects. The mandibular glands of the queen honeybee produce a substance which inhibits ovarian growth in the worker bees as long as they are obtaining it by occasionally licking her body. If an accident happens to the queen or if she is removed from the hive, the ovaries of the workers (neuter-type females) commence developing. Details of the endocrinology have not been worked out, but it is clear that the spread of a chemical through the hive by contact with the queen inhibits ovarian development in the workers. The "queen substance" (9-oxodecenoic acid) has been isolated and synthesized (Engelmann, 1970). Luscher (1972) postulates a dual pheromone control of caste differentiation in primitive termites of the genera *Kalotermes* and *Zootermopsis*. One of these pheromones, which may be juvenile hormone, is given off by the reproductives while the other, an antijuvenile substance, is released by the soldiers.

Echinoderms

In the starfishes, sexual maturation and spawning are regulated by a simple system of hormones (Chaet, 1967; Schuetz, 1969; Kanatani and Shirai, 1972). There are at least two factors: (1) a gonad-stimulating substance (GSS), produced by the radial nerves and (2) a maturation-inducing substance (MIS) synthesized by the follicle cells of the ovary. There may also be a gonad-inhibiting substance formed by the radial nerves (Chaet, 1967). The gonad-stimulating substance is a simple protein of molecular weight about 2000; the follicular hormone is the purine 1-methyl adenine.

With few exceptions, the starfishes are dioecious, shedding their eggs and sperm into the sea. It has long been known that their spawning reactions are synchronized and that the presence of eggs or sperm in the water stimulates other individuals to spawn. The endocrinological explanation of the mass spawning, however, is recent. The evidence is based on *in vivo* and *in vitro* studies, histological examinations and observations of behavior. The controls are relatively simple; a neurosecretion released by the radial nerves (presumably in relation to environmental cues) acts on the follicle cells to induce the synthesis of 1-methyl adenine, which triggers ovulation, release of gametes and reproductive behavior. The addition of this purine to the family of hormones and its synthesis in the gonads, which usually form steroid hormones, is a finding of considerable interest.

Several species of arctic and antarctic asteroids brood their young. This curious adaptation may be carried out in several ways, but it is most frequently performed by the female rising on the tips of her arms while brooding a batch of large yolky eggs between her body and the substratum. The brooding reaction may be induced in either male or female *Henricia nipponica* by the injection of 1-methyl adenine. It is thought that the purine, produced by the ovarian follicle, acts at the level of the nervous system (Kanatani and Shirai, 1972).

VERTEBRATE CONTROLS

In the vertebrates the adenohypophysis, through its gonadotropic hormones, links the neurosecretory centers of the brain to the endocrine tissues of the gonads. The neurosecretory centers, gonadotropins and gonadal hormones form a delicately balanced system for the timing of reproduction, the development of the gonads and the differentiation of the secondary sexual characters; these various hormones together with the nervous system itself regulate the behavior associated with reproduction (Fig. 23.6).

The neurosecretory link between the external environment and reproductive

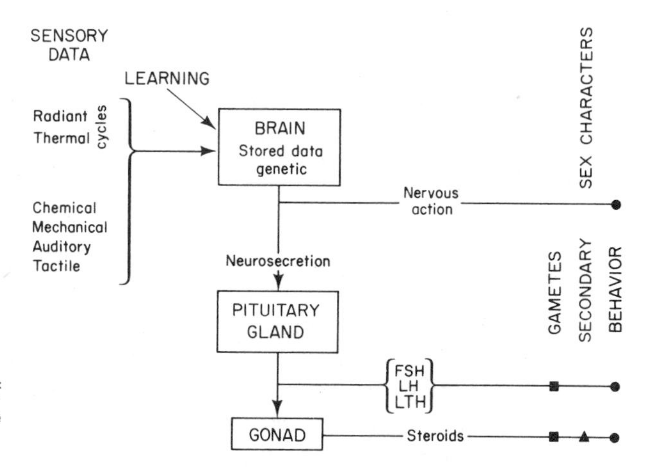

Fig. 23.6 Generalized scheme of the reproductive controls in the vertebrates.

processes is evidently primitive and was probably established early in animal evolution (Chapter 2). It seems probable that the adenohypophysial link between brain and target organs was also present in the most primitive vertebrates. Studies of existing vertebrates indicate that its significance in the regulation of reproduction has increased during phylogeny. In some primitive jawless forms (Agnatha: Petromyzontia), hypophysectomy has a much less drastic effect on reproduction than it has in the gnathostomes. In the male lamprey this operation delays, but does not eliminate, the production of sperm; in the female, ovarian growth is retarded but the developing oocytes do not degenerate (Dodd, 1972). In all higher vertebrates, removal of the pituitary leads to cessation of spermatogenesis and atresia of developing ova (Hoar, 1965a, b). Throughout the gnathostomes the pituitary is in complete command of gametogenesis. The studies of hypophysectomized lampreys suggest that during vertebrate phylogeny the anterior pituitary first exerted a metabolic effect on the gonads (vitellogenesis) and that its gametogenetic activities were secondary, as they evidently are in the insects.

Endocrine tissues are found in the gonads of all vertebrates. From the most primitive to the most advanced forms they produce a series of steroid hormones whose synthesis has evidently been pituitary-regulated from a very early stage in vertebrate phylogeny. This concert of hormones from pituitary and gonads performs together to regulate reproductive processes, and, while the pituitary gonadotropins stimulate the gonadal tissues, the gonadal steroids through a feedback mechanism serve to check the activity of the gonadotropic cells of the pituitary.

The Gonadotropins

Hypophysectomy is always followed by a regression of reproductive functions. In all classes, except the cyclostomes, gametogenesis is blocked at the spermatogonial or oogonial stages. The gametogenetic cells may continue to divide by mitosis, but maturation does not take place. Vitellogenesis ceases or is not initiated; yolk-filled ova degenerate and the yolk is removed, often with the formation of a corpus luteum-like structure. Further, the secondary sex characters fail to develop in hypophysectomized animals of all classes, including cyclostomes. Since these characters are known to be under the influence of the endocrine tissues of the gonads, it is apparent that the gonadotropins control hormone production of the gonad.

There are two distinct gonadotropins in mammals. One of these, the FOLLICLE STIMULATING HORMONE, FSH, is primarily gametogenetic; it initiates the development of ovarian follicles and may also be concerned with spermatogenesis. The main target of the other hormone, the LUTEINIZING HORMONE, LH, is the steroid-producing tissues of ovary or testis. In detail, these controls are somewhat more complex with an interplay between the two hormones both in the regulation of gametogenesis and in steroidogenesis. Thus, the full development of an ovarian follicle requires LH in association with FSH while the action of LH on the interstitial tissues of the testis (where it is also called INTERSTITIAL CELL STIMULATING HORMONE, ICSH), may require the synergistic action of FSH.

The regulation of spermatogenesis is still not fully understood (Steinberger, 1971). Early work indicated a regulation by FSH, either alone or in association with LH. However, it is now recognized that androgen will sometimes maintain spermatogenesis in hypophysectomized animals; this would suggest that LH regulates spermatogenesis by stimulating androgen production in the cells of Leydig or Sertoli. The problem is still not resolved, but it seems likely that both LH and FSH are required to maintain spermatogenesis for long periods in hypophysectomized mammals.

FSH and LH are glycoproteins of molecular weight about 30,000. The molecules are made up of two subunits (polypeptide chains), designated α and β, each with a molecular weight of around 15,000. About 15 per cent of the content is carbohydrate. FSH and LH share a similar chemistry with the thyroid-stimulating hormone TSH. The chemistry of these three factors is being rapidly elucidated with amino acid sequences established for some of the factors in certain species (Papkoff, 1972). The biological activity resides in the β subunit. It seems likely that these three hormones have been derived during phylogeny from a common glycoprotein molecule of some remote ancestral vertebrate.

Based on mammalian studies, the pioneer comparative endocrinologists assumed the presence of two gonadotropins in submammalian vertebrates. It is now reasonably clear that the poikilotherms (probably also the birds) possess a single GONADOTROPIC HORMONE, GTH, and that this varies biochemically in different classes and species. Among the poikilotherms, the teleost fishes and reptiles have been more carefully examined. GTH from teleost fishes is similar to mammalian LH while reptile GTH is similar to FSH (Burzawa-Gerard and Fontaine, 1972; Licht, 1972). Although mammalian gonadotropin is active in the sub-mammalian forms, GTH from the more primitive vertebrates is often inactive in mammals; carp GTH is inactive in rats but lungfish GTH appears to be active in mice. The many curious species differences which have been recorded are best explained by assuming the evolution of a family of glycoprotein hormones (gonadotropic and thyrotropic) from a single molecule. This may have performed gametogenetic, steroidogenetic and thyrotropic functions by specific interactions with different target organs—possibly by way of the adenylcyclase link (Chapter 2; Burzawa-Gerard and Fontaine, 1972). Among present-day poikilotherms there is a distinct separation between GTH and TSH, but the relegation of GTH functions to two different biochemical factors (FSH and LH) is only characteristic of the mammals.

Although the pattern of phylogeny is beginning to emerge, many of the details are still obscure, while the expression of gonadotropic controls in relation to environmental changes has been examined in only a few species. Licht (1972), for example, emphasizes the importance of temperature in regulating the responses of lizards to FSH (Fig. 23.7); some of the earlier work may be virtually meaningless because of lack of temperature controls.

A third factor, the lactogenic hormone PROLACTIN, is also a gonadotropin in some vertebrates. Astwood (1941), who did the pioneer work, found it to be essential for the maintenance of corpora lutea in the rat, and for this reason it has been named LUTEOTROPIN, LTH. Luteotropic activity seems to be confined to some rodents and has not been found in several other mammals (rabbit, monkey, man) nor in any of the lower vertebrates (Nicoll and Bern, 1971). A more detailed discussion of prolactin is reserved for a later section.

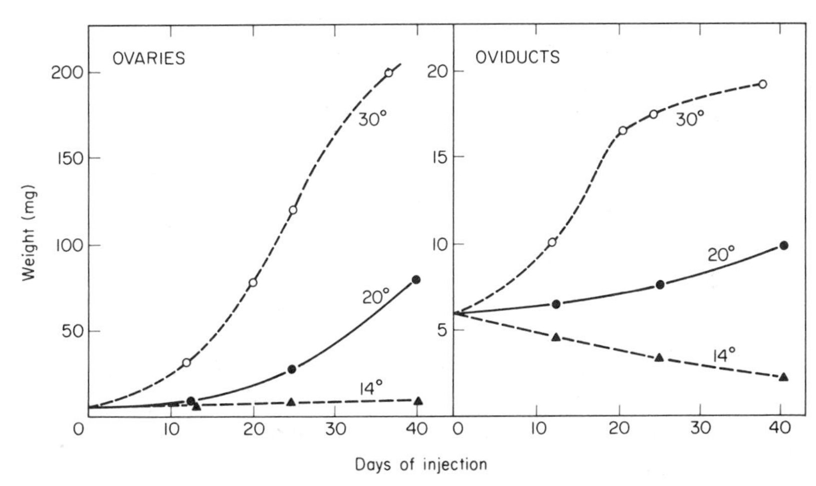

Fig. 23.7 Responses of the ovary and oviducts of adult lizards *Xantusia vigilis* to injections of ovine FSH (at different body temperatures). Subcutaneous injections of 10 μg NIH-FSH-S8 were given every other day, and samples of 3–6 individuals autopsied at the times shown. All pituitaries were intact; uninjected controls at 20 and 30°C showed no detectable ovarian growth during the experimental period. [From Licht, 1972: Gen. Comp. Endocrinol., Suppl. *3*: 484.]

The Gonadal Steroids

Representatives from all classes of living vertebrates have now been successfully gonadectomized. The data are consistent; the operation is followed not only by a suppression of gamete production but also by a regression of secondary sex characters or a failure to develop them. There are only a few recorded exceptions. In weaver finches (*Euplectes* spp), development of the dimorphic plumage seems to be controlled by the pituitary gonadotropins rather than by the gonadal hormones (Witschi, 1961; Hall *et al.*, 1965). This, however, is an exception; among all classes of vertebrates the control of secondary sex characters is normally the direct responsibility of the gonadal hormones. This was probably their first function in phylogeny. In many animals, normal sexual behavior also depends on the gonads but this, and certain other functions associated with reproduction, are much more variable from species to species.

The gonadal hormones are steroid compounds. Several members of this biologically important group of chemicals were identified in Chapter 2 as adrenocortical hormones (Fig. 2.21). Those concerned with vertebrate reproduction are the ANDROGENS from the Leydig and Sertoli cells of the testis, the ESTROGENS from the ovarian follicle and PROGESTERONE (GESTOGENS or PROGESTINS) from the corpus luteum and certain other structures in the mammal. These compounds are all interrelated. Many of the steps in their biosynthesis have been established. Cholesterol, synthesized from acetate, is the parent substance. Through several intermediate steps this gives rise to progesterone and thence to the androgens of which testosterone is physiologically the most important. In turn, the biogenesis of the estrogens is from the androgens with testosterone the major link and estradiol-17β the first step. The adrenocortical steroids are linked to the same chains. Many of these reactions are reversible; the interrelationships are indicated in Fig. 23.8.

An understanding of the biochemical relationships of these substances is essential to the interpretation of many of the physiological findings. The identification of a particular steroid in the blood or tissues of an animal is no evidence that it acts as a hormone in this species; it may be present as a step in the biogenesis of the active endocrine. Moreover, animals may be treated with exogenous steroids which are quickly metabolized to something rather different. Considerable caution is required in interpreting physiological data both from steroid hormone treatments and from chemical assays of hormone levels. Androgens are believed to be the active male hormones in fish as in higher vertebrates, and yet the level of testosterone is sometimes higher in the spawning female salmon than in the male. In the male it may be serving as a physiological androgen; in the female as a precursor of the estrogens. The gonads of many animals, both male and female, contain several androgens, estrogens and gestogens. Differences in secretion are probably quantitative rather than qualitative.

The metabolic chains leading to estrogens and androgens are phylogenetically

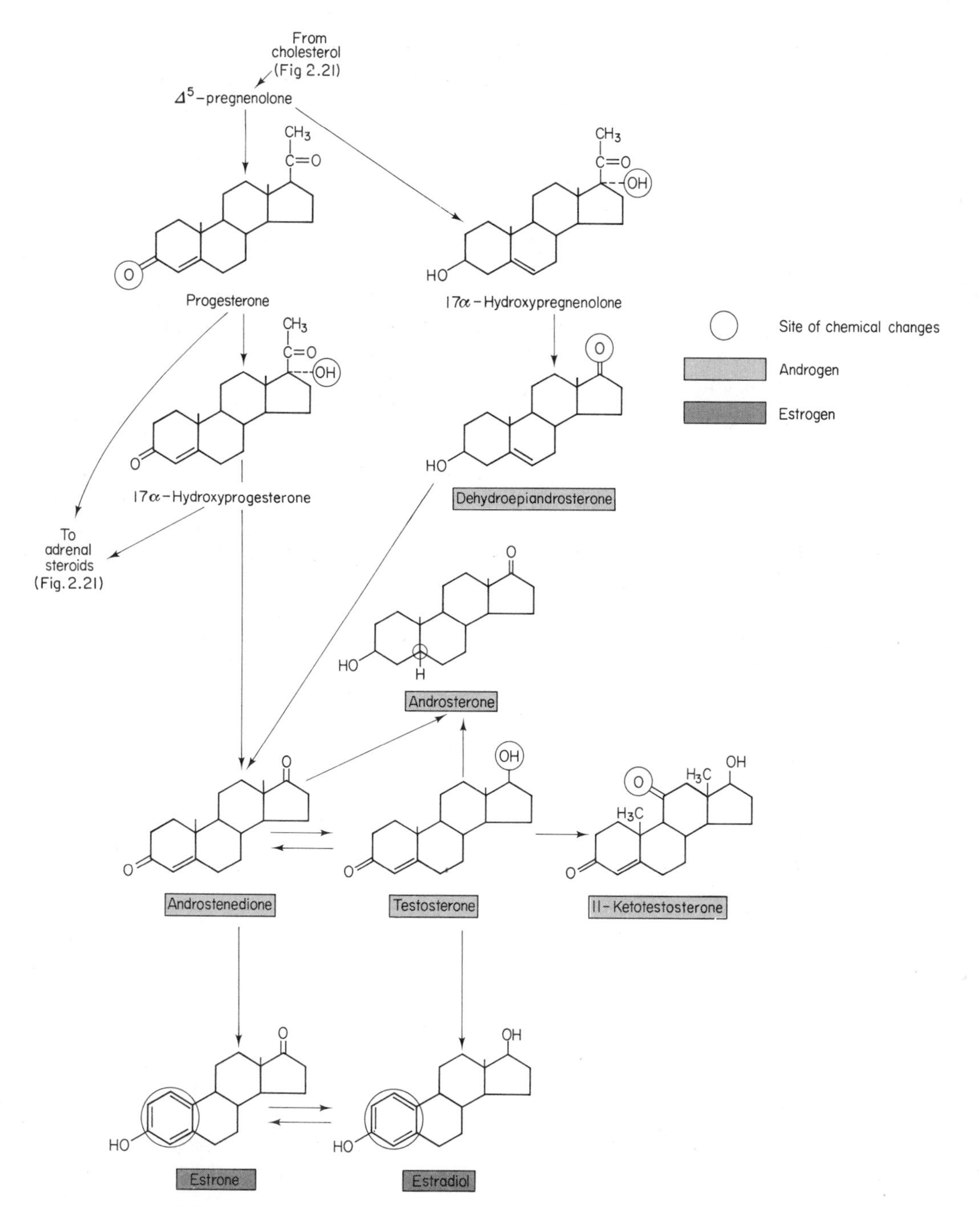

Fig. 23.8 Some important gonadal steroids and their biochemical interrelationships.

very ancient. Both progesterone, at the beginning of the chain, and the estrogens at the end of the chain have been identified in several of the invertebrates (echinoderms, lamellibranchs) as well as in the lower vertebrates. Estrogens are well known in the plant world (Bickoff, 1963).

Androgens The androgens are steroids, characterized biochemically by the presence of methyl groups at C-10 and C-13; they lack a side chain at C-17 and have oxygen substitutions at C-3 and C-17. They are sometimes referred to as the C-19 steroids because of the methyl group in the C-10 position (Fig. 2.21); this is lacking in the estrogens.

One example will suffice to illustrate the type of evidence adduced for a testicular control of secondary sex characters by androgens. The male stickleback *Gasterosteus aculeatus* builds a nest of algae or plant fragments. The strands of building material are piled neatly together at one spot and glued there with a mucous secretion which the fish presses onto the edges of the nest from his urogenital opening as he passes over it in a characteristic motion. The source of this secretion is the kidney which acquires this capacity only during the breeding season. The brush border segments of the kidney tubules increase in height and become packed with secretory granules at the onset of the breeding season. Immature males, adult males in the non-breeding condition and females at all times have brush border tubules which measure about 10 μ in height while the cells lining these tubules in the breeding males range from 35 to 40 μ. Figure 23.9a shows the effects of castration on the kidney tubules of the mature male. Over a period of about 30 days the tubules gradually return to the

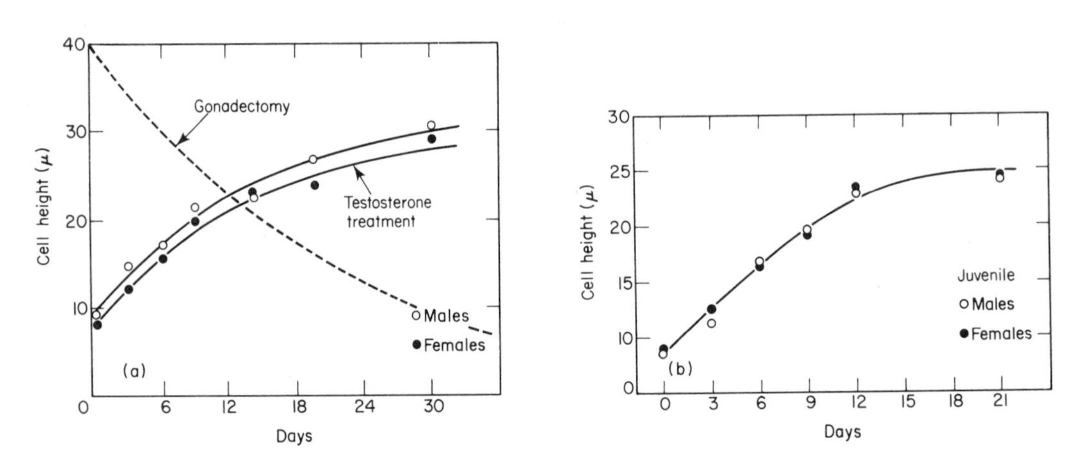

Fig. 23.9 Effects of testicular hormone on the brush border segment of the mesonephric tubules (secondary sex character) in sticklebacks. (a) Castration of sexually mature males (broken line) and treatment of fish gonadectomized for 30 days or more with methyl testosterone (solid lines). (b) Juvenile fish treated with methyl testosterone.

resting condition, and at the same time the animals lose their brilliant breeding colors and cease to build nests. This is a secondary sex character which can be precisely quantified, and the gonad is obviously essential for its expression.

When gonadectomized adult sticklebacks are treated with an androgen, the brush border tubules again assume their glandular characteristics. The reaction can be elicited in females as readily as in males and in juveniles of both sexes (Fig. 23.9b). The data are in accordance with the hypothesis that this secondary sex character is dependent on the testis and that the endocrine factor is an androgen. The literature records many similar examples from all classes of vertebrates. Testosterone is considered to be biologically the most important androgen, but comparative studies are not yet sufficiently extensive to generalize for all groups (Hoar, 1965b). 11-ketotestosterone is a principal androgen in teleost fishes (Idler and Truscott, 1963).

Estrogens The estrogens are C-18 steroids which differ from the androgens in that they lack the C-19 carbon. They are also characterized by the presence of an aromatic ring (*A* or *I*). Two steroids, equilin and equilenin, found in mare's urine, also have the *B* ring partially or totally converted to the aromatic form (Fig. 23.10). Their estrogenic activity is relatively weak. Estradiol-17β, estrone and estriol (Fig. 23.8) are most frequently isolated from animal sources. Estradiol-17β is the most potent of the group and is usually assumed to be the physiologically active substance. It has been found in several invertebrates as well as in all groups of vertebrates (Hisaw, 1963; Hoar, 1965b) and is evidently an ancient biological substance. Several of the estrogens usually associated with animal tissues are also found in plants. Others which are not known to be present in animals are also of regular occurrence in certain plants. Structurally, they may vary considerably from the

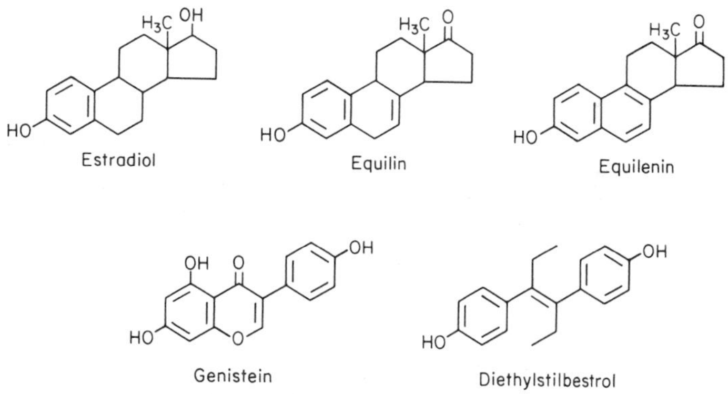

Estradiol

Equilin

Equilenin

Genistein

Diethylstilbestrol

Fig. 23.10 Structure of estradiol and two unusual animal estrogens found in mare's urine (*equilin* and *equilenin*), a plant estrogen *genistein*, and a synthetic estrogen *diethylstilbestrol*. Stilbestrol is the most active of the group.

animal steroids (Fig. 23.10). Their presence is sometimes of physiological significance in animal nutrition. Antigonadal effects are known in domestic animals (Bickoff, 1963; Leavitt and Wright, 1965).

Progesterone Progesterone and certain biologically related substances (gestogens) are similar to the androgens in containing both the C-18 and the C-19 positions; in addition, they have an oxy- or hydroxy-group on C-20 while the terminal carbon (C-21) is a methyl group. The key position of progesterone in steroid biogenesis is shown in Fig. 23.8. Since it is an early step in the sequence which leads both to the cortical steroids and to the gonadal steroids, it is not surprising to find it in many animals where no endocrine activity has yet been ascribed to it.

Progesterone, arising primarily in the corpora lutea, has long been recognized as an indispensable hormone concerned with mammalian viviparity. The phylogeny of this hormone, however, as well as the endocrine status of the corpora lutea, are speculative. Current evidence suggests that the progestins may have cooperated in the hormonal regulation of vertebrate reproduction for a very long time. They have been detected in the blood and tissues of many submammalian vertebrates and have been shown to be physiologically active in a few of them. Some of the amphibians provide nice examples. Following the injection of gonadotropin, a progesterone-like steroid appears in the blood of "winter" (non-breeding) toads *Bufo bufo* (Thornton, 1972); it causes the maturation of ova and the release of jelly from the oviduct. These two events are so timed that the oviducts are prepared to cover the mature ova with protective jelly as they journey to the outside world (Thornton and Evennett, 1969). Maturation and ovulation in the frog *Rana pipiens* also seem to depend on a progestin (Schuetz, 1972). Further, progesterone has been shown to exert several different effects on the reproductive tracts of birds and reptiles. In association with estrogen, its main target appears to be the oviduct (van Tienhoven, 1968). In birds, it is probably synthesized in the wall of the follicle or the interstitial tissues, since post-ovulatory corpora lutea are not formed in this class of vertebrates (Brambell, 1956); the same is likely in the amphibia. It should also be noted that pre-ovulatory progesterone is present as a hormone in some mammals (Parkes and Deansley, 1966; van Tienhoven, 1968). In viviparous reptiles, post-ovulatory corpora lutea are well developed, glandular in appearance, and active in the secretion of progesterone (Yaron, 1972). This is not the whole story, however. Neither the removal of these bodies nor ovariectomy interferes with normal gestation and parturition. Thus, progesterone is not essential for the maintenance of pregnancy in reptiles; the same is probably true of the viviparous elasmobranchs (*Torpedo,* for example) which also form suggestive looking post-ovulatory corpora lutea (Hoar, 1965b; Dodd, 1972).

In summary, progesterone, one of the key compounds in steroid synthesis, has acquired some hormonal functions in the submammalian vertebrates. These functions are largely associated with the physiology of the oviduct. In many lower forms, this hormone is elaborated by the wall of the follicle or the interstitial tissues but in some groups—particularly reptiles and mammals—the synthesizing activities are concentrated in a glandular corpus luteum.

Hisaw (1959, 1963) long maintained that the corpus luteum was added to the endocrine system in the mammals and that the presence of progesterone in submammalian groups was incidental to its key position as a step in steroid synthesis. This argument is still basically sound. Estrogens are the principal female hormones produced during the follicular phase of ovarian activity; they control the development of secondary sex characters and prepare the animal for ovulation and fertilization. Although more recent work shows that progesterone has assumed hormonal functions in some of these preparatory activities, the corpus luteum as an endocrine organ, responsive to pituitary regulation and with feedback relationships, is characteristic of mammalian viviparity. A gradual shift in the synthesizing activities of the ovary from estrogens to progestins is characteristic of the normal mammalian reproductive cycle (Fig. 23.11) and has also probably occurred during vertebrate phylogeny. In the frog, for example, both the meiotic maturation of oocytes and ovulation are inhibited by estrogens but stimulated by progestins formed later in the ovarian cycle (Schuetz, 1972). Viviparity in many of the lower forms is of the

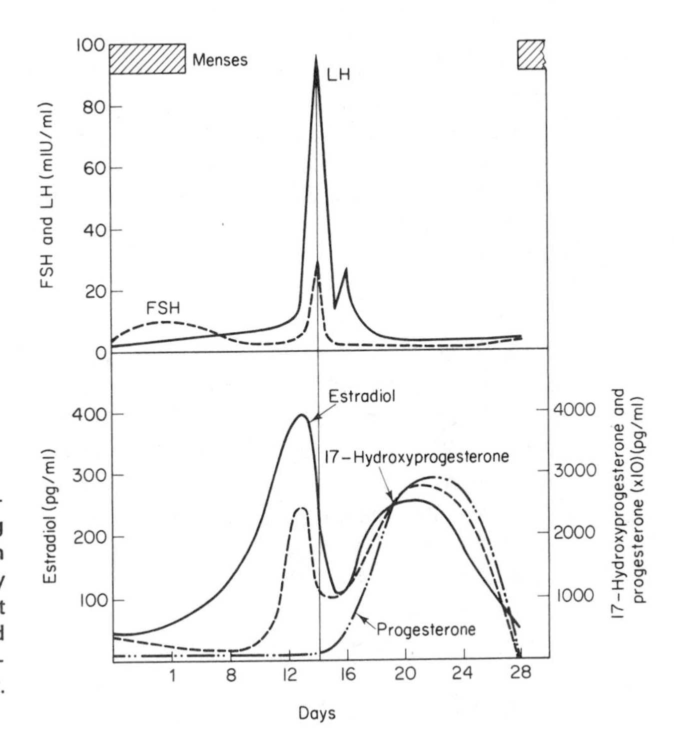

Fig. 23.11 Fluctuations of hormones in the blood serum during the normal menstrual cycle in women. Ovulation is indicated by the vertical line near the midpoint of the cycle. [From Odell and Moyer, 1971: Physiology of Reproduction. C. V. Mosby Co., St. Louis, Mo.]

ovoviviparous type in which yolky eggs are merely housed within the mother and do not depend on extensive elaboration of maternal tissues. In some forms, ovarian or oviducal secretions are a source of nourishment for the young, and progesterone may first have entered the endocrine family as a regulator of these secretions. It would be surprising if no phylogenetic steps for the hormone progesterone could be found among the lower vertebrates and it seems reasonable to look upon these oviduct reactions as preadaptations for its role in mammalian reproduction. The viviparous reptiles should hold many clues in this intriguing biological puzzle.

In mammals, the ovarian follicle enters a distinct luteal phase following ovulation. Many classical experiments have shown that the corpus luteum so formed is essential to the proliferation of the uterine tissues, the establishment of the maternal-fetal connections and the continuance of gestation. In some mammals the placenta gradually takes over the functions of endocrine production (estrogens, progesterone and gonadotropins) and serves as its own chemical regulator (page 743).

Regulation of Breeding Cycles

It is not usually valid to attribute a single function to any one of the several hormones just described. They very often act in a cooperative manner. Sometimes their action is that of an antagonist, sometimes that of a synergist; an independent action in isolation from other factors is much less frequent. Several examples will illustrate this point.

Differentiation of the secondary sex characters is usually entirely controlled by gonadal steroids. The male accessory glands of a castrated and hypophysectomized rat regress to the juvenile condition but can be fully restored with androgen (Turner and Bagnara, 1971); within limits, the response to androgen is proportional to the dose. The literature abounds in comparable examples recorded for all the major groups of vertebrates. There are, however, a few notable exceptions. Numerous reproductive phenomena have been shown to depend on an interaction of pituitary and gonadal hormones. In male sticklebacks nest building and sexual behavior appear much sooner and more regularly when castrates (treated with androgens) are maintained under photoperiod regimes known to stimulate gonadotropic activity (Hoar, 1965a). Presumably, gonadotropin is acting synergistically with the androgen in regulating the breeding behavior of the fish. Again, the inhibitory action of gonadal steroids on the pituitary has frequently been demonstrated in many ways. It is, for example, apparent in the histological changes of the gonadotropic cells and areas of the pituitary following castration or treatment with exogenous steroids. In the first type of experiment the gonadotropic area enlarges; in the second, it decreases in size. The mammalian endocrinologist has recorded many instances of synergistic as well as inhibitory interaction of gonadal and gonadotropic hormones. Several examples have already been noted; further details are recorded in many recent texts

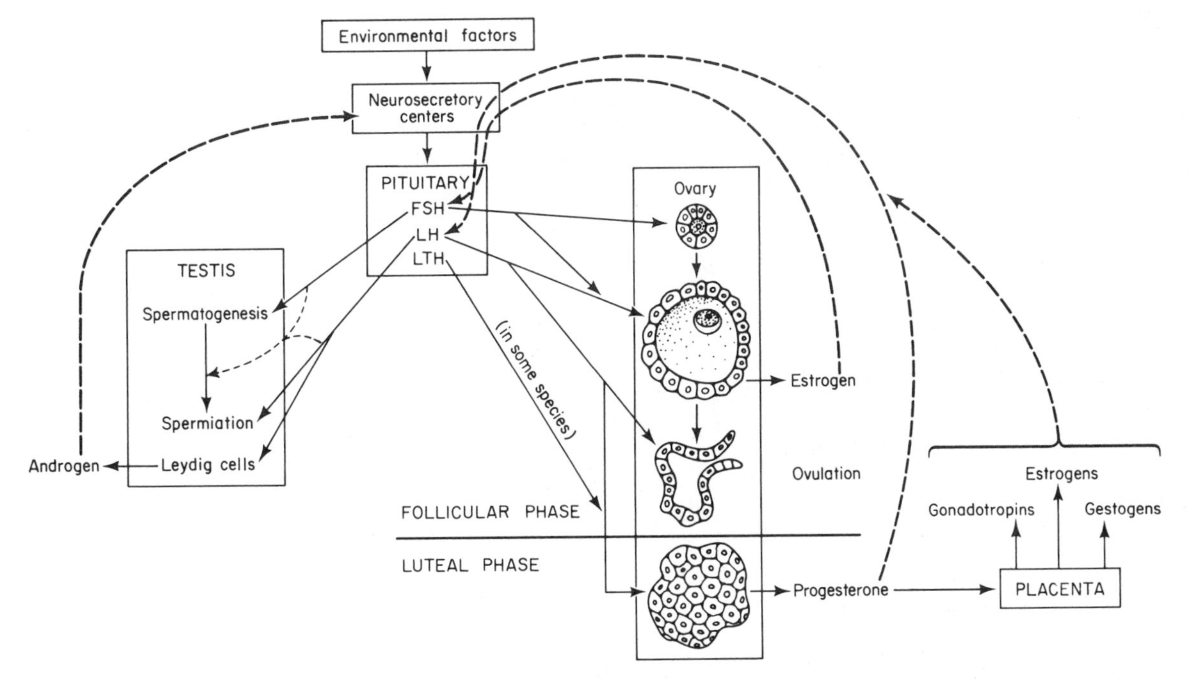

Fig. 23.12 Endocrine regulation of gonadal maturation in vertebrates. Continuous lines, stimulation. Heavy broken lines, inhibition. Light broken lines, probable interaction of FSH and LH. The luteal phase is present only in mammals. Lower vertebrates possess a single gonadotropin (GH).

(Williams, 1968; Turner and Bagnara, 1971). These interactions are stressed as a warning in the examination of several of the simplified diagrams which follow.

Breeding cycles Figure 23.12 summarizes the hormonal regulation of reproductive processes in the vertebrates. In some species (Pacific salmon *Oncorhynchus*, for example), gametogenesis occurs only once in the life of the animal, and the single period of reproduction culminates in death. Usually, however, vertebrate reproductive processes are recurring and cyclical. If the cycle is a seasonal one, environmental changes trigger the neurosecretory centers at the most advantageous seasons for the development and growth of the young. If reproductive activity is continuously cyclical or almost so, as is the case with many homeothermic animals and some of the tropical poikilotherms, then the controls are usually endogenous and depend on feedback mechanisms to the neurosecretory centers and perhaps also on neural pacemakers.

There is an evident parallel between the ovarian and testicular controls in the submammalian groups where the ovary shows only the follicular phase (Fig. 23.12). A true luteal phase is present only in the mammals. It is associated with viviparity and has been built onto the ovarian follicular phase with the development of a new endocrine gland (corpus luteum) which synthesizes progesterone in large amounts, under the control of the pituitary. A concomitant responsiveness of uterine and associated tissues to progesterone is a necessary part of the functional system. In some mammals the placenta also acquires endocrine responsibilities and takes over from both pituitary and gonads the production of hormones concerned with gestation.

Estrous cycle in mammals These endocrine modifications associated with viviparity and parental care in the mammals form a significant part of the many evolutionary adaptations required for this mode of life (Amoroso, 1960). Successful fertilization, implantation and subsequent nourishment of the developing embryo, the birth of the young at an appropriate stage and its sustenance by secretions of the mammary glands depend on a precisely timed sequence of cyclical endocrine changes and tissue responses.

The sexual cycles of vertebrates are extremely variable (Asdell, 1964; Parkes, 1960a; Barrington, 1963). Usually, the female will only receive the male during a relatively brief period of ESTRUS or heat, but in our own species receptivity may occur throughout the cycle. Again, ovulation is normally spontaneous and occurs during estrus (Fig. 23.13), but in some animals, such as the rabbit, ferret and mink which are INDUCED OVULATORS, the stimulation of coitus acting through the hypothalamo-hypophysial pathways with the release of LH or FSH/LH is required for the discharge of the ovum from the ovary. Fertilization usually takes place within a few hours of coitus, but in certain bats, copulation occurs in the autumn and fertilization is delayed until the spring. Implantation occurs after about one week of development in human, rabbit, mouse and some other mammals when the dividing

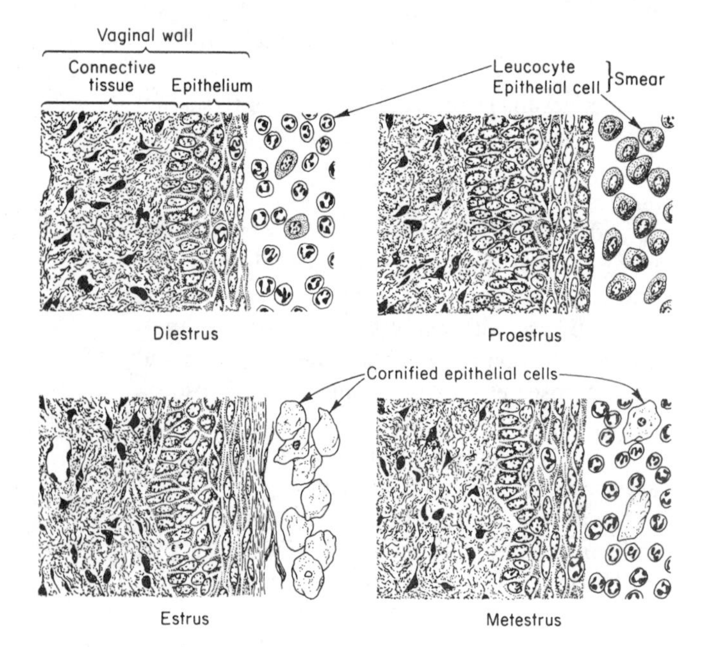

Fig. 23.13 Sections through the rat's vaginal wall during different stages of the estrous cycle with the cells (at right) which characterize vaginal smears during each of the corresponding stages. [From Turner and Bagnara, 1971: General Endocrinology, 5th ed. Saunders, Philadelphia.]

mass of cells has attained the status of a BLASTOCYST. In the pig, dog and cat, implantation is delayed for about two weeks, while in the marten, weasel and badger, mating takes place during the summer but implantation of the blastocyst occurs late in the winter (DELAYED IMPLANTATION). This diversity emphasizes the plasticity of organisms in adjusting reproductive processes to the conditions of the environment and the limitations of the species. Wherever the processes have been carefully studied, they have been found to be highly adaptive in the evolutionary sense (Chapter 22).

Details of the mammalian sexual cycle are best known in the laboratory rodents and in the human. Rats and mice have a cycle of four to five days. Ovulation occurs during the period of heat (estrus) lasting 9 to 15 hours; estrus is under the influence of FSH and estrogen. Since the female will only receive the male during this period, ovulation and fertilization are well coordinated. During estrus, the uterus becomes enlarged and edematous in preparation for implantation; the vaginal mucosa proliferates and the superficial epithelial layers are squamous and cornified in preparation for copulation. Vaginal smears made at this time contain characteristic squamous cells and serve as a means for ready diagnosis of the stage in the estrous cycle (Fig. 23.13; van Tienhoven, 1968).

Subsequent events depend on sexual contact with a male. In the absence of copulation the cycle is almost completely follicular. During METESTRUS (10 to 14 hours), however, a small corpus luteum forms and some progesterone is secreted. This stage is characterized by a diminished vascularity of the uterus, while the vaginal smears show leucocytes along with cornified cells, indicating a thinning of the mucosa and migration of leucocytes through it. The corpora lutea regress during DIESTRUS (60 to 70 hour period), the uteri are relatively small and anemic; cells in the vaginal smears are almost entirely leucocytes. PROESTRUS, lasting about 12 hours, is a preparation for the next estrus. Degeneration of the old corpora lutea continues, but new follicles are growing rapidly, and the uterus is again becoming distended with fluid; vaginal smears contain nucleated (not cornified) epithelial cells which may be detached singly or in sheets. The changes are shown diagrammatically in Fig. 23.13. The classical description for the rat is that of Long and Evans (1922). Rats and mice are POLYESTROUS, and this cycle of events occurs in unmated animals every four to six days. Many animals have only one estrous cycle during the year, and a prolonged period of ANESTRUS (diestrous period) occurs between the metestrus and proestrus. The silver fox, for example, has a single period of heat lasting about 5 days during late winter; it is said to be MONESTROUS. Sheep and cattle are seasonally polyestrous with several periods of heat during one season of the year; several interesting variations seem to have appeared during domestication (Barrington, 1963).

In unmated rats and mice the luteal phase does not produce the progestational changes usually associated with progesterone. These may, however, be artificially induced in the absence of pregnancy by mating females with sterile males or by stimulating the cervix uteri with various mechanical or electrical means. The response is called PSEUDOPREGNANCY and evidently depends on afferent stimuli to the hypothalamic-pituitary axis, as indicated by the failure of such experiments in anaesthetized animals or in animals following destruction of the uterine nerves. Pseudopregnancy lasts for about 13 days in the rat (pregnancy is of about 21 days duration), and during this period the corpora lutea remain active and the uteri undergo changes similar to those seen in pregnancy. Hormone changes and the associated proliferation of the mammary glands are similar to those of the pregnant animal. This pseudopregnant phase occurs as a normal part of the cycle in many unmated animals (guinea pigs) and is not affected by sterile matings or similar stimulation of the reproductive tract.

Menstrual cycles In man, apes and monkeys the progestational phase is terminated by MENSTRUATION during which the inner portion of the endometrium or mucous membrane of the uterus collapses and is discharged with a certain amount of bleeding. As indicated in Fig. 23.14, this occurs at a time of rapid decline in levels of hormones which were responsible for the proliferation of the endometrium, the development of its glands and highly vascular condition. In some other animals (dogs, for example) there may be some bleeding from the hyperemic uterus during estrus, but this is not comparable to menstruation (a terminal event), and no sloughing of the endometrium takes place; the menstrual cycle is characteristic of the higher primates (Asdell, 1964).

This cycle depends on an extremely delicate balance of hypothalamic factors, pituitary and gonadal hormones. It may be easily disrupted to prevent ovum development (FSH dependent) or to suppress or trigger ovulation (dependent on a LH/FSH surge). Thus, man has achieved a measure of control over his fertility by

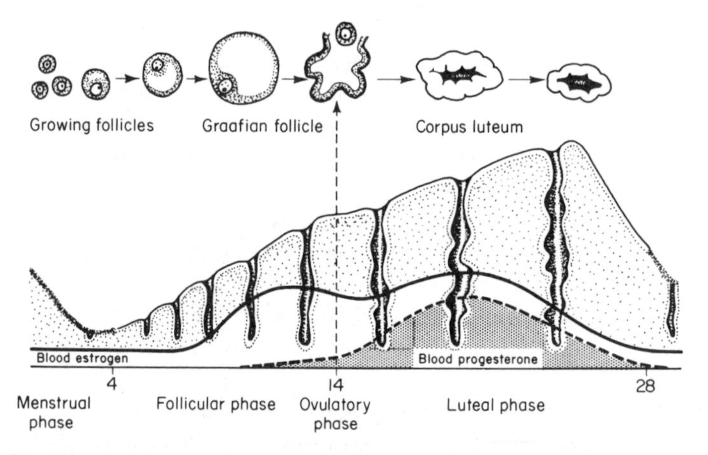

Fig. 23.14 Diagram showing changes in the endometrium, the ovaries, and the circulating ovarian hormones during the menstrual cycle. [From Turner and Bagnara, 1971: General Endocrinology, 5th ed. Saunders, Philadelphia.]

the use of hormones and drugs. Contraceptive "pills" contain estrogen and progesterone (taken together or in sequence) or progesterone alone. Through "feedback mechanisms," these substances completely suppress FSH (hence follicle development) while the LH blood levels are depressed or vary erratically (Odell and Moyer, 1971). The normal blood levels of steroids are shown in Fig. 23.11. The pharmaceutical problems generated by the demand for steroids—both for the treatment of adrenocortical pathologies and for the massive production of contraceptive pills—were first met by turning to a plant steroid *diosgenin,* which proved to be an economical source of starting material for the synthesis of progesterone. Diosgenin is mainly extracted from a type of wild yam *Dioscorea* native to Mexico. Fertility "pills" depend on the use of FSH (usually extracted from menopausal urine) and LH (usually of placental origin). These hormones are proteins and must be injected, but there are drugs, such as clomiphene citrate, which may be taken orally and probably act by triggering a surge of gonadotropic releasing factor GRF from the hypothalamus (Williams, 1968).

Other Hormones Concerned with Vertebrate Reproduction

The placental hormones In many mammals the placenta is not only an organ of embryonic attachment, nourishment, respiration and excretion but also an endocrine gland which produces a variety of hormones concerned with gestation and parturition. The extent to which the placenta assumes responsibilities for steroid and gonadotropic synthesis varies greatly in different animal species (van Tienhoven, 1968). In rats, mice, rabbits and hamsters, removal of the ovaries at any time during pregnancy results in abortion, and it is evident that the ovary is the major source of the steroid hormones of pregnancy. In the guinea pig, cat, dog and ewe, however, ovaries may be removed (without causing abortion) after mid pregnancy; in humans, monkeys and mares they may be removed at an even earlier stage. Progesterone will usually maintain pregnancy in ovariectomized animals which characteristically abort their embryos after ovariectomy. However, the situation is somewhat variable, and it is recognized that the placenta is sometimes a major source of estrogen (mouse and rabbit) and sometimes a source of both estrogen and progesterone (man, some monkeys, mare). On the basis of the earlier discussion of steroid synthesis, it is not surprising to find that a number of different steroids have been isolated from placental tissues.

Mammals also vary greatly in their dependence on the pituitary gonadotropins. Although the hypophysis is always essential for ovulation and implantation, its secretions can evidently be dispensed with during the latter part of pregnancy in several different species (mouse, rat, guinea pig, monkey). In others (rabbit, cat, dog) hypophysectomy leads to abortion at any time during pregnancy. Thus, in some mammals (guinea pig, sheep, monkey), neither pituitary nor gonad is essential to the maintenance of the later stages of pregnancy. The placenta has assumed the total responsibility for the production of gonadal steroids. In the higher primates and the

Equidae, the placenta seems to be unique in its capacities to synthesize pituitary-like protein hormones. The human placenta is the site of production of a chorionic gonadotropin HCG, a chorionic thyrotropin HCT, and a chorionic growth hormone prolactin HCGP (Odell and Moyer, 1971). These factors are presumably essential regulators in either the fetus or the mother.

The two placental gonadotropins most familiar to the endocrinologist are HUMAN CHORIONIC GONADOTROPIN (HCG) isolated from pregnancy urine of women and PREGNANT MARE SERUM (PMS) obtained from the blood serum of the pregnant mare. These have often proved useful in experimental work because they can be readily isolated in quantity and have been commercially available for some years. Since they are used frequently as substitutes for the pituitary fractions, it should be noted that they are not the same biochemically, even though both types are glycoproteins. They mimic the pituitary gonadotropins, but their actions are somewhat different.

HCG is secreted by the chorionic villi of the placenta (embryonic origin) and appears in large quantities in the urine during the first two months of pregnancy; it continues to be secreted in lesser amounts until parturition. Experimentally, its action in the mammal is usually similar to that of LH. PMS is of uterine (endometrial) origin and often has an action similar to a mixture of FSH and LH, with the predominant effect depending on dosage. Unlike HCG, FSH, and LH the uterine factor is not excreted by the mare but remains in the blood; the amounts are particularly high from the fortieth to the hundred-and-twentieth day of pregnancy (Turner and Bagnara, 1971).

Relaxin Mammals must enter the outer world through a relatively narrow bony ring, the pelvis. This requirement of mammalian anatomy has imposed certain restrictions and compromises during the evolution of viviparity. In all mammals the size of the pelvis sets an upper limit on the growth of the young in the uterus. In some habitats, such as those exploited by burrowing and flying species, there may be definite advantages in restricting the size of the pelvis, but any tendency in this direction imposes a disadvantage which must be compensated for at parturition.

In 1926, Hisaw isolated a substance with a pronounced effect on the connective tissues of the symphysis pubis and the sacroiliac joints and ligaments of the estrogen-primed guinea pig. He named the substance RELAXIN because of its apparent function in relaxing the pelvis and softening the cervical mucosa in preparation for parturition. Hormonally-induced tissue changes which facilitate stretching of the cervix and pelvis are now recognized in many mammals (cow, pig, human, rat). Relaxin has not been completely characterized, but it is known to be a polypeptide which acts in association with estrogen and, sometimes, growth hormone as well (Steinetz *et al.*, 1965). It has been isolated from the blood and reproductive tissues (ovary, placenta) of several mammals and has also been identified in some birds and fish. Several functions have been ascribed to it, in addition to those involving a splitting and softening of collagenous fibers and associated tissues, but its comparative physiology is not well known (Steinetz *et al.*, 1959; van Tienhoven, 1968).

The comparative physiologist rarely finds that evolutionary processes have provided a single solution to any problem. This is certainly true of the pelvic adaptations of mammals. The mole represents an extreme situation in which the cartilaginous pubic arch is resorbed during early life; the two halves then grow together dorsal to the digestive and reproductive tracts to form a secondary symphysis. Thus, the adult reproductive tract is entirely outside and ventral to the pelvis. This solution seems to be independent of the endocrine system and has been acquired through genetic changes. In the pocket gopher there are also drastic changes in the morphology of the pelvis, but these occur only in the female and are under the control of estrogen. The juvenile pelvis is complete in both sexes, but at the time of the first estrus there is a resorption of pubic bone leaving the pelvis open ventrally. It remains intact in the male.

Prolactin Prolactin is a member of the family of adenohypophysial hormones. Highly purified fractions from mammals (mol wt about 25,000) are proteins consisting of a single peptide chain which, unlike the gonadotropins FSH and LH, is not bound to carbohydrate. Mammalian prolactin is biochemically similar to growth hormone STH; preparations from human pituitaries appear to be identical.

As the name suggests, prolactin is concerned with lactation in mammals; its first demonstrated action was that of initiating lactation in fully developed mammary glands. At about the same time (1932), it was also shown to stimulate the pigeon's crop glands which produce a cheesy secretion by desquamation of the epithelium; this material, called pigeon's "milk," serves as nourishment for the young squabs (Riddle, 1963). Since that time, prolactin-like substances have been identified in the pituitaries of many different animals at all levels in vertebrate phylogeny, male as well as female (Bern and Nicoll, 1968b), while numerous functions, other than those concerned with lactation, have been ascribed to it (Nicoll and Bern, 1971). Moreover, studies of the endocrinology of lactation have shown that prolactin is only one of several hormonal factors involved in milk production. These two important considerations, the complexity of hormones which regulate lactation and the comparative endocrinology of prolactin can be conveniently separated in the following discussion.

Lactogenic or mammotropic effects Three different processes are involved in the production and delivery of milk to suckling animals. The first of these is the development of the mammary glands at the time of puberty or sexual maturity; the second involves their further proliferation and the secretion of milk, while the third is the evacuation or delivery of milk to the nursing young. The complex of interacting hormones concerned with these processes is shown diagrammatically in Fig. 23.15.

Mammary gland development at puberty, like other secondary sex characters of the female, is under the control of estrogen with some cooperation from the growth hormone somatotropin. The further proliferation of the glands during gestation, with the subsequent production of milk, is a highly complex process which requires the simultaneous activation of metabolic pathways concerned with protein, carbohydrate and fat synthesis, the mobilization of calcium and phosphorus and the regulation of

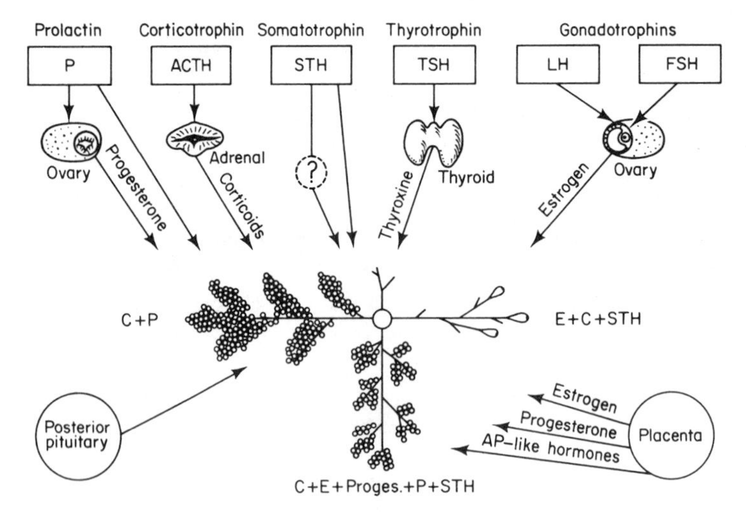

Fig. 23.15 Diagram showing the action of hormones on the growth of the mammary glands and lactation. In the diagram of the gland: upper, rudimentary gland; right, prepubertal and pubertal gland; lower, prolactational gland of pregnancy; left, lactating gland; C, corticosteroids; E, estrogens. [From Turner and Bagnara, 1971: General Endocrinology, 5th ed. Saunders, Philadelphia. Based on Lyons, Li, and Johnson, 1958: Rec. Progr. Hormone Res., *14*: 246.]

specific levels of electrolytes and water. The milk secreting cell is a very complex machine (Linzell and Peaker, 1971). It is not surprising to find that lactogenesis is regulated to some extent by all the metabolic hormones as well as the reproductive hormones. The research of the past quarter of a century has displaced prolactin from its position as the central regulator of lactation to a membership in a complex endocrine committee. Detailed discussion will be left to textbooks of mammalian endocrinology (Bloom and Fawcett, 1968; Williams, 1968; Turner and Bagnara, 1971).

The third process involved in milk production depends on nervous as well as on endocrine factors. Milk removal requires afferent stimuli from the nipple. An anaesthetized mother can only supply the milk stored in the cistern and major ducts. Release of milk from the main portions of the mammary apparatus depends on contractions of the myoepithelial or "basket cells" which surround the alveoli. Their contractions are normally initiated by the hormones of the posterior pituitary. Suckling provides afferent stimuli to the supraoptic and paraventicular nuclei of the brain; these lead to the release of oxytocin which, in turn, brings about contraction of the myoepithelial cells surrounding the alveoli. The same afferent stimuli acting through the hypothalamo-hypophysial pathways trigger release of prolactin to stimulate milk secretion by the gland cells. Thus, prolactin stimulates milk secretion (lactogenesis), a process which also requires the regulatory activities of different metabolic hormones; oxytocin is responsible for milk ejection.

Comparative physiology of prolactin The varied and curious functions attributed to prolactin form one of the most interesting chapters in comparative endocrinology. Prolactin has been termed a "jack-of-all-trades" with more than 80 different hormonal activities ascribed to it (Nicoll and Bern, 1971). There are three classical techniques for the bioassay of prolactin (Zarrow *et al.*, 1964): (1) the increase in interducal hydrostatic pressure following introduction of the hormone into the major mammary duct of the estrogen- and progesterone-primed rabbit, (2) stimulation of the crop glands of the pigeon, and (3) induction of a "water drive" in the hypophysectomized newt *Diemictylus viridescens* (Chapter 24). Assays based on the growth stimulation in amphibian larvae and hydromineral regulation in certain teleosts have also been used. Mammalian prolactin is positive in all these assays, but prolactin preparations from most fish lack not only the mammotropic and luteotropic effects, but also fail to stimulate tadpole growth and pigeon crop glands (Bern and Nicoll, 1968a, b). These differences probably reflect fundamental changes during the evolution of the prolactin molecule and cannot be attributed simply to the species-specificity which is recognized in all proteins. Nicoll and Bern (1971) group the many functions of prolactin into five major categories: reproduction, osmoregulation, growth, integument, and a group of actions in which prolactin has a synergistic effect with the steroid hormones.

The first functions of prolactin to be recognized were those associated with reproduction. Mammogenesis is not alone in this category. Its luteotropic action in rodents has already been mentioned. In addition, prolactin has been associated with the development of incubation or brood patches, incubating behavior and parental care in some birds; these functions may also be related to estrogens and progesterone. Among the amphibians, prolactin plays a part in preparing the immature *Diemictylus* for its aquatic life and may also be involved in regulating the secretory activity of the oviducal glands in *Bufo arenarum*. In the discus fish *Symphysodon discus*, prolactin is claimed to stimulate the formation of epidermal secretions which serve to nourish the young and to stimulate fanning in a closely related species and in the wrasse *Crenilabrus ocellatus*. These several examples provide support for Hisaw's (1963, 1966) observation of a widespread action of prolactin in processes which pertain to the well-being of young animals. There are, however, several other functions which show no such obvious relationship.

The important and widespread action of prolactin in regulating the balance of water and salt has already been discussed (Chapter 11). Although there is currently no evidence of a single prolactin function at the cellular and biochemical level, Nicoll and Bern (1971) suggest that further studies of hydromineral regulation may prove to be the most rewarding line of inquiry in this connection.

In many different test situations, prolactin stimulates somatic growth and general metabolic processes associated with it. Particular target organs and cells (male sex accessories, sebaceous glands, feathers) may also be stimulated. However, it is often difficult to decide whether the prolactin molecule is acting specifically on some basic process associated with growth, or whether these effects are byproducts of some other action, or if the effects are related to the biochemical similarities between the pituitary growth hormone and prolactin. Although two distinct factors are recognized in vertebrates other than man, they are very similar chemically and may have evolved from a single parent protein. The amphibians have provided the most interesting experimental material for the analysis of the distinct growth promoting effects of these two substances. Larvae of several amphibian species show accelerated growth and fail to metamorphose when injected with mammalian prolactin; mammalian somatotropin is ineffective in similar tests. In direct contrast, mammalian somatotropin stimulates growth of postmetamorphic amphibians while prolactin is inactive (Frye *et al.*, 1972). The meaning of these differences must wait a more complete understanding of the functions of somatotropin and prolactin—particularly at the cellular level.

A specific prolactin effect on skin was first noted in some of Pickford's studies of *Fundulus heteroclitus*. A gradual loss in pigmentation followed hypophysectomy and this loss was shown to be caused by the disappearance of melanin rather than changes in the distribution of the melanophores. Prolactin proved to be specific in the restoration of melanin to the ghosts of the depigmented melanophores (Kosto *et al.*, 1959). Subsequent work has implicated prolactin in several other processes concerned with the physiology of integumentary structures; epidermal mucus cells in fish and amphibians, epidermal sloughing in reptiles, feather growth in birds, sebaceous glands and hair maturation in mammals (Ball, 1969; Nicoll and Bern, 1971).

Finally, it should be noted that prolactin often acts synergistically with the steroids—either adrenocortical or gonadal. The list of these synergistic effects, compiled by Nicoll and Bern (1971), is already a very long one.

Retrospect

The endocrinology of reproduction is a many-sided subject. The pituitary and gonadal hormones of vertebrates have been emphasized in this chapter. Hypothalamic releasing factors and pineal physiology were considered in Chapter 2; the environmental cues which trigger reproduction were noted in chapters on environmental physiology and sense organs; the action of hormones on reproductive behavior was mentioned in Chapter 21. It is now evident that reproductive processes are integrated in a similar manner throughout all vertebrate classes. From the Agnatha to the primates, neurosecretory centers in the brain trigger a release of gonadotropins from the adenohypophysis while the gonadotropins, in their turn, regulate the steroid-secreting tissues of the gonads. It seems probable that this basic machinery has remained substantially unchanged since the origin of the vertebrates, some 500 million years ago.

Generalizations can only be cautiously extended from this broad statement. It is probable that pituitary hormones always start the changes associated with the reproductive cycle. This is true of gametogenesis (except perhaps in the cyclostomes) and perhaps also of some aspects of presexual behavior. Subsequently, the pituitary delegates some or all of its responsibilities to the gonadal hormones or, in some of the mammals, to the extragonadal tissues of the placenta. Thus, the gonads are often concerned with the later events of reproduction (such as the differentiation of secondary sex characters and sexual behavior) while the pituitary is more directly involved in the earlier phases (such as gametogenesis and pre-breeding migrations). It should be noted, however, that a strictly independent action of pituitary and gonadal hormones is unlikely in the intact animal. Secretion of the gonadal hormones is stimulated by the pituitary; secretion of the pituitary gonadotropins is inhibited by the gonadal hormones. Within this broad framework of reproductive controls, interrelationships of the hormones and their actions on specific target organs are almost as variable as the animals themselves. The evolutionary processes have made use of these chemicals in ways which are just as curious and diversified as the morphological correlates of sex and reproduction. Students of reproductive endocrinology may confidently expect many variants of the general rule as the investigations are extended to include more and more species.

24

GROWTH AND DEVELOPMENT

The analysis of early stages of growth and differentiation is traditionally the prerogative of the embryologist. Mechanisms regulating postembryonic development come under the purview of the physiologist. Experimental biologists in many areas of research have analyzed the mathematics of growth and described it graphically. Mammalian physiologists and biochemists have been particularly active in studies of the pituitary growth hormone, somatotropin, and the growth effects of thyroxine, corticosteroids, insulin and diet. General and cell physiologists have analyzed cell division, measured growth, studied effects of various stimulants and inhibitors on regeneration and tissue differentiation and probed into the cytological details of protein synthesis and tissue organization. The comparative physiologist has made his major contribution in studies of metamorphosis. There is now a rich literature on these many different aspects of growth (Needham, 1964a): only two topics (regeneration and metamorphosis) will be considered here.

REGENERATION

All animals have some capacity to restore tissues or body parts which have been lost through normal physiological processes or destroyed accidentally. A complete *Amoeba* may grow from a fragment representing $\frac{1}{80}$ th of the original animal provided that the nucleus is included in the fragment. Only $\frac{1}{200}$th of a *Hydra* or $\frac{1}{280}$th of a planarian is necessary to regenerate a completely new animal. Some of the primitive oligochaete worms can regenerate from a single segment. This potential for extensive regeneration from very small fragments is common in other lower phyla. Sometimes, under unfavorable conditions, certain sponges are reduced to an amorphous mass which then acquires its original structure when conditions improve. Many protozoans reproduce asexually by binary fission; hydroids and planarians may also reproduce by fission and regenerate the missing parts. In the more advanced phyla, missing appendages are often regenerated and the outer coverings of cuticle, chitin or cornified epidermis are periodically replaced. Salamanders replace missing legs and some of the lizards can restore a lost tail. However, regeneration, strictly speaking, occurs rarely among the vertebrates and extensive damage is repaired by the formation of SCAR TISSUE that is largely white fibrous connective tissue. Specialized tissues, with few exceptions (liver, gonads in some forms), are not regenerated in most vertebrates. Regeneration in man is limited to the regular cell renewal of our epithelial surfaces. The germinal layers of the epidermis and the mucous membranes of the digestive tract are continuously proliferating to replace cells which are always being rubbed off. Holocrine glands such as the sebaceous glands secrete by accumulating a load of secretory materials in the cytoplasm and then disintegrating.

In their monograph, Vorontsova and Liosner (1960) consider three categories of regeneration: PHYSIOLOGICAL REGENERATION which is a part of the normal and regular functioning of such organs as sebaceous glands, mucous membranes and the outer layers of the skin; REPARATIVE REGENERATION which is provoked by wounding or traumatic destruction, and ASEXUAL REPRODUCTION which is a natural process involving the isolation of a part of the animal and its transformation into a daughter organism. The distinction is useful since the three processes have quite different functions. It should be noted, however, that all three are, in reality, "physiological," that CELL RENEWAL (Leblond and Walker, 1956) is probably a more appropriate term than physiological regeneration and that the word REGENERATION is usually reserved for reparative and post-traumatic processes. It is this latter phenomenon which is considered here.

Reparative Processes

The recorded descriptions of regeneration in all groups of animals are now voluminous. The earlier literature was reviewed by Vorontsova and Liosner (1960). There are several recent symposia and monographs (Kiortsis and Trampusch, 1965; Goss, 1969; Goss and Hay, 1970; Rose, 1970; Adiyodi, 1972).

The sequence of events Needham (1952) describes six morphological events which can be conveniently separated for descriptive purposes. The first is WOUND CLOSURE which in lower forms involves only a contraction of neighboring tissues and a stretching of the surrounding cells over the wound; in higher forms the vascular fluids clot (Chapter 6) and thus form a basis for the later processes of repair. Wound closure is followed by DEMOLITION and DEFENCE which, in the higher vertebrates, begins with a triple response (page 443) caused by the release of toxic substances from injured tissues and expressed as increased dilatation of blood vessels, collection of fluid and eventually the removal of the damaged tissues through autolysis and phagocytosis. The vascular reactions are observed only in molluscs, arthropods and vertebrates; lower forms probably depend entirely on phagocytosis.

Several different processes are associated with the healing which follows demolition and defence. The first of these is often a DEDIFFERENTIATION of tissues to provide indifferent cells for subsequent regenerative processes. Although dedifferentiation has been described in lower forms and may occur to some degree in all cases, it is most characteristic of the vertebrates and has been extensively studied in regenerating amphibian appendages. In lower forms (hydroids, for example), nondifferentiated or pluripotent cells (NEOBLASTS) are always present and these migrate into the area of injury from nearby tissues. In the more advanced phyla, the origin of indifferent cells for regenerating tissues has long been debated. However, it now seems that the new tissues arise from dedifferentiated cells of different types in the region of the injury (Hill, 1970). Needham (1952) suggests that the development of the highly efficient vertebrate circulatory system reduced the importance of maintaining a stock of undifferentiated migrant cells. The process of dedifferentiation brings to an end the regressive phase of regeneration; this is followed by the progressive phases: formation of the BLASTEMA or regeneration bud, its GROWTH and subsequent DIFFERENTIATION into the regenerated structure. The blastema or regeneration bud is composed of a mass of dedifferentiated or immigrant cells; in the amphibians, the blastema is fully established prior to a sudden initiation of mitotic activity which heralds the new growth. The intense cellular proliferation which follows produces a mass of relatively small cells that subsequently increase in size and become somewhat separated as intercellular spaces appear during differentiation. The mitotic rate declines as the regenerating structure continues to differentiate and becomes functional.

These then are the morphological events which characterize regeneration. Explanations of the causal mechanisms have been persistently sought for more than half a century. Experimental embryologists have been particularly active in the search for factors which induce regeneration, the determination of events within the differentiating blastema, gradients in regenerative capacity and metabolism in organisms or parts of organisms, and the inductive action of different tissue transplants (Balinsky, 1970b; Goss and Hay, 1970; Berrill, 1971). Medical physiologists have investigated the posttraumatic blood clotting, vascular responses and inflammatory reactions which precede repair (Montagna and Billingham, 1964; Schilling, 1968; Ross, 1969). Comparative physiologists have recently demonstrated the commanding position of the neurosecretory system in regeneration among some lower forms.

Invertebrate hormones and regeneration Although hydroids, planarians, and annelids have spectacular capacities of regeneration, this is not true of all invertebrates. Nematodes (like most internal parasites), are almost devoid of regenerative power; chaetognaths, rotifers, copepods and cladocerans possess only limited abilities to restore lost parts (Ghirardelli, 1965). These differences are not well understood. There are, however, several observations which may be pertinent. It has, for example, been pointed out that most animals which reproduce by budding or fission (AGAMIC REPRODUCTION) readily regenerate missing parts. These animals maintain a reserve of pluripotent cells throughout life, and, in this characteristic, differ sharply from those groups in which germ cells are set aside at a very early stage in development. Thus, the ability to regenerate lost parts is minimal in the chaetognaths where the germ cells are segregated in the two-blastomere stage; the future of one of these blastomeres is purely somatic while the other gives rise to primordial germ cells. There appears to be some connection between the power to regenerate and the reserve of totipotent or pluripotent cells. Another relationship that has been frequently emphasized is the decline in growth—both new and regenerative—with sexual maturity. This inverse relationship between growth and reproduction has already been noted (Chapter 23) and seems to be one of the general principles of animal life (Clark, 1962; Highnam and Hill, 1969).

The regenerative capacities of several polychaete worms have been carefully investigated and the findings reviewed by Clark and his associates (Clark *et al.*, 1962; Clark, 1966). *Nereis diversicolor* grows by adding new segments until it has produced 40 or 50 of them; thereafter new segments are only added slowly and growth is mostly due to increase in segment size. When the worms have about 90 segments, growth by either means ceases. Segment proliferation is regulated by hormones produced in the supraesophageal ganglia. If these ganglia are removed in young individuals, segment proliferation ceases but is resumed following implantation of ganglia from other young worms. Moreover, old worms which have ceased to grow can be induced to grow by implanting ganglia from young, growing worms but the brains of old worms do not stimulate segmentation in other old worms.

Worms less than 60 segments in length have remarkable capacities for regeneration and this, like normal growth, depends on the supraesophageal ganglia. Amputation of a part of the worm stimulates the neurosecretory cells to elaborate a substance or substances which provoke segment proliferation. The potency of ganglia to stimulate regeneration has been shown to reach a maximum about three days after the trauma and thereafter to decline. Regeneration capacity is gradually lost as the animals become older so that mature worms show little tendency to grow additional segments following injury (Golding, 1967). However, old, sexually mature worms which normally do not regenerate can be induced to proliferate segments by implanting ganglia from young regenerating worms. It is not yet known whether the GROWTH-PROMOTING and the REGENERATION-PROMOTING HORMONES are identical or distinct (Jenkin, 1970).

These studies have implicated the neuroendocrine system in the control of regeneration in *Nereis*. The voluminous literature points toward comparable controls in several other groups ranging from hydroids to vertebrates. Injury acting directly on nerves or through the release of cellular products presumably stimulates neurosecretory cells to produce hormones which trigger the regeneration processes. This may be a universal component of the regulatory mechanisms concerned with regeneration in multicellular animals. Even if this proves to be the case, there are still many unanswered questions in the analysis of regeneration.

Biological significance of regeneration Many workers have noted that the capacity to regenerate is primitive in both the phylogenetic and the ontogenetic sense (Needham, 1964b). Some lower animals regenerate completely after the loss of large portions of their bodies and younger individuals usually respond much more readily than older mature ones. These principles apply widely throughout the animal world and numerous examples will be found in the monographs already cited. If it is true that the ability to regenerate lost parts is primitive, then it seems curious that such an important capacity should have been partially lost during the evolution of many higher animals and some of the very lowly ones as well. Needham (1952, 1964b) discusses the phylogenetic implications. There are really two questions: Is the ability to regenerate lost parts really adaptive in an evolutionary sense? If so, why has it been almost totally lost in numerous animals?

In the biology of regeneration, there is much to support an argument for natural selection. Sedentary worms regenerate their exposed heads better than their tails while the errant species restore tails better than other parts; many arthropods have evolved special autotomy planes where appendages break off easily, wounds are closed by special devices and regeneration is rapid; the more sluggish salamanders regenerate legs readily but more active vertebrates have lost this capacity completely; lizards not only regenerate their vulnerable tails but the lost tails writhe to distract predators. The list of examples could be extended. There is ample evidence that this capacity has been subject to natural selection. The reasons for losing it are more difficult to analyze. Like most biological properties, there have probably been compromises between competing pressures. Thus, copepods, rotifers and many parasites seem to have specialized in rapid reproduction and a speedy population turnover rather than in the repair of damaged individuals. The vertebrates have evolved complex neuromuscular systems and highly adaptive behavior; they either escape with minor injuries or are captured and killed. Where regeneration becomes trivial or has only a neutral value, it seems to have been lost. Thus, Needham (1964b) argues that the ability to regenerate lost parts is both pristine and adaptive in the evolutionary sense; the ability "has persisted wherever its value was not outweighed by other biological considerations, by evolutionary developments with competing advantages, or by proximate difficulties."

METAMORPHOSIS

Metamorphosis is a profound postembryonic reorganization of tissues and processes which usually prepares an animal for life in a different habitat. In many animals (lamellibranchs, barnacles, echinoderms) the larval stages are highly motile or planktonic while the adults are either sessile or have a restricted distribution in relation to specialized habitats and food requirements. Planktonic larvae of fishes often drift from the spawning areas to distant feeding grounds, while the adults actively migrate back to the spawning grounds at sexual maturity. Larval forms of many insects are relatively sedentary and feed actively where they hatch; the adults often exploit a different source of food and may fly about to mate and disperse the species. Amphibian metamorphosis permits escape from a habitat essential for reproduction and early development, but at the same time quite unsuitable for the adult and liable to be radically changed at certain seasons.

These larval stages and the consequent problems of metamorphosis hold a justifiably important place in biological thought and theory. They form the basis of the distribution of many species and are just as significant to their survival and success as are the adult forms. In some cases, larval forms may have provided the evolutionary material for radically new phylogenetic lines. Sexual maturity sometimes occurs in larval forms (NEOTENY), and it seems not unreasonable to assume that certain groups of animals have evolved through the omission of highly specialized adult stages and the adaptation of the more youthful and generalized larval features to new and different ways of life (PAEDOMORPHOSIS). This line or argument has been particularly helpful in speculation concerning the ancestry of insects and vertebrates (Garstang, 1922; De Beer, 1958)—to quote Garstang (1951):

> *Now look at* Ammocoetes *there, reclining in the mud,*
> *Preparing thyroid-extract to secure his tiny food:*
> *If just a touch of sunshine more should make his gonads grow,*
> *The Lancelet's claims to ancestry would get a nasty blow!*

The primitive endocrine system probably assumed responsibilities for the regulation of growth, tissue regeneration and metamorphosis at an early stage in phylogeny. However, at present, the established factual information is almost completely restricted to polychaetes, crustaceans, insects and vertebrates. The present discussion is confined to some aspects of the endocrinology of growth and metamorphosis in these groups.

STOLONIZATION AND EPITOKY IN POLYCHAETE WORMS

Most polychaetes are sluggish marine animals which live a sedentary life under rocks, in burrows or in characteristic tubes. The sexes are separate and their reproductive success can be largely attributed to a specialized pelagic phase, that is preceded by profound structural modifications for swimming, and is characterized by a synchronized mass spawning at the surface of the sea (Clark, 1961; Barrington, 1967). In most of the syllids and in the eunicid palolo worms, reproduction is preceded by an asexual division (STOLONIZATION) in which one or more posterior subdivisions are formed—each with a new, somewhat modified jawless head and with mature reproductive organs. These fragments separate from the parent worm and subsequently swim together in the pelagic swarming. Stolonization is considered a primitive condition (Clark, 1961).

In the nereids, stolonization is absent and sexual maturity is usually associated with a metamorphosis of the posterior portion of the worm. In some species the entire tail region is modified; in others only the middle region is altered. The changes which transform a sedentary worm into an active swimmer are both morphological and biochemical: elongation of swimming setae, enlargement and increased vascularization of parapodia, histolysis of certain body wall muscles and formation of powerful swimming musculature, specialization of sense organs, differentiation of genital ducts, development of gonads. The transformed worm is called a HETERONEREIS; the process is EPITOKY (less frequently EPIGAMY); the highly modified posterior region is the EPITOKE. In a few of the nereids (particularly brackish and freshwater species), a metamorphosis does not occur; reproduction is often viviparous, hermaphroditic or parthenogenetic; sometimes there is a downstream migration of sexually mature worms. The heteronereid phase and pelagic spawning have been lost in habitats where they would be ineffective.

These metamorphic events are regulated by neurosecretory hormones. A premature transformation follows removal of the nereid brain (cerebral or suprapharyngeal ganglia). This precocious change may be prevented by implanting brains from juvenile nereids of either sex and any species; brains from metamorphosed worms are ineffective. These, and many similar experiments, show that the cerebral ganglia of young worms produce a JUVENILE or MOLT-INHIBITING HORMONE which prevents the heteronereid transformation. Production of hormone gradually declines as the worms mature; epitoky follows as a consequence. Biochemically, this factor is probably identical with the gonad-inhibiting hormone described in the previous chapter. Sexual metamorphosis is biologically linked to gonad maturation and the two processes appear to be regulated by the same mechanism. Stolonization is also regulated by a juvenile hormone but this is not secreted by the brain; its exact source is unknown. Moreover, it is not yet known whether the growth-promoting and regeneration hormone of polychaetes is, in fact, the juvenile hormone; they may be identical but the answers to these questions must wait purification of the actual hormone(s). There are excellent summaries of the literature (Barrington, 1967; Highnam and Hill, 1969; Jenkin, 1970).

ARTHROPOD GROWTH AND METAMORPHOSIS

The "growth" of an arthropod is not a continuous process but occurs in a series of disjunct steps. The rigid exoskeleton is an excellent solution to the problem of protection but precludes a continuous increase in size. Growth of arthropods depends on the periodic discarding of the old shell (MOLT or ECDYSIS) with a decided expansion of the body before the hardening of a new one. This size increase is caused by the rapid uptake of water (crustaceans) or air (insects); actual growth, which in the strict sense involves synthesis of new protoplasm, occurs throughout the intermolt. The metabolism of an arthropod is markedly cyclical as a consequence of this sequence of changes associated with molting. Metamorphosis, where it occurs, is regulated in the same general way as molting, and the two phenomena may be considered together.

Molting in Crustaceans

Metamorphic changes in most crustaceans are less dramatic than those of insects and are probably regulated quite differently; the endocrinology remains speculative (Costlow, 1968). Molting, on the other hand, has been scientifically investigated for almost half a century, and its endocrinology, together with the associated metabolism of growth, is now known in considerable detail.

The molt or ecdysis during which a crustacean withdraws from its old shell and acquires a new one is a conspicuous event which seems to divide the animal's life neatly into periods of molt and intermolt. Studies of metabolism, however, show that the cycle is one of almost continuous activity during which structures such as the integument and the hepatopancreas are constantly showing gradual but measurable changes (Yamaoka and Scheer, 1970). It is customary to divide the cycle into the PREMOLT (proecdysis), MOLT (ecdysis), POSTMOLT (metecdysis) and INTERMOLT; the latter is referred to as DIECDYSIS if it is very short or imperceptible as in animals which molt regularly throughout the year, or ANECDYSIS if the intermolt is prolonged as in the crayfish which molts seasonally. There are several different variants of the molting cycle (Carlisle and Knowles, 1959).

The premolt is a period of active metabolism during which calcium is withdrawn from the exoskeleton and stored in the hepatopancreas, blood or gastroliths (concretions formed in the anterior wall of the cardiac stomach of some crustaceans such as the crayfish). Glycogen accumulates in the hepatopancreas, the epithelial and subepithelial connective tissues, and the new exoskeleton starts to form near the end of proecdysis. Concomitant changes also occur in the lipids and proteins, with the net result that organic and mineral reserves are withdrawn from the old exoskeleton and mobilized for the development of a new one. The extent of the resorption varies with species and the conditions of the molt; as much as 79 per cent of the organic and 18 per cent of the inorganic matter may be withdrawn from the carapace of the crab *Carcinus,* although the values are often considerably lower than this (Passano, 1960). During the later stages of premolt in the crayfish, there is a marked decline in the activity of hepatopancreatic enzymes involved in the pentose oxidative cycle. This pathway is strongly active during intermolt and the cyclical changes that it shows are again evidence of profound alterations in physiological chemistry during the different stages of molting (Highnam and Hill, 1969).

During the molt there is a rapid uptake of water, amounting to about 70 per cent of the premolt body weight in *Carcinus.* The influx of water is largely through the lining of the digestive tract and is dependent on both osmotic and hydrostatic factors (Passano, 1960). The osmotic pressure of the blood rises to about 105 per cent of the sea water at the beginning of the molt; the gastric lining breaks down and becomes permeable, the animal drinks water and thus increases the hydrostatic pressure in the gut. Swelling produced by the influx of water breaks the old carapace and molt or exuviation follows. The water uptake continues for a time after the animal emerges from its shell. During the postmolt period the shell hardens and the true tissue growth of the organism commences. The latter may continue through the period of intermolt, or there may be a period of relative metabolic equilibrium if this period is an anecdysis.

Hormonal regulation of the molt cycle depends on the eyestalks (*X*-organ, sinus gland) and the *Y*-organ (Fig. 20.20). It has been known for more than 50 years that removal of the eyestalks during the intermolt will initiate a molt. Furthermore, it was demonstrated that removal of the *Y*-organs completely eliminates the molting processes, while implants of *Y*-organs frequently stimulate a molt. Figure 24.1 illustrates the results of experiments of this type. The eyestalks contain a neurohemal organ (sinus gland) which receives neurosecretory fibers from the *X*-organ (medulla terminalis *X*-organ or MTGX).

It is now agreed that the *X*-organ produces a MOLT-INHIBITING HORMONE which is stored in the sinus gland and passes to the *Y*-organ to inhibit its activity. When the level of molt-inhibiting hormone falls, the *Y*-organ becomes fully active and secretes crustacean ECDYSONE (the molting hormone) which stimulates the molt. A MOLT-ACCELERATING HORMONE has also been described but its existence remains speculative. Experimental evidence for the presence of molting hormones has been summarized (Carlisle and Knowles, 1959; Passano, 1960; Highnam and Hill, 1969).

In some crustaceans (*Homarus, Cancer pagurus*) molting cycles are continuous until death, and the animals may reach a very large size. In other species (*Maia, Carcinus, Callinectes, Pachygrapsus*) a definite adult size is fixed at the final molt, and the animals continue to live for some time thereafter. Mechanisms determining the final molt seem to vary; in *Maia* it is said to be related to a degeneration of the *Y*-gland; in *Carcinus* the *X*-organ-sinus-gland complex seems to produce excessive amounts of molt-inhibiting hormone (Carlisle and Knowles, 1959). It appears that,

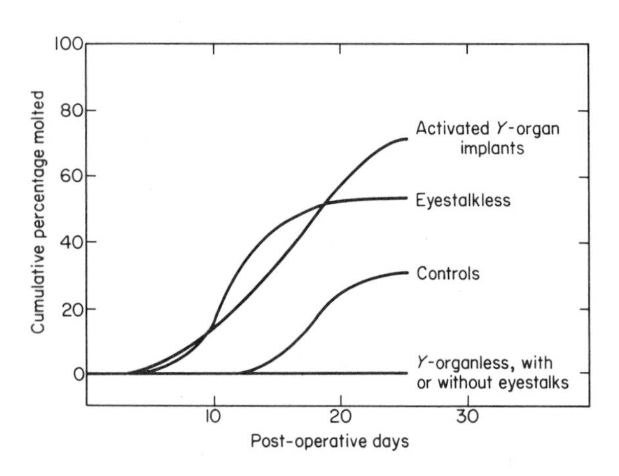

Fig. 24.1 Molting in juvenile *Carcinus maenas*. Note some spontaneous molting in the controls, suppression of molting in the absence of the *Y*-organ and a stimulation of molting with removal of the eyestalks. The group with "activated *Y*-organ implants" received two to four *Y*-organs from eyestalkless donors. [From Passano, 1961: Am. Zoologist, *1*: 92.]

in this as in many other phenomena, the available mechanisms have been used in different ways during evolution to attain the same ends.

It is obvious that the mechanisms regulating reproduction must be related to those concerned with molting. Females do not molt while they are carrying developing young. Moreover, in most of the Malacostraca, copulation is only possible between a freshly molted female and a male with hard integument. These integrating mechanisms have not yet been worked out in detail.

Insect Metamorphosis and Diapause

Several lines of endocrine research have originated from the investigations of insect physiologists. When the Polish scientist Kopéc found evidence for chemical regulation of metamorphosis in caterpillars of the moth *Porthetria* (*Lymantria*), he not only initiated work on the endocrinology of invertebrates but also implicated the nervous tissues in processes of chemical integration. Much of the basic work on neurosecretion stems from these pioneer studies of insect brain hormones. The presence of pheromones may also be cited as an important area of endocrinology which was first explored by students of insect physiology.

Insects are such hardy experimental animals that they are extremely useful in endocrine research. Kopéc (1922 and earlier) found that his animals lived for a considerable period after the removal of the brain but that the caterpillars failed to pupate unless the brain had been present for a period of about 10 days following the final larval molt. If the extirpation was carried out before this "critical period" had elapsed, pupation never occurred; if, on the contrary, the brain was removed 10 days or more after this molt, then pupation was normal although the animals were brainless. He also noted that a tight ligature posterior to the thorax had no effect on pupation if 10 days had elapsed since the final molt; but if the ligature was applied before 10 days, only the portions of the body anterior to the ligature would pupate. He concluded from these and many other convincing experiments that the brain was the source of a factor responsible for pupation and that within 10 days after the final larval molt enough of this material had diffused from the brain into the tissues to effect their transformation. His conclusions are now supported by comparable experiments on many species of insects. Several other experiments are cited in the following sections. The history of insect endocrinology has been adequately reviewed (Gilbert, 1964; Gabe *et al.*, 1964) and will not be detailed here.

Patterns of insect metamorphosis There are three general patterns of metamorphosis in insects. The AMETABOLA, a primitive group represented by the springtails (Collembola), undergo a series of postembryonic molts without any particular change in general appearance; they lack a proper metamorphosis. The HEMIMETABOLA are said to show an incomplete or gradual metamorphosis since the young or NYMPHS have the same general appearance as the adults but lack wings which gradually appear in a series of nymphal molts. The developmental stages are called INSTARS; the full expression of adult characters appears with the final molt. Familiar examples are grasshoppers and cockroaches (Orthoptera), bugs such as *Rhodnius* (Hemiptera), dragonflies (Odonata) and earwigs (Dermaptera). In the third group (HOLOMETABOLA), development is said to show a complete metamorphosis; the larvae are very different in appearance from the adult, and pupation occurs between the last larval instar and the adult stage. Familiar examples are the true flies (Diptera), the moths and butterflies (Lepidoptera), the ants, wasps and bees (Hymenoptera) and the beetles (Coleoptera). Larval stages are commonly called maggots, caterpillars and grubs. In each case this series of worm-like stages, that bear no resemblance to the adult, is followed by a molt in which the last instar is transformed into a PUPA or CHRYSALIS, a superficially quiescent, non-feeding stage, during which there is an active transformation of tissues leading to the emergence of the adult at final molting (Whitten, 1968).

Diapause This is a period of suppressed development characteristic of the life histories of many insects and mites. It is a physiological mechanism for survival during adverse conditions and confers the same biological advantages as comparable stages of dormancy found in other groups of animals: encystment in the protozoans, drought resistant eggs of the brine shrimp *Artemia,* aestivation in the lungfish, delayed implantation in some mammals and hibernation in the homeotherms (page 360). In each case, the rate of metabolism is depressed and growth is either completely arrested or very slow. Both onset of dormancy and resumption of development are usually abrupt; the diapause is followed by metamorphosis, rapid growth or reproductive activity. Many examples are detailed in an excellent monograph by Lees (1955); there are several good reviews of the physiology and endocrinology (Beck, 1968; Highnam and Hill, 1969).

Like so many biological phenomena, diapause is a spectrum of similar conditions rather than a single well-defined state. In a relatively constant environment, such as that of the laboratory, some species (the blowfly *Lucilia sericata,* the red spider mite *Metatetranychus ulmi*) have been reared for years and found to enter diapause only when the environment is appropriately altered. Their diapause is said to be FACULTATIVE. In other species, diapause is OBLIGATORY and occurs as a regular event at a specific time in the annual life cycle. Even in closely related species and subspecies or in different populations of the same species, both facultative and obligatory diapause may occur as well as non-diapausing forms. The European corn borer *Ostrinia nubilalis* spends the winter in diapause—as adults in more northern latitudes where only one generation occurs each year (UNIVOLTINE life cycle) but as full grown caterpillars inside old corn stalks in more southern regions where a second generation reaches the prepupal larval stage by late summer or autumn. The mosquito *Aedes canadensis* also has a univoltine life cycle, but in this case diapause occurs in the eggs that are laid in the autumn. *Aedes dorsalis* produces several generations during the year (MULTIVOLTINE species), but only the eggs laid in the fall enter diapause. In the first species of mosquito, diapause is obligatory; in the second, it is facultative.

In Quebec, the spruce sawfly *Gilpinia polytoma* has one generation per year with an obligatory diapause but in Connecticut, several hundred miles to the south, there are three generations per year with a predominantly facultative diapause; an intermediate condition occurs in areas between these two extremes. Subspecies of migratory locust *Locusta migratoria* develop without diapause in the tropics but show diapause in the northerly limits of their range. Lees gives many other examples.

Thus, the stage in the life cycle at which diapause takes place varies in different species; it may be characteristic of egg, larva, pupa or adult. In adults there is typically a failure in development of reproductive organs and a hypertrophy of fat bodies and storage structures while, in the earlier stages, growth is arrested. Occasionally diapause may occur during more than one stage in a life cycle. The grasshopper *Pardalophora apiculata* has a two-year life cycle and spends the first winter as a diapause egg and the second as a late nymph. In some insects there is a QUIESCENCE characterized by slow growth and irregular development rather than a true diapause. These conditions have their parallels in the state of torpor or narcosis and true hibernation among the homeotherms. Lees cites several examples: the alder fly *Sialis lutaria* and the dragonfly *Anax imperator*, which have two-year life cycles, spend the first winter in quiescence as half-grown larvae and the second in the true diapause of the last larval stadium.

This spectrum of conditions found in closely related insect species suggests that diapause has evolved many times and emphasizes once more the opportunism of phylogenetic processes.

The phenomenon is hormonally regulated. In species with facultative diapause, activity of the endocrine organs is environmentally triggered; when the diapause is obligatory, genetically established endogenous rhythms may play a dominant role. As in many other environmentally controlled physiological processes, photoperiod provides the most universal and reliable cue; in northern latitudes where winter diapause is common, short days usually initiate the change. Temperature, moisture, available oxygen, quality and quantity of food may also be involved. Temperature, in particular, often modifies the photoperiod effect and, in some cases, completely overrides it. The literature now contains numerous examples of the way in which several environmental cues have been woven together to form reliable controls which meet the survival requirements of different species (Clements, 1963; Beck, 1968).

Endocrine controls It has now been abundantly demonstrated that the brain produces only one of several substances concerned with the regulation of metamorphosis in insects. The several links in this regulatory chain are shown in Fig. 24.2. They are basically the same in all groups, whether the metamorphosis is complete or incomplete.

The neurosecretory cells in the pars intercerebralis of the brain produce a tropic substance (ECDYSIOTROPIN OR THORACOTROPIN) which passes by axon transport to a neurohemal organ, the CORPUS CARDIACUM (CORPORA CARDIACA), located in close proximity to the brain. Under appropriate stimulation, the ecdysiotropic hormone is released into the blood and activates the ECDYSIAL OR PROTHORACIC GLAND to secrete

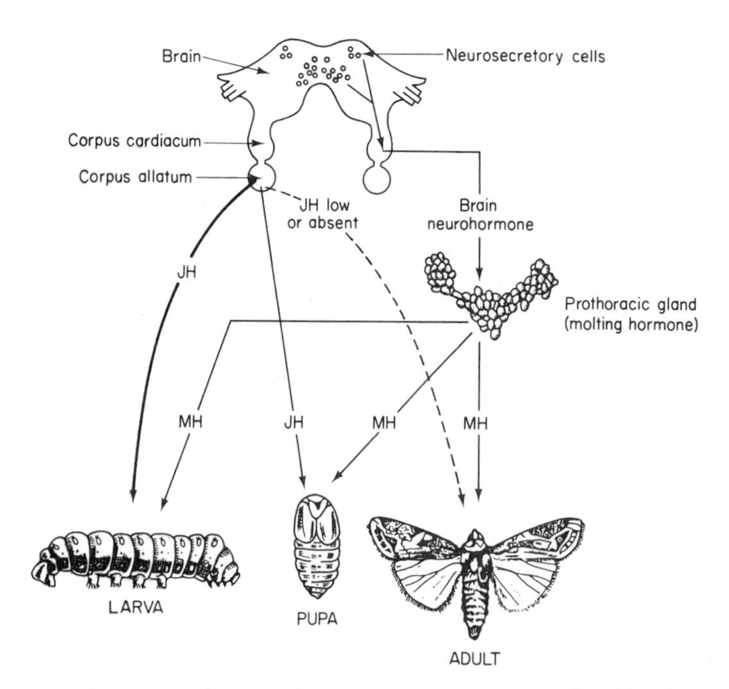

Fig. 24.2 Neural and endocrine control of growth and molting in a moth. Neurosecretory cells in the brain release a principle which stimulates the prothoracic glands to secrete molting hormone (MH or ecdysone). Juvenile hormone (JH) arises from the corpora allata and promotes the retention of larval characters. Adult differentiation occurs when MH acts in the absence of JH. [From Turner and Bagnara, 1971: General Endocrinology, 5th ed. Saunders, Philadelphia.]

ECDYSONE (also called the molting hormone, the prothoracic gland hormone, PGH and the growth and differentiation hormone GD). Ecdysone acts directly on the chromosomal mechanisms concerned with molting. During larval stages, the action of ecdysone is checked by the juvenile hormone JH secreted by the CORPORA ALLATA. The presence of JH ensures the development of another larval form; during the last instar, the corpora allata become inactive so that little or no JH is secreted and the adult transformation occurs.

Evidence for this chain of controls has been found in many different experiments involving several groups of insects. Wigglesworth's (1936, 1964) experiments on *Rhodnius* will serve as a single example, but many others are summarized in the literature already cited. The molt-initiating stimulus differs in various species of insects. In *Rhodnius* the mechanical distension of the body resulting from a full meal of blood activates the neurosecretory centers of the brain and triggers the chain of hormonal activity. *Rhodnius* fails to molt if given only small meals of blood; a large meal with body distension is requisite to growth. Further, the neurosecretory cells require a certain period of time to elaborate enough hormone to initiate a molt. If the brain (source of thoracotropic hormone) is removed very shortly after the meal of blood, the animal will never molt, although it may continue to live for more than a year. If, however, a week or more has elapsed since the engorgement with blood, removal of the brain and corpora cardiaca has no effect on the subsequent molt. In this case, sufficient hormone has diffused into the tissues to effect transformation.

The long head of *Rhodnius* is singularly convenient for surgical work. Animals may be decapitated somewhat anteriorly to remove only the protocerebrum (source of the ecdysiotropic hormone), while a more posterior section will also remove the corpora allata (source of juvenile hormone). Decapitated animals have been joined with capillary tubes or fused together parabiotically as indicated in Fig. 24.3 to demonstrate several features of their endocrinology. When a fourth-instar nymph is decapitated immediately after feeding and combined with a nymph which has been decapitated *anteriorly* several days after feeding (Fig. 24.3b), both molt together to form the next nymphal stage, since the anterior decapitation did not injure the corpora allata which is the source of the juvenile hormone. If, however, the experiment is done in the same way except for a more *posterior* decapitation of the second individual, both nymphs molt into adults, thus omitting further larval stages because the source of juvenile hormone was removed (Fig. 24.3c). Comparable experiments can produce supernumerary nymphal stages or a reversion of adults to nymph-like forms. The importance of the corpora allata may also be shown by direct extirpation of the glands (leading to premature adult molts) or implantation of allata (producing extra nymphal instars).

This pattern of controls has been described in several groups of insects. However,

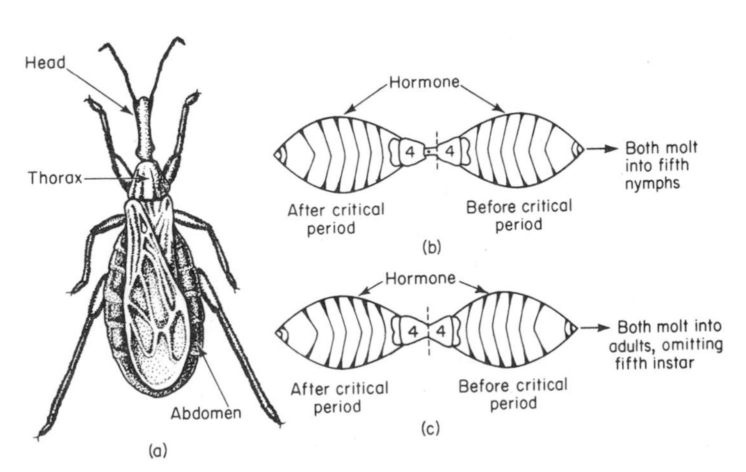

Fig. 24.3 (a) The bloodsucking bug *Rhodnius prolixus*. (b) and (c) Diagrams illustrating some of Wigglesworth's (1936) endocrinological experiments outlined in the text. Broken line, the level of decapitation. Black dot, the presence of the corpus allatum. The joining of two animals to create a common blood system is called PARABIOSIS; the individuals are like Siamese twins.

it is not universal. Variations in both morphological and physiological details have been described (Gilbert, 1964; Highnam and Hill, 1969) and are to be expected, for evolutionary processes have very often attained similar ends by somewhat devious routes. Comparative endocrinology is replete with examples of variable usage of a group of hormones to achieve some specific regulatory function.

Diapause is a special stage in the hormonally regulated growth sequence of some insects and mites. Environmental changes are responsible for both the cessation of growth and the reactivation of development, but these changes always operate through the neuroendocrine system. No peculiar organs have been associated with the production of diapause hormones; on the contrary, diapause is regulated through an adaptation of the endocrine controls already described for other metamorphic processes. Additional complexity is found in both the environmental cues which trigger the processes and in the varied nature of the stages and processes which are temporarily suspended. Diapause is frequently triggered by one environmental change and "broken" by another. In many lepidopterans such as the giant cecropia silkworm *Hyalaphora cecropia* (Williams, 1952) and the European corn borer *Ostrinia nubilalis* (Beck, 1968), diapause is initiated in the autumn by decreasing photoperiods and comes to an end only after an exposure to the chilling winter temperatures. The cecropia silkworms have a pupal diapause. Williams performed many interesting experiments with their isolated abdomens after sealing them off anteriorly with coverslips. The prothoracic glands are removed by this operation and the abdomens may remain in diapause for a very long time. If, however, they are implanted with glands from active pupae, or if a diapausing abdomen is joined parabiotically to an activated (chilled) pupa, metamorphosis of the isolated abdomen occurs and, in some cases, these preparations may lay eggs. Regulation by chemicals produced in the anterior part of the body and triggered by low temperature was thus established. An exposure to temperatures of 3° to 5°C for a period of one to two months is required for activation of diapausing cecropia pupae. The aedine mosquitoes have an egg diapause and also require chilling temperatures for activation—as long as six months in *Aedes stimulans*. In this case there is an additional factor since the eggs will not hatch until they have been submerged in deoxygenated water (Clements, 1963). Many other examples are given in the literature previously cited.

A complexity in diapause endocrinology can be anticipated from the varied nature of the processes regulated. Larval and pupal diapause have been traced to an interruption in the cyclical events responsible for the production and action of ecdysone. Thus, the primary links in regulation are the brain, corpora cardiaca and prothoracic glands. Diapause in adult insects (adult or reproductive diapause) is marked by reproductive dormancy and controlled through neurosecretory centers in the brain. These centers are inactivated in response to environmental stimuli; as a consequence, the corpora allata suspend production of gonadotropic hormone (Chapter 23) and ovarian development ceases.

Egg and embryo diapause have posed some of the most puzzling physiological problems. These processes are maternally regulated. The endocrine system of the female, in response to environmental cues, produces substances which act in the ovary and hence regulate developmental processes which operate in eggs or embryos many months later. Details are different in various insects. The oriental silkworm *Bombyx mori* has been intensively studied by Japanese workers (bibliography in Wigglesworth, 1964, 1970). A special "diapause hormone" has been extracted from the subesophageal ganglion of *Bombyx* pupae; this when injected into female pupae leads to the production of diapause eggs by the resulting moth even in insects which would normally produce non-diapausing eggs. Details of the regulation are being actively investigated but are, at present, incompletely explained.

Biochemistry of insect hormones P. Karlson and his associates have been responsible for the chemical isolation and identification of ecdysone (review by Gilbert, 1964). This is one of the great achievements of twentieth-century biochemistry since such fantastically small amounts of the hormone are present in a single insect. The first sample of 25 mg of crystalline material came from 500 kg of silkworm (*Bombyx*) pupae. In a later isolation a somewhat greater yield of 250 mg was obtained from four tons (wet weight) of animals. In 1963, Karlson and his associates, using

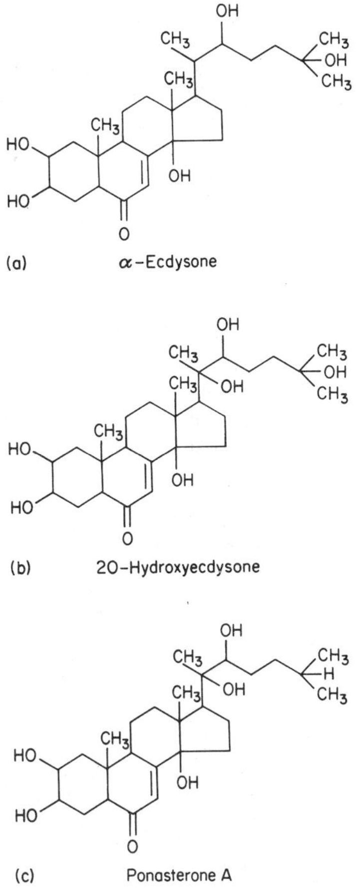

(a) α-Ecdysone

(b) 20-Hydroxyecdysone

(c) Ponasterone A

Fig. 24.4 Some steroid compounds from arthropods and plants. (a) α-ecdysone first extracted from silkworm pupae; (b) 20-hydroxyecdysone which is a molting hormone of crustaceans (crustecdysone) and also occurs in insects (β-ecdysone); (c) a plant steroid with ecdysone activity.

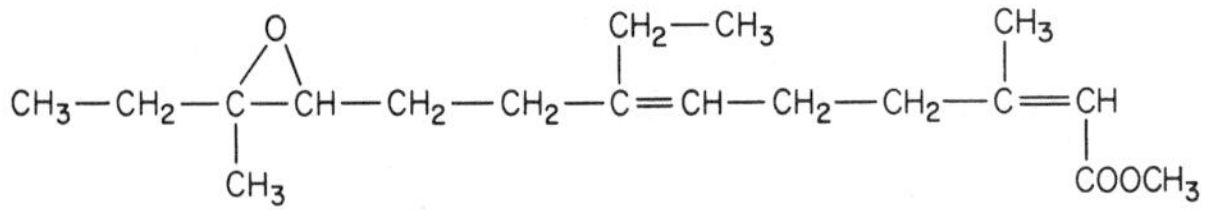

(a) Methyl 10,11-epoxy-7-ethyl-3,11-dimethyl-2,6-tridecadienoate

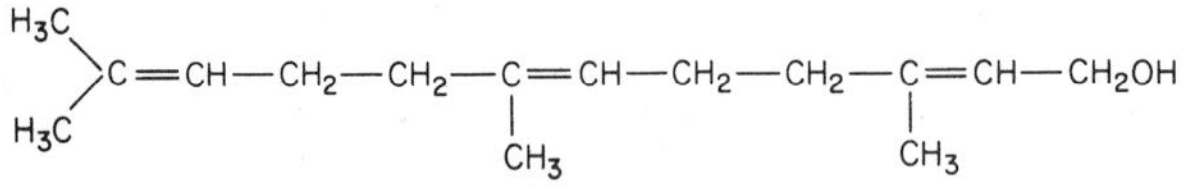

(b) Farnesol

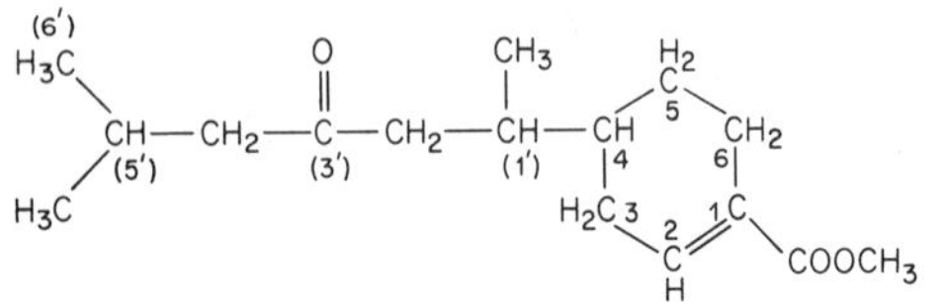

(c) Methyl 4-[1′,5′-dimethyl-3′-oxo] hexyl cyclohex-1-enoate

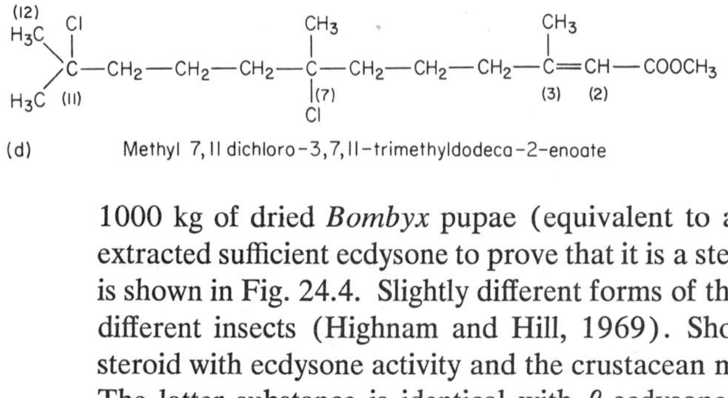

(d) Methyl 7,11 dichloro-3,7,11-trimethyldodeca-2-enoate

Fig. 24.5 Some compounds with juvenile hormone activity. (a) Structure of juvenile hormone as given by Röller *et al.* (1967); (b) farnesol first isolated from the feces of mealworms; (c) "paper factor" isolated from the balsam fir *Abies balsamea*; (d) synthetic substance which very actively mimics JH.

1000 kg of dried *Bombyx* pupae (equivalent to about four tons of fresh insects), extracted sufficient ecdysone to prove that it is a steroid. The structure of α-ecdysone is shown in Fig. 24.4. Slightly different forms of the molecule have been identified in different insects (Highnam and Hill, 1969). Shown also in this figure is a plant steroid with ecdysone activity and the crustacean molting hormone CRUSTECDYSONE. The latter substance is identical with β-ecdysone found along with α-ecdysone in *Bombyx* and some other insects. Thus, it is not surprising that molting in crustaceans can be induced by insect molting hormone (Krishnakumaran and Schneiderman, 1969).

Fig. 24.5 shows the structure of juvenile hormone as characterized by Röller and his associates (1967, 1968). Several other substances with similar biological activity are shown in the same figure. The terpene farnesol, isolated from the excreta of mealworms, was the first substance shown to have JH activity and thought to be the secretion product of the corpora allata. However, farnesol and its derivatives are much less active than the chemical isolated by Röller and now recognized as the true hormone. Several different JH mimics have been extracted from plants as well as animals; in addition, the chemist has added a number of extremely potent synthetic analogues. The possibility of using these substances as insecticides is a very real one

(Williams, 1967; Bowers, 1969; Highnam and Hill, 1969; Sehnal, 1971). Minute amounts may indefinitely prolong larval development, or induce abnormal final molts, or cause adults to lay inviable eggs, or block embryonic development. They have the advantage over other insecticides of being highly specific, active in very small amounts and not likely to contaminate the environment.

The action of hormones at the molecular or cellular level is a subject of great biological interest. The insect hormones, particularly ecdysone, have been central to much of the most productive work. There are several informative reviews of the history and current status of thinking (Highnam and Hill, 1969; Lezzi, 1970; Sehnal, 1971). Three hypotheses have been considered: (1) direct involvement of hormones in specific enzyme systems, (2) an action on membrane permeability (cell surface or other), and (3) the activation or suppression of specific genes. The latter theory is the prevailing one and now supported by an impressive body of experimental evidence. Karlson and his associates did the pioneer work using nuclear preparations of the midge fly *Chironomus* (Karlson and Sekeris, 1964; Beermann and Clever, 1964). Specific chromosomal "puffs" appear at the time of molting or experimentally in response to ecdysone. These "puffs" are considered to be the sites of specific mRNA formation; the RNA, in turn, is the key link in protein synthesis. These studies have included isolated chromosomes as well as separated nuclei and cells. In other productive lines of research, ecdysone has been injected into the larvae of blowflies *Calliphora* and its effects studied on several enzyme systems, particularly dopa decarboxylase—an enzyme concerned with cuticular tanning. The obvious questions are whether the observed effects are at the level of the enzyme reaction or at the point of RNA transcription which must control the synthesis of the enzyme, or whether the hormone influences the intracellular ionic environment which may in turn control the activities of the genes. The hypothesis of differential gene activation and deactivation by the molting and juvenile hormones satisfies most of the accumulated facts, but further investigation will be required to settle the controversy concerning molecular details (Laufer and Calvet, 1972; Lezzi and Gilbert, 1972; Sekeris, 1972).

GROWTH, MOLTING, AND METAMORPHOSIS IN THE VERTEBRATES

Juvenile vertebrates are usually similar in appearance to the adult. Their postembryonic development lacks the disjunct molt-intermolt sequences and the spectacular metamorphoses of the arthropods. Nevertheless, progress from birth to maturity is regulated in accordance with environmental conditions and the biology of the species; there are often phases of slow and more rapid growth. Temperature, season, food availability, and reproductive development are particularly important while some animals, when living in crowded conditions, produce species-specific chemicals (pheromones) which adjust the rate of development of the population (Akin, 1966; Yu and Perlmutter, 1970). The pituitary gland, responding through the hypothalamus to environmental cues, provides the important coordinating factors. Although several metabolic hormones (insulin, thyroxine, gonadal and adrenal steroids) affect growth, pituitary somatotropin and prolactin are the specific growth hormones.

Growth Hormone

The significance of the vertebrate pituitary in the regulation of growth was convincingly demonstrated during the first two decades of the twentieth century. Aschner (1912) first described arrested growth in the hypophysectomized dog; his studies were soon followed by a classical series of experiments on the white rat (Smith, 1930). Since that time, growth hormone (somatotropin, STH) has been highly purified, studied biochemically and used in many physiological experiments.

Li (1969) established the amino acid sequence of human somatotropin. There are 188 amino acid residues and two disulfide bridges; the molecular weight is 21,500. Other mammalian somatotropins have comparable molecular weights or appear to be diamers of units of this size. Details of the biochemistry and physiology of the mammalian factors, as well as speculation concerning the mode of action at the cellular level, will be found in textbooks of mammalian endocrinology (Knobil, 1961; Williams, 1968; Turner and Bagnara, 1971). The comparative endocrinologist has found his major area of excitement in studies of the close affinities of this factor with prolactin, in the marked species specificity of STH, and in the temporal changes in sensitivity of both factors during development.

The close affinities of growth hormone and prolactin suggest that they have evolved from a single parent chemical. Carefully purified preparations often display both growth-promoting and lactogenic effects. Indeed, it has not been possible to isolate a separate prolactin from the human pituitary where two distinct factors seem questionable (Li, 1969). Separate biochemical entities have been characterized from cattle, pigs and sheep, but there is considerable overlap in their growth-promoting and lactogenic effects. The prolactins regularly stimulate growth of several different target organs (Nicoll and Bern, 1971) and may be as effective as STH in accelerating somatic growth; for example, juvenile lizards grow much more rapidly when injected with either factor (Licht and Hoyer, 1968) while hypophysectomized chicks and pigeons respond as well to prolactin as to growth hormone (Jenkin, 1970; Turner and Bagnara, 1971).

Species specificities of growth hormones have long been recognized and details will be found in the literature cited. The phylogenetic correlation is slight. Although primates do not respond to growth hormone from other mammalian classes, the rat responds to all extracts tested except those from teleost fishes; it does, however, respond to extracts from sturgeons and paddlefishes (Hayashida and Lagios, 1969)—members of the Chondrostei, a group considered more primitive than the teleosts (Romer, 1970). These specificities are well established, although it is always wise to be cautious in drawing conclusions from negative data obtained in testing protein hormones because of their marked antigenic tendencies; failure of the hypophysectomized guinea pig to respond to any growth hormone including its own may be due to rapid production of antibodies (Knobil, 1961).

The sensitivity of target organs often shows curious changes during development and these changes may also relate to the biochemical similarities of prolactin and STH. During fetal life, growth is independent of the mammalian pituitary (Knobil, 1961; Jenkin, 1970); young rats are almost a month old before their growth is substantially reduced by hypophysectomy. Thus, the sensitivity of the target organs gradually develops during ontogeny; there may also have been changes during phylogeny since some of the lower forms (elasmobranchs, axolotls), if well fed, grow normally after removal of the pituitary (Pickford and Atz, 1957; Hoar, 1965a). The amphibians are interesting in this connection because of the variety of growth-regulating controls demonstrated and because of the importance of this group in discussions of the evolution of the land habit. Prolactin is the major growth factor in several, but not all, species of larval anurans but acts only at very high doses in adults where STH shows strong growth-promoting effects (Cohen *et al.*, 1972; Frye *et al.*, 1972). In at least one larval anuran (*Rana temporaria*), STH is more effective than prolactin while in the larval toad, *Bufo bufo*, neither hormone affects growth. Premetamorphosed newts *Taricha torosa* show accelerated growth when treated with mammalian STH but not with prolactin. These variable effects found among amphibians may be entirely fortuitous. However, the present evidence, although scanty, suggests that a single metabolic hormone in ancestral forms acquired a variety of regulatory functions associated with somatic growth and the differentiation of specific target organs; in certain forms, these functions have been parceled out to separate factors that have evolved from the parent compound.

Molting in Vertebrates

Molting is characteristic of many of the terrestrial vertebrates. Amphibians and reptiles periodically shed the outer layers of their skin while birds and mammals often alter their plumage or pelage with the season. Concomitant color changes are characteristic of some species. The correlation of molt with season, often also with reproduction, suggests an involvement of the endocrine system. In many cases where the molt is associated with reproduction, pituitary gonadotropins and/or gonadal steroids have been definitely implicated (Jenkin, 1970). Seasonal color changes (morphological color changes) of some of the northern birds and mammals are pituitary-regulated and can be modified by an appropriate manipulation of the photoperiod. In the varying hare *Lepus*, for example, the pituitary gonadotropins are said to induce shedding of white winter fur and the growth of a brown coat. Only the gonadotropins seem to be involved since the reaction can be induced in both castrated and thyroidectomized animals (Lyman, 1943).

In general, molting or changes in the pigmentation of plumage and pelage are regulated by endocrines but, as in other physiological processes, the precise role of the specific hormones is often different in the various groups. In some birds (domestic fowl, pigeon), thyroidectomy inhibits molting while thyroxine stimulates it; in others (crows, jackdaws), the epidermis is relatively insensitive to thyroid manipulation (Gorbman and Bern, 1962; Jenkin, 1970). Thyroid hormone, perhaps more than any other factor, is frequently associated with morphogenetic and epidermal changes in the lower vertebrates. The plumage and color changes associated with reproduction in many birds are regulated by the gonadotropins and gonadal steroids (particularly progesterone) as well as by thyroxine. In some lizards, prolactin is also connected with the epidermal sloughing, characteristic of molting (Chiu *et al.*, 1970; Nicoll and Bern, 1971).

Details of molting endocrinology in the vertebrates are best understood in some of the Amphibia. A relationship between thyroid state and skin structure was noted many years ago, and more recently the corticosteroids have also been associated with these processes (Jenkin, 1970; Gordon *et al.*, 1972). The Amphibia show two distinct stages in molting, and these seem to be subject to different endocrinological controls. There is first an epidermal proliferation with a cornification of the outer layers; this is followed by the secretion of mucus and the shedding and eating of the detached outer layers of skin (slough). Hypophysectomy inhibits both processes but the tropic hormones involved are quite different in urodeles and amphibians. Among the urodeles, the first process (proliferation and cornification) seems independent of thyroid hormone; it may continue in thyroidectomized salamanders to produce layers of unusually thick epidermis, since the second process (epidermal shedding and sloughing) is inhibited or completely suppressed by this operation. Among the anurans, molting continues in the absence of the thyroid; the entire process is regulated by ACTH and the adrenal steroids. Hypophysectomized toads *Bufo bufo* show the proliferation and cornification reaction with formation of a slough when treated with ACTH or corticosteroids. Corticosterone is more active than aldosterone, while cortisol is relatively inactive. Factors such as thyroxine, somatotropin, prolactin and the neurohypophysial octapeptides have no effect on this particular response of the hypophysectomized toad (Jørgensen and Larsen, 1964). The literature on amphibian molting has been summarized by Jenkin (1970).

Vertebrate Metamorphosis

The amphibians provide the best-known examples of metamorphosis among the vertebrates. The young usually hatch as fish-like tadpoles in fresh water and, after a period of aquatic life, metamorphose into terrestrial adults that use lungs instead of gills for respiration and show several profound biochemical changes, associated with a change of diet and restrictions in the availability of water (Frieden, 1961, 1963; Cohen, 1966). Some of these changes are reversed when the adults return again to the aquatic habitat at the time of spawning.

Some fishes show equally significant metamorphic changes (Barrington, 1961,

1968b). The ammocoete larvae of the lamprey live a sedentary life in the mud of freshwater streams, feeding on small food particles which they trap in a sticky ciliated pharyngeal groove; the sticky mucus is secreted by the subpharyngeal gland (endostyle) and discharged into the pharynx. This feeding technique is similar to that of the protochordates. After some years there is a metamorphosis which includes the formation of eyes, the loss of the endostyle and the development of a rasping circular mouth. The lamprey then swims forth to feed on other fishes and sometimes to migrate into the ocean, a habitat that makes very different physiological demands on mechanisms concerned with water and electrolyte balance. Several species of teleost likewise change from a freshwater to a marine habitat at a specific time and in association with definite biochemical and metabolic changes. The smolt transformation of salmonids is the most familiar example. However, in the salmonids, as in most teleosts, the post-embryonic changes in morphology are slight compared with those of the cyclostomes. Metamorphic changes are also characteristic of the Atlantic eel *Anguilla* which, as larvae or leptocephali, are carried by ocean currents from the spawning grounds of the Sargasso sea into the fresh waters from rivers in the North Atlantic. There is a distinct metamorphosis from the leptocephalus to the elver which enters the rivers and again, after a number of years when fully grown, from the yellow stage to the silver stage, which migrates to the ocean on the spawning journey. The examples could be multiplied; the flatfish (Heterosomata) changes from a symmetrical larva to the laterally compressed adult form with both eyes on one side; the mudskipper *Periophthalmus* adopts a semiterrestrial life at metamorphosis.

Wald (1960b) argues that every metamorphosis invites a second metamorphosis. The juveniles which undergo changes necessary for life in an entirely different habitat (such as the change from water to land, or from fresh to salt water) must often show an equally radical transformation when they return to their original habitat at spawning time. Several of these physiological and biochemical changes have already been described (visual pigments, hemoglobins and the transport capacity of the blood, electrophoretic mobility of plasma proteins, excretion and electrolyte balance, alterations in digestive enzymes and in several of the metabolic pathways). They will not be detailed further; in general, the change is appropriate to the physiological demands of the altered habitat and these demands were considered at length in Part Two. Excellent reviews of the biochemistry of vertebrate metamorphosis are available (Wald, 1960b; Cohen, 1966; Deuchar, 1966; Frieden, 1968).

The endocrinology of vertebrate metamorphosis has been studied for more than half a century. In one of the pioneer studies of thyroid physiology, Gudernatsch (1912) showed the acceleration of tadpole development that follows thyroid feeding. Since that time the thyroid has been most frequently implicated in the endocrinological control of metamorphosis. Among fishes, it has been noted that some of the cells of the endostyle are transformed into thyroid follicles when the ammocoete changes to the form of an adult lamprey and that the thyroid shows evidence of increased activity at certain critical stages in the transformation of the eel, in the metamorphosis of flatfishes and in the smoltification of salmonids. In many cases, however, it is not yet clear whether the thyroid is regulating the metamorphic processes. Attempts to hasten the metamorphosis of the ammocoete, to induce the assumption of semiterrestrial life in *Periophthalmus*, or to increase the salinity resistance of young salmonids have been uniformly unsuccessful. The thyroid may be only one of several metabolic hormones activated during these critical periods; thyroid changes may be associated with the metamorphosis rather than causative. In any case, the pituitary which regulates the thyroid must form the first link between environmental triggers and the growth and metamorphic processes.

As already indicated, correlations between thyroid activity and amphibian metamorphosis are long-standing and convincing. Metamorphosis from tadpole to adult is always associated with a surge of pituitary thyrotropic and thyroid activity. Amphibians occupy a key position in the phylogeny of terrestrial vertebrates. Their endocrinology is of particular significance in considerations of biochemical evolution. There are always two distinct parts to an endocrine mechanism since the responsiveness of the target organs is just as significant as the hormone itself. The tailed amphibians show a curiously interesting variety in tissue responsiveness to thyroid stimulation. Metamorphosis in many of the Caudata, as in the Anura, may be accelerated with thyroid hormone. Some salamanders, however, have a facultative metamorphosis. Axolotls, *Ambystoma tigrinum*, in many localities metamorphose regularly, losing their external gills and assuming a terrestrial existence. In certain regions (some high lakes of the Rocky Mountains), however, the larval characteristics are retained throughout life and the animals become sexually mature as neotenous adults. *"They cling to youth perpetual and rear a tadpole brood"* (Garstang, 1951). Thyroid hormone stimulates metamorphosis in these neotenous races. In nature, *Ambystoma tigrinum* can have the best of two worlds.

The pituitary-thyroid control is not the only endocrine mechanism concerned with amphibian metamorphosis. The growth-promoting action of prolactin on anuran tadpoles has been noted. This hormone appears to prolong larval life (Cohen, *et al.*, 1972), stimulating the growth of larval structures such as the tail. Toward the end of larval life, rising levels of TSH and STH override the prolactin effect and lead to metamorphosis and adult growth (Bern and Nicoll, 1968b). Thyroxine has been referred to as an amphibian "land-drive hormone" while prolactin is a "water-drive factor" (Grant and Cooper, 1965). The action of prolactin in triggering the water-drive (second metamorphosis) of *Diemictylus* (*Triturus*) *viridescens* was described in the previous chapter. At this point it is of interest to add that migration and the alteration of skin texture, induced by prolactin or prolactin-like substances, represent only the first stage of the transformation. Although gills do not appear, there is a functional restoration of the lateral line, development of a keeled, swimming tail, changes in pigmentation, and biochemical changes in visual pigments, nitrogen excretion and electrolyte balance. Indications are that the pituitary-thyroid machinery, melanophore-stimulating hormones and neurohypophysial factors are involved in these later phases (Grant, 1961), but details of the endocrinology are still uncertain.

Some urodeles (the perennibranchs *Proteus* and *Necturus*) are genetically incapable of losing their external gills and assuming the terrestrial form (Dent, 1968). Aquatic life is obligatory. During phylogeny, genetic changes in the responsiveness of tissues must have preceded or must have been associated with the development of endocrine controls. Only negative results have been obtained with thyroid treatment of the perennibranchs—to use Garstang's words once more:

> *They do not even contemplate a change to suit the weather,*
> *But live as tadpoles, breed as tadpoles, tadpoles altogether!*

REFERENCES

ABBOTT, B. C., F. LANG, and I. PARNAS. 1969. Physiological properties of the heart and cardiac ganglion of *Limulus polyphemus*. Comp. Biochem. Physiol. **28:** 149–158.

ABBOTT, B. C., and J. LOWY. 1958. Contraction in molluscan smooth muscle. J. Physiol. **141:** 385–397.

ACHER, M., J. CHAUVET, and M-T. CHAUVET. 1972. Identification de deux nouvelles hormones neurohypophysaires, la valitocine (val[8]-oxytocine) et l'aspartocine (asn[4]-oxytocine) chez un poisson sélacien, l'anguillat *(Squalus acanthias)*. C. R. Acad. Sci. Paris. **274D:** 313–316.

ADAM, H. 1963. Structure and histochemistry of the alimentary canal. *In* Brodal and Fänge (1963): 256–288.

ADAMS, E. 1959. Poisons. Sci. Am. **201 (5):** 76–84.

ADAMS, W. E. 1958. The comparative morphology of the carotid body and carotid sinus. C. C Thomas, Springfield, Ill. 272 p.

ADIYODI, K. G., and R. G. ADIYODI. 1970. Endocrine control of reproduction in decapod crustacea. Biol. Rev. **45:** 121–165.

ADIYODI, R. G. 1972. Wound healing and regeneration in the crab *Paratelphusa hydrodromus*. Int. Rev. Cytol. **32:** 257–289.

ADLER, J. 1966. Chemotaxis in bacteria. Science **153:** 708–716.

ADOLPH, E. F. 1947. Physiology of man in the desert. Interscience, New York. 357 p.

ADOLPH, E. F. 1957. Ontogeny of physiological regulations in the rat. Quart Rev. Biol. **32:** 89–137.

ADOLPH, E. F. 1961. Early concepts of physiological regulations. Physiol Rev. **41:** 737–770.

ADOLPH, E. F. 1973. Physiological adaptations to hypoxia in infant mammals. Am. Zool. **13:** 469–473.

ADRIAN, E. D. 1928. The basis of sensation. Christophers, London. 122 p.

ADRIAN, E. D., and F. J. J. BUYTENDIJK. 1931. Potential changes in the isolated brain stem of the goldfish. J. Physiol. **71:** 121–135.

ADRIAN, E. D., and Y. ZOTTERMAN. 1926. The impulses produced by sensory nerve-endings. Part 2. The response of a single end-organ. J. Physiol. **61:** 151–171.

AHLBERT, I. 1969. The organization of the cone cells in the retinae of four teleosts with different feeding habits *(Perca fluviatilis* L. *Lucioperca lucioperca* L. *Acerina cernua* L. and *Coregonus albula* L.*)* Ark. Zool. **22:** 445–481.

AHLQUIST, R. P. 1948. A study of adrenotropic receptors. Am. J. Physiol. **153:** 586–600.

AHSAN, S. N. 1966. Effects of temperature and light on the spermatogenetic activity of the lake chub, *Couesius plumbeus* (Agassiz). Can. J. Zool. **44:** 161–171.

AIDLEY, D. J. 1967. The excitation of insect skeletal muscles. Adv. Insect Physiol. **4:** 1–31.

AIDLEY, D. J. 1971. The physiology of excitable cells. Cambridge U. P., London. 468 p.

Aiello, E. 1970. Nervous and chemical stimulation of gill cilia in bivalve molluscs. Physiol. Zool. **43:** 60–70.

Akin, G. C. 1966. Self-inhibition of growth in *Rana pipiens* tadpoles. Physiol. Zool. **39:** 341–356.

Alderdice, D. F. 1972. Factor combinations. *In* M. O. Kinne (ed.), Marine ecology. **1(3):** 1659–1722. Wiley-Interscience, London.

Alexander, R. McN. 1966. Physical aspects of swimbladder function. Biol. Rev. **41:** 141–176.

Alexander, R. McN. 1972. The energetics of vertical migration by fish. Symp. Soc. Exp. Biol. **26:** 273–294.

Alexandrowicz, J. S. 1951. Muscle receptor organs in the abdomen of *Homarus vulgaris* and *Palinurus vulgaris*. Quart. Rev. Microscop. Sci. **92:** 163–200.

Ali, M. A. 1959. The ocular structure, retinomotor and photobehavioral responses of juvenile Pacific salmon. Can. J. Zool. **37:** 965–996.

Allen, R. D. 1961. Ameboid movement. *In* Brachet and Mirsky, **2:** 135–216.

Allen, R. D. 1962. Amoeboid movement. Sci. Am. **206 (2):** 112–122.

Allen, R. D. 1968. Differences of a fundamental nature among several types of amoeboid movement. Symp. Soc. Exp. Biol. **22:** 151–168.

Allen, R. D., and N. Kamiya. 1964. Primitive motile systems in cell biology. Academic Press, New York. 642 p.

Alliott, E., and J. Bocquet. 1967. Présence d'un système enzymatique chitinolytique dans le tube digestif d'un Sélacien: *Scylliorhinus canicula*. C. R. Soc. Biol. **161:** 840–845.

Amoore, J. E. 1970. Computer correlation of molecular shape and odor: a model for structure-activity relationships. *In* Wolstenholme and Knight (1970): 293–312.

Amoroso, E. C. 1960. Viviparity in fishes. Symp. Zool. Soc. London. **1:** 153–181.

Andersen, B., and H. H. Ussing. 1960. Active Transport. *In* Florkin and Mason, **2:** 371–402.

Andersen, H. T. 1961. Physiological adjustments to prolonged diving in the American alligator *Alligator mississippiensis*. Acta Physiol. Scand. **53:** 23–45.

Andersen, H. T. 1966. Physiological adaptations in diving vertebrates. Physiol. Rev. **46:** 212–243.

Anderson, J. F. 1966. The excreta of spiders. Comp. Biochem. Physiol. **17:** 973–982.

Andreassen, J. 1966. Cilia in nematodes? Science **152:** 231.

Andrew, W. 1959. Textbook of comparative histology. Oxford University Press, New York. 652 p.

Anfinsen, C. B. 1959. The molecular basis of evolution. John Wiley and Sons, New York. 228 p.

Angell James, J. E., and M. de B. Daly, 1972. Some mechanisms involved in the cardiovascular adaptations to diving. Symp. Soc. Exp. Biol. **26:** 313–341.

Annison, E. F., and D. Lewis. 1959. Metabolism in the rumen. Methuen, London. 184 p.

Anthony, E. H. 1961. Survival of goldfish in presence of carbon monoxide. J. Exp. Biol. **38:** 109–125.

Antonini, E. 1965. Interrelationship between structure and function in hemoglobin and myoglobin. Physiol. Rev. **45:** 123–170.

Arden, G. B. 1969. The excitation of photoreceptors. Prog. Biophys. and Mol. Biol. **19:** 373–421.

Arena, V. 1971. Ionizing radiation and life. Mosby, St. Louis, Mo. 543 p.

Arnon, D. I. 1960. The role of light in photosynthesis. Sci. Am. **203 (5):** 105–118.

Arnon, D. I. 1966. Photosynthetic phosphorylation: facts and concepts. *In* Goodwin, **2:** 461–503.

Arnon, D. I., H. T. Tsujimoto and B. D. McSwain. 1965. Photosynthetic phosphorylation and electron transport. Nature **207:** 1367–72.

Arturson, G. 1970. Glomerular permeability to dextrans. *In* C. Crone, and N. A. Lassen (eds.). Capillary permeability. pp. 520–530. Academic Press, New York.

Asahina, E. 1969. Frost resistance in insects. Adv. Insect Physiol **6:** 1–49.

Aschner, B. 1912. Über die Funktion der Hypophyse. Pflüger's Arch. **146:** 1–146.

Asdell, S. A. 1964. Patterns of mammalian reproduction. 2nd ed. Comstock, Ithaca, New York. 670 p.

Asher, J. H., Jr., and G. W. Nace. 1971. The genetic structure and evolutionary fate of parthenogenetic amphibian populations as determined by Markovian analysis. Am. Zool. **11:** 381–398.

Astwood, E. B. 1941. The regulation of corpus luteum function by hypophysial luteotrophin. Endocrinology, **28:** 309–320.

Atkinson, D. E. 1966. Regulation of enzyme activity. Ann. Rev. Biochem. **35:** 85–124.

Atwood, H. L. 1967. Crustacean neuromuscular mechanisms. Am. Zool. **7:** 527–552.

Atwood, H. L. 1968. Peripheral inhibition in crustacean muscle. Experientia. **24:** 753–763.

Auerbach, A. A., and M. L. V. Bennett. 1969. A rectifying electrotonic synapse in the central nervous system of a vertebrate. J. Gen. Physiol. **53:** 211–237.

Austin, C. R. 1965. Fertilization. Prentice-Hall, Englewood Cliffs, N.J. 145 p.

Austin, C. R., and A. Walton. 1960. Fertilisation. *In* Parkes. **1(2):** 310–416.

AWAPARA, J., and J. W. SIMPSON. 1967. Comparative Physiology: Metabolism. Ann. Rev. Physiol. **29:** 87–112.

BAAS-BECKING, L. G. M. 1928. On organisms living in concentrated brine. Tijdschr. der Ned. Dierkundige. Ver. Ser. 3. **1:** 6–9.

BADMAN, D. G. 1971. Nitrogen excretion in two species of pulmonate land snail. Comp. Biochem. Physiol. **38A:** 663–673.

BAERENDS, G. P. 1971. The ethological analysis of fish behavior. *In* Hoar and Randall, **6:** 279–370.

BAGGERMAN, B. 1972. Photoperiodic responses in the stickleback and their control by a daily rhythm of photosensitivity. Gen. Comp. Endocrinol. Suppl. **3:** 466–476.

BAGNARA, J. T., and M. E. HADLEY, 1973. Chromatophores and color change. Prentice-Hall, Englewood Cliffs, N.J. 202 p.

BAHL, K. N. 1947. Excretion in the Oligochaeta. Biol. Rev. **22:** 109–147.

BAIRD, I. L. 1970. The anatomy of the reptilian ear. *In* Gans, **2:** 193–275.

BAKER, P. F. 1966. The nerve axon. Sci. Am. **214(3):** 74–82.

BAKER, P. F., A. L. HODGKIN, and T. I. SHAW. 1962. Replacement of axoplasm of giant nerve fibres with artificial solutions. J. Physiol. **164:** 330–354.

BALDWIN, E. 1964. An introduction to comparative biochemistry. 4th ed. University Press, Cambridge. 179 p.

BALDWIN, E. 1967. Dynamic aspects of biochemistry. 5th ed. University Press, Cambridge. 466 p.

BALINSKY, B. I. 1970a. An introduction to embryology. 3rd ed. Saunders, Philadelphia. 725 p.

BALINSKY, J. B. 1970b. Nitrogen metabolism in amphibians. *In* J. W. Campbell, 519–637.

BALINSKY, J. B., M. M. CRAGG, and E. BALDWIN. 1961. The adaptation of amphibian waste nitrogen excretion to dehydration. Comp. Biochem. Physiol. **3:** 236–244.

BALL, J. N. 1969. Prolactin (fish prolactin or paralactin) and growth hormone. *In* Hoar and Randall, **2:** 207–240.

BANTING, F. G., and C. H. BEST. 1922. The internal secretion of the pancreas. J. Lab. Clin. Med. **7:** 251–266.

BARBER, V. C. 1968. The structure of mollusc statocysts, with particular reference to cephalopods. Symp. Zool. Soc. London. **23:** 37–62.

BARBER, V. C., and P. N. DILLY. 1969. Some aspects of the fine structure of the statocysts of the molluscs *Pecten* and *Pterotrachea*. Z. Zellforsch. **94.** 462–478.

BARCROFT, J. 1934. Features in the architecture of physiological function. University Press, Cambridge. 368 p.

BARDACH, J. E., and J. ATEMA. 1971. The sense of taste in fish. Handbook of Sensory Physiology. (Springer-Verlag). **4(2):** 293–336.

BARLOW, H. B., and T. J. OSTWALD. 1972. Pecten of the pigeon's eye as an inter-ocular eye shade. Nature, New Biol. **236:** 88–90.

BARNES, T. C. 1940. Experiments on Ligia in Bermuda. VII. Further effects of sodium, ammonium and magnesium. Biol. Bull. **78:** 35–41.

BARNETT, A. J. G., and R. L. REID. 1961. Reactions in the rumen. Edward Arnold, London 252 p.

BARNETT, L. B. 1970. Seasonal changes in temperature acclimatization of the house sparrow *Passer domesticus*. Comp. Biochem. Physiol. **33:** 559–578.

BARR, L., and M. ALPERN. 1963. Photosensitivity of the frog iris. J. Gen. Physiol. **46:** 1249–1265.

BARRETT, F. M., and W. G. FRIEND. 1970. Uric acid synthesis in *Rhodnius prolixus*. J. Insect Physiol. **16:** 121–129.

BARRETT, R., P. F. A. MADERSON, and R. M. MESZLER. 1970. The pit organs of snakes. *In* Gans, **2:** 277–300.

BARRINGTON, E. J. W. 1942. Blood sugar and the follicles of Langerhans in the ammocoete larva. J. Exp. Biol. **19:** 45–55.

BARRINGTON, E. J. W. 1957. The alimentary canal and digestion. *In* M. E. Brown, **1:** 109–161.

BARRINGTON, E. J. W. 1959. Some endocrinological aspects of the protochordata. *In* Gorbman (1959): 250–265.

BARRINGTON, E. J. W. 1961. Metamorphic processes in fishes and lampreys. Am. Zool. **1:** 97–106.

BARRINGTON, E. J. W. 1962. Digestive enzymes. Adv. Comp. Physiol. Biochem. **1:** 1–65.

BARRINGTON, E. J. W. 1963. An introduction to general and comparative endocrinology. Clarendon Press, Oxford. 387 p.

BARRINGTON, E. J. W. 1964. Hormones and evolution. English Universities Press, London. 154 p.

BARRINGTON, E. J. W. 1967. Invertebrate structure and function. Nelson, London. 549 p.

BARRINGTON, E. J. W. 1968a. Phylogenetic perspectives in vertebrate evolution. *In* Barrington and Jørgensen, (1968): 1–46.

BARRINGTON, E. J. W. 1968b. Metamorphosis in lower chordates. *In* Etkin and Gilbert (1968): 223–270.

BARRINGTON, E. J. W. 1972. The pancreas and intestine. *In* Hardisty and Potter, **2:** 135–169.

BARRINGTON, E. J. W., and C. B. JØRGENSEN. 1968. Perspectives in endocrinology. Academic Press, London. 583 p.

BARRINGTON, E. J. W., and M. SAGE. 1972. The endostyle and thyroid gland. *In* Hardisty and Potter. **2:** 105–134.

BARRY, J. M. 1964. Molecular biology: Genes and the chemical control of living cells. Prentice Hall, Inc., Englewood Cliffs, N.J. 139 p.

BARTHOLOMEW, G. A. 1964. The roles of physiology and behaviour in the maintenance of homeostasis in the desert environment. Symp. Soc. Exp. Biol. **18:** 7–29.

BARTHOLOMEW, G. A. 1972. Body temperature and energy metabolism. *In* Gordon *et al.* 1972: 198–368.

BARTHOLOMEW, G. A., and R. C. LASIEWSKI. 1965. Heating and cooling rates, heart rate and simulated diving in the Galapagos marine iguana. Comp. Biochem. Physiol. **16:** 573–582.

BASSHAM, J. A. 1962. The path of carbon in photosynthesis. Sci. Am. **206(6):** 88–100.

BASU, S. P. 1959. Active respiration of fish in relation to ambient concentration of oxygen and carbon dioxide. J. Fish. Res. Bd. Canada **16:** 175–212.

BATTLE, H. I. 1926. Effects of extreme temperatures on muscle and nerve tissue in marine fishes. Trans. Roy. Soc. Canada Ser. III. **20:** 127–143.

BATTLE, H. I. 1929. A note on lethal temperature in connection with skate reflexes. Contrib. Can. Biol. and Fish., N.S. **4:** 497–500.

BAYLISS, L. E. 1960. Principles of general physiology. Vol. II. Longmans, Green and Co., London. 848 p.

BAYLISS, W. M. 1920. Principles of general physiology. 3rd ed. Longmans Ltd., London. 862 p.

BAYLISS, W. M., and E. H. STARLING. 1902. The mechanism of pancreatic secretion. J. Physiol. **28:** 325–353.

BEACH, F. A. 1958. Evolutionary aspects of psychoendocrinology. *In* A. Roe, and G. G. Simpson (eds.), Behavior and evolution. pp. 81–102. Yale U. P., New Haven, Conn.

BEADLE, G. W. 1948. The genes of men and molds. Sci. Am. **179. (3):** 30–39.

BEADLE, G. W. and E. L. TATUM. 1941. Genetic control of biochemical reactions in *Neurospora.* Proc. Nat. Acad. Sci. **27:** 499–506.

BEADLE, L. C. 1934. Osmotic regulation in *Gunda ulvae.* J. Exp. Biol. **11:** 382–396.

BEAMENT, J. W. L. 1964. The active transport and passive movement of water in insects. Adv. Insect Physiol. **2:** 67–129.

BEAMENT, J. W. L. 1965. The active transport of water: evidence, models and mechanisms. Symp. Soc. Exp. Biol. **19:** 273–298.

BEAMISH, F. W. H. 1964. Respiration of fishes with special emphasis on standard oxygen consumption. Can J. Zool. **42:** 161–194; 847–856.

BEATON, G. H., and E. W. MCHENRY (eds.) 1964. Nutrition. Vol. 1. Academic Press, New York. 547 p.

BEATTY, D. D. 1969a. Visual pigment changes in juvenile kokanee salmon in response to thyroid hormone. Vision Res. **9:** 855–864.

BEATTY, D. D. 1969b. Visual pigments of the burbot, *Lota lota,* and seasonal changes in their proportions. Vision Res. **9:** 1173–1183.

BECK, S. D. 1960. Insects and the length of the day. Sci. Am. **202 (2):** 109–118.

BECK, S. D. 1963. Animal photoperiodism. Holt, Rinehart and Winston, New York. 124 p.

BECK, S. D. 1968. Insect photoperiodism. Academic Press, New York, 288 p.

BEERMANN, W., and U. CLEVER, 1964. Chromosome puffs. Sci. Am. **210 (4):** 50–58.

BEETS, M. G. J. 1971. Olfactory responses and molecular structure. Handbook of Sensory Physiology (Springer-Verlag), **4 (1):** 257–321.

BEIDLER, L. M. 1970. Physiological properties of mammalian taste receptors. *In* Wolstenholme and Knight, (1970): 51–70.

BÉKÉSY, G. VON. 1956. Current status of theories of hearing. Science. **123:** 779–783.

BÉKÉSY, G. VON. 1957. The ear. Sci. Am. **197 (2):** 66–78.

BĚLEHRÁDEK, J. 1930. Temperature coefficients in biology. Biol. Rev., **5:** 30–58.

BELL, G. E., J. N. DAVIDSON and D. EMSLIE-SMITH. 1972. Textbook of physiology and biochemistry. E. & S. Livingstone, Edinburgh. 1160 p.

BELL, J. B. G., G. P. VINSON, and D. LACY. 1971. Studies of the structure and function of mammalian testis III. *In vitro* steroidogenesis by the seminiferous tubules of rat testis. Proc. Roy. Soc. London B **176:** 433–443.

BENDALL, J. R. 1969. Muscles, molecules and movement. Heinemann, London. 219 p.

BENEDICT, F. G. 1938. Vital energetics. Carnegie Inst. Washington, D.C. 215 p.

BENEDICT, F. G., and R. C. LEE. 1937. Lipogenesis in the animal body, with special reference to the physiology of the goose. Carnegie Inst. Wash. Publ. No. **489:** 1–232.

BENNETT, M. V. L. 1966. Physiology of electrotonic functions. Ann. N.Y. Acad. Sci. **137:** 509–539.

BENNETT, M. V. L. 1970. Comparative physiology: electric organs. Ann. Rev. Physiol. **32:** 471–528.

BENNETT, M. V. L. 1971a. Electric organs. *In* Hoar and Randall (eds.) **5:** 347–491.

BENNETT, M. V. L. 1971b. Electroreception. *In* Hoar and Randall, **5:** 493–574.

BENNETT, M. V. L., C. D. PAPPAS, E. ALJURE, and Y. NAKAJIMA. 1967. Physiology and ultrastructure of electrotonic junctions. II. Spinal and medullary electromotor nuclei in mormyrid fish. J. Neurophysiol. **30:** 180–208. *see also* pp. 209–300.

BENNETT, R., JR., and H. I. NAKADA. 1968. Comparative carbohydrate metabolism of marine molluscs—I. The intermediary metabolism of *Mytilus californianus* and *Haliotus rufescens.* Comp. Biochem. Physiol. **24:** 787–797.

BENNETT, T. P., and E. FRIEDEN. 1962. Metamorphosis and biochemical adaptation in amphibia. *In* Florkin and Mason, **4:** 483–556.

BENNETT, T. P., and E. FRIEDEN. 1966. Modern topics in biochemistry. Macmillan, New York. 186 p.

BENSON, G. K., and J. G. PHILLIPS (eds.). 1970. Hormones and the environment. Mem. Soc. Endocrinol. **18:** 1–629.

BENTLEY, P. J. 1966. The physiology of the urinary bladder of amphibia. Biol. Rev. **41:** 275–316.

BENTLEY, P. J. 1971. Endocrines and osmoregulation. Springer-Verlag, Heidelberg. 300 p.

BERG, T., and J. B. STEEN. 1965. Physiological mechanisms for aerial respiration in the eel. Comp. Biochem. Physiol. **15:** 469–484.

BERG, T., and J. B. STEEN. 1966. Regulation of ventilation in eels exposed to air. Comp. Biochem. Physiol. **18:** 511–516.

BERG, T., and J. B. STEEN. 1968. The mechanism of oxygen concentration in the swim-bladder of the eel. J. Physiol. **195:** 631–638.

BERGEIJK, W. A. VAN. 1966. Evolution of the sense of hearing in vertebrates. Am. Zool. **6:** 371–377.

BERN, H. A. 1966. On the production of hormones by neurons and the role of neurosecretion in neuroendocrine mechanisms. Symp. Soc. Exp. Biol. **20:** 325–344.

BERN, H. A. 1969. Urophysis and caudal neurosecretory system. *In* Hoar and Randall, **2:** 399–418.

BERN, H. A. 1972. Comparative endocrinology—the state of the field and the art. Gen. Comp. Endocrinol. Supp. **3:** 751–761.

BERN, H. A., and C. S. NICOLL. 1968a. The taxonomic specificity of prolactins. *In* C. Gaul (ed.), Progress in endocrinology pp. 433–439. Excerpta Medica Found. Amsterdam.

BERN, H. A., and C. S. NICOLL. 1968b. The comparative endocrinology of prolactin. Recent Prog. Hormone Research, **24:** 681–720.

BERNARD, C. 1957. An introduction to the study of experimental medicine. Translation by H. C. Greene, Dover, New York. 226 p.

BERNFELD, P. 1962. Polysaccharidases. *In* Florkin and Mason, **3:** 355–425.

BERNHARD, C. G., (ed.) 1966. The functional organization of the compound eye. Pergamon, Oxford. 591 p.

BERNSTEIN, J. 1902. Untersuchungen zur Thermodynamik der bioelektrischen Ströme. I. Pflüger's Arch. **92:** 521–562.

BERNSTEIN, J. J. 1970. Anatomy and physiology of the central nervous system. *In* Hoar and Randall, **4:** 1–90.

BEROZA, M., and E. F. KNIPLING. 1972. Gypsy moth control with the sex attractant pheromone. Science. **177:** 19–27.

BERRIDGE, M. J. 1970. Osmoregulation in terrestrial arthropods. *In* Florkin and Scheer, **5:** 287–319.

BERRIDGE, M. J., and J. L. OSCHMAN. 1972. Transporting epithelia. Academic Press, New York. 91 p.

BERRILL, N. J. 1929. Digestion in ascidians and the influence of temperature. J. Exp. Biol. **6:** 275–292.

BERRILL, N. J. 1971. Developmental biology. McGraw-Hill, New York. 535 p.

BERT, P. 1943. Barometric pressure. (La pression barometrique, 1878). Translation by M. A. Hitchcock, College Book Co., Columbus, Ohio. 1055 p.

BEST, C. H. 1959. A Canadian trail of medical research. J. Endocrinol. **19:** i–xvii.

BHOWMICK, D. W. 1967. Electron microscopy of *Trichamoeba villosa* and amoeboid movement. Exp. Cell. Research **45:** 570–89.

BICKOFF, E. M. 1963. Estrogen-like substances in plants. *Proc. 22nd Ann. Biol. Colloq., Oregon State Univ.,* Corvallis, 1961: 93–118.

BIDDER, A. M. 1966. Feeding and digestion in cephalopods. *In* Wilbur and Yonge, **2:** 97–124.

BIGGS, R., and R. G. MACFARLANE. 1962. Human blood coagulation and its disorders. 3rd ed. Blackwell, Oxford. 474 p.

BIRCH, M. P., C. G. CARRE, and G. H. SATCHELL. 1969. Venous return in the trunk of the Port Jackson shark, *Heterodontus portusjacksoni.* J. Zool. Lond., **159:** 31–49.

BIRKLE, D., L. G. TILNEY, and K. R. PORTER. 1966. Microtubules and pigment migration in the melanophores of *Fundulus heteroclitus* L. Protoplasma. **61:** 322–345.

BISHAI, H. M. 1960. The effect of gas content of water on larval and young fish. Z. Wiss. Zool. **163:** 37–64.

BISHOP, S. H., and J. W. CAMPBELL. 1965. Arginine and urea biosynthesis in the earthworm *Lumbricus terrestris.* Comp. Biochem. Physiol. **15:** 51–71.

BITTNER, G. D. 1968. Differentiation of nerve terminals in crayfish opener muscle and its functional significance. J. Gen. Physiol. **51:** 731–758.

BLACK, E. C. 1940. The transport of oxygen by the blood of freshwater fish. Biol. Bull. **79:** 215–229.

BLACK, E. C., F. E. J. FRY, and V. S. BLACK. 1954. The influence of carbon dioxide on the utilization of oxygen by some freshwater fish. Can. J. Zool. **32:** 408–420.

BLACK, E. C., A. C. ROBERTSON, and R. R. PARKER. 1961. Some aspects of carbohydrate metabolism in fish. *In* A. W. Martin (ed.) Comparative physiology of carbohydrate metabolism in heterothermic animals. pp. 89–124. Univ. Washington Press, Seattle.

BLAIR-WEST, J. R., J. P. COGHLAN, D. A. DENTON, J. F. NELSON, E. ORCHARD, B. A. SCOGGINS, R. D. WRIGHT, K. MYERS, and C. L. JUNQUEIRA. 1968. Physiological, morphological and behavioural adaptation to a sodium deficient environment by wild native Australian and introduced species of animals. Nature, **217:** 922–928.

BLAXTER, J. H. S., and F. G. T. HOLLIDAY. 1963. The behaviour and physiology of herring and other clupeids. Adv. Marine Biol., **1:** 261–393.

BLAŽKA, P. 1958. The anaerobic metabolism of fish. Physiol. Zool. **31:** 117–128.

BLES, E. J. 1929. Arcella. A study in cell physiology. Quart. J. Microscop. Sci. **72:** 527–648.

BLISS, C. I. 1967. Statistics in biology. Vol. 1. McGraw-Hill, New York. 558 p.

BLOOM, W., and D. W. FAWCETT. 1968. A textbook of histology. 9th ed. Saunders, Philadelphia. 858 p.

BLOS, M. S., H. F. BLUM, and W. F. LOOMIS. 1968. Vitamin D, sunlight and natural selection. Science. **159:** 652–653.

BLUM, H. F. 1941. Photodynamic action and diseases caused by light. Reinhold, New York. 300 p.

BLUM, H. F. 1945. The physiological effects of sunlight on man. Physiol. Rev. **25:** 483–530.

BLUM, H. F. 1955. Sunburn. *In* Hollaender, **2:** 487–528.

BLUM, H. F. 1961. Does the melanin pigment of human skin have adaptive value? Quart. Rev. Biol. **36:** 50–63.

BLUM, J. J. (ed.). 1970. Biogenic amines as physiological regulators. Prentice-Hall, Englewood Cliffs, N.J. 360 p.

BLUM, M. S. 1969. Alarm pheromones. Ann. Rev. Entomol. **14:** 57–80.

BLUM, M. S., R. E. DOOLITTLE, and M. BEROZA. 1971. Alarm pheromones: utilization in evaluation of olfactory theories. J. Insect Physiol. **17:** 2351–2361.

BODIAN, D. 1962. The generalized vertebrate neuron. Science. **137:** 323–326.

BOETTIGER, E. G. 1957. The machinery of insect flight. *In* Scheer (1957): 117–142.

BOETTIGER, E. G. 1961. Cellular processes in transmission and reception. Bull. Inst. Cellular Biol. Univ. Connecticut (mimeogr.) **3 (7):** 1–8.

BONE, Q. 1958. Synaptic relations in the atrial nervous system of amphioxus. Quart. J. Microscop. Sci., **99:** 243–261.

BONE, Q. 1960. The central nervous system of amphioxus. J. Comp. Neurol. **115:** 27–64.

BONE, Q. 1961. The organization of the atrial nervous system of amphioxus [*Branchiostoma lanceolatum* (Pallas)]. Phil. Trans. Roy. Soc. London. **243B:** 241–269.

BONE, Q. 1966. On the function of the two types of myotonal muscle fibre in elasmobranch fish. J. Mar. Biol. Ass. U.K. **46:** 321–349.

BONE, Q. 1972. Buoyancy and hydrodynamic factors in integument in the castor oil fish, *Ruvettus pretiosus* (Pisces: Gempylidae). Copeia (1972) **1:** 78–87.

BONHAG, P. F. 1958. Ovarian structure and vitellogenesis in insects. Ann. Rev. Entomol. **3:** 137–160.

BONNER, J. T. 1958. The evolution of development. University Press, Cambridge. 103 p.

BOOLOOTIAN, R. A., and A. C. GIESE. 1959. Clotting of echinoderm coelomic fluid. J. Exp. Zool. **140:** 207–229.

BOONE, E., and L. G. M. BAAS-BECKING. 1931. Salt effects on eggs and nauplii of *Artemia salina.* L. J. Gen. Physiol. **14:** 753–763.

BOURNE, G. H. (ed.). 1960. The structure and function of muscle. Vols. 1–3. Academic Press, New York.

BOURNE, G. H. (ed.). 1968–72. The structure and function of nervous tissue. Vols. 1–6. Academic Press, New York.

BOWEN, H. J. M. 1966. Trace elements in biochemistry. Academic Press, London. 241 p.

BOWEN, S. T., H. G. LEBHERZ, M. POON, V. H. S. CHOW, and T. A. GRIGLIATTI. 1969. The hemoglobins of *Artemia salina*—I. Determination of phenotype by genotype and environment. Comp. Biochem. Physiol **31:** 733–747.

BOWERS, W. C. 1969. Juvenile hormone: activity of aromatic terpenoid ethers. Science **164:** 323–325.

BOYD, W. C. 1950. Genetics and the races of man. Little, Brown and Company, Boston. 453 p.

BOYDEN, A. 1942. Systematic serology: a critical appreciation. Physiol. Zool. **15:** 109–145.

BOYDEN, A. 1963. Precipitin testing and classification. Systematic Zool., **12:** 1–7.

BOYDEN, S. V. 1965. Natural antibodies and the immune response. Adv. Immunol. **5:** 1–28.

BOYER, P. D. (ed.). 1970. The enzymes. Vol. 1. 3rd ed. Academic Press, New York. 559 p.

BOYER, S. H. 1963. Papers on human genetics. Prentice-Hall, Englewood Cliffs, N.J. 305 p.

BRACHET, J., and A. E. MIRSKY (eds.) 1961. The cell. Vols. 1–5. Academic Press, New York.

BRAMBELL, F. W. R. 1956. Ovarian changes. *In* Parkes, Vol. **1(1):** 397–542.

BRAND, T. VON. 1946. Anaerobiosis in invertebrates. Biodynamica, Normandy, Missouri. 328 p.

BRAND, T. VON. 1966. Biochemistry of parasites. Academic Press, New York. 429 p.

BRAUN, G., G. KÜMMEL, and J. A. MANGOS. 1966. Studies on the ultrastructure and function of a primitive excretory organ, the protonephridium of the rotifer *Asplanchna priodonta.* Pflüger's Arch. **289:** 144–154.

BRAWN, V. M. 1969. Buoyancy of Atlantic and Pacific herring. J. Fish. Res. Bd. Canada. **26:** 2077–91.

BRAZIER, M. A. B. 1968. The electrical activity of the nervous system. 3rd ed. Pitman Medical Publishing Co., London. 317 p.

BRAZIER, M. A. B. 1969. Electrical activity of the nerve cell. *In* Bourne, **2:** 393–408.

BRESNICK, E., and A. SCHWARTZ. 1968. Functional dynamics of the cell. Academic Press, New York. 482 p.

BRETT, J. R. 1956. Some principles in the thermal requirements of fishes. Quart. Rev. Biol. **31:** 75–87.

BRETT, J. R. 1958. Implications and assessments of environmental stress. *In* P. A. Larkin, (ed.) The investigation of fish-power problems. H. R. MacMillan Lect. in Fish. pp. 69–83. Univ. British Columbia, Vancouver.

BRETT, J. R. 1962. Some considerations in the study of respiratory metabolism in fish, particularly salmon. J. Fish. Res. Bd. Canada. **19:** 1025–1038.

BRETT, J. R. 1963. The energy required for swimming by young sockeye salmon with a comparison of the drag force on a dead fish. Trans. Roy. Soc. Canada, Ser. IV, **1:** 441–457.

BRETT, J. R. 1971. Energetic responses of salmon to temperature. A study of some of the thermal relations in the physiology and freshwater ecology of sockeye salmon *(Oncorhynchus nerka)*. Am. Zool. **11:** 99–113.

BRETT, J. R. 1972. The metabolic demand for oxygen in fish, particularly salmonids, and a comparison with other vertebrates. Respir. Physiol. **14:** 151–170.

BRIGGS, W. R., and H. V. RICE. 1972. Phytochrome: chemical and physical properties and mechanism of action. Ann. Rev. Plant Physiol. **23:** 293–334.

BRITTEN, R. J., and E. H. DAVIDSON. 1969. Gene regulation for higher cells: a theory. Science. **165:** 349–357.

BRODAL, A., and R. FÄNGE. 1963. The biology of *Myxine*. Universitetsforlaget, Oslo, Norway. 588 p.

BRODY, S. 1945. Bioenergetics and growth. Reinhold, New York. 1023 p.

BROWN, F., and W. D. STEIN. 1960. Balance of water, electrolytes, and nonelectrolytes. *In* Florkin and Mason, **2:** 403–470.

BROWN, F. A. Jr. 1957. The rhythmic nature of life. *In* Scheer (1957): 287–304.

BROWN, F. A., JR. 1972. The "clocks" timing biological rhythms. Am. Scient. **60:** 756–766.

BROWN, F. A., JR., J. W. HASTINGS, and J. D. PALMER. 1970. The biological clock: two views. Academic Press, New York. 94 p.

BROWN, J. J., R. FRASER, A. F. LEVER, and J. J. S. ROBERTSON. 1968. Renin and angiotensin in the control of water and electrolyte balance: relation to aldosterone. *In* U. H. T. James (ed.), Recent advances in endocrinology. 8th ed. pp. 271–292. Churchill, London.

BROWN, M. E. (ed.) 1957. The physiology of fishes. Vols. 1-2. Academic Press, New York.

BROWN, P. K., and G. WALD. 1964. Visual pigments in single rods and cones of the human retina. Science. **144:** 45–52.

BÜCHERL, W., E. E. BUCKLEY, and V. DEULOFEU. 1968. Venomous animals and their venoms. I. Venomous Vertebrates. Academic Press, New York. 707 p.

BUCK, J. B. 1948. The anatomy and physiology of the light organ in fireflies. Ann. N.Y. Acad. Sci. **49:** 397–482.

BUCK, J. B., and E. BUCK. 1968. Mechanism of rhythmic synchronous flashing of fireflies. Science. **159:** 1319–1327.

BUCK, J. B., and E. BUCK. 1972. Photic signalling in the firefly *Photinus greeni*. Biol. Bull. **142:** 195–205.

BUCKMAN, M. T., and G. T. PEAKE. 1973. Osmolar control of prolactin secretion in man. Science. **181:** 755–757.

BUDDENBROCK, W. VON, and I. MOLLER-RACKE. 1953. Über den Lichtsinn von *Pecten*. Pubbl. Staz. Zool. Napoli, **24:** 217–245.

BULL, H. O. 1957. Behavior: conditioned responses. *In* M. E. Brown, **2:** 211–228.

BULLER, A. J. 1970. The neural control of the contractile mechanism in skeletal muscle. Endeavour, **29:** 107–111.

BULLOCK, T. H. 1952. The invertebrate neuron junction. Symp. Quant. Biol. **17:** 267–273.

BULLOCK, T. H. 1957. Neuronal integrative mechnisms. *In* Scheer (1957): 1–20.

BULLOCK, T. H. 1958. Evolution of neurophysiological mechanisms. *In* A. Roe and G. G. Simpson (eds.), Behavior and Evolution. pp. 165–177. Yale U. P., New Haven, Conn.

BULLOCK, T. H. 1959a. Initiation of nerve impulses in receptor and central neurons. *In* Oncley (1959): 504–514.

BULLOCK, T. H. 1959b. Neuron doctrine and electrophysiology. Science, **129:** 997–1002.

BULLOCK, T. H. 1961. The origins of patterned nervous discharge. Behaviour, **17:** 48–59.

BULLOCK, T. H., and F. P. J. DIECKE. 1956. Properties of an infra-red receptor. J. Physiol. **134:** 47–87.

BULLOCK, T. H., and G. A. HORRIDGE. 1965. Structure and function of the nervous systems of invertebrates. Vols. 1 and 2. Freeman, San Francisco. 1719 p.

BUNGE, R. P. 1968. Glial cells and the central myelin sheath. Physiol. Rev. **48:** 197–251.

BÜNNING, E. 1967. The physiological clock. 2nd ed. Springer-Verlag, New York. 167 p.

BURGER, J. W. 1962. Further studies on the function of the rectal gland in the spiny dogfish. Physiol. Zool. **35:** 205–217.

BURKHARDT, D. 1962. Spectral sensitivity and other response characteristics of single visual cells in the arthropod eye. Symp. Soc. Exp. Biol. **16:** 86–109.

BURN, J. H., D. J. FINNEY, and L. G. GOODWIN. 1950. Biological standardization. 2nd ed. University Press, Oxford. 440 p.

BURNETT, A. L., and N. A. DIEHL. 1964. The nervous system of *Hydra*. J. Exp. Zool. **157:** 217–250.

BURNSTOCK, G. 1969. Evolution of the autonomic innervation of visceral and cardiovascular systems in vertebrates. Pharm. Rev. **21:** 247–324.

BURNSTOCK, G. 1972. Purinergic nerves. Pharm. Rev. **24:** 509–581.

BURNSTOCK, G., M. E. HOLMAN, and C. L. PROSSER. 1963. Electrophysiology of smooth muscle. Physiol. Rev. **43:** 482–527.

BURSELL, E. 1967. The excretion of nitrogen in insects. Adv. Insect Physiol. **4:** 33–67.

BURTON, A. C., and O. G. EDHOLM. 1955. Man in a cold environment. Edward Arnold, London. 273 p.

BURTON, R. F. 1969. Buffers in the blood of the snail, *Helix pomatia* L. Comp. Biochem. Physiol. **29:** 919–930.

BURTT, E. T., and W. T. CATTON. 1962. Resolving power of the compound eye. Symp. Soc. Exp. Biol. **16:** 72–85.

BURTT, E. T., and W. T. CATTON. 1966. Image formation and sensory transmission in the compound eye. Adv. Insect Physiol. **3:** 1–52.

BURZAWA-GERARD, E., and Y. A. FONTAINE. 1972. The gonadotropins of lower vertebrates. Gen. Comp. Endocrinol. **3:** 715–728.

BUSNEL, R. G. (ed.). 1963. Acoustic behaviour of animals. Elsevier, Amsterdam. 933 p.

BUTLER, P. J., and E. W. TAYLOR. 1973. The effect of hypercapnic hypoxia, accompanied by different levels of lung ventilation, on heart rate in the duck. Resp. Physiol. **19:** 176–187.

BUTLER, W. L., and R. J. DOWNES. 1960. Light and plant development. Sci. Am. **203(6):** 56–63.

BUXTON, P. A. 1923. Animal life in deserts. Arnold, London. 176 p.

CAHN, P. D. (ed.) 1967. Lateral line detectors. Indiana U.P. Bloomington. 496 p.

CALVIN, M. 1962a. Evolution of photosynthetic mechanisms. Persp. Biol. Med. **5:** 147–172.

CALVIN, M. 1962b. The path of carbon in photosynthesis. Science **135:** 879–889.

CALVIN, M. 1967. Chemical evolution. Evol. Biol. **1:** 1–25.

CALVIN, M. 1969. Chemical evolution. Clarendon Press, Oxford. 278 p.

CAMERON, J. N., and T. A. MECKLENBURG. 1973. Aerial gas exchange in the coconut crab, *Birgus latro* with some notes on *Gecarcoidea lalandii*. Resp. Physiol. **19:** 245–261.

CAMERON, J. N., and D. J. RANDALL. 1972. The effect of increased ambient CO_2 on arterial CO_2 tension, CO_2 content and pH in rainbow trout. J. Exp. Biol. **57:** 673–680.

CAMPBELL, G. 1970. Autonomic nervous system. *In* Hoar and Randall, **4:** 109–132.

CAMPBELL, G., and G. BURNSTOCK. 1968. Comparative physiology of gastrointestinal motility. Handbook of Physiology. (Am. Physiol. Soc.). Sect. 6, **4:** 2213–2266.

CAMPBELL, J. W., and K. V. SPEEG. 1969. Ammonia and the biochemical deposition of calcium carbonate. Nature **224:** 725–726.

CAMPBELL, J. W. (ed.). 1970. Comparative biochemistry of nitrogen metabolism. Vols. 1–2. Academic Press, New York. 916 p.

CANNON, W. B. 1929. Organization for physiological homeostasis. Physiol. Rev. **9:** 399–431.

CANNON, W. B. 1939. The wisdom of the body. Norton, New York. 312 p.

CANNON, W. B., and J. E. URIDIL. 1921. Some effects on the denervated heart of stimulating the nerves to the liver. Am. J. Physiol. **58:** 353–364.

CAPRÉOL, S. V., and L. E. SUTHERLAND. 1968. Comparative morphology of juxtaglomerular cells. I. Juxtaglomerular cells in fish. Can. J. Zool. **46:** 249–256.

CARDELL, R. R., JR., S. BADENHAUSEN, and K. R. PORTER. 1967. Intestinal triglyceride absorption in the rat. J. Cell Biol. **34:** 123–155.

CAREY, F. G. 1973. Fishes with warm bodies. Sci. Am. **228(2):** 36–44.

CAREY, F. G., and J. M. TEAL. 1966. Heat conservation in tuna fish muscle. Proc. Nat. Acad. Sci. U.S. **56:** 1464–1469.

CARLISLE, D. B. 1968. Vanadium and other metals in ascidians. Proc. Roy. Soc. B., **171:** 31–41.

CARLISLE, D. B., and F. KNOWLES. 1959. Endocrine control in crustaceans. University Press, Cambridge. 120 p.

CARLSON, A. D. 1969. Neural control of firefly luminescence. Adv. Insect Physiol. **6:** 51–96.

CARPENTER, R. E. 1968. Salt and water metabolism in the marine fish-eating bat, *Pizonyx vivesi*. Comp. Biochem. Physiol. **24:** 951–964.

CARRIKER, M. R., D. VAN ZANDT, and G. CHARLTON. 1967. Gastropod *Urosalpinx:* pH of accessory boring organ while boring. Science **158:** 920–922.

CARTER, G. S. 1931. Aquatic and aerial respiration in animals. Biol. Rev. **6:** 1–35.

CARTER, G. S. 1957. Air breathing. *In* M. E. Brown, **1:** 65–79.

CARTER, G. S. 1961. A general zoology of the invertebrates. 4th ed. Sidgwick and Jackson, London. 421 p.

CARTHY, J. D. 1958. An introduction to the behaviour of invertebrates. Allen & Unwin, London. 380 p.

CASE, J. 1966. Sensory mechanisms. Macmillan, New York. 113 p.

CASTILLA, J. C. 1972. Avoidance behaviour of *Asterias rubens* to extracts of *Mytilus edulis,* solutions of bacteriological peptone, and selected amino acids. Marine Biol. **15:** 236–245.

CASTILLA, J. C., and D. J. CRISP. 1970. Responses of *Asterias rubens* to olfactory stimuli. J. Mar. Biol. Assoc. U.K. **50:** 829–847.

CASTILLO, J. DEL, W. C. DE MELLO, and T. MORALES. 1967. The initiation of action potentials in the somatic musculature of *Ascaris lumbricoides.* J. Exp. Biol. **46:** 263–279.

CATTELL, MCK. 1936. The physiological effects of pressure. Biol. Rev. **11:** 441–476.

CATTON, W. T. 1970. Mechanoreceptor function. Physiol. Rev. **50:** 297–318.

CHAET, A. B. 1967. Gamete releasing and shedding substance in sea-stars. Symp. Zool. Soc. London. **20:** 13–24.

CHAFFEE, R. R. J. and J. C. ROBERTS. 1971. Temperature acclimation in birds and mammals. Ann. Rev. Physiol. **33:** 155–202.

CHAGAS, C., and A. PAES DE CARVALHO (eds.) 1961. Bioelectrogenesis. Elsevier, Amsterdam. 413 p.

CHAN, D. K. O. 1972. Hormonal regulation of calcium balance in teleost fishes. Gen. Comp. Endocrinol. Suppl. **3:** 411–420.

CHAPMAN, G. 1958. The hydrostatic skeleton in the invertebrates. Biol. Rev. **33:** 338–371.

CHAUDHURI, C. R., and I. B. CHATTERJEE, 1969. L-ascorbic acid synthesis in birds: phylogenetic trends. Science **164:** 435–436.

CHEFURKA, W. 1965. Intermediary metabolism of carbohydrates in insects. *In* Rockstein, **2:** 581–667.

CHEN, T. (ed.). 1967–72. Research in protozoology. Vols. 1–4. Pergamon, Oxford.

CHESTER JONES, I. 1957. The adrenal cortex. University Press, Cambridge. 316 p.

CHESTER JONES, I., D. K. O. CHAN, I. W. HENDERSON, and J. N. BALL. 1969. The adrenocortical steroids, adrenocorticotropin, and the corpuscles of Stannius. *In* Hoar and Randall, **2:** 321–376.

CHESTER JONES, I. and J. G. PHILLIPS. 1960. Adrenocorticosteroids in fish. Symp. Zool. Soc. London. **1:** 17–32.

CHEW, R. M. 1961. Water metabolism of desert-inhabiting vertebrates. Biol. Rev. **36:** 1–31.

CHIU, K. W., B. LOFTS, and H. W. TSUI. 1970. The effect of testosterone on the sloughing cycle and epidermal glands of the female gecko, *Gekko gecko* L. Gen. Comp. Endocrinol. **15:** 12–19.

CLARK, A. M., and V. J. CRISTOFALO. 1961. Some effects of oxygen on the insects *Anagasta kuehniella* and *Tenebrio molitor.* Physiol. Zool. **34:** 55–61.

CLARK, E. 1959. Functional hermaphroditism and self-fertilization in a Serranid fish. Science **129:** 215–216.

CLARK, R. B. 1961. The origin and formation of the heteronereis. Biol. Rev. **36:** 199–236.

CLARK, R. B. 1962. The hormonal control of growth and reproduction in polychaetes, and its evolutionary implications. Mem. Soc. Endocrinol. **12:** 323–327.

CLARK, R. B. 1966. The integrative action of a worm's brain. Symp. Soc. Exp. Biol. **20:** 345–379.

CLARK, R. B., M. E. CLARK, and R. J. G. RUSTON. 1962. The endocrinology of regeneration in some errant polychaetes. Mem. Soc. Endocrinol. **12:** 275–286.

CLARKE, F. W. 1924. The data of geochemistry. 5th ed. U.S. Geol. Survey Bull. **770:** 1–841.

CLAYTON, R. K. 1964. Phototaxis in microorganisms. *In* Giese, **2:** 51–77.

CLAYTON, R. K. 1971. Light and living matter. Vol. 2. McGraw-Hill, New York. 243 p.

CLEGG, P. C. 1969. Introduction to mechanisms of hormone action. Heinemann, London. 99 p.

CLEGG, P. C., and A. G. CLEGG. 1969. Hormones, cells and organisms. Heinemann, London. 214 p.

CLEMENT, P. 1968. Ultrastructures d'un rotifère: *Notommata copeus.* I. La cellule-flamme. Hypothèses physiologiques. Z. Zellforsch. **89:** 478–498.

CLEMENTS, A. N. 1963. The physiology of mosquitoes. Pergamon, Oxford. 393 p.

CLONEY, R., and E. FLOREY. 1968. Ultrastructure of cephalopod chromatophore organs. Z. Zellforsch. **89:** 250–280.

CLOUDSLEY-THOMPSON, J. L. 1972. Temperature regulation in desert reptiles. Symp. Zool. Soc. London. **31:** 39–59.

COBB, J. L. S., and M. S. LAVERACK. 1967. Neuromuscular systems in echinoderms. Symp. Zool. Soc. London. **20:** 25–51.

COËRS, C. 1967. Structure and organization of the myoneural junction. Int. Rev. Cytol. **22:** 239–267.

COHEN, A. I. 1963. Vertebrate retinal cells and their organization. Biol. Rev. **38:** 427–459.

COHEN, D. S., J. A. GREENBERG, P. LICHT, H. A. BERN, and R. D. ZIPSER. 1972. Growth and inhibition of metamorphosis in the newt *Taricha torosa* by mammalian hypophysial and placental hormones. Gen. Comp. Endocrinol. **18:** 384–390.

COHEN, M. J., and S. DIJKGRAAF. 1961. Mechanoreception. *In* Waterman, **2:** 65–108.

COHEN, M. J., and A. HESS. 1967. Fine structural differences in "fast" and "slow" muscle fibers of the crab. Am. J. Anat. **121:** 285–304.

COHEN, P. P. 1966. Biochemical aspects of metamorphosis: transition from ammoniotelism to ureotelism. Harvey Lectures **60:** 119–154.

COHEN, P. P., and G. W. BROWN, JR. 1960. Ammonia metabolism and urea biosynthesis. *In* Florkin and Mason, **2:** 161–244.

COLE, K. S. 1968. Membranes, ions and impulses. Univ. California Press, Berkeley. 569 p.

COLES, G. C. 1967. Modified carbohydrate metabolism in the tropical swamp worm *Alma emeni*. Nature **216:** 685–686.

COLLIP, J. B. 1925. The extraction of a parathyroid hormone which will prevent or control parathyroid tetany and which regulates the level of blood calcium. J. Biol. Chem. **63:** 395–438.

COMBES, B. 1964. Excretory function of the liver. In Rouiller, Ch. (ed.) The Liver: Vol. **2:** 1–35. Academic Press, New York.

COMLINE, R. S., I. A. SILVER, and D. H. STEVEN. 1968. Physiological anatomy of the ruminant stomach. Handbook of Physiology (Am. Physiol. Soc.) Sect. 6, **5:** 2647–2671.

COMROE, J. H., JR. 1964. The peripheral chemoreceptors. Handbook of Physiology. (Am. Physiol. Soc.) Sect. 3, **1:** 557–583.

CONNER, R. L. 1967. Transport phenomena in Protozoa. *In* Florkin and Scheer, **1:** 309–350.

CONTE, F. P. 1969. Salt secretion. *In* Hoar and Randall, **1:** 241–292.

CONTE, F. P., S. R. HOOTMAN, and P. J. HARRIS. 1972. Neck organ of *Artemia salina* nauplii. J. Comp. Physiol. **80:** 239–246.

COOK, W. F. 1968. The detection of renin in juxtaglomerular cells. J. Physiol. **194:** 73*P*–74*P*.

COOPER, E. L. 1969. Specific tissue graft rejection in earthworms. Science **166:** 1414–1415.

COPELAND, D. E. 1968. Fine structures of the carbon monoxide secreting tissue in the float of Portuguese man-of-war (*Physalia physalis* L.) Biol. Bull. **135:** 486–500.

COPP, D. H. 1969a. The ultimobranchial glands and calcium regulation. *In* Hoar and Randall, **2:** 377–398.

COPP, D. H. 1969b. Calcitonin and parathyroid hormone. Ann. Rev. Pharmacol. **9:** 327–344.

COPP, D. H. 1969c. Endocrine control of calcium homeostasis. J. Endocrinol. **43:** 137–161.

CORMIER, M. J., and J. R. TOTTER. 1968. Bioluminescence: enzymic aspects. *In* Giese, **4:** 315–353.

CORNER, E. D. S., E. J. DENTON, and G. R. FORSTER. 1969. On the buoyancy of some deep-sea sharks. Proc. Roy. Soc. B, **171:** 415–429.

CORRIGAN, J. J. 1969. D-Amino acids in animals. Science **164:** 142–149.

CORTELYOU, J. R. (ed.) 1967. Comparative aspects of parathyroid function. Am. Zool. **7:** 822–895.

COSTLOW, J. D. JR. 1968. Metamorphosis in crustaceans. *In* Etkin and Gilbert (1968): 3–41.

COUTEAU, J., and J. H. CORRIOL. 1971. Physiological aspects of deep sea diving. Endeavour **30:** 70–76.

COWEY, C. B., J. ADRON, and A. BLAIR. 1970. Studies on the nutrition of marine flatfish. The essential amino acid requirements of plaice and sole. J. Mar. Biol. Assoc. U.K. **50:** 87–95.

COWLES, R. B. 1958. Possible origin of dermal temperature regulation. Evolution, **12:** 347–357.

COWLES, R. B. 1959. Some ecological factors bearing on the origin and evolution of pigment in the human skin. Am. Naturalist **93:** 283–293.

COWLES, R. B., W. J. HAMILTON III, and F. HEPPNER. 1967. Black pigmentation: Adaptation for concealment or heat conservation. Science **158:** 1340–41.

CRAIG, R. 1960. The physiology of excretion in the insect. Ann. Rev. Entomol. **5:** 53–68.

CRANE, F. L. and H. LOW. 1966. Quinones in energy-coupling systems. Physiol. Rev. **46:** 662–695.

CRANE, R. K. 1960. Intestinal absorption of sugars. Physiol. Rev. **40:** 789–825.

CRANE, R. K. 1968. Absorption of sugars. Handbook of Physiology. (Am. Physiol. Soc.) Sect. 6, **3:** 1323–1351.

CRANEFIELD, P. F. 1965. The atrioventricular node and the ventricular conducting system in the nonmammalian vertebrate heart. Ann. N.Y. Acad Sci. **127:** 145–150.

CREUTZBERG, F. 1961. On the orientation of migrating elvers *(Anguilla anguilla* Turt.*)* in a tidal area. Netherl. J. Sea Res. **1:** 257–338.

CRICK, F. H. C. 1962. The genetic code. Sci. Am. **207 (4):** 66–74.

CROGHAN, P. C. 1958a. The osmotic and ionic regulation of *Artemia salina* (L.). J. Exp. Biol. **35:** 219–233.

CROGHAN, P. C. 1958b. The mechanism of osmotic regulation in *Artemia salina* (L.). J. Exp. Biol. **35:** 234–242 and 243–249.

CROGHAN, P. C. 1958c. Ionic fluxes in *Artemia salina* (L.). J. Exp. Biol. **35:** 425–436.

CROZIER, W. J. 1924–25. On biological oxidations

as function of temperature. J. Gen. Physiol. **7:** 189–216 also **7:** 571–579.

CSAPO, A. 1960. Molecular structure and function of smooth muscle. *In* Bourne, **1:** 229–264.

CUELLAR, O. 1971. Reproduction and the mechanism of meiotic restitution in the parthenogenetic lizard. *Cnemidophorus uniparens.* J. Morph. **133:** 139–166.

CURRAN, P. F., and S. G. SCHULTZ. 1968. Transport across membranes: general principles. Handbook of Physiology. (Am. Physiol. Soc.) Sect. 6, **3:** 1217–1243.

CURTIS, H. 1968. The marvelous animals. Natural History Press, Garden City, New York. 189 p.

CUSHING, J. E. 1970. Immunology of fish. *In* Hoar and Randall, **4:** 465–500.

CUTHBERT, A. W. (ed.) 1970. A symposium on calcium and cellular function. Macmillan, London. 301 p.

DADD, R. H. 1966. Beeswax in the nutrition of the wax moth, *Galleria mellonella* (L.). J. Insect Physiol. **12:** 1479–92.

DALES, R. P. 1969. Respiration and energy metabolism in annelids. *In* Florkin and Scheer, **4:** 93–109.

D'ANCONA, U. 1949. Ermafroditismo ed intersessualità nei Teleostei. Experientia, **5:** 381–389.

D'AOUST, B. G. 1970. The role of lactic acid in gas secretion in the teleost swimbladder. Comp. Biochem. Physiol. **32:** 637–668.

DARTNALL, H. J. A., and K. TANSLEY. 1963. Physiology of vision: retinal structure and visual pigments. Ann. Rev. Physiol. **25:** 433–458.

DARWIN, C. 1877. The effects of cross and self fertilization in the vegetable kingdom. Appleton, New York. 482 p.

DAVENPORT, H. W. 1958. The ABC of acid-base chemistry. 4th ed. Univ. Chicago Press, Chicago. 86 p.

DAVEY, K. G. 1965. Reproduction in the insects. Freeman, San Francisco. 96 p.

DAVIES, F., and E. T. B. FRANCIS. 1946. The conducting system of the vertebrate heart. Biol. Rev. **21:** 173–188.

DAVIES, J. T. 1971. Olfactory theories. Handbook of Sensory Physiology (Springer-Verlag) **4(1):** 322–350.

DAVIES, R. E. 1963. A molecular theory of muscle contraction: calcium-dependent contractions with hydrogen bond formation plus ATP-dependent extensions of part of the myosin-actin cross-bridges. Nature **199:** 1068–1074.

DAVIS, B. D. 1961a. The teleonomic significance of biosynthetic control mechanisms. Cold Spring Harbor Symp. Quant. Biol. **26:** 1–10.

DAVIS, D. D. 1961b. Origin of the mammalian feeding mechanism. Am. Zool. **1:** 229–234.

DAVIS, H. 1965. A model for transducer action in the cochlea. Cold Spring Harbor Symp. Quant. Biol. **30:** 181–190.

DAVSON, H. 1970. A textbook of general physiology. Vols. 1–2. 4th ed. Churchill, London. 1694 p.

DAVSON, H. 1972. The physiology of the eye. 3rd ed. Churchill Livingstone, Edinburgh and London. 643 p.

DAVSON, H. and M. G. EGGLETON (eds.) 1968. Principles of human physiology. 14th ed. Churchill, London. 1668 p.

DAWKINS, M. J. R., and D. HALL. 1965. The production of heat by fat. Sci. Am. **213 (2):** 62–67.

DAWSON, A. B. 1951. Functional and degenerate or rudimentary glomeruli in the kidney of two species of Australian frog. Anat. Record **109:** 417–429.

DE BEER, G. 1958. Embryos and ancestors. 3rd ed. Clarendon Press, Oxford. 197 p.

DE BRUYN, P. P. H. 1947. Theories of amoeboid movement. Quart. Rev. Biol. **22:** 1–24.

DEHNEL, P. A., and T. H. CAREFOOT. 1965. Ion regulation in two species of intertidal crabs. Comp. Biochem. Physiol. **15:** 377–397.

DEJOURS, P., J. ARMAND, and G. VERRIEST. 1968. Carbon dioxide dissociation curves of water and gas exchange in water-breathers. Resp. Physiol. **5:** 23–33.

DEMSKI, L. S. 1973. Feeding and aggressive behavior evoked by hypothalamic stimulation in a cichlid fish. Comp. Biochem. Physiol. **44A:** 685–692.

DENISON, R. H. 1956. A review of the habitat of the earliest vertebrates. Fieldiana: Geol. (Chicago Nat. Hist. Mus.) **11:** 359–457.

DENT, J. N. 1968. Survey of amphibian metamorphosis. *In* Etkin and Gilbert (1968): 271–311.

DENTON, D. A. 1965. Evolutionary aspects of the emergence of aldosterone secretion and salt appetite. Physiol Rev. **45:** 245–295.

DENTON, E. J. 1961. The buoyancy of fish and cephalopods. Progr. Biophys. **11:** 177–234.

DENTON, E. J. 1963. Buoyancy mechanisms of sea creatures. Endeavour **22:** 3–8.

DENTON, E. J. 1970. On the organization of reflecting surfaces in some marine animals. Phil. Trans. Roy. Soc. London B **258:** 285–313.

DENTON, E. J. 1971. Reflectors in fishes. Sci. Am. **224(1):** 64–72.

DENTON, E. J., and J. B. GILPIN-BROWN. 1966. On the buoyancy of the pearly nautilus. J. Mar. Biol. Ass. U.K. **46:** 723–759.

DENTON, E. J., and J. B. GILPIN-BROWN. 1971. Further observations on the buoyancy of *Spirula.* J. Mar. Biol. Ass. U.K. **51:** 363–373.

DE ROBERTIS, E. 1959. Submicroscopic morphology of the synapse. Int. Rev. Cytol. **8:** 61–96.

De Robertis, E. 1971: Molecular biology of synaptic receptors. Science **171:** 963–971.

De Reuck, A. V. S., and J. Knight (eds.). 1965. Color vision. Ciba Foundation Symp. Churchill, London. 382 p.

De Reuck, A. V. S., and J. Knight (eds.). 1966. Touch, heat and pain. Ciba Foundation Symp. Churchill, London. 389 p.

Dethier, V. G. 1962. Chemoreceptor mechanisms in insects. Symp. Soc. Exp. Biol. **16:** 180–196.

Dethier, V. G. 1963. The physiology of insect senses. Methuen, London. 266 p.

Deuchar, E. M. 1966. Biochemical aspects of amphibian development. Methuen, London. 206 p.

DeVries, A. L. 1971. Freezing resistance in fishes. *In* Hoar and Randall, **6:** 157–190.

Dewey, V. C. 1967. Lipid composition, nutrition, and metabolism. *In* Florkin and Scheer, **1:** 161–274.

Diamond, I. T., and W. C. Hall. 1969. Evolution of neocortex. Science **164:** 251–262.

Diamond, J. 1970. The Mauthner cell. *In* Hoar and Randall, **5:** 265–346.

Diamond, J. M., and J. M. Tormey. 1966a. Role of long extracellular channels in fluid transport across epithelia. Nature **210:** 817–820.

Diamond, J. M., and J. M. Tormey. 1966b. Studies on the structural basis of water transport across epithelial membranes. Fed. Proc. **25:** 1458–1463.

Dijkgraaf, S. 1963. The functioning and significance of the lateral-line organs. Biol. Rev. **38:** 51–105.

Dijkgraaf, S. 1967. Biological significance of the lateral line organs. *In* Cahn (1967): 83–95.

Dijkgraaf, S. 1968. Electroreception in the catfish, *Ameiurus nebulosus.* Experientia, **24:** 187–188.

Dill, P. A. 1971. Perception of polarized light by yearling sockeye salmon *(Oncorhynchus nerka).* J. Fish. Res. Bd. Canada **28:** 1319–1322.

Dingle, H. 1972. Migration strategies of insects. Science **175:** 1327–1335.

Dixon, G. H. 1966. Mechanisms of protein evolution. Essays in Biochem. **2:** 147–204.

Dobzhansky, T. 1955. Evolution, genetics and man. Wiley, New York. 398 p.

Dobzhansky, T. 1962. Mankind evolving. Yale Univ. Press, New Haven. 381 p.

Dodd, J. M. 1972. The endocrine regulation of gametogenesis and gonad maturation in fishes. Gen. Comp. Endocrinol. Suppl. **3:** 675–687.

Dodd, J. M., P. J. Evennett, and C. K. Goddard. 1960. Reproductive endocrinology in cyclostomes and elasmobranchs. Symp. Zool. Soc. London, **1:** 77–103.

Dogiel, V. A. (revised by J. I. Poljanskij and E. M. Chejsin). 1965. General protozoology. 2nd ed. Clarendon Press, Oxford. 747 p.

Donaldson, E. M., and J. R. McBride. 1967. The effects of hypophysectomy in the rainbow trout *Salmo gairdnerii* (Rich.) with special reference to the pituitary-interrenal axis. Gen. Comp. Endocrinol. **9:** 93–101.

Donnan, F. G. 1927. Concerning the applicability of thermodynamics to the phenomena of life. J. Gen. Physiol. **8:** 685–688.

Doolittle, R. F., and D. M. Surgenor. 1962. Blood coagulation in fish. Am. J. Physiol. **203:** 964–970.

Dowben, R. M. 1969a. General physiology. Harper and Row, New York. 619 p.

Dowben, R. M. (ed.). 1969b. Biological membranes. Churchill, London. 303 p.

Dowling, J. E. 1968. Synaptic organization of the frog retina: an electron microscopic analysis comparing the retinas of frogs and primates. Proc. Roy. Soc. London B **170:** 205–228.

Dowling, J. E., and B. B. Boycott, 1966. Organization of the primate retina: electron microscopy. Proc. Roy. Soc. London B **166:** 80–111.

Drew, C. E. 1961. Profound hypothermia in cardiac surgery. Brit. Med. Bull. **17:** 37–42.

Drummond, G. I. 1967. Muscle metabolism. Fortschritte Zool. **18:** 359–429.

Drummond, J. C., and A. Wilbraham. 1939. The Englishman's food. Jonathan Cape, London. 574 p.

Dubois, R. 1885. Fonction photogénique des pyrophores. C. R. Soc. Biol. Paris. **37:** 559–562.

Dubois, R. 1887. Note sur la fonction photogénique chez les Pholades. C. R. Soc. Biol. Paris **39:** 564–566.

Ducrocq, A. 1957. The origins of life. (Translated by A. Brown). Elek Books, London. 213 p.

Duerr, F. G. 1968. Excretion of ammonia and urea in seven species of marine prosobranch snails. Comp. Biochem. Physiol. **26:** 1051–1059.

Dunning, D. C. 1968. Warning sounds of moths. Z. Tierpsychol. **25:** 129–138.

Dunson, W. A., and A. M. Taub. 1967 Extrarenal salt excretion in sea snakes *(Laticauda).* Am. J. Physiol. **213:** 975–982.

de Duve, C., and R. Wattiaux. 1966. Functions of lysosomes. Ann. Rev. Physiol. **28:** 435–492.

Eakin, R. E. 1963. An approach to the evolution of metabolism. Proc. Nat. Acad. Sci. USA. **49:** 360–366.

Eakin, R. M. 1965. Evolution of photoreceptors. Cold Spring Harbor Symp. Quant. Biol. **30:** 363–370.

Eakin, R. M. 1968. Evolution of photoreceptors. Evol. Biol. **2:** 194–242.

Eaton, J. E. Jr., and E. Frieden. 1969. Primary

mechanism of thyroid hormone control of amphibian metamorphosis. Gen. Comp. Endocrinol. Suppl. **2:** 398–407.

EBASHI, S., and M. ENDO, 1968. Calcium ion and muscle contraction. Progr. Biophys. **18:** 123–183.

ECCLES, J. C. 1957. The physiology of nerve cells. The Johns Hopkins Press, Baltimore. 270 p.

ECCLES, J. C. 1964. The physiology of synapses. Springer-Verlag, Heidelberg. 316 p.

ECCLES, J. C. 1965. The synapse. Sci. Am. **212 (1):** 56–66.

ECCLES, J. C. 1968. Postsynaptic inhibition in the central nervous system. *In* von Euler *et al.,* (1968): 291–308.

ECK, R. V., and M. O. DAYHOFF. 1966. Evolution of the structure of ferredoxin based on living relics of primitive amino acid sequences. Science **152:** 363–366.

EDNEY, E. B. 1954. Woodlice and the land habitat. Biol. Rev. **29:** 185–219.

EDNEY, E. B. 1957. The water relations of terrestrial arthropods. University Press, Cambridge. 109 p.

EDWARDS, J. G. 1928–35. Studies on aglomerular and glomerular kidneys. Am. J. Anat. **42:** 75–108; also Am. J. Physiol. **86:** 383–398.

EFFORD, I. E., and K. TSUMURA. 1973. Uptake of dissolved glucose and glycine by *Pisidium,* a freshwater bivalve. Can J. Zool. **51:** 825–832.

EIBL-EIBESFELDT, I. 1970. Ethology—the biology of behavior. Holt, Rinehart and Winston, New York. 530 p.

EISENSTEIN, E. M. 1972. Learning and memory in isolated insect ganglia. Adv. Insect Physiol. **9:** 111–181.

ELLIOTT, G. F. 1967. Variations of the contractile apparatus in smooth and striated muscle. *In* Stracher (1967): 171–184.

ELLIOTT, K. A. C. 1970. Binding and metabolism of gamma-aminobutyric acid and other physiologically active amino acids in the brain. *In* Blum (1970): 253–273.

ELLIOTT, T. R. 1905. The action of adrenalin. J. Physiol. **32:** 401–467.

ELLIS, W. G. 1937. The water and electrolyte exchange of *Nereis diversicolor* (Müller). J. Exp. Biol. **14:** 340–350.

ELLS, H. A. 1969. Physiology of the vinegar eel, *Turbatrix aceti* (Nematoda). II. Enzymes of the glycolytic sequence. Comp. Biochem. Physiol. **29:** 689–701.

ELLS, H. A., and C. P. READ. 1961. Physiology of the vinegar eel, *Turbatrix aceti* (Nematoda). I. Observations on respiratory metabolism. Biol. Bull. **120:** 326–336.

ELSNER, R. 1969. Cardiovascular adjustments to diving. *In* H. T. Andersen (ed.). The biology of marine mammals. pp. 117–145. Academic Press, New York.

ELYAKOVA, L. A. 1972. Distribution of cellulases and chitinases in marine invertebrates. Comp. Biochem. Physiol. **43B:** 67–70.

ENDEAN, R. 1972. Aspects of molluscan pharmacology. *In* Florkin and Scheer, **7:** 421–466.

ENDERS, A. C. (ed.). 1963. Delayed implantation. Chicago U.P. Chicago. 318 p.

ENGLE, R. L., and K. R. WOODS. 1960. Comparative biochemistry and embryology. *In* F. W. Putnam, (ed.) The Plasma Proteins. Vol. 2. pp. 183–265. Academic Press, New York.

ENGLEMANN, F. 1970, The physiology of insect reproduction. Pergamon Press, Oxford. 307 p.

ENRIGHT, J. T., and W. M. HAMNER. 1967. Vertical diurnal migration and endogenous rhythmicity. Science **157:** 937–941.

EPPLE, A. 1969. The endocrine pancreas. *In* Hoar and Randall, **2:** 275–319.

ETKIN, W. 1941. On the control of growth and activity of the pars intermedia of the pituitary by the hypothalamus in the tadpole. J. Exp. Zool. **86:** 113–139.

ETKIN, W. 1962. Hypothalamic inhibition of the pars intermedia activity in the frog. Gen. Comp. Endocrinol. Suppl. **1:** 148–159.

ETKIN, W., and L. I. GILBERT (eds.). 1968. Metamorphosis. Appleton-Century-Crofts, New York. 459 p.

EULER, U. S. VON. 1946. A specific sympathomimetic ergone in adrenergic nerve fibres (sympathin) and its relation to adrenaline and noradrenaline. Acta Physiol. Scand. **12:** 73–97.

EULER, U. S. VON, S. SKOGLUND, and U. SÖDERBERG. (eds.). 1968. Structure and function of inhibitory neuronal mechanisms. Pergamon, Oxford. 563 p.

EXNER, S. 1891. Die Physiologie der facettirten Augen von Krebsen und Insekten. F. Deuticke, Leipzig. 206 p.

EYZAGUIRRE, C. 1969. Physiology of the nervous system. Year Book Medical Pub., Chicago. 216 p.

EYZAGUIRRE, C., and H. KOYANO. 1965. Origin of sensory discharges in carotid body receptors. Cold Spring Harbor Symp. Quant. Biol. **30:** 227–231.

FABRICIUS, E. 1950. Heterogeneous stimulus summation in the release of spawning activities in fish. Rept. Inst. Freshwater Res. Drottningholm **31:** 57–99.

FAGERLUND, U. H. M. 1967. Plasma cortisol concentration in relation to stress in adult sockeye salmon during the fresh water stage of their life. Gen. Comp. Physiol. **8:** 197–207.

FAHRENBACK, W. H. 1967. Fine structure of fast

and slow crustacean muscles. J. Cell. Biol. **35:** 69–80.

FAIN, G. L., and J. E. DOWLING. 1973. Intracellular recordings from single rods and cones in the mudpuppy retina. Science **180:** 1178–1181.

FAIRBAIRN, D. 1970. Biochemical adaptation and loss of genetic capacity in helminth parasites. Biol. Rev. **45:** 29–72.

FALKMER, S. 1972. Insulin production in vertebrates and invertebrates. Gen. Comp. Endocrinol. Suppl. **3:** 184–191.

FALKMER, S., and G. J. PATENT. 1972. Comparative and embryological aspects of the pancreatic islets. Handbook of Physiology. (Am. Physiol. Soc.) Sect. 7, **1:** 1–23.

FÄNGE, R. 1953. The mechanisms of gas transport in the euphysoclist swimbladder. Acta Physiol. Scand. **30:** Suppl. **110:** 1–133.

FÄNGE, R. 1962. Pharmacology of poikilothermic vertebrates and invertebrates. Pharm. Rev. **14:** 281–316.

FÄNGE, R. 1966. Physiology of the swimbladder. Physiol. Rev. **46:** 299–322.

FÄNGE, R., and K. FUGELLI, 1962. Osmoregulation in Chimaeroid fishes. Nature **196:** 689.

FARMER, G. J., and F. W. H. BEAMISH. 1969. Oxygen consumption of *Tilapia nilotica* in relation to swimming speed and salinity. J. Fish. Res. Bd. Canada **26:** 2807–2821.

FARNER, D. S., and J. R. KING (eds.). 1971–73. Avian biology. Vols. 1–3. Academic Press, New York.

FARNER, D. S., and R. A. LEWIS. 1971. Photoperiodism and reproductive cycles in birds. *In* Giese, **6:** 325–370.

FEIGL, E., and B. FOLKOW. 1963. Cardiovascular responses in "diving" and during brain stimulation in ducks. Acta Physiol. Scand. **57:** 90–110.

FENWICK, J. C. 1970. The pineal organ. *In* Hoar and Randall, **4:** 91–108.

FERGUSON, D. R. 1969. The genetic distribution of vasopressins in the peccary *(Tayassu angulatus)* and the warthog *(Phacochoerus aethiopicus)*. Gen. Comp. Endocrinol. **12:** 609–613.

FERGUSON, J. C. 1969. Feeding, digestion, and nutrition in Echinodermata. *In* Florkin and Scheer, **3:** 71–100.

FERLUND, P., and L. JOSEFSSON. 1972. Crustacean color-change hormone: amino acid sequence and chemical synthesis. Science **177:** 173–175.

FICK, A. 1872. Ueber die messung des Blutquantums in den Herzventrikeln. Sitzungsb. d. physik.-med. ges. sch. z. Würzburg. p. XVI, 1870, Verhandl. Physikal.-Medicin. Gesellsch. Würzburg. Neue Folge. **2:** XVI.

FINGERMAN, M. 1959. The physiology of chromatophores. Int. Rev. Cytol. **8:** 175–210.

FINGERMAN, M. 1963. The control of chromatophores. Pergamon Press, Oxford. 184 p.

FINGERMAN, M. 1969. Cellular aspects of the control of physiological color changes in crustaceans. Am. Zool. **9:** 443–452.

FINGERMAN, M. 1970. Comparative physiology: chromatophores. Ann. Rev. Physiol. **32:** 345–372.

FINNEY, D. J. 1964. Statistical method in biological assay. 2nd ed. Charles Griffin & Co. Ltd. London. 668 p.

FISCHMAN, D. A. 1967. An electron microscope study of myofibril formation in embryonic chick skeletal muscle. J. Cell. Biol. **32:** 557–575.

FISHER, A. E. 1964. Chemical stimulation of the brain. Sci. Am. **210 (6):** 60–68.

FISHER, H. D. 1963. Delayed implantation. *In* Enders (1963): 127–128.

FISHER, J. W. 1972. Erythropoietin: pharmacology, biogenesis and control of production. Pharm. Rev. **24:** 459–508.

FISHER, K. C. 1958. An approach to the organ and cellular physiology of adaptation to temperature in fish and small mammals. *In* Prosser (1958): 3–49.

FISHER, K. C. (ed.). 1967. International symposium on natural mammalian hibernation. Oliver & Boyd, London. 535 p.

FLETCHER, C. R. 1970. The metabolism of iodine by a polychaete. Comp. Biochem. Physiol **35:** 105–123.

FLOCK, Å. 1965. Transducing mechanisms in the lateral line canal organ receptors. Cold Spring Harbor Symp. Quant. Biol. **30:** 133–145.

FLOCK, Å. 1971. The lateral line organ mechanoreceptors. *In* Hoar and Randall, **5:** 241–263.

FLOOD, P. R. 1968. Structure of the segmental trunk muscle in *Amphioxus*. Z. Zellforsch. **84:** 389–416.

FLORES, G., and E. FRIEDEN. 1968. Induction and survival of hemoglobin-less and erythrocyteless tadpoles and young bullfrogs. Science **159:** 101–103.

FLOREY, E. 1966. An introduction to general and comparative physiology. Saunders, Philadelphia. 713 p.

FLOREY, E. 1967. Neurotransmitters and modulators in the animal kingdom. Fed. Proc. **26:** 1164–1178.

FLOREY, E. 1969. Ultrastructure and function of cephalopod chromatophores. Am. Zool. **9:** 429–442.

FLOREY, E., and M. E. KRIEBEL. 1969. Electrical and mechanical responses of chromatophore muscle fibers of the squid, *Loligo opalescens*, to nerve stimulation and drugs. Z. vergl. Physiol. **65:** 98–130.

FLOREY, E., and B. WOODCOCK. 1968. Presynaptic excitatory action of glutamate applied to crab nerve-muscle preparations. Comp. Biochem. Physiol. **26:** 651–661.

FLORKIN, M. 1934. La fonction respiratoire du "milieu intérieur" dans la série animale. Ann. Physiol. Physicochim. Biol. **10:** 599–684.

FLORKIN, M. 1960. Blood chemistry. *In* Waterman, **1:** 141–159.

FLORKIN, M. 1966. Nitrogen metabolism. *In* Wilbur and Yonge, **2:** 309–351.

FLORKIN, M. 1969a. Respiratory proteins and oxygen transport. *In* Florkin and Scheer, **4:** 111–134.

FLORKIN, M. 1969b. Nitrogen metabolism. *In* Florkin and Scheer, **4:** 147–162.

FLORKIN, M., and H. S. MASON. 1960–1964. Comparative biochemistry. Vols. 1–6. Academic Press, New York.

FLORKIN, M., and B. T. SCHEER. 1967–1972. Chemical zoology. Vols. 1–7. Academic Press, New York.

FOGG, G. E. 1972. Photosynthesis. 2nd ed. English Univ. Press, London. 116 p.

FOLK, G. E. 1966. Introduction to environmental physiology. Lea & Febiger, Philadelphia. 308 p.

FONTAINE, A. R. 1965. The feeding mechanisms of the ophiuroid *Ophiocomina nigra.* J. Mar. Biol. Assoc. U.K. **45:** 373–385.

FORREST, H. S. 1962. Pteridines: Structure and metabolism. *In* Florkin and Mason, **4:** 615–641.

FORSTER, R. P. 1948. Use of thin kidney slices and isolated renal tubules for direct study of cellular transport kinetics. Science **108:** 65–67.

FORSTER, R. P. 1961. Kidney cells. *In* Brachet and Mirsky, **5:** 89–161.

FORSTER, R. P., and L. GOLDSTEIN. 1969. Formation of excretory products. *In* Hoar and Randall, **1:** 313–350.

FOX, D. L. 1966. Pigmentation of molluscs. *In* Wilbur and Yonge, **2:** 249–274.

FOX, H. M. 1955. The effect of oxygen on the concentration of haem in invertebrates. Proc. Roy. Soc. London B, **143:** 203–214.

FOX, H. M., and E. A. PHEAR, 1953. Factors influencing haemoglobin synthesis by *Daphnia.* Proc. Roy. Soc. London B. **141:** 179–189.

FOX, H. M., and G. VEVERS. 1960. The nature of animal colours. Sidgwick and Jackson, London. 246 p.

FOX, S. W. 1960. How did life begin? Science **132:** 200–208.

FOX, S. W. (ed.) 1965. The origins of prebiological systems. Academic Press, New York. 482 p.

FOXON, G. E. H. 1955. Problems of the double circulation in vertebrates. Biol. Rev. **30:** 196–228.

FOXON, G. E. H. 1964. Blood and respiration. *In* Moore, J. A. (ed.). Physiology of the amphibia. pp. 151–209. Academic Press, New York.

FRAENKEL, G. S., and D. L. GUNN. 1940. The orientation of animals. Clarendon Press, Oxford. 352 p.

FRAZER, A. C. 1946. Absorption of triglyceride fat from the intestine. Physiol. Rev. **26:** 103–119.

FREE, J. B., and J. SIMPSON. 1968. The alerting pheromones of the honeybee. Z. vergl. Physiol. **61:** 361–365.

FRIDBERG, G., and H. A. BERN. 1968. The urophysis and the caudal neurosecretory system of fishes. Biol. Rev. **43:** 175–199.

FRIEDEN, E. 1961. Biochemical adaptation and anuran metamorphosis. Am. Zool. **1:** 115–149.

FRIEDEN, E. 1963. The chemistry of amphibian metamorphosis. Sci. Am. **209 (5):** 110–118.

FRIEDEN, E. 1968. Biochemistry of amphibian metamorphosis. *In* Etkin and Gilbert (1968): 349–398.

FRIEDEN, E. 1972. The chemical elements of life. Sci. Am. **227 (1):** 52–60.

FRIEDEN, E., and H. LIPNER. 1971. Biochemical endocrinology of the vertebrates. Prentice-Hall, Englewood Cliffs, N.J. 164 p.

FRIEDMAN, S. 1967. The control of trehalose synthesis in the blowfly, *Phormia regina* Meig. J. Insect Physiol. **13:** 397–405.

FRIEDMANN, H., and J. Kern. 1956. The problem of cerophagy or wax-eating in the honey-guides. Quart. Rev. Biol. **31:** 19–30.

FRISCH, K. VON. 1911. Beiträge zur Physiologie der Pigmentzellen in der Fischhaut. Pflüger's Arch. ges. Physiol. **138:** 318–387.

FRISCH, K. VON. 1950. Die Sonne als Kompass in Leben der Bienen. Experientia, **6:** 210–221.

FRISCH, K. VON. 1971. Bees, their vision, chemical senses, and language. 2nd ed. Cornell U.P., Ithaca N.Y. 157 p.

FRUTON, J. S. 1972. Molecules and life: historical essays on the interplay of chemistry and biology. Wiley-Interscience, New York. 579 p.

FRUTON, J. S., and S. SIMMONDS. 1958. General biochemistry. 2nd ed. Wiley, New York. 1077 p.

FRY, F. E. J. 1947. Effects of the environment on animal activity. Univ. Toronto. Stud. Biol. Ser. No. 55; Pub. Ontario Fish. Res. Lab. No. **68:** 1–62.

FRY, F. E. J. 1957. The aquatic respiration of fish. In M. E. Brown, **1:** 1–63.

FRY, F. E. J. 1958. Temperature compensation. Ann. Rev. Physiol. **20:** 207–224.

FRY, F. E. J. 1967. Responses of vertebrate poikilotherms to temperature. *In* A. H. Rose (ed.) Thermobiology. pp. 375–409. Academic Press, New York.

FRY, F. E. J. 1971. The effect of environmental

factors on the physiology of fish. *In* Hoar and Randall, **6:** 1–98.

FRYE, B. E., P. S. BROWN, and B. W. SNYDER. 1972. Effects of prolactin and somatotropin on growth and metamorphosis of amphibians. Gen. Comp. Endocrinol. Suppl. **3:** 209–220.

FUHRMAN, F. A. 1967. Tetrodotoxin. Sci. Am. **217 (2):** 60–71.

FUJII, R. 1969. Chromatophores and pigments. *In* Hoar and Randall, **3:** 307–353.

FUJII, R., and R. R. NOVALES. 1969. Cellular aspects of the control of physiological color changes in fishes. Am. Zool. **9:** 453–463.

FULTON, J. F. 1955. (ed.). A textbook of physiology. 17th ed. Saunders. 1275 p.

FULTON, J. F. 1966. Selected readings in the history of physiology. 2nd ed. Completed by L. G. Wilson. C. C. Thomas, Springfield, Ill. 492 p.

FURSHPAN, E. J., and D. D. POTTER. 1959. Transmission at the giant motor synapses of the crayfish. J. Physiol. **145:** 289–325.

FURUKAWA, T., and E. J. FURSHPAN. 1963. Two inhibitory mechanisms in the Mauthner neurons of goldfish. J. Neurophysiol. **26:** 140–176.

GABE, M., P. KARLSON, and J. ROCHE. 1964. Hormones in invertebrates. *In* Florkin and Mason, **6:** 245–298.

GABRIEL, M. L., and S. FOGEL. 1955. Great experiments in biology. Prentice-Hall, Englewood Cliffs, N.J. 317 p.

GAFFRON, H. 1944. Photosynthesis, photoreduction and dark reduction of carbon dioxide in certain algae. Biol. Rev. **19:** 1–20.

GAFFRON, H. 1960. The origin of life. Persp. Biol. Med. **3:** 163–212.

GALAMBOS, R. 1961. The glia-neural theory of brain function. Proc. Nat. Acad. Sci. U.S. **47:** 129–136.

GALTSOFF, P. S. 1938. Physiology of reproduction in *Ostrea virginica*. II. Stimulation of spawning in the female oyster. Biol. Bull. **75:** 286–307.

GALTSOFF, P. S. 1961. Physiology of reproduction in molluscs. Am. Zool. **1:** 273–289.

GAMOW, R. I., and J. F. HARRIS. 1973. The infrared receptors of snakes. Sci. Am. **228 (5):** 94–100.

GANDOLFI, G. 1969. A chemical sex attractant in the guppy *Poecilia reticulata* Peters (Pisces, Poeciliidae). Monitore Zool. Ital. **3:** 89–98.

GANNON, B. J., G. R. CAMPBELL, and D. J. RANDALL. 1973. Scanning electron microscopy of vascular casts for the study of vessel connections in a complex vascular bed—the trout gill. *In* C. J. Arceneaux, (ed.) 31st Ann. Proc. Electron Microscopy Soc. New Orleans, La.

GANS, C. 1969–70. Biology of the Reptilia. Vols. 1–3. Academic Press, New York.

GANS, C., H. J. DEJONGH, and J. FARBER. 1969. Bullfrog *(Rana catesbeiana)* ventilation: how does the frog breathe? Science **163:** 1223–1225.

GARSTANG, W. 1922. The theory of recapitulation: a critical re-statement of the biogenetic law. J. Linn. Soc. (Zool.) **35:** 81–101.

GARSTANG, W. 1951. Larval forms with other zoological verses. Blackwell, Oxford. 76 p.

GAUNT, R., J. J. CHART, and A. A. RENZI. 1965. Inhibitors of adrenal cortical function. Ergeb. Physiol. Biol. Chem. Exp. Pharmacol. **56:** 114–172.

GEORGE, J. C., and A. J. BERGER. 1966. Avian myology. Academic Press, New York. 500 p.

GEREN, B. B. 1954. The formation of the Schwann cell surface of myelin in the peripheral nerves of chick embryos. Exp. Cell. Res. **7:** 558–562.

GERGELY, J. 1966. Contractile proteins. Ann. Rev. Biochem. **35 (II):** 691–722.

GERNANDT, B. E. 1959. Vestibular mechanisms. Handbook of Physiology, (Am. Physiol. Soc.). Sect. 1, **1:** 549–564.

GHIRARDELLI, E. 1965. Regeneration in the chaetognaths. *In* Kiortsis and Trampusch (1965): 272–277.

GHIRETTI, F. 1966a. Molluscan hemocyanins. *In* Wilbur and Yonge, **2:** 233–248.

GHIRETTI, F. 1966b. Respiration. *In* Wilbur and Yonge, **2:** 175–208.

GHIRETTI, F. 1968. Physiology and biochemistry of haemocyanins. Academic Press, London. 133 p.

GHISELIN, M. 1969. The evolution of hermaphroditism among animals. Quart. Rev. Biol. **44:** 189–201.

GIBBONS, I. R. 1965. Chemical dissection of cilia. Arch. Biol. (Liège) **76:** 317–352.

GIBSON, D., and G. H. DIXON. 1969. Chymotrypsin-like proteases from the sea anemone, *Metridium senile*. Nature **222:** 753–756.

GIESE, A. C. 1950. Action of ultraviolet radiation on protoplasm. Physiol. Rev. **30:** 431–458.

GIESE, A. C. (ed.) 1964–72. Photophysiology. Vols. 1–7. Academic Press, New York.

GIESE, A. C. 1971. Photosensitization by natural pigments. *In* Giese, **6:** 77–129.

GIESE, A. C. 1973. Cell physiology. 4th ed. Saunders, Philadelphia. 741 p.

GILBERT, L. I. 1964. Physiology of growth and development. *In* Rockstein, **1:** 149–225.

GILBERT, L. I. 1967. Lipid metabolism and function in insects. Adv. Insect Physiol. **4:** 69–211.

GILGAN, M. W., and D. R. IDLER. 1967. The conversion of androstenedione to testosterone by some lobster *(Homarus americanus* Milne Edwards) tissues. Gen. Comp. Endocrinol. **9:** 319–324.

GILLIS, J. M. 1969. The site and action of calcium

in producing contraction in striated muscle. J. Physiol. **200:** 849–864.
GILMOUR, D. 1965. The metabolism of insects. Freeman, San Francisco. 195 p.
GILPIN-BROWN, J. B. 1972. Buoyancy mechanisms of cephalopods in relation to pressure. Symp. Soc. Exp. Biol. **26:** 251–259.
GINSBURG, M. 1968. Production, release, transportation and elimination of neurohypophysial hormones. *In* O. Eichler *et al.* (eds.) Handb. Exp. Pharm. New Series, **23:** 286–371. Springer-Verlag, Heidelberg.
GLASS, G. B. J. 1968. Introduction to gastrointestinal physiology. Prentice-Hall, Englewood Cliffs, N.J. 207 p.
GODDARD, C. K., P. I. NICOL, and J. F. WILLIAMS. 1964. The effect of albumen gland homogenate on the blood sugar of *Helix aspersa* Müller. Comp. Biochem. Physiol. **11:** 351–366.
GODDARD, C. K., and A. W. MARTIN. 1966. Carbohydrate metabolism. *In* Wilbur and Yonge, **2:** 275–308.
GOH, S. L. 1971. Mechanism of water and salt absorption in the *in vitro* locust rectum. M.Sc. Thesis, Univ. Brit. Col. Vancouver. 92 p.
GOLDBARD, G. A., J. R. SAUER, and R. R. MILLS. 1970. Hormonal control of excretion in the American cockroach, II. Preliminary purification of a diuretic and antidiuretic hormone. Comp. Gen. Pharmac. **1:** 82–86.
GOLDBLATT, H. 1947. The renal origin of hypertension. Physiol. Rev. **27:** 120–165.
GOLDING, D. W. 1967. Endocrinology, regeneration and maturation in *Nereis*. Biol. Bull. **133:** 567–577.
GOLDING, D. W. 1972. Studies in the comparative neuroendocrinology of polychaete reproduction. Gen. Comp. Endocrinol. Suppl. **3:** 580–590.
GOLDSMITH, T. H. 1964. The visual system of insects. *In* Rockstein, **1:** 397–462.
GOLDSMITH, T. H., and G. D. BERNARD. 1973. *In* M. Rockstein, (ed.), The physiology of Insecta. Vol. 2. 2nd ed. *In press*, quoted from Prosser (1973). Academic Press, New York.
GOLDSTEIN, L. and R. P. FORSTER 1970. Nitrogen excretion in fishes. *In* J. W. Campbell, **2:** 495–518.
GOOD, R. A., A. M. GABRIELSEN, B. POLLARA, H. GEWURZ, and J. FINSTAD. 1971. Phylogenetic development of the lymphoid tissue and immunologic capacity among the lower vertebrates. *In* B. Cinader, (ed.), Regulation of antibody response. 2nd ed. pp. 212–231. C. C. Thomas, Springfield, Ill.
GOODMAN, L. S., and A. GILMAN (eds.). 1970. The pharmacological basis of therapeutics. 4th ed. Macmillan, New York. 1794 p.
GOODRICH, E. S. 1945. The study of nephridia and genital ducts since 1895. Quart. J. Microscop. Sci. **86:** 113–392.
GOODWIN, T. W. 1962. Carotenoids: structure, distribution and function. *In* Florkin and Mason, **4:** 643–675.
GOODWIN, T. W. (ed.). 1966. Biochemistry of chloroplasts. Vols. 1 and 2. Academic Press, London. 476 p and 776 p.
GOODWIN, T. W. 1971. Pigments—Arthropoda. *In* Florkin and Scheer, **6:** 279–306.
GORBMAN, A. (ed.). 1959. Comparative endocrinology. Wiley, New York. 746 p.
GORBMAN, A., and H. A. BERN. 1962. A textbook of comparative endocrinology. Wiley, New York. 468 p.
GORBMAN, A., and C. W. CREASER. 1942. Accumulation of radio-active iodine by the endostyle of larval lampreys and the problem of homology of the thyroid. J. Exp. Zool. **89:** 391–401.
GORDON, M. (ed.). 1959. Pigment cell biology. Academic Press, New York. 647 p.
GORDON, M. S. 1962. Osmotic regulation in the green toad *(Bufo viridis)*. J. Exp. Biol. **39:** 261–270.
GORDON, M. S. 1968. Oxygen consumption of red and white muscles from tuna fishes. Science **159:** 87–90.
GORDON, M. S., G. A. BARTHOLOMEW, A. D. GRINNELL, C. B. JØRGENSEN, and F. N. WHITE. 1972. Animal physiology. 2nd ed. Macmillan, New York. 592 p.
GORDON, M. S., K. SCHMIDT-NIELSEN, and H. M. KELLY. 1961. Osmotic regulation in the crab-eating frog *(Rana cancrivora)*. J. Exp. Biol. **38:** 659–678.
GORDON, M. S., and V. A. TUCKER. 1965. Osmotic regulation in the tadpoles of the crab-eating frog *(Rana cancrivora)*. J. Exp. Biol. **42:** 437–445.
GORDON, M. S., and V. A. TUCKER. 1968. Further observations on the physiology of salinity adaptation in the crab-eating frog *(Rana cancrivora)*. J. Exp. Biol. **49:** 185–193.
GOSS, R. J. 1969. Principles of regeneration. Academic Press, New York. 287 p.
GOSS, R. J., and E. D. HAY (eds.) 1970. Metazoan regeneration. Am. Zool. **10:** 90–186.
GOSSELIN, R. E. 1961. The cilioexcitatory activity of serotonin. J. Cell. Comp. Physiol. **58:** 17–26.
GOSSELIN, R. E. 1967. Kinetics of pinocytosis. Fed. Proc. **26:** 987–993.
GOSTAN, G. 1965. Cytophysiologie de l'excrétion chez les mollusques pulmonés. Ann. Biol. **4:** 481–494.
GOTTFRIED, H., and R. I. DORFMAN. 1970. Steroids of invertebrates. Gen. Comp. Endocrinol. **15:** 101–142.

Gottschalk, C. W. 1960. Osmotic concentration and dilution in the mammalian nephron. Circulation **21:** 861–868.

Gould, E., N. C. Negus, and A. Novick. 1964. Evidence for echolocation in shrews. J. Exp. Zool. **156:** 19–38.

Granit, R. 1955. Receptors and sensory perception. Yale U.P. New Haven, Conn. 369 p.

Grant, W. C., Jr. 1961. Special aspects of the metamorphic process: second metamorphosis. Am. Zool. **1:** 163–171.

Grant, W. C., Jr. and G. Cooper, IV. 1965. Behavioral and integumentary changes associated with induced metamorphosis in *Diemictylus*. Biol. Bull. **129:** 510–522.

Grant, W. C., F. J. Hendler, and P. M. Banks. 1969. Studies of blood-sugar regulation in the little skate *Raja erinacea*. Physiol. Zool. **42:** 231–247.

Gratzer, W. B., and A. C. Allison. 1960. Multiple haemoglobins. Biol. Rev. **35:** 459–506.

Graubard, M. 1964. Circulation and respiration. Harcourt Brace & World, Inc., New York. 278 p.

Gray, I. E. 1957. A comparative study of the gill area of crabs. Biol. Bull. **112:** 34–42.

Gray, J. 1928. Ciliary movement. Cambridge U. P., London. 162 p.

Gray, J. S. 1950. Pulmonary ventilation and its physiological regulation. C. C. Thomas, Springfield, Ill. 82 p.

Green, D. E. 1962. Structure and function of subcellular particles. Comp. Biochem. Physiol. **4:** 81–122.

Greene, C. W. 1926. The physiology of the spawning migration. Physiol. Rev. **6:** 201–241.

Greenwood, P. H. 1963. A history of fishes (revision of J. R. Norman). Benn, London. 398 p.

Greep, R. O. 1963. Parathyroid glands. *In* U. S. von Euler and H. Heller (eds.). **1:** 325–370. Academic Press, New York.

Grégoire, C. 1970. Haemolymph coagulation in arthropods. Symp. Zool. Soc. London **27:** 45–74.

Grégoire, C., and H. J. Tagnon. 1962. Blood coagulation. *In* Florkin and Mason, **4:** 435–482.

Gregory, R. P. F. 1971. Biochemistry of photosynthesis. Wiley-Interscience, London. 202 p.

Greville, G. D., and P. K. Tubbs, 1968. The catabolism of long chain fatty acids in mammalian tissues. Essays in Biochem. **4:** 155–212.

Griffin, D. R. 1958. Listening in the dark. Yale U. P., New Haven, Conn. 413 p.

Griffin, D. R. 1959. Echoes of bats and men. Anchor Books, Garden City, N.Y. 156 p.

Griffin, D. R. 1962. Comparative studies of the orientation sounds of bats. Symp. Zool. Soc. London **7:** 61–72.

Grimstone, A. V., A. M. Mullinger, and J. A. Ramsay. 1968. Further studies on the rectal complex of the mealworm *Tenebrio molitor* L. (Coleoptera, Tenebrionidae). Phil. Trans. Roy. Soc. London B **253:** 343–382.

Grinnell, A. D. 1969. Comparative physiology of hearing. Ann. Rev. Physiol. **31:** 545–580.

Grisolia, S., and J. Kennedy. 1966. On specific dynamic action, turnover, and protein synthesis. Persp. Biol. Med. **9:** 578–585.

Gross, W. J. 1955. Aspects of osmotic regulation in crabs showing the terrestrial habit. Am. Naturalist, **89:** 205–222.

Gross, W. J. 1957. A behavioral mechanism for osmotic regulation in a semi-terrestrial crab. Biol. Bull. **113:** 268–274.

Gross, W. J. 1961. Osmotic tolerance and regulation in crabs from a hypersaline lagoon. Biol. Bull. **121:** 290–301.

Gruber, S. A., and D. W. Ewer. 1962. Observations on the myo-neural physiology of the polyclad, *Planocera gilchristi*. J. Exp. Biol. **39:** 459–477.

Grundfest, H. 1959. Evolution of conduction in the nervous system. *In* A. D. Bass (ed.), Evolution of nervous control from primitive organisms to man. pp. 43–86. A.A.A.S. Pub. 52, Washington, D.C.

Grundfest, H. 1960. Electric fishes. Sci. Am. **203 (4):** 115–124.

Grundfest, H. 1961. Ionic mechanisms in electrogenesis. Ann. N.Y. Acad. Sci. **94:** 405–457.

Grundfest, H. 1965a. Evolution of electrophysiological varieties among sensory receptor systems. *In* Pringle (1965b): 107–138.

Grundfest, H. 1965b. Electrophysiology and pharmacology of different components of bioelectric transducers. Cold Spring Harbor Symp. Quant. Biol. **30:** 1–14.

Grundfest, H. 1966. Comparative electrobiology of excitable membranes. Adv. Comp. Physiol. Biochem. **2:** 1–116.

Grundfest, H. 1969. Synaptic and ephaptic transmission. *In* Bourne, **2:** 463–491.

Grundfest, H., and M. V. L. Bennett. 1961. Electrophysiology of marine electric fishes. *In* Chagas and Paes De Carvalho (1961): 57–101.

Gudernatsch, J. F. 1912. Feeding experiments on tadpoles. Arch. Entwicklungsmech. Organ. **35:** 457–483.

Guillemin, R., and R. Burgus. 1972. The hormones of the hypothalamus Sci. Am. **227(5):** 24–33.

Gulick, W. L. 1971. Hearing. Oxford U. P. New York. 258 p.

Gunter, G., L. L. Sulya, and B. E. Box. 1961. Some evolutionary patterns in fishes' blood. Biol. Bull. **121:** 302–306.

GUTH, L. 1971. Degeneration and regeneration of taste buds. Handbook of Sensory Physiology (Springer-Verlag), **4(2):** 63–74.

GUTHRIE, D. M., and J. R. BANKS. 1970a. Observations on the function and physiological properties of a fast paramyosin muscle–the notochord of *Amphioxus (Branchiostoma lanceolatum).* J. Exp. Biol. **52:** 125–138.

GUTHRIE, D. M., and J. R. BANKS. 1970b. Observations on the electrical and mechanical properties of the myotomes of the lancet *(Branchiostoma lanceolatum).* J. Exp. Biol. **52:** 401–417.

GUTHRIE, R. D., and J. HONEYMAN. 1968. An introduction to the chemistry of carbohydrates. 3rd ed. Clarendon Press. Oxford. 144 p.

GUYTON, A. C. 1971. Textbook of medical physiology. 4th ed. Saunders, Philadelphia. 1032 p.

HADENFELDT, D. 1929. Das Nervensystem von *Stylochoplana maculata* und *Notoplana atomata.* Z. Wiss. Zool. **133:** 586–638.

HADLEY, M. E., and J. M. GOLDMAN. 1969. Physiological color changes in reptiles. Am. Zool. **9:** 489–504.

HADLEY, M. E., and J. M. GOLDMAN. 1970. Adrenergic receptors and geographic variation in *Rana pipiens* chromatophore responses. Am. J. Physiol. **219:** 72–77.

HAGGAR, R. A., and M. L. BARR. 1950. Quantitative data on the size of synaptic end-bulbs in the cat's spinal cord. J. Comp. Neurol. **93:** 17–36.

HAGIWARA, S., and K. NAKA. 1964. The initiation of spike potential in barnacle muscle fibers under low intracellular Ca^{++}. J. Gen. Physiol. **48:** 141–162 and 163–179.

HAGIWARA, S., and S. NAKAJIMA. 1966. Differences in Na and Ca spikes as examined by application of tetrodotoxin, procaine, and magnesium ions. J. Gen. Physiol. **49:** 793–806.

HAGIWARA, S., and A. WATANABE. 1956. Discharges of motoneurons of cicada. J. Cell. Comp. Physiol. **47:** 415–428.

HAHN, W. E., and D. COPELAND. 1966. Carbon monoxide concentrations and the effects of aminopterin on its production in the gas bladder of *Physalia physalis.* Comp. Biochem. Physiol. **18:** 201–207.

HALDANE, J. S. 1927. Respiration. Yale U.P., New Haven, Conn. 427 p.

HALDANE, J. S., and J. G. Priestley. 1935. Respiration. Oxford U. P., London. 493 p.

HALL, P. F., C. L. RALPH, and D. L. GRINWICH. 1965. On the locus of action of interstitial cell-stimulating hormone (ISCH or LH) on feather pigmentation of African weaver birds. Gen. Comp. Endocrinol. **5:** 552–557.

HAM, A. W. 1969. Histology. 6th ed. Lippincott, Philadelphia. 1037 p.

HAMILTON, J. A. R., and F. J. ANDREW. 1954. An investigation of the effect of Baker dam on downstream-migrant salmon. Bull. Int. Pacific Salmon Fish. Comm. **6:** 1–73.

HAMMEN, C. S. 1966. Carbon dioxide fixation in marine invertebrates—V. Rate and pathway in the oyster. Comp. Biochem. Physiol. **17:** 289–296.

HAMMOND, J., JR., 1954. Light regulation of hormone secretion. Vitamins and Hormones, **12:** 157–206.

HANSON, J., and J. LOWY. 1960. Structure and function of the contractile apparatus in the muscles of invertebrate animals. *In* Bourne, **1:** 265–335.

HANYU, I., and M. A. ALI. 1963. Flicker fusion frequency of electroretinogram in light-adapted goldfish at various temperatures. Science **140:** 662–663.

HARA, T. J. 1970. An electrophysiological basis for olfactory discrimination in homing salmon: A review. J. Fish. Res. Bd. Canada **27:** 565–586.

HARA, T. J. 1971. Chemoreception. *In* Hoar and Randall, **5:** 79–120.

HARDISTY, M. W., and I. C. POTTER. 1971–72. The biology of lampreys. Vols. 1-2. Academic Press, New York.

HARDY, J. D. 1961. Physiology of temperature regulation. Physiol. Rev. **41:** 521–606.

HARGENS, A. R. 1972. Freezing resistance in polar fishes. Science **176:** 184–186.

HARNDEN, D. G. 1968. Digestive carbohydrases of *Balanus nubilis* (Darwin, 1854). Comp. Biochem. Physiol. (1968), **25:** 303–310.

HARPER, H. A. 1973. Review of physiological chemistry. 14th ed. Lange Medical Publications, Los Altos, California. 545 p.

HARRINGTON, R. W., JR. 1963. Twenty-four-hour rhythms of internal self-fertilization and of oviposition by hermaphrodites of *Rivulus marmoratus.* Physiol. Zool. **36:** 325–341.

HARRINGTON, R. W., JR., and K. D. KALLMAN. 1968. The homozygosity of clones of the self-fertilizing hermaphroditic fish *Rivulus marmoratus* Poey (Cyprinodontidae, Atheriniformes). Am. Naturalist **102:** 337–343.

HARRIS, R. A., C. H. WILLIAMS, M. CALDWELL, and D. E. GREEN. 1969. Energized configurations of heart mitochondria *in situ.* Science **165:** 700–702.

HART, J. S. 1957. Climatic and temperature induced changes in the energetics of homeotherms. Rev. Can. Biol. **16:** 133–174.

HART, J. S. 1964a. Insulative and metabolic adaptations to cold in vertebrates. Symp. Soc. Exp. Biol. **18:** 31–48.

HART, J. S. 1964b. Geography and season: mammals and birds. Handbook of Physiology. (Am. Physiol. Soc.) **4:** 295–321.

HART, J. S. (ed.). 1969. Proceedings of the International Symposium on Altitude and Cold. Fed. Proc. **28:** 917–1321.

HART, J. S., O. HEROUX, W. H. COTTLE, and C. A. MILLS. 1961. The influence of climate on metabolic and thermal responses of infant caribou. Can. J. Zool. **39:** 845–856.

HARTENSTEIN, R. 1968. Nitrogen metabolism in the terrestrial isopod, *Oniscus asellus.* Am. Zool. **8:** 507–519.

HARTENSTEIN, R. 1970. Nitrogen metabolism in the non-insect arthropods. *In* J. W. Campbell, **2:** 299–385.

HARTLINE, H. K. 1969. Visual receptors and retinal interaction. Science **164:** 270–278.

HARVEY, E. N. 1914. On the chemical nature of the luminous material of the firefly. Science **40:** 33–34.

HARVEY, E. N. 1952. Bioluminescence. Academic Press, New York. 649 p.

HARVEY, E. N. 1960. Bioluminescence. *In* Florkin and Mason, **2:** 545–591.

HARVEY, H. H., and S. B. SMITH. 1961. Supersaturation of the water supply and occurrence of gas bubble disease at Cultus lake trout hatchery. Can. Fish. Culturist **30:** 39–47.

HASKELL, P. T. 1964. Sound production. *In* Rockstein, **1:** 563–608.

HASLEWOOD, G. A. D. 1968. Evolution and bile salts. Handbook of Physiology. (Am. Physiol. Soc.) Sect. 6, **5:** 2375–2390.

HAUGAARD, N. 1968. Cellular mechanisms of oxygen toxicity. Physiol. Rev. **48:** 311–373.

HAUROWITZ, F. 1965. Antibody formation. Physiol. Rev. **45:** 1–47.

HAWKING, F. 1970. The clock of the malaria parasite. Sci. Am. **222(6):** 123–131.

HAYASHIDA, T., and M. D. LAGIOS. 1969. Fish growth hormone: a biological, immunochemical, and ultrastructural study of sturgeon and paddlefish pituitaries. Gen. Comp. Endocrinol. **13:** 403–411.

HAYWARD, J. S., C. P. LYMAN, and C. R. TAYLOR. 1965. The possible role of brown fat as a source of heat during arousal from hibernation. Ann. N.Y. Acad. Sci. **131:** 441–446.

HAZEL, J. R. 1972. The effect of temperature acclimation upon succinic dehydrogenase activity from the epaxial muscle of the common goldfish *(Carassius auratus* L.*)* Comp. Biochem. Physiol. **43B:** 837–861, 863–882.

HAZELWOOD, D. H., W. T. W. POTTS, and W. R. FLEMING. 1970. Further studies on the sodium and water metabolism of the fresh-water medusa, *Craspedacusta sowerbyi.* Z. vergl. Physiol. **67:** 186–191.

HEALEY, E. G. 1957. The nervous system. *In* M. E. Brown, **2:** 1–119.

HEATH, J. E., and P. A. ADAMS. 1967. Regulation of heat production by large moths. J. Exp. Biol. **47:** 21–33.

HECHT, H. H. 1965. Comparative physiological and morphological aspects of pacemaker tissues. Ann. N.Y. Acad. Sci. **127:** 49–83.

HECHT, S. 1937. Rods, cones, and the chemical basis of vision. Physiol. Rev. **17:** 239–290.

HECHT, S., S. SHLAER, and M. H. PIRENNE. 1942. Energy, quanta, and vision. J. Gen. Physiol. **25:** 819–840.

HEILBRUNN, L. V. 1940. The action of calcium on muscle protoplasm. Physiol. Zool. **13:** 88–94.

HEILBRUNN, L. V. 1952. An outline of general physiology. 3rd ed. Saunders, Philadelphia. 818 p.

HEILBRUNN, L. V., and F. J. WIERCINSKI. 1947. The action of various cations on muscle protoplasm. J. Cell. Comp. Physiol. **29:** 15–32.

HELLER, H. 1972. The effect of neurohypophyseal hormones on the female reproductive tract of lower vertebrates. Gen. Comp. Endocrinol. Suppl. **3:** 703–714.

HEMMINGSEN, A. M. 1950. The relation of standard (basal) energy metabolism to total fresh weight of living organisms. Rep. Steno Hosp. Copenhagen **4:** 1–58.

HEMMINGSEN, A. M. 1960. Energy metabolism as related to body size and respiratory surfaces, and its evolution. Rep. Steno Hosp. Copenhagen **9** (pt. 2): 1–110.

HENDERSON, L. M., R. K. GHOLSON, and C. E. DALGLIESH. 1962. Metabolism of aromatic amino acids. *In* Florkin and Mason, **4:** 245–342.

HENDLER, W. W. 1971. Biological membrane ultrastructure. Physiol. Rev. **51:** 66–97.

HENSEL, H. 1966. Classes of receptor units predominantly related to thermal stimuli. *In* De Reuck and Knight (1966): 275–290.

HENSON, O. W., JR. 1970. The ear and audition. In Wimsatt, 2: 181–263.

HERTER, K. 1947. Vergleichende Physiologie der Tiere. I. Stoff- und Energiewechsel. Gruyter, Berlin. 148 p.

HESS, A. 1970. Vertebrate slow muscle fibers. Physiol Rev. **50:** 40–62.

HEYMANS, C., and E. NEILL. 1958. Reflexogenic areas of the cardiovascular system. Churchill, London. 271 p.

HICKMAN, C. P., JR., and B. F. TRUMP. 1969. The Kidney. *In* Hoar and Randall, **1:** 91–239.

HIGHNAM, K. C., and L. HILL. 1969. The comparative endocrinology of the invertebrates. Edward Arnold, London. 270 p.

HILL, R. 1965. The biochemists' green mansions:

the photosynthetic electron-transport chain in plants. Essays Biochem. **1:** 121–151.

HILL, R. B., and J. H. WELSH. 1966. Heart, circulation and blood cells. *In* Wilbur and Yonge, **2:** 125–174.

HILL, S. D. 1970. Origin of the regeneration blastema in polychaete annelids. Am. Zool. **10:** 101–112.

HILLE, B. 1968. Pharmacological modifications of the sodium channels of frog nerve. J. Gen. Physiol. **51:** 199–219.

HINDE, R. A. 1970. Animal behavior. 2nd ed. McGraw-Hill, New York. 876 p.

HINKE, J. A. M. 1961. The measurement of sodium and potassium activities in the squid axon by means of cation-selective glass micro-electrodes. J. Physiol. **156:** 314–335.

HINTON, H. E. 1968. Spiracular gills. Adv. Insect Physiol. **5:** 65–162.

HIRSCH, J. G. 1965. Phagocytosis. Ann. Rev. Microbiol. **19:** 339–350.

HISAW, F. L. 1959. Endocrine adaptations of the mammalian estrous cycle and gestation. *In* Gorbman (1959): 533–552.

HISAW, F. L. 1963. Endocrines and the evolution of viviparity among the vertebrates. Proc. 22nd Ann. Biol. Colloq., Oregon State Univ., Corvallis, 1961: 119–138.

HOAGLAND, H. 1932. Impulses from sensory nerves of catfish. Proc. Nat. Acad. Sci. U.S., **18:** 701–705.

HOAGLAND, H. 1933. Electrical responses from the lateral line nerves of catfish. J. Gen. Physiol. **16:** 695–732 and **17:** 77–82, 195–209.

HOAR, W. S. 1955a. Seasonal variations in the resistance of goldfish to temperature. Trans. Roy. Soc. Can. Ser. III, **49:** 25–34.

HOAR, W. S. 1955b. Reproduction in teleost fish. Mem. Soc. Endocrinol. **4:** 5–24.

HOAR, W. S. 1965a. The endocrine system as a chemical link between the organism and its environment. Trans. Roy. Soc. Can. Ser. IV, **3:** 175–200.

HOAR, W. S. 1965b. Comparative physiology: hormones and reproduction in fishes. Ann. Rev. Physiol. **27:** 51–70.

HOAR, W. S. 1966. Hormonal activities of the pars distalis in cyclostomes, fish and amphibians. *In* G. W. Harris and B. T. Donovan (eds.) The pituitary gland. **1:** 242–294. Butterworths, London.

HOAR, W. S. 1967. Environmental physiology of animals. Trans. Roy. Soc. Can. Ser. IV. **5:** 127–153.

HOAR, W. S. 1969. Reproduction. *In* Hoar and Randall, **3:** 1–72.

HOAR, W. S., and H. A. BERN (eds.). 1972. Progress in comparative endocrinology. Gen. Comp. Endocrinol. Suppl. **3:** 405–458.

HOAR, W. S., and J. G. EALES. 1963. The thyroid gland and low-temperature resistance of goldfish. Can. J. Zool. **41:** 653–669.

HOAR, W. S., and C. P. HICKMAN, JR. 1975. A laboratory companion for general and comparative physiology, 2nd ed., Prentice-Hall, Englewood Cliffs, N.J. 296 p.

HOAR, W. S., and D. J. RANDALL (eds.) 1969–71. Fish Physiology. Vols. 1–6; Academic Press, New York.

HOCHACHKA, P. W., J. FIELDS, and T. MUSTAFA. 1973. Animal life without oxygen: basic biochemical mechanisms. Am. Zool. **13:** 543–555.

HOCHACHKA, P. W. and T. MUSTAFA. 1972. Invertebrate facultative anaerobiosis. Science **178:** 1056–1060.

HOCHACHKA, P. W., and G. N. SOMERO. 1971. Biochemical adaptation to the environment. *In* Hoar and Randall, **6:** 99–156.

HOCHACHKA, P. W. and G. N. SOMERO. 1973. Strategies of biochemical adaptation. Saunders, Philadelphia. 358 p.

HOCK, R. J. 1960. Seasonal variations in physiologic functions of Arctic ground squirrels and black bears. Bull. Mus. Comp. Zool. Harvard **124:** 155–171.

HOCK, R. J. 1970. The physiology of high altitude. Sci. Am. **222 (2):** 52–62.

HODGKIN, A. L. 1964. The conduction of the nerve impulse. Liverpool U. P., Liverpool. 108 p.

HODGKIN, A. L., and A. F. HUXLEY, 1952. Movement of sodium and potassium ions during nervous activity. Cold Spring Harbor Symp. Quant. Biol. **17:** 43–52.

HODGKIN, A. L., A. F. HUXLEY, and B. KATZ. 1952. Measurement of current-voltage relations in the membrane of the giant axon of *Loligo*. J. Physiol. **116:** 424–448.

HODGKIN, A. L., and B. KATZ. 1949. The effect of sodium ions on the electrical activity of the giant axon of the squid. J. Physiol. **108:** 37–77.

HODGSON, E. S., and K. D. ROEDER. 1956. Electrophysiological studies of arthropod chemoreception. J. Cell. Comp. Physiol. **48:** 51–75.

HOFFMAN, R. A. 1964. Terrestrial animals in the cold: hibernators. Handbook of Physiology. (Am. Physiol. Soc.) **4:** 379–403.

HOFFMANN-BERLING, H. 1960. Other mechanisms producing movement. *In* Florkin and Mason, **2:** 341–370.

HOGBEN, L. 1942. Chromatic behaviour. Proc. Roy. Soc. London B **131:** 111–136.

HOGBEN, L., and D. SLOME. 1931. The pigmentary effector system. VI. The dual character of endocrine co-ordination in amphibian colour change. Proc. Roy. Soc. London B **108:** 10–53.

HOLETON, G. F. 1971. Oxygen uptake and transport by rainbow trout during exposure to carbon monoxide. J. Exp. Biol. **54:** 239–254.

HOLETON, G. F. 1972. Gas exchange in fish with and without hemoglobin. Resp. Physiol. **14:** 142–150.

HOLETON, G. F., and D. J. RANDALL. 1967. The effect of hypoxia upon the partial pressure of gases in the blood and water afferent and efferent to the gills of the rainbow trout. J. Exp. Biol. **46:** 317–327.

HOLLAENDER, A. (ed.). 1954–56. Radiation biology. Vols. 1–3. McGraw-Hill, New York.

HOLLANDÉ, A. 1952. Classe des Eugléniens. *In* Grassé. P. (ed.), Traité de Zoologie. Vol. 1: 238–284.

HÖLLDOBLER, B. 1971. Homing in the harvester ant *Pogonomyrmex badius.* Science **171:** 1149–1151.

HOLLIDAY, F. G. T. 1969. The effects of salinity on the eggs and larvae of teleosts. *In* Hoar and Randall, **1:** 293–311.

HOLMES, R. S., C. J. MASTERS and E. C. WEBB. 1968. A comparative study of vertebrate esterase multiplicity. Comp. Biochem. Physiol. **26:** 837–852.

HOLMES, W. N., and R. L. MCBEAN. 1964. Some aspects of electrolyte excretion in the green turtle, *Chelonia mydas mydas.* J. Exp. Biol. **41:** 81–90.

HOLMES, W. N., J. G. PHILLIPS, and I. CHESTER JONES. 1963. Adrenocortical factors associated with adaptation of vertebrates to marine environments. Recent Progr. Hormone Res. **19:** 619–672.

HOLST, E. VON, and U. VON SAINT-PAUL. 1962. Electrically controlled behavior. Sci. Am. **206 (3):** 50–59.

HOLTER. H. 1961. How things get into cells. Sci. Am. **205 (3):** 167–180.

HOLWILL, M. E. J. 1966. Physical aspects of flagellar movement. Physiol. Rev. **46:** 696–785.

HOLZ, G. G. 1964. Nutrition and metabolism of ciliates. *In* Hutner (1964): 199–242.

HORIUCHI, S., and C. E. LANE. 1966. Carbohydrases of the crystalline style and hepatopancreas of *Strombus gigas* Linné. Comp. Biochem. Physiol. **17:** 1189–1197.

HORNE, F. R. 1969. Purine excretion in five scorpions, a uropygid and a centipede. Biol. Bull. **137:** 155–160.

HORNE, F., and V. BOONKOOM. 1970. The distribution of the ornithine cycle enzymes in twelve gastropods. Comp. Biochem. Physiol. **32:** 141–153.

HORRIDGE, G. A. 1954. Observations on the nerve fibres of *Aurellia aurita.* Quart. J. Microscop. Sci. **95:** 85–92.

HORRIDGE, G. A. 1963. Integrative action of the nervous system. Ann. Rev. Physiol. **25:** 523–544.

HORRIDGE, G. A. 1965a. Relation between nerves and cilia in ctenophores. Am. Zool. **5:** 357–375.

HORRIDGE, G. A. 1965b. Non-motile sensory cilia and neuromuscular junctions in a ctenophore independent effector organ. Proc. Roy. Soc. London B **162:** 333–350.

HORRIDGE, G. A. 1965c. Macrocilia with numerous shafts from the lips of the ctenophore *Beroë.* Proc. Roy. Soc. London B. **162:** 351–364.

HORRIDGE, G. A. 1966a. Some recently discovered underwater vibration receptors in invertebrates. in marine science. pp. 395–405. Reprinted with permission of MacMillan Publishing Co., Inc. © George Allen & Unwin Ltd. 1966.

HORRIDGE, G. A. 1966b. The retina of the locust. *In* Bernhard (1966): 513–541.

HORRIDGE, G. A. 1968a. The origins of the nervous system. *In* Bourne, **1:** 1–31.

HORRIDGE, G. A. 1968b. Interneurons. Freeman, San Francisco. 436 p.

HORRIDGE, G. A. 1971a. Primitive examples of gravity receptors and their evolution. *In* S. A. Gordon and M. J. Cohen (eds.), Gravity and the organism. pp. 203–221. Chicago U. P., Chicago.

HORRIDGE, G. A. 1971b. Alternatives to superposition images in clear-zone compound eyes. Proc. Roy. Soc. London B, **179:** 97–124.

HORRIDGE, G. A. 1974. Insect vision. *In* CSIRO (Supp.) The Insects of Australia. Melbourne U.P., Melbourne, Australia.

HORRIDGE, G. A., and B. MACKAY. 1964. Neurociliary synapses in *Pleurobrachia* (Ctenophora). Quart. J. Microscop. Sci. **105:** 163–174.

HOUSE, H. L. 1965. Insect nutrition. *In* Rockstein, **2:** 769–813.

HOUSTON, A. H., and J. A. MADDEN. 1968. Environmental temperature and plasma electrolyte regulation in the carp, *Cyprinus carpio.* Nature **217:** 969–970.

HOWELL, B. J., F. W. BAUMGARDNER, K. BONDI, and H. RAHN. 1970. Acid-base regulation in cold-blooded vertebrates as a function of body temperature. Am. J. Physiol. **218:** 600–606.

HOYLE, G. 1957. Comparative physiology of the nervous control of muscular contraction. Cambridge U. P., London. 147 p.

HOYLE, G. 1962a. Neuromuscular physiology. Adv. Comp. Physiol. Biochem. **1:** 177–216.

HOYLE, G. 1962b. Comparative physiology of conduction in nerve and muscle. Am. Zool. **2:** 5–25.

HOYLE, G. 1965. Neural control of skeletal muscle. *In* Rockstein, **2:** 407–449.

HOYLE, G. 1967. Diversity of striated muscle. Am. Zool. **7:** 435–449.

HOYLE, G. 1969. Comparative aspects of muscle. Ann. Rev. Physiol. **31:** 43–84.

HOYLE, G. 1970. How is muscle turned on and off? Sci. Am. **222(4):** 84–93.

HOYLE, G., and P. A. MCNEILL. 1968. Correlated physiological and ultrastructural studies on specialized muscles. J. Exp. Zool. **167:** 487–522.

HUENNEKENS, F. M., and H. R. WHITELEY. 1960. Phosphoric acid anhydrides and other energy-rich compounds. *In* Florkin and Mason, **1:** 107–180.

HUFF, C. G. 1940. Immunity in invertebrates. Physiol. Rev. **20:** 68–88.

HUGGINS, A. K., and K. A. MUNDAY. 1968. Crustacean metabolism. Adv. Comp. Physiol. Biochem **3:** 271–378.

HUGHES, G. M. 1963. Comparative physiology of vertebrate respiration. Heinemann, London. 145 p.

HUGHES, G. M. 1966. The dimensions of fish gills in relation to their function. J. Exp. Biol. **45:** 177–195.

HUGHES, G. M. 1967a. Evolution between air and water. *In* Development of the Lung. Ciba Foundation Symposium (1967): 64–84. Churchill, London.

HUGHES, G. M. 1967b. Further studies on the electrophysiological anatomy of the left and right giant cells in *Aplysia*. J. Exp. Biol. **46:** 169–193.

HUGHES, G. M., and C. M. BALLINTIJN. 1965. The muscular basis of the respiratory pumps in the dogfish *(Scyliorhinus canicula)*. J. Exp. Biol. **43:** 363–383.

HUGHES, G. M., and C. M. BALLINTIJN. 1968. Electromyography of the respiratory muscles and gill water flow in the dragonet. J. Exp. Biol. **49:** 583–602.

HUGHES, G. M., and G. SHELTON. 1958. The mechanism of gill ventilation in three freshwater teleosts. J. Exp. Biol. **35:** 807–823.

HUGHES, G. M., and G. SHELTON. 1962. Respiratory mechanisms and their nervous control in fish. Adv. Comp. Physiol. Biochem. **1:** 275–364.

HULT, J. E. 1969. Nitrogenous waste products and excretory enzymes in the marine polychaete *Cirriformia spirabrancha*. Comp. Biochem. Physiol. **31:** 15–24.

HUNGATE, R. E. 1966. The rumen and its microbes. Academic Press, New York. 533 p.

HUNTSMAN, A. G., and M. I. SPARKS. 1924. Limiting factors for marine animals. 3. Relative resistance to high temperatures. Contr. Can. Biol. N.S. **2:** 95–114.

HURTADO, A. 1964. Animals in high altitudes: resident man. Handbook of Physiology. (Am. Physiol. Soc.) **4:** 843–860.

HUTCHINSON, G. E. 1957. A treatise on limnology. Vol. I. John Wiley & Sons, New York. 1015 p.

HUTCHINSON, V. H., H. G. DOWLING, and A. VINEGAR. 1966. Thermoregulation in a brooding female Indian python, *Python molurus bivittatus*. Science **151:** 694–696.

HUTNER, S. H. (ed.). 1964. Biochemistry and physiology of Protozoa. Vol. 3. Academic Press, New York. 616 p.

HUTNER, S. H., and A. LWOFF (eds.). 1955. Biochemistry and physiology of protozoa. Vol. 2. Academic Press, New York. 388 p.

HUXLEY, A. F. 1957. Muscle structure and theories of contraction. Progr. Biophys. **7:** 255–318.

HUXLEY, A. F., and R. E. TAYLOR. 1958. Local activation of striated muscle fibres. J. Physiol. **144:** 426–441.

HUXLEY, H. E. 1958. The contraction of muscle. Sci. Am. **199 (5):** 66–86.

HUXLEY, H. E. 1960. Muscle cells. *In* Brachet and Mirsky, **4:** 365–481.

HUXLEY, H. E. 1965. The mechanism of muscular contraction. Sci. Am. **213 (6):** 18–27.

HUXLEY, H. E. 1969. The mechanism of muscular contraction. Science **164:** 1356–1366.

HUXLEY, J. S. 1932. Problems of relative growth. Methuen, London. 276 p.

HYDÉN, H. 1967. RNA in brain cells. *In* Quarton *et al.* (1967): 248–266.

HYMAN, L. H. 1940–67. The invertebrates. Vols. 1–6. McGraw-Hill, New York.

IDLER, D. R. 1971. Some comparative aspects of corticosteroid metabolism. *In* V. H. T. James and L. Martini (eds.), Hormonal steroids. Proc. 3rd Internat. Congr. Hormonal Steroids. Hamburg. 1970. pp. 14–28. Excerpta Medica, Amsterdam. 1063 p.

IDLER, D. R., G. B. SANGALANG, and M. WEISBART. 1971. Are corticosteroids present in the blood of all fish? *In* V. H. T. James and L. Martini (eds.), Hormonal steroids. Proc. 3rd Internat. Congr. Hormonal Steroids. Hamburg. 1970. pp. 983–989. Excerpta Medica, Amsterdam. 1063 p.

IDLER, D. R., G. B. SANGALANG, and B. TRUSCOTT. 1972. Corticosteroids in the South American lungfish. Gen. Comp. Endocrinol. Suppl. **3:** 238–244.

IDLER, D. R., and B. TRUSCOTT. 1963. *In vivo* metabolism of steroid hormones by sockeye salmon. Can. J. Biochem. Physiol. **41:** 875–887.

INOUÉ, S., and H. SATO. 1967. Cell motility by labile association of molecules. *In* Stracher (1967): 259–292.

IRVING, L. 1939. Respiration in diving mammals. Physiol. Rev. **19:** 112–134.

IRVING, L. 1964. Comparative anatomy and physiology of gas transport mechanisms. Handbook of Physiology. (Am. Physiol. Soc.) Sect 3, **1:** 177–212.

IRVING, L. 1972. Arctic life of birds and mammals. Springer-Verlag, Heidelberg. 192 p.

IRVING, L., and J. S. HART. 1957. The metabolism and insulation of seals as bare-skinned mammals in cold water. Can. J. Zool. **35:** 497–511.

IRVING. L., K. SCHMIDT-NIELSEN, and N. S. B. ABRAHAMSEN. 1957. On the melting points of animal fats in cold climates. Physiol. Zool. **30:** 93–105.

JACOBS, M. H., H. N. GLASSMAN, and A. K. PARPART. 1950. Hemolysis and zoological relationship. J. Exp. Zool. **113:** 277–300.

JACOBS, W. 1954. Fliegen. Schwimmen. Schweben. Springer-Verlag, Berlin. 136 p.

JACOBSON, M. 1972. Insect sex pheromones. Academic Press, New York. 381 p.

JAHN, T. L. 1964. Relative motion in *Amoeba proteus*. *In* Allen and Kamiya (1964): 279–302.

JAHN, T. L., and E. C. BOVEE. 1964. Protoplasmic movements and locomotion of Protozoa. *In* Hutner, **3:** 61–129.

JAHN, T. L., and E. C. BOVEE. 1967. Motile behavior of protozoa. *In* Chen, **1:** 41–200.

JAHN, T. L., and E. C. BOVEE. 1969. Protoplasmic movements within cells. Physiol. Rev. **49:** 793–862.

JAHN, T. L. and R. A. RINALDI. 1959. Protoplasmic movement in the foraminiferan, *Allogromia laticollaris;* and a theory of its mechanism. Biol. Bull. **117:** 100–118.

JANDER, R., K. DAUMER, and T. H. WATERMAN. 1963. Polarized light orientation by two Hawaiian decapod cephalopods. Z. vergl. Physiol. **46:** 383–394.

JANSEN, J., P. ANDERSEN, and J. K. S. JANSEN. 1964. On the structure and innervation of the parietal muscle of the hagfish *(Myxine glutinosa)*. Acta Morph. Néerl. Scand. **5:** 329–338.

JANSKÝ, L. (ed.) 1971. Nonshivering thermogenesis. Academia, Prague. 310 p.

JANSKÝ, L. and J. S. HART. 1968. Cardiac output and organ blood flow in warm- and cold-acclimated rats exposed to cold. Can. J. Physiol. Pharmacol. **46:** 653–663.

JAYNES, J. 1969. The historical origins of "ethology" and "comparative psychology." Anim. Behav. **17:** 601–606.

JENKIN, P. M. 1970. Control of growth and metamorphosis. Pergamon, Oxford. 383 p.

JENNINGS, H. S. 1923. The behavior of lower organisms. Columbia U. P. New York. 366 p.

JENNINGS, J. B. 1962. Further studies on feeding and digestion in triclad Turbellaria. Biol. Bull. **123:** 571–581.

JENNINGS, J. B. 1965. Feeding, digestion and assimilation in animals. Pergamon, Oxford. 228 p.

JENNINGS, J. B. 1968. Nutrition and digestion. *In* Florkin and Scheer, **2:** 303–326.

JENNINGS, J. B., and R. GIBSON. 1969. Observations on the nutrition of seven species of rhynchocoelan worms. Biol. Bull. **136:** 405–433.

JEUNIAUX, C. 1961. Chitinase: an addition to the list of hydrolases in the digestive tract of vertebrates. Nature **192:** 131–136.

JHA, R. K. 1965. The nerve elements in silver-stained preparations of *Cordylophora*. Am. Zool. **5:** 431–438.

JHA, R. K., and G. O. MACKIE. 1967. The recognition, distribution and ultrastructure of hydrozoan nerve elements. J. Morph. **123:** 43–62.

JÖBSIS, F. F., and M. J. O'CONNOR. 1966. Calcium release and reabsorption in the sartorius muscle of the toad. Biochem. Biophys. Res. Comm. **25:** 246–252.

JOHANNES, R. E., S. J. COWARD, and K. L. WEBB. 1969. Are dissolved amino acids an energy source for marine invertebrates? Comp. Biochem. Physiol. **29:** 283–288.

JOHANSEN, K. 1960. Circulation in the hagfish, *Myxine glutinosa* L. Biol. Bull. **118:** 289–295.

JOHANSEN, K. 1965. Cardiovascular dynamics in fishes, amphibians, and reptiles. Ann. N.Y. Acad. Sci. **127:** 414–442.

JOHANSEN, K. 1970. Air breathing in fishes. *In* Hoar and Randall, **4:** 361–411.

JOHANSEN, K. 1971. Comparative physiology: Gas exchange and circulation in fishes. Ann. Rev. Physiol. **33:** 569–612.

JOHANSEN, K., C. LENFANT, and T. A. MECKLENBURG. 1970. Respiration in the crab, *Cancer magister*. Z. vergl. Physiol. **70:** 1–19.

JOHANSEN, K., and D. HANSON. 1968. Functional anatomy of the hearts of lungfishes and amphibians. Am. Zool. **8:** 191–210.

JOHANSEN, K., and A. W. MARTIN. 1962. Circulation in the cephalopod, *Octopus dofleini*. Comp. Biochem. Physiol. **5:** 161–176.

JOHANSEN, K., and A. W. MARTIN. 1965. Comparative aspects of cardiovascular function in vertebrates. Handbook of Physiology. (Am. Physiol. Soc.) Sect. 2, **3:** 2583–2614.

JOHANSEN, K., and A. W. MARTIN. 1966. Circulation in a giant earthworm, *Glossoscolex giganteus*. II. Respiratory properties of the blood and some patterns of gas exchange. J. Exp. Biol. **45:** 165–172.

JOHANSEN, P. H. 1967. The role of the pituitary in the resistance of the goldfish *(Carassius auratus* L.) to a high temperature. Can. J. Zool. **45:** 329–345.

JOHANSEN, P. H. 1968. The role of the pituitary of

the goldfish in the development of resistance to repeated high temperature exposures. Can. J. Zool. **46:** 805–806.

JOHNELS, A. G. 1956. On the origin of the electric organs in *Malapterurus*. Quart. J. Microscop. Sci. **97:** 455–464.

JOHNSON, B. E., F. DANIELS, JR., and I. A. MAGNUS. 1968. Responses of human skin to ultraviolet light. *In* Giese, **4:** 139–202.

JOHNSON, F. H., and H. EYRING. 1970. The kinetic basis of pressure effects in biology and chemistry. *In* Zimmerman (1970): 1–44.

JOHNSON, F. H., H. EYRING, and M. J. POLISSAR. 1954. The kinetic basis of molecular biology. Wiley, New York. 874 p.

JOHNSON, F. H., and Y. HANEDA (eds.). 1966. Bioluminescence in progress. Princeton U. P., Princeton, N.J. 649 p.

JOHNSON, F. H., and O. SHIMOMURA. 1968. The chemistry of luminescence in coelenterates. *In* Florkin and Scheer, **2:** 233–261.

JOHNSON, F. H., and O. SHIMOMURA. 1972. Enzymatic and nonenzymatic bioluminescence. *In* Giese, **7:** 275–334.

JOHNSON, L. B., and M. I. GROSSMAN. 1968. Secretin: the enterogastrone released by acid in the duodenum. Am. J. Physiol, **215:** 885–888.

JONES, D. R. 1972. Anaerobiosis and the oxygen debt in an anuran amphibian, *Rana esculenta* (L.) J. Comp. Physiol. **77:** 356–382.

JONES, D. R., and K. JOHANSEN. 1972. The blood vascular system of birds. *In* Farner and King, **2:** 157–285.

JONES, D. R., and G. SHELTON. 1964. Factors influencing submergence and the heart rate in the frog. J. Exp. Biol. **41:** 417–431.

JONES, F. R. H., and N. B. MARSHALL, 1953. The structure and function of the teleostean swimbladder. Biol. Rev. **28:** 16–83.

JONES, J. C. 1964. The circulatory system of insects. *In* Rockstein, **1:** 1–107.

JONES, J. D. 1961. Aspects of respiration in *Planorbis corneus* L. and *Lymnaea stagnalis* L. (Gastropoda: Pulmonata). Comp. Biochem. Physiol. **4:** 1–29.

JONES, J. D. 1963. The functions of the respiratory pigments of invertebrates. Problems in Biology (G. A. Kerkut, ed.) **1:** 9–90.

JONES, J. D. 1964. The role of haemoglobin in the aquatic pulmonate, *Planorbis corneus*. Comp. Biochem. Physiol. **12:** 283–295.

JONES, J. D. 1971. Comparative physiology of respiration. Arnold, London. 202 p.

JONES, H. D. 1970. Hydrostatic pressures within the heart and pericardium of *Patella vulgata* L. Comp. Biochem. Physiol. **34:** 263–272.

JONES, H. D. 1971. Circulatory pressures in *Helix pomatia* L. Comp. Biochem. Physiol. **39A:** 289–295.

JOOSSE, J. 1972. Endocrinology of reproduction in molluscs. Gen. Comp. Endocrinol. Suppl. **3:** 591–601.

JØRGENSEN, C. B. 1966. Biology of suspension feeding. Pergamon, Oxford. 357 p.

JØRGENSEN, C. B., and L. O. LARSEN. 1963. Neurohypophysial relationships. Symp. Zool. Soc. London **9:** 59–82.

JORGENSEN, C. B., and L. O. LARSEN. 1964. Further observations on molting and its hormonal control in *Bufo bufo* (L.). Gen. Comp. Endocrinol. **4:** 389–400.

JOSEPHSON, R. K. 1966. Neuromuscular transmission in a sea anemone. J. Exp. Biol. **45:** 305–319.

JOSEPHSON, R. K., and G. O. MACKIE. 1965. Multiple pacemakers and the behaviour of the hydroid *Tubularia*. J. Exp. Biol. **43:** 293–332.

JOSEPHSON, R. K., and J. UHRICH. 1969. Inhibition of pacemaker systems in the hydroid *Tubularia*. J. Exp. Biol. **50:** 1–14.

JUNGREIS, A. M. 1970. The effects of long-term starvation and acclimation temperature on glucose regulation and nitrogen anabolism in the frog, *Rana pipiens*. II. Summer animals. Comp. Biochem. Physiol. **32:** 433–444.

KAHL, M. P. 1963. Thermoregulation in the wood stork, with special reference to the role of the legs. Physiol. Zool. **36:** 141–151.

KAHLSON, G., and E. ROSENGREN. 1971. Biogenesis and physiology of histamine. Arnold, London. 318 p.

KALCKAR, H. M. 1969. Biological phosphorylations. Prentice-Hall, Englewood Cliffs, N.J. 735 p.

KALLMAN, K. D. 1962a. Gynogenesis in the teleost, *Mollienesia formosa* (Girard), with a discussion of the detection of parthenogenesis in vertebrates by tissue transplantation. J. Genetics **58:** 7–24.

KALLMAN, K. D. 1962b. Population genetics of the gynogenetic teleost, *Mollienesia formosa* (Girard). Evolution **16:** 497–504.

KAMEMOTO, F. I., and R. E. TULLIS 1972. Hydromineral regulation in decapod crustacea. Gen. Comp. Endocrinol. Suppl. **3:** 299–307.

KANATANI, H., and H. SHIRAI. 1972. On the maturation-inducing substance produced in starfish gonad by neural substance. Gen. Comp. Endocrinol. Suppl. **3:** 571–579.

KANDEL, E. R., and I. KUPFERMANN. 1970. The functional organization of invertebrate ganglia. Ann. Rev. Physiol. **32:** 193–258.

KARLSON, P. 1960. Pheromones. Ergeb. Biol. **22:** 212–225.

KARLSON, P., H. HOFFMEISTER, W. HOPPE, and R.

HUBER. 1963. Zur. Chemie des Ecdysons. Justus Liebigs Annl. Or. Chem. **662:** 1–20.

KARLSON, P., and M. LÜSCHER. 1959. Pheromones: a new term for a class of biologically active substances. Nature **183:** 55–56.

KARLSON, P., and C. E. SEKERIS. 1964. Biochemistry of insect metamorphosis. *In* Florkin and Mason, **6:** 221–243.

KATSUKI, Y. 1965. Comparative neurophysiology of hearing. Physiol. Rev. **45:** 380–423.

KATZ, A. M. 1970. Contractile proteins of the heart. Physiol. Rev. **50:** 63–158.

KATZ, B. 1950. Depolarization of sensory terminals and the initiation of impulses in the muscle spindle. J. Physiol. **111:** 261–282.

KATZ, B. 1959. Nature of the nerve impulse. *In* J. E. Oncley (ed.). Biophysical science—a study program. pp. 466–474. Wiley, New York.

KATZ, B. 1966. Nerve, muscle, and synapse. McGraw-Hill, New York. 193 p.

KATZ, B. 1969. The release of neural transmitter substance. Liverpool U. P., Liverpool. 60 p.

KATZ, B. 1971. Quantal mechanism of neural transmitter release. Science **173:** 123–126.

KELLER, H. U., and E. Sorkin. 1968. Chemotaxis of leucocytes. Experientia **24:** 641–652.

KELLOGG, W. N. 1961. Porpoises and sonar. University Press, Chicago. 177 p.

KENDEIGH, S. C. 1939. The relation of metabolism to the development of temperature regulation in birds. J. Exp. Zool. **82:** 419–438.

KENNEDY, D. 1966. The comparative physiology of invertebrate central neurons. Adv. Comp. Physiol. Biochem. **2:** 117–184.

KENNEDY, D. 1967. Small systems of nerve cells. Sci. Am. **216 (5):** 44–52.

KENNEDY, D., A. I. SILVERSTON, and M. P. REMLER. 1969. Analysis of restricted neural networks. Science **164:** 1488–1496.

KEOSIN, J. 1966. The Origin of Life. 2nd ed. Reinhold, New York.

KERKUT, G. A. 1963. The invertebrata. 4th ed. of Borradaile, L. A. *et al.* Cambridge U. P., London. 820 p.

KERKUT, G. A., L. C. BROWN, and R. J. WALKER. 1969a. Cholinergic IPSP by stimulation of the electrogenic sodium pump. Nature **223:** 864–865.

KERKUT, G. A., N. HORN, and R. J. WALKER. 1969b. Long-lasting synaptic inhibition and its transmitter in the snail *Helix aspersa.* Comp. Biochem. Physiol. **30:** 1065–1074.

KETTLEWELL, H. B. D. 1961. The phenomenon of industrial melanism in Lepidoptera. Ann. Rev. Entomol. **6:** 245–262.

KEYNES, R. D. 1957. Electric organs. *In* M. E. Brown, **2:** 323–343.

KEYNES, R. D., and G. W. MAISEL. 1954. The energy requirement for sodium extrusion from frog muscle. Proc. Roy. Soc. London B, **142:** 383–392.

KEYS, A. B. 1931. The heart-gill preparation of the eel and its perfusion for the study of a natural membrane *in situ.* Z. vergl. Physiol. **15:** 352–363 *also* 364–388.

KIDDER, G. W. 1967. Nitrogen: distribution, nutrition and metabolism. *In* Florkin and Scheer, **1:** 93–159.

KIDDER, G. W., and V. C. DEWEY. 1945. Studies on the biochemistry of *Tetrahymena.* V. The chemical nature of factors I and III. Arch. Biochem. **8:** 293–301.

KING, J. R., and D. S. FARNER. 1961. Energy metabolism, thermoregulation and body temperature. *In* Marshall, **2:** 215–288.

KINOSITA, H. 1963. Electrophoretic theory of pigment migration within fish melanophore. Ann. N.Y. Acad. Sci. **100:** 992–1004.

KINOSITA, H., and A. MURAKAMI. 1967. Control of ciliary motion. Physiol. Rev. **47:** 53–82.

KIORTSIS, V., and H. A. L. TRAMPUSCH. 1965. Regeneration in animals and related problems. North-Holland Publishing Co., Amsterdam. 568 p.

KIRIM, S. M. M. (ed.). 1972. The prostaglandins. Wiley-Interscience, New York. 327 p.

KIRSCHNER, L. B. 1967. Comparative Physiology: Invertebrate excretory organs. Ann. Rev. Physiol. **29:** 169–196.

KIRSCHNER, L. B. 1970. The study of NaCl transport in aquatic animals. Am. Zool., **10:** 365–376.

KITCHING, J. A. 1952. Contractile vacuoles. Symp. Soc. Exp. Biol. **6:** 145–165.

KITCHING, J. A. 1954. Osmoregulation and ionic regulation in animals without kidneys. Symp. Soc. Exp. Biol. **8:** 63–75.

KITCHING, J. A. 1967. Contractile vacuoles, ionic regulation, and excretion. *In* Chen, **1:** 308–336.

KLEEREKOPER, H. 1969. Olfaction in fishes. Indiana U. P., Bloomington, Indiana. 222 p.

KLEIBER, M. 1961. The fire of life. Wiley, New York. 454 p.

KLEINHOLZ, L. H. 1938. Studies in reptilian colour changes. J. Exp. Biol. **15:** 474–499.

KLEINHOLZ, L. H. 1961. Pigmentary effectors. *In* Waterman, **2:** 133–169.

KLOTZ, I. M., T. A. KLOTZ, and H. A. FIESS. 1957. The nature of the active site of hemerythrin. Arch. Biochem. Biophys. **68:** 284–299.

KNOBIL, E. 1961. The pituitary growth hormone: some physiological considerations. *In* M. X. Zarrow (ed.), Growth in living systems. pp. 353–381. Basic Books, New York.

KNOX, W. E. 1951. Two mechanisms which increase *in vivo* the liver tryptophan peroxidase activity: specific enzyme adaptation and stimulation of the pituitary-adrenal system. Brit. J. Exp. Path. **32:** 462–469.

KOLLER, G. 1929. Die innere Sekretion bei wirbellosen Tieren. Biol. Rev., **4:** 269–306.

KONISHI, J., and C. P. HICKMAN, JR. 1964. Temperature acclimation in the central nervous system of the rainbow trout *(Salmo gairdnerii)*. Comp. Biochem. Physiol. **13:** 433–442.

KOOYMAN, G. L. 1972. Deep diving behaviour and effects of pressure in reptiles, birds, and mammals. Symp. Soc. Exp. Biol. **26:** 295–312.

KOOYMAN, G. L. 1973. Respiratory adaptations in marine mammals. Am. Zool. **13:** 457–468.

KOPEĆ, S. 1922. Studies on the necessity of the brain for the inception of insect metamorphosis. Biol. Bull. **42:** 323–342.

KORNBERG, H. L., and H. A. KREBS. 1957. Synthesis of cell constituents from C_2-units by a modified tricarboxylic acid cycle. Nature **179:** 988–991.

KORRINGA, P. 1957. Lunar periodicity. Mem. Geol. Soc. Am. **67 (1):** 917–934.

KOSTO, B., G. E. PICKFORD, and M. FOSTER. 1959. Further studies of the hormonal induction of melanogenesis in the killifish, *Fundulus heteroclitus*. Endocrinol. **65:** 860–881.

KOVÁCS, G., and M. PAPP. 1972. XIIth European conference on animal blood groups and biochemical polymorphism. Junk, The Hague. 606 p.

KRAVITZ, E. A., L. L. IVERSEN, M. OTUSKA, and Z. W. HALL. 1968. Gamma-aminobutyric acid in the lobster nervous system; release from inhibitory nerves and uptake into nerve-muscle preparations. *In* U. S. von Euler, S. Skoglund, and U. Söderberg (eds.), Structure and function of inhibitory neuronal mechanisms. pp. 371–376. Pergamon, Oxford.

KREBS, H. A. 1935. Metabolism of amino-acids. III. Deamination of amino-acids. Biochem. J. **29:** 1620–1644.

KREBS, H. A. 1970. The history of the tricarboxylic acid cycle. Persp. Biol. Med. **14:** 154–170.

KREBS, H. A., and K. HENSELEIT. 1932. Untersuchungen über die Harnstoffbildung im Tierkörper. Z. Physiol. Chem. **210:** 33–66.

KREBS, H. A.. and H. L. KORNBERG. 1957. Energy transformations in living matter. Ergeb. Physiol. **49:** 212–298.

KRIEBEL, M. E. 1968a. Pacemaker properties of tunicate heart cells. Biol. Bull. **135:** 166–173.

KRIEBEL, M. E. 1968b. Cholinoceptive and adrenoceptive properties of the tunicate heart pacemaker. Biol Bull. **135:** 174–180.

KRIEBEL, M. E., M. V. L. BENNETT, S. G. WAXMAN, and G. P. PAPPAS. 1969. Oculomotor neurons in fish: electrotonic coupling and multiple sites of impulse initiation. Science **166:** 520–524.

KRIJGSMAN, B. J. 1952. Contractile and pacemaker mechanisms in the heart of arthropods. Biol. Rev. **27:** 320–346.

KRIJGSMAN, B. J., and G. A. DIVARIS. 1955. Contractile and pacemaker mechanisms of the heart of molluscs. Biol. Rev. **30:** 1–39.

KRISHNAKUMARAN, A., and H. A. SCHNEIDERMAN. 1969. Induction of molting in Crustacea by an insect molting hormone. Gen. Comp. Endocrinol. **12:** 515–518.

KRISHNAMURTHY, V. G., and H. A. BERN. 1973. Juxtaglomerular cell changes in the euryhaline freshwater fish *Tilapia mossambica* during adaptation to sea water. Acta Zool. **54:** 9–14.

KROGH, A. 1914. The quantitative relation between temperature and standard metabolism in animals. Int. Z. physik.-chem. Biol. **1:** 491–508.

KROGH, A. 1929. The anatomy and physiology of capillaries. Yale U. P. New Haven, Conn. 422 p. (reprinted, Hafner, New York 1959).

KROGH, A. 1939. Osmotic regulation in aquatic animals. Cambridge U. P., London. 242 p.

KROGH, A. 1941. The comparative physiology of respiratory mechanisms. Univ. Pennsylvania Press, Philadelphia. 172 p.

KUDO, R. R. 1966. Protozoology. 5th ed. C. C Thomas, Springfield, Ill. 1174 p.

KUHN, W., A. RAMEL, H. J. KUHN, and E. MARTI. 1963. The filling mechanism of the swimbladder. Experientia, **19:** 497–511.

KUIPER, J. W. 1962. The optics of the compound eye. Symp. Soc. Exp. Biol. **16:** 58–71.

KUTCHAI, H. and J. B. STEEN. 1971. The permeability of the swimbladder. Comp. Biochem. Physiol. **39A:** 119–123.

LAM, T. J. 1972. Prolactin and hydromineral regulation in fishes. Gen. Comp. Endocrinol. Suppl. **3:** 328–338.

LAMBERTSEN, C. J. (ed.). 1971. Underwater physiology. Academic Press, New York. 575 p.

LAMMENS, J. J. 1967. Growth and reproduction in a tidal flat population of *Macoma balthica* (L.). Netherlands J. Sea Res. **3:** 315–382.

LANG, F., A. SUTTERLIN, and C. L. PROSSER. 1970. Electrical and mechanical properties of the closer muscle of the Alaskan king crab *Paralithodes camtschatica*. Comp. Biochem. Physiol. **32:** 615–628.

LANGE, R. 1963. The osmotic functions of amino acids and taurine in the mussel, *Mytilus edulis*. Comp. Biochem. Physiol. **10:** 173–179.

LANGLEY, J. N. 1921. The autonomic nervous system. Heffer, Cambridge. 80 p.

LANGLEY, P. A. 1967. Experimental evidence for a hormonal control of digestion in the tsetse fly, *Glossina moristans* Westwood: a study of the larva, pupa, and teneral adult fly. J. Insect Physiol. **13:** 1921–31.

LANMAN, J. T. 1970. Delayed implantation of the blastocyst: An explanation of its effect on the developing embryo. Am. J. Obstet. Gynec. **106:** 463–468.

LANYON, W. E., and W. N. TAVOLGA. 1960. Animal sounds and communication. Am. Inst. Biol. Sc. Washington, D.C. 443 p.

LARSEN, L. O., and B. ROTHWELL. 1972. Adenohypophysis. *In* Hardisty and Potter, **2:** 1–67.

LASHLEY, K. S. 1950. In search of the engram. Symp. Soc. Exp. Biol. **4:** 454–482.

LASKOWSKI, M., SR. 1972. The poly(dA-dT) of crab. Prog. Nucleic Acad. Res. and Mol. Biol. **12:** 161–188.

LAUFER, H., and J. P. CALVET. 1972. Hormonal effects on chromosomal puffs and insect development. Gen. Comp. Endocrinol. Suppl. **3:** 137–148.

LAURENS, H. 1933. The physiological effects of radiant energy. The Chemical Catalog Co. Inc. New York. 610 p.

LAVERACK, M. S. 1963. The physiology of earthworms. Pergamon Press, London. 206 p.

LEA, D. E. 1962. Actions of radiations on living cells. 2nd ed. University Press, Cambridge. 416 p.

LEAVITT, W. W., and P. A. WRIGHT. 1965. The plant estrogen coumestrol, as an agent affecting hypophysial gonadotropic function. J. exp. Zool. **160:** 319–328.

LEBLOND, C. P., and B. E. WALKER. 1956. Renewal of cell populations. Physiol. Rev. **36:** 255–276.

LEDERIS, K. 1970. Active substances in the caudal neurosecretory system of bony fishes. Mem. Soc. Endocrinol. **18:** 465–484.

LEDERIS, K. 1972. Recent progress in research on the urophysis. Gen Comp. Endocrinol. Suppl. **3:** 339–344.

LEE, T. W., and J. W. CAMPBELL. 1965. Uric acid synthesis in the terrestrial snail, *Otala lactea.* Comp. Biochem. Physiol. **15:** 457–468.

LEES, A. D. 1955. The physiology of diapause in insects. University Press, Cambridge. 151 p.

LEES, A. D. 1960. Some aspects of animal photoperiodism. Cold Spring Harbor Symp. Quant. Biol. **25:** 261–268.

LEES, A. D. 1968. Photoperiodism in insects. *In* Giese, **4:** 47–137.

LEHMANN, H., and R. G. HUNTSMAN. 1961. Why are red cells the shape they are? The evolution of the human red cell. *In* R. G. Macfarlane and A. H. T. Robb-Smith (eds.), Functions of the blood. 73–148. Blackwell, Oxford.

LEHNINGER, A. L. 1962. Water uptake and extrusion by mitochondria in relation to oxidative phosphorylation. Physiol. Rev. **42:** 467–517.

LEHNINGER, A. L. 1964. The mitochondrion. Benjamin, New York. 263 p.

LEHNINGER, A. L. 1970. Biochemistry. Worth, New York. 833 p.

LEHNINGER, A. L. 1971. Bioenergetics. 2nd ed. Benjamin. Menlo Park, California. 245 p.

LEMBERG, M. R. 1969. Cytochrome oxidase. Physiol. Rev. **49:** 48–121.

LENFANT, C., and K. JOHANSEN. 1965. Gas transport by hemocyanin–containing blood of the cephalopod *Octopus dofleini.* Am. J. Physiol. **209:** 991–998.

LENFANT, C., and K. JOHANSEN. 1966. Respiratory function in the elasmobranch *Squalus suckleyi* G. Resp. Physiol. **1:** 13–29.

LENFANT, C., K. JOHANSEN, and G. C. GRIGG. 1966. Respiratory properties of blood and pattern of gas exchange in the lungfish *Neoceratodus foresteri.* (Krefft). Resp. Physiol. **2:** 1-21.

LENHOFF, H. M. 1968. Chemical perspectives on the feeding response, digestion and nutrition of selected coelenterates. *In* Florkin and Scheer, **2:** 157–221.

LENHOFF, H. M. 1969. pH Profile of a Peptide Receptor. Comp. Biochem. Physiol. **28:** 571–586.

LENTZ, T. L. 1968. Primitive nervous systems. Yale U. P., New Haven, Conn. 148 p.

LEVEY, R. H. 1964. The thymus hormone. Sci. Am. **211 (1):** 66–77.

LEVIN, R. L. 1969. The living barrier. Heinemann, London. 170 p.

LEVINSON, H. Z., and A. NAVON. 1969. Ascorbic acid and unsaturated fatty acids in the nutrition of the Egyptian cotton leafworm, *Prodenia litura.* J. Insect Physiol. **15:** 591–595.

LEWIS, R. W., and J. W. LENTFER. 1967. The Vitamin A content of polar bear liver: range and variability. Comp. Biochem. and Physiol. **22:** 923–926.

LEZZI, M. 1970. Differential gene activation in isolated chromosomes. Int. Rev. Cytol. **29:** 127–168.

LEZZI, M., and L. I. GILBERT. 1972. Hormonal control of gene activity in polytene chromosomes. Gen. Comp. Endocrinol. Suppl. **3:** 159–167.

LI, C. H. 1969. Comparative chemistry of pituitary lactogenic hormones. Gen. Comp. Endocrinol. Suppl. **2:** 1–9.

LICHT, P. 1972. Environmental physiology of reptilian breeding cycles: role of temperature. Gen. Comp. Endocrinol. Suppl. **3:** 477–488.

LICHT, P., and H. HOYER. 1968. Somatotropic effects of exogenous prolactin and growth hormone in juvenile lizards *(Lacerta s. sicula)*. Gen. Comp. Endocrin. **11:** 338–346.

LILEY, N. R. 1969. Hormones and reproductive behavior in fishes. *In* Hoar and Randall, **3:** 73–116.

LILLY, D. M. 1967. Growth factors in protozoa. *In* Florkin and Scheer, **1:** 275–307.

LINDBERG, O. (ed.). 1970. Brown adipose tissue. Elsevier, New York. 337 p.

LINDSTEDT, K. J. 1971. Biphasic feeding response in a sea anemone: control by asparagine and glutathione. Science **173:** 333–334.

LINDSTEDT, K. J., L. MUSCATINE, and H. M. LENHOFF. 1968. Valine activation of feeding in the sea anemone *Boloceroides*. Comp. Biochem. Physiol. **26:** 567–572.

LINZELL, J. L., and M. PEAKER. 1971. Mechanism of milk secretion. Physiol. Rev. **51:** 564–597.

LIORET, C., and A. MOYSE. 1963. Acid metabolism: the citric acid cycle and other cycles. *In* Florkin and Mason, **5:** 203–306.

LISK, R. D., L. A. CIACCIO, and L. A. REUTER. 1972. Neural receptor mechanisms for estrogens and progesterone and the regulation of ovulation and sex-related behavior in mammals. Gen. Comp. Endocrinol. Suppl. **3:** 553–564.

LISSMANN, H. W. 1958. On the function and evolution of electric organs in fish. J. Exp. Biol. **35:** 156–191.

LISSMANN, H. W. 1961. Ecological studies on gymnotids. *In* Chagas and Paes de Carvalho (1961), 215–226.

LISSMANN, H. W. 1963. Electric location by fishes. Sci. Am. **208 (3):** 50–59.

LISSMAN, H. W., and K. E. MACHIN. 1958. The mechanism of object location in *Gymnarchus niloticus* and similar fish. J. Exp. Biol. **35:** 451–493.

LLOYD, J. E. 1966. Studies on the flash communication system in *Photinus* fireflies. Michigan Univ. Misc. Pub. Museum Zoology. **130:** 1–95.

LOCKE, M. 1964. The structure and formation of the integument in insects. *In* Rockstein **3:** 379–470.

LOCKWOOD, A. P. M. 1962. The osmoregulation of Crustacea. Biol. Rev. **37:** 257–305.

LOCKWOOD, A. P. M. 1964. Animal body fluids and their regulation. Harvard U. P., Cambridge, Mass. 177 p.

LOCKWOOD, A. P. M. 1971. Membranes of animal cells. Arnold, London. 71 p.

LOEB, J. 1918. Forced movements, tropisms, and animal conduct. Lippincott, Philadelphia.

LOEWENSTEIN, W. R. 1960. Biological transducers. Sci. Am. **203 (2):** 98–108.

LOEWENSTEIN, W. R. 1961. Excitation and inactivation in a receptor membrane. Ann. N.Y. Acad. Sci. **94:** 510–534.

LOEWENSTEIN, W. R. 1966. Permeability of membrane junctions. Ann. N.Y. Acad Sci. **137:** 441–472.

LOEWENSTEIN, W. R. 1970. Intercellular communication. Sci. Am. **222 (5):** 79–86.

LOFTS, B. 1970. Animal photoperiodism. Arnold, London. 64 p.

LOFTS, B. 1972. The sertoli cell. Gen. Comp. Endocrinol. Suppl. **3:** 636–648.

LONG, J. A., and H. M. EVANS. 1922. The estrous cycle of the rat and its associated phenomena. Mem. Univ. Calif. **6:** 1–148.

LOOMIS, W. F. 1967. Skin-pigment regulation of Vitamin-D biosynthesis in man. Science **157:** 501–506.

LOOMIS, W. F. 1970. Rickets. Sci. Am. **223 (6):** 77–91.

LOWENSTEIN, O. 1967. The concept of the acousticolateral system. *In* Cahn (1967): 3–12.

LOWENSTEIN, O. 1971. The labyrinth. *In* Hoar and Randall, **5:** 207–240.

LOWY, J., and B. M. MILLMAN. 1962. Mechanical properties of smooth muscle of cephalopod molluscs. J. Physiol. **160:** 353–363.

LOWY, J., and B. M. MILLMAN. 1963. The contractile mechanism of the anterior byssus retractor muscle of *Mytilus edulis* Phil. Trans. Roy. Soc. London B. **246:** 105–148.

LOWY, J., B. M. MILLMAN, and J. HANSON. 1964. Structure and function in smooth tonic muscles of lamellibranch molluscs. Proc. Roy. Soc. London B. **160:** 525–536.

LUCE, G. G. 1971. Biological rhythms in human and animal physiology. Dover Publications, New York. 183 p.

LUNDBERG, A. 1958. Electrophysiology of salivary glands. Physiol. Rev. **38:** 21–40.

LÜSCHER, M. 1972. Environmental control of juvenile hormone (JH) secretion and caste differentiation in termites. Gen. Comp. Endocrinol. Suppl. **3:** 509–514.

LWOFF, A. 1947. Some aspects of the problem of growth factors for protozoa. Ann. Rev. Microbiol. **1:** 101–114.

LWOFF, A. (ed.). 1951. Biochemistry and physiology of protozoa. Vol. I. Academic Press, New York. 434 p.

LYMAN, C. P. 1943. Control of coat color in the varying hare *Lepus americanus* Erxleben. Bull. Mus. Comp. Zool. Harvard **93:** 393–461.

LYMAN, C. P., and P. O. CHATFIELD. 1955. Physiology of hibernation in mammals. Physiol. Rev. **35:** 403–425.

LYMAN, C. P., and A. R. DAWE (eds.). 1960. Mammalian hibernation. Bull. Mus. Comp. Zool. Harvard **124:** 1–549.

LYONS, W. R., C. H. LI, and R. E. JOHNSON. 1958. The hormonal control of mammary growth and lactation. Rec. Progr. Hormone Res. **14:** 219–254.

MACKIE, G. O. 1960. The structure of the nervous system in *Velella*. Quart. J. Microscop. Sci. **101:** 119–131.

MACKIE, G. O. 1965. Conduction in the nerve-free epithelia of siphonophores. Am. Zool. **5:** 439–453.

MACKIE, G. O. 1970. Neuroid conduction and the evolution of conducting tissues. Quart Rev. Biol. **45:** 319–332.

MACKLIN, M. 1967. Osmotic regulation in Hydra. Sodium and calcium localization and source of electrical potential. J. Cell. Physiol. **70:** 191–196.

MACKLIN, M., T. ROMA, and K. DRAKE. 1973. Water excretion in *Hydra*. Science **179:** 194–195.

MADDRELL, S. H. P. 1963. Excretion in the blood-sucking bug *Rhodnius prolixus*. Stål. I. the control of diuresis. J. Exp. Biol. **40:** 247–256.

MADDRELL, S. H. P. 1964. Excretion in the blood-sucking bug, *Rhodnius prolixus*. Stal. III. the control of the release of the diuretic hormone. J. Exp. Biol. **41:** 459–472.

MAETZ, J. 1963. Physiological aspects of neurohypophysial function in fishes with some reference to the amphibia. Symp. Zool. Soc. London, **9:** 107–140.

MAETZ, J. 1968. Salt and water metabolism. *In* Barrington and Jϕrgensen (1968): 47–162.

MAETZ, J., and F. GARCÍA-ROMEU. 1964. The mechanism of sodium and chloride uptake by the gills of a fresh-water fish, *Carassius auratus*. III. Evidence for NH_4^+/Na^+ and HCO_3^-/Cl^- exchanges. J. Gen. Physiol. **47:** 1209–1227.

MALOIY, G. M. O. 1972. Comparative physiology of desert animals. Symp. Zool. Soc. London **31:** 1–413.

MANGUM, C. P., and R. P. DALES. 1965. Products of haem synthesis in polychaetes. Comp. Biochem. Physiol. **15:** 237–257.

MANNING, A. 1967. An introduction to animal behaviour. Edward Arnold, London. 208 p.

MANSOUR, T. E. 1967. Effect of hormones on carbohydrate metabolism of invertebrates. Fed. Proc. **26:** 1179–1185.

MANWELL, C. 1958. On the evolution of hemoglobin. Respiratory properties of the hemoglobin of the California hagfish, *Polistotrema stouti*. Biol. Bull. **115:** 227–238.

MANWELL, C. 1960. Comparative physiology: blood pigments. Ann. Rev. Physiol. **22:** 191–244.

MANWELL, C. 1963. The blood proteins of cyclostomes. A study in phylogenetic and ontogenetic biochemistry. *In* Brodal and Fänge (1963): 372–458.

MANWELL, C. 1964. Chemistry, genetics and function of invertebrate respiratory pigments–configurational changes and allosteric effects. *In* F. Dichens and E. Neil (eds.), Oxygen in the organism. 49–119. Macmillan, New York.

MANWELL, C. 1966. Sea cucumber sibling species: polypeptide chain types and oxygen equilibrium of hemoglobin. Science **152:** 1393–1395.

MAREN, T. H. 1967a. Carbonic anhydrase: Chemistry, physiology, and inhibition. Physiol. Rev. **47:** 595–781.

MAREN, T. H. 1967b. Carbonic anhydrase in the animal kingdom. Fed. Proc. **26:** 1097–1103.

MARKS, V., and E. SAMOLS. 1968. Glucose homeostasis. *In* V. H. T. James, (ed.), Recent advances in endocrinology. 111–138. Churchill, London.

MARLER, P., and W. J. HAMILTON III. 1966. Mechanisms of animal behavior. Wiley, New York. 771 p.

MARSHALL, A. J. (ed.). 1961. Biology and comparative physiology of birds. Vols. 1 and 2. Academic Press, New York.

MARSHALL, E. K. 1934. The comparative physiology of the kidney in relation to theories of renal secretion. Physiol. Rev. **14:** 133–159.

MARSHALL, E. K., and H. W. SMITH. 1930. The glomerular development of the vertebrate kidney in relation to habitat. Biol. Bull. **59:** 135–153.

MARSHALL, N. B. 1954. Aspects of deep sea biology. Hutchinson, London. 380 p.

MARSHALL, N. B. 1960. Swimbladder structure of deep-sea fishes in relation to their systematics and biology. Discovery Rep. **31:** 1–122.

MARSHALL, N. B. 1972. Swimbladder organization and depth ranges of deep-sea teleosts. Symp. Soc. Exp. Biol. **26:** 261–272.

MARSLAND, D. A. 1944. Mechanism of pigment displacement in unicellular chromatophores. Biol. Bull. **87:** 252–261.

MARSLAND, D. A. 1956. Protoplasmic contractility in relation to gel structure: Temperature-pressure experiments on cytokinesis and amoeboid movement. Internat. Rev. Cytol. **5:** 199–227.

MARSLAND, D. A. 1958. Cells at high pressure. Sci Am. **199 (4):** 36–43.

MARTIN, A. W. 1957. Recent advances in knowledge of invertebrate renal function. *In* Scheer (1957): 247–276.

MARTIN, A. W., and F. M. HARRISON. 1966. Excretion. *In* Wilbur and Yonge **2:** 353–386.

MARTIN, A. W., and K. JOHANSEN, 1965. Adaptations of the circulation in invertebrate animals. Handbook of Physiology. (Am. Physiol. Soc.) Sect. 2, **3:** 2545–2581.

MARTOJA, M. 1972. Endocrinology of Mollusca. *In* Florkin and Scheer, **7:** 349–392.

MASLIN, T. P. 1971. Parthenogenesis in reptiles. Am. Zool. **11:** 361–380.

MATTHEWS, B. H. C. 1931. The response of a single end organ. J. Physiol. **71:** 64–110.

MATTHEWS, D. M. 1967. Absorption of water-soluble vitamins. Brit. Med. Bull. **23:** 258–262.

MATTHEWS, P. B. C. 1972. Mammalian muscle receptors and their central actions. Arnold, London. 630 p.

MATTHEWS, S. A. 1931. Observations on pigment migration within the fish melanophore. J. Exp. Zool. **58:** 471–486.

MAYNARD, D. M. 1955. Activity in a crustacean ganglion. II. Pattern and interaction in burst formation. Biol. Bull. **109:** 420–436.

MAYNARD, D. M. 1960. Circulation and heart function. *In* Waterman, **1:** 161–226.

MAYNARD, D. M. 1961. Thoracic neurosecretory structures in Brachyura. I. Gross anatomy. Biol. Bull. **121:** 316–329.

MEGLITSCH, P. A. 1972. Invertebrate zoology. 2nd ed. Oxford U. P. New York. 834 p.

MELLON, D. 1968. The physiology of sense organs. Freeman, San Francisco. 107 p.

MENAKER, M. 1972. Nonvisual light reception. Sci. Am. **226 (3):** 22–29.

MENDELSON, M. 1971. Oscillator neurons in crustacean ganglia. Science **171:** 1170–1173.

MERCER, E. H. 1959. An electron microscopic study of *Amoeba proteus*. Proc. Roy. Soc. London B **150:** 216–232.

METCALFE, J., H. BARTELS, and W. MOLL. 1967. Gas exchange in the pregnant uterus. Physiol. Rev. **47:** 782–838.

METZ, C. B., and A. MONROY. 1967–69. Fertilization. Vols. 1–2. Academic Press, New York.

MICHAELIS, L., and M. L. MENTEN. 1913. Der Kinetik der Invertinwirkung. Biochem. Zeit. **49:** 333–369.

MILEDI, R. 1969. Transmitter action in the giant synapse of the squid. Nature **223:** 1284–86.

MILEDI, R., and C. R. SLATER. 1966. The action of calcium on neuronal synapses in the squid. J. Physiol. **184:** 473–498.

MILES, F. A. 1969. Excitable cells. Heinemann, London. 147 p.

MILL, J. P., and M. F. KNAPP. 1970. The fine structure of obliquely striated body wall muscles in the earthworm *Lumbricus terrestris* Linn. J. Cell Sci. **7:** 233–261.

MILLER, J. F. A. P., and D. OSOBA. 1967. Current concepts of the immunological function of the thymus. Physiol. Rev. **47:** 437–520.

MILLER, P. L. 1964. Respiration—aerial gas transport. *In* Rockstein, **2:** 557–615.

MILLER, P. L. 1966a. The regulation of breathing in insects. Adv. Insect Physiol. **3:** 279–354.

MILLER, R. L. 1966b. Chemotaxis during fertilization in the hydroid *Campanularia*. J. exp. Zool. **162.** 23–44.

MILLER, S. L. 1953. A production of amino acids under possible primitive earth conditions. Science **117:** 528–529.

MILLER, S. L. 1955. Production of some organic compounds under possible primitive earth conditions. J. Am. Chem. Soc. **77:** 2351–2361.

MILLER, S. L., and L. E. ORGEL. 1974. The origins of life on the earth. Prentice-Hall, Englewood Cliffs N. J. 229 p.

MILLER, S. L., and H. C. UREY. 1959. Organic compound synthesis on the primitive earth. Science **130:** 245–251.

MILLER, W. H., F. RATLIFF, and H. K. HARTLINE. 1961. How cells receive stimuli. Sci. Am. **205 (3):** 233–238.

MILLMAN, B. M. 1967. Mechanism of contraction in molluscan muscle. Am. Zool. **7:** 583–591.

MILLOTT, N. 1960. The photosensitivity of sea urchins. Symp. Comp. Biol. (Academic Press), **1:** 279–293.

MILLOTT, N. 1968. The dermal light sense. Symp. Zool. Soc. London. **23:** 1–36.

MILLS, J. N. 1966. Human circadian rhythms. Physiol. Rev. **46:** 128–171.

MILLS, R. P., and R. C. KING. 1965. The pericardial cells of *Drosophila melanogaster*. Quart, J. Microscop. Sci. **106:** 261–8.

MILLS, R. R. 1967. Hormonal control of excretion in the American cockroach. I. Release of a diuretic hormone from the terminal abdominal ganglion. J. Exp. Biol. **46:** 35–41.

MILNE, L. J., and M. MILNE. 1959. Photosensitivity of invertebrates. Handbook of Physiology (Am. Physiol. Soc. Sect. 1, **1:** 621–645.

MILNE, L. J., and M. MILNE. 1962. The senses of animals and men. Atheneum, New York. 305 p.

MIZOGAMI, S., M. OGURI, H. SOKABE and H. NISHIMURA. 1968. Presence of renin in the glomerular and aglomerular kidney of marine teleosts. Am. J. Physiol. **215:** 991–994.

MOEHRES, F. P. 1960. Sonic orientation of bats and other animals. Symp. Zool. Soc. London. **3:** 57–66.

MOIR, R. J. 1968. Ruminant digestion and evolution. Handbook of Physiology (Am. Physiol. Soc.). Sect. 6, **5:** 2673–2694.

MONCRIEFF, R. W. 1967. The chemical senses. 3rd ed. Leonard Hill, London. 760 p.

MONROE, E. A., and S. E. MONROE. 1968. Origin of iridescent colors on the indigo snake. Science **159:** 97–98.

MONROY, A. 1965. Chemistry and physiology of

fertilization. Holt, Rinehart and Winston. New York. 150 p.
MONTAGNA, W. 1961. Skin and integument and pigment cells. *In* Brachet and Mirsky (1961) **5:** 267–322.
MONTAGNA, W., and R. E. BILLINGHAM. 1964. Wound healing. Adv. Biol. Skin. **5:** 1–254. Pergamon, Oxford.
MORALES, M. F. 1959. Mechanisms of muscle contraction. *In* Oncley (1959): 426–432.
MORDUE, W. 1969. Hormonal control of Malpighian tube and rectal function in the desert locust, *Schistocerca gregaria.* J. Insect Physiol. **15:** 273–285.
MORDUE, W. 1972. Hormones and excretion in locusts. Gen. Comp. Endocrinol. Suppl. **3:** 289–298.
MORIN, J. G., and J. W. HASTINGS. 1971. Energy transfer in a bioluminescent system. J. Cell. Physiol. **77:** 313–318.
MORRIS, R. 1965. Studies on salt and water balance in *Myxine glutinosa* (L.). J. exp. Biol. **42:** 359–71.
MORRISON, P., and F. A. RYSER. 1962. Metabolism and body temperature in a small hibernator, the meadow jumping mouse, *Zapus hudsonius.* J. Cell. Comp. Physiol. **60:** 169–180.
MORTON, J. E. 1952. The role of the crystalline style. Proc. Malac. Soc. London. **29:** 85–92.
MORTON, J. E. 1958. Molluscs. Hutchinson Univ. Library. London. 232 p.
MORTON, J. E. 1960. The functions of the gut in ciliary feeders. Biol. Rev. **35:** 92–140.
MORTON, J. E. 1967. Guts. Arnold, London. 58 p.
MORTON, R. A. 1971. Ubiquinones, plastoquinones and vitamins K. Biol. Rev. **46:** 47–96.
MOTAIS, R., and F. GARCÍA-ROMEU. 1972. Transport mechanisms in the teleostean gill and amphibian skin. Ann. Rev. Physiol. **34:** 141–176.
MOTT, J. C. 1957. The cardiovascular system. *In* M. E. Brown, **1:** 81–108.
MOULTON, D. J. 1971. The olfactory pigment. Handbook of Sensory Physiology (Springer-Verlag), 4 **(1):** 59–74.
MOUNIB, M. S., and J. S. EISAN. 1972. Fixation of carbon dioxide by the testes of rabbit and fish. Comp. Biochem. Physiol. **43B:** 393–401.
MOUNTCASTLE, V. B. (ed). 1968. Medical physiology. 12th ed. Vol. 1. Mosby, St. Louis, Mo. 1054 p.
MUIR, B. S., and J. I. KENDALL. 1968. Structural modifications in the gills of tunas and some other oceanic fishes. Copeia (1968) **2:** 388–398.
MÜLLER, J. 1826. Zur vergleichenden Physiologie des Gesichtssinnes des Menschen und der Thiere. C. Cnobloch, Leipzig. 462 p.
MÜLLER, M. 1967. Digestion. *In* Florkin and Scheer, **1:** 351–380.
MULLONEY, B. 1970. Structure of the giant fibers of earthworms. Science **168:** 994–996.
MUNTZ, W. R. A. 1964. Vision in frogs. Sci. Am. **210 (3):** 111–119.
MUNZ, F. W. 1971. Vision: visual pigments. *In* Hoar and Randall, **5:** 1–32.
MURRAY, R. G. 1969. Cell types in rabbit taste buds. *In* Pfaffmann (1969): 311–344.
MURRAY, R. G. 1971. Ultrastructure of taste receptors. Handbook of Sensory Physiology (Springer-Verlag), **4(2):** 31–50.
MURRAY, R. W. 1962. Temperature receptors. Adv. Comp. Physiol. Biochem. **1:** 117–175. *Also* Symp. Soc. Exp. Biol. **16:** 245–266.
MURRAY, R. W. 1966. Nerve membrane properties and thermal stimuli. *In* De Reuck and Knight (1966): 164–185.
MURRAY, R. W. 1971. Temperature receptors. *In* Hoar and Randall, **5:** 121–133.
MURRISH, D. E., and K. SCHMIDT-NIELSEN. 1970. Exhaled air temperature and water conservation in lizards. Resp. Physiol. **10:** 151–158.
MCAFEE, D. A., and P. GREENGARD. 1972. Adenosine 3′5′-monophosphate: electrophysiological evidence for a role in synaptic transmission. Science **178:** 310–312.
MCAFEE, D. A., M. SCHORDERET, and P. GREENGARD. 1971. Adenosine 3′,5′-monophosphate in nervous tissue: increase associated with synaptic transmission. Science **171:** 1156–1158.
MCALEAR, J. H., N. S. MILBURN, and G. B. CHAPMAN. 1958. The fine structure of Schwann cells, nodes of Ranvier and Schmidt-Lanterman incisures in the central nervous system of the crab, *Cancer irroratus.* J. Ultrastruct. Res. **2:** 171–176.
MCBEAN, R. L., and L. GOLDSTEIN. 1967. Ornithine—urea cycle activity in *Xenopus laevis*: adaptation in saline. Science **157:** 931–932.
MCCRONE, J. D. 1969. Spider venoms: biochemical aspects. Am. Zool. **9:** 153–156.
MCELROY, W. D. 1971. Cell physiology and biochemistry. 3rd ed. Prentice-Hall, Inc. Englewood Cliffs, N. J. 152 p.
MCELROY, W. D., and B. GLASS (ed.). 1961. A symposium on light and life. Johns Hopkins Press, Baltimore. 924 p.
MCELROY, W. D., and H. H. SELIGER. 1961. Mechanisms of bioluminescent reactions. In McElroy and Glass (1961): 219–257.
MCELROY, W. D., and H. H. SELIGER. 1962. Biological luminescence. Sci. Am. **207 (6):** 76–89.
MACFARLANE, R. G. 1964. An enzyme cascade in

the blood clotting mechanisms, and its function as a biochemical amplifier. Nature **202:** 498–499.

McGilvery, R. W. 1970. Biochemistry. Saunders, Philadelphia. 769 p.

McGuire, J., and G. Moellmann. 1972. Cytochalasin B: effects on microfilaments and movement of melanin granules within melanocytes. Science **175:** 642–644.

McInerney, J. E. 1964. Salinity preference: an orientation mechanism in salmon migration. J. Fish. Res. Bd. Canada, **21:** 995–1018.

McLeese, D. W. 1956. Effects of temperature, salinity and oxygen on the survival of the American lobster. J. Fish. Res. Bd. Canada, **13:** 247–272.

McLennan, H. 1970. Synaptic transmission. 2nd ed. Saunders, Philadelphia. 178 p.

McMahon, B. R. 1969. A functional analysis of the aquatic and aerial respiratory movements of the African lungfish, *Protopterus aethiopicus*, with reference to the evolution of the lung-ventilation mechanism in vertebrates. J. Exp. Biol. **51:** 407–430.

McMillan, I. K. R., and E. S. Machell. 1961. The technique of induced hypothermia. Brit. Med. Bull. **17:** 32–36.

MacNichol, E. F., Jr., 1958. Subthreshold excitation processes in the eye of *Limulus*. Exp. Cell. Res. Supp. **5:** 411–425.

Naisse, J. 1966. Contrôle endocrinien de la différenciation sexuelle chez l'Insecte *Lampyris noctiluca*. Arch. Biol. **77:** 139–201.

Needham, A. E. 1952. Regeneration and wound-healing. Methuen, London. 152 p.

Needham, A. E. 1964a. The growth process in animals. Pitman, London. 522 p.

Needham, A. E. 1964b. Biological considerations of wound healing. Adv. Biol. Skin, **5:** 1–38.

Needham, A. E. 1965. The uniqueness of biological materials. Pergamon, Oxford. 593 p.

Needham, A. E. 1970a. Haemostatic mechanisms in the invertebrates. Symp. Zool. Soc. London **27:** 19–44.

Needham, A. E. 1970b. Nitrogen metabolism in Annelida. *In* J. W. Campbell, **1:** 207–297.

Needham, A. E. 1974. The significance of zoochromes. Springer-Verlag, Heidelberg. 429 p.

Needham, J. 1931. Chemical embryology. Vol. 2. 615–1253. University Press, Cambridge.

Needham, J. 1942. Biochemistry and morphogenesis. University Press, Cambridge. 785 p.

Nelsen, O. E. 1953. Comparative embryology of the vertebrates. McGraw-Hill, New York. 982 p.

Neuhaus, O. W., and J. E. Halver. 1969. Fish in research. Academic Press, New York. 311 p.

Neurath, H. 1964. Protein-digesting enzymes. Sci. Am. **211 (6):** 68–79.

Neurath, H., K. A. Walsh, and W. P. Winter. 1967. Evolution of structure and function of proteases. Science **158:** 1638–1644.

Newell, R. C. 1970. Biology of intertidal animals. Logos Press, London. 555 p.

Newell, R. E. 1971. The global circulation of atmospheric pollutants. Sci. Am. **244 (1):** 32–42.

Newth, D. R., and D. M. Ross. 1955. On the reaction to light of *Myxine glutinosa* L. J. Exp. Biol. **32:** 4–21.

Nicol, J. A. C. 1952. Autonomic nervous systems in lower chordates. Biol. Rev. **27:** 1–49.

Nicol, J. A. C. 1960. The regulation of light emission in animals. Biol. Rev. **35:** 1–42.

Nicol, J. A. C. 1961. The tapetum in *Scyliorhinus canicula*. J. Mar. Biol. Ass. U.K. **41:** 271–277.

Nicol, J. A. C. 1962. Animal luminescence. Adv. Comp. Physiol. Biochem. **1:** 217–273.

Nicol, J. A. C. 1963. Some aspects of photoreception and vision in fishes. Adv. Marine Biol. **1:** 171–208.

Nicol, J. A. C. 1964. Special effectors. *In* Wilbur and Yonge, **1:** 353–381.

Nicol, J. A. C. 1965. Migration of chorioidal tapetal pigment in the spur dog *Squalus acanthias*. J. Mar. Biol. Ass. U.K. **45:** 405–427.

Nicol, J. A. C. 1967a. The biology of marine animals. 2nd ed. Pitman, London. 699 p.

Nicol, J. A. C. 1967b. The luminescence of fishes. Symp. Zool. Soc. London, **19:** 27–55.

Nicol, J. A. C. 1969. Bioluminescence. *In* Hoar and Randall, **3:** 355–400.

Nicol, J. A. C., and H. J. Arnott. 1972. Riboflavin in the eyes of gars (Lepisosteidae). Can. J. Zool. **50:** 1211–1214.

Nicol, J. A. C., and H. J. Arnott. 1973. Tapeta lucida in bony fishes (Actinopterygii): a survey. Can. J. Zool **51:** 69–81.

Nicoll, C. S., and H. A. Bern. 1971. On the actions of prolactin among the vertebrates: is there a common denominator? *In* G. E. W. Wolstenholme and J. Knight (eds.), Ciba Foundation Symposium on Lactogenic Hormones. 299–324. Churchill, London.

Niel, C. B. van. 1935. Photosynthesis of bacteria. Cold Spring Harbor Symp. Quant. Biol. **3:** 138–150.

Niel, C. B. van. 1954. The chemoautotrophic and photosynthetic bacteria. Ann. Rev. Microbiol. **8:** 105–132.

Nilsson, A. 1970. Gastrointestinal hormones in the holocephalian fish *Chimaera monstrosa* (L.). Comp. Biochem. Physiol. **32:** 387–390.

Nilsson, A., and R. Fänge. 1970. Digestive proteases in the cyclostome *Myxine glutinosa* (L.). Comp. Biochem. Physiol. **32:** 237–250.

Nilsson, S. 1971. Adrenergic innervation and drug

responses of the oval sphincter in the swim-bladder of the cod (*Gadus morhua*). Acta Physiol. Scand. **83:** 446–453.

NIRENBERG, M. W. 1963. The genetic code: II. Sci. Am. **208 (3):** 80–94.

NOBLE-NESBITT, J. 1970. Water balance in the firebrat, *Thermobia domestica* (Packard). The site of uptake of water from the atmosphere. J. Exp. Biol. **52:** 193–200.

NOSSAL, G. J. V. 1964. How cells make antibodies. Sci. Am. **211 (6):** 106–115.

NOVÁK, V. J. A. 1964. The phylogenetic origin of neurosecretion. Gen. Comp. Endocrinol. **4:** 696–703.

NOVALES, R. R., and R. FUJII. 1970. A melanin-dispersing effect of cyclic adenosine monophosphate on *Fundulus* melanophores. J. Cell. Physiol. **75:** 133–136.

NOVICK, A. 1959. Acoustic orientation in the cave swiftlet. Biol. Bull. **117:** 497–503.

NOVIKOFF, M. M. 1953. Regularity of form in organisms. Systematic Zool. **2:** 57–62.

ODELL, W. D., and D. L. MOYER. 1971. Physiology of Reproduction. Mosby, St. Louis, Mo. 152 p.

O'DELL, R., and B. SCHMIDT-NIELSEN. 1960. Concentrating ability and kidney structure. Fed. Proc. **19:** 366.

OGAWA, M. 1968. Osmotic and ionic regulation in goldfish following removal of the corpuscles of Stannius or the pituitary gland. Can. J. Zool. **46:** 669–676.

OGAWA, M., M. OGURI, H. SOKABE, and H. NISHIMURA. 1972. Juxtaglomerular apparatus in the vertebrates. Gen. Comp. Endocrinol. Suppl. **3:** 374–381.

OHNO, S., U. WOLF, and N. B. ATKIN. 1968. Evolution from fish to mammals by gene duplication. Hereditas **59:** 169–187.

OKON, E. E. 1970. The effect of environmental temperature on the production of ultrasounds by isolated non-handled albino mouse pups. J. Zool. Lond. **162:** 71–83.

OLIVER, J. H., Jr. (ed.). 1971. Parthenogenesis. Am. Zool. **11:** 239–398.

OLSEN, M. W., and S. J. MARSDEN. 1954. Natural parthenogenesis in turkey eggs. Science **120:** 545–546.

ONCLEY, J. L. (ed.). 1959. Biophysical science—a study program. Wiley, New York. 568 p.

OORD, A. VAN DEN. 1964. The absence of cholesterol synthesis in the crab, *Cancer pagurus* L. Comp. Biochem. Physiol. **13:** 461–467.

OPARIN, A. I. 1936. The origin of life. Translation by S. Margulius (1953). Dover Publications, New York. 270p.

ORGAN, A. E., E. C. BOVEE, T. L. JAHN, D. WIGG, and J. R. FONSECA. 1968. The mechanism of the nephridial apparatus of *Paramecium multimicronucleatum*. I. Expulsion of water from the vesicle. J. Cell. Biol. **37:** 139–145.

ORGAN, A. E., E. C. BOVEE, and T. L. JAHN. 1969. The mechanism of the nephridial apparatus of *Paramecium multimicronucleatum* II. The filling of the vesicle by action of the ampullae. J. Cell. Biol. **40:** 389–394.

OTTOSON, D. 1970. Electrical signs of olfactory transducer action. *In* Wolstenholme and Knight (1970): 343–356.

OU, L. C., and S. M. TENNEY. 1970. Properties of mitochondria from hearts of cattle acclimatized to high altitude. Resp. Physiol. **8:** 151–159.

OVERTON, E. 1902. Beiträge zur allgemeinen Muskel- und Nervenphysiologie. Pflüger's Arch. Ges. Physiol. **92:** 346–386.

OWEN, G. 1956. Observations on the stomach and digestive diverticula of the Lamellibranchia. Quart. J. Microscop. Sci. **97:** 541–567.

OWEN, G. 1966. Digestion. *In* Wilbur and Yonge, **2:** 53–96.

PAGÉ, E. 1957. Body composition and fat deposition in rats acclimated to cold. Rev. Can. Biol. **16:** 269–278.

PAGE, R. M. 1968. Phototropism in fungi. *In* Giese, **3:** 65–90.

PALMER, M. F., and G. CHAPMAN. 1970. The state of oxygenation of haemoglobin in the blood of living *Tubifex* (Annelida). J. Zool. Lond. **161:** 203–209.

PANTIN, C. F. A. 1935. The nerve net of the Actinozoa. J. Exp. Biol. **12:** 119–164.

PANTIN, C. F. A. 1950. Behaviour patterns in lower invertebrates. Symp. Soc. Exp. Biol. **4:** 175–195.

PANTIN, C. F. A. 1952. The elementary nervous system. Proc. Roy. Soc. London B, **140:** 147–168.

PANTIN, C. F. A. 1956a. The origin of the nervous system. Pubbl. Staz. Zool. Napoli, **28:** 171–181.

PANTIN, C. F. A. 1956b. Comparative physiology of muscle. Brit. Med. Bull. **12:** 199–202.

PANTIN, C. F. A. 1965. Capabilities of the coelenterate behavior machine. Am. Zool. **5:** 581–589.

PAPERMASTER, B. W., R. M. CONDIE, and R. A. GOOD. 1962. Immune response in the California hagfish. Nature **196:** 355–357.

PAPERMASTER, B. W., R. M. CONDIE, J. K. FINSTAD, R. A. GOOD, and A. E. GABRIELSEN. 1963. Phylogenetic development of adaptive immunity. Fed. Proc. **22:** 1152–1155.

PAPI, F. 1969. Light emission, sex attraction and male flash dialogues in a firefly, *Luciola lusitanica* (Charp.) Monitore Zool. Ital. **3:** 135–184.

PAPKOFF, H. 1972. Subunit interrelationships among the pituitary glycoprotein hormones. Gen. Comp. Endocrinol. Suppl. **3:** 609–616.

PAPPAS, G. D., and D. P. PURPURA. 1972. Structure and function of synapses. Raven Press, New York. 308 p.

PÁRDUCZ, B. 1967. Ciliary movement and coordination in ciliates. Int. Rev. Cytol. **21:** 91–128.

PARKER, G. A. 1970. Sperm competition and its evolutionary consequences in the insects. Biol. Rev. **45:** 525–567.

PARKER, G. H. 1919. The elementary nervous system. Lippincott, Philadelphia. 229 p.

PARKER, G. H. 1922. Smell, taste, and allied senses in the vertebrates. Lippincott, Philadelphia. 192 p.

PARKER, G. H. 1948. Animal colour changes and their neurohumors. London. 377 p.

PARKES, A. S. (ed.). 1956. Marshall's physiology of reproduction. 3rd ed. Vol. 1(1). Longmans, Green and Co., London. 688 p.

PARKES, A. S. (ed.). 1960a. Marshall's physiology of reproduction. 3rd ed. Vol. 1(2): Longmans, Green and Co. London. 877 p.

PARKES, A. S. 1960b. The biology of spermatozoa and artificial insemination. *In* Parkes Vol. 1(2): 161–263.

PARKES, A. S., and R. DEANSLY. 1966. The ovarian hormones. *In* A. S. Parkes (ed.). Marshall's physiology of reproduction. **3:** 570–828.

PARRY, G. 1966. Osmotic adaptation in fishes. Biol. Rev. **41:** 392–444.

PARSONS, T. R., and W. PARSONS. 1923. Observations on the transport of carbon dioxide in the blood of some marine invertebrates. J. Gen. Physiol. **6:** 153–166.

PARSONS, T. S. 1970. The nose and Jacobson's organ. *In* Gans, **2:** 99–191.

PASSANO, L. M. 1960. Molting and its control. *In* Waterman (1960), **1:** 473–536.

PASSANO, L. M., and C. B. MCCULLOUGH. 1965. Co-ordinating systems and behaviour in *Hydra.* II. The rhythmic potential system. J. Exp. Biol. **42:** 205–231.

PASTAN, I. 1972. Cyclic AMP. Sci. Am. **227 (2):** 97–105.

PAX, R. A. 1969. The abdominal cardiac nerves and cardio-regulation in *Limulus polyphemus.* Comp. Biochem. Physiol. **28:** 293–305.

PAYNE, R. S. 1971. Acoustic location of prey by barn owls. *(Tyto alba).* J. Exp. Biol. **54:** 535–573.

PEACHEY, L. D. 1968. Muscle. Ann. Rev. Physiol. **30:** 401–440.

PEARSON, O. P. 1960. Torpidity in birds. Bull. Mus. Comp. Zool. Harvard, **124:** 93–103.

PEDLER, C. 1969. Rods and cones—a new approach. Int. Rev. Gen. Exp. Zool. **4:** 219–274.

PENGELLEY, E. T., and S. J. ASMUNDSON. 1971. Annual biological clocks. Sci. Am. **224 (4):** 72–79.

PENGELLEY, E. T., and K. C. FISHER. 1963. The effect of temperature and photoperiod on the yearly hibernating behavior of captive golden-mantled ground squirrels *(Citellus lateralis tescorum).* Can. J. Zool. **41:** 1103–1120.

PEQUIGNAT, E. 1966. "Skin digestion" and epidermal absorption in irregular and regular urchins and their probable relation to the outflow of spherule-coelomocytes. Nature **210:** 397–399.

PERKINS, E. B. 1928. Color changes in crustaceans, especially in *Palaemonetes.* J. Exp. Zool. **50:** 71–105.

PERKS, A. M. 1969. The neurohypophysis. *In* Hoar and Randall, **2:** 111–205.

PERKS, A. M., and E. VIZSOLYI. 1973. Studies of the neurohypophysis in foetal mammals. *In* K. S. Cromline, K. W. Cross, G. S. Dawes, and P. W. Nathanielsz (eds.). Foetal and neonatal physiology: proceedings of the Sir Joseph Barcroft symposium. pp. 430–438. Cambridge U. P., Cambridge.

PERRY, S. V. 1960. Muscular contraction. *In* Florkin and Mason, **2:** 245–340.

PERUTZ, M. F. 1964. The hemoglobin molecule. Sci. Am. **211 (5):** 64–76.

PERUTZ, M. F. 1969. Structure and function of hemoglobin. Harvey Lectures **63:** 213–261.

PETTUS, D. 1958. Water relationships in *Natrix sipedon,* Copeia (1958), **3:** 207–211.

PFAFFMANN, C. (ed.). 1969. Olfaction and taste. Rockefeller U. P. New York. 648 p.

PFAFFMANN, C. 1971. Sensory reception of olfactory cues. Biol. Reprod. **4:** 327–343.

PFEIFFER, W. 1962. The fright reaction of fish. Biol. Rev. **37:** 495–511.

PFEIFFER, W. 1963. Alarm substances. Experientia, **19:** 113–123.

PFEIFFER, W. 1966. Die Verbreitung der Schreckreaktion bei Kaulquappen und die Herkunft des Schreckstoffes. Z. vergl. Physiol. **52:** 79–98.

PFENNING, N. 1967. Photosynthetic bacteria. Ann. Rev. Microbiol. **21:** 285–324.

PHILLIPS, D. M. 1968. Exceptions to the prevailing pattern of tubules ((9+9+2) in the sperm flagella of certain insect species. J. Cell. Biol. **40:** 28–43.

PHILLIPS, J. E. 1964. Rectal absorption in the desert locust, *Schistocerca gregaria* Fôrskal. J. Exp. Biol. **41:** 15–80.

PHILLIPS, J. E. 1969. Osmotic regulation and rectal absorption in the blowfly *Calliphora erythrocephala.* Can. J. Zool. **47:** 851–863.

PHILLIPS, J. E. 1970. Apparent water transport by insect excretory systems. Am. Zool. **10:** 413–436.

PHILLIS, J. W. 1970. The pharmacology of synapses. Pergamon, Oxford. 358 p.

PICK, J. 1970. The autonomic nervous system. Lippincott, Philadelphia. 483 p.

PICKFORD, G. E., and J. W. ATZ. 1957. The physiology of the pituitary gland of fishes. N.Y. Zoological Society, New York. 613 p.

PICKFORD, G. E., and F. B. GRANT. 1967. Serum osmolarity in the coelacanth, *Latimeria chalumnae:* Urea retention and ion regulation. Science **155:** 568–570.

PIKE, J. E. 1971. Prostaglandins. Sci. Am. **225 (5):** 84–92.

PINSKER, H., and E. R. KANDEL. 1969. Synaptic activation of an electrogenic sodium pump. Science **163:** 931–935.

PITELKA, D. R. 1969. Fibrillar systems in Protozoa. *In* Chen **3:** 279–388.

PITMAN, R. M. 1971. Transmitter substances in insects: a review. Comp. Gen. Pharmacol. **2:** 347–371.

PITTENDRIGH, C. S., and D. M. MINIS. 1964. The entrainment of circadian oscillations by light and their role in photoperiodic clocks. Am. Naturalist **98:** 261–294.

PLASS, G. N. 1959. Carbon dioxide and climate. Sci. Am. **201 (1):** 41–47.

PLATE, L. 1891. Studien über opisthopneumone Lungenschnecken. Zool. Jb. Abt. Anat. u. Ontog. **4:** 505–630.

PLATZER, E. G., and J. R. ADAMS. 1967. The life history of a dracunculoid *Philonema oncorhynchi,* in *Oncorhynchus nerka.* Can. J. Zool. **45:** 31–43.

POLLER, L. (ed.) 1969. Recent advances in blood coagulation. Churchill, London. 362 p.

POLYA, J. B., and A. J. WIRZ. 1965. Studies on carbonic anhydrase. II. Occurrence of the enzyme in some invertebrates. Enzymologia **29:** 27–37.

PON, N. G. 1964. Expressions of the pentose phosphate cycle. *In* Florkin and Mason, **7:** 1–92.

PORTER, K. R., and G. E. PALADE. 1967. Studies on the endoplasmic reticulum. III. Its form and distribution in striated muscle cells. J. Biophys. Biochem. Cytol. **3:** 269–310.

PORTER, R. 1970. Breathing: Hering-Breuer centenary symposium. Ciba Foundation Symposium. Churchill, London. 402 p.

PORTER, R., and J. KNIGHT (eds.), 1971, High altitude physiology. Ciba Foundation Symposium. Churchill, London. 196 p.

POTTS, W. T. W. 1967. Excretion in the molluscs. Biol. Rev. **42:** 1–41.

POTTS, W. T. W. 1968. Osmotic and ionic regulation. Ann. Rev. Physiol. **30:** 73–104.

POTTS, W. T. W., and G. PARRY. 1964a. Osmotic and ionic regulation in animals. Pergamon, London. 423 p.

POTTS, W. T. W., and G. PARRY. 1964b. Sodium and chloride balance in the prawn, *Palaemonetes varians.* J. Exp. Biol. **41:** 591–601.

PRAKASH, R. 1957. Structure, development and phylogeny of the impulse conducting (connecting) tissue of the vertebrate heart. J. Anat. Soc. India, **6:** 30–39.

PRECHT, H. 1958. Concepts of the temperature adaptation of unchanging reaction systems of cold-blooded animals. *In* Prosser (1958): 50–78.

PRECHT, H., J. CHRISTOPHERSEN, and H. HENSEL. 1955. Temperatur und Leben. Springer-Verlag, Berlin. 514 p.

PRICHARD, R. K., and P. J. SCHOFIELD. 1968a. Phosphoenolpyruvate carboxykinase in the adult liver fluke, *Fasciola hepatica.* Comp. Biochem. Physiol. **24,** 773–786.

PRICHARD, R. K., and P. J. SCHOFIELD. 1968b. A comparative study of the tricarboxylic acid cycle enzymes in *Fasciola hepatica* and rat liver. Comp. Biochem. Physiol. **25:** 1005–1019.

PRICHARD, R. K., and P. J. SCHOFIELD. 1969. The glyoxylate cycle, fructose-1, 6-diphosphatase and glyconeogenesis in *Fasciola hepatica.* Comp. Biochem. Physiol. **29:** 581–590.

PRINCE, J. H. 1956. Comparative anatomy of the eye. C. C. Thomas, Springfield, Illinois. 418 p.

PRINGLE, J. W. S. 1949. The excitation and contraction of the flight muscles of insects. J. Physiol. **108:** 226–232.

PRINGLE, J. W. S. 1954. The mechanism of myogenic rhythm of certain insect striated muscles. J. Physiol. **124:** 269–291.

PRINGLE, J. W. S. 1957. Insect flight. University Press, Cambridge. 132 p.

PRINGLE, J. W. S. 1965a. Locomotion: flight. *In* Rockstein **2:** 283–329.

PRINGLE, J. W. S. (ed.). 1965b. Essays on physiological evolution. Pergamon, Oxford. 364 p.

PRINGLE, J. W. S. 1967. The contractile mechanism of insect fibrillar muscle. Progr. Biophys. Mol. Biol. **17:** 1–60.

PROSSER, C. L. 1933. Action potentials in the nervous system of the crayfish. II. Responses to illumination of the eye and caudal ganglion. J. Cell. Comp. Physiol. **4:** 363–377.

PROSSER, C. L. 1955. Physiological variation in animals. Biol. Rev. **30:** 229–262.

PROSSER, C. L. (ed.). 1958. Physiological adaptation. American Physiological Society. Washington, D.C. 185 p.

PROSSER, C. L. 1960. The comparative physiology of activation of muscles, with particular attention to smooth muscles. *In* Bourne, **2:** 387–434.

PROSSER, C. L. (ed.). 1967. Molecular mechanisms of temperature adaptation. Am. Assoc. Sci. Washington, D.C. 390 p.

PROSSER, C. L. 1969. Principles and general concepts of adaptation. Environ. Research, **2:** 404–416.

PROSSER, C. L. (ed.). 1973. Comparative animal physiology. 3rd ed. Saunders, Philadelphia. 966 p.

PROSSER, C. L., and F. A. BROWN. 1961. Comparative animal physiology. 2nd ed. Saunders, Philadelphia. 688 p.

PSCHEIDT, G. R. 1963. Biochemical correlates with phyletic division of the nervous system. T. Theor. Biol. **5:** 52–56.

PUMPHREY, R. J. 1940. Hearing in insects. Biol. Rev. **15:** 107–132.

PUMPHREY, R. J. 1950. Hearing. Symp. Soc. Exp. Biol. **4:** 3–18.

PUMPHREY, R. J. 1961. Sensory organs: Hearing *In* Marshall (1961), **2:** 69–86.

PURPLE, R. L., and F. A. DODGE. 1965. Interaction of excitation and inhibition in the eccentric cell of the eye of *Limulus.* Cold Spring Harbor Symp. Quant. Biol. **30:** 529–537.

PYE, J. D. 1968. Hearing in bats. *In* A. V. S. De Reuck, and J. Knight (eds.). Hearing mechanisms in the vertebrates. pp. 66–88. Ciba Foundation Symposium, Churchill, London.

QUARTON, G. C., T. MELNECHUK, and F. O. SCHMITT, (eds.). 1967. The neurosciences. Rockefeller U. P., New York. 962 p.

QUTOB, Z. 1962. The swimbladder of fishes as a pressure receptor. Arch. Néerl. Zool. **15:** 1–67.

RABINOWITCH, E. I. 1948. Photosynthesis. Sci. Am. **179 (2):** 25–35.

RACE, R. R., and R. SANGER. 1968. Blood groups in man. 5th ed. Blackwell, Oxford. 599 p.

RAHN, H. 1967. Gas transport from the external environment to the cell. *In* A. V. S. De Reuck, and R. Porter (eds.), Development of the lung. Ciba Found. Symp. pp. 3–29. Churchill, London.

RAHN, H. and F. W. BAUMGARDNER. 1972. Temperature and acid-base regulation in fish. Resp. Physiol. **14:** 171–182.

RALL, D. P., P. SCHWAB and C. G. ZUBROD. 1961. Alteration of plasma proteins at metamorphosis in the lamprey *(Petromyzon marinus dosatus).* Science **133:** 279–280.

RALPH, C. L. 1967. Recent developments in invertebrate endocrinology. Am. Zool. **7:** 145–160.

RAMSAY, J. A. 1955. The excretory system of the stick insect, *Dixippus morosus* (Orthoptera, Phasmidae). J. Exp. Biol. **32:** 183–199.

RAMSAY, J. A. 1956. Excretion by the Malpighian tubules of the stick insect, *(Dixippus morosus)* (Orthoptera, Phasmidae): calcium, magnesium, chloride, phosphate, and hydrogen ions. J. Exp. Biol. **33:** 697–708.

RAMSAY, J. A. 1958. Excretion by the Malpighian tubules of the stick insect, *Dixippus morosus* (Orthoptera, Phasmidae): amino acids, sugars and urea. J. Exp. Biol. **35:** 871–891.

RAMSAY, J. A. 1964. The rectal complex of the mealworm *Tenebrio molitor,* L. (Coleoptera, Tenebrionidae). Phil. Trans. Roy. Soc. London. B **248:** 279–314.

RAMSAY, J. A. 1968. A physiological approach to the lower animals. 2nd ed. University Press, Cambridge. 150 p.

RANDALL, D. J. 1968. Functional morphology of the heart in fishes. Am. Zool. **8:** 179–189.

RANDALL, D. J. 1970a. The circulatory system. *In* Hoar and Randall, **4:** 133–172.

RANDALL, D. J. 1970b. Gas exchange in fish. *In* Hoar and Randall, **4:** 253–292.

RANDALL, D. J., and J. N. CAMERON. 1973. Respiratory control of arterial pH as temperature changes in rainbow trout *Salmo gairdneri.* Am. J. Physiol. **225:** 997–1002.

RANDALL, D. J., and D. R. JONES. 1973. The effect of deafferentation of the pseudobranch on the respiratory response to hypoxia and hyperoxia in the trout *(Salmo gairdneri).* Resp. Physiol. **17:** 291–301.

RAO, G. M. M. 1968. Oxygen consumption of rainbow trout *(Salmo gairdneri)* in relation to activity and salinity. Can. J. Zool. **46:** 781–786.

RAO, K. P. 1958. Oxygen consumption as a function of size and salinity in *Metapenaeus monoceros* Fab. from marine and brackish-water environments. J. Exp. Biol. **35:** 307–313.

RASMUSSEN, H., and M. M. PECHET. 1970. Calcitonin. Sci. Am. **223 (4):** 42–50.

RAVEN, C. P. 1961. Oogenesis: the storage of developmental information. Pergamon, Oxford. 274 p.

RAVEN, C. P. 1966. An outline of developmental physiology. 3rd ed. Pergamon, Oxford. 230 p.

READ, C. P. 1968. Intermediary metabolism of flatworms. *In* Florkin and Scheer **2:** 327–357.

READ, K. R. H. 1966. Molluscan hemoglobin and myoglobin. *In* Wilbur and Yonge **2:** 209–232.

READ, K. R. H. 1968. The myoglobins of the gastropod molluscs *Busycon contrarium* Conrad, *Lunatia heros* Say, *Littorina littorea* L. and *Siphonaria gigas* Sowerby. Comp. Biochem. Physiol **25,** 81–94.

READ, L. J. 1968. Urea and trimethylamine oxide levels in elasmobranch embryos. Biol. Bull **135:** 537–547.

REDFIELD, A. C. 1934. The haemocyanins. Biol. Rev. **9:** 175–212.

REDMOND, J. R. 1968. The respiratory function of hemocyanin. *In* F. Ghiretti (ed.), Biochemistry and physiology of haemocyanins. pp 4–23. Academic Press, New York.

REEVES, R. B. 1972. An amidazole alphastat hypothesis for vertebrate acid-base regulation; tissue carbon dioxide content and body temperature in bullfrogs. Resp. Physiol. **14:** 219–236.

Reid, R. G. B. 1968. The distribution of digestive tract enzymes in lamellibranchiate bivalves. Comp. Biochem. Physiol. **24:** 727–744.

Reimer, A. A. 1971. Chemical control of feeding behavior in *Palythoa* (Zoanthidea, Coelenterata). Comp. Biochem. Physiol. **40A:** 19–38.

Reiter, R. J. (ed.). 1970. Comparative endocrinology of the pineal. Am. Zool. **10:** 187–267.

Reiter, R. J. 1973. Comparative physiology: pineal gland. Ann. Rev. Physiol. **35:** 305–328.

Riddick, D. H. 1968. Contractile vacuoles in the amoeba, *Pelomyxa carolinensis*. Am. J. Physiol. **215:** 736–740.

Riddle, O. 1963. Prolactin in vertebrate function and organization. J. Nat. Cancer. Inst. **31:** 1039–1110.

Ridgway, E. B. and C. C. Ashley. 1967. Calcium transients in single muscle fibers. Biochem. Biophys. Res. Comm. **29:** 229–234.

Riedesel, M. L. 1960. The internal environment during hibernation. Bull. Mus. Comp. Zool. Harvard **124:** 421–435.

Riegel, J. A. 1972. Comparative physiology of renal excretion. Oliver and Boyd, Edinburgh. 204 p.

Riegel, J. A., and L. B. Kirschner. 1960. The excretion of inulin and glucose by the crayfish antennal gland. Biol. Bull. **118:** 296–307.

Rigg, G. B., and L. A. Swain. 1941. Pressure-composition relationships of the gas in the marine brown alga, *Nereocystis luetkeana*. Plant Physiol., **16:** 361–371.

Riggs, A. 1965. Functional properties of hemoglobins. Physiol. Rev. **45:** 619–673.

Riggs, A. 1970. Properties of fish hemoglobins. *In* Hoar and Randall, **4:** 209–252.

Riley, J. P., and R. Chester. 1971. Introduction to marine chemistry. Academic Press, London. 465 p.

Rittenberg, S. C. 1969. The roles of exogenous organic matter in the physiology of chemolithotropic bacteria. Adv. Microbial Physiol. **3:** 159–196.

Roberts, B. L. 1969. The buoyancy and locomotory movements of electric rays. J. Mar. Biol. Ass. U.K. **49:** 621–640.

Robertson, J. D. 1957. The habitat of the early vertebrates. Biol. Rev. **32:** 156–187.

Robertson, J. D. 1960. Studies of the chemical composition of muscle tissues. J. Exp. Biol. **37:** 879–888.

Robertson, J. N. 1967. Fundamentals of acid-base regulation. Blackwell, Oxford. 109 p.

Robison, G. A., R. W. Butcher, and E. W. Sutherland. 1971. Cyclic AMP. Academic Press, New York. 531 p.

Robson, E. A., and R. K. Josephson. 1969. Neuromuscular properties of mesenteries from the sea-anemone *Metridium*. J. Exp. Biol. **50:** 151–168.

Rockstein, M. 1964. The physiology of Insecta. Vols. 1–3. Academic Press, New York.

Rockstein, M. 1971. The distribution of phosphoarginine and phosphocreatine in marine invertebrates. Biol. Bull. **141:** 167–175.

Roeder, K. D. 1951. Movements of the thorax and potential changes in the thoracic muscles of insects during flight. Biol. Bull. **100:** 95–106.

Roeder, K. D. (ed.). 1953. Insect physiology. Wiley, New York. 1100 p.

Roeder, K. D. 1966. Auditory system of noctuid moths. Science **154:** 1515–1521.

Roeder, K. D., and A. E. Treat. 1957. Ultrasonic reception by the tympanic organ of noctuid moths. J. Exp. Zool. **134:** 127–157.

Rohrlich, S. T., and K. R. Porter. 1972. Fine structural observations relating to the production of color by the iridophores of a lizard, *Anolis carolinensis*. J. Cell. Biol. **53:** 38–52.

Röller, H., and K. H. Dahm. 1968. The chemistry and biology of juvenile hormone. Recent Prog. Hormone Research **24:** 651–680.

Röller, H., K. H. Dahm, C. C. Sweeley, and B. M. Trost. 1967. The structure of the juvenile hormone. Angew. Chem. Internat. Edit. **6:** 179–180.

Romer, A. S. 1946. The early evolution of fishes. Quart. Rev. Biol. **21:** 33–69.

Romer, A. S. 1963. The "ancient history" of bone. Ann. N.Y. Acad. Sci., **109:** 168–176.

Romer, A. S. 1970. The vertebrate body. 4th ed. Saunders, Philadelphia. 601 p.

Root, R. 1931. The respiratory function of the blood of marine fishes. Biol. Bull. **61:** 427–456.

Roots, B. I., and C. L. Prosser. 1962. Temperature acclimation and the nervous system of fish. J. Exp. Biol. **39:** 617–629.

Rose, S. M. 1970. Regeneration. Appleton-Century-Crofts, New York. 264 p.

Rose, W. C. 1938. The nutritive significance of the amino acids. Physiol. Rev. **18:** 109–136.

Rose, W. C. 1949. Amino acid requirements of man. Fed. Proc. **8:** 546–552.

Rosen, W. G. 1962. Cellular chemotropism and chemotaxis. Quart. Rev. Biol. **37:** 242–259.

Rosenberg, T. 1948. On accumulation and active transport in biological systems: I. Thermodynamic considerations. Acta Chem. Scand. **2:** 14–33.

Rosenbluth, J. 1967. Obliquely striated muscle. III. Contraction mechanism of *Ascaris* body muscle. J. Cell. Biol. **34:** 15–33.

Rosenfeld, I., and O. A. Beath. 1964. Selenium. Academic Press, New York. 411 p.

Rosenthal, G. M. 1957. The role of moisture and temperature in the local distribution of the

plethodontid salamander *Aneides lugubris.* Univ. Calif. Pub. Zool. **54:** 371–420.
ROSENZWEIG, M. R., E. L. BENNETT, and M. C. DIAMOND. 1972. Brain changes in response to experience. Sci. Am. **226 (2):** 22–30.
ROSS, D. M. 1965. Complex and modifiable behavior patterns in *Calliactis* and *Stomphia.* Am. Zool. **5:** 573–580.
ROSS, D. M., and L. SUTTON. 1964. The swimming response of the sea anemone *Stomphia coccinea* to electrical stimulation. J. Exp. Biol. **41:** 731–749.
ROSS, R. 1969. Wound healing. Sci. Am. **220 (6):** 40–50.
ROSSI, A. C. 1969. Chemical signals and nest-building in two species of *Colisa* (Pisces, Anabantidae). Monitore Zool. Ital. **3:** 225–237.
ROTHSCHILD, M. 1965. Fleas. Sci. Am. **213 (6):** 44–53.
ROTHSCHILD, M., and B. FORD. 1964. Breeding of the rabbit flea *Spilopsyllus cuniculi* (Dale) controlled by the reproductive hormones of the host. Nature **201:** 103–104.
ROTHSCHILD, M., B. FORD, and M. HUGHES. 1970. Maturation of the male rabbit flea *(Spilopsyllus cuniculi)* and the oriental rat flea *(Xenopsylla chaeopis):* some effects of mammalian hormones on development and impregnation. Trans. Zool. Soc. London **32:** 105–188.
ROTHSCHILD, N. M. V. 1956. Fertilization. Methuen, London. 170 p.
ROTHSTEIN, M. 1968. Nematode biochemistry—IX. Lack of sterol biosynthesis in free-living nematodes. Comp. Biochem. Physiol. **27:** 309–317.
ROTHSTEIN, M., and H. MAYOH. 1966. Nematode biochemistry—VIII. Malate synthetase. Comp. Biochem. Physiol. **17:** 1181–1188.
ROTRUCK, J. T., A. L. POPE, H. E. GANTHER, A. B. SWANSON, D. G. HAFEMAN, and W. G. HOEKSTRA. 1973. Selenium: biochemical role as a component of glutathione peroxidase. Science, **179:** 588–590.
ROWLEY, D. 1962. Phagocytosis. Adv. Immunol. **2:** 241–264.
RUBEY, W. W. 1951. Geologic history of sea water. Bull. Geol. Soc. Am. **62:** 1111–1148.
RUBIN, R. P. 1970. The role of calcium in the release of neurotransmitter substances and hormones. Pharm. Rev. **22:** 389–428.
RUBNER, M. 1968. The laws of energy consumption in nutrition. U.S. Army. Research Inst. Environ. Med. Natick, Mass., U.S.A. 371 p. (Trans.)
RUCH, T C., and H. D. PATTON (eds.). 1965. Physiology and biophysics. Saunders, Philadelphia. 1242 p.
RUCK, P. 1962. On photoreceptor mechanisms of retinula cells. Biol. Bull. **123:** 618–634.
RÜEGG, J. C. 1961. On the tropomyosin–paramyosin system in relation to the viscous tone of lamellibranch "catch" muscle. Proc. Roy. Soc. London B, **154:** 224–249.
RUSHFORTH, N. B. 1965. Inhibition of contraction responses of *Hydra.* Am. Zool. **5:** 505–513.
RUSHMER, R. F., and O. A. SMITH, JR. 1959. Cardiac control. Physiol. Rev., **39:** 41–68.
RUSHTON, W. A. H. 1962. Visual pigments in man. Liverpool U. P., Liverpool. 38 p. *See also:* Symp. Soc. Exp. Biol. **16:** 12–31 *and* Sci. Am. **207 (5):** 120–132.
RUSSELL, F. E. 1969. Poisons and venoms. *In* Hoar and Randall, **3:** 401–449.
RUSSELL, F. E., and P. R. SAUNDERS (eds.). 1967. Animal toxins. Pergamon, Oxford. 428 p.
RUSSELL, I. J. 1971. The role of the lateral-line efferent system in *Xenopus laevis.* J. Exp. Biol. **54:** 621–641.
RUUD, J. T. 1954. Vertebrates without erythrocytes and blood pigment. Nature **173:** 848–850.
RYLEY, J. F. 1967. Carbohydrates and respiration. *In* Florkin and Scheer **1:** 55–92.
SACKTOR, B. 1965. Energetics and respiratory metabolism of muscular contraction. *In* Rockstein **2:** 483–580.
SAGE, M. 1973. The evolution of thyroidal function in fishes. Am. Zool. **13:** 899–905.
SALT, G. 1970. The cellular defence reactions of insects. Cambridge U. P. London. 118 p.
SALT, G W., and E. ZEUTHEN. 1960. The respiratory system. *In* Marshall, **1:** 363–409.
SALT, R. W. 1959. Role of glycerol in the cold-hardening of *Bracon cephi* (Gahan). Can. J. Zool. **37:** 59–69
SALT, R W. 1964. Terrestrial animals in cold: arthropods. Handbook of Physiol. (Am. Physiol. Soc) **4:** 349–359.
SANDOW, A. 1965. Excitation–contraction coupling in skeletal muscle. Pharmacol. Rev. **17:** 265–320
SATCHELL, G. H. 1959. Respiratory reflexes in the dogfish. J. Exp. Biol. **36:** 62–71.
SATCHELL, G. H. 1960. The reflex co-ordination of the heart beat with respiration in the dogfish. J. Exp. Biol. **37:** 719–731.
SATCHELL, G. H., and H. K. WAY. 1962. Pharyngeal proprioceptors in the dogfish *Squalus acanthias* L. J. Exp. Biol. **39:** 243–250.
SATIR, P. 1961. Cilia. Sci. Am. **204 (2):** 108–116.
SATIR, P. 1965. Studies of cilia. II. Examination of the distal region of the ciliary shaft and the role of the filaments in motility. J. Cell Biol. **26:** 805–834.
SAUNDERS, R. L. 1961. The irrigation of the gills in fishes. I. Studies of the mechanism of branchial irrigation. Can. J. Zool. **39:** 637–653.
SAUNDERS, R. L. 1962. The irrigation of the gills in

fishes. II. Efficiency of oxygen uptake in relation to respiratory flow activity and concentration of oxygen and carbon dioxide. Can J. Zool. **40:** 817–862

SAWYER, W. H. 1967. Evolution of antidiuretic hormones and their functions. Amer J. Med. **42:** 678–686.

SAWYER, W. H. 1968. Phylogenetic aspects of the neurohypophysial hormones *In* O. Eichler, *et al.* (eds), Handb. Exp. Pharm. N. S., **23:** 717–747. Springer-Verlag, Heidelberg.

SAWYER, W. H. 1972. Neurohypophysial hormones and water and sodium excretion in the African lungfish. Gen. Comp. Endocrinol. Suppl. **3:** 345–349.

SAZ, H. L. 1969. Carbohydrate and energy metabolism of nematodes and Acanthocephala. *In* Florkin and Scheer, **3:** 329–360.

SCHACHTER, M. 1969. Kallikreins and kinins. Physiol. Rev. **49:** 509–547.

SCHALLY, A. V., A. ARIMURA, and A. J KASTIN. 1973. Hypothalamic regulatory hormones. Science **179:** 341–350.

SCHARRER, B. 1936. Über "Drüsen-Nervenzellen" im Gehirn von *Nereis virens* Sars. Zool. Ang. **113:** 299–302.

SCHARRER, B. 1970. General principles of neuroendocrine communication. *In* F. O. Schmidt (ed.). The neurosciences: second study program. pp. 519–529. Rockefeller U. P., New York.

SCHARRER, E. 1965. The final common path in neuroendocrine integration. Arch. Anat. Microscop. **54:** 359–370.

SCHARRER, E., and B. SCHARRER. 1963. Neuroendocrinology. Columbia U. P. New York. 289 p.

SCHEER, B T. (ed.). 1957. Recent advances in invertebrate physiology. Univ. Oregon Pub., Eugene, Ore. 304 p.

SCHEER, B. T. 1960. The neuroendocrine system of arthropods. Vitamins and Hormones **18:** 141–204.

SCHEER, B. T. 1969. Carbohydrates and carbohydrate metabolism: Annelida, Sipunculida, Echiurida. *In* Florkin and Scheer, **4:** 135–145.

SCHERBA, G. 1962. Mound temperature of the ant *Formica ulkei* Emery. Am. Midland Nat. **67:** 373–385.

SCHILLING, J. A. 1968. Wound healing. Physiol. Rev. **48:** 374–423.

SCHLIEPER, C. 1958. Physiologie des Brackwassers. Die Binnengewässer **22:** 217–330.

SCHMIDT-NIELSEN, B., and L. E. DAVIS. 1968. Fluid transport and tubular intercellular spaces in reptilian kidneys. Science **159:** 1105–1108.

SCHMIDT-NIELSEN, B., and C. R. SCHRAUGER. 1963. *Amoeba proteus:* studying the contractile vacuole by micropuncture. Science **139:** 606–607.

SCHMIDT-NIELSEN, K. 1959. The physiology of the camel. Sci. Am. **201 (6):** 140–151.

SCHMIDT-NIELSEN, K. 1964. Desert animals. Clarendon Press, Oxford. 277 p.

SCHMIDT-NIELSEN, K. 1971. How birds breathe. Sci. Am. **225 (6):** 72–79.

SCHMIDT-NIELSEN, K. 1972. How animals work. Cambridge U. P., London. 114 p.

SCHMIDT-NIELSEN, K., F. R. HAINSWORTH, and D. E MURRISH. 1970. Counter-current heat exchange in the respiratory passages: effect on water and heat balance. Resp. Physiol. **9:** 263–276.

SCHMIDT-NIELSEN, K., and B. SCHMIDT-NIELSEN. 1952. Water metabolism of desert animals. Physiol. Rev **32:** 135–166.

SCHMIDT-NIELSEN, K., C. R. TAYLOR, and A. SHIKOLNIK. 1972. Desert snails: problems of survival. Symp. Zool. Soc. Lond. **31:** 1–13.

SCHMITT, O. H. 1959. Biological transducers and coding. *In* Oncley (1959): 492–503.

SCHNEIDER, D. 1969. Insect olfaction: deciphering system for chemical messages. Science **163:** 1031–1037.

SCHNEIDER, D., and R. A. STEINBRECHT. 1968. Checklist of insect olfactory sensilla. Symp. Zool. Soc. Lond. **23:** 279–297.

SCHOFFENIELS, E., and R. GILLES. 1970. Osmoregulation in aquatic arthropods. *In* Florkin and Scheer **5:** 255–286.

SCHOFFENIELS, E., and R. GILLES. 1972. Ionoregulation and osmoregulation in Mollusca. *In* Florkin and Scheer, **7:** 393–420.

SCHOLANDER, P. F. 1940. Experimental investigations on the respiratory function in diving mammals and birds. Hvalrådets Skr. **22:** 1–131.

SCHOLANDER, P. F. 1955. Evolution of climatic adaptation in homeotherms. Evolution **9:** 15–26.

SCHOLANDER, P. F. 1958. Studies on man exposed to cold. Fed. Proc. **17:** 1054–1057.

SCHOLANDER, P. F., E. BRADSTREET, and W. F. GAREY. 1962. Lactic acid response in the grunion. Comp. Biochem. Physiol. **6:** 201–203.

SCHOLANDER, P. F., L. VAN DAM, J. W. KANWISHER, H. T. HAMMEL, and M. S. GORDON. 1957. Supercooling and osmoregulation in Arctic fish. J. Cell. Comp. Physiol. **49:** 5–24.

SCHOLES, J. 1965. Discontinuity of the excitation process in locust visual cells. Cold Spring Harbor Symp. Quant. Biol. **30:** 517–527.

SCHUETZ, A. W. 1969. Induction of oocyte shedding and meiotic maturation in *Pisaster ocraceus:* kinetic aspects of radial nerve factor and ovarian factor induced changes. Biol. Bull: **137:** 524–534.

SCHUETZ, A. W. 1972. Estrogens and ovarian follicular function in *Rana pipiens*. Gen. Comp. Endocrinol **18:** 32–36.

SCHULTZ, R. J. 1971. Special adaptive problems associated with unisexual fishes. Am. Zool. **11:** 351–360.

SCHULTZ, S. G. and P. F. CURRAN. 1970. Coupled transport of sodium and organic solutes. Physiol. Rev. **50:** 637–718.

SCHULTZE-WESTRUM, T. G. 1969. Social communication by chemical signals in flying phalangers *(Petaurus breviceps papuanus). In* Pfaffmann (1969): 268–277.

SCHWANZARA, S. A. 1967. The visual pigments of freshwater fishes. Vision Res. **7:** 121–148.

SCHWARTZKOPFF, J. 1963. Vergleichende Physiologie des Gehörs und der Lautäusserungen. Fortschritte Zool. **15:** 214–336.

SEBEOK, T. A. (ed.). 1968. Animal communication. Indiana U. P., Bloomington. 686 p.

SEDALLIAN, A. 1968. Recherche des germes chitinolytiques dans le tube digestif d'un Sélacien *(Scyliorhinus canicula)*. C. R. Soc. Biol. Paris **162:** 949–951.

SEEGERS, W. H. 1969. Blood clotting mechanisms: three basic reactions. Ann. Rev. Physiol. **31:** 269–294.

SEGAAR, J., and R. NIEUWENHUYS. 1963. New ethophysiological experiments with male *Gasterosteus aculeatus,* with anatomical comment. Behaviour, **11:** 331–346.

SEHNAL, F. 1971. Endocrines of arthropods. *In* Florkin and Scheer, **6B:** 307–345.

SEKERIS, C. E. 1972. Ecdysone and RNA synthesis in target organs. Gen. Comp. Endocrinol. Suppl. **3:** 149–158.

SELWOOD, L. 1968. Interrelationships between developing oocytes and ovarian tissues in the chiton *Sypharochiton septentriones* (Ashby) (Mollusca, Polyplacophora). J. Morph. **125:** 71–104.

SELYE, H. 1949. Textbook of endocrinology. 2nd ed. Acta Endocrinol. Inc. Montreal. 914 p.

SELYE, H. 1956. The stress of life. McGraw-Hill, New York. 324 p.

SELYE, H. 1961. Nonspecific resistance. Ergeb. Pathol. **41:** 208–241.

SERVENTY, D. L. 1971. Biology of desert birds. *In* Farner and King, **1:** 287–339.

SHALER, R. 1972. An eagle's eye: quality of the retinal image. Science **176:** 920–922.

SHARMAN, G. B., and P. J. BERGER. 1969. Embryonic diapause in marsupials. Adv. Rep. Physiol. **4:** 211–240.

SHAW, J. 1960. The mechanisms of osmoregulation. *In* Florkin and Mason, **2:** 471–518.

SHAW, J., and R. H. STOBBART. 1972. The water balance and osmoregulatory physiology of the desert locust *(Schistocerca gregaria)* and other desert and xeric arthropods. Symp. Zool. Soc. London, **31:** 15–38.

SHAW, S. R. 1969a. Optics of arthropod compound eye. Science **165:** 88–90.

SHAW, S. R. 1969b. Interreceptor coupling in ommatidia of drone honeybee and locust compound eyes. Vision Res. **9:** 999–1029.

SHELTON, G. 1970. Regulation of breathing. *In* Hoar and Randall **4:** 293–359.

SHERRINGTON, C. 1929. Some functional problems attaching to convergence. Proc. Roy. Soc. London **B 105:** 332–362.

SHERRINGTON, C. 1947. The integrative action of the nervous system. 2nd ed. Yale University Press. New Haven. 413 p.

SHIMOMURA, O., and F. H. JOHNSON. 1972. Structure of the light-emitting moiety of aequorin. Biochemistry, **11:** 1602–1608.

SHIMOMURA, O., F. H. JOHNSON, and T. MASUGI. 1969. Cypridina bioluminescence: light-emitting oxyluciferin-luciferase complex. Science **164:** 1299–1300.

SHROPSHIRE, W., JR. 1972. Phytochrome, a photochromic sensor. *In* Giese, **7:** 33–72.

SHULMAN, R. G., S. OGAWA, K. WÜTHRICH, T. YAMANE, J. PEISACH, and W. E. BLUMBERG. 1969. The absence of "heme-heme" interactions in hemoglobin. Science **165:** 251–257.

SIBAOKA, T. 1966. Action potentials in plant organs. Symp. Soc. Exp. Biol. **20:** 49–73.

SIMPSON, L., H. A. BERN, and R. S. NISHIOKA. 1966. Survey of the evidence for neurosecretion in gastropod molluscs. Am. Zool. **6:** 123–138.

SINDERMANN, C. J., and D. F. MAIRS. 1959. A major blood group system in Atlantic sea herring. Copeia (1959) **3:** 228–232.

SINGER, C. 1959. A short history of scientific ideas to 1900. Oxford U. P.. London. 525 p.

SINSHEIMER, R. L. 1962 Single-stranded DNA. Sci. Am. **207 (1):** 109–116.

SJÖSTRAND, F. S. 1959. The ultrastructure of the retinal receptors of the vertebrate eye. Ergeb. Biol. **21:** 128–160.

SKADHAUGE, E. 1972. Salt and water excretion in xerophilic birds. Symp. Zool. Soc. London **31:** 113–131.

SLEIGH, M. A. 1962. The biology of cilia and flagella. Pergamon, Oxford. 242 p.

SLEIGH, M. A. 1968. Patterns of ciliary beating. Symp. Soc. Exp. Biol. **22:** 131–168.

SLEIGH, M. A. 1969. Coordination of the rhythm of beat in some ciliary systems. Int. Rev. Cytol. **25:** 31–54.

SLIFER, E. H. 1970. The structure of arthropod chemoreceptors. Ann. Rev. Entomol. **15:** 121–142.

SLIJPER, E. J. 1962. Whales. Hutchinson, London. 475 p.

SMARSH, A., H. H. CHAUNCEY, and M. R. CARRIKER. 1969. Carbonic anhydrase in the accessory boring organ of the gastropod, *Urosalpinx*. Am. Zool. **9:** 967–982.

SMITH, A. D. 1972. Subcellular localization of noradrenaline in sympathetic neurons. Pharm. Rev. **24:** 435–457.

SMITH, A. U. 1954. Effects of low temperatures on living cells and tissues. In R. J. C. Harris, (ed.), Biological applications of freezing and drying. pp. 1–62. Academic Press, New York.

SMITH, A. U. 1958. The resistance of animals to cooling and freezing. Biol. Rev. **33:** 197–253.

SMITH, D. S. 1965. The flight muscles of insects. Sci. Am. **212 (6):** 77–88.

SMITH, D. S. 1966. The organization and function of the sarcoplasmic reticulum and T-system in muscle cells. Progr. Biophys. **16:** 107–142.

SMITH, H. W. 1951. The kidney. Oxford U. P. New York. 1049 p.

SMITH, H. W. 1953. From fish to philosopher. Little, Brown and Company, Boston. 264 p.

SMITH, H. W. 1956. Principles of renal physiology. Oxford U. P., New York. 237 p.

SMITH, P. E. 1930. Hypophysectomy and a replacement therapy in the rat. Am. J. Anat. **45:** 205–273.

SMITH, R. E., and B. A. HORWITZ. 1969. Brown fat and thermogenesis. Physiol. Rev. **49:** 330–425.

SMITH, R. I. 1958. On reproductive pattern as a specific characteristic among nereid polychaetes. Syst. Zool. **7:** 60–73.

SMITH, R. T., R. A. GOOD, and P. A. MIESCHER. 1967. Ontogeny of immunity. Univ. Florida Press, Gainsville. 208 p.

SNIPES, C. A. 1968. Effects of growth hormone and insulin on amino acid and protein metabolism. Quart. Rev. Biol. **43:** 127–147.

SOKABE, H., and T. NAKAJIMA. 1972. Chemical structure and role of angiotensins in the vertebrates. Gen. Comp. Endocrinol. Suppl. **3:** 382–392.

SOLOMON, A. K. 1962. Pumps in the living cell. Sci. Am. **207 (2):** 100–108.

SOMERO, G. N. 1969. Enzymic mechanisms of temperature compensation: immediate and evolutionary effects of temperature on enzymes of aquatic poikilotherms. Am. Nat. **103:** 517–530.

SOMERO, G. N., and P. W. HOCHACHKA. 1971. Biochemical adaptation to the environment. Am. Zool. **11:** 159–167.

SONGDAHL, J. H., and V. H. HUTCHISON. 1972. The effect of photoperiod, parietalectomy and eye enucleation on oxygen consumption in the blue granite lizard *Sceloporus cyanogenys*. Herpetologica **28:** 148–156.

SOUTHWICK, E. E., and J. N. MUGAAS. 1971. A hypothetical homeotherm: the honeybee hive. Comp. Biochem. Physiol. **40A:** 935–944.

SPEEG, K. V., JR., and J. W. CAMPBELL. 1968. Formation and volatilization of ammonia gas by terrestrial snails. Am. J. Physiol. **214:** 1392–1402.

SPIKES, J. D. 1968. Photodynamic action. *In* Giese, **3:** 33–64.

SPIKES, J. D., and R. LIVINGSTON. 1969. The molecular biology of photodynamic action: sensitized photooxidations in biological systems. Adv. Radiation Biol. **3:** 29–121.

STARKEY, R. L. (CONVENER). 1962. Symposium on autrophy. Bact. Rev. **26:** 142–175.

STARLING, E. H. 1918. The law of the heart. Longmans, Green and Co. London. 27 p.

STEELE, J. E. 1963. The site of action of insect hyperglycemic hormone. Gen. Comp. Endocrinol. **3:** 46–52.

STEEN, J. B. 1970. The swim bladder as a hydrostatic organ. *In* Hoar and Randall, **4:** 413–443.

STEEN, J. B. 1971. Comparative physiology of respiratory mechanisms. Academic Press, New York. 182 p.

STEEN, J. B., and A. KRUYSSE. 1964. The respiratory function of teleostean gills. Comp. Biochem. Physiol. **12:** 127–142.

STEFANELLI, A. 1951. The Mauthnerian apparatus in the Ichthyopsida; its nature and function and correlated problem of neurohistogenesis. Quart. Rev. Biol. **26:** 17–34.

STEIN, W. D. 1967. The movement of molecules across cell membranes. Academic Press, New York. 369 p.

STEINBERGER, E. 1971. Hormonal control of mammalian spermatogenesis. Physiol. Rev. **51:** 1–22.

STEINBRECHT, R. A. 1969. Comparative morphology of olfactory receptors. *In* Pfaffmann (1969): 3–21.

STEINETZ, B. G., V. L. BEACH, and R. L. KROC. 1959. The physiology of relaxin in laboratory animals. *In* C. W. Lloyd, (ed.), Recent progress in the endocrinology of reproduction. pp. 389–427. Academic Press, New York.

STEINETZ, B. G., J. P. MANNING, M. BUTTER, and V. BEACH. 1965. Relationships of growth hormone, steroids and relaxin in the transformation of pubic joint cartilage to ligament in hypophysectomized mice. Endocrinology **76:** 876–882.

STEPHENS, G. C. 1963. Uptake of organic material by aquatic invertebrates. II. Accumulation of amino-acids by the bamboo worm, *Clymenella torquata*. Comp. Biochem. Physiol. **10:** 191–202.

STERBA, G. 1972. Neuro- and gliasecretion. *In* Hardisty and Potter, **2:** 69–89.

STERLING, K., M. A. BRENNER, and V. F. SALDANHA. 1973. Conversion of thyroxine to triiodothyronine by cultured human cells. Science **179:** 1000–1001.

STERNBERG, J. 1963. Autointoxication and some stress phenomena. Ann. Rev. Entomol. **8:** 19–38.

STEVEN, D. M. 1963. The dermal light sense. Biol. Rev. **38:** 204–240.

STRACHER, A. (chairman). 1967. Proceedings of a symposium on the contractile process. J. Gen. Physiol. **50(6):** 1–299.

SUEOKA, N. 1965. On the evolution of informational molecules. *In* V. Bryson and H. J. Vogel (eds.), Evolving genes and chromosomes. pp. 479–496. Academic Press, New York.

SULLIVAN, M. 1967. The message of the genes. Basic Books, New York. 198 p.

SUMNER, J. B. 1951. Urease. *In* J. B. Sumner, and K. Myrback (eds.), The enzymes. Vol. 1. Part 2, pp. 873–892. Academic Press, New York.

SUOMALAINEN, E. 1962. Significance of parthenogenesis in the evolution of insects. Ann. Rev. Ent. **7:** 349–366.

SUOMALAINEN, P. (ed.). 1964. Mammalian hibernation. Ann. Acad. Sci. Fennicae, Ser. A (IV) **71:** 1–453.

SUTTERLIN, A. M., and R. L. SAUNDERS. 1969. Proprioceptors in the gills of teleosts. Can. J. Zool. **47:** 1209–1212.

SVERDRUP, H. U., M. W. JOHNSON, and R. H. FLEMING. 1942. The oceans. Prentice-Hall, Englewood Cliffs, N.J. 1087 p.

SWAN, L. W. 1961. The ecology of the high Himalayas. Sci. Am. **205 (4):** 68–78.

SWEENEY, B. M. 1960. The photosynthetic rhythm in single cells of *Gonyaulax polyedra.* Cold Spring Harbor Symp. Quant. Biol. **25:** 145–148.

SWEENEY, B. M. 1969. Rhythmic phenomena in plants. Academic Press, New York. 147 p.

SWENSON, M. J. (ed.). 1970. Dukes' physiology of domestic animals. 8th ed. Cornell U. P., Ithaca, New York. 1463 p.

SZENT-GYÖRGYI, A. 1949. Free-energy relations and contraction of actomyosin. Biol. Bull. **96:** 140–161.

TAKAGI, S. F. 1971. Degeneration and regeneration of the olfactory epithelium. Handbook of Sensory Physiology (Springer-Verlag), **4(1):** 75–94.

TANSLEY, K. 1965. Vision in Vertebrates. Chapman and Hall, London. 132 p.

TASAKI, I. 1954. Nerve impulses in individual auditory nerve fibers of guinea pig. J. Neurophysiol. **17:** 97–122.

TASAKI, I. 1968. Nerve excitation. C. C. Thomas, Springfield, Ill. 201 p.

TATA, J. R. 1969. The action of thyroid hormones. Gen. Comp. Endocrinol. Suppl. **2:** 385–397.

TAUC, L. 1966. Physiology of the nervous system. *In* Wilbur and Yonge, **2:** 387–454.

TAUC, L. 1967. Transmission in invertebrate and vertebrate ganglia. Physiol. Rev. **47:** 521–593.

TAUC, L., and H. M. GERSCHENFELD. 1961. Cholinergic transmission mechanisms for both excitation and inhibition in molluscan central synapses. Nature **192:** 366–367.

TAVOLGA, W. N. 1971. Sound production and detection. *In* Hoar and Randall, **5:** 135–205.

TAYLOR, A. G. 1969. The direct uptake of amino acids and other small molecules from sea water by *Nereis virens* Sars. Comp. Biochem. Physiol. **29** (1). 243–250.

TAYLOR, C. R. 1972. The desert gazelle: a paradox resolved. Proc. Zool. Soc. Lond. **31:** 215–227.

TAYLOR, E. W. and P. J. BUTLER. 1973. The behavior and physiological responses of the shore crab *Carcinus maenas* during changes in environmental oxygen tension. Netherl. J. Sea Res. **7:** 496–505.

TAYLOR, E. W., P. J. BUTLER, and P. J. SHERLOCK. 1973. The respiratory and cardiovascular changes associated with the emersion response of *Carcinus maenas* (L.) during environmental hypoxia, at three different temperatures. J. Comp. Physiol. **86:** 95–115.

TEMIN, H. M. 1972. RNA-directed DNA synthesis. Sci. Am. **226 (1):** 25–33.

TERWILLIGER, R. C. and K. R. H. READ. 1969. The radular muscle myoglobins of the amphineuran mollusc, *Acanthopleura tranulata* Gmelin. Comp. Biochem. Physiol. **29:** 551–560.

THESLEFF, S. 1961. Nervous control of chemosensitivity in muscle. Ann. N.Y. Acad. Sci. **94:** 535-546.

THIMANN, K. V. 1963. The life of bacteria. 2nd ed. Macmillan, New York, 909 p.

THIMANN, K. V., and G. M. CURRY. 1960. Phototropism and phototaxis. *In* Florkin and Mason, **1:** 243–309.

THOAI, N. VAN, and Y. ROBIN. 1969. Guanidine compounds and phosphagens. *In* Florkin and Scheer, **4:** 163–203.

THOMPSON, M. L. 1955. Relative efficiency of pigment and horny layer thickness in protecting the skin of Europeans and Africans against solar ultraviolet radiation. J. Physiol. **127:** 236–246.

THOMSON, R. H. 1962. Melanins. *In* Florkin and Mason, **3:** 727–753.

THORNTON, V. F. 1972. A progesterone-like factor detected by bioassay in the blood of the toad *(Bufo bufo)* shortly before induced ovulation. Gen. Comp. Endocrinol. **18:** 133–139.

THORNTON, V. F., and P. J. EVENNETT. 1969. Endocrine control of oocyte maturation and oviducal jelly release in the toad, *Bufo bufo* (L.). Gen. Comp. Endocrinol. **13:** 268–274.

THORPE, W. H. 1950. The concepts of learning and their relation to those of instinct. Symp. Soc. Exp. Biol. **4:** 387–408.

THORPE, W. H. 1958. Ethology as a new branch of biology. *In* A. A. Buzzati-Traverso (ed.), Perspectives in marine biology. pp. 411–428, Univ. California Press, Berkeley.

THORPE, W. H. 1963. Learning and instinct in animals. 2nd ed. Methuen, London. 558 p.

TIEGS, O. W. 1955. The flight muscles of insects—their anatomy and histology; with some observations on the structure of skeletal muscle in general. Phil. Trans. Roy. Soc. London B, **238:** 221–348.

TIENHOVEN, A. VAN. 1968. Reproductive physiology of vertebrates. Saunders, Philadelphia. 498 p.

TINBERGEN, N. 1951. The study of instinct. Clarendon Press, Oxford. 228 p.

TINBERGEN, N. 1963. On aims and methods of ethology. Z. Tierpsychol. **20:** 410–433.

TOEWS, D. P. 1969. Respiration and circulation in *Amphiuma tridactylum*. Ph.D. Thesis, Univ. of Brit. Col., Vancouver. 109 p.

TOMBES, A. S. 1970. An introduction to invertebrate endocrinology. Academic Press, New York. 217 p.

TOMLINSON, J. 1966. The advantages of hermaphroditism and parthenogenesis. J. Theoret. Biol. **11:** 54–58.

TREGEAR, R. T. (1967). The oscillation of insect flight muscle. Current Topics Bioenergetics **2:** 269–286.

TREHERNE, J. E. 1967. Gut absorption. Ann. Rev. Entomol. **12:** 43–58.

TRIBUKAIT, B. 1963. Der Einfluss chronischer Hypoxie entsprechend 1,000–8,000 m Höhe auf die Erythropoiese der Ratte. Acta Physiol. Scand., **57:** 1–25.

TRUEMAN, E. R. 1966a. Bivalve mollusks: fluid dynamics of burrowing. Science **152:** 523–525.

TRUEMAN, E. R. 1966b. The mechanism of burrowing in the polychaete worm, *Arenicola marina* (L.). Biol. Bull. **131:** 369–377.

TRUEMAN, E. R. 1967. The dynamics of burrowing in *Ensis* (Bivalva). Proc. Roy. Soc. London B **166:** 459–476.

TRUEX, R. C., and M. Q. SMYTHE. 1965. Comparative morphology of the cardiac conduction tissue in animals. Ann. N.Y. Acad. Sci. **127:** 19–33.

TRYSTAD, O. 1968. The lipid-mobilizing effect of some pituitary gland preparations. Acta Endocrinol., **58:** 277–294; 295–317.

TSUKUDA, H. 1960. Heat and cold tolerance in relation to body size in the guppy, *Lebistes reticulatus*. J. Inst. Polytech., Osaka City Univ. D, **11:** 55–62.

TSUYUKI, H., E. ROBERTS, and R. E. A. GADD. 1962. Muscle proteins of Pacific salmon *(Oncorhynchus)*. Can. J. Biochem. Physiol. 40: 929–936.

TUCKER, D. 1971. Nonolfactory responses from the nasal cavity: Jacobson's organ and the trigeminal system. Handbook of Sensory Physiology (Springer-Verlag), **4(1):** 151–181.

TUCKER, V. A. 1968. Respiratory physiology of house sparrows in relation to high-altitude flight. J. Exp. Biol. **48:** 55–66.

TURNER, C. D. 1966. General endocrinology. 4th ed. Saunders, Philadelphia. 579 p.

TURNER, C. D., and J. T. BAGNARA. 1971. General endocrinology. 5th ed. Saunders, Philadelphia. 659 p.

TURNER, C. L. 1937. Reproductive cycles and superfetation in poeciliid fishes. Biol. Bull. **72:** 145–164.

TYLER, A., and B. T. SCHEER. 1945. Natural heteroagglutinins in the serum of the spiny lobster, *Panulirus interruptus*. II. Chemical and antigenic relation to blood proteins. Biol. Bull. **89:** 193–200.

UDE, J. 1964. Untersuchungen zur Neurosekretion bei Dendrocoelum lacteum Oerst. (Platyhelminthes-Turbellaria). Zeit. wiss. Zool. **170:** 223–255.

ULLRICH, K. J. 1960. Function of the collecting ducts. Circulation **21:** 869–874.

UMMINGER, B. L. 1971. Osmoregulatory overcompensation in the goldfish, *Carassius auratus*, at temperatures near freezing. Copeia (1971) **4:** 686–691.

UMMINGER, B. L. 1972. Physiological studies on supercooled killifish *(Fundulus heteroclitus)*. J. Exp. Zool. **181:** 217–222.

UNDERWOOD, G. 1968. Some suggestions concerning vertebrate visual cells. Vision Res. **8:** 483–488.

UNDERWOOD, E. J. 1971. Trace elements in human and animal nutrition. 3rd ed. Academic Press, New York. 543 p.

UNSWORTH, B. R., J. B. BALINSKY, and E. M. CROOK. 1969. Evidence for direct excretion of blood ammonia by an ammoniotelic amphibian. Comp. Biochem. Physiol. **31:** 373–377.

UREY, H. C. 1952. The planets, their origin and development. Yale U. P., New Haven, Conn. 245 p.

URIST, M. R. 1963. The regulation of calcium and other ions in the serums of hagfish and lampreys. Ann. N.Y. Acad. Sci. **109:** 294–311.

USHERWOOD, P. N. R. 1967. Insect neuromuscular mechanisms. Am. Zool. **7:** 553–582.

VARELA, F. G., and W. WIITANEN. 1970. The optics of the compound eye of the honeybee (*Apis mellifera*). J. Gen. Physiol. **55:** 336–358.

VERNBERG, F. J., and W. B. VERNBERG. 1970. The animal and the environment. Holt, Rinehart and Winston, New York. 398 p.

VERWORN, M. 1899. General physiology. Macmillan, London. 615 p.

VERWORN, M. 1913. Irritability. Yale U. P., New Haven, Conn. 264 p.

VINNIKOV, J. A. 1965. Principles of structural, chemical and functional organization of sensory receptors. Cold Spring Harbor Symp. Quant. Biol. **30:** 293–299.

VITÉ, J. P., and J. A. A. RENWICK. 1971. Population aggregating pheromone in the bark beetle, *Ips grandicollis*. J. Insect Physiol. **17:** 1699–1704.

VIZSOLYI, E. 1972. Aspects of neurohypophysial physiology during fetal development and pregnancy in the fur seal *Callorhinus ursinus*. Ph.D. thesis, Univ. Brit. Col., Vancouver. 198 p.

VIZSOLYI, E., and A. M. PERKS. 1969. New neurohypophysial principle in foetal mammals. Nature **223:** 1169–1171.

VLIEGENTHART, J. F. G., and D. H. G. VERSTEEG. 1967. The evolution of vertebrate neurohypophysial hormones in relation to the genetic code. J. Endocrinol. **38:** 3–12.

VONK, H. J. 1937. The specificity and collaboration of digestive enzymes in metazoa. Biol. Rev. **12:** 245–284.

VONK, H. J. 1960. Digestion and metabolism. *In* Waterman, **1:** 291–316.

VONK, H. J. 1962. Emulgators in the digestive fluids of invertebrates. Arch. Int. Physiol. Biochem. **70:** 67–85.

VONK, H. J. 1969. The properties of some emulsifiers in the digestive fluids of invertebrates. Comp. Biochem. Physiol **29:** 361–372.

VORONTSOVA, M. A., and LIOSNER, L. D. 1960. Asexual propagation and regeneration. Pergamon Press, Oxford. 489 p.

WAGGE, L. E. 1955. Amoebocytes. Int. Rev. Cytol. **4:** 31–78.

WAGNER, A. F., and K. FOLKERS. 1964. Vitamins and coenzymes. Wiley, New York. 532 p.

WALD, G. 1959. Life and light. Sci. Am. **201 (4):** 92–108.

WALD, G. 1960a. The distribution and evolution of visual systems. *In* Florkin and Mason, **1:** 311–345.

WALD, G. 1960b. The significance of vertebrate metamorphosis. Circulation **21:** 916–938.

WALD, G. 1961. The molecular organization of visual systems. *In* McElroy and Glass (1961): 724–753.

WALD, G. 1968. Molecular basis of visual excitation. Science **162:** 230–239.

WALKER, J. G. 1970. Oxygen poisoning in the annelid *Tubifex tubifex*. I. Response to oxygen exposure. Biol. Bull. **138:** 235–244.

WALL, B. J. 1967. Evidence for antidiuretic control of rectal water absorption in the cockroach *Periplaneta americana* L. J. Insect Physiol. **13:** 565–578.

WALL, B. J., and C. L. RALPH. 1964. Evidence for hormonal regulation of Malpighian tubule excretion in the insect, *Periplaneta americana* L. Gen. Comp. Endocrinol. **4:** 452–456.

WALLS, G. L. 1942. The vertebrate eye and its adaptive radiation. Cranbrook Inst. Sci. Bull. 19, Bloomfield Hills, Michigan. 785 p.

WALTMAN, B. 1966. Electrical properties and fine structure of the ampullary canals of Lorenzini. Acta Physiol. Scand. 66, Suppl. **264:** 1–60.

WARBURG, M. R. 1972. Water economy and thermal balance of Israeli and Australian amphibia from xeric habitats. Symp. Zool. Soc. London **31:** 79–111.

WARING, H. 1963. Color change mechanisms of cold-blooded vertebrates. Academic Press, New York. 266 p.

WATANABE, A., S. OBARA, T. AKIYAMA, and K. YUMOTO. 1967. Electrical properties of the pacemaker neurons in the heart ganglion of a stomatopod *Squilla oratoria*. J. Gen. Physiol. **50:** 813–838.

WATERMAN, T. H. 1960–61. The physiology of the crustacea. Vols. 1 and 2. Academic Press, New York.

WATERMAN, T. H., and R. B. FORWARD, JR. 1972. Field demonstration of polarotaxis in the fish *Zenarchopterus*. J. Exp. Zool. **180:** 33–54.

WATERMAN, T. H., and K. W. HORCH. 1966. Mechanism of polarized light perception. Science **154:** 467–475.

WATSON, J. D. 1968. The double helix. Atheneum, New York. 226 p.

WATSON, J. D. 1970. Molecular biology of the gene. 2nd ed. Benjamin, New York. 662 p.

WEBB, S. J. 1965. Bound water in biological integrity. C. C. Thomas, Springfield, Ill. 187 p.

WEBER, G. (ed.). 1972. Advances in enzyme regulation. Vol. 10. Pergamon Press, Oxford. 463 p.

WEBER, H. H. 1958. The motility of muscle and cells. Harvard U. P., Cambridge. 69 p.

WEBER, H. H. 1960. Chemical reactions during contraction and relaxation. *In* D. Nachmansohn (ed.), Molecular biology. pp. 1–16. Academic Press, New York.

WEEL, P. B. VAN. 1974. "Hepatopancreas"? Comp. Biochem. Physiol. **47A:** 1–9.

WEISBART, M., and D. R. IDLER. 1970. Re-examination of the presence of corticosteroids in two cyclostomes, the Atlantic hagfish *(Myxine glutinosa* L.*)* and the sea lamprey *(Petromyzon marinus* L.) J. Endocrinol. **46:** 29–43.

WELLS, G. P. 1949. Respiratory movements of *Arenicola marina* L: intermittent irrigation of the tube, and intermittent aerial respiration. J. Mar. Biol. Assoc. U.K., **28:** 447–464.

WELLS, G. P. 1950. Spontaneous activity cycles in polychaete worms. Symp. Soc. Exp. Biol. **4:** 127–142.

WELLS, M. J. 1962. Brain and behaviour of cephalopods. Heinemann, London. 171 p.

WELLS, M. J. 1966. The brain and behavior of cephalopods. *In* Wilbur and Yonge, **2:** 547–590.

WELLS, M. J., and J. WELLS. 1959. Hormonal control of sexual maturity in *Octopus.* J. Exp. Biol. **36:** 1–33.

WELLS, M. J., and J. WELLS. 1969. Pituitary analogue in the octopus. Nature **222:** 293–294.

WELLS, N. A. 1935. Variations in the respiratory metabolism of the Pacific killifish *Fundulus parvipinnis* due to size, season and continued constant temperature. Physiol. Zool. **8:** 318–336.

WELSH, J. H. 1934. The caudal photoreceptor and responses of the crayfish to light. J. Cell. Comp. Physiol. **4:** 379–388.

WELSH, J. H. 1970. Phylogenetic aspects of the distribution of biogenic amines. *In* Blum (1970): 75–94.

WENDELAAR BONGA, S. E. 1972. Neuroendocrine involvement in osmoregulation in a freshwater mollusc, *Lymnaea stagnalis.* Gen. Comp. Endocrinol. Suppl. **3:** 308–316.

WENNER, A. M., P. H. WELLS, and D. L. JOHNSON. 1969. Honey bee recruitment to food sources: olfaction or language. Science **164:** 84–86.

WERNER, G., and V. B. MOUNTCASTLE. 1965. Neural activity in mechanoreceptive cutaneous afferents: stimulus-response relations, Weber functions, and information transmission. J. Neurophysiol. **28:** 359–397.

WERSÄLL, J., A. FLOCK, and P. G. LUNDQUIST. 1965. Structural basis for directional sensitivity in cochlear and vestibular sensory receptors. Cold Spring Harbor Symp. Quant. Biol. **30:** 115–132.

WHITE, A., P. HANDLER, and E. L. SMITH. 1968. Principles of biochemistry. 4th ed. McGraw-Hill, New York. 1187 p.

WHITE, M. J. D. 1973. Animal cytology and evolution. 3rd ed. Cambridge U. P., London. 961 p.

WHITFORD, W G. and V. H. HUTCHINSON. 1963. Cutaneous and pulmonary gas exchange in the spotted salamander, *Ambystoma maculatum.* Biol. Bull. **124:** 344–354.

WHITTEN, J. 1968. Metamorphic changes in insects. *In* Etkin and Gilbert (1968): 43–105.

WIERSMA, C. A. G. 1961. The neuromuscular system. Reflexes and the central nervous system. *In* Waterman, **2:** 191–279.

WIESER, W. 1972. Oxygen consumption and ammonia excretion in *Ligia beaudiana* M.-E. Comp. Biochem. Physiol. **43A:** 869–876.

WIESER, W., and G. SCHWEIZER. 1970. A re-examination of the excretion of nitrogen by terrestrial isopods. J. Exp. Biol. **52:** 267–274.

WIGG. D., E. C. BOVEE, and T. L. JAHN. 1967. The evacuation mechanism of the water expulsion vesicle ("contractile vacuole") of *Amoeba proteus.* J. Protozool. **14:** 104–108.

WIGGLESWORTH, V. B. 1936. The function of the corpus allatum in the growth and reproduction of *Rhodnius prolixus* (Hemiptera). Quart. J. Microscop. Sci. **79:** 91–121.

WIGGLESWORTH, V. B. 1964. The hormonal regulation of growth and reproduction in insects. Adv. Insect Physiol., **2:** 247–336.

WIGGLESWORTH, V. B. 1970. Insect hormones. Freeman, San Francisco. 159 p.

WIGGLESWORTH, V. B. 1972. The principles of insect physiology. 7th ed. Chapman & Hall, London. 827 p.

WILBER, C. G. 1957. Physiological regulations and the origin of human types. Human Biology **29:** 329–336.

WILBUR, K. M. 1964. Shell formation and regeneration. *In* Wilbur and Yonge **1:** 243–282.

WILBUR, K. M., and C. M. YONGE (eds.). 1964–1966. Physiology of Mollusca. Vols. 1–2. Academic Press, New York.

WILKIE, D. R. 1956. The mechanical properties of muscle. Brit. Med. Bull. **12:** 177–182.

WILKIE, D. R. 1968. Muscle. Arnold, London. 64 p.

WILLIAMS, C. M. 1952. The physiology of insect diapause. Biol. Bull. **103:** 120–138.

WILLIAMS, C. M. 1967. Third-generation pesticides. Sci. Am. **217 (1):** 13–17.

WILLIAMS, R. H. (ed.). 1968. Textbook of endocrinology. 4th ed. Saunders, Philadelphia. 1258 p.

WILLIAMS, R. J. 1970. Freezing tolerance in *Mytilus edulis.* Comp. Biochem. Physiol. **35:** 145–161.

WILLMER, E. N. 1934. Some observations on the respiration of certain tropical freshwater fishes. J. Exp. Biol. **11:** 283–306.

WILLOWS, A. O. D. 1967. Behavioral acts elicited by stimulation of single, identifiable brain cells. Science **157:** 570–574.

WILSON, E. O. 1965. Chemical communication in social insects. Science **149:** 1064–1071.

WILSON, R. P., R. O. ANDERSON, and R. A. BLOOMFIELD. 1969. Ammonia toxicity in selected fishes. Comp. Biochem. Physiol. **28:** 107–118.

WILSON, R. P., M. E. MUHRER, and R. A. BLOOMFIELD. 1968. Comparative ammonia toxicity. Comp. Biochem. Physiol. **25:** 295–301.

WIMSATT, W. A. (ed.). 1970. Biology of bats. Vols. 1 and 2. Academic Press, New York.

WINGSTRAND, K. G., and O. MUNK. 1965. The pecten oculi of the pigeon with particular regard to its function. Biol. Skr. Dan. Vid. Selsk. **14 (3):** 1–64.

WIRZ, H. 1961. Newer concepts of renal mechanism in relation to water and electrolyte excretion. *In* C. P. Stewart and Th. Stengers. (eds.), Water and electrolyte metabolism. 100–108. Elsevier, Amsterdam.

WISKEMANN, A. 1969. Effects of ultraviolet light on the skin. *In* C. P. Swanson (ed.), An introduction to photobiology. 81–98. Prentice-Hall, Englewood Cliffs, N. J.

WITHROW, R. B. (ed.). 1959. Photoperiodism and related phenomena in plants and animals. Amer. Assoc. Adv. Sci. Publ. 55, Washington, D.C. 903 p.

WITSCHI, E. 1961. Sex and secondary sexual characters. *In* Marshall, **2:** 115–168.

WITTENBERG, J. B. 1960. The source of carbon monoxide in the float of the Portuguese man-of-war, *Physalia physalis* L. J. Exp. Biol. **37:** 698–705.

WITTENBERG, J. B. 1961. The secretion of oxygen into the swim-bladder of fish. I. The transport of molecular oxygen. J. Gen. Physiol. **44:** 521–526.

WOLBARSHT, M. L., and S. S. YEANDLE. 1967. Visual processes in the *Limulus* eye. Ann. Rev. Physiol. **29:** 513–542.

WOLKEN, J. J. 1958. Studies of photoreceptor structures. Ann. N.Y. Acad. Sci. **74:** 164–181.

WOLKEN, J. J. 1971. Invertebrate photoreceptors. Academic Press, New York. 179 p.

WOLSTENHOLME, G. E. W., and J. BIRCH (eds.). 1971. Neurohypophysial hormones. Ciba Foundation Study Group No. 39. Churchill, London. 146 p.

WOLSTENHOLME, G. E. W., and J. KNIGHT (eds.). 1970. Taste and smell in vertebrates. Ciba Foundation Symposium. Churchill, London. 402 p.

WOLVEKAMP, H. P. 1961. The evolution of oxygen transport. *In* Macfarlane and Robb-Smith (1961): 1–72.

WOLVEKAMP, H. P., and T. H. WATERMAN. 1960. Respiration. *In* Waterman, **1:** 35–100.

WOODRUFF, G. N. 1971. Dopamine receptors: a review. Comp. Gen. Pharmac. **2:** 439–455.

WOODS, F. H. 1931. History of the germ cells in *Sphaerium striatum* (Lam.). J. Morph. **51:** 545–595.

WOOTTON, J. A., and L. D. WRIGHT. 1962. A comparative study of sterol biosynthesis in Annelida. Comp. Biochem. Physiol. **5:** 253–264.

WOURMS, J. P. 1967. Annual fishes. *In* F. W. Wilt, and N. K. Wessels, (eds.), Methods in developmental biology. pp. 123–137. Thomas Y. Crowell Co., New York.

WRIGHT, R. H. 1964. Odor and molecular vibration: far infrared spectra of some perfume chemicals. Ann. N.Y. Acad. Sci. **116:** 552–558.

WRIGHT, R. H., and R. E. BURGESS. 1970. Specific physicochemical mechanisms of olfactory stimulation. *In* Wolstenholme and Knight (1970): 325–342.

WURTMAN, R. J., J. AXELROD, and D. E. KELLY. 1968. The pineal. Academic Press, New York. 199 p.

WYATT, G. R. 1967. The biochemistry of sugars and polysaccharides in insects. Adv. Insect Physiol. **4:** 287–360.

YAMAOKA, L. H., and B. T. SCHEER. 1970. Chemistry of growth and development in crustaceans. *In* Florkin and Scheer, **5A:** 321–341.

YARON, Z. 1972. Endocrine aspects of gestation in viviparous reptiles. Gen. Comp. Endocrinol. Suppl. **3:** 663–674.

YONGE, C. M. 1928. Feeding mechanisms in the invertebrates. Biol. Rev. **3:** 21–76.

YONGE, C. M. 1937. Evolution and adaptation in the digestive system of the metazoa. Biol. Rev. **12:** 87–115.

YOST, H. T. 1972. Cellular physiology. Prentice-Hall, Englewood Cliffs, N.J. 925 p.

YOUNG, J. Z. 1935. The photoreceptors of lampreys. 1. Light-sensitive fibres in the lateral line nerves. J. Exp. Biol. **12:** 229–238.

YOUNG, J. Z. 1939. Fused neurons and synaptic contacts in the giant nerve fibres of cephalopods. Phil. Trans. Roy. Soc. London B, **229:** 465–505.

YOUNG, J. Z. 1957. The life of mammals. Clarendon Press, Oxford. 820 p.

YOUNG, J. Z. 1961. Learning and discrimination in the octopus. Biol. Rev. **36:** 32–96.

YOUNG, J. Z. 1963. The number and sizes of nerve cells in *Octopus*. Proc. Zool. Soc. London **140:** 229–254.

YOUNG, J. Z. 1964. A model of the brain. Clarendon Press, Oxford. 348 p.

YOUNG, J. Z. 1971. The anatomy of the nervous system of *Octopus vulgaris*. Clarendon Press, Oxford. 690 p.

YOUNG, M. 1969. The molecular basis of muscle contraction. Ann. Rev. Biochem. **38:** 913–950.

YOUNG, R. W. 1970. Visual cells. Sci. Am. **223 (4):** 81–91.

Yousef, M. K., S. M. Horvath and R. W. Bullard (eds.). 1972. Physiological adaptations. Academic Press, New York. 258 p.

Yu, M., and A. Perlmutter. 1970. Growth inhibiting factors in the zebrafish, *Brachydanio rerio* and the blue gourami, *Trichogaster trichopterus*. Growth **34:** 153–175.

Zandee, D. I. 1966. Metabolism in the crayfish *Astacus astacus* (L). II. The energy-yielding metabolism. Arch. Int. Physiol. Biochem. **74:** 45–57.

Zandee, D. I. 1967. The absence of cholesterol synthesis in *Sepia officinalis* L. Arch. Int. Physiol. Biochem. **75:** 487–491.

Zarrow, M. X., J. M. Yochim, and J. L. McCarthy. 1964. Experimental endocrinology. Academic Press, New York. 519 p.

Zenkevitch, L. A. (translated by S. Botcharskaya). 1963. Biology of the seas of the U.S.S.R. Allen & Unwin, London. 955 p.

Zeuthen, E. 1953. Oxygen uptake as related to body size in organisms. Quart. Rev. Biol. **28:** 1–12.

Zeuthen, E. 1955. Comparative physiology (respiration). Ann. Rev. Physiol. **17:** 459–482.

Zimmerman, A. M. (ed.). 1970. High pressure effects on cellular processes. Academic Press, New York. 324 p.

Zimmerman, A. M. 1971. High-pressure studies in cell biology. Internat. Rev. Cytol. **30:** 1–47.

Zoond, A., and J. Eyre. 1934. Studies in reptilian colour response. Phil. Trans. Roy. Soc. London B, **223:** 27–55.

Zotterman, Y. 1959. Thermal sensations. Handbook of Physiology (Am. Physiol. Soc.) Sect. 1, **1:** 431–458.

Zuckerkandl, E. 1965. The evolution of hemoglobin. Sci. Am. **212 (5):** 110–118.

Zuckerman, S. 1962. The ovary. Vols. 1–2. Academic Press, New York and London.

Zwicky, K. T. 1968. Innervation and pharmacology of the heart of *Urodacus,* a scorpion. Comp. Biochem. Physiol. **24:** 799–808.

INDEX